Early Flowers and Angiosperm Evolution

The recent discovery of diverse fossil flowers and floral organs in Cretaceous strata has revealed astonishing details about the structural and systematic diversity of early angiosperms. Exploring the rich fossil evidence that has been accumulated over the past three decades, this unique study follows the evolutionary history of flowering plants from their earliest phases in obscurity to their dominance in modern vegetation.

The book provides comprehensive biological and geological background information, before moving on to summarise the fossil record in detail. Including previously unpublished results based on research into Early and Late Cretaceous fossil floras from Europe and North America, the authors draw together direct palaeontological evidence of the pattern of angiosperm evolution through time.

Synthesising palaeobotanical data with information from living plants, this book explores the latest research in the field and highlights connections with phylogenetic systematics as well as the structure and the biology of extant angiosperms.

Else Marie Friis is in the Department of Palaeobotany at the Swedish Museum of Natural History. Her research interests include Cretaceous flowers and other fossil reproductive structures, with particular focus on the origin and early diversification of angiosperms and related seed plants.

Peter R. Crane is in the School of Forestry and Environmental Studies at Yale University. His research interests include large-scale patterns and processes of plant evolution and integrated palaeobotanical and neobotanical studies of plant diversity and evolution.

Kaj Raunsgaard Pedersen is in the Department of Geology at the University of Aarhus. His research interests include integrated palynological and palaeobotanical studies of Mesozoic seed plants with particular focus on Cretaceous reproductive structures and flowering plant evolution.

Early Flowers and Angiosperm Evolution

Else Marie Friis
Swedish Museum of Natural History, Stockholm

Peter R. Crane
School of Forestry & Environmental Studies, Yale University

Kaj Raunsgaard Pedersen
Department of Geology, University of Aarhus

Line drawings by Pollyanna von Knorring

CAMBRIDGE
UNIVERSITY PRESS

CAMBRIDGE UNIVERSITY PRESS
Cambridge, New York, Melbourne, Madrid, Cape Town, Singapore,
São Paulo, Delhi, Tokyo, Mexico City

Cambridge University Press
The Edinburgh Building, Cambridge CB2 8RU, UK

Published in the United States of America by Cambridge University Press, New York

www.cambridge.org
Information on this title: www.cambridge.org/9780521592833

First published 2011

Printed in the United Kingdom at the University Press, Cambridge

A catalogue record for this publication is available from the British Library

Library of Congress Cataloging-in-Publication Data

Friis, Else Marie.
Early flowers and angiosperm evolution / Else Marie Friis, Peter R. Crane,
Kaj Raunsgaard Pedersen.
p. cm.
ISBN 978-0-521-59283-3 (Hardback)
1. Angiosperms, Fossil. 2. Angiosperms–Evolution. I. Crane, Peter R.
II. Pedersen, Kaj Raunsgaard. III. Title.
QE980.F75 2011
561–dc22

2011001815

ISBN 978-0-521-59283-3 Hardback

Contents

Preface

Developments in the study of fossil and living plants over the past few decades have greatly clarified many aspects of early angiosperm evolution. Explicit phylogenetic analyses, facilitated by the development of computer technology and based on both morphological and molecular data, have renewed interest in the relationships of angiosperms to other plants, the patterns of relationship among major groups of angiosperms, and the processes that have generated angiosperm diversity at both microevolutionary and macroevolutionary scales. At the same time, a rapid accumulation of new information on the structure and biology of many key groups of living angiosperms has catalysed comparative studies and brought to light many previously unrecognised features that provide new perspectives on angiosperm evolution.

Palaeobotanical studies have also been central in revitalising research on early angiosperm evolution and have advanced significantly our understanding of early angiosperm history. In particular, the discovery of diverse and exquisitely preserved fossil flowers and floral organs from the Cretaceous has yielded detailed information on the structural and systematic diversity of early angiosperms. These data complement the information available from living plants, and are also invaluable for testing evolutionary hypotheses based on extant taxa against palaeobotanical and stratigraphic evidence. The recognition of fossil pollen grains *in situ* within flowers has also provided new possibilities for interpreting the record of dispersed fossil pollen. Only a few decades ago the abundant occurrence of fossil angiosperm flowers in Cretaceous strata was unimagined, but today there is a rich floral record, much of which still remains to be analysed in detail. The key breakthrough was the recognition that numerous small fossil flowers, which are generally not visible to collectors in the field, can be extracted from Cretaceous sediments by using bulk-sieving techniques and studied with scanning electron microscopy (SEM), and now also with synchrotron X-ray microtomography (SXRTM). These techniques, modified from standard approaches to Cenozoic fossil floras in Europe, and pioneered in the Late Cretaceous of Scania, Sweden, have now yielded diverse angiosperm flowers from many new fossil floras (mesofossil floras) discovered in Lower and Upper Cretaceous strata in Europe, North America, Asia, New Zealand and Antarctica.

In this book we provide a synthesis and overview of current data and ideas on the major patterns of angiosperm evolution, focusing especially on the early evolution of the group. Our emphasis is on the new information from the fossil record that has accumulated over the past three decades and how this relates to recent findings on the phylogenetic systematics, structure and biology of extant angiosperms. Central to this synthesis of the palaeobotanical data is its integration with information from living plants and the presentation of previously unpublished results based on our research with Early and mid-Cretaceous fossil floras from eastern North America and Portugal.

Chapters 1 to 4 provide the background to information and ideas discussed in more detail later in the book. Chapter 1 introduces recent developments in angiosperm palaeobotany, molecular systematics and studies of the flowers of living plants, and briefly considers some of the ways in which these advances are changing our perspective on early angiosperm evolution. Major features of angiosperm structure and biology are also reviewed along with previous ideas on the origin and early evolution of angiosperms and their flowers, as well as the rise of angiosperms to ecological dominance. Chapter 2 provides an overview of the nature of the angiosperm fossil record. Chapter 3 briefly outlines changes in palaeogeography and climate since the Early Cretaceous, as an introduction to the changing world in which angiosperm diversification took place. Chapter 4 briefly discusses the stratigraphic framework and occurrence of the angiosperm fossils considered in this book and provides a review of the key fossil localities.

Chapter 5 places angiosperms in context with respect to other groups of extant and extinct seed plants and focuses especially on those plants that have been thought to be closely related to angiosperms. In particular, we highlight new palaeobotanical data on the Gnetales and the potentially related extinct Bennettitales and Erdtmanithecales. Chapters 6 and 7 review the development of ideas concerning seed plant and angiosperm phylogeny.

The core of this book, Chapters 8–15, summarises in a phylogenetic framework the fossil record of angiosperms with particular emphasis on floral structures known from the Cretaceous. Brief mention is also made of key records from the Early Cenozoic.

In Chapter 16 we consider major patterns in the structural diversification of angiosperm flowers based on current phylogenetic hypotheses and evidence from the fossil record.

Chapters 17–20 consider the biological and ecological consequences of angiosperm diversification, including the nature of vegetational change during the Cretaceous and the evolution of interactions with pollinators and dispersers. Through these interactions, the diversification of flowering plants has been inextricably linked with diversification in the animal world, as well as with the origin of modern ecosystems.

All photographs and plates are by Else Marie Friis and Kaj Raunsgaard Pedersen, and all line drawings by Pollyanna von Knorring, unless otherwise specified; the maps were drawn by Wieslaw Smolinsky. We thank all staff members of the Department of Palaeobotany, Swedish Museum of Natural History, Stockholm, for much help and encouragement during the production of this book. We are also grateful for the support and patience of our families in the process of completing this work. We are also deeply grateful for stimulating discussions with many colleagues around the world over many years. We have greatly appreciated and benefited from their advice and friendship.

Major support for this work was obtained from the Swedish Museum of Natural History and the Swedish Natural Science Research Council. Additional substantial support was provided by the Carlsberg Foundation, the United States National Science Foundation, The Field Museum, The Royal Botanic Gardens, Kew, The University of Chicago, Yale University and The University of Aarhus. We also acknowledge support from the TOMCAT Beamline at the Swiss Light Source, Paul Scherrer Institute, Switzerland, the Asian–Swedish Research Link Programme, and the WCU program of the National Research Foundation of Korea.

1

Introduction to angiosperms

The phylogenetic diversification and ecological radiation of angiosperms (flowering plants) that took place in the Early Cretaceous, between about 135 and 65 million years ago, was one of the major biotic upheavals in the history of life. It had dramatic consequences for the composition and subsequent evolution of terrestrial ecosystems. Ancient Mesozoic vegetation, which was dominated by ferns, conifers, ginkgos and cycads, as well as Bennettitales and other groups of extinct seed plants, was eventually almost entirely replaced by more modern ecosystems dominated by angiosperms. Since the Early Cretaceous, high diversification rates have generated more than 350 000 extant angiosperm species. Today there are more living species of angiosperms than all other groups of land plants combined.

In their rise to ecological dominance angiosperms have exhibited extraordinary developmental and evolutionary plasticity. This has resulted in overwhelming morphological diversity and a great variety of adaptive types. Angiosperms are far more diverse in vegetative form and in the structure of their reproductive organs than any other group of land plants.

In growth habit angiosperms range from minute free-floating aquatics less than 1 mm long (*Wolffia*) and moss-like plants of fast-flowing water (Podostemaceae), to herbs, epiphytes, lianas, shrubs and tall trees. The massive *Eucalyptus* trees of Tasmania reach heights of more than 100 m. In floral morphology the structural diversity of angiosperms ranges from the minute staminate flowers of *Hedyosmum*, which comprise only a single stamen, to the giant blossoms of *Rafflesia*, up to 90 cm in diameter, or to the slender flowers of *Aristolochia* that may be up to 1 m long (Endress, 1994b).

Structural diversity is matched by ecological diversity: angiosperms occupy an astonishing range of habitats from deserts to freshwater swamps, and from tropical rain forests to the Antarctic Peninsula. Except for boreal conifer forests and moss–lichen tundra, angiosperms dominate the vegetation of all the major terrestrial biomes. They are also the only group of land plants that have reinvaded the marine habitat (e.g. *Zostera*). As a result of their ecological dominance angiosperms account for the majority of terrestrial primary productivity, play a significant role in major geochemical cycles and influence atmospheric composition and climate. They comprise the autotrophic foundation on which almost all terrestrial ecosystems are built.

The diversity of angiosperms also includes great variety in physiology and biochemistry. Most angiosperms are autotrophs, but there are also diverse parasites and saprophytes. Some angiosperms supplement their intake of nitrogen through trapping and digesting animals. A much larger number are involved in elaborate symbiotic associations with a wide range of fungi and bacteria. Several groups of angiosperms have evolved modifications to the basic processes of photosynthesis, which confer significant ecological advantages in arid or hot environments. Angiosperms also produce a bewildering variety of biochemical compounds, many of which appear to be important in interactions with other organisms ranging from fungi to vertebrates. As food, as a source of raw materials, and as key participants in global ecological systems, angiosperms are fundamental to human life and survival.

1.1 PHYLOGENETIC POSITION OF ANGIOSPERMS

1.1.1 Anthophytes and alternative patterns

For almost 150 years, attempts to understand the origin and diversification of flowering plants were hindered by uncertain relationships among the vast array of extant angiosperms, as well as the apparently insuperable morphological 'gap' between angiosperms and other seed plants (gymnosperms). More generally, progress in studying the patterns and larger-scale processes of biological evolution was retarded by the absence of an appropriate methodological framework in which to develop, and choose among, competing phylogenetic hypotheses. As a result, it was difficult to evaluate the great variety of ideas that had been published on angiosperm origins, and that had implicated almost all groups of fossil and living gymnosperms as potential angiosperm ancestors (Crane, 1985a).

The role of the plant fossil record in contributing to an understanding of angiosperm origins and early evolution was also uncertain. Some authors assumed that palaeobotanical data were either unavailable or uninformative as a source of useful information bearing on angiosperm evolution (e.g. Stebbins, 1974); others argued that only the fossil record could provide reliable data for interpreting angiosperm evolution (e.g. Hughes, 1976, 1994).

Over the past few decades the development of phylogenetic systematics (cladistics) has stimulated intense discussion of the philosophical and methodological bases for reconstructing phylogenetic patterns. Many of the crucial theoretical and other issues that formerly blocked progress in phylogenetics have now been wholly or partly resolved, clearing the way for new research that has made rapid progress. The development of computer software for powerful numerical cladistic analyses of large complex datasets, as well as techniques for rapidly amplifying and sequencing nucleic acids, have also reactivated interest in phylogenetic reconstruction as a key component of modern evolutionary biology and as one of the central goals of contemporary plant systematics. In the process, the nature of what the palaeobotanical record can, and cannot, contribute to studies of angiosperm evolution has been clarified (Crane *et al.*, 2004).

The diversity of land plants can be viewed as a nested set of four major groups: land plants (embryophytes), vascular plants (tracheophytes), seed plants (spermatophytes) and flowering plants (angiosperms) (Figure 1.1). In this context, and in cladistic terms, resolving the question of angiosperm origin requires recognising and defining the major groups of seed plants, determining their phylogenetic interrelationships, and thus establishing the group with which flowering plants share a most recent common ancestor.

Since the early 1980s (Hill and Crane, 1982; Crane, 1985a) a series of cladistic analyses using parsimony have been conducted to investigate the phylogenetic position of angiosperms among living and fossil seed plants based on syntheses of mainly morphological data (Doyle and Donoghue, 1986, 1992, 1993; Loconte and Stevenson, 1990; Nixon *et al.*, 1994; Rothwell and Serbet, 1994). All of these studies identify angiosperms as a strongly supported clade, and indeed as one of the most strongly supported monophyletic groups in the plant kingdom. Cladistic studies

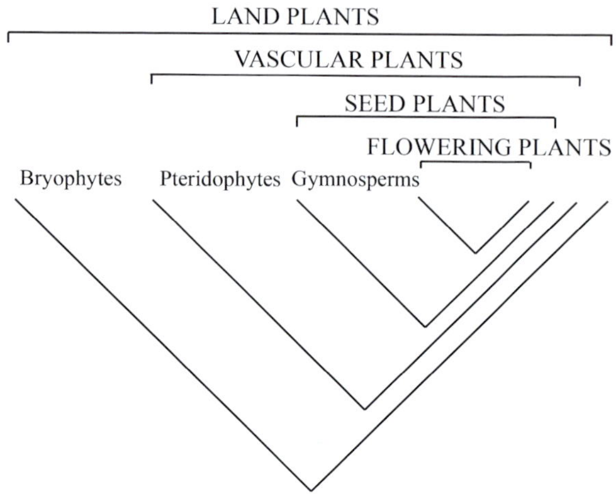

Figure 1.1 Cladistic relationships among land plants (embryophytes) showing the nested positions of three monophyletic groups: vascular plants (tracheophytes), seed plants (spermatophytes) and flowering plants (angiosperms). Under this interpretation bryophytes, pteridophytes and gymnosperms are defined by exclusion (paraphyletic). For example, gymnosperms are those seed plants that are not angiosperms. Adapted from Crane (1985a).

that include a significant representation of extinct seed plants also recognise that the seed ferns (Pteridospermales) as traditionally defined constitute a highly unnatural assemblage of seed plants of diverse relationships (Crane, 1985a). Most studies also confirm the Gnetales as a monophyletic group (but see Nixon *et al.*, 1994).

Morphology-based phylogenetic analyses of extant land plants have generally placed the Gnetales as the closest living relatives to angiosperms (Figure 1.2), and analyses incorporating extinct plants show that among fossils the Bennettitales are also closely related (Figure 1.3), supporting earlier hypotheses of angiosperm and seed plant relationships (Bessey, 1897, 1915; Hallier, 1901; Arber and Parkin, 1907). The angiosperms, Gnetales and Bennettitales, sometimes together with the extinct *Pentoxylon* plant, form a monophyletic group, usually referred to as the anthophytes (e.g. Doyle and Donoghue, 1986), in recognition of their shared possession of flower-like reproductive structures (Figure 1.3). An area of disagreement among these cladistic analyses based on morphological data concerns the resolution of relationships among anthophytes (Figure 1.3). This disagreement arises mainly as a result of uncertainties over the homologies among seed plant reproductive structures, including angiosperm flowers.

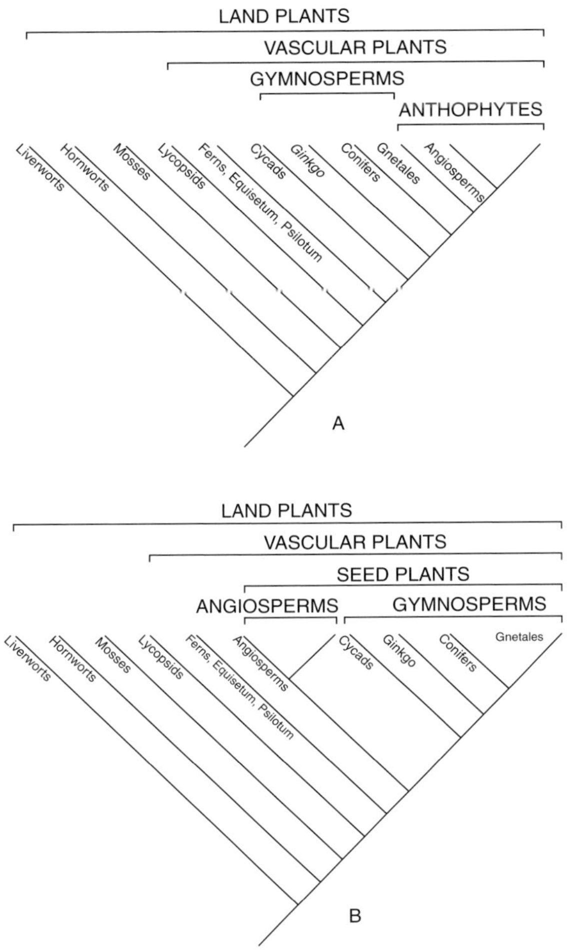

Figure 1.2 Alternative cladistic relationships among extant land plants. (A) Cladogram showing angiosperms nested among other seed plants and forming a monophyletic group with Gnetales (anthophyte clade); under this hypothesis extant gymnosperms are paraphyletic. (B) Cladogram showing angiosperms as sister group to all other seed plants; under this hypothesis extant gymnosperms are monophyletic. (A) and (B) synthesised from several sources.

The relationships of anthophytes to other seed plants are also uncertain. In several analyses anthophytes are nested among two or more of the so-called Mesozoic seed fern groups (Glossopteridales, Caytoniales, Corystospermales, Peltaspermales), often with Caytoniales as the sister taxon to the anthophytes (e.g. Crane, 1985a; Doyle and Donoghue, 1986, 1993). Again, many of the difficulties behind these different interpretations arise from ongoing uncertainties about how key fossil taxa should be compared with angiosperms and also with each other. These difficulties are further exacerbated by imperfect knowledge of important extant and extinct plants. There is still much more work to be done to develop a more satisfactory comparative dataset on which future integrated cladistic analyses of living and fossil seed plants can be based.

Recent analyses based on morphological data give phylogenetic patterns very similar to those from 25 years ago (Doyle, 2006; Hilton and Bateman, 2006; Friis *et al.*, 2007), but so far this consistency has not translated into broad confidence in the results. A particular difficulty is that recent analyses based on molecular data support alternative interpretations of relationships among extant seed plants. Some of these analyses suggest that angiosperms are the sister group to a clade comprising all other extant seed plants (cycads, *Ginkgo*, conifers, Gnetales) (Figure 1.2), which would make all living gymnosperms (but not necessarily all fossil and living gymnosperms) monophyletic. Other analyses are especially radical in supporting the inclusion of Gnetales within conifers (Hansen *et al.*, 1999; Qiu *et al.*, 1999; Bowe *et al.*, 2000; Chaw *et al.*, 2000; Burleigh and Mathews, 2004). Conifers, as they are traditionally defined, would therefore be paraphyletic. Still other analyses indicate monophyly for conifers with Gnetales as sister to conifers (Burleigh and Mathews, 2004), or suggest that angiosperms and cycads may be sister taxa (Mathews, 2010). Taken together current molecular data do not appear to be sufficient for the unambiguous identification of relationships among extant seed plants (Rydin *et al.*, 2002; Friis *et al.*, 2009a; Mathews, 2010). Recent phylogenetic analyses that bear on the question of angiosperm origin, including those relevant for ideas about the age of angiosperms are reviewed in more detail in Chapter 6.

1.1.2 Relationships among living angiosperms

Another factor that has contributed to uncertainties about phylogenetic patterns among major groups of seed plants has been the question of how best to represent the enormous diversity of angiosperms in higher-level analyses. Until recently this problem has been further compounded by the very poor resolution of phylogenetic relationships near the base of the angiosperm clade.

In the 1980s important initial progress was made in the development of the phylogenetic framework for angiosperm evolution by overturning the traditional view that angiosperms comprise two distinct lineages; dicotyledons (dicots) and monocotyledons (monocots). Early cladistic

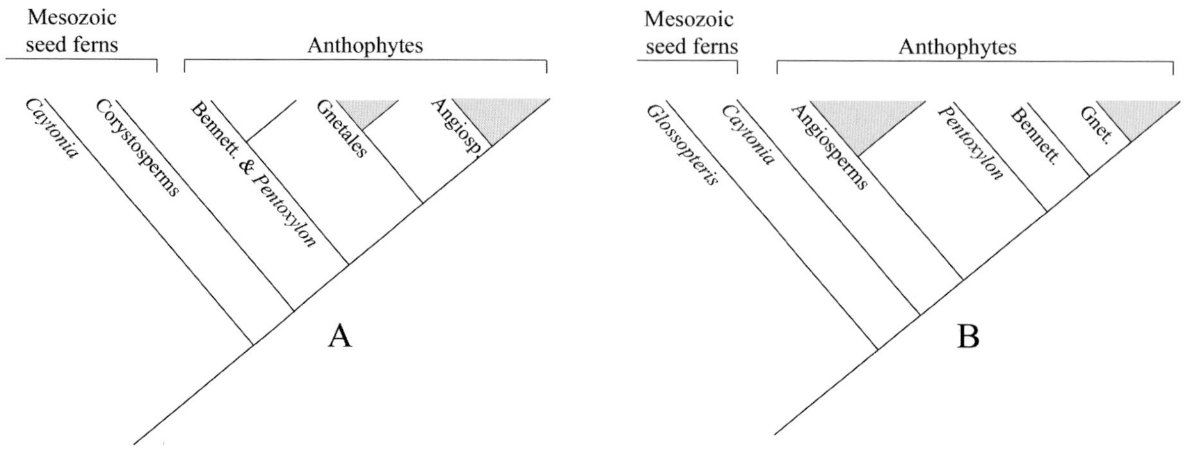

Figure 1.3 Alternative cladistic relationships among anthophytes. (A) Cladogram showing angiosperms nested among other anthophytes. (B) Cladogram showing angiosperms as sister group to all other anthophytes. (A) adapted from Crane (1985a); (B) adapted from Doyle and Donoghue (1992). For other models of relationships, see Chapter 6.

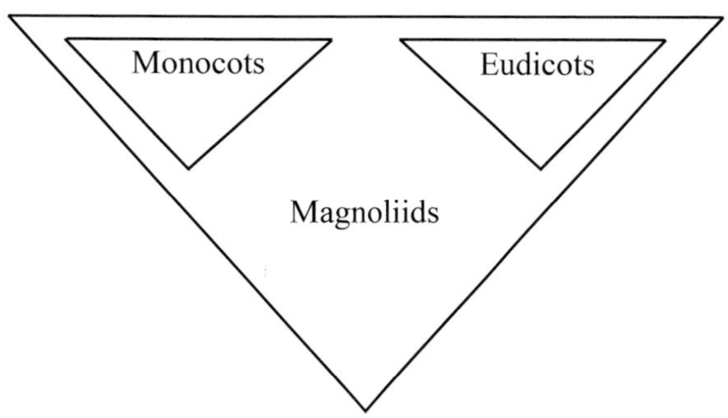

Figure 1.4 Cladistic relationship of the two major groups of angiosperms (monocots and eudicots), both monophyletic, whereas magnoliids *sensu lato* are paraphyletic, in effect the residue of angiosperms after monocots and eudicots have been removed. Magnoliids include the ANITA lineages, Chloranthaceae, *Ceratophyllum* and eumagnoliids (see Figure 1.5).

assessments based on morphological and molecular data identified a paraphyletic basal grade of dicots (broadly equivalent to subclass Magnoliidae of previous classifications such as that of Takhtajan, 1980) in which two major monophyletic groups, monocotyledons (monocots) and eudicotyledons (eudicots), were embedded (Figure 1.4). Eudicots (Chapters 12–15) are defined morphologically by the presence of triaperturate pollen (Donoghue and Doyle, 1989a, 1989b). Such pollen contrasts with the fundamentally monoaperturate pollen of magnoliids, monocots and some gymnosperms. Monocots (Chapter 11) are also strongly supported as a monophyletic group based on molecular data (Chase *et al.*, 2000), the single cotyledon and other features (Herendeen and Crane, 1995).

Although extant magnoliids account for only about 3% of living angiosperm species they are very diverse in habit, vegetative form and floral structure and biology (Chapter 10). Variation in the number and arrangement of floral parts is especially extreme (Endress, 1994b). Large, multipartite and bisexual flowers, small, simple and frequently unisexual forms, and a variety of other kinds of flowers, are widespread at this level of angiosperm evolution. Until recently, comparative studies of extant magnoliids had been impeded by poor knowledge of many of the key groups. However, as a result of recent detailed morphological and anatomical research these plants are now sufficiently well understood to incorporate into phylogenetic studies in a meaningful way (e.g. Endress, 1987, 1990; Endress and Hufford, 1989; Igersheim and Endress, 1997).

The recognition of magnoliids as a paraphyletic group simplifies and focuses attempts to precisely root the angiosperm tree, which is essential for clarifying both the patterns and processes of morphological evolution within the group. Extensive sampling of magnoliids for phylogenetic analyses based on molecular data has achieved good coverage of extant diversity at this level and has begun to produce consistent resolution of relationships (Figure 1.5). These analyses identify Amborellaceae, Nymphaeales and Austrobaileyales (including Austrobaileyaceae, Schisandraceae, Trimeniaceae) as the earliest diverging lineages at the base of the angiosperm phylogenetic tree (Qiu *et al.*,

1999, 2000; Soltis *et al.*, 1999, 2002; APGII, 2003; APGIII, 2009). This basal grade of angiosperms has been referred to as the ANITA grade; from *Amborella*, Nymphaeaceae, *Illicium* (Schisandraceae), Trimeniaceae and Austrobaileyaceae (Qiu *et al.*, 1999). The remaining magnoliids, except for Chloranthaceae and *Ceratophyllum*, form a monophyletic group composed of Magnoliales, Laurales, Piperales and Canellales (APGIII, 2009). This clade was termed eumagnoliids by Soltis *et al.* (2000b) and magnoliids by the APGII (2003). In this book, to avoid confusion, and to maintain the broader conventional use of the informal term magnoliid, we refer to the basal grade of angiosperms as the ANITA grade and to the Magnoliales, Laurales, Piperales and Canellales clade as eumagnoliids. The phylogenetic framework of angiosperm evolution that has been established based on molecular phylogenetics is briefly reviewed in Chapter 7.

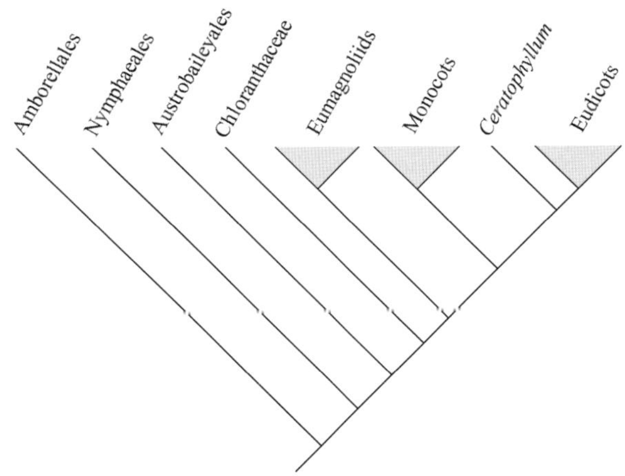

Figure 1.5 Cladistic relationships among major groups of angiosperms. Note the position of monocots, eudicots and eumagnoliids nested among ANITA grade plants (*Amborella*, Nymphaeales, and Austrobaileyales) and other magnoliids (Chloranthaceae and *Ceratophyllum*).

1.1.3 Origin of the angiosperm flower

Pre-cladistic hypotheses of flowering plant phylogeny were inextricably linked to concepts of the primitive angiosperm flower. They can be divided broadly into two contrasting sets of ideas: the Euanthial Theory, also known as the Ranalian Theory (Bessey, 1894, 1896, 1897, 1915; Hallier, 1900, 1901, 1902, 1912; Arber and Parkin, 1907), and the Pseudanthial Theory (Wettstein, 1907).

Under the Euanthial Theory the angiosperm flower is interpreted as a simple, bisexual, uniaxial system bearing spirally arranged lateral leaf-like appendages (bracts and sporophylls) potentially homologous with the strobilus of the Cycadales or Bennettitales (Chapter 6). According to the Euanthial Theory the insect-pollinated flowers of Magnoliaceae and related families are most similar to the angiosperm floral archetype, which would have had numerous helically arranged parts. More simple flowers, for example, those of Piperales, or wind-pollinated trees such as Betulaceae, Juglandaceae and Myricaceae, are considered to be derived from the basic *Magnolia*-type by simplification, reduction and fusion of parts.

Under the Pseudanthial Theory the angiosperm flower is interpreted as a compound, pluriaxial (multiaxial) structure potentially homologous to the cone of conifers or Gnetales (Chapter 6). According to this interpretation the angiosperm flower is composed fundamentally of a primary axis bearing secondary axes, with both orders of branching bearing lateral appendages. According to the classical Pseudanthial Theory the wind-pollinated flowers of angiosperms such as the Piperales, as well as those of the Betulaceae, Juglandaceae and Myricaceae, are most similar to the angiosperm floral archetype. In these plants the flowers are typically small, simple, unisexual and aggregated on elongated inflorescence axes. The gynoecium is unilocular and contains a single orthotropous or anatropous ovule. Large, bisexual, insect-pollinated flowers, such as those of *Magnolia* and its allies, are considered derived. They are interpreted as pseudanthia formed by the aggregation of unisexual floral units.

With the recognition that Betulaceae, Juglandaceae and Myricaceae occupy a relatively derived position within the angiosperm clade, many authors in the 1980s implicitly accepted the Euanthial Theory. However, since then the situation has become more complicated. Many early fossils, and several of the earliest diverging angiosperm lineages (e.g. Amborellaceae, Trimeniaceae, Hydatellaceae, Chloranthaceae), have flowers that are very simple. It seems unlikely that they can be accounted for simply by pervasive patterns of reduction from larger flowers with more numerous parts. A comprehensive theory of the origin and early evolution of angiosperm flowers, which accounts for all the relevant variation is yet to emerge. Identifying the likely basic condition for floral features in angiosperms, and establishing their homology with reproductive structures of other seed plants, will require integration of a fully resolved pattern of relationships at the magnoliid grade with significantly improved information on early angiosperm fossils and their close relatives among other seed plants.

1.2 CHARACTERISTIC FEATURES OF ANGIOSPERMS

There is currently no support from phylogenetic analyses based on morphological or molecular data for angiosperm polyphyly. The extraordinary diversity of angiosperms, and especially the bewildering variety among their flowers, has sometimes led to the suggestion that they are polyphyletic. However, a clear suite of characters (autapomorphies) unites angiosperms as a well-defined monophyletic group and distinguishes them from other seed plants. Several other potential defining features are not universally present in the group, for example vessel elements with scalariform or simple perforation plates (section 1.2.1), and tectate columellate pollen (section 1.2.2). Full resolution of relationships at the magnoliid grade based on both morphological and molecular data, and also including fossils, will be necessary to clarify the precise pattern of evolution of these features. Other characteristic angiosperm features, such as broad leaves with reticulate, open venation, or ovules with more than one envelope surrounding the nucellus, also occur in non-angiospermous plants.

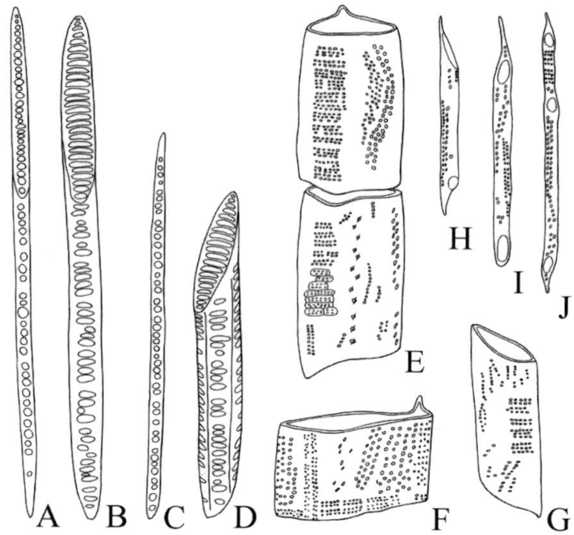

Figure 1.6 Tracheids and vessels of angiosperms. (A, B) Tracheids from a vessel-less angiosperm showing circular pits (A) and scalariform pits (B). (C) Tracheid from a vessel-bearing angiosperm showing circular pits. (D) Vessel element with scalariform perforation plate. (E–G) Wide vessel elements with simple perforation plates: note two continuous vessel elements in (E). (H–J) Narrow vessel elements with simple perforation plates. (A–D) redrawn from Carlquist (1988); (E–J) redrawn from Esau (1977).

1.2.1 Vegetative features

Vascular system. Two principal types of stem vascular system are present in angiosperms. In most dicots the stems have longitudinal strands of primary vascular tissue arranged in a ring around the central pith to form a eustele. This is the basic condition in most seed plants. In most woody seed plants, including most dicots, a cambium, which produces secondary xylem to the inside and secondary phloem to the outside, develops between the primary xylem and the primary phloem and extends between the bundles to form a continuous cylinder producing secondary tissue. However, in herbaceous dicots cambial development is often partially or completely suppressed. In monocots, and in a few dicots, vascular bundles are scattered throughout the stem to form an atactostele. A cambial zone may develop in each bundle early in its development, but all trace of it is usually lost at maturity (Sporne, 1974).

Monocots do not produce typical wood (secondary xylem), but in some groups (e.g. many palms) a different form of secondary growth occurs in which a zone of dividing cells (primary thickening meristem) develops in the parenchyma tissue beneath the apical meristem. In other monocots, a zone of dividing cells forms a thickening ring that produces new parenchyma as well as secondary bundles. In the context of seed plants as a whole it is clear that in monocots the capacity to produce typical secondary tissues has been lost, apparently in an irreversible way.

The cellular structure of angiosperm vascular tissue is significantly modified compared with that in other seed plants. In the phloem, a unique feature of angiosperms, seen in both dicots and monocots, is that the companion cells accompanying the sieve tubes are typically derived developmentally from the same initial cell as the sieve elements. In the xylem, a characteristic modification seen in almost all angiosperms is the presence of vessels composed of vessel elements (Figure 1.6), which are connected by scalariform or simple perforation plates to form long tubes.

In some highly specialised groups (e.g. certain Cactaceae and aquatic plants) secondary loss of vessels has clearly occurred. However, there are also vessel-less taxa among the ANITA grade angiosperms (Amborellaceae) and eumagnoliids (Winteraceae), as well as among the earliest diverging groups of eudicots (Trochodendraceae). In these cases it is more difficult to decide whether the lack of vessels is plesiomorphic or reflects secondary loss. The absence of vessels in

Amborella is especially interesting given its position as the sister group to all other angiosperms in phylogenetic analyses based on molecular data.

Arguments based on parsimony analyses of basal angiosperms (prior to the recognition of the ANITA grade) suggest that vessels in angiosperms originated only once, and that the vessel-less condition is secondary (Donoghue and Doyle, 1989b). However, based on functional considerations, Carlquist (1996a) argued that vessels originated several times, and there is also the complication that the distinction between tracheids and vessel elements is not always straightforward based on light microscope studies alone (Carlquist and Schneider, 2002). Whichever interpretation is correct, it is clear that vessel elements evolved from tracheids. Vessel elements differ from tracheids in having perforated end walls (perforation plates) and are also generally wider in diameter. Among angiosperms, ontogenetic and phylogenetic studies indicate a broad evolutionary trend from long narrow vessel elements with long oblique scalariform perforation plates to short, broad elements with transverse, simple perforation plates. The changes associated with this general trend occurred repeatedly within many separate lineages (Carlquist, 1996a).

In non-angiosperm seed plants, vessels are known with certainty only in the Gnetales. Vessels in the Gnetales differ from those of angiosperms in having foraminate perforation plates with circular perforations, and thus the phylogenetic significance of this character as a link to angiosperms, where scalariform perforation plates are thought to be basic, is uncertain (Muhammad and Sattler, 1982; Carlquist, 1996b). On the whole, differences in vessel element and tracheid characters support an independent origin of the vessels in the two groups. In particular, vessels of Gnetales often have pits with a thick central torus, suspended by margo strands; a feature that is unknown in the angiosperms (Carlquist, 1996b). Vessel-element-like cells have been observed in Permian wood from China of possible gigantopteridalean affinity (Li *et al.*, 1996) and have also been reported from the leaf bundles of a bennettitalean plant (Krassilov, 1984). The significance of these discoveries remains uncertain.

Leaves. Angiosperm leaves show great variety in form, ranging from the minute scale-like leaves of *Casuarina* and the linear leaves of grasses, to the giant peltate floating leaves of *Victoria* (Nymphaeaceae) and the large oblong leaves of *Musa* (Musaceae). Despite this diversity, Hickey (1978) recognised four features typical of most angiosperm leaves: intercalary growth during most of the leaf blade extension; a hierarchical system of successively thinner veins (usually three or more vein orders); freely ending veinlets; and vein anastomoses between two or more orders of veins to form a reticulate pattern (Figure 1.7). Hickey and Wolfe (1975) also consider stipules as typical of angiosperms although they are not common among monocots.

In monocots the main veins are usually more or less parallel, and the leaf is differentiated into a sheath and a blade. Some monocots (e.g. *Dioscorea*) have distinctly petiolate leaves with an expanded blade and more complex venation. In dicots the main veins are typically arranged in a pinnate or palmate pattern, and the leaf is differentiated into a petiole and a blade. Compound leaves are also common in many groups of dicots. Leaves with reticulate venation, sometimes with freely ending veinlets, occur scattered in non-angiosperm plants, for example in dipteridaceous ferns (Figure 1.8) and the enigmatic Palaeozoic Gigantopteridales (Figure 1.9), but outside angiosperms, among living seed plants, broad, petiolate leaves with several discrete orders of venation, freely ending veinlets, and a reticulate pattern of venation occur only in Gnetales (*Gnetum*, Figure 1.9).

Stomatal anatomy can be used to distinguish different groups among seed plants. Within angiosperms two main stomatal types, as well as more complex variations, have been recognised (Fryns-Claessens and Van Cotthem, 1973). The most common developmental pattern is the mesogenous, or syndetocheilic, type, in which guard cells and their adjacent cells are formed from the same epidermal initial cell. Cells adjacent to the guard cells may be undifferentiated, but typically they are readily distinguished from the surrounding epidermal cells as two subsidiary cells that are parallel to the guard cells. The perigenous, or haplocheilic, pattern of stomatal development consists of guard cells and adjacent cells (undifferentiated or differentiated into subsidiary cells) that are formed from different epidermal initials. This developmental pattern is widely distributed among monocots (Tomlinson, 1974).

Among non-angiosperm seed plants, mesogenous stomata (inferred from the arrangement of the subsidiary cells parallel to the guard cells: paracytic) are known in extant Gnetales (*Gnetum*, *Welwitschia*), and also in the extinct Bennettitales and Erdtmanithecales. This is one of the features that has been used to define the anthophyte clade (Chapter 5). Stomatal characters have not contributed substantially to establishing large-scale phylogenetic patterns among angiosperms, but have proved useful in systematically more restricted studies (e.g. Baranova, 1972).

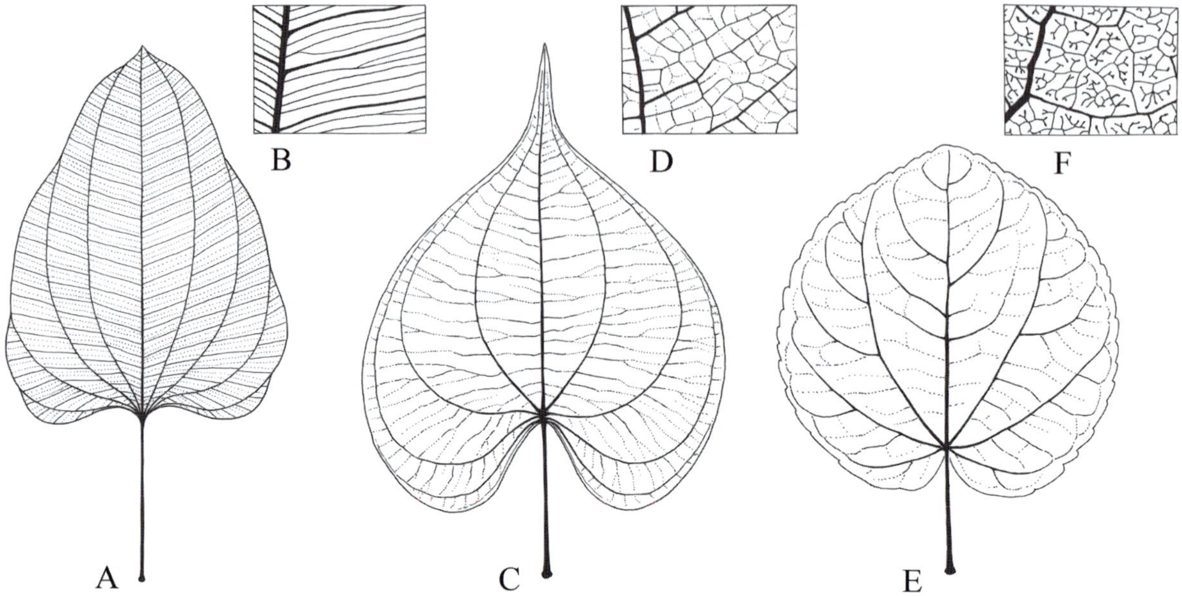

Figure 1.7 Leaves in angiosperms. (A, B) Leaf (A) and detail of venation (B) in *Stemona cochinchinensis* (Stemonaceae, monocot). (C, D) Leaf (C) and detail of venation (D) in *Dioscorea hemsleyi* (Dioscoreaceae, monocot). (E, F) Leaf (E) and detail of venation (F) in *Cercidiphyllum japonicum* (Cercidiphyllaceae, eudicot). Based on herbarium specimens in the Swedish Museum of Natural History.

Stomatal details have also been widely used in studies of fossil angiosperm leaves, especially in comparing fossil leaves with those of living plants (e.g. Dilcher, 1974).

1.2.2 Reproductive features

Angiospermy. Most of the unique defining characters of angiosperms are related to their characteristic and complex reproductive system. The most important feature, which has also given its name to the group, is angiospermy: the enclosure of the ovules in carpels. Endress and Igersheim (2000a) showed that in many angiosperms at the ANITA grade, which now appear to be the earliest diverging lineages in the group, carpel closure is achieved solely by secretion and not by postgenital fusion of the carpel wall itself. They also showed that a combination of secretion and postgenital fusion in the closure of the carpels is common among other early diverging angiosperm lineages.

Endress and Igersheim (2000a) recognise four different types of angiospermy among magnoliids (Figure 1.10): Type 1 carpel closure formed by secretion only, without any postgenital fusion of the carpel wall, as seen in Amborellaceae, some Nymphaeales (e.g. *Cabomba*, *Brasenia*), Trimeniaceae, Schisandraceae and Austrobaileyaceae at the ANITA grade and also in Chloranthaceae; Type 2 carpel closure formed partly by secretion in the centre and partly by postgenital fusion in the periphery leaving a completely unfused canal up to the stigma, as seen in Schisandraceae, Annonaceae, Myristicaceae and Canellaceae; Type 3 carpel closure formed by postgenital fusion over the entire periphery of the carpel wall with secretion in an inner unfused canal that ends below the stigma, as in many Magnoliaceae; and Type 4 carpel closure by complete postgenital fusion of the carpel combined with a pollen-tube-transmitting tissue, as seen for example in Degeneriaceae, Laurales, and some Winteraceae among magnoliids, as well as in most monocots and eudicots.

The closed carpel protects the developing ovules and in most cases also protects the seeds during their development. Enclosure also excludes the pollen grains from direct contact with the ovule. Instead of germinating inside the micropyle, which is typical of non-angiosperm seed plants, pollen grains germinate on a specialised portion of the carpel: the stigma (Figure 1.11). Pollen tubes then grow from the stigma through the pollen-transmitting tract of the carpel and into the embryo sac within the ovule, where the male gametes are released. Growth of the pollen tube through the sporophytic carpellary

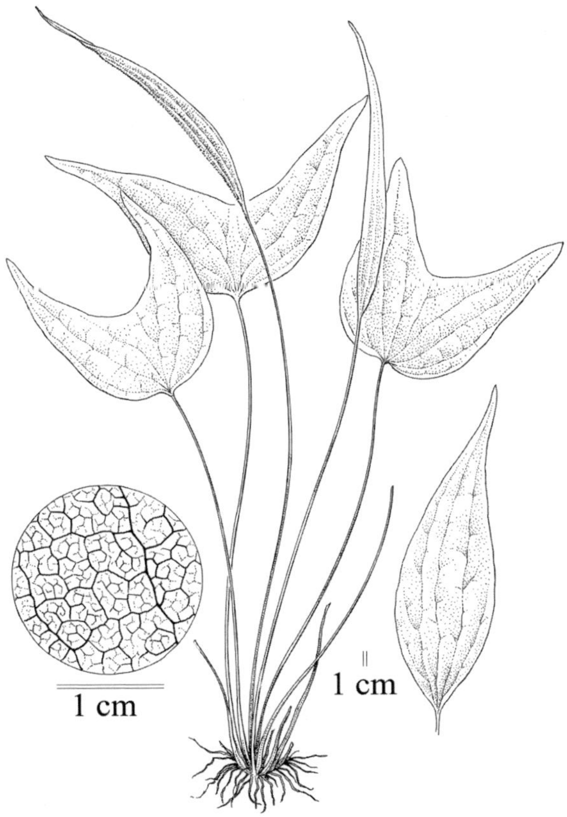

Figure 1.8 *Cheiropleuria bicuspis*, an extant fern (Dipteridaceae) with reticulate venation and freely ending veinlets. Based on herbarium specimens in the Swedish Museum of Natural History.

tissue, or through a secretion from this sporophytic tissue, creates the possibility for sophisticated interactions between gametophytes and sporophytes, and in some cases the growth of pollen with a particular genetic profile (e.g. pollen from the parent plant) may be blocked. Such mechanisms can help to ensure outbreeding, and to the extent that this generates additional variability or facilitates reproductive isolation it may be an important factor in the evolutionary success of angiosperms (Heslop-Harrison, 1983). Male gametophyte competition, expressed by differences in growth rates among the pollen tubes, may also be another advantageous trait made possible by the closed carpel (Mulcahy, 1979). The formation of syncarpous ovaries with a common stigmatic area from which pollen tubes may access ovules in different carpels may further enhance pollen competition (Endress, 1982).

Flowers. The angiosperm flower is formed by female organs (carpels) and male organs (stamens) that are often surrounded by a perianth (Figures 1.12, 1.13). The perianth may be undifferentiated, consisting of parts that are all of one kind (tepals), or it may be differentiated into an outer zone of sepals forming the calyx, and an inner zone of petals forming the corolla. Most angiosperm flowers are bisexual (hermaphroditic) with both carpels and stamens in the same flower. However, unisexual flowers, in which the stamens and carpels are separated into staminate and pistillate flowers, occur in many different angiosperm lineages. In bisexual flowers the stamens always occur below or around the carpels. *Lacandonia* (Triuridaceae) is exceptional in that this pattern is apparently reversed, with stamens occurring in the centre of the flower and surrounded by carpels (but see also Rudall and Bateman, 2010).

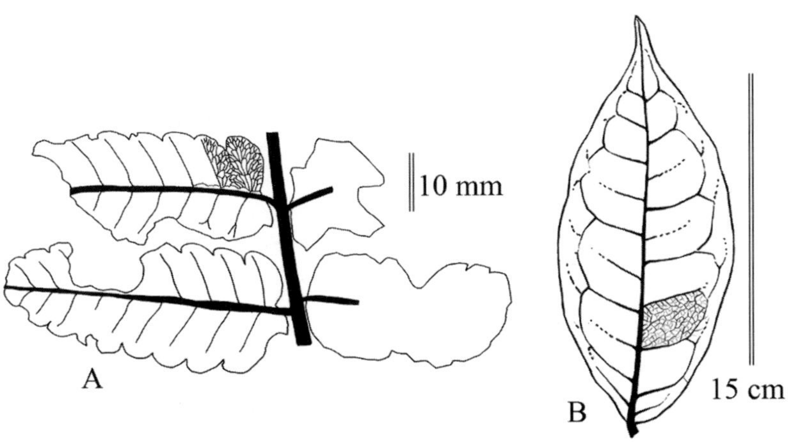

Figure 1.9 (A) Leaf of *Gigantonoclea lagrelii* (Gigantopteridales), a Palaeozoic plant (probable seed plant) with reticulate venation. (B) Leaf of extant *Gnetum gnemon* (Gnetales) with reticulate venation and freely ending veinlets. (A) redrawn from Glasspool *et al.* (2004); (B) based on herbarium specimens in the Swedish Museum of Natural History.

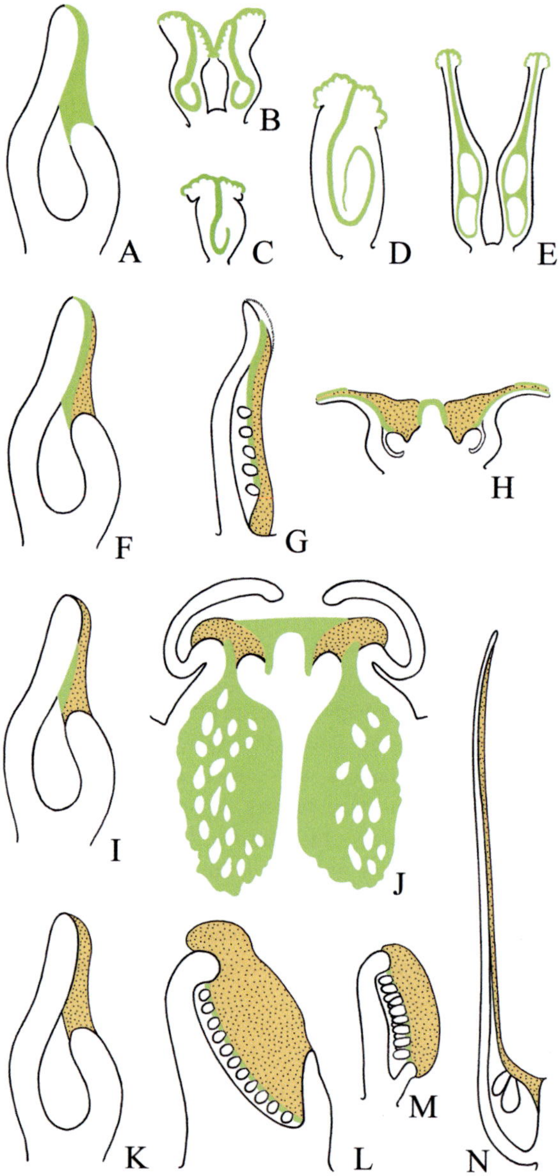

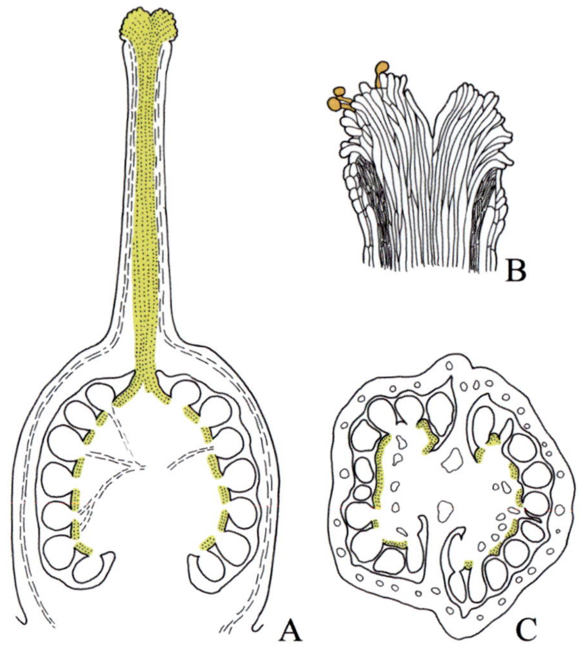

Figure 1.11 Pollen–carpel interactions in angiosperms (*Lycopersicon esculentum*, Solanaceae). (A) Longitudinal section of ovary showing location of pollen-tube-transmitting tissue (yellow stippling) through which the pollen tubes grow *en route* to the ovules. (B) Longitudinal section of apical part of style showing pollen grains on the stigmatic surface, where they germinate and each produces a single pollen tube; pollen grain and pollen tube shown in orange. (C) Transverse section of ovary with two locules, showing location of pollen-tube-transmitting tissue (yellow stippling) adjacent to the locules in which the ovules occur. Redrawn from Endress (1994b).

Figure 1.10 The four main types of angiospermy (carpel closure) among magnoliid angiosperms, illustrated in median longitudinal section. (A–E) Angiospermy type 1 closure formed by secretion; schematic section (A) with ovules omitted; *Amborella*, Amborellaceae (B); *Chloranthus*, Chloranthaceae (C); *Trimenia*, Trimeniaceae (D); *Cabomba*, Nymphaeaceae (E). (F–H) Angiospermy type 2 closure formed partly by secretion in a continuous internal canal and partly by postgenital fusion in the periphery; schematic section (F) with ovules omitted; *Asimina*, Annonaceae (G); *Illicium*, Schisandraceae (H). (I, J) Angiospermy type 3 closure formed by a combination of types 1 and 2 with fusion in the periphery of the carpel wall and secretion in the inner parts, but without a continuous canal; schematic section (I) with ovules omitted; *Nymphaea*, Nymphaeaceae (J). (K–N) Angiospermy type 4 closure formed entirely by postgenital fusion; schematic section (K) with ovules omitted; *Degeneria*, Degeneriaceae (L); *Tasmannia*, Winteraceae (M); *Liriodendron*, Magnoliaceae (N). Green areas show secretion; orange areas show postgenital fusion. Based on Endress and Igersheim (2000a).

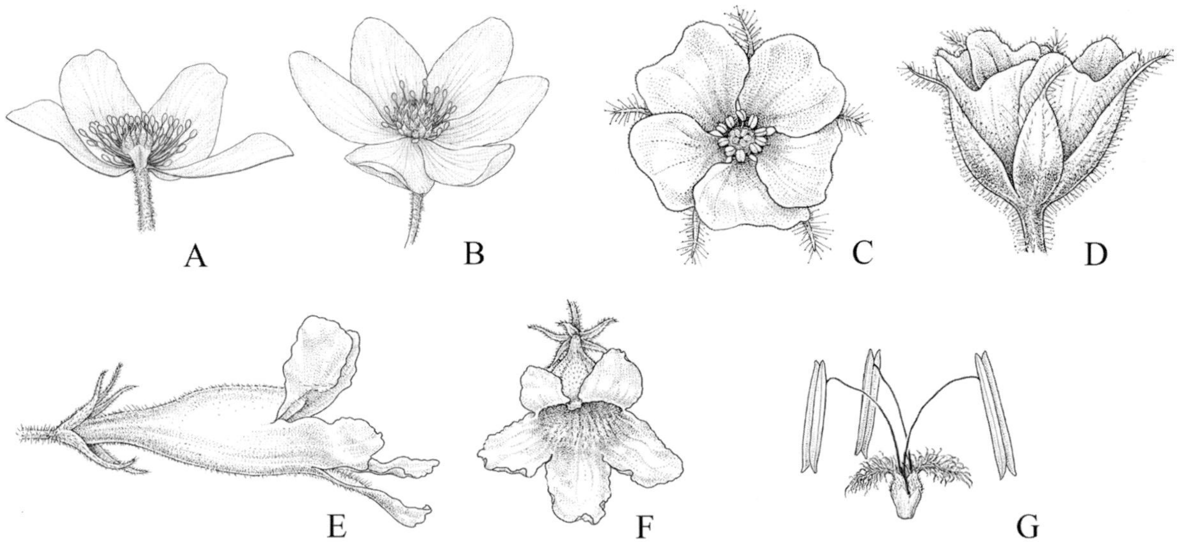

Figure 1.12 The diversity of flowers in angiosperms. (A, B) Section and oblique view of open, actinomorphic flower with undifferentiated perianth and exposed carpels and stamens (eudicot, insect-pollinated). (C, D) Apical and lateral view of open, actinomorphic flower with prominent calyx and corolla (rosid, insect-pollinated). (E, F) Lateral and apical view of zygomorphic flower with carpels and stamens hidden within the corolla tube (asterid, insect-pollinated). (G) Grass flower with reduced perianth, long stamen filaments, exposed anthers and two feathery stigmas (monocot, wind-pollinated).

The type and number of floral organs present in a flower, and the details of their structure and function, varies considerably among different groups of angiosperms. This extreme variation reflects phylogeny as well as ecology (Endress, 1994b) and has been of considerable importance in angiosperm classification since Linnaeus (1735, 1758). It is also of great significance for pollination and other aspects of angiosperm reproductive biology.

Among extant seed plants bisexual reproductive structures occur only in angiosperms, but were also present in some extinct Bennettitales (Chapter 5). Male flowers of *Welwitschia* (Gnetales) also have a non-functional ovule inside the whorl of pollen-producing organs.

Number of floral parts and floral symmetry. There is great variation in number of floral parts in the flowers of different groups of angiosperms and this variation is especially marked among magnoliids (Endress, 1994b). In general, however, flowers with parts in threes, or in simple multiples of three, are especially common among monocots, magnoliids and early-diverging lineages of eudicots. Flowers with sepals, petals and stamens in fives, or simple multiples of five, and a gynoecium consisting of two carpels, are characteristic of most eudicots. Dimerous flowers are also widespread among early-diverging lineages of eudicots (Drinnan *et al.*, 1994).

Radial symmetry (actinomorphic) is the most common condition for angiosperm flowers (Figures 1.12, 1.13), and is widespread among flowers of basal angiosperms, as well as many monocots and eudicots. In many derived dicots and monocots there are also other symmetry patterns often associated with specialised pollination systems, of which monosymmetry (zygomorphy) is the most common (Figures 1.12, 1.13). This predominates in several large and very diverse groups such as legumes and orchids.

Perianth. In most angiosperm flowers the outermost organs (sepals), collectively termed the calyx, are green and photosynthetic. They protect the flower during development in the bud and are generally interpreted as modified bracts or leaves (Endress, 1994b). The organs internal to the calyx (petals), collectively termed the corolla, are frequently larger, showy and coloured. Petals are generally involved in the attraction of pollinators. There is morphological and ontogenetic evidence that most petals are modified stamens (Endress, 1994b). In many flowers, especially among magnoliids, there is no modification of the perianth into a distinct calyx and

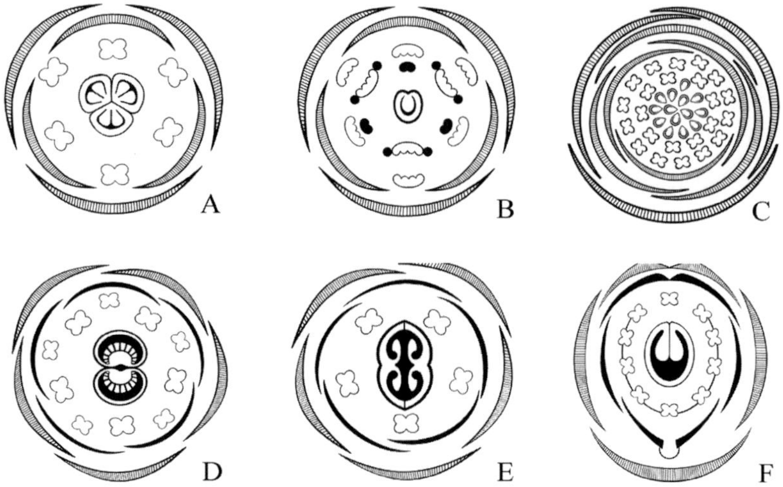

Figure 1.13 The diversity of floral architecture in angiosperms summarised in floral diagrams showing variation in the number and arrangement of floral parts. Sepals and tepals shaded; petals solid. (A) Actinomorphic, trimerous flower of Cabombaceae (ANITA grade angiosperm) with syncarpous gynoecium of three carpels. (B) Actinomorphic trimerous flower of Lauraceae (eumagnoliid) with gynoecium consisting of a single carpel; staminal glands and staminodes in black. (C) Multipartite flower of Magnoliaceae (eumagnoliid) with an apocarpous gynoecium composed of numerous carpels. (D) Actinomorphic, pentamerous flower of Saxifragaceae (core eudicot) with a bicarpellate, syncarpous gynoecium. (E) Actinomorphic, pentamerous flower of Pittosporaceae (core eudicot) with a bicarpellate, syncarpous, unilocular gynoecium. (F) Zygomorphic, pentamerous flower of Fabaceae (core eudicot) with a monocarpellate gynoecium; note fusion of parts in the calyx, corolla and androecium.

corolla and the individual perianth elements are termed tepals. In relatively derived groups of eudicots, and in many other groups of angiosperms, the calyx, corolla, or both are often fused into a tube (synsepalous, sympetalous) and this is frequently associated with specialised insect pollination. In these sympetalous plants the stamens may be fused to the corolla tube.

Androecium. Stamens are the pollen-producing organs of the flower and together with sterile stamens (staminodes) are collectively termed the androecium. They are a characteristic angiosperm feature and their homology with the pollen-producing structures of other seed plants is not well understood (Chapter 6).

Stamens are rather uniform in construction in all angiosperms (Endress, 1996b), and typically they are clearly differentiated into a fertile anther and a sterile filament (Figure 1.14). Frequently the filament is thread-like, but it may also be swollen, flattened (laminar) or modified in other ways. It is diagnostic of angiosperms that the anther is tetrasporangiate, consisting of four sporangia (pollen sacs) arranged in two lateral pairs, the thecae. Sterile tissue (the connective) is usually developed between the thecae and may expand above the anther to form a sterile protrusion. The presence of a sterile protrusion is particularly common among the stamens of magnoliids and early-diverging lineages of eudicots (Endress and Hufford, 1989; Hufford and Endress, 1989). Dehiscence of the anthers is usually by longitudinal slits, or more rarely by pores or valves. Valvate dehiscence is often developed in stamens in which the thecae are embedded in the connective and where there is a connective protrusion. Valvate dehiscence is also generally associated with insect pollination (Endress, 1996b).

Pollen. Pollen grains are produced in the pollen sacs in groups of four by the meiotic division of the pollen mother cells. In the prepollen of extinct early seed plants from the Palaeozoic, the aperture is on the proximal surface with respect to the tetrad in which the grains were formed, as in the spores of lycopsids and ferns. However, the pollen of most non-angiosperm seed plants is characterised by a single germination aperture (monoaperturate) that may be slit-like (monocolpate/monosulcate) or pore-like (monoporate), and that is located at the distal pole. More rarely, pollen grains of both angiosperms and non-angiosperms may be inaperturate.

Pollen grains of angiosperms are more diverse in aperture configuration and wall structure than those of other seed plants. The only character recognised so far that

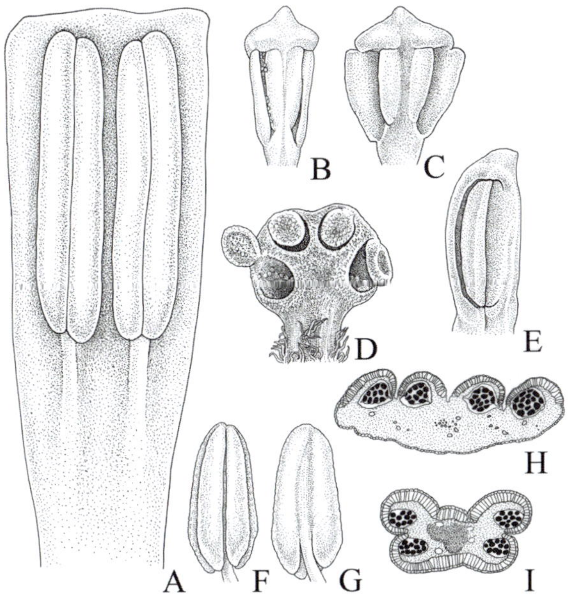

Figure 1.14 Angiosperm stamens. (A) *Nuphar lutea* (Nymphaeaceae), ventral view of laminar stamen showing four protruding pollen sacs (tetrasporangiate) in two pairs (dithecate). (B, C) Stamens of *Artabotrys uncinatus* (Annonaceae) showing a short filament that grades into the tetrasporangiate anther without a joint, valvate dehiscence of the anther, and a massive apical expansion of the connective; (B) indehisced stamen, (C) dehisced stamen showing each theca opening by two longitudinal valves. (D) *Cinnamomum camphora* (Lauraceae), lateral view of stamen showing two thecae; each of the four pollen sacs dehisces by a single flap-like valve. (E) *Dicoryphe viticoides* (Hamamelidaceae), lateral view of stamen showing one theca dehiscing by a single longitudinal valve. (F, G) Stamen of *Genista pilosa* (Fabaceae) showing ventral (F) and dorsal (G) view of tetrasporangiate anther and slender filament; dehiscence is by longitudinal slits. (H) *Talauma singapurensis* (Magnoliaceae), cross-section of laminar stamen showing four protruding pollen sacs on ventral surface. I. *Magnolia denudata* (Magnoliaceae), cross-section of anther showing four pollen sacs in two lateral thecae separated by extensive sterile tissue. (A) drawn from SEM-micrograph in Hufford (1996); (B) drawn from SEM-micrograph in Endress and Hufford (1989); (C–E) drawn from SEM-micrographs in Endress (1994b); (F, G) drawn from SEM-micrographs in Endress and Stumpf (1991); (H–I) drawn from LM-micrographs in Canright (1952).

appears to unite angiosperm pollen, and distinguish it from pollen of other seed plants, is the lack of a distinct laminated endexine. In non-angiosperm seed plants this layer is well developed, both under the apertures and in non-apertural areas of the grains. In angiosperms the endexine is generally better developed under the aperture, but over most of the grain it is typically thin and granular to faintly laminar, and may even be absent (e.g. Foreman and Sampson, 1987). The ektexine of most angiosperms is stratified and consists of three layers (Figure 1.15). The outer layer (tectum) may form a solid roof, may have openings of various sizes (e.g. foveolate or reticulate), or may be absent or strongly reduced (atectate). The middle layer (infratectal layer) is composed of columellae or granulae, and the inner layer (pedium, foot layer) is solid. Among living plants, pollen grains with a reticulate tectum and columellate infratectal layer occur only in angiosperms and this feature is widely used to recognise dispersed fossil pollen grains as being of probable angiosperm affinity (Chapter 6).

Monoaperturate pollen grains are characteristic of most groups of living seed plants (e.g. cycads, *Ginkgo*), but among angiosperms they are typical of only magnoliids and monocots (Figure 1.15). Pollen grains of eudicot angiosperms are distinguished from those of all other seed plants by the presence of three or more apertures that are generally equally distributed around the equator in a longitudinal orientation (tricolpate, tricolporate, triporate) (Figure 1.15). Some monocots (e.g. *Areca* and *Sclerosperma*, Arecaceae) (Harley, 1997) also have radially symmetrical triporate pollen with apertures arranged in an equatorial or subequatorial position, and thus the higher-level systematic affinities of fossil dispersed triporate pollen grains may not always be easy to establish. Inaperturate pollen grains occur in several different lineages among magnoliids and monocots, and in rare cases also within eudicots. Pollen grains with many pores in an equatorial or global arrangement occur sporadically in many different groups of magnoliids, monocots and eudicots and appear to have evolved many times and in several different ways.

Gynoecium. The female organ of the flower, the gynoecium, is a unique feature of angiosperms. It is differentiated into an ovary that encloses the ovules (angiospermy) and a stigma that receives the pollen grains. Typically there is also a style that raises the stigma above the ovary. The gynoecium is formed from one or more carpels. The origin of the carpel and its homology with structures in other seed plants is a key unresolved question in angiosperm evolution (Chapter 6).

Two main types of carpel development are sometimes distinguished: ascidiate (peltate) and plicate (epeltate). However, the situation is more complex when the full range of developmental sequences exhibited by angiosperms are considered. In early carpel development ascidiate carpels are urn-shaped, but with one side of the urn (generally

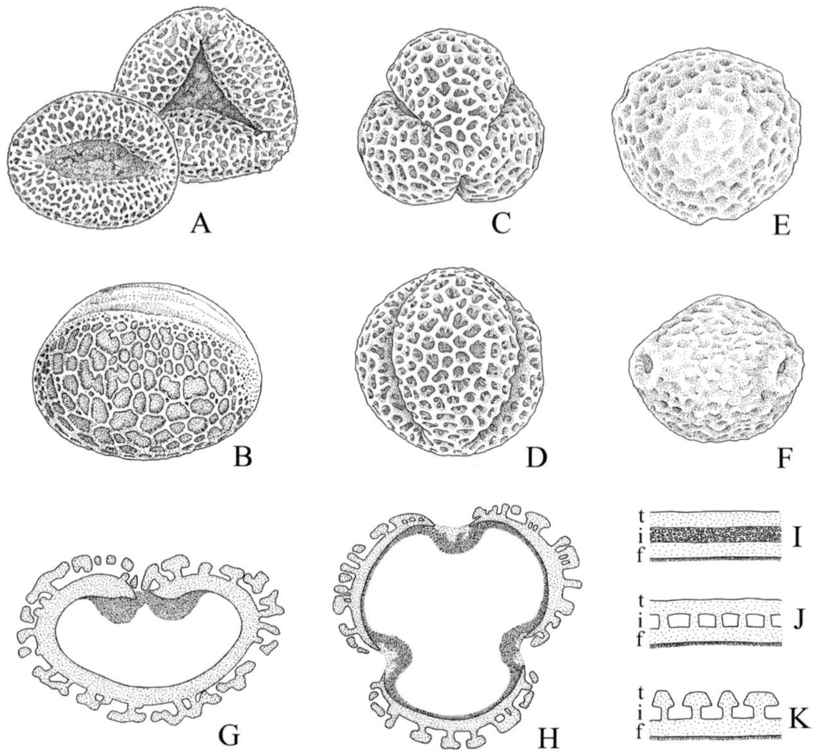

Figure 1.15 Angiosperm pollen. (A, B, G) Monoaperturate pollen of magnoliids and monocots. (C–F, H) Triaperturate pollen of eudicots. (A) Pollen grains of *Ascarina* (Chloranthaceae), polar view (distal) showing reticulate tectum; one specimen with a simple colpus and one specimen with a triradiate colpus (trichotomocolpate). (B) *Puya ferruginea* (Bromeliaceae), equatorial view of grain showing coarsely reticulate tectum and a long simple colpus at distal pole. (C, D) *Euphorbia fulgens* (Euphorbiaceae), tricolpate grains seen in polar (C) and equatorial view (D). (E, F) *Zelkova abelicea* (Ulmaceae), triporate grains seen in polar (E) and equatorial view (F). (G) Cross-section of monoaperturate pollen (fossil Chloranthaceae) through the polar axis showing ektexine (lighter layer) and endexine (darker layer); endexine thick and restricted to the aperture region. (H) Cross-section of tricolpate grain (fossil Platanaceae) through the equatorial axis showing ektexine (lighter layer) and endexine (darker layer); endexine thick and not restricted to aperture regions. (I–K) Three common types of pollen wall stratification showing variation in the architecture of the ektexine, endexine (below) shaded: (I) tectate–granular type with solid tectum and foot layer separated by a granular infratectal layer; (J) tectate–columellate type with solid tectum and foot layer separated by a columellar infratectal layer; (K) semi-tectate–columellate type with perforated tectum and solid foot layer separated by a columellar infratectal layer; t, tectum; i, infratectal layer; f, foot layer. (A, G, H) drawn from material in the collections of the Swedish Museum of Natural History; (B) drawn from SEM-micrograph in Halbritter and Hesse (1993); (C, D) drawn from SEM-micrographs in El-Ghazaly and Chaudhary (1993); (E, F) drawn from SEM-micrographs in Stafford (1995).

dorsal) better developed than the other. In non-ascidiate carpels the primordium is initially C-shaped, but further growth often results in a form resembling an infolded leaf blade. Ascidiate early carpel development predominates among ANITA-grade angiosperms, but also occurs in many other groups. According to Endress and Igersheim (2000a) there is no clear evolutionary trend in carpel form among the early lineages of angiosperms.

Carpels may be free (apocarpous) or fused (syncarpous). The syncarpous condition is the most widespread in angiosperms and occurs in more than 80% of extant angiosperm species (Endress, 1982). Apocarpous gynoecia are particularly common among magnoliids and monocots as well as early-diverging lineages of eudicots (e.g. Dahlgren *et al.*, 1985; Endress, 1994b; Igersheim and Endress, 1997). However, apocarpous gynoecia also occur sporadically in more derived groups.

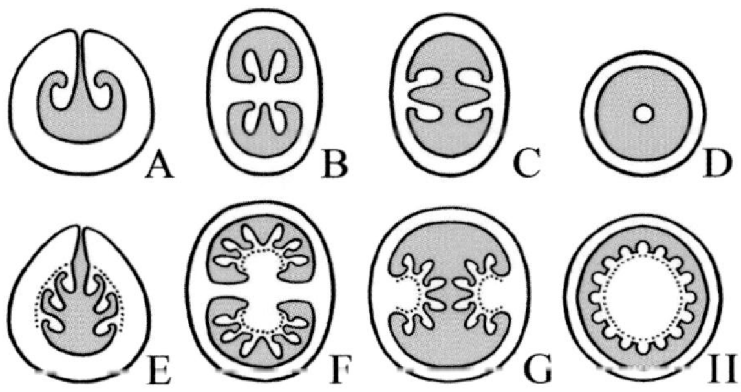

Figure 1.16 Main types of placentation in apocarpous (A, E) and syncarpous (B–D, F–H) ovaries as seen in transverse section. (A) Placentae linear, ovules in two series, one on either side of the ventral suture. (B) Placentae axile, ovules in two series. (C) Placentae parietal, ovules in two series. (D) Placenta free central with a single ovule. (E) Placentae laminar-diffuse, ovules irregularly arranged or in several series. (F) Placentae axile, ovules irregularly arranged or in several series on inner surface of the carpel. (G) Placentae parietal, ovules irregularly arranged or in several series. (H) Placentae free, central ovules irregularly arranged or in several series. Redrawn from Endress (1994b).

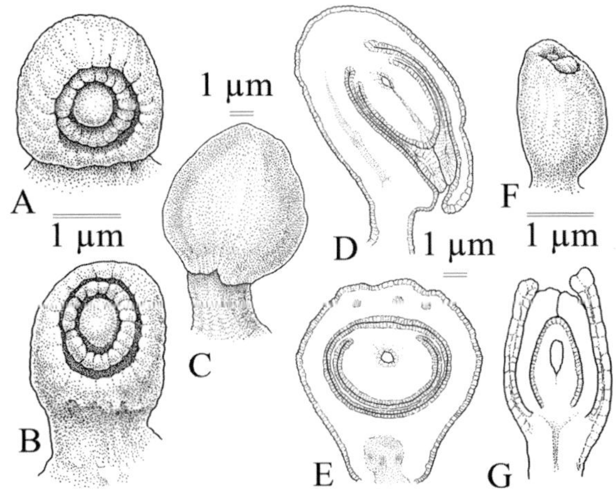

Figure 1.17 Angiosperm ovules. (A, B) Young bitegmic, anatropous ovule showing initiation of the two integuments in different views (*Canella alba*, Canellaceae). (C) Young bitegmic, anatropous ovule showing outer, lobed, integument and distinct funicle (*Liriodendron tulipifera*, Magnoliaceae). (D, E) Longitudinal (D) and transverse section (E) of bitegmic, anatropous ovule (*Magnolia officinalis*, Magnoliaceae). (F, G) Bitegmic, orthotropous ovule (*Saururus cernuus*, Saururaceae) in surface view (F) and longitudinal section (G). (A–E) drawn from Igersheim and Endress (1997); (F, G) drawn from Igersheim and Endress (1998).

The ovary may have a single cavity (locule) that contains the ovule, but frequently it is divided by partitions (septae) into several locules (Figure 1.16). The ovules are borne on one or more placentae within the locule or locules, and where there are many ovules the placentae may be linear or diffuse. The most common ovule type in angiosperms, including among the earliest diverging extant lineages, is anatropous and bitegmic (Figure 1.17). This is most likely the basic condition for angiosperms as a whole. The nucellus is generally prominent and well developed (crassinucellar). In a few angiosperms, the ovules are orthotropous (Figure 1.17). The number of integuments may also be reduced to one, or rarely three integuments may be present. The nucellus may also be poorly developed (tenuinucellar).

In most non-angiospermous seed plants ovules are orthotropous, but recurved ovules, similar to the anatropous condition, also occur in some conifers. Most non-angiosperm seed plants have a single integument, but in Gnetales and some conifers there are additional envelopes surrounding the nucellus. Studies of ovule development show that the additional envelopes of Gnetales do not develop in the same way as the angiosperm integument (Endress, 1996a), but the homologies of these additional envelopes with structures in angiosperms remain uncertain.

Pollination. Insects and a variety of other animals play an important role in pollen transfer in angiosperms, and it is generally believed that the bisexual angiosperm flower, as well as the enormous floral diversity exhibited by the group, is intimately linked to co-evolutionary interactions between angiosperms and their pollinators. Insect pollination does occur in other seed plants, but it is only in angiosperms that it is exploited so consistently, and in conjunction with such diverse and elaborate modifications of the reproductive structures (Chapter 17).

The bisexual flower that is characteristic of angiosperms may have evolved by facilitating greater efficiency in pollination, including the more efficient utilisation of insects, by ensuring that pollen may be deposited on the stigma of a flower at the same time as a new load of pollen is collected, and that only one set of protective and attractive organs (sepals, petals, nectaries) is required. Many of the specific structural and other features of flowers appear to operate to minimise the probability of self-pollination, while maximising the probability of outcrossing. Separating the staminate and pistillate flowers on different parts of the same individual is one obvious mechanism (monoecy, dicliny), but more often these mechanisms involve structural or developmental modifications within the same flower, including, most commonly, maturation of the pollen- and ovule-producing organs at different times (dichogamy). In extreme cases staminate and pistillate flowers may be separated onto different individuals (dioecy).

Gametophytes. The male gametophyte of angiosperms consists of only two or three cells, whereas in other seed plants it typically contains four, five or six, or occasionally as many as 40, cells (Friedman, 1993). Following pollination the male gametophyte germinates to produce a pollen tube that grows through the carpellary tissue or its secretions. Male gametes pass through the pollen tube to the ovule that contains the female gametophyte, where fertilisation takes place.

The female gametophyte (embryo sac) of angiosperms is also reduced in comparison with the female gametophyte of other seed plants. Typically it consists of only seven cells and eight nuclei at maturity (Friedman, 2001). There are three cells at the micropylar end (the egg and two synergids), three antipodal cells at the opposite end, and two polar nuclei in the central cell between the egg cell and the antipodal cells. These two polar nuclei often fuse to form a diploid secondary nucleus prior to fertilisation.

Although the eight-nucleate female gametophyte is the most common condition in angiosperms, other patterns also occur among early-diverging angiosperm lineages. The *Amborella*-type gametophyte has seven cells and nine nuclei, whereas the *Nuphar/Schisandra*-type gametophyte has four cells and four nuclei (Friedman and Ryerson, 2009). However, in all cases no archegonia are developed and there is only one egg. Archegonia are also lacking in the more extensively developed female gametophytes of *Gnetum* and *Welwitschia* (Gnetales). In all other seed plants the female gametophyte is always well developed, often with several clearly differentiated archegonia and eggs.

Double fertilisation and endosperm development. In angiosperms, fertilisation involves the release of two male gametes from the pollen tube into the female gametophyte. Both gametes unite with nuclei of the female gametophyte in a process termed double fertilisation. One gamete unites with the egg nucleus while the other unites with the diploid product resulting from the fusion of the two polar nuclei. Fusion of a male gamete with the nucleus formed by the fusion of the two polar nuclei results in a triploid nucleus, which usually divides in parallel with embryo development to form a nourishing tissue: the endosperm.

Endosperm formation has long been regarded as a unique feature that distinguishes angiosperms from all other seed plants, but a comparable double fertilisation process, which also results in the formation of nutritive tissue, also occurs in species of *Ephedra*, *Welwitschia* and *Gnetum* (Gnetales). In *Ephedra* the second male gamete fuses with the sister nucleus of the egg and develops into an additional embryo (Friedman, 1992). In *Ephedra*, however, only one female nucleus is involved in the formation of the second embryo, and the product is diploid, in contrast to the triploid endosperm of angiosperms.

Fruits and seeds. In some angiosperms the unit of dispersal is the seed. In others it is the mature carpel (fruit) and enclosure of the ovules has clearly provided further possibilities for structural modification beyond what is possible in other extant seed plants. Angiosperm fruits and seeds exhibit enormous morphological diversity, which largely reflects great variation in modes of dispersal (Chapter 18). Some fruits and seeds are frequently modified in extraordinary ways, apparently to enhance their dispersal by wind or water (abiotic) or by animals (biotic).

Fruits composed of numerous free follicles with ventral dehiscence have often been regarded as basic among angiosperms and have sometimes been inferred to be the predominant condition among magnoliids. However, a survey of magnoliid fruits shows that less than 1% of the species have follicles. Most magnoliid angiosperms have dry or fleshy, indehiscent fruits (Endress, 1996c).

1.3 TIMING OF ANGIOSPERM DIVERSIFICATION

1.3.1 Pre-Cretaceous angiosperms?

The fossil record strongly suggests that the first major diversification of angiosperms took place during the Cretaceous period and that differentiation into many of the major clades that are recognised among extant flowering plants also occurred through this interval. However, in discussing the origin of angiosperms, or the origin of any other clade, it is important to specify exactly what is meant by the time of origin and to differentiate this clearly from the time of diversification of the extant group. The time of origin can be viewed either as the time at which the lineage leading to angiosperms diverged from the most closely related fossil or extant taxon, or it can be viewed as the time at which a plant in the angiosperm lineage had assembled all the defining features (synapomorphies) that today unite the group (Chapter 6). This distinction is important, since the fossil record of other major plant groups indicates that their synapomorphies were acquired sequentially.

Progymnosperms and protracheophytes, for example, exhibit some, but not all, of the features that define extant seed plants and extant vascular plants (Kenrick and Crane, 1997; Crane *et al.*, 2004). A similar pattern might be expected for angiosperms. The time of diversification is the time at which differentiation into different lineages within the crown group (the clade containing all extant taxa and with all the characteristic defining features of the extant group) began (Doyle and Donoghue, 1992).

Phylogenetic hypotheses that propose a sister group relationship between angiosperms and Gnetales or Bennettitales (or both) imply that the lineage leading to angiosperms may have diverged from related groups as early as the Triassic. Both the Bennettitales and the Gnetales occur in Triassic sediments (Chapter 5), and other possible anthophytes are also diverse in the Triassic (e.g. the Crinopolles-complex, Chapter 6) and Jurassic (e.g. Erdtmanithecales, Chapter 5). Alternative phylogenetic hypotheses based on molecular data, which suggest that angiosperms may be the sister group to all other seed plants, imply still greater antiquity for the lineage leading to flowering plants (back to the mid-Palaeozoic). It is therefore possible that the angiosperm lineage, although not necessarily the angiosperm crown group, may have had a long pre-Cretaceous history.

A pre-Cretaceous origin of angiosperms has frequently been suggested; such speculations go back to Darwin (Friedman, 2009). The Upland Theory (Axelrod, 1952) suggests that angiosperms evolved as early as the Permian and then subsequently diversified during the Triassic and Jurassic in upland regions remote from the sedimentary basins, and in habitats where fossilisation potential was minimal. According to this theory, environmental changes during the Early Cretaceous promoted angiosperm dispersal into lowland regions where their preservation was enhanced. Under this interpretation the apparent diversification of angiosperms seen in the Early Cretaceous palaeobotanical record is a function of changing ecology that enhanced their preservation potential and does not accurately reflect their initial evolutionary diversification (Axelrod, 1952, 1960, 1970).

Currently, the Upland Theory, in its most extreme form, receives very little support. The first appearance of different angiosperm subgroups is broadly consistent with predicted first appearances inferred from phylogenetic studies of extant plants. This orderly sequence of increased complexity and diversity of different angiosperm organs, with different preservational characteristics, through the Early and mid-Cretaceous, is not consistent with a sudden ecological influx of an already highly diversified group from a remote source. Furthermore, extensive studies of rich, diverse and well-preserved dispersed pollen floras from Triassic and Jurassic sequences from all over the world have failed to identify any of the very distinctive triaperturate pollen that is characteristic of eudicots.

Based on the available evidence, it seems unlikely that eudicots, or angiosperms as a whole, were diverse prior to the Early Cretaceous. There is no direct evidence for any significant diversification of crown group angiosperms prior to the Cretaceous, and the first appearance of plants that are presumed to have all the synapomorphies of the extant angiosperms is also likely to have been in the Early Cretaceous, which is when characteristic angiosperm carpels, fruits, pollen and stamens appear in the fossil record for the first time. In our view, the pattern of diversification seen in the Early and mid-Cretaceous palaeobotanical record most likely reflects both the origin and initial diversification of the angiosperm crown group with reasonable accuracy. Estimates for the greater antiquity of certain derived angiosperm subclades based on molecular clock assumptions (Martin *et al.*, 1989; Troitsky *et al.*, 1991) have a variety of inherent problems and should be treated with caution. Several claims of pre-Cretaceous angiosperm macrofossils have been published, but all are questionable and none has been documented in satisfactory detail (Chapter 6). In several cases it has been demonstrated convincingly that the fossils were misplaced stratigraphically, or misinterpreted systematically (Hughes, 1976). In other cases the fossils are not sufficiently well preserved to document their critical systematic features and thus to securely establish their relationships.

1.3.2 The fossil record

The earliest fossils with distinctive angiosperm features are dispersed pollen grains from the earliest Cretaceous (Hauterivian) of China, Israel and England (Hughes, 1994; Brenner, 1996; Zhang, 1999). In earlier (Valanginian) strata putative angiosperm pollen grains are extremely rare and not diverse (Chapter 16). Their relationship to extant angiosperms is much less secure. The grains from the Hauterivian are referred to the angiosperms based on their small size and reticulate–columellate wall structure, although details of pollen wall structure have not yet been investigated with transmission electron microscopy.

All of these earliest grains are either inaperturate or based on a monoaperturate ground plan. The other features of the plants that produced these very early pollen grains are unknown, but their similarity to pollen of living forms implies that the plants that produced them possessed all of the characters diagnostic of extant angiosperms. As we learn more about these plants the expectation is that they will exhibit typical angiosperm features, such that if we were to see them today we would clearly recognise them as members of the angiosperm clade.

In the succeeding Barremian and Aptian stages of the Early Cretaceous, angiosperm pollen grains become more diverse and more abundant. The oldest fossil floras that contain well-preserved angiosperm flowers, carpels and stamens are also interpreted to be of Late Barremian – Early Aptian age (Friis *et al.*, 1994b, 1997a). The first major diversification of angiosperms, as reflected in the fossil record by global occurrences of dispersed pollen and by the first recorded appearance of distinctive angiosperm leaves and reproductive organs, therefore took place during the Hauterivian–Aptian interval.

Resolving the exact timing of the first appearance of angiosperms, and attempting to establish in detail the initial patterns of their geographic spread and increasing diversity, are impeded by difficulties in the precise stratigraphic correlation of the critical plant-bearing sequences (Chapter 4). There is, however, in many parts of the world, a distinct orderly sequence of increasing abundance, diversity and complexity of angiosperm fossils through the Early Cretaceous. This was demonstrated clearly for dispersed pollen and leaves from the Early and mid-Cretaceous Potomac Group of eastern North America (Doyle and Hickey, 1976) and is supported by similar palynological sequences in other areas, including some with better stratigraphic age control. Comparable sequences of leaf floras from elsewhere in the Early Cretaceous are unfortunately less common, and frequently the fossil floras of greatest interest are not precisely dated. However, the data that are available, especially from North America, are consistent with the scheme outlined for the Potomac Group. A comparable sequence of increasing diversity and complexity has now been demonstrated for angiosperm reproductive organs from Early and Late Cretaceous sediments based on studies of mesofossils (Chapter 16).

The earliest angiosperm pollen grains indicate the presence of plants at the magnoliid grade and early monocots. One of the most frequently cited genera of early angiosperm pollen is *Clavatipollenites*, which was first described from the Barremian of southern England (Couper, 1958). Pollen grains assigned to *Clavatipollenites* include monocolpate and trichotomocolpate forms with a finely reticulate tectum, and often with supratectal ornamentation (Chapter 9). Some of these grains are closely similar to pollen of extant *Ascarina* of the magnoliid family Chloranthaceae (Couper, 1958, 1960; Walker and Walker, 1984).

Tricolpate pollen typical of eudicots is first reported from around the Barremian–Aptian boundary (e.g. Hughes, 1994), and subsequently, during the Aptian and Albian, there is a rapid appearance of diverse triaperturate grains. Tricolporoidate and triporate pollen enter the fossil record for the first time during the later part of the Albian and during the mid-Cenomanian, respectively (Doyle and Hickey, 1976; Hickey and Doyle, 1977). The eudicot affinity of early tricolpate pollen has been confirmed by its discovery *in situ* within Late Aptian–Albian and Cenomanian flowers of ranunculalean, platanaceous and buxaceous affinity (Chapter 12).

The earliest leaf fossils from the Potomac Group sequence are Late Barremian or Early Aptian in age. At this level angiosperm leaves are rare, occur only sporadically and comprise only a few morphotypes. They are characterised by their small size and simple, generally elliptical, ovate or obovate shape. Usually the petiole is poorly differentiated, and the venation is generally disorganised, with irregularly branched veins and poor differentiation between tertiary and higher-order veins (low-rank venation; Doyle and Hickey, 1976). The character combinations seen in these early leaf fossils indicate relationships with magnoliid angiosperms (Upchurch, 1984) and monocots (Doyle, 1973).

During the Albian and Cenomanian of the Potomac Group, and elsewhere, the structural and taxonomic diversity of angiosperm leaves, and their abundance, increases considerably. Simple leaves with relatively disorganised patterns of venation become more diverse, probably reflecting in part diversification at the magnoliid grade. However, new leaf forms also appear that include simple, palmate, pinnatifid and pinnately compound leaves, most of which generally have a more ordered pattern of venation. Branching and spacing of the veins is often regular, and veins of tertiary and higher orders are well differentiated. Palmately veined leaves, peltate leaves and lobate–cordate–ovate leaves also occur for the first time during the early part of the Albian. Pinnatifid and truly pinnate compound leaves are first recorded from the later part of the Albian (Doyle and Hickey, 1976). The leaf architecture of some of these Albian–Cenomanian leaf types suggests a relationship with several groups of eudicots, such as the Platanaceae and

Trochodendrales (Upchurch and Wolfe, 1987; Crane *et al.*, 1993). These groups are among the earliest diverging lineages of the eudicot clade according to recent molecular phylogenetic analyses.

Numerous angiosperm reproductive structures have been discovered from Early Cretaceous floras in the western Portuguese Basin and in the Potomac Group of eastern North America. Although the precise stratigraphic position of these floras is not fully established, the oldest are of probable Late Barremian or Early Aptian age (Friis *et al.*, 1994b, 1997a) (Chapter 4). Flowers are small, and both bisexual and unisexual forms are present. Flowers with few parts and a poorly elaborated perianth predominate, and most appear to be of magnoliid or monocot affinity. Probable eudicots, represented by simple forms with tricolpate pollen, are also present in these early floras, but are less common and much less diverse. During the later part of the Aptian and Albian the diversity of angiosperm reproductive organs increases. Where their relationships to extant taxa are clear, most of these Aptian–Albian taxa are related to early-diverging lineages of angiosperms among magnoliids, monocots or eudicots (Chapters 8–12).

During the Late Cretaceous more elaborate and more complex flowers appear, including forms with a differentiated perianth and synorganisation of floral parts (e.g. forms with sympetalous corolla and syncarpous gynoecium). These flowers reflect more complex and more integrated patterns of floral construction and the development of more sophisticated pollination systems. By the end of the Cretaceous most of the major lineages of extant angiosperms, and many extant angiosperm families, were already well established (Chapters 8–15).

1.4 RISE TO ECOLOGICAL DOMINANCE

Through the mid-Cretaceous, soon after their first appearance in the fossil record, angiosperms increased dramatically in their importance in fossil assemblages. This apparent ecological expansion continued alongside the phylogenetic diversification of the group through the Late Cretaceous and into the Cenozoic. Understanding the details of this explosive angiosperm diversification and the potential interaction of angiosperms with other elements of Cretaceous vegetation and relevant animal groups (herbivores, pollinators, dispersal vectors) is critical for meaningful understanding of the origin of modern terrestrial ecosystems. The changing palaeogeography and palaeoclimate of the Cretaceous and Cenozoic also undoubtedly had a significant impact and needs to be considered in understanding the development of angiosperm-dominated vegetation and how this relates to other aspects of global environmental change.

1.4.1 Increasing angiosperm diversity and abundance

In attempting to estimate the rate and magnitude of angiosperm expansion, and associated vegetational changes during the Cretaceous, it is crucial to understand the taphonomic biases that influence the representation of different plant groups in fossil assemblages. It is also important to distinguish between measures of diversity and abundance as different, but complementary, indices of ecological importance (Wing *et al.*, 1993). In most of the Early Cretaceous fossil floras in which they first appear, angiosperms are neither diverse nor abundant. They are a relatively minor component of the overall fossil assemblage. This is especially true of macrofossil floras: there are very few pre-Albian macrofossil assemblages in which angiosperm leaves are well represented. Angiosperms are also typically a minor component of pre-Albian palynofloras or mesofossil floras. However, in some mesofossil assemblages (e.g. Famalicão, Vale de Agua, Torres Vedras and Buarcos in western Portugal; Chapter 4) angiosperms may be locally dominant as judged by both diversity and abundance, indicating that the initial diversification of the group was already under way and that angiosperms were already ecologically important in selected environments.

The most complete information on Early Cretaceous angiosperms is from middle palaeolatitudes in the Northern Hemisphere, but to judge from palynofloras elsewhere in the world angiosperms were also diverse and abundant at low palaeolatitudes. Evidence from North America indicates that angiosperms only became ecologically significant at the highest latitudes by the earliest Late Cretaceous (Wing *et al.*, 1993). Quantitative assessments based on both macrofossil floras and palynofloras indicate that by the latest Cretaceous, angiosperms typically accounted for 40% and 60%, respectively, of the species in individual floras (Lidgard and Crane, 1990) and also had achieved approximately equivalent levels of abundance (Lupia *et al.*, 1999).

Estimates based on macrofossil floras (diversity and abundance) suggest that angiosperms were already dominant in some environments at mid-palaeolatitudes by the Cenomanian (Lupia, 1999). However, diversity estimates based on palynofloras give lower values for angiosperms throughout the Cretaceous (Lidgard and Crane, 1990). These contrasting patterns may suggest that angiosperms are especially well represented in the environments sampled by macrofossil floras, whereas they are perhaps relatively under-represented across the broader landscapes sampled by palynofloras. Detailed studies of individual fossil assemblages from the Campanian and Maastrichtian indicate that in some areas angiosperms, although diverse, were relatively low in abundance compared with ferns in the same vegetation (Wing *et al.*, 1993). However, in other assemblages it is evident that angiosperms were dominant by any measure (Wing and Boucher, 1998; Lupia *et al.*, 1999). These differences may reflect different kinds of vegetation developed under different climatic regimes. They may also reflect different successional situations.

Taken as a whole, the expansion in the abundance and diversity of angiosperms through the Late Cretaceous seems to have been accompanied by a decline in the abundance and diversity of certain kinds of ferns, cycads, ginkgos, Bennettitales, and other groups of typical Mesozoic seed plants. Nevertheless, it is clear that the Late Cretaceous was a period of active evolution for some of these ancient groups (e.g. some ferns) even as their overall prominence in global vegetation declined. The Late Cretaceous also represents an active phase in the evolution and origin of modern conifers, which is reflected in significant turnover in the composition of conifer assemblages and the first appearance and diversification of certain extant groups. Conifers as a whole do not appear to undergo a significant decline as angiosperms diversify.

1.4.2 Co-evolution with animals: herbivory, dispersal and pollination

Throughout their evolutionary history angiosperms have influenced, and been influenced by, the changing ecology of contemporaneous organisms, including especially through interactions with the evolving insect and vertebrate fauna. In some cases these interactions have been significant but diffuse, resulting in broad effects on both animals and plants. In other cases these interactions have resulted in highly specific interactions at the level of individual species. In broad terms there is no doubt that extant angiosperms look the way they do today because of interactions with herbivores, dispersal agents and pollinators. It is also clear that the form and behaviour of many animals has been profoundly influenced by the plants on which they feed.

The transition to angiosperm-dominated ecosystems modified the foundations of terrestrial food webs and undoubtedly created new challenges and new opportunities for animal life. Protection from herbivores also undoubtedly underpins much of the diversity in angiosperm secondary metabolites, as well as a great variety of physical defences ranging from spines to anatomical characteristics and particular modes of growth. Similarly, the changed form, structure and chemistry of the angiosperm plant body, relative to that of more ancient groups of plants, undoubtedly had an effect on the evolution of behaviour, masticatory apparatus, mode of digestion and other features in insects and vertebrates.

Details of the changing ecological interactions among different groups of organisms are rarely amenable to direct observation in the fossil record, and inferences of cause and effect depend primarily on coincidence in the timing of palaeontological patterns. Judged at the family level there is no major increase in insect diversity coincident with the Cretaceous diversification of angiosperms (Labandeira and Sepkoski, 1993), but this masks significant changes that occurred at other levels such as subfamily and genus/species level as well as total species diversity (Grimaldi, 1999). Insect groups that were well established long before angiosperms appear to have diversified significantly as they exploited angiosperms for food and other purposes. Many flower-visiting groups of insects also clearly show their first significant diversification in the Cretaceous, including butterflies (Lepidoptera), hoverflies (Syrphidae), bee flies (Bombyliidae) and bees (Apoidea/Apidae), as well as other major flower-visiting insect groups (Grimaldi, 1999) (Chapter 17).

Lepidoptera appear in the fossil record only slightly before angiosperms in the Late Jurassic, but ditrysian butterfly families, which comprise about 98% of all extant species of Lepidoptera, apparently first radiated in the Late Cretaceous when angiosperms were well established (Grimaldi, 1999). These butterflies are significant today, at different stages of their life cycle, as both herbivores and pollinators of angiosperms. Bees are also first known from the Late Cretaceous (Grimaldi, 1999), but today they are diverse and extremely important pollinators of angiosperm flowers. The diversification of ants also occurred during the Cretaceous (Carpenter, 1992). Today ants play an important role in angiosperm seed dispersal.

Since the Cretaceous there have been significant changes in the vertebrate fauna, which again have influenced, and been influenced by, changes in plants and vegetation (Chapters 17, 18). During the Early Cretaceous herbivores were generally large and the disturbance they created may have favoured the evolution of weedy life cycles and small propagules (Wing and Tiffney, 1987a, b). Large dinosaurs continued to dominate the Late Cretaceous fauna of the Southern Hemisphere while communities of smaller dinosaurs dominated Late Cretaceous faunas in the Northern Hemisphere. This, in turn, may have created differential disturbance pressures and contributed to the marked current vegetational differences in the two hemispheres (Novacek, 1999).

With the disappearance of generalised dinosaur herbivores at the Cretaceous–Cenozoic boundary, the nature of ecological interactions between plants and vertebrates changed dramatically, as did the selective pressures on angiosperms. Angiosperms are thought to have increased in stature along with an increase in propagule size and the development of increasingly closed plant communities. These changes were also accompanied by the diversification of small frugivorous birds and mammals (Wing and Tiffney, 1987a, b) (Chapter 18). Changes in palaeogeography and climate from the Cretaceous into the Cenozoic, and during the Cenozoic, must also have played an important role in shaping vegetation structure and animal–plant interactions (Eriksson *et al.*, 2000a). The kinds of interactions established in closed environments in the earliest Cenozoic still persist today in those kinds of habitats. However, as grasslands and savannahs developed, perhaps in response to global changes in climate, the new, open conditions were again exploited by large generalised herbivores (mammals). This in turn created further possibilities for diversification in new groups of opportunistic angiosperms (Collinson and Hooker, 1987; Wing and Tiffney, 1987a).

The changing interactions between angiosperms, insects and vertebrates through time provide fascinating avenues for further research. Modern ecosystems are, in large part, a function of these interactions, which were played out through the Cretaceous, and then later through the Cenozoic, against a background of changing geography and changing climate. Taken together these processes, in which angiosperms played a central role, helped create the biological diversity of our modern world.

2

The nature of the angiosperm fossil record

The main sources of information on the timing and pattern of early angiosperm diversification have traditionally been studies of dispersed pollen and detached leaves (e.g. Doyle and Hickey, 1976; Hughes, 1976; Hickey and Doyle, 1977; Hughes *et al.*, 1994). However, since about 1980 discoveries of well-preserved flowers and dispersed floral organs from sediments of Early and Late Cretaceous age have provided a large amount of new and previously unsuspected information, especially on the structure and systematic relationships of early flowering plants.

In this chapter we briefly consider the nature of the angiosperm fossil record, particularly in relation to studies of angiosperms from the Cretaceous. We review the difficulties introduced by the incompleteness of the fossil record, especially the biases that result from aspects of the biology of different kinds of plants and the particular circumstances under which fossil assemblages are preserved. We also discuss the most common types of angiosperm fossils, their preservation, and the implications of the nature of the plant fossil record for environmental and evolutionary interpretations based on palaeobotanical data. More extensive treatments of the nature and depositional aspects of the plant fossil record are given by Krassilov (1975b) and Behrensmeyer and Hook (1992), who also provide an entry into the substantial literature on this topic.

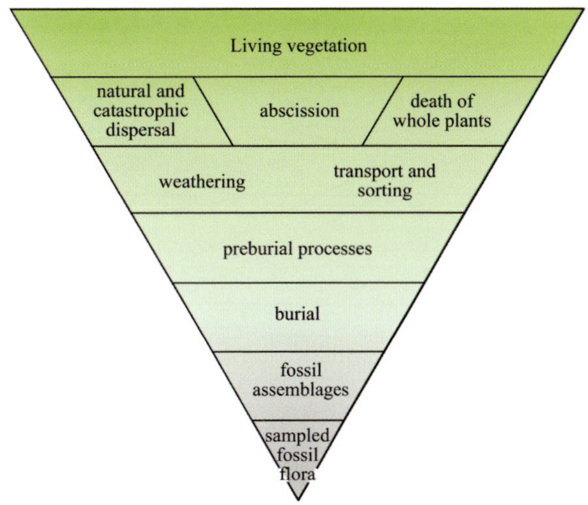

Figure 2.1 Key processes responsible for the formation of plant fossil assemblages. The complexity of these processes often precludes straightforward extrapolations from the floral assemblages to the nature of the original, living, source vegetation. The sampled fossil flora is also just a subset of the total fossil assemblages.

2.1 UNDERSTANDING THE PLANT FOSSIL RECORD

Only a small fraction of the Earth's vegetation will ever leave evidence of its existence in the fossil record, and the processes responsible for generating plant fossil assemblages are diverse and complex (Figure 2.1). Nevertheless, increasingly sophisticated studies of post-mortem processes and fossilisation potential in plants have clarified many of the biases of representation and recognition inherent in the palaeobotanical record. They have also significantly improved our understanding of the possibilities and limits of using plant fossils to reconstruct past vegetation, to better understand the biology of ancient plants, and to develop and test hypotheses of angiosperm evolution (e.g. Burnham, 1989; Gastaldo, 1989; Behrensmeyer *et al.*, 2000).

Typically, only fragments of plants growing close to depositional basins are incorporated into accumulating sediments by wind or water transport, and even then only those fragments entombed in such a way that normal processes of decomposition are halted or reduced have the potential to survive as fossils. In general, long-distance transport, especially of larger plant parts, is rare, and this implies that plants from drier or upland areas, far from lowland lakes, rivers or the coast, have significantly lower fossilisation potential than plants from lowland areas that are close to many different kinds of depositional environments in which they may be preserved (Figure 2.2). There

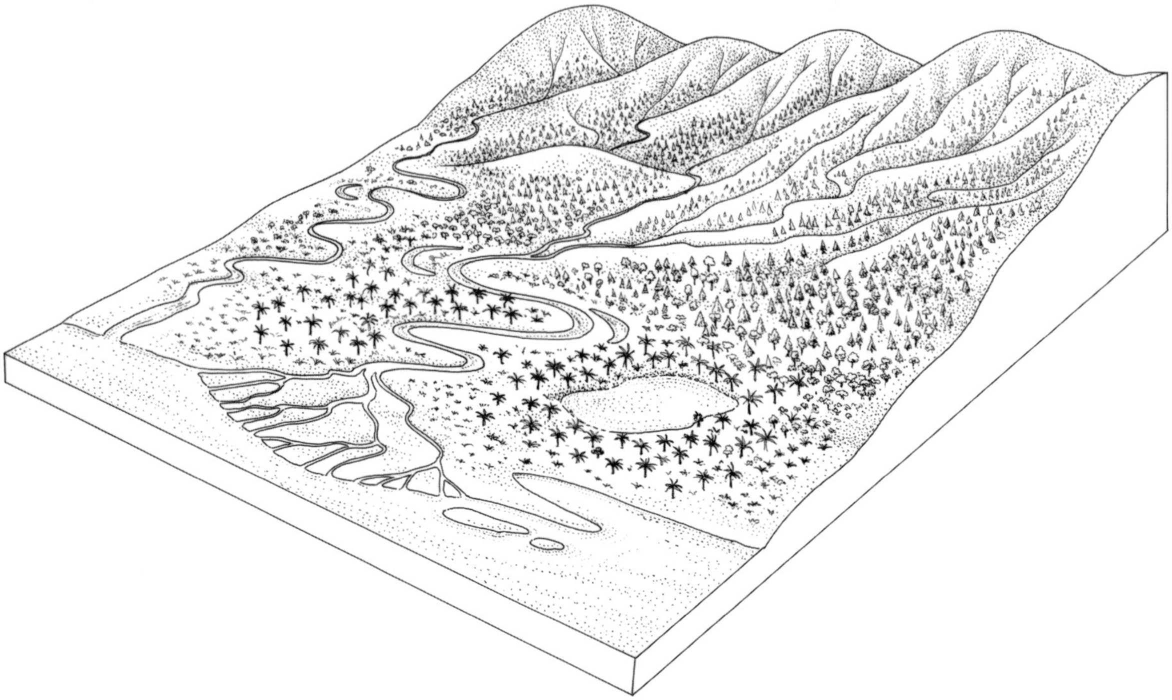

Figure 2.2 Diagrammatic summary of some of the main depositional environments in which plant fossil assemblages may be preserved. Most plant fossils are preserved in lowland environments, typically on the flood plain where sediments are accumulating, for example in small lakes, abandoned river channels (e.g. ox-bow lakes) or deltas. The source vegetation reflected in these assemblages is largely that of the lowlands, with much lower representation from the uplands. Preservation of plant fossil assemblages in upland environments is less common because these are mainly regions of net erosion rather than net deposition. Preservation of plant fossil assemblages is also less common in marine deposits. There are, however, important exceptions to these generalisations, especially in the Cenozoic, where both upland (e.g. the Florissant Lake Basin, Colorado, USA) and marine deposits (e.g. the London Clay, southern England) have yielded important and very informative fossil plant assemblages.

are, however, many exceptions to these generalisations. For example, some plant parts are particularly well adapted for transport over long distances, such as the pollen of wind-pollinated plants or fruits that are water-dispersed. Similarly, particularly resistant plant parts such as tough leaves or woody fruits or stems may survive extensive transport prior to fossilisation (Figure 2.3).

As a further generalisation, potential fossil assemblages that are deposited at, or close to, sea level also have a greater probability of surviving erosion over long periods of geological time. However, again, in detail, the situation is complex. For example, not all lowland environments offer good conditions for fossilisation. In many wet tropical lowlands decomposition rates are high and plant material is either not preserved at all, or may be strongly degraded prior to, and during, fossilisation. Equally, informative fossils from both tropical and temperate areas may be preserved in upland lake basins. Such ancient lakes may provide a great deal of useful palaeobotanical data, particularly if they occur, as is often the case, in association with the rapid deposition of fine sediments (e.g. ash) released through volcanism. Nevertheless, because the chances of erosion of inland or upland basins increase with age, such deposits are most common and best known from Cenozoic strata.

A further general limitation of the plant fossil record is that whole plants, or even large parts of plants, are preserved only infrequently. In the relatively few cases where this occurs, the depositional environment is usually exceptional (e.g. quiet lake basins) and the plants are of small stature or aquatics (Figure 2.4). Much more commonly, plant fossils are dispersed parts of plants that have been shed as part of the natural life cycle: pollen dispersed at

Figure 2.3 Modern depositional environment in the Mobile Delta of southern Alabama, USA, in which plants are being preserved. (A) Overbank deposits colonised by herbaceous vegetation and containing abundant allochthonous and autochthonous plant remains. (B) Sediment sample from overbank deposit showing plant remains including leaves of trees transported from upstream in the delta system.

Figure 2.4 Young shoot of *Archaefructus* preserved in fine-grained lake sediments from the Early Cretaceous (Early to early Late Aptian) Yixian Formation, China. The inflorescence axis and leaves are still attached to the stem. Specimen in the collections of the Nanjing Institute of Geology and Palaeontology. Photograph C. Pott.

pollination; aborted flowers, floral parts and stamens shed after anthesis; fruits and seeds dispersed at time of maturity; and leaves and twigs shed through the life of a plant, or especially at the time of defoliation in the autumn or at the onset of the dry season. Under normal conditions, the time of shedding of the various parts is generally not synchronous, and therefore the different organs of the same plant species may not always occur together. Different plant parts may also be sorted and concentrated in different ways during transport. Plant fragments accumulated after a storm, volcanic activity, or other catastrophic events may include parts that are not normally shed from the plant. Such specimens may also furnish spectacular and rare examples of attachment between different plant organs (Figure 2.5).

In addition to environmental factors, the nature of the plants, or of the detached organs themselves, also strongly influences their fossilisation potential. Herbaceous plants, for example, tend to be strongly under-represented in the fossil record. Their leaves typically lack an abscission layer and generally die, wither and decompose in place. They are therefore less easily transported and less frequently incorporated into depositional environments. Herbaceous plants also contain few woody tissues and their low stature may further reduce their preservation potential, for example by reducing the possibilities for long-distance transport of wind-dispersed pollen. However, pollen, as well as the fruits and seeds of herbaceous plants, may also be transported by water, and these structures provide the most common form of representation in the fossil record for herbaceous plants.

Figure 2.5 Fossil branch of *Ulmus okanaganensis* from the Early Cenozoic (Eocene) McAbee locality, British Columbia, Canada, showing twig, leaves and young fruits in organic connection. From Denk and Dillhoff (2005); specimen in the collections of the Burke Museum, Seattle, Washington, UWBM. Photograph T. Denk.

Figure 2.6 Polished transection of a permineralised palm stem (provenance unknown) showing an outer zone of adventitious roots surrounding a stem with scattered vascular bundles. Specimen in the collections of the Swedish Museum of Natural History, Stockholm. Photograph C. Pott.

2.1.1 Fossil wood

Wood is generally resistant to both physical and biotic degradation. As a result, trunks and large fragments of wood are commonly preserved throughout the plant fossil record (Figure 2.6). Most of the published records of angiosperm wood from the Cretaceous are based on permineralised specimens (e.g. Stopes and Fujii, 1910; Stopes, 1915; Spackman, 1948; Mädel, 1960; Page, 1968; Wheeler *et al.*, 1987; Poole and Francis, 1999; Cantrill and Poole, 2002; Poole, 2002), but fragments of charcoalified wood are also recovered together with flowers, fruits and seeds in mesofossil floras (section 2.1.3). These charcoal fragments have provided a new source of data on early angiosperm wood (Herendeen, 1991a, b).

Specimens of permineralised wood are generally larger than charcoalified fragments and often have both primary and secondary tissue preserved (Figure 2.7). They may also be preserved in growth position and show growth rings or other features that are useful for ecological interpretations. Fine anatomical details of cell walls, for example pitting, which are often of systematic importance, may also be preserved in permineralised specimens. However, such details are also exceptionally well preserved in charcoalified material. An important limitation to the study of permineralised wood is that it is rarely preserved in the same beds as other plant fossils, except where there has been a mineralisation of more or less *in situ* plant assemblages (for example a peat or a soil litter horizon). As a result, it is often difficult to link dispersed permineralised wood with other parts of the original plant. The main limitation of charcoalified wood is that it is always fragmentary, because of the way it forms and because it is brittle. It is generally preserved in small pieces only a few millimetres to a few centimetres long. As a result, the information that charcoal provides on wood gross structure is often very incomplete.

The systematic determination of fossil angiosperm wood may be problematic even with well-preserved specimens (Herendeen *et al.*, 1999b). Furthermore, these difficulties generally increase with the age of the specimens, because few wood characters are reliably diagnostic above the family level, and because of the increased likelihood of mosaic character combinations in extinct taxa. Comparative wood anatomy of extant plants is relatively well known, but there is also considerable structural variation among species that are taxonomically closely related, as well as significant convergence in wood anatomy among taxa that are only distantly related. Taken together, the effect of these different factors means that fossil wood has so far provided relatively little information on the systematic dimensions of angiosperm diversification through the Early and mid-Cretaceous. Studies of fossil angiosperm wood from the latest Cretaceous and Cenozoic are of much

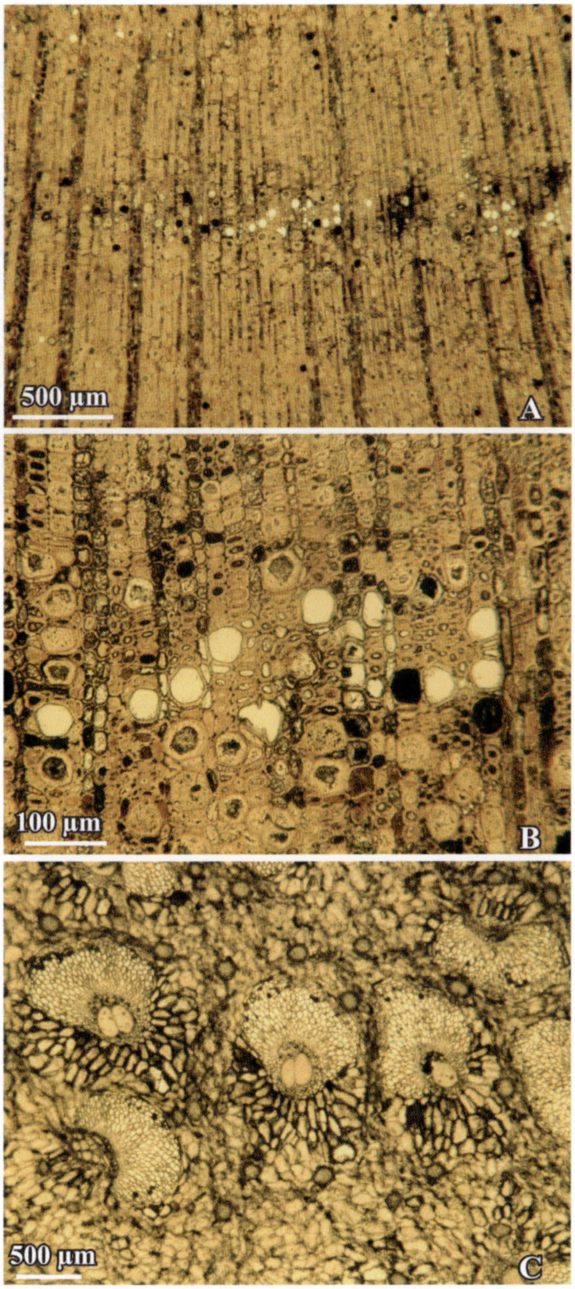

Figure 2.7 Fossil angiosperm wood. (A, B) Permineralised wood of the fossil angiosperm *Aptiana radiata* from the Early Cretaceous (Aptian) of southern England showing secondary xylem with vessels concentrated along the growth rings (ring porous). (C) Permineralised wood of a fossil palm from the Cenozoic of the West Indies (provenance unknown), showing vascular bundles scattered in parenchymatous ground tissue. Specimens in the collection of the Swedish Museum of Natural History, Stockholm.

greater importance for elucidating evolutionary patterns and broad changes in the nature of vegetation. For example, Cantrill and Poole (2005) used a long stratigraphic series of wood fossils from Antarctica to characterise the evolution of woody vegetation in the region from the Cretaceous into Early Cenozoic. Wood also preserves useful climatic and other environmental signals (Creber and Chaloner, 1985). In particular, growth rings indicate changing growth rates through the growing season. Such seasonality generally results from the slowing or cessation of growth in regions with cold winters, but it can also result from alternating wet and dry seasons, as occur in many parts of the tropics.

Larger pieces of fossil angiosperm wood are surprisingly sparse during the Early and mid-Cretaceous, especially compared with the record during the latest Cretaceous and Cenozoic. Only rarely (e.g. Thayn *et al.*, 1985) have larger fragments of angiosperm wood been described from the Early Cretaceous and in some cases these have subsequently been shown to have been stratigraphically misplaced. For example, the sources of the few large specimens of angiosperm wood, up to about 30 cm in diameter, reported from the Early Cretaceous (Aptian) of England (Stopes, 1912, 1915) have long been questioned (Harris, 1956; Hughes, 1976). Subsequent searches for other specimens in the same strata have failed, and recent reinvestigation of the five wood types originally described has shown that only one, *Aptiana radiata* (Figure 2.7), is of probable Early Cretaceous age (Crawley, 2001). Three other taxa (*Cantia arborescence*, *Hythia elgarii*, *Sabulia scottii*) are Paleocene, while the stratigraphic position of the fifth taxon, *Woburnia porosa*, is still uncertain, but also most likely Paleocene (Crawley, 2001).

2.1.2 Fossil leaves

Prior to the expansion of studies of fossil pollen in the second half of the twentieth century, and the more recent discovery of fossil flowers, fossil leaves were the most important source of information on early angiosperm history. Investigations of several important Cretaceous leaf floras published in the late nineteenth century and the early twentieth century greatly influenced early ideas on angiosperm origin and evolution. Most significant were leaf floras from the Potomac Group and elsewhere in the mid-Cretaceous of North America, leaf floras from the Early and mid-Cretaceous of the western Portuguese Basin, and several mid- to Late Cretaceous leaf floras from western Greenland (Chapter 4).

Figure 2.8 Fossil angiosperm leaves from the Cretaceous. (A) Leaf impression of *Credneria integerrima* (Platanaceae) from the Late Cretaceous (Santonian) of Quedlinburg, Germany. (B) Leaf compression of *Artocarpus dicksonii* (possible Moraceae) from the Late Cretaceous of Igdlokunguak, Disco Island, West Greenland. Specimens in the collections of the Swedish Museum of Natural History, Stockholm. Photograph Y. Arremo.

Fossil leaf assemblages are strongly biased against the representation of herbaceous taxa (section 2.1). They also typically represent rather local vegetation growing close to the depositional basin, although longer-distance transport of leaves does occasionally occur in major river systems (Gastaldo, 1989). However, in contrast to fossil pollen or spores, once deposition has occurred and partial decomposition has been initiated, fossil leaves are unlikely to be reworked through secondary transport and deposition.

Fossil leaves are usually preserved as impressions or compressions in clays or silts, or as impressions in sandstones (Figure 2.8). In early palaeobotanical studies, ideas on the systematic relationships of fossil angiosperm leaves were typically based solely on gross morphology and major patterns of venation. In many cases the resulting assessments were based on superficial resemblances to modern plants, and the fossils were assigned either to extant genera (e.g. *Populus*, *Sassafras*, *Vitis*) or to extinct genera considered closely related to extant forms (e.g. *Menispermites*, *Sapindopsis*). This practice resulted in considerable confusion about the systematic diversification of angiosperms through the Cretaceous. One result was the widespread belief that by the mid-Cretaceous angiosperms had already attained a substantially modern level of diversification with most modern lineages well established.

More recent and more detailed studies of fossil angiosperm leaves have included detailed investigation of fine venation (Hickey and Wolfe, 1975) and epidermal structure (as preserved in the cuticle; Dilcher, 1974). These studies have shown that many of the earlier systematic determinations of both Cretaceous and Cenozoic leaf fossils were incorrect. It is now clear that most of leaf fossils from the Cretaceous should not be referred to extant genera and that the supposed modern aspect of angiosperms in mid-Cretaceous floras is only superficial.

Modern studies of Cretaceous fossil leaves, which include detailed analyses of venation patterns and cuticular structure, have provided useful evidence of the systematic relationships of Cretaceous angiosperms and have contributed significantly to an improved understanding of early angiosperm evolution (Hickey and Doyle, 1977; Upchurch, 1984; Upchurch and Dilcher, 1990). Such work has also been greatly aided by improved awareness and increased documentation of the systematic importance of leaf architecture among extant plants (Hickey and Wolfe, 1975). Further, studies of fossil leaves have been of great importance in attempting to infer ancient climates because of well-established correlations between particular key characters and climatic parameters that are largely, although not completely, independent of systematic relationships (Chapter 3).

2.1.3 Fossil reproductive structures

Well-preserved angiosperm reproductive organs, including complete flowers and flower buds, as well as dispersed stamens, fruits and seeds, provide an important source of information on Cretaceous angiosperms. Such structures are relatively rich in features of potential systematic value, and because the equivalent structures in living plants are often well documented, at least in broad terms, systematic comparisons can be relatively straightforward. Flowers and other reproductive structures can also provide important clues to aspects of the biology of ancient angiosperms such as pollination and dispersal.

Mesofossils. The most common fossil angiosperm flowers, and those that have been studied most intensively from the Cretaceous, are typically very small and are generally visible in sediment samples only as dark spots, or dark laminae (Figure 2.9). Individual fossils cannot be distinguished with the naked eye until they are isolated from the rock (Figure 2.10). Fossils of this kind are referred to as mesofossils because they are intermediate in size between the fossil pollen and spores preserved in palynofloras (microfossils) and leaves and other structures preserved in macrofossil assemblages.

In most cases mesofossils (and the mesofossil floras comprising them) are recovered from relatively soft, unconsolidated rocks that were generally deposited in floodplain settings and that have not been deeply buried or extensively weathered. Sediment samples can therefore be disaggregated relatively easily. Most often the fossils are extracted by simple sieving and panning in water over a fine-mesh sieve. The extracted plant debris is then cleaned in bulk with hydrofluoric and hydrochloric acid, washed thoroughly in water, dried in air, sorted under a binocular microscope and examined for morphological and structural details (Figures 2.10, 2.11).

The recognition that exquisitely well-preserved fossil flowers are widespread during the Cretaceous revolutionised the study of the early angiosperm fossil record. In addition, because mesofossils can be extremely abundant such fossil assemblages provide valuable additional evidence of Cretaceous angiosperm diversity and abundance that can supplement the information provided by macro fossils (mostly leaves) and microfossils (mostly pollen). External structures and many internal details of mesofossils are most easily studied by using scanning electron microscopy (SEM) (Figure 2.12). Further details and higher resolution can be obtained by serial sectioning. Standard serial sectioning is carried out on specimens that have first been embedded in a hard resin or plastic. Thin sections are cut about 3 μm thick with a glass or diamond knife, and examined for major structural features using light microscopy (LM). Ultrathin sections can also be prepared for transmission electron microscopy (TEM).

Most recently the application of synchrotron-radiation X-ray tomographic microscopy (SRXTM) and associated phase contrast approaches (PCXTM) has provided an alternative, non-destructive, method for studying mesofossils that reveals internal features with high resolution (e.g. Friis *et al.*, 2007). The application of this method has proven to be ideal for charcoalified mesofossils. It has greatly improved the quality of information retrievable from such material and revealed astonishingly fine details at the cellular level. The method has now been applied successfully to a wide range of mesofossil material (Figures 2.13, 2.14). To a large extent it replaces the need for serial sectioning, except in those cases where resolution is required at the ultrastructural level, for example to determine exine stratification in the wall of *in situ* pollen grains. Synchrotron-radiation X-ray tomographic microscopy also has the added advantage that the information is captured digitally and is readily converted into three-dimensional reconstructions.

In many cases fossil flowers and other reproductive structures preserved as mesofossils are charcoalified, as a result of natural fires in the source vegetation that probably consumed both living plants and the leaf and other litter on the soil surface. Charcoalified specimens are typically three-dimensional and have a black, highly reflective surface. Internal cellular structures studied in thin sections of fossils from the Åsen locality, Sweden (Chapter 4) showed homogenisation of the cell walls, characteristic of

Figure 2.9 The Åsen locality in Scania, southern Sweden, exposing soft, kaolinitic limnic and fluviatile sediments with rich mesofossil assemblages that contain abundant well-preserved flowers and other small plant fossils. (A) One of the authors (E.M. Friis) in the field. (B) Detail of fluviatile sediments showing scattered and laminar concentrations of coaly material; red bar = 10 cm.

charcoalified plant tissues (Friis and Pedersen, 2000). Based on comparison with cell walls in experimentally charred plant material (Scott and Jones, 1991) it is estimated that fossilisation took place under low to moderate temperatures, between 240 °C and 370 °C. Charcoalification provides three-dimensional rigidity and resistance to the plant tissue. In many cases it also prevents the fossils from collapsing after burial in the sediments.

Charcoalified flowers may have all floral parts present, including sepals and petals, stamens and styles (Figure 2.12). Anatomical details are generally well preserved. Pollen grains may be preserved *in situ* in the anthers with their external morphological and sculptural details more or less intact, but ultrastructural details of the pollen wall are often lost in charcoalified material. Delicate details of indumentum and other cellular features are often beautifully preserved, even though the cuticle usually cannot be isolated. The main disadvantage of charcoalified preservation is that the specimens are brittle and fragile. Most well-preserved charcoalified fossil assemblages have probably undergone relatively little transport.

Lignitised to partly coalified reproductive organs are frequently preserved as mesofossils in association with charcoalified specimens. In some mesofossil assemblages all of the material is lignitised and charcoal is absent. Lignitised fossils are usually very easily distinguished from charcoalified ones because they are typically brownish in colour with non-reflective, or only slightly reflective, surfaces (Figure 2.10). Lignitised specimens sometimes have much of their original three-dimensional structure

Figure 2.10 Sorting trays with Cretaceous mesofossil assemblages with different preservation. (A) Charcoalified and lignitised plant mesofossils from the Late Cretaceous (Early Campanian) Neuse River locality, North Carolina, USA. (B) Lignitised plant mesofossils from the Early Cretaceous (Late Berriasian-Valanginian) Carl Nielsen A/S locality, Bornholm, Denmark. Squares of sorting tray, 1 cm.

Figure 2.11 Picking mesofossils under the stereo microscope.

preserved, but often they are more or less compressed. In these cases the original three-dimensional form may be difficult to reconstruct, particularly for complex structures such as flowers. However, lignitised flowers may have all floral parts preserved along with cuticles and important cellular details. Pollen grains *in situ* within lignitised anthers are also typically better preserved than in charcoalified material and their ultrastructural details can be studied with transmission electron microscopy. Often, cuticles can be isolated from lignitised material, and this may be important for comparison with extant taxa. Cuticles may also be helpful in establishing connections between the leaves, fruits and other parts of the same plant species that are preserved separately.

Both lignitised and charcoalified organs have often shrunk during preservation. Lignitised material may have been exposed to some desiccation before it was fossilised, and charcoalification also drives off water and other cell wall components. Experimental studies also show that the charring process results in shrinkage (Figure 2.15), and that the degree of shrinkage may vary among different plant parts (Harris, 1981; Lupia, 1995). Shrinkage is often linked to the degree of lignification of the cells: soft tissues typically shrink more than hard tissues (Figure 2.15).

Compressions. Prior to the discovery of flowers preserved as charcoal, angiosperm reproductive organs from the Cretaceous were mostly reported as scattered compressions or impressions in fossil floras dominated by leaf macrofossils. They were studied by using the same approach that had been applied successfully to compressed angiosperm flowers and inflorescences from Early Cenozoic fossil floras (Dilcher, 1979). Fossil flowers preserved in this way are typically strongly compressed, with details that are difficult to study, especially where cuticles are not preserved. Often, however, they are less fragmentary than mesofossils and such material includes the most complete inflorescence axes currently known from the Cretaceous, as well as rare instances of reproductive structures attached to leafy shoots (Figure 2.4). Thus, while mesofossils provide detailed information on floral structure and organisation, macrofossils may provide additional information on inflorescence structure, the form of the whole plant and the opportunity to link leaves and reproductive structures as part of the same plant species (e.g. Dilcher and Crane, 1984a). Fossil angiosperms from the Crato Formation of Brazil (Mohr and Friis, 2000) and from the Yixian Formation of China (Sun *et al.*, 2002; Friis *et al.*, 2003a; Leng and Friis, 2003) provide good examples. Some of the larger reproductive organs known in Cretaceous sediments (e.g. *Archaeanthus*; Dilcher and Crane, 1984a) have also been described from compressions in macrofossil floras. Occasionally, for example in sandstones, impressions may also provide evidence of the three-dimensional form of relatively large angiosperm reproductive structures (e.g. *Lesqueria*; Crane and Dilcher, 1984).

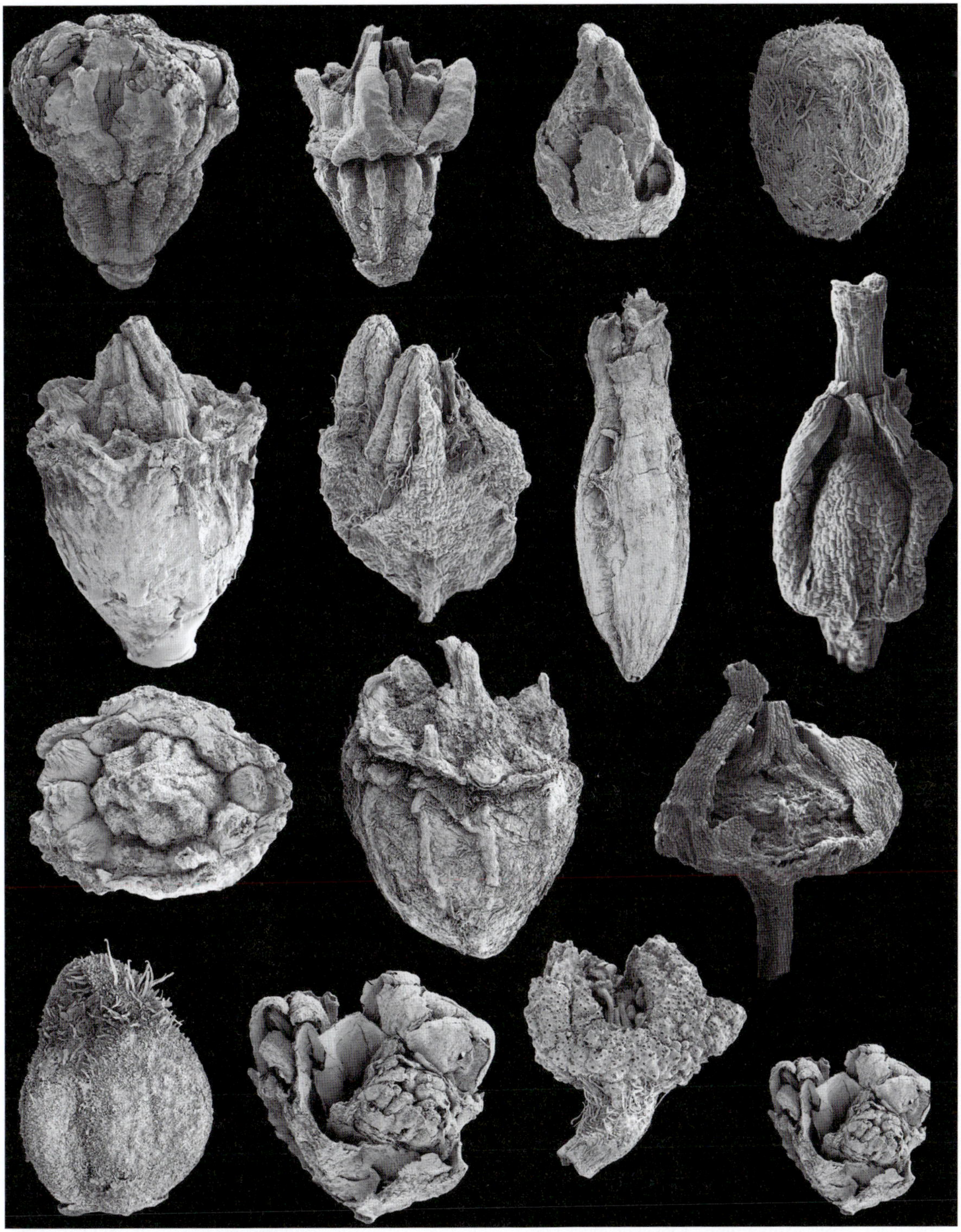

Figure 2.12 A variety of charcoalified mesofossils (flowers) from the Late Cretaceous of Sweden and Portugal showing named and unnamed taxa with different kind of organs preserved; all flowers between about 0.5 and 2 mm long. SEM images of specimens in the collection of the Swedish Museum of Natural History, Stockholm.

Figure 2.13 (A) Charcoalified flower mounted on stub for X-ray tomographic microscopy. (B) Scanning electron micrograph of the same flower. Specimen in the collection of the Swedish Museum of Natural History, Stockholm.

Permineralisations. Permineralised fossils, in which carbonates, silicates or iron compounds have precipitated within the plant tissue, often have excellent preservation and potentially yield information on morphology, gross organisation and anatomy. However, angiosperm reproductive structures preserved in this way are relatively rare in Cretaceous sediments. A few permineralised angiosperm reproductive organs have been described from a Late Cretaceous (Turonian–Santonian) marine sequence from Hokkaido, Japan (e.g. *Cretovarium japonicum*, Stopes and Fujii, 1910; *Protomonimia kasai-nakajhongii*, Nishida and Nishida, 1988; *Elsemaria kokubunii*, Nishida, 1994b). Cellular details are well preserved and the fossils provide important information on the organisation and structure of the floral components (Figure 2.16). Other permineralised Late Cretaceous angiosperm reproductive organs with many diagnostic details preserved are known from the Late Santonian – Early Campanian of British Columbia, Canada (e.g. Delevoryas and Mickle, 1995) and also from the Maastrichtian Deccan Intertrappean Beds, India (e.g. Sahni, 1943; Shukla, 1944; Verma, 1958; Chitaley and Patel, 1975). The Deccan Intertrappean Beds are especially rich in permineralised angiosperm reproductive structures, and these assemblages include several different flowers. From the Early and mid-Cretaceous permineralised angiosperm reproductive structures are rare; they include material from Western Australia (Dettmann *et al.*, 2009) and East Siberia (Samylina, 1960).

Amber. Occasionally fossil flowers and other reproductive structures occur in amber (Figure 2.17). Flowers preserved in such fossil resin, which may be of both conifer and angiosperm origin, are known mainly from the Cenozoic (e.g. Conwentz, 1886; Poinar, 1992; Grimaldi, 1996). Amber that contains fossil insects occurs at several levels in the Cretaceous, but so far the only convincing report of a Cretaceous flower preserved in this way is a single specimen from the Late Cretaceous of New Jersey (Grimaldi, 2000). Floral remains have also been reported from Burmese amber of the Hukawng Valley in northern Myanmar (Poinar and Chambers, 2005, 2008; Poinar *et al.*, 2007). However, the age of the Burmese amber is controversial. The Late Albian age assignment (Cruickshank and Ko, 2003) appears much too old, judging from the organisation of the flowers described so far. In our view, the age of this material is more likely Late Cretaceous or even Early Cenozoic (Chapter 4).

In situ pollen. Pollen grains found *in situ* in charcoalified and lignitised flowers provide an important link between dispersed pollen and mesofossils. In several cases pollen grains *in situ* within flowers have provided information on the systematic affinities of pollen grains that were previously only known as dispersed specimens (e.g. Friis, 1983; Drinnan *et al.*, 1991; Friis *et al.*, 2000a, 2003b;

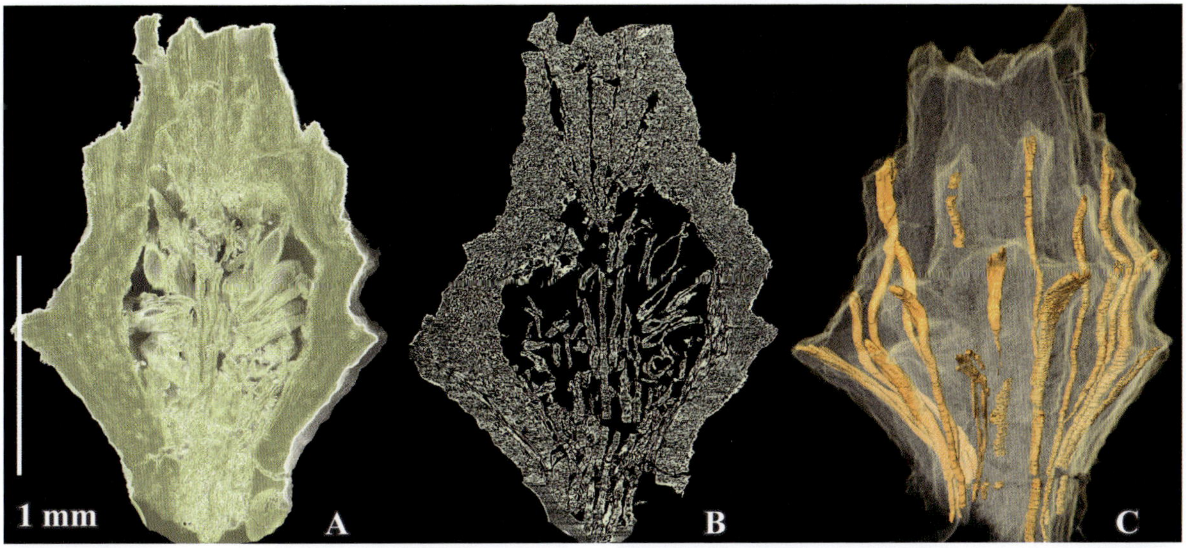

Figure 2.14 Synchrotron-radiation X-ray tomographic microscopy (SRXTM) applied to the fossil flower *Monetianthus mirus*, showing various digitally reconstructed sections. (A) Longitudinal section of locules and seeds showing internal structures in 3-D. (B) Longitudinal 2-D section. (C) Transparent view of flower (voltex) with the vascular bundles marked. Specimen in the collection of the Swedish Museum of Natural History, Stockholm.

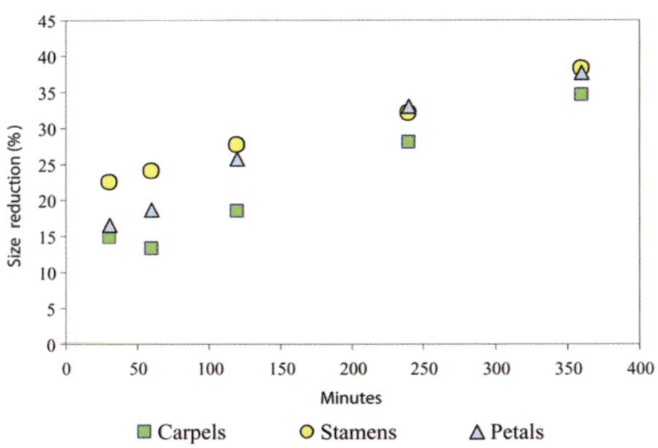

Figure 2.15 Plant organs shrink to various degrees when burnt and transformed to charcoal. Experimental studies of floral organs show the extent of differential size change for stamens, carpels and petals from extant plants that have been exposed to temperatures between 325 and 350 °C for 30–360 minutes. Based on data from Lupia (1995).

Schönenberger *et al.*, 2001b). The occurrence of similar pollen grains within anthers and on stigmatic surfaces also provides evidence that can help to connect the different dispersed parts (e.g. carpels and stamens) of the same plant species (Figure 2.18). With regard to pollen dispersal and understanding the palynological record, it is also significant that some pollen grains preserved *in situ* in flowers from mesofossil floras are not detected in the dispersed palynoflora from the same sediments.

2.1.4 Fossil pollen

By far the most abundant angiosperm fossils in Cretaceous and younger rocks are dispersed pollen grains that have been shed from flowers. The walls of pollen grains and spores are composed of sporopollenin, a carotenoid polymer that is extremely resistant to the physical and chemical changes involved in the fossilisation process. This wall is also resistant to strong acids. Pollen and spores may therefore be isolated from sediments that are not strongly oxidised by using standard palynological preparation techniques that use hydrofluoric, hydrochloric and other acids to dissolve away the mineral matrix. Fossil spores and pollen may occur in huge quantities even in a small sample. A few grams or cuttings of core material from drillings are sufficient for a palynological study in most cases. Assemblages of spores and pollen and other small acid-resistant fossils are usually referred to as microfloras or microfossil floras/palynofloras.

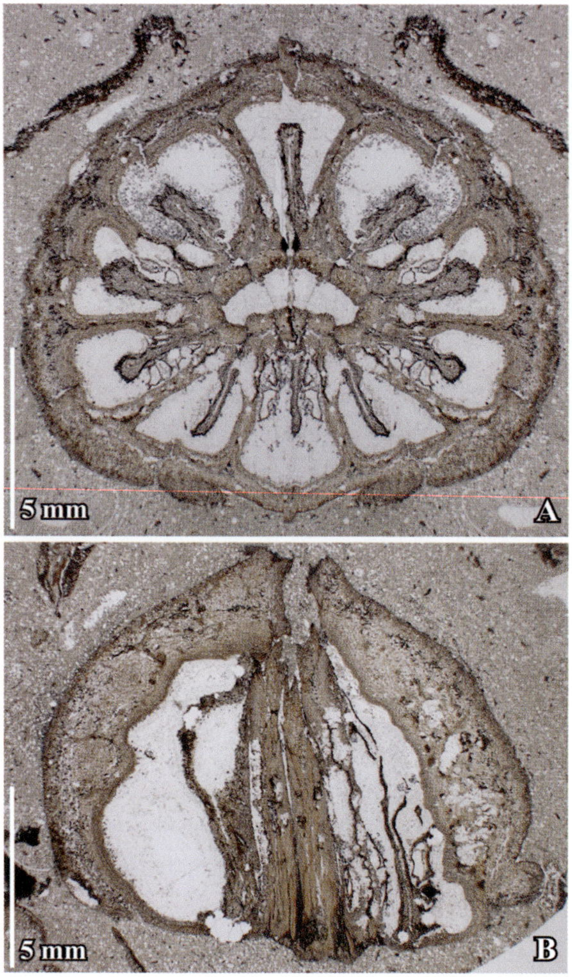

Figure 2.16 *Elsemaria kokubunii*, a permineralised angiosperm reproductive structure from the Late Cretaceous (Coniacian-Santonian) of Hokkaido, Japan, in transverse section (A) and longitudinal section (B). Material in the collections of the Laboratory of Phylogenetic Botany, Faculty of Science, Chiba University. Photograph H. Nishida.

Figure 2.17 Fossil flower preserved in Baltic amber; flower about 2 mm in diameter. Specimen at the Department of Geology, University of Aarhus.

Wind-pollinated taxa typically have much higher pollen production than taxa with biotic pollination and they are therefore generally over-represented in dispersed palynofloras. The concentration of wind-pollinated grains declines markedly away from their source, but nevertheless pollen grains of wind-pollinated taxa may be transported over long distances and incorporated into sediments far from the original vegetation. For example, transport of *Betula* pollen from southwestern Russia to Scandinavia, more than 2000 km, occurs in sufficient quantities to cause allergenic symptoms in sensitive Scandinavians before the local flowering of *Betula*. Pollen is transported into high-elevation air-masses and deposited by turbulence or rainfall (Hjelmroos, 1991). The resistant pollen wall, combined with production in large quantities, and the possibilities of long-distance dispersal by both wind and water, result in fossil pollen grains occurring in many different kinds of terrestrial depositional environments. Fossil pollen and spores from land plants are also commonly transported by wind and water into sediments deposited in marine conditions relatively close to the shore.

The morphological features of some angiosperm pollen grains can provide a broad indication of the likely mode of pollination in the parent plant. Pollen grains of wind-pollinated plants are often smooth and around 20–40 μm in diameter (Whitehead, 1969, 1983; Crane, 1986). Larger, strongly sculptured or coarsely reticulate pollen grains are often associated with animal pollination (Fægri and Iversen, 1964). There are, however, many exceptions to these generalisations, and by itself information from the morphology of pollen grains is generally not sufficient for inferring modes of pollination.

Palynological characters are useful in establishing the systematic affinities of dispersed pollen grains or spores in terms of major plant groups, but are often not diagnostic at

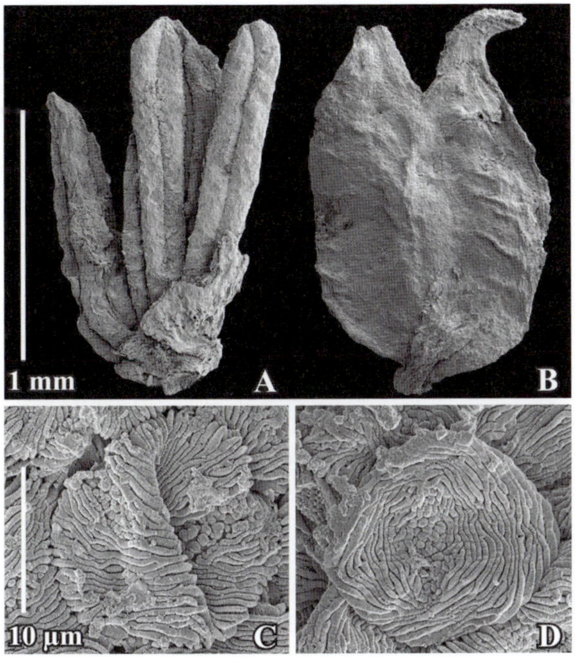

Figure 2.18 Coalified flowers of *Lusistemon striatus* (A, staminate) and *Lusicarpus planatus* (B, pistillate) from the Early Cretaceous (Late Aptian-Early Albian) of Portugal, linked together by the occurrence of similar tricolpate-striate pollen grains *in situ* in stamens (C, D) and adhering to the stigmatic surface (not shown). Material in the collection of the Swedish Museum of Natural History, Stockholm.

the level of order, family or genus. This is particularly the case for pollen grains from Early Cenozoic and older deposits. Studies of the pollen morphology of extant plants have demonstrated that closely similar grains can be produced in groups that are widely separated systematically, and also that closely related taxa can produce distinctly different pollen. Nevertheless the palynological diversity of extant angiosperms has been relatively well surveyed and provides a useful basis for comparisons with fossil material.

Several different approaches have been used to name angiosperm pollen and other dispersed organs, depending on the focus of the study and the type of organs available. In research aimed at understanding phylogeny and evolutionary patterns in fossil plants it is customary to adopt methods and taxon-concepts that are as close as possible to those used for extant taxa. However, in many palynological–stratigraphical applications a pragmatic approach has often been to use a morphographic system (Potonié, 1956), which may result in artificial genera that are highly heterogeneous in terms of plant systematics. A further refinement of the morphographic system that excludes Linnean binomial nomenclature entirely has also been used in a limited number of studies with a strong stratigraphic focus (e.g. Hughes, 1986, 1994; Penny, 1991).

Most species of dispersed pollen have been established based on light microscopy (LM). This method is usually adequate to observe critical information on aperture configuration and wall structure, but optical resolution is usually not sufficient to provide information on supratectal ornamentation and other fine details that can be observed with modern scanning electron microscopy (SEM). Such characters may be of considerable taxonomic importance, especially in studying pollen grains of the earliest angiosperms (see discussion of *Clavatipollenites hughesii*, Chapter 9). Also, in the past, important type specimens for dispersed taxa studied by LM were often embedded in a fluid or solid medium in sealed preparations, which makes them difficult or impossible to access for re-examination and comparison using SEM. One consequence has been a tendency to adopt a very broad definition of dispersed pollen taxa, with the result that many of the genera in current use for dispersed pollen may be unnatural and difficult to use for detailed interpretations of diversification patterns.

The need to refine concepts of pollen taxa based on light microscopy was one of the motivations for Hughes' suggestion (e.g. Hughes, 1986, 1994) that the Linnean nomenclatural system should be abandoned for studies of dispersed pollen taxa. The alternative data-handling method that Hughes developed and recommended offered a partial solution to the problem of large, weakly defined and unnatural taxa, and maintained a strong link between the original sample of fossils studied and the reference system used. However, it also had disadvantages in terms of further divorcing palynology from mainstream palaeobotany and plant systematics and by overemphasising a view of the palynological record that presumed gradualistic evolution. It is also clear that whatever nomenclatural system is used, the requirement for detailed comparison of newly discovered fossil pollen grains with previously described material remains the same.

One major advantage of the pollen and spore record over the record of other organs, at least with current levels of knowledge, is that palynomorphs have been studied throughout the Mesozoic and Cenozoic from deposits all over the world. While much of this work has been carried out for purposes of geological exploration it nevertheless provides a vast wealth of data on the distribution of different kinds of plants in time and space. It thus provides a surprisingly well-documented basis to begin to understand angiosperm diversification in a broad temporal, as well as palaeogeographic and palaeoclimatic, context (Crane and Lidgard, 1989; Lupia *et al.*, 1999).

Another important and useful attribute of the fossil record of palynomorphs is that because fossil pollen and spores are often extracted from sediment samples in large numbers the relative abundance of different taxa is more readily quantified than for mesofossils and macrofossils. Fossil spore/pollen assemblages, therefore, provide opportunities to assess the nature of vegetational changes based on patterns of abundance in addition to patterns of diversity. These two different measures provide important complementary information on the nature of vegetational change during the Cretaceous (Lupia *et al.*, 1999).

2.2 THE ADEQUACY OF THE ANGIOSPERM FOSSIL RECORD

Notwithstanding the inherent limitations of the fossil record, it is important to emphasise that in many cases the morphological and anatomical characters that can be assessed in fossil material are precisely comparable to those that can be obtained from samples of extant taxa. There is therefore a common basis for comparison and for determining systematic relationships. It is also self-evident that current knowledge of plant evolution would be considerably poorer without the information provided by almost two centuries of palaeobotanical research. Studies of the fossil record have advanced our understanding of many aspects of plant evolution. They also have much to contribute to elucidating the evolutionary history of angiosperms. Studies of structurally preserved angiosperm flowers are only in their initial phase, and the Cretaceous record of angiosperm flowers is likely to be expanded considerably in the future.

For plants as a whole, the fossil record has provided a broad chronological overview of the development of major clades (e.g. Kenrick and Crane, 1997; Doyle, 2006; Hilton and Bateman, 2006) and different kinds of vegetation (e.g. Johnson, 2002). It has proved of great practical value in stratigraphic studies and geological exploration. Further, the discovery and description of fossil plants has dramatically expanded our knowledge of plant diversity (Crane *et al.*, 2004). When integrated with information from extant taxa this has provided new insights into the origin of many of the characteristic features of living plants. Taken together, both experience and analysis (e.g. Benton *et al.*, 2000; Benton, 2001) show that the fossil record is useful for a whole range of purposes, from illustrating broad patterns in the history of life, to reconstructing the nature of vegetational and climatic changes, to contributing to the dating of sedimentary rocks.

3

The environmental context of early angiosperm evolution

The global environment during the earliest phases of angiosperm diversification was radically different from that of today. The geography and distribution of continents was unlike that of our modern world and the Cretaceous was also a time of generally higher sea levels and higher global temperatures. There were low thermal gradients from the equator to the poles (DeConto *et al.*, 2000a; Gale, 2000; MacLeod *et al.*, 2000), patterns of rainfall were quite different (Parrish, 1987) and there is no unequivocal evidence of polar ice caps during the Cretaceous and Early Cenozoic (Gale, 2000). The Cretaceous is often considered the classic example of 'greenhouse' or 'supergreenhouse' conditions in Earth history. These unusual conditions compared with today profoundly influenced the ecology and evolution of life on land as angiosperms underwent their initial radiation and then diversified to become the dominant primary producers in most terrestrial ecosystems. Our modern world provides a poor analogue for the environmental backdrop against which more than half of the evolutionary history of angiosperms unfolded. In this chapter we provide a brief overview of changes in palaeogeography and palaeoclimate through the Cretaceous period, a time interval of 80 million years (Figure 3.1). More detailed accounts can be found elsewhere (e.g. Skelton, 2003b). The aim here is to provide a short summary that places patterns of early angiosperm evolution in their environmental context.

3.1 PALAEOGEOGRAPHY

During the later Palaeozoic and Triassic a single supercontinent, Pangaea, was formed through the coalescence of all pre-existing major continental masses. The southern part, Gondwana, consisted of South America, Africa, Apulia (present-day Italy), Arabia, Australia, New Zealand, Antarctica, India, and Madagascar. The northern part, Laurasia, consisted of many smaller and larger continents with North America, Greenland, Europe, Iberia, Siberia, Kazakhstan, Kolyma and China as its main components. Pangaea persisted into the early Mesozoic, but began to break up during the Triassic and Jurassic through a series of processes that continued into the Cretaceous (Figures 3.2–3.5) (Smith *et al.*, 1981; Scotese *et al.*, 1988).

3.1.1 Geologic and geographic changes during the Cretaceous

The breakup of Pangaea continued and intensified through the later Mesozoic (Smith *et al.*, 1981; Scotese *et al.*, 1988). As a result, the Cretaceous period is characterised by an extensive reorganisation and redistribution of tectonic plates, which resulted in major changes in the distribution of land and sea (Figures 3.3–3.5). This, in turn, had significant effects on climate. The extensive rifting and mantle plume formation associated with high rates of tectonic activity caused large-scale volcanic eruptions during the Cretaceous (Gale, 2000). These events created massive oceanic plateaux in the Pacific, Atlantic and Indian Oceans. Total oceanic crust production increased by about 50%–100% over previous levels (e.g. Larson, 1991; Hallam, 1994).

Intense volcanism also resulted in the high concentrations of atmospheric carbon dioxide that perhaps contributed to Cretaceous greenhouse conditions. Repeated oceanic disoxic and anoxic events during the mid-Cretaceous, which are reflected in the deposition of organic-rich marine black shales, may also be linked to the volcanic outgassing of carbon dioxide (Hallam, 1994). Orogenic activity was extensive, particularly during the Late Cretaceous. It resulted in several intracontinental orogenic events including the uplift of the South China block in Asia, and the initiation of the Andean and Rocky Mountain uplifts in the Americas (Gale, 2000).

The changing palaeogeography of the Cretaceous resulted not only from plate movements, but also from marked fluctuations in sea level, which themselves may have been driven by tectonic events (Figure 3.5). The Cretaceous period is characterised by generally high sea levels that resulted in widespread inundation of continental

Period	Epoch	Stage	Ma
			65.5
Cretaceous	Late	Maastrichtian	
			70.6
		Campanian	
			83.5
		Santonian	
			85.8
		Coniacian	
			89.3
		Turonian	
			93.5
		Cenomanian	
			99.6
	Early	Albian	
			112.0
		Aptian	
			125.0
		Barremian	
			130.0
		Hauterivian	
			136.4
		Valanginian	
			140.2
		Berriasian	
			145.5

Figure 3.1 Geological timescale for the Cretaceous period showing the boundaries between the stages in million years (Ma). Note the long duration of the Early Cretaceous epoch (45.9 Ma) compared to the duration of the entire Late Cretaceous epoch (34.1 Ma). Based on Gradstein *et al.* (2004).

margins. In some cases continental interiors were also extensively flooded during the major transgressions that occurred through the Early and Late Cretaceous (Figure 3.6). Especially high sea levels occurred during the Late Cenomanian – Early Turonian as well as the Late Campanian (Gale, 2000). Estimates of absolute eustatic levels vary considerably (Skelton, 2003a), but they were clearly very high compared with the present day. According to Hallam (1992) the maximum Cretaceous sea level may have been about 250 m above present levels: higher than at any other time since the beginning of the Mesozoic.

The overview of Cretaceous palaeogeography and plate movements provided for Laurasia and Gondwana in the following sections is based on palaeocoastal maps (Smith *et al.*, 1994) and plate reconstructions (Smith *et al.*, 1981; Scotese *et al.*, 1988) as well as other research referred to in the text. These studies provide a broad global framework for understanding the history of the continents, but the detailed geology of many regions remains poorly known. Precise reconstructions of palaeogeography, coastlines, sea-level changes and timing of plate movements are therefore problematic in many areas and are the focus of continuing geological research. In this review we focus only on the large-scale events to provide a framework for discussions of climatic change and angiosperm diversification.

3.1.2 Early and mid-Cretaceous

Relatively low global sea level is inferred for the Early Berriasian, but subsequently, in the mid-Berriasian, a long-term transgression was initiated that lasted for most of the Cretaceous. This was interrupted only intermittently by short-term regressions. The first of these brief regressive phases occurred in the latest Berriasian and earliest Valanginian (Haq *et al.*, 1987; Gale, 2000). Short regressive phases also occurred in the Late Barremian – Early Aptian, the mid-Aptian and the Late Aptian – Early Albian (Gale, 2000). During these phases of lower sea level, anoxic facies developed extensively in the ocean basins. Sea level continued to rise through the mid- and Late Albian and Cenomanian and reached its Cretaceous maximum in the latest Cenomanian – earliest Turonian, an interval during which orogenic activity also intensified (Gale, 2000).

Separation of the southern Gondwana continents (Australia, Antarctica, New Zealand, India, Madagascar) from equatorial Gondwana (Africa, Apulia, Arabia, and South America) had already commenced during the latest

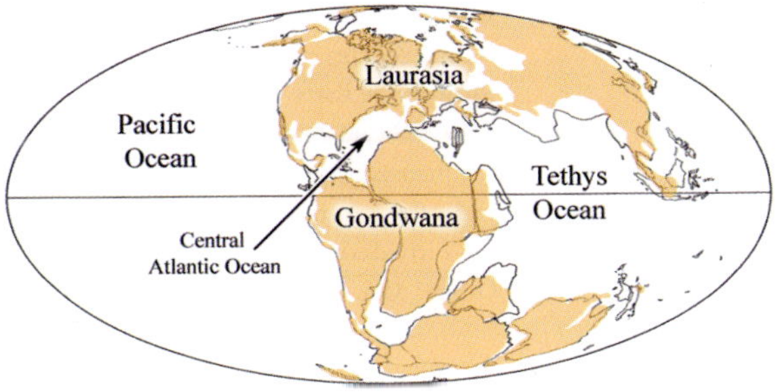

Figure 3.2 Palaeogeographic reconstruction of continental positions during the Early Cretaceous showing the two main continent masses, Laurasia and Gondwana, and the Pacific, Central Atlantic, and Tethys oceans. Light brown areas indicate emergent land surface. Map based on Smith *et al.* (1994).

Jurassic. This continued through the earliest Cretaceous. One effect was that the proto-Indian Ocean between Africa and Antarctica–Madagascar–India widened considerably, but land connection between southern and equatorial Gondwana was probably maintained via the Patagonia–Grahamland land bridge. This land bridge had been disrupted by the formation of a narrow seaway by the Aptian, but whether this key land connection between the southern and equatorial Gondwana continents was subsequently re-established is unclear (Parrish, 1987). Land connection between Antarctica and Australia still existed during the Cenomanian, but rifting, which had already been initiated early in the Cretaceous, caused the two continents to move slowly apart during the Late Cretaceous (Scotese *et al.*, 1988).

Another effect of the Gondwana breakup was that India and Madagascar separated from the southern Gondwana landmasses and began to move north. The timing and course of this northward movement is not fully established (Scotese *et al.*, 1988), but apparently the separation from Antarctica was initiated during the Hauterivian–Barremian (Wilford and Brown, 1994), and the isolation of India–Madagascar from Antarctica was complete by the mid-Cretaceous.

The separation of South America and Africa also began in the earliest Cretaceous, and marine waters extended northward into the rift zone between Argentina and South Africa. As the separation of the two continents continued, marine waters spread further north, and by the end of the Aptian the developing South Atlantic Ocean reached as far north as Gabon, although land connections between South America and Africa were still maintained further north. During the Albian, separation of South America and Africa was almost completed and the opening of the South Atlantic, with connection to the Tethys and the Central Atlantic, was fully established by the Cenomanian. However, the marine connection over a broad elevated zone, the Rio Grande Rise – Walvis Ridge complex in the central South Atlantic, was very shallow during the Albian–Cenomanian. According to Parrish (1987) there is no evidence that large parts of this zone were above sea level, but the extent of potential connection, for example via ephemeral islands, is hard to clarify. The last date at which significant faunal or floral exchange between Africa and South America could have occurred therefore remains uncertain and there is evidence that some exchange, most likely by dispersal, remained possible until well into the Cenozoic (e.g. Flynn and Wyss, 1998).

In the earliest Cretaceous the central Atlantic Ocean separating Gondwana and western Laurasia began to widen and the Tethys expanded further westwards, although it continued to be constricted and narrow between the Iberian Peninsula and Africa. The Tethys also extended north to form a narrow connection to the Arctic Ocean (the Turgai Strait) that persisted throughout the Cretaceous (Figure 3.6). Land connection between North and South America apparently was also maintained into the Early Cretaceous. Large parts of Europe were also inundated by the ocean.

In North America the establishment of the Western Interior Seaway was initiated by an extension of a southerly embayment of the Arctic Ocean that spread from the north towards the south (Figure 3.6). During the mid-Cretaceous, further expansion of this embayment from the Arctic connected with a northward-expanding embayment from the Gulf of Mexico. With this connection established, the North Atlantic, the Tethys and the Arctic Ocean were linked, and there was also connection through the developing South Atlantic into the waters around Antarctica.

3.1.3 Late Cretaceous

After the extremely high sea levels of the latest Cenomanian and Early Turonian a major regressive phase took place in the Late Turonian in conjunction with a very marked lowering of sea level. However, a new transgressive phase began in the Coniacian and culminated in the Campanian with the breakdown of barriers between the Pacific and the developing Atlantic and Indian Oceans (Huber and Watkins, 1992; Gale, 2000). During the

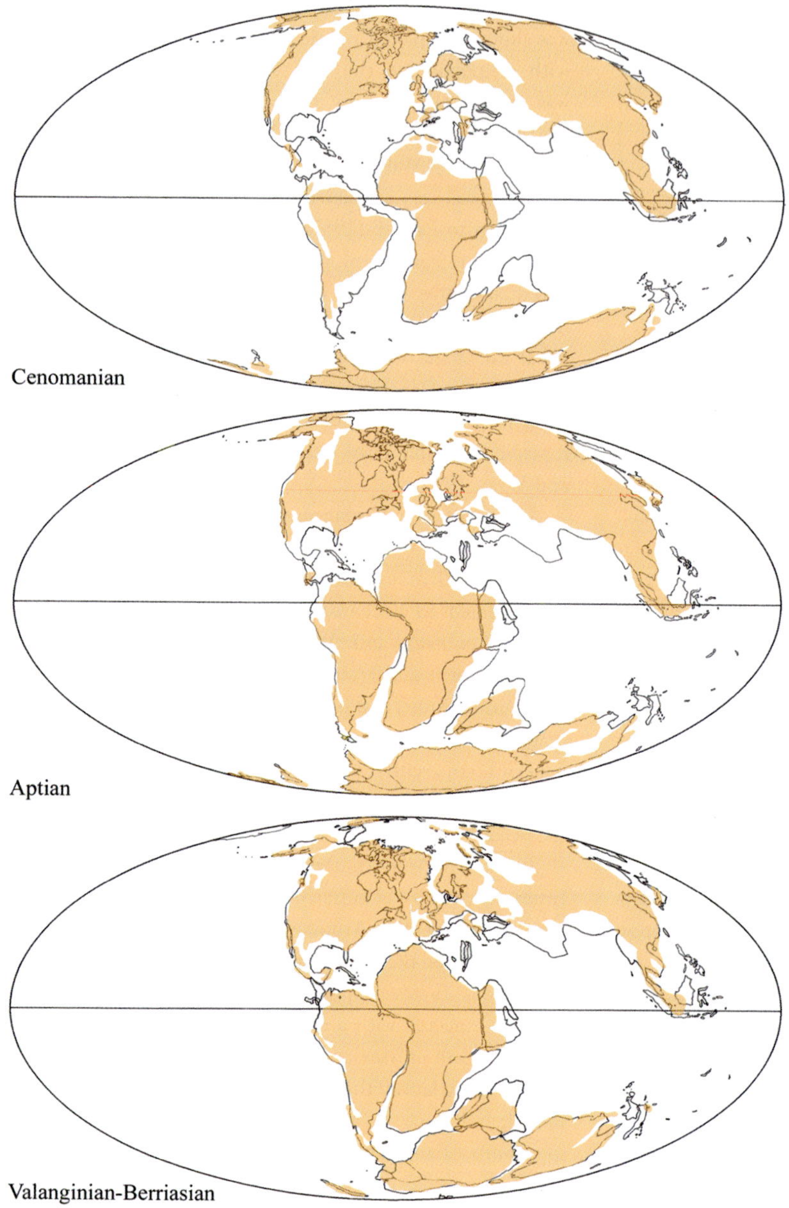

Figure 3.3 Palaeogeographic reconstruction of continental positions through the Early Cretaceous. Light brown areas indicate emergent land surface. Note the increasing separation of Africa and India from Antarctica and Australia and the early development of the Atlantic Ocean. Maps based on Smith *et al.* (1994).

Campanian it is estimated that the total global land area was about 20% less than at present (DeConto *et al.*, 2000a). The last major regressive phase of the Cretaceous started in the Late Campanian and continued through the Maastrichtian with a minimum sea level stand occurring at around the Cretaceous–Palaeogene boundary. An estimated fall in sea level of perhaps 50–100 m in the Maastrichtian coincides with a decrease in global temperature (Hallam, 1992). The later part of the Cretaceous was also marked by intensified orogenic uplift (Gale, 2000). By the Campanian–Maastrichtian both the Andes and the Rocky Mountains are thought to have achieved considerable heights (Gale, 2000), which would also have had an impact on regional climate.

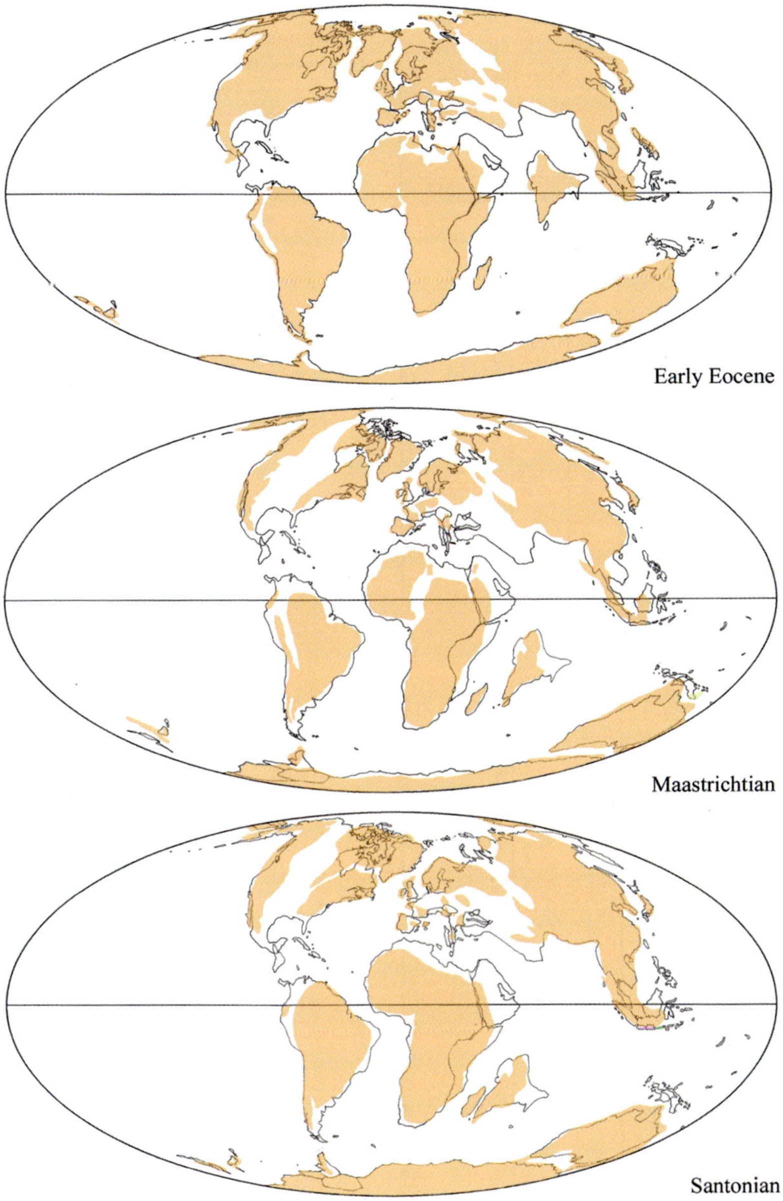

Figure 3.4 Palaeogeographic reconstruction of continental positions through the Late Cretaceous into the Early Cenozoic (Early Eocene). Light brown areas indicate emergent land surface. Note the development of the Atlantic Ocean, the closure of the Mediterranean from the much wider Tethys, the northward movement of India and the separation of Australia from Antarctica. Maps based on Smith *et al.* (1994).

After the Cenomanian, Australia became isolated from Antarctica, but similarities in fossil marsupial faunas suggest that some level of connection between the two landmasses persisted as late as the Eocene (Marshall, 1980). Antarctica and South America may also have been linked by land connections as a result of uplift in the southern Andes and the Antarctic Peninsula, although again definitive evidence on the timing of connections and separations is not available (Parrish, 1987).

Madagascar became isolated as the northward movement of India continued throughout the Cretaceous. This further northward movement apparently also initiated one of the most extensive volcanic events of the Cretaceous, which led to the formation of the Deccan Traps. These

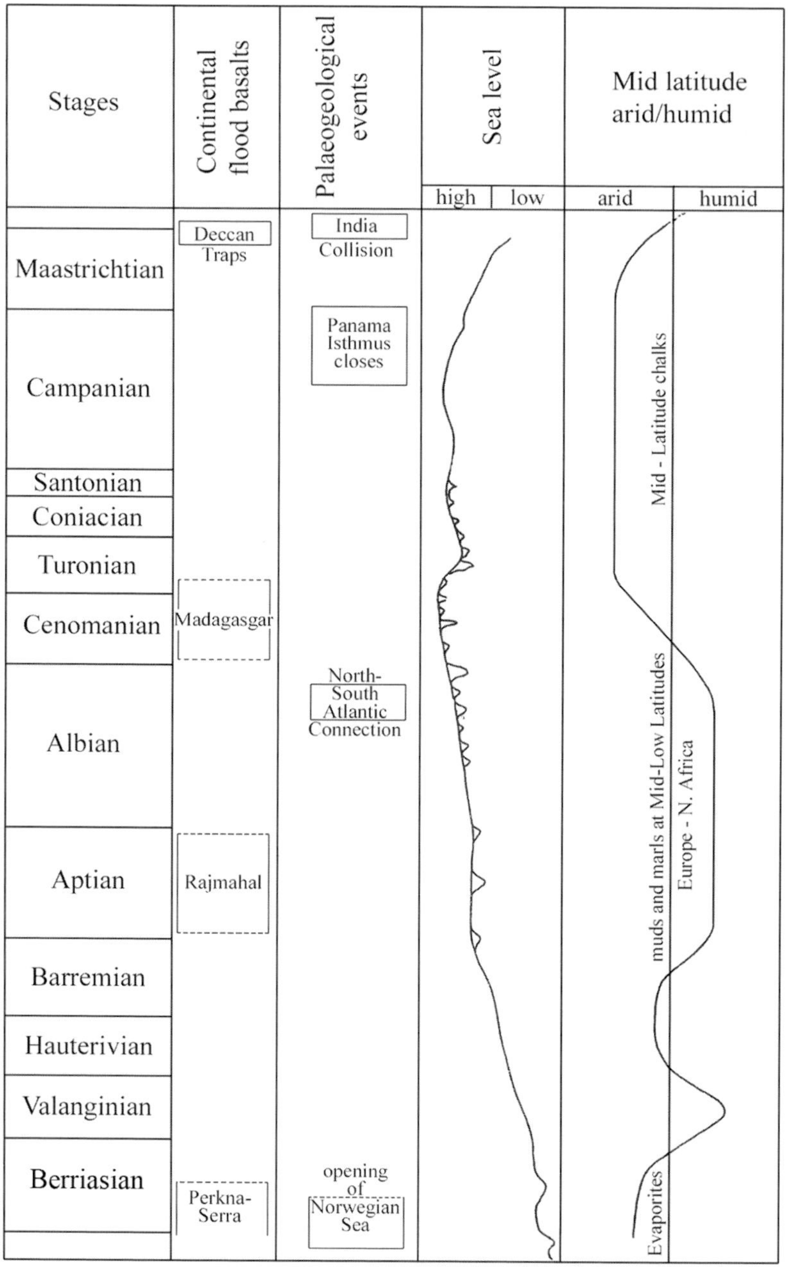

Figure 3.5 Summary of major changes in key environmental parameters during the Cretaceous. Based on Gale (2000).

were apparently formed over a narrow time interval spanning the Cretaceous–Palaeogene boundary and were associated with India passing over a hotspot as it moved northward towards Asia (Jaeger *et al.*, 1989; Gale, 2000).

The separation of South America and Africa also continued during the Late Cretaceous as the South Atlantic broadened. A seaway separating northwestern Africa from the rest of the continent was also established and persisted throughout the Late Cretaceous. Apulia (Italy) was finally

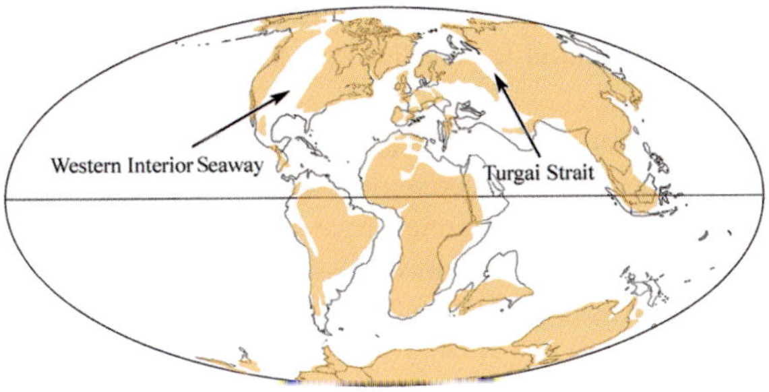

Figure 3.6 Palaeogeographic reconstruction of continental positions during the Cenomanian showing the two major intercontinental seaways created by especially high sea levels: the mid-continental Western Interior Seaway in North America and the Turgai Strait separating Europe and Asia. Map based on Smith *et al.* (1994).

separated from Africa and was connected with southern Europe by the Santonian–Campanian.

The relatively high sea levels of most of the Turonian–Maastrichtian continued to inundate large parts of Europe. High sea levels also ensured that the Turgai seaway separating Europe and Asia, as well as the Western Interior Seaway separating eastern and western North America, persisted throughout the Late Cretaceous. High sea levels also separated Europe from Greenland and North America through most of the Late Cretaceous (Parrish, 1987).

The relative positions of North and South America have remained almost unchanged since the latest Cretaceous (Parrish, 1987; Hallam, 1994). For most of the Cretaceous they appear to have been separated, although connection between North America and South America may have been partly established during the Campanian and Maastrichtian by a volcanic arc in the Central American – Caribbean region (Archibald, 1996).

3.2 PALAEOCLIMATE

The tectonic, orogenic and sea-level changes that occurred through the Cretaceous and Cenozoic had significant and lasting effects on global climate, primarily through their effect on patterns of oceanic and atmospheric circulation, but also indirectly through their effect on atmospheric composition.

The nature of global and regional climate change during the Cretaceous and Cenozoic has received a great deal of research attention, especially in recent decades, based both on palaeontological and geological evidence. There has also been interest in the retrospective application of physical models of climate change developed primarily for predicting climate change in the future (Barron *et al.*, 1981; DeConto *et al.*, 2000b).

3.2.1 Palaeontological indicators of climate

The most straightforward and most commonly used methods for inferring palaeoclimate are based on the premise that the climatic tolerances of extinct groups of plants and animals were fundamentally the same as those of their nearest living relatives. This technique has been employed particularly successfully for reconstructing climate during the later part of the Cenozoic, and over these relatively short timescales (a few thousands to a few millions of years) potential errors as a result of misidentifications or marked changes in climatic tolerance in certain lineages are probably relatively minor. However, over longer timescales (tens of million of years) the potential problems introduced by systematic misidentifications, or by evolutionary change in ecological characteristics, increase. Nevertheless, analyses of nearest living relatives have been widely and successfully used to provide a crude indication of the climatic conditions prevailing in the past, as far back as the Early Cenozoic.

An alternative approach to inferring ancient climates using palaeontological data, which seeks to be independent of systematic determinations, is based on correlations between broad climatic parameters and the structural features of fossil plants (e.g. tree rings, leaf physiognomy) or animals (e.g. coral reefs).

From studies of modern vegetation it is clear that there is a strong correlation between leaf physiognomy and the climate under which the vegetation is growing (Wolfe, 1979). Based on this correlation, and building on the early work of Bailey and Sinnott (1916), fossil leaves have been widely used to infer palaeoclimate (especially temperature and precipitation) and palaeoenvironment (e.g. Dilcher, 1973; Upchurch and Wolfe, 1987; Wolfe and Upchurch, 1987). Six physiognomic features of particular importance are discussed by Wolfe and Spicer (1999): leaf margin; leaf size; leaf apex; leaf length; leaf length:width ratio; and leaf shape.

In broad terms, large leaves are characteristic of plants of humid vegetation or low light intensities, for example leaves of plants growing in ever-wet rain forest or the understory of evergreen stratified forest. In contrast, small leaves are characteristic of vegetation from colder or drier climates. Similarly, entire leaf margins are typical of vegetation from warmer and drier climates, whereas leaves with serrate margins are more common in cooler, more humid climates. Thick and tough leaves characterise many, largely evergreen and xerophytic plants; leaves of deciduous plants in mesic climates are often thin. Drip-tips are almost exclusively associated with evergreen leaves and are common in the understory vegetation of humid evergreen forests. Compound leaves are broadly associated with deciduous vegetation and are particularly common among plants of successional or disturbed vegetation. Palmately veined leaves are also more common in successional or disturbed vegetation, and are characteristic of many lianas.

Modern efforts to generate temperature and precipitation estimates from assemblages of fossil leaves seek to quantify one or more of these generalisations based on extant vegetation and then extrapolate into the fossil record. A sophisticated methodology, Climate–Leaf Analysis Multivariate Program (CLAMP), has been developed by Wolfe (1990, 1993) and Wolfe and Spicer (1999) to describe the complex interrelationships among different aspects of leaf physiognomy of angiosperms and environmental parameters. This approach is now widely used to obtain results that are comparable among fossil sites. The core of CLAMP is a large database of leaf character states in woody dicots, from a broad spectrum of modern vegetation, and their corresponding climatic parameters.

In a similar way to gross form, epidermal features of fossil leaves also reflect the conditions under which the plants grew, and correlation between epidermal features and environment also applies to the non-angiosperm components of a fossil flora. For example, stomatal density is used increasingly to estimate concentrations of atmospheric carbon dioxide. With regard to more direct estimates of climate, potentially significant features include the thickness of the cuticle, the presence of epicuticular waxes, the distribution and position of stomata, and the presence of trichomes, papillae and glands. In broad terms, a thin-walled epidermis, thin cuticles, evenly distributed and non-sunken stomata, and a moderate or absent indumentum, are all features characteristic of plants growing under humid conditions. Epidermal features that tend to characterise plants from xeric environments (Figures 3.7, 3.8) include a thick-walled epidermis and hypodermis, a dense, waxy, cuticle with cutinisation extending to the inner periclinal walls of the epidermis, a thick (5 μm or more) shiny reflective cuticle, sunken stomata, stomatal crypts, papillate subsidiary cells, stomata clustered or restricted to certain areas, presence of papillae and trichomes, and the presence of salt glands (Watson and Alvin, 1996). So far these parameters have not been quantified in a similar way to the morphological features incorporated into the CLAMP database.

Many Early Cretaceous plants show strong indications of having grown under xeric conditions. For example, leaves of Bennettitales often have stomatal pits overhung by papillae, stomata confined to narrow grooves, numerous trichomes, possible salt glands and probably a large amount of wax on the upper cuticle (Watson and Alvin, 1996). Similarly, leaves of Cheirolepidiaceae, a characteristic group of extinct conifers including *Frenelopsis* and *Pseudofrenelopsis*, are extremely reduced (Figure 3.8). Photosynthesis evidently mainly occurred in the stems. The epidermis has a very thick cuticle and is often papillate with sunken stomata. Marginal hairs on the leaves of many species (Figure 3.8) have been compared with similar hairs on the leaves of the extant cupressaceous conifer *Tetraclinis articulata*. In *Tetraclinis*, which grows in semi-arid environments, these hairs play a role in condensing moisture from the air during the night (Pons, 1979). Wood of Cheirolepidiaceae has growth rings of uneven width, a feature characteristic of trees growing in savannah and similar environments with irregular seasonally dry climate. Seeds of Cheirolepidiaceae are also protected, perhaps against desiccation, by the enclosing cone scale (Watson and Alvin, 1996).

Like the physiognomic characteristics of leaves, the presence of growth rings in fossil wood (Figure 3.9) is widely used as an indication of precipitation and seasonality in ancient climates. Rings are formed by an alternation between intervals that are favourable for plant growth, alternating with intervals of halted or reduced plant growth due to either low temperatures or drought (Francis, 1984). Other environmental inferences are also possible from wood anatomy (for a review see Herendeen *et al.*, 1999b), but the use of wood characters to deduce palaeoecological and palaeoclimatic parameters is complex (Carlquist, 1988) and the results may not be definitive. Nevertheless, information from wood, in conjunction with other lines of evidence to help infer palaeoclimate, can be valuable.

During the Cretaceous, changes in floristic diversity and changes in the relative proportions of different groups of plants in different fossil floras have also been used to generate crude inferences about climate. For example, the presence of rich floras in high-latitude regions of both hemispheres during the Late Cretaceous and Early Cenozoic (e.g. Nathorst, 1890; Koch, 1964; Kvaček *et al.*,

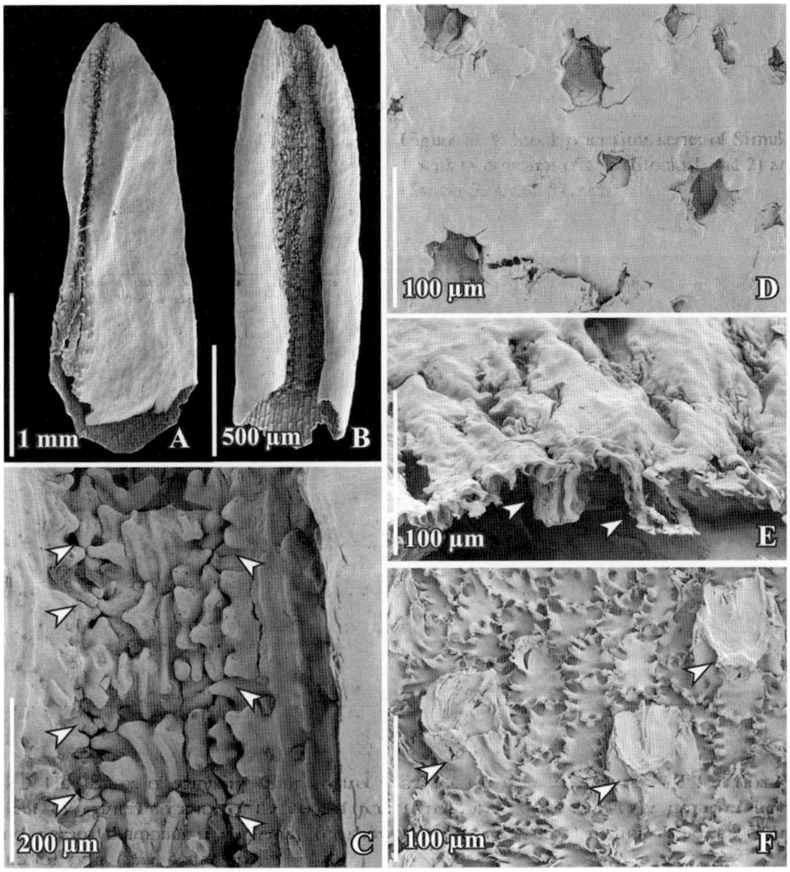

Figure 3.7 SEM micrographs of leaf fragments from the Early Cretaceous (Berriasian) Skyttegård locality, Bornholm, Denmark, showing xeromorphic epidermal features. Only the cuticle with imprints of epidermal cells is preserved. (A, B) Two different leaf fragments with a thick shiny, reflective cuticle and stomata crowded in a narrow sunken groove covered with papillae and hairs. (C) Detail of leaf in 3.7B showing stomata covered by papillae (arrows). (D–F) Epidermal features of leaf with deeply sunken stomata (arrows) shown from outer (D) and inner (F) surface of cuticle and in section (E). Specimens in the collections of the Swedish Museum of Natural History.

1994; Cantrill and Poole, 2002) has been used to suggest warmer climates and the probable absence of polar ice caps at this time (see also Gale, 2000). In a similar way, most living bryophytes, lycopsids, horsetails and ferns are characteristic of relatively humid conditions; the diversity of these plants, or abundance of their spores, has often been used to infer a humid climate (e.g. Schrank and Nesterova, 1993). Conversely, among seed plants, Bennettitales, Cheirolepidiaceae and ephedroids (certain Gnetales) are generally thought to favour xeric conditions. In particular, the occurrence of *Classopollis* (pollen of Cheirolepidiaceae, Figure 3.8) and ephedroid pollen has often been used as an indication of drier climates (Figure 3.10).

There are, however, many complications to almost all methods that use plants to infer palaeoclimate. For example, during the Early Cretaceous certain very characteristic ferns have distinct xerophytic features. Especially prominent are *Onychiopsis* and *Weichselia*, which have small, reduced pinnules and sporangia that are deeply embedded in a protective tissue (Figure 3.11). Similarly, although leaf physiognomic methods are independent of some of the complications of systematics they still require consistent delimitation of systematic units. Further, certain key physiognomic features (e.g. leaves with serrate margins) also appear to be concentrated in certain clades of extant angiosperms. Nevertheless, climatic reconstruction based on fossil plants can be useful when used with restraint. Very specific conclusions about ancient climates, based on only few lines of evidence, need to be treated with caution.

3.2.2 Geological indicators of climate

There is a well-documented broad correlation between climate and the distribution of different types of

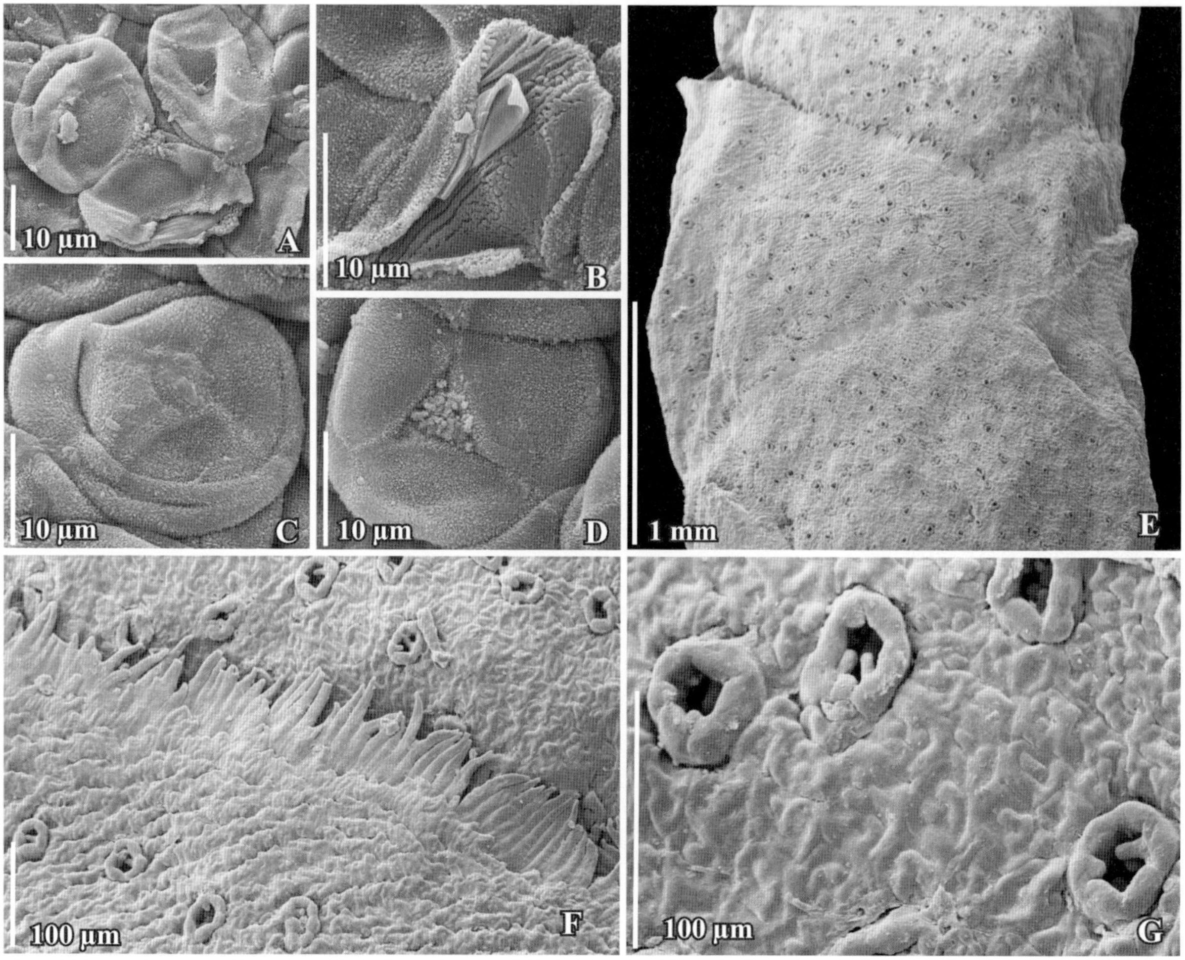

Figure 3.8 SEM micrographs of pollen and twig of Cheirolepidiaceae from the Early Cretaceous Vale de Agua and Buarcos localities, Portugal. (A–D) *Classopollis* pollen in tetrad (A), internal view (B), distal (C) and proximal (D) view. (E–G) *Pseudofrenelopsis* twig showing strongly reduced leaves with fringed margins (E, F) and sunken stomata surrounded by a raised rim with papillae protruding over the stomatal pit (F, G). Specimens in the collections of the Swedish Museum of Natural History, Stockholm.

sedimentary deposit. Distributions of various sediment types have therefore been used to infer palaeoclimate, especially during the Palaeozoic and Mesozoic where extrapolation based on the climatic tolerances of extant relatives of fossil organisms is more problematic than for the Cenozoic. Some of these geological indicators are particularly useful in terrestrial sediments; others are applicable only to marine deposits. The presence of ironstones, coal beds, bauxite, laterite and kaolinite in a sedimentary sequence is generally used as an indicator of warm and/or humid climates. The presence of charcoal, smectite and evaporites are generally taken to indicate dry conditions.

Coal beds form almost exclusively in freshwater swamps. Extensive occurrences of coal beds in a geological sequence are commonly used to infer humid climates. The formation of substantial peat deposits occurs today mainly in warm temperate to tropical regions in which high precipitation is distributed evenly throughout the year, but peat is also formed under colder climatic conditions and occurs abundantly at high latitudes, particularly in the Northern Hemisphere (Ziegler *et al.*, 1987).

Bauxite, laterite and kaolinite are weathering products formed in terrestrial environments under warm and humid climates. Bauxite and laterite are typically formed

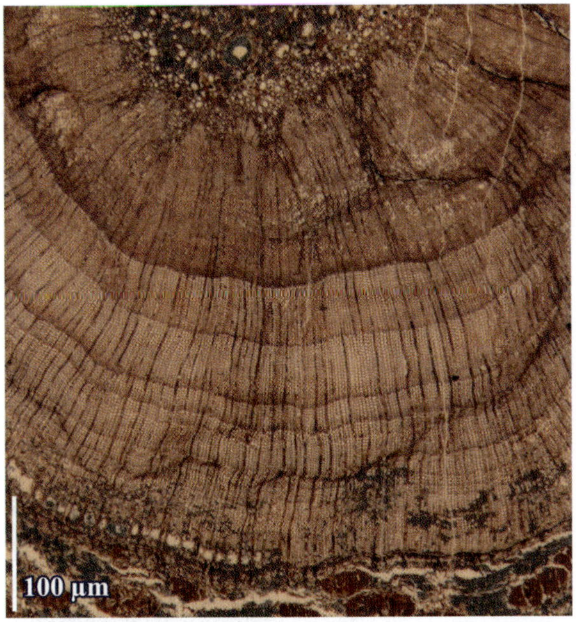

Figure 3.9 Permineralised wood from the Late Cretaceous of Hokkaido, Japan, showing distinct growth rings indicating climatic seasonality. Specimen in the collections of the Swedish Museum of Natural History, Stockholm.

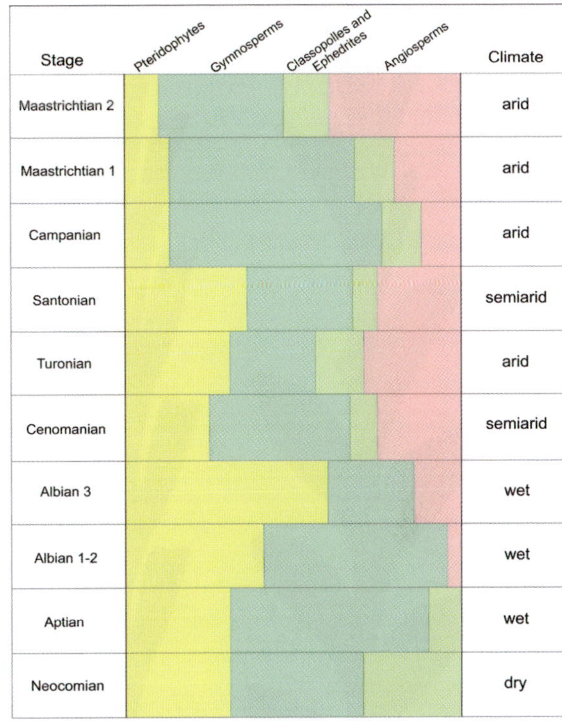

Figure 3.10 Changes in the composition of palynofloras and inferred changes in climate based on the changing abundance of different kinds of spores and pollen in Kazakhstan through the Cretaceous. Inferred climatic changes reflect interpretations based on the presumed ecological preferences of different groups of land plants. Adapted from Schrank and Nesterova (1993).

in better-drained soils, whereas kaolinite is characteristic of wetter conditions (Hallam, 1984, 1985). Bauxite and laterite are also restricted to terrestrial deposits. Kaolinite is also commonly deposited in marine environments and its presence or absence in marine sediments is used to infer humid or arid climates, respectively, in the source area. Among the other minerals formed under humid conditions that have been used as indicators of palaeoclimate are siderite and goethite. Smectite is another weathering product that is formed in terrestrial environments, but under strong seasonal variation in humidity (Hallam, 1985).

The formation of charcoal is clearly related to the presence of wildfires (e.g. Harris, 1981). Today high frequencies of wildfire mainly occur in areas with strong seasonal climate where a growing season with sparse to abundant precipitation is followed by a dry period of several months (Bond and van Wilgen, 1996). Abundant occurrences of charcoal in sediments may thus indicate seasonal drought, although extensive wildfires may occur also in more humid areas, such as rainforests, following periods without rain (Bond and van Wilgen, 1996).

The presence of evaporites provides probably the most unambiguous geological indication of a dry climate. Evaporites may form in terrestrial environments, but these are rarely preserved in geological sequences. Most fossil evaporite deposits are of marine origin, where they form in shallow basins with high evaporation and low precipitation. Today marine evaporites are mainly formed at latitudes between 15° and 35° in both hemispheres (Sellwood and Price, 1994). In a similar way, carbonate deposits are formed mainly in tropical waters where there is both inorganic carbonate precipitation and high organic calcium carbonate production. Pure carbonate deposits are characteristic of areas dominated by warm and dry climate. Low inputs of clastic material are also often a result of low precipitation and less intense weathering in the land areas surrounding the depositional basin.

Geological indicators generally provide rather coarse estimates of palaeoclimate, but they can often be assessed over large geographical areas based on information in the geological literature. Potentially more precise temperature

Figure 3.11 The extinct fern *Onychiopsis psilotoides* from the earliest Cretaceous of Bornholm, Denmark, showing fertile (A) and sterile (B) portions of fronds. On fertile fronds the sporangia are completely enclosed by a strongly thickened, overlapping leaf lamina. Specimens in the collections of the Swedish Museum of Natural History, Stockholm.

curves, or measures of other environmental parameters such as atmospheric composition, may be obtained using several kinds of isotope data. For example, oxygen isotope studies, usually performed on marine carbonates and calcareous shells of marine organisms, record the ratio of ^{16}O to ^{18}O ($\delta^{18}O$ values). To a large extent this ratio is determined by the temperature of the seawater in which the carbonate was precipitated (Gale, 2000). Similarly, the concentration of carbon dioxide in the atmosphere is reflected in the ratio of ^{12}C to ^{13}C ($\delta^{13}C$ values) preserved in inorganic and biogenetic carbonates, as well as in carbon accumulated in fossil plants and coal beds.

3.3 CLIMATE CHANGE DURING THE CRETACEOUS

Based on both palaeontological and geologic evidence, it is widely accepted that the Cretaceous was a period of elevated global warmth with estimated average sea surface temperatures 6–14 °C higher than at present (Kerr, 1984). Especially high sea surface temperatures have been inferred for the Turonian with temperatures up to 35 °C at low palaeolatitudes (Forster *et al.*, 2007). One effect of these elevated temperatures was that during the Cretaceous, unlike today, deep ocean water was warm and saline. This had an impact on both ocean circulation and land temperatures (Gale, 2000).

Further indications of global warmth during the Cretaceous come from the presence of rich floras at high palaeolatitudes, as well as inferences based on climate models. Both suggest that extensive polar ice caps did not exist during the Cretaceous (Upchurch and Wolfe, 1987; Barron *et al.*, 1994; Gale, 2000) and if true this means that cycles of marine transgression and regression during the Cretaceous were mainly of tectonic origin. Polar temperatures appear to have been moderate, and equator to pole thermal gradients were apparently low. Modelling also suggests higher temperatures than today in the tropics and that these higher temperatures were substantially more pronounced over land than over the oceans (Barron *et al.*, 1994; Gale, 2000). Temperatures apparently remained high into the Early Cenozoic. Based on a leaf-size index and the percentage of entire-margined species, Upchurch and Wolfe (1987) recorded mean annual temperatures in North America (low to mid palaeolatitudes) for Albian to Eocene floras between about 21°C and 30°C, with the highest temperatures recorded for the Eocene.

In addition to generally higher temperatures, the Cretaceous period is also characterised by marked variability in both temperature and humidity. Both these features have been explained partly by greenhouse conditions associated with high atmospheric carbon dioxide. Cretaceous concentrations of atmospheric CO_2 are estimated at eight to ten times higher than today (Berner, 1991; Barron *et al.*, 1994; Gale, 2000). These high carbon dioxide levels have been linked to extensive volcanism associated with widespread rifting, particularly the substantial seafloor spreading that occurred in the Atlantic, Indian, and Pacific Oceans during the mid- and Late Cretaceous.

The overall warming trend in Cretaceous climate began in the earliest Cretaceous and continued through the

Cretaceous until about the Campanian. There is a marked temperature peak in the Aptian, and maximum temperatures were obtained through the Cenomanian–Campanian interval. A decline in temperature started in the Campanian–Maastrichtian and continued through the Paleocene until temperatures rose again during the Eocene. Associated with the overall warmth of Cretaceous climates, arid or semi-arid conditions were apparently widespread, particularly at low and mid-palaeolatitudes. This seems to have been especially the case during the earliest part of the Early Cretaceous, and in the Late Cretaceous. The Barremian–Albian time interval has been inferred to include more humid intervals. Dry conditions are reflected in the widespread occurrence of charcoal in both Early and Late Cretaceous sediments at mid-palaeolatitudes and may reflect strong seasonality in precipitation. The initial development of more humid conditions in palaeoequatorial regions is indicated by the presence of diverse palm pollen in palaeoequatorial regions during the Late Cretaceous, but conditions suitable for the development of wet tropical forest may not have existed until close to the end of the Cretaceous and were perhaps not widespread until the Early Cenozoic.

3.3.1 Early and mid-Cretaceous

Generalisations about the nature of climatic change through the Early and mid-Cretaceous are complicated by significant regional and palaeolatitudinal variation, but by the latest Jurassic – earliest Cretaceous, both modelling and geological evidence suggests a combination of warmth and low precipitation across low and mid-palaeolatitude continents (Parrish *et al.*, 1982; Ziegler *et al.*, 1987) (Figure 3.12). Palynological assemblages from the Berriasian–Hauterivian of West Africa and Brazil include common pteridophyte spores and bisaccate pollen, which are sometimes thought to indicate more humid conditions (Herngreen *et al.*, 1996), but the climate was probably dominantly warm and dry. Mid-latitude palynological assemblages from this time interval include many xerophytic elements consistent with arid or semi-arid conditions (Herngreen *et al.*, 1996). Dry conditions in North America and Europe for the earliest Cretaceous were also inferred by Hallam (1984, 1985) based on a low content of terrestrial organic matter in the sediments. The amount of terrestrial organic matter is higher in Valanginian sediments in North America and Europe and it has been suggested that a climatic change to more humid conditions took place in these regions during the Valanginian (Gale, 2000). At this time a temperate humid climate is inferred for the northernmost high-palaeolatitude regions and the southern continents based on the presence of palynofloras dominated by abundant pteridophyte spores and saccate conifer pollen.

Beginning in the Hauterivian–Barremian and continuing until the mid-Aptian (Ruffell and Batten, 1990; Gale, 2000) the western European region shows evidence of a more intense arid or semi-arid phase. Similarly an increase in the diversity of ephedroid pollen may indicate that the climate became drier at low palaeolatitudes during the Barremian–Aptian (Herngreen *et al.*, 1996). A thick sequence of evaporites, formed in the northern part of the opening South Atlantic in the Aptian–Albian, also indicates the prevalence of arid conditions at low palaeolatitudes (Doyle *et al.*, 1982). An arid or seasonally dry climate for this region is also supported by the occurrence of numerous and diverse ephedroid pollen grains in a zone extending from Peru through Brazil (Osborn *et al.*, 1993) across West Africa and into southern China during the later part of the Early Cretaceous.

In Laurasia there are several indications from mid-palaeolatitude regions of intervals with more humid climate. For example, Rat (1989) suggested a warm and more humid climate during the Aptian–Middle Albian in the Iberian area based on geological evidence. Similarly, based on the composition of palynological assemblages, Schrank and Nesterova (1993) inferred relatively humid conditions for the Aptian–Albian of Kazakhstan (Figure 3.10). Late Aptian – Early Albian bauxite and coal deposits at mid-palaeolatitudes also indicate increased humidity during this time interval.

At high palaeolatitudes more humid conditions are indicated through the later part of the Early Cretaceous by abundant siliciclastic sediments and the presence of many coal beds. Barremian coals are reported from Spitsbergen; the Barremian–Aptian Kome Formation from Greenland has several coal seams up to 1 m thick (Pedersen, 1976). Palynological data indicate generally cool and humid conditions in southern high-palaeolatitude regions, but semi-arid conditions and marked seasonality are also inferred for some areas during the Aptian (Herngreen *et al.*, 1996).

Compared with conditions earlier in the Cretaceous, increasingly humid conditions are inferred over broad areas during the Late Albian – Cenomanian based on geological data such as distribution of coals and ironstones (Parrish *et al.*, 1982; Parrish, 1987; Gale, 2000). This is supported by palynological data from mid- and high palaeolatitudes that show diverse assemblages of pteridophyte pollen and bisaccate conifer pollen (Herngreen *et al.*, 1996). Studies of

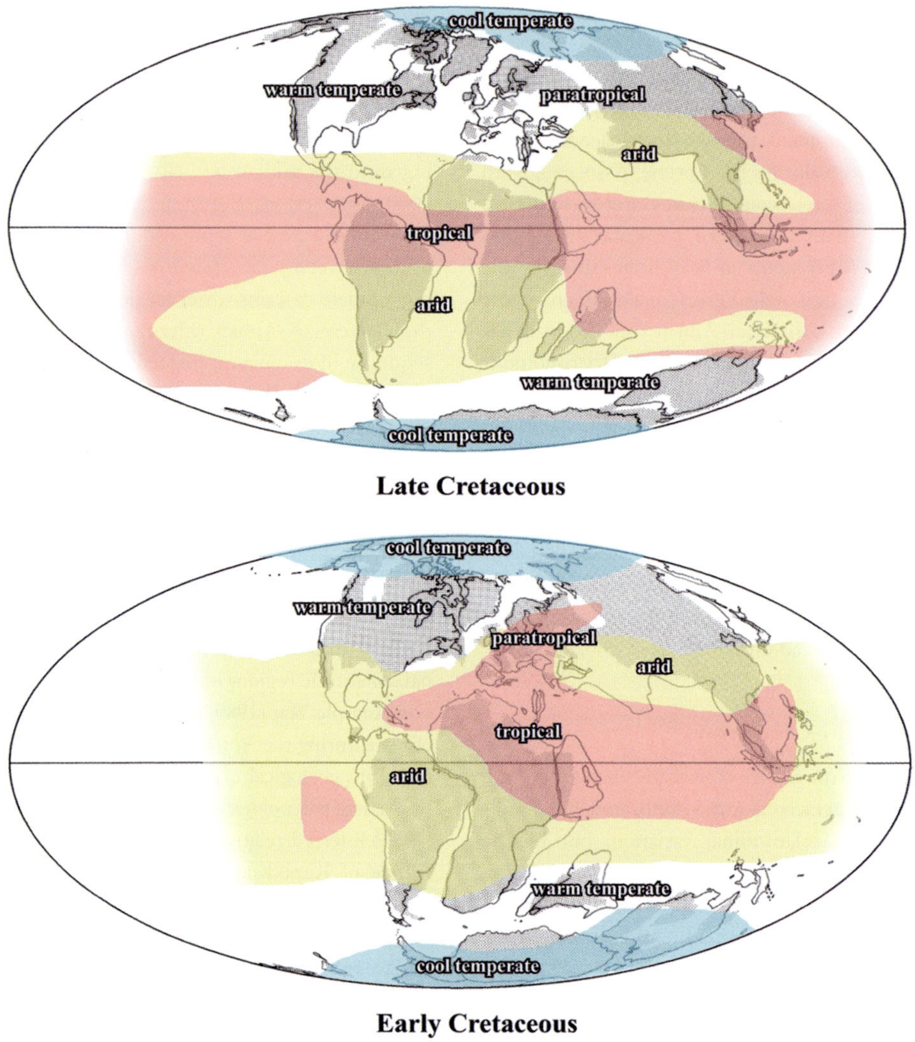

Figure 3.12 Broadly defined climatic zones in the Early and Late Cretaceous. Information from the PALEOMAP Project (http://www.scotese.com/). Map based on Smith *et al.* (1994).

palynological assemblages also indicate that more humid conditions were established in some low-palaeolatitude coastal areas; Parrish *et al.* (1982) suggested an increase in humidity in the regions bordering the Atlantic Ocean as a result of the opening of the South Atlantic and expansion of the North Atlantic. In inland areas palynological assemblages at low palaeolatitudes remained characterised by abundant xerophytic elements, such as *Classopollis*, a variety of ephedroid pollen, and numerous elater-bearing pollen taxa (Schrank and Nesterova, 1993; Herngreen *et al.*, 1996; Dino *et al.*, 1999). Arid or semi-arid conditions were also inferred for parts of Northern Africa and Central Asia based on the presence of evaporites (Parrish *et al.*, 1982).

3.3.2 Late Cretaceous

Following a sea-surface temperature maximum in the Late Cenomanian – Early Turonian (Forster *et al.*, 2007) a cooling phase lasting for about 200 000 years during the Middle Turonian has been suggested based on isotopic data. This has been interpreted as a glaciation event (Bornemann *et al.*, 2008), but the existence of polar ice caps during the Cretaceous remains controversial and sea surface temperatures up to 32 °C have been estimated for high palaeolatitudes in the Southern Hemisphere during the Turonian (Bice *et al.*, 2003). High temperatures were apparently maintained in high-palaeolatitude regions through the Santonian, but were followed by a temperature decline through the Campanian–Maastrichtian into the earliest Cenozoic. In low- and mid-palaeolatitude regions the decrease in temperature started earlier and was under way by the Late Turonian (Gale, 2000). This is consistent with the generally much more restricted distribution of evaporites by the end of the Cretaceous compared with the Early and mid-Cretaceous, and the increase in widespread coals and ironstones at mid- and higher palaeolatitudes. Hallam (1984) concluded that humid zones became more extensive in the Late Cretaceous and related this to the increased influence of epicontinental seas and further separation of the continents (Hallam, 1985).

The formation of the permanent connection between the South and North Atlantic in the Late Albian – Cenomanian resulted in new patterns of ocean circulation and apparently higher humidity at least in some of the coastal areas of the Atlantic region (Parrish *et al.*, 1982). In these low palaeolatitudes, high abundances of palm pollen and pollen of other angiosperms indicate more humid conditions from the Turonian and onwards. Xerophytic elements, however, continued to be important in some areas including northern Africa, suggesting dry phases in this region until about the Santonian (Schrank and Nesterova, 1993). This was followed by a humid phase in the Campanian–Maastrichtian as indicated by an increasing variety of palm pollen including *Spinozonocolpites–Nypa* and *Mauritiidites* grains. Both of these pollen types are thought to indicate wet tropical coastal vegetation (Schrank, 1987). This interpretation is also supported by the presence of coal-bearing strata from the Maastrichtian of northern Somalia (Schrank, 1994a) and Nigeria (van Hoeken-Klinkenberg, 1964).

In mid-palaeolatitudes the presence of abundant charcoal through the Late Cretaceous of Europe and North America may indicate that dry, or at least seasonally dry, conditions prevailed in these areas. Warm, dry climates are also supported by the abundant occurrence of carbonate deposits, such as the chalk of southern England and adjacent areas. In eastern North America and Europe many palynological assemblages are dominated by pollen of the so-called Normapolles complex. In some assemblages pollen of this kind may account for more than 80% of all angiosperm grains (Pacltová, 1981). Upchurch and Wolfe (1987) suggested that these plants grew under warm, seasonally dry climate. In Kazakhstan dry phases are also inferred for the Turonian and Campanian–Maastrichtian, whereas the Santonian is believed to have been relatively humid based on the relative occurrence of xerophytic versus more humid elements in palynological assemblages (Schrank and Nesterova, 1993) (Figure 3.10). A CLAMP analysis of the Early Campanian Grünbach Flora, Austria, indicates seasonal climate, with longer humid and warm to hot seasons alternating with short dry periods (Herman and Kvaček, 2002).

At high palaeolatitudes in the Northern Hemisphere the climate during the Late Cretaceous has been inferred in broad terms to be warm temperate to subtropical with high precipitation, based on the presence of abundant fern spores and saccate conifer pollen (Herngreen *et al.*, 1996). This is supported by studies of leaf physiognomy that indicate mean annual temperatures during the Coniacian of about 12–13 °C and summer temperatures of about 25 °C (Spicer *et al.*, 1994b). At high palaeolatitudes in the Southern Hemisphere a very humid and cool climate, with little or no seasonality, is indicated based on the abundant occurrence of saccate conifer pollen and pollen of *Nothofagites*, comparable to pollen of modern *Nothofagus*.

In the late Maastrichtian there is clear palaeontological and other evidence for substantial cooling of the climate. The interval immediately prior to the Cretaceous–Palaeogene boundary was also characterised by intensified climatic and eustatic fluctuations, progressive extinctions in both marine and terrestrial organisms, and widespread significant volcanic activity, especially associated with the massive extrusion of basalts on the Deccan Plateau, India (MacLeod *et al.*, 2000; Pickering, 2000). Superimposed upon these significant environmental changes the impact of the Chicxulub bolide perhaps resulted in further climatic perturbation and further levels of extinction (Alvarez *et al.*, 1980).

3.4 IMPLICATIONS FOR ANGIOSPERM DIVERSIFICATION

The palaeogeographic and climatic changes that took place during the Cretaceous undoubtedly had a profound influence on the first 80 million years or more of angiosperm diversification. Most importantly, through this interval, climatic patterns, both temperature and rainfall, remained very different from those of today, undermining the notion that any particular modern biome was the cradle of angiosperm diversity. In addition, through much of the Cretaceous high sea levels exacerbated the isolation of land masses, concentrations of atmospheric carbon dioxide remained high and ultimately the creation of significant new mountain chains was initiated. However, an especially pervasive and important trend through the Cretaceous was that the continents became more dispersed, creating further possibilities for geographic isolation. During the Cenozoic, continental dispersion became still more pronounced and at the same time global climates became more strongly differentiated as a result of pronounced climatic deterioration at the poles. The more recent evolution of angiosperms has therefore been in a world that gradually became increasingly finely divided in terms of geography, altitude and climate. These physical changes contributed significantly to patterns of isolation and diversification in different angiosperm lineages. Some of these patterns have no doubt been obscured by subsequent migration, extinction and other events, but many are still reflected in the patterns of distribution among living angiosperm groups.

4

Stratigraphic framework and key areas for Cretaceous angiosperms

In this chapter we review the geology and stratigraphic setting of selected localities and areas that are of special importance for research on early angiosperm fossils. We also provide a stratigraphic overview and discussion of the possible age of the major angiosperm floras discussed in this book. Among the key areas considered are the Estremadura and Beira Litoral regions of Portugal, southern Sweden, and eastern North America, where most of our own work has been concentrated, but we also include other areas that have yielded important fossil material, much of which is considered later in this book.

4.1 THE STRATIGRAPHIC FRAMEWORK

Establishing a well-corroborated and detailed stratigraphic framework to assess the timing and pattern of angiosperm diversification is not a straightforward task. With few exceptions, angiosperms are land plants and most diverse assemblages of fossil angiosperm leaves and reproductive structures occur in terrestrial sediments deposited close to where the plants grew. Dispersed angiosperm organs, especially pollen, may be transported by wind or water into marine deposits, but the terrestrial input into marine sediments generally decreases with increasing distance from the land. The quality of preservation of palynomorphs and other plant parts in fully marine sediments is also generally poorer than in terrestrial sediments. As a result, although angiosperm fossils do occur in marine deposits, they are more abundant, and usually better preserved, in terrestrial sediments.

The chronostratigraphic units (stages) used for the Cretaceous are all based on marine strata defined based on stratotype localities and sections in Europe (France, Switzerland, The Netherlands) (Gradstein *et al.*, 2004). The stratigraphic position of marine sequences outside the stratotype areas is usually established by direct or indirect lithological or palaeontological correlation to the type sections. Many marine organisms, particularly pelagic forms or forms with pelagic larvae, may be dispersed over wide geographic areas and are therefore especially valuable for such correlation. However, correlation between different marine depositional basins may be problematic. During the Early Cretaceous, for instance, the South Atlantic and the Tethys were separated and apparently had little or no exchange of marine organisms. As a result, precise correlation between the two areas is difficult. For many Cenozoic sequences, and for some Cretaceous sequences, palaeomagnetism provides an important means of linking stratigraphic sequences in the marine and terrestrial realms. However, this technique is less useful in the dating of strata that are important for the initial diversification of angiosperms because there is a large uninterrupted phase of normal polarity through the critical mid-Cretaceous interval.

Correlation of terrestrial sediments is much more problematic than for sediments deposited under marine conditions because the individual sedimentary units are generally much less laterally extensive than in the marine realm. Biostratigraphic correlation is also more difficult because the microfossils that occur in terrestrial sediments are generally much less widespread than those used in correlating marine deposits. In a few instances volcanic ashes in Cretaceous terrestrial sediments have been dated by using isotopic techniques. These provide very useful absolute dates, including those for some key fossil occurrences, but currently they are sparsely distributed, particularly in some of the sections most important for early angiosperm fossils. Typically, correlation in the terrestrial realm is indirect. Age is often established by comparing the spores and pollen in a sample to palynological assemblages in marine strata of known age, or in other terrestrial sediments that are dated independently by marine intercalations or by the presence of volcanic beds.

Dispersed spores and pollen may be found in large quantities in both terrestrial and marine sediments, but because many dispersed palynomorph taxa are very broadly defined (Chapter 2) and wide-ranging the resulting stratigraphic resolution is generally not precise. Very few of the

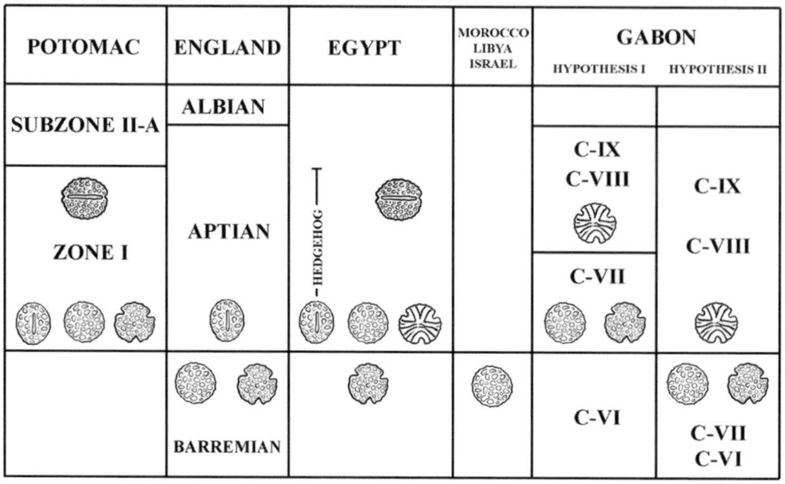

Figure 4.1 Palynological correlation of Early Cretaceous palynological assemblages from the Potomac Group, USA, England, Egypt, Morocco, Libya, Israel and Gabon showing two hypotheses for correlation to Gabon. The correlation is based on five different types of pollen and illustrates the difficulties in exact correlations over longer distances. Redrawn from Doyle (1992).

key sections containing early angiosperm fossils (e.g. Gabon, Egypt, Potomac Group, Portugal, southeastern England) have been precisely dated based on independent evidence. There is some limited marine and absolute date (isotopic) control, but correlations are to a large extent based on the occurrence of selected angiosperm taxa and what is known about their stratigraphic range in southern England, where there is marine control in some parts of the section (Doyle, 1992). An example of the difficulties of correlation is given in Figure 4.1, which shows correlation of Early Cretaceous strata from Gabon to Morocco, Egypt, England and eastern North America (Potomac Group) based on five different types of pollen. All types do not occur in all areas and detailed SEM investigation still remains to be done in order to establish that putatively similar pollen grains from the various areas are really conspecific.

4.2 KEY AREAS FOR CRETACEOUS ANGIOSPERMS

The first comprehensive reports on early angiosperm fossils appeared in the later part of the nineteenth century and were based on leaf floras collected at Early Cretaceous localities in Portugal (Saporta, 1894), the Potomac Group of eastern North America (Ward, 1888, 1905; Fontaine, 1889), the Western Interior of North America (Lesquereux, 1868, 1874, 1883, 1892; Newberry, 1868; Berry, 1911) and western Greenland and Spitsbergen (Heer, 1868, 1882; Saporta, 1877). These studies greatly influenced early views on angiosperm origin and diversification (Darwin in a letter to Hooker 1879, cf. Friedman, 2009), and for a long time, subsequent work focused on these same geographical regions. More detailed investigations of the classic leaf floras from Portugal (Teixeira, 1946, 1947, 1948, 1950, 1952) and from the Potomac Group (Doyle and Hickey, 1976; Hickey and Doyle, 1977) were published along with studies of new leaf floras collected in the same areas.

Beginning in the 1950s and 1960s there are series of important studies on the pollen and spores of Early Cretaceous rocks from southern England (e.g. Couper, 1958; Kemp, 1968; Hughes, 1994). These proved crucial in linking early angiosperm pollen with the age of reasonably well dated sediments. Subsequently, the Potomac Group also became the focus for studies of dispersed angiosperm pollen (Brenner, 1963; Doyle, 1969; Doyle and Robbins, 1977) as well as pioneering attempts to study and integrate the parallel stratigraphic patterns of leaves and pollen (Doyle and Hickey, 1976; Hickey and Doyle, 1977). More recently still, as a result of new field work by ourselves and our colleagues, both the Portuguese and the Potomac Group localities have become the focus for studies of fossil angiosperm flowers, fruits and seeds (e.g. Crane *et al.*, 1986, 1994; Friis *et al.*, 1986, 1994a, b, 1999, 2006a).

Most of the recently investigated sites from Portugal and North America are surface exposures, often of very limited vertical and lateral extent. Precise correlation among localities is therefore difficult. More continuous vertical sequences have been studied through palynological investigations of samples from drillings, and this has been especially helpful in the case of the Potomac Group (Doyle and Robbins, 1977). Numerous such boreholes, made all over the world during the twentieth century as part of the search for oil and water, have contributed many new palynofloras. These are especially important for the information they provide on Cretaceous vegetation in areas of the world where macrofossils are sparse or unknown. Sections that have been studied palynologically from outcrops, and also from boreholes, and that are of particular significance for unravelling the early history of angiosperms, are known from Israel, England, North and Western Africa, and Brazil (see references below). However, informative Cretaceous palynofloras have also been obtained from many other parts of the world, ranging from Peru to Papua New Guinea, and from Arctic Canada and Greenland to Antarctica. Taken together the scattered, but extensive, information on Cretaceous palynofloras is a unique resource for the study of angiosperm evolution that still remains to be fully exploited.

4.3 EUROPE

4.3.1 Portugal: Early Cretaceous localities

Some of the oldest and most diverse angiosperm macro- and mesofossil floras have been reported from Lower Cretaceous sediments in the Lusitanian Basin of western Portugal, where there are also important angiosperm floras from Upper Cretaceous strata. The classic Early Cretaceous leaf floras of Cercal, Buarcos-para-Tavarede and Nazaré described by Saporta (1894) include several angiosperm taxa that are preserved mainly as impressions, or more rarely as compressions. These floras were later restudied by Teixeira, who also included several new macrofossil floras from both Lower and Upper Cretaceous strata (Teixeira, 1945, 1946, 1947, 1948, 1950, 1952). Palynofloras with angiosperm pollen have been reported from numerous horizons throughout the Portuguese Cretaceous; most are from the Late Cretaceous (see below), but there are also some Early Cretaceous palynofloras (Groot and Groot, 1962; Hasenboehler, 1981; Pais and Reyre, 1981; Trincão, 1990; Heimhofer *et al.*, 2005, 2007). Rich mesofossil floras with angiosperm flowers, fruits and seeds have now also been discovered in both Lower and Upper Cretaceous strata of the Western Portuguese Basin (see below).

Cretaceous sediments in Portugal occur in the Algarve region and in the Western Portuguese (Lusitanian) Basin. In the Algarve, the sediments are of Early to mid-Cretaceous age and are distributed in a narrow belt close to the coast (Rey, 1983). These sediments are predominantly marine, and no angiosperm macro- or mesofossil floras have been reported from this region. In the Lusitanian Basin the Cretaceous sequence includes terrestrial as well as marine sediments that range from earliest Cretaceous (Berriasian) to latest Cretaceous (Maastrichtian) in age (Ribeiro *et al.*, 1980). The oldest sediments occur in the southern and central parts of the basin; the youngest sediments are in the northern part (Figures 4.2, 4.3).

The Lusitanian Basin is one of several rift basins associated with the extension and final opening of the North Atlantic Ocean. Major extensional rift phases occur from the Late Triassic to the earliest Cretaceous along with extensive regional subsidence (Rasmussen *et al.*, 1998). Sedimentation was influenced by changes in sea level and the resulting fluctuation of the Cretaceous coastline across the area. Detailed geological and palaeontological studies of the Early Cretaceous sequence in the Lusitanian Basin have been carried out for the Estremadura region (Rey, 1972, 1979, 1982, 1992, 1993; Rey *et al.*, 2006), and more detailed investigations have been initiated for Cretaceous sediments further north in the basin (Dinis and Trincão, 1995; Dinis, 2001; Dinis *et al.*, 2002; Rey *et al.*, 2006). Marine conditions predominated in the southern and western parts of the basin, whereas terrestrial conditions prevailed towards its eastern and northern margins. Despite the presence of marine sediments in some parts of the basin the exact dating of the Cretaceous floras from Portugal is problematic, primarily owing to the lack of marine control in the areas where plant fossils are most abundant. These difficulties are further exacerbated by the paucity of well-dated floras from other areas that can be used as a stratigraphic reference for the Portuguese floras, and by the fragmentation of the depositional basin caused by the tectonic activity in the area. In addition, most outcrops are rather small and correlation is difficult because it is rarely possible to trace horizons laterally over longer distances.

Terrestrial sediments of Early Cretaceous age that contain abundant plant material are particularly extensive in the areas around Torres Vedras and Runa, around Cós-Juncal, Nazaré and Leiria, and around Figueira da Foz. The Early Cretaceous sequence in these areas consists mainly of clastic sediments deposited in fluviatile to limnic

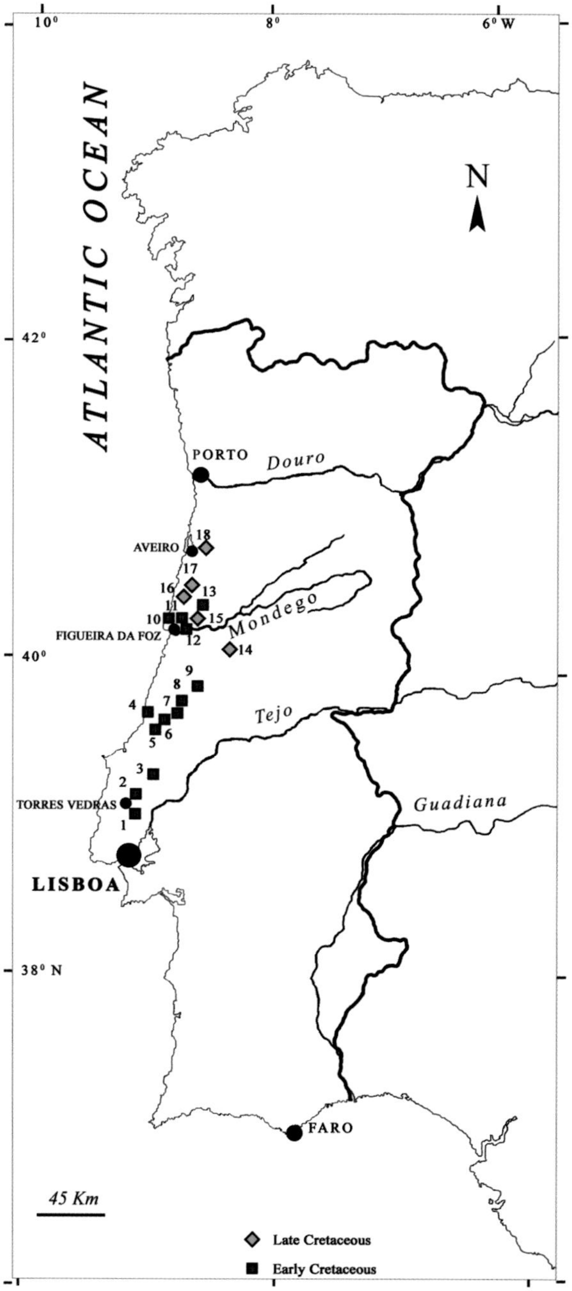

Figure 4.2 Major localities with Cretaceous macro- and mesofossil floras from Portugal. 1, Catefica; 2, Torres Vedras (NE of Forte de Forca); 3, Cercal; 4, Nazaré; 5, Juncal/Vale Paino; 6, Juncal; 7, Vale de Agua/Vale Farelo; 8, Vale de Agua; 9, Famalicão; 10, Buarcos; 11, Tavarede; 12, Vila Verde; 13, Ança; 14, Vila Flor; 15, Vila Verde de Tentugal; 16, Mira; 17, Presa; 18, Esgueira. From Friis *et al.* (2010a).

environments. Lower Cretaceous strata in the Torres Vedras – Runa area were previously assigned to the 'Grés de Torres Vedras' beds (Zbyszewski *et al.*, 1955), but are now subdivided formally into several formations that range from Berriasian to Albian in age (Figure 4.3) (Rey, 1972, 1993; Rey *et al.*, 2006). Mesofossil assemblages collected in the Almargem Formation from the Torres Vedras locality are from the lower part of the formation, thought to be of Late Barremian – Early Aptian age, and are currently the oldest of the Portuguese mesofossil floras (see below).

Further north in the basin there is a hiatus in deposition that appears to extend from the Berriasian–Valanginian to the Late Aptian. Angiosperm remains are known from strata above this hiatus. In the Cós-Juncal–Nazaré–Leiria area these plant-bearing horizons were previously assigned to the 'Complexos gresosos de Nazaré e de Cós-Juncal' beds, and are in places overlain by marine sediments of Cenomanian age (França and Zbyszewski, 1963). In the Figueira da Foz region the plant-bearing horizons were assigned to the 'Arenitos de Carrascal' beds (Rocha *et al.*, 1981). Both the 'Complexos gresosos de Nazaré e de Cós-Juncal' and the 'Arenitos de Carrascal' are now included in the Figueira da Foz Formation, which is subdivided into a number of formal members (Figure 4.3) (Dinis, 1999, 2001; Dinis *et al.*, 2002). The mesofossil floras from these regions are from the basalmost members, the Calvaria and the Famalicão members. The age of the Figueira da Foz Formation was first interpreted as Late Aptian (Dinis *et al.*, 2002), but palynological correlation between the Algarve area and the Lusitanian basin suggest a slightly younger age. Rey *et al.* (2006) suggested a Late Aptian – Early Albian age for the Calvaria and Famalicão members and this is accepted here.

The Torres Vedras – Runa area. Leaf floras have been described from several localities around Torres Vedras (Saporta, 1894; Teixeira, 1948). These floras are mainly dominated by ferns and conifers. Only fragmentary angiosperm remains have been reported. These leaf floras have not been restudied and a more detailed evaluation of the angiosperm component is currently lacking. The floras in the area most likely come from several different stratigraphic levels.

Torres Vedras. Several mesofossil floras have been recovered from the Torres Vedras area. The most diverse is the assemblage collected from a large open clay pit east of Forte de Forca on the road towards Sarge (Figure 4.4). Although it is not the only flora from this region we refer to it as the Torres Vedras flora. The locality has been destroyed by town development and is no longer available

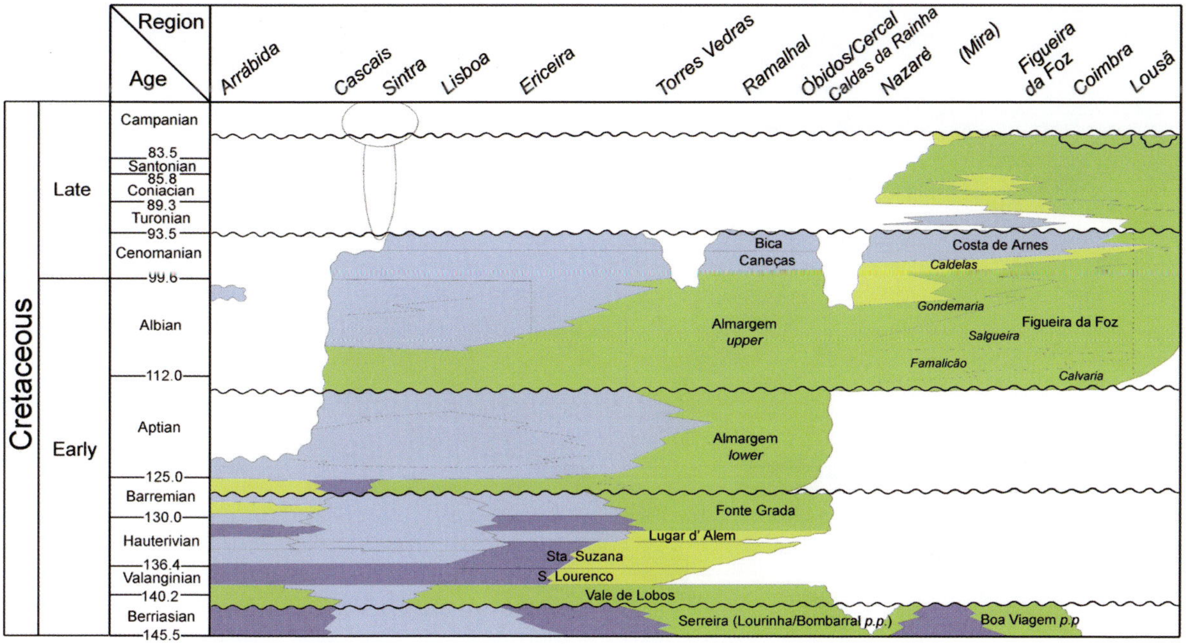

Figure 4.3 Overview of Cretaceous lithostratigraphic units in the Lusitanian Basin of western Portugal. Green, fluviatile to deltaic clastic sediments; yellow, coastal to estuarine clastic sediments; light blue, marine carbonate deposits; dark blue, coastal clastic and carbonate deposits. Redrawn from Rey *et al.* (2006).

for collection. The sediments exposed in the clay pit included light yellowish and purple silts and clays as well as thin horizons of greyish clays, silts and sands containing coalified material. The mesofossil flora is rich in angiosperm remains and is dominated by non-eudicot angiosperms including *Hedyosmum*-like flowers (Chloranthaceae) and *Mayoa* (Araceae) (Friis *et al.*, 2006a). Only two different kinds of eudicot pollen have currently been identified in the mesofossil flora. There are also remains of many non-angiospermous seed plants, including small seeds related to the Bennettitales–Erdtmanithecales–Gnetales complex (Friis *et al.*, 2009a). The age is Late Barremian – Early Aptian (see above).

Catefica. The Catefica locality is a road cut between the villages of Catefica and Mugideira, about 4 km south of Torres Vedras on the western margin of the Runa Basin (Figure 4.5). The Cretaceous strata at Catefica consist of light cross-bedded sands with some darker clay beds and lenses that are rich in plant fragments. The mesofossil flora from Catefica is very diverse and includes well-preserved flowers, fruits, seeds, dispersed stamens with pollen *in situ* (Friis *et al.*, 1994b, 1999), twigs of cheirolepidiaceous conifers, many non-angiosperm seed plants (Friis *et al.*, 2009a), a variety of ferns, numerous fragments of thalloid liverworts, and many shoots and megaspores of Selaginellaceae. The age of this flora is not completely clear. It shares several taxa with the Torres Vedras mesofossil flora, and the strata also appear to be from the lower part of the Almargem Formation, indicating a Late Barremian – Early Aptian age. However, there are also similarities to the mesofossil floras from the Cos-Juncal–Leiria area and we therefore give a broader age range (Late Barremian – Aptian) for the Catefica flora.

Cercal. The Cercal leaf flora was collected from a small road cut close to the village of Cercal approximately 28 km northeast of Torres Vedras and 35 km south of Alcobaça. The road cut exposes a lens 4.5 m thick of light brown lacustrine clay containing a leaf flora composed mainly of ferns and conifers, but also with some angiosperm taxa (Saporta, 1894; Teixeira, 1947, 1948). Fossil leaves can still be collected although the exposure is heavily overgrown and tree roots penetrate the sediments (Figure 4.6) resulting in fragmentation of the bedding planes. The leaf fossils are predominantly fine impressions with no organic

Figure 4.4 The Early Cretaceous Torres Vedras locality (Late Barremian – Early Aptian), Portugal. Diverse mesofossil assemblages have been extracted from thin coaly horizons in the basal part of the exposure. The locality has now been destroyed by urban development.

Figure 4.5 The Early Cretaceous (Late Barremian – Aptian) Catefica locality, Portugal, showing cross-bedded sands and gravels with subordinate dark organic horizons that have yielded mesofossils.

Figure 4.6 The Early Cretaceous Cercal locality, Portugal, that has yielded the classic Cercal impression flora. The finely laminated beds are now strongly broken up by modern tree roots. In the picture are E.M. Friis and P.R. Crane (1989).

material preserved. Several of the taxa are probable aquatics. The plant-bearing sequence is part of the Almargem Formation (Rey, 1972), but the age of the Cercal flora has not been bracketed with more precision. The leaf flora is rather unusual and shares few elements with other Early Cretaceous leaf floras from Portugal. There are no palynomorphs preserved in the clay and the sediments yielded no mesofossils.

Nazaré. The Nazaré leaf flora is from terrestrial sediments exposed in the outskirts of the fishing village of Nazaré (Saporta, 1894; Teixeira, 1948). The fossils occur in clay layers intercalated among coarser sediments and belong to the Figueira da Foz Formation, which in its upper part is probably of Late Albian age (Dinis, 1999). The majority of the plant fossils are angiosperm leaves (Saporta, 1894; Teixeira, 1948), and Saporta indicated that this flora had a younger aspect than the flora from Buarcos-para-Tavarede. About 12 m above the clay beds with angiosperm leaves is another thick sequence with fossil plants. This sequence is dominated by cheirolepidiaceous conifers and is overlain by marine sediments (Saporta, 1894). Mesofossil floras have not been recovered from this locality.

Famalicão. This locality is a deep clay pit on the outskirts of the village of Famalicão, SSE of Leiria. Most of the exposed sequence is rather uniform, consisting of clays that are mainly reddish or greenish in colour without visible organic remains, and belongs to the Calvaria and Famalicão members of the Figueira da Foz Formation (Dinis, 1999, 2001). In the deepest part of the clay pit, below the Calvaria Member, a dark grey clay horizon yielded an extremely rich mesofossil flora. Thousands of specimens were extracted from a sample of less than a kilogram. The flora is also extremely diverse, with more than one hundred angiosperm taxa recognised so far (Friis *et al.*, 1994b, 1997a; Friis and Pedersen, 2011; Eriksson *et al.*, 2000a, b). The stratigraphic position below the Calvaria Member indicates a Late Aptian age. Unfortunately, the mesofossil-bearing strata are no longer accessible.

Vale de Agua. The Vale de Agua assemblage was collected in a large complex of clay pits situated close to the small village of Vale de Agua, near Juncal, southwest of Batalha (Figure 4.7). Sediments outcropping in the clay pits are mainly reddish clays, but horizons with grey organic-rich clay have yielded a large number of angiosperm flowers, fruits, seeds and dispersed stamens, as well as abundant shoots of cheirolepidiaceous conifers (Friis *et al.*, 1994b, 1997a, 1999, 2000a, 2000b, 2001, 2009b; Friis and Pedersen, 2011; von Balthazar *et al.*, 2005; Pedersen *et al.*, 2007). The mesofossil floras from this pit complex were extracted from sediments belonging to the basal part of the Figueira da Foz Formation indicating a Late Aptian – Early Albian age.

Figure 4.7 The Early Cretaceous (Late Aptian – Early Albian) Vale de Agua locality, Portugal. The locality is a complex of clay pits with thin darker and grey horizons that contain mesofossils.

Juncal. This locality is an old clay pit close to the village of Juncal south of Vale de Agua and Batalha. A thin horizon of light brownish to grey clay containing a relatively rich macrofossil flora was exposed in 1989, but the layers are no longer accessible due to road construction in the area. The macrofossil flora collected at the locality includes a variety of angiosperm leaves and flowers as well as ferns and rare conifers, all of which are preserved as impressions (Friis *et al.*, 1994b). The bedding planes are irregular and complete leaf impressions are rare. A small mesofossil flora containing angiosperm fruits and seeds has been extracted from grey clay beds at the locality. The fruits and seeds identified so far from Juncal are conspecific with taxa from the Vale de Agua complex of clay pits and the sediments are most likely referable to the basal part of the Figueira da Foz Formation of Late Aptian – Early Albian age. Another small mesofossil flora has been collected in the Juncal/Vale Painho clay pit, which is situated only about 1 km south-west of the old Juncal clay pit. The sediments at the two localities are similar in appearance, but the fossils included in the Juncal/Vale Painho flora are exclusively non-angiospermous plants and much older, belonging to the Bombarral formation, which is of possible Berriasian age (Mendes *et al.*, 2008a).

Figueira da Foz region. Several fossil floras have been discovered in an area extending from Buarcos to Tavarede. These floras include classic leaf floras and rich assemblages of mesofossils. In this region the Early Cretaceous sequence of the Figueira da Foz Formation is several hundred metres thick and consists mainly of sandy sediments with intercalating beds of clay (Rocha *et al.*, 1981; Dinis, 2001). The Figueira da Foz Formation, in this area, unconformably overlays Jurassic sediments, and in some places is followed above by marine sediments of Cenomanian age. The plant fossils from Buarcos and Tavarede occur near the base of the sequence about 10–15 m above the Jurassic strata (Rocha *et al.*, 1981). According to Dinis (2001) this part of the sequence belongs to the lowermost member (Calvaria Member) of the Figueira da Foz Formation.

The classic leaf flora from Buarcos-para-Tavarede described by Saporta (1894) and Teixeira (1948) was collected close to the old Buarcos–Tavarede road from sections that are no longer available for study. The leaf fossils were preserved in three clay lenses intercalated in a sequence of sands and gravels. The flora includes a rich assemblage of twigs and leaves of ferns and conifers, in addition to leaves of angiosperms. The position of the leaf flora in relation to the strata yielding the Buarcos mesofossil flora (below) is uncertain, but it may also be in the basal part of the Figueira da Foz Formation (Late Aptian – Early Albian).

Buarcos. The Buarcos–Tavarede area is now urbanised and most of the exposures that yielded the classic leaf floras have disappeared. A locality in the town of Buarcos that has yielded well-preserved mesofossils could still be collected until 2001, but has now also disappeared. This exposure was a partly overgrown road cut along the old road between Buarcos and Tavarede that exposed a sequence of coarse, cross-bedded sands with intercalated layers of silt and clay (Figure 4.8). A rich mesofossil flora with angiosperm flowers, fruits, seeds and anthers was extracted from dark organic-rich clay and silt layers at this locality, together with abundant fragments of Cheirolepidiaceae, other conifers, ephedroid seeds and other seeds of the Bennettitales–Erdtmanithecales–Gnetales complex (Friis *et al.*, 1997a, 1999, 2000b, 2009a; Rydin *et al.*, 2006a; Friis and Pedersen, 2011). The flora occurs in the basal part of the Figueira da Foz Formation (Late Aptian – Early Albian).

Figure 4.8 The Early Cretaceous (Late Aptian – Early Albian) Buarcos locality, Portugal. Mesofossils occur in the small organic-rich deposit in the middle of the exposure (asterisk). The locality has now disappeared under urban development.

Villa Verde. Mesofossil assemblages with well-preserved plant remains have been collected in the basal part of the Figueira da Foz Formation (Late Aptian – Early Albian) about 1 km NNE of the village of Villa Verde (northeast of Figueira da Foz). One of the assemblages (Villa Verde 2) is remarkable in containing several well-preserved monocot flowers and inflorescences (Friis *et al.*, 2010b). Another assemblage (Villa Verde 1) is less diverse and includes reproductive organs of Erdtmanithecales (Friis and Pedersen, 1996).

There are several other localities with mesofossil floras that have yielded angiosperm remains from the Early Cretaceous of Portugal, but that remain to be studied in more detail. There are also many mesofossil floras from the earliest Cretaceous that have not yielded angiosperm remains. These additional localities are not considered here.

4.3.2 Portugal, Late Cretaceous localities

The youngest Cretaceous sediments in Portugal are known from the northern part of the Lusitanian Basin in the Beira Litoral region. The sediments were deposited in a near-shore environment and consist of alternating terrestrial and marine facies that range from Cenomanian to Campanian–Maastrichtian in age. Some of these sediments are very rich in organic material and contain many plant fossils. Leaf floras have been reported from the eastern part of the basin at the Vila Flor locality and near Buçaco (Teixeira, 1950). Palynological studies of the Vila Flor sediments have revealed well-preserved Normapolles type pollen and indicate a Late Cenomanian age for the flora (Kedves and Diniz, 1981). The age of the Buçaco flora is uncertain. Other leaf floras have been reported from localities in the area around Mira and Presa between Aveiro and Figueira da Foz. There are also floras in the vicinity of Aveiro, including the flora of Vila Verde de Tentugal and the flora of Esgueira (Saporta, 1894; Lima, 1900; Teixeira, 1946; Lauverjat and Pons, 1978; Pons and Broutin, 1978). Rich microfossil floras have also been described from several outcrops and boreholes in the Aveiro area and around Mira and Presa. These floras contain a variety of Normapolles pollen (Diniz, 1967; Kedves and Diniz, 1967; Diniz *et al.*, 1974; Kedves and Pittau, 1979; Medus, 1981; Batten, 1986a), as well as diverse mesofossil floras with exquisitely preserved fossil flowers, including Normapolles flowers. Studies of charophytes (Gutierrez and Lauverjat, 1978) and marine intercalations indicate a Campanian–Maastrichtian age for the Upper Cretaceous strata in the northern part of the Lusitanian Basin in the Beira Litoral region. The leaf floras have not been resampled or restudied recently and here we consider only two of the key mesofossil floras from the latest Cretaceous: Mira and Esgueira.

Mira. The Mira locality is an old clay pit in the southern outskirts of the village of Mira, about 25 km south of Aveiro (Figure 4.9). Material could be collected at this locality until 1990, but the pit is now flooded and collecting is no longer possible. The sediments exposed at the Mira locality comprise mostly dark clays and silts, but greenish clays with abundant pyrite and shells of invertebrate fossils also occurred in the southern part of the pit. A rich mesoflora with well-preserved angiosperm flowers, fruits, seeds, and stamens was extracted from the Mira sediments. The fossils are mainly preserved as charcoal, but lignitised fossils are common in some samples. The angiosperm fossils include a variety of Normapolles flowers of fagalean affinity (Schönenberger *et al.*, 2001b; Friis *et al.*, 2003b, 2006b) as well as many other eudicots and some monocots (Friis *et al.*, 1992, 2006a, 2010a). Sediments in the Mira area are referred to the 'Argilas de Vagos' and 'Conglomerado de Mira'. Marine horizons in the sedimentary sequence indicate a Santonian? to Maastrichtian age for the whole sequence; the plant-bearing sediments are indicated as being of Campanian–Maastrichtian age on geological maps of the area (Barbosa, 1981; Friis *et al.*, 1992). Marine sediments exposed close to Mira, and apparently more or less contemporaneous with the Mira plant beds (Friis *et al.*, 1992), are dated as being of Late Campanian age (Beauvais *et al.*, 1975). Because of the current uncertain stratigraphic placement we give the broader Campanian–Maastrichtian age range for the Mira mesofossils.

Figure 4.9 The Late Cretaceous (Campanian–Maastrichtian) Mira locality, Portugal. The locality is an old, now abandoned and water-filled, clay pit that has yielded diverse mesofossils.

Figure 4.10 The Late Cretaceous (Campanian–Maastrichtian) Esgueira locality, Portugal, exposed during extensive road building. The sediments are finely laminated and interpreted as being deposited in a near-coastal, possibly tidal flat environment.

Esgueira (Olho de Agua). Alternating layers of light brown to grey sands, silts and clays containing numerous plants fossils were exposed in Esgueira, close to Olho de Agua in the outskirts of Aveiro. Collection was facilitated by extensive road building in the area around 1989–1990 (Figure 4.10), but this has now removed or covered all of the exposures, and the most productive sites are no longer available for collecting. The sediments are referred to the 'Arenitos e argilas de Aveiro' and are thought to be Campanian–Maastrichtian in age (Friis *et al.*, 1992). Twigs of the cheirolepidiaceous conifer *Frenelopsis oligostomata* occur abundantly on some bedding planes and have been described in detail together with male cones containing *Classopollis* pollen (Pons and Broutin, 1978). Other horizons are rich in angiosperm fossils, and both leaf fossils and mesofossils have been collected. Among the angiosperm fossils *Esgueiria adenocarpa*, a small epigynous flower of combretaceous affinity, is extremely common in some samples (Friis *et al.*, 1992). There are also several Normapolles-producing genera as well many other eudicots (Friis *et al.*, 2003b, 2010a). Angiosperm wood from the Esgueira locality comprises forms assigned to *Icacinoxylon* and *Paraphyllanthoxylon* as well as several undescribed forms (Herendeen, 1991a). The depositional environment for the Esgueira locality has been interpreted as a near-coastal, possibly tidal flat, situation (Pons *et al.*, 1980; Ribeiro *et al.*, 1980; Batten and Maclennan, 1984). The occurrence of tidal conditions may also be supported by high δC^{13} values for the plant remains from Esgueira and Mira, indicating that the plants grew under xeric conditions with osmotic or saline stress (Bocherens *et al.*, 1993).

4.3.3 Spain

In Spain, Early Cretaceous continental sediments with rich fossil-bearing horizons occur in the eastern provinces of Asturias, Burgos, Cuenca, La Rioja, Lérida and Teruel (for a useful review see Diéguez *et al.*, 2010). The presence of a few lignite horizons indicates occasional humid conditions (e.g. Gierlowski-Kordesch and Janofske, 1989). The fossil floras from the Early Cretaceous of Spain are interesting in providing important evidence of early angiosperms, mostly from fossil pollen (e.g. Villanueva-Amadoz *et al.*, 2010).

One of the most intensively studied fossil assemblages from the Early Cretaceous of Spain is a lacustrine lithographic limestone assigned to the La Huérguina Limestone Formation from the Las Hoyas Sub-Basin in Serranía de Cuenca (Cuenca Province). This fossil assemblage has been studied mainly for its rich and diverse vertebrate fauna, which includes early birds, dinosaurs, crocodilians, turtles, amphibians and fish (for references see Meléndez, 1995). The Las Hoyas sequence is up to 300 m thick and was deposited in a sub-basin of the Iberian Basin during a period of internal rifting and a high rate of subsidence in the Late Barremian (Fregenal and Meléndez, 1995; Diéguez and Meléndez, 2000). The plant fossils include typical Early Cretaceous xeromorphic elements with a dominance of ferns such as *Weichselia* and *Onychiopsis* (Diéguez and Meléndez, 2000). There are also a variety of extinct conifers and probable Gnetales, as well as angiosperm leaves and tricolpate pollen (Diéguez *et al.*, 1995). Detailed documentation of these remains would be of considerable interest. Plant remains extracted from the La Huérguina Limestone Formation at Uña, immediately north of Las Hoyas, are dominated by cheirolepidiaceous conifers (Gomez *et al.*, 2001).

Localities in the Sierra del Montsec of Lérida Province have also provided rich fossil assemblages of vertebrates, insects and plants preserved in a limnic lithographic limestone. The limestone is of Early Cretaceous age, but the age range given varies from earliest Cretaceous (Barale *et al.*, 1984) to Hauterivian–Barremian (Diéguez *et al.*, 2010). The macrofossil flora from the Montsec area is apparently much more diverse than that of Las Hoyas and has been studied for more than a century (e.g. Barale *et al.*, 1984). The fossil plants are mostly impressions or compressions of leaves and twigs, often with well-preserved cellular details. The plants are typical xeromorphic Early Cretaceous forms and include many ferns such as *Weichselia* and *Onychiopsis*, as well as extinct conifers such as *Frenelopsis* and *Brachyphyllum*. The flora is similar to other Hauterivian–Barremian floras from Europe. The flora has also yielded the enigmatic fossils described as *Montsechites (Ranunculus) ferreri* (Blanc-Louvel, 1984) and *Montsechia vidali*, which also occur in the Las Hoyas assemblage.

Plant assemblages of Albian age from Teruel Province, also eastern Spain, are rich in gymnosperms (Gomez *et al.*, 2001) and also contain a variety of angiosperm leaves (Sender *et al.*, 2005).

4.3.4 The Czech Republic

Rich Cretaceous floras containing angiosperm fossils have long been known from the Bohemian Massif. These floras occur in two main basins, the Bohemian Cretaceous Basin and the South Bohemian Basin, and they range from Cenomanian to Santonian in age (Knobloch and Mai, 1986). The earliest descriptions of Cretaceous floras from Bohemia appeared in the last part of the nineteenth century and the early part of twentieth century, and included mostly leaf floras from the Middle to Late Cenomanian Peruc–Korycany Formation (e.g. Velenovský, 1882, 1883, 1884, 1889; Velenovský and Viniklář, 1926, 1927, 1929). Later studies added many new leaf floras both from the Peruc–Korycany Formation and from younger strata. Numerous mesofossil floras with abundant angiosperm fruits and seeds have also been discovered from the Bohemian Massif (Knobloch, 1964, 1971, 1977, 1985; Knobloch and Mai, 1983, 1984, 1986; Kvaček, 1992; Eklund and Kvaček, 1998; Kvaček and Eklund, 2003; Kvaček and Friis, 2010).

The major leaf floras are from the Bohemian Cretaceous Basin, which extends from about Brno to north of Prague, with extensions into Germany and Poland (Figure 4.11). The sedimentary sequence is limnic to brackish with marine intercalations, and ranges in age from Cenomanian to Santonian. The Cenomanian Peruc–Korycany Formation includes the richest plant fossil assemblages, and numerous localities have been discovered. Kvaček (1998) gives a useful summary of the localities and the stratigraphy of these Cenomanian floras. The classic locality of Vyšerice (now Vyšehořovice), which has yielded most of the leaf fossils described from the Peruc–Korycany Formation, is now a natural monument and the exposures heavily weathered and overgrown (Kvaček, 1999). Abundant plant fossils can still be collected at the Pecínov Quarry, in which the Peruc–Korycany Formation is well exposed. The sedimentary sequence consists of mudstones, sandstones and conglomerates interpreted as having been deposited in a near-shore environment with fluvial to marsh and tidal conditions (Kvaček, 1998). Other important localities for leaves and mesofossil floras are Brník and Hloubětín-Hutě. The Peruc–Korycany sequence is well dated, based on overlying marine sediments with molluscs of Late Cenomanian – Turonian age. A Cenomanian age is also supported by palynological evidence (Pacltová, 1971).

The Cenomanian macrofossil floras from the Peruc–Korycany Formation are rich in angiosperms and other seed plants. The non-angiosperm component was studied by Velenovský (1885) and has recently been restudied by J. Kvaček (1998). New studies of the angiosperm

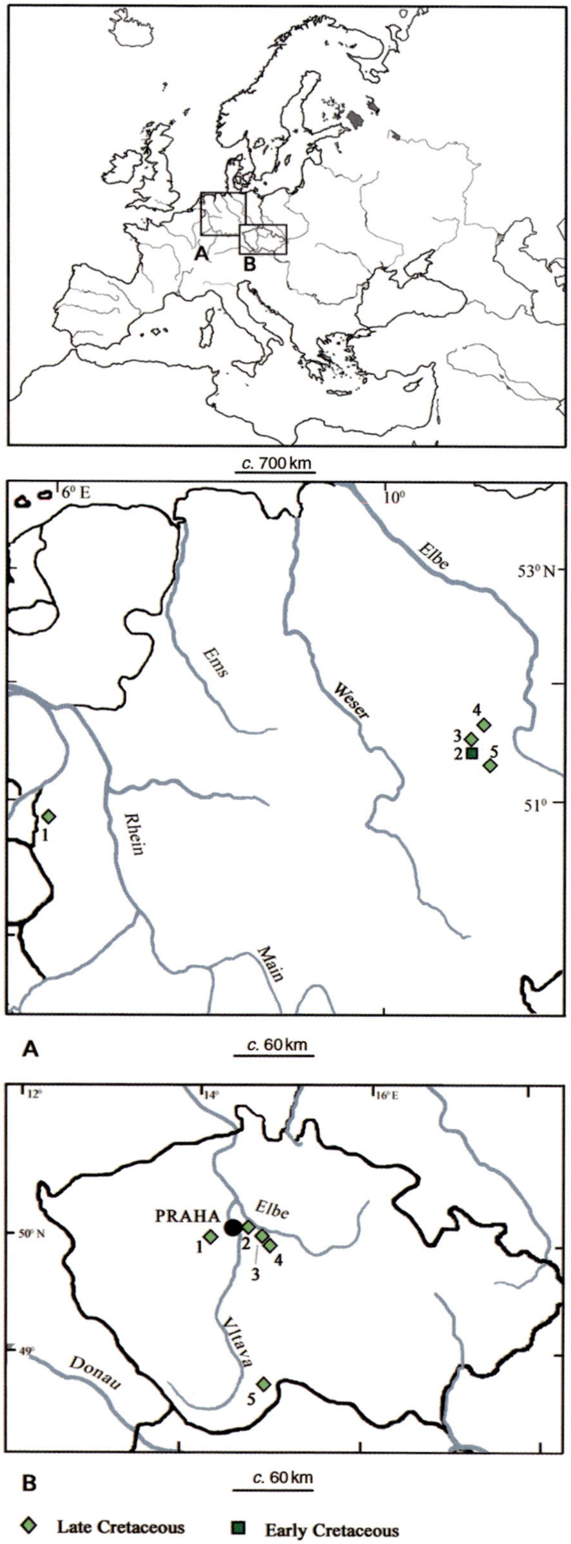

Figure 4.11 Late Cretaceous localities with important mesofossil floras from Central Europe. (A) Localities in the Netherlands and Germany: 1, Aachen; 2, 3, Quedlinburg; 4, Walbeck; 5, Eisleben. (B) Localities in the Czech Republic: 1, Pecínov; 2, Hloubětín-Hutě; 3, Vyšehořovice; 4, Brník; 5, Klikov.

component have been initiated by J. Kvaček and some angiosperm leaves were also studied by Z. Kvaček. Of particular interest are floras that include both leaf fossils and reproductive material preserved as macro- and mesofossils (Eklund and Kvaček, 1998; Kvaček and Eklund, 2003; Kvaček and Friis, 2010).

The South Bohemian Basin has yielded mesofossil floras that are rich in angiosperm remains, and many assemblages have been described from the Late Turonian – Santonian Klikov Formation (Klikov-Schichtenfolge). The Klikov Formation is up to 340 m thick and consists of alternating sandstones, mudstones and claystones deposited in a fluvio-lacustrine freshwater environment (Zetter *et al.*, 2002). Some of the mesofossil floras are very diverse and almost 100 different taxa of angiosperm fruits and seeds have been described (e.g. Knobloch, 1964; Knobloch and Mai, 1984, 1986) as well as a diversity of angiosperm leaves (Němejc, 1961). Prominent angiosperm components in the Klikov sequence are fruits and flowers related to the Normapolles complex (Knobloch and Mai, 1986; Friis and Crane, 1989; Heřmanová *et al.*, 2011). Dispersed pollen grains referred to the Normapolles complex are extremely common in parts of this sedimentary sequence and Pacltová (1981) noted abundances of Normapolles up to 67% of the total palynological assemblage (corresponding to about 80% of all angiosperm grains). The Klikov Formation also extends into Lower Austria and well-preserved pollen assemblages were described from the Gmünd at the Austrian–Czech border (Zetter *et al.*, 2002).

4.3.5 Austria

In Austria, Late Cretaceous floras with well-preserved angiosperm leaves and reproductive organs are known from the Gosau Group sediments of the Eastern Alps near Saltzburg and Klagenfurt, and from flysch deposits in the Wien–Sievering and Langenzersdorf area (Goth, 1986; Knobloch and Mai, 1986; Herman and Kvaček, 2002; Kvaček and Herman, 2004).

Grünbach. In the Gosau Group, a well-preserved leaf flora has been collected from the Grünbach Formation at the village of Grünbach in the Neue Welt Basin about 50 km south-southeast of Vienna. The formation is a coal-bearing sequence consisting of coal seams deposited in freshwater swamp environments with interbedded near-shore marine sediments. The Grünback Formation is dated as Early Campanian, based on marine intercalations and its position between two fully marine formations, the Late Santonian Maiersdorf Formation and the Late Campanian–Maastrichtian Piesting Formation. The Grünbach flora comprises about 60 different taxa, of which about 70% are angiosperms (Herman and Kvaček, 2002).

Gmünd. Pollen assemblages extracted from the Klikov Formation near the small town of Gmünd at the Czech–Austrian border include a considerable variety of well-preserved Normapolles pollen (Zetter *et al.*, 2002). The sediments at this locality are dated as Santonian based on palynological correlation (Zetter *et al.*, 2002).

4.3.6 Germany

Terrestrial sediments of Cretaceous age are known from several places in Germany, and well-preserved fossil floras have been reported from both Lower and Upper Cretaceous strata (Figure 4.11). Plant assemblages of Berriasian–Valanginian age without angiosperm fossils include the classic Wealden-coal flora from Bückeburg, Germany (Osterwald and Obernkirchen members of the Bückeburg Formation; Pelzer and Wilde, 1987), and other floras from northwestern Germany studied by Dunker (1846), Schenk (1871), Benda (1961) and Riegel *et al.* (1986; see also Mägdefrau, 1968). Angiosperms are also not found in the classic Hauterivian–Barremian assemblages from the Quedlinburg area described by Richter (1906, 1909) and Daber (1968, 1990; see also Mägdefrau, 1968). Floras with angiosperm fossils are known only for the mid- and Late Cretaceous, ranging in age from Late Albian – Early Cenomanian to the Maastrichtian. They include leaf floras, such as the classic Quedlinburg flora (Richter, 1905; Rüffle, 1968; Rüffle and Knappe, 1988; Tschan *et al.*, 2008), as well as numerous mesofossil floras recovered both from outcrops and from boreholes (Vangerow, 1954; Knobloch and Mai, 1986). Recently, a mesofossil flora with angiosperm reproductive organs has also been discovered from mid-Cretaceous karst infillings (see below; Drozdzewski *et al.*, 1998; Viehofen *et al.*, 2008). Several of the Late Cretaceous floras from Germany are relatively well dated by marine intercalations.

Karst infillings in the Rhenish Massif. Rich mesofossil assemblages containing angiosperm reproductive structures have been discovered in karst infillings in Upper Devonian reef limestones near Wülfrath and Wuppertal in the Rhenish Massif. Caves and shafts are common in this limestone and several are infilled with Cretaceous sediments containing plant fossils. The Prangenhaus and Rohdenhaus localities have provided particularly rich and well-preserved mesofossil floras. The sediments are predominantly light grey clays with abundant charcoal and were apparently washed into the karst shaft over a very short time interval (Drozdzewski *et al.*, 1998; Viehofen *et al.*, 2008). Based on palynological evidence the infilling is dated as mid-Cretaceous (Late Albian – Early Cenomanian; Viehofen *et al.*, 2008) but more precise dating has not been possible. The mesofossils are exquisitely preserved and include many fragments of ferns and conifers as well as inflorescence fragments and flowers. One of the most common taxa is *Mauldinia angustiloba*, an extinct member of the extant angiosperm family Lauraceae (Viehofen *et al.*, 2008).

Aachen. The first study of a Cretaceous mesofossil flora containing angiosperm fruits and seeds was by Vangerow (1954). The flora occurs in the basal clay (Basiston or Hergenrath Beds) of the Aachen Formation, which outcrops in the area around Aachen in the German–Dutch–Belgian border zone. The depositional environment is interpreted as near-shore lagoonal with variable limnic to marine influence (Knobloch and Mai, 1986; Batten *et al.*, 1988). The clay rests on weathered Palaeozoic rocks and is overlain by marine beds of Campanian age. Based on superimposed marine beds and palynological studies the Aachen Formation is dated as of Santonian age (Batten *et al.*, 1988).

The mesofossil assemblage is dominated by angiosperm remains and includes several taxa related to the Normapolles complex (Knobloch and Mai, 1986; Friis and Crane, 1989). Palynological studies of the Aachen Formation also identified a considerable variety of Normapolles taxa (Batten *et al.*, 1988). A rich assemblage of megaspores of selaginellaceous and isoetalean affinity also occurs among the mesofossils (Dijstra, 1949; Batten, 1988). A diverse leaf flora comprising several angiosperm taxa was described in a very early palaeobotanical study of the clay beds at Aachen (Debey, 1848), as well as in subsequent work (Kräusel, 1922, 1923). A variety of petrified wood has also been reported from the Aachen Formation. Most of this wood is coniferous, but the material also includes several angiosperm taxa (Meijer, 2000).

Quedlinburg–Blankenberg area. Two classic macrofossil floras have been described from Cretaceous strata in the Hartz region near the towns of Quedlinburg and Blankenberg. Cretaceous sediments in this area consist mainly of well-sorted sandstone and intercalated clay beds. They were deposited close to the Cretaceous sea, perhaps associated with a coastal dune system. One flora is from the Early Cretaceous (Hauterivian–Barremian) and is particularly well known for the presence of the isoetalean fossil *Nathorstiana* and the fossil fern *Hausmannia*. No angiosperms are known from this Early Cretaceous flora (Richter, 1906, 1909; Mägdefrau, 1968). The other flora from the Quedlinburg sandstone is much younger and of Late Cretaceous age. It comprises abundant angiosperm leaf fossils, some angiosperm reproductive organs and many conifers. The fossils in the Quedlinburg sandstone are mostly preserved as impressions and moulds (Figure 2.8). Beautifully preserved compression fossils are also recorded from associated clay horizons in the same area (e.g. Rüffle and Knappe, 1988; Rüffle, 1995). Most conspicuous in the flora are leaves of *Credneria*, an extinct genus of probable platanaceous affinity (Richter, 1905; Tschan *et al.*, 2008). The Upper Cretaceous plant-bearing strata are part of the Heidelberg-Schichten, dated as mid- to Late Santonian based on overlaying marine sediments and palynological studies (Krutzsch, 1966b; Knobloch and Mai, 1986).

Mesofossils have also been recovered in the Quedlinburg area. A small flora has been extracted from clay beds collected at the Altenburg locality and includes about 15 taxa. Most are angiosperm fruits and seeds (Knobloch and Mai, 1986), but the flora has also yielded platanoid inflorescences with pollen *in situ* (Pacltová, 1982; Knobloch and Mai, 1986; Tschan *et al.*, 2008).

Eisleben. The Eisleben locality is about 40 km southeast of Quedlinburg and has long been known for angiosperm fossils that occur in layers of grey coaly clay. The plant-bearing sequence ranges from Maastrichtian to Paleocene and was deposited unconformably on top of Permian (Zechstein) strata (Knobloch and Mai, 1986). Mesofossils are known from the basal part of the sequence (Maastrichtian) and include a diverse assemblage of angiosperms, many of which have been assigned to modern families and genera (Knobloch and Mai, 1986).

Walbeck. Leaf and mesofossil floras have been recovered from the Walbeck area, about 50 km north of Quedlinburg. The sediments are mostly sandy with darker clay and coaly layers. They were deposited in a deltaic or near-shore environment and are dated as Late Maastrichtian in age based on palynological studies (Knobloch and Mai, 1986). The mesofossil flora from Walbeck contains diverse angiosperms; as for the Eisleben mesofossil flora, many of them have been assigned to modern families and genera (Knobloch and Mai, 1986).

4.3.7 Great Britain

Early Cretaceous terrestrial, near-shore and marine sediments from southern and eastern parts of England have provided numerous macrofossil floras, a few mesofossil assemblages and rich palynological assemblages. Several of the sequences are conformable and include strata from the Berriasian to Aptian or Albian. Early Cretaceous sediments are also present further north in England, but in this region the sequence is much thinner and with more marine influence. A useful overview of the most important boreholes and outcrops that have been studied for early angiosperm pollen is given by Hughes (1994) and a review of the stratigraphical and geological data for the Wealden area of southeast England is given by Allen (1990).

The Early Cretaceous macrofossil flora of southeast England has been intensively studied for almost 200 years and is particularly important for the concept of a 'Wealden flora'. An overview of the flora and its history was compiled by Seward in two extensive catalogues of the Wealden fossil plants housed at the British Museum (Natural History) (now The Natural History Museum) (Seward, 1894, 1895). Stopes (1915) later added a catalogue of the Aptian plant fossils from this area. More recent studies of the macrofossil flora have primarily been carried out by Watson (e.g. Watson and Sincock, 1992; Watson *et al.*, 2001; Watson and Cusack, 2005). In addition to the macrofossil floras, several mesofossil floras have also been studied from the Early Cretaceous of southeast England (Batten, 1998). None of the British Early Cretaceous macro- or mesofossil floras has yielded undisputed angiosperms, but the palynological assemblages, mainly from the younger part of the sequence, include numerous angiosperm pollen taxa that are of central importance for understanding the early angiosperm radiation (e.g. Couper, 1958; Kemp, 1968; Hughes *et al.*, 1979, 1991; Hughes and McDougall, 1987, 1990; Hughes, 1994). These palynological assemblages are also the main reference floras for calibrating the stratigraphy

Stages	Lithostratigraphic units	Beds
Aptian	Lower Greensand (part)	
Barremian	Upper Weald Clay	Weald Clay
Hauterivian	Lower Weald Clay	
Valanginian	Upper Tunbridge Wells Sands	Hastings Bed
	Grinstead Clay	
	Lower Tunbridge Wells Sands	
	Wadhurst Clay	
	Ashdown Beds	
	Fairlight Clay	
Berriasian	Durlston Beds	Purbeck
	Cinder Beds	

Figure 4.12 Stratigraphy of the Wealden strata of southern England. Based on Watson and Sincock (1992).

of terrestrial sequences in other parts of the world where marine intercalations are lacking.

Most of the Cretaceous macrofossil floras and all mesofossil assemblages from southern England have been described from so-called Wealden sediments (for a summary see Watson, 1969; Watson and Sincock, 1992; Batten, 1998). The stratigraphic term 'Wealden' was first applied to terrestrial sediments outcropping in the anticline of the Weald, southeast England. These sediments are Late Berriasian to Barremian in age and comprise the Hasting Beds (Fairlight Clay, Ashdown Beds, Wadhurst Clay, Lower and Upper Tunbridge Wells Sands, and Grinstead Clay) and the Weald Clay (Lower and Upper) (Figure 4.12). Subsequently, the term Wealden has been used in other areas, particularly in northern Europe, for Lower Cretaceous sedimentary sequences that are typically older than the British sequence in the Weald (Allen, 1990).

The Wealden sequence in southeast England is conformably overlain by marine glauconitic sands (Greensand) of Aptian age and marine blue clay (Gault Clay) of Albian age. The Wealden sequence in turn overlays the Berriasian Cinder Beds and Durlston Beds, which are part of the mainly marine Purbeck sequence that is of latest Jurassic to earliest Cretaceous age (Allen, 1990). Palaeoenvironmental studies indicate that the Wealden sequence was deposited in a coastal flood plain under warm climatic conditions with marked seasonality (Batten, 1975, 1998; Allen, 1976, 1981).

Angiosperm wood has been described from the Aptian Greensand of southern England (Stopes, 1912, 1915) but locality information is imprecise or lacking and the records have for a long time been controversial (Chapter 9). According to Crawley (2001) only one of the specimens is of probable Aptian age, but there is no locality information associated with this fossil (section 2.1.1).

The only other possible angiosperm macrofossil described from the Early Cretaceous of southern England is *Bevhalstia* (Chapter 9) collected from outcrops of the Weald Clay Formation in Sussex (Keymer Tileworks, Rudgwick Brickworks) and Surrey (Clockhouse and Smokejacks Brickworks) (Hill, 1996). However, there are important descriptions of angiosperm pollen grains from several outcrops and boreholes. The following summary is mainly based on Hughes (1994) and Hill (1996) and focuses especially on those successions in southern England that are particularly important for stratigraphic correlation and for understanding patterns of early angiosperm diversification.

Isle of Wight. Early Cretaceous sediments can be studied at several coastal exposures on the Isle of Wight. The sequence includes terrestrial sediments of mid- to Late Barremian age that are conformably overlain by the Lower Greensand, which is marine and of Aptian age (Hughes, 1994). The Barremian sequence is about 150 m thick. Parts of the sequence are barren owing to oxidation, but other parts contain a well-preserved and diverse palynoflora. The Aptian sequence is rich in pollen, and is less fully marine than the Aptian of the Warlingham Borehole further north (see below). The dating of the Barremian part of the sequence is based on its stratigraphic position below the marine Aptian sequence, and its content of monoaperturate angiosperm pollen.

A notable macrofossil site on the Isle of Wright is the Hanover Point locality, which has yielded petrified trunks of conifers and Bennettitales (e.g. Watson and Sincock, 1992). Another plant fossil site about 200 m southeast of Hanover Point has yielded the notable 'Pine Raft' logs, which are rich in cheirolepidiaceous wood preserved either as charcoal or as permineralisations (Alvin *et al.*, 1981).

Worbarrow Bay, Dorset. A coastal exposure along Worbarrow Bay in Dorset shows an almost complete sequence of Lower Cretaceous sediments a few hundred metres thick (Hughes and McDougall, 1987, 1989; Hughes, 1994). This sequence conformably overlies beds of Late Jurassic age and is apparently conformable throughout the Early Cretaceous from the Berriasian to Albian. The major part of the sequence consists of Hauterivian–Barremian strata and several angiosperm pollen types have been reported from this interval. Stratigraphic correlation is based on spores and pollen, but there is no marine control on the critical Hauterivian–Barremian part of the sequence. Hughes (1994) noted that the Cretaceous sequence at Worbarrow was atypical owing to its marginal position in the Cretaceous depositional basin.

Warlingham Borehole, Surrey. This borehole was drilled south of London in 1954 and large parts of it are cored. The sequence is a key stratigraphic reference point for other Early Cretaceous strata that include early angiosperms because of the presence of marine intercalations that provide more secure dating. The sequence extends from the Berriasian to the Early Aptian, but the Hauterivian portion is disturbed by faulting (Hughes, 1994). The Hauterivian/Barremian boundary has been established based on marine dinocysts, and the Barremian strata are overlain by the glauconitic marine Lower Greensand (Aptian). Other boundaries in the section are less certain (Hughes, 1994). Numerous well-preserved angiosperm pollen grains have been recorded from the Warlingham Borehole, particularly from the Barremian part of the section (Hughes, 1994).

Smokejacks Brickworks, Surrey. Sediments exposed in this clay pit consist of variously coloured mudstones with silty ironstone concretions, small siltstone lenses and phosphatic nodules (Hill, 1996; Batten *et al.*, 1998). Based on lithological correlation and palaeontological evidence the sediments are believed to be part of the Upper Weald Clay, which is of Barremian age (Batten *et al.*, 1998). The sequence is rich in fossils, predominantly insects and various invertebrates, as well as dinosaurs and other vertebrate remains (Jarzembowski, 1991), but has also yielded a variety of plant macrofossils and mesofossils (Hill, 1996; Batten *et al.*, 1998). One of the plant fossils, described as *Bevhalstia pebja*, was interpreted as a possible angiosperm (Hill, 1996), but otherwise all plant fossils from this locality are non-angiospermous. They include a variety of lycopsids, ferns, horsetails and cheirolepidiaceous conifers (Batten, 1998).

Clockhouse Brickworks, Surrey. Sediments exposed in this clay pit comprise about 35 m of mudstones with intercalated layers of more silty and sandy material that were deposited in a shallow marine to marginal lacustrine environment. The mudstone sequence contains many fossils such as ostracods, fish and reptiles as well as some plant remains, most importantly fronds of *Weichselia* and fragments of *Bevhalstia pebja* (Hill, 1996). The sequence is assigned to the Lower Weald Clay Formation (Hauterivian) and is overlain by sandstone referred to the Upper Weald Clay Formation (Barremian) (Batten, 1998).

Keymer Tileworks, Sussex. Keymer Tileworks is the type locality for the enigmatic *Bevhalstia pebja* (Hill, 1996). The sediments exposed at the locality are from near the top of the Lower Weald Clay Formation, which is of Hauterivian age.

Rudgwick Brickworks, Sussex. This locality exposes sediments referred to the Upper Weald Clay Formation, which is of Barremian age. The locality is significant in yielding fossil remains of *Bevhalstia pebja* (Hill, 1996).

Kingsclere Borehole, Berkshire. This borehole was drilled in 1934 and includes several cores that have yielded abundant palynological material. Two stratigraphic levels were particularly important and productive, one from the Late Hauterivian and one from the Barremian/Aptian boundary (Hughes, 1994). No marine fossils were reported from the Hauterivian–Barremian/Aptian interval and the stratigraphic position was established based on correlation with the Warlingham borehole, in which marine dinocysts of latest Hauterivian age have been reported (Hughes and McDougall, 1987; Hughes *et al.*, 1991). Several important pollen records are known from the Kingsclere Borehole, including a variety of well-preserved tectate columellate and monocolpate pollen grains from the Late Hauterivian part of the sequence (Hughes *et al.*, 1991). The type specimen of *Clavatipollenites hughesii* is also recorded from the Barremian part of the sequence in this borehole (Couper, 1958; Hughes *et al.*, 1991).

4.3.8 Sweden

Cretaceous sediments occur in several areas of Scania, southern Sweden, where they were deposited close to the fluctuating Cretaceous coastline (Figure 4.13). They are mainly of marine origin and contain few or no land-derived fossils. Rich angiosperm assemblages have, however, been discovered in a kaolinitic deposit of terrestrial origin in northeast Scania in the Lake Ivösjön area. Early studies of these deposits by Grönwall (1915) and Lundegren (1931, 1934) described their lithology and noted the abundance of plant fragments, particularly wood and larger trunks up to about 10 m long. The plant-bearing sequence also includes a rich and well-preserved pollen flora (e.g. Ross, 1949; Skarby, 1968, 1986; Skarby *et al.*, 1990).

Åsen. The plant-bearing sequence in the Scanian kaolinites is best exposed in old kaolin quarries at the Åsen locality (Figures 2.9, 4.14). The fluviatile–lacustrine sequence at Åsen is about 20 m thick and comprises kaolinitic clays, silts and sands of lacustrine and fluviatile origin. A marked weathering horizon divides the sequence into a lower and an upper unit. Palynological studies indicate a mid Late Cretaceous age (approximately Campanian) for the fluviatile–lacustrine sequence (Skarby, 1964, 1968). In places the fluviatile–lacustrine sequence is overlain by marine glauconitic sands, which have been dated as late Early Campanian based on the belemnite *Belemnellocamax mammillatus* (Christensen, 1975). A magnetic reversal at the weathering horizon is identical with chron 33R, which can be correlated with zone A in the Gubbio sequence of Italy. This indicates an Early Campanian age for the upper unit; the age of the lower unit is probably latest Santonian. A Late Santonian – Early Campanian age for the whole sequence is also in accordance with inferences based on the dispersed megaspore assemblages in the Åsen samples (Koppelhus and Batten, 1989). There are some differences in composition of plant assemblages from the lower and the upper unit, but they also have many taxa in common, indicating that there is no significant hiatus of deposition at the weathering horizon.

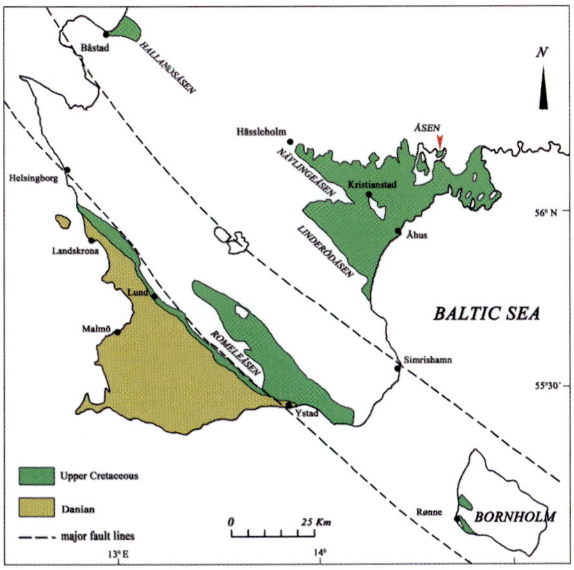

Figure 4.13 Map showing distribution of Cretaceous and Early Cenozoic (Danian) sediments in Scania, southern Sweden, and the position of the Åsen locality (red arrow). All Cretaceous sediments are marine except for the deposits in the northernmost part around Åsen.

Plant fragments occur at all levels in the sequence and include an immense diversity of forms. There are some differences in both composition and preservation among samples throughout the sequence, but generally all samples include rich and well-preserved fossils. Except for conifer cones, twigs and wood most fossils are tiny, typically a few millimetres or less, and consist mainly of angiosperm flowers, fruits and seeds. Most of the plant

Figure 4.14 The Late Cretaceous (Late Santonian – Early Campanian) Åsen locality exposing fluviatile and limnic sediments with abundant plant mesofossils. The weathering horizon marked by asterisks indicates the Santonian–Campanian boundary. The clay pit has now been filled with domestic refuse.

fossils are three-dimensionally preserved as charcoal, and charcoalified fossils occur at all levels in the sequence. In addition, the Åsen assemblages also include lignitised compression fossils. These are most common in a clay gyttja horizon of the lower sedimentary sequence, and twigs and cones of taxodiaceous conifers are particularly abundant at this level (Srinivasan and Friis, 1989). Angiosperms occur at all levels and include many systematically informative floral structures (Friis and Skarby, 1981, 1982; Friis, 1983, 1984, 1985a, 1985c, 1990; Friis *et al.*, 1988, 2006b; Friis and Crane, 1989; Eklund *et al.*, 1997; Schönenberger and Friis, 2001). Larger wood pieces and trunks are particularly common at the weathering horizon. All of these larger specimens that have been studied are coniferous (Nykvist, 1957). Angiosperm wood has only been identified from smaller wood fragments (Herendeen, 1991a).

4.3.9 Other parts of Europe

There are numerous reports of Cretaceous macro-, meso- and microfloras containing angiosperms from other parts of Europe. Most of these are of Late Cretaceous age and typically they are less diverse, and have been less intensively studied, than the floras mentioned above. From the island of Bornholm, Denmark, a rich mesofossil flora of Late Berriasian – Valanginian age was discovered from the Carl Nielsen A/S gravel pit (Figure 4.15). The flora is rich in non-angiospermous seed plants (e.g. Pedersen *et al.*, 1989b) and is particularly rich in fragments of the extinct fern *Onychiopsis* (Friis and Pedersen, 1990). No angiosperms have been discovered in the flora. This is currently the most diverse and best-preserved mesofossil flora from the earliest Cretaceous.

In Italy the discovery of dispersed angiosperm pollen from the Valanginian of southern Tuscany is probably one of the most important records from Europe (Trevisan, 1988). However, because the discovery remains to be documented in detail, its full significance is not yet clear. The sediments studied from Tuscany are of marine origin and also include dinoflagellate cysts, which are an important independent tool for dating these sediments and the early angiosperms that they contain. Further studies of these sequences could be of considerable importance for strengthening the stratigraphic context in which early angiosperm evolution is interpreted. Macrofossil floras with angiosperm remains are also reported from the Middle Albian and from the Cenomanian of southern Italy (Bravi *et al.*, 2004, 2010). In the Middle Albian flora the angiosperm component includes a fragment of a possible herbaceous plant with attached fruits and lobed leaves co-occurring with various twigs of conifers (Bravi *et al.*, 2010).

Figure 4.15 The Carl Nielsen A/S locality, Bornholm, Denmark, exposing Early Cretaceous (Late Berriasian–Valanginian) brackish to limnic sediments. A thin peat horizon in the top of the exposure has yielded a rich mesofossil flora dominated by *Onychiopsis* and non-angiosperm seed plants.

Recently, Late Cretaceous mesofossils have also been discovered from Italy and Romania. The age of the Italian mesofossil flora is latest Cretaceous or earliest Cenozoic. The Romanian mesofossil flora is of latest Cretaceous (Maastrichtian) age. A preliminary account was given for the Romanian mesofossil flora from the Budurone locality (Lindfors *et al.*, 2010). It is important as the first mesofossil flora from the Cretaceous to be found associated with vertebrate fossils. The plant-bearing sediments at this locality are deposited in a flood-plain environment and are part of the continental Densuş–Ciula Formation that is exposed in the northwestern part of Haţeg Basin (Grigorescu, 1992).

4.4 EASTERN NORTH AMERICA

Numerous Cretaceous floras with angiosperm fossils have been reported from the Atlantic Coastal Plain of eastern North America from Georgia in the south to Martha's Vineyard, Massachusetts, in the north. The Cretaceous sedimentary sequence in this region consists predominantly of clays, silts, and sands deposited in terrestrial environments. Most palaeobotanical attention has been focused in Virginia, Maryland and

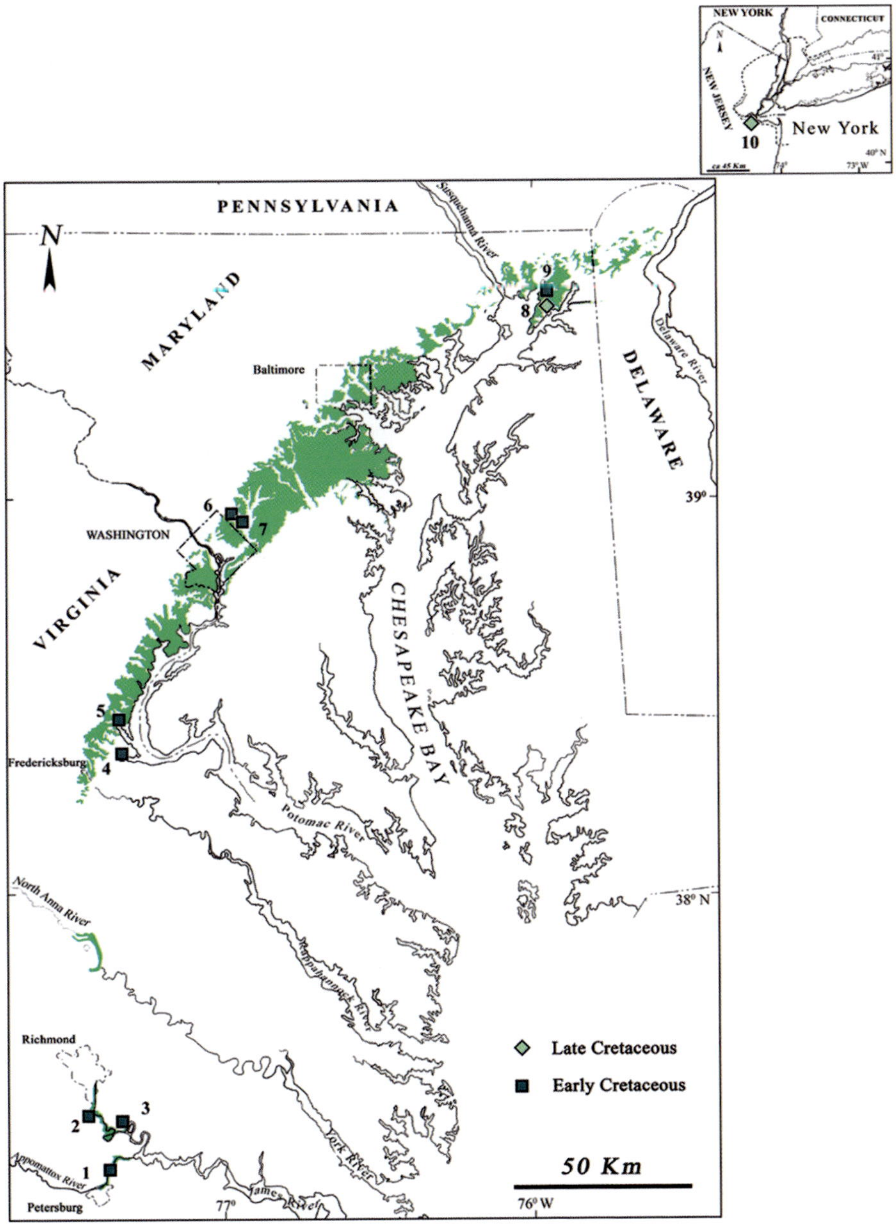

Figure 4.16 Map showing the Potomac Group outcrop belt (green) and the location of the most important plant fossil localities in Virginia, Maryland and New Jersey, USA. 1, Puddledock; 2, Drewry's Bluff; 3, Dutch Gap; 4, Bank near Brooke; 5, Quantico; 6, Kenilworth; 7, West Brothers; 8, Mauldin Mountain; 9, Bull Mountain; 10, Old Crossman. Map based on Glaser (1969).

New Jersey where the Cretaceous deposits have traditionally been subdivided into the Potomac Group sequence, followed above by the Raritan Formation and the Magothy Formation. The Potomac Group is a fluviatile–limnic sequence deposited by major river systems that drained the Appalachians during the later part of the Early Cretaceous (Jordan, 1983). It has an extensive distribution mostly underlying the Middle Atlantic Coastal Plain, but with scattered exposures along a NNE–SSW belt (Figure 4.16). The Potomac Group is traditionally subdivided into three formations (Patuxent, Arundel and

Patapsco), although sometimes the sediments are referred to a single formation, the Potomac Formation, and the Patuxent, Arundel and Patapsco Formations are referred to as facies (e.g. Benson, 2006; Lipka *et al.*, 2006). Because of the extensive literature that refers to the Potomac Group plant fossils we have retained the term in this book.

In Maryland and Delaware, Potomac Group sediments are overlain by more sandy deposits that have sometimes been referred to the younger Raritan Formation, which outcrops mainly in New Jersey, but these sediments are now known to be older than those of the type Raritan Formation (Doyle and Robbins, 1977). They have sometimes been treated as part of the Patapsco Formation (Wolfe and Pakiser, 1971) or separated informally as the 'Maryland Raritan' or the Elk Neck Beds within the Potomac Group (Doyle and Robbins, 1977; Hickey and Doyle, 1977).

The Potomac Group is the oldest of the Cretaceous sequences on the Atlantic Coastal Plain and is exposed in a narrow belt from Virginia in the south to Delaware in the north (Figure 4.16). The sediments are deposited exclusively in terrestrial environments with the exception of a single interval of brackish origin encountered in a deep well from the eastern shore of Maryland (Doyle, 1969). The age of the Potomac Group sequence is poorly controlled. The sediments unconformably overlay the crystalline basement rocks or earlier Mesozoic sediments, and over much of their outcrop they are unconformably overlain by Cenozoic deposits. Superimposed younger Cretaceous strata (Elk Neck Beds) are reported in Maryland, but these are also terrestrial. More precise dating of the Potomac Group sediments therefore relies almost entirely on palynological correlation with better-dated deposits elsewhere (Doyle and Robbins, 1977; Doyle, 1992). Higher in the sequence, the Woodbridge Clay Member of the New Jersey Raritan Formation is dated by marine molluscs as of mid-Cenomanian age (Doyle, 1969; Wolfe and Pakiser, 1971).

Within the Potomac Group, correlation based on lithology is problematic. The Patuxent Formation is predominantly sandy and gravelly, the Arundel Formation (Arundel Clay) is predominantly highly organic clay, and the Patapsco and Raritan Formations are more heterogeneous (Doyle and Hickey, 1976; Hickey and Doyle, 1977), but the Potomac Group formations/facies are difficult to separate in the field (Doyle, 1969). Extreme lateral variation in lithology (typical of sediments deposited in fluvial systems), combined with limited outcrops, makes traditional internal stratigraphic subdivision and correlation difficult. Based on palynological studies of surface samples and borehole material, an informal palynological zonation was suggested by Brenner (1963) for the Potomac Group sequence with Zone I corresponding to Patuxent Formation and Arundel Formation and Zone II (divided into Subzone II-A, Subzone II-B-1, and Subzone II-B-2) corresponding to the Patapsco Formation. This zonation was refined by studies of angiosperm pollen from two wells near Baltimore (Doyle and Robbins, 1977). It was also extended to younger strata (Doyle and Hickey, 1976; Doyle and Robbins, 1977) with the addition of Subzone II-C and Zone III, corresponding in part to the Elk Neck Beds ('Maryland Raritan'), and Zone IV, corresponding to the lower members of the Raritan Formation of New Jersey (Farringdon Sand, Woodbridge Clay) (Doyle and Hickey, 1976; Doyle and Robbins, 1977; Hickey and Doyle, 1977).

Correlation of the Potomac Group sediments with marine-based stages in Europe is based mainly on long-distance comparison with palynofloras from southern and eastern England, Portugal and Africa. Some of these European and Africa palynofloras are themselves from predominantly non-marine sequences, which introduce additional uncertainty into the dating. However, in general the Potomac Group sequence (including the Elk Neck Beds) is thought to extend from the Late Barremian or earliest Aptian to the latest Albian or earliest Cenomanian. The ages of the boundaries between the zones and subzones are not fully established. According to Doyle (1992) the base of Zone I may be of earliest Aptian age, but a Late Barremian age cannot be ruled out. The upper part of Zone I may be of mid-Aptian age. Subzone II-A may be of latest Aptian or Early Albian age, while the base of Zone II-B correlates with Early or Middle Albian. Previous studies (Doyle and Robbins, 1977) indicated a Late Albian age for Zone II-C and an Early Cenomanian age for Zone III. The base of Zone IV is considered to be of Middle Cenomanian age, and the top of Zone IV may be latest Cenomanian or Early Turonian (Doyle and Robbins, 1977).

4.4.1 Virginia

Several classic leaf floras have been described from the Potomac Group of Virginia, including the floras from Aquia Creek, Bank near Brooke, Drewry's Bluff, Dutch Gap, Fredericksburg, Mount Vernon, Quantico and Widewater (Fontaine, 1889, summarised in Doyle and Hickey, 1976). Sediments of the Potomac Group sequence are distributed in Virginia in a narrow belt from Petersburg in the south to Washington, D.C., in the north with exposures along the James, Appomattox, North Anna, Rappahannock and Potomac rivers as well as in several clay pits and other

Figure 4.17 The Early Cretaceous (Early Aptian) Dutch Gap locality at James River, Virginia, USA, exposing Potomac Group deposits of the Patuxent Formation/facies comprising cross-bedded sands with laminated, organic-rich silts containing mesofossils. The outcrop is now obscured with concrete for erosion protection.

temporary exposures (Figure 4.16). The oldest plant assemblages in the sequence, from the Late Barremian or earliest Aptian, occur in the south and are especially well exposed along the James River. In the following sections we describe some of the localities that have yielded informative angiosperm fossils including reproductive structures.

Bank near Brooke. This locality is an overgrown north-facing bank close to the Richmond, Fredericksburg and Potomac railway line, which exposes a sedimentary sequence of sandstones with intercalated claystones and siltstones. A sketch of the sedimentary facies at this locality is given in Hickey and Doyle (1977). The sediments belong to the Patapsco Formation of the Potomac Group and are referred to the lower part of the pollen Zone II-B of Early to Middle Albian age (Doyle and Hickey, 1976). Leaf macrofossils occur abundantly in several horizons and are dominated by *Sapindopsis* leaves (Fontaine, 1889; Hickey and Doyle, 1977; Crane *et al.*, 1993). Most of the sediments at the locality are too oxidised to yield mesofossils, but mesofossils extracted from a small pocket of less oxidised material include the earliest unequivocal platanoid reproductive structures known from the Potomac Group (Crane *et al.*, 1993). These platanoid reproductive organs are believed to come from the same plants that produced the associated *Sapindopsis* leaves.

Drewry's Bluff. The Drewry's Bluff plant assemblages are from a high riverbank on the James River that exposes about 20 m of the Arundel Formation facies, comprising mostly fluviatile gravels, sands, silts and clays. Based on palynological studies, the Drewry's Bluff sequence is referred to the middle part of pollen Zone I and thought to be of Aptian age. Reworked clayballs occur in the basal part of the sequence. Palynological studies indicate that these older blocks are of Late Barremian or Early Aptian age, corresponding to the basal part of pollen Zone I (Brenner, 1967).

Macrofossils have been described from several horizons in the Drewry's Bluff sequence and comprise mostly ferns, bennettitaleans and conifers (Fontaine, 1889). Angiosperm leaves have also been reported from Drewry's Bluff, but are rare (Doyle and Hickey, 1976). The best-studied macrofossil assemblage is that from the Drewry's Bluff leaf bed exposed to the north of the main section (now obscured by slumping of the river bank), which includes ferns and several kinds of angiosperm leaves as well as rare gnetalean fossils (Upchurch and Doyle, 1981; Crane and Upchurch, 1987). Mesofossil floras obtained from some of the sandy layers exposed at Drewry's Bluff and from the clayballs include fruits, seeds and stamens of angiosperms, as well as reproductive organs of diverse non-angiospermous seed plants (Brenner, 1967; Pedersen *et al.*, 1993; Rydin *et al.*, 2006a).

Dutch Gap. The Dutch Gap plant assemblages occur on the bank of the James River, which exposes a sedimentary sequence of mostly cross-bedded light sand with laminated, organic-rich silt and clay beds (Figure 4.17). Much of the former exposure, especially the lower part of the sequence, is now covered by large blocks put in place as erosion protection. The depositional environment is interpreted as fluviatile and deltaic, possibly with some tidal influence (Upchurch and Doyle, 1981). The sequence is assigned to the Patuxent Formation, and based on palynological studies is referred to the basal part of pollen Zone I corresponding

Figure 4.18 The Early Cretaceous (Early–Middle Albian) Puddledock locality, Virginia, USA, a large gravel and sand pit exposing cross-bedded sand and gravel with subordinate organic-rich clay horizons that contain mesofossils.

to the Early Aptian. Leaf macrofossils have been recovered from several horizons at the Dutch Gap locality and include species of Bennettitales, Cheirolepidiaceae and other conifers, as well as very rare angiosperms (Fontaine, 1889; Ward, 1905; Doyle and Hickey, 1976; Upchurch and Doyle, 1981). Mesofossil floras have also been recovered from several horizons at this locality and include diverse angiosperm fruits and seeds as well as reproductive organs and twigs of non-angiosperm seed plants (Pedersen *et al.*, 1993; Rydin *et al.*, 2006a).

Puddledock (Tarmac Lone Star Industries). This locality is a large sand and gravel pit southwest of Hopewell close to the Appomattox River. Several clay and silt horizons in the pit have yielded well-preserved mesofossils (Figure 4.18). In particular, a dark clay horizon in the basal part of the pit has yielded rich assemblages of angiosperm flowers, fruits and seeds (Crane *et al.*, 1994; Friis *et al.*, 1994a, 1995, 1997a; von Balthazar *et al.*, 2007, 2008). Based on palynological studies, the sediments at the Tarmac Lone Star gravel pit were correlated with the basal part of Subzone II-B (Christopher in Dischinger, 1987), which indicates an Early to Middle Albian age. This correlation has also been confirmed by a preliminary palynological analysis of one Puddledock sample (J. A. Doyle, personal communication).

4.4.2 Maryland and Washington, D.C.

The Potomac Group outcrop broadens northward from around Richmond, Virginia, and is at its widest in the vicinity of Washington, D.C., and Baltimore. It becomes narrower further north and peters out in the vicinity of the Elk Neck Peninsula at the head of the Chesapeake Bay (Figure 4.16). Many of the old localities studied in and around Washington, D.C., by Fontaine (1889) are no longer available because of urbanisation, but some of the localities studied more recently by Brenner (1963) and Doyle and Hickey (1976) still exist.

Kenilworth (Bladensburg). The Kenilworth locality is an old clay pit immediately north of Washington, D.C. The presence of angiosperms at this locality has been documented by palynological studies (Brenner, 1963) and the sediments have also yielded some well-preserved fruits and seeds (Friis *et al.*, 1997a). The sedimentary sequence is referred to the Patapsco Formation and consists of about 20 m of predominantly red and grey clays and silts of terrestrial origin. Based on dispersed spores and pollen, Brenner (1963) placed the basal part of the exposure in Subzone II-A, which may be either Early Albian or Late Aptian in age. Angiosperm fruits and seeds extracted from the Kenilworth locality come from a level more than 15 m above the Subzone II-A samples of

Brenner (1963). The palynoflora at this level is more characteristic of palynological Subzone II-B (J. A. Doyle, personal communication), and this basal position in Subzone II-B indicates an Early–Middle Albian age.

West Brothers. The locality is an abandoned clay pit situated close to Washington, D.C., in Prince George's County, Maryland. The Cretaceous strata consist of a sequence approximately 10 m thick of fluviatile clays, silts, and cross-bedded sands interpreted as the fill of abandoned channels (Hickey, 1984). A macrofossil flora of compression leaves was reported from the locality (Upchurch, 1984) and associated well-preserved flowers, fruits and seeds occur abundantly in the clay (Friis *et al.*, 1988; Crane *et al.*, 1989b; Drinnan *et al.*, 1991). The sequence is referred to the Patapsco Formation, and based on palynological studies it is assigned to the upper part of Subzone II-B equivalent to about Late Albian (Doyle and Hickey, 1976).

Mauldin Mountain. The locality occurs in Elk Neck State Park on the Elk Neck Peninsula near the top of a coastal cliff overlooking Chesapeake Bay. About 20 metres of grey clay and silt of terrestrial origin are exposed. This locality has yielded leaf fossils together with a rich mesofossil flora of angiosperm flowers, fruits and seeds (Drinnan *et al.*, 1990, 1991; Pedersen *et al.*, 1991). The sequence is referred to the Elk Neck Beds based on geological evidence. Palynological analyses place the Mauldin Mountain flora in the lowermost part of Zone III (Drinnan *et al.*, 1990), which is probably of earliest Cenomanian age. Preliminary results from studies of the Mauldin Mountain leaf flora recognise 26 species of angiosperms dominated by foliage with magnoliid-grade architecture (Wolfe, 1987).

Bull Mountain. The Bull Mountain locality is a coastal cliff on the western shore of the Elk Neck Peninsula, a few kilometres south of Mauldin Mountain. The sediments exposed at the locality are dark grey clays belonging to the Elk Neck Beds (Pedersen *et al.*, 1994). According to Doyle and Hickey (1976) the sequence belongs to Subzone II-C and is probably Late Albian in age (Doyle and Hickey, 1976). The clay contains angiosperm leaves (Hickey and Doyle, 1977) and a diverse mesofossil flora with many fragments of angiosperm reproductive organs (Pedersen *et al.*, 1994).

4.4.3 New Jersey

Sediments of Late Cretaceous age from New Jersey have long been known to contain angiosperm fossils (Newberry, 1895; Berry, 1909) and leaf floras, cones, and wood remains have all been described previously from this area. Palynofloras have also been described by Christopher (1979).

Old Crossman Clay Pit. Clay deposits exposed in the Old Crossman Clay Pit in Sayreville, New Jersey, contain a rich mesoflora of three-dimensionally preserved plant fragments. The clay is referable to the Raritan or Magothy Formation and, based on an unpublished palynological study by Brenner, it is interpreted as of Late Turonian age (e.g. Nixon and Crepet, 1993). The sediments at the Old Crossman locality are thought to represent the Amboy Fire Clay. Many of the mesofossils observed in the Old Crossman flora have also been recorded from other, slightly younger, floras in North America (Santonian–Campanian: Neuse River and Allon floras) and there is therefore the possibility that the Old Crossman flora is slightly younger (perhaps Santonian or Campanian). The fossil plant remains are mainly preserved as charcoal and are often exceptionally well preserved. Angiosperm flowers, fruits and seeds predominate (Crepet *et al.*, 1992, 2005; Nixon and Crepet, 1993; Crepet and Nixon, 1994, 1998a, b; Crepet, 1996; Gandolfo, 1998; Gandolfo *et al.*, 1998a, b, 2002, 2004; Hermsen *et al.*, 2003; Martínez-Millán *et al.*, 2009) but the flora also includes well-preserved conifers and fragments of ferns (Gandolfo *et al.*, 1997).

4.4.4 Massachusetts

The Late Cretaceous deposits that outcrop in New Jersey also extend through Long Island (Hollick, 1894, 1912) and into southern Massachusetts, with several fossil localities on Martha's Vineyard (Hollick, 1894, 1902, 1906).

Gay Head. The Cretaceous sequence at the Gay Head locality of Martha's Vineyard is of particular interest because in addition to macrofossils (e.g. Hollick, 1906) mesofossils have also been recovered, including a suite of small angiosperm flowers (Tiffney, 1977b). The Cretaceous sediments at Gay Head comprise sandy clays with iron nodules and concretions distorted by glacial activity. More than a hundred different taxa of macrofossils were reported by Hollick (1906) of which about 85 are angiosperms. The angiosperm leaves have not been revised since Hollick's studies. Currently only a single flower has been described from the mesofossil flora, but it is well preserved with the three-dimensional form intact and with *in situ* pollen (Tiffney, 1977b). According to Tiffney (1977b) the age of the Martha's Vineyard flora is earliest Campanian based on palynological studies, but details of the palynological studies have not been published.

4.4.5 North Carolina

Cretaceous strata known from the Coastal Plain of North Carolina are mostly sands and clays deposited in deltaic to shelf environments. They range in age from Coniacian to Maastrichtian (Owens and Sohl, 1989). Lignite horizons, trunks, and fossil leaves have been reported from several localities, mainly along the Cape Fear and Neuse rivers (Berry, 1907, 1910a). In the 1980s a rich mesofossil assemblage was discovered at the Neuse River Cut-Off.

Neuse River. The Neuse River Cut-Off, southwest of Goldsboro in Wayne County, exposes a plant-bearing sequence consisting of sand, silt and clay with subordinate lignite beds. According to Owens and Sohl (1989) the sequence was deposited in a delta front environment. Fossil plants recovered from this site include lignitised angiosperm flowers, fruits and seeds, which occur abundantly in the sedimentary sequence (Friis, 1988; Friis *et al.*, 1988; Frumin and Friis, 1996; Mickle, 1996; Eklund, 2000), as well as many conifer shoots, including the unusual frond-like shots of *Androvettia* (Hueber and Watson, 1988). Well-preserved angiosperm leaves have also been recovered from the locality, but have not yet been studied. Large petrified trunks that have been washed out of the sediments are scattered over the riverbed. The sediments have been assigned to the basal part of the Black Creek Formation, which corresponds to the Tar Heel Formation, Black Creek Group, of Owens and Sohl (1989). This part of the sequence is thought to be of Early Campanian age based on palynological studies (Owens and Sohl, 1989) or it may extend down into the Santonian (see references in Raubeson and Gensel, 1991).

4.4.6 Georgia

Although both Early and Late Cretaceous sediments are present in Georgia only few plant macrofossils have been recorded (Berry, 1910b). However, several mesofossil floras with rich and well-preserved assemblages of angiosperm reproductive organs have been discovered in sediments of Late Cretaceous age that outcrop in clay pits, road cuts and riverbanks in central and western Georgia (Crane and Herendeen, 1996).

Upatoi Creek. Several plant-bearing horizons with well-preserved angiosperm mesofossils have been discovered along Upatoi Creek, which forms the boundary between Chattahoochee County and Muscoge County in central western Georgia. The plant fossils occur in clay lenses interspersed with coarse to fine sands and silts, and were deposited in a terrestrial environment. The sedimentary sequence belongs to the lower part of the Eutaw Formation, which is thought to be of Coniacian age (Magallón-Puebla *et al.*, 1997). The material is preserved as three-dimensional charcoalifications and lignitised fossils. The plant fossil assemblage is diverse and very well preserved. Among the fossils from the Upatoi Creek locality are well-preserved staminate and pistillate flowers of platanaceous affinity (Magallón-Puebla *et al.*, 1997).

Allon Quarry. A carbonaceous clay lens exposed in the Allon Quarry of the Atlanta Sand and Supply Company at Gaillard, Crawford County, has yielded numerous well-preserved angiosperm reproductive organs (Herendeen *et al.*, 1995, 1999a; Keller *et al.*, 1996; Magallón-Puebla *et al.*, 1996). Sediments at this locality have been referred to the Buffalo Creek Member of the Gaillard Formation and were probably deposited in a small flood-plain pond. Based on palynological evidence, the plant-bearing sediments at the Allon Quarry were initially correlated with the earliest Campanian (Huddlestun and Hetrick, 1991), but later investigations indicate that they are perhaps slightly older, of Late Santonian age (Konopka *et al.*, 1997). The mesoflora comprises a rich assemblage of flowers, fruits, seeds, stamens and twigs preserved as three-dimensional fossils, both charcoalified and lignified (Herendeen *et al.*, 1999a). Angiosperm reproductive organs predominate in the mesoflora (Herendeen *et al.*, 1995; Crane and Herendeen, 1996; Keller *et al.*, 1996; Magallón-Puebla *et al.*, 1996; Sims *et al.*, 1998), but there are also well-preserved conifers and remains of mosses (Konopka *et al.*, 1997, 1998).

A similar mesofossil flora is known from Whitewater Creek, approximately 25 km south of the Allon locality. Several taxa are shared between the two localities (Crane and Herendeen, 1996).

4.4.7. Alabama

Late Cretaceous sediments assigned to the Tuscaloosa Formation outcrop extensively in western Alabama and have yielded extensive macrofossil floras, especially from the Shirley's Mill locality in Fayette County (Berry, 1919). Over most of its area the Tuscaloosa Formation unconformably overlies Palaeozoic rocks, but to the east it overlies Early Cretaceous sediments. To the west the formation is overlain by sediments of the Eutaw Formation. The Tuscaloosa Formation consists predominantly of light-coloured, cross-bedded, sometimes very coarse sands with several interbedded clay horizons and thin lignitic layers (Berry, 1919). The Tuscaloosa Formation is Middle to Late Cenomanian in age (Christopher, 1982). The flora is dominated by angiosperm leaves, but also contains a variety of conifers and a few other non-angiospermous seed plants as well as some ferns and lycopsids (Berry, 1919).

4.5 WESTERN INTERIOR OF THE UNITED STATES AND CANADA

Beginning in the later part of the Early Cretaceous and continuing through the Late Cretaceous, an epicontinental seaway extended from the Gulf of Mexico to the Arctic Ocean (Chapter 3). Fossil plants are preserved in many of the continental clastic sediments deposited along the margins of this Western Interior Seaway. Most of these fossil floras are of Albian age or younger and are important in providing information on the mid-Cretaceous angiosperm radiation and also on floristic changes over the Cretaceous–Palaeogene (K–T) boundary (e.g. Nichols and Johnson, 2008).

4.5.1 Kansas and Nebraska

The classic fossil flora of the Dakota Formation is one of the most extensively collected of all mid-Cretaceous plant assemblages. The first descriptions of these plant fossils appeared more than a hundred years ago (e.g. Lesquereux, 1868, 1874, 1892; Newberry, 1868) and subsequently the flora has been the focus of many studies (Dilcher, 1979, 1989; Retallack and Dilcher, 1981c; Basinger and Dilcher, 1984; Crane and Dilcher, 1984; Dilcher and Crane, 1984a; Dilcher and Kovach, 1986; Upchurch and Dilcher, 1990). The Dakota Formation outcrops over a broad area of the Western Interior of North America from Utah and Colorado in the west to Iowa in the east, and from Minnesota in the north to Kansas in the south, but most of the classic floras are from Central Kansas and Nebraska, where the plant fossils occur in fluviatile and lacustrine sediments deposited on a deltaic, lagoonal coastal plain.

The following summary of the stratigraphy and depositional environment of the Dakota Formation is based mostly on summaries by Retallack and Dilcher (1981c) and Dilcher and Crane (1984a). The Dakota Formation sediments are predominantly sandstones, siltstones and clay, and include coaly layers particularly in the upper part of the formation. Also in the upper part of the sequence there are several marine intercalations. The variable and complex depositional situation results in considerable lateral variation, which makes correlation within the basin problematic.

Two members of the Dakota Formation are distinguished: the Terra Cotta Clay Member below and the Janssen Clay Member above. The entire formation is dated as latest Albian – earliest Cenomanian by its stratigraphic position above the marine Late Albian Kiowa Shale and below the Early? Cenomanian marine Graneros Shale, which is dated based on marine invertebrates and radiometric age determinations. Brenner *et al.* (2000) suggested an Early Cenomanian age, for the Janssen Clay Member. Palynological studies of the Dakota Formation also indicate an Early Cenomanian age, probably equivalent to the palynological Zone III of the Potomac Group zonation (see Dilcher and Crane, 1984a; Hu *et al.*, 2008b). However, diachronism for the Dakota Formation caused by the northwardly transgressing sea, and resulting in part in slightly older ages, cannot be ruled out (Hu *et al.*, 2008b).

Most of the macrofossils are from sandstones of the Janssen Clay Member. The flora is dominated by angiosperm leaves preserved as high-quality impressions in fine- to coarse-grained sandstones (Lesquereux, 1892). Reproductive structures are relatively rare, but include isolated petals (Hollick, 1903) as well as occasional angiosperm infructescences and fruits (Dilcher, 1979; Crane and Dilcher, 1984). In the late 1970s well-preserved compression floras were also discovered in the Dakota Formation and these have provided important information on angiosperms from the mid-Cretaceous (Dilcher, 1979, 1989; Retallack and Dilcher, 1981c; Crane and Dilcher, 1984; Dilcher and Crane, 1984a; Dilcher and Kovach, 1986; Wang and Dilcher, 2006).

Linnenberger's Ranch. The Linnenberger's Ranch locality is exposed in a low cliff near Bunker Hill, Russell County, in central Kansas. The sediments exposed are brownish-grey clays with narrow sandy horizons, and are referred to the Janssen Clay Member of the Dakota Formation. Fossil plants are preserved as compressions (Retallack and Dilcher, 1981c). The macrofossil flora is dominated by angiosperm leaves; the angiosperm element of the associated pollen flora accounts for about 25% of the dispersed palynoflora (Dilcher and Crane, 1984a). The Linnenberger's Ranch plant fossil assemblage is interpreted as having been deposited in an abandoned channel on the distal flanks of a levee system (Retallack and Dilcher, 1981a, c).

Hoisington. The Hoisington locality from central Kansas is situated about 50 km south of the Linnenberger's Ranch locality. Sediments at this locality also belong to the uppermost part of the Dakota Formation (Janssen Clay Member) and mainly comprise shales deposited in a near-coastal environment (Retallack and Dilcher, 1981a). The plant fossils here are often more or less complete, but preserved as impressions in a light, fine-grained clay. The assemblage is dominated by angiosperm leaves of *Liriophyllum*, which are associated with *Archaeanthus*-like floral structures (Dilcher and Crane, 1984a) and there are several other angiosperm reproductive structures

(Dilcher and Kovach, 1986) as well as various aquatic plants (Wang and Dilcher, 2006) including the water fern *Marsilea* (Skog and Dilcher, 1992).

Braun Ranch. The Braun Ranch locality in central Kansas is situated about 60 km south of the Nebraska–Kansas border southeast of Concordia, Kansas. The fossil flora is dominated by the leaves and reproductive structures of platanoids along with associated *Sequoia*-like conifer shoots (Retallack and Dilcher, 1981c). The fossils occur in laminated clayey shales interpreted as having been deposited in a coastal lake within the flood-plain forest (Retallack and Dilcher, 1981a).

Rose Creek. The Rose Creek locality occurs about 10 km south of Fairbury and about 2 km south of Rose Creek in Jefferson County, southeastern Nebraska, near the border with Kansas. The flora is dominated by angiosperms and includes the suite of well-preserved pentamerous flowers ('Rose Creek flower') described by Basinger and Dilcher (1984) as well as many angiosperm leaf taxa (Upchurch and Dilcher, 1990). The plant fossils occur in mudstones penetrated by fossil roots and also contain brackish-water bivalves. The Rose Creek locality has been assigned to the upper part of the Janssen Clay Member and is interpreted to have been derived from vegetation growing in a brackish water swamp (Retallack and Dilcher, 1981c; Upchurch and Dilcher, 1990).

4.5.2 Texas

There are several Cretaceous floras in Texas from both the Early and the Late Cretaceous. Early Cretaceous floras occur in the Glenrose Formation of central Texas, while Late Cretaceous floras occur in the Woodbine Formation of northeast Texas.

Glenrose. Plant-bearing strata at the Glenrose locality along the Paluxy River in central Texas are mainly dolomitic limestones belonging to the predominantly marine Glenrose Formation (Fontaine, 1893; Watson, 1977; Watson and Fischer, 1984; Srinivasan, 1992). The flora is species-poor and the plants, which are mainly extinct Cheirolepidiaceae and other conifers with xeromorphic features, indicate a vegetation growing under extreme water stress (Watson, 1977; Watson and Fischer, 1984). The plants probably grew in a tidal marsh environment characterised by high salinity and were washed out into the marine basin (Upchurch and Doyle, 1981). The palynoflora includes a high proportion of *Classopollis* pollen (produced by Cheirolepidiaceae), which are often especially abundant in near-shore sediments. There are no angiosperm remains among the macrofossils, but angiosperm pollen grains, including a few tricolpate forms, have been reported. Based on palynological evidence an Early Albian age was tentatively suggested for the plant-bearing sequence (Upchurch and Doyle, 1981).

Arthur's Bluff. The Arthur's Bluff locality is located on the south bank of the Red River in Lamar Country, northeast Texas. The sediments exposed here are green-grey siltstones, yellowish clayey sands and loose sandstones belonging to the Woodbine Formation. The macrofossil flora is dominated by leaves of angiosperms, but also includes a few conifer twigs and leaves of Bennettitales (Berry, 1912, 1922), as well as the first pollen organ described for the extinct seed plant group Erdtmanithecales (Pedersen *et al.*, 1989b).

Several localities from the Woodbine Formation of northeast Texas have yielded a variety of fossil plants described by Knowlton (1901), Berry (1912, 1922), MacNeal (1958), Retallack and Dilcher (1981c) and others. The Woodbine Formation consists mostly of light sands with occasional coal beds, layers of lignitic clay and many organic-rich lenses. The age of the formation is established as Cenomanian based on its intermediate position between the Late Albian – Early Cenomanian marine Grayson Marl and the Late Cenomanian – Early Turonian marine Eagle Ford Shale (MacNeal, 1958). The Woodbine Formation is similar in stratigraphic position to the Dakota Formation and the preservation of some of the fossil plants in sandstone is also very similar to those from the Dakota Formation (MacNeal, 1958). Like the flora of the Dakota Formation, the flora of the Woodbine Formation is dominated by angiosperm leaves and the two assemblages are further linked by the occurrence of the angiosperm floral structure *Lesqueria elocata* (Crane and Dilcher, 1984).

4.5.3 Colorado, Montana, North and South Dakota and New Mexico

Numerous Cretaceous fossils have been discovered in Colorado, Montana, North and South Dakota and New Mexico. Of particular importance are floras from the Cretaceous–Palaeocene (K–T) boundary recovered from the Denver, Raton, Hell Creek and Fort Union formations (Nichols and Johnson, 2008).

Plant-bearing beds of the Raton Formation occur in southern Colorado and extend into northern New Mexico.

The Raton Formation includes the iridium-rich boundary clay corresponding to the Cretaceous–Palaeogene boundary. Vegetational changes through the latest Cretaceous and Early Paleocene have been assessed from both palynofloras (Orth *et al.*, 1981) and leaf macrofossils (Wolfe and Upchurch, 1987). Analyses based on both kinds of data record significant levels of plant extinctions at the K–T boundary (Nichols and Johnson, 2008).

The Denver Formation, exposed in the Denver Basin (centred south of Denver, Colorado) has recently been the subject of detailed palynological and palaeobotanical studies as part of a multidisciplinary research programme to establish the position of the K–T boundary (Barclay *et al.*, 2003; Hicks *et al.*, 2003; Johnson *et al.*, 2003; Raynolds and Johnson, 2003). Age determinations based on isotopic methods, magnetostratigraphy and biostratigraphy show that the K–T boundary occurs within the Denver Formation and associated palynofloras and leaf fossils provide an opportunity to assess vegetational changes during this interval based on well-constrained and extensive sampling. At the Bijon site in Arapohoe County, Colorado, the K–T boundary claystone is 3 cm thick and coincides with an iridium anomaly and shocked minerals. There is a 21% level of extinction of Cretaceous palynomorphs across the boundary (Barclay *et al.*, 2003). Significant floral changes across the K–T boundary are also recorded among the fossil leaf floras at the 149 localities studied (Johnson *et al.*, 2003; Nichols and Johnson, 2008).

In western North Dakota, northwestern South Dakota and eastern Montana, plants from the K–T boundary are found in the Hell Creek Formation and the overlying Fort Union Formation in the Williston Basin. This region has provided the most extensive sample of macrofossils across the K–T boundary: 353 leaf morphospecies from a collection of over 22 000 specimens from more than 161 localities (Wilf and Johnson, 2004; Nichols and Johnson, 2008). The Hell Creek Formation is exclusively Late Cretaceous (Maastrichtian) in age, whereas the Fort Union is largely Paleocene, except locally where its most basal sediments (*c.* 2 m) are Cretaceous (Pearson *et al.*, 2001; Nichols and Johnson, 2002). The age of the Hell Creek Formation and Cretaceous–Palaeogene boundary section is well constrained by magnetostratigraphy and biostratigraphy. The extensive sampling of Hell Creek and Fort Union leaf macrofossils, combined with palynological data, provides the best available insight into the nature of vegetational change at the K–T boundary.

4.5.4 Alberta and British Columbia, Canada

Rich floras with angiosperm remains have been discovered in sediments of the Horseshoe Canyon Formation of southern Alberta, Canada, particularly at localities around Drumheller, and include diverse palynofloras (Jarzen, 1982), leaf fossils preserved as compressions/impressions as well as exceptionally preserved permineralised fruits, seeds, wood, twigs and cones representing diverse assemblages of angiosperms as well as conifers, other seed plants, ferns, lycopsids and horsetails (Serbet and Stockey, 1991; Aulenback, 2009). The Horseshoe Canyon Formation is part of the Edmonton Group and is of Late Cretaceous (Late Campanian – Maastrichtian) age. It has been studied in detail for its important dinosaur fauna (e.g. Béland and Russell, 1978; Russell, 1983, 1989). The study of the fossil plants has been initiated relatively recently and offers considerable potential for detailed studies of latest Cretaceous plants and plant communities.

Permineralised fossils also occur in the Haslam Formation deposited in the Nanaimo Basin on Vancouver Island, British Columbia. This formation is a marine shale and includes a number of calcified plant fossils preserved in carbonate nodules or as isolated calcareous petrifactions (Ludvigsen and Beard, 1997). The formation is rich in marine invertebrates, including ammonites, and is dated as Early Campanian (Rothwell and Stockey, 2002). Detailed studies of the plant fossils show excellent cellular preservation. Among the plant fossils described so far are a single angiosperm fruit described from Brannan Creek (Delevoryas and Mickle, 1995) and two different cones of Bennettitales described from the Brannan Lake locality (Rothwell and Stockey, 2002; Stockey and Rothwell, 2003).

4.6 ALASKA

Numerous Cretaceous macrofossil floras containing angiosperm leaves have been reported from the North Slope of Alaska (Smiley, 1969; Spicer, 1987; Spicer *et al.*, 1994a). Systematic descriptions and comparisons of most of the plant fossils preserved in these floras remain to be done but these assemblages are particularly important for assessing the palaeoclimate and palaeoenvironment in the Arctic during the Cretaceous. They indicate significantly warmer conditions in the Arctic than occur today. In the Colville Basin the sediments are predominantly shallow marine to coal-bearing coastal plain deposits ranging from Middle Albian to Maastrichtian in age. The Nanushuk Group is of Albian to Cenomanian age and has yielded extensive macrofossil collections from the Kukpowiuk, Corwin and Chandler formations. The Coville Group has yielded macrofossils from the Prince Creek Formation, which is probably of Coniacian to Campanian–Maastrichtian age (Spicer, 1987). Angiosperms are reported in fossil floras from the latest Albian–Cenomanian and onwards (Smiley, 1969).

4.7 GREENLAND

Terrestrial sediments of Early and Late Cretaceous age that contain plant fossils are exposed at several localities in central West Greenland, including Disko Island, the Nuussuaq (Nûgssuaq) Peninsula, Upernivik Island and the Svartenhuk Peninsula (Koch, 1964; Pedersen, 1976; Dam *et al.*, 2009). Plant fossils are preserved mainly as compressions and impressions of leaves and twigs, but there are also some compressed reproductive structures.

Leaf floras from Greenland were first studied by Heer (e.g. 1882, 1883a, b, c) and Nathorst (1890) and later re-examined by Seward (1926, 1935), Seward and Conway (1939) and Boyd (1992). Angiosperm remains occur abundantly in the Late Cretaceous floras and include several probable tropical or subtropical elements. The discovery of Cretaceous floras from the polar regions that are rich in angiosperm fossils strongly influenced early theories on angiosperm origin and diversification, and raised the idea (now discredited) that angiosperms originated in the Arctic regions and migrated southward during the mid-Cretaceous (Heer, 1868; Seward and Conway, 1935). Brief summaries of the stratigraphy and occurrence of the floras, as well as their floristic composition, were given by Koch (1964) and Pedersen (1976). Koch (1964) pointed out that some earlier reports on the floras contain confusing mistakes. The exact stratigraphic position from which all the collections were made is not settled, and in some cases fossils from stratigraphically separate horizons have been mixed up in the descriptions. Critical re-evaluation of the collections and their stratigraphy is therefore needed.

Terrestrial sediments in central West Greenland include a variable sequence of shales, sandstones and coal layers (Dam *et al.*, 2009). The age of the whole sedimentary sequence is not established with certainty, but marine intercalations in the younger part of the sequence, together with palynological investigations, allow more precise dating for some of the Cretaceous floras. The oldest of the Cretaceous floras from this region is the Kome flora, now thought to be Albian (pre-Late Albian) although the exact age is still not fully established (Dam *et al.*, 2009). The flora occurs in the Kome Formation at Nuussuaq and Svartenhuk. This formation consists of dark shales and sandstones with coal seams up to 1 m thick (Pedersen, 1976). The flora is dominated by ferns, bennettitaleans, and conifers; angiosperms are extremely rare. Many of the distinctive angiosperm leaves from Upernivik, such as *Platanus*, which were earlier described as elements of the Kome flora by Seward (1926), are from younger strata (Koch, 1964).

The most diverse of the Cretaceous floras including angiosperms are the Atane and Paatuut (Patoot) floras, both of which contain abundant and diverse angiosperm remains: both occur in sediments of the Atane Formation. The Atane Formation consists mainly of shales and sandstones and has numerous coal layers up to about 1 m thick. The age of the Atane Formation is not fully established, but it appears to extend from the latest Albian into the Santonian or earliest Campanian (Pedersen and Pulvertaft, 1992; Dam *et al.*, 2000). The Atane flora is known from several localities at Disko, Nuussuaq, and in the Umanak Fjord area, and is from the lower part of the formation. According to Dam *et al.* (2009) the Paatuut flora is probably Middle Turonian – Early Campanian age.

4.8 ISRAEL, JORDAN AND LEBANON

Important Early Cretaceous palynofloras, which include some of the oldest known angiosperm pollen, have been reported from Israel; macrofossil floras of Albian to Turonian age have also been described from the area (Figure 4.19).

Early Cretaceous terrestrial sediments with intercalations of shallow marine to more open marine deposits are referred to the Kurnub Group. These sediments occur in southern Israel in an area extending from the coastal plain to northern Negev. This sequence unconformably overlies Jurassic sediments and includes shales, sandstones, limestone and dolomite. Shales predominate to the west; sands predominate to the east (Brenner, 1996). The region has been tectonically active since the Late Jurassic, and extensive tectonic movements, together with changing coastlines, make lithological correlation difficult. Only a few enigmatic macrofossils have been reported from the Kurnub Group (Krassilov and Dobruskina, 1995), but palynological studies have been carried out on core material from several boreholes, two of which were particularly rich in palynomorphs: the Kokhav 2 well and the Zohar 1 well.

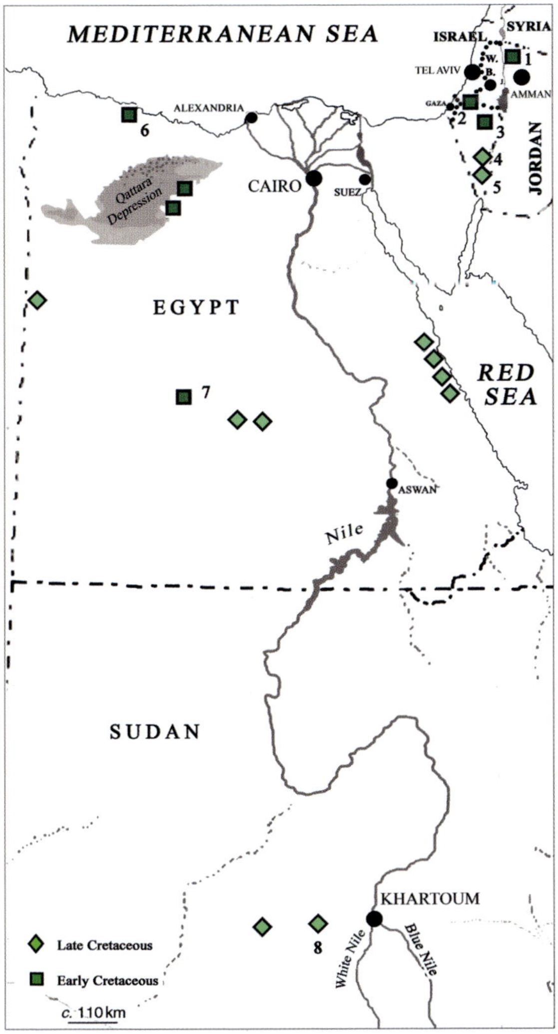

Figure 4.19 Map of the Middle East and northeastern Africa showing the position of important Cretaceous localities. Jordan: 1, Mahis. Israel: 2, Kokhav well; 3, Zohar well; 4, Qetura; 5. Gerofit. Egypt: 6, Mersa Matruh; 7. Mawhoub West. Sudan: 8, Jebel (Djebel) Mudaha. Other important localities, mainly for palynological correlations, are indicated by stratigraphic symbols, but are not mentioned in the text.

The Kurnub sequence is also known from Jordan (see below) where it has yielded well-preserved angiosperm leaves.

The Kokhav 2 well. This well, on the coastal plain of Israel, penetrated the Early Cretaceous Helez Formation of Valanginian to Barremian age that unconformably overlies Jurassic (Oxfordian) sediments. The Helez Formation consists of terrestrial to marine shales, sandstones, limestones and dolomites deposited in a near-shore environment. It is subdivided into eight lithological members numbered H_M–H_{1-7}. Angiosperm pollen is found at several levels, with the oldest occurrences in the Lower Sand Member (H_1) dated as Late Valanginian to Early Hauterivian based on the presence of Valanginian foraminiferans and ostracods in the underlying limestone, and the occurrence of Early Hauterivian ammonites in the overlying dolomite. Angiosperm pollen is extremely rare at this horizon, constituting fewer than 2 grains per thousand recorded palynomorphs. In total 15 grains were assigned to the angiosperms by Brenner (1996); if this assignment is accepted then these are among the earliest angiosperms recorded.

The Zohar 1 well. The Early Cretaceous sequence in the Zohar 1 well, drilled in the northern Negev, comprises strata divided into four formations: the Zeweira, Dragot, Malhatta and Uza. The Zeweira Formation is at the base and unconformably overlies Jurassic rocks. A marine horizon, which is dated by ammonites and molluscs as of Early Aptian age, is present about 40 m above the base of the formation, but there are no stratigraphic markers for the basal part of the sequence. Based on the stratigraphic position below the marine horizon, and on palynological evidence, Brenner and Bickoff (1992) inferred a Late Barremian age for the basal part of the sequence and an Early Aptian age for the layer immediately below the marine horizon. Several distinct monocolpate angiosperm pollen grains occur in the Barremian–Aptian part of the sequence. Tricolpate pollen grains first appear together with *Afropollis* and *Pennipollis* (*Brenneripollis*), immediately below the marine horizon. Above, the Dragot Formation is dated as Late Aptian – Early Albian or Early Albian. The Dragot Formation has yielded the tetrads of the winteraceous pollen *Walkeripollis* (Walker *et al.*, 1983).

Negev. The fossil flora of the Ora Shales Formation in southern Israel is important as an example of Cretaceous leaf macrofossil assemblage from the northern Gondwana realm. The fossils occur in the upper member of the formation and are dated as of Early Turonian age based on underlying marine faunas. Fossil material has been collected from two localities (Gerofit and Qetura) both in the southern Negev. A first account of the flora reported *Equisetum*, *Brachyphyllum*-type conifers, fragmentary ferns and eight species of angiosperms (Dobruskina, 1996, 1997; Krassilov and Dobruskina, 1998). More recently a comprehensive account of the Turonian Negev flora has documented diverse angiosperms including many possible monocots and aquatic plants as well as diverse floral structures (Krassilov, 2004; Krassilov *et al.*, 2005).

Jordan. The oldest angiosperm fossils from Jordan are from the Jarash Formation of the Kurnub Group. This formation is characterised by a basal marine transgressive phase followed by a regressive fluviatile phase and is probably of Albian age (Taylor *et al.*, 2008). *Cabomba*-like leaf fossils were found in the fluviatile sediments at the Mahis locality associated with some ferns (Taylor *et al.*, 2008).

Several younger plant macrofossil assemblages approximately contemporaneous with those from Gerofit and Qetura localities in the Negev have also been described from Jordan. These floras occur in continental sandstones and are interpreted as of Cenomanian age. The plant assemblage is dominated by leaves of dicotyledons, which account for 19 of the 22 species recorded (Bender and Mädler, 1969).

Lebanon. Plant macrofossils from the Late Cretaceous, including a peculiar *Sapindopsis*-type leaf and an impression flower (*Nupharanthus cretacea*), have been discovered in a lithographic limestone where they co-occur with fossil fish and other faunal elements (Dilcher and Basson, 1990; Krassilov and Bacchia, 2000). The plant fossils come from two quarries near the village of Nammoura, about 20 km northeast of Beirut. The beds are of marine origin and Cenomanian in age. The plant fossils are preserved as impressions, sometimes with poorly preserved patches of organic material (Krassilov and Bacchia, 2000).

4.9 NORTH AFRICA

4.9.1 Egypt and Sudan

Terrestrial and marine deposits of Cretaceous age are widely distributed in northeast Africa (Figure 4.19). They range in age from the earliest Cretaceous to the Maastrichtian. They are known from numerous exposures and also from boreholes drilled for oil and water. Intensive studies of the palynofloras, particularly from Egypt and Sudan, have documented rich angiosperm assemblages from the Late Barremian – Aptian and onwards (e.g. Schrank, 1982, 1983, 1992, 1994b; Penny, 1991, 1992; Schrank and Nesterova, 1993; Schrank and Ibrahim, 1995; Ibrahim and Schrank, 1996; Ibrahim, 2002; Schrank and Rüffle, 2003). A summary of the non-marine palynological assemblages in Egypt and northern Sudan, and their correlation, includes 61 localities that have yielded palynological material (Schrank, 1992).

A north–south section from northern Egypt to the Abyad Basin in northern Sudan shows a marked decrease in the thickness of Cretaceous strata. A similar decrease in thickness is shown in a section from west to east from the Libyan border to the Red Sea. Marine intercalations are rare in these Early Cretaceous sequences, and also decrease from north to south away from the main axis of the Tethys. In northern Egypt marine sediments predominate through the Cretaceous. In southern Egypt and Sudan, Early and mid-Cretaceous sediments are predominantly continental sandstones, while marine shales dominate the sedimentary sequence from the later part of the Cretaceous. The continental sandstones in the south were previously included in the so-called Nubian Sandstone, a thick sequence ranging at least from the Jurassic to the mid-Cretaceous. These sandstones also extend west into Chad and Libya.

Although much more scattered than the palynofloras, diverse macrofossil assemblages with angiosperms also occur in Cretaceous strata from Egypt and Sudan (Schrank, 1999). The latest Jurassic and earliest Cretaceous floras are difficult to distinguish. They include xeromorphic elements that are widespread and also known from other areas, such as the fern *Weichselia*, the cheirolepidiaceous conifers *Frenelopsis* and *Pseudofrenelopsis*, and conifers of probable araucariaceous affinity. The earliest angiosperm leaf floras are from the Sabaya Formation and are generally interpreted as of Albian age, but could range downward into the Aptian (Schrank and Mahmoud, 1998). The leaves of the Sabaya Formation are, however, much more similar to Zone I leaves of the Potomac Group sequence, which may suggest that they are older. Schrank (1999) also indicated that the formation may be incorrectly dated and suggested that an older age was possible.

Two boreholes are of particular interest for the study of Early Cretaceous angiosperms: the Mawhoub West 2 borehole in central Egypt and the Mersa Matruh 1 borehole in northern Egypt.

The Mawhoub West 2 borehole. This borehole was drilled for water in the Western Desert of Egypt in the Dakhla Basin. It is 639 m deep, penetrating mainly sands and shales referred to the Nubian Group (Schrank, 1982, 1983). Two samples from the basal part of the borehole are thought to belong to the Abu Ballas Formation, or they may be slightly older. Samples from this borehole include only spores and pollen, and age determinations are based on palynological correlation to palynofloras from other regions. Fossiliferous strata regarded as equivalent to the Abu Ballas Formation are exposed 200 km southwest of the drilling site. Macrofossils from this exposure indicate a probable Aptian age (Schrank, 1983). An Aptian or Early Albian age has also been suggested for the basal samples of the Mawhoub borehole based on the occurrence of *Afropollis* and *Asteropollis* pollen grains (Schrank, 1983). The age of the two samples near the base of the Mawhoub West borehole could therefore be older than Aptian – Early Albian. Schrank (1983) noted that an Early Albian age was unlikely owing to the common occurrence of *Ephedripites*, which is less common in Albian strata from the Sahara (Reyre, 1973). In addition, tricolpate pollen is not found in the oldest of the Mawhoub samples and only few examples are known from the younger strata. This would also point to an Aptian or older age.

The Mersa Matruh 1 borehole. This borehole was drilled in the North West Desert of Egypt and penetrated about 300 m of Lower Cretaceous strata (Penny, 1991, 1992). The sedimentary sequence is continuous and coarsens upwards from predominantly shales at the base into coarser sandstones towards the top. The shaly part of the sequence is referred to the Matruh Shale and the overlying coarser sediments are referred to the Kharita Member of the Burg el Arab Formation. The basal part of the Matruh Shale is of marine origin and the depositional environment for the whole sequence is interpreted as changing from open marine to coastal conditions as the basin was filled by a prograding delta. Dinoflagellates and foraminifers indicate a Neocomian (Berriasian–Hauterivian) to Early Barremian age for the basal, marine sequence, but there is no marine control higher in the sequence. The non-marine part of the sequence is thought to range from the mid-Barremian to mid-Albian based on correlation with palynological successions from West Africa, Brazil, Europe and North America. The Mersa Matruh sequence was divided into nine informal palynological sections. At the base, Section 1, which comprises the marine part of the sequence, has no angiosperm pollen. Several angiosperm taxa are reported from Section 2, which represents the beginning of the terrestrial part of the sequence and is of possible Late Barremian age based on palynological correlation. Sections 3–9 are more diverse in angiosperm pollen and considered as covering the earliest Aptian to Middle Albian (Penny, 1991, 1992).

Jebel (Djebel) Mudaha, Sudan. This locality is situated about 30 km southwest of Khartoum and exposes a sequence of cross-bedded sandstones ('Nubian Sandstone') with intercalated hard claystone that contains well-preserved palynological assemblages as well as angiosperm fruits and leaves (Schrank, 1992; Schrank and Rüffle, 2003). The Jebel Mudaha flora was previously thought to be of Cenozoic age (Vaudois-Miéja and Lejal-Nicol, 1987b, 1988), but is now considered to be Late Cretaceous (?Turonian to Early Senonian) based on it stratigraphic position and palynological correlation (Schrank and Rüffle, 2003). The composition and morphology of the leaf flora has been used to infer that the sediments were deposited in an area dominated by warm, seasonally dry climate (Schrank and Rüffle, 2003).

4.9.2 Tunisia and Libya

Several Late Jurassic – Early Cretaceous floras are reported from fluvio-deltaic and aeolian sediments in southern Tunisia and Libya (Barale *et al.*, 1997, 1998; Barale and Ouaja, 2001, 2002; Mohr *et al.*, 2006). The floras are of Gondwanan type similar to contemporaneous floras from Madagascar and include mainly bryophytes, horsetails, ferns and conifers. From Tunisia two Early Cretaceous floras of Late Aptian – Early Albian age, from the Bir el Karma and Foum el Hassan localities, include several angiosperm leaves that indicate herbaceous and aquatic habitat (Barale and Ouaja, 2001). The Bir el Karma flora includes the peculiar fossil *Klitzschophyllites* that was probably also an aquatic plant. *Klitzschophyllites* had a wide distribution and is also known from the Crato flora of Brazil as well as other Gondwanan floras (Mohr *et al.*, 2006).

4.10 WEST AFRICA AND BRAZIL

For most of the Mesozoic Era Africa and South America were part of a continuous landmass, that started to divide in the Late Jurassic – Early Cretaceous. Intensive drilling for oil and gas, both offshore and onshore, has provided an important basis for understanding the evolution and history of the developing rift valley and associated vegetational changes during the Cretaceous. Well-preserved palynological assemblages from Lower Cretaceous strata include abundant angiosperm pollen as well as ephedroid and elater-bearing pollen that proliferated together with angiosperms during the late Early to mid-Cretaceous at low palaeolatitudes (e.g. Jardiné and Magloire, 1965; Boltenhagen, 1967; Regali *et al.*, 1974; Herngreen, 1975; Doyle *et al.*, 1977, 1982; Lima, 1978, 1979; Dejax, 1987; Osborn *et al.*, 1993). Except for the rich Crato flora (Mohr *et al.*, 2007) macrofossils are rare throughout the area.

The sediments preserved in the marginal basins of Africa and South America record the progressive separation of the two continents. To the south the separation was characterised by a marked rift-valley phase with spreading, while in the north the separation of the continents was characterised by transform movements. In the southern basins terrestrial, fluvial and lacustrine sediments occur towards the base, but are replaced by shallow marine, lagoonal deposits and more fully marine sediments above. Non-marine sediments were deposited during the pre-Albian rift-valley phase, and sedimentation was influenced by strong vertical tectonism. This resulted in an extremely thick sedimentary sequence that in Gabon reaches a thickness of up to 7000 m. The clastic non-marine sequence is overlain by a sequence of evaporites that mark the first influence of marine conditions by the Late Aptian. This is then followed by a sequence of carbonate sediments indicating more fully marine conditions.

The marine transgression reached Gabon–Congo–Angola and southern Brazil sometime in the Late Aptian and progressed towards the north during the Late Aptian and Early Albian. The marine sediments contain marine fauna and marine microorganisms (dinoflagellates, nanoplankton, foraminiferans) that have been used in correlating and dating the strata. However, their spatiotemporal distribution is not fully understood; as noted by Doyle (1982), different organisms suggest different ages. The most serious problem is correlation to the European marine areas where the Cretaceous stages are defined, since Tethyan elements are not known from the Early Cretaceous of the South Atlantic before it opened to the north, which took place in the Albian.

Based on palynological studies the complete Early to mid-Cretaceous sequence in West Africa has been subdivided into 13 zones (C-I – C-XIII) and 23 subzones (Jardiné *et al.*, 1974; Doyle *et al.*, 1977, 1982). Other local zonations have been used in Brazil (see discussion in Doyle *et al.*, 1982; Regali and Viana, 1989). Strong tectonic activity during the rift-valley phase creates special difficulties for correlation. In addition, dating of the thick non-marine sequence faces the same problems as in many other areas. Correlation of the local biostratigraphic zones with those of other classical areas, such as England and North America, is also impeded by distance and the associated differences in geographic and environmental conditions.

Late Cretaceous sediments with well-preserved palynomorphs are also reported from both West Africa and Brazil, but in the following sections we focus on the Early Cretaceous sequences that are particularly important for understanding early angiosperm diversification.

4.10.1 West Africa

In West Africa, Early Cretaceous angiosperms have been reported over a wide geographical area from Angola in the south to Senegal in the north. In Gabon there is an almost complete Early Cretaceous sequence that has been intensively studied (Doyle *et al.*, 1977, 1982).

Gabon. The Cocobeach System was defined in the Eastern Basin of North Gabon (see references in Doyle *et al.*, 1982). It includes an extremely thick non-marine sequence (Cocobeach Sequence, up to 7000 m) of clastic sediments deposited in the rift-valley phase of the Africa – South America separation, as well as an overlying evaporite sequence (Gamba Sequence) that reflects the first marine influence in the rifting zone. This sequence is overlain by the marine Madiéla and Cap Lopez Series. The base of the marine Madiéla Series is dated tentatively as Late Aptian and the top as Late Albian/Early Cenomanian, based on marine invertebrates and foraminiferans. However, correlation of the marine sequence to the European stratigraphic stages is not certain and the faunal dating is still approximate (Doyle *et al.*, 1982). The Madiéla Series is composed of carbonates whereas the Cap Lopez sequence is clastic. This facies change has been interpreted as reflecting a drop in sea level.

Angiosperm pollen has been recorded from the terrestrial as well as the marine part of the sequence. The base of the terrestrial Cocobeach Sequence is unspecified as Neocomian (Doyle *et al.*, 1977). Based on the overlying marine sequences and palynological correlation, the top of the Cocobeach Sequence is dated as Late Aptian. The first angiosperm pollen grains occur in the middle part of the Cocobeach Sequence (pollen zone C-V). This part of the sequence, and the following pollen zone C-VI, were dated as Barremian. Above, pollen zones C-VII to C-XI were referred to the Aptian, mainly based on long-distance correlation with angiosperm assemblages from southern England and the Potomac Group (Doyle *et al.*, 1982). Doyle (1992) later suggested a possible Late Barremian age for pollen zone C-VII based partly on correlation with non-marine sediments in Egypt (dating of the Egyptian sequence was made in part by correlation to Gabon, England and the Potomac Group sequence; see above).

The Cocobeach Sequence is rich in early angiosperm pollen. Several distinct pollen types occur abundantly in the dispersed palynofloras, including the enigmatic *Afropollis* (Chapter 9), which has its first appearance in the Late Barremian (Doyle *et al.*, 1982).

Congo. The Early Cretaceous sequence in Congo is also extensive and is generally equivalent to that of Gabon although it is less complete with a hiatus corresponding to the interval C-VII to C-VIII. Among early angiosperms (possibly Barremian) recorded from Congo are several species of *Stellatopollis*, a distinctive angiosperm pollen grain with a crotonoid tectum (Dejax, 1987).

Other West African Cretaceous occurrences. Owing to differences in their tectonic development the depositional history of the basins further to the north (Ghana, Liberia, Guinea Bissau, Nigeria and Senegal) is different from that further south. The extensive pre-Albian rift-valley infillings are generally lacking and in the northern part marine influence from the Tethys occurs from around the Barremian and onwards. Several important palynological studies document the changing vegetation in the region, including studies of Aptian–Maastrichtian samples from offshore Ghana (Atta-Peters and Salami, 2006) and Cretaceous assemblages from Senegal and the Côte d'Ivoire (Jardiné and Magloire, 1965) and from Nigeria (van Hoeken-Klinkenberg, 1964; Jan du Chêne *et al.*, 1978a, b). From West Africa there are only scattered reports of macrofossil floras. These include Early Cretaceous assemblages without angiosperm remains from northern Cameroun (Dejax and Brunet, 1996), as well as fruit and seed floras from the Maastrichtian of Nigeria (Chester, 1955) and from the Late Campanian – Maastrichtian of Senegal (Monteillet and Lappartient, 1981).

4.10.2 Brazil

In eastern Brazil Early Cretaceous sediments occur in several separate basins mostly along or close to the Atlantic coast (Figure 4.20). The basins were separated by basement massifs and in some periods marine waters transgressed into the central parts, e.g. via the Tucano Seaway. The Cretaceous sequences have been studied from outcrops, as well as from many land and offshore drillings conducted for oil and gas prospecting in the area. Palynological studies have documented rich and diverse palynofloras, including angiosperm assemblages, comparable to those of West Africa (Regali *et al.*, 1974; Herngreen, 1975; Herngreen and Chlonova, 1981; Doyle *et al.*, 1982; Regali and Viana, 1989; Dino *et al.*, 1999). In addition to angiosperm pollen there is also an important low-palaeolatitude occurrence of

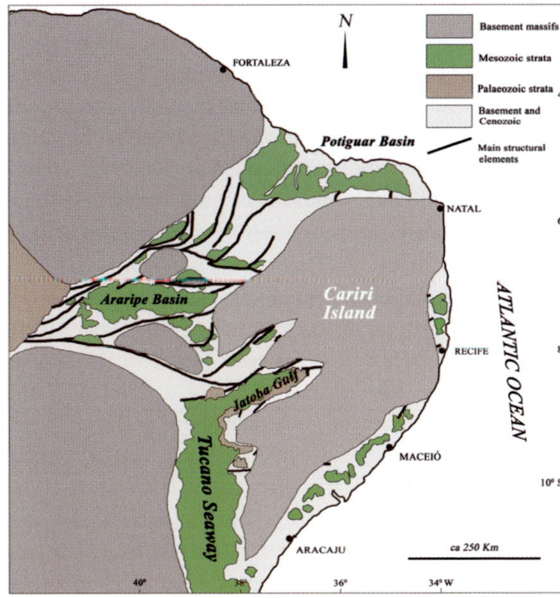

Figure 4.20 Map of northeastern Brazil showing the Tucano Seaway and Mesozoic sedimentary basins. The Crato flora (Late Aptian – Early Albian) is from the Araripe Basin. Redrawn from Martill (2007).

Early Cretaceous angiosperm macrofossils in the Araripe Basin (e.g. Martill *et al.*, 1993; Mohr *et al.*, 2007).

Southeastern Brazil. During the Early Cretaceous the southern part of the area was geographically close to Congo and Gabon and the sedimentary sequence corresponds to that of Gabon in comprising a thick non-marine sequence followed by evaporites and then by fully marine sediments (Doyle *et al.*, 1982; Regali, 1989). Several different systems of stratigraphic zonation have been applied locally to the various basins, and correlation among the local units may be problematic (Herngreen, 1973). According to the system used by Petrobras (Regali, 1989) the Late Jurassic – Early Cretaceous sequence of southeast Brazil can be subdivided into various ostracod and palynomorph zones that are correlated with six local chronostratigraphic stages (Dom João, Rio da Serra, Aratu, Buracica, Jiquiá and Alagoas stages). Based mainly on ostracods and pollen data the local stages are correlated with the West African sequence and the international standard scale ranging from the Oxfordian to the Early/Middle Albian.

Sediments from the latest Jurassic (Dom João stage) consist mostly of red beds and are apparently completely barren of palynomorphs. They are followed by a sequence of fine clastic sediments with well-preserved pollen and spores, but with very few distinct stratigraphic markers. This part of the sequence, which is broadly referred to the Berriasian–Valanginian–Hauterivian interval, corresponds to the Rio da Serra stage and the lower part of the Aratu stage. This is overlain by a sedimentary sequence dominated by coarser clastic sediments of fluvio-lacustrine origin that also contain well-preserved palynomorphs. These sediments are dated as Barremian (corresponding to the upper part of the Aratu stage through to the middle part of the Jiquiá stage) based on correlation with West Africa. The first occurrence of angiosperm pollen in this area is reported at the base of this sequence and includes the characteristic tectate and monoaperturate pollen of *Tucanopollis crisopolensis*. Further up in the sequence, at the base of the Late Barremian, tricolpate pollen is recorded in the area for the first time.

Northeastern Brazil. The Cretaceous basins in the northern part of the area were geographically close to Ghana in the Early Cretaceous and preserve a similar sequence of sediments with the non-marine rift-valley phase missing. The Early Cretaceous sequence starts with evaporites followed by fully marine sediments. The sequence is referred to the Alagoas stage and the base of the marine sequence dated as Early Albian based on foraminiferans. The sequence is most complete in the Maranhão continental shelf area; in other areas it is incomplete with a pronounced hiatus in the Late Aptian – Early Albian (Regali, 1989). A macrofossil flora was reported from the Maranhão area that includes elements also known from the Araripe Basin (Duarte and Santos, 1993).

The Araripe Basin. The Araripe Basin occurs inland between the southern and northern Cretaceous basins (Figure 4.20). It contains a thick sedimentary sequence of several hundred metres of mainly continental Jurassic and Cretaceous sediments. Outcrops in the uppermost formations have yielded a rich and diverse fauna as well as many unique plant macrofossils. The fauna in particular has been intensively studied for almost two hundred years (since Spix and Martius, 1823–1831). The richest fossiliferous horizons for plant fossils are in the Crato Formation, that consists of about 30 m organic-rich mudstones and laminated micritic plattenkalk limestones (Martill *et al.*, 2007). Above the Crato Formation is an evaporite sequence assigned to the Ipubi Formation. This is followed by the Santana Formation, which also contains many fossils, particularly many fish and other spectacularly preserved vertebrates. The uppermost unit in the sequence is the Exu and Abaiara formations. The definition of the Araripe Group and its formations is relatively new. In earlier literature circumscription of the Santana Formation was broader, which may cause some confusion. The age of the sedimentary sequence in the Araripe Basin is not fully established, owing to the lack of good stratigraphic markers. Bertou (1990) placed the Aptian–Albian boundary within the Crato Formation, while Doyle *et al.* (1982) correlated the sequence (Santana Formation *sensu lato*) with the pollen zone C-IX established for Gabon and dated it as Late Aptian. Based on palynological data the Exu Formation was assigned to the Albian (Maisey, 1991). Batten (2007) argued for a Late Aptian or perhaps Early Albian age, and in this book we refer to the Crato Formation as being of Late Aptian – Early Albian age, although the age for this formation, as for many other terrestrial sequences, may change with more precise dating.

The Araripe Group sediments have yielded many palynomorphs, including a wide variety of ephedroid grains (Lima, 1978, 1979, 1980; Osborn *et al.*, 1993). The plant macrofossil flora includes a diverse assemblage of pteridophytes and non-angiospermous seed plants including Gnetales. Angiosperm macrofossils are diverse and preserved as impressions, sometimes with whole plants comprising roots, stems, leaves and reproductive organs organically connected (e.g. Mohr and Friis, 2000; Mohr and Rydin, 2002; Mohr and Eklund, 2003; Mohr and Bernardes-de-Oliveira, 2004; Mohr *et al.*, 2006, 2007).

4.11 ASIA

Cretaceous floras are widespread in Asia and a vast literature has accumulated during the past century on Cretaceous floras from Central and East Asia, particularly for regions that were formerly included in the Soviet Union. In most cases information on localities and stratigraphy from Central and East Asia is published in Russian and it can also be difficult to access. There is now renewed focus on the angiosperm palaeobotany of Central and eastern Asia and more detailed accounts of the geological setting and major fossil floras are under way. In some cases previous floral concepts are redefined and the stratigraphy refined (L. Golovneva, personal communication, 2010). It is therefore not possible here to give a comprehensive account of this important region and the focus is mainly on areas where angiosperm reproductive structures have been found (Figure 4.21).

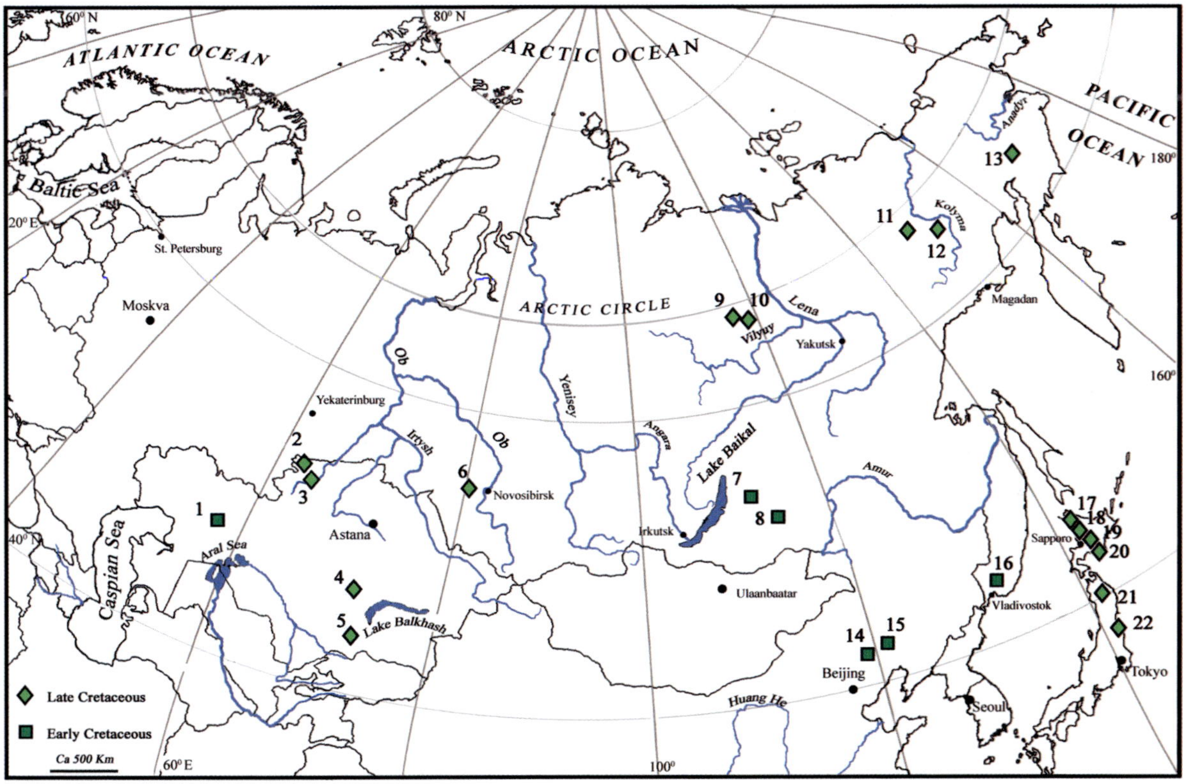

Figure 4.21 Map showing localities with major Cretaceous floras in Central and eastern Asia containing angiosperm reproductive organs. 1–5, Kazakhstan: 1, Karatsche-Tau and Kyzyl-Shen; 2, Kachar; 3, Sarbay; 4, Taldysay; 5, Kysul-Zhar (Karatau). 6, West Siberia, Russia: Chulym–Yenisei area. 7, 8, Transbaikalia, Russia: 7, Baisa; 8, Semion Valley. 9, 10, East Siberia, Lena–Vilyuy River Basin: 9, Tyung River; 10, Linda River. 11–13, Northeastern Russia: 11, Zyrianka: 12, Kolyma River; 13, Grebenka River-Eliseevskoye. 14, 15, Liaoning Province, China: 14, Dawangzhangzi; 15, Beipiao. 16, The Far East of Russia: Bolshoy Kamen. 17–20, Hokkaido, Japan: 17, Obirashibe; 18, Sankebetsu; 19, Mikasa; 20, Hidaka-Monbetsu. 21, 22, Honshu, Japan: 21, Gokurakuzawa; 22, Kamikitaba.

India was part of the southern Gondwanan landmass during the Early Cretaceous and is treated together with this region in Section 4.12.3.

4.11.1 Kazakhstan

Terrestrial sediments of Cretaceous age are widespread in Kazakhstan, and many macrofossil floras with angiosperm leaves have been described (Vakhrameev, 1952, 1991; Vakhrameev and Krassilov, 1979; Shilin, 1986). Only a few macrofossil floras with angiosperm remains are known from the Early Cretaceous; knowledge of vegetational changes in the region during this phase of angiosperm radiation is based mainly on dispersed spores and pollen (Herngreen and Chlonova, 1981; Papulov, 1990; Papulov *et al.*, 1990; Schrank and Nesterova, 1993). Dry climatic conditions prevailed during the earliest Cretaceous (Berriasian–Barremian), resulting in the formation of extensive red beds and evaporites. Conifers generally account for more than half of palynological assemblages; pollen of cheirolepidiaceous conifers (*Classopollis*) was particularly abundant. *Classopollis* constitutes up to 80% of some palynological assemblages from these deposits (Vakhrameev, 1952; Schrank and Nesterova, 1993). During the later part of the Early Cretaceous the climate became more humid and the remaining part of the Cretaceous was characterised by fluctuating humid and dry climates (Schrank and Nesterova, 1993).

Angiosperms are reported from the latest Early Cretaceous onwards. Leaf floras that include angiosperms come from a broad stratigraphic interval that extends from the latest Albian to the Maastrichtian and into the Cenozoic (Shilin, 1986). These floras include impression fossils as well as better-preserved compressions. An enumeration and description of the most important localities for leaf floras was given by Shilin (1986) together with floral lists and descriptions of new taxa. In rare cases the macrofossil floras also contain angiosperm reproductive organs (Vakhrameev and Krassilov, 1979; Krassilov *et al.*, 1983), and recently several mesofossil floras with three-dimensionally preserved angiosperm reproductive organs have been described from northwestern Kazakhstan (e.g. Frumin and Friis, 1996; Hvalj, 2001).

Kyzylshen suite. Several angiosperm leaves including two species of *Nelumbites* were reported from sediments of the Kyzylshen suite in western Kazakhstan. These sediments also include the oldest angiosperm reproductive organs known from Kazakhstan. The apocarpous, follicular fruiting structure *Hyrcantha karatscheensis* was collected at the Karatsche-Tau locality and the inflorescence *Caspiocarpus paniculiger* at the Kyzyl-Shen locality south of Karatsche-Tau (Vakhrameev and Krassilov, 1979; Krassilov *et al.*, 1983). These two localities are situated approximately halfway between the Caspian and Aral Seas. The age of the plant-bearing sediments is indicated as Middle Albian, but there is no further information on the locality or the basis for the dating of the fossils.

Sarbay Iron Quarry. The rich Sarbay mesofossil flora has been extracted from sediments collected in the Sarbay Iron Quarry near the town of Rudnyy in the Kustanay Region of northwestern Kazakhstan. The Cretaceous sediments in the Sarbay area belong to two sedimentary suites, the Novokozyrevsky suite (Novokozyrevsky Formation) and the Shetirgiz suite (Shet-Irgiz Formation), which are interrelated facies and probably contemporaneous (Golovneva and Oskolski, 2007). The Sarbay fossils come from the Shetirgiz suite. The sedimentary sequence is dated, based on palynological studies and stratigraphic correlation, as Cenomanian – Early Turonian (Levina *et al.*, 1990). The Sarbay flora is important as the first substantial mesoflora containing angiosperm reproductive organs discovered from the Cretaceous of Asia. The sediments are predominantly unconsolidated kaolinitic silts with some clays and sands that were clearly deposited in a terrestrial environment. Sedimentological investigations suggest wet subtropical–tropical conditions (H. Friis in Frumin and Friis, 1999). The mesoflora includes a rich assemblage of flowers, fruits, seeds and twigs. The material is well preserved, mostly as slightly compressed lignitic fossils, but there are also a few charcoalified specimens. Angiosperms are the most common component of the flora with several magnoliids related to extant Schisandraceae, Lauraceae and Magnoliaceae as well as many eudicots (Frumina *et al.*, 1995; Frumin and Friis, 1996, 1999; Frumin *et al.*, 2004). The mesofossil flora includes about 50 different types of angiosperm. In contrast, a macrofossil flora from the same locality, and probably contemporaneous with the mesoflora, includes ferns, conifers and only about seven different angiosperm leaves, as well as a few angiosperm reproductive organs (Shilin, 1986; Krassilov and Shilin, 1995).

Kachar. Well-preserved angiosperm mesofossil floras have been recovered in a large magnetite quarry about 45 km NNE of the Sarbay site. The plant-bearing beds at this locality belong to the Novokozyrevsky Formation and are also Cenomanian–Turonian in age. The mesofossil flora is diverse in angiosperm fruits and seeds and shares many taxa with the Sarbay flora. The mesofossils are carefully documented in a doctoral thesis (Hvalj, 2001), but only a single taxon has been formally described (Golovneva and Oskolski, 2007).

Taldysay. Several compression/impression fossils of angiosperm reproductive organs have been reported from the Taldysay locality on the Sarysu River in south-Central Kazakhstan (Krassilov and Shilin, 1995). No information on the sediments and stratigraphy was given with the description of the angiosperms, but Vakhrameev (1991) indicates a Santonian – Early Campanian age for the Taldysay locality based on overlaying marine beds of Late Campanian – Maastrichtian age.

Kysyl-zhar (Karatau). A macrofossil flora with many angiosperm leaves, including several platanoid taxa, has been described from the northwestern slopes of the Karatau range in southern Kazakhstan. The flora is variously referred to as Karatau or Kysul-zhar. In addition to the leaves the flora also includes isolated platanoid fruits described as *Platanocarpus*. The flora was compared to the Dakota and Peruc floras (Jarmolenko, 1935). A Turonian age was later suggested by Shilin (1986). This flora should not be confused with the Late Jurassic Karatau Lagerstätten, which, in addition to numerous insects and other faunal elements, also includes an interesting macrofossil flora (Doludenko and Orlovskaya, 1976).

4.11.2 Transbaikalia

In the Transbaikalia area continental sediments of Early Cretaceous age are widespread and also include many macrofossils. The sediments are deposited in several small separate basins. Two major floras (Turga and Kuta) have been recognised (Bugdaeva, 1984; Vakhrameev, 1991). The Turga flora includes pteridophytes, a variety of ginkgos and czekanowskialean plants, cycads and many conifers, but also some angiosperms. The age of the sequence was first estimated as earliest Cretaceous, but this is uncertain owing to the lack of marine intercalations or other stratigraphic markers (Vakhrameev, 1991). It is probably of Barremian–Aptian age (see below). The Kuta flora is younger and thought to be Aptian – Early Albian in age. It occurs in strata overlaying the Turga suite. The sediments are organic-rich with many coal beds, and the flora is particularly rich in ferns, ginkgos and conifers, but no angiosperms have been reported from this flora (Vakhrameev, 1991).

Plants assigned to the Turga flora occur at the Semion Valley locality, situated about 40 km south of the city of Chita. This locality exposes a lacustrine and alluvial sequence characterised by paper shale deposits. Several enigmatic reproductive organs of non-angiospermous seed plants such as *Semionogyna, Meeusella* and *Baikalophyllum* (Bugdaeva, 1984; Krassilov and Bugdaeva, 1988b) were described from this locality.

Plants assigned to the Turga flora were also recovered at the Baisa locality at the Vitim River, and include a small angiosperm leaf described as *Dicotylophyllum pusillum* (Vakhrameev and Kotova, 1977). Pollen grains of *Asteropollis* were also reported together with the fossil leaves. From the same locality Krassilov and Bugdaeva (1982) described small achene-like fossils (*Baisia hirsuta*) interpreted as a possible transitional stage to angiosperms (proangiosperm). The probable gnetalean seed-bearing structure *Eoantha zherikhinii* (Chapter 5) is another important fossil from the Baisa locality (Krassilov, 1986). The locality has also yielded well-preserved insects with pollen in the gut (Krassilov *et al.*, 2003).

4.11.3 Siberia, Northeastern Russia and the Far East of Russia

Numerous macrofossil and microfossil floras of Cretaceous age have been reported from Siberia, Northeastern Russia and the Far East of Russia (see references in Herngreen and Chlonova, 1981; Zaklinskaya, 1981; Herman and Lebedev, 1991; Vakhrameev, 1991; Herman, 1994; Herman and Spicer, 1995; Golovneva, 1998, 2007, 2008; Kirillova *et al.*, 2000; Maslova and Herman, 2006; Hofmann and Zetter, 2007; Tekleva and Krassilov, 2009). The depositional environment, particularly in the eastern regions, is controlled by strong tectonic activity with a major plate reorganisation in the earliest Cretaceous (Hauterivian), the formation of the East Asian volcanic belt in the Late Cretaceous and considerable elevation of the continental margin in the latest Cretaceous (Kirillova *et al.*, 2000). Tuffaceous beds and other volcanogenic sediments are extensive in some places. In the Early and Late Cretaceous wide spread paralic swamps were formed in the coastal regions, resulting in very thick coal-bearing sequences. Coals were also formed further to the west in inland basins (Kirillova *et al.*, 2000). These coal formations are among the most extensive known anywhere in the world.

West Siberia, Chulym–Yenisei area. Cretaceous sediments including abundant angiosperm remains are known from West Siberia, where they occur mainly in the southeastern part of the West Siberian Plain in strata deposited along the margin of the epicontinental Turgai Strait. Most important are the Cenomanian floras of the Simonovo Formation, which are widespread in the Chulym–Yenisei area with collections from localities along the Chulym, Kiya and Kem' rivers including abundant platanoid leaves assigned to *Platanus* and *Sapindopsis* (Golovneva, 2007, and references therein). In addition to platanoid leaves the Simonovo Formation also includes several angiosperm reproductive structures such as *Cathiaria*, *Freyantha* and *Friisicarpus*. The Kas flora from Turonian strata also includes platanoids and the reproductive structure *Kasicarpa* (Golovneva, 2008).

East Siberia, Lena–Vilyuy River Basin. Numerous Jurassic and Early to mid-Cretaceous floras are reported from the Lena–Vilyuy River Basin in East Siberia, also sometimes referred to as the Linda–Vilyuy Basin. The Vilyuy and Linda rivers are tributaries of the Lena River. Outcrops yielding fossil plants are exposed along these rivers as well as their tributaries. In preliminary reports of the fossil floras Vakhrameev (1958) did not describe any angiosperms. Subsequently, several angiosperm remains, including a variety of reproductive organs, have been reported. At one locality on the Tyung River, sediments assigned to the basal part of the Cenomanian Timerdyakhskaya Formation have yielded small inflorescences (Maslova and Golovneva, 2000a) and pistillate heads of uncertain affinity (Krassilov and Golovneva, 2004) that may be related to Platanaceae. Other platanoid reproductive structures have been described from slightly younger strata on the Linda River, provisionally dated as Coniacian (Maslova and Goloneva, 2000a).

Northeastern Russia. Cretaceous strata with angiosperm macrofossil floras are abundant in the vast region of Northeastern Russia (e.g. Golovneva, 1998). The region is characterised by very different geological conditions from west to east, from strata deposited in isolated continental basins in the west, to depositional environments influenced by extensive volcanic activity in the middle, to continental and coastal near-marine deposits in the east (Herman, 1990; Herman and Lebedev, 1991). The eastern area has yielded particularly rich plant-bearing strata, which in several cases can be dated by marine intercalations. Some of these floras occur in sediments exposed along the Anadyr River and its tributaries the Grebenka, Chukotskaya and Bystraya rivers; other floras are from coastal exposures. The flora collected at the Eliseevskoye locality on the Grebenka River is from sediments assigned to the Krivorechenskaya Formation that is dated as latest Albian – Early Cenomanian. It is one of the most diverse mid-Cretaceous floras from this part of Russia and includes a variety of angiosperm leaves as well as platanoid reproductive structures (e.g. Maslova and Herman, 2004). Floras from northwestern Kamchatka are generally younger, ranging in age from the Turonian to the Campanian (Herman and Lebedev, 1991). The youngest floras are from the Ugolnaya Bay and Amaam Lagoon, where the plant-bearing strata extend from the Campanian into the Maastrichtian (Herman and Lebedev, 1991; Herman and Spicer, 1995).

Several Early Cretaceous floras have also been reported from plant-bearing beds exposed along the Kolyma River and its tributaries, such as the Zyrianka River and Sugoy River (Samylina, 1968, 1976). The Zyrianka coal field is situated in the middle part of the Kolyma River area and the Omsukschan coal field in the upper reaches of the Kolyma River between the Sugoy and Balygychan rivers. Plant fossil localities in the Omsukschan coal field, on the Kolyma River bank between the Sugoy and Balygychan rivers, are divided into Omsukschan, Toptan and Zaria suites or stratofloras. The flora of the Omsukschan suite consists of abundant ferns and many ginkgoalean and czekanowskialean plants, as well as cycads and Bennettitales, but there are also a few problematic angiosperm remains. The flora from the Toptan suite is less diverse, but with several distinct angiosperm leaves as well as the enigmatic reproductive structure *Kennella*, which is also reported from the Omsukschan suite (Samylina, 1960, 1968, 1976). Exposures along the Zyrianka River include a small flora with angiosperm leaves and permineralised reproductive organs described as *Ranunculaecarpus quinquiecarpellatus* and *'Aralia' kolymensis*. The age of these fossils is given as undefined Albian in Samylina (1968), but in Samylina (1976) it is correlated with the Early to mid-Albian part of the Omsukschan suite.

Samylina (1974) recognised a number of stratofloras from this vast region, including mainly floras from the Early and mid-Cretaceous. Later, Golovneva (1998) emended the stratoflora concept to include also the whole of the Late Cretaceous. Each stratoflora is a summary of contemporaneous floras over the entire area. They are useful as a framework for discussing vegetational changes at a broad scale, but according to Lina Golovneva (personal communication, 2010) radically different geological/environmental conditions in the various subregions cause great facies-related differences that may be mistaken for differences in age, and vegetational patterns are complex.

Far East of Russia. Early Cretaceous floras are widespread in the southern Primorye region of the Far East of Russia. The earliest floras from the Berriasian–Valanginian are preserved in near-coastal marine sediments. Younger floras come from a non-marine or partly paralic coal-bearing sequence of Barremian–Aptian age and from a volcanomictic sequence of Albian age (Krassilov, 1967; Krassilov and Volynets, 2008). Reproductive organs of uncertain affinity (*Nyssidium orientale*; Samylina, 1961 and *Onoana nicanica*; Krassilov, 1967) have been reported from the Barremian–Aptian part of the sequence. Indisputable angiosperm remains are known only from the Albian, where they are relatively rare, but recently two new fruiting structures have been reported from the Ussuriysk Bay locality near the village of Bolshoy Kamen. The sediments at this locality are tuffaceous and belong to the Kangaus Formation, dated as Early to Middle Albian (Krassilov and Volynets, 2008).

4.11.4 Mongolia

Cretaceous deposits of continental origin are widespread in Mongolia and have long been known for their rich and well-preserved vertebrate faunas, which include spectacular dinosaur fossils from both Lower and Upper Cretaceous strata (e.g. Khand *et al.*, 2000). Plant fossils are less common, but are known from many localities, particularly in Lower Cretaceous sediments (Krassilov, 1982;

Vakhrameev, 1991). The Cretaceous sequence in Mongolia is deposited in many isolated intracratonic basins and is strongly influenced by extensive tectonic and volcanic activity (Khand *et al.*, 2000). Together with the lack of marine sediments, this makes correlations and dating problematic (Van Itterbeeck *et al.*, 2004; Nichols *et al.*, 2006). None of the Early Cretaceous macrofossil floras from Mongolia includes angiosperm remains, but angiosperm pollen such as *Asteropollis*, *Clavatipollenites* and *Tricolpites* has been reported from the Early Cretaceous of southeastern Mongolia (Nichols *et al.*, 1997, 2006). These sediments were dated as Albian (Nichols *et al.*, 2006) based on the presence of *Asteropollis*, but *Asteropollis* is also reported from older sediments and an Aptian or Late Barremian – Aptian age cannot be ruled out.

Early Cretaceous macrofossil floras from Mongolia show considerable similarity to the broadly coeval floras in northeastern China assigned to the Jehol Biota. There is particularly close similarity between the plant assemblages described from the Gurvan Eren Mountain range in western Mongolia and the assemblages from the Yixian flora of northeastern China, which is now dated as Early Aptian – early Late Aptian.

4.11.5 China

Continental sediments of Cretaceous age are widespread in China and the depositional environment in many regions is influenced by extensive tectonic and volcanic activity (Yang *et al.*, 1986; Chen, 2000). In western and northwestern China a variety of dark and red terrestrial sediments, which often contain gypsum and other evaporites, were deposited during the Early Cretaceous in large inland basins remote from any marine influence (Yang *et al.*, 1986; Chen, 2003). In eastern and northeastern China terrestrial Early Cretaceous sediments were deposited in many smaller shallow basins, which include extensive volcanic and pyroclastic intercalations. In some places the sedimentary sequence also includes coal beds and gypsum horizons. Marine influence is recorded in the northeastern part of the region where an embayment of the Pacific Ocean extended into eastern Heilongjiang in the Early Cretaceous (Yang *et al.*, 1986; Chen, 2000). In southern China terrestrial sediments intercalate with marine and paralic deposits. However, the lack of marine intercalations in most areas of China creates considerable difficulties for the establishment of stratigraphic boundaries in most terrestrial sequences (Yang *et al.*, 1986; Chen, 2003; Zhou, 2006).

Macrofossil floras and diverse palynological assemblages have been reported from numerous localities in China (see references in Sun *et al.*, 1995; Chen, 2003). Two major vegetational types can be recognised for the Early Cretaceous: a southern type, which is comparable to coeval floras from Europe and the Outer Zone of Japan, and a northern type, which is similar to coeval floras from Siberia, Mongolia and the Inner Zone of Japan (Chen, 2003). Both types are characterised by a variety of pteridophytes, cycads, bennettitaleans and ginkgophytes (Li, 1995; Ohana and Kimura, 1995). Angiosperm fossils are rare, but have been reported from the northeastern provinces Heilongjiang (Sun *et al.*, 1993; Sun and Dilcher, 1997; Yang, 2003) and Liaoning (e.g. Sun *et al.*, 1998, 2001; Leng and Friis, 2003; Ji *et al.*, 2004) as well as Hong Kong (Zhou *et al.*, 1990). Recently, a minute angiosperm leaf has also been discovered from the Anhui Province (Li, 2003). Angiosperm pollen fossils from Early Cretaceous strata are more common and include forms such as *Clavatipollenites* and *Asteropollis* (Li and Liu, 1994).

Late Cretaceous macrofossil floras with angiosperm leaves have been reported from Northeast, Southeast, and Southwest China (Li, 1995). Taken as a whole, studies of these Late Cretaceous floras indicate some provinciality, but also many similarities to floras from other parts of the Northern Hemisphere, and particularly from North America. None of the Late Cretaceous floras has so far been studied in detail. Locality information is included here only for the Early Cretaceous floras containing angiosperm remains. There are also extensive red bed deposits from the Late Cretaceous in the central parts of China.

Heilongjiang. Angiosperm fossils from Heilongjiang Province were reported from the upper beds of the Chengzihe Formation, close to the city of Jixi. A marine horizon lower down in the Chengzihe Formation, which is dated as Valanginian–Hauterivian based on the occurrence of dinocysts, gives a maximum age for the plant-bearing beds. Based on the marine occurrence in the sequence, and also on indirect correlation to southern Primorye and Israel, it was concluded that the plant material is probably of Late Hauterivian or Early Barremian age (Sun *et al.*, 1993; Sun and Dilcher, 1997, 2002). However, according to Ohana and Kimura (1995) the age of the Chengzihe Formation is controversial. Based on marine intercalations, Sha *et al.* (2003) suggested a mainly Aptian age for the Chengzihe Formation with the basal part perhaps ranging into the Barremian. Angiosperm fossils similar to those described by Sun and Dilcher from the Chengzihe Formation were

Figure 4.22 The Early Cretaceous (Early Aptian – early Late Aptian) Sihetun locality, Liaoning, northeastern China, exposing the Yixian Formation with finely laminated lacustrine shales and mudstones intercalated with volcanic tuffs. The locality is famous for its exquisitely preserved fauna, including complete specimens of the early bird *Confuciusornis*.

also reported from the Muling Formation (mid-Barremian – Early Aptian) in the Jixi Basin by Yang (2003).

Jilin. Angiosperm remains reported from the Jilin Province (Tao and Zhang, 1992) are poorly preserved and show few details to securely document their angiospermous affinity. Details of the locality or stratigraphic setting of this material have not yet been published.

Liaoning and Inner Mongolia. The Jehol Group, which contains rich fossil assemblages ascribed to the Jehol Biota, is widely distributed in the western part of the Liaoning Province and Inner Mongolia (Figures 4.21, 4.22). The fossil-bearing strata have been excavated extensively, mainly for their exquisitely preserved fauna, which includes some of the most spectacular discoveries of early birds such as *Confuciusornis* and *Liaoningornis* and small, feathered dinosaurs including *Sinosauropteryx* (e.g. Hou *et al.*, 1996; Chen *et al.*, 1998; Zhou *et al.*, 2003). However, there are also many plant macrofossils (e.g. Wu, 1999; Sun *et al.*, 2001). Fossil assemblages equivalent to the Jehol Biota are widely distributed in eastern Asia and extend from Transbaikal and Mongolia in the west to Korea and Japan in the east (Chang *et al.*, 2003).

The Jehol Group is divided into the Yixian and Jiufotang formations. The Dabeigou Formation occuring in Hebei has sometimes also been included as a basal formation. However, the age and correlation of the Dabeigou Formation is uncertain, and it has also been suggested that this formation is equivalent to the Yixian Formation in Liaoning (Zhou, 2006). From a palaeobotanical point of view the Yixian Formation is of particular interest. It contains the early angiosperm *Archaefructus* as well as a number of other important and intriguing plant fossils of angiosperm and probable gnetalean affinity. The Yixian sequence is up to about 2700 m thick and consists mainly of weakly laminated to finely bedded clastic sandstones and shales deposited in a low-energy lacustrine environment in a seasonally semi-arid climate. Numerous intercalated beds of tuffs and tuffaceous sandstones, as well as basalt and andesite horizons, document extensive volcanic activity in the area (Chang *et al.*, 2003; Chen, 2003; Zhou *et al.*, 2003).

The Yixian Formation is divided into four units: the Lujiatun, Jianshangou, Dawangzhangzi and Jingangshan beds. The age of the Jehol Group and the Yixian Formation has been much debated and the fossil-bearing horizons have been referred to either the Late Jurassic or the Early Cretaceous (see references in Barrett, 2000; Zhou *et al.*, 2003). There are no marine intercalations in the sequence and most age estimates are based on biostratigraphic correlations, often to areas that also suffer from a lack of secure age determination and detailed stratigraphic resolution. However, several recent chronological studies using isotope dating have been carried out for various parts of the sequence (Smith *et al.*, 1995; Swisher *et al.*, 1999; Wang *et al.*, 1999, 2001a; He *et al.*, 2004, 2006) and there is now a general consensus that the Yixian Formation is of mid–Early Cretaceous age. $^{40}Ar/^{39}Ar$ dates indicate that the duration of the Yixian Formation deposition was relatively short, lasting from about 125 Ma to 120 Ma, and that the basal Lujiatun and Jianshangou formations were probably deposited approximately at the same time (He *et al.*, 2006; Zhou, 2006). In the Gradstein *et al.* (2004) timescale, this time span corresponds to Early to early–Late Aptian. Fossils plants occur throughout the Jehol Group sequence, but in contrast to the exceptional preservation of the faunal elements the plant fossils rarely have anatomical details preserved. The plant tissues are oxidised and often destroyed by pyrite framboids and microcrystalline deposits (Leng and Yang, 2003). Some of the plant fossils are preserved as whole plants with roots, stems, leaves and reproductive structures in organic connection (Sun *et al.*, 2002). Rich assemblages of plant fossils have been described from the villages of Huangbaijiegou, Jianshangou and Sihetun in the vicinity of Beipiao city and from Dawangzhangzi village in Lingyuan County. Much of the material is collected by local collectors and the exact location not always clear. Most of the fossils are from the Jianshangou beds, while some are from the younger beds. The material includes diverse pteridophytes and non-angiospermous seed plants (e.g. Wu, 1999, 2003; Sun *et al.*, 2001; Rydin *et al.*, 2006b). Angiosperms have also been reported from this area, but several of these fossils are almost certainly related to Gnetales rather than to angiosperms (Chapter 5). Only the reproductive axes of *Archaefructus* and *Sinocarpus*, as well as a few other fossils, show unequivocal angiosperm features (Sun *et al.*, 1998, 2002; Leng and Friis, 2003, 2006; Leng *et al.*, 2003).

4.11.6 Japan

Depositional environments in Japan during the Cretaceous were strongly influenced by its close proximity to the subduction zone between the East Asian continent and the proto-Pacific oceanic plates. Relief is believed to have been high and erosion intensive, resulting in high sedimentation rates with thick and often coarse sedimentary sequences deposited in many separate tectonic basins (Okada and Sakai, 2000). Cretaceous floras occur both in marine and terrestrial sediments. Continental deposits with coal beds are reported for the northern regions (Kimura and Ohana, 1992; Ohana and Kimura, 1995). Southern Japan was connected with China until the opening of the Sea of Japan in the Late Cenozoic, and the floras of the two areas share many common elements.

Early Cretaceous floras, particularly in the south of Japan, are characterised by many xeromorphic elements including ferns such as *Weichselia*, *Onychiopsis* and many Cheirolepidiaceae as well as a variety of cycads, ginkgophytes, Bennettitales and some conifers (Kimura, 2000). In the north, Early Cretaceous floras are characterised by diverse Ginkgoales and Czekanowskiales, but there are also many ferns, cycads and Bennettitales. Currently only a few angiosperm fossils have been described from the Early Cretaceous of Japan, including the early Late Albian seed of *Stopesia* (Yamada and Kato, 2008), assigned to the Trimeniaceae, and the Albian wood of *Icacinoxylon* (Takahashi and Suzuki, 2003).

From the Late Cretaceous, however, there are two important groups of localities in Northeast Japan with angiosperm remains. Permineralised plant fragments from Hokkaido, including reproductive organs and wood of angiosperms, have been studied for 100 years (Stopes and Fujii, 1910; Nishida, 1985, 1994b; Nishida and Nishida, 1988; Nishida *et al.*, 1993; Ohana *et al.*, 1999). More recently, well-preserved mesofossil floras broadly equivalent in age to the Hokkaido material have been discovered in the Futaba Group sequence of Honshu (e.g. Takahashi *et al.*, 1999b, 2008a).

Hokkaido. The permineralised plant fossils from Hokkaido occur in calcium carbonate nodules in a marine sequence assigned to the Upper Yezo Group, which is dated as Coniacian–Campanian based on ammonites and other marine fossils. Some of the nodules are apparently older and of mid- to Late Turonian age. The nodules are usually found along river outcrops, where they are washed out of the sediments. They may be of considerable size and even large stems may be preserved (Stopes and Fujii, 1910). The nodules have yielded a rich flora of ferns and conifers. Bennettitales, cycads and *Ginkgo* are also present along with several enigmatic seed plants (Nishida, 1991). Recently described angiosperm fossils are from several different localities and of slightly different ages. Nodules from the Obirashibe River, western Hokkaido (Coniacian–Santonian), yielded the eudicot capsular fruit of *Elsemaria* (Nishida, 1994b); samples from Sankebetsu River, also in western Hokkaido, yielded the possible magnoliid reproductive structure of *Protomonimia*, for which associated ammonites indicated a Coniacian–Santonian age (Nishida and Nishida, 1988). A sample from the Hidaka–Monbetsu River, southeastern Hokkaido, yielded the possible magnoliid reproductive structure of *Hidakanthus* of Coniacian age (Nishida *et al.*, 1996), and the nodule containing *Keraocarpon* was collected at the Kumaoizawa locality, in the city of Mikasa, and dated as Coniacian–Santonian (Ohana *et al.*, 1999). Other angiosperm fossils remain to be described (Nishida, 1991).

Honshu. The mesofossil floras from the Futuba Group were discovered from localities in northeastern Honshu and are the first such floras to be described from eastern

Asia. These floras range in age from Early Coniacian to Early Santonian based on marine invertebrates that occur within the sequence (Ando *et al.*, 1995). The most intensively studied of the several mesofossil floras is the Kamikitaba assemblage from the Ashizawa Formation, which is of Early Coniacian age (e.g. Takahashi *et al.*, 1999a, b, 2001, 2008a). The flora is dominated by conifers and angiosperms, with Schizaeaceae and Selaginellaceae among the relatively minor spore-producing component. Another flora with well-preserved angiosperm mesofossils is from the Gokurakuzawa locality. The flora is collected in sediments of the Tamayama Formation (Early Santonian) exposed along the Kobisa River (Takahashi *et al.*, 2007).

4.12 SOUTHERN GONDWANA AND INDIA

The depositional environment in many parts of southern Gondwana during the Cretaceous was highly influenced by tectonic activity related to the breakup of Gondwana. Major events included rifting of Australia and Antarctica and proto-Pacific subduction. This resulted in the accumulation of thick sedimentary sequences in some areas, while erosion and lack of sedimentation prevailed in other areas. The most complete younger Cretaceous sequence in southern Gondwana is probably from the Antarctic Peninsula and southern South America.

4.12.1 Southern Africa

Cretaceous sediments of continental origin are not extensive in southern Africa and are mainly restricted to scattered basins along the eastern and southern coast of South Africa. Reviews of the fossil floras are provided by Anderson and Anderson (1985; see also Drinnan and Crane, 1989) who report scattered macrofossil assemblages from the Kirkwood Formation (Middle to Late Valanginian), the Mngazana Formation (Late Valanginian), and the Makatini Formation (Barremian – Late Aptian). From the Kirkwood Formation there is also an occurrence of amber that has yielded leaf fragments with well-preserved cuticles (Gomez *et al.*, 2002). All of these Early Cretaceous fossil assemblages are dominated by ferns and Bennettitales, and *Brachyphyllum*-like conifers were reported from the Kirkwood amber. Angiosperms, based on pollen or meso/macrofossils, have not been recorded so far from the Early Cretaceous, but diverse angiosperm pollen grains are described from Cenomanian palynofloras at a Deep Sea Drilling Project (DSDP) site off the southern tip of South Africa (McLachlan and Pieterse, 1978). A few macrofossil floras are preserved from the Late Cretaceous in lake sediments deposited in the vents of kimberlite pipes, but there are no published records of angiosperms in these floras (Anderson and Anderson, 1985). Angiosperms are mentioned to occur in the Cenomanian–Coniacian Orapa kimberlite in central Botswana (Rayner, 1993), but the occurrence was not documented. Another macrofossil flora including undetermined angiosperm remains was reported from Namibia in a volcanic dome at the Brukkaros locality. The plant-bearing sediments at this locality are probably Late Cretaceous, but the precise age is not established (Kelber *et al.*, 1992/93).

4.12.2 Madagascar

Cretaceous continental sediments occur in several basins in Madagascar, but the only information on fossil plants comes from pollen and spores, for example from the Majunga Basin, which contains continental sediments ranging from Early Aptian to Campanian in age (Doyle *et al.*, 1982; Krause and Hartman, 1996). One palynological sample dated as probably Early Cenomanian contains angiosperm pollen including *Afropollis*, tricolpates and tricolporates. The most detailed study of Cretaceous angiosperms from Madagascar is based on pollen from a sandy carbonaceous siltstone that outcrops along a stream on the northeastern flank of the Manamana mountain in the Morondova Basin, southwest Madagascar (Zavada, 2003). The siltstone contains marine molluscs and was dated, based on ammonites, as Early Cenomanian. About 16 different kinds of angiosperm pollen were described, comprising mainly monocolpate and dicolpate grains with single occurrences of tricolpate and tricolporate grains. Angiosperms are a significant component of the diversity of the flora (32%), but in terms of abundance the flora is dominated by spores and gymnosperm pollen (Zavada, 2003).

4.12.3 India

Several classic fossil floras from India that were once thought to be of Jurassic age are now assigned to the Early Cretaceous. These include the floras of the Rajmahal Hills in western Bengal, northeastern India (Oldham and Morris, 1863; Zeba-Bano *et al.*, 1979) that yielded the original material of Pentoxylales (e.g. Sahni, 1948), and the fossil flora of Kachchh in Gujarat State, western India (Feistmantel, 1876; Bose and Banerji, 1984). These and other 'Upper Gondwana' floras from India are probably of Berriasian to Barremian age, and perhaps in some cases extend up into the Albian (for a review see Drinnan and Crane, 1989). However, all of these fossil floras are devoid of angiosperms.

Little is known about mid- and Late Cretaceous floras from India except for the intensively studied permineralised fossils from the Deccan Intertrappean Beds. These are lacustrine and fluviatile sediments intercalated in flood basalt lava flows of the vast Deccan Traps. The Deccan Traps, which constitute the largest volcanic outpouring preserved on land anywhere in the world, were formed in association with hotspot development as India moved northward. Sediments preserved between the basalts are often silicified. These cherts are especially rich in angiosperm reproductive structures and the assemblages include several different flowers as well as a variety of fruits and seeds (e.g. Sahni, 1934; Sahni and Rode, 1937; Chitaley, 1956; Prakash, 1960; Chitaley and Sheikh, 1973; Chitaley and Patel, 1975; Nambudiri and Tidwell, 1978; Patil and Singh, 1978; Patil and Upadhye, 1984; Bonde, 2005).

Plant fossils have been collected from many localities at different levels in the Intertrappean Beds. The most important discoveries have been from the area around the village of Mohgaon Kalan (Mohgaonkalan) in the Chhindwara district of Madhya Pradesh, but there are many other localities with well-preserved plant fossils from the Madhya region. Permineralised plants have also been reported from Vicarabad in Hyderabad, Rajahmundry in South India, and regions near Bombay (Prakash, 1960).

The age of the plant fossils from the Deccan cherts is not conclusively established, but combined palaeontological evidence, magnetostratigraphy and isotopic age determinations indicate that at least part of the sequence is of latest Cretaceous (Maastrichtian) age (Jaeger *et al.*, 1989; Allègre *et al.*, 1999; Hofmann *et al.*, 2000).

4.12.4 Australia

Thick, continuous sequences of Lower Cretaceous sediments occur in Australia and contain important macrofossil floras, particularly from the Otway and Gippsland basins in Victoria and from the Maryborough and Artesian (including the Carpentaria, Eromanga and Surat subbasins) basins of Queensland (Douglas, 1986, 1994; Drinnan and Crane, 1989; McLoughlin *et al.*, 1995, 2002, 2010; Dettmann *et al.*, 2009). The earliest floras are from Berriasian–Hauterivian strata (McLoughlin *et al.*, 2002) but the Lower Cretaceous sequence and plant-bearing beds continue into the earliest Late Cretaceous. Angiosperms are rare in these Early to mid-Cretaceous floras and are mostly known from dispersed pollen. Angiosperm macrofossils have been reported from only two localities so far: one from Victoria (Koonwarra Fossil Beds) and one from Queensland (Winton Formation).

Lower Cretaceous strata are also reported from Western Australia. The sequence ranges from Berriasian to Albian, but plant macrofossils are only known from the Hauterivian part of the sequence and no angiosperms have been reported (McLoughlin, 1996).

Late Cretaceous sediments are more restricted in Australia and current knowledge of Late Cretaceous vegetation on the continent is based on studies of dispersed pollen (e.g. Dettmann, 1973, 1986, 1994; Dettmann and Jarzen, 1996). Late Cretaceous macrofossil floras have not been described. In the following we give a brief overview of the two formations that include angiosperm macrofossils. Other floras are reviewed elsewhere (Drinnan and Crane, 1989; Douglas, 1994).

Koonwarra Fossil Bed. The earliest record of angiosperms from Australia is from the Koonwarra Fossil Bed in the Gippsland Basin (Taylor and Hickey, 1990). The fossil-bearing sequence is exposed on the South Gippsland Highway approximately 140 km southeast of Melbourne, Victoria, and comprises 8 m of fine-grained and finely laminated mudstone deposited in a small freshwater basin (Drinnan and Chambers, 1985; Dettmann, 1986). The sequence belongs to the Korumburra Group, which is predominantly of fluviatile origin. There are no marine intercalations, and fission track studies of volcanogenic apatites in sediments immediately above the fossil-bearing sequence indicate an Aptian age. The exposed fossil-bearing sediments are barren of palynomorphs, but palynological studies of possible correlated horizons from boreholes drilled close to the fossil site support a Barremian or Aptian age, although a slightly younger age cannot be ruled out (Dettmann, 1986). The Koonwarra Fossil Bed include a diverse fauna, which is particularly rich in fossil insects and fish. There is also a rich assemblage of plant macrofossils, typically preserved as leaf impressions with no or little organic material. The flora includes a single angiosperm fossil (Taylor and Hickey, 1990) but is otherwise characterised by many ginkgoalean plants and araucariaceous conifers as well as other kinds of seed plants and ferns (Drinnan and Chambers, 1986).

Winton Formation. The Winton Formation is a plant-bearing continental sequence consisting of volcanoclastic sediments with intercalated horizons of siltstone and coal beds (McLoughlin *et al.*, 1995, 2010; Dettmann *et al.*, 2009). The formation is widely distributed in southwestern Queensland, northwestern New South Wales and northeastern South Australia and is thought to have been deposited in fluvial to lacustrine environment. It conformably overlays a volcanoclastic near-coastal sequence (Marckunda Formation) that is dated as Late Albian based on its marine fauna. Dettmann *et al.* (2009) suggest a Late Albian age for the Winton Formation, while a latest Albian – Early Cenomanian age was suggested based on palynological studies (see references in McLoughlin *et al.*, 2010). Numerous plant fossils have been collected over many years from many different localities in the Eromanga Basin, and much of the published material is in the collections of the Queensland Museum. The fossils occur as permineralisations in blocks of silicified fine-grained sediments or as impression/compressions in siltstone or fine-grained lithic sandstones (McLoughlin *et al.*, 1995, 2010; Pole and Douglas, 1999; Pole, 2000). The most abundant plant fossils are leafy shoots, twigs, cones and wood of conifers, but there are also diverse angiosperm leaves. The angiosperm floral structure *Lovellea wintonensis* is also from the Winton Formation (Dettmann *et al.*, 2009). This fossil was collected in western Queensland at the Lovelle Downs Station, about 50 km WNW of Winton.

4.12.5 Antarctica

A thick sequence of Cretaceous – Early Cenozoic sediments was deposited on the Antarctic Peninsula microcontinent in an environment strongly influenced by tectonic and volcanic activities related to the breakup of Gondwana. The Cretaceous sequence occurs in two well-defined basins, the fore-arc basin exposed at Alexander Island and the James Ross Basin (a subbasin of the more extensive Larsen Basin) exposed in the northeastern part of the Antarctic Peninsula and nearby islands. The Cretaceous sequence of the James Ross Basin is volcanoclastic and more than 5 km thick, consisting mainly of marine sediments deposited in a back-arc setting (see references in Crame *et al.*, 2004; Hayes *et al.*, 2006; Kennedy *et al.*, 2007). Plant fossils have been recovered from many localities (Truswell, 1991; Cantrill and Poole, 2002). Outside the main fore-arc and back-arc basins are a series of deposits that accumulated on the arc. These are presumably associated with small basins, and correlation among these deposits is difficult due to tectonic juxtaposition and limited outcrops (David J. Cantrill, personal communication, 2009).

Although more or less all Cretaceous stages are represented on the Antarctic Peninsula, and there are some important macrofossil floras from the Early Cretaceous (Cantrill, 1995, 1996, 2000; Cantrill and Nichols, 1996), most of the plant fossils are from the Late Cretaceous. Late Cretaceous assemblages of permineralised wood are particularly abundant and have been reported from many localities (e.g. Chapman and Smellie, 1992; Poole and Francis, 1999; Poole *et al.*, 2000a, b; Poole and Gottwald, 2001; Cantrill and Poole, 2002; Poole and Cantrill, 2006). The wood is typically well preserved and often systematically informative. There are also several leaf floras from the Late Cretaceous (e.g. Zastawniak, 1994; Zhou and Li, 1994; Cantrill and Poole, 2002; Hayes *et al.*, 2006), and the first mesofossil floras discovered in the Southern Hemisphere are from the Antarctic Peninsula (Eklund, 2003; Eklund *et al.*, 2004a). Angiosperms have also been described from a number of Cretaceous palynofloras (e.g. Dettmann and Thomson, 1987; Askin, 1990; Truswell, 1991; Chapman and Smellie, 1992; Dutra and Batten, 2000).

Alexander Island. Cretaceous strata exposed at Alexander Island were deposited in the fore-arc basin. The most important plant macrofossil assemblages are from the Triton Point Member of the Neptune Glacier Formation at the Coal Nunatak, Citadel Bastion and Titan Nunataks localities on Alexander Island (Cantrill, 1995, 1996; Hayes *et al.*, 2006). The plant-bearing sequence consists of sandstones and siliceous siltstones. It is part of the predominantly marine Fossil Bluff Group and is dated as Late Albian based on the marine fauna below and above the plant-bearing beds (Moncrieff and Kelly, 1993). The plant assemblages were collected at different stratigraphic levels in the Triton Point Member and Cantrill and Nichols (1996) recognised at least five different associations. Although the plant fossils are preserved only as impressions, they document that a diverse community of angiosperms thrived in the area during the Albian, and that this community also included several herbaceous and perhaps also aquatic angiosperms (Cantrill, 1996). Another important plant fossil from this sequence is the youngest report of *Pentoxylon* (Howe and Cantrill, 2001), a seed plant probably related to the Bennettitales–Erdtmanithecales–Gnetales group (Chapter 5).

Antarctic Peninsula, Table Nunatak. Table Nunatak is an isolated exposure on the Antarctic Peninsula, immediately east of Kenyon Peninsula and is the most southerly outcrop of the Larsen Basin. The Cretaceous sequence at this locality is about 62 m thick and consists of fine-grained mudstones, siltstones and sandstones with fine laminae containing a diverse mesofossil flora of charcoalified plant debris that also includes small flowers (Eklund, 2003; Eklund *et al.*, 2004a). The sequence is interpreted as shallow-marine and is dated based on dinoflagellate cysts to the Late Santonian.

James Ross, Vega, Seymour, and Snow Hill islands. Abundant fossil wood occurs in the upper part of the sequence assigned to the Santa Marta, Snow Hill Island and López de Bertodano formations, which are dated as latest Coniacian–Campanian, latest Campanian – earliest Maastrichtian, and ?Late Maastrichtian, respectively (Crame *et al.*, 2004). The wood is thought to have been washed into the James Ross Basin from vegetation growing on the Antarctic Peninsula (Francis, 1986; Poole and Francis, 1999; Poole *et al.*, 2000c). On James Ross Island fossil wood has been collected from several outcrops at different stratigraphic levels of the Santa Marta Formation. Wood collected from Vega Island is from the Snow Hill Island Formation (Poole and Gottwald, 2001) and fossil wood from Seymour Island was described from the López de Bertodano Formation (Poole *et al.*, 2000c; Poole and Gottwald, 2001).

Macrofossil floras containing angiosperm leaves have also been discovered at several Late Cretaceous outcrops on James Ross Island. The oldest assemblage is from the Hidden Lake Formation referred to the Coniacian (Hayes *et al.*, 2006). Another flora is from the younger Santa Marta Formation (Hayes *et al.*, 2006). A small macrofossil flora that included only conifer remains was reported from Vega Island collected from the López de Bertodano Formation (Césari *et al.*, 2001).

South Shetland Islands. Fossil floras occur abundantly on the South Shetland Islands and include leaves, fossil wood and diverse palynological assemblages. King George Island, the largest of the islands, has yielded most of the plant fossils from the area (reviewed by Dutra and Batten, 2000). Isotopic dating indicates a Late Campanian to probably Early Maastrichtian age for the plant-bearing horizons. The sediments are continental and thought to have been deposited in a major braided river system in an area with considerable topography (Dutra and Batten, 2000). Four localities with plant fossils were discussed in more detail by Dutra and Batten: three on the Fildes Peninsula at Half Point, Price Point and Skua Bay first studied by Zhou and Li (1994) and one near Admiralty Bay at Zamek Hill first studied by Zastawniak (1994). In addition to the leaf assemblages, which include ferns, conifers and some angiosperms, the sediments have also yielded diverse palynological assemblages (Dutra and Batten, 2000).

Fossil wood (Poole and Gottwald, 2001) and leaves (Rees and Smellie, 1989) collected from Livingston Island in sediments of the Williams Point Beds are broadly referred to as of Cenomanian – Early Campanian age (Chapman and Smellie, 1992). According to David J. Cantrill (personal communication, 2009) these beds are more likely of Santonian age.

4.12.6 New Zealand

Angiosperm fossils from the Cretaceous of New Zealand have so far received relatively little attention. They are known mostly from dispersed palynological assemblages (e.g. Couper, 1953; McQueen, 1956; McIntyre, 1968; Raine, 1984; Vajda *et al.*, 2001; Raine *et al.*, 2006), but several leaf floras have also been described (Pole, 1998; Pole and Douglas, 1999; Pole and Vajda, 2009; Pole and Philippe, 2010) and recently a mesofossil flora has been reported by Cantrill *et al.* (2011). Angiosperms are first encountered in the Late Albian of the Clarence Series in the northeast of the South Island (Couper, 1953, 1960) after which they steadily become more diverse and more abundant.

A leaf flora from Pitt Island, situated about 700 km east of New Zealand, was described from part of the Tupuangi Formation thought to be of Turonian age. The flora is dominated by conifers, but also includes some angiosperm leaves, and charcoal occurs abundantly in some horizons (Pole and Philippe, 2010).

Plant macrofossils are known mainly from the South Island and they have been particularly important in the study of floristic turnover at the K–T boundary (Vajda *et al.*, 2001; Pole and Vajda, 2009). One flora, from the Pakawau Bush Road locality, preserves compression fossils in the non-marine Rakopi Formation of the Pakawau Group. The Rakopi Formation is dated as latest Cretaceous (Campanian or Maastrichtian) based on palynological and geological evidence (Kennedy *et al.*, 2003). The flora is dominated by angiosperms with minor components of ferns and conifers. It also includes the only Cretaceous flower described from New Zealand (Kennedy *et al.*, 2002, 2003). Another fossil flora from the latest Cretaceous (Late Maastrichtian, Cave Stream locality) is dominated by conifers and is devoid of angiosperms (Pole and Vajda, 2009).

4.12.7 Southern South America

Several Cretaceous floras have been described from sedimentary basins in southern South America (Argentina, Chile) (for reviews see Drinnan and Crane, 1989; Archangelsky *et al.*, 2009). The deposits are strongly influenced by volcanic and tectonic activity and the plant fossils are frequently preserved in a tuffaceous succession deposited in a fluviatile and lacustrine environment. Several of the plant fossils show strongly xeromorphic traits that in some cases have been interpreted as adaptation to volcanic stress (Archangelsky *et al.*, 1995). Earlier reports of angiosperms in these Cretaceous floras were scattered, but recent, intensified studies have resulted in important discoveries, particularly from Argentina. In Argentina fossil floras occur in five Cretaceous basins, from north to south: the San Luis Basin, the Neuquén Basin, the San Jorge Gulf Basin, the Deseado Massif Basin and the Austral Basin (Archangelsky *et al.*, 2009). Rich palynological assemblages from the Early Cretaceous have been reported from all basins (Archangelsky, 1980; Archangelsky *et al.*, 2009) and there are also several informative macrofossil floras with angiosperm leaves. Most are from the Early and mid-Cretaceous. Younger (post-Cenomanian) macrofossil floras are currently known only from the Austral Basin and the Neuquén Basin where they extend into the Coniacian. The earliest angiosperms reported are pollen grains, assigned to *Clavatipollenites*, from the Barremian (Archangelsky, 1980). Tricolpate pollen grains are known from the Middle Albian and onwards (Volkheimer and Salas, 1975).

The most intensively studied Early Cretaceous plant assemblages from southern South America, referred to as the Ticó flora, have been collected from several localities in the Deseado Massif Basin in Santa Cruz Province, where they occur at different stratigraphic levels (Archangelsky, 1963, 1967, 2001; Archangelsky *et al.*, 1986; Archangelsky and Taylor, 1986, 1993; Romero and Archangelsky, 1986). The lower fossil assemblages are from the Ticó Anfiteatro Formation, dated as early Late Aptian. The upper fossil assemblages are from the Punta del Barco Formation of Late Aptian age. The Ticó flora is dominated by cycadophytes and conifers together with diverse ferns and unusual seed plants, but it also includes some angiosperms. Another flora from the Santa Cruz Province with angiosperm leaves has been reported from the Bajo de los Corrales locality (Passalia *et al.*, 2001), which is dated as mid-Cretaceous (Late Albian – Early Cenomanian; Archangelsky *et al.*, 2009).

5

Angiosperms in context: extant and fossil seed plants

Over more than a hundred years, a vast literature has accumulated on the origin of angiosperms, but different authors have approached the question in different ways. Many studies have focused solely on extant angiosperms, seeking to establish which groups are most primitive, largely based on ideas about the evolution of angiosperm flowers. Others have sought to identify one or more ancestral groups from which angiosperms may have been derived, often based on ideas about how specific angiosperm features (e.g. flower, carpel, stamen) may have evolved from the reproductive structures of other fossil seed plants (e.g. Gaussen, 1946; Melville, 1962, 1963; Retallack and Dilcher, 1981b; Krassilov, 1997). Palaeobotanical data have also been used to constrain the time and place of angiosperm origin (e.g. Doyle, 1969; Doyle and Hickey, 1976; Hughes, 1976), to expand the known diversity within the seed plant clade and also to identify putative 'proangiosperms' (e.g. Thomas, 1925). Some extreme views (Hughes, 1976, 1994) have argued that palaeobotanical data alone, divorced almost entirely from studies of living plants, hold the key to understanding angiosperm origins.

A modern approach to understanding the origin of angiosperms reduces fundamentally to recognising the major groups of land plants and understanding how they are phylogenetically related (Hill and Crane, 1982; Crane, 1985a). This approach focuses on developing a well-corroborated hypothesis of relationships among living and fossil seed plants. It implies that the pattern of relationships will inform ideas on how the characteristic features of angiosperms may have evolved, as well as the nature of the evolutionary transitions and processes that may have been important in angiosperm origins.

Many previous studies have examined seed plant relationships (Chapter 6). At this point, to be useful, further phylogenetic analysis would need to take a new approach. It would also require a new and critical synthesis of data from across all groups of living and fossil seed plants. This is beyond the scope of this book. However, the Gnetales and several groups of fossil seed plants have figured prominently in discussions of angiosperm origins (Chapter 6) and warrant further consideration here. For some of these groups there is also new and potentially important information. In this chapter we place angiosperms in the broader context of land plant evolution with a brief overview of several potentially closely related groups of seed plants.

5.1 ANGIOSPERMS AMONG EXTANT AND FOSSIL SEED PLANTS

Understanding the phylogenetic position of angiosperms requires knowledge of relationships among the major groups of seed plants, which in turn are influenced in part by the position of seed plants themselves in the broader context of land plant evolution. Among extant vascular plants, sphenopsids (*Equisetum*), ferns (including *Psilotum*) and seed plants are more closely related to each other than any of these groups is to lycopsids (Figure 5.1). This group of tracheophytes has been termed the euphyllophytes (Kenrick and Crane, 1997).

Phylogenetic studies (Crane, 1985a; Doyle and Donoghue, 1986, 1992; Nixon *et al.*, 1994; Doyle, 1996, 2006; Hilton and Bateman, 2006) resolve living and fossil seed plants (spermatophytes) as a monophyletic group nested within a more inclusive clade (lignophytes *sensu* Doyle and Donoghue, 1986; Figure 5.1) defined by the presence of a bifacial cambium producing typical seed-plant secondary xylem and secondary phloem. Progymnosperms make up a grade of lignophytes that are not seed plants (i.e. that have retained free-sporing reproduction). Based on phylogenetic analyses of molecular sequence data from extant vascular plants it also seems likely that sphenopsids and ferns are sister taxa (moniliforms; Kenrick and Crane, 1997) to the exclusion of seed plants (Qiu and Palmer, 1999; Pryer *et al.*, 2001), but morphological evidence for this is currently limited (Kenrick and Crane, 1997; Pryer *et al.*, 2004).

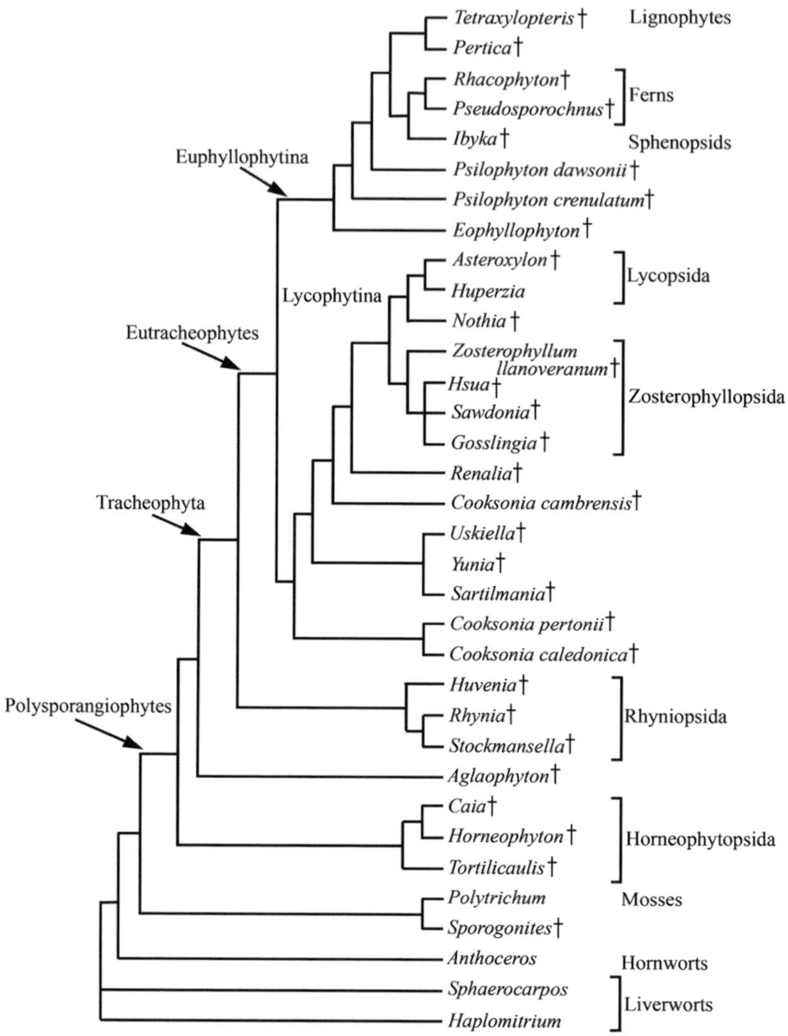

Figure 5.1 Phylogenetic analysis of living and fossil plants showing the position of euphyllophytes and lignophytes in the broader context of land plant evolution. Only *Huperzia*, *Polytrichium*, *Anthoceros*, *Sphaerocarpos* and *Haplomitrium* are extant. All other genera are based on fossils (†). Lignophytes and their subgroup, seed plants, are represented here by *Tetraxylopteris*, a Devonian progymnosperm. Based on Kenrick and Crane (1997).

Despite some consensus on these aspects of vascular plant relationships significant uncertainties remain. For example, few integrated neobotanical–palaeobotanical studies have focused specifically on phylogenetic questions at the level at which the major groups of euphyllophytes diverge, and it is uncertain whether aneurophytalean or archaeopteridalean progymnosperms are the sister group to the seed plant clade (Stein and Beck, 1987). Exactly how this and other relationships at the trimerophyte and progymnosperm grades are resolved will have implications for understanding phylogenetic patterns among early seed plants.

Among seed plants themselves, there are only five extant groups: cycads, *Ginkgo*, conifers, Gnetales and angiosperms. Angiosperms stand out in terms of their extraordinary systematic, structural and ecological diversity, which is all the more remarkable given their relatively recent appearance in the fossil record. The other four groups are all relatively stereotyped in their morphology and relatively low in species diversity, but have well-established fossil records that extend back to the late Palaeozoic or early Mesozoic.

Extant cycads comprise approximately 300 species and there is broad agreement regarding the monophyly of the group and the pattern of relationships among the genera (Stevenson, 1992). Cycads have a reliable fossil record that begins in the Early Permian (Crane, 1988) and they were diverse and widespread during the Mesozoic (e.g. Harris, 1964). The diversity of cycads may have decreased during the mid-Cretaceous (Lidgard and Crane, 1988), perhaps in parallel with the diversification and ecological expansion of angiosperms (Chapter 19). Extant cycads are relictual, in the sense that they are less diverse and more narrowly distributed today than they were during the Mesozoic. However, the fact that all genera (except *Cycas*) are restricted to a single continental area (Johnson and Wilson, 1990) suggests that most genera and species may be of relatively recent origin. Most of the diversity of extant cycads has probably evolved in parallel with angiosperm diversification and since the major tropical landmasses became separated in the Late Cretaceous. Cycads have not figured prominently in discussions of angiosperm origins but a recent analysis of seed plant relationships using duplicate gene rooting resolves cycads and angiosperms as more closely related to each other than either is to any other living group (Mathews, 2010).

Ginkgo biloba, native to China, is the single extant species of Ginkgoales. Putative Ginkgoales have been described from the Permian (i.e. *Trichopitys*), but the relationships of these fossil plants to extant *Ginkgo* are not established securely (Meyen, 1988). Fossil plants that are unequivocally related to extant *Ginkgo* are known from the Middle Jurassic (Zhou, 1997; Zhou and Zheng, 2003), and throughout the Mesozoic fossils were widely distributed in both the Northern and Southern Hemispheres, especially at high palaeolatitudes (Chapter 19). *Ginkgo* remained widespread in the Northern Hemisphere into the Late Cenozoic. The currently restricted geographic distribution clearly reflects regional extinction associated with climatic deterioration (Tralau, 1967; Zhou, 1997) and perhaps loss of its seed dispersers (Tiffney, 1984). In phylogenetic analyses based on both morphological and molecular data, *Ginkgo* is generally resolved as closely related to conifers.

Extant conifers comprise approximately 550 extant species distributed among about 52 extant genera. Conifers are generally considered monophyletic (e.g. Crane, 1985a; Doyle, 1996) although it is difficult to identify unequivocal morphological or anatomical synapomorphies (Crane, 1985a; Rothwell and Serbet, 1994) and the monophyly of the group has been challenged by several studies of seed

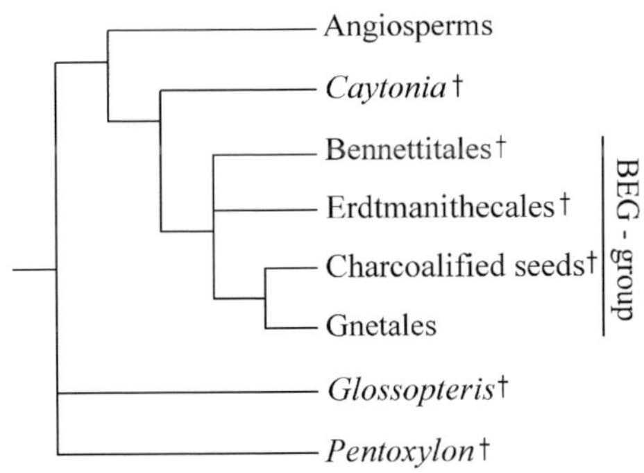

Figure 5.2 Relationships among members of the Bennettitales–Erdtmanithecales–Gnetales (BEG) group and their position in the broader anthophyte/glossophyte context. Fossil taxa indicated by †. Based on Friis *et al.* (2007).

plant phylogeny based on molecular data. These place Gnetales inside conifers close to Pinaceae (e.g. Hansen *et al.*, 1999; Qiu *et al.*, 1999; Bowe *et al.*, 2000; Chaw *et al.*, 2000; Burleigh and Mathews, 2004; Braukmann *et al.*, 2009) or other groups (see discussion in Mathews, 2009, 2010), thus making conifers paraphyletic (Chapter 6). Conifers have a fossil record that extends back to the Late Pennsylvanian (Scott and Chaloner, 1983; Mapes and Rothwell, 1984). The group is diverse and widespread in the Permian, and remains important through the Mesozoic and Cenozoic to the Recent.

Extant Gnetales comprise three genera: *Gnetum*, *Welwitschia* and *Ephedra*. Despite the diversity of habit and form among these genera most phylogenetic analyses, based on both morphological and molecular data, resolve the group as monophyletic (section 5.3). Distinctive pollen that is characteristic of the Gnetales is first known from the Permian (Wilson, 1962), and attains its greatest diversity in the Late Triassic and Early to mid-Cretaceous (Crane, 1996). In recent years the formerly sparse macro- and mesofossil record of the Gnetales has been significantly expanded (section 5.3). The Gnetales have played a prominent role in discussions of angiosperm origin based on phylogenetic systematics and have often been regarded as the closest living relatives of angiosperms (Chapter 6).

Beyond these four extant groups there is much greater diversity of extinct seed plants in the fossil record (Figure 5.2). The earliest seed plants are known from the Late Devonian (e.g. Fairon-Demaret and Scheckler, 1987; Rothwell *et al.*, 1989) and through the Palaeozoic the diversity of seed plants expands dramatically. Diversification continued through the Mesozoic with the first appearance of further new groups. Information on extinct seed plants is therefore necessary to place extant seed plants, including angiosperms, in a broader context. The five groups of living seed plants are not representative of the total diversity in the seed plant clade.

In broad terms, the prospects for obtaining useful information from studies of fossil seed plants are limited by the constraints of palaeontological preservation and the need to assemble reliable 'whole' plant reconstructions from the often fragmentary and physically separated stems, leaves and reproductive structures encountered in most palaeobotanical assemblages (Chapter 2). Specimens that show physical attachment between different plant organs are rare. As a result, most reconstructions of fossil plants, especially in compression floras, are based on anatomical or other structural similarities and/or on evidence of repeated co-occurrence (association evidence). Current concepts of most higher taxa of extinct seed plants are therefore based on a large number of often loosely associated, dispersed, plant fossil organs, with a few partly reconstructed fossil plants at their core (Crane *et al.*, 2004). As a result, precise delimitation of fossil groups is often unsatisfactory, and most phylogenetic studies have, either explicitly or implicitly, leaned heavily on the relatively small number of more or less convincingly reconstructed 'whole' fossil plants. All of the major groups of extinct seed plants that have featured prominently in phylogenetic analyses of seed plant relationships (e.g. Crane, 1985a; Rothwell and Serbet, 1994; Doyle, 2006; Hilton and Bateman, 2006; Friis *et al.*, 2007; Rothwell *et al.*, 2009) are of this kind.

Here we review several groups of mainly Mesozoic seed plants to illustrate the extent and limitations of current knowledge, as well as aspects of the structure and biology of some of the important plants in pre-angiosperm vegetation. We focus especially on those groups that have figured prominently in previous discussions of angiosperm origins. Gnetales, Bennettitales and Pentoxylales have all been important in various manifestations of the 'anthophyte' hypothesis (Chapter 6) and also show interesting similarities to angiosperms in reproductive and vegetative features. The Erdtmanithecales, a recently recognised group, also appear to be part of this complex (Friis *et al.*, 2007). Caytoniales, corystosperms, glossopterids and, to a lesser extent, peltasperms have been implicated as potential relatives of angiosperms in several studies (Chapter 6) and have been particularly important in terms of developing ideas about the origin of the carpel and bitegmic ovule.

5.2 BENNETTITALES–ERDTMANITHECALES–GNETALES (BEG) GROUP

A recent development in studies of seed plant relationships, related to, but distinct from, the results of phylogenetic analyses, has been the recognition of the Bennettitales–Erdtmanithecales–Gnetales (BEG) group based on the very distinctive structure of their seeds (Friis *et al.*, 2007, 2009a). In Bennettitales, Erdtmanithecales and Gnetales the seeds are rounded, but generally somewhat angular in cross-section. Usually they are three- or four-angled, but sometimes they may be five-, six- or only two-sided. The seeds are distinctive in having a thin membranous integument that is completely enclosed by one, or sometimes two, seed envelopes, except for a small opening at the micropylar region. The integument is fused to the innermost envelope only at the base and at the apex the integument extends into a long micropylar tube that forms a long narrow micropylar canal. The micropylar tube extends well beyond the seed envelope(s). In mature seeds the micropylar canal is open only apically. It is closed further down, either by cellular growth of the inner epidermis of the micropylar tube or by secretion. In extant *Ephedra* micropylar closure is by secretion, but in *Welwitschia* and *Gnetum* it is by cellular growth. Berridge (1911) noted that the micropylar closing mechanism for Bennettitales is very similar to that in *Gnetum* and the same mechanism apparently also occurred in *Cycadeoidea morierei* (Friis *et al.*, 2009a) as well as in some dispersed seeds of uncertain relationships, such as *Buarcospermum* and *Lignierispermum* (section 5.5).

Adding Erdtmanithecales and one of the dispersed seeds (unnamed square seeds) to a phylogenetic analysis based on the seed plant matrix of Hilton and Bateman (2006), but adjusting the scoring of associated characters, supports the significance of the distinctive BEG seed morphology, and it groups Bennettitales, Erdtmanithecales and Gnetales in a single clade (Figure 5.2) (Friis *et al.*, 2007). In the context of seed plants as a whole, other features that appear to unite the BEG group are paracytic stomata and distinct tectate pollen wall ultrastructure with a more or less homogeneous ektexine, a granular infratectal layer, a foot layer that is often thin, and a thick laminate endexine (Friis *et al.*, 2009a).

Although recognition of the BEG group is controversial (Rothwell *et al.*, 2009) emerging evidence from Bennettitales and other fossils, based mainly on new levels of detail provided by synchrotron X-ray microtomography, appear to strengthen the BEG link further. Ongoing evaluation, based on more material and new analyses, will be needed. Here we survey our current knowledge on the fossil record of the BEG group, treating first Gnetales and then Erdtmanithecales and Bennettitales. We also review emerging information on a large number of isolated seeds that have the same organisational ground plan as seeds of the BEG group, but that cannot currently be included in any of the formally defined groups. These include seeds assigned to the extinct genera *Buarcospermum*, *Lignierispermum*, *Lobospermum*, *Raunsgaardispermum* and *Rugonella*, described from the Early Cretaceous of Portugal and eastern North America (Mendes *et al.*, 2008b; Friis *et al.*, 2009a).

5.3 GNETALES

Gnetales comprise three extant genera: *Ephedra* with approximately 35–45 species of shrubs, climbers, or small trees distributed in the arid and semi-arid regions of the world (Kubitzki, 1990b); *Gnetum*, a pantropical genus of mainly lianas with about 30 extant species (Kubitzki, 1990a); and *Welwitschia mirabilis*, endemic to the Namib Desert (Kubitzki, 1990c). There is clear palaeontological evidence that the three extant genera are the relictual extant remnants of a group that was once more diverse (Crane, 1988; Crane and Lidgard, 1989; Lidgard and Crane, 1990). The most recent monograph of *Ephedra* is that of Stapf (1889), and for *Gnetum* that of Markgraf (1926), but recent renewed interest in Gnetales over the past few decades has resulted in several studies that begin to address relationships among species within the genera (Ickert-Bond and Wojciechowski, 2004; Rydin *et al.*, 2004; Won and Renner, 2006).

Despite the obvious morphological differences among the three extant genera, there is strong evidence that they comprise a monophyletic group based on their shared possession of opposite and decussate, or sometimes whorled, phyllotaxis, the presence of vessels with end plates apparently derived from circular bordered pits, multiple axillary buds, circular bordered pits in the protoxylem and female and male reproductive organs clustered in compound strobili (Figure 5.3), sometimes with male and female parts close together. Pollen-producing organs of Gnetales are distinctive and generally grouped together in synangia of two to four sporangia with apical dehiscence (Figures 5.3, 5.4). Sporangia are borne on a stout stalk, the antherophore, which may be rounded or flattened (Mundry and Stützel, 2004).

A further potential synapomorphy of Gnetales are the anastomoses in venation that occur in the cotyledons and leaves of *Welwitschia* and *Gnetum*, which may have been obscured in *Ephedra* by extreme leaf reduction (e.g. Crane and Upchurch, 1987; Rydin *et al.*, 2003). Similarly, a potential synapomorphy that is of particular palaeobotanical significance is the characteristic ribbed pollen of extant *Ephedra* (Figures 5.3, 5.5) and *Welwitschia*. This has apparently been modified in *Gnetum*, where the pollen has fine spines. The pollen of all three genera also has a granular infratectal layer (e.g. El-Ghazaly and Rowley, 1997; Yao *et al.*, 2004).

Further distinctive features of Gnetales are the presence of an additional layer (apparently formed from bracts) around the nucellus, the very thin integument, the long protruding micropylar tube formed from the apex of the integument (Figure 5.3) and the distinct closure of the micropylar canal after pollination (Pearson, 1929; Crane, 1985a; Doyle and Donoghue, 1986; Doyle, 1996; Endress, 1996a; Friis *et al.*, 2009a). Although it was once thought to be characteristic of Gnetales, it is becoming clear that this suite of features defines a more inclusive group that also includes Bennettitales, Erdtmanithecales and perhaps other taxa (Friis *et al.*, 2009a).

Cladistic analyses of Gnetales, based on morphological data, suggest that *Ephedra* is sister to *Welwitschia* and *Gnetum*, which are united by the shared possession of anastomoses in foliar venation, paracytic stomata (Figure 5.6), a four-nucleate male gametophyte, tetrasporic megagametophyte development, absence of archegonia, cellular embryogeny, and the presence of a 'feeder' in the embryo. This pattern of relationships is also strongly supported both by molecular data (e.g. Hamby and Zimmer, 1992; Chase *et al.*, 1993; Doyle *et al.*, 1994; Price, 1996) and combined analyses of morphological and molecular data (Albert *et al.*, 1994; Doyle *et al.*, 1994). The one alternative possibility that has been suggested is that Gnetales are paraphyletic, with *Gnetum* the sister group to angiosperms (Nixon *et al.*, 1994), but this pattern of relationships has not been supported by subsequent analyses based on morphological or molecular data.

Because most hypotheses of relationships among anthophytes resolve the Gnetales as monophyletic, the derived similarities of *Gnetum–Welwitschia* and angiosperms (e.g. reticulate-veined leaves, reduction of male gametophyte, tetrasporic female gametophyte lacking archegonia and with free nuclei serving as eggs, cellular early embryogeny) are most parsimoniously interpreted as convergence rather than as potentially homologous features. However, an interesting

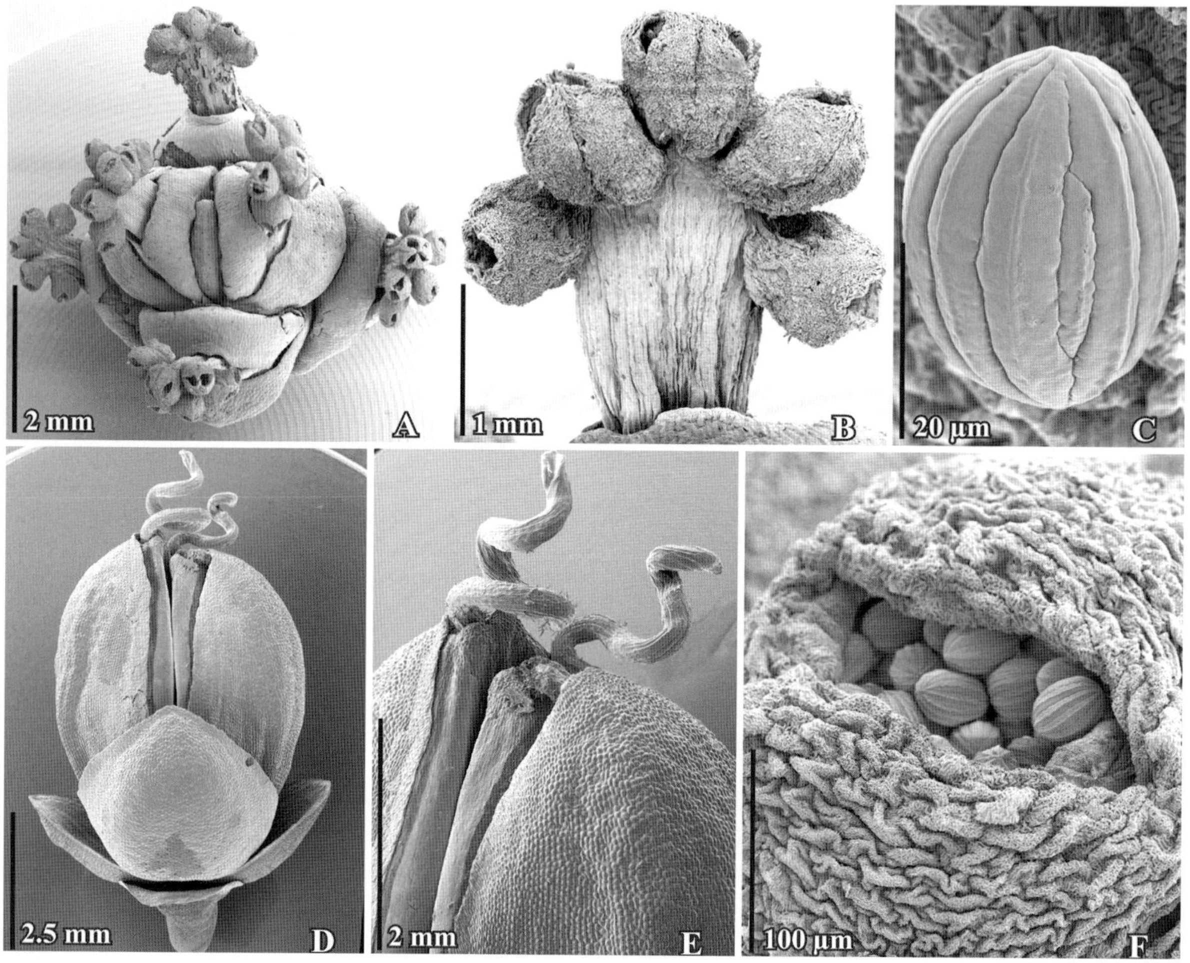

Figure 5.3 Reproductive structures of extant *Ephedra* (Gnetales). (A–C, F) *Ephedra americana*; (A) pollen-producing cone showing six antherophores with their associated bracts; (B) detail of an antherophore showing five synangia with apical dehiscence slits; (C) polyplicate pollen; (F) detail of synangium showing one sporangium with apical dehiscence exposing the pollen. (D, E) *Ephedra intermedia*, (D) ovulate cone showing two seeds surrounded by four pairs of opposite and decussate leaf-like bracts; (E) detail of apical part of two seeds from D showing the uppermost pair of leaf-like bracts; note the micropylar tubes extending out of the tips of the seed envelope. All SEM images of critical-point-dried specimens.

issue is to what extent these similarities reflect the expression of conserved underlying developmental potential. Leaves of *Gnetum*, for example, are so similar to those of angiosperms that they are routinely confused in major herbaria when only sterile material is available. A related question is which features of *Ephedra* are general for Gnetales as a whole, and which reflect more recent specialisation? A case in point is the presence of an inner papillate lining in the micropylar region of the seed envelope. Among extant Gnetales this is diagnostic of *Ephedra*, but it also occurs in the fossil seed *Raunsgaardispermum*, which appears not to be gnetalean based on the non-ephedroid pollen found in the micropyle (Mendes *et al.*, 2008b).

5.3.1 Temporal and spatial patterns of gnetalean radiation

The best insights into the distribution of Gnetales in space and time are provided by dispersed fossil pollen. This requires accepting that ribbed fossil pollen grains

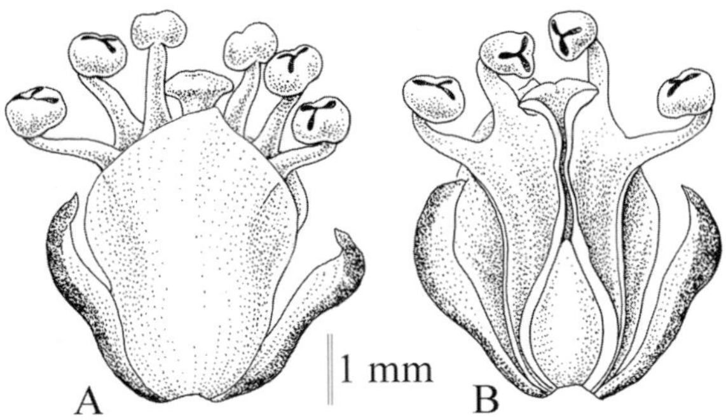

Figure 5.4 Pollen-producing structure of *Welwitschia mirabilis.* (A) Complete reproductive structure showing lateral bracts, the anterior–posterior bract pair, protruding micropyle and protruding antherophore; (B) longitudinal section through reproductive structure showing aborted ovule in the centre with its nucellus and long micropyle with expanded funnel-shaped apex. Based on specimens in the collections of the Swedish Museum of Natural History, Stockholm; from Friis *et al.* (2006a).

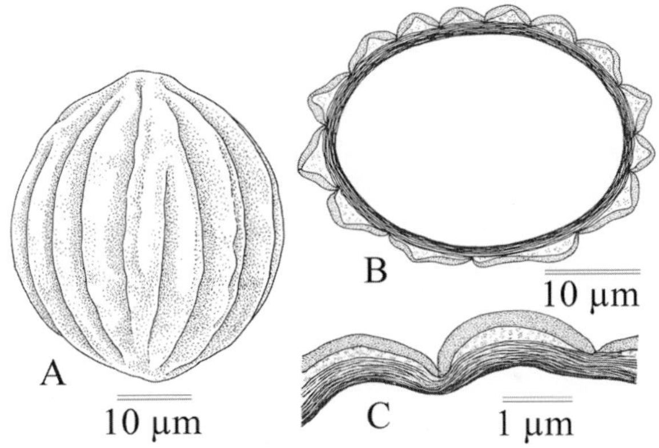

Figure 5.5 Pollen morphology and ultrastructure of extant *Ephedra foliata* showing polyplicate tectum (A), as well as granular infratectal layer and thick laminate endexine (B–C). Redrawn from El-Ghazaly and Rowley (1997).

(ephedroid pollen) resembling the characteristic ribbed pollen of extant *Ephedra* and *Welwitschia* are referable, without supporting evidence from other parts of these fossil plants, to the gnetalean clade. This assumption is reasonable given the distinctiveness of the grains, but there is also reason for caution. Similar striate pollen occurs in a few angiosperms (e.g. Araceae), and there is one report of such grains (*Mayoa portugallica* closely similar to pollen of extant *Holochlamys*, Araceae; Friis *et al.*, 2004) from the Early Cretaceous (Chapter 11). In this case the angiosperm affinity is secure, based on ultrastructural details of the pollen wall (i.e. the lack of a distinct endexine seen in Gnetales), but the absence of such information for most dispersed pollen grains needs to be borne in mind. As far as we are aware, dispersed fossil pollen comparable to the echinate grains of extant *Gnetum* has not yet been reported.

There are scattered reports of fossil ephedroid pollen from the latest Palaeozoic (Permian) (Wilson, 1962). There are also intriguing similarities between some ephedroid grains and Permian glossopterid-type pollen with coarse ribs on the body of the grain in which the sacci are reduced or absent (see for example pollen of *Kendostrobus*; Surange and Chandra, 1975). An improved understanding of these glossopterid grains relative to those of extant *Ephedra* and *Welwitschia*, especially ultrastructural studies of the pollen wall, would be helpful to establish whether these similarities are significant or merely superficial.

Ephedroid pollen grains become more common during the Triassic, and are widespread in fossil palynofloras from the Northern Hemisphere (Chaloner, 1969; Traverse, 1988). During the Jurassic they become generally less common, but the record remains continuous into the Early and mid-Cretaceous and becomes more diverse and abundant, especially in low-palaeolatitude areas (Stover, 1964; Brenner, 1976; Lima, 1980; Muller, 1984; Lidgard and Crane, 1988; Crane and Lidgard, 1989; Osborn *et al.*, 1993). A compilation of probable gnetalean polyplicate fossil pollen from northern Gondwana by Dilcher *et al.* (2005) reported more than 70 different taxa from the Early and mid-Cretaceous of northern Gondwanan South America. Cretaceous ephedroid grains are mainly simple, polyplicate forms assigned to genera such as *Ephedripites*, *Equisetosporites*, *Gnetaceaepollenites*, *Jugella*, *Regalipollenites*, *Singhia*, *Steevesipollenites* and *Welwitschiapites.* Most are inaperturate, as in extant *Ephedra*. Only *Jugella* has a distinct colpus, as seen in *Welwitschia*. In addition to these simple polyplicate ephedroid grains, morphologically more complex forms, the Elaterates, are also diverse in palaeoequatorial regions during the mid-Cretaceous. Elaterates may also be related to Gnetales, but are considered separately below.

Ultrastructural studies of ephedroid pollen from Brazil show that the external morphological similarities to extant *Ephedra* and *Welwitschia* also extend to details of the pollen wall. As in the pollen of the two extant genera, the fossil grains have a distinct granular infratectum and a thick, laminar endexine (Osborn *et al.*, 1993). Ephedroid pollen grains have also been found inside the micropyle of Early Cretaceous seeds that are clearly related to extant *Ephedra*.

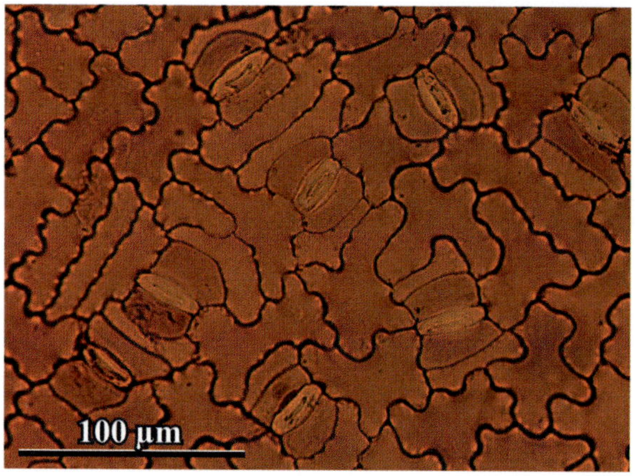

Figure 5.6 Cuticle preparation from leaf of *Gnetum edule* showing outlines of epidermal cells and stomata with two subsidiary cells parallel to the guard cells (paracytic). Preparation in the collections of the Swedish Museum of Natural History, Stockholm.

The diversity and abundance of ephedroid grains expands rapidly during the Early and mid-Cretaceous, at about the same time and in the same place (low palaeolatitudes) as the initial marked expansion of the diversity and abundance of angiosperm pollen (Crane and Lidgard, 1989; Lidgard and Crane, 1990; Lupia *et al.*, 2000). However, while angiosperm pollen continues to expand at middle and high palaeolatitudes, ephedroid pollen is never especially prominent in palynofloras from mid- or high palaeolatitudes. At low palaeolatitudes the diversity and abundance of ephedroid grains declines precipitously toward the end of the Cenomanian, after which it is generally only a minor component of later Cretaceous and Cenozoic palynofloras from these areas (Crane and Lidgard, 1989; Lidgard and Crane, 1990).

In addition to broad temporal and spatial parallels in the increasing abundance and diversity of ephedroid and angiosperm pollen during the Early and mid-Cretaceous, more localised patterns of co-occurrence also suggest that Early and mid-Cretaceous Gnetales and angiosperms occurred in similar habitats. In the Early Cretaceous of southern England, Phase 1 of Hughes (1994) (Late Hauterivian) is marked by a sudden influx of *Ephedripites*-type pollen together with the appearance of diverse small monocolpate, tectate, grains of probable angiosperm affinity (Hughes, 1994). Preliminary analyses of Early to mid-Cretaceous pollen floras from low palaeolatitudes also show that at a regional level the abundance of ephedroid pollen is highest where angiosperm pollen is also very prominent. Further work to clarify these patterns of co-occurrence would be of great interest.

More direct evidence of a close ecological association between Early Cretaceous angiosperms and Gnetales is provided by macro- and mesofossil assemblages in which angiosperms and Gnetales co-occur. In the Potomac Group of eastern North America, the Drewry's Bluff Leaf Bed includes the gnetalean macrofossils *Drewria potomacensis* together with remains of ferns and angiosperm leaves. The Drewry's Bluff Leaf Bed has been interpreted as a more or less *in situ* assemblage of early-successional vegetation (Crane and Upchurch, 1987) implying that in this instance both angiosperms and Gnetales were early colonisers of disturbed habitats. *Drewria* and other macrofossil remains (section 5.3.3) further indicate that Cretaceous Gnetales were probably small herbaceous or shrubby plants. Two other Early Cretaceous macrofossil floras in which gnetalean and angiosperm fossils co-occur are the Crato flora of Brazil (Mohr and Friis, 2000; Rydin *et al.*, 2003) and Jehol flora of northeastern China (Wu, 1999; Sun *et al.*, 2001; Leng *et al.*, 2003; Rydin *et al.*, 2006b). Gnetalean seeds also occur together with angiosperm reproductive structures in Early Cretaceous mesofossil floras from Portugal and eastern North America (Rydin *et al.*, 2006a).

5.3.2 Elaterates

A group of dispersed polyplicate pollen referred to as the Elaterates (Elaterate Complex) may also have been produced by Gnetales or related plants in the BEG group, but their systematic position is uncertain. Their probable relationship to Gnetales, or to the BEG group more broadly, is based on ultrastructural details of the pollen wall as well as morphological intermediates that link them with more normal ephedroid grains (Crane, 1996; Dino *et al.*, 1999). Elaterate grains are diverse and are placed in several distinct genera such as *Alaticolpites*, *Elateroplicites*, *Elaterocolpites*, *Elateropollenites*, *Elaterosporites*, *Galeacornea*, *Regalipollenites*, *Senegalosporites* and *Sofrepites* (e.g. Jardiné, 1967; Herngreen, 1973; Dino *et al.*, 1999).

Elaterate pollen has a more elaborate morphology than typical ephedroid grains. In some there are relatively few exine ribs but they are greatly expanded in thickness. In others, the ribs are expanded into horn-like projections. In other forms the exine ribs are almost completely detached from the body of the grains to form long elater-like structures (Figure 5.7) analogous to those of spores of *Equisetum*.

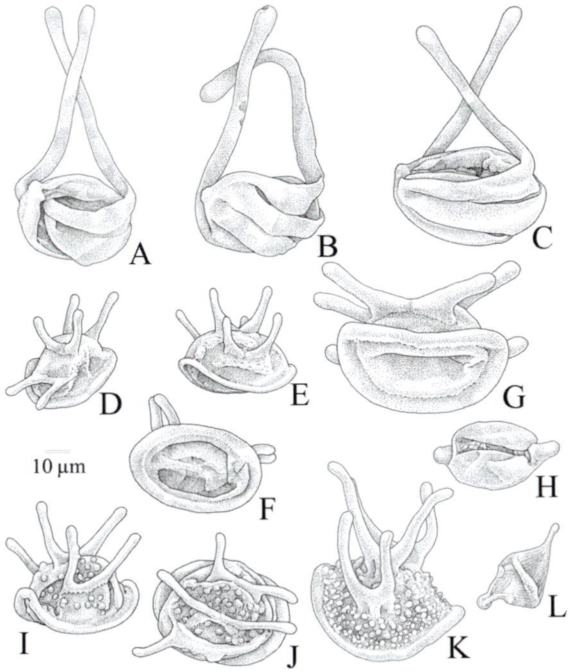

Figure 5.7 Elaterates pollen of possible gnetalean affinity from the mid-Cretaceous (Albian–Cenomanian) of South America. (A–C) *Elateroplicites africaensis*: (D–F) *Elaterosporites klaszii*; (G) aff. *Elaterosporites klaszii*; (H) *Sofrepites legouxiae*, (I, J) *Elaterosporites verrucatus*; (K) *Elaterosporites protensus*; (L) *Elateropollenites jardinei*. Drawn from SEM micrographs in Dino *et al.* (1999).

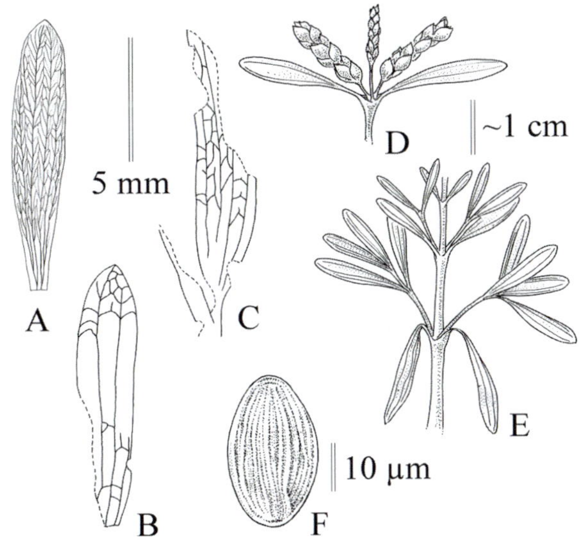

Figure 5.8 Extant *Welwitschia* and fossil *Drewria*. (A) Cotyledon of *Welwitschia mirabilis* showing chevron-shaped pattern of cross-veins. (B–E) *Drewria potomacensis* from the Early Cretaceous (Aptian) Drewry's Bluff Leaf Bed, Virginia, USA: (B, C), detail of venation in leaves; (D) reconstruction of reproductive shoot showing three cones with the terminal cone smaller than the two lateral cones; (E) reconstruction of vegetative shoot showing opposite – decussate arrangement of leaves and axillary branching. (F) Polyplicate ephedroid pollen associated with *Drewria potomacensis*. (A) redrawn from Martens (1971); (B–F) redrawn from Crane and Upchurch (1987).

The size and complexity of these grains may indicate that their parent plants were insect-pollinated (Chapter 17).

Like that of more conventional ephedroid pollen, the diversity and abundance of elaterate grains also increased dramatically in palaeoequatorial regions in the Early Albian (Chapter 19), but elaterate grains are rare in post-Cenomanian sediments and disappeared from the fossil record around the Coniacian. They are best known mainly from West Africa and South America, but they are recorded from Peru, through Brazil and West Africa, to East Africa and New Guinea. They are characteristic of these areas at this time and give their name to the Elaterates Province (*sensu* Herngreen *et al.*, 1996).

5.3.3 Gnetalean mesofossils and macrofossils

Apart from dispersed pollen grains the fossil record of Gnetales is relatively sparse compared with that of other groups of living seed plants. However, it has recently been greatly expanded by new macro- and mesofossil evidence. The most reliable records are from the Cretaceous, where the presence of diverse and widespread Gnetales is now confirmed by excellent data from both macrofossils and mesofossils. Pre-Cretaceous macrofossil records of Gnetales are important (e.g. Crane 1988), but are less common, less well understood and less securely associated with the group.

Drewria potomacensis was among the first gnetalean macrofossils to be recognised, based on a collection of small macrofossils from the Drewry's Bluff Leaf Bed (Early to mid-Aptian) in the Potomac Group of eastern North America (Crane and Upchurch, 1987). The fossils are impressions and thin compressions. Cellular details are not preserved. *Drewria* is reconstructed as a herb or small shrub with slender stems, bearing pairs of opposite and decussate leaves at swollen nodes (Figure 5.8). Branching is monopodial and axillary. The pairs of leaves have sheathing bases. Each leaf is oblong with a distinctive pattern of reticulate venation. Two veins enter the base of the leaf and dichotomise almost immediately, resulting in four major veins, which run more or less parallel for the entire length of the leaf. In some leaves a further dichotomy close to the leaf base results in two additional veins, which run just inside the leaf margin. The longitudinal veins are linked into a reticulum by a series of cross-veins, which form distinctive inverted chevrons. The overall pattern of leaf venation is very similar to that in the cotyledons of extant *Welwitschia* (Figure 5.8) (Rodin, 1953).

Reproductive structures of *Drewria* are attached to the stems and consist of dichasia that are borne either terminally, or in the axil of a leaf. Each dichasium comprises three short, loose spikes. The two lateral spikes of the dichasia bear presumed seeds with associated small bracts. The precise arrangement of the seeds and the bracts is not clear, and the preservation is inadequate to yield cellular details or information on pollen *in situ*. Seeds do not appear to be present on the central spike, indicating that this perhaps bore pollen-producing structures, but again details and pollen grains are unknown. Dispersed pollen sacs, which occur in the same bed as the *Drewria* material, contain typical polyplicate pollen of the *Welwitschia*-type with an elongated colpus.

Dispersed gnetalean seeds have not been recovered from the Drewry's Bluff Leaf Bed but are known from other levels at the Drewry's Bluff locality. Small seeds of *Ephedra drewriensis* have been described from clay balls lower in the sequence (Rydin *et al.*, 2006a) and these are very similar to seeds of *Ephedra portugallica* from the Buarcos locality, Portugal (Rydin *et al.*, 2004, 2006a). Both species are also similar to seeds of *Ephedrispermum lusitanicum* from the Buarcos and Torres Vedras localities, Portugal. The Drewry's Bluff clay balls are of Early Aptian age, or perhaps slightly older. The mesofossil flora at the Buarcos locality is Late Aptian – Early Albian in age; the mesofossil assemblages from Torres Vedras are Late Barremian – Early Aptian.

Seeds of all three species are small, up to about 1.2–1.4 mm long, four-angled and with a pointed micropylar region. All have a sclerenchymatic seed envelope surrounding a thin integument. The integument is extended into a long micropylar tube that contains polyplicate ephedroid pollen. The outer sclerenchyma layer is extended to support the micropylar tube and is composed of cells with slightly wavy walls. The three species are distinguished mainly in details of their shape, pollen grains and sclerenchyma cells.

In *Ephedra drewriensis* and *E. portugallica* the sclerenchymatic seed envelope has a distinct papillate zone on its inner surface in the micropylar region. Polyplicate ephedroid pollen grains have been observed at the tip of the micropyle as well as inside it. The pollen grains inside the micropyles are distorted in a very characteristic way with striations of the pollen wall perpendicular to their long axis. This is very similar to what occurs in extant *Ephedra*, where the pollen wall is shed at pollen germination and curls up in a characteristic manner (El-Ghazaly and Rowley, 1997). According to El-Ghazaly and Rowley (1997) this is unique to *Ephedra*. Together with the characteristic papillae in the micropylar region, this indicates that plants very similar to modern *Ephedra*, and with its unique pollen germination, were already present in the Early Cretaceous (Rydin *et al.*, 2006a). *Ephedrispermum lusitanicum* differs from *Ephedra drewriensis* and *E. portugallica* in having much smaller pollen grains in the micropylar tube. It was not assigned to the extant genus because it apparently lacks the papillate zone on the seed envelope.

Several fossils from the Early Cretaceous Crato flora, Araripe Basin, Brazil, show opposite branching, reduced or absent leaves and compact cone-like reproductive structures, all features reminiscent of gnetalean plants (Dilcher *et al.*, 2005; Mohr *et al.*, 2007). Pollen assemblages from the Early Cretaceous of the Araripe Basin also contain significant amounts of ephedroid grains (Pons *et al.*, 1990; Osborn *et al.*, 1993). Fossil ephedroid pollen has also been reported *in situ* in gnetalean cones, but descriptions of this material have not yet been published (e.g. Mohr *et al.*, 2007). The gnetalean fossils from the Crato flora that have been formally described are all thought to be most closely related to extant *Welwitschia* than the other two extant genera. These include two different kinds of seedlings, *Cratonia cotyledon* and *Priscowelwitschia austroamericana*, as well as a reproductive shoot with putative pollen cones, *Welwitschiostrobus murili*. There are also isolated leaves, *Welwitschiophyllum brasiliense* (Dilcher *et al.*, 2005). These leaves are narrowly triangular to linear with more or less parallel primary veins and a thick, leathery lamina.

Cratonia cotyledon is an unusual plant (Rydin *et al.*, 2003) preserved as a seedling with two large, ovate cotyledons, about 4 cm long and up to about 2 cm wide. The cotyledons have an entire leaf margin and a distinct pattern of venation in which about 20 parallel primary veins are linked into a reticulum by secondary cross-veins that form distinctive inverted chevrons. The cotyledons are born on a root that is laterally branched and below the cotyledons is a thick feeder. Both the peculiar chevron cross-veins and the feeder

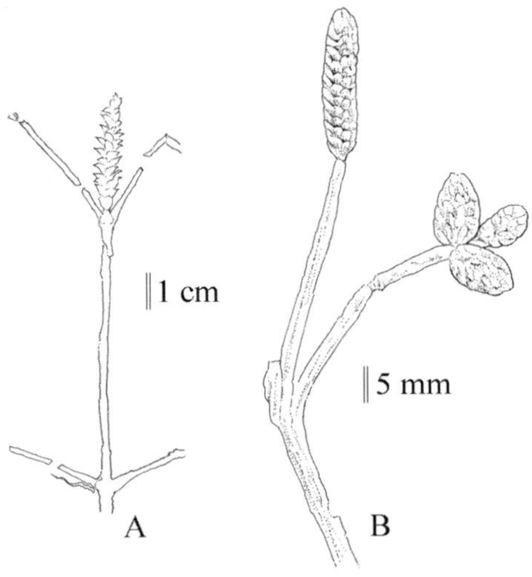

Figure 5.9 *Welwitschiostrobus murili*, gnetalean fossil from the Early Cretaceous (Late Aptian – Early Albian) Crato flora, Brazil. (A) Shoot showing axillary branches and a single terminal cone. (B) Shoot showing a single terminal cone and three cones on a lateral branch; bracts on the terminal cone are in an opposite–decussate arrangement. Drawn from photographs in Dilcher *et al.* (2005).

are characters that place *Cratonia* in the *Gnetum*-*Welwitschia* clade, probably on the *Welwitschia* lineage (Rydin *et al.*, 2003).

Priscowelwitschia (*Welwitschiella*) *austroamericana* (Dilcher *et al.*, 2005) is similar to *Cratonia cotyledon*. It also has two cotyledons and primary veins linked into a reticulum by chevron-shaped secondary veins. The cotyledons are narrower than those of *Cratonia* and the number of primary veins is smaller. The root is not preserved and whether this fossil is a seedling is less clear than for *Cratonia*.

Welwitschiostrobus murili (Figure 5.9) comprises terminal or axillary cones borne on slender axes that have a finely striate surface (Dilcher *et al.*, 2005). The cones are elongated, about 2 cm long and 0.65 cm wide, with bracts aligned in four rows and a four-angled cross-section that reflects the opposite and decussate arrangement of the bracts. There is also a distinct median keel on the bracts. There are no indications of antherophores or sporangia and no pollen was found. The cones were interpreted as pollen cones based on similarity between the fossils and the pollen cones of extant *Welwitschia*.

In addition to the discoveries from Brazil, diverse macrofossils of gnetalean affinity are also known from Central and eastern Asia, several of which were first reported as angiosperms. Re-evaluation of these fossils and studies of new material have resulted in the recognition of diverse assemblages of gnetalean fossils, primarily of ephedroid affinity (Tao and Zhang, 1992; Sun *et al.*, 2001; Tao and Yang, 2003; Yang *et al.*, 2005; Rydin *et al.*, 2006b; Rydin and Friis, 2010). Especially important are records from the Early Cretaceous (Early Aptian – early Late Aptian) Yixian Formation in northeastern China, but similar occurrences are also known from plant assemblages of roughly equivalent age in Mongolia. In both areas the preservation of the fossils provides relatively good morphological information, but there are no details of anatomical structure, epidermal features or pollen.

Gnetalean fossils from northeastern China and Mongolia include shoots with cone-like reproductive structures that have been described as *Alloephedra*, *Cyperacites* sp., *Eragrosites changii*, *Erenia stenoptera*, *Liaoxia cheniae*, *Potamogeton*-like, and *Sparganium*-like, as well as fossils with winged seeds described as *Chaoyangia liangii* and *Gurvanella dictyoptera* (Krassilov, 1982; Cao *et al.*, 1998; Duan, 1998; Tao and Yang, 2003). The Yixian Formation alone includes ten or more different species that are thought to be closely related to *Ephedra*, although in some instances species limits are not clear.

The relationship of *Ephedra archaeorhytidosperma* from the Yixian Formation to extant *Ephedra* is especially secure. The fossils have distinctive ephedroid vegetative morphology and small seed cones with rugulate seeds. These seeds are very similar to those of extant *Ephedra rhytidosperma* (Yang *et al.*, 2005) and fossil *Eoantha zherikhinii* (see below). This characteristic rugulate (wrinkled) seed surface also occurs in other seeds assigned to the BEG group such as *Rugonella*, the unnamed square seeds (section 5.5) and *Erdtmanispermum* (section 5.4). *Problematospermum* also has the same characteristic surface and may also belong to the BEG group (section 5.5).

The genus *Liaoxia* was also established from the Yixian Formation based on shoots with attached reproductive structures (Cao *et al.*, 1997, 1998). The type species, *Liaoxia cheniae*, was described originally as a monocot. It has slender, profusely branched stems. Branching is monopodial and axillary, and the leaves are attached with opposite and decussate phyllotaxis (Figure 5.10). Leaves are narrow and elongated, with entire margins and a thin lamina. Leaf venation consists of two to three longitudinal and parallel first-order veins. Reproductive structures are terminal or axial spikes consisting of two to six pairs of opposite and decussate sterile bracts.

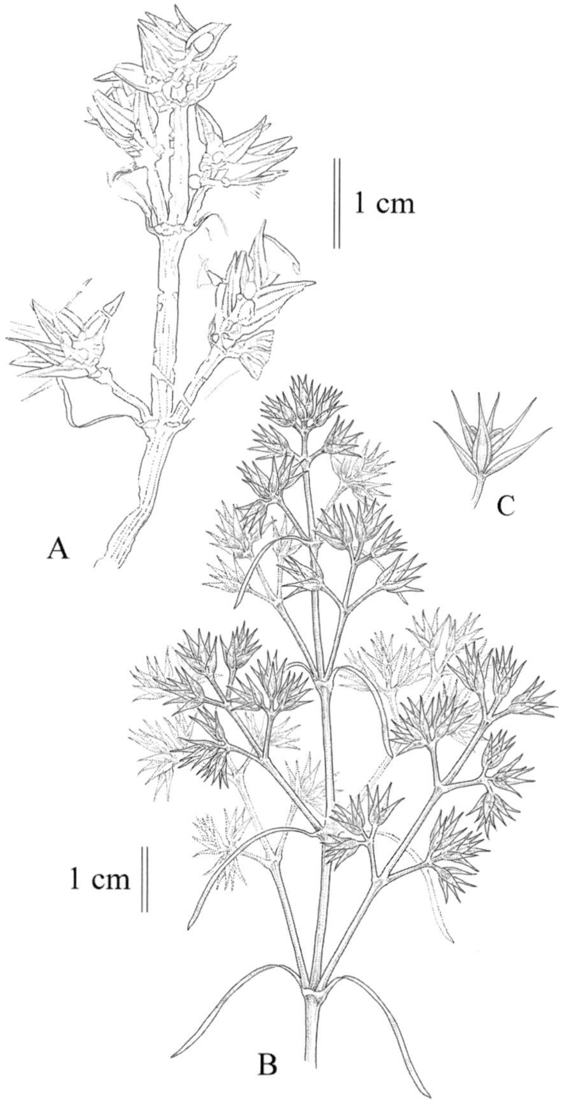

Figure 5.10 *Liaoxia cheniae*, an ephedroid fossil from the Early Cretaceous Yixian Formation (Early Aptian – early Late Aptian), Liaoning, northeastern China. (A) Reproductive shoot showing opposite leaves and opposite axillary branches bearing clusters of seeds. (B, C) Reconstruction of the *Liaoxia* plant; overview (B) and detail of cone (C). From Rydin *et al.* (2006b).

Re-examination of *Liaoxia cheniae* and new material from the Yixian Formation by Rydin *et al.* (2006b) suggests that the reproductive structures are ovulate and also shows that *Liaoxia* shares several important features with extant and fossil Gnetales. In particular, *Liaoxia* is similar to species of extant *Ephedra* in several respects. This is also consistent with previous suggestions (Guo and Wu, 2000; Sun *et al.*, 2001) that had transferred the species to the genus *Ephedrites*. *Ephedrites*, however, is not a suitable name for ephedroid fossils since it was established for a Cenozoic plant of probable loranthaceous affinity (Rydin *et al.*, 2006b). Later authors transferred *Liaoxia cheniae* to the extant genus *Ephedra* (Liu *et al.*, 2008), but critical features of *Ephedra*, such as an inner papillate lining of the seed envelope and polyplicate pollen, are not known for the fossil. Plants of *Liaoxia cheniae* appear similar in habit to *Drewria potomacensis* from the Drewry's Bluff Leaf Bed (see above), but the leaves are narrower with a different pattern of venation.

Another plant from the Yixian Formation, China, originally described as *Eragrosites changii* and compared to extant Poaceae (Cao *et al.*, 1997, 1998), is very similar to *Liaoxia cheniae*. It was included in the same genus under the name *Liaoxia changii* by Rydin *et al.* (2006b). It apparently has leafless stems. The reproductive structures are cone-like and may be pedunculate or sessile. The fossil is poorly preserved, but shows distinct nodes and an opposite–decussate arrangement of the bracts in the reproductive structures. There are no traces of seeds or sporangia and it is unknown whether the cones were pollen-producing or ovule-bearing (Rydin *et al.*, 2006b).

Several other species of *Liaoxia* (*L. acutiformis*, *L. elongata*, *L. longibractea* and *L. robusta*) have also been described from the Yixian Formation (Rydin *et al.*, 2006b). The species are separated by differences in cone size and shape as well as leaf and bract morphology. Similar species reported from the Yixian Formation (Sun *et al.*, 2001) have been assigned to *Ephedrites* (*E. antiquus*, *E. guozhongiana* and *E? elegans*). Some of these specimens are ovulate, whereas others apparently bore pollen-producing structures. It is also likely that some are male and female plants of the same fossil species.

A further possible ephedroid macrofossil is *Alloephedra xingxuei* from the Early Cretaceous Dalazi Formation of Jilin Province, northeastern China (Tao and Yang, 2003). *Alloephedra* was described as a small, profusely branched herb or shrub, with distinct nodes and opposite–decussate branching. Leaves are small and membranous. Cones were described as ovulate.

Several fossil plants described from the Early Cretaceous of Mongolia, such as *Cyperacites* sp., *Potamogeton*-like fossil and *Sparganium*-like fossil (Krassilov, 1982), show the same jointed stems and condensed spike-like cones seen in the material from the Yixian Formation and are probably also of ephedroid relationship. *Vitimantha crypta*, from the Early Cretaceous of Transbaikalia, may also be of ephedroid affinity. *Vitimantha* consists of striate stems with bracts and perhaps reproductive structures crowded at the node. Polyplicate ephedroid pollen is attached to the surface of some of these structures, but it is uncertain whether they were produced by the same plant (Krassilov and Bugdaeva, 2000).

Eoantha zherikhinii was described based on two isolated, compressed and fragmentary ovulate structures from the Early Cretaceous (Barremian–Aptian) Lake Baikal area of Transbaikalia (Krassilov, 1986). Three further specimens were described later from the same locality, one attached to a narrow, striate stem (Krassilov and Bugdaeva, 2000). Because of the fragmentary and strongly compressed nature of the fossils the organisation of the *Eoantha* reproductive structure is not completely clear. However, individual reproductive units are apparently arranged at the nodes and one specimen that is compressed from the apex shows four seeds radiating from the central axis in a whorled or decussate arrangement (Figure 5.11). The seeds are surrounded by long narrow bracts.

Seeds of *Eoantha* are small, about 2.2 mm long, broadly ovoid, apparently angular in cross-section and with a pointed apex. The seed surface is wrinkled with irregular transverse ridges. Maceration of the seeds shows that they are orthotropous with a distinct megaspore membrane. The integument is extended into a long micropyle. Several polyplicate, *Ephedripites*-type, pollen grains are enclosed at the base of the micropylar tube at the apex of the nucellus (Krassilov, 1986).

Seeds of *Eoantha* are very similar to those of some *Ephedra* species, such as extant *Ephedra rhytidosperma* and fossil *Ephedra archaeorhytidosperma*. They are also similar to several unassigned BEG-type seeds (section 5.5) such as *Rugonella* and the unnamed square seeds. In these seeds it is the sclerenchyma cells of the outer seed envelope that form the wrinkled surface and enclose the integument. The presence of a distinct, relatively well-developed megaspore membrane distinguishes *Eoantha* from extant Gnetales, but a megaspore membrane is also present in seeds of the putatively related Erdtmanithecales (section 5.4). Co-occuring with *Eoantha*, and identified as the possible vegetative parts of the *Eoantha* plant (Krassilov and Bugdaeva, 2000), are small vegetative axes of ephedroid appearance, but with very long leaves, described as *Praeherba spathulate*.

Baisianthus ramosus is another putative ephedroid plant from the Lake Baikal area. It co-occurs with *Eoantha*

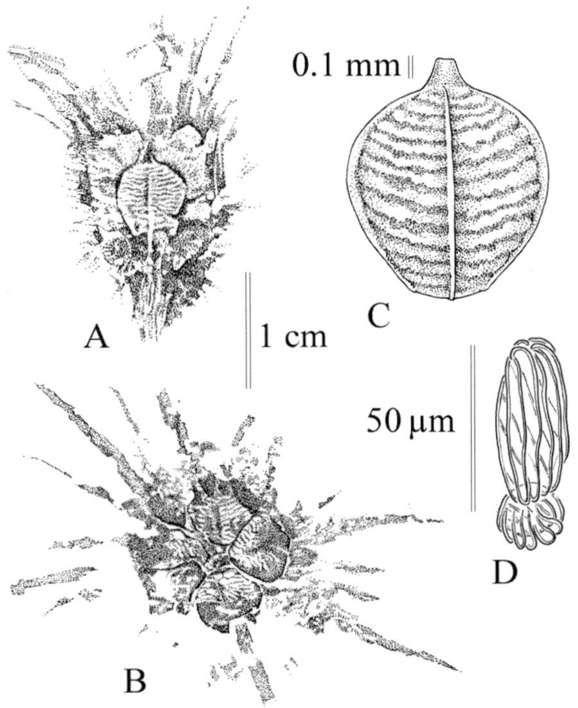

Figure 5.11 *Eoantha zherikhinii*, an ephedroid plant from the Early Cretaceous (Barremian–Aptian) of Transbaikalia, Russia; compressed reproductive structures showing a possible whorl of rugulate seeds in (A) lateral and (B) apical view; reconstruction of (C) seed and (D) polyplicate pollen grains found inside the integument. Drawn from photographs and line drawings in Krassilov (1986).

(Krassilov and Bugdaeva, 2000). *Baisianthus* has slender, striate stems with opposite–decussate phyllotaxis and leaves borne at the nodes separated by long internodes. The leaves are long and narrow. The reproductive structures are small axial spikes that are borne in whorls at the nodes. They consist of bracts supporting clusters of sporangia that are sometimes separate or more usually organised in two to four bisporangiate synangia. Pollen is monocolpate with a thick pollen wall that has a solid tectum, a granular infratectal layer, no foot layer and a thick endexine (Tekleva and Krassilov, 2009). The preservation of the fossils does not allow detailed reconstruction of the reproductive structure and how it is organised. Together with the monocolpate, non-plicate nature of pollen, this leaves some uncertainty as to the exact systematic position of *Baisianthus*.

Most of the probable gnetalean macrofossils from eastern Asia have vegetative features resembling those of extant *Ephedra*. However, since the reproductive structures are

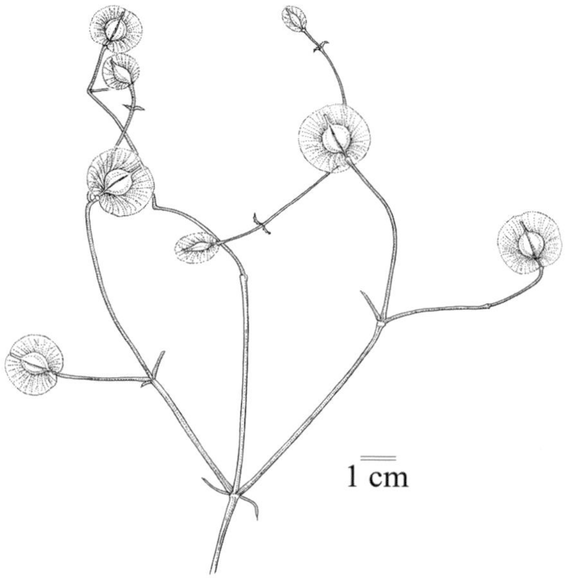

Figure 5.12 Reconstruction of the *Gurvanella* (*Chaoyangia*) plant from the Early Cretaceous Yixian Formation (Early Aptian – early Late Aptian), Liaoning, northeastern China, showing opposite leaves and axillary branches with terminal winged seeds. Based on Duan (1998) and personal observations.

not known in detail it is possible that some of them are more closely related to *Welwitschia*, and have perhaps retained some ephedroid features, as is apparently the case for some of the Crato fossils (e.g. *Welwitschiostrobus murili*). The fossil *Gurvanella* (*Chaoyangia*) may be another example. It appears to combine possible welwitschioid reproductive structures with ephedroid vegetative morphology.

Gurvanella was first described from Mongolia based on dispersed winged seeds assigned to *Gurvanella dictyoptera* (Krassilov, 1982). Identical seeds were described subsequently as *Chaoyangia liangii* (Figure 5.12) from the Yixian Formation, where they occur both isolated and also attached to vegetative shoots (Duan, 1998). The Yixian fossils were interpreted as an angiosperm, but the shoots have distinct ephedroid features with jointed stems, slender branches and a branching pattern very similar to that of *Liaoxia* and *Ephedra*. The stems are finely striate and bear elongated narrow leaves in an opposite–decussate arrangement.

Reproductive structures of *Gurvanella* were interpreted by Duan (1998) as tricarpellate hairy fruits with long styles. However, even though the reproductive structures are not well-preserved there is no evidence of carpels. The reproductive structures appear to be winged seeds with anastomosing veins running outwards, up to the margin of the wing. The seeds resemble those of extant *Welwitschia*, in which the seed, with its thin integument and long micropylar tube, is surrounded by a membranous wing formed by a pair of bracts. The tip of the integument projects between the bracts at the apex of the wing. Another enigmatic fossil, *Erenia stenoptera*, first described from the Early Cretaceous of Mongolia (Krassilov, 1982) and also occurring in the Yixian Formation, may belong to the same complex of plants as *Gurvanella*.

Other macrofossils from the mid-Cretaceous that may be attributable to the Gnetales have been described, but all remain to be investigated in detail. Some specimens of *Conospermites hakeaefolius* from the Cenomanian of Czech Republic (Velenovský and Viniklář, 1926; see Crane, 1988) resemble leaves of *Drewria* in shape and pattern of venation, while *Cyperacites potomacensis* (Berry, 1911) resembles *Drewria* in the dichasial arrangement of the reproductive structures. *Casuarina covilli* (Ward, 1895; Crane, 1988) from the Potomac Group has a whorled arrangement of parts. *Phyllotheca wonthaggiensis* from the Early Cretaceous Koonwarra Flora of Victoria, Australia, is currently poorly known (Drinnan and Chambers, 1986; Krassilov *et al.*, 1998), but also deserves further study as a potential gnetalean.

5.4 ERDTMANITHECALES

The order Erdtmanithecales was established to accommodate fossil material linked to the widespread and distinctive genus of Mesozoic pollen grains *Eucommiidites* (Friis and Pedersen, 1996). Subsequent work has expanded previous knowledge of the seeds of the group and three genera of pollen organs (*Bayeritheca*, *Erdtmanitheca*, *Eucommiitheca*) have been described. Phylogenetic analyses suggest that the Erdtmanithecales are part of a group that also includes Bennettitales and Gnetales (section 5.2).

The pollen genus *Eucommiidites* was first described from the Early Jurassic of Scania, southern Sweden (Erdtman, 1948). It is now known to be widespread in Triassic to Lower Cretaceous rocks throughout the Northern Hemisphere (Pedersen *et al.*, 1989b). Based on the presence of three elongated apertures, Erdtman (1948) initially interpreted *Eucommiidites* as a tricolpate angiosperm pollen grain. However, later it became clear that the unequal development and positioning of the apertures reflected bilateral symmetry (Kuyl *et al.*, 1955; Couper, 1956). *Eucommiidites* was then reinterpreted as a gymnosperm

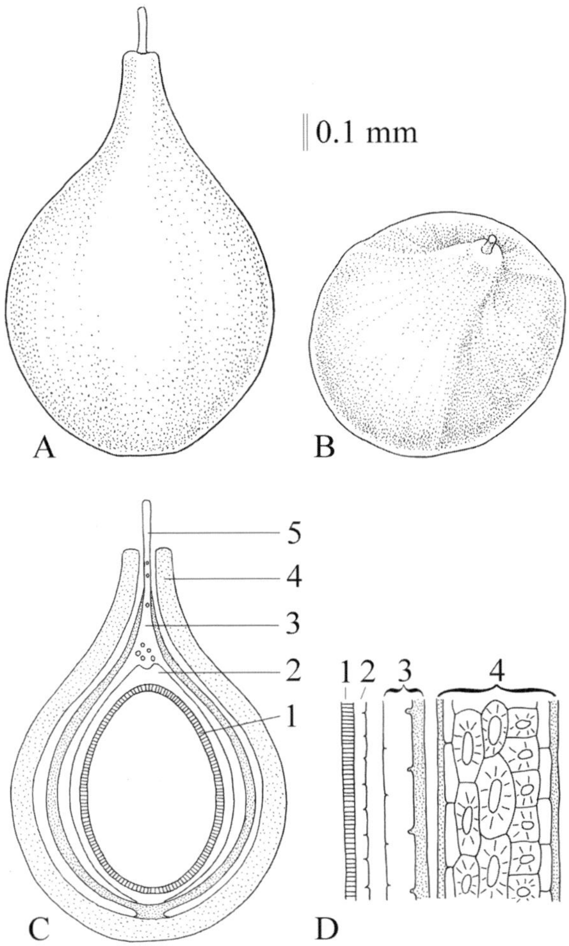

Figure 5.13 Seeds of *Erdtmanispermum balticum* (Erdtmanithecales) from the Early Cretaceous (Late Berriasian – Valanginian) Carl Nielsen A/S locality, Bornholm, Denmark; schematic drawing of seeds showing (A, B) external morphology, (C) internal structures and (D) anatomy of the seed wall highlighting the cuticles of the various layers. 1, Megaspore membrane; 2, nucellus; 3, integument (cuticles of inner and outer epidermis indicated); 4, seed envelope with sclerenchyma layer; 5, micropylar tube (extension of integument).

grain with a single distal colpus flanked by two subsidiary lateral colpi (Couper, 1956, 1958). The gymnosperm rather than angiosperm status of *Eucommiidites* pollen has since been confirmed by its presence in the micropyles of Early Jurassic and Early Cretaceous seeds (Hughes, 1961b; Brenner, 1963, 1967; Reymanówna, 1968; Pedersen *et al.*, 1989b). A typical, well-developed, laminate endexine (characteristic of pollen of non-angiosperm seed plants) has also been documented by TEM studies for both *in situ* and isolated *Eucommiidites* pollen grains (e.g. Pedersen *et al.*, 1989b; Batten and Dutta, 1997).

The most completely understood of the *Eucommiidites*-containing seeds is *Erdtmanispermum balticum* (Figures 5.13, 5.16), which is based on large numbers of small, dispersed seeds recovered from the Early Cretaceous (Late Berriasian – Valanginian) of Bornholm, Denmark (Pedersen *et al.*, 1989b). The seeds are ovoid with a distinct attachment scar at the base, but the structure on which they were borne is unknown. They are weakly triangular in cross-section and gradually taper into an elongated apical tip. The nucellus apparently has a thin cuticle and internally there are remains of a granular megaspore membrane. The nucellus is surrounded by a thin integument, which is extended at the apex into a long, narrow micropylar tube. The integument is enclosed in a thicker seed envelope of sclerified cells. This envelope gives the seed its weakly triangular shape and also shows a tendency to split into three valves. Typical *Eucommiidites* pollen occurs consistently in the micropyles of *Erdtmanispermum balticum* seeds (Pedersen *et al.*, 1989b). The pollen has a smooth outer surface with a foveolate tectum, a granular infratectal layer and a thick foot layer. The endexine is thick and laminate.

Seeds of *Erdtmanispermum* (*E. juncalense*) have also been described from the earliest Cretaceous (probably Berriasian) of Portugal (Mendes *et al.*, 2008a). Pollen has not been observed in the micropyle and they are distinguished from the Bornholm seeds by their slightly more rugulate surface. Other seeds linked to *Eucommiidites* pollen are *Allicospermum retimirum* from the Early Jurassic of Poland (Reymanówna, 1968), and Early Cretaceous seeds assigned to *Spermatites pettensis* from southern England (Hughes, 1961b) and to *Spermatites patuxensis* from the Potomac Group of eastern North America (Brenner, 1967). All have *Eucommiidites* pollen in the micropyle; although knowledge of these seeds is less detailed than for *Erdtmanispermum*, they also have an integument extended into a long, narrow micropylar tube and an outer seed envelope.

Pollen-producing structures with *Eucommiidites* pollen *in situ* have been described from the mid-Cretaceous of Texas, USA (Pedersen *et al.*, 1989b), the Early Cretaceous of Portugal (Friis and Pedersen, 1996; Mendes *et al.*, 2010) and the Late Cretaceous of Bohemia in the Czech Republic (Kvaček and Pacltová, 2001). They are assigned to three extinct genera, *Erdtmanitheca*, *Eucommiitheca* and *Bayeritheca*.

Erdtmanitheca texensis (Figures 5.14, 5.15) from the Late Cretaceous (Early Cenomanian) Arthur's Bluff locality,

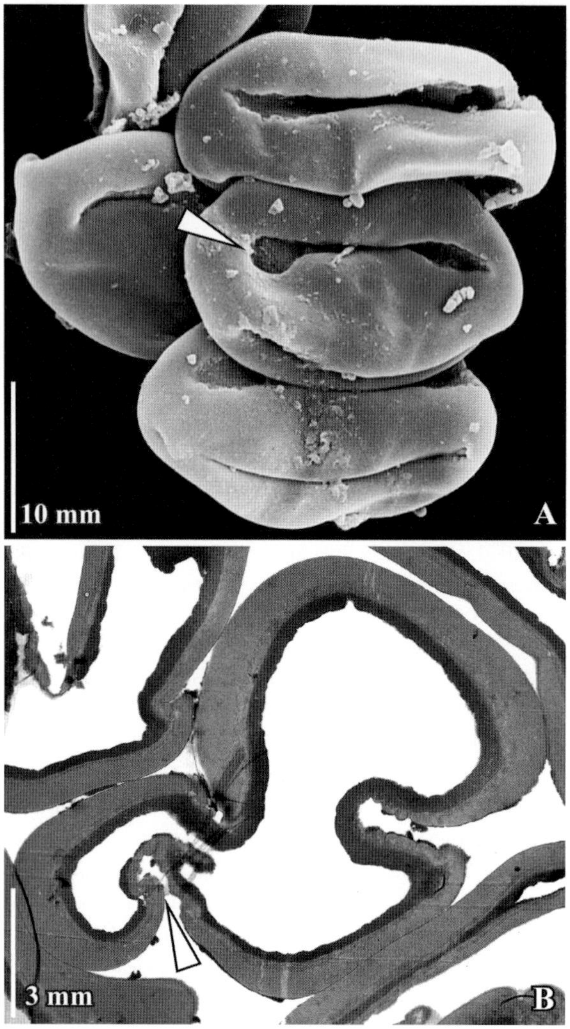

Figure 5.14 *Eucommiidites*-type pollen found *in situ* in pollen organs. (A) SEM image of several pollen grains of *Erdtmanitheca texensis* from the Late Cretaceous (Cenomanian) of Texas, USA, showing main colpus (arrow head) and lateral furrows. (B) TEM image of *Erdtmanitheca portucalensis* from the Early Cretaceous (Late Aptian – Early Albian) of Portugal showing thick ektexine, granular infratectal layer and thick, darkly staining endexine; main colpus indicated by arrow head. (A) From Pedersen *et al.* (1989b); (B) from Mendes *et al.* (2010).

Texas, USA (Pedersen *et al.*, 1989b), is a more or less spherical cone-like structure consisting of numerous (*c.* 300) radiating and densely crowded pollen-producing units. Each unit consists of a peltate, hexagonal head with about ten long and narrow sporangia below that contain *Eucommiidites* pollen. It is unknown how the sporangia are arranged or borne, but they appear to be parallel to the presumed stalk of each unit, and perhaps longitudinally fused into some kind of synangiate structure. Pollen grains are of the typical *Eucommiidites* type with a solid tectum and granular infratectal layer. The foot layer is extremely thin or lacking.

Erdtmanitheca portucalensis from the Early Cretaceous (Late Aptian – Early Albian) of Portugal (Mendes *et al.*, 2010) is a pollen organ very similar to *E. texensis*. It has the same spherical cone-like shape with densely crowded pollen-producing units radiating from a central receptacle (Figure 5.15). There are about ten sporangia per unit and in this case the sporangia are clearly synangiate and open by longitudinal slits. *E. portucalensis* is distinguished from *E. texensis* mainly in having fewer and more loosely arranged pollen-producing units (*c.* 100–150) and the pollen grains have a distinct foot layer and prominent endexine (Figure 5.14).

Eucommiitheca hirsuta, also from the Early Cretaceous (Late Aptian–Early Albian) of Portgual comprises fragments of pollen organs as well as dispersed pollen-producing units (Friis and Pedersen, 1996). The stalks of the pollen-producing units, and the axes to which they are attached, are densely hairy. The units are arranged in an opposite–decussate pattern along the main axis. Each consists of a short stalk with a hexagonal peltate head and a whorl of numerous paired sporangia beneath. The sporangia are arranged around the distal part of the stalk in two whorls (Figure 5.15). Each sporangium dehisces by a longitudinal slit and the dehiscence slits of each sporangium in a pair are aligned facing each other. The sporangia contain characteristic *Eucommiidites* pollen.

Eucommiitheca hirsuta is very similar to *Sturiella langeri*, which was interpreted originally as a very small, perhaps bisexual, bennettitalean flower based on a single specimen (now missing) from the Late Triassic of Lunz, Austria (Kräusel, 1948; Crane, 1988). *Sturiella* consists of an axis bearing approximately 20 lateral units. Each unit consists of a flattened, sub-hemispherical head. Each of these units was interpreted by Kräusel (1948) as a mass of interseminal scales with a peripheral 'perianth' of 25–30 bracts. Some of the 'bracts' were thought to have a distinct bulge on their inner surface, which Kräusel (1948) interpreted as a sporangium, although no pollen was isolated. In light of the structure of *Eucommiitheca hirsuta* the 'bracts' seem more likely to be equivalent to the sporangia of *Eucommiitheca hirsuta* (Friis and Pedersen, 1996).

Bayeritheca hughesii from the Late Cretaceous (Middle Cenomanian) of Bohemia, Czech Republic (Kvaček and Pacltová, 2001), is an elongated cone-like structure bearing

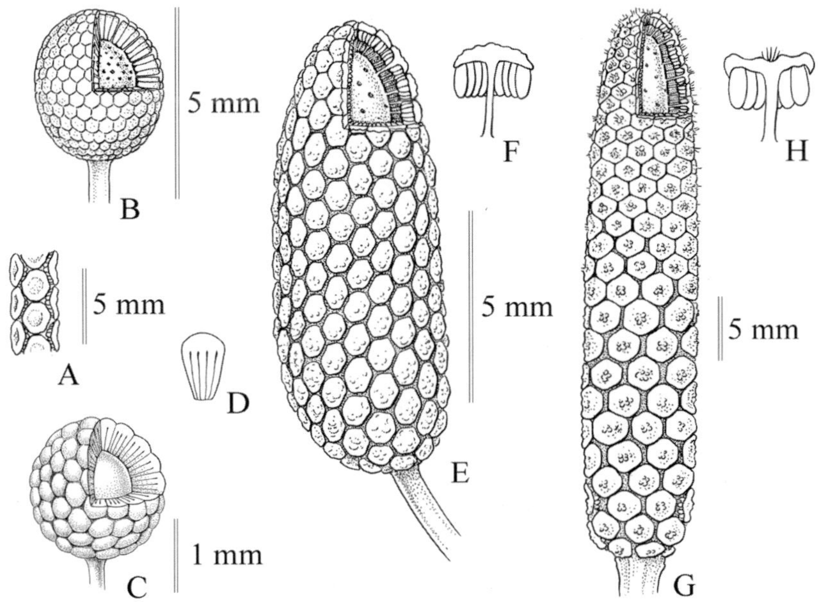

Figure 5.15 Reconstructions of pollen organs of Erdtmanithecales. (A) *Eucommiitheca hirsuta* from the Early Cretaceous (Late Aptian – Early Albian) of Portugal with sporangia borne on pollen-producing units that are opposite–decussate on an elongated reproductive axis. (B) *Erdtmanitheca texensis* from the Late Cretaceous (Cenomanian) of Texas, USA, with sporangia borne on pollen-producing units in a spherical head. (C, D) *Erdtmanitheca portucalensis* from the Early Cretaceous (Late Aptian – Early Albian) of Portugal with sporangia borne in a spherical head (C) with loosely arranged pollen-producing units, each unit peltate with several sporangia (D). (E, F) Undescribed pollen organ of probable Erdtmanithecales from the Early Cretaceous Yixian Formation (Early Aptian – early Late Aptian) of Liaoning, northeastern China, showing (E) elongated reproductive axis and (F) peltate pollen-producing units. (G, H) *Bayeritheca hughesii* from the Late Cretaceous (Cenomanian) of the Czech Republic with sporangia borne in an elongated reproductive axis (G) with densely clustered, peltate pollen-producing units (H). (A) Based on Friis and Pedersen (1996); (B) based on Pedersen *et al.* (1989b); (C, D) based on Mendes *et al.* (2010); (E, F) based on unpublished material in the Institute of Geology and Palaeontology, Nanjing; (G, H) based on Kvaček and Pacltová (2001).

about 180–200 lateral pollen-producing units attached to a massive main axis (Figure 5.15). The lateral units are apparently arranged in whorls. Each unit consists of a stalk with a roughly hexagonal peltate head with several pollen sacs on their underside. These are interpreted as united into synangia. The form of the synangia, and the number of pollen sacs they contain, is unknown. Each pollen sac contains masses of *Eucommiidites* pollen. Stomata are described as paracytic. Macerations also reveal the presence of resin bodies in the tissues of the lateral units.

The general organisation of the lateral pollen-producing units in *Bayeritheca*, *Erdtmanitheca* and *Eucommiitheca* also occurs in several other fossils from the Mesozoic and Permian. Fossils similar to *Bayeritheca* are known from the Early Cretaceous of northeastern China (Figure 5.15; Zhou Zhiyan, personal communication), but have no pollen preserved. *Aegianthus sibiricus*, from the Jurassic of Ust-Balej, East Siberia, comprises similar pollen organs with lateral units that have peltate, polygonal heads and pendulous pollen sacs on their lower surface. *Aegianthus* was originally described as *Kaidacarpum sibiricum* by Heer (1876), and assigned to Gnetales by Krassilov and Bugdaeva (1988a). However, stomata are anomocytic. Pollen from the pollen sacs is monocolpate, and also lacks the characteristic features of ephedroid or *Eucommiidites* grains. Krassilov and Bugdaeva (1988a) linked *Aegianthus* with the associated seeds of *Heerala*, bracts of *Angarolepis* and dispersed leaves of *Cadmisega*. *Kendostrobus*, from the Permian glossopterid-dominated fossil assemblage of India, presents a similar situation. It resembles *Bayeritheca*, but the pollen is monocolpate with faint striae (Surange and Chandra, 1974a, 1975).

A systematic position for Erdtmanithecales close to Bennettitales and Gnetales has been suggested for *Eucommiidites*-producing plants (e.g. Hughes, 1961b;

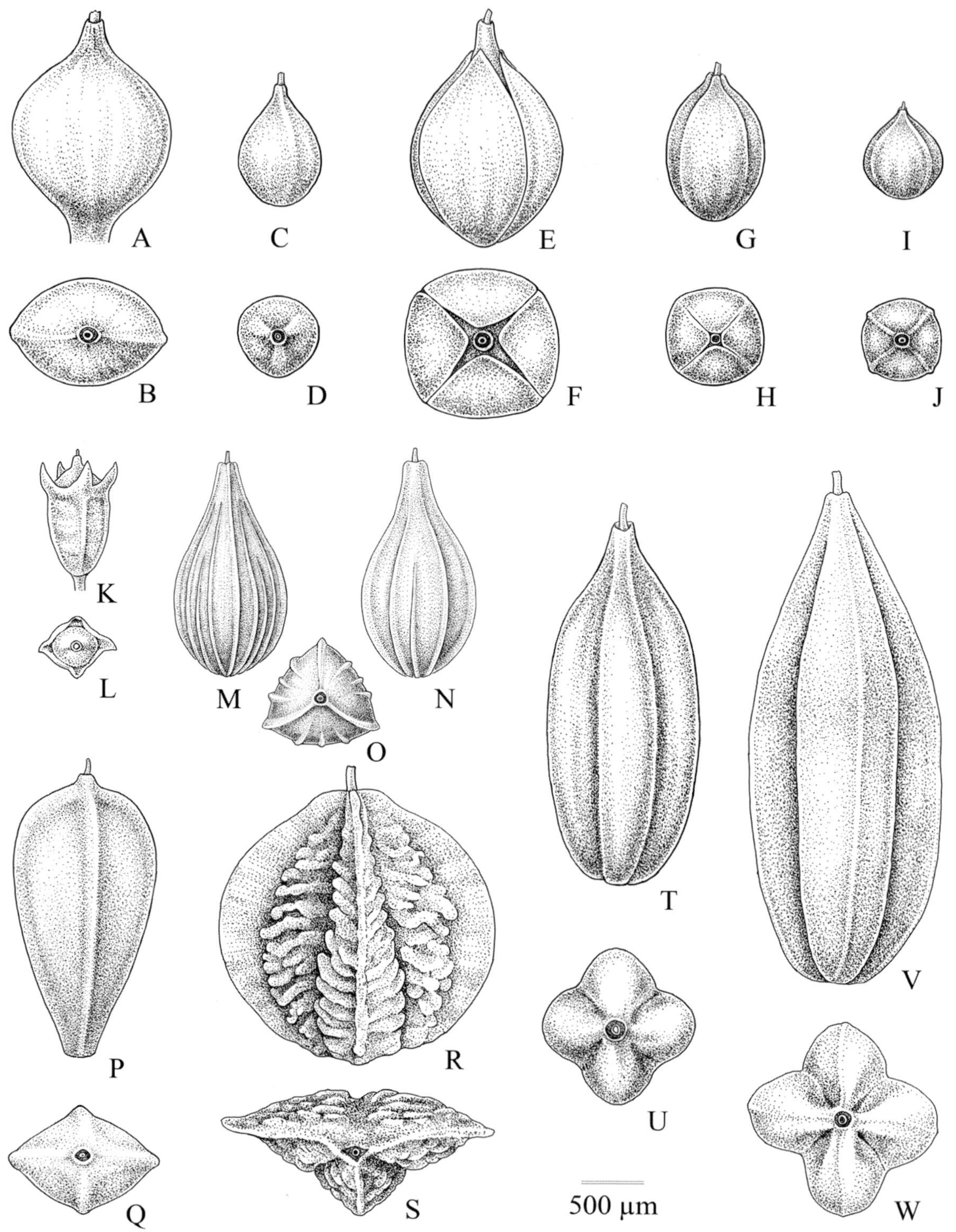

Figure 5.16 Reconstructions of BEG-type seeds from the Early Cretaceous of Denmark, Portugal and eastern North America, in lateral and apical view. (A, B) *Raunsgaardispermum lusitanicum*, Juncal/Vale Painho locality, Portugal. (C, D) *Erdtmanispermum*

Friis and Pedersen, 1996; Friis *et al.*, 2007). Similarities with Gnetales and Bennetittales include the granular infratectal layer in the pollen wall and the presence of an additional seed envelope surrounding the nucellus and integument. There is also the possible homology between the two lateral colpi of *Eucommiidites* and the grooves between the longitudinal ridges in pollen of *Ephedra* and *Welwitschia*. The relative simplicity of the pollen-producing units, and their occurrence in *Eucommiitheca hirsuta* in opposite–decussate pairs, may also be significant similarities with Gnetales.

5.5 UNASSIGNED DISPERSED SEEDS OF THE BEG GROUP

In addition to dispersed seeds that can clearly be assigned to Gnetales or Erdtmanithecales, mesofossil floras from the Early Cretaceous of Portugal and eastern North America have yielded many other seeds that are identical in their general organisation (Figure 5.16). None of these seeds has been linked to other reproductive structures and only *Raunsgaardispermum* has pollen preserved in the micropyle. Nevertheless, based on their distinctive structure these seeds are thought to have been produced by plants phylogenetically closely related to Gnetales, Erdtmanithecales and Bennettitales (Friis *et al.*, 2009a). Seven species of these dispersed seeds have been described and attributed to five extinct genera, *Buarcospermum*, *Lignierispermum*, *Lobospermum*, *Raunsgaardispermum* and *Rugonella* (Mendes *et al.*, 2008b; Friis *et al.*, 2009a). Several others remain to be formally described and named, including the characteristic square seeds illustrated by Friis *et al.* (2007). There are also dispersed seeds from compression floras that are probably related to the BEG group. *Raunsgaardispermum* is from the earliest Early Cretaceous (probably Berriasian) and there are several different seeds from the earliest Cretaceous of Bornholm (Late Berriasian – Valanginian), but most of the unassigned dispersed seeds are from the later part of the Early Cretaceous (Late Barremian – Middle Albian) where they co-occur with angiosperm fossils of various kinds (Friis *et al.*, 2007).

All of the dispersed seeds that we treat in this section are united with each other, as well as with seeds of Gnetales, Erdtmanithecales and Bennettitales, by similarities in morphology, organisation and anatomy. All have a thin integument surrounded by an outer seed envelope. The integument is extended apically into a long, narrow micropylar tube that forms the micropylar canal. At the tip the micropylar canal is open and the micropylar tube consists of a single layer of cells formed by its outer epidermis. Slightly further down there is an additional cell layer, the inner epidermis, and in all specimens the micropylar canal becomes closed in the middle. The most common closing mechanism is by cells that enlarge radially toward the centre of the canal. The seeds are distinguished from each other by differences in symmetry and details of the seed wall. Some species are more like the seeds of ephedroid gnetaleans (e.g. *Rugonella*, unnamed square seeds); others are more like the seeds of Bennettitales and *Gnetum* (e.g. *Buarcospermum*, *Lignierispermum*).

Buarcospermum tetragonium is based on small seeds from the Buarcos, Catefica and Torres Vedras localities, Portugal, as well as from the Puddledock locality, Virginia, USA (Friis *et al.*, 2009a). The seeds are broadly ovoid with a pointed micropylar region. They are up to about 2.2 mm long, radially symmetrical and four-angled in cross-section with multicellular closure of the micropyle (Figures 5.16–5.18). The seed envelope is composed mainly of chevron-shaped sclerenchyma cells. Four longitudinal, wing-like crests radiate from the inner part of the seed envelope to the surface.

Seeds of *Lignierispermum maroneae* from the Puddledock locality, Virginia, USA, and from the Catefica locality, Portugal, are also four-angled in cross-section. The seed envelope has an inner layer of transversely aligned fibre cells that form four narrow, wing-like crests. These are most pronounced near the apex, but also extend to the base. In the apical region the areas between the crests are filled by larger cells that radiate from the crest towards the outside (Figures 5.16–5.18).

Caption for Figure 5.16 (*cont.*)
balticum, Carl Nielsen A/S locality, Bornholm, Denmark. (E, F) *Buarcospermum tetragonium*, Buarcos locality, Portugal. (G, H) *Ephedra drewriensis*, Drewry's Bluff locality, Virginia, USA. (I, J) *Ephedrispermum lusitanicum*, Torres Vedras locality, Portugal. (K, L) 'Square seeds', Torres Vedras locality, Portugal. (M–O) Undescribed triangular seed, Puddledock locality, Virginia, USA. (P, Q) *Lignierispermum maroneae*, Puddledock locality, Virginia, USA. (R, S) *Rugonella trigonospermum*, Puddledock locality, Virginia, USA. (T, U) *Lobospermum stampanonii*, Puddledock locality, Virginia, USA. (V, W) *Lobospermum glabrum*, Famalicão locality, Portugal. From Friis *et al.* (2009a).

Figure 5.17 SEM micrographs and SXRTM images of BEG-type seeds from (A–D, I) the Early Cretaceous of Portugal and (E–H, J) the Potomac Group, USA. (A–D) *Buarcospermum tetragonium* in (A) lateral and (B) apical view and cross-section through

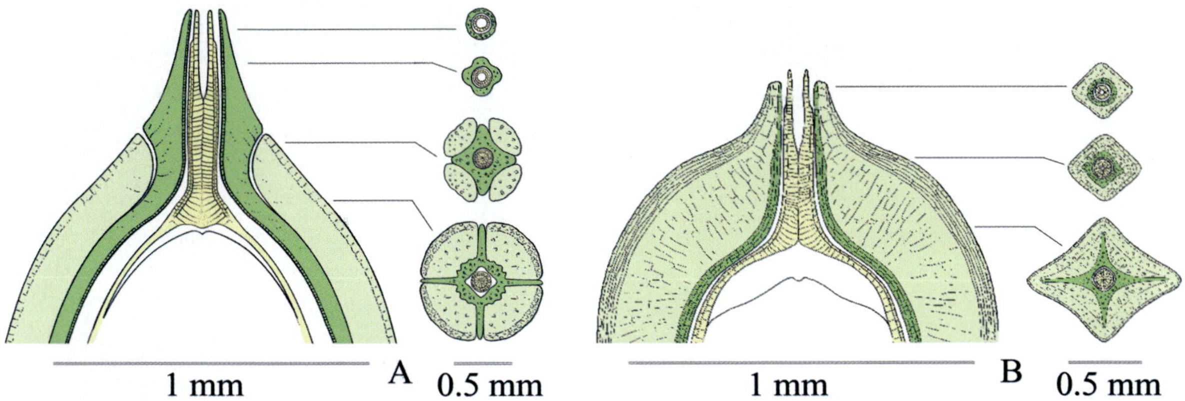

Figure 5.18 Schematic longitudinal and cross-sections of (A) *Buarcospermum tetragonium* and (B) *Lignierispermum maroneae* showing integument (yellow) with closure of long micropylar tube as well as inner and outer tissues of seed envelope (darker and lighter green). Modified from Friis *et al.* (2009a).

Rugonella trigonospermum is currently known only from the Puddledock locality, Virginia, USA (Friis *et al.*, 2009a). The seeds are small, up to about 2.5 mm long, slightly winged and with a triangular-lobed cross-section. The envelope is rugose and transversely ribbed (Figures 5.16, 5.17).

Lobospermum includes three species, *L. stampanonii*, *L. glabrum* and *L. rugosum*, all from either Early Cretaceous localities in Portugal or the Puddledock locality, Virginia, USA (Friis *et al.*, 2009a). The seeds are up to about 3.8 mm long and distinctly four-lobed in cross-section (Figure 5.17). The seed envelope has an inner layer of long, chevron-shaped fibres and an outer layer of longitudinally aligned sclerenchyma cells. The outer surface of the seed envelope is smooth in *Lobospermum stampanonii* and *L. glabrum*, but irregularly rugose in *L. rugosum*.

A group of distinctive four-angled, rugulose seeds (Figure 5.19), which occur in all of the Early Cretaceous mesofossils floras from Portugal and in many of the Early Cretaceous mesofossil floras from the Potomac Group of eastern North America, still remain to be formally described. The seeds are small, and vary from about 0.5 mm to about 1.8 mm in length. Some of the seeds have four, pointed, tepal-like structures extending apically from their four corners, whereas others lack these projections. Remains of a stalk have been observed in several specimens. The integument is thin and extended into a long, narrow micropylar tube. The micropylar canal is open near the apex, but closed further down by radiating cells. The seed envelope is four-valved, often with sharp edges, and is ornamented by distinct transverse ridges formed from transversely aligned sclerenchyma cells. The seeds are similar to those of the putative ephedroid *Eoantha*-plant from Transbaikalia (section 5.3.3). A similar rugulate seed envelope also occurs in other Early Cretaceous seeds, such as *Ephedra archaeorhytidosperma*, *Erdtmanispermum juncalense*, *Lobospermum rugosum* and *Rugonella trigonospermum*, as well as in some Bennettitales and in the seeds of *Problematospermum* (see below).

Raunsgaardispermum lusitanicum from the earliest Cretaceous (probably Berriasian) of Portugal (Mendes *et al.*, 2008b) includes small, bisymmetrical seeds, about 2 mm long, that are ovoid to broadly fusiform in lateral view with a pointed apex (Figure 5.16). In cross-section the seeds are circular to broadly elliptical. The seed envelope is

Caption for Figure 5.17 (*cont.*)
the micropylar area showing (C) micropylar tube surrounded by envelope, and (D) lower level at which the micropyle is closed by inward growth of the cells of the micropylar tube. (E–G) *Lignierispermum maroneae* in (E) lateral and (G) apical view and longitudinal section through the micropylar area showing the micropylar tube, which is (F) open at the apex and closed below. (H) *Rugonella trigonospermum* in lateral view. (I) *Lobospermum glabrum* in lateral view. (J) *Lobospermum rugosum* in lateral view. Specimens in the collections of the Swedish Museum of Natural History (A–D, I) and the Field Museum, Chicago (E–H, J).

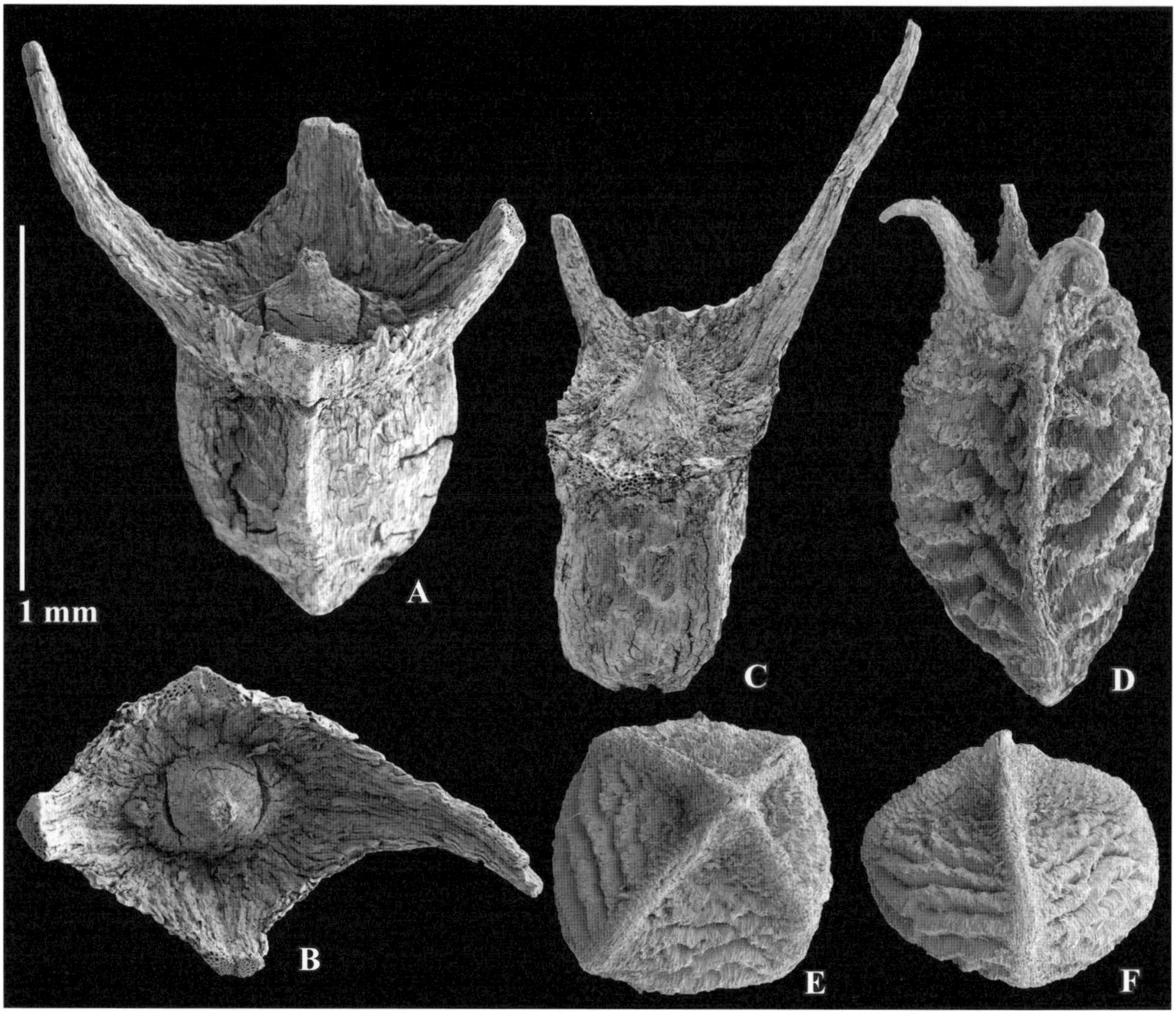

Figure 5.19 Unnamed four-angled BEG-type seeds from (A, B) the Early Cretaceous of Portugal and (C–F) the Potomac Group, USA. Specimens in the collections of the Swedish Museum of Natural History (A, B) and the Field Museum, Chicago (C–F).

two-valved with distinct vascular bundles that extend from the base to near the apex. *Raunsgaardispermum* is distinct from the other dispersed seeds so far described in having an inner papillate lining in the apical region of the seed envelope, a character that is otherwise unique to *Ephedra*. Pollen grains occur in the micropyle and also on the seed surface. However, the pollen is monocolpate, non-plicate, and has a continuous psilate to perforate tectum. Similar pollen is unknown for Gnetales, but occurs in Bennettitales and also in *Pentoxylon*.

Dispersed seeds similar to those described above are known so far only in Early Cretaceous mesofloras from Portugal, Bornholm, Denmark, and the Potomac Group of eastern North Amercia. However, judging from the similarities to *Eoantha*, and also to seeds that occur in compression floras from elsewhere, this complex of isolated seeds was much more widespread during the Early Cretaceous. In particular, there are several seeds from Early Cretaceous floras in Central Asia and China that are very similar to those of the BEG group.

One such group of enigmatic, isolated, potentially BEG-related small seeds are the distinctive forms with hair-like appendages known from several Mesozoic floras (Figure 5.20). These were first described from the ?Late

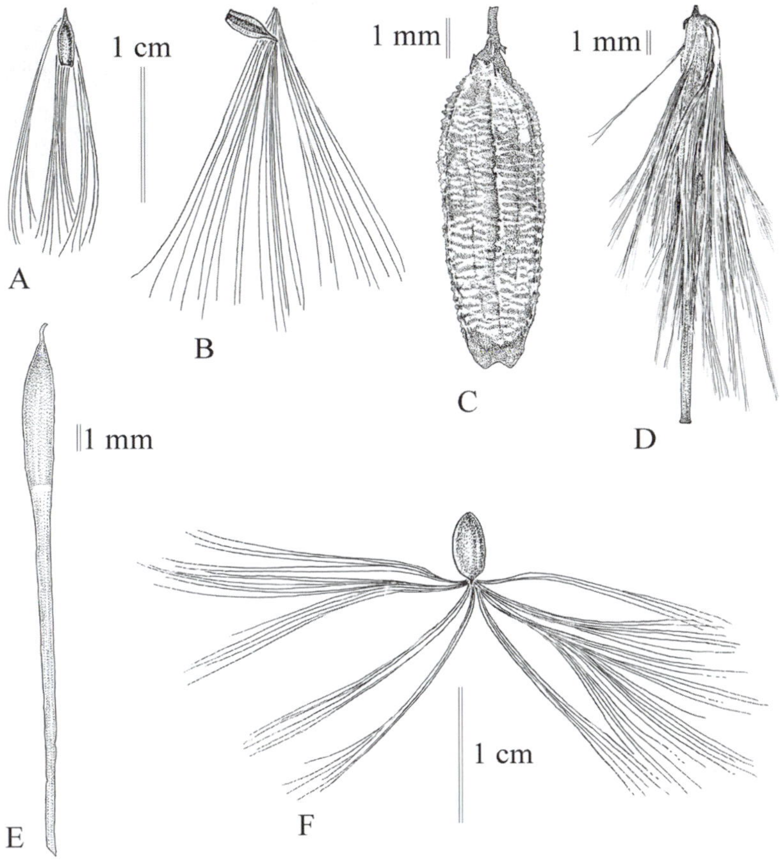

Figure 5.20 Seeds of (A–D, F) *Problematospermum ovale* and (E) *Problematospermum beipiaoense* from (A, B, E) the Early Cretaceous of Liaoning and (C, D, F) the Jurassic of Inner Mongolia, both China, showing rugulate, transversely ribbed seeds surrounded by 'hairy' appendages (possible 'tubular cells'); in our interpretation the seeds are borne on a long stalk (seen in D, E) and have an elongated apical micropyle (seen in A, C–E; see text for alternative interpretation); seeds are sometimes found isolated from the hairy tuft (C). (A, B, F) Drawn from photographs in Sun *et al.* (2001); (C–E) drawn from photographs in Wang *et al.* (2010).

Jurassic flora of the Karatau (Tien Shan) mountain range, Kazakhstan, and assigned to the fossil genus *Problematospermum* with two species, *P. elongatum* and *P. ovale* (Turutanova-Ketova, 1930). Specimens of *Problematospermum ovale* from Karatau were also later described by Krassilov (1973). *Typhaera fusiformis* from the Early Cretaceous flora of Gurvan-Eren, Mongolia (Krassilov, 1982), is similar to *Problematospermum* in several respects, but it is unclear how these taxa should be compared.

Seeds of *Problematospermum* are elongated, elliptical and about 5–8 mm long with a rugulate surface. They were described as having a long tube apically ('micropylar' tube or 'style' in Krassilov, 1973). Often there is a conspicuous pappus-like covering of fine bristles associated with the seeds. These bristles may extend for up to 20 mm beyond the seed proper. Macerations of the seeds showed several layers of tissue. Krassilov (1973) suggested affinity with seeds of Bennettitales, but also compared them with fruits of the extant angiosperm family Asteraceae. Leaves of Bennettitales are a dominant element in the Karatau flora, but there is no evidence of the presence of angiosperms.

Problematospermum (*P. ovale* and *P. beipiaoense*) is also known from the Yixian Formation of northeast China (Figure 5.20). Sun *et al.* (2001) tentatively suggested a relationship with Bennettitales or Gnetales. In these seeds, and also in those illustrated by Krassilov (1973) and those described by Wang *et al.* (2010) from the Middle Jurassic Jiulongshan Formation of western Liaoning and eastern Inner Mongolia, the attachment of the hairs is generally not clear. In some specimens the hairs are clustered into a poorly defined fascicle along the long tube. The similarity of the seed body to that of other rugulate BEG-type seeds, including seeds of Bennettitales, may suggest that the long tube is homologous with the long stalk of some bennettitalean seeds. In this case, the hairs may be homologous to the 'tubular layer': the elongated, isolated cells that surround the seed stalk and seeds in bennettitalean reproductive structures (section 5.6). If this interpretation is correct then *Problematospermum* may be a relatively rare example of a bennettitalean seed that is known to have been dispersed as an isolated unit. This may also be the case for *Cycadeoidea morierei*, where seeds apparently separated easily from each other (Friis *et al.*, 2009a).

5.6 BENNETTITALES (CYCADEOIDALES)

The Bennettitales are a group of extinct seed plants with cycad-like leaves that are diverse, abundant and widespread in Triassic, Jurassic and Cretaceous macrofossil floras (e.g. Harris, 1932a, 1969; Crane, 1985a, 1987; Watson and Sincock, 1992). Leaves of Bennettitales differ from those of true cycads, and most other seed plants, in having characteristically cutinised paracytic stomata. These distinctive paracytic (or syndetocheilic) stomata, combined with the characteristic interseminal scales of the ovule-bearing structures, have led to the widespread acceptance of the Bennettitales as a natural group (e.g. Harris, 1976).

When first recognised, the Bennettitales were interpreted as closely related to cycads (Wieland, 1906, 1916; Chamberlain, 1913, 1920) based mainly on their cycad-like foliage and the cycad-like habit of some late-occurring and probably relatively specialised genera (e.g. *Cycadeoidea*). However, it was soon recognised that the two groups were probably only distantly related (Arber and Parkin, 1907; Scott, 1923; Florin, 1933; Arnold, 1953). For more than a hundred years Bennettitales have also been thought to be closely related to Gnetales (Arber and Parkin, 1907, 1908). Particular similarities of the ovule and elongated micropyle (Berridge, 1911; Thoday, 1911), as well as the presence of paracytic stomata (Martens, 1971), support this suggestion, as does also the possible occurrence of vessels in some Bennettitales (Krassilov, 1984) and the possible equivalent of a 'feeder' in *Cycadeoidea* embryos (p. 158 in Crepet, 1974; Crane, 1985a).

In broad terms, most cladistic analyses based on morphological data suggest a close relationship between Bennettitales, Gnetales and angiosperms (Crane, 1985a; Doyle and Donoghue, 1986; Doyle, 2006; Hilton and Bateman, 2006) (Chapter 6). This has received further support with the recognition of fundamental similarities in the structure of the seeds in Bennettitales and Gnetales (Friis *et al.*, 2007, 2009a). Based on this distinctive seed structure, as well as other features, the Erdtmanithecales (section 5.4) also appear to be closely related to Bennettitales and Gnetales (Friis *et al.*, 2007).

The Bennettitales were diverse in habit (Figure 5.21). They ranged from unbranched, pachycaul forms (e.g. *Cycadeoidea*, *Monanthesia*) to forms that were more slender and more highly branched, and that apparently exhibited axillary branching (*Wielandiella angustifolia* plant, Nathorst, 1909b; Harris, 1932a; *Williamsonia leckenbyi* plant, Harris, 1969; *Ischnophyton*, Delevoryas and Hope, 1976). The anatomy of highly branched forms is unknown, but pachycaul species have a stem with a large pith surrounded by a single broad ring of vascular tissue with both secondary xylem and secondary phloem. The wood is separated into many segments by broad rays, but individual segments have only uniseriate or biseriate rays. Leaf-traces do not girdle the stem as in cycads (Crepet, 1974).

Leaves of Bennettitales vary considerably in size, but all are basically pinnate with a well-developed rachis that gives rise to numerous fine, parallel and occasionally dichotomous veins in the lamina or lamina segments. A distinctive feature of most bennettitalean leaves is that the laminae are inserted onto the rachis close to the adaxial surface. The overall form of bennettitalean leaves may be simple (*Nilssoniopteris*), but more typically it is pinnately compound with distinct pinnae. Even in *Nilssoniopteris* there are often small pinnate laminae. In *Eoginkgoites* the leaf appears palmate, but again the organisation is pinnate with several pairs of pinnae crowded together at the apex of the petiole, resulting in a pseudopalmate arrangement (Ash, 1976a, 1977). In *Dictyozamites* the pinnule venation is reticulate.

Bennettitalean reproductive structures are typically organised into either unisexual or bisexual flower-like reproductive structures (Figures 5.22, 5.23). These are the most flower-like reproductive structures seen outside angiosperms themselves. They vary greatly in size. Typically each consists of bracts that surround the ovulate and pollen-producing organs. When bisexual the pollen-producing structures are attached below the ovules but both structures are quite different from the characteristic carpels and stamens of angiosperms. Therefore, although these reproductive structures resemble the flowers of angiosperms in their basic architecture, because the details differ, the bisexual structure of the two groups is very

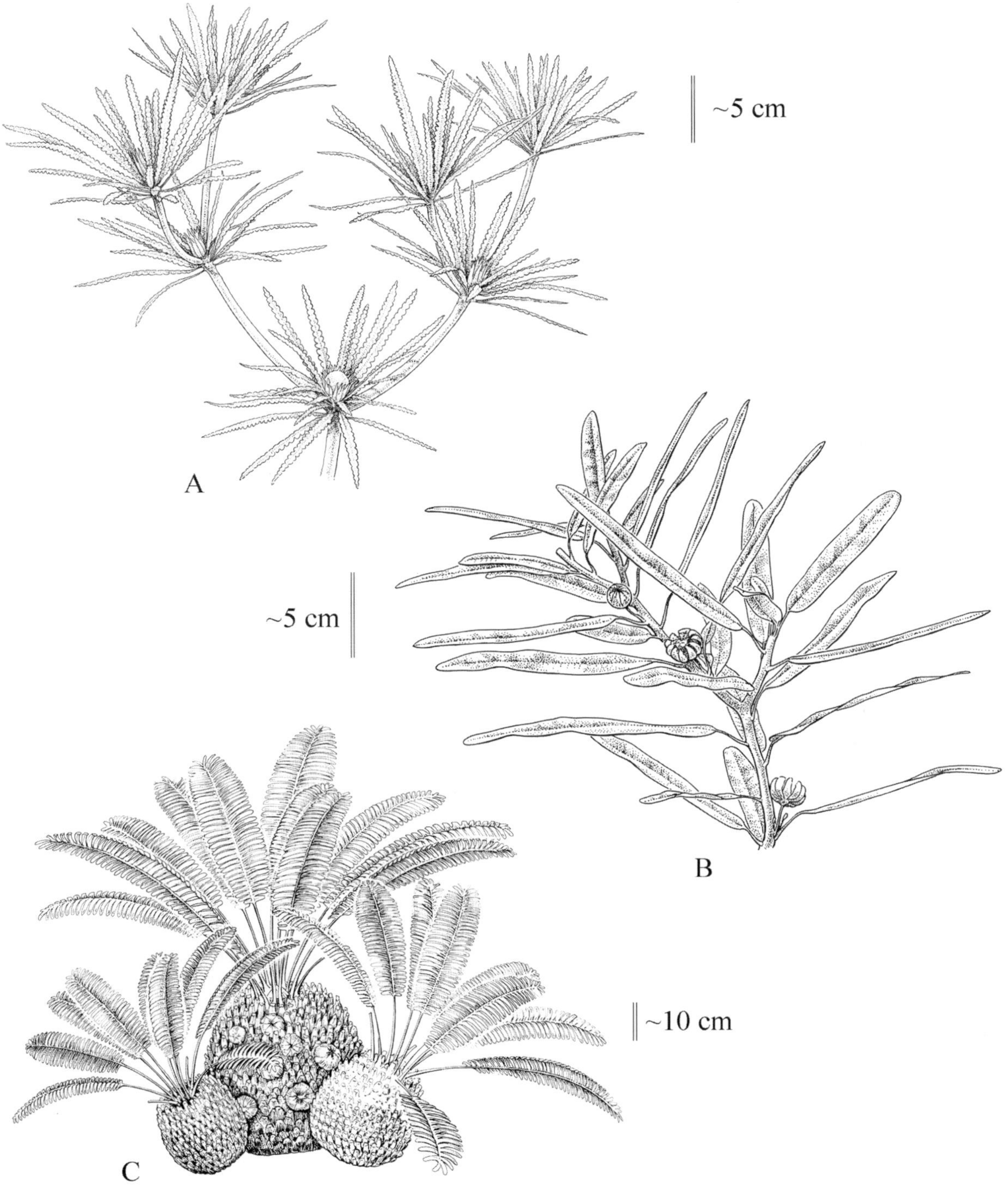

Figure 5.21 Reconstructions of the growth habit of Mesozoic Bennettitales. (A) *Wielandiella* from the Late Triassic of southern Sweden. (B) *Williamsoniella* from the Middle Jurassic of Yorkshire, England. (C) *Cycadeoidea* from the Early Cretaceous of Europe and North America. (A, B) Based on models in the Field Museum, Chicago; (C) redrawn from Watson and Sincock (1992).

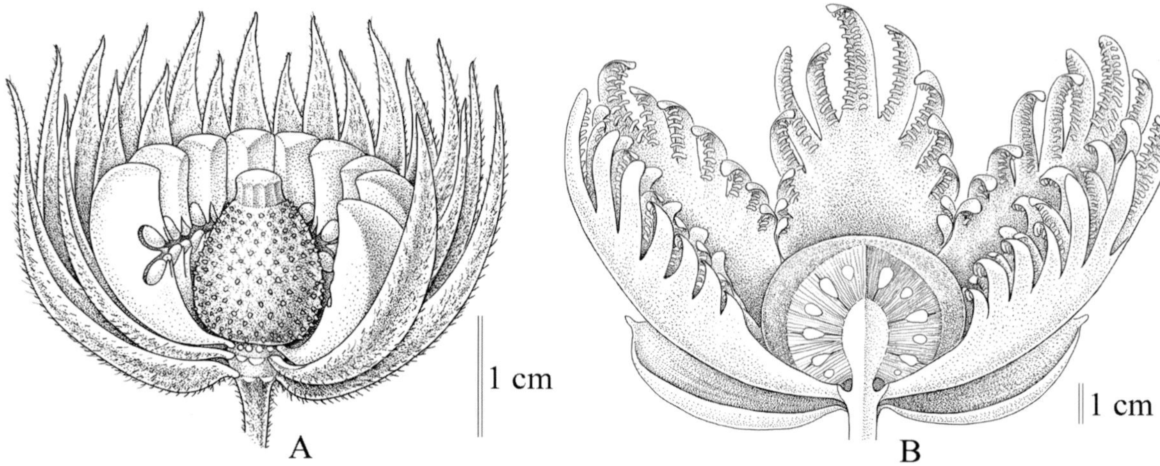

Figure 5.22 Reconstruction of two bisexual bennettitalean reproductive structures. (A) *Williamsoniella* from the Middle Jurassic of Yorkshire, England, showing the central ovule-bearing axis, surrounded by wedge-shaped pollen-producing structures, and sterile bracts. (B) Bennettitalean reproductive structure from Lunz, Austria, composed of *Cycadolepis wettsteinii* scale leaves, *Haitingeria krasseri* pollen organs, and *Bennetticarpus wettsteinii* ovulate organs/seed cones. (A) From Friis *et al.* (2006a); (B) from Pott *et al.* (2010).

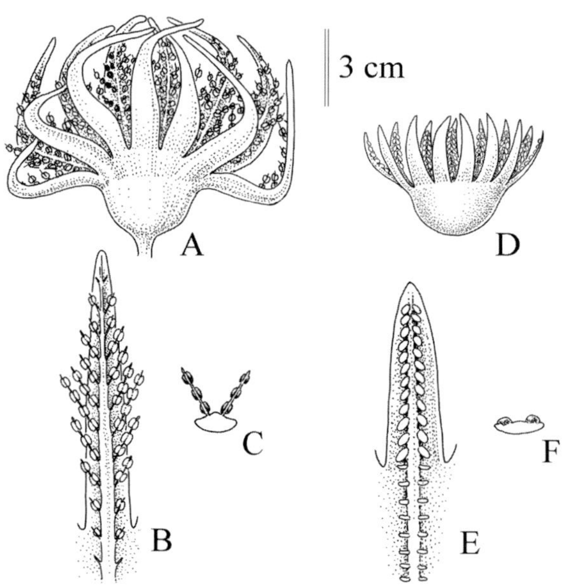

Figure 5.23 Reconstructions of pollen-producing structures of Bennettitales. (A–C) *Weltrichia* (*Williamsonia*) *spectabilis* (A) with synangia on branched structures that are each apparently adaxial to the bract (B, C). (D–F) *Weltrichia* (*Williamsonia*) *whitbiensis* (D) with synangia in two rows (E, F). Based on specimens in the collections of the Swedish Museum of Natural History, Stockholm, and Nathorst (1911).

unlikely to be homologous. The pollen-producing structure of *Welwitschia* (Gnetales), which has an aborted ovule in the centre, also shows this same arrangement where the pollen organs are borne proximally in relation to the ovule.

Bennettitalean ovule-producing structures consist of ovules and interseminal scales borne laterally on a swollen axis, the receptacle (Figure 5.24). Ovules are orthotropous and generally small, less than 7 mm long, and frequently only 1–2 mm long. They are usually radially symmetrical. Bennettitalean ovules are usually described as having a single integument (Rothwell *et al.*, 2009) but in our view most bennettitalean ovules are enclosed by an additional (second) layer to the outside: the envelope. This envelope surrounds the integument (Figures 5.24, 5.25) (Friis *et al.*, 2009a). The integument is fused to the nucellus for most of its length but extended at the apex into a long micropylar tube. Where preservation is good the micropylar canal is often seen to be occluded in its lower part by expanded cells from the lining of the micropylar tube. Each ovule is supplied by a single vascular trace that extends to a pad of vascular tissue in the base of the seed, but no vasculature is known in the integument. In many cases, the ovules, which each consist of nucellus, integument and envelope, are borne on a long stalk. The megaspore membrane is thin, or not resistant to maceration, but based on studies of macerated compression

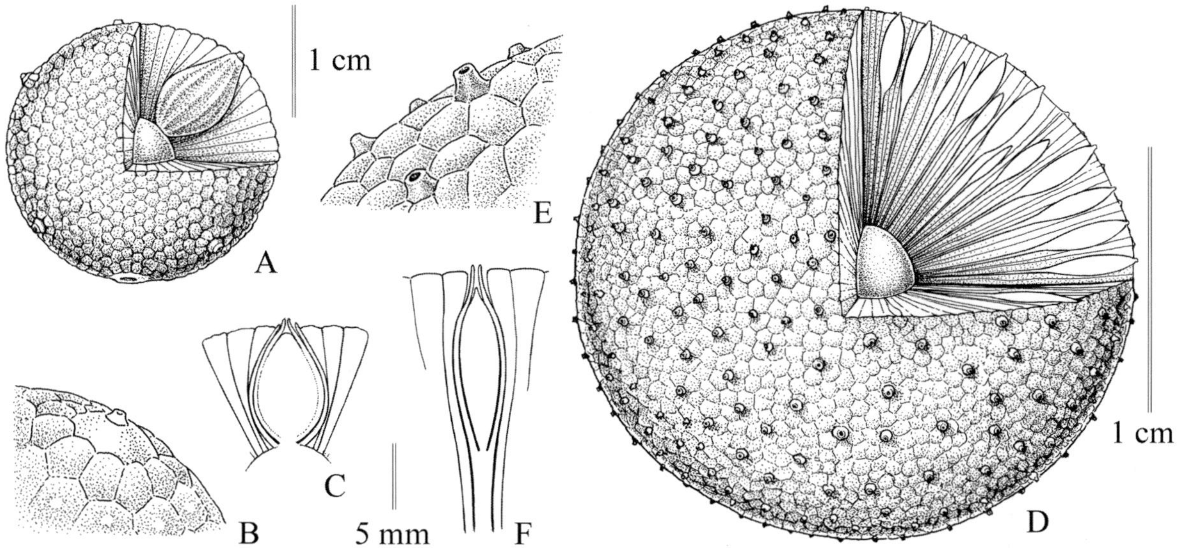

Figure 5.24 Reconstructions and schematic sections of the more or less spherical ovule-bearing structures of two bennettitalean plants. (A–C) *Vardekloeftia* from the Late Triassic of East Greenland, consisting of a spherical head with few, relatively large seeds and many interseminal scales (A, B). (C) Section of seed and interseminal scales; note two layers surrounding the nucellus. (D–F) *Williamsonia*, showing (D) spherical head with many small seeds, each borne on a long stalk and interspersed between interseminal scales; (E) tips of seeds projecting from the surface of the ovule-bearing structure; (F) section of seed and interseminal scales, showing a single layer around the nucellus as in the interpretation of Stockey and Rothwell (2003). (A–C) Based on Pedersen *et al.* (1989a); (D–F) based on information in Stockey and Rothwell (2003).

specimens the nucellus and inner surface of the integument is well cutinised (Harris, 1954, 1969).

The envelope that surrounds the integument in bennettitalean ovules is seen in both permineralised and compression material (Figures 5.24, 5.25). Among compression fossils it is seen especially clearly in *Bennetticarpus crossospermus*, but is also obvious in *Vardekloeftia* (Harris, 1932a; Crane, 1985a; Pedersen *et al.*, 1989a). It can also be seen as a layer surrounding the micropylar tube in more typical Bennettitales preserved as compressions, such as *Williamsonia himas* (fig. 60K in Harris, 1969), *Bennettites tylosus* (pl. 15, fig. 6 in Harris, 1932a) and *Williamsonia leckenbyi* (fig. 57F in Harris, 1969).

In permineralised material there was considerable controversy early in the twentieth century over whether the bennettitalean ovules had one or two integuments (Solms-Laubach, 1891; Berridge, 1911; Lignier, 1911; Stopes, 1918). Re-examination of the classic material of *Cycadeoidea morierei* described by Lignier (1894) using synchrotron X-ray microtomography shows that what has generally been referred to as integument in permineralised bennettitalean ovules is, in fact, the envelope (Friis *et al.*, 2009a). The integument itself is thin and extended apically into a long micropylar tube (Friis *et al.*, 2009a). Based on these observations bennettitalean ovules therefore could be considered to be fundamentally bitegmic, with an integument surrounding the nucellus that is itself surrounded by the envelope (Friis *et al.*, 2009a). They are also directly comparable to seeds of extant and fossil Gnetales (section 5.3; Figure 5.25) and Erdtmanithecales (section 5.4). Although this view has been questioned (Rothwell *et al.*, 2009), and more material is currently under investigation by ourselves and others, all of the specimens that we have examined so far support the 'integument plus envelope' interpretation.

In all bennettitalean ovule-bearing structures the ovules are crowded together and embedded in a mass of sterile interseminal scales. Neither the ovules nor the interseminal scales are subtended by other structures. Like the ovules, each interseminal scale is supplied by a single vascular bundle (but see Lignier, 1894). The structure of *Bennetticarpus crossospermus* provides evidence of the

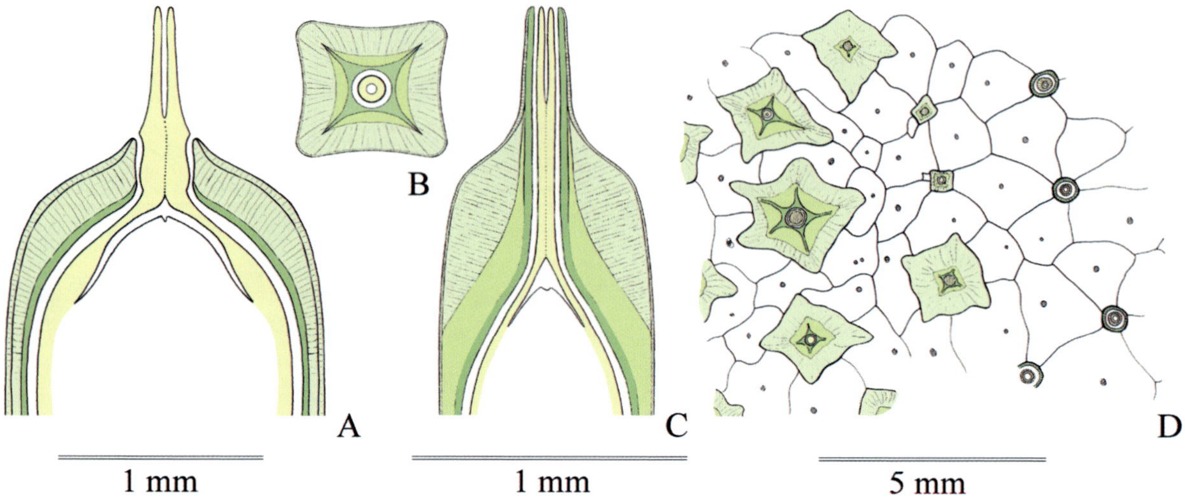

Figure 5.25 Schematic sections of the apical portion of the seed in (A, B) extant *Gnetum* and (C, D) extinct *Cycadeoidea morierei* in (A, C) longitudinal and (B, D) cross-sections, showing integument (yellow) with long micropyle surrounded by seed envelope (green). From Friis *et al.* (2009a).

homology between ovules and interseminal scales as suggested earlier by Seward (1912, 1917; see also Delevoryas, 1968). It has an elongated micropylar tube that passes through the centre of a thickly cutinised structure very similar to the flattened cutinised apices of the interseminal scales. It is also significant that interseminal scales in many Bennettitales (e.g. *Bennettites litchi*; Harris, 1969; *Williamsonia cynthiae*; Watson and Sincock, 1992) have a raised or otherwise differentiated region at the centre (Crane and Herendeen, 2009). Ovules and interseminal scales are therefore most likely homologous; the scales forming from seed initials by deviation from normal ovule development (Harris, 1932a). Together, the interseminal scales may have provided some protection for the developing ovules and at maturity their massed, interlocking, heavily cutinised, heads would have helped protect the mature seeds. Internally the interseminal scales may have been fleshy at maturity and were perhaps attractive to potential animal dispersers (Crane and Kenrick, 1997; Chapter 18).

Like the ovule-bearing structures, the pollen-producing structures of most Bennettitales were aggregated into flower-like forms with bracts at the base (Figure 5.23). Pollen sacs are generally borne in bivalved synangia, but the synangia, or in some cases the structures on which they appear to be borne, are attached to the bracts in various ways. In no case are the morphological details completely clear or well understood. In some cases synangia are borne on fertile appendages, in others they are apparently sessile on the adaxial surface of the bracts. The synangia open like a clam and each half of the synangium contains up to 20 or 30 pollen sacs embedded in its surface. Pollen is monocolpate, boat-shaped, and more or less smooth-walled. Based on studies of *Cycadeoidea* (Taylor, 1973; Osborn and Taylor, 1994) and *Leguminanthus* (Ward *et al.*, 1989) the pollen wall has a granular infratectal layer.

There is ample evidence that pollen grains entered the micropylar canal, even though this has been questioned for some taxa based on recognition of putative pollen tubes in *Williamsonia bockii* (Stockey and Rothwell, 2003). Pollen is clearly seen inside the micropylar tube in *Vardekloeftia* (Pedersen *et al.*, 1989a) and also, for example, in *Williamsonia leckenbyi* (fig. 57A in Harris, 1969) and *Williamsonia bryonyae* (fig. 98B in Watson and Sincock, 1992). Female gametophyte development was apparently monosporic, developing from the innermost of a linear tetrad of megaspores (*Cycadeoidea*, Crepet and Delevoryas, 1972).

5.6.1 Permineralised material

Permineralised material of Bennettitales is generally assigned to genera such as *Cycadeoidea*, *Monanthesia* and *Bennettites*. Such material has been studied for almost two hundred years and provides some of the most detailed information available on the habit and structure of

Bennettitales. Most of the permineralised bennettitalean material, perhaps all of it, is of latest Jurassic or Cretaceous age, which is relatively late in the history of the group. This cautions against regarding these fossils as necessarily typical of the Bennettitales as a whole.

Watson and Sincock (1992) recognised three informal groups among the many reproductive structures described from permineralised material. In all cases, the flower-like reproductive structures are surrounded by bracts apparently arranged in a spiral. Each reproductive structure with its bracts is embedded in an armour of persistent petiole bases.

The *Cycadeoidea saxbyana* group includes *Monanthesia magnifica* (Wieland, 1934; Delevoryas, 1959). *Monanthesia* was described from the Late Cretaceous Mesaverde Formation of New Mexico, USA, but is now believed to come from rocks of Early Cretaceous age (Tidwell *et al.*, 1981; Watson and Sincock, 1992). Similar specimens are also known from elsewhere, including *Cycadeoidea fisherae* from the Wealden Formation of southern England (Watson and Sincock, 1992). Plants of this kind had tall, parallel-sided trunks and produced relatively small flower-like reproductive structures in the axil of almost every leaf. In some specimens the leaf bases subtending the reproductive structures become abruptly smaller some distance up the stem.

Plants of the *Cycadeoidea gibsoniana/wielandii* group are similar to those of the *C. saxbyana* group in having small reproductive structures, but these occur scattered on the trunks. Sometimes they are sparsely distributed; in other cases they are more densely arranged and may occur in the axil of nearly every leaf as in *C. saxbyana*. The trunk in this group is more like that of the typical short, hemispherical *C. dacotensis* (Watson and Lydon, 2004).

Plants of the *Cycadeoidea reichenbachiana/dacotensis* group have the typical *Cycadeoidea* habit, like that of the *C. gibsoniana/wielandii* group, but the reproductive structures are much larger. They also have a taller, more conical receptacle. Reproductive structures were also relatively sparse, scattered on the trunk (Watson and Sincock, 1992), and are known to be bisexual (e.g. Wieland, 1906; Delevoryas, 1963, 1968; Crepet, 1974). The pollen-producing organs are fused below into a shallow cup. Each is compressed laterally and is apparently pinnate with two rows of fertile appendages. Each appendage has a row of kidney-shaped, bivalved synangia of elongated pollen sacs. Distally the pollen-producing structures are incurved with their adaxial surface against the conical receptacle. The receptacle itself is covered with ovules and interseminal scales. The flowers may never have opened, as Wieland (1906) supposed, and may perhaps have been self-pollinated (Delevoryas, 1963, 1968; Crepet, 1972, 1974). However, there is also evidence that the flowers were attractive to insects (Crepet, 1974).

Among other important records of permineralised bennettitlean reproductive structures from Europe and North America are those from the Early Cretaceous of France (*Cycadeoidea* (*Bennettites*) *morierei*, Lignier, 1894; Friis *et al.*, 2009a), Scotland (*Williamsonia scotica*, Seward, 1912, 1917), Poland (*Cycadeoidea* sp., Reymanówna, 1960), the Late Cretaceous of British Columbia, Canada (*Cycadeoidea maccafferyi*, Rothwell and Stockey, 2002; *Williamsonia bockii*, Stockey and Rothwell, 2003).

Permineralised fossils from other areas have also contributed important information to our understanding of diversity and biology of the Bennettitales. Of particular importance is *Cycadeoidella japonica* (Nishida, 1994a) from the Late Cretaceous of Japan and specimens from the classic Early Cretaceous silicified flora from the Rajmahal Hills, India, including *Williamsonia harrisiana* (Bose, 1968), *W. sewardiana* (Sahni, 1932) and *Amarjolia dactylota* (Bose *et al.*, 1984a). *Williamsonia sewardiana* is thought to have been borne on *Bucklandia indica* stems, which also bore leaves of *Ptilophyllum* cf. *cutchense*. Stems and leaves of *Williamsonia harrisiana* and *Amarjolia dactylota* are not known. *Amarjolia* is known to have been bisexual with about 12 pollen-producing structures that are folded into contact with the receptacle (Bose *et al.*, 1984a). Also important from the Rajmahal Hills flora is the bennettitalean pollen-producing structure *Weltrichia santalensis*, the pollen-producing units of which have an interesting bipartite structure (e.g. Sitholey and Bose, 1953; Sharma, 1969). It is unknown whether *Weltrichia santalensis* corresponds to *Williamsonia harrisiana*, *Williamsonia sewardiana* or some other bennettitalean.

5.6.2 Compression material

Bennettitalean reproductive structures known from compressions or impressions are particularly well represented in the Middle Jurassic flora of Yorkshire, England. These have provided important information on Bennettitales other than those typically preserved as petrifactions. The flora includes several species of ovulate structure (Harris, 1969). The organisation of the ovule-bearing structures in *Williamsonia hildae* and *W. leckenbyi* is similar to that in most other Bennettitales with a dome-shaped receptacle surrounded by prominent bracts. *W. gigas* differs in the presence of a sterile 'corona' formed from the elongated receptacle and poorly developed interseminal scales (Harris, 1969). *W. himas* has not so far been linked with a corresponding pollen-producing structure, but *W. gigas*, *W. hildae* and *W. leckenbyi* are more completely known as whole plants.

The pollen-producing structures of all three *Williamsonia* species are basically similar and assigned to the genus *Weltrichia*. *Weltrichia sol* (linked with *Williamsonia gigas* and *Zamites gigas* leaves) is cup-like with bracts fused basally, but free above. The inner surface of the cup bears large numbers of 'resinous sacs', which Harris (1969) interpreted as the remains of nectaries (Chapter 17). The fertile appendages projecting from the bracts have been interpreted as pinnae, but their morphological interpretation is uncertain. Each appendage bore bivalved synangia with many microsporangia similar to those seen in *Cycadeoidea*. Pollen is monocolpate and psilate.

Also from the Middle Jurassic of Yorkshire there are two very similar species of *Williamsoniella*, *W. coronata* and *W. ligneri* (Harris, 1944, 1969, 1974; Cridland, 1957). Both species have relatively small and bisexual structures. *Williamsoniella coronata* is the better understood of the two species. It has a whorl of bracts surrounding a whorl of about 12 laterally compressed pollen-producing structures, each of which resembes an orange segment. Each segment has two to three pairs of short fertile appendages on the inner (adaxial) surface bearing two-valved synangium containing several pollen sacs. The ovulate structures have numerous ovules and interseminal scales arranged around the central receptacle that is expanded at the apex into a terminal sterile corona. Pollen is monocolpate and verrucate with a homogeneous ektexine (Zavialova *et al.*, 2009).

From the Triassic Lunz flora, Austria, the best-known bennettitalean is a possible bisexual structure composed of *Cycadolepis wettsteinii* scale leaves, *Haitingeria krasseri* pollen organs and *Bennetticarpus wettsteinii* ovule-bearing structures (Figure 5.22). The different dispersed organs were linked based on similarities in gross morphology and epidermal anatomy (Pott *et al.*, 2010). The association seems secure, but the reconstructed bisexual nature of the structure, although it is possible, is not confirmed by organically connected organs. The ovulate-bearing organ, *Bennetticarpus wettsteinii*, is more similar to *Vardekloeftia*, which is of about the same age, than to more typical bennettitalean reproductive structures from the Jurassic and Cretaceous. *Haitingeria krasseri* also differs from the pollen-producing structures of younger bennettitaleans in having free sporangia rather than sporangia borne in bivalved synangia.

Other reproductive structures from the Lunz flora that have been assigned to the Bennettitales are *Leguminanthus siliquosus* and *Westersheimia pramelreuthensis*, but in both cases the fossils are strongly compressed and do not provide sufficient characters for a detailed systematic analysis.

There are also interesting forms from Late Triassic and Early Jurassic of East Greenland and Scania, southern Sweden. *Wielandiella angustifolia* (Figure 5.21) is reported from both areas (Nathorst, 1909a; Harris, 1932a). It was described as having an ovule-bearing reproductive structure with a distal sterile corona, but the structure is poorly understood and requires restudy. *Vardekloeftia sulcata* is an ovulate structure also reported from both areas (Harris, 1932a; Pedersen *et al.*, 1989a). There are no associated or attached bracts. Each head is composed of numerous interseminal scales with relatively few, relatively large, ovules. The ovules have an outer cutinised layer, the envelope, that surrounds the integument and with an apical opening through which the micropyle projects (Figure 5.24).

5.7 PENTOXYLALES

The Pentoxylales are a small, distinctive group of seed plants known from the Middle Jurassic and Early Cretaceous with reports in Australia, Antarctica, India and New Zealand (Sahni, 1948; Harris, 1962; White, 1981; Bose *et al.*, 1984b, 1985; Drinnan and Chambers, 1985, 1986; McLoughlin and Drinnan, 1995; Howe and Cantrill, 2001; see Crane, 1985a, 1988 for reconstructions of the isolated organs). They have an unusual arrangement of five vascular segments in the stems (*Pentoxylon*), distinctive spherical–ellipsoidal to elongated ovule-bearing structures (*Carnoconites*) and flower-like aggregations of pollen-producing structures (*Sahnia*). The most detailed information is based on silicified material from the Early Cretaceous of the Rajmahal Hills, India. Pioneering work by Sahni (1948) linked the ovulate heads (*Carnoconites*) with the stem (*Pentoxylon*) and the leaves (*Nipaniophyllum*) based on association evidence and structural similarity. Microsporangiate organs (*Sahnia*) were described later by Vishnu-Mittre (1953). Knowledge of the *Pentoxylon* plant was reviewed by Rao (1976, 1981) and more recent studies of the Indian material were undertaken by Bose *et al.* (1984b, 1985). The relationships of the *Pentoxylon* plant to other seed plants have been regarded as enigmatic (Andrews, 1961; Stewart, 1983; Bose *et al.*, 1985) but most cladistic analyses based on morphological data support earlier suggestions by Ehrendorfer (1976) that the group is closely related to the Bennettitales (Crane, 1985a) or to the Bennettitales and Gnetales (Doyle and Donoghue, 1986).

Reports of *Pentoxylon* from outside India are based on compressed ovulate heads and leaves from the Late Jurassic of the North Island, New Zealand (Harris, 1962, 1983b), and a variety of different organs from Victoria (Drinnan and Chambers, 1985, 1986), New South Wales (White, 1981), and Queensland (McLoughlin and Drinnan, 1995), Australia. The different organs of the *Pentoxylon* plant are also known from the Early Cretaceous (Albian) Triton Point Formation of Alexander Island, Antarctica, where the material includes ovule-bearing structures (*Carnoconites cranwelli*) with associated leaves (*Taeniopteris daintreei*) and stems (*Pentoxylon* sp.). The Alexander Island occurrence suggests that *Pentoxylon* plants were perhaps small shrubs growing in the understory of flood-plain forests (Howe and Cantrill, 2001).

Ovules are sessile, orthotropous, and aggregated into compact ovoid (*Carnoconites compactum*, *C. cranwelliae*) or elongated heads (*C. laxum*). No bracts or other structures are associated with the heads, and there are no interseminal scales as in Bennettitales. Ovulate heads are borne terminally on short shoots. Ovules of *Carnoconites compactum* are flattened. They have generally been interpreted as unitegmic with the integument free from the nucellus except at the chalaza. However, there is a distinct beak at the apex of the nucellus. It is possible that as in Bennettitales, as well as Erdtmanithecales, Gnetales and the dispersed BEG seeds, the integument is thin and fused to the nucellus for most of its length. What has been described as integument may be an envelope. The 'sclerotesta' is bicarinate and distinctly flattened in transverse section with a distinct layer of dark cells towards the outside. The remaining outer part forms a thick, fleshy 'sarcotesta'. Each ovule is supplied by a single vascular bundle. No other vascular tissue has been reported in the ovules.

Pollen-producing structures (*Sahnia*) are borne aggregated in dense clusters at the end of short shoots around a broad, conical, sterile receptacle. They are apparently in a whorl, or low helix, with deciduous bracts below and to the outside. According to Vishnu-Mittre (1953), the pollen-producing structures are fused proximally to form a shallow cup, but according to Rao (1981), they are free at the base. Each pollen-producing structure bears 10–20 spirally arranged, stalked unilocular sporangia singly or in groups of two or four. Sporangia are pyriform and thick-walled. Pollen is monocolpate, non-saccate, and probably with a granular infratectal layer (Osborn *et al.*, 1991). Monocolpate pollen occurs in the micropyles of *Carnoconites* ovules.

5.8 OTHER PALAEOZOIC AND MESOZOIC SEED PLANTS

In pre-cladistic treatments several groups of extinct seed plants (hydraspermans, lyginopterids, medullosans, Callistophytales) were often grouped together as seed ferns (pteridosperms), and this concept has sometimes been further expanded to include glossopterids, peltasperms, corystosperms and Caytoniales (the so-called Mesozoic seed ferns or Mesozoic pteridosperms). It is now clear that seed ferns defined in this way are of diverse relationships and do not form a monophyletic group (Figure 5.2, Chapter 6). The concept is therefore unhelpful in discussions of seed plant phylogeny and angiosperm relationships. Most groups included in the so-called Mesozoic seed ferns have at some point figured as potential ancestors to angiosperms and are discussed only briefly here.

5.8.1 Corystospermales and potentially related plants

The Corystospermaceae was established for leaves, microsporophylls and megasporophylls from the Middle Triassic Molteno Formation of South Africa (Thomas, 1933; Townrow, 1962; see also Anderson and Anderson, 1983, 1985, 1989). Additional material has come from the Molteno Formation of Zimbabwe (Lacey, 1976), the Late Triassic/Early Jurassic of Tasmania, Australia (Townrow, 1965), the Triassic of New South Wales, Australia (Holmes and Ash, 1979; Holmes, 1987), the Triassic of India (Pant and Basu, 1973, 1979; Srivastava, 1974), the Triassic of Argentina (Archangelsky, 1968; Petriella, 1978, 1981), the Triassic of Antarctica (Taylor, 1996; Axsmith *et al.*, 2000; Taylor and Taylor, 2009) and the Middle Jurassic of Yorkshire (Harris, 1964). Until recently the individual organs assigned to the corystosperms had never been found in organic connection. However, in the past few decades very important new material has been collected from

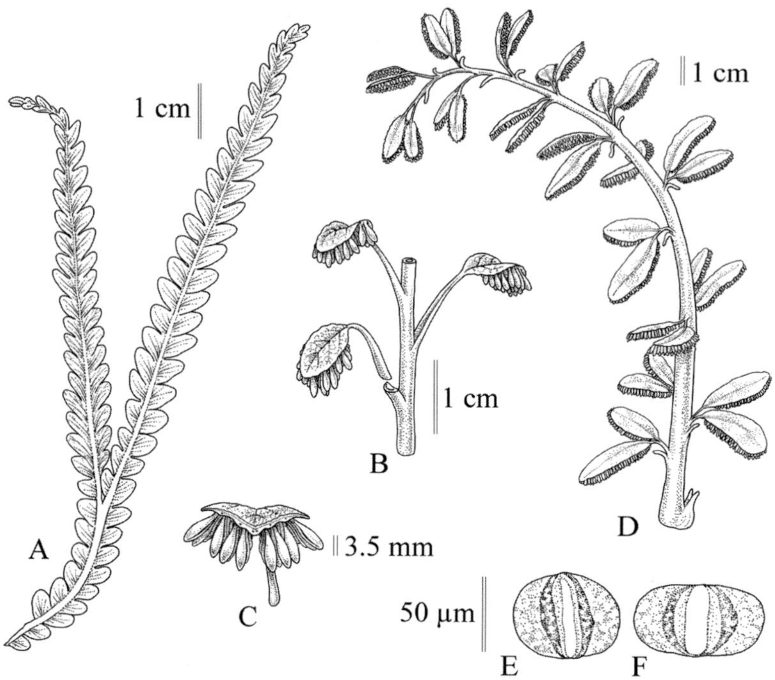

Figure 5.26 Leaves and pollen-producing structures of Corystospermales from the Triassic of (A, D–F) South Africa and (B, C) Antarctica. (A) *Dicroidium odontopteroides* subsp. *obiculoides* leaf. (B, C) Reconstruction of pollen-producing axis of *Pteruchus fremouwensis* showing (B) helical arrangement of pollen-producing structures and (C) attachment of sporangia to the underside of laminar structures. (D) Reconstruction of pollen-producing axis of *Pteruchus matatimajor* showing pairs of sporangium-bearing structures borne in a helical arrangement. (E, F) Saccate pollen grains found associated with *Pteruchus matatimajor* and known *in situ* in other material. (A) Based on Anderson and Anderson (1983); (B, C) based on Yao *et al.* (1992); (D–F) based on Anderson and Anderson (2003).

South Africa and from Antarctica (Figures 5.26, 5.27). Some of this material has documented connection between organs that were previously only known as separated specimens.

Corystosperms were woody plants. Harris (1983a) described the stem of *Pachypteris papillosa* based on compressions, and Archangelsky (1967, 1968) suggested that the permineralised stem *Rhexoxylon* was probably attributable to Triassic corystosperms. Archangelsky also suggested that the ring of vascular segments seen in *Rhexoxylon*, with secondary xylem developed both internal and external to the primary xylem, may indicate a liana-like habitat for some taxa. Other corystosperms may have been small to medium-sized trees. The study of a petrified forest from the Late Triassic of Argentina consisting of *Elchaxylon* stems has documented forest vegetation in which corystosperm trees were monodominant and reached estimated heights of about 13–15 m (Artabe *et al.*, 2007).

Corystosperm leaves have been referred to several genera, including *Dicroidium*, *Johnstonia*, *Pachypteris* and *Xylopteris* (Townrow, 1957, 1965). Such leaves are among the most common fossils in Triassic floras from the Southern Hemisphere. They generally have a forked leaf rachis with a simple or bipinnate arrangement of pinnae (Figure 5.26). Stomata are anomocytic. Accumulating evidence that links leaves with reproductive structures suggests that leaves that are indistinguishable in form bore significantly different kinds of reproductive structure (Axsmith *et al.*, 2000).

Corystosperm ovule-bearing structures were first recognised and linked with *Dicroidium* foliage by Thomas (1933), who recognised three genera: *Umkomasia*, in which the cupules are divided into two lobes by clefts in the plane of branching; *Pilophorosperma*, in which the cupules have a single cleft; and *Spermatocodon*, which has cupules with no conspicuous clefts. Cupules of all three genera contain a single ovule, but this is known to vary in other material. Other corystosperm cupules have been described subsequently, but especially informative compression material assigned to *Umkomasia uniramia* has been described from the Middle to Late Triassic Shackleton Glacier locality, Antarctica (Axsmith *et al.*, 2000).

Interpretation of the morphology of corystosperm ovule-bearing structures has been controversial, but the studies of well-preserved compression material from the

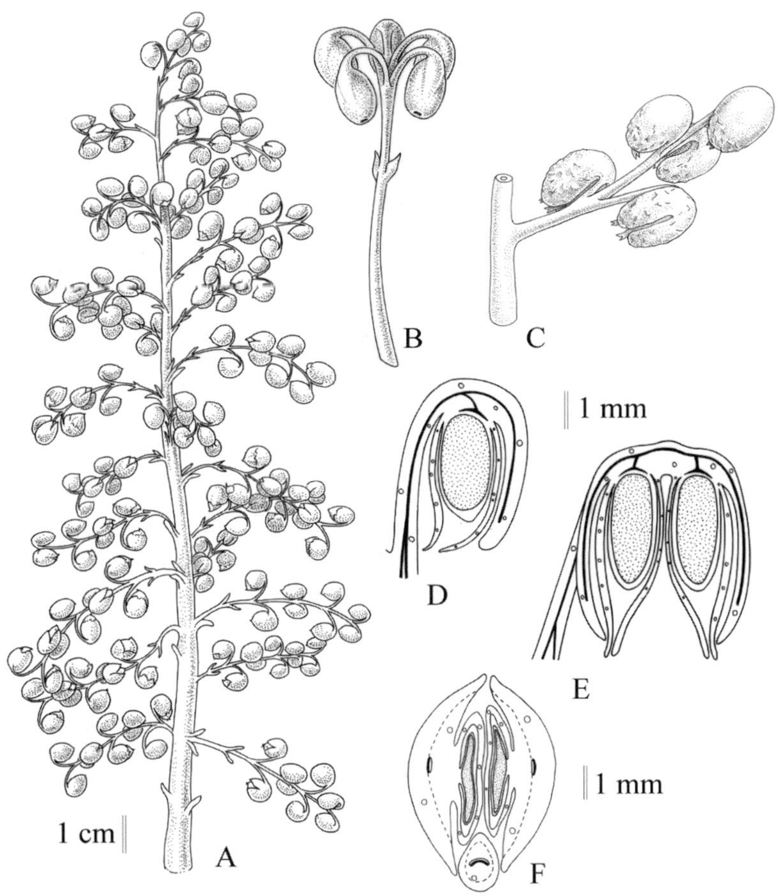

Figure 5.27 Reconstructions of ovulate structures of Corystospermales from (A) the Triassic of South Africa and (B–F) Antarctica. (A) *Umkomasia quadripartita* showing compound branching axis with ovulate cupules and seeds. (B) *Umkomasia uniramia* showing five recurved cupules. (C–F) *Umkomasia resinosa*, reconstruction (C) and schematic section of cupules with seeds in longitudinal (D, E) and transverse (F) sections. (A) Redrawn from Anderson and Anderson (2003); (B) redrawn from Axsmith *et al.* (2000); (C–F) redrawn from Klavins *et al.* (2002).

Triassic of Antarctica show that it is a branch system bearing lateral units that have ovules on their adaxial surfaces. The interpretation of corystosperm reproductive structures as pinnate sporophylls, as suggested by Thomas (1933) and followed by others (Crane, 1985a; Doyle and Donoghue, 1986), now seems unlikely to be correct. *Umkomasia uniramia* from the Middle–Late Triassic Shackleton Glacier locality has ovulate structures borne in groups at the tip of short shoots that also bore *Dicroidium odontopteroides* leaves (Axsmith *et al.*, 2000). Each ovule-bearing structure consists of an elongated axis bearing a pair of opposite bracts and a whorl of five to eight cupules very close to, or at, the apex (Figure 5.27). Each cupule is borne on a recurved stalk and consists of a midrib with two lateral flaps that enclosed the ovule. Ovules are not known, but by analogy with other corystosperms it is likely that the cupules of *Umkomasia uniramia* contained just one, or perhaps two, ovules.

Umkomasia resinosa from the Middle Triassic Fremouw Peak locality has ovule-bearing structures consisting of a determinate axis (anatomically a branch) bearing helically arranged, short-stalked lateral laminar structures that are bilobed and folded to form a cupule enclosing the ovules (Figure 5.27) (Klavins *et al.*, 2002). Each lamina unit bears one or two ovules on its internal surface. Clear preservation of vascular tissue shows that the ovules are borne on the abaxial surface of the cupule. The ovules are small, flattened or irregular in transverse section, with a thin integument that is bifid at the apex. Bisaccate pollen grains, comparable to those produced by *Pteruchus fremouwensis*, occur inside the integument at the apex of the nucellus.

Pollen-producing organs of corystosperms (*Pteruchus*) consist of an axis bearing lateral units (Figure 5.26). The lateral units had a short stalk and expanded distal lamina with groups of pendulous, ellipsoidal pollen sacs on the abaxial surface. Lateral units (rachises, Townrow, 1962) were attached either in a single plane (Townrow, 1962) or spirally (Thomas, 1933; Pant and Basu, 1973, 1979; Srivastava, 1974; Anderson and Anderson, 2003). Townrow (1962) emphasised the leaf-like nature of *Pteruchus*, and following him it has generally been interpreted as a microsporophyll (e.g. Crane, 1985b). However, permineralised material shows that the pollen-producing structure and the seed-bearing axes are fundamentally axial (Klavins *et al.*, 2002; Yao *et al.*, 1995). Pollen sacs dehisced along a single suture to release bisaccate pollen with a single distal colpus. The pollen wall is granular (Taylor *et al.*, 1984; Zavada and Crepet, 1986). Pollen has been described in the micropyle of both compression and permineralised specimens.

In addition to the main group of species that form the better-known core of the corystosperms, several other genera of fossil plants have been described that may be related to this group, or that may provide a link to Caytoniales. Especially important is *Petriellaea triangulata*, described based on permineralised material from the early Middle Triassic Fremouw Formation in the central Transantarctic Mountains, Antarctica (Taylor *et al.*, 1994). *Petriellaea* is an ovule-bearing structure, probably with a central axis bearing lateral units comprising short, forked axes, the tips of which bear single recurved cupules. Each contains five to six orthotropous, triangular ovules. Anatomical details show that the cupule is formed by a reflexed leaf-like structure, which bears sessile ovules on its internal surface in various arrangements. The ovule-bearing surface is interpreted as adaxial. The ovules are orthotropous, and generally distinctly triangular in cross section. The nucellus and integument are fused. Ovules are without vascular bundles except for disc of tracheids at the chalazal end. The pollen, pollen organs and leaves of *Petriellaea triangulata* are unknown. The affinities of *Petriellaea* are uncertain. On the one hand the cupules are similar to those of corystosperms, with which *Petriellaea* co-occurs, but on the other hand the multiple ovules per cupule, and the essentially closed nature of the cupules, resembles the situation in *Caytonia*.

5.8.2 Caytoniales

The Caytoniales range in age from Late Triassic to Late Cretaceous and comprise five genera: *Sagenopteris* (leaves), *Caytonanthus* (pollen-producing structures), *Vitreisporites* (isolated pollen) and *Caytonia* and *Reymanownaea* (ovule-bearing structures). The different organs have never been found organically connected, but are linked by association evidence, similarity of cuticular structure, and the consistent occurrence of *Caytonanthus* pollen in the micropyles of *Caytonia* ovules. Only a few reconstructed *Caytonia* plants are known and for the best of these there is information on leaves, pollen-producing structures, ovule-bearing structures and external features of the stem. No petrified material of *Caytonia* is known.

The most common fossils of Caytoniales are *Sagenopteris* leaves. These are geographically widespread through most of the Mesozoic in the Northern Hemisphere, but most information on the Caytoniales comes from a few localities in East Greenland (Harris, 1933, 1937), eastern Russia (Krassilov, 1977b), Poland (Reymanówna, 1973), Hungary (Barbacka and Bóka, 2000) and most importantly the Middle Jurassic of Yorkshire, England (Thomas, 1925; Harris, 1940a, b, 1941, 1951, 1958, 1964). There are also a few records from the Southern Hemisphere (Rees, 1993).

The stem of the *Caytonia* plant is known only from several small twigs preserved as compressions (Harris, 1940a, 1971). These show that *Caytonia* was a woody shrub or tree. Branching was sympodial, the apical bud ceasing to grow and being succeeded by two lateral buds to give forked branching as also occurred, for example, in some Bennettitales.

Leaves were borne on raised leaf cushions and were apparently alternate. Each leaf has a long petiole and four narrowly elliptical leaflets at the apex (Figure 5.28). The arrangement is superficially palmate, but the fundamental organisation is pinnate with the leaflets attached in two pairs. Each leaflet is narrowly elliptical to ovate with a midrib that becomes less pronounced towards the apex. The fine venation is reticulate, but lacks free-ending veinlets. Stomata are anomocytic. Small leaves with broad stipule-like petiolar flanges and bud scales are also known (Harris, 1940a). The leaves apparently were deciduous, with both leaflets and petioles shed.

The ovule-bearing structures (*Caytonia*) consist of a central axis or rachis bearing lateral units (Figure 5.28), but it is not known how they were borne on the stem. Each lateral unit consists of a short stalk terminated by a small sac-like cupule containing the ovules. The cupules are asymmetrical and reflexed. Each has a small lip on one side above the entrance to the mouth. All the cupules appear

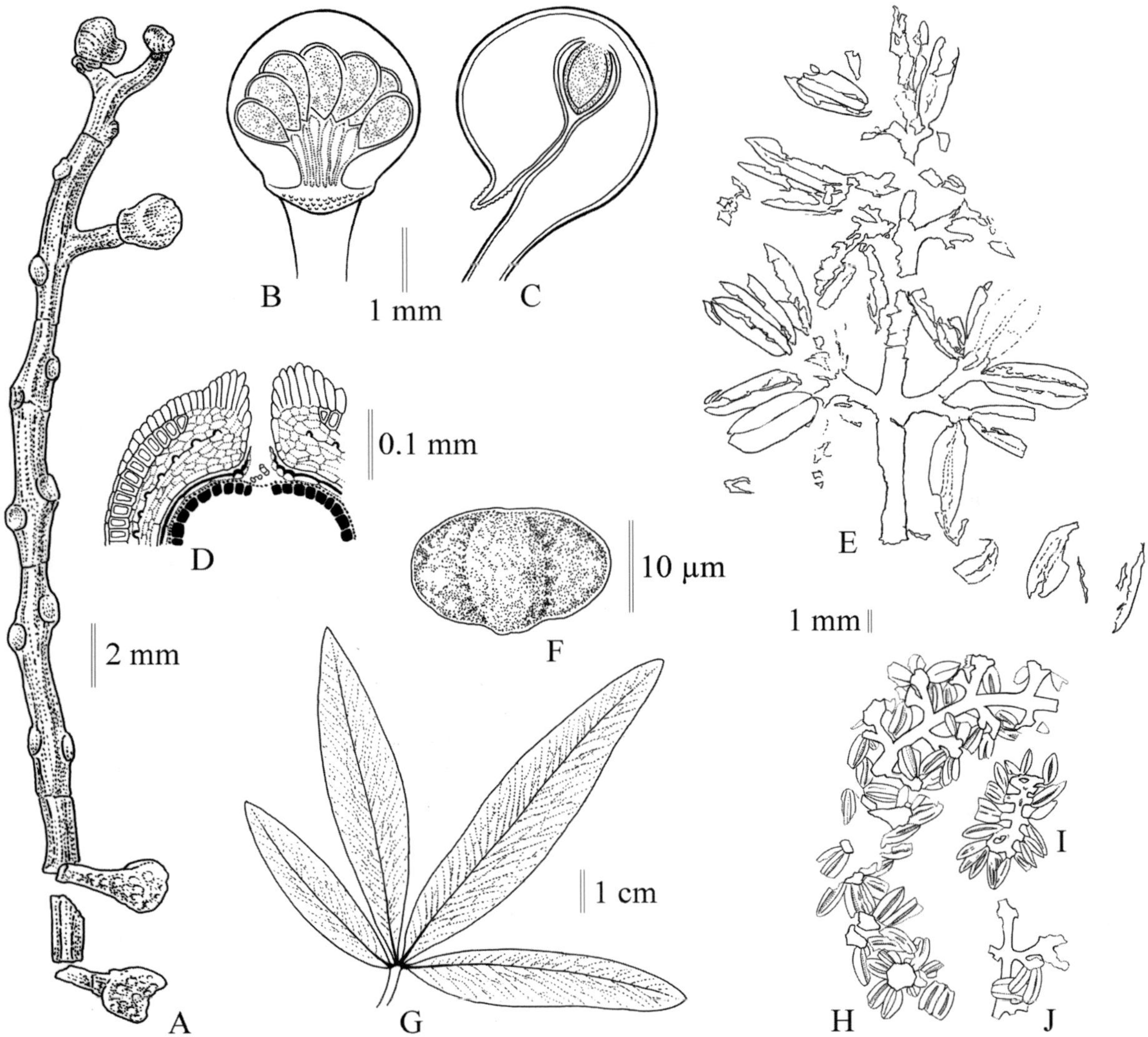

Figure 5.28 Caytoniales. (A) *Caytonia nathorstii*, ovulate axis from the Jurassic of Yorkshire, England. (B, C) Schematic section through *Caytonia* cupule with seeds in (B) dorsiventral and (C) lateral view. (D) Section through micropylar region of seeds of *Caytonia nathorstii* showing cells of seed wall and pollen at top of nucellus. (E) *Caytonanthus*, pollen-producing axis from the Jurassic of Yorkshire, England. (F) Bisaccate *Vitreisporites*-type pollen from *Caytonanthus arberi* sporangium from the Jurassic of Yorkshire, England. (G) *Sagenopteris colpodes*, leaf from the Jurassic of Yorkshire, England. (H–J) *Kachchia navicula*, fragments of putative caytonialean pollen organ from the Early Cretaceous of India. (A) Redrawn from Harris (1964); (B, C) redrawn from Crane (1985a); (D) redrawn from Harris (1958); (E) drawn from specimen in the Natural History Museum, London; (F) drawn from LM image in Pedersen and Friis (1986); (G) drawn from Thomas (1925); (H–J) redrawn from Bose and Banerji (1984).

slightly displaced towards one surface of the axis or rachis, and all are reflexed in the same way, with their lips pointing in the same direction. The cells around the lip are papillate (Thomas, 1925). At maturity the cupules were probably fleshy and berry-like (Harris, 1951) and shed individually. Fragments of cupules and other parts of the *Caytonia* plant have been described from small coprolites (Harris, 1945; Chapter 18). The fleshy tissue of the cupule consists of large rounded cells, branched sclereids and a network of vascular strands (Reymanówna, 1973).

Each cupule contains several small ovules surrounded by a common cuticular membrane. This membrane is continuous with the outer cuticle of the cupule and forms a small packet that contains the seeds (Reymanówna, 1973; Krassilov, 1977b). Micropyles of the ovules are oriented towards the mouth. A series of parallel narrow channels run between the cuticles of the mouth and lead from the mouth of the cupule, just below the lip, to the micropyles of the seeds. The seeds are flattened and apparently unitegmic, with a narrow micropylar slit (Thomas, 1925; Harris, 1940b, 1958). The anatomy of the seeds has been studied in detail by Harris (1940b) and Reymanówna (1973) based on maceration of compression material (Figure 5.28). The integument is supplied by a pair of vascular bundles and is free from the nucellus except at the chalaza. Harris (1958) reported that the integument and nucellus were strongly cutinised, but no acid-resistant megaspore membrane was detected.

Pollen-producing structures (*Caytonanthus*) are flattened with irregular, short lateral units (Figure 5.28). As with the ovule-bearing structures, it is not known how they were attached to the plant. The occurrence of different cuticles on the two different surfaces has been used to infer that the pollen-bearing structures were flattened and leaf-like. Elongated synangia, with four (rarely three or five) pollen sacs were borne on the presumed lower surface. The pollen sacs dehisced longitudinally. The pollen (*Vitreisporites*) is bisaccate with a distal sulcus and alveolate pollen wall with large cavities (Pedersen and Friis, 1986; Zavada and Crepet, 1986). Harris (1933, 1940b) demonstrated *Vitreisporites* pollen in the micropyles of *Caytonia* seeds.

A pollen organ potentially similar to *Caytonanthus*, *Kachchia navicula*, has been described from the Early Cretaceous flora of Kachchh in northwestern India (Bose and Banerji, 1984). The main difference from *Caytonanthus* is that the pollen sacs are not in groups of four in a single synangium (Figure 5.28). Instead, two distinct pollen masses have been seen in macerations of some of the *Kachchia* sporangia, suggesting that they may be bisporangiate synangia. Pollen within the sporangia is immature, but pollen grains attached to macerated fragments of the sporangial walls suggest that the grains are bisaccate. Specimens of *Kachchia navicula* are closely associated with leaves of *Pachypteris specifica*, with which it also shares stomatal similarities. However, *Kachchia* also co-occurs with *Caytonia* and *Sagenopteris* at some of the localities studied by Bose and Banerji (1984).

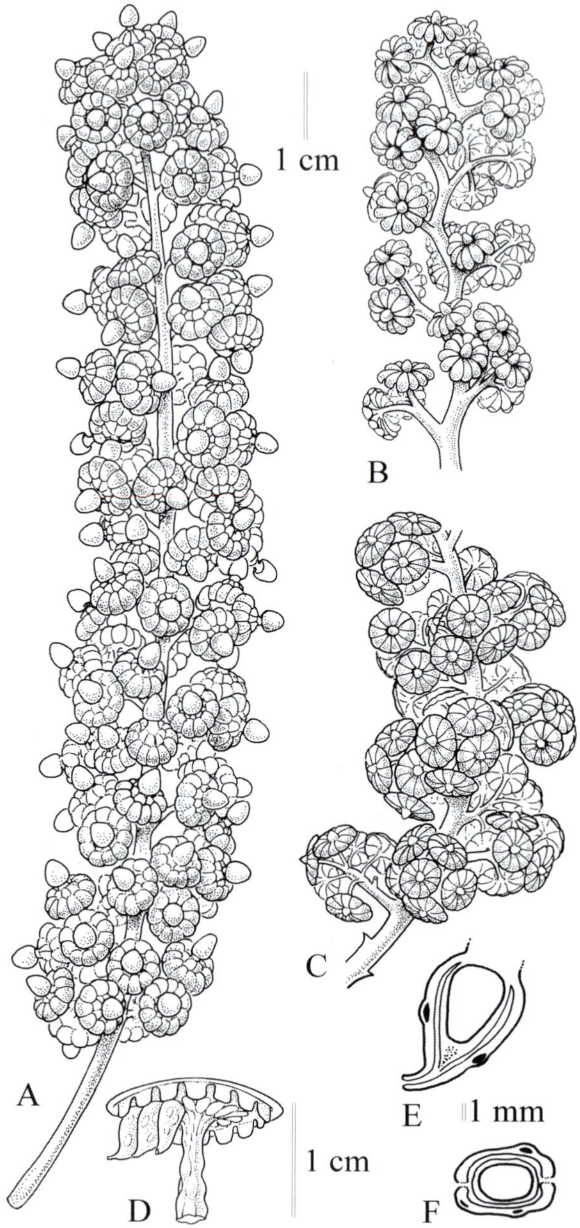

Figure 5.29 Peltaspermales from the Triassic of (A–C) South Africa and (D–F) East Greenland; reconstructions of ovulate axes with stalked peltate ovule-bearing units. In the material from South Africa each peltate head has a prominent central raised area: seeds are borne beneath the peltate rim. In the material from East Greenland the peltate head is less elaborate. (A) *Peltaspermum turbanatum*. (B) *Peltaspermum tridiscum*. (C) *Peltaspermum quindiscum*. (D–F) *Peltaspermum rotula*, reconstruction of (D) peltate head with seeds and (E, F) sections through seed. (A–C) Redrawn from Anderson and Anderson (2003); (D–F) redrawn from Harris (1932b).

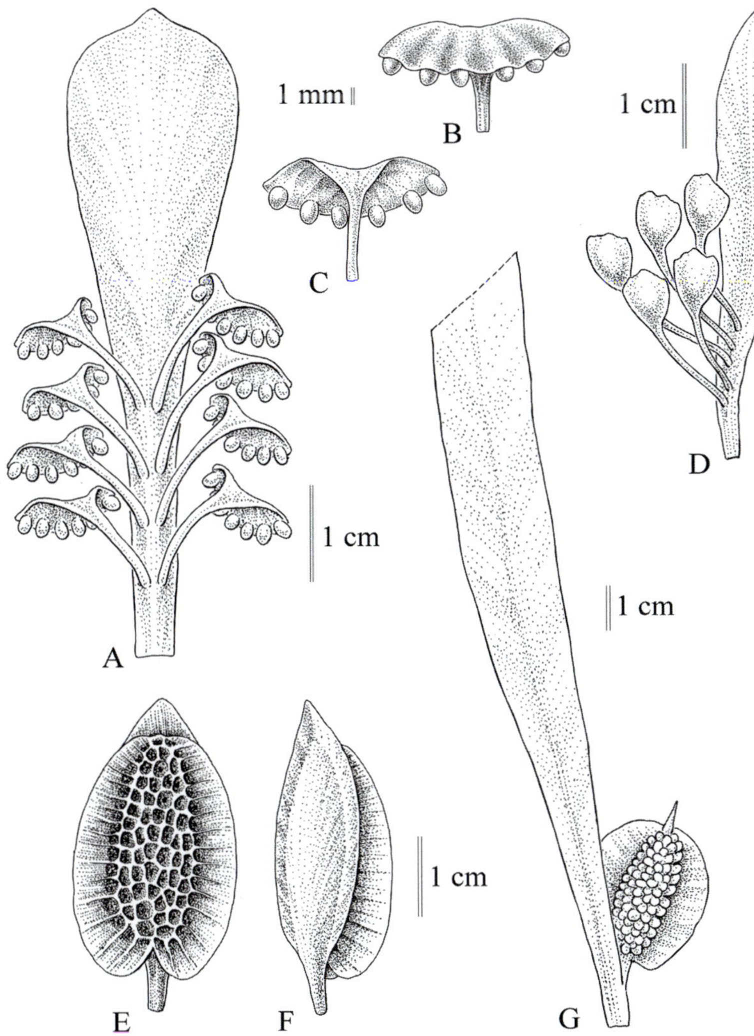

Figure 5.30 Ovulate organs of Glossopteridales from the Permian of (A–D) India and (E–G) South Africa. (A–C) Reconstruction of *Lidgettonia mucronata*, (A) leaf with pairs of ovule-bearing structures and (B, C) details of ovulate unit. (D) *Denkania indica*, leaf bearing uniovulate cupules. (E, F) Reconstruction of *Bifariala intermittens*, a double-winged ovulate structure, in two different views. (G) Reconstruction of *Gladiopomum dutoitides*. (A–C) Based on Surange and Chandra (1974b); (D) based on Surange and Chandra (1971); (E, F) based on Prevec *et al.* (2008); (G) based on Adendorff *et al.* (2002).

5.8.3 Peltaspermales

The Peltaspermales are widely distributed and diverse in Permian and Triassic floras. Phylogenetic analyses that have incorporated peltasperms have focused on the two most thoroughly investigated whole fossil plants. The *Autunia* plant is based on ovule-bearing structures (*Autunia*) that have been linked with leaves (*Lepidopteris*) and pollen-producing structures (*Antevsia*) (Kerp, 1988). A second plant, described by Harris (1932b, 1937) from the Late Triassic of East Greenland, is based on *Peltaspermum* ovule-bearing structures (Figure 5.29) that have been linked with associated *Lepidopteris* leaves, as well as *Antevsia* pollen-producing structures. A variety of other plant fossils may also be relevant to the peltasperm complex, especially from Angaran floras (see Meyen, 1984, 1988), and many new forms have also been described from South Africa (Figure 5.29) (Anderson and Anderson, 2003). However, the defining features of the group remain to be established and very little of this additional material has so far been incorporated into comparative syntheses or detailed phylogenetic analyses.

The most appropriate morphological interpretation of the ovule-bearing and pollen-producing structures of peltasperms is uncertain. The situation is also made difficult by

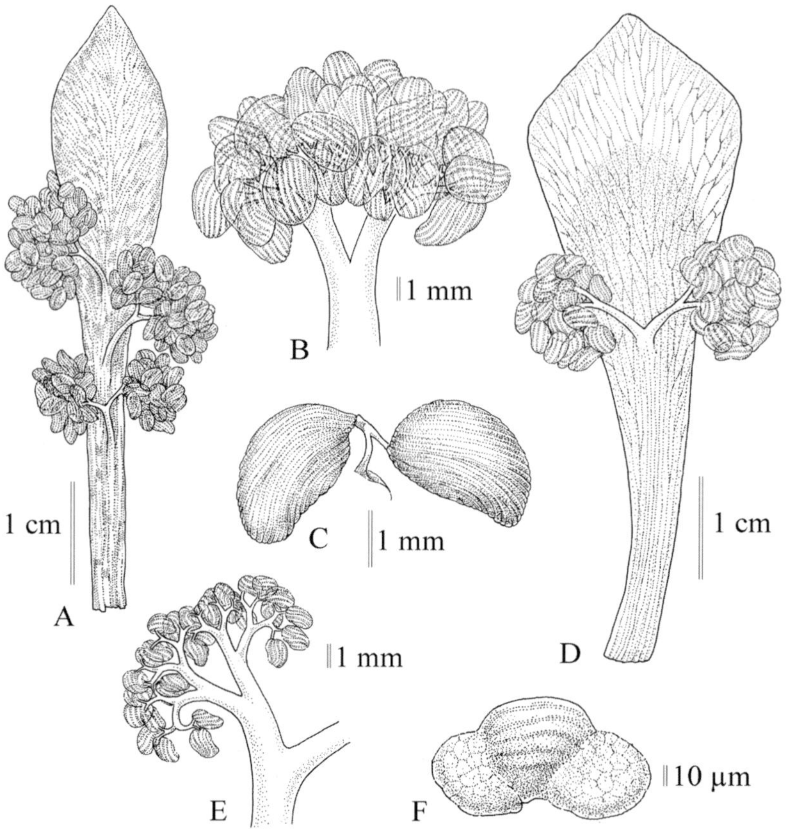

Figure 5.31 Pollen-producing structures of Glossopteridales from the Permian of India. (A–C) *Glossotheca orissiana*; (A, B) leaf with branched clusters of sporangia and (C) isolated sporangia. (D, E) *Eretmonia* sp.; (D) leaf with one pair of branched microsporangia clusters and (E) detail of cluster. (F) Bisaccate and striate glossopterid pollen. (A–C) Based on Surange and Chandra (1974c); (D, E) based on Surange and Chandra (1974a).

the fact that the group is known only from compression material. However, some of the compressed *Antevsia* material from East Greenland that we have examined is exceptionally well preserved. These specimens show a very stout central axis bearing three-dimensionally arranged lateral units. They are not leaf-like and in our view almost certainly consist of a main axis bearing compound lateral units that branch a few times. The ultimate branches terminate in elongate peltate structures that each bear about eight pollen sacs.

5.8.4 Glossopteridales

The Glossopteridales are the dominant group of plants in high-palaeolatitude areas of the Southern Hemisphere during the Permian; they occur abundantly in compression floras from southern South America, southern Africa, India, Antarctica and Australia. Useful reviews of glossopterid material are provided by McLoughlin (1990) for Australia, Pant (1977) and Surange and Chandra (1975) for India, and Anderson and Anderson (1985) for South Africa. Most glossopterid material is preserved as compressions or impressions, but permineralised material has been described from the Permian of Antarctica (Pigg, 1990; Taylor and Taylor, 1992; Pigg and Taylor, 1993; Pigg and Trivett, 1994) and Australia (Gould and Delevoryas, 1977; Nishida *et al.*, 2003).

Glossopterids are characterised by their distinctive tongue-shaped leaves with reticulate venation (*Gangamopteris*, *Glossopteris*). In most taxa the ovule-bearing structures and pollen-producing structures are partly fused to the adaxial surface of a leaf-like bract. Seeds such as *Pterygospermum* and *Platycardia* (e.g. Pant and Nautiyal, 1960; Pant, 1977) are flattened. Pollen (*Protohaploxypinus* and other genera) is saccate and striate. As would be expected from the wide distribution and ubiquitous occurrence of the distinctive leaves, glossopterid pollen is widespread and abundant in Permian palynofloras from Gondwana (Traverse, 1988).

The morphology of glossopterid ovule-bearing structures is complex (Figure 5.30), and detailed morphological interpretation is made more difficult by their generally poor preservation as compressions or impressions. Usually the ovule-bearing structure appears to be attached to the adaxial surface of a leaf-like bract. In some cases this structure appears to be pinnate (e.g. *Lidgettonia*, Thomas, 1958; Surange and Chandra, 1975, 1976) whereas in other cases it appears to be unbranched (Gould and Delevoryas, 1977; Taylor, 1996). However, more complex ovulate reproductive organs are also known in which the ovule-bearing structure appears to be sandwiched between a larger and a smaller leaf-like bract as in *Dictyopteridium feistmanteli*, *Venustostrobus* and *Plumsteadiostrobus* (Surange and Chandra, 1975, 1976) or the structure is double-winged as in *Hirsutum intermittens* (Prevec *et al.*, 2008).

Pollen-producing structures have an organisation similar to that of the simpler ovule-bearing structures (i.e. they are partly fused to a single bract). Pollen sacs are borne on one (*Eretmonia*) or more (*Glossotheca*) pairs of branches that emerge from the mid-rib of the attached leaf-like bract (Figure 5.31).

Until a convincing and comprehensive morphological interpretation of glossopterid reproductive structures is available, it will remain difficult to assess their relationships to other groups of seed plants. This will almost certainly require additional permineralised material, including well-preserved specimens showing attachment to stems (Taylor, 1996).

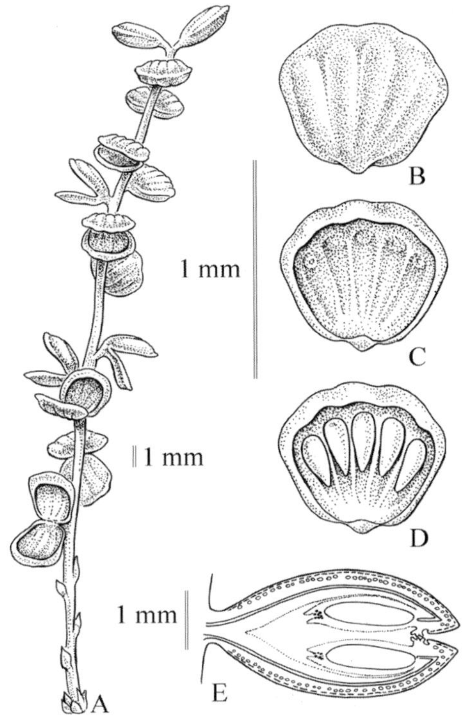

Figure 5.32 *Leptostrobus cancer*, the ovule-bearing structure linked to Czekanowskiales. (A) Axis with valvate ovulate units in a spiral arrangement. (B–D) Reconstruction of ovulate unit seen (B) from outside, (C) from inside with seed scars and (D) with seeds. (E) Reconstructed section through ovulate unit showing the two valves and seeds. (A–D) Based on Schweitzer (1977); (E) redrawn from Harris *et al.* (1974).

5.8.5 Czekanowskiales

The Czekanowskiales are a group of enigmatic plants that are widespread in the Jurassic and Early Cretaceous, especially at high palaeolatitudes in the Northern Hemisphere. They are easily recognised by their distinctive deciduous shoots bearing 11 – 16 elongated leaves (*Czekanowskia*, *Phoenicopsis*, *Solenites*), but their reproductive structures are poorly understood (Crane, 1985a). Ovule-bearing structures of *Leptostrobus* (Figure 5.32) have been linked with *Czekanowskia* or *Solenites* leaves based on association evidence and cuticular similarity. *Leptostrobus* consists of an elongated axis with characteristic bivalved ovulate units, each unit enclosing several seeds. The pollen producing structures, however, have not been identified securely. At some localities the pollen organ *Ixostrobus* has been linked with *Czekanowskia* leaves. However, at other localities, the evidence for linking *Czekanowskia* with *Ixostrobus* is weak. *Ixostrobus* may be attributed equally well to *Desmiophyllum* leaves (Harris *et al.*, 1974). The Czekanowskiales are one of several enigmatic groups of seed plant about which we currently understand relatively little. In particular, the small capsule-like ovulate structures are poorly understood. Nevertheless, they may be important for a more refined understanding of seed plant relationships and the position of angiosperms within the group.

6

Origin and age of angiosperms

A convincing hypothesis of angiosperm origins in the context of a secure understanding of seed plant phylogeny still remains to be achieved. A major impediment is the need for improved information on fossil seed plants, but there are also other challenges, particularly concerning how best to integrate the massive and increasing disparity between what is known about the genomes of living plants and what is known about the morphology and anatomy of extinct and extant taxa. In this chapter we briefly review the development of ideas concerning seed plant phylogeny and outline the status of research in this area. We also review the implications of ideas on relationships for understanding the origin and age of angiosperms.

6.1 HYPOTHESES OF SEED PLANT RELATIONSHIPS

Ideas on relationships among seed plants that developed in the early twentieth century (e.g. Coulter and Chamberlain, 1917; Chamberlain, 1935) focused on gymnosperms (often excluding angiosperms completely) and recognised two groups: cycadopsids and coniferopsids. Cycadopsids included cycads, Bennettitales (Cycadeoidales) and the fossil plants grouped together at that time as seed ferns (e.g. lyginopterids, medullosans). Coniferopsids comprised conifers, cordaites and *Ginkgo*, and in many schemes also included Gnetales. This biphyletic interpretation of seed plant evolution gathered further support with the recognition of Devonian progymnosperms and the hypothesis that aneurophytalean progymnosperms might have given rise to the cycadopsid line, while archaeopteridalean progymnosperms might have given rise to the coniferopsid line (Beck, 1966, 1970, 1971, 1981). As developed by some authors (e.g. Doyle, 1977), this notion of two main lines of seed plant evolution interpreted angiosperms as the culmination of evolutionary elaboration in the cycadopsid line, while Gnetales were viewed in the equivalent position among coniferopsids (Figure 6.1).

Among those who challenged the concept of separate cycadopsid and coniferopsid lineages of seed plants, Rothwell (1981, 1982) noted that the Palaeozoic seed plant *Callistophyton* combined conifer-like reproductive features (especially platyspermic seeds and saccate pollen) with fern-like foliage. Meyen (1984) proposed a more complex pattern of seed plant evolution that recognised two platyspermic lineages (Ginkgoopsida, Pinopsida), and included Gnetales among other radiospermic seed plants in Cycadopsida.

Some of the most influential hypotheses on angiosperm origin mainly focused on the origin of the angiosperm flower, and the early development of ideas in this area was greatly influenced by expanding knowledge of non-angiosperm fossil plants. In particular, the Euanthial Theory (Anthostrobilus or Ranalian Theory) developed by Arber and Parkin (1907), Bessey (1894, 1896, 1897, 1915), Hallier (1900, 1901, 1902, 1912) and others, was strongly influenced by the discovery of bisexual flower-like reproductive structures in the Bennettitales by Wieland (1906, 1911).

Under the Euanthial Theory, as conceived by Arber and Parkin (1907), the angiosperm flower is a single, bisexual axis bearing lateral appendages in a spiral (Figure 6.2). *Magnolia*-like flowers are similar to the basic angiosperm flower, with the small, inconspicuous flowers of the 'Amentiferae', and other angiosperms, interpreted as secondarily reduced. Takhtajan (1969), for example, characterised the basic angiosperm flower as large, insect-pollinated, with numerous, helically arranged, free parts, a perianth of modified bracts and broad, laminar stamens without differentiation into filament and anther. Carpels were large and leaf-like, and the large seeds developed from anatropous bitegmic ovules.

The Gnetales were interpreted as the closest living relatives of angiosperms by Arber and Parkin (1908), with the extinct Bennettitales the sister group to both lineages. Flower-like reproductive structures in all three groups were interpreted as derived from a pro-anthostrobilus, consisting

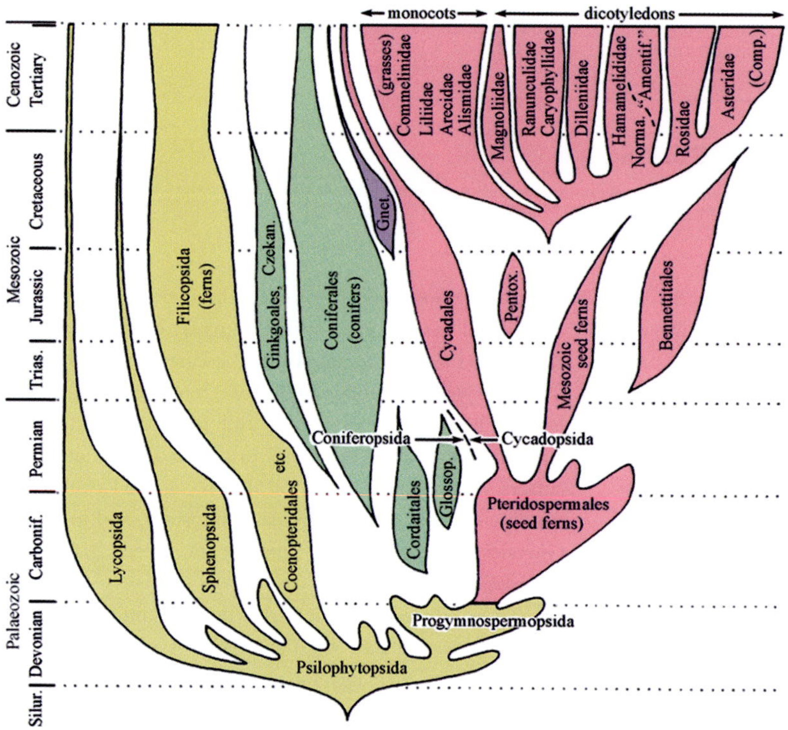

Figure 6.1 Pre-cladistic view of land plant relationships that also summarises the stratigraphic distribution of major groups and generalised temporal changes in their abundance. Note the traditional division of seed plants into Cycadopsida (pink) and Coniferopsida (green), and the implied close relationship among Mesozoic seed ferns. Glossopt., Glossopteridales; Gnet., Gnetales; Pentox., Pentoxylales; Norma., Normapolles pollen group, "Amentif.", 'Amentiferae', Comp., Compositae; Silur., Silurian; Carbonif., Carboniferous; Trias., Triassic. Modified from Doyle (1977).

of pinnate microsporophylls and megasporophylls. In angiosperms, the microsporophylls were thought to have been reduced and simplified to form the characteristic stamen. In Bennettitales the megasporophylls were interpreted as reduced to a single ovule. In Gnetales both micro- and megasporophylls were interpreted as highly reduced.

The Pseudanthial Theory, advanced by Wettstein (1907) and others (Neumeyer, 1924; Janchen, 1950), offered a different interpretation of angiosperm flowers. Under the Pseudanthial Theory (Figure 6.2) the flower was interpreted as a pluriaxial system: essentially a condensed inflorescence consisting of a primary axis and more or less numerous, variously reduced, secondary axes. Both orders of axes bear lateral appendages, but pollen sacs and ovules (i.e. sporangia) are borne terminally on axes (stachysporous *sensu* Lam, 1950). Under this interpretation the stamen is an axis bearing two pairs of sporangia, while the carpel is a structure that combines a bract with the ovule-bearing axis that it subtends.

According to the Pseudanthial Theory, small, simple, unisexual flowers, such as those of Betulaceae, Casuarinaceae, Fagaceae, Juglandaceae and Myricaceae, were interpreted as basic within angiosperms as a whole, and compared to the cones of Gnetales. These families, together with several others that have small, unisexual flowers (e.g. Chloranthaceae, Piperaceae), were grouped by Wettstein (1907) as the Monochlamydeae.

Despite very different interpretations of the structure of the flower, initial formulations of the Euanthial Theory and the Pseudanthial Theory both interpreted the Gnetales as closely related to angiosperms. However, this possibility fell from favour with the recognition that vessels in angiosperms and Gnetales probably originated independently (Thompson, 1918; Bailey, 1944) and was further undermined by interpretations of wood anatomy (Bailey, 1944) and reproductive structures (Eames, 1952) that inferred a close relationship between Gnetales and conifers. Evidence from comparative palynology, wood anatomy and other features also showed clearly that the 'Amentiferae', the angiosperm group to which Gnetales had been compared most closely, occupied a relatively derived position in angiosperm phylogeny, making a direct link to the Gnetales unlikely.

Another factor that led to declining interest in Gnetales (and also Bennettitales) as possible close relatives to angiosperms was the apparent difficulty of deriving an

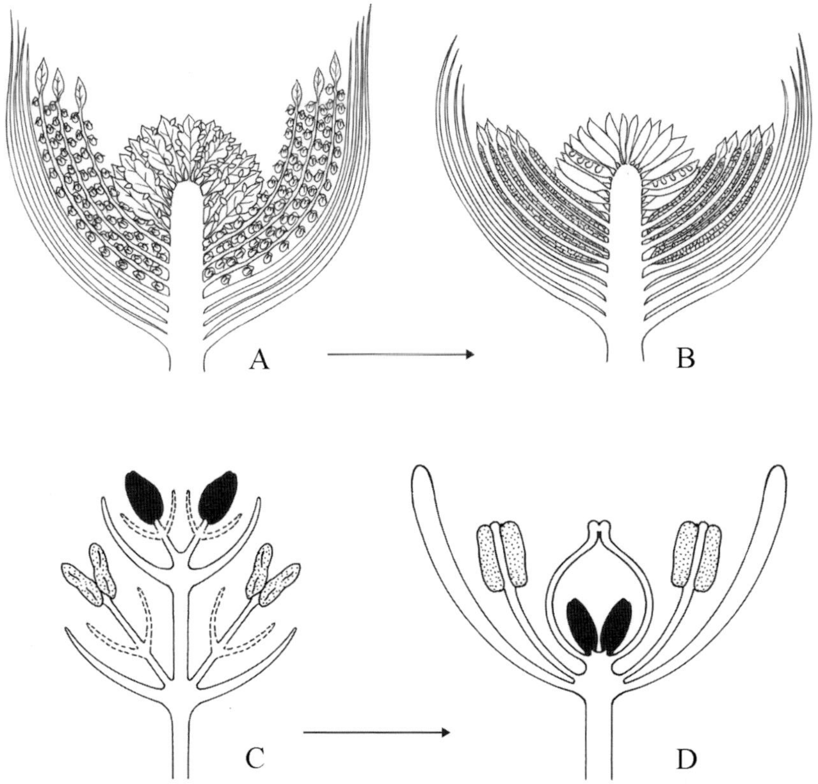

Figure 6.2 (A, B) Euanthial Theory. (A) Uniaxial 'pro-anthostrobilus' of the hypothetical 'Hemiangiospermeæ' as envisaged by Arber and Parkin (1907): a bisexual, flower-like reproductive structure with inner leaf-like megasporophylls (carpels) bearing naked ovules on their margin, in a similar way to extant *Cycas*, surrounded by leaf-like microsporophylls and an outer zone of sterile bracts. (B) *Magnolia*-like angiosperm flower derived from the hypothetical form in (A), with numerous multiovulate carpels, numerous stamens with flattened tips and numerous perianth parts not differentiated into sepals and petals; the flower is a uniaxial structure and all organs of the flower are leaf homologues. (C, D) Pseudanthial Theory. (C) Hypothetical, multiaxial reproductive structure derived from an *Ephedra* cone bearing axillary shoots, each with a terminal ovule or terminal pollen-producing structure; axillary shoots subtended by bracts. (D) Angiosperm flower, derived from the hypothetical form in (B) in which the carpel wall is formed by the bracts of axillary ovule-bearing shoots, stamens are modified axillary shoots, and the perianth is formed by bracts; the flower is multiaxial and the organs of the flower are a mixture of leaf and shoot homologues. (A, B) Drawn from Arber and Parkin (1907); (C, D) redrawn from Firbas (1947).

angiosperm carpel from their simple ovulate reproductive structures. In seeking to explain the origin of the angiosperm carpel, attention focused instead on several fossil groups (glossopterids, corystosperms, Caytoniales) in which the ovule-bearing structures were apparently much more complex.

Other ideas on the homology of the angiosperm flower have combined the Euanthial and Pseudanthial theories, and some authors have also suggested that angiosperm flowers have been derived more than once (Karsten, 1918; Hagerup, 1934; Fagerlind, 1947; Lam, 1950; Burger, 1977; Melville, 1983a, b; Meeuse, 1987). Under these interpretations, not all angiosperm flowers are homologous, and clusters of flowers in some groups are homologous to single flowers in other groups. Lam (1950), for example, divided the angiosperms into phyllosporous and stachysporous groups, whereas Meeuse (1986) in his Anthocorm Theory suggests that there are at least two different kinds of flower (or in Meeuse's terminology FRUs, functional reproductive units): holoanths and anthoids. The holoanths, including for instance the flowers of Magnoliaceae, are modified pluriaxial systems (holoanthocorms), whereas the anthoid flowers, including, for example, the flowers of Chloranthaceae, are modified uniaxial systems. Similarly, Karsten (1918) derived large multipartite flowers such as those of Magnoliaceae from clusters of *Gnetum*-like reproductive units. He derived the small simple flowers of, for example, Casuarinaceae, from an *Ephedra*-like reproductive unit.

Two theories in which the flowers of different groups of angiosperms are interpreted as non-homologous relate the number of radial sectors in the flower to the number of fundamental units from which the flower is constructed. According to Melville (1962, 1963, 1983a, b) the angiosperm flower is composed of several reproductive shoot systems with their subtending bracts (gonophylls) derived from the reproductive structures of glossopterids. The number of gonophylls is equivalent to the number of floral sectors. Similarly, Burger (1977) interpreted most angiosperm flowers as derived from simple flowers such as those in Chloranthaceae. The trimerous flowers of monocots, for example, would be composed of three such units. Under this interpretation most angiosperm flowers are pseudanthia, except for a few relictual forms (e.g. Chloranthaceae) that are uniaxial true flowers.

A logical extension of the idea that not all angiosperm flowers are homologous (e.g. Karsten, 1918; Hagerup, 1934) is that angiosperms may be polyphyletic rather than monophyletic. This view has been advocated by several authors based on both morphological and stratigraphic arguments (e.g. Meeuse, 1972a, b, c; Hughes, 1976; Krassilov, 1977a, 1997). Some of these ideas link different groups of angiosperms with different groups of non-angiosperm seed plants, but frequently such hypotheses have been based on superficial similarities between the reproductive structures of angiosperm and non-angiosperm groups, with little regard to other features.

In cladistic terms, angiosperm polyphyly requires that some subgroup of angiosperms can be shown to be more closely related to one or more plants at the 'gymnosperm' grade than it is to other angiosperms. The question of whether all angiosperm flowers are homologous or not, may or may not be relevant to this issue, but hypotheses of angiosperm polyphyly have not found wide support because they require multiple origins of the many characteristic structural and reproductive novelties of the group. Given current knowledge of living and fossil seed plants, as well as cladistic analyses based on both morphological and molecular data, angiosperm polyphyly seems highly unlikely.

6.1.1 Cladistic hypotheses based on morphology

Cladistic analysis (phylogenetic systematics *sensu* Hennig, 1965) was first used in plant systematics in the late 1960s (Koponen, 1968); in the early 1980s (Parenti, 1980; Bremer and Wanntorp, 1981) cladistic approaches were first applied to broader systematic problems in the evolution of land plants. The first cladistic analysis to focus specifically on relationships among extant seed plants (Hill and Crane, 1982) was devised primarily as an example intended to illustrate the principles of cladistics, rather than as a definitive study of seed plant relationships. Nevertheless, many of the characters first used in that study have been utilised repeatedly in subsequent work.

The first cladistic analysis to incorporate both extant and fossil seed plants (Crane, 1985a) examined relationships among 19 taxa using 38 characters. Results identified a clade in which Bennettitales plus *Pentoxylon* were the sister group of Gnetales plus angiosperms. Within Gnetales, *Gnetum* and *Welwitschia* were sister taxa (figs. 20 and 22 in Crane, 1985a). If the 'cupule' of Bennettitales and *Pentoxylon* were coded as homologous to the 'cupules' of glossopterids, *Caytonia*, corystosperms, and the outer integument of angiosperms, then corystosperms formed the sister group to the Bennettitales–*Pentoxylon*–Gnetales–angiosperm clade, with *Caytonia* as the sister group to corystosperms plus that clade (Figure 6.3). Another result of this early cladistic analysis was that *Lyginopteris* was resolved as a sister taxon to all other seed plants. A large clade (platysperms, Crane, 1985b), characterised by flattened seeds and saccate pollen, was also recognised, which included all the seed plants considered, except *Lyginopteris*, medullosans and cycads (Figure 6.3).

The synthesis of character data and initial phylogenetic analysis by Crane (1985a) was extended by Doyle and Donoghue (1986), who focused on a similar set of 20 taxa. Significant advances included improved coding of characters and greater representation of vegetative features. This study also made use of computer-assisted analyses to evaluate the robustness of results, assess the influence of different character codings and examine the implications of alternative phylogenetic hypotheses. However, the results obtained (Figure 6.4) were very similar to those found previously. Lyginopterids and *Medullosa* were placed as the earliest diverging groups of seed plants, and a large clade defined by the presence of flattened seeds and saccate pollen was also recognised. This corresponded to the platysperm clade of Crane (1985b), except

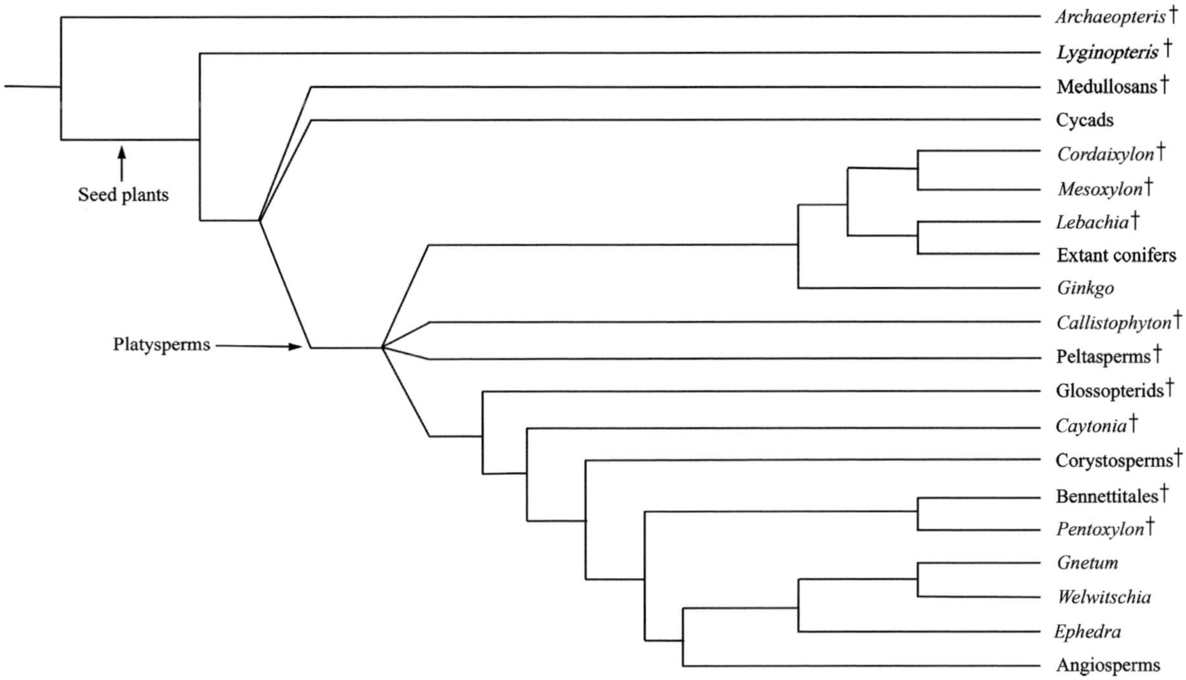

Figure 6.3 Cladistic relationships among seed plants according to the first attempt to incorporate extant and fossil seed plants into the same cladistic analysis. Note that *Archaeopteris* is used as an outgroup, that angiosperms and Gnetales are sister taxa, and that both are sister to Bennettitales and *Pentoxylon*. Fossil taxa indicated by †. Redrawn from Crane (1985a).

for the inclusion of cycads. Relationships within the platysperm clade were similar to those found earlier, and again, as in Crane (1985a, b), angiosperms, Bennettitales, *Pentoxylon* and Gnetales formed a monophyletic group (Figure 6.4). Doyle and Donoghue (1986) termed this group comprising angiosperms and its closest relatives the 'anthophyte clade' based on the 'strong aggregation of sporophylls into flower-like structures' (p. 355 in Doyle and Donoghue, 1986).

Several other analyses, based largely on the dataset in Doyle and Donoghue (1986), were published in the late 1980s. Doyle and Donoghue (1985) focused on relationships between the angiosperms and Gnetales, whereas later (Doyle and Donoghue 1987) they addressed the issue of angiosperm origins. Later revisions and reanalyses resulted in patterns of relationships basically similar to earlier results. However, from the beginning it was clear that trees that were almost equally parsimonious showed significantly different patterns. In some of these topologies angiosperms were linked directly with late Palaeozoic–Mesozoic seed ferns (*Caytonia*, Doyle and Donoghue, 1986; *Caytonia* plus glossopterids, Doyle and Donoghue, 1992), which were then linked to other anthophytes. Such results raised concerns about the monophyly of anthophytes as they were originally circumscribed. In other topologies (termed 'neo-englerian' by Doyle, 1996), which were only slightly different from their most parsimonious trees, anthophytes were linked with coniferopsids (conifers, cordaites, *Ginkgo*) rather than late Palaeozoic–Mesozoic seed ferns, and Gnetales were placed as the earliest diverging group of anthophytes.

The analysis of Rothwell and Serbet (1994) further improved the data matrix of Doyle and Donoghue (1992) in terms of the inclusion of additional taxa (especially a greater range of extant and fossil coniferophytes) as well as the incorporation of new character information for early seed plants, coniferopsids and *Pentoxylon*. The result was similar to the 'neo-englerian' topologies of Doyle and Donoghue (1986, 1992), but an unusual feature was the separation of extant conifers (resolved as the sister group to anthophytes) from the Palaeozoic conifer *Emporia*, cordaites and *Ginkgo*.

A significant theoretical issue recognised by early cladistic studies (Crane, 1985a; Doyle and Donoghue,

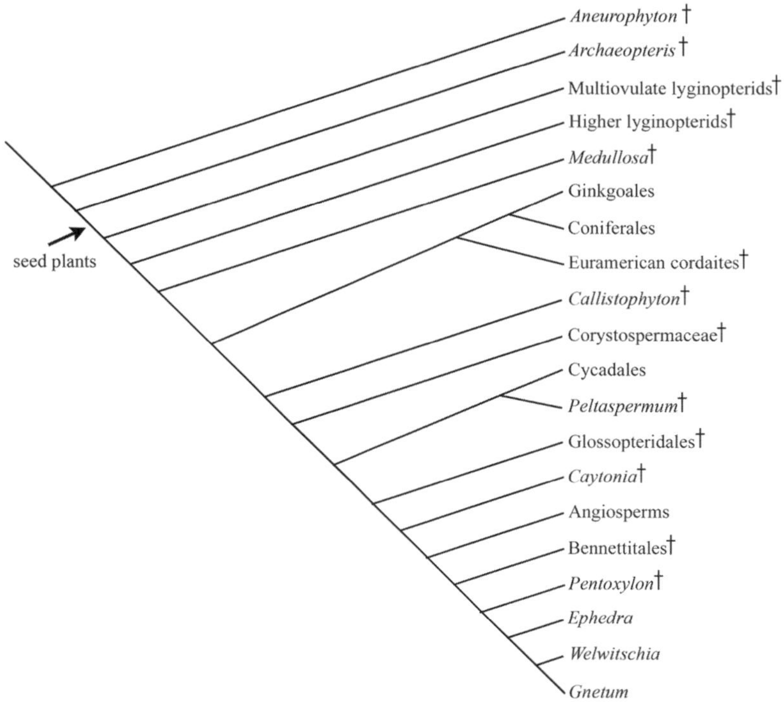

Figure 6.4 Cladistic relationships among seed plants from the early analysis of Doyle and Donoghue (1986). Note that *Aneurophyton* and *Archaeopteris* are used as outgroups and that angiosperms are sister to a clade including Bennettitales, *Pentoxylon* and Gnetales. Fossil taxa indicated by †. Redrawn from Doyle and Donoghue (1986).

1986), but not adequately dealt with in initial analyses, was the question of how very diverse clades should best be represented in large-scale analyses of seed plant relationships. Of particular concern was how angiosperms should be most appropriately included, especially given the very different vegetative and reproductive morphology among those extant taxa that might plausibly be among the early branching lineages in the group. At this time, prior to widespread application of molecular phylogenetics, patterns of relationship at the base of the angiosperm clade were very uncertain.

Analyses by Doyle and Donoghue (1992) and Doyle *et al.* (1994) were the first that tested, rather than assumed, angiosperm monophyly. Doyle *et al.* (1994) included nine angiosperm taxa ('core' Magnoliales, *Austrobaileya*, Chloranthaceae, Winteraceae, eudicots, Aristolochiaceae, Piperales, Nymphaeales, monocots) in morphological analyses based on a modification of the 1992 dataset (Doyle and Donoghue, 1992). In all analyses Gnetales were the closest living relative to angiosperms, and in some trees glossopterids and *Caytonia* were linked directly to angiosperms, with the resulting clade being the sister group to the remaining anthophytes.

Nixon *et al.* (1994) went further than all previous cladistic analyses in terms of breaking down diverse and complex terminal taxa. Although the monophyly of these groups was not seriously questioned, Nixon *et al.* (1994) recognised that some of these groups were sufficiently heterogeneous that different scoring, based on variation within the group, might potentially affect overall patterns of seed plant relationships. Cycads were represented by Cycadaceae, Stangeriaceae and Zamiaceae. Conifers were represented by Cephalotaxaceae, Podocarpaceae, Taxaceae, Taxodiaceae plus Cupressaceae, Araucariaceae and Pinaceae. Angiosperms were represented by *Lilium* (monocot) and *Betula*, *Caltha*, *Chrysolepis*, *Casuarina*, *Dillenia*, *Hamamelis*, *Platanus* and *Trochodendron* (eudicots), as well as a variety of magnoliids (*Ceratophyllum*, *Chloranthus*, *Piper*, *Persea*, Winteraceae, *Nymphaea*, *Magnolia*, *Calycanthus*, *Eupomatia*). The resulting topologies again recognised anthophytes (angiosperms, Bennettitales, Gnetales) as monophyletic. As in previous analyses, the earliest diverging seed plants were *Lyginopteris* and medullosans. Cycads and conifers were resolved as monophyletic, and cycads were placed as sister group to the platysperm clade.

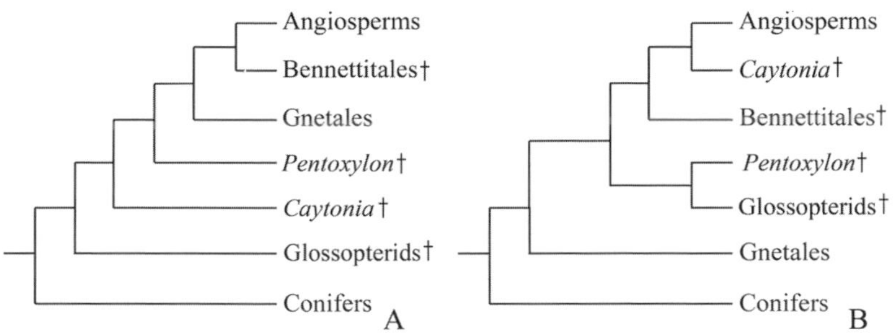

Figure 6.5 Relationships among anthophytes under two different scenarios. Fossil taxa indicated by †. (A) Analysis by Doyle (2006). (B) Analysis by Hilton and Bateman (2006).

Surprisingly, Gnetales were not monophyletic. *Welwitschia* and *Gnetum* formed a clade that was the sister group to angiosperms.

Doyle (1996) responded to some of the criticisms of Nixon *et al.* (1994) and conducted further analyses that incorporated the six major groups of extant conifers (including the Late Carboniferous genus *Emporia*), as well as 11 groups of angiosperms (Nymphaeales, Piperales, Aristolochiaceae, monocots, eudicots, Winteraceae, Laurales, Chloranthaceae, Austrobaileyaceae, *Eupomatia*, Magnoliales). Results linked *Caytonia* with angiosperms as part of a larger clade (termed glossophytes by Doyle, 1996) that also included Gnetales, Bennettitales, glossopterids and *Pentoxylon*.

Analyses of seed plant phylogeny published more recently (Doyle, 2006; Hilton and Bateman, 2006) have continued to recover topologies that are broadly similar to those found in earlier cladistic analyses, especially in terms of the groups that they associate most closely with angiosperms. In the strict consensus of the 21 most parsimonious trees found in the analyses of Hilton and Bateman (2006) *Caytonia* is sister to angiosperms, with Bennettitales sister to both. At the next level out, glossopterids plus *Pentoxylon* are sister to the Bennettitales–*Caytonia*–angiosperm clade and the Gnetales are sister to them all (Figure 6.5). Similarly, in the analysis by Doyle (2006) that included morphological data and fossils, angiosperms are also linked with Gnetales (Figure 6.5), but topologies with Gnetales linked to conifers are almost as parsimonious. In trees that constrain Gnetales and conifers to form a clade Cycadales, *Glossopteris* + *Pentoxylon*, Bennettitales and *Caytonia* are successive sisters to angiosperms.

Friis *et al.* (2007) expanded the Hilton and Bateman dataset to include the Erdtmanithecales and newly recognised fossil seeds similar to those of Gnetales and Bennettitales. This analysis also recognised the anthophyte clade. In particular, it highlighted the Erdtmanithecales (Chapter 5) as another member of a loosely defined anthophyte assemblage (Friis *et al.*, 2007), based on features such as the presence of an additional envelope surrounding the integument and nucellus, the granular infratectal layer in the pollen wall, and the apparently simple microsporophylls. Later work (Friis *et al.*, 2009a) confirmed the close similarity between seeds of Gnetales and Erdtmanithecales, but went further in documenting the diversity of dispersed seeds of this kind as well as the strong similarities to seeds of Bennettitales. A further analysis of seed plant relationships by Rothwell *et al.* (2009) including both extinct and extant seed plants questioned the Bennettitales–Erdtmanithecales–Gnetales link, but otherwise their morphological analysis also supported the anthophyte model.

As anthophytes were originally conceived, the supporting morphological characters included: the presence of an additional integumentary envelope surrounding the integument and nucellus, the possession of a pollen wall with a granular–columellate infratectal layer, at least some stomata with two subsidiary cells parallel to the guard cells, which is inferred to reflect mesogenous (syndetocheilic) development, and the aggregation of the reproductive organs into flower-like structures (Crane, 1985a; Doyle and Donoghue, 1986; Nixon *et al.*, 1994; Rothwell and Serbet, 1994). Other potential defining characters of anthophytes include: scalariform pitting in the secondary xylem, whorled or opposite arrangement of microsporophylls, apical meristems with tunica–corpus organisation, wood with syringal lignin subunits (Mäule reaction), and (secondarily) non-saccate pollen (Crane, 1985a; Doyle and Donoghue, 1986; Nixon *et al.*, 1994; Rothwell and Serbet, 1994). Information on fertilisation in *Ephedra* and *Gnetum* (Friedman, 1990; Friedman and Carmichael, 1996) also raises the possibility that double fertilisation may be a general feature of anthophytes, but the situation in extinct taxa is unknown.

Examined critically, several characters on this list of putative defining features of the anthophytes are unsatisfactory in various respects and highlight general problems with all existing morphological datasets. First, some of these features (e.g. apical meristem) are poorly surveyed among extant plants and unlikely to be preserved, even under ideal circumstances, in most fossil groups. Whether or not the fossil taxa actually possessed these features is open to different interpretations. Second, one of the key features (aggregation of reproductive organs into flower-like reproductive structures) is very poorly defined, raising questions about the basic hypothesis of character homology. Third, the non-saccate pollen character requires acceptance of homology based on the absence of a key feature (sacci). It is obviously tenuous to regard two or more organisms as fundamentally similar in a feature that neither of them possesses.

Nevertheless, despite these and other difficulties, the core of the anthophyte idea has proved remarkably resilient. The same groups, Bennettitales, Gnetales, *Pentoxylon*, and now Erdtmanithecales, together with Caytoniales, corystosperms and glossopterids, are resolved repeatedly in various relationships as the lineages that are most closely related to flowering plants. Other patterns of relationships that introduce other groups (e.g. conifers) are almost equally parsimonious, but there nevertheless appears to be a signal in the morphological dataset that aggregates the familiar list of the same eight groups together. That signal deserves continuing careful consideration and further investigation.

6.1.2 Cladistic hypotheses based on molecular data

Analyses of phylogenetic relationships among living and fossil seed plants initiated in the 1980s, based on structural features, have gradually given way to phylogenetic approaches based on molecular data and focused solely on extant taxa. Early work was based on sequence data either from the chloroplast gene *rbc*L (Hasebe *et al.*, 1992; Chase *et al.*, 1993) or nuclear rRNA sequences (Hamby and Zimmer, 1992). In the Chase *et al.* (1993) study seed plants were rooted with cycads, and Gnetales and angiosperms were resolved as monophyletic sister groups. In the Hasebe *et al.* (1992) analysis the results varied depending on whether the raw sequences, or the sequences converted into amino acids, were analysed. Results also depended on the method of analysis (parsimony, maximum likelihood or neighbour joining). In the Hamby and Zimmer (1992) analysis the resulting topologies also varied with method of analysis (parsimony or neighbour joining). In some analyses Gnetales and angiosperms were sister taxa. In others, Gnetales were sister to angiosperms plus conifers, *Ginkgo* and cycads.

Doyle *et al.* (1994) reanalysed an expanded version of the Hamby and Zimmer (1992) dataset including 71 taxa, and resolved Gnetales and angiosperms as sister groups. In analyses in which 49 of the original taxa were combined into a representative sample of 18 taxa, cycads were resolved as the earliest branching seed plant lineage with conifers and then *Ginkgo*, as successive sister groups to a clade comprising monophyletic Gnetales and monophyletic angiosperms. An identical topology for the different groups of seed plants (but not for groups within the angiosperms) was also supported by combined analyses of morphological and rRNA data.

Thus, in the mid-1990s there seemed to be broad support for the hypothesis that Gnetales are the closest extant relatives of angiosperms based on morphological data (Crane, 1985a, 1985b; Doyle and Donoghue, 1986; Nixon *et al.*, 1994), phylogenetic analyses of sequence data (Hamby and Zimmer, 1992; Doyle *et al.*, 1994), as well as analyses of combined morphological and molecular data (Albert *et al.*, 1994; Doyle *et al.*, 1994). However, this conclusion has been seriously challenged by more recent work.

Beginning in the later 1990s a series of studies based on molecular data have called into question the notion that Gnetales and angiosperms are more closely related to each other than either is to cycads, *Ginkgo* or conifers. Analyses based on 28S (Stefanoviac *et al.*, 1998), cpITS (Goremykin *et al.*, 1996), 18S (Chaw *et al.*, 1997) and *cox*1 (Bowe *et al.*, 2000) sequence data all placed angiosperms as sister to a monophyletic group comprising cycads, *Ginkgo*, Gnetales and conifers. In several of these studies Gnetales were the sister group to conifers, and this result was also supported in phylogenetic studies based on a long sequence of chloroplast DNA (Hansen *et al.*, 1999) as well as MADs box genes (Winter *et al.*, 1999).

Subsequently, phylogenetic analyses using multiple genes and a variety of analytic approaches, have broadly supported these results in resolving gymnosperms and angiosperms as monophyletic sister taxa (e.g. Qiu and Palmer, 1999; Bowe *et al.*, 2000; Chaw *et al.*, 2000; Burleigh and Mathews, 2004; Mathews, 2009, 2010). In addition, these analyses very frequently place Gnetales and conifers together in various ways. Surprisingly, however, patterns of relationship in which Gnetales and conifers are sister taxa are recovered much less frequently than those that place Gnetales within conifers, most often as the sister group to Pinaceae ('gne-pine' trees), but sometimes as sister of cupressophytes (Mathews, 2010, and references therein). In recent analyses based on molecular data there is very little or no support for a close relationship between Gnetales and angiosperms.

On morphological grounds, a close relationship between Gnetales and conifers may not be surprising. It would also not necessarily rule out a close relationship of both conifers and Gnetales with angiosperms, if gymnosperms are not monophyletic. However, gymnosperm monophyly has several potentially problematic implications: it suggests that the lineage leading to angiosperms has been separated from that leading to all other extant seed plants since the Carboniferous and that all supposed apomorphic similarities between angiosperms and specific groups of living seed plants are convergent. It implies, for example, that siphonogamy in angiosperms is not homologous with siphonogamy in conifers and Gnetales.

Positioning Gnetales among conifers is still more surprising and undermines the presumed evolutionary integrity of conifers. Conifers have long been recognised as difficult to define on morphological grounds (e.g. Crane, 1985a), but on the other hand morphological concerns have rarely focused on Pinaceae, which share clear morphological and other features with other conifer families. These features either would have to be convergent, or would have to have been lost in Gnetales under 'gne-pine' and other topologies that place Gnetales inside conifers. Interestingly, more extensive sampling of conifers appears to preserve conifer monophyly (Rai *et al.*, 2008), perhaps providing a potentially instructive example of the importance of adequate systematic sampling in addressing large-scale phylogenetic questions.

Attempts to resolve definitively seed plant relationships using molecular data continue to yield different, sometimes radically different, results that individually appear to be strongly supported. Conclusions vary depending on the data and the analytic techniques employed. For example, a recent duplicate gene rooting approach links angiosperms and cycads, a pattern that has previously received little support, with *Ginkgo* as sister to Gnetales and conifers (Mathews, 2009). Nevertheless, a persistent signal in this and most other molecular datasets places Gnetales and conifers together in some way. That signal, like the equally persistent signal from morphological analyses that links Bennettitales, Gnetales, *Pentoxylon*, Erdtmanithecales, Caytoniales, corystosperms, glossopterids and angiosperms, deserves continuing careful consideration and further investigation. The key question for the future is: can both signals be reconciled or does acceptance of one require that the other be rejected entirely?

6.1.3 Current status of phylogenetic studies

The plethora of phylogenetic analyses of seed plants over the last decades has resulted in a forest of often seemingly contradictory trees, but nevertheless there are still some significant points of agreement.

First, all studies based on morphological evidence indicate that several lineages of early seed plants (hydraspermans, lyginopterids, and probably medullosans) diverged before the common ancestor of all living seed plants. There is therefore evidence for a basal grade of evolution among seed plants that is not represented by any of the five living groups. It is among these taxa, and their relatives at the progymnosperm grade, that answers to questions relating to the origin of seed plants and early seed plant biology must be sought.

Second, as a corollary to recognition of a basal grade of seed plants, all studies that include living and fossil seed plants recognise a clade that is broadly equivalent to the platysperms of Crane (1985b). This group appears to be characterised by the presence of flattened seeds and saccate pollen. Despite concerns about the homology of platyspermy in different groups, this clade appears robust. It contains all five groups of extant seed plants and was perhaps characterised by a more 'modern' seed plant reproductive biology that operated at the hydrasperman, lyginopterid, medullosan grade. For example, the switch from tetrahedral to linear megaspore tetrads may have occurred in the origin of this group. The appearance of pollen grains with sacci may also indicate new modes of pollination that were later transformed several times, with a subsequent repeated loss of the sacci, in the platysperm clade.

Third, all studies that include living and fossil seed plants reinforce the clear heterogeneity of the groups that were formerly grouped together as seed ferns. Seed ferns *sensu lato* include the earliest seed plants, as well as some taxa that diverged very early in seed plant phylogeny, together with other groups of perhaps reproductively more sophisticated seed plants that are more closely related to the five extant lineages. Even among seed ferns within the platysperm clade, patterns of relationship are very diverse.

Fourth, almost all morphological studies, and many molecular studies, resolve cycads, Gnetales and angiosperms as monophyletic groups. Only a single analysis of morphological data (Nixon *et al.*, 1994) separated *Gnetum* and *Welwitschia* from *Ephedra*, and this result did not survive reanalysis of several critical characters (Doyle, 1996). The situation with respect to conifers is much less clear. Many recent analyses based on molecular data have cast doubt on the monophyly of conifers, while others support it.

Fifth, within Gnetales almost all analyses of both molecular and morphological data resolve *Gnetum* and *Welwitschia* as sister taxa to the exclusion of *Ephedra*. The only exception is the early analysis of *rbc*L sequence data (converted to amino acid sequences) by Hasebe *et al.* (1992) that resolves *Gnetum* as the sister group to *Ephedra* plus *Welwitschia*. However, when the same DNA sequences are analysed the normal pattern of relationships is recovered: *Ephedra* is the sister group to *Gnetum* and *Welwitschia*. This may suggest that both the unusual habit of *Welwitschia* and the unusual leaf morphology of *Gnetum* are secondary developments within the group.

Less secure, but also important, is that among the morphological analyses that include both extant and fossil plants, most resolve Bennettitales as relatively closely related to Gnetales, although not always to the exclusion of other groups. This pattern has received further support from documentation of unusual seed structure that indicates a close relationship of Bennettitales, Erdtmanithecales and Gnetales. The implication from the recognition of this BEG group is that if Gnetales and angiosperms are not closely related, then Bennettitales and Erdtmanithecales are not likely to be closely related to angiosperms either. Wherever Gnetales are placed by analyses of molecular data, then its extinct seed plant relatives, must also be placed there.

A further theme among all analyses based on morphological data that include both living and fossil plants is that Gnetales, Bennettitales, *Pentoxylon*, and now Erdtmanithecales, together with a few groups of other extinct seed plants (glossopterids, *Caytonia*, corystosperms) are somehow closely related. The pattern of relationships among these seven groups of seed plants and their relationship to angiosperms remains very uncertain, in large part because the homologies among their reproductive structures are not well understood. Until the key homologies are resolved more securely, this problem will continue to undermine the reliability of phylogenetic results based on morphology.

6.2 ORIGIN OF ANGIOSPERM STRUCTURE

Uncertainty over the relationships of angiosperms to other groups of seed plants precludes specific conclusions about the origin of many aspects of angiosperm structure and biology. Nevertheless, clarification of patterns of relationships among angiosperms themselves, together with palaeobotanical information, permits some conclusions about the origin of some characteristic angiosperm features.

6.2.1 The angiosperm flower

Angiosperm flowers are merely aggregations of their characteristic reproductive organs, and there are clear analogies (and perhaps some homologies) among living and fossil seed plants. In many seed plants, the ovule- and pollen-producing structures are aggregated on a determinate axis and there is an irreversible conversion of the apical meristem from the vegetative phase into the reproductive phase. Reestablishment of the vegetative phase is relatively uncommon. Interestingly, it occurs most conspicuously in the ovule-producing plants of *Cycas*, which has the most leaf-like megasporophylls of any living seed plant. In most cases, both in angiosperms and in other seed plants, the dividing cells of the apical meristem are used up in the production of reproductive organs. The formulation of the ABC model (Figure 6.6), and its gradual extension and modification to different kinds of angiosperm flowers, has been pivotal for understanding floral organ identity and its genetic regulation during angiosperm floral development (e.g. Coen and Meyerowitz, 1991; Kramer and Hodges, 2010).

Despite this apparent clarity, interesting questions still remain concerning the origin of the angiosperm flower. One issue concerns whether the characteristic reproductive organs of angiosperms evolved at the same time, or after, the process of aggregation that resulted in flower-like construction, or whether they evolved their characteristic form first and became aggregated only later (Stuessy, 2004). The latter possibility raises the question of whether

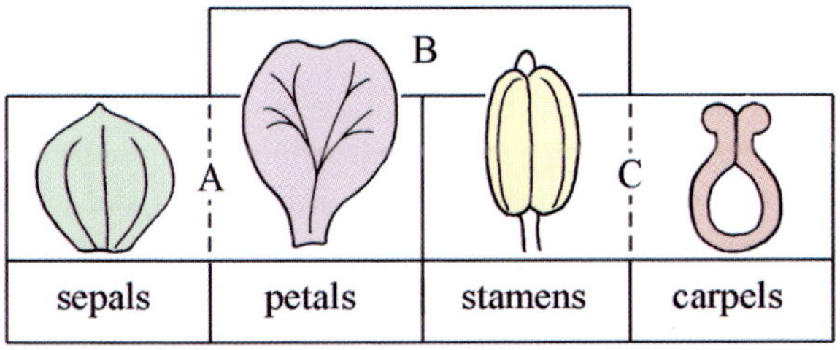

Figure 6.6 ABC-model for floral organ identity and regulation in the development of the angiosperm flower.

angiosperms went through a 'prefloral' phase at an early stage in their evolution.

Flowers of living and fossil angiosperms currently provide no clear evidence of a prefloral phase in angiosperm evolution. Initial suggestions that the organisation of the reproductive structures of *Archaefructus* reflect such a condition (Stuessy, 2004) now seem unlikely (Friis *et al.*, 2003a; Endress and Doyle, 2009), and the reproductive axes of *Archaefructus* can be interpreted straightforwardly as an inflorescence bearing numerous simple unisexual flowers. Interpretations of the floral structure of *Hydatella* are much more ambiguous (Rudall *et al.*, 2007; Rudall and Bateman, 2010). Although *Hydatella* may ultimately be shown to be of profound significance for interpreting the origins of the angiosperm flower, it seems more likely that it will reflect secondary simplification, albeit at a very early stage in angiosperm evolution, perhaps as a result of heterochrony.

A second open question is whether bisexual or unisexual flowers are basic for angiosperms as a whole. Evidence from both extant angiosperms and the Early Cretaceous record suggests that early in angiosperm evolution the production of bisexual or unisexual flowers was not strongly fixed.

6.2.2 The angiosperm stamen

The unique structure of the angiosperm stamen is a diagnostic feature for angiosperms as a whole. Typical angiosperm stamens, which have two pairs of pollen sacs, not only are ubiquitous among those living angiosperms that are placed close to the base of the angiosperm phylogenetic tree, but also occur in the earliest angiosperm flowers. However, although the diagnostic value of the angiosperm stamen is now clearer than it was 20 years ago, the origin of its characteristic structure remains uncertain. It is possible to imagine various scenarios for the evolution of the angiosperm stamen from the microsporangiate structures of other seed plants, but none of these will be wholly convincing without clearer insights into the likely homologies of the key distinctive features of angiosperm stamens. Angiosperm stamens are distinctive in their apparently extreme reduction compared with the more complex microsporangiate structures of the other seed plants, their bilateral symmetry and their dehiscence, which occurs between the sporangia rather than individually through the wall of each sporangium. Recognition of extinct seed plants that show some approach to these conditions in their microsporangiate structures would be of great interest.

Relatively little attention has been given to developing ideas about how the characteristic angiosperm stamen may have evolved. Most studies, including those carried out in a cladistic framework (e.g. Crane, 1985a; Doyle and Donoghue, 1986), either explicitly or implicitly interpret the stamen as a microsporophyll with two pairs of pollen sacs reduced from the more complex forms seen in other seed plants (e.g. cycads). Structurally, angiosperm stamens are unlike the microsporophylls of any living or fossil seed plants. They differ significantly from the microsporophylls of cycads, while their bilateral symmetry is more like the microsporophylls of conifers and *Ginkgo* than the antherophores of Gnetales. Among fossil seed plants, the structures bearing the pollen sacs are generally more complex and often compound in their organisation (bipinnate or branched).

The bulk of the information now available suggests that the pollen-producing structures of peltasperms (*Antevsia*), glossopterids (*Eretmonia*, *Glossotheca*), corystosperms (*Pteruchus*), Pentoxylales (*Sahnia*) and Erdtmanithecales (*Bayeritheca*, *Erdtmanitheca*, *Eucommiitheca*) are most straightforwardly interpreted as axial structures bearing lateral branches or small modified leaves on which the pollen sacs are borne (Chapter 5). It no longer seems tenable to interpret these organs as pinnately organised microsporophylls.

The microsporophylls of *Caytonia* (*Caytonanthus*) are of interest in having four pollen sacs fused together to a synangium and this has led to comparisons with angiosperm stamens (Thomas, 1925). Unfortunately, the major impediment in pursuing this comparison further is the lack of detailed structural information. It is currently unclear whether *Caytonanthus* is a pinnately organised microsporophyll or complex branching structure with relatively simple synangial units. It has generally been interpreted as a microsporophyll, based mainly on Harris' description of the organ as flattened with a differentiated upper and lower cuticle (Harris, 1940b). However, in our view *Caytonanthus* could just as easily be interpreted as an axial structure as a planated microsporophyll and this would also bring it more in line with the pollen organs of corystosperms, to which Caytoniales seem very likely to be closely related.

Another difficult situation is to interpret the microsporangiate structures in the Bennettitales. Unfortunately the anatomical evidence from permineralised material is not clear, but there is no doubt that the pollen-producing structures in this group are complex. In *Bennettistemon amblum* the form suggests a pinnately branched structure fused entirely to the bract (Crane, 1988), whereas in *Weltrichia sol* the pollen sacs are 'borne on short fertile appendages on inner surface of rays' (Harris, 1969). In *W. spectabilis* Harris (1969) reports 'the inner face of rays bearing fertile branches'. In *W. pecten* and *W. whitbiensis* the synangia appear to be borne directly on the 'rays'. In the three-dimensionally preserved material of *W. santalensis* the fertile appendages are 'borne in a double row on the upper (adaxial surface) of microsporophyll segments' (Sitholey and Bose, 1971) and the peculiar inward rolling of the most distal fertile appendages also suggests a morphology more complicated than a simple microsporophyll. However *Cycadeoidea* is interpreted (compare Wieland, 1906, 1916; Delevoryas, 1963, 1965, 1968), it is also clear that the pollen-producing structures are much more than simple microsporophylls. Taken together we believe that the pollen-producing structures of Bennettitales are most likely to be fundamentally axial, with a fertile branch system, of varying degrees of complexity, fused either in part or completely to its corresponding bract. This is a situation not too different from that in glossopterids. Further studies are needed to establish this more securely.

Also difficult to interpret are the pollen-producing structures of Gnetales. The antherophores of extant Gnetales have often been interpreted as a pair of fused structures. However, careful developmental studies show that in both *Welwitschia* and *Ephedra distachya* the antherophore is composed of two fused lateral shoots (Mundry and Stützel, 2004). In *Welwitschia* the apex of each of the two shoots is visible early in development, but later aborts. If this interpretation holds for Gnetales as a whole, it brings the morphology of gnetalean pollen organs into line with that of other key groups of seed plants. It also reconciles the cone-like morphology of the putative early gnetalean pollen organ *Masculostrobus clathratus* (Crane, 1988) with the seemingly very different morphology of the antherophores in the three extant genera.

6.2.3 The angiosperm ovule

A key issue with respect to the origin of angiosperms is the evolution of anatropous, bitegmic ovules from the orthotropous, unitegmic ovules of most other extant and fossil seed plants. In those angiosperms where there is only one integument, it appears to reflect secondary loss. Developmental evidence from angiosperms, including from studies of ovule mutants in *Arabidopsis* (e.g. Gasser *et al.*, 1998; Kelley and Gasser, 2009), indicates that formation of the inner and outer integuments is under different genetic control. However, determining how best to compare the coverings around the ovules in different groups of seed plants encounters various problems.

First, among extant angiosperms there is variation in whether one or two integuments are developed, sometimes even among closely related plants. More problematic from a practical standpoint is that as the seed matures, different layers and different tissues within the covering(s) around the nucellus develop to different extents. Others are crushed and entirely or partly obliterated. At maturity, and without knowledge of development, it can be difficult, in both living and fossil plants, to establish whether the original ovule had one or two integuments.

Second, it is clear that in several groups of extant seed plants a second covering around a unitegmic ovule has arisen *de novo*. Among conifers, for example, the fleshy arils of *Taxus* and Podocarpaceae are apparently independent developments. There are also tendencies toward ovule enclosure in some cycads. In peltasperms, the ovules are partially enclosed by the expanded heads of the ovule-bearing structures. In most cases (e.g. *Peltaspermum rotula*) this structure is peltate and radial, but in other cases it is asymmetric.

Third, there are a variety of structures among early seed plants that surround, partly enclose, or fully enclose the ovules. Among hydraspermans, for example, the ovules are surrounded by sterile cupule lobes, in *Lagenostoma* the seed sits within an obvious cup-like cupule, and in *Calathospermum* and similar forms several ovules are completely or partly enclosed within a cupule. The extent to which these different kinds of structures may have been incorporated developmentally into the apparently unitegmic ovules of plants such as medullosans or Callistophytales deserves further study.

However, notwithstanding these potential difficulties of homology among the seed coverings of early seed plants, the seeds of *Callistophyton* (*Callospermarion*) appear to be

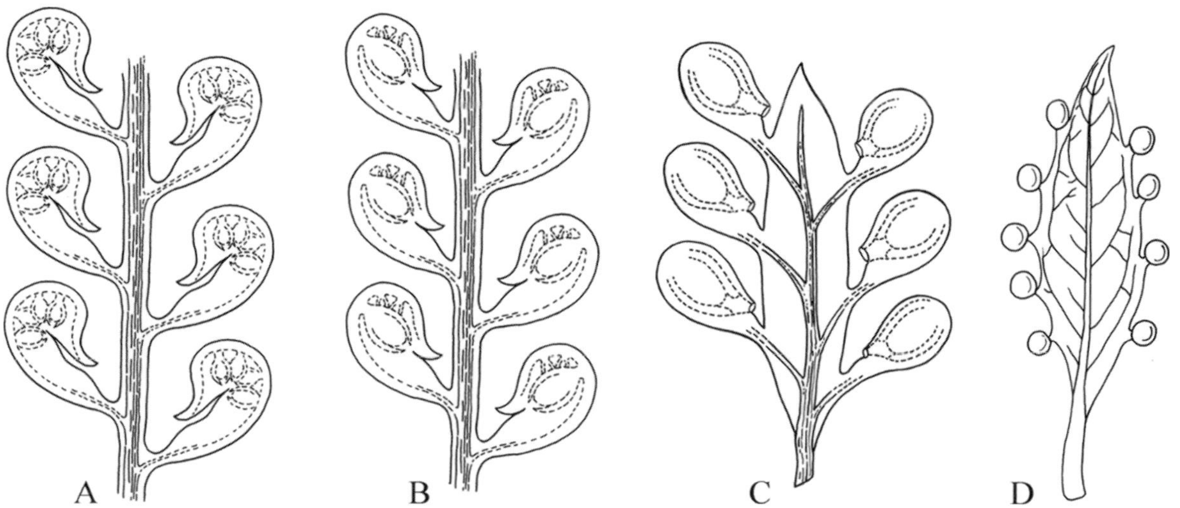

Figure 6.7 Theoretical model for deriving the angiosperm bitegmic ovule and closed carpel from *Caytonia*. (A) *Caytonia* ovule-bearing structures bearing recurved cupules, each containing several seeds. (B) Hypothetical intermediate form; only one ovule in each cupule matures to seed. (C) Hypothecial intermediate form; only a single ovule and seed per cupule, cupule and enclosed ovule correspond to the bitegmic ovule of angiosperms; ovule-bearing axis enlarged and planate. (D) Leafy ovule-bearing structure bearing bitegmic ovules along the margin; this corresponds to an opened plicate (leaf-like) angiosperm carpel. Based on Gaussen (1946).

unitegmic in the same way as those of extant conifers and provide a convenient point of reference for the ovules of later plants from the Permian and Mesozoic. Among glossopterids and corystosperms, as in *Callospermarion*, there is anatomically preserved material. In all cases the ovules of these groups have been interpreted as unitegmic. For Caytoniales, and also for peltasperms, information on seed anatomy is based solely on studies of compression (coalified) material, but again the most detailed studies that have been done have concluded that the seeds of these groups are unitegmic.

The working hypothesis in studies of angiosperm ovule evolution is therefore that the inner integument is comparable to the single integument seen in most other seed plants, and what needs to be explained for a deeper understanding of angiosperm evolution is the origin of the second integument. A variety of suggestions have been made for deriving the second integument from pre-existing structures in other seed plants. Especially popular has been the idea that the second integument may have been derived from the cupule that encloses the ovules in *Caytonia*, as suggested by Gaussen (1946) and several subsequent authors (e.g. Stebbins, 1974; Doyle, 1996), by reduction of the number of ovules in the cupule to one (Figure 6.7). At the same time the recurved cupules seen in both *Caytonia* and corystosperms provide a plausible basis for explaining the anatropous condition of angiosperm ovules. An elaboration of this idea is that the cupules of *Caytonia* and corystosperms, as well as the outer integument of angiosperms, are homologous with the outer layer surrounding the ovules (envelope) in the seeds of Bennettitales (e.g. Crane, 1985a), thereby linking the situation in orthotropous and anatropous forms.

Although these ideas provide a unifying framework for understanding the structures surrounding the ovule and the true integument in several different groups of seed plants, they are not yet supported by detailed morphological and anatomical comparisons. There are also persistent uncertainties about how other aspects of the ovulate structures of angiosperms, *Caytonia*, corystosperms, Bennettitales and other groups should be compared to each other. For example, there is still controversy over whether the cupule-bearing axis of *Caytonia* is a modified leaf (megasporophyll) or a branch bearing several megasporophylls, as seems more likely for corystosperms.

The situation in Bennettitales has also been controversial (section 5.6). In addition to being embedded among interseminal scales, many bennettitalean seeds are also enclosed by an outer envelope around the nucellus and integument. This is seen very clearly in compression fossils such as *Bennettites crossospermus* and *Vardekloeftia*, but has also now been detected in permineralised specimens (Chapter 5). This is almost certainly a general feature of the group even though it may be difficult to detect in some material. Based on comparison with the similar seeds of fossil Gnetales and Erdtmanithecales, the origin of the outer envelope seems most likely to have been from a small number of bracts; in all these taxa, it is the outer envelope that has a sclerenchymatous layer that provides the main protection for the developing embryo. If correct, this would indicate that the bennettitalean ovulate reproductive structure is itself composed of many, often hundreds, of very substantially reduced lateral branches rather than reduced megasporophylls.

Also relevant to the situation in Bennettitales and angiosperms is the morphological interpretation of seeds of *Pentoxylon* (*Carnoconites*). In *Carnoconites* the tightly packed seeds have a thick, probably somewhat fleshy, outer layer surrounding a harder, bicarinate sclerified layer. The standard interpretation is that these tissues are sarcotesta and sclerotesta. However, it seems just as likely that they are the inner and outer parts of an envelope equivalent to that seen in Bennettitales (Chapter 5). Under this interpretation, and if this layer originated from a pair of bracts, each *Carnoconites* ovulate reproductive structure is composed of numerous highly reduced lateral branches.

From the above it is clear that a tendency toward enclosure of the ovules by some kind of additional layer is a common feature of Late Palaeozoic and Mesozoic seed plants. In glossopterids, corystosperms and Caytoniales this is accomplished by complete or partial enclosure by some kind of lamina-like structure. In many cases more than one seed is enclosed (even in some corystosperms). In Bennettitales (and perhaps *Pentoxylon*) there is a more intimate association of an outer layer around a single seed, but here the enclosing structure seems to be homologous to a small number of bracts. In Gnetales and Erdtmanithecales ovule enclosure around a single seed also appears to have been by associated bracts. All these innovations may be important for understanding how the second (outer) integument of angiosperms may have come about. But despite their similar functional attributes, for example additional protection for the ovule(s) and perhaps involvement in dispersal, the extent to which these different situations may have a common developmental or evolutionary basis remains uncertain. The central question, especially in light of the anthophyte concept, is whether the cupules of *Caytonia* and corystosperms can be reconciled with the structure of the envelope in Bennettitales, Erdtmanithecales and Gnetales, or whether these two kinds of covering are completely independent developments.

6.2.4 The angiosperm carpel

The origin of the carpel remains one of the most important persistent questions for a deeper understanding of the origin of angiosperms. Based on extant taxa, it now seems clear that the carpels of early angiosperms contained only one or a small number of ovules. This is also supported by palaeobotanical evidence from early fossil angiosperms, in which carpels also contain one or only a small number of ovules. Studies of extant taxa also show that carpel development in early-diverging lineages of angiosperms is fundamentally ascidiate (e.g. Endress and Doyle, 2009). This current view contrasts with the earlier idea that the primitive carpel is fundamentally a laminar, leaf-like, plicate structure; a view that was also fundamental to the development of the Euanthial Theory.

There is currently no clear indication of the likely origin of the angiosperm carpel from a pre-existing state in living or fossil seed plants. A great variety of ideas have been presented in attempting to explain the origin of the angiosperm carpel, but most of these have not studied the related problem of the origin of the bitegmic ovule. For example, Thomas (1933) suggested that the carpel was homologous to the cupule wall of *Caytonia*, whereas Melville (1962, 1963, 1983a) derived the carpel from the bracts of glossopterids. Gaussen (1946) and later authors (e.g. Doyle, 1996) suggested a derivation of the carpel from the *Caytonia* cupule-bearing axis (megasporophyll) by planation of the rachis and folding over of the cupules (Figure 6.7), resulting in a plicate, rather than an ascidiate, structure.

Variants of the Pseudanthial Theory derive the carpel from subtending bracts of the axes bearing ovules (e.g. Wettstein, 1907). The possibility of enclosing an orthotropous ovule with more than one integument (as in Gnetales) makes it possible to derive both the bitegmic ovule and the carpel in this way. A disadvantage, however, is that unlike the *Caytonia* or corystosperm scenario this does not account for the origin of the anatropous condition.

6.2.5 Vegetative structure of angiosperms

The transition from woody to herbaceous lineages, and *vice versa*, has happened many times in angiosperm phylogeny with significant effects on rates of molecular, and perhaps organismal, evolution (Smith and Donoghue, 2008). In the context of current knowledge of angiosperm phylogeny the basic condition, as in other seed plants, was most likely woody, but with several distinct origins of herbaceous lineages very early in angiosperm evolution (e.g. Nymphaeales, *Ceratophyllum*, Piperales). Most importantly, in monocotyledons, complete loss of the ability to form normal seed plant secondary xylem resulted in fundamental changes to vegetative structure that had profound evolutionary consequences.

Some scenarios of early angiosperm evolution posit a direct link between the development of laminar leaves and the presence of vessels, and highlight the combination of those features as a potential key innovation underpinning angiosperm success. However, the occurrence of vessel-less taxa close to the base of the angiosperm tree introduces new complexity that needs to be taken into account. In particular, the placement of *Amborella* (which lacks vessels) as the sister group to all other angiosperms implies that vessels originated within the group. This possibility, or at least the possibility that the production of vessels was not fully fixed at this level of angiosperm evolution, is further strengthened by the absence of vessels in *Austrobaileya* and some/all Nymphaeales. Vessels are also absent in Chloranthaceae, which remain in an uncertain position close to the point of eudicot–eumagnoliid–monocot divergence.

It is now clear that laminar leaves with several orders of reticulate venation are basic for the angiosperms as a whole. Based on currently accepted phylogenetic patterns, the origin of the monocot leaf, with its characteristic parallel venation, is a secondary modification that occurred early in angiosperm history. The closest similarity to angiosperm leaves in a non-angiosperm group occurs in *Gnetum* (Figure 1.9). Comparison of the development of the leaves of *Gnetum* with that of angiosperms, such as *Amborella*, would be interesting. It would also be interesting to revisit the scenario of leaf evolution elaborated by Hickey and Doyle (1977) in the context of what is now known about angiosperm phylogeny and the molecular biology of angiosperm leaf development.

6.3 THE AGE OF ANGIOSPERMS

6.3.1 Implications of hypotheses of relationships

Although the pattern of relationships among different groups of seed plants remains uncertain, the fact that all

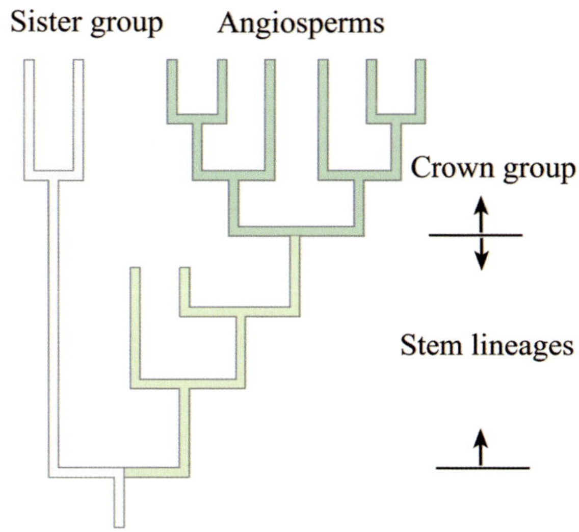

Figure 6.8 Time of origin compared to time of diversification. Crown group angiosperms (green) possess all the apomorphies of the extant taxa. Lineages in the angiosperm stem group (yellow) have acquired some, but not all, of the apomorphies of the angiosperm crown group. Adapted from Doyle and Donoghue (1993).

extant and fossil groups of seed plants, except angiosperms, have a fossil record that extends back at least to the Triassic has significant implications for ideas on angiosperm origins. Because these groups appear to be monophyletic, and because it is therefore unlikely that angiosperms evolved from 'within' one of the groups, the inescapable conclusion is that the lineage leading to angiosperms diverged from its nearest seed plant relatives early in the Mesozoic, or perhaps even earlier. In this context, it is especially important to distinguish the timing of angiosperm divergence (splitting of the angiosperm lineage from the stem lineages of its sister group) from the timing of angiosperm diversification (splitting of the angiosperm crown group possessing all the characteristic angiosperm synapomorphies into extant angiosperm lineages) (Doyle and Donoghue, 1993) (Figure 6.8). In addition, because angiosperms are a clear monophyletic group with a large number of synapomorphies, stem group angiosperms ('proangiosperms') might be expected to possess some, but not all, of the characters seen in extant members of the group. The hypothetical group that includes all extant angiosperms, plus the currently unrecognised 'proangiosperms', has been termed the angiophytes (Doyle and Donoghue, 1993).

6.3.2 Age estimates of angiosperms based on molecular data

With the increasing availability of molecular data there have been many attempts to date the age of angiosperms, and angiosperm subclades, by approaches that seek to calibrate, and then extrapolate, rates of molecular evolution. Pioneering studies using this approach (Ramshaw *et al.*, 1972) utilised protein sequencing of plant mitochondrial cytochrome *c* and generated an age for angiosperms, much older than the presumed Early Cretaceous diversification of the group. These results were interpreted as reflecting a higher rate of molecular evolution within angiosperms (see references in Smith and Donoghue, 2008). Martin *et al.* (1989) used sequences of the glycolytic enzyme glyceraldehyde 3-phosphate dehydrogenase (GAPDH), and an assumption of an approximately constant (clock-like) rate of evolution, to estimate the divergence of monocots and dicots as having occurred in the Carboniferous (319 ± 33 Myr) (see also Martin *et al.*, 1993). Wolfe *et al.* (1989) used a similar approach, but resolved the monocot–dicot divergence at 200 ± 40 Myr. These results were criticised as conflicting with the orderly sequence of appearance of different angiosperm subgroups in the fossil record, based on rich pollen floras from both hemispheres (Crane *et al.*, 1989a). This orderly pattern is also consistent with results from cladistic analyses of extant angiosperms based on molecular data.

More recently there have been renewed attempts to utilise rates of molecular evolution to determine the age of angiosperms (Magallón and Sanderson, 2001; Sanderson and Doyle, 2001; Wikström *et al.*, 2001; Bell *et al.*, 2005) and angiosperm subgroups (Bremer, 2000; Bremer *et al.*, 2004; Anderson *et al.*, 2005) stimulated in part by the development of more sophisticated numerical techniques (Sanderson, 1997; Huelsenbeck *et al.*, 2000; Yoder and Yang, 2000) as well as the increasing availability of extensive molecular sequence data for many angiosperm groups. All of these approaches require the use of fossils, or other geological evidence, as calibration points.

In the analysis of Wikström *et al.* (2001) two Late Cretaceous fagalean fossils (*Protofagacea*, *Antiquacupula*; Chapter 14) were used to fix the divergence time between Cucurbitales and Fagales. This resulted in an age estimate for crown group angiosperms of 179–158 Myr, corresponding approximately to the latest Early Jurassic – Middle Jurassic (~Toarcian–Bajocian) and of 147–131 Myr for the origin of eudicots corresponding to the latest Jurassic – earliest Cretaceous (approximately Tithonian–Hauterivian). Anderson *et al.* (2005), using multiple calibration points, estimated the age of crown group eudicots to about 120 Myr (Early Aptian), which corresponds approximately to the first occurrence of tricolpate pollen in the fossil record. Anderson *et al.* (2005) also suggested that the radiation of eudicots was very rapid particularly around the mid-Cretaceous (Figure 6.9), a conclusion that is roughly in accordance with the fossil data.

6.3.3 Angiosperm age based on palaeobotanical data

There are occasional reports of possible angiosperm pollen in strata of Valanginian age in Italy (Trevisan, 1988), Portugal (Trincão, 1990), Morocco (Gübeli *et al.*, 1984) and Israel (Brenner, 1996). All are somewhat uncertain, either because the systematic relationships of the material are not well established or because the records have not yet been fully documented. Much more secure, both from the standpoint of stratigraphy and the likely robustness of the systematic relationships, are the early assemblages of angiosperm pollen documented from the Hauterivian of southern and eastern England (Hughes *et al.*, 1991; Hughes, 1994) and Israel (Brenner and Bickoff, 1992; Brenner, 1996). Hauterivian assemblages with angiosperms have also been reported from other areas (e.g. China, Zhang, 1999).

All these occurrences are based on isolated organs and none provides information on angiosperm reproductive structures other than pollen. However, some of these early pollen grains are distinctively angiospermous with reticulate–columellate pollen wall structure, a type not reported definitively prior to about the Valanginian–Hauterivian, except for the somewhat similar Crinopolles grains from the Triassic (section 6.4). Leaf floras of this age that contain angiosperms are rare and their precise stratigraphic position is often debatable (Chapter 9). At this level (Valanginian–Hauterivian) angiosperms are neither diverse nor abundant. The pollen–spore assemblages in which they occur are dominated by a variety of fern spores and the pollen of other groups of seed plants. These early angiosperm pollen grains also exhibit little morphological diversity: they are all either inaperturate or monocolpate.

In the Barremian there are many much more reliable reports of angiosperm pollen and by around the end of the Barremian, close to the Late Barremian – Early Albian boundary, there are several records of angiosperm macro- and mesofossils. Most importantly, unequivocal

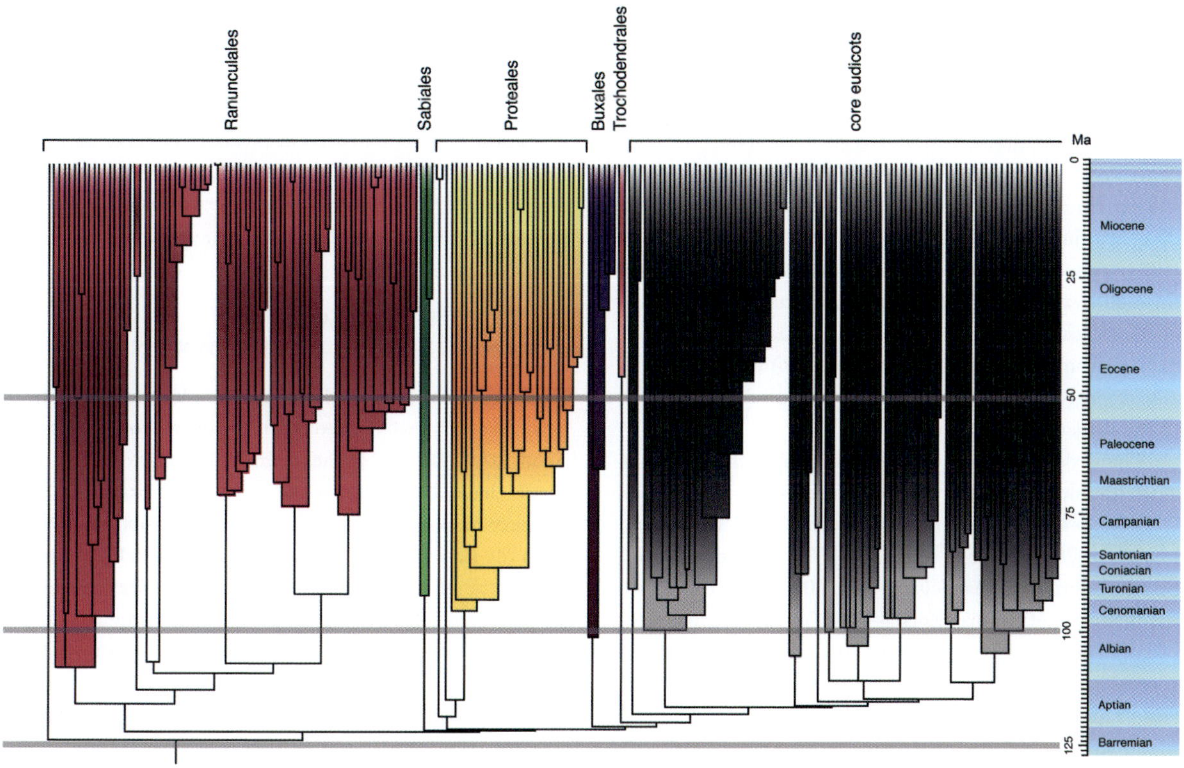

Figure 6.9 Age estimates for different groups of eudicots based on a penalised likelihood analysis of molecular data. From Anderson *et al.* (2005); colour version courtesy of C. N. Anderson.

angiosperm flowers, carpels with pollen on the stigmatic surface, and characteristic angiosperm stamens, are known from the earliest mesofossil floras from Portugal. These are of Late Barremian – Early Aptian age. There are also indisputable angiosperm leaves at this level. However, all kinds of angiosperm fossils (micro-, meso-, and macrofossils) are still rare. Most angiosperm pollen is still either inaperturate or fundamentally monocolpate, but there is greater diversity in exine ornamentation and trichotomocolpate forms are also present. The indication from pollen is that angiosperms are now more diverse, although still within certain limits perhaps broadly equivalent to the palynological diversity that is basic to extant ANITA-grade angiosperms and perhaps basal monocots.

Unequivocal triaperturate pollen indicative of the presence of eudicots appears for the first time (e.g. Hughes, 1994) around the Barremian–Aptian boundary. At first triaperturate pollen is rare and much less abundant than monocolpate angiosperm pollen in the same palynofloras. At this level, angiosperm pollen as a whole is still only a small component of palynofloras compared with the spores and pollen of bryophytes, pteridophytes and other seed plants. Within the angiosperm component, pollen of eudicots is rare, exhibiting low diversity. Furthermore, all of the earliest triaperturate grains have simple tricolpate apertures. There is no evidence of triaperturate grains with more elaborate apertures.

In the Aptian, and also in the succeeding Albian stages, triaperturate pollen steadily becomes more diverse and more abundant, reflecting the initial diversification of eudicots. Meso- and macrofossil floras from the Late Aptian – Albian provide further evidence of angiosperm diversification. Most obviously, in these floras there is additional diversity among eudicots and it is also around this time that there is early evidence of several lineages that diverged close to the base of the eudicot clade. Also at this level there are early flowers that can be attributed securely to extant lineages of eumagnoliids.

By the Early–Late Cretaceous boundary it is very clear that angiosperm diversity, and the vegetational prominence of angiosperms, had increased still further. In mesofossil floras there are diverse eumagnoliids as well as a range of eudicots. By about this stage angiosperms were evidently well established in Cretaceous landscapes. In some places they may have dominated the vegetation. However, evidence from fossil pollen suggests that their hegemony was still far from complete. Based on analyses of palynomorph diversity in different low-palaeolatitude areas, Muller (1984) concluded that the angiosperm 'ecological breakthrough' was not completed until the Turonian. Analyses of the relative abundance of pollen and spores in Cenomanian palynofloras from middle palaeolatitudes show that angiosperms typically account for only about 40%–60% of the dispersed palynomorphs.

The orderly pattern of angiosperm appearance in the stratigraphic record shown from England (e.g. Hughes, 1994), North America (e.g. Brenner, 1963; Doyle and Hickey, 1976) and other regions (e.g. Egypt, Penny, 1991; South America, Archangelsky *et al.*, 2009) constrains both the likely age of angiosperms as a whole, and the likely age of angiosperm subgroups. It is also significant that angiosperms show the same level of organisation and patterns almost simultaneously at an almost global scale.

In summary, there are no convincing reports of fossils with unequivocal angiosperm characteristics in rocks older than the Early Cretaceous, despite the presence of a large number of very intensively studied, and well-understood, palynofloras and macrofossil floras from earlier rocks stretching back 300 million years to the Silurian. Investigations of mesofossil floras from the Jurassic (e.g. Middle Jurassic of Cayton Bay, Yorkshire, England, Crane and Herendeen, 2009), many mesofossil floras from the earliest Cretaceous of Portugal (Berriasian–Hauterivian, Pedersen and Friis, personal observations, and Mendes *et al.*, 2008b) and from Bornholm, Denmark (Pedersen *et al.*, 1989b) have also failed to yield angiosperms. It is thus very unlikely that angiosperms were diverse or abundant in the Jurassic, and this also sets limits on estimates of the likely age of angiosperm clades based on extrapolations from rates of molecular evolution.

The first records of eudicots around the Barremian–Aptian boundary most likely reflect the origin of the group around this time. The morphological distinctiveness of triaperturate grains makes it highly unlikely that they have been overlooked in Hauterivian, Valanginian or older rocks. Their first appearance at around the Barremian–Aptian boundary therefore provides an especially useful datum for molecular clock studies. Following from this, we think it very unlikely that the age of crown group eudicots extends much further back than the Barremian–Aptian boundary, about 125 Myr, and that hyperdiverse groups of angiosperms, such as orchids, Fabaceae, commelinids or core asterids, have a fossil history that is younger. The extraordinary accumulation of diversity in these and other groups of angiosperms has to be explained within these and other temporal constraints, which will continue to be refined by further studies of the palaeobotanical record.

6.4 PRE-CRETACEOUS ANGIOSPERM-LIKE FOSSILS

A variety of fossils from Jurassic and older rocks have been described as pre-Cretaceous angiosperms (for reviews see Scott *et al.*, 1960; Hughes, 1961a, 1976). Some of these fossils have been shown to be angiosperms, but from younger rocks; for example, the palm woods described from the Late Jurassic of Utah (Tidwell *et al.*, 1970) proved later to be of Cenozoic age (Scott *et al.*, 1972). Similarly, *Archaefructus*, originally described as from the Late Jurassic (Sun *et al.*, 1998), has since been shown to be of Aptian age (e.g. Zhou *et al.*, 2003). However, there are several other intriguing fossil plants that have been claimed as pre-Cretaceous angiosperms or that show certain angiosperm-like features. Some of these are briefly reviewed below.

Sanmiguelia is an interesting early Mesozoic plant of unknown affinity that was initially thought to be an angiosperm. *Sanmiguelia lewisii* was first described by Brown (1956), from the Upper Triassic of Colorado, USA. The type material of *Sanmiguelia* is based on large, elliptical leaves with a short stalk and distinct longitudinal plications (Figure 6.10). Superficially these leaves resemble those of extant *Veratrum* (Melanthiaceae) (Figure 6.11) and other monocots, and Brown initially classified *Sanmiguelia* as a palm. Additional material has been described from the Upper Triassic of the western USA (Becker, 1972b; Ash, 1976b; Tidwell *et al.*, 1977), but like the specimens described by Brown these consist entirely of compressions in a highly oxidised, heavily indurated, reddish siltstone or sandstone.

Better-preserved material attributed to *Sanmiguelia*, including specimens preserved as compressions, was described by Cornet (1986, 1989b) from the Late Triassic Sunday Canyon locality in northwestern Texas, USA. At this locality *Sanmiguelia* plants were found in growing position and were quite convincingly linked with separate ovule-bearing and pollen-producing reproductive structures. The ovule-bearing structures, *Axelrodia burgeri*, are head-like aggregations of small partial inflorescences, each consisting of many ovulate units that Cornet described as free carpels with several anatropous ovules. The pollen-producing structures, *Synangispadixis tidwellii*, are elongated, compound, and bore numerous, spirally arranged and paired bisporangiate synangia, termed anthers by Cornet. The synangia apparently dehisced by an apical slit. Pollen found *in situ* is monocolpate, psilate and with tectate–granular wall structure.

The new reconstruction of *Sanmiguelia* with its reproductive structures (Figure 6.10) reveals an intriguing early Mesozoic plant that is unlike any known seed plant.

Figure 6.11 Extant *Veratrum album* from the Swiss Alps with plicate, *Sanmiguelia*-like leaves.

Figure 6.10 Reconstructions of the *Sanmiguelia* plant from the Late Triassic of Texas, USA. (A) Vegetative plant with plicate leaves of *Sanmiguelia lewisii*. (B) Axis with ovulate heads of *Axelrodia burgeri*. (C) Plant with leaves and the pollen-producing structure of *Synangispadixis tidwellii*. (A) Redrawn from Tidwell (1977); (B, C) redrawn from Cornet (1986).

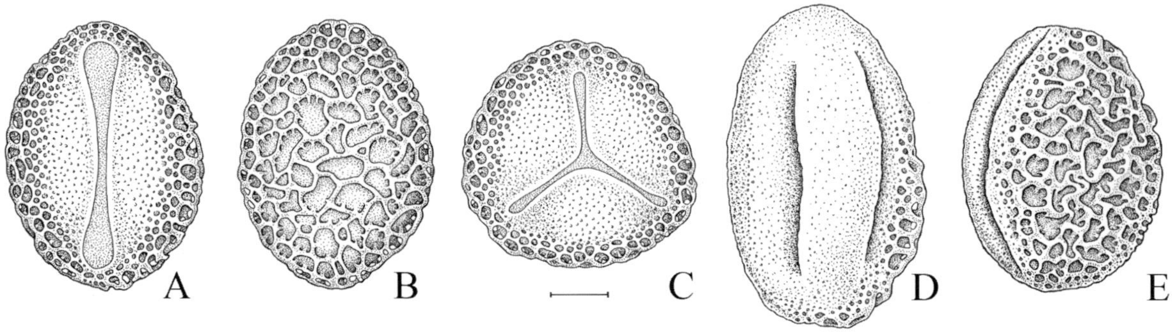

Figure 6.12 Reticulate pollen grains from the Late Triassic of Virginia, USA, assigned to the Crinopolles group. (A–C) *Monocrinopollis* in (A, C) distal and (B) proximal view showing (A) simple monocolpate or (C) trichotomocolpate aperture. (D, E) *Tricrinopollis olsenii* in (D) distal and (E) equatorial view. Scale bar = 10μm. Drawn from SEM-images in Cornet (1989a).

Unfortunately, the fossils are strongly compressed and details of the reproductive structures are not fully understood. Therefore, the angiosperm affinity suggested by Cornet (1986, 1989b) is not documented securely. Key features of this material, and especially the details of the reproductive structures, require much more careful documentation, which is probably not possible with the material currently available. Nevertheless, if they can be substantiated, characters such as the possible tectate–granular pollen wall, the microsporangiate units, the apical dehiscence of the synangia, and the broad leaf blade, may suggest a relationship to anthophytes including Gnetales. Further information on the *Sanmiguelia* plant could prove helpful for resolving some of the remaining questions relating to the relationships of anthophytes.

Another group of interesting fossils that may be important for future consideration of the relationship between angiosperms and anthophytes are pollen grains of the Crinopolles group (Figure 6.12), described from the Late Triassic of the Richmond Basin, Virginia, and the Chinle Formation, Arizona, USA (Cornet, 1989a; Doyle and Hotton, 1991). These grains are perhaps the most intriguing of all the pre-Cretaceous fossils that may have a possible connection to angiosperms. Cornet (1989a) included five genera in the Crinopolles group, of which four, *Tricrinopollis*, *Monocrinopollis*, *Dicrinopollis* and *Zonacrinopollis*, have reticulate exine sculpture. The other genus (*Pentecrinopollis*) has a clavate exine sculpturing. All the reticulate-walled genera are basically monocolpate or dicolpate, sometimes with additional lateral furrows that are perhaps comparable to those of *Eucommiidites* (Erdtmanithecales). Scanning and transmission electron microscopy (Cornet, 1989a; Doyle and Hotton, 1991) show that the reticulate tectum is supported by well-developed columellae on a thin foot layer. The endexine is thick and apparently lacks laminations. Cornet (1989a) argued that the absence of laminations was an angiosperm feature, but a similar thick endexine is not known for angiosperms. Most angiosperms that produce monocolpate grains lack a well-developed endexine (Doyle and Hotton, 1991; Doyle and Donoghue, 1993).

Unfortunately, Crinopolles-type pollen is known only as dispersed grains, and there is no information on other parts of the plants by which they were produced. It is therefore impossible to determine whether these intriguing pollen grains were produced by plants close to angiosperms, for example at an evolutionary stage prior to the loss of endexine, or whether they are unrelated to angiosperms and have acquired reticulate–columellate wall structure convergently. Until more information is available, the possibility of a close relationship of Crinopolles-type pollen to angiosperms remains equivocal.

Angiosperm-like pollen was also described from the Late Jurassic of Normandy, France, and includes crotonoid grains assigned to the dispersed genera *Stellatopollis* and *Multimarginites* (Cornet and Habib, 1992) that are otherwise known only from younger sediments in the Cretaceous. The grains are also similar to grains of extant angiosperms (monocots and Acanthaceae, respectively). Subsequent examinations of the preparations showed that several of the grains had either disintegrated or swollen (Hughes, 1994) and the possibility of contamination either from younger sediments or from the local vegetation cannot be ruled out. In addition, new palynological preparations made from the same strata failed to

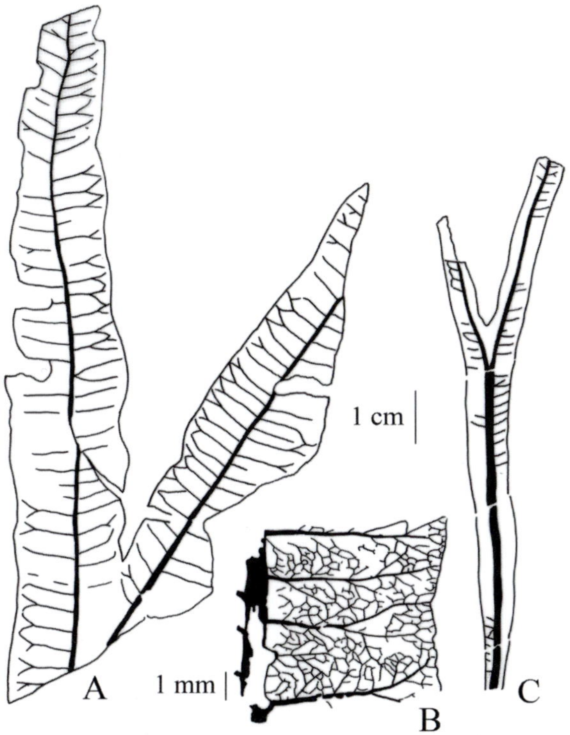

Figure 6.13 Leaves and venation of *Furcula granulifer* from the Late Triassic of East Greenland showing (A, C) forked leaf and (B) details of venation. Redrawn from Harris (1932b).

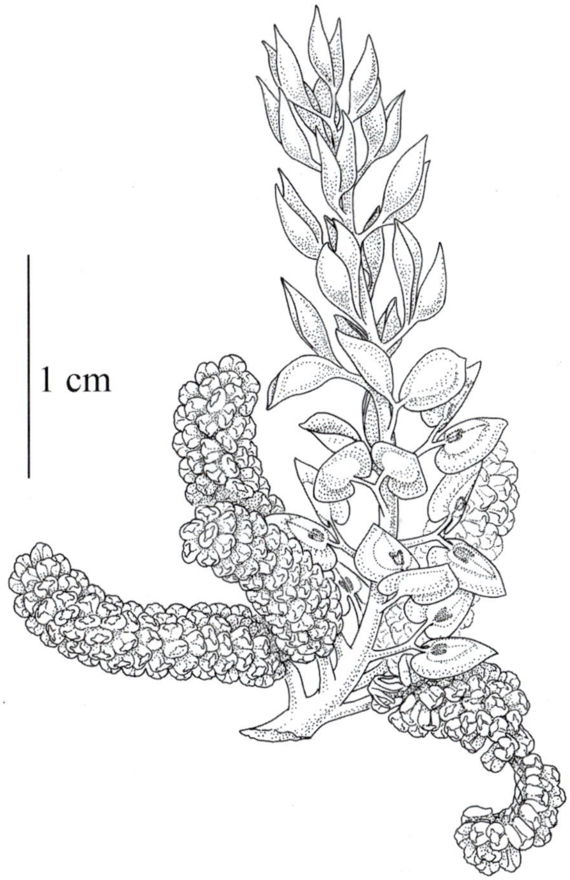

Figure 6.14 Reconstruction of the *Schweitzeria (Irania) hermaphroditica* plant from the Late Triassic of Iran showing ovulate structures above and presumed pollen-producing structures below. Redrawn from Schweitzer (1977).

yield angiosperm pollen (K. R. Pedersen, personal observations). According to Rioult (1966, and personal communication, 2008) the coastal exposures at the Villiers-sur-Mer locality comprise mainly Jurassic sediments, but they are superimposed by Cretaceous soft sediment, and Cretaceous fossils are regularly washed down over the Jurassic strata.

Several Triassic–Jurassic leaves with reticulate venation have received attention from time to time in discussions of pre-Cretaceous angiosperms. These include such fossils as *Furcula*, *Phyllites*, *Marcouia* and *Scoresbya*. *Furcula granulifer* (Harris, 1932b) from the Late Triassic of East Greenland is a distinctive and unusual leaf that is simple proximally, but forked distally (Figure 6.13). The midrib gives off prominent, more or less perpendicular, secondary veins that branch and become less distinct toward the margin. Distally they divide and give rise to finer veins that form a clear reticulate pattern and freely ending veinlets. In *Furcula granulifer* the stomata are syndetocheilic. The reproductive structures of the *Furcula* plant are unknown and its systematic relationships are uncertain. The reticulate venation has prompted comparison with gigantopterids, while the simple forked leaf is reminiscent of some species of *Dicroidium*. Other leaves from the early Mesozoic that exhibit reticulate venation include *Marcouia* (Ash, 1972) and *Scoresbya* (Harris, 1932b). The relationships of all of these fossils are uncertain. *Scoresbya* may be related to Dipteridaceae (Kräusel and Schaarschmidt, 1968), but affinity with some undefined group of gymnosperms has also been suggested (Weber, 1995).

Phyllites from the Middle Jurassic Stonesfield Slate of Oxfordshire, England, was described by Seward (1904). The specimen is a poorly preserved impression, but has palmate acrodromous primary venation similar to that of certain magnoliids and early-diverging groups of eudicots (Doyle and Donoghue, 1993). However, details of the venation and epidermal features are not preserved and the relationship of this intriguing fossil remains equivocal.

Schweitzeria (Irania) hermaphroditica, an enigmatic fossil described from the Late Triassic of the Alborz Mountains, Mazandaran, Iran (Schweitzer, 1977; Shipunov and Sokoloff, 2003), is of interest because of its bisexual organisation. The reproductive structure consists of elongated, presumed pollen-producing axes, arranged in a spiral below a terminal axis bearing ovulate structures (Figure 6.14). The specimens are strongly compressed and preserved in a hard matrix with no cellular details. The ovulate part of the axis is elongate and bears about 20 spirally arranged ovulate units. Each ovulate unit is stalked and bears two follicle-like structures that open at maturity to form a flat circular disc exposing the attachment scars of two seeds. The presumed pollen-producing axes are elongate, thick and with densely crowded, stalked structures that bear several, apparently free, microsporangia.

In gross morphology the arrangement of the ovulate organs in *Schweitzeria (Irania)* is similar to that in *Archaefructus*, which also shows pairs of ovulate structures along the distal part of the reproductive axis (Chapter 9) with the pollen-producing structures below. Schweitzer (1977) emphasised the relevance of *Schweitzeria (Irania)* for understanding the evolution of the angiosperm flower, but he also noted that the uniqueness of the structure prevented its inclusion in any group of living or fossil plants, at least with current knowledge. He also compared the fossil to the reproductive structures of Czekanowskiales, particularly based on the apparently bivalved construction of the ovule-bearing structures (section 5.8.5).

7

Phylogenetic framework and the assignment of fossils to extant groups

Over the past two decades there has been rapid progress in developing a phylogenetic framework within which to interpret many aspects of angiosperm evolution. This progress has been brought about mainly through the application of molecular phylogenetics to an increasingly large sample of flowering plants. In this chapter we briefly review the development of ideas on angiosperm phylogeny and outline the current state of knowledge based on the most recent APGIII classification (2009). Phylogenetic patterns within particular groups are summarised in Chapters 8–15. In this chapter we also briefly discuss the issues that need to be considered in assigning fossils to extant groups. This is essential background to what follows in Chapters 8–15.

7.1 EARLY IDEAS ON ANGIOSPERM PHYLOGENY

The Euanthial (Anthostrobilus) Theory of the angiosperm flower, which interprets the flower as a uniaxial structure (Chapter 6), is at the core of most phylogenetic interpretations of angiosperm evolution developed in the second half of the twentieth century. In their emphasis on Magnoliaceae and their relatives as a starting point for angiosperm evolution, such phylogenetic schemes trace their origin to the pre-Darwinian classifications of De Candolle and Bentham and Hooker, but they have been broadly supported by comparative data from many sources, especially from wood anatomy (Bailey, 1944; Eames, 1961; Dickison, 1975) and palynology (Woodhouse, 1935; Walker, 1974a, b, 1976). The classifications of Hutchinson (1959), Cronquist (1968, 1981), Takhtajan (1969, 1980), Thorne (1976) and Dahlgren (1980) all represent different manifestations, each with their own idiosyncrasies, of the Euanthial paradigm.

Taken as a whole these pre-cladistic classifications of flowering plants, and the many detailed studies on which they were based, were a considerable achievement in terms of synthesising and systematising knowledge of flowering plant diversity, and providing a foundation on which future work could be based. Many of the groups recognised, from genera to orders, have subsequently been shown to be monophyletic and in broad terms the conclusion that magnoliids *sensu lato* (rather than Magnoliaceae and a narrow group of allied families) comprise a basal grade of angiosperms has been borne out by recent phylogenetic studies based on molecular data. However, at a more detailed level the many different classifications developed in the second half of the twentieth century often presented different phylogenetic pictures based on the perceived value of different features as guides to evolutionary relationships. Progress was also frustrated by the lack of a clear basis for choosing among potentially plausible, but sometimes radically different, systematic and evolutionary interpretations.

7.2 PHYLOGENETIC STUDIES OF ANGIOSPERMS BASED ON MOLECULAR DATA

The past few decades have seen significant further progress in resolving relationships among angiosperms using cladistic methodology applied to DNA sequence data. A key early study was that of Chase *et al.* (1993), which presented parsimony analyses for almost 500 species based on DNA sequences from the chloroplast gene *rbc*L. This study corroborated several major lineages corresponding to monophyletic groups recognised by non-molecular methods (e.g. monocots, Caryophyllidae, and many orders and families). However, it also presented a radical challenge to previous ideas, especially in showing that the Hamamelididae and Dilleniidae, at least as recognised in Cronquist's and Takhtajan's influential schemes, were grossly polyphyletic. At a finer level many of the results based on molecular data confirmed previous ideas of relationships, but in other cases the new hypotheses were in conflict with traditional views.

Since the early 1990s, the initial large-scale analysis by Chase *et al.* (1993) has been supplemented and extended by many other studies. Taxon sampling has been further expanded to better represent key groups (almost all previously recognised angiosperm families have now been incorporated) and to develop a balanced initial sample of the full spectrum of angiosperm diversity. At the same time, the sequence data available have been extended so as to combine information from the chloroplast, nuclear and mitochondrial genomes. There is perhaps no other major group of organisms for which taxon sampling is so extensive, and for which so much molecular data has been gathered so quickly, by so many different researchers, in such a coordinated and consistent way.

In 1998 the first concerted effort to convert the new information from phylogenetic studies of angiosperms into a new classification was made by the Angiosperm Phylogeny Group (APG, 1998). This sought to provide a revised suprafamilial classification based on the principle of monophyly. It was also guided by a desire to maintain 'well established and familiar entities', while at the same time seeking to minimise redundancy by avoiding monofamilial orders and monogeneric families (APG 1998). The resulting classification recognised 462 families and 40 orders of flowering plants.

Five years later, the first APG classification was revised based on information from further phylogenetic analyses (APGII, 2003) and this was followed by another update in 2009 (APGIII, 2009). Only a few families are not yet placed and currently only five taxa (Apodanthaceae, Cynomoriaceae, *Gumillea*, *Petenaea*, *Nicobariodendron*) are listed as of uncertain position. In some cases the groups recognised are also supported by comparative data from non-molecular sources, but in many cases comparable phylogenetic analyses of morphological data remain to be completed. A general problem is that the rapidity with which molecular data can be gathered and translated into phylogenetic hypotheses has outstripped the botanical community's capacity to test those hypotheses against patterns in other kinds of data. Although phylogenetic studies based on sequence data from different genes in different genomes (chloroplast, mitochondrial, nuclear) often give broadly consistent results, tests using morphological data are still desirable.

7.3 ANGIOSPERM PHYLOGENY GROUP CLASSIFICATION (APGIII)

The interrelationships of orders and families accepted by the APGIII classification are illustrated in Figure 7.1. The brief comments below provide the framework for the discussion of fossil taxa in Chapters 8 and 10–15. In general, throughout this book, we adopt the APGIII classification with only relatively minor modification.

7.3.1 ANITA grade, Chloranthaceae and Ceratophyllaceae

The concept of the ANITA grade (*Amborella*, Nymphaeales, *Illicium*, Trimeniaceae, Austrobaileyaceae) for the first lineages diverging from the main line of angiosperm evolution was first advanced by Qiu *et al.* (1999) based on phylogenetic analyses of molecular data. The same pattern of relationships has received strong support from subsequent analyses. All recent phylogenetic analyses based on molecular data (e.g. Moore *et al.*, 2007), as well as molecular data combined with morphological data (e.g. Endress and Doyle, 2009), recognise this basal grade of angiosperms.

Amborella trichopoda, the single species of the family Amborellaceae (order Amborellales) is the sister group to all other angiosperms (APGIII, 2009), followed by Nymphaeales (including Cabombaceae, Hydatellaceae, Nymphaeaceae) and then Austrobaileyales, which include Austrobaileyaceae, Schisandraceae (including Illiciaceae) and Trimeniaceae. Qiu *et al.* (1999) coined the term euangiosperms for the clade that excludes the ANITA grade but comprises all other angiosperms (Chloranthaceae, *Ceratophyllum*, monocots, eumagnoliids, eudicots).

The position of Chloranthaceae, a key family in the palaeobotanical record (Chapter 8), is important. It appears to diverge at some point above the ANITA grade, but below the divergence of eumagnoliids (Chapter 8). There is some evidence that Chloranthaceae may be the sister taxon to eumagnoliids (Moore *et al.*, 2007). Similar uncertainties surround the placement of the enigmatic Ceratophyllaceae, which may be the sister group to eudicots (Jansen *et al.*, 2007; Moore *et al.*, 2007). Excluding the ANITA grade, Chloranthaceae and Ceratophyllaceae, core angiosperms comprise three clades: eumagnoliids, monocots and eudicots.

7.3.2 Eumagnoliids

Eumagnoliids (Magnoliids *sensu* APGIII) comprise four orders: Canellales (Canellaceae, Winteraceae), Piperales (Aristolochiaceae, Hydnoraceae, Lactoridaceae, Piperaceae, Saururaceae), Laurales (Atherospermataceae, Calycanthaceae, Gomortegaceae, Hernandiaceae, Lauraceae, Monimiaceae, Siparunaceae), and Magnoliales (Annonaceae, Degeneriaceae, Eupomatiaceae, Himantandraceae, Magnoliaceae, Myristicaceae). Molecular data suggest that Canellales and Piperales are sister taxa while Laurales and Magnoliales are sister taxa (Doyle and Endress, 2000)

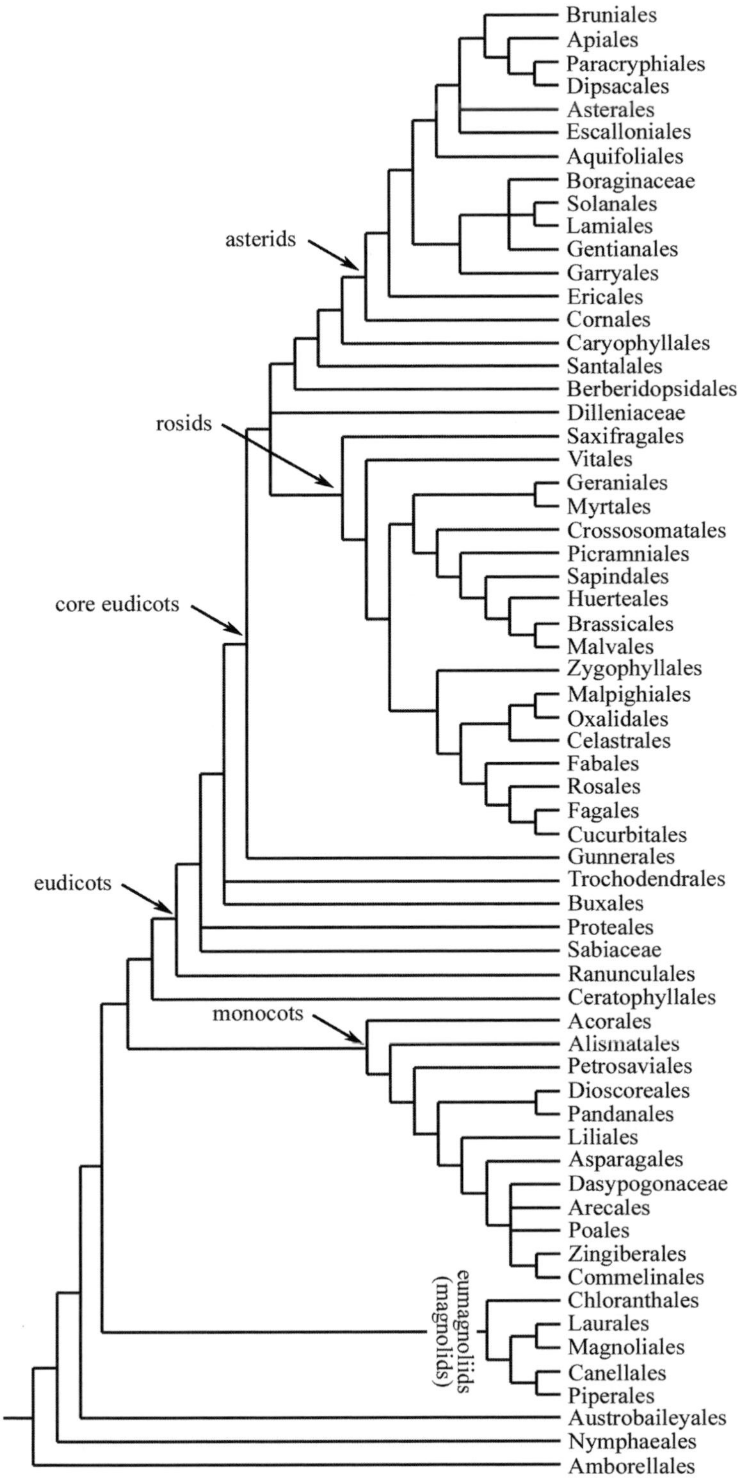

Figure 7.1 Cladistic representation of Angiosperm Phylogeny Group III classification of extant angiosperms at the level of orders (APGIII, 2009). See text for additional explanation.

(APGIII, 2009). Eumagnoliids have an extensive fossil record in the Cenozoic and are also becoming increasingly well known from the Cretaceous (Chapter 10). In the summary cladogram accompanying APGIII, the Chloranthaceae are placed as the sister group to eumagnoliids.

7.3.3 Monocots

Considerable progress has been made with identifying major monophyletic groups among monocots and evaluating the relationships among them. Acorales, comprising the single genus *Acorus*, are the sister group to all other monocots. Alismatales are the sister group to the core monocot clade that comprises Asparagales, Dioscoreales, Liliales, Pandanales, Petrosaviales and commelinids. The fossil record of monocots is extensive in the Cenozoic, but with a few exceptions it is relatively sparse in the Cretaceous. Araceae (Alismatales) as well as Arecaceae and Zingiberales (both commelinids) are among the groups known from both the Cretaceous and the Cenozoic (Chapter 11).

7.3.4 Eudicots

Eudicots include more than 75% of all extant angiosperm species and have an extensive fossil record from the Early Cretaceous onwards (Chapters 12–15). The Ranunculales (Ranunculid clade *sensu* Magallón *et al.*, 1999), comprising Berberidaceae, Circaeasteraceae, Eupteleaceae, Lardizabalaceae, Menispermaceae, Papaveraceae and Ranunculaceae, are the sister group to the main eudicot clade. Within this large group there is a basal grade comprising Buxales, Proteales, Sabiaceae and Trochodendrales. Several of these basal-grade eudicots have a fossil record that begins in the Early Cretaceous (Chapter 12), but which of these groups is closest to the core eudicots is not fully secure. The cladogram presented with the APGIII scheme (Figure 7.1) places Buxales and Trochodendrales in an unresolved polychotomy with core eudicots. Sabiaceae and Proteales (including *Nelumbo*, *Platanus* and Proteaceae) are placed in an unresolved polychotomy with the Buxales–Trochodendrales – core eudicot group.

Core eudicots. Core eudicots are a strongly supported monophyletic group (e.g. Soltis *et al.*, 2000a) and within this clade Gunnerales (*Gunnera*, *Myrothamnus*) appear to be the sister group to all other taxa (Soltis *et al.*, 2003). The uncertain position of some eudicot lineages in previous APG systems has now been resolved with more confidence. The Saxifragales are now placed as the sister group to rosids, while Berberidopsidales, Santalales, and Caryophyllales are placed as successive sister taxa to asterids. The position of Dilleniaceae remains unresolved with respect to rosids, asterids and their associated groups (APGIII, 2009).

The Saxifragales comprise a heterogeneous assemblage of 15 families including Cercidiphyllaceae, Crassulaceae, Hamamelidaceae, Iteaceae, Paeoniaceae and Saxifragaceae. Several of these families are well represented in the Late Cretaceous and Cenozoic fossil record (Chapter 13).

Rosids. Within rosids, the position of Vitales is not fully secure, but in some analyses it is sister to the two main clades of rosids: fabids (broadly equivalent to the former eurosid I group) and malvids (broadly equivalent to the former eurosid II group). Rosids are extensively represented in both the Late Cretaceous and the Cenozoic fossil record (Chapter 14).

Fabids include Zygophyllales (Krameriaceae, Zygophyllaceae), which are sister to the COM clade (Celastrales, Oxalidales, Malpighiales) and the nitrogen-fixing clade (Fabales, Rosales, Cucurbitales, Fagales). Fabales and Rosales are successive sister groups to Fagales plus Cucurbitales (Figure 7.1).

Celastrales and Oxalidales contain two and seven families, respectively, whereas the Malpighiales are larger and much more diverse and include Euphorbiaceae and Malpighiaceae as well as 33 other families.

The Fabales comprise Fabaceae (legumes) plus Polygalaceae, Quillajaceae and Surianaceae. The Rosales comprise Moraceae, Rhamnaceae, Rosaceae, Ulmaceae, Urticaceae and four other families. The fossil record of legumes and certain groups of Rosales is extensive, particularly from the Cenozoic.

The Cucurbitales comprise Anisophylleaceae, Begoniaceae, Cucurbitaceae and four other families, while the Fagales comprise Fagaceae and Nothofagaceae plus four (APGIII, 2009) or five (APGII, 2003) 'core fagalean' families (putative 'Normapolles families'), mainly from the Northern Hemisphere (Betulaceae, Juglandaceae, Myricaceae, Rhoipteleaceae, Ticodendraceae) as well as Casuarinaceae, mainly from the Southern Hemisphere. The Fagales are very important palaeobotanically and are well known from both the Late Cretaceous and the Cenozoic (Chapter 14).

Malvids comprise the Geraniales (three families) and Myrtales (nine families) that form a sister group to the Crossosomatales (seven families), Picramniales (one family), Sapindales (nine families), and Huerteales (three families). These four orders are themselves successive sister groups to Brassicales (Brassicaceae plus 16 other families) plus Malvales (Cistaceae, Dipterocarpaceae, Malvaceae and seven other families).

Asterids. Asterids are the most species-rich clade of eudicots. Although the main components of the group (Cornales, Ericales, lamiids, campanulids) have been clear for some time, it was not until APGIII that the likely position of Berberidopsidales, Santalales, and Caryophyllales as successive sister taxa to asterids was regarded as reasonably reliable. The Berberidopsidales comprise only two families (Aextoxicaceae, Berberidopsidaceae); the Santalales (seven families) and Caryophyllales (34 families) are much more diverse. All three groups have a relatively sparse fossil record in the Late Cretaceous and Early Cenozoic (Chapter 15).

The Cornales (Cornaceae, Hydrangeaceae plus four other families) are sister group to the Ericales plus core asterids. The Ericales comprise a total of 22 families including Actinidiaceae, Clethraceae, Ericaceae, Primulaceae and Theaceae. Both Cornales and Ericales are very well represented in fossil floras from the Late Cretaceous, but the record of core asterids is almost exclusively from the Cenozoic (Chapter 15).

Lamiids (Euasterids I) (Lamiidae *sensu lato* of Magallón *et al.*, 1999) comprise the Garryales (a small order including Eucommiaceae and Garryaceae), which is the sister to four clades: Boraginaceae, Gentianales, Lamiales and Solanales. The cladogram presented with the APGIII scheme (Figure 7.1) resolves Lamiales plus Solanales as sister taxa, but in an unresolved polychotomy with Boraginaceae and Gentianales.

The most difficult of these four groups to circumscribe is the Boraginaceae (APGIII, 2009), the relationships of which to Vahliaceae, Icacinaceae, Metteniusaceae and Oncothecaceae are not fully resolved. From a palaeobotanical point of view the position of the Icacinaceae, which has an extensive fossil record, is especially important. The Gentianales comprise the Apocynaceae, Gelsemiaceae, Gentianaceae and Loganiaceae, as well as the hyper-diverse family Rubiaceae.

The Lamiales–Solanales clade is also very diverse. Solanales comprise Convolvulaceae, Hydroleaceae, Montiniaceae and Sphenocleaceae, as well as the Solanaceae. The Lamiales are about as diverse as the Gentianales and Solanales combined and include several very large groups, such as Acanthaceae, Bignoniaceae, Lamiaceae and Verbenaceae, as well as 19 other families.

Campanulids (Euasterids II) include the Aquifoliales (Aquifoliaceae and four other families), which, in the cladogram presented with APGIII, are placed as the sister to an unresolved polychotomy involving Escalloniales (Escalloniaceae) and Asterales and a clade comprising Bruniales (Bruniaceae, Columelliaceae) and Apiales (Apiaceae, Araliaceae, Pittosporaceae and four smaller families) as successive sister groups to a clade comprising Dipsacales (Adoxaceae, Caprifoliaceae) plus Paracryphiales (one family). The vast bulk of campanulid species diversity is accounted for by the Asteraceae, which, together with Campanulaceae and nine smaller families, comprise the Asterales.

7.4 ANGIOSPERM PHYLOGENY: FUTURE DIRECTIONS

Enormous progress has been made in a very short time towards clarifying the major patterns of angiosperm phylogeny based on molecular data. The broad outlines of the phylogenetic pattern are now clear. These outlines also appear relatively stable to the incorporation of new data, both from additional sampling of extant angiosperm diversity and from the sequencing of additional genes. Nevertheless there is still much to do in order to establish more securely those nodes that currently have only weak support, and also to clarify the phylogenetic position of key taxa for which relationships remain uncertain. Taxa of considerable palaeobotanical interest, for which phylogenetic relationships are somewhat uncertain, include Ceratophyllaceae, Chloranthaceae, Vitaceae and Icacinaceae.

A particular problem concerns the secure resolution of relationships near the base of the angiosperm tree. At this level of angiosperm evolution, rooting of the angiosperm tree is especially important (Zanis *et al.*, 2002) and this is made especially difficult by the apparently large evolutionary gap that separates the angiosperms from their closest relatives (Chapter 6). This gap might be narrowed if the sister group of angiosperms could be identified with certainty from among the currently known groups of extant seed plants, but nevertheless it would probably still be substantial. In these circumstances the potential discovery of relevant extinct taxa could also have a significant effect on the rooting of the angiosperm tree. Similar considerations also apply to the substantial gaps that separate those lineages of extant angiosperms that apparently diverged near the base of the angiosperm tree. Such gaps add to the difficulties of correctly interpreting key structural and biological homologies. These considerations emphasise the importance of improved knowledge of fossil taxa from the earliest phases of angiosperm evolution to improve the available sample of diversity at this level.

More generally, across the whole of angiosperms, a further need is to continue to expand sampling of extant taxa and to strive for even more comprehensive representation of angiosperm diversity in phylogenetic analyses. Almost all angiosperm families are now represented in molecular studies. As the ease of obtaining large amounts of sequence data continues to increase, a key collaborative goal, which appears to be in reach, is to sample at least one representative of all angiosperm genera. Just as the delimitation of orders and families has been revised based on the molecular data currently available, realignment of generic boundaries will be an ongoing task in the next phase of this work over the coming years.

There is also a clear need for increased emphasis on combined morphological and molecular analyses in critical groups. Such work is vital, both to supplement and test the results based on molecular data, and to permit the integration of palaeobotanical data with the information available for extant plants. Until an adequate phylogenetic framework is available based on morphological data the full potential of palaeobotanical discoveries will not be realised and the important morphological tests of hypotheses based on molecular data will remain incomplete.

7.5 ASSIGNMENT OF FOSSILS TO EXTANT GROUPS

The lack of carefully constructed morphological datasets for extant taxa is one significant impediment to integrating knowledge of extant and extinct plant diversity into comprehensive phylogenetic hypotheses. Another is the difficulty of dealing with missing data for fossil taxa. However, there are now many examples where such integration has been attempted (Crane and Manchester, 1982; Crane, 1985a; Kenrick and Crane, 1997; Friis *et al.*, 2009b; Doyle and Endress, 2010) and has proved valuable. An integrated approach is clearly necessary to gain maximum value from both neobotanical and palaeobotanical evidence of plant evolution. However, given the scope of what we seek to cover in Chapters 8–15 (all angiosperms) an explicit phylogenetic approach to all fossils considered in this book is currently impracticable. We are also cautious about the potential confusion that can arise from the premature incorporation of insufficiently well-studied fossils into insufficiently well-scrutinised data matrices for extant taxa.

Our objective in this book is to provide an overview and initial critical analysis of the palaeobotanical record as it currently stands. More rigorous phylogenetic analyses of particular taxa need to follow, as the fossils and the extant taxa to which they need to be compared become sufficiently well understood. We intend that the synthesis provided here will encourage future research of this kind, ideally involving collaboration between neobotanical and palaeobotanical specialists (e.g. Manchester and Donoghue, 1995; Matthews *et al.*, 2001; Schönenberger *et al.*, 2001a). In the interim, knowledge of the relationship between extant taxa and their fossil relatives will inevitably remain imprecise, and in some cases this uncertainty may limit the use of the data to calibrate or test estimates of clade ages based on molecular clock calculations. However, in so far as the overall pattern documented here forms a coherent whole we believe it sets clear limits on reasonable hypotheses for the timing of major events in angiosperm evolution.

At a more detailed level, as far as possible, we adopt an implicitly phylogenetic approach in the chapters that follow. We attempt to give due weight to probable synapomorphies, and where possible we make the distinction between assignment of fossils to a crown group and assignment to a stem group. In many cases fossil taxa show apparently plesiomorphic features compared to extant taxa, suggesting that they are more likely to be part of the stem group.

The distinction between stem groups and crown groups (Chapter 6) is especially important in comparing age estimates based on fossils with age estimates based on molecular data. Whether age estimates based on fossils are an age estimate for a stem group or for the crown group depends on the precise phylogenetic position of the fossil in question. However, estimates based on molecular data are age estimates for crown groups. Again we consider it prudent to adopt a conservative approach. To be considered part of the crown group a fossil must not only have all the synapomorphies of the crown group in all its preserved characters, but in practice it should also possess at least one synapomorphy of a subclade of that group.

The Cretaceous fossil record of angiosperms has grown rapidly over the past 30 years, and continues to expand through further exploration of mesofossil floras that contain well-preserved angiosperm reproductive organs. In the following eight chapters we provide an overview of current knowledge of fossil angiosperms, focusing mainly on the Cretaceous record. The fossils are arranged as far as possible systematically (with the caveats outlined above) and we focus particularly on those fossils that provide the most useful information on phylogenetic diversity and reproductive biology of Cretaceous angiosperms: typically, with current knowledge, these are flowers, fruits and seeds.

8

Fossils near the base of the angiosperm tree

In this chapter we consider fossils that appear related to the extant ANITA lineages: Amborellaceae, Nymphaeales (Hydatellaceae, Cabombaceae, Nymphaeaceae), and Austrobaileyales (Austrobaileyaceae, Schisandraceae including Illiciaceae, Trimeniaceae) (Figure 8.1). These three lineages are recognised in all recent phylogenetic analyses as the first to diverge from the main line of angiosperm evolution and form a series of successive sister groups to all other angiosperms (Chapter 7). We also consider the fossil history of Chloranthaceae and Ceratophyllaceae. Both are usually resolved as separate lineages that diverge above the level of Austrobaileyales, but their position with respect to eumagnoliids, monocots and eudicots is not fully secure.

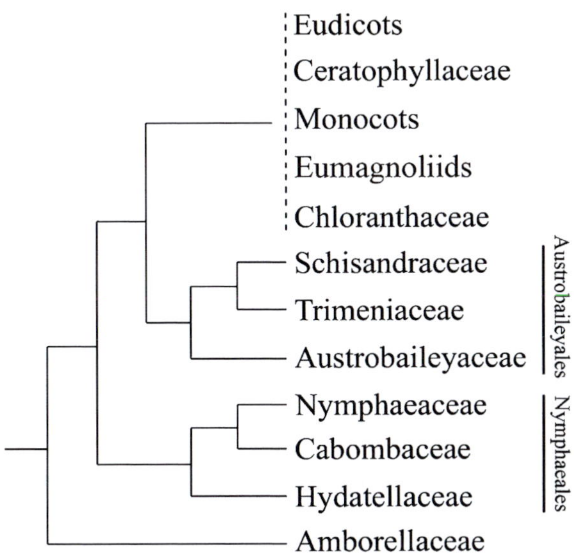

Figure 8.1 Summary of phylogenetic relationships of the ANITA lineages (Amborellaceae, Nymphaeales, Austrobaileyales) and the five other groups of angiosperms (Chloranthaceae, Ceratophyllaceae, eumagnoliids, monocots and eudicots) shown here conservatively as a polychotomy. Relationships among the ANITA lineages based on Saarela *et al.* (2007) and also consistent with APGIII (2009).

There is some indication that Chloranthaceae may be the sister group to eumagnoliids, while *Ceratophyllum* may be the sister group to eudicots (Moore *et al.*, 2007). In Chapter 9 we describe fossil reproductive structures, mainly from the Early Cretaceous, that may also be relevant to this initial phase of angiosperm diversification, but that cannot be assigned to any extant family or order with the information currently available.

8.1 EARLY-DIVERGING ANGIOSPERM LINEAGES AT THE ANITA GRADE

The known fossil record of plants related to the ANITA lineages, as well as to Chloranthaceae, continues to increase rapidly. The relevant fossil material includes leaf compressions and impressions and a variety of dispersed pollen, as well as flowers and floral organs. Some of these fossils can be assigned reliably to extant orders, families or even genera. Ceratophyllaceae also have an extensive fossil record, but are most common in Cenozoic floras. There are only a few, uncertain records of Ceratophyllaceae from the Cretaceous.

An interesting feature of the early angiosperm record is that, although many fossils cannot be assigned easily to extant taxa, there are other cases where the reproductive morphology of extant lineages has changed relatively little since the Early Cretaceous. The most striking examples of such morphological stasis occur in Chloranthaceae, where it is clear that flowers and inflorescences very similar to those of extant *Hedyosmum* were already present in the Early Cretaceous by the Barremian–Aptian, and that flowers and inflorescences very similar to those of extant *Chloranthus* were present around the mid-Cretaceous.

8.1.1 Fossils of uncertain relationships at the ANITA grade

Several fossils from the Early Cretaceous appear to be generally related to extant taxa at the ANITA grade, but

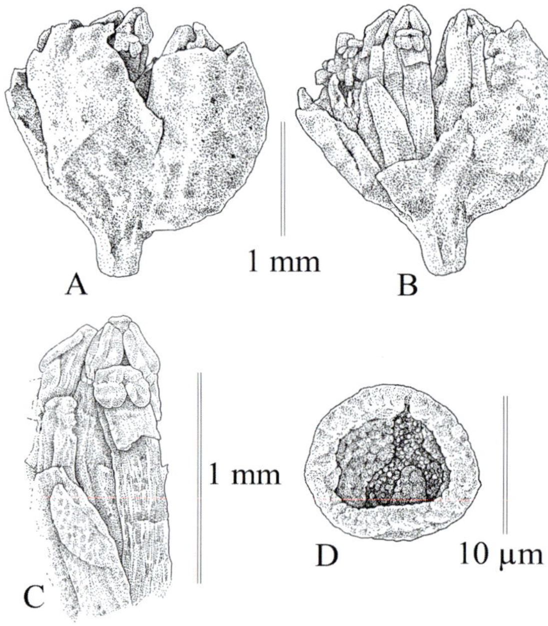

Figure 8.2 Staminate flower from the Early Cretaceous (Late Barremian–Aptian) Catefica locality, Portugal, possibly related to ANITA-grade angiosperms. (A) Flower showing broad inner tepals and the indistinct impression left by a small outer tepal. (B) Inner tepals partly removed, showing multistaminate androecium. (C) Detail of stamens showing flattened filaments. (D) Pollen grain showing circular depression with a distinct trichotomocolpate aperture. Drawn from SEM images in Friis *et al.* (2000b) and material in the collections of the Swedish Museum of Natural History, Stockholm.

cannot be assigned to any of the three extant lineages. One such specimen (Figure 8.2), not yet formally described, is from the Early Cretaceous (Late Barremian – Aptian) Catefica locality, Portugal (Friis *et al.*, 2000a). It is a staminate flower, about 2.3 mm long, with an outer set of smaller, triangular tepals and an inner set of larger, broader tepals. These are followed internally by elongate, narrow structures that may be filaments or perhaps a further set of narrow tepals. The androecium consists of about 10–15 stamens. Filaments are long, broad, flattened, and were probably originally laminar. Anthers are small, triangular, tetrasporangiate and dithecate. Thecae are lateral to slightly abaxial, and are separated by an extensive connective that is expanded apically into a small protrusion. Dehiscence is extrorse. Pollen *in situ* is small, about 15 μm in diameter, and monoaperturate with a distinctly delimited, trichotomocolpate aperture. The pollen apparently has a more or less complete tectum with a slightly verrucate–rugulate surface, except over the aperture, where the surface is granular. There is no evidence of a gynoecium.

The fossil staminate flower is similar to flowers of extant *Amborella* in several respects, but differs in having extrorse, rather than introrse, anther dehiscence. Pollen of *Amborella* also has an irregular, rather than a trichotomocolpate, aperture, and a tectum in which the surface is not verrucate–rugulate but covered with small gemmae. The fossil also resembles flowers of some species of extant *Kadsura* (Schisandraceae), which have a similar arrangement of tepals, anthers with extrorse dehiscence, and pollen with trichotomocolpate apertures. However, pollen of *Kadsura* has three shorter colpi between the diverging arms of the trichotomocolpate aperture and a coarsely reticulate tectum. The fossil flower may be broadly related to ANITA-grade plants, but plesiomorphic in several respects compared to extant *Amborella* and *Kadsura*.

Another flower that is probably related to ANITA-grade angiosperms is *Carpestella lacunata* (Figure 8.3) from the Early Cretaceous (Early to Middle Albian) Puddledock locality, Virginia, USA (von Balthazar *et al.*, 2008). *Carpestella* is based on a single abraded specimen almost certainly preserved at a post-anthetic stage. The specimen is small, about 0.65 mm long and 0.45 mm in diameter, and actinomorphic. Most of the fossil consists of a fragmentary gynoecium, which is covered on the outer surface by helically arranged scars left by detached floral organs. Proximally, there are 15 larger oval scars that are interpreted as scars from tepals. Distally, these are followed by about 60 smaller quadrangular scars, which are interpreted as scars from stamens or staminodes. The gynoecium is syncarpous, consisting of 13 carpels that are radially arranged around a central column in a star-shaped pattern. Carpel flanks are partly free from each other and there are distinct septal slits between the carpels. The apical part of the gynoecium is not preserved and there are no remains of ovules or seeds.

The star-shaped pattern of the gynoecium, the central gynoecial column, and the multipartite perianth and androecium of *Carpestella* are all features shared with flowers of some Nymphaeaceae and *Illicium* (Schisandraceae). The presence of septal slits links the fossil more directly to Nymphaeaceae, but phylogenetic analysis failed to resolve the relationships of the fossil securely. *Carpestella* may represent a separate lineage at the ANITA grade or

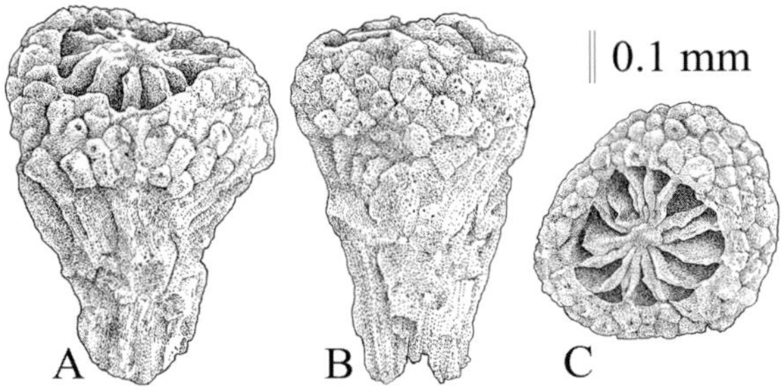

Figure 8.3 Abraded flower of *Carpestella lacunata* from the Early Cretaceous (Early–Middle Albian) Puddledock locality, Virginia, USA. (A, B) Lateral view showing scars of detached floral organs, and (C) apical view showing 13 carpels arranged radially around a central column. Drawn from SEM images in von Balthazar *et al.* (2007).

perhaps an early extinct taxon within Nymphaeaceae (von Balthazar *et al.*, 2008).

8.2 AMBORELLACEAE

Amborella has been assigned traditionally to Laurales, but modern phylogenetic analyses place it in a remarkable, isolated position as the sister group to all other angiosperms (e.g. Soltis *et al.*, 1997; Nandi *et al.*, 1998; Qiu *et al.*, 1999, 2000). The family is monotypic with a single species, *Amborella trichopoda*, endemic to New Caledonia.

Amborella is a woody shrub or small tree with alternate leaves. The wood lacks vessels and also shows many other features that have often been regarded as unspecialised. Flowers are unisexual with five to eight undifferentiated tepals and all floral parts in a spiral arrangement. Staminate flowers have up to 14 tetrasporangiate stamens that produce inaperturate, monoporate or irregular monocolpate pollen with the tectum ornamented with fine spines and small holes (Sampson, 1993, 2000b). Pistillate flowers have five to eight free, short-stalked carpels, each of which contains a single, pendent and almost orthotropous, bitegmic ovule (Philipson, 1993b; Nandi *et al.*, 1998; Endress and Igersheim, 2000b). Individual carpels are strongly ascidiate and carpel closure is by secretion (Endress and Igersheim, 2000b).

The family has no published fossil record. The small staminate flower discussed in section 8.1.1 from the Early Cretaceous Catefica locality, Portugal, shows interesting similarities to flowers of extant *Amborella*, but also differs in several respects and cannot be taken as a reliable record of the family.

8.3 NYMPHAEALES

The Nymphaeales include about 70–90 species of aquatic herb. The eight genera (*Barclaya*, *Brasenia*, *Cabomba*, *Euryale*, *Nuphar*, *Nymphaea*, *Ondinea*, *Victoria*) are sometimes treated as part of a single family, the Nymphaeaceae *sensu lato*, or *Brasenia* and *Cabomba* are separated from a more narrowly defined Nymphaeaceae in a separate family, Cabombaceae. A further family, Hydatellaceae, has recently been recognised as part of the Nymphaeales clade (Saarela *et al.*, 2007).

8.3.1 Hydatellaceae

The Hydatellaceae are a small family of inconspicuous, mainly annual, more or less aquatic, herbs. Until recently, the family was placed among monocots in Poales, but its relationship is now known to be close to the base of the angiosperm phylogenetic tree, most likely as sister group to Nymphaeaceae *sensu lato* (Saarela *et al.*, 2007). This new view of the phylogenetic position of Hydatellaceae is the most dramatic of all the many realignments that have taken place in angiosperm classification as a result of phylogenetic analyses based on molecular data.

The Hydatellaceae comprise a single genus, *Trithuria* (including *Hydatella*), with 11 species from Australia, New Zealand and India (Sokoloff *et al.*, 2008). In their vegetative structure Hydatellaceae plants are diminutive and simple, probably reflecting unique, perhaps paedomorphic, ecological specialisation. Leaves resemble those of monocots, but also juvenile leaves of Nymphaeales (Rudall *et al.*, 2007). Whether the Hydatellaceae have a single cotyledon, or a dicotyledonous embryo as in other Nymphaeales, is not yet known (Rudall *et al.*, 2007). Reproductive structures of Hydatellaceae differ substantially from those of other Nymphaeales, and also most other plants near the base of the angiosperm tree. How these reproductive structures should be interpreted morphologically is not completely clear (Rudall *et al.*, 2007; Rudall and Bateman, 2010), but there is no straightforward trimery in any aspect of floral organisation, as has sometimes been reported.

The reproductive units of *Trithuria* are unisexual or bisexual, consisting of bracts enclosing several carpels and/or stamens (Figure 8.4). It seems likely that each

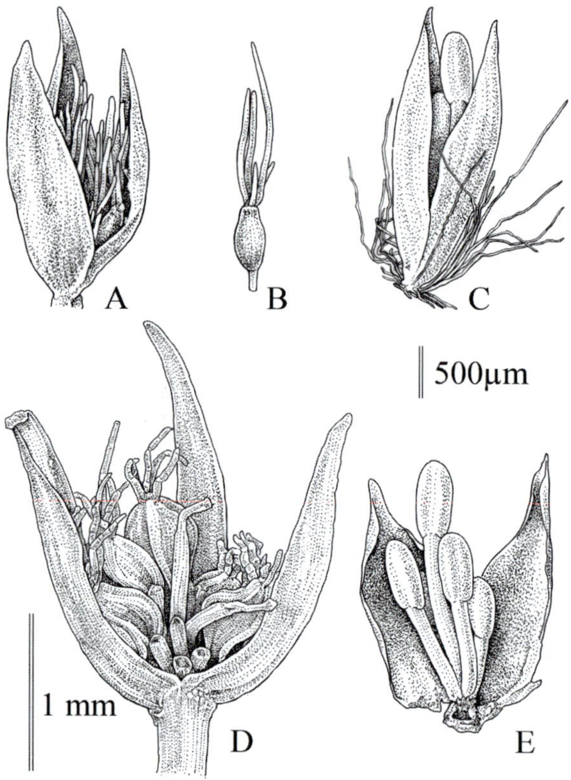

Figure 8.4 Reproductive units of extant *Trithuria australis* (A, B, D, E) and *Trithuria submersa* (C) (Hydatellaceae). (A) Pre-anthetic, pistillate reproductive unit with hair-like stigmatic papillae. (B) Isolated pistil with five stigmatic papillae of unequal length. (C) Pre-anthetic, staminate reproductive unit showing involucral bracts and protruding anthers. (D) Bisexual reproductive unit with three involucral bracts and several mature fruits. (E) Post-anthetic, staminate reproductive unit showing four stamens. Drawn from SEM images in Rudall *et al.* (2007).

reproductive unit is a cluster of unisexual flowers that are very simple, and presumably highly reduced, compared with the more typical flowers of other ANITA-grade angiosperms (Rudall *et al.*, 2007). Stamens are tetrasporangiate, dithecate and basifixed. Pollen is monocolpate, or sometimes trichotomocolpate, with a distinct colpus margin. The pollen wall is tectate–columellate with a perforate tectum and microechinate ornamentation (Remizowa *et al.*, 2008). Each pistil is most reasonably interpreted as an ascidiate carpel containing a single pendulous, anatropous, bitegmic ovule (Rudall *et al.*, 2007). Carpels have unfused margins and an apical tuft of peculiar, jointed stigmatic hairs. Fruits are dehiscent and prominently three-ribbed in some species, or indehiscent and not ribbed in other species. Seeds have a smooth or reticulate surface (Sokoloff *et al.*, 2008).

The Hydatellaceae currently have no fossil record and comparisons with *Archaefructus* (Chapter 9) (Rudall *et al.*, 2007; Saarela *et al.*, 2007; Endress and Doyle, 2009; Rudall and Bateman, 2010) seem premature given the very obvious differences. Nevertheless, the Hydatellaceae are of great interest because they are morphologically distinct from the other extant taxa at the ANITA grade. The family raises new questions about floral evolution in early angiosperm diversification, and further expands the search image for recognising and interpreting early angiosperms in the fossil record.

8.3.2 Cabombaceae and Nymphaeaceae

Cabombaceae and Nymphaeaceae are two families of aquatic herbs of almost worldwide distribution. However, except for *Nymphaea*, all of the genera have a relatively restricted geographic range (Schneider and Williamson, 1993; Williamson and Schneider, 1993). Cabombaceae include the extant genera *Brasenia* and *Cabomba*. The Nymphaeaceae include six genera: *Barclaya*, *Euryale*, *Nuphar*, *Nymphaea*, *Ondinea* and *Victoria*. The two families are sometimes merged into a broad concept of Nymphaeaceae, but APGIII (2009) leaves the limits of the two families open.

Flowers of Cabombaceae and Nymphaeaceae vary greatly in size and complexity, but are all actinomorphic, bisexual and with at least some petaloid tepals. The perianth is generally hypogynous, except in *Nymphaea*, *Euryale* and *Victoria* where the flowers are cup-shaped, and the perianth is epigynous. In *Brasenia* and *Cabomba* stamens are clearly differentiated into a filament and basifixed tetrasporangiate anther. In other Nymphaeaceae stamens are often laminar with the two thecae separated on the adaxial surface. Pollen is typically monocolpate or zonacolpate with ornamentation of various kinds (Sampson, 2000b). The androecium varies from six stamens in *Cabomba* to more than 100 in *Victoria*. In *Cabomba*, which is insect-pollinated, the gynoecium consists of three carpels, each with a long style and a short rounded stigma. In *Brasenia*, which is wind-pollinated, there are 10–14 carpels, each with a short style and a long papillate stigma. In all Nymphaeaceae, but not in *Brasenia* and *Cabomba*, the carpels are generally partly fused to syncarpous. Carpels of Nymphaeales are closed by a secretion in which the ovules are embedded. Ovules are one to numerous in each carpel. They are anatropous, except in *Barclaya*, which has orthotropous ovules. Ovules do not fill out the ovary cavity and in Nymphaeaceae the ovules are not in contact with the inner surface of the ovary (Igersheim and Endress, 1998). In Nymphaeaceae placentation is laminar–diffuse, whereas in Cabombaceae it is ventral, dorsal or lateral. Fruits in multicarpellate forms are generally many-seeded, spongy berries (Schneider and Williamson, 1993).

Seeds of Cabombaceae and Nymphaeaceae are very characteristic. They are typically small, elliptical or ovate, with a smooth or spiny surface. Seeds form from bitegmic ovules and are exotestal, often with the palisade cells in the outer epidermis having strongly undulate anticlinal walls (Corner, 1976). The most distinctive feature of seeds of Cabombaceae and Nymphaeaceae is the presence of a micropylar lid (embryotheca), which is well developed in all genera except for *Barclaya* (Collinson, 1980).

The Cenozoic fossil record of Cabombaceae and Nymphaeaceae is very extensive, as would be expected from their aquatic habit, which places them in, or close to, suitable environments for preservation. The Cretaceous record of Nymphaeaceae is more sparse and composed mainly of poorly preserved leaf impressions. However, there are also seeds and at least one floral structure, the Early Cretaceous *Monetianthus mirus*, that provide unequivocal early documentation of the group (Friis *et al.*, 2009b). Another Early Cretaceous floral structure, *Pluricarpellatia peltata*, is probably also of nymphaealean affinity (Mohr *et al.*, 2008). The interpretation of the organisation of the *Microvictoria svitkoana* flowers and the position of the genus as sister to *Euryale*, *Nymphaea* and *Victoria* suggested by Gandolfo *et al.* (2004), has been questioned (Endress, 2008) (section 9.5.1).

Monetianthus mirus (Figures 2.14, 8.5, 8.6) is known from a single, small, coalified flower with clear nymphaeaceous features described from the Early Cretaceous (Late Aptian – Early Albian) Vale de Agua locality, Portugal (Friis *et al.*, 2001, 2009b). The specimen is fragmentary, but can be reconstructed in some detail by SEM studies combined with synchrotron X-ray microtomography (SXRTM). The flower is actinomorphic, perigynous, about 3 mm long and 2 mm in diameter, and apparently bisexual with the remains of 9 or 10 perianth parts, an androecium of 20 stamens, and a gynoecium of 12 carpels. The ovary is syncarpous with the carpels arranged radially around a central column. The phyllotaxis of the perianth parts and stamens is not completely clear. There are 12 locules, each containing around six anatropous ovules. Placentation is laminar. Ovules do not fill the whole locule cavity. Monocolpate, coarsely reticulate–columellate pollen, about 20 μm in diameter, occurs on the surface of the fossil between the stamen scars.

The arrangement of carpels around a central projection of the floral axis is a unique feature known only in some Nymphaeaceae and *Illicium* (Endress and Igersheim, 2000a). However, flowers of *Illicium* are distinguished from *Monetianthus* in having a superior ovary and an apocarpous gynoecium of spirally arranged carpels; each carpel contains only a single ovule that almost completely fills out the ovary cavity. A relationship of *Monetianthus* with *Illicium*, as suggested by Gandolfo *et al.* (2004), is therefore not supported by the information currently available. The characters of *Monetianthus* clearly place it among Nymphaeaceae and it appears most closely related to those extant genera that have flowers with multicarpellate, partly syncarpous gynoecia and a semi-inferior ovary (Friis *et al.*, 2009b). The presence in the fossil flower of these derived features of Nymphaeaceae *sensu stricto* places *Monetianthus* in crown group Nymphaeaceae. In turn, this implies that Cabombaceae, Hydatellaceae and perhaps other lineages of Nymphaeales had already diverged in the Early Cretaceous.

Pluricarpellatia peltata (Figure 8.7) is an Early Cretaceous nymphaealean plant based on fossils described from the Crato Formation of Brazil (Late Aptian – Early Albian). It comprises roots, leaves and reproductive organs that are attached in unusual whole-plant preservation (Mohr *et al.*, 2008). The peltate leaves are similar to those of some extant Nymphaeales and suggest that the plant was an aquatic. Reproductive structures are not well preserved, but show clusters of free carpels, each of which contains many seeds. With its apocarpous gynoecium, *Pluricarpellatia* may be more closely related to Cabombaceae than to Nymphaeaceae *sensu stricto*.

A variety of leaves and seeds with affinity to Cabombaceae or Nymphaeaceae from the classic Early Cretaceous floras of Portugal (Saporta, 1894; Teixeira, 1945; Friis *et al.*, 2000b) provide further support for the presence of Nymphaeales in the Early Cretaceous. *Brasenia*-like leaves from the Early Cretaceous leaf flora of Buarcos-para-Tavarede (Figure 8.8) have been assigned to the fossil genus *Braseniopsis* (Saporta, 1894), but a more detailed analysis of leaf venation is needed to secure the systematic placement

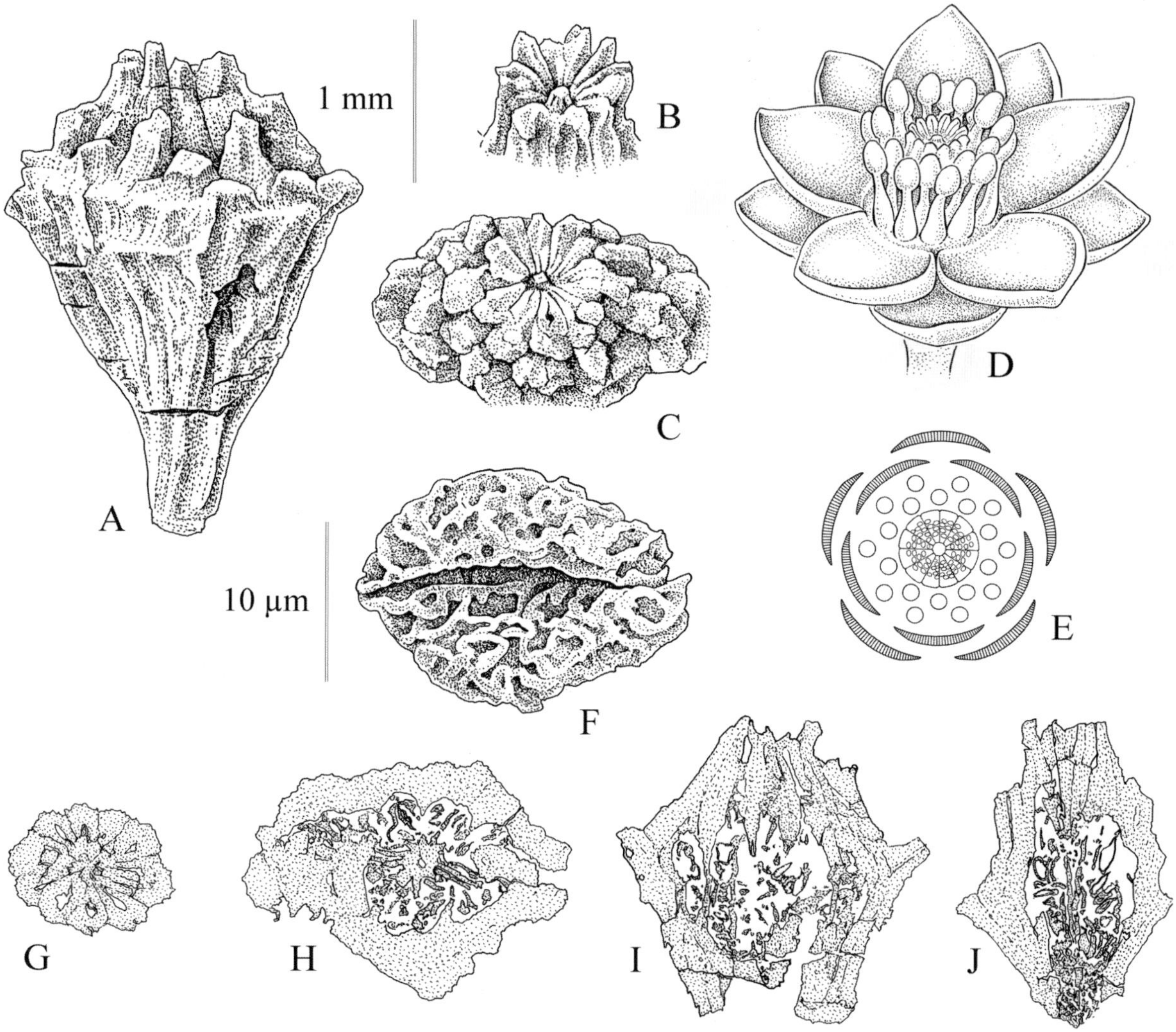

Figure 8.5 *Monetianthus mirus* from the Early Cretaceous (Late Aptian – Early Albian) Vale de Agua locality, Portugal. (A–C) Flower in (A) lateral and (B, C) apical view showing carpels, (B, C) central column and (A, C) scars from stamens and tepals. (D) Reconstruction of flower in oblique view. (E) Floral diagram. (F) Monocolpate, reticulate pollen grain from surface of flower. (G–J) Synchrotron X-ray microtomography (SXRTM) sections through flower showing locules with many ovules in (G, H) transverse and (I, J) longitudinal view. Drawn from SEM and SXRTM images in Friis *et al.* (2001, 2009b).

of these leaves. Other Early Cretaceous fossils of possible nymphaealean affinity are leaves of *Scutifolium jordanicum* from the Albian of Jordan (Taylor *et al.*, 2008). These are small, elliptical to almost circular, centrally peltate leaves with actinodromous to palinactinodromous primary venation. In leaf architecture they are closely similar to leaves of extant *Cabomba*.

The probable nymphaealean affinity of the leaves from the Early Cretaceous of Portugal is supported indirectly by the presence of several probable nymphaealean seeds from several Early Cretaceous localities from Portugal, including the Torres Vedras locality of Late Barremian – Early Aptian age (Figure 8.9). Similar seeds also occur in the clay balls of the Drewry's Bluff locality (Late

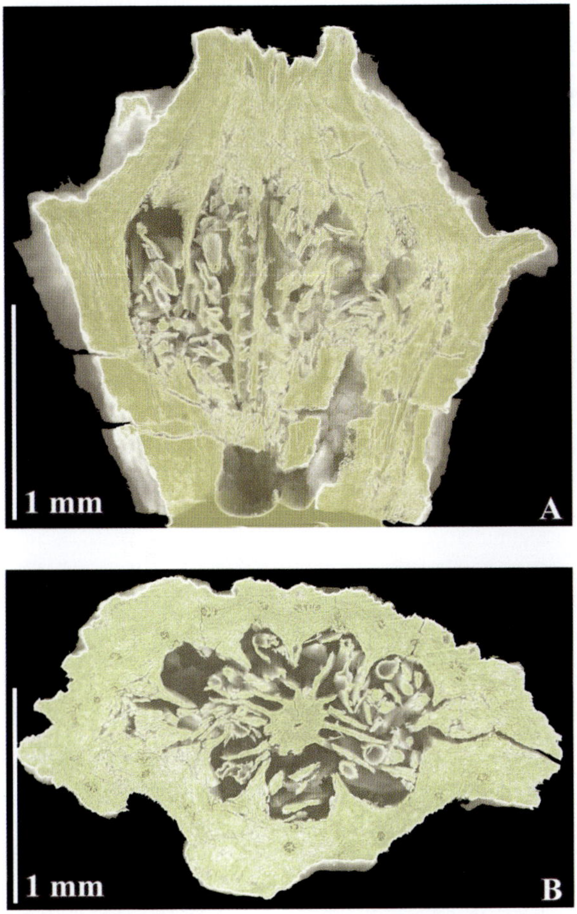

Figure 8.6 *Monetianthus mirus* from the Early Cretaceous (Late Aptian – Early Albian) Vale de Agua locality, Portugal. Synchrotron X-ray microtomography (SXRTM) image of ovary in (A) longitudinal and (B) transverse section showing locules with many ovules that do not fill the locule cavities.

Barremian – Early Aptian) in the Potomac Group of eastern North America (Figure 8.9). These seeds have an exotesta of distinctive tall palisade sclerenchyma with strongly undulate (wavy) anticlinal cell walls. Very similar cells are seen in seeds of extant Nymphaeales. Several of the fossil seeds also have a distinct micropylar lid that strongly suggests affinity with Nymphaeales.

From the Late Cretaceous, there are several records of fossil leaves and seeds of Nymphaeales. Leaves of *Nymphaea mesozoica* from the Late Cretaceous (Turonian) of the southern Negev Desert, Israel (Dobruskina, 1997), are deeply cordate with an entire margin and palmate primary venation. Long petioles are attached perpendicular to the leaf blade as in extant Nymphaeaceae. Still more secure are unequivocal seeds of Nymphaeales described as *Symphaenale futabensis* from the Late Cretaceous (Early Santonian) Gokurakuzawa locality, northeastern Honshu, Japan (Takahashi *et al.*, 2007). These seeds are anatropous, ovoid and about 1 mm long (Figure 8.10). The raphe enters the seed obliquely at a hilum that is adjacent to, but separate from, the micropyle. The micropyle has a distinct plug with a papillate surface. The exotesta is composed of columnar palisade sclerenchyma cells with anticlinal walls that are strongly undulate to the outside, but straight towards the inside. *Symphaenale* is similar to seeds of extant *Brasenia*, but differs in features of the micropyle and hilum, which are more similar to the condition in other Nymphaeales (Takahashi *et al.*, 2007).

Nymphaealean seeds are also reported as *Sabrenia pachyderma* from the Late Cretaceous flora of the Negev Desert, Israel (Krassilov *et al.*, 2005). However, the critical details on the exact position of the hilum and micropyle needed for a secure generic or familiar assignment are lacking. Similarly, the putative nymphaeaceous affinity of seeds described as *Barclayopsis urceolata* from the Maastrichtian and Early Cenozoic (Paleocene) of Central Europe (Knobloch and Mai, 1984, 1986) needs further clarification (Mai, 1985a, 1995).

The earliest record of dispersed pollen assigned to Nymphaeaceae is from the Late Cretaceous (Maastrichtian) of North America. This is based on *Nymphaea*-like pollen from Canada, described as *Zonosulcites scollardensis* and *Z. parvus* (Srivastava, 1969; Muller, 1981). *Nuphar*-like pollen reported from the Teapot Dome locality, Wyoming, USA (Wolfe, 1991), which includes the Cretaceous–Palaeogene boundary, is identical to *Pandaniidites typicus* according to Nichols (1992). This fossil pollen differs from pollen of extant Nymphaeaceae in the configuration of the aperture and is probably related to extant Pandanaceae (Nichols, 1992).

Although the Nymphaeales were evidently established early in the Cretaceous, most extant genera may not have differentiated or become diverse until the Cenozoic. The Cenozoic record of Nymphaeales is very extensive, based mainly on their characteristic seeds. Numerous fossil species are assigned to the extant genera *Brasenia*, *Cabomba*, *Euryale*, *Nymphaea* and *Nuphar*, but the Cenozoic fossil record also includes many species assigned to extinct genera such as *Braseniella*, *Dusembaya*, *Eoeuryale*, *Irtyshenia*, *Nikitinella*, *Palaeoeuryale*, *Palaeonymphaea*, *Protobarclaya*, *Pseudoeuryale*, *Sabrenia*, *Tavdenia* and *Tomskiella* (Reid and Chandler, 1933; Weyland, 1938;

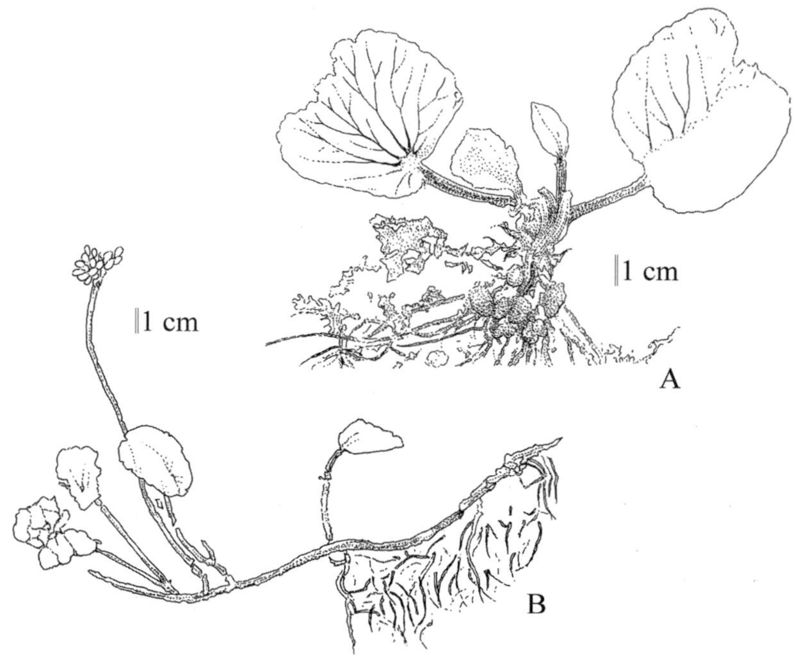

Figure 8.7 *Pluricarpellatia peltata*, a nymphaealean 'whole' plant from the Early Cretaceous (Late Aptian – Early Albian) Crato flora, Brazil. (A) Plant showing attached roots and leaves. (B) Plant showing attached rhizomes, roots, leaves and fruits. (A) Drawn from photograph in Mohr *et al.* (2008); (B) from Friis *et al.* (2007).

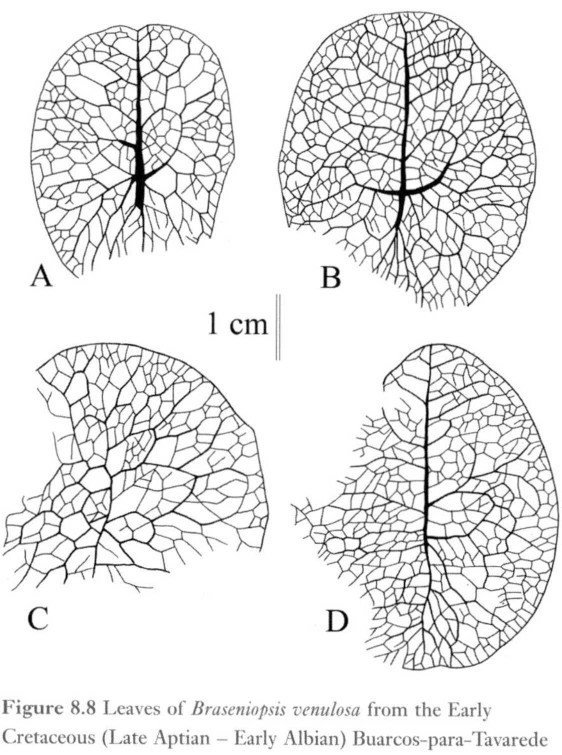

Figure 8.8 Leaves of *Braseniopsis venulosa* from the Early Cretaceous (Late Aptian – Early Albian) Buarcos-para-Tavarede locality, Portugal. Redrawn from Saporta (1894).

Dorofeev, 1973, 1974a; Collinson, 1980; Mai, 1985a, 1995). The family seems to have been more diverse and more widespread during the Cenozoic than it is today. Since the seeds are so easy to recognise, and also fossilise well, their relatively sparse presence in the Cretaceous may indicate that Nymphaeales were less important in aquatic ecosystems at this time. Their subsequent radiation in the Cenozoic may be related to climatic or other environmental changes. The phylogenetic relationships and relatively restricted geographic distributions of *Victoria*, *Euryale*, *Ondinea* and *Barclaya* may also suggest a Cenozoic radiation (Chapter 20).

8.4 AUSTROBAILEYALES

The Austrobaileyales include three families, Austrobaileyaceae, Schisandraceae (including Illiciaceae) and Trimeniaceae. Takhtajan (e.g. 1980) placed Austrobaileyaceae in Laurales, a position that was also supported by early cladistic analyses (Donoghue and Doyle, 1989a). Cronquist (1981) included the family in Magnoliales. In recent molecular analyses Austrobaileyaceae are placed as sister to Schisandraceae (Qiu *et al.*, 1993; Nandi *et al.*, 1998) plus Trimeniaceae (APGII, 2003), both families with which it had not previously been associated.

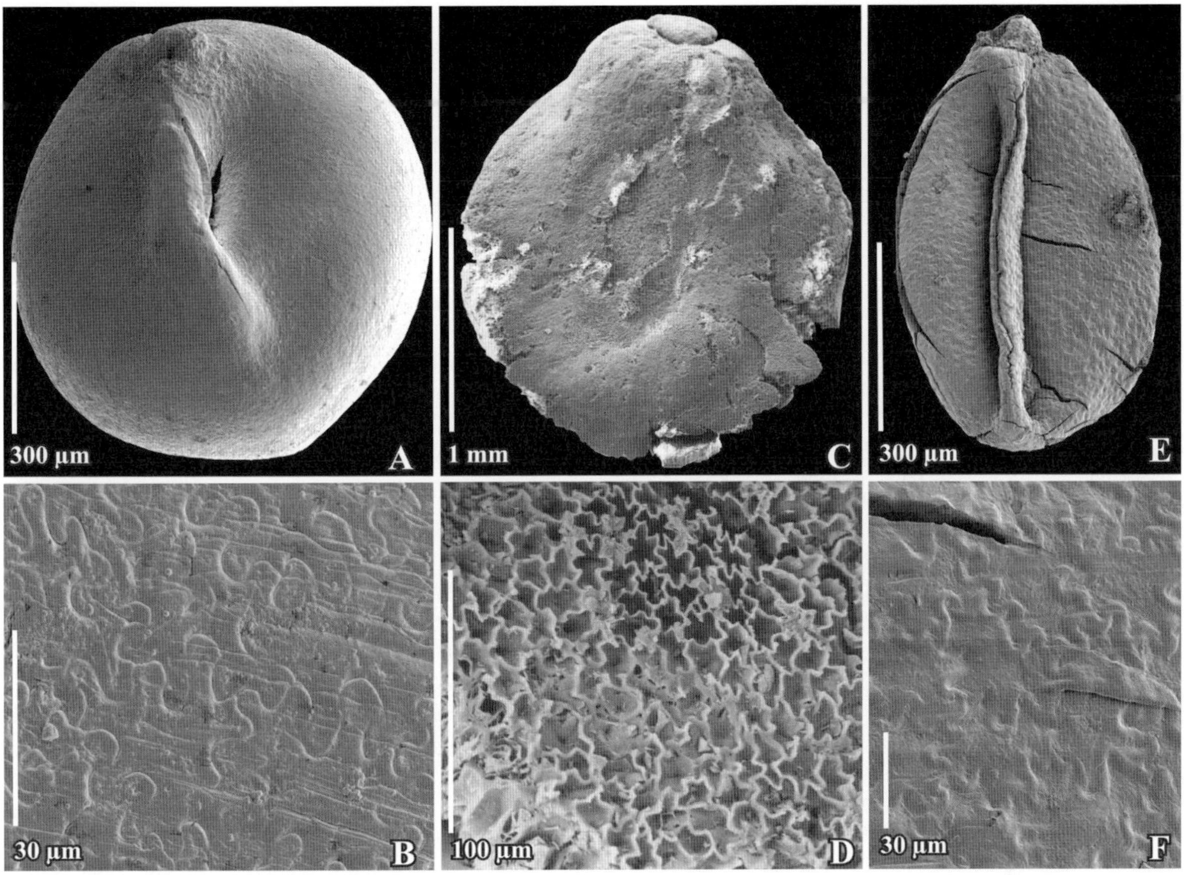

Figure 8.9 SEM pictures of Early Cretaceous seeds from the Potomac Group of eastern North America and from Portugal with features suggesting relationship to Nymphaeales. (A, B) Seed from the clay balls (Late Barremian – Early Aptian) at the Drewry's Bluff locality, Virginia, USA, (A) in lateral view showing raised micropylar area and (B) detail of exotestal cells with wavy outlines. (C, D) Seed from Famalicão locality, Portugal (Late Aptian), (C) in lateral view showing apical micropylar lid and (D) detail of exotestal cells with wavy outlines seen from inside. (E, F) Seed from Torres Vedras locality, Portugal (Late Barremian – Early Aptian) (E) in lateral view showing apical micropylar lid and (F) detail of exotestal cells with wavy outlines. (A, B) Material in the collections of the Field Museum, Chicago; (C–F) material in the collections of the Swedish Museum of Natural History, Stockholm.

8.4.1 Austrobaileyaceae

The Austrobaileyaceae are a monotypic family with a single species, *Austrobaileya scandens*, which is endemic to northern Queensland, Australia (Endress, 1993a). *Austrobaileya* is a liana with opposite, evergreen leaves. Flowers are bisexual with spirally arranged parts. There are 10–13 free, ascidiate carpels, each containing 4–10 anatropous ovules (Igersheim and Endress, 1997). Pollen is monocolpate and reticulate, and is similar in some respects to fossil pollen assigned to *Clavatipollenites* (Endress and Honegger, 1980). The Austrobaileyaceae have no published fossil record.

8.4.2 Schisandraceae (including Illiciaceae)

The Schisandraceae comprise about 85 species in three genera, *Illicium*, *Kadsura* and *Schisandra*. *Kadsura* and *Schisandra* are climbers that are mainly distributed in tropical and temperate parts of eastern Asia. There is one species, *Schisandra glabra*, in southeastern North America. *Illicium* includes evergreen shrubs or small trees. There are about five species in North America with the remaining 30 species in Southeast Asia (Keng, 1993).

Flowers are unisexual in *Kadsura* and *Schisandra*, but bisexual in *Illicium*. Flowers have numerous floral parts in a

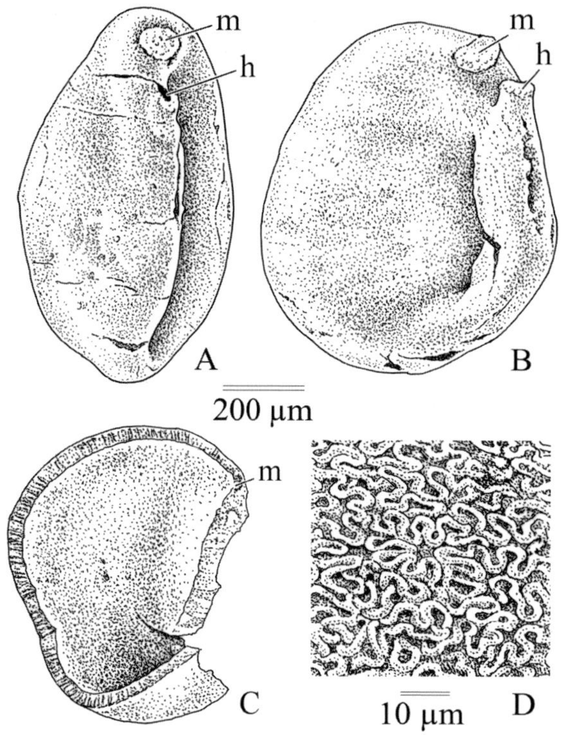

Figure 8.10 Seeds of *Symphaenale futabensis* (Nymphaeales) from the Late Cretaceous (Early Santonian) Gokurakuzawa locality, northeastern Honshu, Japan. (A, B) Lateral view of anatropous seeds showing hilum (h) and micropylar area (m). (C) Broken seed showing columnar palisade sclerenchyma of seed wall. (D) Detail of seed surface showing wavy cell outlines. Drawn from SEM images in Takahashi *et al.* (2007).

spiral arrangement. Perianth parts are in two or three whorls. Staminate flowers of *Kadsura* and *Schisandra* have few to many stamens with tetrasporangiate anthers. Pollen of *Illicium* and some species of *Schisandra* is syntricolpate with three colpi that are fused at the distal pole. Pollen of *Kadsura* and other species of *Schisandra* is hexacolpate with three additional shorter colpi in an equatorial position alternating with the fused colpi. The colpi do not converge at the pole (Praglowski, 1976; Doyle, 2005). The symmetry and homology of this usual aperture arrangement is not well understood (see discussion in Doyle, 2005).

The gynoecium of *Illicium* is apocarpous with up to about 20 carpels that are partly, or completely, plicate (Igersheim and Endress, 1997). They are spirally arranged, but the spiral is very condensed and mature carpels appear whorled. Each carpel contains a single, anatropous, bitegmic ovule that almost completely fills the cavity of the ovary (Igersheim and Endress, 1997). The carpels are closed by a secretory canal as well as by postgenital fusion at their periphery. Fruits are star-shaped. Seeds are similar in several respects to those of extant Nymphaeales. They are exotestal with an outer epidermis of palisade sclerenchyma cells (Figure 8.11). The palisade cells have strongly undulate anticlinal walls towards both the outside and the inside of the seeds. The hilar area is distinct (Oh *et al.*, 2003). The gynoecium of *Kadsura* and *Schisandra* is multicarpellate and apocarpous with 12–300 carpels in a spiral arrangement along a more or less elongated receptacle. Each carpel is closed by secretion and contains two to five, or sometimes more, anatropous to campylotropous, bitegmic ovules that almost completely fill the ovary cavity. Seeds are exotestal (Denk and Oh, 2005).

The earliest fossils that can be assigned to Schisandraceae are seeds of *Illiciospermum*, but otherwise the Cretaceous record of the family is very sparse. It includes only a few leaf impressions and pollen. Nevertheless, the fossil record indicates the presence of at least stem-group Schisandraceae by the mid-Cretaceous (Oh *et al.*, 2003).

Illiciospermum pusillum (Figure 8.11) is based on well-preserved seeds from the Late Cretaceous (Cenomanian–Turonian) of Kazakhstan (Frumin and Friis, 1999). The seeds show a suite of characters linking them to extant *Illicium*. Seeds are anatropous, bitegmic and exotestal. The testa has an outer epidermis of tall, thick-walled palisade sclerenchyma with strongly undulate anticlinal walls and an inner layer of large parenchyma cells. The tegmen is thin and crushed. The most diagnostic feature linking the fossil seeds to those of extant *Illicium* species is the slit-like micropylar opening in the outer integument. This slit is transverse to the plane of symmetry and placed on a slight bulge dorsal to the hilum. *Illiciospermum pusillum* is considerably smaller than seeds of extant *Illicium* (Figure 8.11). The lack of information on other aspects of the plant, including floral morphology and fruits, precludes assignment of the fossils to the modern genus.

Pollen from the Maastrichtian of California described as 'Forma A' and compared to pollen of extant Schisandraceae (Chmura, 1973) is hexacolpate and may be related to *Schisandra* and *Kadsura*. Fossil wood of possible Schisandraceae has been described from two Late Cretaceous localities in Antarctica, the Coniacian–Santonian of James Ross Island and the Late Campanian – Early Maastrichtian of Vega Island (Poole *et al.*, 2000b). Other Cretaceous

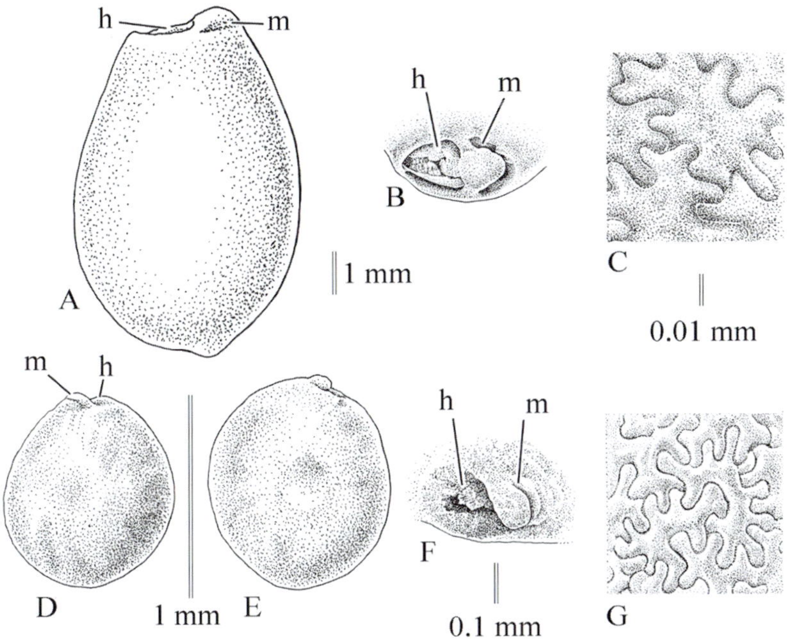

Figure 8.11 Seeds of (A–B) extant *Illicium anisatum*, (C) extant *Illicium floridanum* and (D–G) extinct *Illiciospermum pusillum* from the Late Cretaceous (Cenomanian–Turonian) Sarbay locality, Kazakhstan. (A, B) Lateral (A) and apical (B) views showing hilum (h) and slit-shaped micropylar area (m). (C) Detail of seed surface showing wavy cell outlines. (D, E) Seeds in lateral view. (F) Apical view of seed showing hilum (h) and micropylar area (m). (G) Detail of seed surface showing wavy cell outlines. (A–C) Drawn from SEM images in Oh *et al.* (2003); (D–G) drawn from SEM- images in Frumin and Friis (1999).

fossils assigned to Schisandraceae require re-examination before their systematic affinity can be established with certainty. They include fossil leaves from the Cenomanian of Bohemia, Czech Republic, and the Turonian of Kazakhstan, assigned to the extinct genus and species *Illiciphyllum deletum* (Velenovský, 1884, 1889; Shilin, 1986). Leaves with *Illicium*-like features have also been reported from the mid-Cretaceous (latest Albian – earliest Cenomanian) of Nebraska as *Longstrethia varidentata* (Upchurch and Dilcher, 1990), but they also show features of other early-diverging angiosperm lineages such as Trimeniaceae.

Records of Schisandraceae are more abundant in the Cenozoic than in the Cretaceous and there is reliably determined fossil material from Asia, Europe and North America (Oh *et al.*, 2003). In particular there is an extensive record of leaves and seeds from the Cenozoic of Central Europe assigned to extant *Illicium* (e.g. Kräusel and Weyland, 1950; Jähnichen, 1965; Weyland *et al.*, 1967; Mai, 1970). Most of these possess features that are diagnostic for the extant genus, but some fossils, such as *Illicium germanicum*, *I. fliegelii* and *I. geiseltalensis*, should be excluded (Oh *et al.*, 2003).

Characteristic star-shaped fruits, described as *Illicium avitum* (Tiffney and Barghoorn, 1979), are known from the Early Miocene Brandon Lignite of Vermont, USA. They are very similar to fruits of extant *Illicium*. This material includes several three-dimensional immature gynoecia that have many details of their organisation and anatomy well preserved. A record of *Illicium*-like fruits from the younger Cenozoic of Australia, *Illicites astrocarpa* (Mueller, 1883), has not been confirmed by recent studies and the phylogenetic position of these fossils remains uncertain (Frumin and Friis, 1999). Fossil leaves and seeds of *Schisandra* are known from Asia, Europe and North America with the earliest well-documented records in the Eocene of Europe and North America (Gregor, 1981; Mai and Walther, 1985; Manchester, 1994; Mai, 1995; Denk and Oh, 2005). Fossil leaves of *Kadsura* are also reported from the Late Cenozoic of Europe (Mai, 1995). These records document the once much more widespread distribution of the three extant genera and show that the current disjunct distribution of *Illicium* and *Schisandra* is relictual.

8.4.3 Trimeniaceae

The Trimeniaceae are a small family of five to nine species in one genus, *Trimenia* (including *Piptocalyx*), distributed from eastern Australia to Celebes and the Southwest Pacific. The plants are shrubs, small trees or lianas with opposite leaves.

Flowers are small, simple and mostly bisexual with 2–38 tepals and 7–25 stamens in a spiral arrangement. Stamens are tetrasporangiate and dehisce by longitudinal slits. Pollen is unusually diverse and includes dicolpate and polyporate as well as inaperturate forms. The gynoecium consists of a single carpel closed by secretion and contains a single, pendulous, anatropous and bitegmic ovule (Endress and Sampson, 1983; Sampson and Endress, 1984; Philipson, 1993c).

There are a few putative records of Trimeniaceae from the Cretaceous and Cenozoic, but all are based on dispersed pollen. The oldest potential record is dispersed pollen of *Cretacaeisporites scabratus* from the mid-Cretaceous (Albian–Cenomanian) of Brazil (Muller, 1981) and the Late Cretaceous (Turonian) of West Africa (Herngreen, 1973). Other possible records from the Late Cretaceous are dispersed pollen from the Campanian–Maastrichtian of Australia described as *Periporopollenites fragilis* (Dettmann, 1994). An Early Cenozoic record of *Periporopollenites demarcatus* from Australia has also been referred to Trimeniaceae (Macphail *et al.*, 1994), although a slightly younger record of the same species has been referred to Caryophyllaceae (Blackburn and Sluiter, 1994). None of these dispersed grains has been studied at the ultrastructural level; further study is required to confirm or refute their proposed affinity to Trimeniaceae.

8.5 CHLORANTHACEAE

Previous classifications have often placed Chloranthaceae in Laurales (e.g. Takhtajan, 1980; Donoghue and Doyle, 1989a) or sometimes in Piperales (Cronquist, 1981). However, phylogenetic analyses based on molecular data (Qiu *et al.*, 1999, 2000; Soltis *et al.*, 2000b) separate Chloranthaceae from both of these groups. The precise phylogenetic position of Chloranthaceae has proved difficult to resolve, but the family appears to have diverged at a level above the ANITA grade, and below crown-group eumagnoliids, monocots and eudicots. It is sometimes resolved as the sister group to eumagnoliids (Moore *et al.*, 2007). The position of Chloranthaceae in angiosperm phylogeny, as a lineage that diverged and began to differentiate at a very early stage in angiosperm evolution, is consistent with the antiquity of the group revealed by its very extensive and early fossil record (Friis *et al.*, 1997b) (Chapter 20).

The Chloranthaceae are a well-defined family comprising four genera and about 77 species distributed mainly in tropical regions of the world. They include herbs, shrubs and trees with opposite leaves in a decussate arrangement. Flowers are small and simple. They may be unisexual or bisexual. The flowers lack a perianth, except in *Hedyosmum* (Endress, 1987). Fusion of the carpel wall is not complete in Chloranthaceae. Angiospermy is achieved by an inner secretion that closes the carpels (Endress and Igersheim, 1997). Three genera, *Ascarina*, *Chloranthus* and *Sarcandra*, are restricted to the Old World. *Hedyosmum* has one species in the Old World and about 45 species in the New World (Todzia, 1993).

The Chloranthaceae diverged, and began to differentiate, in the Early Cretaceous, very early in angiosperm history. The oldest fossils that can be assigned to the family with reasonable certainty are dispersed pollen grains of *Asteropollis* from the Hauterivian–Barremian of Central Asia (Vakhrameev, 1991). Still more secure are staminate and pistillate flowers from the Late Barremian – Early Albian mesofossil floras of Portugal. Both dispersed *Asteropollis* pollen and the fossil flowers are very similar to those of extant *Hedyosmum*. By the Aptian–Albian, the common occurrence of *Asteropollis* pollen in palynological assemblages from Europe, Asia, Australia and North America indicates that these *Hedyosmum*-like plants had attained an almost worldwide distribution before the end of the Early Cretaceous.

Dispersed pollen of *Clavatipollenites* has been compared to pollen of extant *Ascarina*. Some of the *Clavatipollenites* grains very closely resemble those of *Ascarina* in details of their morphology and pollen wall ultrastructure (e.g. Couper, 1960; Walker and Walker, 1984). Records of *Clavatipollenites* pollen have therefore often been taken as an indicator of the presence of Chloranthaceae in the fossil record. However, diverse monocolpate or trichotomocolpate pollen grains with different kinds of exine ornamentation have been included in *Clavatipollenites*. Some of these grains were almost certainly produced by Chloranthaceae, and perhaps by plants close to extant *Ascarina*, but the precise circumscription of *Clavatipollenites* is problematic. As a result, the systematic position of many of the grains referred to *Clavatipollenites* is uncertain (Chapter 9).

Hammenia is another dispersed pollen type from the Cretaceous that has been compared to Chloranthaceae. *Hammenia* pollen is stephanocolpate and reticulate, similar to that of some extant *Chloranthus* and *Sarcandra* (Hedlund and Norris, 1968; Walker and Walker, 1984; Ward, 1986). However, *Hammenia* was also compared to *Hedyosmum*. More detailed comparison with extant Chloranthaceae is needed for a secure systematic placement of this pollen type.

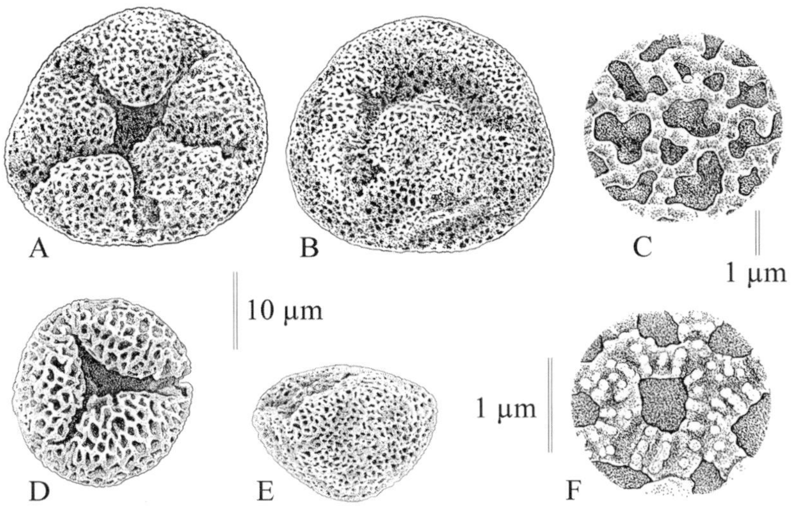

Figure 8.12 Dispersed chloranthaceous pollen from the Early Cretaceous of the USA. (A–C) *Asteropollis asteroides* from the Middle Albian of Oklahoma in (A) distal and (B) proximal view; (C) details of the reticulum. (D–F) *Clavatipollenites*-type pollen from the Middle–Late Albian of Maryland, USA, trichotomocolpate grain in (D) distal and (E) equatorial view; (F) details of reticulum. Drawn from SEM images in Walker and Walker (1984).

Fossil leaves assigned to *Crassidenticulum* and *Densinervum* from the mid-Cretaceous (latest Albian – earliest Cenomanian) Rose Creek locality, Nebraska, USA, closely resemble leaves of extant Chloranthaceae and may represent extinct genera of the family (Upchurch and Dilcher, 1990). The oldest fossils clearly related to *Chloranthus* and *Sarcandra* are from the mid-Cretaceous of Europe (section 8.5.3) and include isolated androecia and fruits from the karst infillings of the Rhenish Massif, Germany (C. Hartkopf-Fröder, A. Viehofen and E. M. Friis, in progress). From the Late Cretaceous of Europe, Asia and North America there are also several records of flowers, dispersed stamens and inflorescences assigned to *Chloranthistemon* (e.g. Eklund *et al.*, 1997).

8.5.1 *Asteropollis*

Dispersed pollen assigned to *Asteropollis* is conspicuous in many Early Cretaceous palynofloras (Figure 8.12). The genus was first described from the Albian or Early Cenomanian of central Alberta, Canada, and includes one broadly circumscribed species, *Asteropollis asteroides* (Hedlund and Norris, 1968). *Asteropollis* pollen is monoaperturate and reticulate with beaded or spiny supratectal ornamentation on the muri. The aperture is irregularly star-shaped and very distinctive, which usually makes the identification of dispersed or *in situ* grains straightforward. The number of aperture arms varies from three to six (trichotomocolpate to hexachotomocolpate). Regularly trichotomocolpate pollen assigned to *Asteropollis* probably does not belong to the genus (Chapter 9). In the pollen wall the ektexine in *Asteropollis* is three-layered, consisting of a reticulate tectum, an infratectal layer of simple columellae and a distinct foot layer. The endexine is thin except under the aperture, where it is strongly thickened. *Asteropollis* pollen closely resembles that of extant *Hedyosmum*, both in details of morphology and in ultrastructure of the pollen wall (Walker and Walker, 1984). The relationship of *Asteropollis* to extant *Hedyosmum* is also confirmed by the discovery of *Asteropollis* pollen *in situ* in stamens and adhering to pistillate flowers that are closely similar to staminate and pistillate flowers of *Hedyosmum*.

Dispersed *Asteropollis* pollen is known from many Early and mid-Cretaceous palynofloras from both the Northern and the Southern Hemisphere. The oldest records are from the ?Hauterivian–Barremian Andakhuduk Formation of Mongolia (Vakhrameev, 1991). Other occurrences from the Early Cretaceous of Asia are from the Barremian–Aptian of the Lake Baikal area (Vakhrameev, 1991) and the Barremian–Aptian of China (e.g. Li, 1995). Early to mid-Cretaceous records from Europe include dispersed and *in situ Asteropollis* pollen from several localities in Portugal, with the oldest occurrences from the Barremian or Aptian (e.g. Groot and Groot, 1962; Pais and Reyre, 1981; Friis *et al.*, 1999). In North Africa *Asteropollis* enters the fossil record in the Barremian–Aptian (e.g. Schrank, 1987) and in North

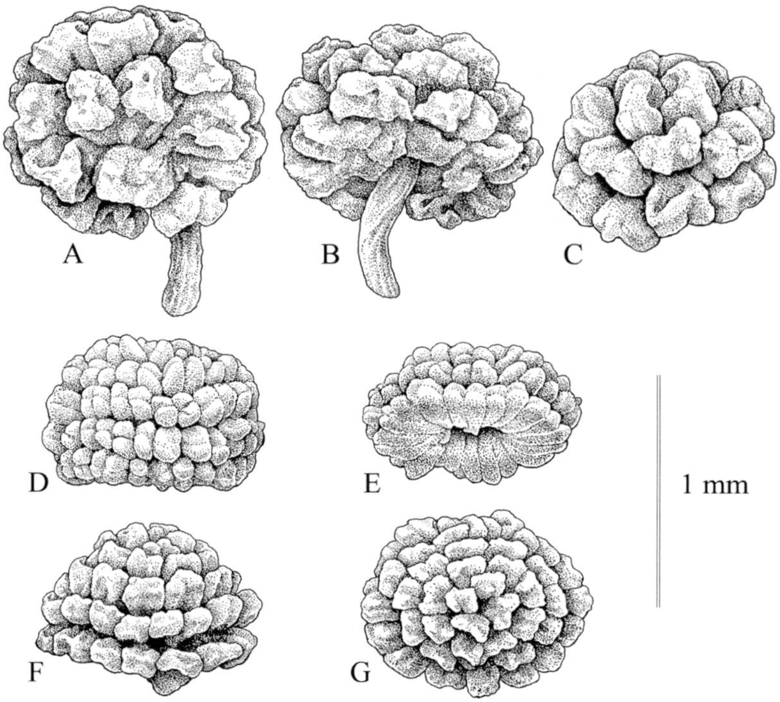

Figure 8.13 Staminate *Hedyosmum*-like inflorescences from the Early Cretaceous of Portugal. (A–C) Inflorescences from the Catefica locality (Late Barremian–Aptian) with stalk preserved in (A) oblique lateral, (B) oblique basal and (C) apical view. (D, E) Inflorescence from the Vale de Agua locality (Late Aptian – Early Albian) in (D) lateral and (E) oblique basal view. (F–G) Inflorescence from the Torres Vedras locality (Late Barremian – Early Aptian) in (F) lateral and (G) apical view. (A, B, D–G) From Friis *et al.* (2006a); (C) from material in the collections of the Swedish Museum of Natural History, Stockholm.

America *Asteropollis* ranges from the Albian to the Turonian (e.g. Hedlund and Norris, 1968; Phillips and Felix, 1971; Doyle *et al.*, 1977; Srivastava, 1977; Ward, 1986). In the Southern Hemisphere the genus is known from the Albian of South America and from the ?Albian to the Turonian of Australia (Dettmann, 1973; Burger, 1980, 1990). Through the Late Cretaceous *Asteropollis* decreased in importance and by the Cenozoic this kind of pollen is sparse.

8.5.2 *Hedyosmum*-like flowers

Staminate and pistillate flowers very similar to those of extant *Hedyosmum* have been recovered from several Early Cretaceous (Late Barremian – Early Albian) mesofossil floras from Portugal. Pistillate flowers are particularly abundant in the Buarcos flora, but are also present in the Famalicão, Vale de Agua and Torres Vedras floras (Friis *et al.*, 1997b). Some of these floras have also yielded well-preserved staminate inflorescences (Friis *et al.*, 2006a). A single pistillate *Hedyosmum*-like flower has also been identified from the Early Cretaceous (Early–Middle Albian) Puddledock flora, Virginia, USA (Friis *et al.*, 1997b).

Staminate flowers (Figures 8.13, 8.14) are naked, generally interpreted as unistaminate, and borne in dense inflorescences. The inflorescences are conical, or almost spherical, and borne on a long stalk. There are no remains of bracts or other supporting organs at the base of the inflorescences. The arrangement of flowers in the inflorescence is whorled and number of stamens per whorl decreases towards the apex. Anthers are basifixed, almost sessile, dithecate and tetrasporangiate. The connective is poorly developed between the pollen sacs, but extends apically into a short sterile extension. Pollen *in situ* is of the *Asteropollis* type. Most staminate flowers are found in inflorescences (Figure 8.13) or in inflorescence fragments (Figure 8.14); more rarely they occur as isolated stamens. Based on the size of the stamens and the morphology of the pollen grains several different taxa can be distinguished, all of which remain to be formally described.

Pistillate flowers (Figure 8.15) are unicarpellate and consist of a distinctly triangular ovary topped by three

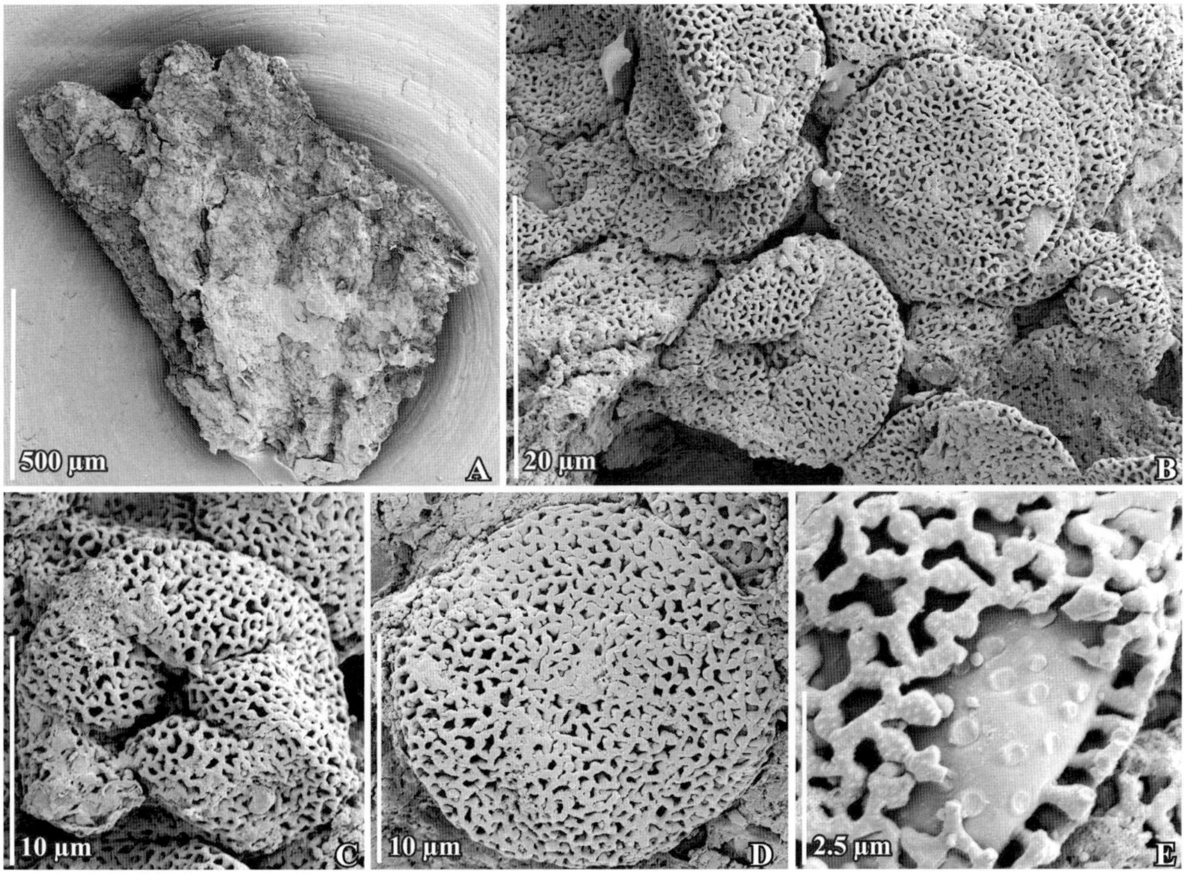

Figure 8.14 SEM images of inflorescence fragment of a *Hedyosmum*-like staminate structure from the Early Cretaceous (Late Aptian – Early Albian) Vale de Agua locality, Portugal, with *Asteropollis*-type pollen *in situ*. (A) Several stamens adhering together. (B) Pollen grains in distal and proximal view. (C) Pollen in distal view showing four-armed, star-shaped, aperture. (D) Pollen in proximal view. (E) Detail of pollen wall with fragmentary reticulum, showing supratectal ornamentation, columellate infratectal layer, and scars from columellae on the surface of the foot layer. Specimen in the collections of the Swedish Museum of Natural History, Stockholm.

membranous tepal-like structures. The edges of the ovary are sharp, while the faces are flat. Outgrowths from the edges of the ovary wall form a narrow zone of tissue around the margin of each face, resulting in three 'windows' on the three surfaces of the fruit that are also characteristic of extant *Hedyosmum*. There is a single massive, sessile stigma, but this is typically broken off to leave a small, rounded scar at the top of the ovary. The ovary contains a single orthotropous ovule. Pollen of the *Asteropollis* type occurs commonly on the surface of the flowers and almost certainly belongs to the same plant.

In all floral features and pollen characters the fossils are almost identical to extant *Hedyosmum*. Staminate inflorescences differ slightly in having more stamens in each whorl, but other diagnostic features of the fossils justify placement of these flowers in the stem group, or perhaps the crown group, of the modern genus. Inclusion of the fossils in phylogenetic analyses of the Chloranthaceae placed the *Asteropollis* plant as sister to modern *Hedyosmum* (Doyle *et al.*, 2003; Eklund *et al.*, 2004b).

8.5.3 *Chloranthistemon*

The genus *Chloranthistemon* (Figures 8.16–8.19) is based on small trilobed androecia from the Late Cretaceous (Late

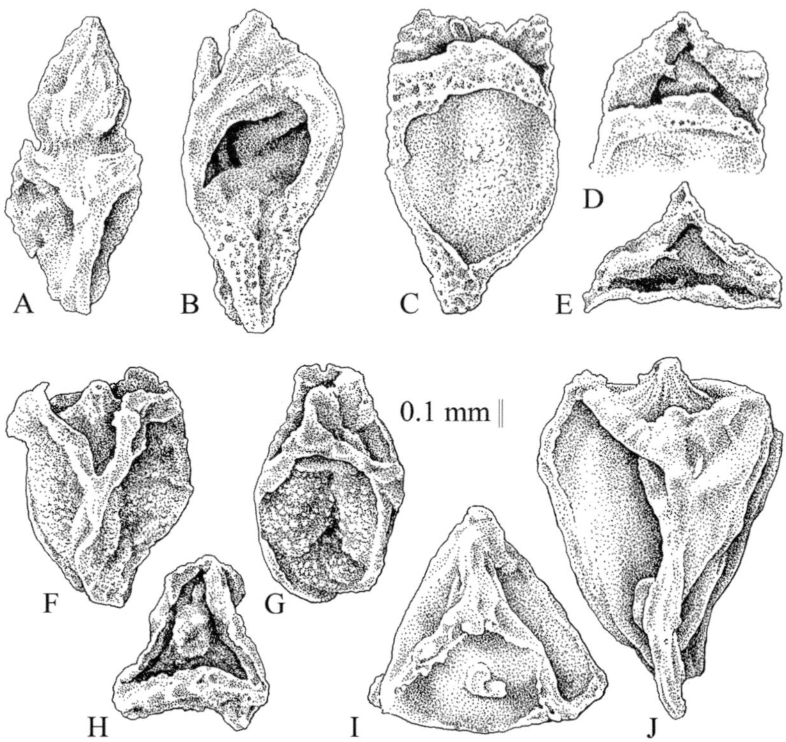

Figure 8.15 Pistillate *Hedyosmum*-like flowers from the Early Cretaceous of Portugal. (A) Flower from the Vale de Agua locality (Late Aptian – Early Albian) in lateral view showing well preserved prominent stigmatic area. (B) Flower from the Buarcos locality (Late Aptian – Early Albian) in lateral view showing tepals at apex. (C–E) Flower from the Buarcos locality (Late Aptian – Early Albian) in (C) lateral and (D, E) apical view showing the triangular shape and the distinctive lateral 'window'. (F–H) Flower from the Torres Vedras locality (Late Barremian – Early Aptian) in (F, G) lateral and (H) apical view showing distinct windows and remains of apical tepals. (I, J) Flower from the Torres Vedras locality (Late Barremian – Early Aptian) in (I) apical and (J) lateral view. (A, C–H) From Friis *et al.* (2006a); (B, I, J) from material in the collections of the Swedish Museum of Natural History, Stockholm.

Santonian – Early Campanian) flora of Åsen, Sweden, assigned to *C. endressii* (Crane *et al.*, 1989b). The original description was based on dispersed androecia. Subsequently, more complete specimens of *C. endressii* were recovered from the Åsen flora, together with an additional species, *Chloranthistemon alatus* (Eklund *et al.*, 1997). Both species have bisexual flowers borne on elongated inflorescence axes in the axils of cup-like bracts in an opposite–decussate arrangement. Flowers are minute and naked, without a perianth. The androecium is attached to the ovary in an abaxial to lateral position. The androecium is broad and laterally flattened and consists of three stamens that are free at the base. Pollen sacs are small; each opens by two laterally hinged valves. The median stamen is dithecate and tetrasporangiate, with the two pairs of pollen sacs borne on its lateral margins. The two lateral stamens are monothecate and bisporangiate, with the pollen sacs borne on their outer margins. Connective tissue is extensive and expands into an apical protrusion. In *C. endressii* the apical extensions of the three stamen lobes are laterally fused to form a massive shield-like structure. In *C. alatus* the apical protrusions are wing-like and free from each other. Pollen observed *in situ* in *C. endressii* flowers is spherical and reticulate, with a spiraperturate aperture that resembles the seam of a tennis ball (Figure 8.17). In *C. alatus* pollen is spherical and tectate–foveolate, with two opposite colpi/furrows arranged perpendicular to each other (Figure 8.17). The gynoecium is monocarpellate and contains a single ovule. At the apex of the ovary there is an indistinct sessile stigma.

In general floral and inflorescence structure *Chloranthistemon* is very similar to flowers and inflorescences of extant species of *Chloranthus*. In particular, the three-lobed androecium attached to the upper part of the ovary wall is unique. Among extant plants this is known only in *Chloranthus*. Most extant *Chloranthus* species have stamen lobes that are fused at the base, but in a few species the lobes are free as in *Chloranthistemon alatus* and *C. endressii*. *Chloranthistemon* therefore falls clearly within crown-group Chloranthaceae. Whether it also falls within crown-group *Chloranthus* is less certain. Pollen in both fossil species differs from that in all extant *Chloranthus* species. Fusion of the

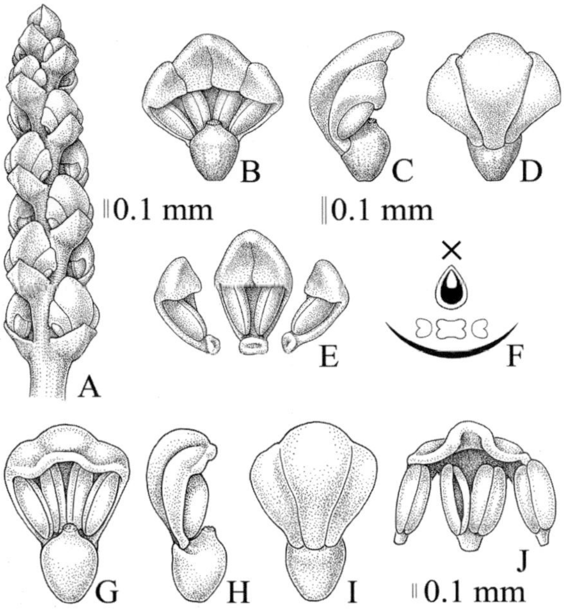

Figure 8.16 Reconstructions of inflorescence and flowers of (A–F) *Chloranthistemon alatus* and (G–J) *Chloranthistemon endressii* from the Late Cretaceous (Late Santonian – Early Campanian) Åsen locality, Sweden. (A) Inflorescence axis showing flowers in a decussate arrangement. (B–D) Flower in (B) adaxial, (C) lateral and (D) abaxial view showing ovary with attached trilobed androecium. (E) Isolated lobes of androecium showing two lateral, monothecate lobes and one median, dithecate lobe. (F) Floral diagram. (G–I) Flower in (G) adaxial, (H) lateral and (I) abaxial view showing ovary with attached trilobed androecium. (J) Isolated trilobed androecium with lobes that are free proximally and fused distally. (A–E, G–J) redrawn from Eklund *et al.* (1997).

connective extensions, as seen in *C. endressii*, is not known in the extant genus.

Chloranthus-like androecia have also been reported from the Late Cretaceous (Turonian) of New Jersey, USA, and described as *Chloranthistemon crossmanensis* (Figure 8.18) (Herendeen *et al.*, 1993). The North American material comprises dispersed androecia, but pollen has not been described. In this species, the androecial lobes are apparently fused at the base, as in many extant species of *Chloranthus*, but in contrast to the situation in *Chloranthistemon endressii* and *C. alatus*.

Chloranthus/Chloranthistemon-like androecia and fruits have also been recovered from the mid-Cretaceous karst infillings at the Prangenhaus and Rohdenhaus localities of the Rhenish Massif, Germany (C. Hartkopf-Fröder, A. Vienofen and E. M. Friis, work in progress) (Figure 8.19). The androecia are trilobed as in *Chloranthistemon* and *Chloranthus* and they also have a fully developed dithecate median lobe and two monothecate lateral lobes. However, in contrast to *Chloranthus* and *Chloranthistemon*, pollen sacs on the lateral lobes are borne on the inner, rather than the outer, margins. The pollen sacs are also in a lateral position, whereas in *Chloranthus* and *Chloranthistemon* they are more or less adaxial.

The combined features of the *Chloranthistemon* flowers, with small pollen sacs, valvate dehiscence of the anthers, extensive sterile tissue between the pollen sacs, and a small sessile stigma, indicate that entomophily was probably characteristic of these early *Chloranthus*-like plants, as is also inferred for extant *Chloranthus* (Endress, 1987).

8.6 CERATOPHYLLACEAE

In previous classifications the Ceratophyllaceae have sometimes been grouped with Nymphaeaceae (e.g. Takhtajan, 1980; Cronquist, 1981), but recent phylogenetic analyses emphasise the isolated position of the family. In some early analyses based on molecular data, and one study based on morphology, *Ceratophyllum* was resolved as the sister group to all other angiosperm families (e.g. Chase *et al.*, 1993; Qiu *et al.*, 1993; Nandi *et al.*, 1998). More recently, the Ceratophyllaceae have sometimes been placed as the sister group to eudicots (Moore *et al.*, 2007).

The Ceratophyllaceae are widely distributed. The family comprises a single genus (*Ceratophyllum*) with six species. All are small, submerged aquatic herbs with small, simple, unisexual flowers that are water-pollinated (Endress, 1994a; Les, 1993). Pollen is dispersed under water and has an extremely thin exine. The fruits are distinctive, easily recognisable achenes, with a woody, often spiny, fruit wall.

Despite the high fossilisation potential of the fruits there are only few published records of possible Ceratophyllaceae from the Cretaceous (Samylina, 1976; Krassilov *et al.*, 2005; Dilcher and Wang, 2009). *Donlesia dakotensis*, described from the mid-Cretaceous (latest Albian – earliest Cenomanian) Dakota Formation by Dilcher and Wang (2009) as an early member of Ceratophyllaceae, is based on fruits with an arrangement of spines that is different from that in all extant and fossil *Ceratophyllum*, and the systematic assignment to the extant family is not fully secure (Chapter 9). Similarly, the nature of the small spiny ovulate structures described as *Ievlevia dorofeevii* from the Early Cretaceous of Siberia, which were

Figure 8.17 *In situ* pollen from *Chloranthistemon* fossils from the Late Cretaceous (Late Santonian–Early Campanian) Åsen locality, Sweden. (A–C) *Chloranthistemon alatus*, grains showing the two colpi/furrows arranged perpendicular to each other. (D, E) *Chloranthistemon endressii*, grains showing spiraperturate aperture. Material in the collections of the Swedish Museum of Natural History, Stockholm.

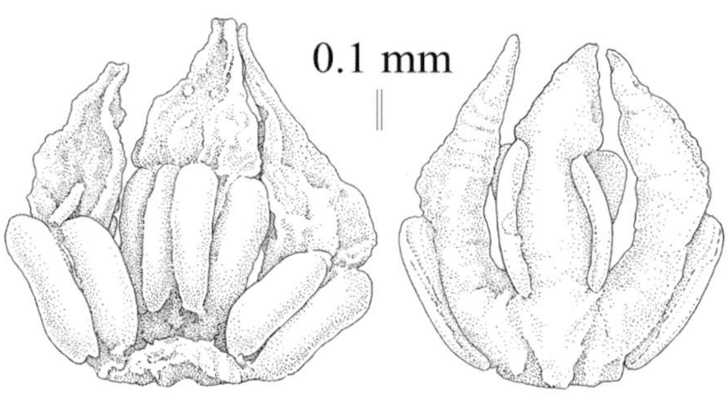

Figure 8.18 Trilobed androecium of *Chloranthistemon crossmanensis* from the Late Cretaceous (Turonian) Old Crossman locality, New Jersey, USA, in (A) adaxial and (B) abaxial view. Drawn from SEM images in Herendeen *et al.* (1993).

compared by Samylina (1976) to *Ceratophyllum*, is not fully understood (Chapter 9).

Another putative early record of Ceratophyllaceae, *Ceratostratiotes cretaceus*, is based on a single impression of a fruit from the Late Cretaceous of the Negev Desert, Israel (Krassilov *et al.*, 2005). The fruit is small, apparently two-valved, and has the remains of a prominent spine at the apex and several spines at the base. The possibility of confusion with seeds of the Bennettitales–Erdtmanithecales–Gnetales group, for *Donlesia, Ievlevia, Ceratostratiotes* and similar material (Chapter 5), needs to be considered (Chapter 9).

Ceratophyllum is common in Cenozoic sediments, where it is represented mainly by the characteristic fruits. At least

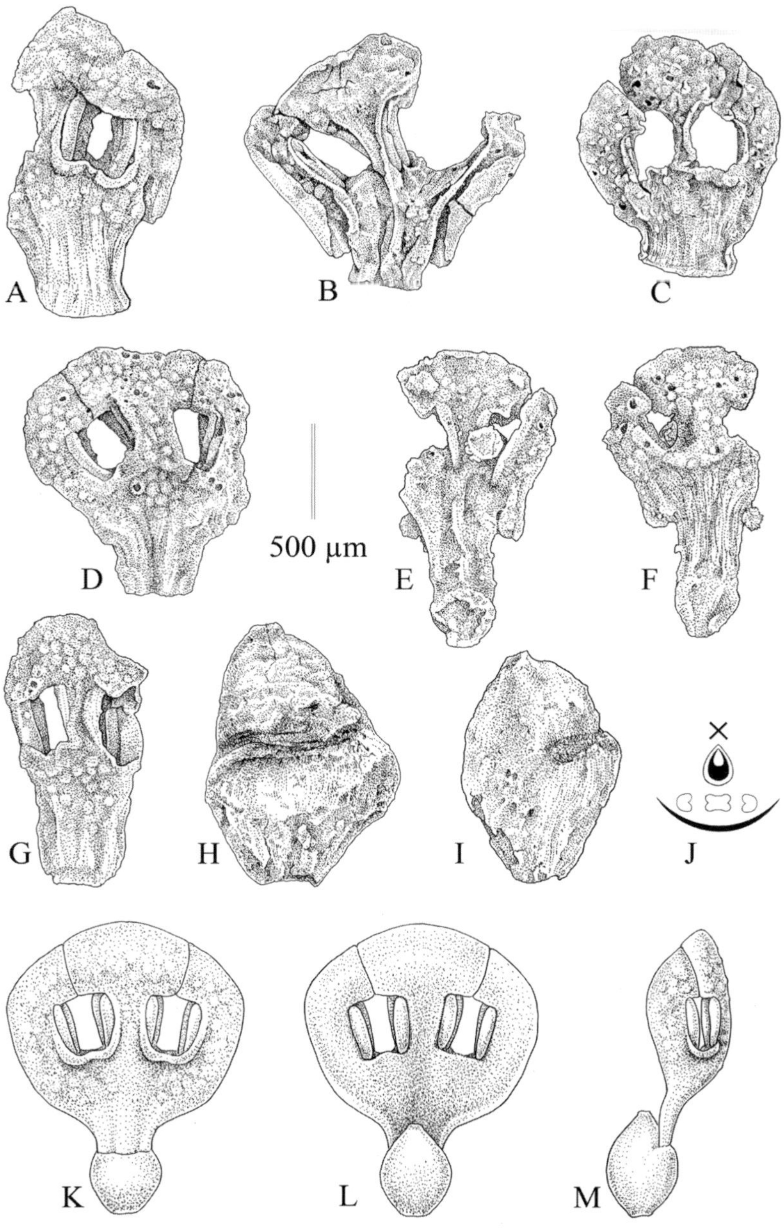

Figure 8.19 Chloranthaceous fossils from the mid-Cretaceous karst infillings of the Rhenish Massif, Germany. (A–G) Trilobed *Chloranthistemon*-like androecia in abaxial view. (C, D) Note that lobes adhere together apically usually leaving two prominent holes between the median and lateral lobes. (H, I) Pistillate structures in (H) abaxial and (I) lateral view showing scar from the attachment of the androecium. (J) Floral diagram. (K–M) Reconstruction of bisexual flower in (K) abaxial, (L) adaxial and (M) lateral view. Drawn from material in the collections of the Geological Survey of North Rhine – Westphalien, Krefeld, Germany.

12 different fossil species have been reported (e.g. Mai, 1985a). In North America, unequivocal fossils of the genus are present from the Paleocene and onwards. In Europe and western Asia they are present from the Oligocene and onwards. Vegetative axes and leaves of *Ceratophyllum* are rarely preserved, but occur associated with dispersed fruits in the Middle Eocene Green River Formation of Wyoming, USA (e.g. Herendeen *et al.*, 1990). Pollen of extant *Ceratophyllum* has a thin wall and very low fossilisation potential and, as expected, there is no record of fossil *Ceratophyllum* pollen.

9

Early fossil angiosperms of uncertain relationships

In this chapter we describe plant fossils, mainly from the Early Cretaceous, that for various reasons cannot be assigned reliably to any extant group of angiosperms. In a few cases the relationship to angiosperms themselves is also uncertain (sections 9.1, 9.7). Some of these fossils are well-preserved floral structures, and the available information is reasonably extensive, but assignment to an extant order or family is precluded because the fossil has characters, or suites of characters, that are difficult to relate to a particular extant angiosperm group. In some instances the lack of relevant comparative data for extant taxa is also a problem. Other fossils considered in this chapter are dispersed floral parts, such as pollen, fruits and seeds. These provide further documentation of diversity among early angiosperms, but are difficult to assign to an extant group because they lack sufficient diagnostic characters.

In addition to the material discussed in this chapter there are many other fossils, especially from the Portugal and Potomac Group mesofossil floras, that remain to be described and for which relationships are not yet evaluated. A comprehensive account of these fossils is beyond the scope of this book. However, many of them probably represent early-diverging lineages of angiosperms of uncertain relationship that would probably also be included here.

9.1 PUTATIVE ANGIOSPERMS

Several fossils with reproductive organs preserved have been described as early angiosperms, but lack sufficient characters to securely document their relationship with angiosperms, or to any other group of seed plants. In another case, spike-like reproductive structures, such as *Eragrosites changii* and *Lioaxia cheniae*, that were originally thought to be angiosperms, have later been assigned to the Gnetales. These are considered in Chapter 5. Other non-angiosperm fossils that have played a key role in the discussions of a possible pre-Cretaceous history of angiosperms are considered in Chapter 6.

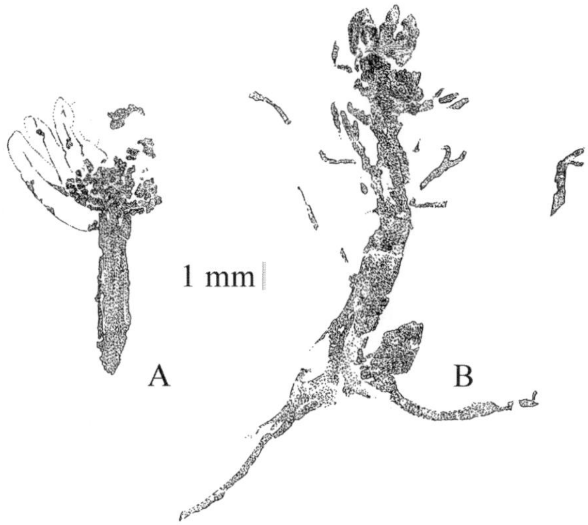

Figure 9.1 *Bevhalstia pebja*, fossil shoot with attached leaves and flower-like structures from the Early Cretaceous (latest Hauterivian to earliest Barremian) Weald Clay of Sussex and Surrey, southern England. Drawn from photographs in Hill (1996).

9.1.1 *Bevhalstia*

Bevhalstia pebja (Figure 9.1) is an early putative angiosperm fossil based on small shoots with attached leaves and flower-like organs from the Early Cretaceous (latest Hauterivian – earliest Barremian) Weald Clay of Sussex and Surrey, southern England. The holotype is from the Keymer Tileworks, Sussex. Other specimens have been collected from the Smokejacks, Rudgwick and Clockhouse brickworks (Hill, 1996).

In most specimens the contrast between fossil and matrix is very low, and thus morphological details are not easily observed. Critical study is also hindered by poor preservation of the compressed parts. Stems are apparently parenchymatous, vascularised and frequently branched. Each branch is subtended by bract-like structures. Leaves attached to the stem are minute, laminar and with two to three broad lobes. Occasionally they are more extensively dissected. Leaf venation is simple, dichotomous and apparently without anastomoses or interconnected veins. Flower-like, presumed reproductive, organs occur at the tips of a few shoots, but all are poorly preserved and details of their structure are unclear. The most prominent feature of these organs is a lobed leaf-like structure with coarse dichotomous venation. This structure is apparently associated with a rounded central body. Hill (1996) suggested that these organs are either specialised terminal buds subtended by laminar leaves, or perhaps radially symmetrical flowers with fused leaf-like tepals, a central gynoecium and a blunt, perhaps stigmatic, apical projection.

Based on the morphology of the shoot fragments, and the apparent soft and parenchymatous nature of the stems, *Bevhalstia* was interpreted as a small, herbaceous plant about 25 cm high. Among seed plants the herbaceous habit is restricted to angiosperms, but is widespread among ferns and lycopsids. However, the leaves and pattern of venation of *Bevhalstia* are unusual among angiosperms. Open dichotomous venation is known only in two eudicot genera (*Circaeaster*, *Kingdonia*, Ranunculales). Leaves or leaflets with this kind of open dichotomous venation do not occur among lycopsids, but are common among some groups of ferns. Hill (1996) dismissed the possibility of an affinity between *Bevhalstia* and ferns because of the lack of circinate vernation in smaller (perhaps younger) leaves. He therefore argued by process of elimination that the herbaceous habit indicated an angiospermous affinity. *Bevhalstia* remains an intriguing plant, but much more information is needed before the presence of flowers, carpels and stamens, and hence its angiosperm affinity, can be established securely.

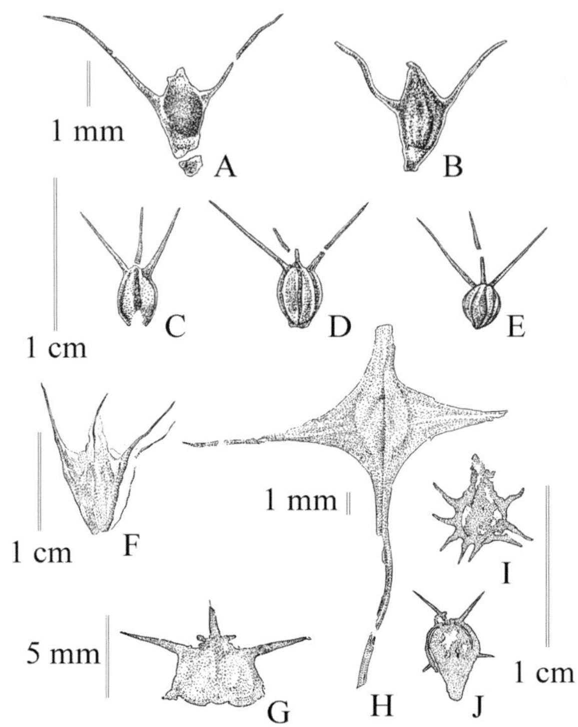

Figure 9.2 Spiny ovulate structures from the Early Cretaceous. (A, B) 'Awned seeds' from the Koonwarra Fossil Bed (Aptian), Victoria, Australia. (C–G) Species of *Beipiaoa* from the Yixian Formation (Early – early Late Aptian), Liaoning Province, northeastern China; (C–E) *Beipiaoa parva* and (F, G) *Beipiaoa spinosa*. (H) *Donlesia dakotensis* from the Acme Brick Co. locality (latest Albian – earliest Cenomanian), Kansas, USA. (I, J) *Ievlevia dorofeevii* from the Toptan flora (Middle Albian), Kolyma River, Northeastern Russia (Samylina, 1976). (A, B) Drawn from Drinnan and Chambers (1986); (C–G) drawn from Sun *et al.* (2001); (H) drawn from Dilcher and Wang (2009); (I, J) drawn from Samylina (1976).

9.1.2 *Lappacarpus*, *Ievlevia*, *Donlesia* and *Beipiaoa*

Several kinds of small dispersed ovulate structures with various spiny or hairy appendages have been reported from Early Cretaceous floras in both the Northern and Southern Hemispheres. Some of these are probably extinct plants related to the BEG group (Chapter 5). Others have been described as angiosperms and linked to extant *Ceratophyllum*. In some of these fossils the spines are straight, sharply pointed and arranged symmetrically. In other material, the spines are slightly curved, blunt and more irregularly arranged.

Lappacarpus aristata and *Ievlevia dorofeevii* were among the first fossils of this kind to be formally named. *Lappacarpus aristata* was described from the Early Cretaceous (Floral Zone D, Albian) of the Yangery Bore 1 in Victoria, Australia (Douglas, 1969). The fossils are small compressions, about 1–2 mm long (without the spines) with an obtriangular outline. Apically they have a short central protrusion and two lateral, straight or slightly curved, spines, about 2–5 mm long. The fossils were compared with the spiny fruits of extant *Acaena anserinifolia* (Rosaceae), but no definite systematic assignment was suggested. Similar spiny structures were also reported from the Early Cretaceous Gemmill's Hill and Koonwarra localities, Victoria, Australia, and referred to as *Hemitrapa*? sp. (Douglas, 1969) or 'Awned seeds' (Figure 9.2) (Drinnan and Chambers, 1986).

Ievlevia dorofeevii (Figure 9.2) was described from the Early Cretaceous (Middle Albian) Toptan flora from the Omsukchan coal field on the Kolyma River (Samylina, 1976). These fossils are small flattened structures that are ovate to elliptical in outline. They are about 4–6 mm long and 2–3 mm wide (without spines). On the seeds that are illustrated there are up to eight spines arranged along the margins in a more or less symmetrical arrangement. Spines are straight or slightly curved, up to about 4 mm long (measured from the illustration). In the original description the fossils were compared to fruits of extant *Ceratophyllum* (Samylina, 1976). They have also been cited, without further comment, as a putative early record of Vitaceae (Collinson *et al.*, 1993).

Donlesia dakotensis (Figure 9.2) is another early fossil based on spiny ovulate organs. It was described from the mid-Cretaceous (latest Albian – earliest Cenomanian) Dakota Formation from two localities (The Acme Brick Co. I and II), Kansas, USA, by Dilcher and Wang (2009) and includes well-preserved fruits characterised by a long stylar spine, a long stalk and four spine-like wings, two lateral and two dorsiventral, placed around the middle of the fruit in a radiating pattern. The fossil was assigned to Ceratophyllaceae, but the authors also emphasised important differences in organisation compared with fruits of the extant genus, such as the presence of four winged spines, arranged at the same level of the fruits in a radial pattern, and the long stalk. The structure is different from that in all extant and fossil *Ceratophyllum* species and the systematic relationship of these fossils to the extant genus is not fully secure.

Fossils similar to *Lappacarpus aristata*, *Ievlevia dorofeevii* and the 'Awned seeds' from the Koonwarra locality are common in the Early Cretaceous (Early Aptian – early Late Aptian) Yixian Formation, China (Figure 9.2). The Chinese material has been described under various names such as *Trapa*? sp. (Wu, 1999), *Beipiaoa spinosa*, *Beipiaoa rotunda* and *Beipiaoa parva* (Sun *et al.*, 2001). Some of these *Beipiaoa* fossils may belong to the same complex of plants as *Lappacarpus* or *Ievlevia*, but the three *Beipiaoa* species are very different from each other structurally and mostly likely they are not closely related. *Beipiaoa spinosa* appears to be a node with bracts, perhaps related to Gnetales.

The possibility that *Lappacarpus*, *Ievlevia* and *Beipiaoa* are not angiosperms, and are perhaps related to the Bennettitales–Erdtmanithecales–Gnetales group, needs to

Figure 9.3 *Montsechites* (*Ranunculus*) *ferreri*, a putative aquatic angiosperm from the Early Cretaceous (Hauterivian–Barremian) lithographic limestone from the Sierra del Montsech, Lérida, Spain. Based on Teixeira (1954).

be considered. Some specimens are similar in general morphology to unnamed square seeds from the Early Cretaceous of Portugal and the Potomac Group of eastern North America (Chapter 5). In this context, it is interesting that *Lappacarpus aristata* is associated with leafy stems (Angiosperm ? sp. 'b' of Douglas, 1969) that were later described as *Leongathia elegans* and assigned to Gnetales (Krassilov *et al.*, 1998).

Small spiny reproductive units also occur in leaf axils of the enigmatic Early Cretaceous plant *Montsechites ferreri* preserved in lithographic limestone from the Hauterivian–Barremian Montsech locality, Lerida Province, Spain (Figure 9.3). *Montsechites ferreri* has been interpreted as an aquatic plant. It comprises leafy shoots with slender stems and strongly dissected leaves. It was first described by Teixeira (1954) who interpreted it as a putative angiosperm and compared it to *Ranunculus*. Subsequently Blanc-Louvel (1984) transferred the fossil to the extant *Ranunculus aquatilis*, but later it was listed as *Ranunculus ferreri* (Barale *et al.*, 1984). In overall leaf morphology the fossil resembles some species of extant *Ranunculus*, but it differs in having small spiny seeds or fruits in the axils of the leaves (Figure 9.3). It is possible that *Montsechites ferreri* is related to the other spiny structures described in this section, but unfortunately most of the *Montsechites* material has few structural details preserved. For the moment only comparisons based on superficial resemblance are possible.

9.1.3 Other isolated seeds of uncertain affinity

Kennella harrisiana was described from the Early Cretaceous (Albian) of the Kolyma Basin, East Siberia. The specimens are impressions of small, probably ovulate structures, about 16 mm long and 5 mm broad (Samylina, 1968). The fossils occur in isolation and also in small clusters. They are ellipsoidal in outline with a pointed apex and ribbed surface. They co-occur with angiosperm leaves assigned to the extinct genus *Dalbergites*. Samylina suggested that *Kennella* was perhaps a fruit resembling other Cretaceous fossils that have been described as fruits (*Quereuxia*, *Nyssidium*). However, critical features for assigning *Kennella* and similar impression fossils to angiosperms are lacking. They could be seeds of some kind of non-angiosperm seed plant.

Onoana californica is an enigmatic fossil first described as a very early angiosperm (Chandler and Axelrod, 1961). The fossil is permineralised and was recovered from the Hauterivian Ono Formation of California, USA. It is a small, obovoid, ovulate structure, about 20 mm long and 15 mm in diameter, with a pitted surface and a thick wall (Figure 9.4). Chandler and Axelrod (1961) interpreted *Onoana* as an endocarp from a drupaceous angiosperm fruit and provisionally assigned it to the angiosperm family Icacinaceae. Inside the thick sclerenchymatous wall is a thin, but distinct, yellow layer. An angiosperm affinity cannot be excluded, but the fossil could also be the seed of a non-angiosperm seed plant with well-developed megaspore membrane (Friis and Crepet, 1987).

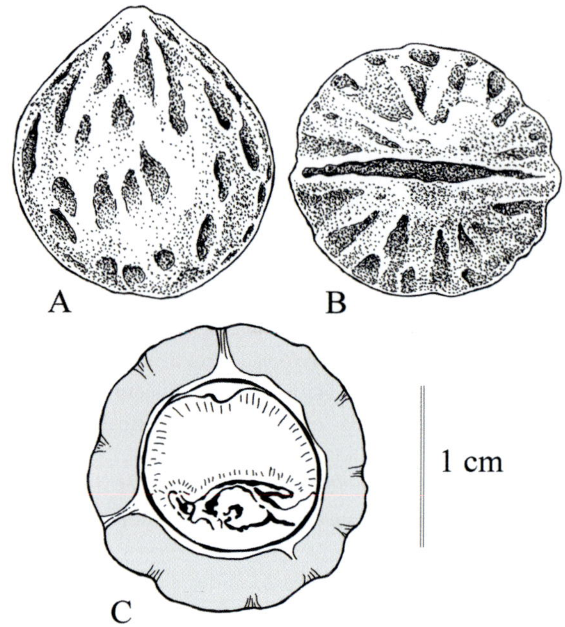

Figure 9.4 *Onoana californica* from the Early Cretaceous (Hauterivian) of California, USA, a fossil seed of uncertain relationship originally described as an angiosperm seen in lateral (A) and apical (B) view, and in transverse section (C). Drawn from photographs in Chandler and Axelrod (1961).

9.2 FOSSIL FLOWERS ATTACHED TO INFLORESCENCES AND STEMS

Most fossil inflorescences are recovered from macrofossil floras preserved as compressions or impressions in a consolidated rock matrix. Occasionally, such inflorescences, or even single flowers, are found attached to axes or leaf-bearing

Figure 9.5 *Archaefructus sinensis* from the Early Cretaceous (Early Aptian – early Late Aptian) Yixian Formation, Liaoning Province, northeastern China. Leafy shoot with reproductive axis bearing carpels above and scars from stamens below. Specimen in the Nanjing Institute of Geology and Palaeontology, photograph Y. Arremo.

shoots. More rarely, they may be attached to whole plants, with roots, stems and leaves all connected. These fossils are rarely informative about details of floral organisation, but they are very valuable in providing insights into the habit of early angiosperms and other aspects of plant architecture.

Early and mid-Cretaceous inflorescences are mostly elongated, simple or compound spikes with small, densely crowded, floral units. Several of these fossils, such as species of *Mauldinia* and *Pragocladus* (Chapter 10), have yielded sufficient information to be placed in a modern family or order. In other cases such fossil inflorescences, which are generally preserved as compressions or impressions, are difficult to study because of poor preservation combined with the dense arrangement of floral units and other parts.

9.2.1 *Archaefructus*

Archaefructus (Figures 2.4, 9.5, 9.6) includes three species, *A. liaoningensis* (Sun *et al.*, 1998), *A. sinensis* (Sun *et al.*, 2002) and *A. eoflora* (Ji *et al.*, 2004), all from the Early Cretaceous (Early Aptian – early Late Aptian) Yixian Formation of Liaoning Province, northeastern China. All specimens of *Archaefructus* are compressions or impressions, but they are sometimes preserved as whole plants with roots, stems, leaves, and reproductive structures attached. This preservation provides excellent information on the general morphology and habit of the plant, but cellular details are absent or only poorly preserved.

Archaefructus has been a focus of considerable interest and also contrasting interpretations. The Jurassic age suggested in the initial publication, combined with the near-whole plant preservation, may have contributed to its particularly high profile in debates about early floral structure and other aspects of early angiosperm evolution. A Jurassic age would have made *Archaefructus* the oldest known angiosperm (Sun *et al.*, 1998), but the Early Aptian – early Late Aptian age now attributed to the fossils brings *Archaefructus* more in line with knowledge of the early angiosperm radiation from other areas.

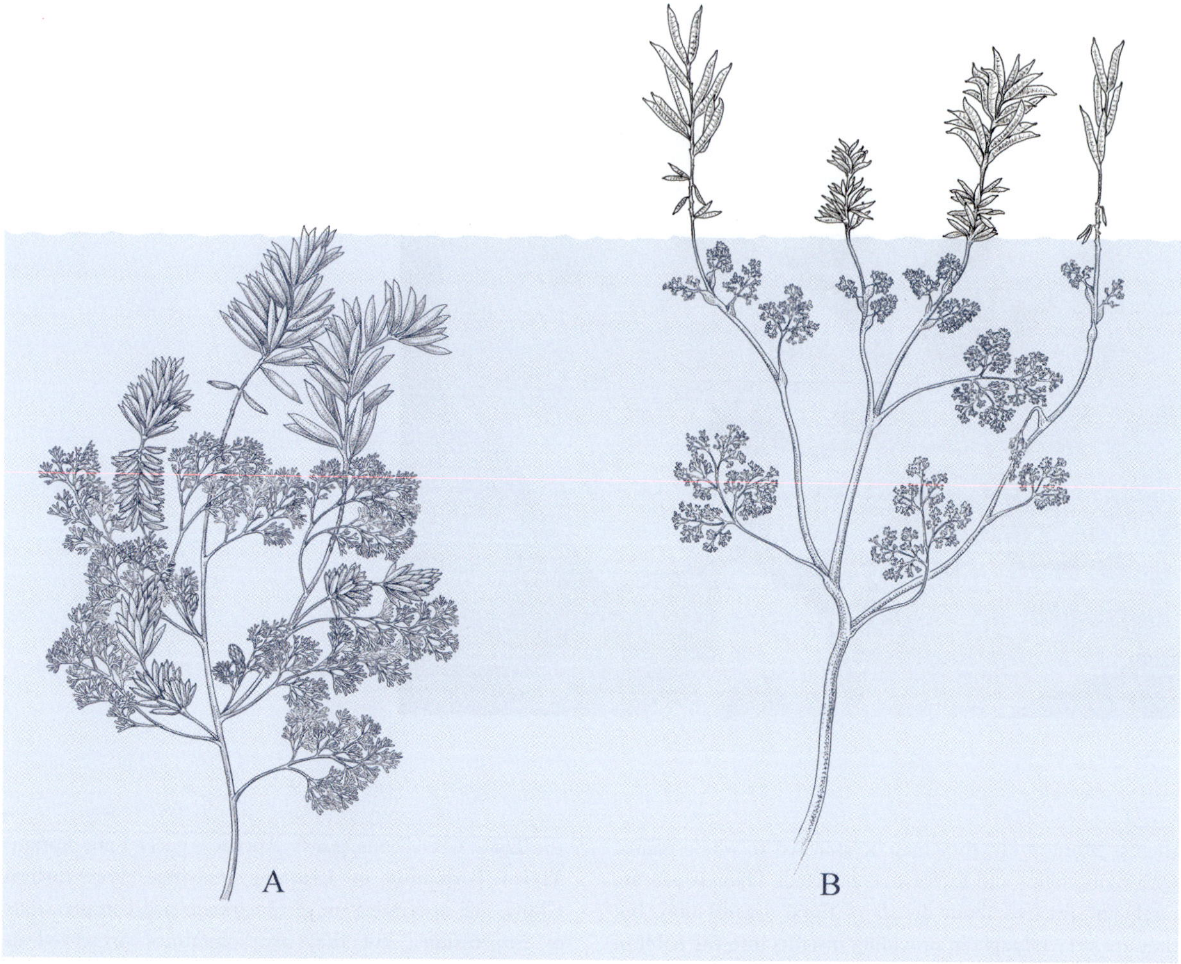

Figure 9.6 Reconstruction of *Archaefructus* under two different interpretations. (A) *Archaefructus* interpreted as a fully submerged aquatic with reproductive structures functioning underwater. (B) *Archaefructus* interpreted as an emergent aquatic with reproductive structures held above the water surface and functioning like those of a normal angiosperm. See text for further explanation. (A) Based on Friis *et al.* (2006a); (B) based on Sun *et al.* (2002).

Archaefructus is an herbaceous plant with axillary branching and strongly dissected leaves (Figure 9.5). It is up to about 50 cm tall. Reproductive axes are elongate and terminal. They are up to about 85 mm long in *A. liaoningensis* and *A. sinensis*, but are more condensed in *A. eoflora* (Sun *et al.*, 1998, 2002; Ji *et al.*, 2004). They consist of numerous ovulate and staminate organs that are typically borne in pairs in a helical or subdecussate arrangement in two different zones along an elongated axis. Ovulate organs are borne distally and the pollen-producing organs are borne proximally on the reproductive axes. In young specimens the reproductive units are condensed and their organisation is difficult to study. In more mature specimens, apparently preserved at fruiting stage, the reproductive units are more widely spaced. No bracts or perianth parts have been observed on the fossils, either subtending the reproductive axes as a whole, or subtending the individual reproductive units.

The ovulate organs of *Archaefructus* are elongate, fusiform, pedicellate, and up to about 1 cm long. They are interpreted as carpels although details of their organisation are poorly known. Each contains few (*A. liaoningensis*, *A. eoflora*) or many (*A. sinensis*) small seeds. The pollen-producing organs are borne in pairs, or in groups of up to four, on a common stalk. They resemble typical angiosperm tetrasporangiate stamens and are described as such although they are too compressed for detailed characterisation and none of them has so far yielded unequivocal pollen (Friis *et al.*, 2003a). *Archaefructus eoflora* is interesting because it appears to deviate from the normal unisexual organisation of the reproductive structures. In the transition region between the pistillate and staminate zones a bisexual flower consisting of a single carpel and two or three stamens was described by Ji *et al.* (2004), but this is difficult to see from the published illustrations.

The morphological interpretation of *Archaefructus* reproductive axes is controversial (Friis *et al.*, 2003a) and contributes to the uncertain systematic relationships of the plant. Sun *et al.* (1998, 2002) interpreted each reproductive axis of *Archaefructus* as a single, multipartite and bisexual flower with an elongated receptacle bearing many stamens and many carpels in a helical arrangement. Carpels were interpreted as follicles. The lack of a perianth or bracts was interpreted as a plesiomorphic feature. Friis *et al.* (2003a) reinterpreted the axes as inflorescences bearing many simple, unisexual flowers. This interpretation was based mainly on the paired arrangement of the carpels and stamens, an organisation that is unknown in multipartite angiosperm flowers. The strongly dissected leaves and general habit were also used to infer that *Archaefructus* was an aquatic plant. Friis *et al.* (2003a) suggested that it perhaps grew completely submerged and interpreted the lack of perianth and bracts as a secondary simplification related to underwater flowering (Figure 9.6).

Archaefructus was first interpreted as an extinct lineage of angiosperm not closely related to any extant angiosperm taxon. A new subclass, Archaemagnoliidae, was established to accommodate the genus (Sun *et al.*, 1998). In a subsequent phylogenetic analysis *Archaefructus* was resolved as sister to all extant angiosperms and placed in a new family, the Archaefructaceae (Sun *et al.*, 2002). Alternatively, in light of the reinterpretation of *Archaefructus* reproductive axes as inflorescences, *Archaefructus* was interpreted as a crown-group angiosperm, perhaps related to Nymphaeales. The unusual floral features were suggested to reflect specialisation for an aquatic habit (Friis *et al.*, 2003a). Recognition of the extant family Hydatellaceae, with its small, simple and naked, unisexual flowers as a new member of Nymphaeales (Saarela *et al.*, 2007) may also provide indirect support for this idea (Endress and Doyle, 2009).

More detailed structural information is needed before the relationships of *Archaefructus* can be determined reliably. It is also important to note that although an angiosperm relationship for *Archaeanthus* seems very probable, its angiospermous affinity has not been fully documented. A possible relationship to *Caytonia* has been suggested (Zhou *et al.*, 2003). *Archaefructus* also shows some similarity with *Dirhopalostachys rostrata* described from the Late Jurassic – Early Cretaceous of the Bureja Basin, in Far Eastern Russia (Krassilov, 1975a) and *Schweitzeria* (*Irania*) *hermaphroditica* from the Late Triassic of Iran (Schweitzer, 1977; Shipunov and Sokoloff, 2003). Both these fossils have elongated reproductive axes with capsule-like structures borne in pairs on short stalks (Chapter 6).

9.2.2 *Xingxueina* and *Caspiocarpus*

Xingxueina heilongjiangensis (Figure 9.7) comprises an elongated spike-like inflorescence associated with small angiosperm leaves from the Early Cretaceous Chengzihe Formation in the Jixi Basin, Heilongjiang Province, northeastern China (Sun and Dilcher, 1997, 2002). The age was originally suggested to be Hauterivian or Early Barremian, but is now considered to be Aptian, based on correlation with marine fossils (Sha *et al.*, 2003). The inflorescence axis is narrowly elongate and about 14 mm long. It consists of numerous small, presumed floral units borne in a helical arrangement. Individual units are densely crowded and there are no details of floral parts or other aspects of their organisation. The proposed angiosperm affinity for this fossil is based on *in situ* pollen. Pollen occurring on the surface of the inflorescence is more or less circular in equatorial outline, about 15–20 μm in diameter, and reticulate. Pollen was described as inaperturate, but some of the grains appear to have a single long colpus that may be slightly extended towards the proximal surface. The relationship of these fossils to a specific group of angiosperms is currently uncertain. *Xingxueina heilongjiangensis* co-occurs with several other angiosperm taxa preserved as leaves. Similar leaves were also reported from the slightly younger Muling Formation also in the Jixi Basin (Yang, 2003).

Caspiocarpus paniculiger (Figure 9.7) is based on a single specimen of a small leafy shoot bearing terminal

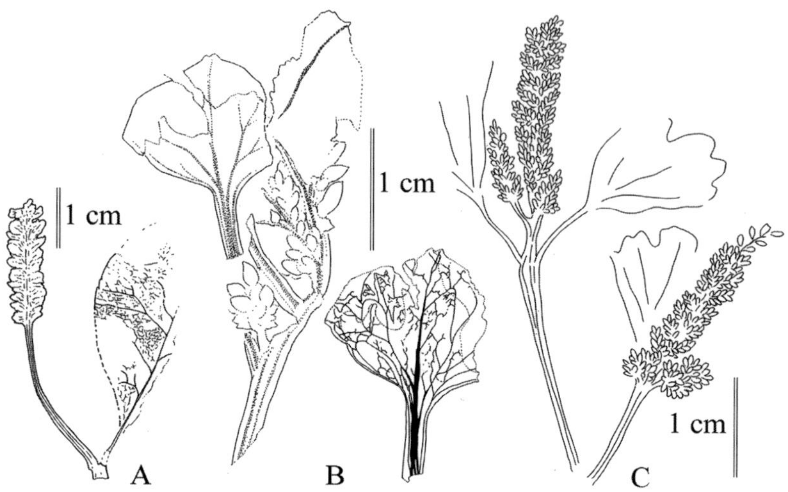

Figure 9.7 Leafy shoots of Early Cretaceous angiosperms with strongly condensed and compressed inflorescences composed of numerous small flowers. (A) *Xingxueina heilongjiangensis* from the Aptian of Heilongjiang Province, northeastern China. (B) The Koonwarra fossil from the Aptian Koonwarra Fossil Bed, Victoria, Australia. (C) *Caspiocarpus paniculiger* from the Middle Albian of Kazakhstan. From Friis *et al.* (2006a).

reproductive axes from the Early Cretaceous (Middle Albian) of Kazakhstan (Vakhrameev and Krassilov, 1979). The whole structure is a few centimetres long and encrusted in iron oxide. Leaves are opposite, slightly lobed, with palmate venation and a thin leaf blade. They are similar to leaves of *Vitiphyllum* described from several Early Cretaceous leaf floras (Fontaine, 1889; Berry, 1911; Doyle and Hickey, 1976), but details of leaf venation have not been illustrated or described. Above the leaves there are three, closely spaced, inflorescence axes, the two lateral axes are shorter, while the terminal axis is longer. The reproductive axes were described as panicles with the secondary axes bearing four to ten fruits arranged in a dense spiral. The fruits were described as follicles with ventral dehiscence and a small number of seeds. However, floral units are minute and closely packed, which together with the poor preservation precludes their detailed study.

The thin leaf blade and overall morphology of *Caspiocarpus paniculiger* indicates that it was probably an herbaceous plant. Krassilov (1997) suggested a relationship to certain Ranunculales based mainly on the form of the leaves and carpels, but poor preservation and equivocal structural information prevents a secure systematic assignment. Despite its poor preservation *Caspiocarpus paniculiger*, like *Archaefructus*, is important as one of the few Early Cretaceous angiosperm fossils in which leaves are attached to axes with reproductive structures. Such fossils provide useful information on the vegetative morphology and growth habit of early angiosperms.

9.2.3 The Koonwarra fossil

The Koonwarra fossil (Figure 9.7) is a small leafy shoot bearing strongly condensed reproductive axes. The specimen is preserved as a thin compression and no cellular details are preserved. It has not yet been formally named. The single specimen was illustrated in an overview of the Early Cretaceous (Aptian) Koonwarra flora from Victoria, Australia, and described as a possible marsilealean fern (Marsileales? indet., Drinnan and Chambers, 1986). Subsequently, more detailed studies have convincingly documented that it is an angiosperm (Taylor and Hickey, 1990).

The fossil is less than 30 mm long and consists of a narrow axis, about 1.4 mm wide, bearing two leaves, each of which has an inflorescence in its axil. The inflorescences are small, elongated, spike-like structures consisting of numerous, densely spaced bracts, bracteoles and probable ovaries. However, the relative positions of the individual parts, and therefore the detailed structure of the inflorescences, are unclear. Taylor and Hickey (1990) suggested that the inflorescences are probably thyrses with a primary axis bearing densely crowded ovate bracts, each with two axillary bracteoles and one ovary. The leaves are alternate. They are simple, with a long, broad petiole and an unlobed, broadly ovate lamina. The lamina has intermediate palmate to pinnate primary venation, and an irregular pattern of poorly differentiated, but clearly reticulate, veins. Small, spherical yellow-brown translucent bodies scattered over all parts of the plant are probably the remains of ethereal oil

Figure 9.8 Inflorescence axes and pollen of (A) *Zlatkocarpus brnikensis* and (B, C) *Zlatkocarpus pragensis* from the Late Cretaceous (Middle Cenomanian) Peruc flora of the Bohemian Basin, Czech Republic. Specimens in the collections of the National Museum, Prague (see also Kvaček and Friis, 2010).

cells. The leaves show a mosaic of features from several different families (e.g. Saururaceae, Piperaceae, Aristolochiaceae, Chloranthaceae, Taylor and Hickey, 1990) and suggest an affinity with those early-diverging lineages of angiosperms that comprise the magnoliids in the broad sense. The relationships of the fossil have not yet been established more precisely.

9.2.4 *Myricanthium*, *Zlatkocarpus* and other inflorescences from the Peruc flora

Spike-like inflorescences/infructescences consisting of elongated main axes, which bear shorter secondary axes in a semi-decussate or helical arrangement, are common in the Late Cretaceous (Middle Cenomanian) Peruc flora of the Bohemian Basin, Czech Republic. Among these fossils are the compound spikes of *Myricanthium amentaceum* first described by Velenovský (1889) based on poorly preserved compression/impression material. Structurally preserved material recovered subsequently from localities that also yield the compression/impression fossils, demonstrates that *Myricanthium*-like fossils include diverse, systematically distinct, taxa (Kvaček, 1992; Eklund and Kvaček, 1998; Kvaček and Eklund, 2003; Kvaček and Friis, 2010). Two species, *Mauldinia bohemica* and *Pragocladus lauroides*, have trimerous, bisexual flowers and can be assigned reliably to the Lauraceae (Chapter 10). Two other species, *Zlatkocarpus pragensis* and *Zlatkocarpus brnikensis*, have simple, unisexual flowers. They closely resemble extant Chloranthaceae, but are currently unassigned.

Other unnamed and unassigned taxa in the *Myricanthium* complex are poorly understood and may be of diverse relationships. For example, spikes of *Myricanthium amentaceum* are compound with secondary axes borne in a semi-decussate manner. Floral units are borne on the secondary axes in a helical arrangement. The preservation does not allow more detailed investigation of the organisation and structure of the individual floral units and it is not known whether they were unisexual or bisexual flowers.

Zlatkocarpus (Figure 9.8) was established for structurally preserved inflorescences that are similar in gross morphology to *Myricanthium amentaceum*, but that have better preservation of the floral units (Kvaček and Friis, 2010). The type species, *Zlatkocarpus brnikensis*, consists of small fragments of spike-like inflorescences. The most complete specimen is about 6.5 mm long and 2.5 mm wide with about 50 densely spaced floral units borne helically along the elongated inflorescence axis. Each floral unit is supported by a small triangular bract that forms a deep floral cup enclosing the basal part of the ovary. There are no indications of stamens. The floral cup appears to be fused to the ovary and is interpreted as a hypanthium in which only the apical rim is free. The ovary is monocarpellate with a small indistinct, sessile stigma near the apex. The fruit wall is thin, wrinkled and may have been parenchymatous. The fruit was apparently a single-seeded berry. Monocolpate, coarsely reticulate, pollen occurs abundantly on the surface of the ovaries. The pollen is about 12–18 µm in equatorial diameter with a heterogeneous reticulum and a columellate infratectal layer. Abundant resin bodies, interpreted as the remains of ethereal oil cells, are embedded in the fruit wall and other tissues.

Zlatkocarpus pragensis (Figure 9.8) was first described as *Myricanthium pragense* by Kvaček and Eklund (2003). It is structurally preserved and has compound inflorescences with secondary axes borne in a semi-decussate to helical arrangement. Floral units are borne helically on the secondary axes as in *Myricanthium amentaceum*. However, *Z. pragensis* differs from *Z. brnikensis* in having a much smaller, bract-like floral cup that does not encircle the ovary completely. Its pollen also has a homogeneous reticulum in contrast to the heterogeneous reticulum seen in pollen associated with *Zlatkocarpus brnikensis* (Kvaček and Friis, 2010).

Individual ovaries/fruits of *Zlatkocarpus* are very similar to those of the more or less contemporaneous *Couperites mauldinensis* from eastern North America (section 9.5.2). However, *Couperites* is only known isolated and there is no information on the inflorescence. Fruits of *Couperites* and *Zlatkocarpus* were both one-seeded fruits with a small sessile stigma and abundant ethereal oil cells in the fruit wall. In both genera the pollen is monocolpate and coarsely reticulate with a columellate infratectal layer. Characters that distinguish the two taxa include the elongate, slightly decurrent stigmatic ridge seen in *Couperites* and differences in the pollen. *Couperites* has pollen with beaded muri, whereas pollen associated with *Zlatkocarpus* has muri that are smooth and lack supratectal ornamentation. Pollen similar to that seen in *Zlatkocarpus* is often referred to the pollen genus *Retimonocolpites*, but the use of this genus is currently very broad and not helpful for understanding the systematics of the *Zlatkocarpus* plants. *Retimonocolpites* pollen has also been linked to fossils of clear monocot (Araceae) affinity (Chapter 11) and to fossil floral structures of *Canrightia resinifera* (Chapter 10) that also show intriguing similarities to *Zlatkocarpus*.

Other structurally preserved spicate inflorescences from the Peruc flora that are similar to *Zlatkocarpus*, such as the 'Spicate Inflorescence with Unilocular Fruits' described by Kvaček and Eklund (2003), may be part of the same complex of plants. They have abundant resin bodies embedded in the various tissues of the inflorescence and they also have reticulate, probably monocolpate pollen. They are distinguished from *Zlatkocarpus* in having larger, cup-shaped supporting bracts. The floral units are currently too poorly understood to allow inclusion in *Zlatkocarpus*.

9.2.5 Inflorescences from the Dakota Formation

Several elongated catkin-like inflorescences have been reported from the mid-Cretaceous (Late Albian – Early Cenomanian) Dakota Formation of central USA (Dilcher, 1979). Several were later described and named; others remain to be formally described. The 'fruit-bearing axis' (Fig. 47 in Dilcher, 1979) was included in *Prisca reynoldsii* (Chapter 10) and the 'elongate axis bearing numerous small florets' (Fig. 48 in Dilcher, 1979) was described as *Caloda delevoryana* (Dilcher and Kovach, 1986).

Caloda delevoryana is based on elongated infructescences up to about 15 cm long that are preserved as compressions. Infructescences comprise small, densely crowded, secondary axes borne helically along the main axis. Each secondary axis terminates in a small swollen receptacle bearing 35–50 free carpels (Figure 9.9). Carpels are stalked, elliptical in outline and about 2 mm long. No style was observed and the stigma is apparently sessile. Poor preservation of the specimens prevents detailed reconstruction of the original fruiting structure; although the flowers are thought to be unisexual the possibility that stamens were shed before fruit maturation cannot be excluded. Broadly similar apocarpous fruiting structures occur in extant plants among eumagnoliids, monocotyledons, and even among eudicots, and the systematic position of *Caloda* with respect to major groups of angiosperms is not yet established (Dilcher and Kovach, 1986).

Other fossil inflorescences from the Dakota Formation (Dilcher, 1979) remain to be studied in detail. The specimen illustrated as 'globose cluster of numerous florets' (Fig. 49 in Dilcher, 1979) provides further evidence of small flowers densely crowded in spike-like inflorescences, but again poor preservation, combined with compression and clustering of the flowers, makes it difficult to obtain detailed information on floral organisation. A different mode of preservation is provided by inflorescences preserved as impressions in sandstone. Some of these, for example *Platanus primaeva* (Figs. 41 and 42 in Dilcher, 1979), are easily referable to extant groups (Chapter 12). Others, for example '*Salix* fruiting catkin' (Fig. 39 in Dilcher, 1979), are less straightforward to interpret, and both morphology and systematic relationships are unclear.

Of particular interest among these undescribed fossils from the Dakota Formation are elongated inflorescences from the Linnenberger's Ranch locality (Fig. 44 in Dilcher, 1979). These inflorescences are preserved as compressions and have numerous, apparently tetramerous, floral units densely crowded on the inflorescence axes. The floral units were interpreted as staminate, and clusters of reticulate

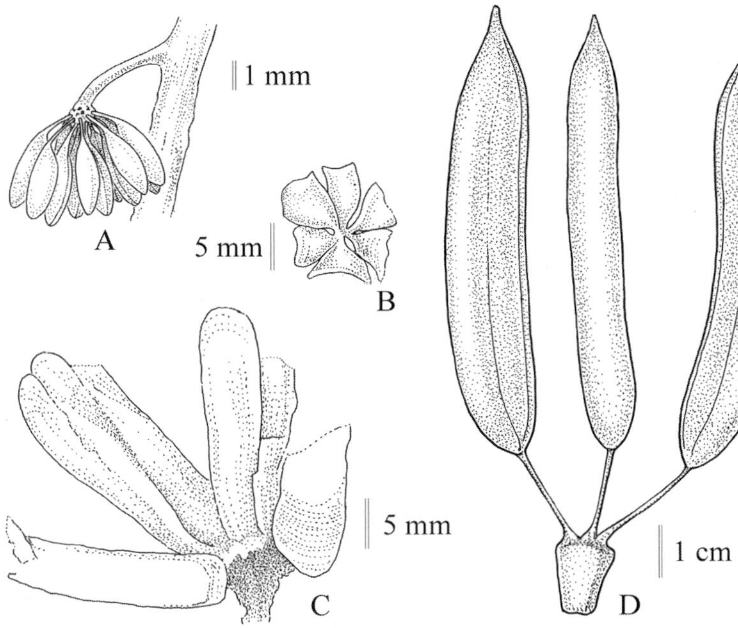

Figure 9.9 Multicarpellate angiosperm reproductive structures of uncertain relationship from the mid-Cretaceous. (A) *Caloda delevoryana* from the Late Albian-Early Cenomanian Dakota Formation, Kansas, USA; reconstruction of part of the main infructescence axis showing a lateral branch with a terminal cluster of small follicles. (B) *Carpolithus conjugatus*, an impression/compression fossil from the Albian of Virginia, USA. (C) Unnamed reproductive structure with apocarpous gynoecium from Virginia, USA. (D) *Triplicarpus purkynei*, a putative fruiting structure from the Late Cretaceous (Middle Cenomanian) Peruc flora of the Bohemian Basin, Czech Republic. (A) Redrawn from Dilcher and Kovach (1986); (B, C) drawn from photographs in Dilcher (1979); (D) redrawn from Velenovský and Viniklář (1926).

pollen grains were isolated from the base of the floral units (Dilcher, 1979). These grains are similar to those assigned to the dispersed pollen genus *Dichastopollenites* (May, 1975). Pollen is apparently zonacolpate with the encircling colpus dividing the pollen in two equal halves. The tectum is reticulate with lumina of two sizes. Zonacolpate pollen of this kind occurs in several monocot taxa, but also in some magnoliids. Tetramerous floral organisation is unusual among monocots, but does occur for instance in Araceae. It is also not common among eumagnoliids, but does occur sporadically in some families (e.g. Saururaceae). Further details of floral structure, confirmation of the link with *Dichastopollenites* pollen, and attempts to link these inflorescences to associated leaves, will be needed for further consideration of the systematic affinities of this interesting material.

9.2.6 Other inflorescences

Several inflorescences of uncertain affinity, together with isolated flowers and fruits, have been reported as compression/impression fossils from the Late Cretaceous (Turonian) of the Negev Desert, Israel (Krassilov *et al.*, 2005). Many of these are too poorly preserved to yield details on floral structure, but they provide important information on Late Cretaceous angiosperms from a region and palaeolatitudinal/palaeoclimatic zone where knowledge is otherwise sparse. Most of the Negev fossils are probably related to eudicots, but Krassilov *et al.* (2005) also suggest that a large proportion are monocots, including those described as *Gerofitia lochii*, *Gerocladus foliosus*, *Gerocladus selaginelloides*, *Gereilatia picnoclada*, *Lochiella setosa*, *Minevronia capitata* and *Qeturocarpus costatus*. In most cases, however, critical systematic evaluation is not possible owing to poor preservation.

From the Lower Cretaceous (Late Aptian – Early Albian) Crato Formation of Brazil several reproductive structures are known that are attached to vegetative axes (Mohr and Friis, 2000; Mohr *et al.*, 2007). In rare cases, whole plants with attached roots, stems, leaves and reproductive structures are present. Some of these fossils possess sufficient characters for a systematic assignment at the family or higher level. *Endressinia brasiliana* is assigned to the Magnoliales (Chapter 10) and *Pluricarpellatia peltata* is assigned to Nymphaeales (Chapter 8). *Araripia florifera* has features suggesting a eumagnoliid affinity (Chapter 10). Other plant fossils are less informative systematically, but provide useful information on habit and diversity of early angiosperms in the palaeoequatorial region. New discoveries can be expected as investigations of the Crato flora progress.

9.3 ISOLATED FLOWERS AND FRUITS PRESERVED AS COMPRESSIONS/ IMPRESSIONS

Several isolated flowers and fruiting structures of uncertain systematic affinity have been recovered from Early and mid-Cretaceous compression/impression floras. Apocarpous fruiting structures with relatively large follicles have been reported from mid- to Late Cretaceous floras of North America, Europe and Asia. Most of these fossils are probably of eumagnoliid affinity, but diagnostic characters are generally lacking, and assignment to specific angiosperm groups remains uncertain. The gynoecia described in this section are almost all from flowers borne terminally on axes, and they are all large compared to other Cretaceous floral or fruiting structures.

From the mid-Cretaceous of North America there are several apocarpous fruiting structures preserved as impressions or compressions from localities in Virginia, New Jersey, Kansas, Texas, USA, and Alberta, Canada (*Carpolithus conjugatus*, *Lesqueria elocata*, *Palaeanthus problematicus*, *Archaeanthus linnenbergeri*, '*Williamsonia*' *recentior*). *Archaeanthus linnenbergeri* can be confidently linked with Magnoliales (Chapter 10), but the relationships of most of the other fossils are more uncertain.

9.3.1 *Lesqueria elocata* and other isolated reproductive structures

Lesqueria elocata is based on several enigmatic fruiting structures preserved as three-dimensional moulds in mid-Cretaceous (latest Albian – earliest Cenomanian) sandstones from the Dakota Formation of central Kansas and the Woodbine Formation of northern Texas, USA (Lesquereux, 1892; Crane and Dilcher, 1984). The preservation is unusual, and although many aspects of their structure remain uncertain it is clear that the fossils are fruits derived from multiparted flowers with an apocarpous gynoecium. The fruiting structure has a particularly unusual receptacle, up to about 56 mm long. The long, stout cylindrical basal part bears 110–150 flap-like laminar appendages, while the expanded conical distal part bears 175–250 tightly packed follicles. Both the laminar appendages and the follicles above them are borne in a helical arrangement, but the transition between the two organ types is abrupt (Crane and Dilcher, 1984). The laminar appendages are all incomplete with their distal portions missing. The preserved portions are about 2 mm long at the base and increase in size to about 20 mm below the gynoecium. Crane and Dilcher (1984) suggested that the appendages were either all parts of a perianth or perhaps graded from a perianth region proximally into an androecial region of laminar stamens distally. Follicles each have a short, stout stalk. They are up to about 24 mm long, with adaxial dehiscence, and two terminal prolongations: one on either side of the suture. Imprints in the fill of some of the follicles indicate that 10–20 seeds were originally present in each.

Lesqueria was first assigned to the Bennettitales as *Williamsonia elocata* (Lesquereux, 1892), but it is clear that the structures attached to the expanded receptacle are carpels, rather than interseminal scales and seeds. The fossils show similarities to extant Magnoliaceae, Annonaceae, Austrobaileyaceae, and other magnoliids. In particular the two prolongations of *Lesqueria* follicles are similar to the two projections at the tips of the carpels of extant *Austrobaileya*, but the fossils cannot be assigned to a specific angiosperm family with the information currently available.

Carpolithus conjugatus (Figure 9.9) is an impression/ compression fossil from the Early Cretaceous (Albian) of Virginia, USA (Fontaine, 1889; Dilcher, 1979). No anatomical details have been described. It appears to consist of three free, dehisced follicle-like carpels, about 10 mm long. There are no seeds or other parts of the plant preserved and the systematic affinity of the fossil to major groups of angiosperms has not been established.

Palaeanthus problematicus from the Amboy Clays (Raritan Formation, mid–Late Cenomanian) of New Jersey, USA, is another possible magnoliid reproductive structure (Dawson, 1886). The specimens are preserved mainly as impressions in grey micaceous silty clay. They consist of compact heads with 50–80 follicle-like structures supported by narrowly triangular bract-like tepals. The suggested systematic affinities of these specimens have varied from *Williamsonia* (Bennettitales) to angiosperms (Asteraceae). The associated leaf flora is dominated by angiosperms and the most straightforward interpretation of *Palaeanthus problematicus* is that it is a multicarpellate, probably magnoliid, angiosperm flower (Crane and Dilcher, 1984).

Triplicarpus purkynei (Figure 9.9) is an unusually large tripartite structure from the Cenomanian of Bohemia, the Czech Republic. It was interpreted as an apocarpous fruit with three long, narrow elliptical carpels. Each carpel is about 90 mm long and pedicellate with a long slender stalk.

The fossil is preserved in a sandstone matrix and seeds or other parts of the plant are not preserved. A systematic affinity close to Annonaceae has been suggested (Velenovský and Viniklář, 1926), but according to J. Kvaček (personal communication, 2010) the preservation of the fossil is poor and details of its structure are uncertain.

Capsulocarpus dakotensis is based on a single specimen preserved as an impression in Dakota Formation sandstone collected southwest of Springfield, Minnesota, USA (Berry, 1939). The specimen is elongated, slightly curved toward the adaxial side, and 47 mm long and 16 mm wide. At one end there is a short point, and at the other a broad stalk. Part and counterpart provide impressions of both the inner and the outer surface. The inner surface has numerous fine transverse striations, probably of the fibres that lined the locule. The outer surface has a few broader oblique longitudinal ribs. The specimen is one half of a large follicle that split cleanly down the midline into two halves prior to preservation. It was preserved at fruiting stage and along the inside of the slightly concave ventral margin are the distinct scars of 20 seed attachments. Assuming that there were a similar number of seeds attached to the other half of the follicle the original fruit contained about 40 seeds. The systematic relationships of this fossil are uncertain, but are probably among magnoliids in the broad sense.

'*Laurus*' *macrocarpa* is based on three specimens preserved as impressions and collected from Dakota Formation sandstone in Kansas and Nebraska, USA (Lesquereux, 1874, 1892; Fig. 34 in Dilcher, 1979). They are up to 19 mm long and up to 17 mm broad. Each appears to be a capsule-like fruit borne on a swollen pedicel. The fruits split into four valves at maturity and each quadrant was supplied by a separate vascular bundle running through the tissue of the pedicel. The number of seeds produced in each fruit is unknown. Fruits of extant Lauraceae have a single seed inside a unilocular ovary and there is no close resemblance between the fossil and extant *Laurus* as implied in the name. The material remains to be studied in detail, but the relationships of these fruits remain obscure.

'*Williamsonia*' *recentior* from the 'Upper Blairmore' flora (mid-Late Albian) of southwestern Alberta, Canada, is preserved as a compression in a hard grey shale (Crane and Dilcher, 1984). The material remains to be studied in detail, but it is clearly not assignable to the Bennettitales. It also occurs in a flora dominated by angiosperm leaves. Specimens are common at the locality (P.R. Crane, personal observation) and show heads of what appear to be about 150–200 curved follicles. The fossil is probably some kind of multicarpellate fruiting structure (Crane and Dilcher, 1984; Dilcher and Crane, 1984a).

9.4 PERMINERALISED FLOWERS

Permineralised angiosperm flowers and other reproductive structures are not common in Cretaceous strata. Most famous are the plant fossils preserved in calcite nodules from the Late Cretaceous of Hokkaido, Japan, which generally have excellent preservation with cellular details intact. These nodules were first studied by Stopes and Fujii (1910) who described the putative monocot flowers of *Cretovarium japonicum*. Another flower, *Elsemaria kokubunii*, has been described as a eudicot (Chapter 13). Three further flowers from Hokkaido, *Protomonimia*, *Hidakanthus* and *Keraocarpon*, have been compared to eumagnoliid angiosperms, but because their relationships are not fully clear they are included in this section. Other occurrences of petrified flowers from the Cretaceous are from the mid-Cretaceous of Australia (Chapter 10), the Late Cretaceous of India (Chapter 11) and the Late Cretaceous of British Columbia, Canada (Chapter 10).

Protomonimia kasai-nakajhongii (Figure 9.10) is a relatively large, multicarpellate fruiting structure described from the Late Cretaceous (Coniacian–Santonian) of Hokkaido, Japan (Nishida and Nishida, 1988). The genus is based on a single specimen with many anatomical details preserved. It is up to about 40 mm in diameter and consists of about 55 sessile carpels arranged on a slightly concave receptacle. Carpels are up to about 17.5 mm long and have numerous oil cells scattered in the carpel wall. Each carpel contains 12–15 ovules. The ovules are bitegmic, anatropous and apparently exotestal with an outer palisade layer of sclerenchyma cells with digitate anticlinal cell walls. The systematic relationships of *Protomonimia* are probably among magnoliids in the broad sense, but its more precise phylogenetic position remains uncertain. The resemblance between the fossil and extant Monimiaceae implied in the name is not strong. In extant Monimiaceae individual carpels consistently contain only a single ovule. Flowers with apocarpous, multicarpellate gynoecia and several ovules per carpel are common among extant Magnoliales, but seeds of Magnoliales are typically endotestal, and a relationship with Magnoliales (Nishida and Nishida, 1988) has not been convincingly demonstrated. Exotestal seeds with digitate cell walls occur in most extant Nymphaeales and also in *Illicium* (Schisandraceae). Dispersed

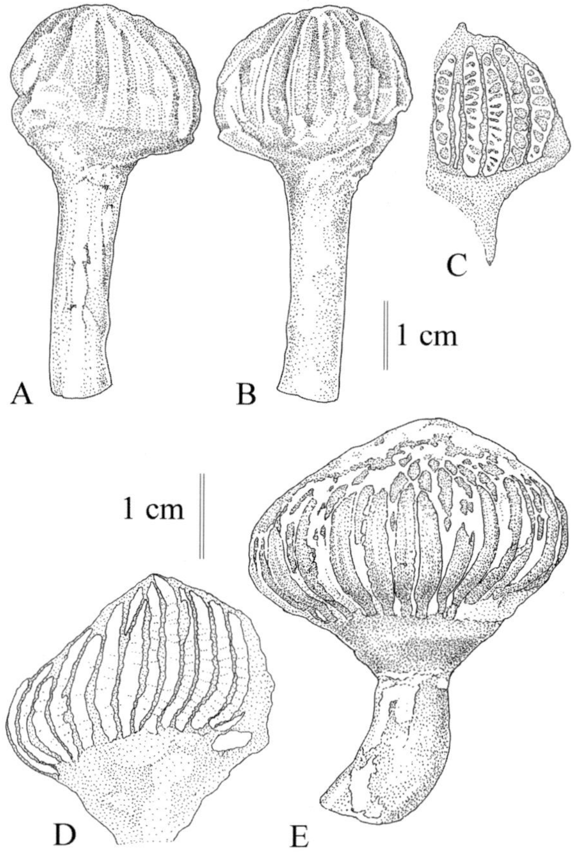

Figure 9.10 Permineralised fruits from the Late Cretaceous (Coniacian–Santonian) of Hokkaido, Japan. (A–C) *Protomonimia kasai-nakajhongii* showing many free carpels in lateral view (A, B), and longitudinal section (C) showing many seeds in each carpel. (D, E) *Hidakanthus shiinae* in lateral view (E) showing many free carpels and longitudinal section (D) showing receptacle and carpels. (A) Drawn from photographs in Nishida and Nishida (1988) (B–E) from Friis *et al.* (2006a).

seeds very similar to those of *Protomonimia* have been recovered from the Early Coniacian Kamikitaba flora, Honshu, Japan (Takahashi *et al.*, 1999b). Also present in the Kamikitaba flora is *Futabanthus asamigawaensis* (Annonaceae, Chapter 10), which is known from a single specimen preserved at the flowering stage (Takahashi *et al.*, 2008b). The possibility of a relationship between *Protomonimia* and *Futabanthus* deserves further consideration as more material becomes available.

Hidakanthus shiinae (Figure 9.10) is another relatively large, multicarpellate fruiting structure described from the Late Cretaceous (Coniacian) of Hokkaido, Japan (Nishida *et al.*, 1996). *Hidakanthus shiinae* is similar to *Protomonimia kasai-nakajhongii* in its general organisation, but the two genera are distinguished from each other by the shape of the receptacle, and also by the number and shape of the carpels. The single specimen of *Hidakanthus shiinae* is up to about 40 mm in diameter with around 170 carpels arranged on a slightly convex receptacle. Carpels are stalked, up to about 20 mm long and 2.6 mm broad. Seeds are not preserved, but some of the carpels show faint impressions indicating that each originally contained several ovules. Anatomical details of the fruit wall are also less well preserved than in *Protomonimia*. Detailed comparison between the two genera, and with extant taxa, is not possible based on the information currently available.

Two other permineralised multicarpellate gynoecia have been described from the Late Cretaceous of Hokkaido, Japan, as *Keraocarpon yasujii* and *Keraocarpon mastoshii* (Ohana *et al.*, 1999). They are similar in general structure to *Protomonimia* and *Hidakanthus* and all three genera are perhaps closely related. *Keraocarpon yasujii* is distinguished by having very numerous carpels, up to about 470; *K. mastoshii* has between 70 and 170 carpels.

Cretovarium japonicum from the Late Cretaceous (Coniacian-Campanian) flora of Hokkaido, Japan (Stopes and Fujii, 1910) is permineralised in calcium carbonate and five individual specimens preserved in a single block were studied in transverse, or oblique transverse, sections. The fossils are small and consist of perianth-like leaves surrounding a trilocular and slightly inferior ovary, about 2 mm in diameter and 3 mm long. Placentation is axile with two separate placentae in each locule, indicating that ovules were borne in two rows. Ovules or seeds were not preserved. Soft tissues in the outer ovary wall, which are developed between the carpels, were interpreted as possible intercarpellary nectaries. Stopes and Fujii (1910) tentatively suggested a relationship to extant Liliaceae, based on the trilocular gynoecium, axile placentation and two rows of ovules per locule. However, the information currently available is not sufficient for accurate systematic assessment and the relationship of *Cretovarium* to monocots remains uncertain.

9.5 ISOLATED ANGIOSPERM MESOFOSSILS

Studies of Cretaceous mesofossil floras have led to the discovery of numerous dispersed fossil flowers, stamens, fruits and seeds. Most of this material is from the Late

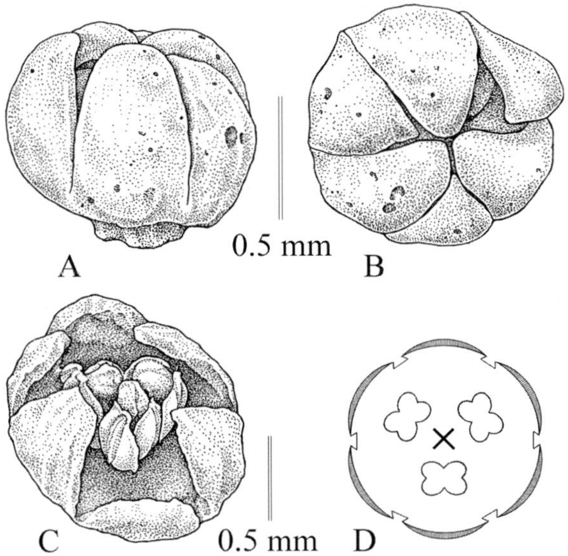

Figure 9.11 *Mabelia archaia*, small, trimerous, staminate flowers from the Late Cretaceous (Turonian) Old Crossman locality, New Jersey, USA, in (A) lateral and (B, C) apical view, showing (A–C) six tepals, (C) three stamens with extrorse dehiscence and (D) floral diagram. Drawn from unpublished specimens.

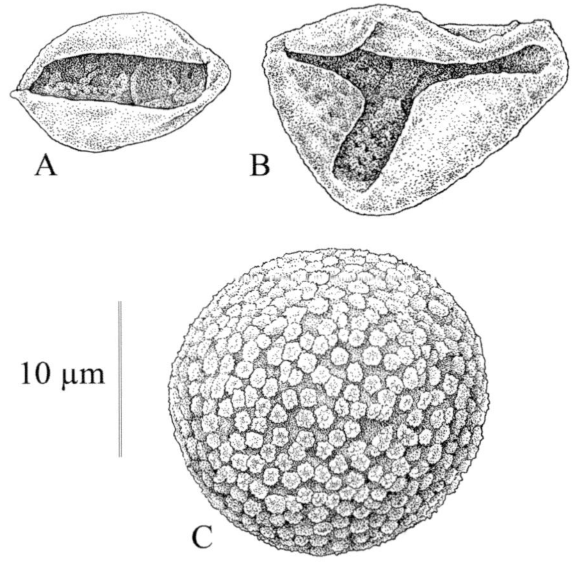

Figure 9.12 Pollen of (A, B) extinct *Mabelia* from the Late Cretaceous (Turonian) Old Crossman locality, New Jersey, USA, and (C) extant *Triuris lutea*. (A) Drawn from SEM-image in Gandolfo *et al.* (2002); (B) drawn from unpublished specimen; (C) drawn from SEM image in Furness *et al.* (2002).

Cretaceous and the majority is clearly of eudicot affinity. However, isolated fossil flowers and other reproductive organs are also abundant in Early Cretaceous mesofossil floras from eastern North America and western Portugal. Several of these Early Cretaceous flowers can be related to the earliest diverging lineages of extant angiosperms (Chapter 8), while others are more closely related to eumagnoliids (Chapter 10), monocots (Chapter 11) or early-diverging lineages of eudicots (Chapter 12). However, there are also many Early and mid-Cretaceous mesofossils that cannot be assigned with certainty to any of the major angiosperm clades, even though they are well preserved and can be characterised in detail. In this section we discuss mesofossils of this kind that have already been formally described.

9.5.1 Isolated flowers

Mabelia (Figure 9.11) and *Nuhliantha* are two extinct genera established based on small trimerous, staminate flowers from the Late Cretaceous (Turonian) Old Crossman locality, New Jersey, USA (Gandolfo *et al.*, 2002). *Mabelia* comprises two species, *M. connatifila* and *M. archaia*. *Nuhliantha* includes a single species, *N. nyanzaiana*. Flowers very similar to *Mabelia* are also known from the Late Cretaceous (Late Santonian) Allon flora, Georgia, USA (Herendeen *et al.*, 1999a). *Mabelia* and *Nuhliantha* are similar in their general floral organisation. Both are actinomorphic, unisexual and have six triangular, leathery tepals that are fused at the base and show valvate aestivation. Bulging structures are scattered over the surface of the tepals, indicating the probable presence of ethereal oil cells. The androecium consists of three stamens, each with a more or less distinct apical extension of the connective. Anthers are dithecate with extrorse, valvate dehiscence. In *M. connatifila* the stamen filaments are fused into a short staminal column, while in *M. archaia* and *Nuhliantha* anthers are sessile. Pollen in the published material is monocolpate and psilate–punctate, but flowers of *Mabelia* with trichotomocolpate pollen have also been observed (our observations; Figure 9.12). *Mabelia* flowers show no indication of pistillate organs; flowers of *Nuhliantha* have a central pistillode.

Both *Mabelia* and *Nuhliantha* were assigned to the extant monocot family Triuridaceae (Gandolfo *et al.*, 1998b, 2002), but currently there is no information on the corresponding pistillate flowers. Pollen of *Mabelia* and

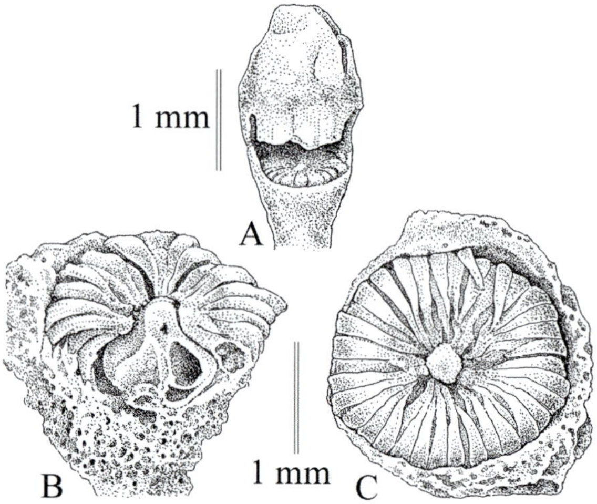

Figure 9.13 *Microvictoria svitkoana* from the Late Cretaceous (Turonian) Old Crossman locality, New Jersey, USA. (A) Flower bud in lateral view with perianth partly removed. (B) Specimen without perianth, showing central structure surrounded by possible pistillate organs interpreted as paracarpels. (C) Apical view of flower with perianth and androecium removed, showing pistillate organs interpreted as paracarpels and central structure. Drawn from SEM images in Gandolfo *et al.* (2004).

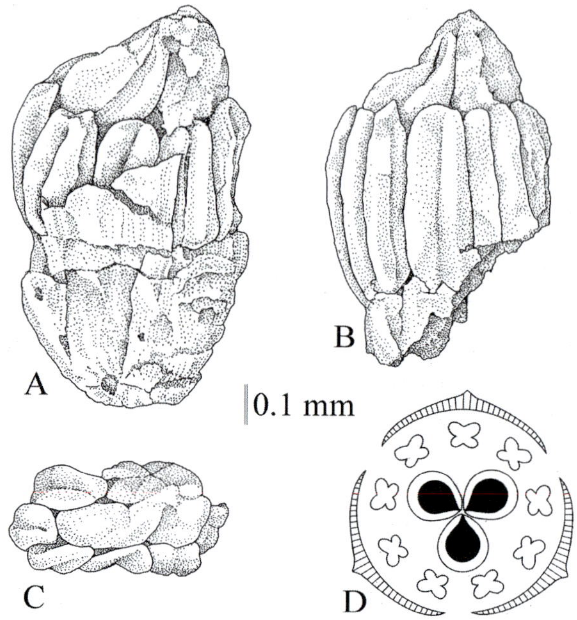

Figure 9.14 Trimerous flower from the Early Cretaceous (Late Barremian – Early Aptian) Torres Vedras locality, Portugal, with superior ovary. (A) Flower in lateral view with perianth parts adpressed to the androecium. (B) Perianth parts removed to show androecium. (C) Apical view of flower showing gynoecium and surrounding stamens. (D) Floral diagram. (A, B, D) From Friis *et al.* (2006a); (C) from material in the collections of the Swedish Museum of Natural History, Stockholm.

Nuhliantha is distinct from that of extant Triuridaceae (Figure 9.12). Assignment of the fossils to Triuridaceae and to monocots is therefore not fully secure. Some of the distinctive features of these fossils occur in magnoliid angiosperms, in particular the fusion of the stamens into an androecial column (e.g. Myristicaceae, Schisandraceae), but the hexamerous perianth would be unusual for both of these families.

Microvictoria svitkoana (Figure 9.13) comprises multipartite flowers also from the Late Cretaceous (Turonian) Old Crossman locality, New Jersey, USA (Gandolfo *et al.*, 2004). The flowers are minute, about 3 mm long and 1.5 mm in diameter, actinomorphic, bisexual and borne on a stout pedicel. Floral parts are numerous and borne on a shallow floral cup. Tepals or bracts cover the outer surface of the floral cup in an imbricate pattern. These are sepaloid below and petaloid above. The androecium is composed of an inner zone of numerous flattened stamens, and an outer zone of flattened and tongue-shaped staminodes. Stamens are laminar, but the organisation of the pollen sacs and their mode of dehiscence are unknown. No pollen was observed. Aspects of gynoecial structure are also not completely clear. According to the descriptions, the gynoecium is formed by paracarpels, several rows of carpellary appendages and several series of carpels around a central column. However, the locules of the carpels and ovules have not been observed.

Cladistic analysis of the relationships of *Microvictoria*, in a combined molecular and morphological data matrix of extant Nymphaeaceae, placed the fossil in an unresolved polychotomy with extant *Euryale*, *Nymphaea* and *Victoria* (Gandolfo *et al.*, 2004). However, given the uncertainties in the interpretation of the gynoecium the proposed relationships may not be fully secure. Based on phyllotaxis and the position and aestivation of tepals, Endress (2008) suggested that *Microvictoria* is more similar to flowers of Austrobaileyales, or certain eumagnoliids, than to flowers of Nymphaeales.

Two different kinds of probable monocot flowers from the Early Cretaceous of Portugal have been illustrated in our previous work and are also included here although they remain to be formally described and named. One of the flower types is represented by a single flower with a superior ovary (Figure 9.14). This flower is from the Torres Vedras flora (Late Barremian – Early Aptian) and is bisexual, about 1 mm long, with nine stamens and an apparently trimerous gynoecium. Fragments of a perianth are closely adpressed to stamens and the arrangement and number of parts is unclear. Removal of the perianth showed that the androecium is also incompletely preserved, but the position of the stamens clearly indicates that the flower had nine stamens. Because the base of the flower is broken it is uncertain whether the stamens occur in one, two or three whorls. The stamens have a broad filament and a dithecate and tetrasporangiate anther with pollen *in situ*. Pollen is monocolpate, and elliptical to circular in outline, with a maximum dimension of about 12–13 μm. The colpus is long extending to the equator. The tectum is reticulate, with lumina of two sizes and low muri supported by short and densely spaced columellae.

The trimerous floral organisation may indicate a monocot affinity for this flower, and several extant monocots have pollen with short, densely spaced columellae. However, trimery combined with monoaperturate pollen also occurs among magnoliids, and the systematic affinity of the flowers remains to be established with certainty.

The other kind of possible monocot flower from Portugal is known from many specimens and from several different floras of Late Barremian – Early Albian age (Catefica, Famalicão, Vale de Agua). Flower morphology and details of pollen show that there are at least two species. They all have the same general organisation with an epigynous, undifferentiated perianth, stamens with massive connective, valvate dehiscence, prominent apical extension of sterile tissue above the pollen sacs, and finely striate pollen. Pollen is dicolpate in one species, and monoaperturate, with either a simple elongate aperture or a triradiate (trichotomocolpate) aperture, in another. The number of perianth parts and the shape of tepals differ between the species. Flowers from the Catefica locality (Figure 9.15) are apparently always trimerous with six tepals in two whorls and six stamens, also in two whorls. Details of gynoecium organisation and ovules/seeds are unknown for all of the material.

The unusual striate pollen and the aperture configuration as well as the trimerous organisation indicate that the fossil flowers are probably an early monocot, but more comparative studies are needed before the material can be placed with confidence in any of the major angiosperm clades.

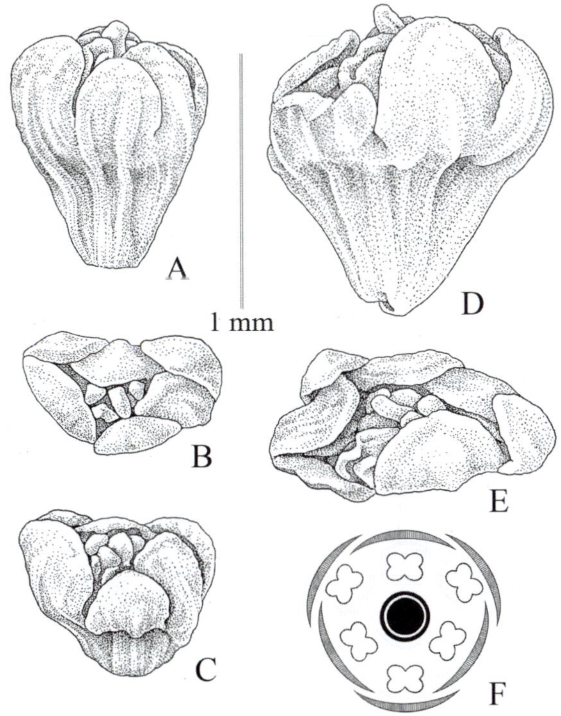

Figure 9.15 Trimerous flowers from the Early Cretaceous (Late Barremian–Aptian) Catefica locality, Portugal, with an inferior ovary. (A–C) Flower in (A) lateral, (B) apical and (C) oblique view, showing epigynous arrangement of tepals. (D, E) Another flower of the same species in (D) lateral and (E) apical view. (F) Floral diagram; note that the details of the gynoecium are unknown. (A, B, D, E) From Friis *et al.* (2006a); (C) from material in the collections of the Swedish Museum of Natural History, Stockholm.

9.5.2 Isolated fruits and seeds

Isolated fruits and seeds are the most abundant angiosperm remains in both Early and Late Cretaceous mesofossil floras. They are usually well-preserved, but often lack diagnostic features critical for detailed systematic analysis. In some cases Early Cretaceous fruits have triaperturate pollen adhering to their stigmatic surfaces and these are almost certainly eudicots. However, where there are monocolpate grains on stigmatic surfaces it is often more difficult to establish whether a given fossil should be placed among magnoliids or monocots. *Anacostia* and *Couperites* are two fossils that fall in this category.

Anacostia (Figure 9.16) is common in Early Cretaceous mesofossil floras from both eastern North America and Portugal (Friis *et al.*, 1997a). The type species, *Anacostia marylandensis*, was described from the Kenilworth locality (Early–Middle Albian), Maryland, USA. Another species, *A. virginiensis*, was described from the Puddledock locality

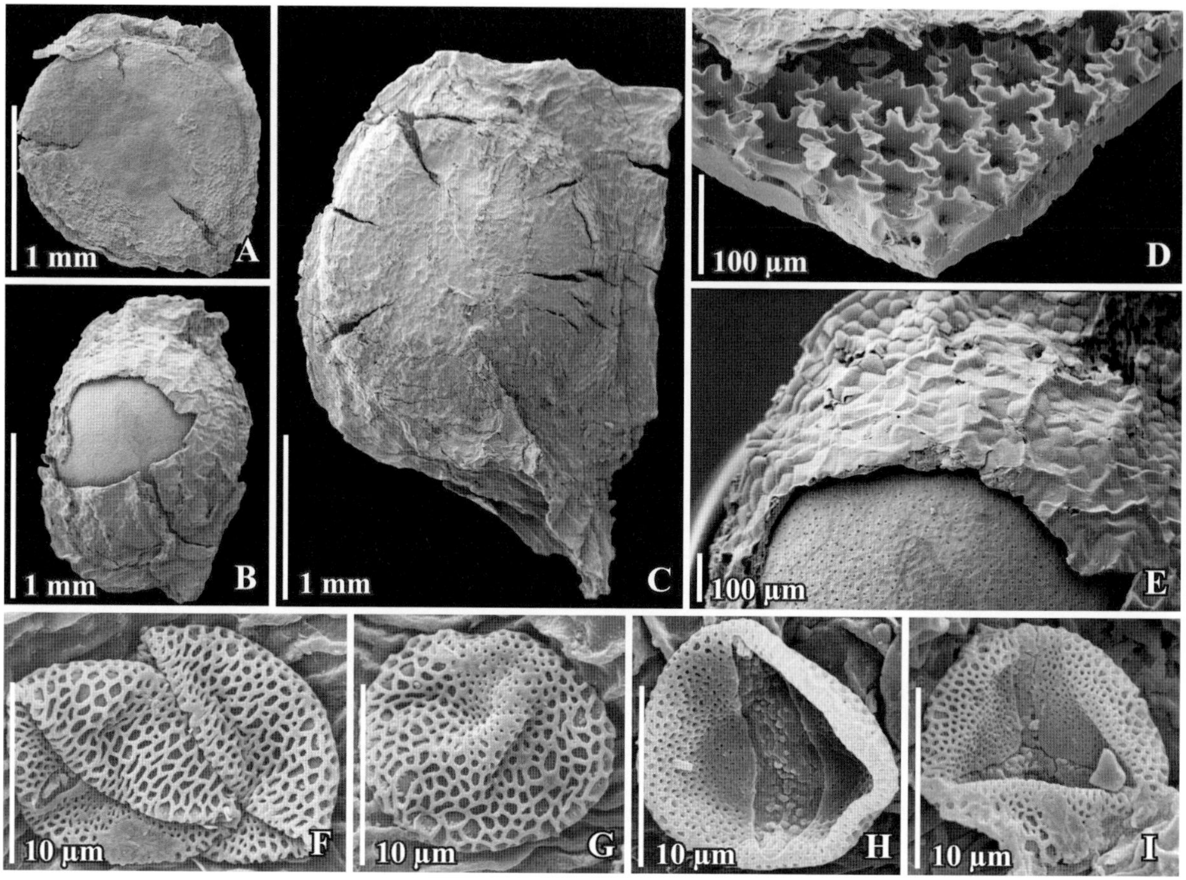

Figure 9.16 Fruits and pollen of *Anacostia* from the Early Cretaceous. (A, I) *Anacostia virginiensis* from the Puddledock locality (Early–Middle Albian), Virginia, USA; (A) fruit and (I) trichotomocolpate pollen. (B, D, E) *Anacostia teixeirae* from the Famalicão locality (Late Aptian), Portugal; (B, E) fruit with part of fruit wall missing showing single seed; (D) inner view of seed wall showing cells with wavy anticlinal walls. (C) Fruit of *Anacostia portugallica* from the Vale de Agua locality (Late Aptian – Early Albian), Portugal. (F–H) Pollen grains on fruit surface of *Anacostia marylandensis* from the Kenilworth locality (Early–Middle Albian), Maryland, USA, similar to dispersed *Similipollis*-type pollen with graded reticulum and monocolpate aperture. Specimens in the collections of the Field Museum, Chicago (A, F–I) and the Swedish Museum of Natural History, Stockholm (B–E).

(Early–Middle Albian), Virginia, USA, and a further two species, *A. portugallica* and *A. teixeirae*, have been reported from several Early Cretaceous localities in Portugal.

Anacostia comprises small unilocular and one-seeded fruiting units. They are laterally flattened and semi-circular in outline with a straight or slightly curved ventral margin. The stigma is indistinct, sessile and positioned at the top of the ventral margin. The fruit wall consists of thin-walled parenchyma cells that are only preserved in some of the charcoalified specimens. The parenchyma layer is covered by the outer epidermis, which has a thick cuticle. In compressed or lignitic specimens where the parenchyma has been lost the epidermis and cuticle form a broad wing-like rim around the seed. The fruitlets were small one-seeded berries. Seeds are anatropous and exotestal with an outer layer of small sclerenchymatic, cuboid to palisade-shaped crystal cells followed towards the inside by a layer of shallow cells with strongly digitate anticlinal walls. These cells are separated from the membranous inner layer by a thin cuticle.

Pollen found adhering to the stigma and fruit surface in many specimens from both the Potomac Group and Portuguese localities is all of the same type (Figure 9.16). The grains are monoaperturate, with a trichotomocolpate aperture or a simple elongated colpus. The aperture margins are distinct and the aperture membrane is granular to verrucate. Pollen is tectate with graded reticulate–foveolate ornamentation. The reticulum is more open around the equator, but lumina decrease in size towards the poles and form a foveolate zone along the aperture margins and over the proximal pole. The tectum and muri are supported by short, densely spaced, columellae. The endexine is moderately thick under the aperture, but is otherwise very thin. The pollen is comparable to the dispersed pollen genus *Similipollis* (section 9.6.4), but similar dispersed grains have also been assigned to various pollen genera such as *Liliacidites* and *Retimonocolpites* (e.g. Walker and Walker, 1986).

Anacostia is linked to associated small multiparted cone-like structures by identical epidermal features and pollen. These multipartite structures indicate that the gynoecium of *Anacostia* was multicarpellate and apocarpous with the fruiting units shed individually. None of these cone-like structures shows the remains of an androecium. A single charcoalified specimen of a similar cone-like structure from the Puddledock locality does show the remains of other organs, most likely stamens and perianth, below the gynoecium. However, further work is needed to determine whether this specimen can be related securely to *Anacostia*.

The systematic position of *Anacostia* remains to be fully established. Monoaperturate pollen with a similar graded reticulum is currently known only for monocots. However, the apocarpous and multiparted gynoecium, together with the structure of the seeds, suggests that *Anacostia* is related to early-diverging lineages at the ANITA grade or to eumagnoliid angiosperms. In particular, extant Schisandraceae (*Kadsura*, *Schisandra*) have similar multicarpellate gynoecia with the individual fruitlets being berries with one or more seeds. However, pollen in Schisandraceae is unique among angiosperms and unlike that associated with *Anacostia* (Chapter 8).

Couperites (Figure 9.17) is another genus of small, unilocular, one-seeded fruiting units. It presents similar challenges to *Anacostia*. *Couperites mauldinensis* was first described from the Early Cenomanian Mauldin Mountain locality, Maryland, USA (Pedersen *et al.*, 1991) and an additional species, *Couperites* sp., was reported from the Early–Middle Albian Puddledock locality, Virginia, USA (Friis *et al.*, 1997b). The fruiting units are small, dispersed, one-seeded berries with anatropous, exotestal seeds. The fruiting units are laterally flattened with a more or less straight ventral margin and a semicircular dorsal margin.

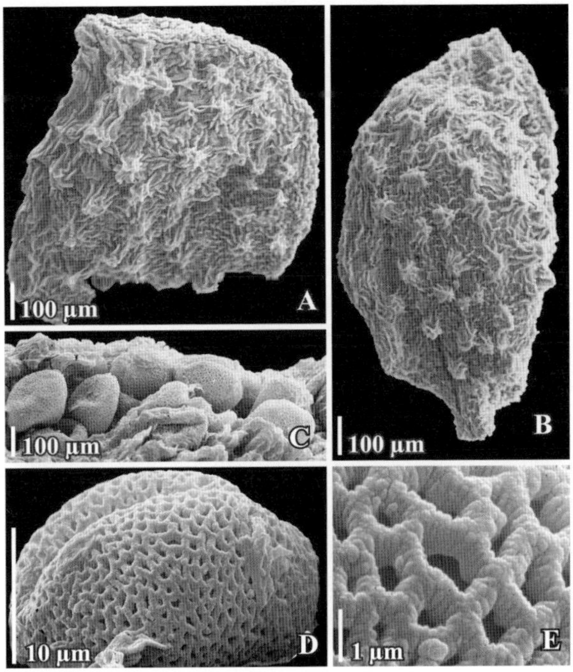

Figure 9.17 *Couperites mauldinensis* from the Late Cretaceous (Early Cenomanian) Mauldin Mountain locality, Maryland, USA. (A, B) Fruits with indistinct apical stigmatic area and fruit wall with blunt spine-like projections; (A) fragmentary specimen and (B) fruit with short stalk. (C) Stigmatic area of fruit fragment showing adhering pollen grains. (D) Monocolpate, reticulate pollen from stigmatic surface. (E) Details of pollen wall with dense reticulum and beaded muri. Specimens in the collections of the Field Museum, Chicago.

All that remains of the fruit wall is a cutinised outer epidermis that shows stout, blunt spine-like projections. Internally, there is a thin amorphous tissue that contains abundant resin bodies. The stigma is sessile, poorly differentiated and slightly decurrent down the ventral margin of the fruit. The single seed in each fruiting unit is anatropous and pendent. The outer epidermis of the seed is hard and composed of short palisade-like cells, and has a black shiny surface. Abundant pollen grains have been found adhering to the fruit wall and around the stigmatic surface. They are monocolpate with a finely reticulate tectum, and with a supratectal ornamentation of fine ribs arranged in two rows along the margin of the muri. These grains are very similar to some dispersed pollen that has been referred to the pollen genus *Clavatipollenites*.

Clavatipollenites-type fossil pollen has been compared to pollen of extant *Ascarina* (Chloranthaceae) (e.g. Couper,

1958; Walker and Walker, 1984). Pedersen *et al.* (1991) also compared *Couperites* to extant Chloranthaceae, but could not place the fossil in the family mainly because of differences in the organisation of the seeds. Later phylogenetic assessments by Doyle *et al.* (2003) and Eklund *et al.* (2004b) placed *Couperites* as a sister to Chloranthaceae, or sometimes embedded in the Chloranthaceae, but the information currently available is too sparse for definitive phylogenetic analysis. While some dispersed *Clavatipollenites*-type pollen is very similar to pollen of extant *Ascarina* (Walker and Walker, 1984) pollen attached to *Couperites* differs from that of the extant genus in having a distinct colpus margin. *Couperites* also differs from all extant Chloranthaceae in having exotestal seeds that develop from anatropous, rather than orthotropous, endotestal ovules. Monocolpate and reticulate pollen with supratectal ornamentation on the muri also occurs in other basal angiosperms and eumagnoliids (Doyle, 2005), as well as in many monocots (Harley, 1997). Abundant resin bodies (presumed fossilised ethereal oil cells) indicate that the fossils are probably not monocots, although oil cells are also known for *Acorus* and thus could have been present in early extinct monocot lineages.

9.6 DISPERSED MONOAPERTURATE POLLEN

Dispersed monoaperturate angiosperm pollen grains are diverse and widely distributed in palynofloras from the Cretaceous. Most of the pollen taxa used to accommodate these dispersed grains were established using light microscope (LM) studies of palynofloras based on material mounted in glycerine or other embedding media. These studies reveal key information on the size, form and aperture configuration of fossil pollen. Unfortunately, however, many taxonomically important features, such as supratectal ornamentation, features of columellae, and ultrastructural details of the pollen wall, cannot be reliably observed at the relatively low magnifications achievable with even the best light microscope. Formal taxa based on LM of dispersed pollen grains are therefore often used in a very broad sense; the current knowledge of some of the more important named genera of dispersed monoaperturate pollen from the Early and mid-Cretaceous is reviewed here.

A further manifestation of very broad generic concepts for fossil pollen is that the number of formally recognised taxa is low. It is now clear that the widespread application of this relatively small number of names has, to a large extent, concealed the true diversity of angiosperms during the early phases of their radiation (Friis *et al.*, 1999). It has also been an impediment to using angiosperm pollen grains for stratigraphic purposes (Hughes, 1994). For example, Hughes (1994) noted that a group of pollen grains recognised in LM as a single species of *Clavatipollenites* could be separated into more than ten distinct species, and perhaps even several distinct genera, by high-resolution SEM studies (see also below). Considerable taxonomic diversity within a group of superficially similar fossil angiosperm pollen was also demonstrated by Penny (1988b) in his study of acolumellate grains from the Early Cretaceous of Egypt. In LM studies different forms could not be securely separated, but SEM studies of numerous specimens documented the presence of 13 well-delimited taxa, as well as several other types that were not sufficiently well characterised to be established as separate taxonomic entities. The future use of fossil pollen to understand taxonomic and phylogenetic and stratigraphic patterns in early angiosperms will therefore require high resolution SEM-studies to delimit natural groups. The discovery of several stratigraphically important pollen types *in situ* in flowers and dispersed stamens, or adhering to the stigmatic surfaces of fruits, has also provided a new tool for understanding how pollen diversity reflects angiosperm diversity more broadly in Early Cretaceous floras.

9.6.1 *Afropollis* and *Schrankipollis*

Afropollis (Figure 9.18) was established by Doyle *et al.* (1982) to accommodate dispersed, spheroidal, acolumellate pollen with a characteristic coarse, loose reticulum that encloses a much smaller main body. The type species, *Afropollis jardinus*, was first described by Brenner (1968) as *Reticulatasporites jardinus* based on dispersed grains from the Albian of Peru. The type material was illustrated by light micrographs and has not been restudied. The generic description was based mainly on material from West Africa studied by LM and SEM (Doyle *et al.*, 1982), which revealed a considerable diversity of new forms. Several new taxa were described (*A.* aff. *jardinus, A. operculatus, A. zonatus, Afropollis* sp. *A, Afropollis* sp. *B*) (Doyle *et al.*, 1982). Further new forms have since been described from the Early Cretaceous of northern Africa and England (Schrank, 1983; Penny, 1989). Grains very similar to typical *Afropollis* were described from the Early Cretaceous of Egypt as *Retimonocolpites mawhoubensis* (Schrank, 1983) and later transferred to the genus *Schrankipollis* (Doyle *et al.*, 1990a).

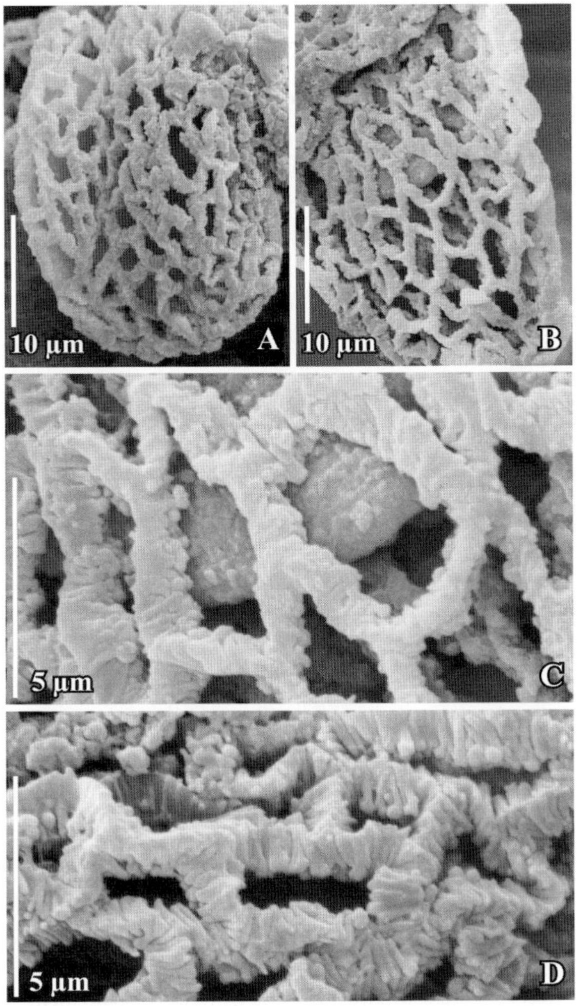

Figure 9.18 *Afropollis* pollen from the Early Cretaceous (Late Aptian – Early Albian) Vale de Agua locality, Portugal, from an isolated pollen sac. (A, B) Whole grains showing loose reticulum around the much smaller central body. (C) Detail of pollen grain in (B) showing loose reticulum and central body. (D) Detail of reticulum showing strongly segmented muri. Specimen in the collections of the Swedish Museum of Natural History, Stockholm.

Afropollis exhibits a variety of aperture configurations, which include inaperturate and zonacolpate arrangements (Doyle *et al.*, 1982) as well as possible monocolpate forms (Schrank, 1983). The ring-furrow in zonacolpate grains is either displaced toward one pole (heteropolar) or divides the grain into two equal hemispheres (isopolar). *Afropollis* pollen is semitectate with a coarse reticulum consisting of an open network of muri. The network of muri is typically separated from the foot layer over most of the surface of the grain. Most species lack columellae and there is usually a granular layer beneath the muri. A further characteristic feature of many *Afropollis* species is muri with a distinctly triangular profile and strong transverse segmentation.

Pollen wall ultrastructure has been studied for *A. operculatus*, *A. zonatus*, and *A.* aff. *jardinus* using TEM (Doyle *et al.*, 1990a). Sections through the ektexine show a thick tectum, a thin infratectal layer of irregular granulae, and a very thin inner layer corresponding to the foot layer. The innermost layer of the pollen wall is darker-staining and is interpreted as an endexine. In *A. operculatus* and *A. zonatus* this innermost darker-staining layer is very thick, and apparently equal in thickness all around the grain. In *A.* aff. *jardinus* this inner layer is thinner, at least in certain parts of the grains, and this presumed endexine also shows some lamination, particularly close to the contact with the foot layer. We have observed similar wall ultrastructure in *in situ Afropollis* grains from the Early Cretaceous of Portugal.

Afropollis is conspicuous in many Early Cretaceous palynofloras and is particularly abundant in Aptian–Albian palynofloras from low palaeolatitudes in Africa and South America. It also occurs at higher latitudes and has been reported from Portugal, southern England and British Columbia. The earliest record of the genus is from the Barremian of southern England (Penny, 1989). *Afropollis* attained its maximum abundance in the Aptian and Albian and became extinct during the Cenomanian.

From the Early Cretaceous (Late Aptian – Early Albian) Vale de Agua locality, Portugal, *Afropollis* grains have been discovered *in situ* in a single isolated pollen sac, but otherwise there is no other fossil evidence of the plants that produced these unique grains. The pollen sac is narrow, elongate and without a distinct dehiscence line. There is no evidence that it was originally part of a pair (as in angiosperms). Unfortunately the fossil is too fragmentary to provide critical information relevant for understanding the systematic position of the *Afropollis*-plant.

The systematic affinity of *Afropollis* has generally been assumed to be angiospermous. Doyle *et al.* (1990a) suggested a relationship with extant Winteraceae, and specifically that *Afropollis*, together with the dispersed pollen genus *Schrankipollis*, were the sister group to the dispersed pollen genus *Walkeripollis* and the extant families Winteraceae and Illiciaceae (Schisandraceae). More recent work makes this interpretation unlikely. Phylogenetic analyses of extant taxa based on molecular data do not support a close relationship between Winteraceae and Schisandraceae. The isolated pollen sac with *Afropollis* pollen is also distinctly different from those of extant Winteraceae or Illiciaceae. The tentative assignment of *Afropollis* to the Nymphaeales ('nymphoids') by Krassilov (1997), based on superficial similarities in aperture configuration, seems similarly unlikely. It is also not supported by other features such as pollen wall ultrastructure.

Pollen wall structure resembling that of *Afropollis* has been described for several monocots. For example, *Aristia dichotoma* (Iridaceae) has similar semitectate and coarsely reticulate pollen with irregular and strongly transversely segmented muri supported by an irregularly columellate infratectal layer borne on a very thin foot layer (Le Thomas *et al.*, 1996). *Aristia* also has a thick, continuous, darker-staining inner layer that resembles the supposed endexine of *Afropollis*. However, in *Aristia* this inner layer is interpreted as an intine and it is not resistant to acetylosis. Generally, intine is not preserved in fossil material, but the possibility that the inner darker-staining layer in *Afropollis* is actually the remains of an unusual kind of intine needs to be considered. In this case, the discrepancy in size between the reticulum and main body of the grain in *Afropollis* may be due to substantial shrinkage of the less resistant intine-like inner layer. Alternatively, and more likely in our opinion, if the inner darker-staining layer in *Afropollis* is actually endexine, then this may indicate that *Afropollis* is more closely related to some (unknown) non-angiosperm seed plant. A thick laminar endexine is unknown in angiosperms. This interpretation is also supported by the unusual pollen sac in which *Afropollis* occurs *in situ*. Dispersed angiosperm pollen sacs typically occur in two pairs, reflecting the characteristic dithecate arrangement of the four pollen sacs in angiosperm stamens. Taken together, the scant evidence currently available supports the possibility that *Afropollis* may have been produced by an interesting, but non-angiosperm, seed plant.

Schrankipollis was established by Doyle *et al.* (1990a) for zonacolpate pollen with a loose, segmented reticulum closely resembling that of *Afropollis* (Doyle *et al.*, 1990a). The type species *Schrankipollis mawhoubensis* from the Early Cretaceous of Egypt was first described as *Retimonocolpites mawhoubensis* (Schrank, 1983). A further species, *Schrankipollis microreticulatus*, from the Potomac Group, originally described as *Schizosporites microreticulatus* (Brenner, 1963), was also included in the genus. *Schrankipollis* is similar to *Afropollis* in having segmented muri that are sharply triangular in cross-section and that are typically detached from the main body of the grain. As in *Afropollis*, the main body of the grain is much smaller than the enveloping reticulum. However, *Schrankipollis* is distinguished from *Afropollis* by having an infratectal layer of very short columellae with rounded bases. Ultrastructural studies of the pollen wall in *S. microreticulatus* (Doyle *et al.*, 1990a) show that the tall sharply triangular muri are supported by a columellar to almost granular infratectum. The inner layer (nexine) is thick and was interpreted as mainly consisting of a foot layer. The structure of the inner part is unclear.

The data currently available are sparse, but suggest that *Schrankipollis* and *Afropollis* are probably closely related. The proposed link between *Schrankipollis–Afropollis* and *Walkeripollis*, as suggested by Doyle *et al.* (1990a), is much less certain. *Walkeripollis* is pollen of presumed Winteraceae with the grains occurring in permanent tetrads (Chapter 10). Pollen of *Schrankipollis* and *Afropollis* always occurs individually (monads). The two taxa are also distinguished from *Walkeripollis* and extant Winteraceae in pollen wall ultrastructure. A winteraceous affinity for *Schrankipollis* and *Afropollis* seems unlikely.

9.6.2 *Asteropollis* and *Clavatipollenites*

The characteristic pollen of *Asteropollis*, with a four- to five-armed aperture, has been found closely associated with *Hedyosmum*-like fruits as well as in *Hedyosmum*-like staminate inflorescences (Chapter 8). This confirms their long-suspected relationship to extant Chloranthaceae. However, other records attributed to *Asteropollis* include grains with more regularly trichotomocolpate apertures. These grains may not be related to *Hedyosmum*, but could perhaps be related to *Ascarina*, another genus of Chloranthaceae, which occasionally has trichotomocolpate pollen. However, reticulate, trichotomocolpate pollen also occurs in other families, both among eumagnoliids and among monocots. The chloranthaceous affinity of those *Asteropollis* grains with a regularly trichotomocolpate aperture is therefore not fully secure.

Dispersed pollen grains assigned to *Clavatipollenites* (Couper, 1958) are among the oldest definitive angiosperm pollen grains to be recognised in dispersed palynofloras, and are often cited as documentation for the antiquity of the family Chloranthaceae. The genus was established based on a single species, *C. hughesii* from the Barremian–Aptian of southern England (Couper, 1958), but has been reported subsequently from Early and Late Cretaceous palynofloras from all over the world. Several new species have been described, with the oldest record from the Hauterivian of Israel (Brenner and Bickoff, 1992). A summary of the stratigraphic and geographic distribution of the genus was given by Muller (1981).

Although the name *Clavatipollenites* is widely applied to Early Cretaceous angiosperm pollen there are problems in defining the genus. As currently used it may include several distinct lineages. The type species, *C. hughesii*, comprises small elliptical to almost spherical, monocolpate pollen with a long, broad colpus and a finely reticulate tectum. However, the holotype is embedded in a solid medium and can only be studied by light microscopy. Details of pollen wall structure and ornamentation are unknown and cannot be studied by SEM and TEM. New preparations were subsequently made from the same sample that yielded the holotype in an attempt to study the species by SEM (e.g. Hughes *et al.*, 1979), but instead of identifying one type of pollen that could have been ascribed to *Clavatipollenites hughesii* ten or more distinct species were recognised, all showing broadly similar general morphology. The various specimens were distinguished from each other mainly by fine structural details that can only be differentiated by using SEM at high magnification. As currently used *Clavatipollenites* is distinguished from other monocolpate and reticulate forms by the indistinct aperture margins and the finely verrucate aperture membrane, but these characters are not completely clear in the original illustrations of *Clavatipollenites hughesii*.

Couper (1958, 1960) was far-sighted in suggesting a chloranthaceous affinity for *Clavatipollenites hughesii*. Since then, detailed studies of fossil pollen from the Potomac Group sequence (Lower Zone I) assigned to *Clavatipollenites hughesii*, and comparative studies of extant Chloranthaceae, have documented considerable ultrastructural similarity between the North American *Clavatipollenites* and modern *Ascarina* (Walker and Walker, 1984).

The only mesofossil evidence of plants producing *Clavatipollenites*-like pollen is provided by the fruiting units *Couperites mauldinensis* (section 9.5.2), but these fossils cannot be included in the Chloranthaceae because of differences in the structure and attachment of the ovules and the distinct aperture margin of the pollen. This may indicate that, although some of the pollen specimens assigned to *Clavatipollenites* represent early Chloranthaceae, others represent plants that are systematically distinct from that family. Considering also the variability in sculptural details observed in *Clavatipollenites*-like grains from southern England, we agree with Hughes (1994) that the use of *Clavatipollenites* as a universal indicator for the presence of Chloranthaceae in the fossil record requires caution.

9.6.3 *Retimonocolpites*, *Brenneripollis* and *Pennipollis*

Retimonocolpites was established by Pierce (1961) for dispersed pollen from the Late Cretaceous (Early Cenomanian) of Minnesota, USA. The type species, *R. dividuus*, was based on a single monocolpate, reticulate pollen grain studied by using LM. Details of pollen wall structure are unknown for the type material. In subsequent palynological studies the genus has often been used as a repository for monocolpate and reticulate pollen grains with a distinct colpus margin and a relatively coarse reticulum that sometimes has supratectal ornamentation of the muri. However, the type species has an aperture that almost encircles the grain, nearly dividing it into two halves. It is thus distinctly different from the normal monocolpate pollen that has typically been attributed to *Retimonocolpites*, including the dispersed grains from the Albian of the Potomac Group assigned to *R. dividuus* and studied by Walker and Walker (1984) using LM, SEM and TEM. These grains are more or less elliptical in equatorial outline, about 30 μm long, and semitectate with an open reticulum. The muri are narrow, low, and ornamented with fine transverse and often discontinuous ridges. Pollen wall ultrastructure shows a thick tectum supported by short columellae, a thin foot layer and no endexine. The dispersed pollen genus *Dichastopollenites* established based on Dakota Sandstone material (May, 1975) also includes reticulate grains that often split in valves.

Other dispersed pollen types assigned to *Retimonocolpites* show considerable variation in the ornamentation of the muri and pollen wall ultrastructure. They most likely represent a heterogeneous assemblage of forms that are widely separated systematically (Friis and Pedersen, 2011). Two species that have been included in *Retimonocolpites* are *R. peroreticulatus* and *R. reticulatus*, first described as species of *Peromonolites* (Brenner, 1963) and later transferred to *Retimonocolpites* by Doyle (in Doyle *et al.*, 1975). However, these two species are distinguished from other reticulate pollen in lacking a columellate infratectal layer. They have therefore been transferred to the dispersed genus *Brenneripollis* and later to *Pennipollis*.

Brenneripollis was established by Juhász and Góczán (1985) to accommodate small to medium-sized monocolpate pollen grains with an irregular reticulum that is often loosely connected to the foot layer. The transfer of *Retimonocolpites peroreticulatus* and *R. reticulatus* to *Brenneripollis* was accepted in several subsequent studies (e.g. Schrank, 1987; Doyle and Hotton, 1991; Doyle, 1992). However, the relationship between these acolumellate grains and *Brenneripollis* has been questioned by Penny (1988b), as well as by Friis *et al.* (1999), because the type species of *Brenneripollis* (*B. pellitus*) is smaller and has a well-developed columellate infratectal layer and a finer reticulum. A new genus, *Pennipollis*, has therefore been established for those dispersed grains that lack a columellate infratectal layer (Friis *et al.*, 2000a); this assignment has now gained general acceptance. *Pennipollis* occurs *in situ* in stamens and on the surfaces of small fruits that have been considered a possible monocot (Chapter 11). Similar acolumellate grains from the Early Cretaceous Mersa Matruh borehole, Egypt, were assigned to the Genusbox 'Retimono' (Penny, 1988b), while columellate grains with finely striate muri, similar to those of *R. dividuus* from the Potomac Group, were assigned to the Genusbox 'Reticoll' (Penny, 1991). Several species originally assigned to *Retimonocolpites* have later been transferred to other dispersed pollen genera.

Other kinds of *Retimonocolpites*-type pollen with smooth muri have been found *in situ* in staminate flowers from the Early Cretaceous that are definitely related to crown group Araceae (Chapter 11). Very similar grains have also been found associated with the Cenomanian inflorescence *Zlatkocarpus* (section 9.2.4) and the Early Cretaceous *Canrightia* flowers (section 10.1.1).

9.6.4 *Liliacidites* and *Similipollis*

Liliacidites was first described by Couper (1953) based on dispersed pollen from the Late Cretaceous and Early Cenozoic of New Zealand. This material comprises monocolpate or trichotomocolpate pollen with a long, broad colpus and a reticulate, semitectate exine. The reticulum is heterogeneous and graded, with smaller lumina around the equatorial ends of the aperture. In the type species, *Liliacidites kaitangataensis*, the muri are clearly beaded or segmented. Among extant plants this kind of graded reticulum is known only in pollen of monocots. It occurs, for example, in several genera of Liliales (Figure 9.19), which also occasionally have beaded muri resembling those of the fossil grains (e.g. Couper, 1953; Walker and Walker, 1984). Dispersed pollen grains with a similar graded reticulum occur in several Early Cretaceous palynofloras (Figure 9.19), with the earliest records from the Barremian–Aptian of Egypt, North America and southern England.

Liliacidites has also been used for a variety of dispersed monocolpate pollen with a different kind of reticulum in which the smaller lumina are concentrated in the polar, rather than equatorial, area of the grains. Grains of this type (Figure 9.19) are now transferred to the genus *Similipollis*. Among the several different pollen grains assigned to *Liliacidites* and studied using LM, SEM and TEM by Walker and Walker (1984) '*Liliacidites* Couper sp. 1' shows the same kind of graded reticulum as observed for the type species of the genus. '*Liliacidites* Couper sp. 2' has a *Similipollis*-type of graded reticulum, and '*Liliacidites*' *minutus* Brenner has an ungraded reticulum.

Similipollis was established by Góczán and Juhász (1984) for medium-sized to large, tectate–reticulate, monoaperturate pollen with an oval to almost circular equatorial outline. The aperture is an elongated colpus that extends to the equator. The reticulum is heterogeneous and graded. Smaller lumina occur on the distal surface around the aperture and over the proximal pole. Larger lumina occur around the equator. The type species, *Similipollis varireticulatus*, described from the Late Albian of Hungary, is characterised by a simple aperture (Góczán and Juhász, 1984). Pollen described from the Potomac Group that has a similar graded reticulum includes forms with a simple colpus as well as trichotomocolpate forms (Figure 9.19) (Walker and Walker, 1986). Similar pollen with both monocolpate and trichotomocolpate aperture configurations has also been observed on the fruits of *Anacostia* (Figure 9.16). Rather similar pollen has been reported from modern palms (Harley, 1997).

9.6.5 *Stellatopollis*

Stellatopollis is one of the most distinctive Early to mid-Cretaceous pollen types. It was established to accommodate large monocolpate and reticulate pollen with large triangular or elliptical clavate supratectal projections that are borne on the muri in a characteristic star-shaped pattern (Doyle *et al.*, 1975). This sculpture is sometimes referred to as crotonoid because it is particularly well developed in pollen of the extant genus *Croton* and other Euphorbiaceae. Similar exine sculpture also occurs in the eudicots *Sarcococca* and *Pachysandra*, two genera of extant Buxaceae, a family that diverges from other eudicots early in the evolution of the group (Chapter 12).

The muri in *Stellatopollis* are low and supported by short columellae; the foot layer is thick. Endexine is present under the aperture, but not in non-apertural areas. The

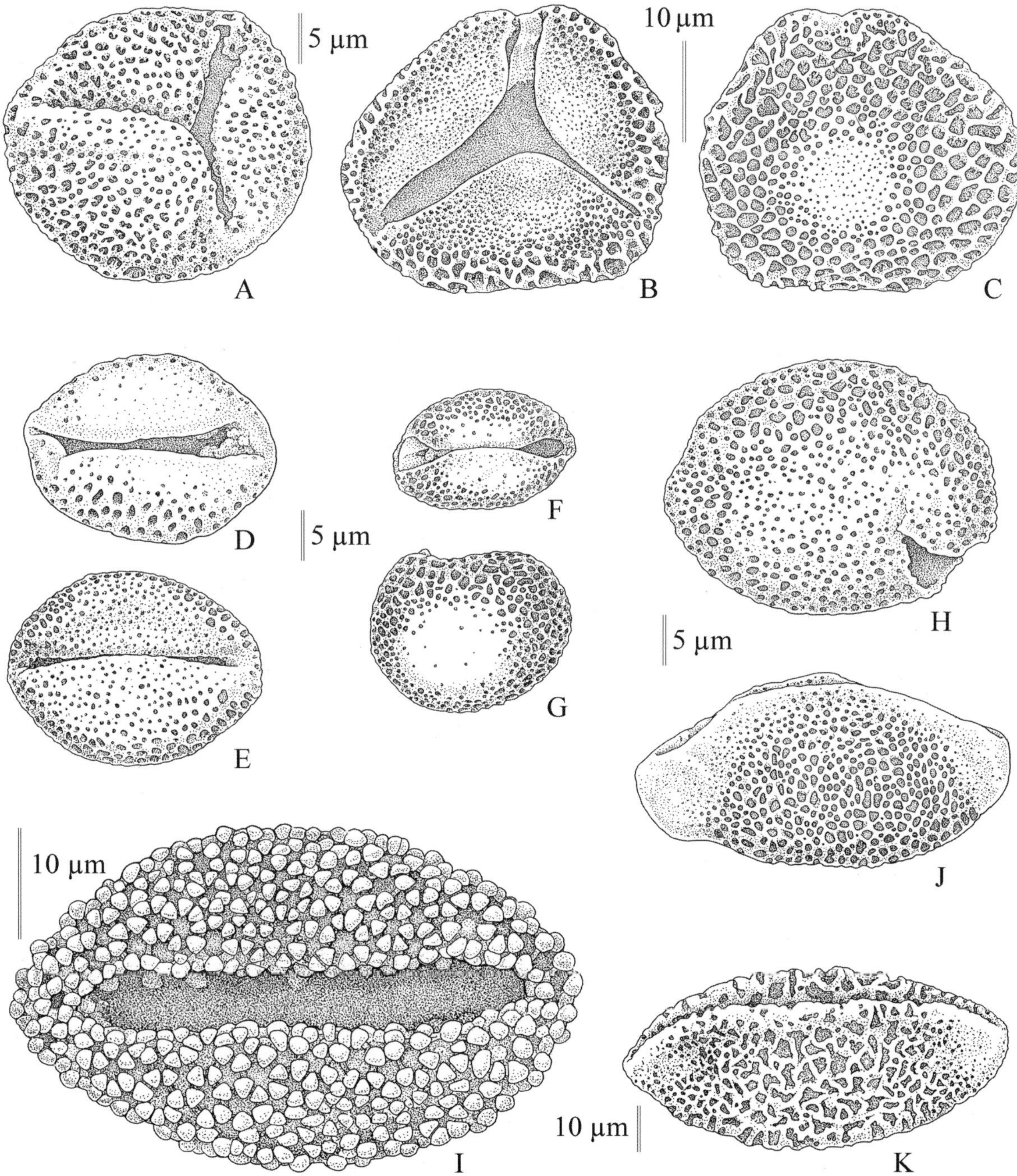

Figure 9.19 (A–H) *Similipollis*-type pollen showing graded reticulum with smaller lumina towards the aperture margin and proximal pole, (J) *Liliacidites*-type pollen showing graded reticulum with smaller lumina towards the equator of the grain, (I) *Stellatopollis barghoornii* from the Early Cretaceous of the Potomac Group, northeastern North America, and (K) pollen of extant *Hemerocallis*, Liliaceae. (A) Trichotomocolpate pollen in distal view (monocotyledonous pollen of Walker and Walker, 1986). (B, C) Trichotomocolpate pollen in (B) distal and (C) proximal view (*Liliacidites* sp. 2 of Walker and Walker, 1984). (D–H) Monocolpate pollen in (D–F) distal and (G, H) proximal view (monocotyledonous pollen of Walker and Walker, 1986). (I) Monocolpate pollen of *Stellatopollis barghoornii* showing crotonoid tectum ornamentation. (J) Monocolpate (*Liliacidites* sp. 1 pollen of Walker and Walker, 1984). (K) Monocolpate pollen of extant *Hemerocallis* showing graded reticulum. (A–F) Drawn from SEM images in Walker and Walker (1986); (G–K) drawn from SEM images in Walker and Walker (1984).

type species of *Stellatopollis*, *S. barghoornii* (Figure 9.19), was described from the Potomac Group where it is reported from the base of Subzone II-B (Albian) through to Zone III (Early Cenomanian) (Doyle *et al.*, 1975). Pollen of the *Stellatopollis*-type has been reported from several Early to mid-Cretaceous palynofloras from North America, Portugal, Egypt, West Africa and Brazil that range in age from Barremian to Cenomanian (Doyle *et al.*, 1975; Walker and Walker, 1984; Schrank, 1992; Hughes, 1994). From Egypt *Stellatopollis*-type grains were described as *S. bituberensis* (Penny, 1986) and as biorecords of the Genusbox 'Superret' (Penny, 1991). A much earlier occurrence of *Stellatopollis*-type pollen (*S. pocockii*) was described from the Late Jurassic (Oxfordian) of Normandy, France (Cornet and Habib, 1992), but this record is problematic (Chapter 6).

The monocolpate aperture of *Stellatopollis* indicates an affinity with magnoliids or perhaps monocots. Crotonoid sculpture is known both among eumagnoliids (*Daphnandra*, *Nemuaron*, Atherospermaceae) and among monocotyledons (*Lilium*, Liliaceae). However, a plausible systematic assignment of *Stellatopollis* to a higher taxon within the angiosperms has not been suggested to date. Nevertheless, it is unlikely that there is a close systematic relationship between these monoaperturate Cretaceous pollen grains and those eudicots that exhibit crotonoid sculpture (e.g. Buxaceae).

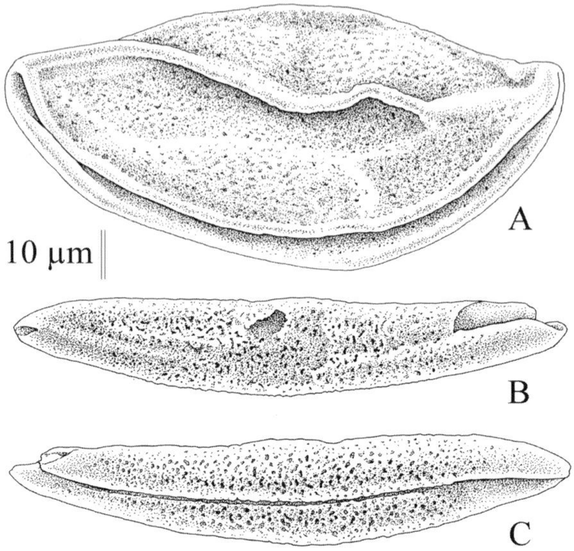

Figure 9.20 *Lethomasites fossulatus* pollen from the Early Cretaceous Potomac Group of eastern North America. (A) Boat-shaped grain folded over the aperture. (B, C) Strongly folded grains. Drawn from SEM images in Ward *et al.* (1989).

9.6.6 *Transitoripollis* and *Tucanopollis*

Transitoripollis is based on dispersed pollen from the Early Cretaceous (Albian) of Hungary. The type species, *Transitoripollis anuliculcatus*, and several other species, were first described from this area (Góczán and Juhász, 1984). Grains are monocolpate, with a broad colpus that does not reach the equator. The tectum is continuous, finely verrucate to microechinate, and occasionally finely punctate. The infratectal layer is granular and there is a thick foot layer. The endexine is thin in non-apertural regions. Other early angiosperm pollen with a similar morphology and wall structure has been described as *Tucanopollis*.

Tucanopollis is based on dispersed pollen grains from the Early Cretaceous (Barremian–Aptian) of Brazil assigned to *Tucanopollis crisopolensis* (Regali, 1989). These grains have a continuous tectum with finely verrucate to microechinate supratectal sculpture and a broad open aperture with a granular aperture membrane. The shape of the aperture varies among the specimens originally figured, from being a relatively narrow slit-like colpus to almost circular. The aperture of the type species is intermediate in shape and has a broadly elliptical outline. Pollen of *Tucanopollis* resembles that of *Transitoripollis* in all essential features, but none of the *Transitoripollis* pollen grains has a circular aperture. Both *Transitoripollis* and *Tucanopollis* are common in Early Cretaceous palynofloras and have an almost worldwide distribution. The earliest occurrence of *Tucanopollis* is from the Barremian of the Southern Hemisphere, where it has been reported from Africa (Doyle and Hotton, 1991) and South America (Regali, 1989). Very similar pollen was also reported from the Hauterivian and Barremian of southern England as CfA Hauterivian-*cactisulc* and CfA Barremian-*ring* (Hughes, 1994).

Pollen grains of this general type are also known in extant Piperales (e.g. Smith and Stockey, 2007a); Doyle and Hotton (1991) suggested that *Tucanopollis* was probably produced by early Piperales. Fossil pollen grains resembling *Transitoripollis* and *Tucanopollis* in both morphology and wall ultrastructure have been observed on Early Cretaceous fruits of *Appomattoxia* from eastern North America and Portugal (section 10.5.1) that may also be related to extant Piperales. However, the piperalean affinity of *Transitoripollis* and *Tucanopollis* is not fully secure. It is possible that the various dispersed grains assigned to *Transitoripollis* and *Tucanopollis* were produced by a variety of distinctly different early angiosperms.

9.6.7 *Lethomasites*

Lethomasites (Figure 9.20) was established for unusually large, dispersed, monocolpate pollen from the basal part of the Potomac Group sequence, eastern North America (Ward *et al.*, 1989). The genus includes a single species, *Lethomasites fossulatus*. The grains have a maximum dimension up to 127 µm. They are tectate with a continuous tectum perforated by small foveolae or fossulae. The tectal perforations are slightly graded, from almost psilate at the equatorial ends of the colpus, to more prominently perforate in the middle of the grains. Ultrastructural studies of the pollen wall show a thick granular infratectal layer of about the same thickness as the tectum. A very thin non-laminar layer inside the granular layer was interpreted as a possible foot layer. Endexine is apparently lacking. Pollen wall ultrastructure, and the large size of the pollen grains, suggests affinity with core Magnoliales, including the families Annonaceae, Degeneriaceae, Magnoliaceae and Myristicaceae, but none of the extant taxa shows the precise combination of features seen in the fossils (Ward *et al.*, 1989; Doyle and Hotton, 1991). The graded tectum of *Lethomasites* is also broadly similar to that of fossil *Liliacidites* and many extant monocots, and a monocot affinity for *Lethomasites* cannot be ruled out (Ward *et al.*, 1989).

9.7 FOSSIL LEAVES OF UNCERTAIN RELATIONSHIPS

Detailed treatment of the many angiosperm leaves described from the Early Cretaceous is beyond the scope of this book, but the focus here mainly on reproductive structures provides a useful context in which further work on the systematic relationships of early angiosperm leaves might be attempted. In this section we briefly discuss only a few Early Cretaceous leaves of uncertain systematic position that have figured in discussions of early monocot diversification.

Leaves related to various magnoliids or eudicots are known from many levels through the later part of the Early Cretaceous (Figure 9.21) and they are usually easy to recognise. Recognition of monocots based on fossil leaves is less straightforward, given the potential for confusion with 'parallel-veined' leaves of other kinds of plants such as extant conifers (e.g. *Agathis, Araucaria, Podocarpus*), the leaves of extinct, perhaps conifer-like, seed plants (e.g. *Podozamites*), or even poorly preserved stems (Doyle, 1973). Some ferns may also have leaves and a pattern of venation that could be confused with monocot leaves.

Trifurcatia flabellata, from the Early Cretaceous (Late Aptian – early Albian) Crato Formation of northern Brazil, now reassigned as one of three species of the extinct genus *Klitzschophyllites*, illustrates some of the potential difficulties (Mohr and Rydin, 2002; Mohr *et al.*, 2006).

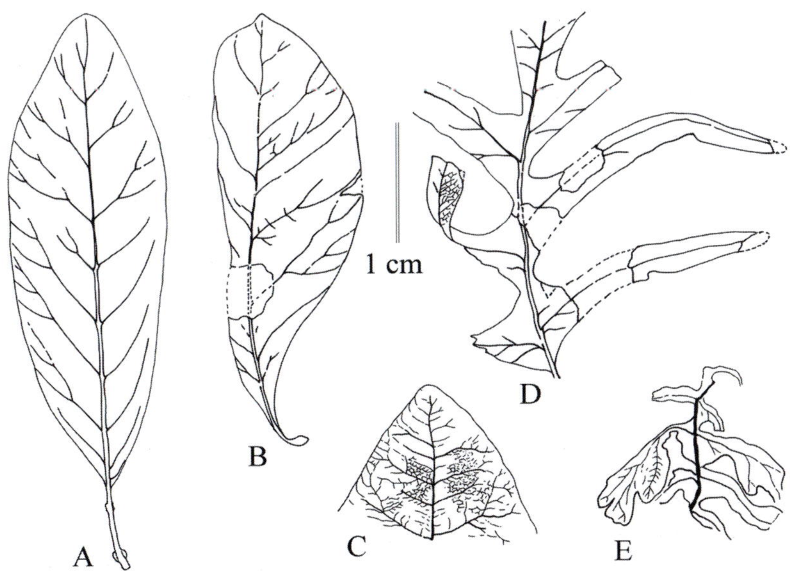

Figure 9.21 Early Cretaceous (Aptian) leaves from the Jixi locality, Heilongjiang Province, China, related to magnoliid or eudicot angiosperms. (A, B) *Asiatifolium elegans*. (C) *Shenkuoa caloneura*. (D) *Jixia pinnatipartita*. (E) *Jixia chenzihensis*. Redrawn from Sun and Dilcher (2002).

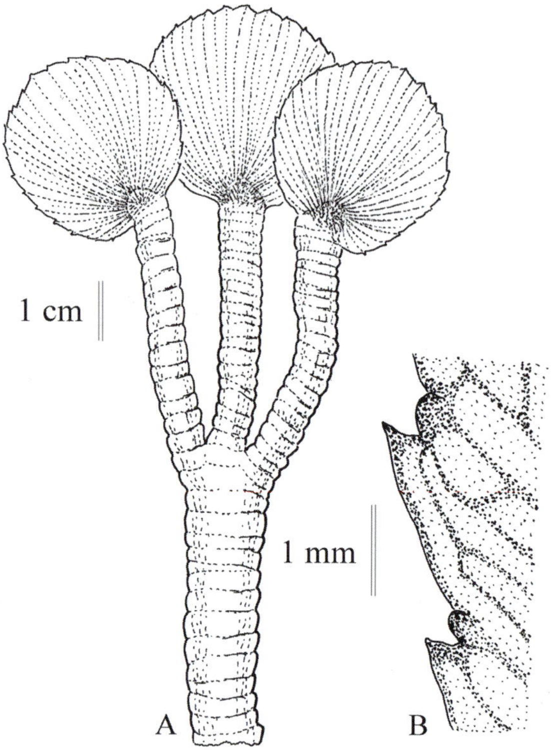

Figure 9.22 *Klitzschophyllites (Trifurcatia) flabellata* from the Early Cretaceous (Late Aptian – Early Albian) Crato Formation, Brazil. (A) Axis with numerous leaf scars bearing three serrate leaves. (B) Details of serrate margin showing bulging structures (probable vegetative buds) between the teeth. Drawn from photograph in Mohr and Rydin (2002).

Specimens assigned to *Klitzschophyllites* from Africa and Brazil were described as having thick leathery leaves and the *Klitzschophyllites* plants were interpreted as xeromorphic monocots perhaps adapted to saline conditions (Mohr *et al.*, 2006). *Trifurcatia flabellata* consists of stout axes bearing apparently rather thick, broadly oval to orbiculate leaves with a serrate margin (Figure 9.22). About 20–25 main veins diverge from the leaf base. These terminate at the base of the spinose teeth or sometimes between them, or they may branch to form secondary veins that also extend to the leaf margin. Tertiary veins form a reticulum between the primary and secondary veins. All of the primary and secondary veins terminate in a fine fimbrial vein that runs around the leaf just inside the leaf margin. Small bulging structures, described as glands, occur along the leaf margin.

Leaves from the Early Cretaceous flora of Cercal, Portugal, first described as *Protorrhipis choffatii* (Saporta, 1894), were

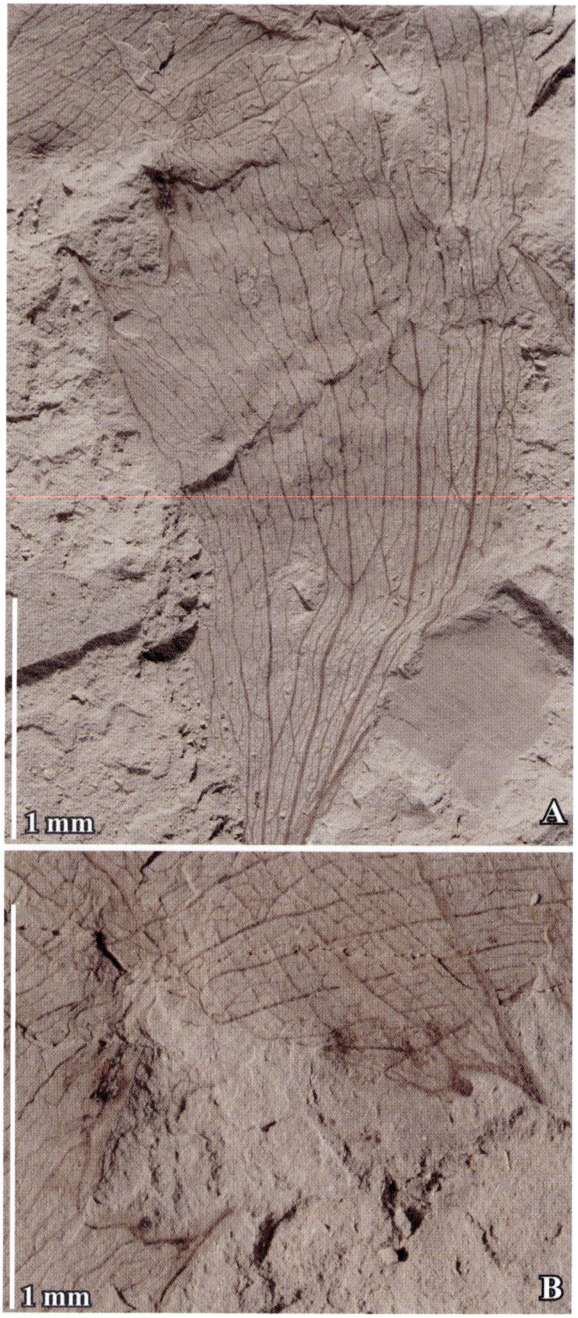

Figure 9.23 Leaves of *Protorrhipis choffatii* from the Early Cretaceous (?Aptian) Cercal locality, Portugal, showing (A) venation pattern and (B) serrate margin and bulging structures between the teeth. Specimens in the collections of the Swedish Museum of Natural History, Stockholm.

also transferred to *Klitzschophyllites* by Mohr *et al.* (2006). However, our examination of *Protorrhipis choffatii* collected at the type locality shows thin leaves with an unusual reticulate venation pattern that lacks freely ending veinlets and that is unlike that of angiosperms (Figure 9.23). The material is preserved only as impression, and details are difficult to study, but the leaf-blade has small discrete scars, or remains of small rounded structures, scattered on the veins. These could be the remains of sporangia and the bulging structures along the leaf margin could be vegetative buds rather than glands. Under this interpretation *Protorrhipis choffatii* is a fern comparable to, and perhaps closely related to, the extant aquatic fern *Ceratopteris*, and in particular, extant *Ceratopteris cornuta* and *Ceratopteris pteridoides*. *Ceratopteris* has solitary sporangia scattered over the leaf-blade and on the veins and also has vegetative buds that arise along the leaf-margin (Friis *et al.*, 2010a).

Acaciaephyllum from the Potomac Group of eastern North America is an Early Cretaceous leaf that has long occupied a prominent position in discussions of early monocot diversification (Doyle, 1973; Gandolfo *et al.*, 2000). We consider this genus in Chapter 11, but several other fossil leaves of possible monocot relationship described from the Potomac Group sequence are less convincing (see also Gandolfo *et al.*, 2000). For example, *Plantaginopsis marylandica* from Zone I of the Potomac Group, as illustrated by Ward (1905), shows a rosette-like plant with parallel-veined leaves and associated ellipsoidal reproductive structures. Some of the leaves show serrate margins. According to Doyle (1973) the most informative specimen is lost. *Alismaphyllum victor-masonii*, from Subzone II-B of the Potomac Group, is based on a poorly preserved impression showing a leaf with campylodromous venation and well-developed basal lobes that Ward (1895) assigned initially to *Sagittaria*. Doyle (1973) was unable to exclude a dicotyledonous affinity for this form. Also regarded by Doyle (1973) as interesting leaf fossils from the Early Cretaceous, but taxa that are equivocal in terms of their possible relationship to monocotyledons, are *Alismacites primaevus* from Portugal (Saporta, 1894; Teixeira, 1948) as well as *Pandanophyllum* (Kryshtofovich, 1929) and *Caricopsis* (Samylina, 1960) from eastern and northeastern Russia, respectively.

Another suite of leaves with possible monocot affinities, but that are also currently of uncertain relationship, are from the Early Cretaceous (Early – early Late Aptian) Yixian flora of northeast China. This complex includes leaves originally described as *Orchidites linearifolius* (Wu, 1999), *Orchidites lancifolius* (Wu, 1999), *Potamogeton jeholensis* (Yabe and Endo, 1935) (later referred to *Ranunculus jeholensis* by Miki, 1964) and 'Monocotyledon Leaf' (Cao *et al.*, 1998). All of these species have morphologically similar, narrowly elliptical leaves with parallel venation. They were reinterpreted by Sun *et al.* (2000, 2001) as conifers and included in *Liaoningocladus baii*. However, the paracytic stomata documented for *Liaoningocladus* are unusual for conifers. A possible relationship to angiosperms, and specifically monocots, requires further consideration as more complete or better-preserved material becomes available (Leng *et al.*, 2003).

10

Early fossils of eumagnoliids

Different applications of the magnoliid concept, and different conclusions as to which families should be included in the group, are a potential source of confusion in understanding early angiosperm evolution (Chapters 7, 8). However, recent phylogenetic studies have greatly clarified the situation and provide a firm basis for interpreting fossil material related to this important group of angiosperms. In this chapter we provide a brief outline of the current classification of eumagnoliids, and a review of those fossils, mainly from the Cretaceous, that can be assigned to the group. The fossil history of eumagnoliids is extensive, particularly from the Cenozoic, but there are also well-preserved and informative fossils from the Cretaceous. The early fossil record of eumagnoliids continues to increase rapidly as new palaeobotanical discoveries are made.

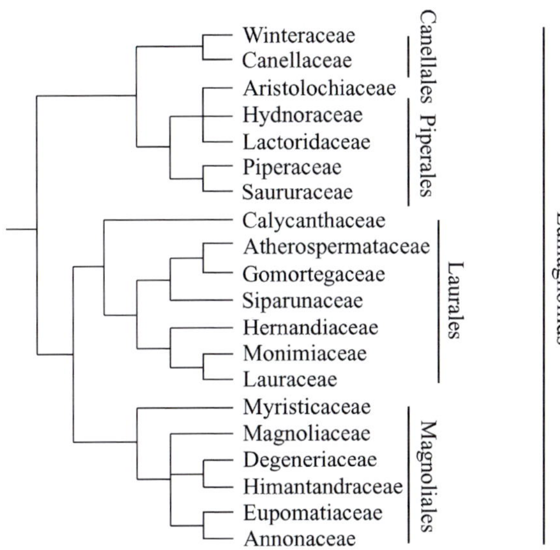

Figure 10.1 Summary of phylogenetic relationships of extant families and orders of the eumagnoliid clade based on Soltis *et al.* (2005). Circumscription of families follows APGIII (2009).

10.1 CLASSIFICATION OF EUMAGNOLIIDS

In the classification of Takhtajan (e.g. 1969) subclass Magnoliidae included six orders (Magnoliales, Laurales, Piperales, Aristolochiales, Rafflesiales, Nymphaeales), but it is now clear from phylogenetic analyses based on molecular data that this group is not monophyletic (e.g. Soltis *et al.*, 2005; APGIII, 2009). Several families are now recognised to comprise the ANITA grade; others have been placed elsewhere in the angiosperm tree. Four orders, Magnoliales, Laurales, Piperales and Canellales (Zanis *et al.*, 2003; APGIII, 2009), which were at the core of the former subclass Magnoliidae, comprise a monophyletic group, termed eumagnoliids by Soltis *et al.* (2000b) and magnoliids by APGIII (2009). Here we follow Soltis *et al.* (2000b) and refer to Magnoliales, Laurales, Piperales and Canellales as eumagnoliids. We use magnoliids informally in the broader conventional sense (Chapter 7). Magnoliales and Laurales form a clade that is sister to a clade comprising Piperales plus Canellales.

The current circumscription of the eumagnoliid clade, and the pattern of relationships within it (Figure 10.1), is based on results from phylogenetic analyses of molecular data. Chloranthaceae are sometimes resolved as the sister group of eumagnoliids (Chapters 7, 8). Morphological support for the monophyly of eumagnoliids, and some of its constituent subgroups above the family level, is not strong. Most eumagnoliids, except Piperales, are woody plants, and many have flowers with an undifferentiated perianth arranged on a trimerous plan. In addition, in many eumagnoliids, but not all, stamens and carpels are numerous and free. Carpels are plicate, ascidiate or intermediate between plicate and ascidiate (Endress and Igersheim, 2000a; Soltis *et al.*, 2005).

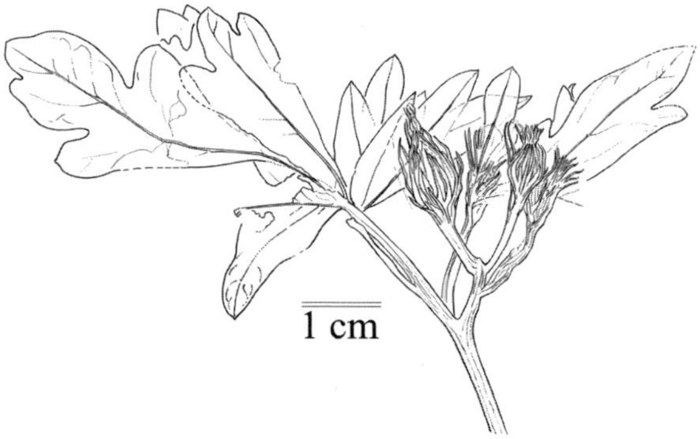

Figure 10.2 *Araripia florifera* from the Early Cretaceous (Late Aptian – Early Albian) Crato flora, Brazil. Redrawn from Mohr and Eklund (2003).

10.1.1 Fossil eumagnoliids of uncertain relationships

Several Cretaceous fossils are almost certainly related to eumagnoliids but cannot be included in extant orders or families, either because of incomplete knowledge of the fossils or because of uncertainty about the phylogenetic significance of the features that they exhibit. One example is the compression fossil *Araripia* in which the critical diagnostic features needed to evaluate its phylogenetic position are not known. Three other examples are *Detrusandra*, *Cronquistiflora* and *Canrightia*. In these cases the material is well preserved, but the combination of features that the fossils exhibit suggests that these taxa are representatives of extinct lineages or sub-lineages among the eumagnoliids or close to the root of the eumagnoliids. Other taxa (e.g. species of *Microvictoria, Protomonimia, Hidakanthus, Keraocarpon*) may also belong to the eumagnoliids, but their systematic position is currently more uncertain; they are discussed in Chapter 9.

Araripia florifera (Figure 10.2), described from the Lower Cretaceous (Upper Aptian – Lower Albian) Crato Formation, Brazil, consists of a branching axis bearing several floral structures and leaves (Mohr and Eklund, 2003). Floral structures are about 20 mm long, pedicellate, and apparently preserved as buds. The preservation precludes detailed assessment of floral organisation, but flowers were clearly multipartite with bracts and undifferentiated perianth parts apparently helically arranged on a cup-shaped floral receptacle. Leaves are small, about 25 mm long, and are apparently decussately arranged. Each leaf is irregularly three-lobed with pinnate primary venation.

Araripia is unusual in having both reproductive structures and vegetative parts preserved on the same branching system. Aspects of floral structure are similar to those of some extant Laurales, particularly certain Calycanthaceae, Idiospermaceae and Hernandiaceae (Mohr and Eklund, 2003), but because many critical floral features are unknown, including especially the nature of androecium and gynoecium, the precise systematic position of *Araripia* is not yet clear.

Detrusandra mystagoga (Figure 10.3) is based on well-preserved floral material (Taxon 'A' of Crepet and Nixon, 1994) from the Late Cretaceous (Turonian) Old Crossman locality, New Jersey, USA (Crepet and Nixon, 1998b). Flowers are pedicellate with a distinct receptacular cup that bears numerous floral parts in a spiral arrangement. Imbricate bracts cover the outer surface of the floral cup. Broad ovate tepals are borne on the rim. The androecium is borne on the upper part of the inner surface of the floral cup, and consists of numerous, flattened, linear–lanceolate stamens. Stamens are strongly incurved and each has a long apical extension of the connective. There are four pollen sacs in two pairs arranged parallel to each other on the adaxial surface of each stamen. Pollen grains found *in situ* are spherical and reticulate. In the published illustrations pollen appears to have more than one colpus, but details of aperture configuration or wall structure were not reported. In the character matrix developed to assess phylogenetic relationships of the fossil material, pollen was scored as monocolpate or monocolpate-derived. The gynoecium consists of about five, free, apparently plicate carpels borne at the base of the floral cup. Individual carpels have a sessile bilobed stigma and several ovules borne in two rows along the ventral margin. The carpels are surrounded by numerous dorsiventrally flattened structures interpreted as pistillodes (Crepet and Nixon, 1998b).

Detrusandra mystagoga shares floral features with several different extant eumagnoliids that have cup-shaped flowers. There are particular similarities with certain Magnoliales and Laurales (Calycanthaceae). However, the character combination is unlike that in the flowers of any extant taxon and phylogenetic analyses were inconclusive (Crepet and Nixon, 1998b). *Detrusandra* is treated here as unassigned with respect to extant families and orders of eumagnoliids, although it is most likely a member of the eumagnoliid clade.

Cronquistiflora sayrevillensis (Figure 10.3) is another probable eumagnoliid flower from the Late Cretaceous (Turonian) Old Crossman locality, New Jersey, USA (Taxon 'B' of Crepet and Nixon, 1994, 1998b). Flowers have a slightly concave floral cup, which is covered on the outer surface with spirally arranged and imbricate bracts. Stamen morphology and organisation of the androecium are unknown. Isolated pollen grains are present on the

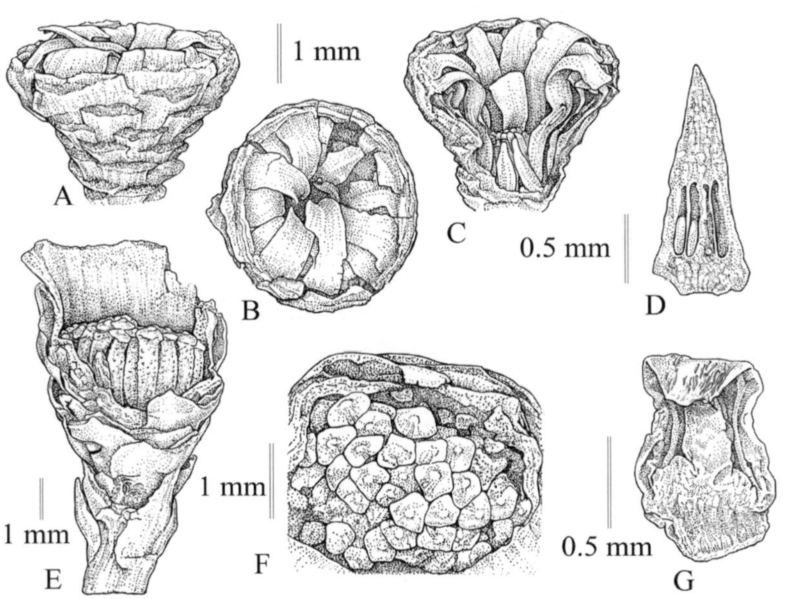

Figure 10.3 Fossil flowers of uncertain relationship among eumagnoliids from the Late Cretaceous (Turonian) Old Crossman locality, New Jersey, USA. (A–D) *Detrusandra mystagoga*; (A) lateral and (B) apical views showing scars of imbricate bracts on the outer surface of the floral cup and inwardly arching tepals borne on the rim; (C) broken flower showing inner part of floral cup with stamens and inwardly arching tepals; (D) isolated stamen showing two pairs of pollen sacs embedded in the adaxial surface and long triangular extension of the connective. (E–G) *Cronquistiflora sayrevillensis*; (E) lateral view of broken flower showing imbricate bracts on the outer surface and carpels borne at the base of the floral cup; (F) apical view showing tightly packed heads of carpels; (G) isolated stamen showing widely separated pairs of pollen sacs. Drawn from SEM images in Crepet and Nixon (1998b).

surface of one flower. From the illustrations the pollen appears to have two colpi, although this is not completely clear. In the character matrix developed to assess the phylogenetic relationships of the fossil material, pollen was scored as monocolpate or monocolpate-derived. Similar pollen grains occur *in situ* in isolated laminar, tetrasporangiate stamens that co-occur with the flowers. The gynoecium consists of numerous free carpels borne in a spiral arrangement at the base of the floral cup. Individual carpels are more or less linear and expanded distally into a distinct peltate stigma. They appear plicate, but were reported as ascidiate. The carpel wall shows densely spaced spherical protrusions that probably represent ethereal oil cells. Each carpel contains many ovules arranged in two rows along the ventral suture. Ovules appear orthotropous and bitegmic, with the micropyle formed from the inner integument, although the nature of the ovules is not clear from the illustrations and description (Crepet and Nixon, 1998b).

Cronquistiflora shares features with flowers of several different extant eumagnoliids. In particular, the flowers closely resemble those of *Eupomatia* (Eupomatiaceae, Magnoliales), which have a similar cup-like receptacle and numerous spirally arranged carpels, each containing many ovules. However, the ovules in *Eupomatia* are anatropous. The character combination of the *Cronquistiflora* flowers is unique and the phylogenetic position of the genus among eumagnoliids is uncertain (Crepet and Nixon, 1998b). Like *Detrusandra*, it may represent an extinct lineage close to Magnoliales or Laurales.

Canrightia resinifera (Figure 10.4) is based on flowers, fruits and seeds that occur abundantly in Early Cretaceous (Aptian–Early Albian) floras from Portugal (Friis and Pedersen, 2011). The flowers are small, about 1–2 mm long, and bisexual with a single whorl of tepals. The tepals are fused laterally to form a hypanthium and also fused to the ovary wall. The gynoecium is formed from two to five carpels and the ovary is semi-inferior, syncarpous and unilocular. There are two to five orthotropous and pendent ovules, one ovule for each carpel. The ovules are unusual in being endotestal-endotegmic with a well-developed endothelium. The endotesta consists of cubic cells with lignified infillings that show abundant imprints of crystals. The endotegmen consists of a single layer of cells that in anthetic flowers are developed into a prominent endothelium. The presence of an endothelium is unusual among extant angiosperms at the magnoliid level (only known for extant *Lactoris*) and has never been described before for fossil angiosperms (Friis and Pedersen, 2011).

The ovary wall is characterised by abundant resin bodies or empty cavities. Resin bodies and the cavities are interpreted as remnants of oil cells. The fruit wall is parenchymatic without hard tissue and the fruits are interpreted as berries.

Pollen grains are found in large quantities on the surface of the fruit, particularly in the stigmatic area. They are

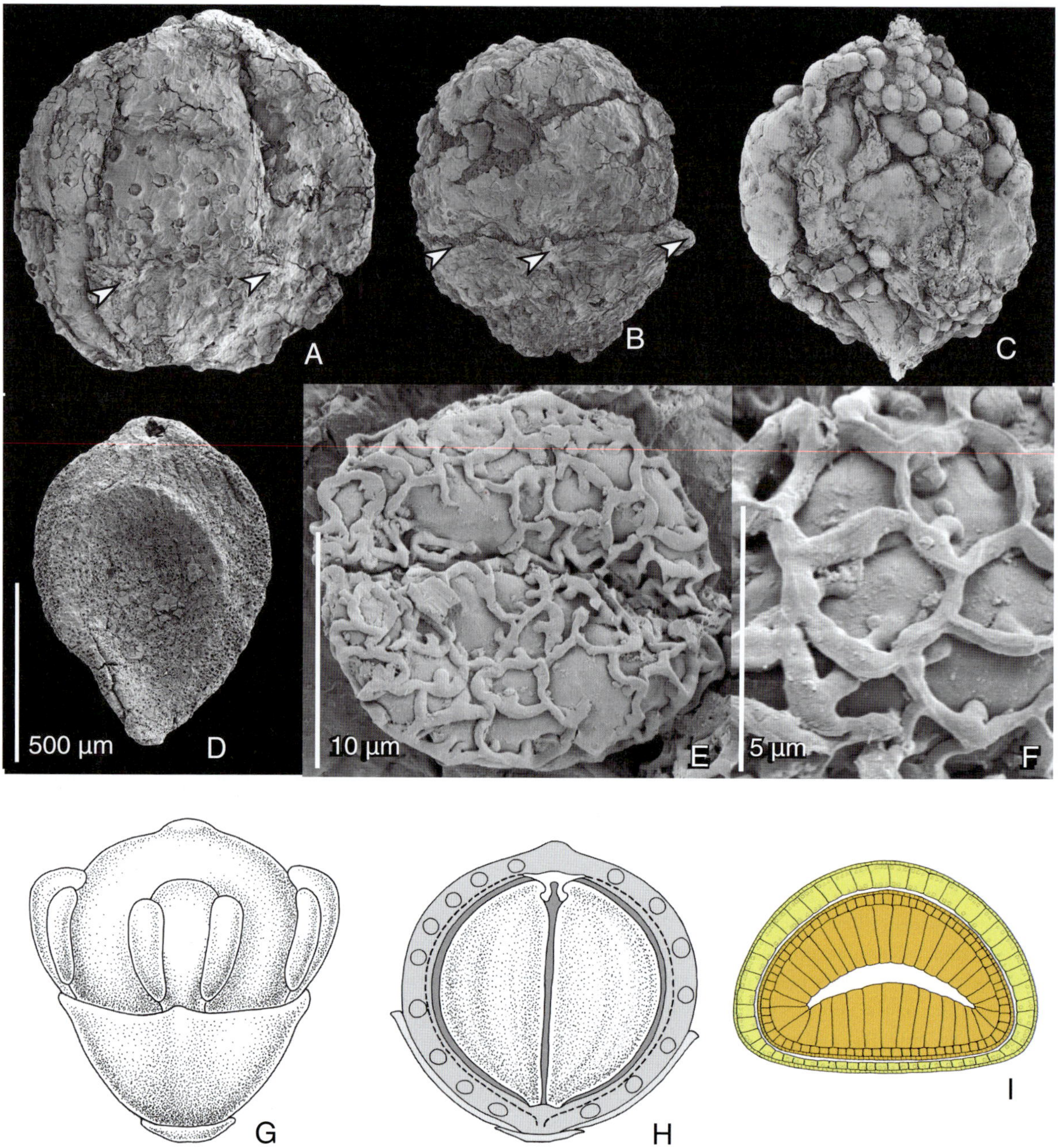

Figure 10.4 SEM images and reconstructions of *Canrightia resinifera* from the Early Cretaceous of Portugal; (A–B) from the Catefica locality (Late Barremian–Aptian), (C–F) from the Famalicão locality (Late Aptian). (A, B) Lateral view of flowers with remains of hypanthium and scars from stamens (arrowheads). C. Flower with abundant resin bodies in the ovary wall. D. Isolated ovule showing heteropyle-like opening at chalaza and surface of endotesta with fine imprints of crystals. E. Monoaperturate, reticulate pollen from fruit surface. F. Detail of pollen showing coarse reticulum and scattered columellae. G. Reconstruction of flower. H. Schematic longitudinal section of flower showing oil-cells in ovary wall and pendent, orthotropous ovules. I. Transection of ovule showing outer integument (light yellow) with crystalliferous endotesta and inner integument (darker yellow) with large cells of endotegmen/endothelium. From Friis and Pedersen (2011).

of the broadly defined *Retimonocolpites*-type and have a long extended colpus, elliptical to circular outline, and are about 16.5–20.5 μm in diameter. The pollen wall is semitectate reticulate with a homobrochate reticulum and columellate infratectal layer. The reticulum is coarse and loosely attached to the foot layer. Columellae supporting the smooth muri are long and widely spaced. Similar pollen has also been observed in laminar stamens that have a strongly expanded apical connective.

The systematic position of *Canrightia* is uncertain. The floral and ovule structure is unique combining features of several magnoliid lineages. Character optimisation indicates that the genus is most closely related to a clade including *Ceratophyllum* and Chloranthaceae, but *Canrightia* also shows close similarity to some members of the eumagnoliid clade that includes Piperales and Canellales, and there are also several features in common with the Magnoliaceae, particularly *Liriodendron*, in seed wall structure (Friis and Pedersen, 2011).

The *Canrightia* fossils are closely similar to the floral/fruiting structures of the extinct genus *Zlatkocarpus* (section 9.2.4), and both have very similar *Retimonocolpites*-type pollen. However, *Canrightia* has clearly a syncarpous ovary, while the ovary of *Zlatkocarpus* appears to be monomerous. We have placed *Canrightia* in this chapter together with the eumagnoliid angiosperms to emphasise the possible link between some chloranthoid angiosperms and the eumagnoliids.

10.2 MAGNOLIALES

The APGIII classification includes six families in the Magnoliales: Annonaceae, Degeneriaceae, Eupomatiaceae, Himantandraceae, Magnoliaceae and Myristicaceae (Figure 10.1). Other families previously placed in Magnoliales, such as the Canellaceae, Winteraceae (e.g. Takhtajan, 1969; Cronquist, 1981) and Lactoridaceae (e.g. Cronquist, 1981) are now placed in the Canellales (Canellaceae, Winteraceae) and Piperales (Lactoridaceae). The Magnoliales, as presently circumscribed, are a group that is well supported by molecular-based phylogenetic studies. There is also good support for the clade from non-molecular characters (Doyle and Endress, 2000; Endress and Doyle, 2009). DNA sequence data resolve Myristicaceae as sister to two subclades: one comprising Annonaceae and Eupomatiaceae, and another in which the Magnoliaceae are sister to Himantandraceae plus Degeneriaceae (Qiu *et al.*, 1999). This topology is also supported by combined molecular and structural data (Doyle and Endress, 2000; Sauquet *et al.*, 2003). All extant Magnoliales are trees or shrubs. Flowers are typically large and bisexual with an apocarpous gynoecium of plicate carpels (intermediate between ascidiate and plicate in Myristicaceae, Doyle and Endress, 2000).

The Magnoliales have a rich fossil record in the Cenozoic. The Cretaceous record is less extensive, but is significant in documenting the presence of the order in the mid- and Late Cretaceous based on reproductive organs. In the Cenozoic, seeds and leaves of Magnoliaceae (e.g. *Liriodendron*, *Magnolia*) and Annonaceae (e.g. *Anonaspermum*) are particularly abundant.

10.2.1 Fossil Magnoliales of uncertain relationships

Several multipartite floral structures from the Cretaceous are probably the reproductive structures of extinct Magnoliales, but cannot be placed securely in any of the extant families. Among these fossils is *Endressinia* from the Early Cretaceous of Brazil.

Endressinia brasiliana (Figure 10.5) is based on a small branching system with attached leaves and flowers from the Lower Cretaceous (Upper Aptian – Lower Albian) Crato Formation, Brazil. *Endressinia* has an unusual preservation. Vegetative parts are mostly preserved as compressions, but the floral parts are partly mineralised and three-dimensionally preserved. Flowers are present as buds as well as later-stage flowers that are more open. Flowers are borne solitarily on long pedicels. Each is about 4–5 mm long and 6–7 mm in diameter with a flattened, slightly concave receptacle that bears a few larger tepals at the rim. In the centre there are numerous free carpels in a spiral arrangement. Between the tepals and the gynoecium there are numerous, narrow, flattened organs borne in a dense spiral. The nature of these organs is not fully understood because they are densely packed and because of poor preservation, particularly to the outside. Anther thecae or pollen were not observed and it is uncertain whether these structures are part of the perianth or androecium. However, the organs facing the gynoecium are better preserved and show distinct stalked and multicellular glandular structures along their margins. These organs are interpreted as staminodes similar to the staminodes of extant Himantandraceae and Eupomatiaceae. There are up to about 20 narrowly ovate and plicate carpels (Mohr and Bernardes-de-Oliveira, 2004).

Based on the multipartite flowers, the glandular staminodes and the plicate carpels, *Endressiana* is attributed

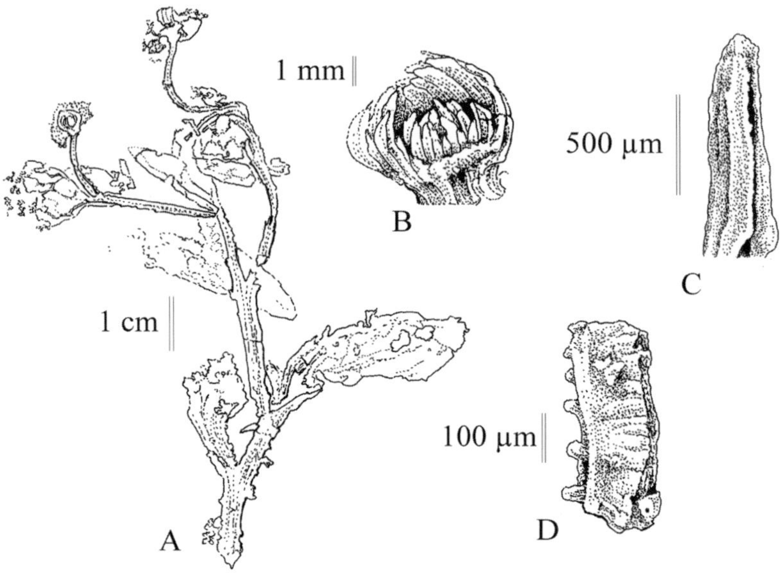

Figure 10.5 *Endressinia brasiliana* from the Early Cretaceous (Late Aptian – Early Albian) Crato flora, Brazil. (A) Branching system showing attached leaves and solitary flowers. (B) Detail of flower showing numerous tepals and a mass of organs in the centre comprising staminodes and carpels. (C) Carpel isolated from flower. (D) Fragment of staminode showing glandular margin. From Friis *et al.* (2006a).

to Magnoliales. There are particular similarities to extant Eupomatiaceae and Himantandraceae, but with the current information *Endressiana* cannot be placed confidently in either the Annonaceae–Eupomatiaceae group or the Magnoliaceae–Himantandraceae–Degeneriaceae group.

10.2.2 Myristicaceae

The Myristicaceae are a pantropical family of trees and shrubs that occur in lowland rain forests. The family includes about 400 species in about 20 genera. Flowers are usually small, unisexual, and mostly with three perianth parts in a single whorl. Staminate flowers have few to many stamens that are usually fused into a columnar synandrium. Most of the floral diversity in the family is in the androecium (Sauquet, 2003). Pistillate flowers have a single carpel that is intermediate between ascidiate and plicate. Fruits are one-seeded berries. Seeds have a distinct aril and extensive ruminate endosperm (Heywood *et al.*, 2007).

The fossil record of Myristicaceae is meagre and currently restricted to the Cenozoic. A report of putative myristicaceous wood from the Paleocene of the Sahara (Boureau, 1950) requires re-evaluation, but recognition of a well-preserved ruminate seed from the Early Eocene London Clay, southern England (Doyle *et al.*, 2008b) confirms the presence of the myristicaceous lineage by this time, even though it is unclear whether this fossil falls within the stem group or the crown group of the family. The Myristicaceae are also known based on fossil wood from the ?Miocene of Chile (Schöning and Bandel, 2004) as well as ruminate seeds from the Middle Miocene of Germany (Gregor, 1977). Several species of dispersed pollen may also indicate the presence of Myristicaceae with the oldest pollen record from the Late Eocene of Africa (Muller, 1981). The Late Cretaceous flowers of *Mabelia connatifila*, *M. archaia* and *Nuhliantha nyanzaiana* described from the Late Cretaceous (Turonian) Old Crossman locality, New Jersey, USA, as members of the monocot family Triuridaceae (Gandolfo *et al.*, 2002) have some distinctive features that resemble those of extant Myristicaceae. These include the leathery perianth and stamens that are fused into a synandrium. *Mabelia* and *Nuhliantha* are currently known only from staminate flowers. More material is needed to establish the systematic position of these fossils with certainty (Chapter 9).

10.2.3 Annonaceae

The Annonaceae are a large and diverse family with about 2300 species and 128 genera. They include shrubs, trees and climbers that are mainly distributed in tropical regions of the world. Two genera extend into temperate regions. Flowers are typically bisexual. The perianth is trimerous, or rarely tetramerous, with two whorls of differentiated perianth parts. The androecium usually has many stamens

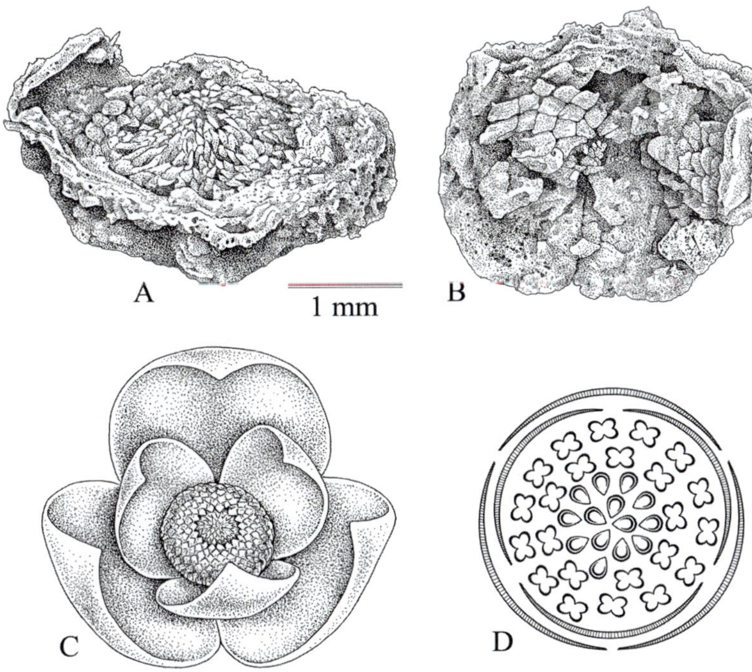

Figure 10.6 *Futabanthus asamigawaensis*, a flower related to Annonaceae from the Late Cretaceous (Early Coniacian) Kamikitaba locality, northeastern Honshu, Japan. (A, B) Lateral and apical views of flower showing numerous stamens surrounding many smaller, free, carpels in the centre. (C) Reconstruction of flower. (D) Floral diagram. (A–C) drawn from SEM images (A, B) and reconstruction in Takahashi *et al.* (2008b).

and staminodes in an irregular arrangement (Kessler, 1993b). Pollen grains are variable in size, shape, aperture configuration and wall structure. They are shed in monads, or rarely in tetrads, and are monocolpate to inaperturate with a granular or columellar infratectal layer (Doyle and Le Thomas, 1994, 1996). The gynoecium is usually apocarpous with many free, plicate carpels, but syncarpous, paracarpous or unicarpellate gynoecia also occur in the family. Fruits are fleshy or sometimes woody, with each fruitlet containing one or several seeds (Kessler, 1993b). Seeds are often arillate with a fibrous seed coat and pronounced rumination of the endosperm, which typically forms densely spaced transverse ridges on the endosperm surface.

The distinctive ruminate seeds of Annonaceae occur extensively in Early Cenozoic floras from Europe and have also been reported from the Early Cenozoic of Egypt and Pakistan (Reid and Chandler, 1933; Chandler, 1954, 1961; Tiffney and McClammer, 1988). They are typically described as extinct species of modern genera, such as *Alphonsea*, *Annona*, *Asimina* and *Uvaria*, or are assigned to genera such as *Anonaspermum* (Reid and Chandler, 1933; Chandler, 1961; Mai and Walther, 1985), which recognises the relationship to the family, but does not require assignment to an extant genus. The diversity of annonaceous seeds in Early Cenozoic fruit and seed floras, for example the Early Eocene flora from the London Clay, southern England, clearly establishes that the family had already undergone significant differentiation by this time.

Records of Annonaceae from the Cretaceous are sparse, but ruminate seeds from the Maastrichtian of Nigeria and Senegal (Chester, 1955; Monteillet and Lappartient, 1981) and dispersed pollen of *Foveomorphomonocolpites humbertoides* from the Maastrichtian and Paleocene of Colombia (compared to pollen of extant *Malmea*) (Sole de Porta, 1971; Muller, 1981), provide secure evidence of the group in the Late Cretaceous. Also important is the informative floral structure *Futabanthus asamigawaensis* from the Late Cretaceous of Japan (Takahashi *et al.*, 2008b).

Futabanthus asamigawaensis (Figure 10.6) is based on a single flower from the Late Cretaceous (Early Coniacian) Kamikitaba locality, northeastern Honshu, Japan (Takahashi *et al.*, 2008b). The flower is small, bisexual, actinomorphic, and borne on a short stalk. Perianth parts are free and borne in at least two cycles at the rim of a flattened and disc-shaped floral receptacle. The androecium is conspicuous and multipartite with about 90–100 stamens. Stamens are stout, have no clear differentiation into anther and filament, and are curved toward the centre of the flower. Anthers extend for almost the full length of the stamens. Each stamen has a prominent, flattened apical extension of the connective. Pollen is unknown. The gynoecium is superior, apocarpous and formed from 100–120 carpels borne in the centre of the flower on a slightly raised conical projection.

The organisation and structure of *Futabanthus* clearly indicates a relationship with extant Annonaceae. Based on stamen morphology and features of the androecium it seems likely that *Futabanthus asamigawaensis* falls within the crown group of Annonaceae (Takahashi *et al.*, 2008b).

10.2.4 Eupomatiaceae

The Eupomatiaceae are a small family comprising a single genus, *Eupomatia*, with three species. All are shrubs or small trees in the tropical rain forests of New Guinea and eastern Australia. Flowers are large, showy, bisexual and multipartite, with spirally arranged parts and an inferior ovary. Perianth parts are lacking, but two fused bracts protect the floral parts in bud. Pollinators are attracted by the modified androecium, which consists of numerous stamens to the outside and several series of petaloid staminodes towards the centre. Stamens have a short, broad filament and are tetrasporangiate. Stamens and staminodes are united at the base. Pollen grains are zonacolpate. The gynoecium consists of numerous plicate carpels that are fused at the base and completely embedded in the receptacle. Each carpel contains two to several anatropous ovules (Endress, 1993e; Heywood *et al.*, 2007).

A small dicolpate (zonacolpate) pollen grain from the Maastrichtian of California described as 'Pollen Forma E' (Chmura, 1973) currently provides the only potential fossil occurrence of Eupomatiaceae. However, the specimen was studied using LM only and the relationship of the grain to Eupomatiaceae is not fully established.

The fossil plant *Endressinia brasiliana* has some features that suggest a close relationship to extant Eupomatiaceae, but the gynoecium is superior and carpels are free (section 10.2.1). Flowers of *Cronquistiflora* and *Detrusandra* also share some features with Eupomatiaceae (section 10.1.1).

10.2.5 Magnoliaceae

The Magnoliaceae comprise about 220 species of trees or shrubs with alternate leaves. The species have been divided traditionally into about seven to ten genera. They are distributed mainly in temperate to tropical regions of Southeast and East Asia and North America, but with some taxa also in the West Indies and Central and South America. The family is subdivided into two subfamilies: Liriodendroideae, including only *Liriodendron*, and Magnolioideae, containing all other genera (sometimes merged into a broadly defined concept of *Magnolia*, Kim *et al.*, 2001). Leaves of *Liriodendron* are distinctively bilobed (Figure 10.7). Flowers are bisexual, or more rarely unisexual, with numerous floral parts that are helically arranged. The perianth parts are undifferentiated and frequently appear to be in whorls. Stamens are tetrasporangiate. Pollen grains are monocolpate and tectate with a granular infratectal layer. The gynoecium consists of many free plicate carpels, each containing two to several anatropous ovules. Fruits are apocarpous and fruitlets are woody. In Magnolioideae fruitlets are dehiscent and open in several different ways. Fruitlets of *Liriodendron* are indehiscent and winged. Seeds are endotestal with a sclerified endotesta and a distinct heteropyle (Tiffney, 1977a; Nooteboom, 1993; Heywood *et al.*, 2007).

The fossil record of Magnoliaceae is extensive. The earliest fossil remains are from the mid-Cretaceous, but fossil Magnoliaceae are particularly common in Cenozoic floras from Europe, Asia and North America. The family is especially well known from the distinctive leaves of *Liriodendron*, as well as wood and dispersed seeds. Dispersed pollen of Magnoliaceae is not easy to recognise and is rarely recorded (Muller, 1981). Several groups of species in the family have distributions that are disjunct between eastern North America and eastern Asia and evidently once had a more continuous range. *Liriodendron*, for example, which now has one species in Eastern North America and one species in China, was widespread during the Cenozoic ranging from eastern and Central Asia to Europe and North America. Cenozoic records of Magnoliaceae include fossils assigned to the extant genera *Liriodendron*, *Magnolia*, *Talauma* and *Manglietia*, as well as fossil genera such as *Liriodendroxylon*, *Magnoliaceoxylon* and *Magnolioxylon* for fossil wood (Wheeler *et al.*, 1977; Scott and Wheeler, 1982; Cevallos-Ferriz and Stockey, 1990), and *Magnoliaespermum* for fossil seeds (Reid and Chandler, 1933; Chandler, 1964; Dorofeev, 1970, 1974b; Tiffney, 1977a; Dorofeev, 1983; Friis, 1985b; Dorofeev, 1988; Manchester, 1994; Mai, 1995).

Except for fossil seeds assigned to *Manglietia* and *Liriodendron* from the latest Cretaceous (Maastrichtian) of Germany (Knobloch and Mai, 1986), all Cretaceous taxa that have been studied in detail appear to represent extinct genera within the family.

Figure 10.7 Leaves and flowers of extant *Liriodendron tulipifera*.

The Late Cretaceous record of Magnoliaceae is particularly rich in fossils that may be closely related to *Liriodendron* rather than to extant Magnolioideae. For example, leaves with distinctive *Liriodendron*-type morphology assigned to genera such as *Liriodendrites*, *Liriodendropsis* and *Liriophyllum* are reported in many Late Cretaceous floras from North American and Europe (Figure 10.8). Other probable liriodendroid fossils include fruits of *Archaeanthus* and associated floral parts and leaves, seeds of *Liriodendroidea*, and seeds and fruits of *Padragkutia*. Fruits and seeds of *Litocarpon* from the Santonian–Campanian of Canada also show a mosaic of fruit and seed characters that may indicate a relationship to liriodendroid Magnoliaceae. In contrast, the relationship to Magnoliaceae of *Magnoliaestrobus gilmouri*, from the Late Cretaceous Kardlok locality of western Greenland, is very uncertain. The fossil is an elongated reproductive axis, about 9 cm long, with crowded, apparently helically arranged, lateral units. Seward and Conway (1935) suggested that these units correspond to fruitlets of extant *Magnolia*. However, the preservation is poor and it is not possible to determine whether this specimen is a single floral structure, an inflorescence, or perhaps the cone of some non-angiosperm seed plant.

Archaeanthus linnenbergeri (Figure 10.8) is the most informative of several North America multicarpellate structures that are probably related to extant Magnoliaceae. *Archaeanthus* was described from the Dakota Formation (latest Albian – earliest Cenomanian) of central Kansas, USA, and includes fossils that are all preserved at a post-anthetic stage (Dilcher and Crane, 1984a). The most complete specimens are elongated receptacles, up to about 130 mm long, with attached follicular fruitlets. The gynoecial region extends for most of the length of the receptacle and consists of 100–130 loosely spaced follicles borne in a helical arrangement. Helically arranged scars of different sizes below the gynoecial region indicate the presence of a multistaminate androecium and a perianth consisting of six to nine inner tepals and three outer tepals. A short distance below the base of the receptacle is a prominent scar indicating the attachment of a stipular bud scale. The follicles are stalked, elliptical in outline, up to about 38 mm long, with a distinct adaxial suture. About 100 ovules may have been present in

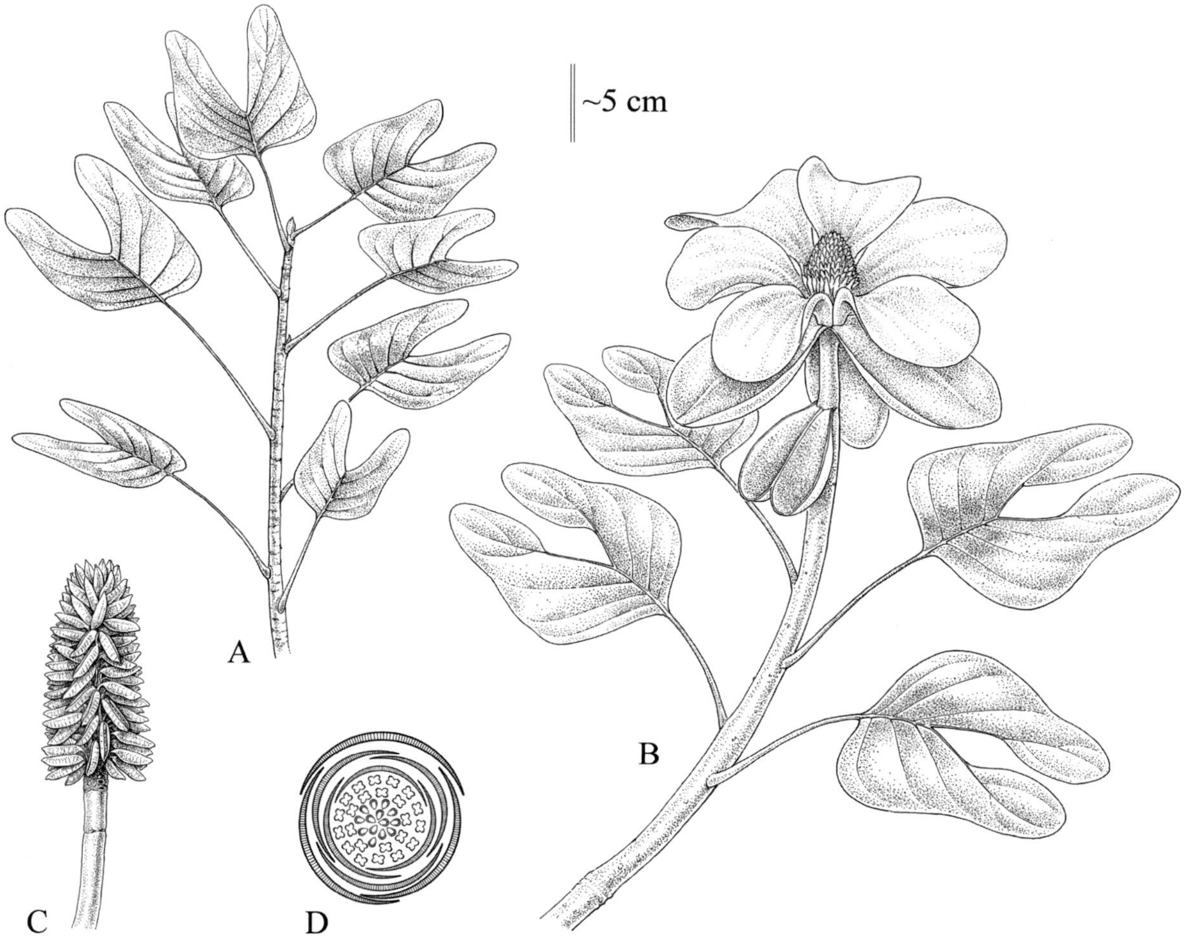

Figure 10.8 Cretaceous leaves and reproductive structures related to Magnoliaceae. (A) Reconstruction of a shoot of the *Liriodendrites bradacii* plant from the Late Cretaceous (Late Maastrichtian) Hell Creek Formation, North Dakota, USA. (B–D) Reconstructions (B, C) and floral diagram (D) of *Archaeanthus linnenbergeri* from the Dakota Formation (latest Albian – earliest Cenomanian) of central Kansas, USA; (B) shoot with alternate leaves and terminal flower showing bilobed stipular bud scale, inner and outer tepals, stamens and carpels; (C) elongated fruiting axis showing numerous follicles, scars of tepals and stamens, and scar of stipular bud scale. (A) Redrawn from Johnson (1996); (B, C) redrawn from Dilcher and Crane (1984a).

immature follicles. Mature follicles contain 10–18 seeds that were shed through the opening of the adaxial suture. The organisation of the seeds and their wall structure is unknown. Only the cuticles of the integuments are preserved.

Archaeanthus linnenbergeri fruiting structures were recovered in association with dispersed leaves, tepal-like structures, and stipular bud scales thought to have been produced by the same plant. These different organs all contain distinctive resin bodies that are probably the remains of ethereal oil cells (Dilcher and Crane, 1984a). Associated fossil leaves (*Liriophyllum kansense*) have an unusual, deeply bilobed morphology that is rare among modern angiosperms. The leaf morphology and venation is similar, but not identical, to that of extant *Liriodendron*. In *Liriophyllum*, unlike in extant *Liriodendron*, the midrib branches into two strong veins at the base of a much deeper sinus. These two strong veins form the inner margin of the sinus in its lower part.

Tepal-like structures associated with *Archaeanthus linnenbergeri* include smaller elliptical forms (*Archaepetala beekeri*) thought to be inner tepals, and larger obovate forms with a clear basal attachment scar (*Archaepetala obscura*) thought to be the outer tepals. The stipular bud scales (*Kalymmanthus walkeri*) are bilobed as in extant *Liriodendron* and some Magnoliaceae. The structure of *Archaeanthus* indicates a close relationship with extant Magnoliaceae. In particular, the arrangement of the perianth and androecium, as inferred from the scars and associated organs, is closely similar to that of certain extant species of *Magnolia*. There are also strong similarities with *Liriodendron*, including the unusual form of the leaves. The occurrence of probable ethereal oil cells also supports a eumagnoliid affinity for *Archaeanthus*. However, *Archaeanthus* differs from all extant Magnoliaceae in the numerous ovules and seeds per carpel (Dilcher and Crane, 1984a; Igersheim and Endress, 1997). Abscission of the fruitlets at maturity is known in extant Magnoliaceae only for *Liriodendron*, which has indehiscent nutlets. It seems likely that *Archaeanthus* falls within the stem group of the extant family or the stem group of extant *Liriodendron*.

Fossils leaves and tepal-like structures similar to those associated with *Archaeanthus* have also been reported from other localities in North America. These include a large tepal-like fossil from the Dakota Formation assigned to *Magnolia* (Hollick, 1903) and unequivocal smaller tepals with distinctive resin bodies, which had previously been misinterpreted as syconia of *Ficus* (*'Ficus' neurocarpa*, Hollick, 1903). There are also leaves of another species of *Liriophyllum*, *L. populoides*, from the Early Cretaceous (Late Albian) Kassler Sandstone of the South Platte Formation, Colorado, USA (Dilcher and Crane, 1984a).

Liriodendroidea (Figure 10.9) was established by Knobloch and Mai (1984) for dispersed seeds closely resembling those of extant *Liriodendron*, but differing in details of seed structure. Originally two species from the Late Cretaceous of Central Europe were assigned to the genus: the type species *Liriodendroidea germanica* from the Maastrichtian of Eisleben, Germany, and *L. protogea* from the Santonian of Quedlinburg, Germany (Knobloch and Mai, 1984, 1986). Subsequently four species of *Liriodendroidea* (*L. alata*, *L. asiatica*, *L. costata*, *L. tenuitesta*) were described from the Late Cretaceous (Cenomanian–Turonian) of Kazakhstan, along with two further species (*L. carolinensis*, *L. latirapha*) from the Late Cretaceous (Campanian) Neuse River locality, North Carolina, USA (Frumin and Friis, 1996, 1999).

The *Liriodendroidea* material from Asia and North America is abundant and well preserved. It has yielded detailed information on the organisation and structure of the seeds. The fossils are similar to seeds of *Liriodendron* in their anatropous organisation, the presence of a distinct heteropyle, and the structure of the endotestal seed coat. Cells of the endotesta are distinct, palisade-shaped sclerenchyma containing fibrous lignifications and cubic crystals. However, the fossils differ from seeds of extant *Liriodendron* in having a distinct lateral wing surrounding the seed body. The wing is clearly visible only in well-preserved specimens, and although it was not observed in the original description of the German material it is present in the type material (Frumin and Friis, 1999). *Liriodendroidea* is further distinguished from *Liriodendron* in having several seeds per carpel.

Padragkutia was established for small fruits and seeds from the Late Cretaceous of Central Europe that are closely similar to those of modern *Liriodendron* and extinct *Liriodendroidea* (Knobloch and Mai, 1984), but that differ in having a marked hook-shaped extension of the endotesta around the heteropyle. *Padragkutia* also has up to 17 seeds in each fruitlet, in contrast to one or two in extant *Liriodendron*. The genus includes two species, *Padragkutia haasii* from the Late Turonian–Santonian Klikov-sequence of the Czech Republic, and the Late Santonian – Early Campanian of Balkony-Waldes, Hungary (Knobloch and Mai, 1984), and *Padragkutia edelenyi*, known only from the Balkony-Waldes locality (Knobloch and Mai, 1986). The seeds are closely similar to those of *Liriodendroidea*, but lack the diagnostic wing.

Litocarpon beardii (Figure 10.10) is a multicarpellate fruit of probable magnoliaceous affinity from the Late Cretaceous (Late Santonian – Early Campanian) Haslam Formation, Vancouver Island, British Columbia, Canada (Delevoryas and Mickle, 1995). The species is based on a single, incomplete, petrified specimen with many details of organisation, as well as fruit and seed structure, preserved. The specimen has 19 fruitlets borne in a helical arrangement along a slender receptacle. The base is lacking and there is no information on other floral parts. Fruitlets are about 4–5 cm long and up to 1.3 cm wide with dorsal dehiscence. The fruit wall is expanded ventrally and laterally into a distinct wing that extends distally beyond the locule. The fruit is preserved at fruiting stage and seeds have been shed from several of the open fruitlets. Fruitlets with seeds still *in situ* have up to five seeds. Seeds are anatropous, borne along the ventral suture, and winged.

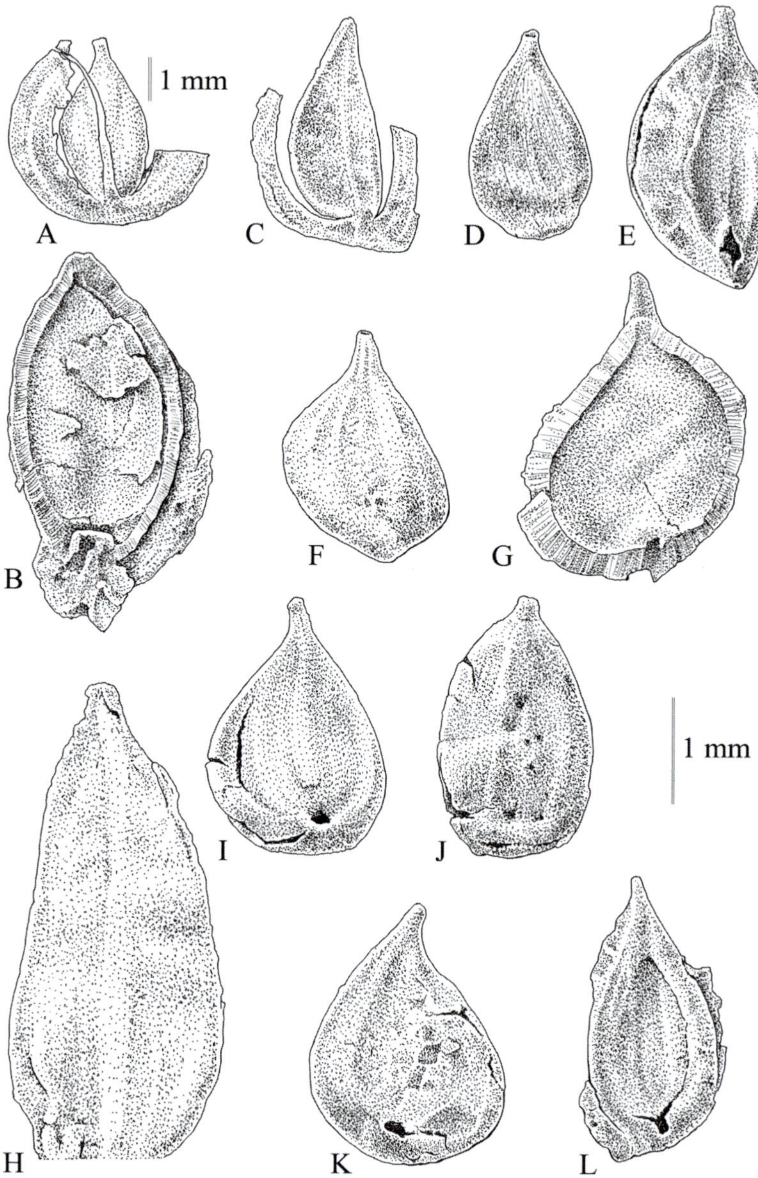

Figure 10.9 Seeds of *Liriodendroidea* from the Late Cretaceous (Cenomanian–Turonian) Sarbay locality, Kazakhstan (A–G) and from the Late Cretaceous (Early Campanian) Neuse River locality, North Carolina, USA (H–L). (A, B) *Liriodendroidea alata* (A) in ventral view showing remains of the wing, and (B) broken specimen showing palisade cells of the fractured seed. (C, D) *Liriodendroidea tenuitesta* (C) in ventral view showing remains of wing, and (D) dorsal view with wing missing. (E) *Liriodendroidea costata* in ventral view showing distinct heteropyle. (F, G) *Liriodendroidea asiatica* (F) in ventral view and (G) broken specimen showing inside of seed. (H–J) *Liriodendroidea carolinensis*; (H) large specimen in ventral view; (I) smaller specimens in ventral view showing heteropyle; (J) dorsal view. (K, L) *Liriodendroidea latirapha*, seeds in ventral view showing heteropyle. (A, H–L) Drawn from SEM images in Frumin and Friis (1996); (B–G) drawn from SEM images in Frumin and Friis (1999).

They are up to 8.5 mm long and 8.5 mm wide. The cells of the seed wall are not fully preserved, but there is an outer layer of parenchyma cells followed internally by a sclerenchymatic layer. The inner cells of the parenchyma layer form an irregular pattern resembling the 'fingerprint pattern' of modern Magnoliaceae. The cells of the sclerenchyma layer are indistinct.

Litocarpon is closely similar to extant Magnoliaceae in the size and arrangement of the fruitlets. Fruitlets of extant *Liriodendron* have a similar ventral and lateral wing (Delevoryas and Mickle, 1995). However, in the modern genus the fruitlets are indehiscent nuts enclosing one or two wingless seeds. The seed coat also appears different from that of extant *Liriodendron*, as well as the extinct genus *Liriodendroidea*, but the preservation of the seeds precludes detailed comparison. *Liriodendroidea* does, however, have winged seeds as in *Litocarpon*, and it is probable that *Litocarpon* is also an early *Liriodendron*-like plant.

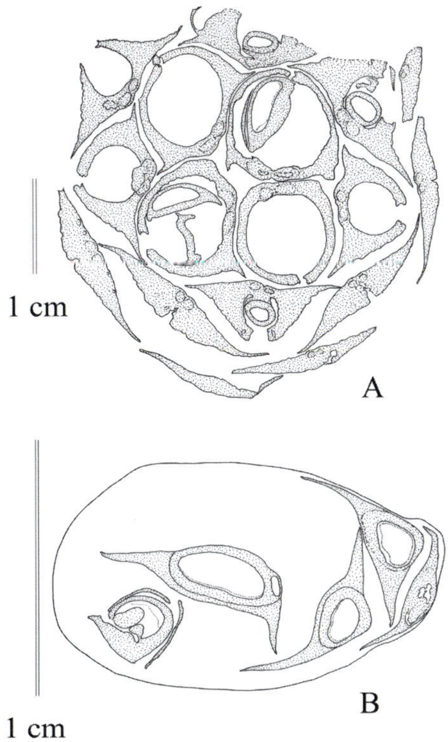

Figure 10.10 *Litocarpon beardii*, a petrified multicarpellate fruit of probable magnoliaceous affinity from the Late Cretaceous (Late Santonian – Early Campanian) Haslam Formation of Vancouver Island, British Columbia, Canada. (A) Transverse section through group of follicles showing both follicles and seeds with lateral wings. (B) Transverse section near the base of a single follicle showing shape of locule and several winged seeds. Drawn from photographs in Delevoryas and Mickle (1995).

10.2.6 Himantandraceae

The Himantandraceae are a small family with one genus, *Galbulimia*, and two species in the tropical rain forests of northwestern Australia and the Papuan region. Flowers are large, showy, multipartite and bisexual with spirally arranged floral parts. Perianth parts are absent, but the floral parts are protected in bud by two closed bracts. Stamens are tetrasporangiate with a strongly expanded apical sterile extension. Anther dehiscence is valvate by two laterally hinged valves. Pollen grains are monocolpate and atectate. The apocarpous gynoecium consists of 7–28 free, plicate carpels, each containing one or two anatropous ovules (Endress, 1993d).

Currently no fossils have been assigned to the Himantandraceae, but several Cretaceous flowers (*Cronquistiflora*, *Detrusandra*, *Endressinia*) show some characters seen in the family (sections 10.1.1, 10.2.1).

10.2.7 Degeneriaceae

The Degeneriaceae include a single genus, *Degeneria*, with two species of large tree endemic to Fiji. Flowers are large, multipartite and bisexual. The perianth is differentiated into an outer whorl of three sepals and several whorls of petals. The androecium consists of many laminar stamens. Pollen grains are monocolpate with a continuous tectum. The gynoecium consists of one, or rarely two, plicate carpels with numerous anatropous ovules (Cronquist, 1981).

Degeneria has long been considered an archetypical 'primitive angiosperm' because of the laminar stamens and carpels that were thought to be unsealed. However, it is now known that the carpels are postgenitally fused, similar to the condition in other Magnoliales (Endress and Igersheim, 2000a). No fossils of Degeneriaceae have so far been recorded.

10.3 LAURALES

In the APGIII classification (2009) the Laurales include the families Atherospermataceae, Calycanthaceae, Gomortegaceae, Hernandiaceae, Lauraceae, Monimiaceae and Siparunaceae. A similar circumscription of the order was suggested by Renner (1999) based on evidence from both molecular and morphological data. The Calycanthaceae are sister to the remaining six families, which are grouped in two clades: Hernandiaceae (Lauraceae–Monimiaceae) and Siparunaceae (Atherospermataceae–Gomortegaceae) (Renner, 1999; Renner and Chanderbali, 2000). There are few defining morphological characters for the order as a whole, but epigynous flowers with a fleshy receptacle, in which the gynoecium is more or less deeply embedded, are a key potential synapomorphy (Renner, 1999). Families assigned previously to Laurales, such as Amborellaceae, Trimeniaceae and Chloranthaceae, are now recognised to be distinct from this group. They are placed instead among those lineages that diverged at an earlier stage from the main line of angiosperm evolution (Chapter 8).

The fossil record of Laurales is extensive. However, many of the fossils assigned to the order are leaves or wood, often with too few diagnostic features for full systematic evaluation. Fossil flowers and dispersed reproductive organs, which generally provide more systematically informative characters, are less common, although studies of the Cretaceous mesofossil floras have considerably expanded the record of early Laurales. Reports of dispersed lauralean pollen grains are rare. This is to be expected given the near-absence of sporopollenin from the pollen wall of extant taxa and therefore the inherently low fossilisation potential of these grains.

10.3.1 Fossil Laurales of uncertain relationships

The oldest floral remains that can be assigned with confidence to Laurales are from the Early Cretaceous (Early–Middle Albian) Puddledock flora, Virginia, USA (Friis *et al.*, 1994; Crane *et al.*, 1994; von Balthazar *et al.*, 2007). Two taxa under formal description (von Balthazar *et al.*, 2011) are mentioned here as Laurales A and Laurales B.

Laurales A (Figure 10.11) is based on floral fragments with a distinct trimerous organisation as well as other features indicating a relationship to Laurales (von Balthazar *et al.*, 2011). However, the available material is too fragmentary for assignment to a particular family. The perianth comprises six tepals in two whorls. The outer whorl consists of three broad, five-veined, tepals. The inner whorl consists of three tepals that are narrower and three-veined. The androecium apparently consists of three stamens, each with an associated pair of staminal appendages that may have been nectariferous. Only the basal parts of the filaments are preserved and anthers have not been observed. The ovary is also not preserved in any of the specimens.

Laurales B, another lauralean flower from Puddledock, also shows clear similarities to members of Lauraceae (von Balthazar *et al.*, 2011). The flowers are small, bisexual and actinomorphic with a trimerous organisation. There are six free tepals in two alternating whorls and six stamens in two whorls. Staminal appendages are apparently associated with inner whorls of staminodes. The anthers are apparently bisporangiate and open by apically hinged valves. The gynoecium is formed from a single carpel and the ovary is superior with a single pendent ovule. In addition to these two floral structures there are several other lauralean structures from the Puddledock flora that are currently undescribed, as well as *Potomacanthus*, attributed to extant Lauraceae (section 10.3.6).

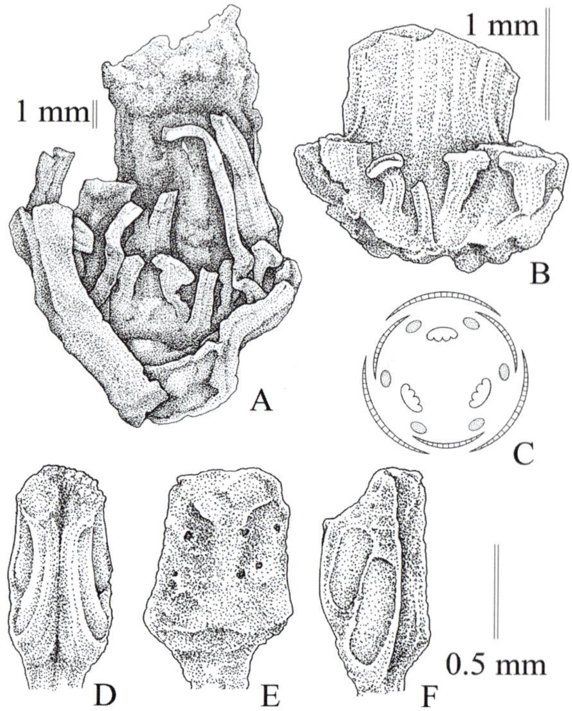

Figure 10.11 Early Cretaceous fossils related to extant Laurales from the Early Cretaceous (Early–Middle Albian) Puddledock locality, Virginia, USA. (A–C) *Laurales A*, (A, B) flower fragments showing stamen filaments and staminodal appendages and (C) floral diagram. (D–F) Dispersed tetrasporangiate stamen in different views showing an elongated filament that grades into a basifixed anther with apically hinged valves. Redrawn from Crane *et al.* (1994).

A further early record of Laurales, also from the Puddledock flora, is a small, dispersed stamen consisting of an elongated filament, which grades above into a basifixed anther (Figure 10.11). The anther is dithecate and tetrasporangiate with the pollen sacs arranged in two pairs. The median pollen sac of each pair is slightly shorter than the lateral pollen sac, and it is also slightly offset to a more distal position. Dehiscence is by four separate, apically hinged valves, one for each pollen sac. Anthers with apically hinged valves are characteristic of several lauralean families, and occur also in the eudicot order Ranunculales (Berberidaceae). However, tetrasporangiate and four-valved anthers are only known for Laurales, and within the order anthers of this type are restricted to Lauraceae (Eklund, 1999). It is likely that this fossil is referable to extant Lauraceae, but because other details of the floral organisation are not known its relationships, at the family level, cannot be established with certainty.

Laurales were also established at an early stage in the Southern Hemisphere, where the group is represented by permineralised flowers of *Lovellea wintonensis* (Dettmann *et al.*, 2009) from the mid-Cretaceous (Late Albian – Early Cenomanian) of Western Australia. *Lovellea* is based on several specimens with well-preserved cellular structure. The flowers are about 11 15 mm long and 12–16 mm wide with a cup-shaped hypanthium. Perianth and androecium are multiparted and borne on the rim of the hypanthium. There are about 16 triangular tepals and 16 stamens. Anthers are basifixed and dithecate with introrse dehiscence. Embedded in the floral cup are about 40 carpels that are fused to the hypanthium wall, but free above. Each carpel contains a single ovule. The fossils were compared to members of the Laurales and in the phylogenetic analyses of Dettmann *et al.* (2009) were resolved as most closely related to *Gomortega* (Gomortegaceae). *Lovellea* also shares several features with Monimiaceae (Dettmann *et al.*, 2009) and its precise position among Laurales remains to be established.

10.3.2 Atherospermataceae

The Atherospermataceae are a small family of shrubs and trees with about 16 species in seven genera. The family is restricted to temperate regions of the Southern Hemisphere (Renner *et al.*, 2000). Flowers are bisexual or unisexual, with bisporangiate anthers that open by two apically hinged valves. Pollen grains are dicolpate or zonacolpate with the apertures extending over the distal and proximal poles (Sampson, 1975). The gynoecium consists of several to numerous carpels, each containing a single ovule (Philipson, 1993a).

The fossil record of Atherospermataceae is based on pollen, wood and leaves. Fossil flowers or other reproductive structures are unknown. Currently known fossils are mainly from the Gondwanan region, but there are also records from Egypt and Europe that indicate a wider distribution of the family in the past. The earliest well-documented record of Atherospermataceae is wood described as *Laurelites jamesrossii* from the Campanian of the Seymour Island, Antarctica, which is closely similar to wood of extant *Laureliopsis* (Poole and Francis, 1999). A possible older record includes pollen similar to that of extant *Laurelia*. This record of Coniacian age is from a borehole on the Kerguelen Plateau (Barbara Mohr, personal communication in Renner *et al.*, 2000).

Other Cretaceous records of Atherospermataceae include wood from the Late Cretaceous of South Africa assigned to the genus *Protoatherospermoxylon* (Mädel, 1960; Poole and Gottwald, 2001), as well as possible leaves from the Late Cretaceous (Santonian) of Germany. Leaves described as *Protohedycarya ilicoides* were first assigned to Monimiaceae (Rüffle, 1965), but later transferred to Atherospermataceae (Knappe and Rüffle, 1975). However, the proposed relationships of this species need further study.

The Cenozoic record of Atherospermataceae is more extensive and includes leaves of *Laurelia* from Argentina and Seymour Island, Antarctica (Dusén, 1908), wood from Antarctica (Poole and Gottwald, 2001), Patagonia (Nishida, 1984; Nishida *et al.*, 1988), Egypt (Kräusel, 1939) and Europe (Gottwald, 1992), and fossil pollen from New Zealand (Couper, 1953, 1960) and Tasmania (Carpenter *et al.*, 1994).

10.3.3 Calycanthaceae

The Calycanthaceae are a small family that includes about eight species and four genera of shrubs and small trees with a relictual distribution. *Sinocalycanthus* and *Chimonanthus* are restricted to southern and eastern China, *Idiospermum* is restricted to northeastern Australia, and *Calycanthus* is restricted to North America (Kubitzki, 1993a). Flowers are relatively large, showy, bisexual and multipartite. Bracts grade into the perianth parts in a spiral arrangement on the outside of a deep cup-shaped receptacle. The androecium is arranged in a dense spiral on the rim of the floral cup. Anthers are almost sessile, and are tetrasporangiate with longitudinal and extrorse dehiscence. There are also prominent apical extensions of the connective. Pollen grains are dicolpate. Numerous staminodes are present inside the fertile anthers on the inner surface of the floral cup. The gynoecium consists of one to many free, plicate carpels that are sunken into the floral cup. Each carpel contains a single anatropous ovule (Kubitzki, 1993a).

The fossil record of Calycanthaceae is sparse. Two floral structures from the Cretaceous of North America, *Virginianthus calycanthoides* and *Jerseyanthus calycanthoides*, are similar to flowers of Calycanthaceae, but both show features that are not known among the three extant genera. Additional information is needed for more precise phylogenetic placement. From the Cenozoic, fossil fruits assigned to the genus *Calycanthus* are reported from the Middle Miocene of Germany (Mai, 1987b).

Virginianthus calycanthoides (Figure 10.12) from the Early Cretaceous (Early–Middle Albian) Puddledock flora,

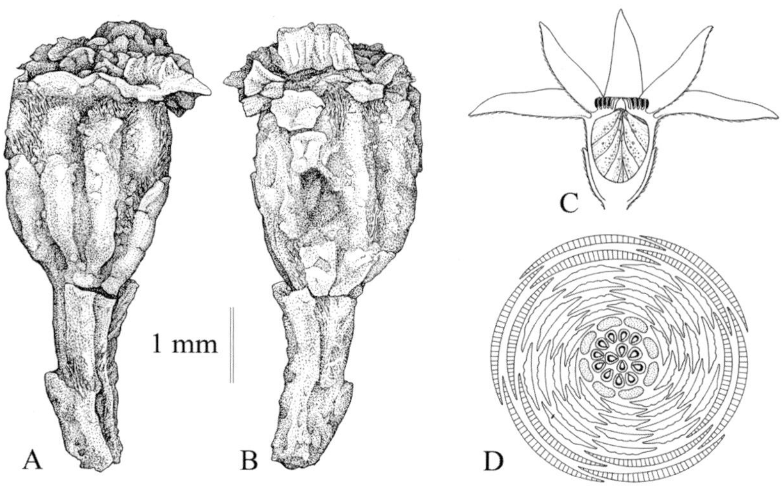

Figure 10.12 *Virginianthus calycanthoides*, a flower related to extant Calycanthaceae from the Early Cretaceous (Early–Middle Albian) Puddledock locality, Virginia, USA; (A, B) fragmentary flower in different lateral views showing tepals and stamens borne around the rim of the floral cup; (C) schematic longitudinal section through flower showing stamens on the rim of the floral cup and carpels inside; (D) floral diagram. Redrawn from Friis *et al.* (1994a).

Virginia, USA, is based on a single flower with good structural details that allow detailed comparison with extant taxa (Friis *et al.*, 1994a). The flower is bisexual, 2–3 mm long, and consists of a deep, cup-like hypanthium bearing numerous spirally arranged tepals and stamens in its apical part. The androecium consists of 30–40 laminar stamens. Anthers have two pairs of abaxial pollen sacs embedded in the connective tissue. Sterile tissue is extensive on the adaxial side of the stamens and extends apically into a broadly triangular expansion. Anther dehiscence is valvate by laterally hinged valves. Pollen grains *in situ* in the pollen sacs are monocolpate with a long slit-like colpus that reaches to the equator. The exine consists of a coarse reticulum supported by low, widely spaced columellae, and a distinct foot layer. The endexine is very thin or absent in non-apertural regions, but well-developed under the aperture. Inserted between the fertile stamens and the gynoecium are numerous sterile organs that may be staminodes. The inner surface of the floral cup has 18–26 free carpels borne in three to four series. Small spherical secretory cells occur in most floral tissues including the hypanthium, tepals, stamens and carpels.

In most floral features *Virginianthus* is closely similar to flowers of extant Calycanthaceae. It differs, however, in details of the stamens and pollen: most obviously the embedded pollen sacs of the stamens and the monocolpate pollen grains. These contrast with the protruding pollen sacs and dicolpate pollen grains of all extant Calycanthaceae and preclude assigning the fossil to a particular extant genus (Friis *et al.*, 1994a). A phylogenetic analysis that included both *Jerseyanthus* (see below) and *Virginianthus* placed *Virginianthus* as unresolved among Laurales (Crepet *et al.*, 2005). It most likely falls somewhere within the stem group of Calycanthaceae.

Jerseyanthus calycanthoides (Figure 10.13) from the Late Cretaceous (Turonian) Old Crossman locality, New Jersey, USA, is based on several well-preserved floral structures (Crepet *et al.*, 2005). Flowers are bisexual and multipartite, up to about 2.8 mm long and 3 mm in diameter, and borne on a slender pedicel. The receptacle is cup-shaped and covered on the outside by 30–50 helically arranged tepals. Proximally the tepals are short, triangular and bract-like. Distally they are longer and petal-like. The androecium consists of an outer series of staminodes and an inner series of fertile stamens. Anthers are dithecate, probably tetrasporangiate, and almost sessile. The apical extension of the connective is distinctly fan-shaped and has small stalked and globose structures at the apex. Pollen grains were described as dicolpate (disulcate), but they are strongly folded and the aperture configuration is not clear from the illustrations. The gynoecium consists of about 24 free carpels borne at the bottom of the floral cup. Carpels have a distinct ventral slit. There is no information on ovules or seeds. The flowers are covered with a dense indumentum of simple hairs.

Preliminary phylogenetic analyses (Crepet *et al.*, 2005) place *Jerseyanthus* in the Calycanthaceae as sister to extant *Calycanthus*. *Jerseyanthus* differs, however, from extant Calycanthaceae in its more numerous floral parts, the presence of an outer series of staminodes, and the fan-shaped apical extension of each anther.

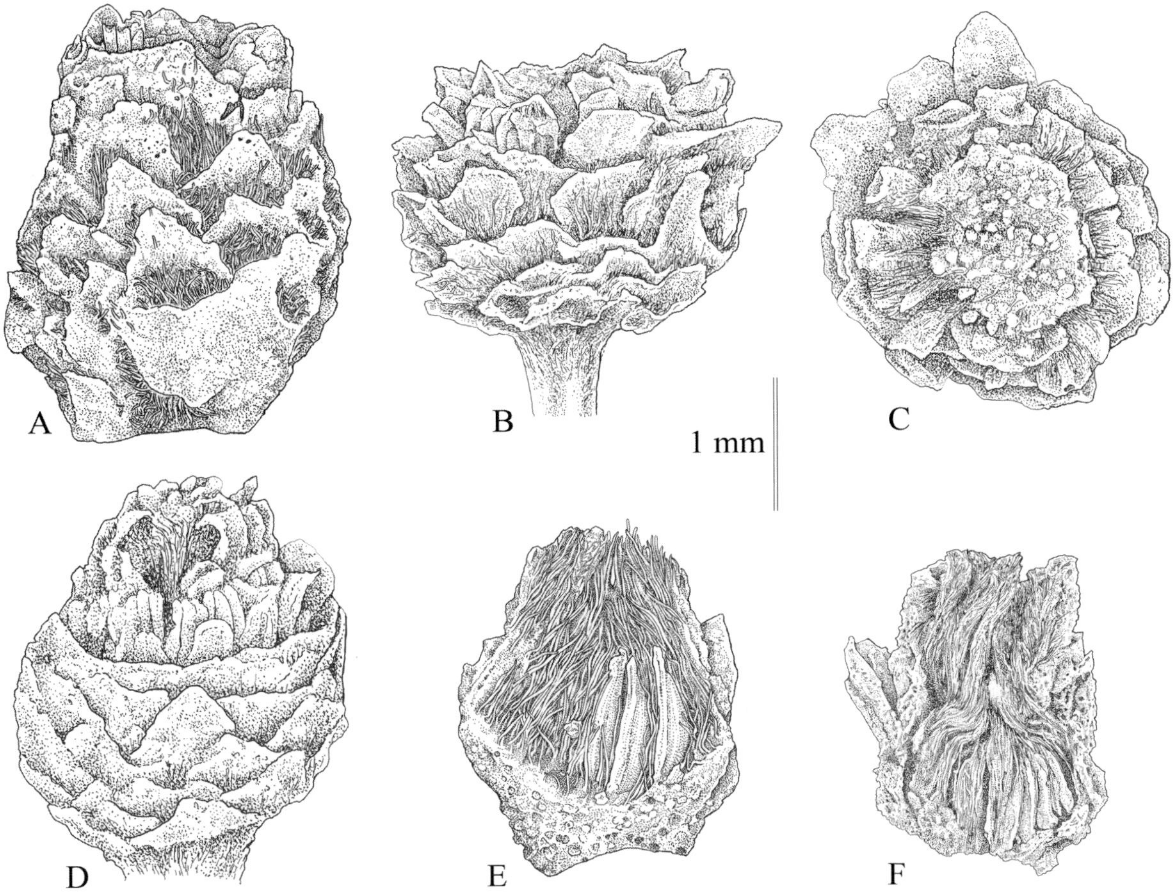

Figure 10.13 *Jerseyanthus calycanthoides*, a flower related to extant Calycanthaceae from the Late Cretaceous (Turonian) Old Crossman locality, New Jersey, USA. (A, B, D) Flowers in lateral view showing imbricate tepals on the surface of the floral cup. (C) Apical view of flower. (E, F) Fragmentary flowers showing inside of floral cup with carpels and densely packed trichomes. Drawn from SEM images in Crepet *et al.* (2005).

10.3.4 Gomortegaceae

The Gomortegaceae comprise a single species, *Gomortega nitida*, restricted to southern Chile. *Gomortega* forms large trees with opposite–decussate leaves. Flowers are epigynous and bisexual with an irregular number of floral parts. Stamens are bisporangiate and anthers open by two apically hinged valves. Pollen grains are inaperturate. There are two or three free carpels, each containing a single ovule (Kubitzki, 1993f).

Currently, the Gomortegaceae have no reported fossil record, but *Lovellea wintonensis*, described by Dettmann *et al.* (2009), shares many features with flowers of extant *Gomortega* and could possibly represent an early member of the family (section 10.3.1).

10.3.5 Hernandiaceae

The Hernandiaceae are a pantropical family comprising about 60 species in five genera, *Hazomalania*, *Hernandia*, *Illigera*, *Gyrocarpus* and *Sparattanthelium*. They range in habit from trees and shrubs to vines and they have alternate leaves. Flowers are small, unisexual or bisexual, with few parts arranged in whorls. Stamens are bisporangiate and anther dehiscence is by two apically hinged valves. Pollen grains are inaperturate with an extremely thin sporopollenin exine. The gynoecium consists of a single carpel and a single ovule (Kubitzki, 1993d).

The only potential fossil Hernandiaceae so far recorded are leaves from the Upper Miocene of Venezuela assigned

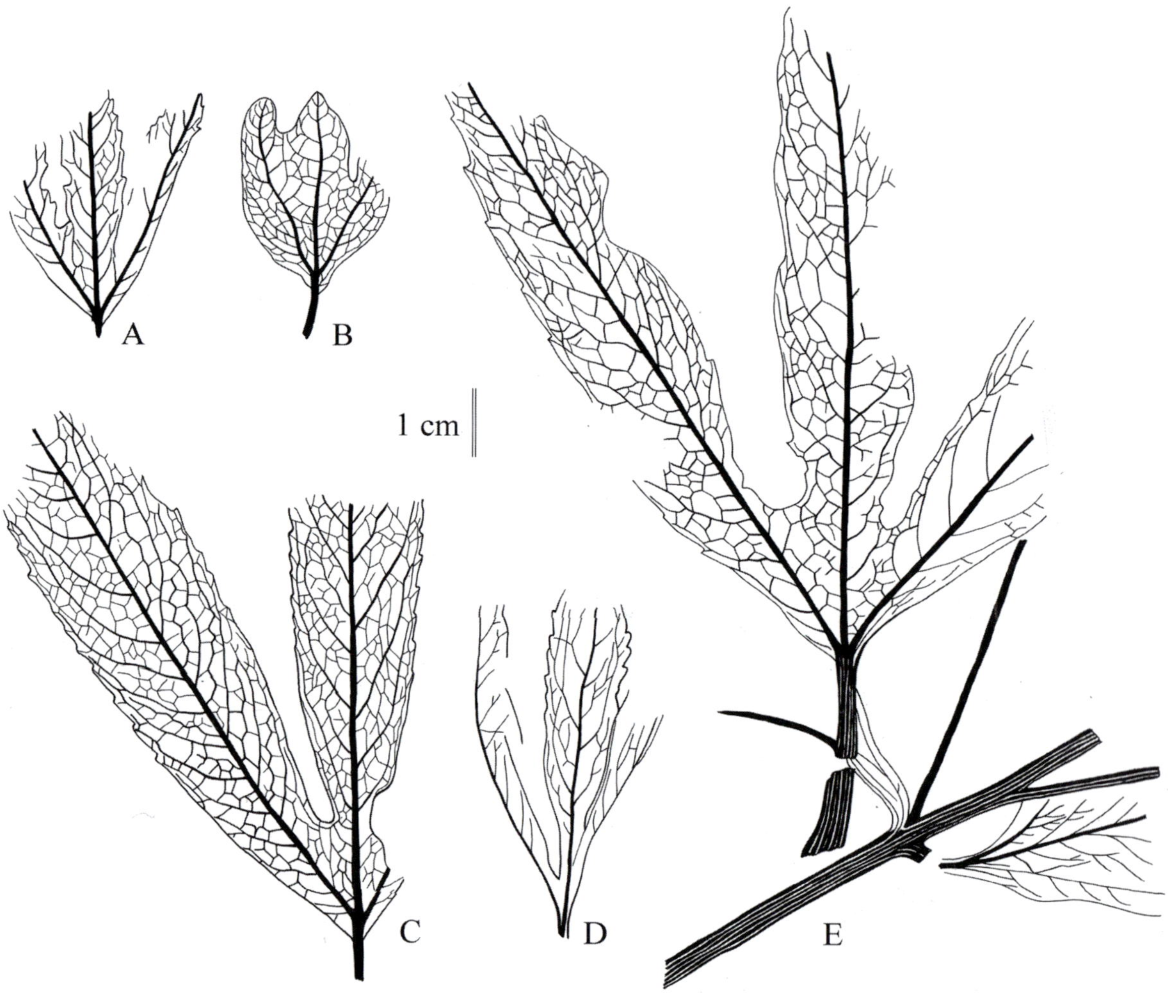

Figure 10.14 Leaves possibly related to extant Laurales from the Early Cretaceous (Late Aptian – Early Albian) Buarcos-para-Tavarede locality, Portugal. Redrawn from Saporta (1894).

to the extant genus *Gyrocarpus* (Berry, 1937), but this determination is in need of restudy.

10.3.6 Lauraceae

The Lauraceae are a large family comprising between 2500 and 3500 species in about 50 genera. They are mostly trees or shrubs with alternate to opposite leaves. The family is largely pantropical, with some taxa extending into temperate regions. Flowers are small, typically bisexual and actinomorphic, and have a distinct trimerous organisation in the perianth and androecium. The perianth consists of two whorls of undifferentiated tepals. Most of the floral variation is within the androecium. Typically the androecium consists of four whorls of three stamens and staminodes. Stamens are tetra- or bisporangiate with valvate dehiscence by two or four apically hinged valves. Sometimes stamens have paired glandular appendages at the base of the filament. Pollen grains are usually inaperturate with an extremely thin sporopollenin exine. The gynoecium consists of a single carpel containing a single, anatropous, bitegmic ovule (Rohwer, 1993a). A preliminary phylogenetic analysis of the family indicates an early division into a Gondwanan group and a Laurasian – South American group (Rohwer, 2000).

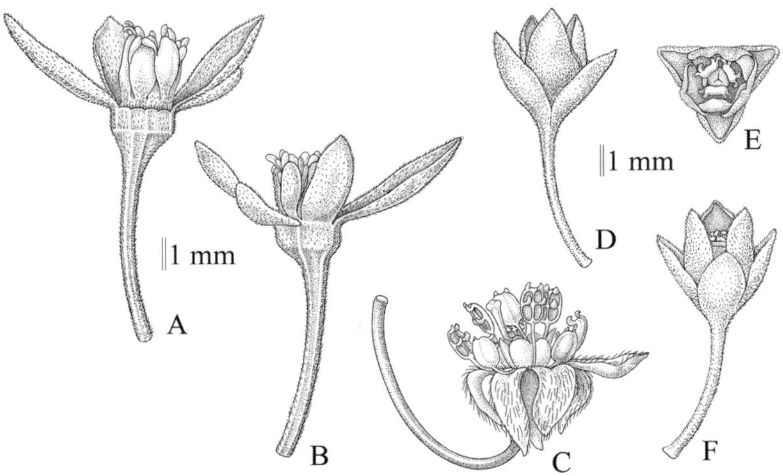

Figure 10.15 Flowers of Lauraceae from the Eocene–Oligocene Baltic amber. (A, B) *Cinnamomum felixii*. (C) *Cinnamomum prototypum*. (D–F) *Trianthera eusideroxyloides*. Redrawn from Conwentz (1886).

Most of the fossil diversity recorded for Laurales falls within Lauraceae. There is a very rich fossil record, particularly from the Cenozoic. Many of these records are based on fossil wood and leaves, but they provide good evidence that Lauraceae were diverse and widespread in the Cenozoic. There are also probable reliable reports of lauraceous leaves from the Cretaceous, for example from the Late Cretaceous of the Czech Republic (Kvaček, 1992) and perhaps also from the Early Cretaceous of Portugal (Figure 10.14). However, most Cretaceous records are problematic and frequently poorly documented. Re-examination of putative lauraceous leaves from the mid-Cretaceous of North America shows that many leaf fossils previously assigned to the extant genus *Sassafras* are more likely related to Platanaceae (e.g. Doyle and Hickey, 1976). In general, reliable assignment of fossil leaves to Lauraceae usually requires micromorphological studies of epidermal features as well as detailed studies of venation (Kvaček, 1992).

Fossil wood with distinct lauraceous features has been described from the Cretaceous of both Gondwana and Laurasia (Gothan, 1908; Herendeen, 1991b; Poole *et al.*, 2000c). *Sassafrasoxylon gottwaldii* from the Late Cretaceous (Santonian–Maastrichtian) of James Ross Island and Seymour Island, Antarctica, is described as having wood characters indistinguishable from those of modern *Sassafras*, a genus now restricted to the Northern Hemisphere (Poole *et al.*, 2000c).

Well-preserved flowers from the Cenozoic and the Cretaceous are important components of the fossil record of Lauraceae and can be assigned unequivocally to the family. From these fossils it is clear that variation in floral morphology among fossil Lauraceae is mainly associated with the androecium, as in the extant family. Lauraceous flowers from the Cenozoic (not discussed in detail here) include compressions from the Middle Eocene Claiborne Formation of Tennessee, USA, described as *Androglandula tennessensis* (Taylor, 1988), as well as compression fossils from the Miocene of Rott, Germany, described as *Litseopsis rottensis* and *Lindera rottensis* (Weyland, 1938). Also well documented are the three-dimensionally preserved flowers of *Cinnamomum prototypum*, *C. felixii* and *Trianthera eusideroxyloides* that are preserved as inclusions in the Eocene–Oligocene Baltic amber (Figure 10.15) (Conwentz, 1886).

Potomacanthus lobatus (Figure 10.16), a small flower from the Early Cretaceous (Early–Middle Albian) Puddledock locality, Virginia, USA, provides the earliest unequivocal record of Lauraceae (von Balthazar *et al.*, 2007). The flower is bisexual with a trimerous perianth and androecium. There are two whorls of equally sized tepals and two whorls of fertile stamens. Anthers are bisporangiate and dehisced by two apically hinged valves. The gynoecium consists of one carpel with a single, pendent ovule. The carpel is ascidiate in the ovary, but plicate in the style.

The trimerous organisation of the flower, the valvate anther dehiscence and the structure of the gynoecium strongly indicate a relationship to Lauraceae. However, the fossil differs from extant taxa in the simple androecium, which has only two whorls of stamens and lacks staminodes

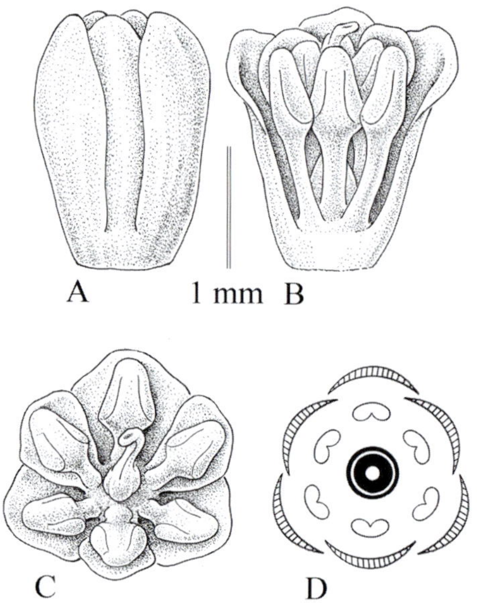

Figure 10.16 *Potomacanthus lobatus*, reconstructions of a small flower of Lauraceae from the Early Cretaceous (Early–Middle Albian) Puddledock locality, Virginia, USA. (A) External view of flower showing tepals of equal length. (B) Internal view of flower with tepals removed showing stamens and central gynoecium. (C) Apical view of flower showing six stamens and central gynoecium. (D) Floral diagram. From von Balthazar *et al.* (2007).

(von Balthazar *et al.*, 2007). *Laurales A* and *Laurales B* as well as several other undescribed structures from the Puddledock flora may also belong to the Lauraceae (section 10.3.1).

Mauldinia (Figures 10.17, 10.18) is an important extinct genus that is widespread in mid-Cretaceous floras of the Northern Hemisphere. It is the most thoroughly documented fossil member of the Lauraceae from the Cretaceous. The genus is based on dispersed flowers and fragments of inflorescences from the Early Cenomanian Mauldin Mountain flora, Maryland, USA, described as *Mauldinia mirabilis* (Drinnan *et al.*, 1990). Subsequently, three other species of *Mauldinia* have been described from the mid- and Late Cretaceous including *M. bohemica* from the Middle Cenomanian Peruc-Korycany Formation of the Czech Republic (Eklund and Kvaček, 1998), *M. hirsuta* from the Cenomanian–Turonian Sarbay flora, Kazakhstan (Frumin *et al.*, 2004) and *M. angustiloba* from the mid-Cretaceous karst infillings in the Rhenish Massif, Germany (Viehofen *et al.*, 2008). A further unnamed species, *Mauldinia* sp., is present in the Late Santonian, Allon flora, Georgia, USA (Herendeen *et al.*, 1999a).

The inflorescence of *Mauldinia* is unique and distinguishes the genus from all other extant and fossil Lauraceae. Inflorescences are compound with bilobed partial inflorescences (lateral units) arranged helically along an elongated central axis. The bilobed lateral units are flattened, cladode-like, and irregular in outline with various bracts or leaf-like scales along the margins and on the surfaces. These bracts or scales are particularly distinct in *M. bohemica* and Eklund and Kvaček (1998) suggested that the bilobed lateral unit might correspond to a homocladic thyrse with flowers and scales arranged in a strongly condensed dichasial–monochasial branching system. Flowers are sessile on the lateral units with one flower in a median position immediately below the separation of the lobes and up to three flowers on each lobe. *Mauldinia mirabilis*, *M. hirsuta* and *M. angustiloba* typically have five flowers on each bilobed unit (Figures 10.17, 10.18); *M. bohemica* has up to seven flowers.

All *Mauldinia* species have the same basic inflorescence organisation and similar floral structure. They differ from each other only in relatively minor morphological details. Flowers are bisexual, actinomorphic and trimerous. They closely resemble those of extant Lauraceae. The perianth consists of an outer whorl of shorter tepals and an inner whorl of longer tepals. The androecium consists of nine fertile stamens in three whorls, plus an additional inner whorl of three sterile staminodes. Stamens of the inner whorl each have an associated pair of appendages with clavate–sagittate heads. Anthers are bilocular and dehisced by two elongate valves that are attached distally. The gynoecium is unicarpellate with a single anatropous and pendant ovule. Pollen grains have not been observed in any of the species. Both flowers and inflorescences have an indumentum of simple, stiff trichomes, which is particularly pronounced in *M. hirsuta*. Secretory cells are present in all tissues and in many cases the lumina of these cells are filled with resin. The secretory cells are probably the remains of ethereal oil cells.

Wood anatomy studied from charcoalified inflorescence axes of *Mauldinia mirabilis* documents that *Mauldinia* had wood of the *Paraphyllanthoxylon* type (Herendeen, 1991b). *Paraphyllanthoxylon* wood is widespread in the Late Cretaceous of North America (Wheeler *et al.*, 1977; Herendeen, 1991b).

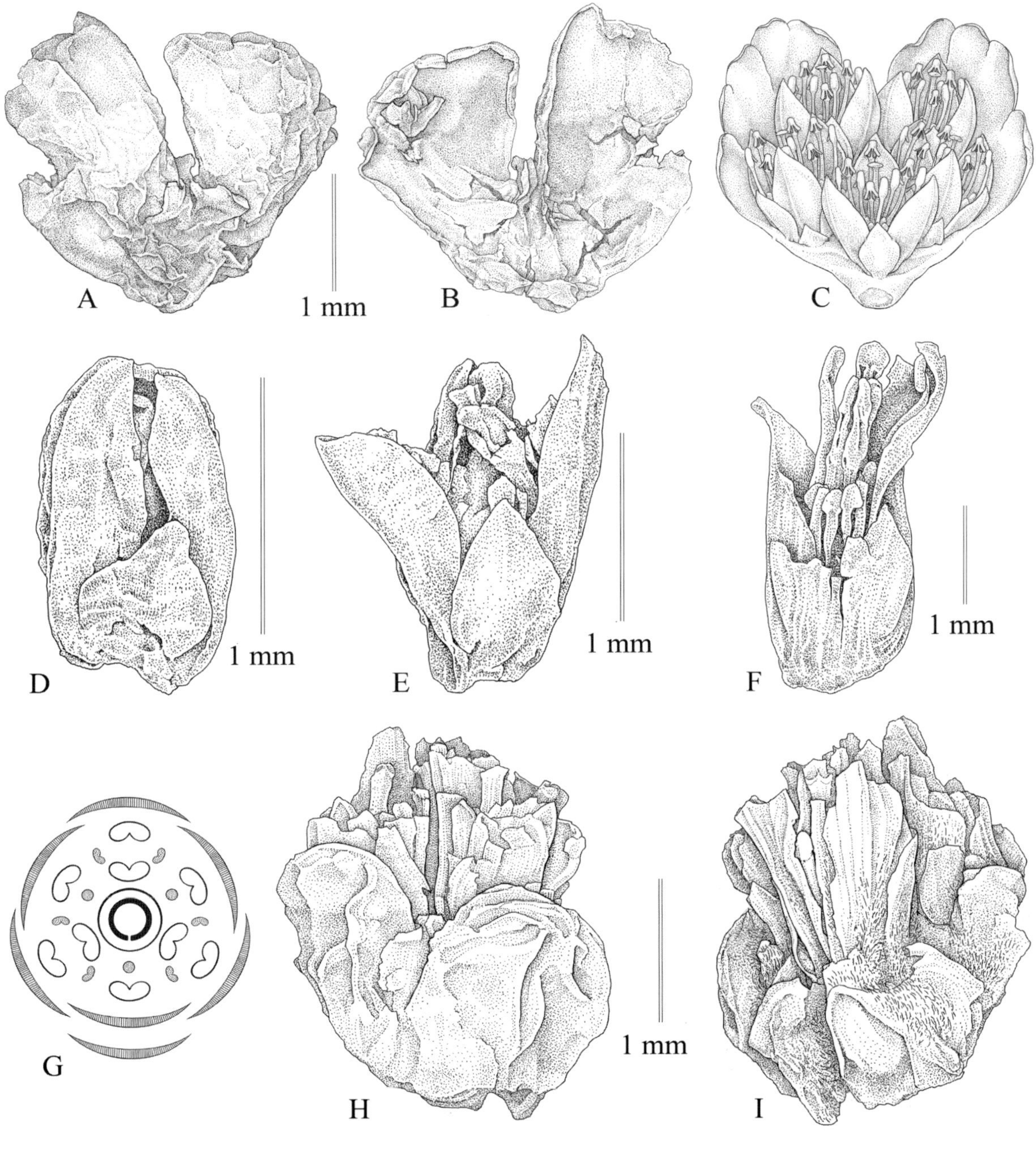

Figure 10.17 Inflorescence units and flowers of (A–G) *Mauldinia mirabilis* from the Late Cretaceous (Cenomanian) Mauldin Mountain locality, Maryland, USA, and (H, I) *Mauldinia hirsuta* from the Late Cretaceous (Cenomanian–Turonian) Sarbay locality, Kazakhstan. (A, B) Empty bilobed lateral inflorescence units in (A) dorsal and (B) ventral view. (C) Reconstruction of inflorescence unit showing five flowers on the adaxial surface. (D, E) Isolated flowers showing short outer tepals and longer inner tepals. (F) Isolated, fragmentary flower showing stamens and staminal appendages. (G) Floral diagram. (H, I) Bilobed lateral inflorescence unit with attached remains of flowers in (H) dorsal and (I) ventral view. (A, B) Drawn from material in the collections of the Field Museum, Chicago; (C) from Friis *et al.* (2006a); (D–F) drawn from SEM images in Drinnan *et al.* (1990); (H, I) drawn from SEM images in Frumin *et al.* (2004).

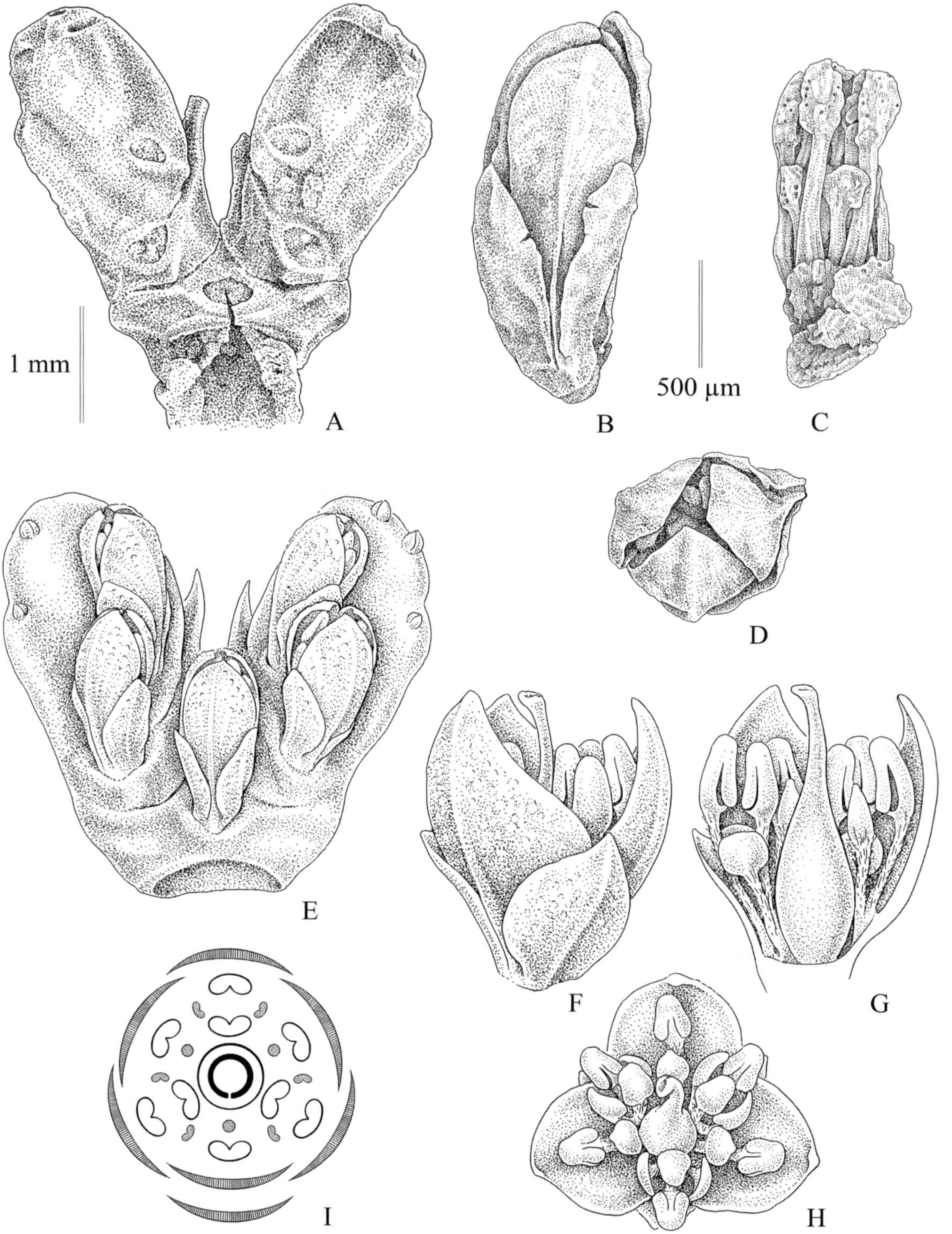

Figure 10.18 *Mauldinia angustiloba* from the mid-Cretaceous karst infillings at the Prangenhaus Quarry, Rhenish Massif, Germany. (A) Empty, bilobed inflorescence units showing scars from five flowers, one in the centre and two on each lobe. (B, D) Isolated flower in (B) lateral and (D) apical view with perianth preserved, showing trimerous perianth with shorter outer tepals and longer inner tepals. (C) Abraded flower revealing the androecium. (E) Reconstruction of inflorescence unit showing five flowers. (F–H) Reconstruction of isolated flower (F) in lateral view, (G) with tepals partly removed and (H) in apical view. (I) Floral diagram. (A–D) Drawn from SEM images in Viehofen *et al.* (2008); (E–I) from Viehofen *et al.* (2008).

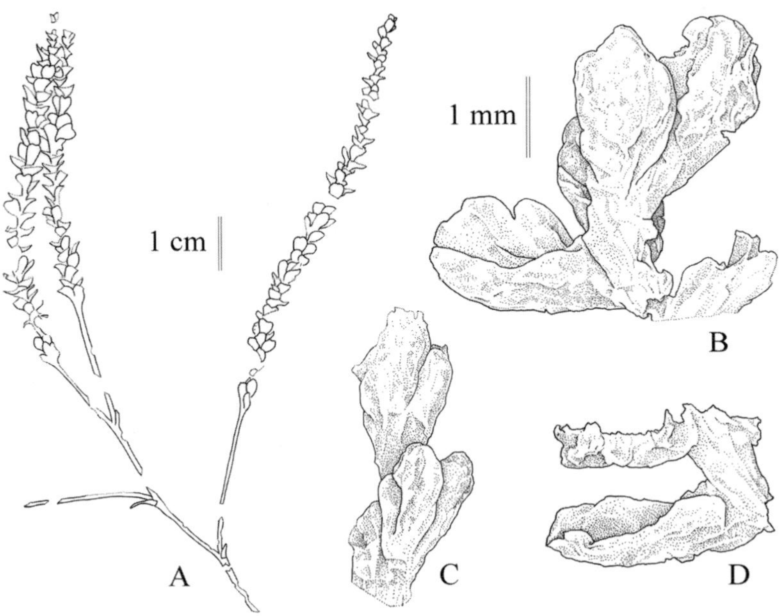

Figure 10.19 (A) Inflorescence axes and (B–D) fragments of lobed lateral units of *Prisca reynoldsii* from the Dakota Formation (latest Albian – earliest Cenomanian), Kansas, USA. Redrawn from Retallack and Dilcher (1981c).

Inflorescence axes of *Prisca reynoldsii* (Figure 10.19) from the Dakota Formation (latest Albian – earliest Cenomanian) of Kansas, USA (Retallack and Dilcher, 1981c), are perhaps identical to *Mauldinia*, but the type species and other specimens from the type locality are too poorly preserved, compared to the *Mauldinia* fossils, for detailed comparison (Drinnan *et al.*, 1990). *Prisca* is associated with entire-margined elliptical leaves described as *Magnoliaephyllum* sp. These leaves are very similar to those of '*Eucalyptus*' *geinitzii* associated with *Mauldinia bohemica* from the Czech Republic (Eklund and Kvaček, 1998). Fragmentary inflorescences probably also related to *Mauldinia* have been recovered from the Early Cenomanian of Utah, USA, and of the Arthur's Bluff locality, Texas, USA (personal observations).

Pragocladus lauroides (Figure 10.20) is another lauraceous fossil from the Peruc-Korycany Formation of Bohemia, Czech Republic (Kvaček and Eklund, 2003). It co-occurs with *Mauldinia bohemica* and shares many floral features with *Mauldinia*. The two taxa are mainly distinguished by differences in inflorescence structure. Inflorescences of *Pragocladus* are compound spikes consisting of a long main axis with many secondary lateral axes in a helical arrangement. Secondary axes bear small tertiary inflorescence units, also in a helical arrangement. Each of these tertiary units bears two to four flowers. Flowers are small, about 1.0–2.3 mm long, and trimerous with two whorls of perianth parts of unequal length, the inner being about twice as long as the outer. The organisation of the androecium is not fully understood, but apparently there are three whorls of three stamens. The stamens and staminal appendages are also very similar to those of *Mauldinia*. Stamens have bilocular anthers that dehisced by two distally attached elongated valves. Flowers are thought to be bisexual, as in *Mauldinia*, but the fossils are strongly compressed and internal structures not clear. The close similarity to *Mauldinia* indicates that *Pragocladus lauroides* also belongs to the Lauraceae (Kvaček and Eklund, 2003).

Neusenia tetrasporangiata (Figure 10.21) is based on a well-preserved flower from the Late Cretaceous (Early Campanian) Neuse River locality, North Carolina, USA (Eklund, 1999). The flower is small, about 2.5 mm long, pedicellate, and isolated. It is bisexual, actinomorphic, and trimerous with a perianth of six tepals in two whorls. Tepals of the outer whorl are slightly shorter than those of the inner whorl. Tepals are fused at the base to form a short floral cup. The androecium consists of nine fertile stamens in three outer whorls, and three staminodes in an inner fourth whorl. Stamens of the third whorl have paired glandular appendages. Fertile stamens have tetrasporangiate anthers that dehisce through four apically hinged valves. The gynoecium is unicarpellate.

Neusenia is clearly assignable to the Lauraceae, and is particularly similar to extant *Neocinnamomum* in the tribe

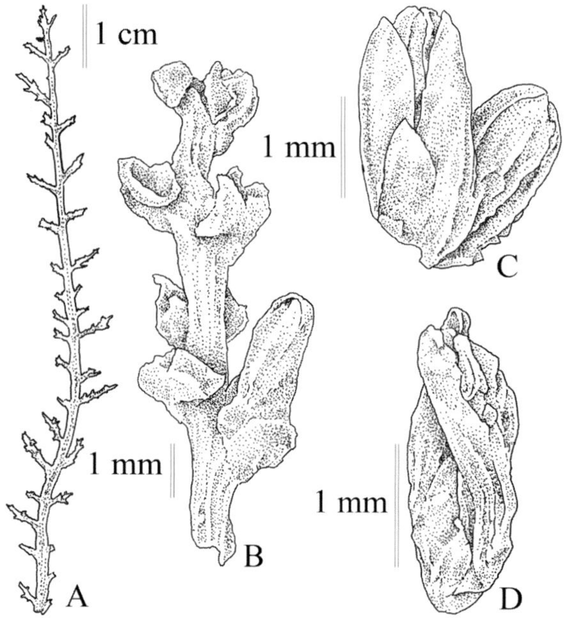

Figure 10.20 Inflorescence fragments and flowers of *Pragocladus lauroides* from the Late Cretaceous (Middle Cenomanian) Peruc-Korycany Formation, Czech Republic. (A) Compound inflorescence with main axis and secondary axes in a helical arrangement. (B) Fragment of secondary axis with remains of tertiary inflorescence units. (C) Two flowers from an inflorescence unit. (D) Single isolated flower showing short outer tepals, longer inner tepals and anther visible between the tepal tips. Drawn from photographs in Kvaček and Eklund (2003).

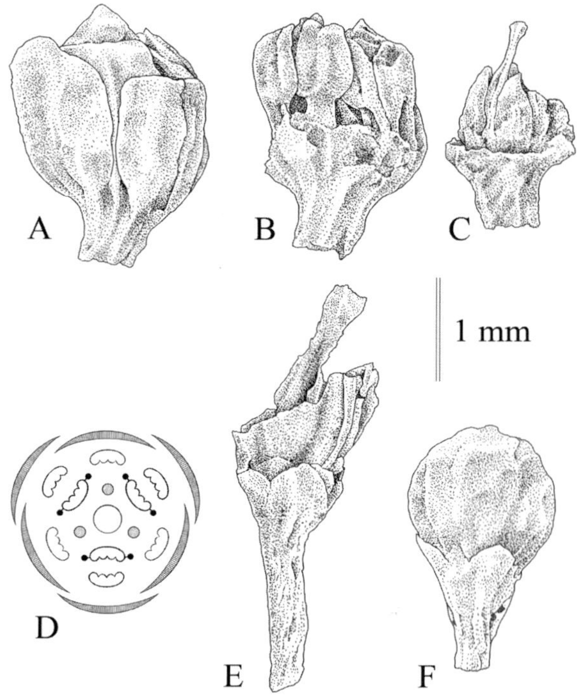

Figure 10.21 Lauraceous flowers from the Late Cretaceous (Early Campanian) Neuse River locality, North Carolina, USA. (A–D) *Neusenia tetrasporangiata*: (A) flower with perianth preserved; (B) same flower with perianth removed showing stamens; (C) same flower with perianth and most of androecium removed showing long style; (D) floral diagram. (E, F) Lauraceae (Taxon B): (E) fragment of flower with outer perianth parts preserved; (F) fruit showing persistent outer perianth parts. Drawn from SEM images and line drawing in Eklund (2000).

Perseeae (Eklund, 1999). *Neusenia* also resembles several other fossil flowers of Lauraceae that have tetrasporangiate anthers, such as *Cinnamomum felixii* and *C. prototypum* from the Baltic amber (Figure 10.15).

Co-occurring with *Neusenia* are two other flowers: Lauraceae Taxon A and Lauraceae Taxon B (Figure 10.21). Both are incompletely preserved, but are distinctly lauraceous in their organisation (Eklund, 1999). Lauraceae Taxon A is distinguished from *Neusenia* in having sessile flowers arranged in a dense inflorescence and having bisporangiate anthers. Flowers of Lauraceae Taxon B are pedicellate and similar to *Neusenia*, but have very short outer tepals and longer inner tepals. The androecium is incompletely preserved and anthers are unknown. Associated with the lauraceous flowers in the Neuse River flora are possible lauraceous fruits described as *Grexlupus carolinensis* (Mickle, 1996).

Lauranthus futabensis (Figure 10.22) is based on a single, isolated, but well-preserved flower from the Late Cretaceous (Early Coniacian) flora of the Kamikitaba locality, northeastern Honshu, Japan (Takahashi *et al.*, 2001). The flower is small, about 1.7 mm in diameter, with a trimerous perianth and androecium. The perianth is undifferentiated with tepal lobes of equal size, arranged in two whorls. The androecium consists of three whorls. Anthers are dithecate and tetrasporangiate, and dehisce by four apically hinged valves arranged in pairs above each other. Anther dehiscence of the two outer whorls is introrse and those of the inner whorl latrorse to extrorse. Staminal appendages were not reported in the initial publication, but have subsequently been shown to be present. However, there is no evidence of a gynoecium. It is unknown whether

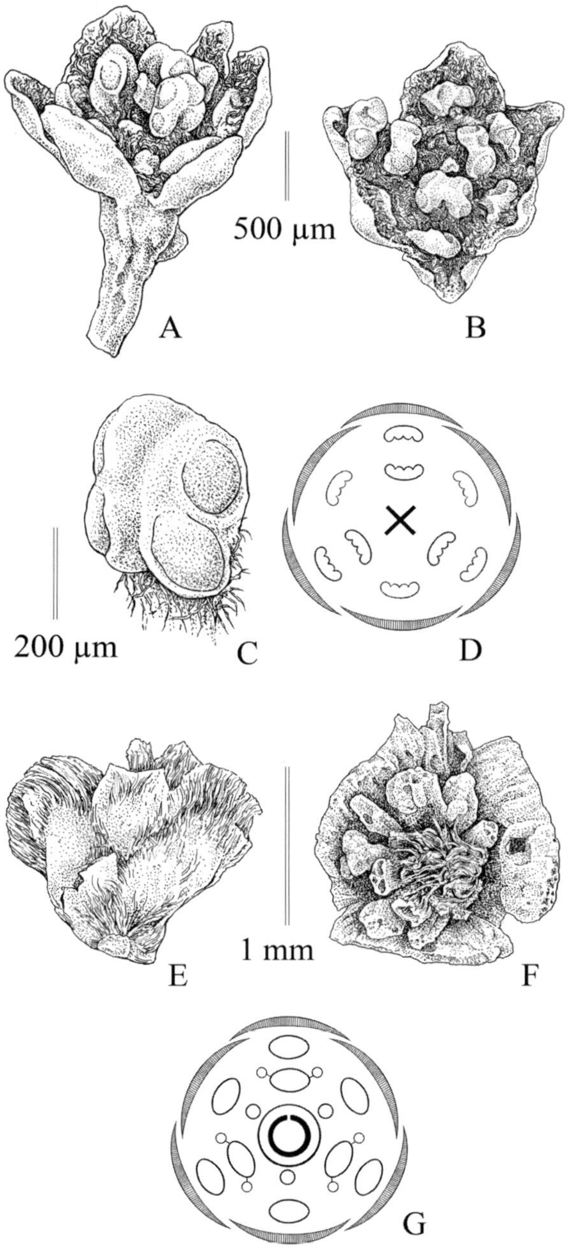

Figure 10.22 (A–D) *Lauranthus futabensis*, a lauraceous flower from the Late Cretaceous (Early Coniacian) Kamikitaba locality, northeastern Honshu, Japan, in (A) lateral and (B) apical view; (C) details of tetrasporangiate stamen and (D) floral diagram. (E–G) *Perseanthus crossmanensis*, a lauraceous flower from the Late Cretaceous (Turonian) Old Crossman locality, New Jersey, USA, in (E) lateral and (F) apical view and (G) floral diagram. (A–C) Drawn from SEM images in Takahashi *et al.* (2001); (E–G) drawn from images in Herendeen *et al.* (1994).

the gynoecium was lost or the flower was unisexual. The general trimerous organisation of *Lauranthus futabensis* combined with characters of the androecium strongly indicates a position within the Lauraceae. Within the family its systematic position is uncertain.

Perseanthus crossmanensis (Figure 10.22) from the Late Cretaceous (Turonian) Old Crossman locality, New Jersey, USA, is based on a single, incomplete, detached flower (Herendeen *et al.*, 1994). The flower is trimerous and very similar in general organisation to the flowers of *Mauldinia*. It is distinguished from *Mauldinia* mainly by the shorter floral tube and shorter staminal appendages. Floral parts are also known to vary in length and form among the different *Mauldinia* species. Further investigation of the possible relationship between *Perseanthus* and *Mauldinia* will require information on how the flowers were borne in the material from New Jersey.

10.3.7 Monimiaceae

As traditionally circumscribed, the Monimiaceae included a heterogeneous assemblage of genera that has now been separated into three families: Atherospermataceae, Monimiaceae *sensu stricto* and Siparunaceae (Renner, 1998). The Monimiaceae *sensu stricto* comprise 11 genera and about 260 species, mainly with a pantropical distribution. Based on phylogenetic analyses of DNA data two major clades are recognised within the family (Renner, 1998), one comprising *Peumus*, *Monimia* and *Palmeria*, and the other comprising the remaining eight genera (*Hortonia*, *Hedycarya*, *Xymalos*, *Tambourissa*, *Hennecartia*, *Mollinedia*, *Kibara*, *Wilkiea*). Except for *Hortonia*, which has bisexual flowers, all Monimiaceae have unisexual flowers. Flowers vary considerably in form and number of floral parts. The receptacle is flat to deeply urceolate and may split at anthesis to form a perianth-like structure. A perianth may be absent, but where present it is tetramerous and arranged in a spiral or whorls. The androecium typically consists of numerous stamens (up to about 1800 in *Tambourissa*) distributed over the surface of the receptacle (Endress and Lorence, 1983; Philipson, 1993a). Pollen grains are in monads, or more rarely in tetrads, typically with a very poorly developed exine without columellae or foot layer (Sampson, 2000b). The gynoecium is apocarpous with one to many carpels in a superior, or rarely inferior, position. Each carpel contains a single anatropous ovule. Carpels mature into drupaceous fruits (Philipson, 1993a).

The fossil record of Monimiaceae includes pollen, leaves and wood. Fossil flowers or other reproductive structures of Monimiaceae are unknown. The oldest fossils that can be assigned reliably to the family are three species of *Hedycaryoxylon* wood from the Late Cretaceous including *H. subaffine* from the Early Senonian of Germany (Süss, 1960), *H. hortonioides* from South Africa (Mädel, 1960; Poole and Gottwald, 2001) and *H. tambourissoides* from the Campanian of James Ross Island, Antarctica (Poole and Gottwald, 2001). Wood of *Hedycaryoxylon* is also recorded from the Cenozoic of Germany (Poole and Gottwald, 2001). Leaf remains described as *Protohedycarya ilicoides* (Heer) Rüffle and assigned to the Monimiaceae may be related to Atherospermataceae although their relationship has not been documented with certainty (section 10.3.2.). A further species of fossil leaf assigned to *Protohedycarya*, *P. pseudoquercifolia*, from the Santonian of Quedlinburg (Rüffle and Knappe, 1988), is probably not related to Monimiaceae (Renner, 1998).

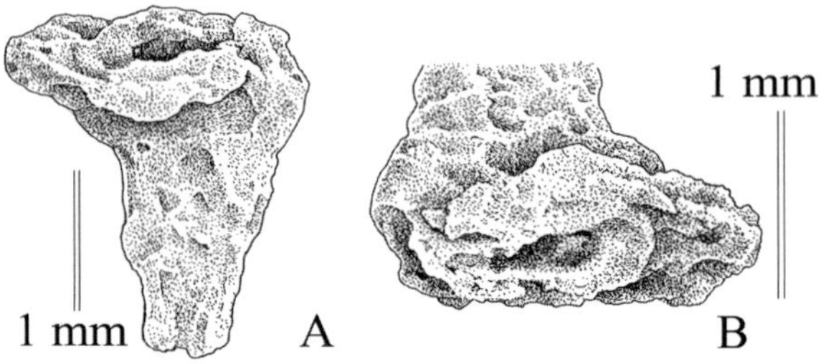

Figure 10.23 Floral structure from the Late Cretaceous (Late Santonian) Table Nunatak locality, Antarctica, of possible relationship with Siparunaceae (Laurales), seen in (A) lateral and (B) apical view, showing a cup-shaped hypanthium covered by a flattened roof with a central opening. Drawn from SEM-images in Eklund *et al.* (2003).

10.3.8 Siparunaceae

The Siparunaceae comprise about 70 species in two genera, *Glossocalyx* and *Siparuna*. They are shrubs or trees with opposite–decussate leaves. They occur mainly in tropical regions of the Southern Hemisphere with *Glossocalyx* in West Africa and *Siparuna* in Central and South America. Flowers are small, unisexual and closed by a roof (velum) inside the perianth that may be either distinct (*Siparuna*) or obscure (*Glossocalyx*) (Philipson, 1993a; Renner, 1999). Staminate flowers have one to numerous stamens. Anthers are bisporangiate and dehisce by a single apically hinged valve (Endress and Hufford, 1989). Pollen grains are inaperturate, spherical and echinate (Pignal *et al.*, 1999). Pistillate flowers have few to many carpels each containing a single unitegmic ovule (Philipson, 1993a; Renner, 1999).

The fossil record of Siparunaceae is meagre and there are currently no fossils that can be placed in the family with certainty. The fossil flowers described as 'Flower Type 1' from the Late Cretaceous (Late Santonian) Table Nunatak Formation, Antarctic Peninsula (Eklund, 2003), are superficially very similar to flowers of extant *Siparuna*. They consist of a cup-shaped hypanthium that is closed by a flattened roof with a central opening (Figure 10.23). Remains of floral parts on the rim of the hypanthium indicate the presence of four tepals. Poor preservation of the fossil precludes detailed comparison with flowers of extant taxa, but nevertheless a similar floral roof with a relatively large central perforation is known for staminate flowers of *Siparuna*. This unusual construction is apparently an adaptation to pollination by gall midges (Eklund, 2003).

10.4 CANELLALES

The Canellales are a well-supported clade consisting of Canellaceae and Winteraceae (APGIII, 2009). In several earlier classifications (e.g. Takhtajan, 1969; Cronquist, 1981) both families were placed in Magnoliales. In other classifications (e.g. Takhtajan, 1997) they were placed in different orders.

10.4.1 Canellaceae

The Canellaceae comprise about 16 species in five genera: *Canella*, *Cinnamodendron*, *Cinnamosma*, *Pleodendron* and *Warburgia*. All are trees and shrubs distributed in tropical Africa, Madagascar, tropical South America and the Caribbean. Flowers are bisexual with a basic trimerous or pentamerous organisation. Perianth parts are differentiated, with the inner parts sometimes forming a corolla tube. Anthers are fused to form a tube. Pollen grains are monoaperturate with a monocolpate or trichotomocolpate aperture. The tectum is continuous or more rarely reticulate. The gynoecium is formed from two to six fused plicate carpels, each with up to about 30 campylotropous ovules (Kubitzki, 1993e; Igersheim and Endress, 1997).

The fossil record of Canellaceae includes a single equivocal record of dispersed pollen from the Early Cenozoic (Oligocene) of Puerto Rico assigned to the extant genus *Pleodendron* (Graham and Jarzen, 1969; Muller, 1981).

10.4.2 Winteraceae

The Winteraceae are a small family of about 65 species in four to nine genera. They are small shrubs and trees with a clearly relictual distribution in montane or cooler rain forests of South and Central America, Madagascar, New Zealand, Australia and New Caledonia (Vink, 1993; Endress, 1994b; Heywood *et al.*, 2007). Flowers are bisexual, or in *Tasmannia* unisexual, and they exhibit considerable variability and irregularity in the form and number of floral parts. Carpels are plicate, free or sometimes fused (Vink, 1993; Endress, 1994b). Pollen grains are monoaperturate and distinctive in being dispersed in permanent tetrads (Praglowski, 1979; Sampson, 2000a). All Winteraceae have vessel-less wood, a feature that has generally been regarded as the basic condition in angiosperms and has been a major focus in discussions of the systematic position of the family (e.g. Carlquist, 1983).

Fossils that can be assigned unequivocally to the Winteraceae are rare. So far no reproductive structure with diagnostic features of Winteraceae has been identified in the Cretaceous fossil record. Knowledge of the history of the family is based mainly on the recognition of the distinctive pollen tetrads (Doyle, 2000), which indicate that the family had a much wider distribution in the Cretaceous and in the Cenozoic than it has today. Most fossils are from the Gondwanan region, but there are also records from Laurasia (Doyle, 2000).

The oldest fossils that can be linked reliably to Winteraceae are pollen tetrads from the Early Cretaceous assigned to the extinct genus *Walkeripollis*. These are thought to represent stem group Winteraceae, while fossils that can be assigned to crown group Winteraceae are known only from the Late Cretaceous onwards (Doyle, 2000). A winteraceous affinity has also been suggested for two kinds of Early to mid-Cretaceous dispersed pollen, *Afropollis* and *Schrankipollis*, but these have never been observed in tetrads. *Afropollis* also has features that indicate that it may not be an angiosperm (section 9.6.1).

Walkeripollis (Figure 10.24) is a fossil genus based on dispersed pollen from the Early Cretaceous (Late Barremian or Early Aptian) of Gabon. The type species *W. gabonensis* (Doyle *et al.*, 1990a) has pollen arranged in permanent tetrahedral tetrads. The individual grains are also distinctive in having a circular to elongate distal aperture surrounded by an annulus that is formed by a thickening of the exine. The pollen wall is finely reticulate with segmented muri supported by short, stubby columellae. The foot layer is distinct and the endexine, strongly thickened under the aperture, is absent or poorly developed in non-apertural regions (Doyle *et al.*, 1990a). The genus also includes *Walkeripollis* sp. A. (Doyle *et al.*, 1990a) from the Early Cretaceous (Late Aptian – Early Albian) of Israel (Walker *et al.*, 1983). *Walkeripollis* sp. A. is distinguished from *Walkeripollis gabonensis* by its non-segmented muri and a somewhat coarser reticulum.

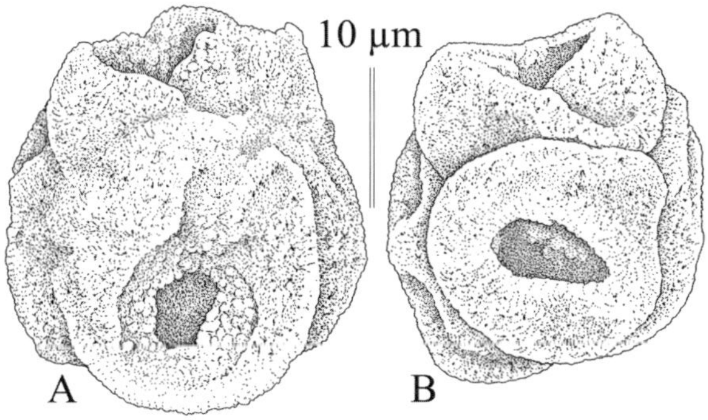

Figure 10.24 Pollen tetrad of *Walkeripollis*, a dispersed pollen type related to extant Winteraceae from the Early Cretaceous (Late Barremian – Early Aptian) of Gabon, showing four adhering pollen grains, each with a single distal pore. Drawn from SEM images in Doyle *et al.* (1990a).

Both *Walkeripollis gabonensis* and *Walkeripollis* sp. A. differ from modern Winteraceae in several respects. According to Doyle (2000) they most likely represent a stem lineage leading to crown group Winteraceae. Dispersed pollen from the Cretaceous that can be assigned to crown group Winteraceae includes species of the extinct genus *Pseudowinterapollis* from the Late Cretaceous (Campanian–Maastrichtian) of Australia that show similarities to pollen of extant *Pseudowintera*, *Belliolum* and *Bubbia* (Dettmann, 1994).

The Cenozoic record of Winteraceae is more extensive than that in the Cretaceous. Dispersed pollen tetrads are known from New Zealand (Couper, 1960; Mildenhall and Crosbie, 1979), Australia (Macphail *et al.*, 1994), South Africa (Coetzee and Muller, 1984; Coetzee and Praglowski, 1987) and South America (Barreda, 1997). The presence of Winteraceae in Antarctica is also indicated by a single pollen record from the Early Cenozoic of Seymour Island (Cranwell, 1959) as well as wood described as *Winteroxylon jamesrossii* from the Late Cretaceous (Late Santonian – Early Campanian) of James Ross Island, Antarctic Peninsula (Poole and Francis, 2000). Wood from the Late Cretaceous (Maastrichtian) of California compared to the Winteraceae

(Page, 1979) may be more closely related to wood of Trochodendraceae (Poole and Francis, 2000). The genus *Winteroxylon* was established based on fossil wood from the Eocene of Germany (Gottwald, 1992).

A fossil leaf described as 'New genus A' from the mid-Cretaceous of Nebraska, USA (Upchurch and Dilcher, 1990) provides further evidence that Winteraceae may also have had a Laurasian range during the Cretaceous and Cenozoic. Other leaf fossils are from the Southern Hemisphere and require re-examination. They include *Drimys antarctica* from the Paleocene of Seymour Island (Dusén, 1908) and *D. patagonica* from the Early Miocene of Argentina (Berry, 1937).

10.5 PIPERALES

The circumscription and relationships of Piperales have been much debated. Traditionally the Piperales have included two families: Piperaceae and Saururaceae. In some previous classifications the order also included Chloranthaceae, but the proposed relationship between Chloranthaceae and Piperales has received little support from recent phylogenetic analyses, but see Friis and Pedersen (2011). Instead two other families, Aristolochiaceae and Lactoridaceae, were placed close to Piperaceae and Saururaceae. Initially Nymphaeales and monocots were included together with these four families in the 'paleoherbs' *sensu* Donoghue and Doyle (1989b). However, in the APGIII classification Nymphaeales and monocots are placed elsewhere, while Piperales comprise Aristolochiaceae, Lactoridaceae, Piperaceae and Saururaceae, together with the holoparastic family Hydnoraceae. In previous classifications Lactoridaceae have been variously placed in Magnoliales (Cronquist, 1981), Laurales (Takhtajan, 1980), Piperales (Walker and Walker, 1984) or a separate order, Lactoridales (Dahlgren, 1983). Aristolochiaceae have typically been placed in a separate order, Aristolochiales (Takhtajan, 1980; Cronquist, 1981).

10.5.1 Fossil Piperales of uncertain relationships

The fossil record of Piperales is scattered, but all families except Hydnoraceae are known from the Cenozoic. There are several dispersed monocolpate, tectate pollen types in Cretaceous sediments, such as *Tucanopollis* and *Transitoripollis*, that may be related to Piperales, but the additional information critical for establishing the relationships of these grains more precisely is currently lacking (Chapter 9). Among Cretaceous mesofossils, fruits of *Appomattoxia* share key features with Piperales and may be an early member of the clade.

Appomattoxia (Figure 10.25) was first reported from the Early Cretaceous (Early–Middle Albian) Puddledock locality, Virginia, USA, based on small spiny fruits described as *Appomattoxia ancistrophora* (Friis *et al.*, 1995). Subsequently, *Appomattoxia* has also been found in Early Cretaceous (Late Barremian – Early Aptian) assemblages from Torres Vedras, Portugal (Friis *et al.*, 2010a). Fruits are unicarpellate with a distinct, sessile stigma and a single, apically attached, orthotropous seed. The fruit wall is thin, typically consisting of only the outer cuticle, which shows distinct imprints of the large epidermal cells. Each epidermal cell bears a central, elongated unicellular spine with a distinctive curved tip. Pollen grains occur abundantly on the stigmatic surface. They are monocolpate and tectate with finely spinulate ornamentation and granular to columellate infrastructure. The foot layer is thick. The endexine is well developed only under the broad aperture. These pollen grains closely resemble dispersed pollen assigned to the genera *Transitoripollis* and *Tucanopollis*, which are particularly common in Barremian–Aptian palynofloras from palaeoequatorial regions (Chapter 9).

Fruit, seed and pollen characters suggest that *Appomattoxia* may be closely related to extant Piperales, but in Piperales the gynoecium is typically formed from three or more carpels and the seeds are basally attached. However, unicarpellate gynoecia do occur in some extant Piperaceae. *Appomattoxia* also shows some similarities to extant Chloranthaceae, but is distinct in seed and pollen characters. Superficially similar spiny fruits occur in the extant genus *Circaeaster* (Circaeasteraceae), but *Circaeaster* has tricolpate pollen, and details of the fruit wall also differ from that of *Appomattoxia* (Friis *et al.*, 1995).

10.5.2 Aristolochiaceae

The Aristolochiaceae comprise about 275 species in 12 genera. They are mainly tropical vines, or more rarely shrubs or herbs, but some taxa are present in temperate regions. The flowers are large, solitary and often strongly zygomorphic. They are bisexual with a trimerous perianth and androecium, but exhibit considerable morphological and organisational variation. Some of the largest flowers may be up to one metre long (Huber, 1993; Endress, 1994b). The androecium consists typically of six to 12 stamens, but many species have as many as 40 stamens. Stamens are usually fused with the gynoecium into a

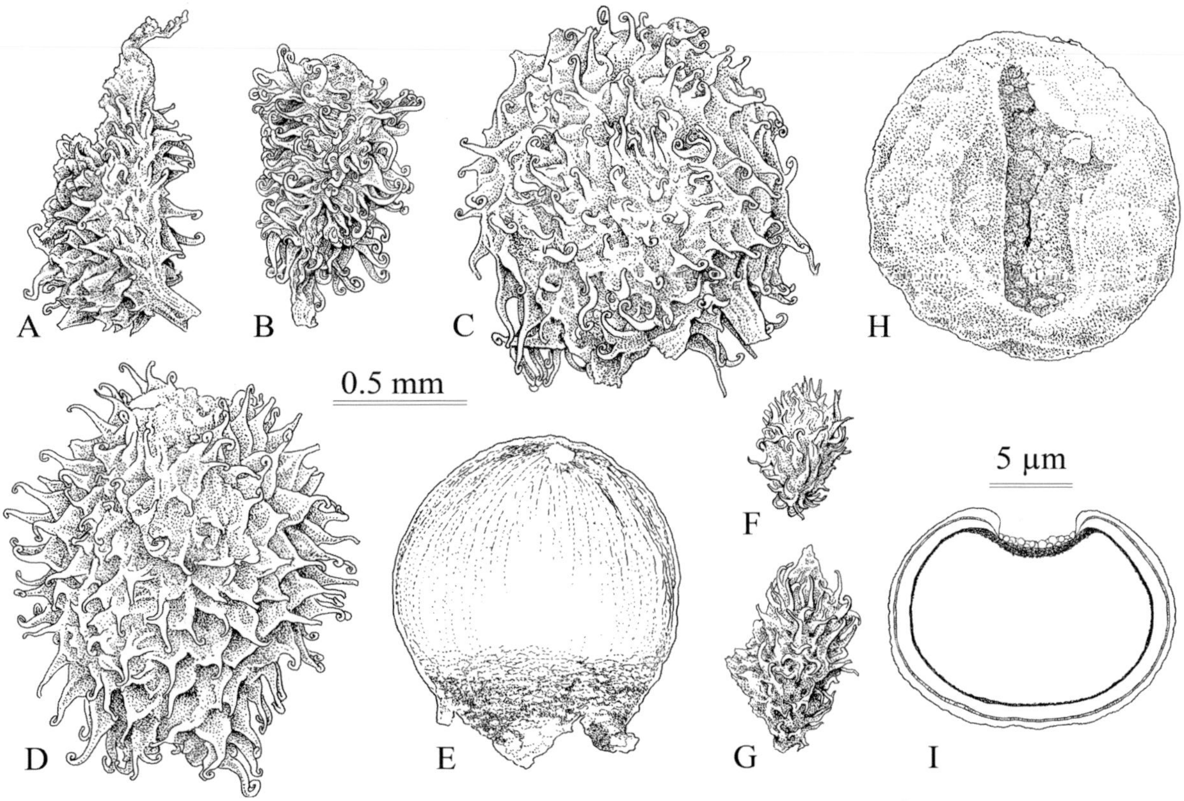

Figure 10.25 Early Cretaceous fruits of *Appomattoxia* possibly related to Piperales. (A–E, H, I) Fruits of *Appomattoxia ancistrophora* from the Early Cretaceous (Early–Middle Albian) Puddledock locality, Virginia, USA, (A, B) Young fruits; (C, D) larger fruits; (E) seed; (H) monocolpate pollen grain from fruit surface; (I) section through pollen grain showing thicker endexine in the apertural region. (F, G) *Appomattoxia* sp. from the Early Cretaceous (Late Barremian – Early Aptian) Torres Vedras locality, Portugal. (A–D, H) Drawn from SEM images in Friis *et al.* (1995); (E) drawn from LM image in Friis *et al.* (1995); (F, G) drawn from SEM images in Friis *et al.* (2006a).

gynostemium. The gynoecium consists of four to six carpels that are usually fused. The fruits are dry, indehiscent or dehiscent, with few to many anatropous seeds.

The Cretaceous record of Aristolochiaceae includes putative leaf impressions from the Late Cretaceous (Early Santonian) of Kamchatka described as *Aristolochites kamchaticus* (Herman and Lebedev, 1991). Another early record of the family includes wood from the Deccan Intertrappean Beds (Late Cretaceous/Early Cenozoic) of India described as *Aristolochioxylon prakashii* (Kulkarni and Patil, 1977). Leaf fossils assigned to the extant genus *Aristolochia* are reported from Early Cenozoic (Eocene–Oligocene) floras from North America (MacGinitie, 1953, 1969, 1974) and Late Cenozoic (Miocene–Pliocene) floras from Abkhazia, Georgia (Kolakovsky, 1957, 1964), Ukraine (Pimenova, 1954) and Poland (Czeczott, 1951). All fossil records of the family probably require critical reassessment.

10.5.3 Lactoridaceae

The Lactoridaceae are a monotypic family with a single species, *Lactoris fernandeziana*, restricted to Masatierra and Masatuera in the Juan Fernandez Islands. *Lactoris* is a shrubby plant with small alternate leaves. Flowers are trimerous, bisexual or unisexual, and show considerable variation in the distribution of stamens, staminodes and pistils, even on the same plant. Fertile stamens are tetrasporangiate. The gynoecium is formed from three basally connate carpels, each containing several anatropous ovules. Pollen grains are shed in permanent tetrads. They are monocolpate with a granular infratectal layer (Cronquist, 1981; Kubitzki, 1993g).

The fossil record of Lactoridaceae is poor and equivocal. Probably the most reliable record is provided by dispersed pollen tetrads of *Lactoripollenites africanus* from the Late Cretaceous (Turonian–Campanian) of Southwest Africa that resemble pollen of extant *Lactoris* in size and shape (Zavada and Benzon, 1987). Other pollen grains reported from the Early and mid-Cretaceous of North America as lactoridaceous (Zavada and Taylor, 1986) are distinguished from extant *Lactoris* pollen in being shed in monads. They are probably related to some other angiosperm group.

10.5.4 Piperaceae

The Piperaceae comprise about 2000 species in five genera: *Macropiper*, *Peperomia*, *Piper*, *Sarcorhachis* and *Zippelia*. The family includes herbs, lianas, shrubs and trees distributed in tropical and subtropical regions of the world. Frequently they grow as epiphytes. Two genera, *Piper* and *Peperomia*, have a pantropical distribution, while *Zippelia* is restricted to southeastern Asia and Malesia, *Macropiper* is restricted to the South Pacific, and *Sarcorhachis* is restricted to the neotropics. Flowers are minute and crowded into dense inflorescences. They are unisexual or bisexual, simple and few-parted, and lack a perianth. Stamens are tetrasporangiate or rarely bisporangiate. Pollen grains are monocolpate, or rarely inaperturate, with a continuous tectum and granular infratectal layer. The gynoecium consists of three or four carpels, or more rarely, a single carpel. Each carpel contains a single orthotropous ovule (Cronquist, 1981; Tebbs, 1993).

Several fossil leaves from Cretaceous and Cenozoic strata have been assigned to the extant genus *Piper* (e.g. Hollick, 1930; Knowlton, 1930), but detailed studies of their venation and epidermal structure are lacking and their systematic relationships have not been documented securely. Small seeds with a reticulate surface pattern, described as *Peperomia sibirica*, are reported from the Late Cenozoic (Miocene) of Siberia (Dorofeev, 1963). Seeds of *Peperomia* have also been reported from the Cenozoic of Europe (Mai, 1995).

10.5.5 Saururaceae

The Saururaceae are a small family comprising five species in four genera: *Anemopsis*, *Houttuynia*, *Gymnotheca* and *Saururus*. All are herbaceous plants and the family has a relictual distribution in Asia and North America. *Saururus* has one species in eastern North America and another species in eastern Asia; *Anemopsis* is restricted to western North America, and *Houttuynia* and *Gymnotheca* occur only in eastern Asia. Flowers are minute and crowded in dense inflorescences that are sometimes subtended by showy bracts. Flowers are bisexual, simple, few-parted, and lack a perianth. Stamens are tetrasporangiate. Pollen grains are monocolpate, or sometimes trichotomocolpate, with a continuous tectum and granular infratectal layer (Smith and Stockey, 2007a). The gynoecium consists of three or four carpels that are free or connate. Each carpel contains two to several orthotropous ovules (Cronquist, 1981; Wu and Kubitzki, 1993b).

The presence of Saururaceae in Cenozoic floras from Europe and Asia is documented by numerous reliable records of fruits and seeds assigned to the extinct species *Saururus biloba*, which is known in Europe from the Eocene to the Miocene (Friis, 1985b; Mai, 1995) and in Asia from the Oligocene and Miocene (e.g. Dorofeev, 1963). Well-preserved permineralised inflorescence axes with attached flowers and fruits have also been described as *Saururus tuckerae* from the Middle Eocene Princeton Chert of British Columbia, Canada (Smith and Stockey, 2007b). The only potential Cretaceous occurrence of Saururaceae is a petrified stem from the Late Cretaceous of Hokkaido, Japan, described as *Saururopsis niponensis* (Stopes and Fujii, 1910). The relationships of this material require further study.

10.5.6 Hydnoraceae

The Hydnoraceae are a small family with about ten species in two genera, *Hydnora* and *Prosopanche*, occuring in dry regions of western Africa and Madagascar (*Hydnora*), and South America from Paraguay to Patagonia (*Prosopanche*). Both are parasites with leafless rhizome-like underground parts. Flowers open just above the surface of the ground. They are bisexual with inferior ovary and three to five fleshy perianth lobes. The androecium consists of three to five fused stamens with numerous pollen sacs. Pollen grains are monocolpate in *Hydnora* and dicolpate or trichotomocolpate in *Prosopanche*. The gynoecium is syncarpous, formed from three or four carpels with numerous ovules (Cronquist, 1981). The Hydnoraceae have no published fossil record.

11

Fossils of monocots

Monocots (monocotyledons) are a major clade of angiosperms that have been recognised as a natural group since John Ray in the early eighteenth century (Ray, 1703). The name refers to the single cotyledon, a feature found in all members of the group. Monocots also lack the ability to produce secondary xylem and phloem in the same way as other angiosperms and most other seed plants. Nevertheless, they are hugely diverse in habit and form. Monocots include aquatic herbs and tall palms as well as Arctic grasses and tropical epiphytes. Taken together, monocots account for a little over a fifth of angiosperm species. A recent attempt to develop a global checklist of all monocots at the species level suggests that about 70 000 extant species of monocot are currently known (The Board of Trustees of the Royal Botanic Gardens, 2008). In this chapter we provide a brief overview of monocot classification and consider the fossil history of the group, focusing particularly on the earliest records and other evidence of monocots from the Cretaceous.

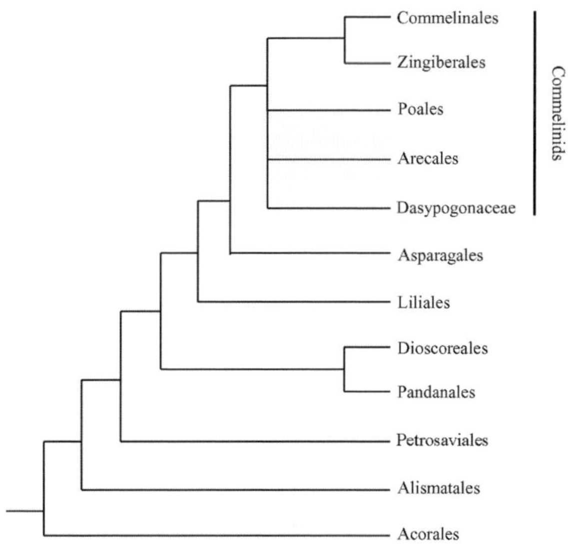

Figure 11.1 Summary of phylogenetic relationships among extant orders of monocots based on APGIII (2009).

11.1 CLASSIFICATION OF MONOCOTS

Rapid progress in the development of a phylogenetic classification of monocots over the past 25 years was stimulated in large part by the important synthetic work of Dahlgren *et al.* (1985). This was taken forward in a series of symposia (Rudall *et al.*, 1995; Wilson and Morrison, 2000; Columbus *et al.*, 2006) as well many individual research contributions. Sampling for phylogenetic analyses based on molecular data has been especially intensive and a broad consensus has emerged on the major clades of monocots (Figure 11.1), as well as many aspects of phylogenetic pattern within these groups (e.g. Chase, 2004). There is also agreement on some aspects of the relationships among these clades.

The current APGIII classification (2009) recognises 11 orders of monocots, plus Dasypogonaceae, which is assigned to commelinids, but is of uncertain ordinal-level relationship within this group. The Acorales, then Alismatales, followed by Petrosaviales, and Pandanales plus Dioscoreales, are successive sister groups to the remaining orders. The bulk of monocot diversity resides in this remaining group, in which Liliales, then Asparagales, are sister group to the commelinids (Dasypogonaceae, Arecales, Poales, Commelinales, Zingiberales) (Figure 11.1).

Monocots are supported as a monophyletic group by strong evidence from many phylogenetic studies based on molecular data (Chase, 2004; Chase *et al.*, 2006). The single cotyledon is an important morphological synapomorphy although this also occurs in some eudicots (Rudall *et al.*, 2007). The presence of proteinaceous inclusions in sieve tube plastids is also widely cited as a defining feature of monocots, although it also occurs in certain eumagnoliids (Aristolochiaceae, Dahlgren *et al.*, 1985). Cladistic studies based on non-molecular data list a variety of other characters that potentially support monocot monophyly (Dahlgren *et al.*, 1985). However, secure interpretation of these features as monocot synapomorphies requires more detailed study.

Although the monophyly of monocots is well established, exactly how the group is related to Chloranthaceae, eumagnoliids, eudicots, Ceratophyllaceae and the three lineages at the ANITA grade remains uncertain. Early cladistic work based on morphology (Donoghue and Doyle, 1989a; Doyle and Donoghue, 1992, 1993) tended to associate monocots with predominantly herbaceous eumagnoliids (Lactoridaceae, Aristolochiaceae, Piperaceae, Saururaceae), as well as Nymphaeaceae, and placed these groups together as 'paleoherbs'. This echoed some aspects of earlier ideas by Burger (1977). However, based on data from molecular phylogenetics, most of these families (except Nymphaeaceae) are now placed within Piperales and allied to Canellales. This makes a relationship between monocots and most of these herbaceous eumagnoliids unlikely.

11.2 FOSSIL EVIDENCE OF MONOCOT DIVERSIFICATION

The evolutionary distinctiveness of monocots has generally been taken to indicate that the group diverged very early in angiosperm evolution. Compared with the fossil record of eumagnoliids, eudicots, or even some ANITA-grade angiosperms (e.g. Nymphaeales), the Cretaceous record of monocots has been regarded as poor (for reviews see Doyle, 1973; Daghlian, 1981; Herendeen and Crane, 1995; Gandolfo *et al.*, 2000; Greenwood and Conran, 2000; Stockey, 2006; Friis *et al.*, 2010b). However, new discoveries from the Early Cretaceous show that monocots were more diverse in the early phases of angiosperm evolution than has previously been appreciated. Future research on Early Cretaceous mesofossil floras seems likely to further expand the early monocot fossil record.

The apparently sparse fossil record of monocots may partly reflect the group's fundamentally herbaceous habit (Chapters 2, 20). Unlike the leaves of trees, leaves of herbaceous plants are not generally shed in large numbers. Typically, they die, wither and rot while still attached to the parent plant and therefore the probability of transport into fossil assemblages is low. There are also difficulties of recognition. Monocot leaves can be hard to distinguish from other kinds of structures, such as compressed stems or dicot leaves that have parallel veins, especially when details of venation and cuticles are not preserved. Combined with difficulties in discriminating among the leaves produced by different species of monocot, this further complicates the recognition of monocot diversity in macrofossil assemblages. In general, monocots tend to be best represented in macrofossil plant assemblages that are preserved *in situ*, such as ash falls (Wing *et al.*, 1993) or silicified peats (Erwin and Stockey, 1994; Stockey *et al.*, 1997). The most extensive record of fossil monocots from the Cretaceous comes from the Deccan Intertrappean Beds of India (e.g. Prakash, 1954, 1960; Chitaley, 1960; Chitaley and Patil, 1971; Nambudiri and Tidwell, 1978; Bonde, 1986, 1995, 1996, 2000; Bonde and Kumaran, 1993).

Monocots are better represented in the fossil record by fruits and seeds, especially in the Cenozoic where these can generally be related relatively easily to extant taxa at the level of families and genera. Fruits and seeds produced by monocots characteristic of wet or damp habitats, which were probably close to potential sites of deposition, are very well represented in the Cenozoic and very well known. However, there are no morphological and anatomical features that specifically distinguish monocot seeds and fruits from those of other angiosperm groups. Consequently, in the early angiosperm fossil record at a stage before the differentiation of most extant families and genera, and before the origin of their most distinctive features, monocot seeds and fruits may be difficult to recognise.

The situation with pollen is similar. Monocot pollen is hard to distinguish from the pollen of some magnoliids, and the difficulties of recognition are perhaps further exacerbated by problems of representation. Widespread entomophily among most monocotyledons generally results in relatively low pollen production. Only a few groups of monocots are wind-pollinated, for example Poaceae, Cyperaceae and Juncaceae, and produce large quantities of pollen. Taken as a whole, the biology and ecology of monocots, combined with the nature of the plant fossil record, makes unravelling the history of this group of angiosperms especially challenging.

While some supposed pre-Cretaceous angiosperms (e.g. *Sanmiguelia*) have sometimes been compared with monocots (Chapter 6) the earliest unequivocal records of the group are several distinctive fossils from the Early Cretaceous that are securely referable to extant Araceae (section 11.5.2). Others fossils of this age may also be early monocots, but more information is needed to establish their systematic affinities securely (Chapter 9).

11.3 PUTATIVE EARLY MONOCOT FOSSILS

11.3.1 Dispersed leaves

In his initial critical evaluation of the early fossil record of monocotyledons, Doyle (1973) highlighted three criteria as

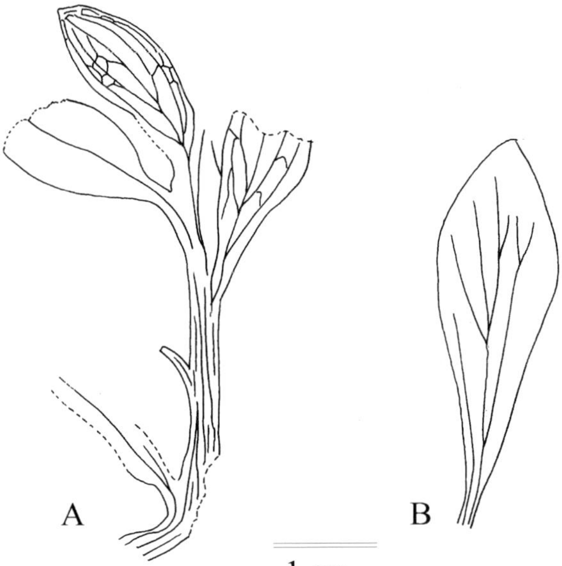

Figure 11.2 *Acaciaephyllum spatulatum*, a putative early monocot shoot with attached leaves from the Early Cretaceous (Early Aptian) Dutch Gap locality, Virginia, USA. (A) Shoot showing attached leaves. (B) Isolated leaf showing poorly organised venation. Redrawn from Doyle and Hickey (1976).

guides for separating potential leaves of monocots from those of similar plants: the presence of a hierarchy of vein orders of the parallel veins, the presence of higher-order cross-veins connecting the parallel veins, and convergence of the leaf veins towards the leaf apex. None of these criteria are definitive by themselves, but they nevertheless provide a basis for assessing the possible monocot relationship of enigmatic material. As noted by Doyle (1973), most putative monocot leaves from Early Cretaceous floras either do not meet these criteria or are too poorly preserved for detailed systematic evaluation. An intriguing large leaf fossil from the latest Albian (?) of Pingzhou Island, Hong Kong, described as *Amesoneuron* sp., may be an early monocot, but the age is currently uncertain and there is no information from the cuticle (Zhou *et al.*, 1990).

Acaciaephyllum (Figure 11.2) has figured prominently in discussions of early monocot diversification. The genus was described from the Dutch Gap locality in the Potomac Group of eastern North America by Fontaine (1889), who included four species in the genus based on seven specimens. Only one specimen has venation that is preserved sufficiently well to provide a basis for comparison with living plants (Doyle, 1973; Gandolfo *et al.*, 2000). The specimen was restudied and re-illustrated by Doyle (1973) and Hickey and Doyle (1977). It comprises a herbaceous stem bearing three alternate leaves and two possible leaf bases. The leaves have sheathing leaf bases. There is a distinct, but relatively weakly developed midvein that gives off apically converging acrodromous secondary veins that are linked by cross-veins. In places there is a delicate reticulum formed by the secondary and tertiary veins. Doyle (1973) and subsequent authors accepted *Acaciaephyllum* as probable evidence of early monocots, but this conclusion was not supported by Gandolfo *et al.* (2000). In addition to *Acaciaephyllum*, Doyle (1973) reviewed several other fossil leaves of possible monocot relationship, all of which were considered less convincing (Chapter 9).

From the Late Cretaceous (Maastrichtian) Deccan Intertrappean Beds, India, several petrified leaves and rhizomes of unequivocal monocot relationship, but of uncertain affinity within the group, are reported. These include fossils such as *Aerophyllites intertrappea* (Chitaley and Patil, 1970) and *Aerorhizos harrissii* (Chitaley, 1968), which occur together with a range of other fossil material of certain monocot relationship.

A significant record from earlier in the Cretaceous is from the Late Cretaceous (Turonian) flora of the southern Negev Desert, Israel. This flora is unusual for its age in the relatively high proportion of putative monocots that it includes. Eight different kinds of monocot leaves have been described (Krassilov *et al.*, 2005) and although their preservation does not allow detailed comparison with modern plants, their monocot affinity appears uncontroversial. Several are leaves of probable water plants. Leaves of *Potamogetophyllum mite* were compared to floating leaves of extant *Potamogeton*, leaves of *Pontederites eichhornioides* have features that may indicate an affinity with *Pontederia* and *Eichornia*, and leaves of *Limnobiophyllum dentatum* may be related to Araceae (Krassilov *et al.*, 2005).

Leaves of *Potamogetophyllum mite* are small, up to about 8 mm long and lanceolate, with an entire margin, a mucronate apex and a long slender petiole. The venation is simple, with the central midrib connected to two lateral veins by densely spaced tertiary venation. Leaves of *Pontederites eichhornioides* are elliptical, up to about 11 cm long, and grade basally into an inflated petiole. Venation is simple, with fine cross-veins connecting the thicker primary veins to the more slender secondary and tertiary veins. *Limnobiophyllum dentatum* is described as having minute, reniform fronds, about 4 mm long and 6 mm wide, borne on a stipe about 6 mm long (Krassilov *et al.*, 2005). These leaf types are also known from other areas, associated with other organs of probable Araceae (section 11.5.2).

Among other possible monocot leaves from the Negev Desert, *Plumafolium bipartitum*, with its bifid leaf structure, parallel venation and long ribbon-like petiole, is distinct from any modern taxon. Krassilov *et al.* (2005) suggested possible affinity with Cyclanthaceae or Arecaceae. *Typhacites negevensis* has ribbon-shaped leaves with parallel venation of thicker primary veins interspaced by thinner secondary and tertiary veins. *Geonomites* sp. and *Quturea fimbriata* were both regarded as probable Arecaceae (Krassilov *et al.*, 2005). *Geonomites* sp. is fragmentary, but *Quturea fimbriata* is based on pinnate, palm-like leaves with a stout petiole that is basally expanded into a semicircular sheath. The rachis tapers to a pointed apex and bears small pinnae. The leaves are similar to those of modern palms, such as *Ravenea*, but are much smaller (Krassilov *et al.*, 2005).

11.3.2 Dispersed pollen

Pollen of most extant monocots is either monocolpate or trichotomocolpate, and may be difficult to distinguish from the pollen of some ANITA-grade angiosperms, Chloranthaceae and eumagnoliids. Several other aperture types, such as monoporate, dicolpate or triporate (e.g. Harley and Baker, 2001; Harley and Dransfield, 2003), as well as inaperturate forms, also occur among monocots. Although these pollen types are relatively uncommon, and are presumably derived from monocolpate forms, they may also be difficult to distinguish from pollen of other angiosperms, even including the pollen of some eudicots.

From the Late Cretaceous there are numerous records of dispersed pollen assigned to the monocots. Several of these are sufficiently characteristic to be assigned to extant families or orders, such as *Spinizonocolpites*, related to extant *Nypa* (Arecaceae). However, recognition of monocot pollen in the Early Cretaceous is more problematic and in most cases supporting evidence from other reproductive structures is desirable for secure determination.

One set of pollen features that appears to be unique for monocots is the presence of a graded reticulum, in which there are finer and coarser lumina on different parts of the grain, combined with a poorly developed endexine and basically monoaperturate pollen (Doyle, 1973; Walker and Walker, 1984; Harley, 1997). Walker and Walker (1984) also suggested other features that might be useful for identifying monocot pollen, such as laterally extended columellae that give the muri a 'frilled' appearance, and regular and clearly polygonal lumina. Much of the dispersed pollen assigned to the genera *Liliacidites* and *Similipollis* (Chapter 9), which is widely distributed in the Early Cretaceous, shows this combination of features. However, a complication is that pollen grains with a *Similipollis*-type reticulum occur on stigmas of *Anacostia*, an Early Cretaceous fruiting structure that also shows similarities to fruits of early-diverging angiosperm lineages such as *Kadsura* and *Schisandra* (Schisandraceae). It is therefore not certain that all *Similipollis*-type dispersed pollen represents early monocots (Chapter 9).

Another feature among monoaperturate pollen that appears to be restricted to some monocots is the acolumellate pollen wall structure seen in some reticulate grains, where the columellae are replaced by a thin granular layer. This occurs in some extant Araceae (Grayum, 1992), as well as pollen of fossil Araceae (*Mayoa*) from the Early Cretaceous of Portugal (section 11.5.2) and *Pennipollis*, a putative early monocot (section 11.5.1; Figure 11.3). Other fossil monocots, such as Araceae sp. A, have a granular inner lining of the muri and very few, scattered columellae (section 11.5.2).

11.3.3 Reproductive structures

Reproductive structures of early monocots are difficult to recognise because they usually cannot be assigned to specific extant groups. At a general level there are no key characters that distinguish them from equivalent structures in other early angiosperms. Several reproductive structures of possible monocot affinity, for which additional information is needed for a more secure systematic placement, are treated in Chapter 9. Among these are the Early Cretaceous fruiting structures of *Anacostia*, known from North America (*A. marylandensis*, *A. virginiensis*) and Portugal (*A. portugallica*, *A. teixeirae*). Fruit and seed characters of *Anacostia* suggest a relationship to certain magnoliids; however, pollen features point towards an affinity with monocots. Also treated in Chapter 9 are *Mabelia* and *Nuhliantha*, two putative

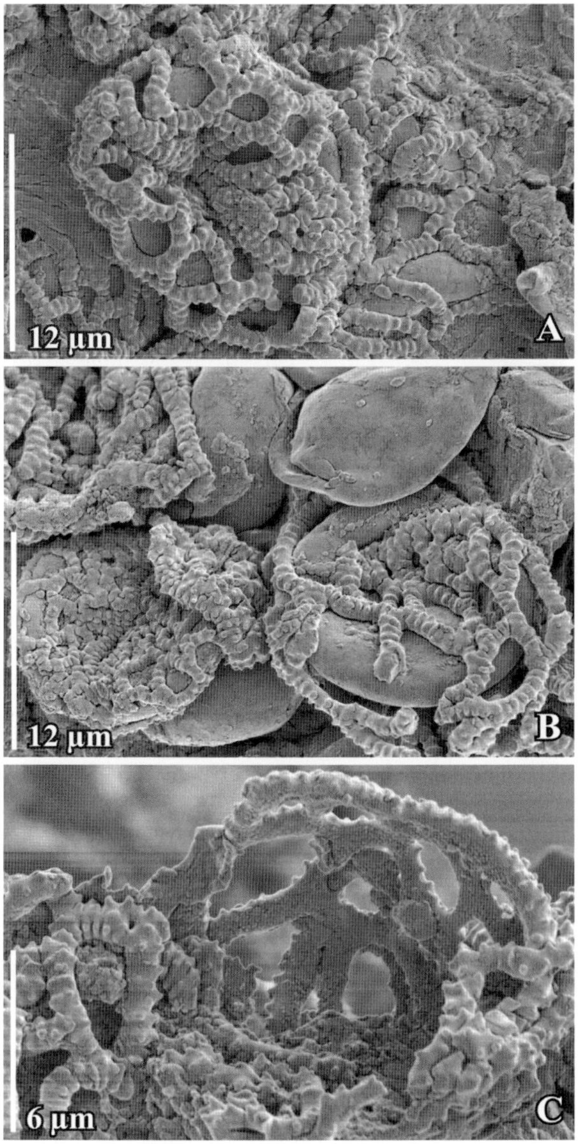

Figure 11.3 Pollen grains of the *Pennipollis*-type in a coprolite from the Early Cretaceous (Late Aptian – Early Albian) Vale de Agua locality, Portugal. (A) Pollen grains showing aperture and coarse reticulum with spiny ornamentation. (B) Group of pollen grains, some with reticulum still attached, others naked. (C) Reticulum that has come loose from the main body of the grain, revealing details of the inner surface and showing fine granulae of the infratectal layer that provide the connection to the foot layer. Material in the collections of the Swedish Museum of Natural History, Stockholm.

monocots from the Late Cretaceous (Turonian) Old Crossman locality, New Jersey, USA. In some features they resemble members of extant Triuridaceae, but assignment of these fossils to the monocots is not fully secure. Two other trimerous flowers from the Early Cretaceous of Portugal may represent some of the oldest monocots, but detailed description and analysis of these fossils remains to be completed and they are here treated in Chapter 9. *Cretovarium japonicum*, the first angiosperm reproductive structure to be described from the Late Cretaceous (Coniacian–Campanian) flora of Hokkaido, Japan (Stopes and Fujii, 1910), is also a putative monocot, discussed in Chapter 9. There are also many putative monocot flowers in the Late Cretaceous floras of Portugal, but these remain to be studied and formally described.

A variety of monocot floral and fruiting structures have been described from the Late Cretaceous (Maastrichtian) Deccan Intertrappean Beds, India. These include palm fruits, seeds and stems (section 11.10.1) as well as fruits and stems related to *Musa* (section 11.10.4). But there are also inflorescences and flowers for which the relationships are less clear. These include *Shuklanthus* and *Viracarpon*, as well as *Tricoccites trigonum* and *Deccananthus savitrii*.

Shuklanthus is based on an inflorescence with flowers; *Viracarpon* includes infructescences with fruits. The two genera are very similar in their organisation and anatomy and probably represent different developmental stages of the same kind of plant (Chitaley and Patil, 1971). *Shuklanthus superbum* (Figure 11.4) was first described from Madhya Pradesh, India (Verma, 1958). It is a fragment of a racemose inflorescence about 47 mm long. The axis is angular at the nodes and rounded in the internodal regions, and consists of numerous small, densely packed flowers borne on short stalks. Flowers are in six longitudinal rows, apparently with three flowers at each node. Flowers are pistillate with six tepals. The gynoecium comprises six apocarpous or partly fused carpels arranged around a stout central extension of the receptacle that widens apically (Verma, 1958). Each carpel is opposite a tepal and has a single ovule that completely fills the locule. The anatomy of the inflorescence axis is clearly monocotyledonous with a broad cortex and a central ground tissue with many closely spaced vascular bundles.

Viracarpon hexaspermum (Figure 11.4) and *V. elongatum* were formally described by Sahni (1944) from Mohgaon

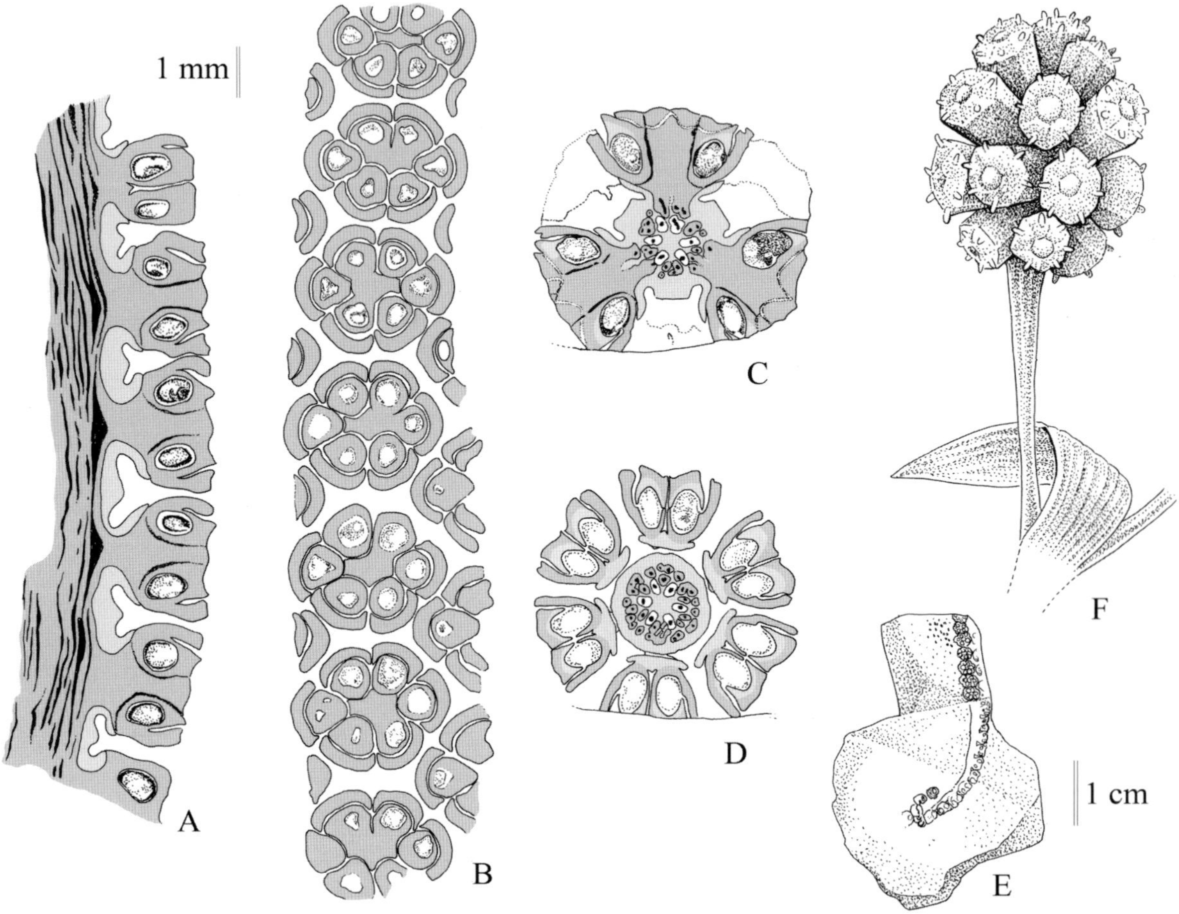

Figure 11.4 Monocot flowers and fruits of uncertain relationship from the Late Cretaceous (Maastrichtian) Deccan Intertrappean Beds, India. (A–E) Inflorescence and flowers of *Shuklanthus superbum* from Madhya Pradesh showing (A) longitudinal–radial, (B) tangential and (C, D) transverse sections (C at node, D between nodes) of inflorescence with (C) flowers in alternating whorls of three; (E) inflorescence in rock. (F) Reconstruction of infructescence of *Viracarpon hexaspermum* from Mohgaon Kalan showing infructescence with fruits. (A–D, F) Redrawn from Verma (1958); (E) redrawn from Nambudiri and Tidwell (1978).

Kalan, India. Further material of *Viracarpon*, including two new species, *V. chitaleyi* and *V. sahnii*, were added later from other areas in the Deccan Intertrappean Beds (Chitaley, 1954, 1958; Chitaley *et al.*, 1969; Patil, 1972; Nambudiri and Tidwell, 1978). Fruits of *Viracarpon* are arranged along the infructescence axis in six rows in a manner similar to the placement of flowers of *Shuklanthus* along the inflorescence axis. Each fruit is formed from six carpels that are free for part of their length and grouped around a massive central tissue. Individual fruitlets are drupes with a single seed (Nambudiri and Tidwell, 1978). Infructescences of *Viracarpon hexaspermum* and *V. sahnii* (included in *V. hexaspermum* by Nambudiri and Tidwell, 1978) are broadly elliptical in outline with perhaps less than 30 fruits. *Viracarpon elongatum* and *V. chitaleyi* have more elongated infructescences that are more comparable to the elongated inflorescences of *Shuklanthus superbum*. Chitaley and Patil (1971) suggested that *S. superbum* should be treated as a synonym of *V. elongatum*.

Verma (1958) indicated a relationship for *Shuklanthus superbum* to Scheuchzeriaceae or related alismatalean

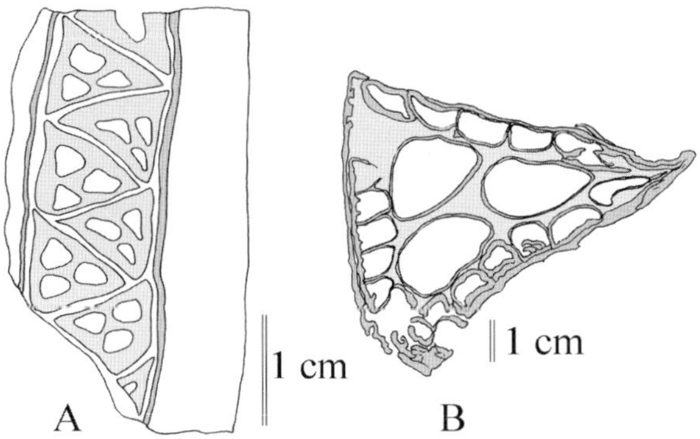

Figure 11.5 *Tricoccites trigonum*, a probable monocot fruiting structure from the Late Cretaceous (Maastrichtian) Deccan Intertrappean Beds, Mohgaon Kalan, India. (A) Tangential section of infructescence with crowded, trilocular fruits. (B) Transverse section of fruit showing the three locules and fruit wall with air chambers. Drawn from photographs in Chitaley (1956).

families. Chitaley (1954) suggested relationship with Pandanaceae or perhaps Arecaceae, which was supported by Nambudiri and Tidwell (1978). Nambudiri and Tidwell suggested that *Viracarpon* is closest to extant Pandanaceae in most floral and fruit characters, but also that some features, such as characters of the embryo, were more comparable to those of certain extant Araceae.

Tricoccites trigonum (Figure 11.5), another fossil fruiting structure from the Deccan Intertrappean Beds, India, has been compared to fruits of extant Pandanaceae and palms (e.g. Sahni and Rode, 1937; Chitaley, 1956). The species is based on fragments of elongate infructescences as well as isolated fruits from Mohgaon Kalan. The infructescence is enveloped by several ensheathing leaves. It is at least 14 cm long, but complete specimens have not been found. Fruits are trilocular drupes, about 30 mm long and 30 mm in diameter, with a smooth wall or with faint longitudinal ribs. There is a single seed in each locule. The fruits are sharply triangular in transverse section and have a distinctive wall consisting of an outer layer of parenchymatous cells and various fibrous and sclerenchymatic tissues internally. Regularly spaced air chambers give this layer a characteristic, coarsely pitted appearance in section and most likely indicate water dispersal. *Tricoccites trigonum* has sometimes been included in *Palmocarpon*, but it is distinguished by the distinctive construction of the fruit wall (Chitaley, 1956). *Tricoccites* was first compared to fruits of Euphorbiaceae (Rode, 1933), but has later been regarded as a monocot and compared to Arecaceae and Pandanaceae. So far, however, it has not proved possible to link the fossil convincingly with a specific family or genus.

Deccananthus savitrii is a possible monocot flower based on a single specimen, also from the Mohgaon Kalan (Chitaley and Kate, 1974). The flower is trimerous with two whorls of perianth parts and apparently two whorls of stamens. Anthers are tetrasporangiate and contain small trichotomocolpate pollen. The gynoecium is trilocular with a single style topped by a solid and simple stigma. There is no information on ovules or seeds. Based on the trichotomocolpate pollen and the floral organisation Chitaley and Kate (1974) indicated a possible relationship with palms, but the information currently available does not allow for a more detailed systematic analysis.

11.4 ACORALES

The Acorales (Acoraceae) comprise a single genus, *Acorus*, with two to four species that are widely distributed in both the tropics and temperate regions, mainly in the Northern Hemisphere (Mayo *et al.*, 1997; Bogner and Mayo, 1998). All species are rhizomatous herbs with long, narrow, parallel-veined leaves. Erect spadix-like inflorescences contain numerous bisexual, trimerous flowers of the basic monocot type. The perianth is composed of two whorls, each of three tepals. The androecium also comprises two whorls, each with three stamens. Anthers open by longitudinal slits. Pollen is monocolpate to subporate with a psilate to foveolate exine (Grayum, 1992). The gynoecium is syncarpous, with two to three locules and a minute, apical, sessile stigma. Each locule contains several pendulous, anatropous, bitegmic ovules that are borne apically (Bogner and Mayo, 1998).

There is strong support from molecular phylogenetics that *Acorus* is the sister group to all other monocots (e.g. Chase *et al.*, 2000; Chase, 2004) although some authors regard this as controversial (Stevenson *et al.*, 2000). The presence of ethereal oil cells in *Acorus*, which are otherwise unknown in other monocots, may reflect retention of a plesiomorphic feature from magnoliid precursors.

The Acoraceae are not well represented in the fossil record and several fossils previously assigned to the family are now excluded (Stockey, 2006). Specimens described as *Acorus brachystachys* from the Eocene of Spitsbergen (Heer, 1870) are now assigned to the extinct eudicot genus *Nordenskioeldia* (Kvaček in Bogner, 2001). Another Eocene fossil, *Acorites heeri*, comprising inflorescences preserved as compressions from the Wilcox flora of southeastern USA, first described as *Acorus heeri* and later transferred to the extinct genus *Acorites* (Crepet, 1978), is regarded as perhaps close to *Acorus* (Bogner *et al.*, 2005).

Small monocolpate pollen with a short colpus and a continuous tectum from the Late Barremian – Early Aptian Torres Vedras flora shows close similarity to pollen of extant *Acorus* L. (Friis *et al.*, 2010a), but ultrastructural studies of the pollen wall are needed before the fossils can be securely placed systematically.

11.5 ALISMATALES

The Alismatales are a well-supported monophyletic group based on molecular data (Chase *et al.*, 2000). In the APGIII classification (2009) the order comprises 13 families, several of which are well represented in the Cenozoic fossil record. Analyses of relationships within the group based on molecular phylogenetics recognise four main clades: Araceae (including Lemnaceae), Alismataceae (including Limnocharitaceae), hydrocharitoids (Hydrocharitaceae, Najadaceae, now included in Hydrocharitaceae, and Butomaceae), and potamogetonoids (comprising Potamogetonaceae and the rest of the order) (Les *et al.*, 1997). These families have also been treated as closely related in morphology-based classifications (e.g. Dahlgren *et al.*, 1985). Also placed within Alismatales in the APGIII system is Tofieldiaceae. The extant Tofieldiaceae comprise four genera and about 27 species and as far as we are aware there is no reported fossil record of this group.

11.5.1 Putative fossil Alismatales

Among the rich mesofossils recovered from the Early Cretaceous one distinctive and relatively common group of reproductive organs stands out as being of possible monocot affinities: *Pennistemon* stamens and *Pennicarpus* fruits, which are linked by the occurrence of *Pennipollis*-type pollen in the anthers and on the fruit surfaces.

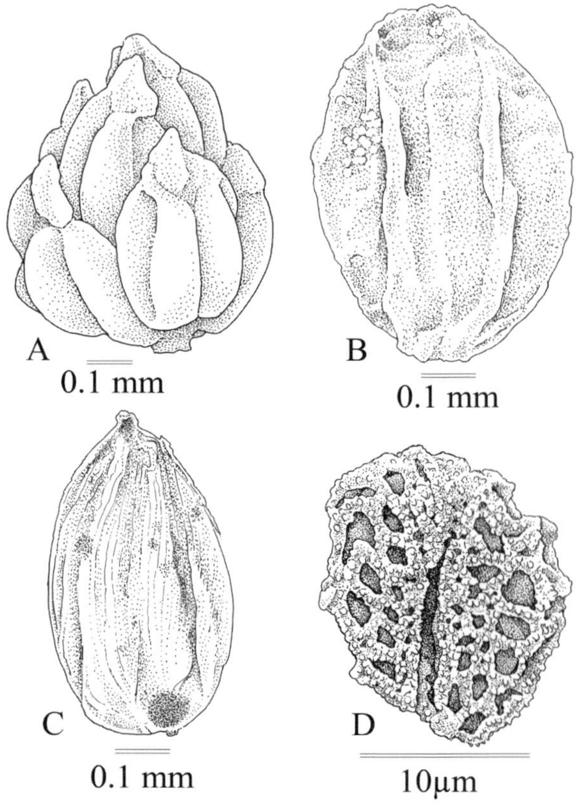

Figure 11.6 *Pennistemon* plant from the Early Cretaceous (Late Aptian – Early Albian) Vale de Agua locality, Portugal. (A) Inflorescence fragment of *Pennistemon portugallicus* showing several stamens in a spiral arrangement. (B, C) *Pennicarpus tenuis*: (B) fruit and (C) seed. (D) Monocolpate, acolumellate pollen grain from *Pennistemon portugallicus*. (A) Redrawn from Friis *et al.* (2000a); (B–D) drawn from SEM and LM images in Friis *et al.* (2000a).

Pennistemon portugallicus (Figure 11.6) is based on a small fragment of a reproductive axis with helically arranged stamens, as well as several isolated stamens, first described from the Early Cretaceous (Late Aptian – Early Albian) Vale de Agua locality, Portugal (Friis *et al.*, 2000a). Stamens are small, about 0.5 mm long, and differentiated into a short filament and a dithecate, tetrasporangiate anther. There is a short apical extension of the connective. Anther dehiscence is extrorse along longitudinal slits. Pollen *in situ* in the stamens is monocolpate. The colpus is long and slit-like with clearly delimited margins, and extends for the full length of the grain. The pollen is almost spherical, about 14–19 μm in diameter, and has a very coarse and loosely attached reticulum without columellae. The reticulum is almost free from the smooth central body. Muri are narrow, ornamented by distinct spines in one to several rows, and have a very thin granular lining on the inner surface. The wall of the central body is formed mainly by a thick foot layer. Endexine is well developed only under the aperture. Pollen of this kind also occurs abundantly in coprolites from the Vale de Agua locality (Figure 11.3), indicating that the flowers were visited by insects.

Similar pollen, with a loosely attached reticulum, acolumellate tectum and spiny muri, occurs abundantly in Early and mid-Cretaceous palynofloras. It is reported from the Barremian through to the Cenomanian, and perhaps extends to the Turonian (Friis *et al.*, 2000a). This kind of pollen is particularly characteristic of Early Cretaceous palynofloras from West Africa and Egypt, but is also common in the Early Cretaceous of Portugal, England and North America. From Egypt, 13 different taxa in this complex were distinguished from the Mersa Matruh borehole in sediments ranging in age from the Late Barremian to the Early Albian (Penny, 1988b). In these samples *Pennipollis* pollen is most diverse and abundant in the Aptian. Different taxa are distinguished based on pollen size, distribution and size of the lumen of the reticulum, configuration of the muri, and supratectal ornamentation. They were assigned to various biorecords with the prefix 'Retimono' (e.g. Retimono-Necklace).

Pennipollis pollen was first described from the Potomac Group palynofloras as *Peromonolites peroreticulatus* and *P. reticulatus* (Brenner, 1963), but later referred to the dispersed pollen genera *Retimonocolpites* (Doyle *et al.*, 1975) and *Brenneripollis* (Juhász and Góczán, 1985). However, the type species of both *Retimonocolpites* and *Brenneripollis* are distinctly different in having a columellate infratectal layer (Chapter 9). A new genus, *Pennipollis*, was therefore suggested for dispersed forms from the Potomac Group and elsewhere that have a loosely attached reticulum supported by a granular infratectal layer (Friis *et al.*, 2000a).

Minute and strongly flattened unilocular fruits described as *Pennicarpus tenuis* (Figures 11.6) are associated with the *Pennistemon* fossils at the Vale de Agua locality, as well as at other Early Cretaceous localities in Portugal, and were most likely produced by the same kind of plant (Friis *et al.*, 2000a). Fruits are about 0.6 mm long with widely spaced longitudinal ridges. The fruit wall is thin and the stigmatic area is indistinct. Each fruit contains a single seed that is apparently basally attached and orthotropous (or semi-orthotropous) with the chalaza slightly displaced. There is no trace of a raphe, but an anatropous organisation cannot be completely ruled out. Cell walls of the seeds are digitate in the micropylar area, but are otherwise straight. Pollen grains observed on the surface of the fruits are similar to those found *in situ* within *Pennistemon* stamens.

Pollen from extant angiosperms with a structure comparable to that of *Pennipollis* has only been reported from extant monocots. Especially similar is pollen of certain Araceae and other Alismatales, although among extant taxa there is no pollen that matches all the details seen in the fossils. Features of the staminate axis and the stamens are also consistent with an assignment to probable early monocots, but the information currently available for the *Pennipollis* plant is insufficient for precise systematic assessment. *Pennipollis* pollen has also been suggested to be similar to pollen of extant *Ascarina* (Chloranthaceae) (e.g. Hesse and Zetter, 2007), but *Ascarina* pollen is very clearly columellate and the reticulum is not as coarse.

11.5.2 Araceae

The Araceae comprise 109 genera and more than 3300 species. The family is virtually cosmopolitan in distribution, but is most diverse across the tropics (Mayo *et al.*, 1997). The Araceae exhibit great morphological diversity within their herbaceous ground plan and range from gigantic tropical geophytes (e.g. *Amorphophallus*) and epiphytes (e.g. *Philodendron*) to temperate herbs (e.g. *Arum*) and diminutive aquatics (e.g. *Lemna, Wolffia*). Leaves are generally differentiated into a petiole and an expanded blade of various shapes.

Flowers of Araceae are borne in inflorescences consisting of a spadix subtended by a variously modified spathe (Figure 11.7). Each spadix consists of numerous, tightly packed unisexual or bisexual flowers. Flowers are often trimerous or dimerous, and the tepals (when free) are in two whorls, or the flowers are naked. Stamens also occur in one or two whorls and may be free or fused in various ways. In subfamily Aroideae flowers are unisexual and lack a perianth. Staminate flowers consist of only one or two stamens. Pollen of Araceae is very

Figure 11.7 Inflorescences of extant *Zantedeschia aethiopica* (Araceae), each with an elongated spadix subtended by a white spathe. Photographed in the tropical greenhouse of the Bergius Botanic Garden, Stockholm.

variable, both in aperture configuration (monocolpate, dicolpate, zonacolpate, periporate or inaperturate) and sculpture (psilate to reticulate, striate or spinose). The gynoecium is syncarpous, typically one to three locular with an inconspicuous style. Ovules are borne in a variety of ways and range from orthotropous to anatropous (Mayo *et al.*, 1997).

Fossil pollen and leaves of Araceae. Araceae have an extensive fossil record extending from the Early Cretaceous to Recent. According to the survey by Hesse and Zetter (2007) the pollen record reaches its zenith in the Paleocene–Eocene, and this may correspond to the maximum extent of warm, humid, rainforest-type vegetation around the Eocene thermal maximum. Pollen assigned to the extinct form *Proxapertites operculatus* is particularly abundant during this time interval. This pollen type has been compared to pollen of extant *Monstera* (subfamily Monstereae) and *Gonatopus* (subfamily Zamioculcadoideae), but its precise systematic affinity is not yet established. It is also possible that the fossil taxon includes more than one lineage within the family. Hesse and Zetter (2007) suggested that a number of polyplicate pollen grains that have traditionally been regarded as ephedroid, and assigned to the extinct genus *Ephedrites*, should be included in the extant araceous genus *Spathiphyllum*. Another distinctive araceous pollen type from the Paleocene–Eocene comprises spiny, porate pollen similar to that found *in situ* in flowers of the extinct *Limnobiophyllum* plant.

Limnobiophyllum is an aquatic plant known from the Cenozoic of Europe and North America based on ovate to reniform leaves resembling those of extant *Spirodella* in the subfamily Lemnoideae (e.g. Kvaček, 1995; Stockey *et al.*, 1997). *Limnobiophyllum scutatum*, described from the Paleocene Paskapoo Formation, Alberta, Canada, is particularly informative with details of flowers, pollen and leaf anatomy preserved (Stockey *et al.*, 1997). These details establish unequivocally the relationship of this plant to the Araceae, but its exact position with respect to extant genera remains uncertain (Collinson *et al.*, 2002). Occurrences of *Pandaniidites* pollen similar to the *in situ Limnobiophyllum* pollen (e.g. Jarzen, 1983) in Maastrichtian sediments indicate that *Limnobiophyllum* was probably also part of the Late Cretaceous vegetation in North America. There is also a record of *Limnobiophyllum dentatum* from the Late Cretaceous (Turonian) of Negev Desert, Israel (Krassilov *et al.*, 2005) (section 11.3.1). Other unequivocal araceous fossils from the Palaeogene include leaves from several localities in North America (Dilcher and Daghlian, 1977; Hickey, 1977; Mayo *et al.*, 1997; Bogner *et al.*, 2007).

The Cretaceous record of Araceae is sparse compared with that from the Cenozoic, but it is now clear that the family has a long history that extends back to the Early Cretaceous. By the latest Cretaceous and Early Cenozoic the family was diverse with several subfamilies represented based on both vegetative and reproductive structures. The earliest fossils that can be assigned to Araceae are striate pollen grains of *Mayoa portugallica* from the Early Cretaceous of Portugal. Other striate pollen from the Early Cretaceous, such as species assigned to the dispersed pollen genus *Jugella* (Mtchedlishvili and Shakhmoundes, 1973), may also be related to Araceae, but based on light microscopy alone these cannot be distinguished easily from pollen of Gnetales (*Ephedra*, *Welwitschia*). There are also unequivocal fossil flowers assignable to Araceae from the Early Cretaceous (Friis *et al.*, 2010b). One of these has *Retimonocolpites*-type

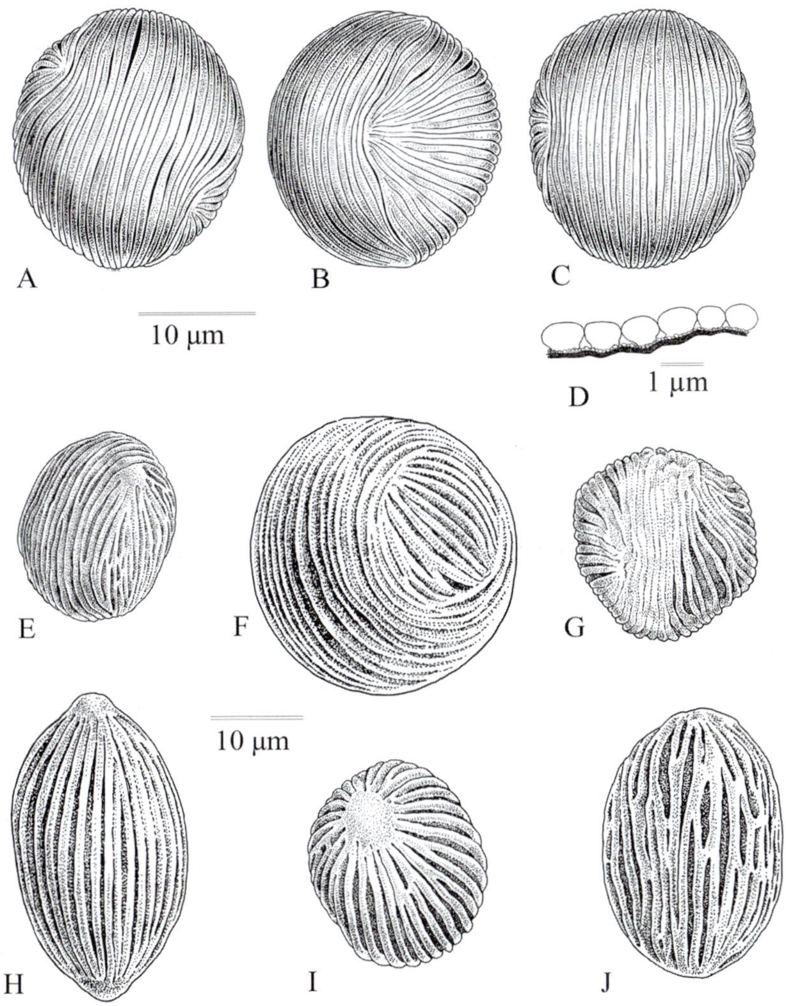

Figure 11.8 (A–D) *Mayoa portugallica* pollen from the Early Cretaceous (Late Barremian – Early Aptian), Torres Vedras locality, Portugal; (A–C) shape reconstructed and (D) section of pollen wall. (E–J) Pollen of extant Araceae: (E, F) *Holochlamys guineensis*; (G) *Holochlamys beccari*; (H, I) *Spathiphyllum kalbreyeri*; (J) *Spathiphyllum laeve*. (A, B) From Friis *et al.* (2004); (E) drawn from SEM image in Hesse *et al.* (2000); (F–J) drawn from SEM images in Grayum (1992).

pollen *in situ* and it is possible that some of the many different species of *Retimonocolpites*-type pollen from the Early and Late Cretaceous are also araceous.

Several unequivocal aroid leaf fossils are known from the Late Cretaceous. Leaves similar to the orontioid genus *Lysichiton* are known from the Early Campanian Grünbach locality of Lower Austria (Kvaček and Herman, 2004; Bogner *et al.*, 2007), leaves of *Orontium* were reported from the Maastrichtian of Mexico (Bogner *et al.*, 2007), and leaves of another orontioid genus, *Symplocarpus*, were reported from the Maastrichtian of North Dakota and from the Paleocene of Colorado (Bogner *et al.*, 2007). An aerial stem with root scars and remains of leaf sheaths of a monsteroid plant described as *Rhodospathodendron tomlinsonii* is known from the Maastrichtian Deccan Intertrappean Beds, India (Bonde, 2000). *Pistia*-like leaves assigned to the extinct genus *Cobbania* are known from the Late Cretaceous (Campanian–Maastrichtian) of North America and Asia (Stockey *et al.*, 2007).

Mayoa portugallica (Figures 11.8, 11.9) from the Early Cretaceous (Late Barremian – Early Aptian) of Torres Vedras, Portugal, is a small fossil of uncertain morphology with masses of pollen adhering to an irregular structure (Friis *et al.*, 2004). The pollen is striate (polyplicate), inaperturate, elliptical in outline and about 20 μm long. The pollen wall is characteristically striate with closely spaced, straight ribs that are separated by narrow grooves. The ribs are arranged in two sets that cross each other in a characteristic way at angles ranging between 45° and 90°. The ribs are solid in cross-section and about 1 μm high. They are

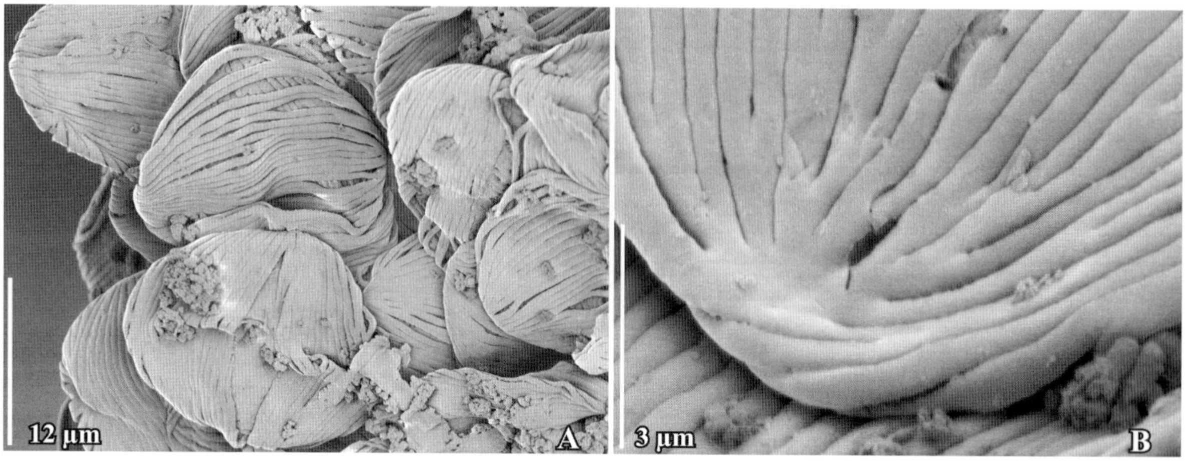

Figure 11.9 *Mayoa portugallica* pollen from the Early Cretaceous (Late Barremian – Early Aptian) Torres Vedras locality, Portugal. (A) Group of pollen grains. (B) Detail of pollen showing the cross-wise striations. SEM micrographs of material in the collections of the Swedish Museum of Natural History, Stockholm.

supported by a very thin, granular or rarely weakly columellate, infratectal layer and a thin, granular, evenly distributed endexine. In many specimens the whole pollen wall is split between the ribs and the pollen is only kept together at the points on the grains where the ribs diverge.

Mayoa pollen grains are strikingly similar to those of extant *Holochlamys*, tribe Spathiphylleae, subfamily Monsteroideae, in every respect. *Holochlamys* has inaperturate, striate pollen with ribs that form a crossed pattern as in the fossil grains (Figure 11.8). Pollen of extant *Spathiphyllum* is also very similar, but the ribs are arranged in a single direction only (Figure 11.8). The pollen wall also has a thin granular to slightly columellate infratectal layer, and a thin, evenly distributed endexine.

Proxapertites-type pollen (Figure 11.10) is another Early Cretaceous pollen type thought to belong to Araceae. Pollen is zonate–aperturate and has been found in an elongated pollen cluster that contains entire and broken pollen grains mixed together indicating that the cluster is a coprolite. The specimen (S136702) is from sample 149 from the Early–Middle Albian Puddledock locality, eastern North America, and not, as reported earlier (Friis *et al.*, 2004), from sample 149 from the Torres Vedras locality, Portugal. Pollen is tectate–acolumellate, circular to elliptical in the plane of the aperture, and about 30–35 μm in diameter. Grains are characterised by an aperture that extends almost around their full circumference and that separates the pollen into two shallow and almost equal, dome-shaped halves. The pollen wall is about 2.5 μm thick with a psilate–punctate tectum. Perforations of the tectum increase in size towards the inside of the exine. The infratectal layer is extremely thin and consists of faintly developed granulae that are visible on the inner surface of the tectum. The foot layer is homogeneous. Fossil pollen of this kind is most similar to pollen in the two araceous subfamilies, Monstereae (e.g. *Monstera*) and Zamioculcadeae (e.g. *Gonatopus*).

Fossil floral structures and seeds of Araceae. The records of Araceae based on fossil floral structures and seeds are mainly restricted to the Cenozoic. In the Neogene, seeds of Araceae are diverse in the lignite floras of Europe and Siberia, and in some floras seeds of *Pistia* are particularly common (Friis, 1985b). In the Palaeogene, there is an aroid inflorescence preserved in Baltic amber (Bogner, 1976) and there are also permineralised flowers, fruits and seeds from the Princeton Chert, British Columbia, Canada (Cevallos-Ferriz and Stockey, 1988; Smith and Stockey, 2003; Stockey, 2006). In the Cretaceous, the record is sparse, but unequivocal araceous fossils include several floral structures from the Early Cretaceous of Portugal, seeds from the Campanian—Maastrichtian of Portugal, and *Albertarum pueri* from the Late Campanian of Canada (Bogner *et al.*, 2005).

Several kinds of fossil floral structures of unequivocal araceous affinity have been discovered from the Early

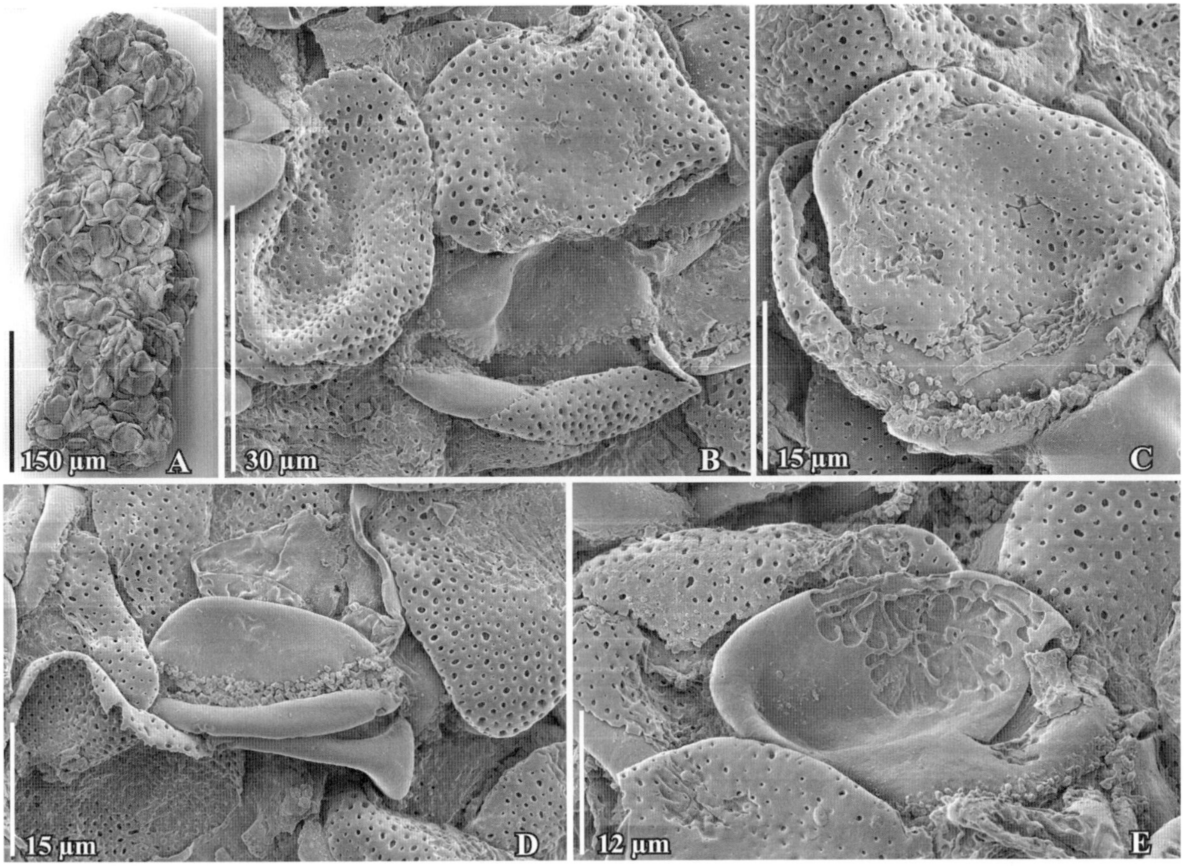

Figure 11.10 *Proxapertites* type pollen from a coprolite-like pollen clump from the Early Cretaceous (Early–Middle Albian) Puddledock locality, Virginia, USA. (A) Pollen clump. (B) Group of pollen grains showing punctate tectum and near-circular aperture extending almost completely around the grain. (C) Pollen grain with punctate tectum and almost circular aperture. (D) Group of fragmentary pollen grains showing one naked grain where tectum has loosened and fragments of grains that consist only of tectum. (E) Naked pollen grain with small grooves in the foot layer. SEM micrographs of material in the collections of the Field Museum, Chicago.

Cretaceous (Late Aptian – Early Albian) Vila Verde 2 locality, Portugal, and remain to be formally described and named. One of these (Figures 11.11, 11.12), referred to as Araceae sp. A by Friis *et al.* (2010b), comprises small fragments of inflorescence axes with stamens. Isolated stamens are also known. Specimens are mostly preserved as charcoal with their original three-dimensional shape intact. The small staminate flowers consist of a single, or perhaps sometimes two, stamens that are densely packed around the inflorescence axis in a spiral arrangement. Stamens have a very short stalk with a bulky tetrasporangiate anther. The four pollen sacs are widely separated from each other by an extensive connective. The surface of the anthers bulges irregularly and there are also remains of a secretion, suggesting that the anthers were probably osmophoric. Anthers produced monocolpate, reticulate pollen that sometimes occurs in strands. The reticulum is coarse with a dimorphic lumen and is only loosely attached to the foot layer by a few, scattered columellae. Pollen is similar to dispersed pollen assigned to *Retimonocolpites* and sometimes to *Liliacidites*. Similar *Retimonocolpites* pollen has also been found in a variety of isolated stamens from other Early Cretaceous localities in Portugal. These stamens are often bulky, and probably also osmophoric. It is likely that at least some of them are also stamens of early Araceae.

The fossil inflorescences and staminate flowers are closely similar to those of certain extant Araceae in the subfamily Aroideae that have simple, naked staminate

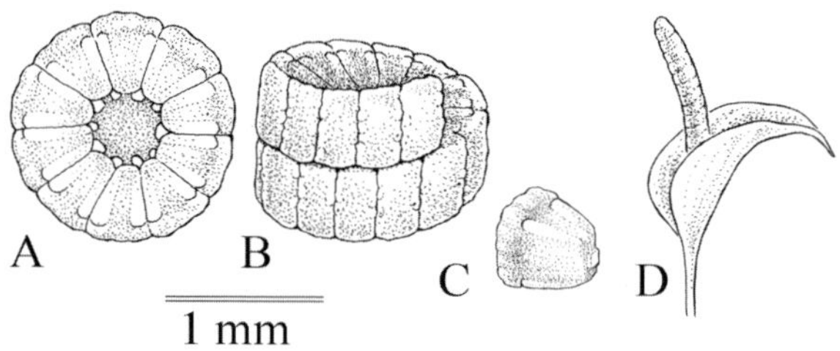

Figure 11.11 Reconstruction of aroid flower and inflorescence from the Early Cretaceous (Late Aptian – Early Albian) Vila Verde 2 locality, Portugal. (A) Apical view of fragmentary inflorescence showing ten unistaminate flowers, (B) fragment of inflorescence in lateral view showing arrangement of unistaminate flowers, (C) isolated staminate flower/stamen, and (D) reconstruction of inflorescence showing spadix and presumed spathe. Based on material in the collections of the Swedish Museum of Natural History, Stockholm.

flowers (Cabrera *et al.*, 2008) and that consist only of one or two stamens. The fossils differ from extant Aroideae only in their reticulate pollen. Except for *Calla*, which is sometimes included in the Aroideae, extant members of the subfamily are characterised by atectate pollen. In all other Araceae, pollen grains are tectate. The fossil aroid may exhibit the plesiomorphic condition for the pollen wall in Araceae, but the very loose reticulum may indicate an intermediate condition between normal reticulate and atectate pollen. Together with the presence of *Mayoa* (subfamily Monsteroideae) in the Late Barremian – Early Aptian it is clear that crown group Araceae were well established and diverse already by the end of the Early Cretaceous.

A second floral structure that can be assigned to Araceae from the Vila Verde 2 locality, Portugal, was referred to by Friis *et al.* (2010b) as Araceae sp. B. This fossil is a larger inflorescence fragment, also with densely spaced flowers that are arranged helically around the inflorescence axis (Figure 11.13). This fossil is lignitised and slightly compressed, and details of floral organisation are less distinct than in the charcoalified fossils. Flowers also have a distinct perianth (perigonate) of four tepals. They were apparently bisexual with four free stamens and a central gynoecium. Pollen *in situ* is periporate and semitectate–reticulate with a coarse, homogeneous reticulum, psilate muri and long, scattered columellae. This kind of pollen has not been described from any dispersed Early Cretaceous palynoflora, but has been found *in situ* in isolated stamens from other Early Cretaceous floras in Portugal. In extant Araceae similar periporate pollen occurs in subfamily Pothoideae. The fossil flowers are also similar to those of extant Pothoideae in being perigonate and bisexual. A further possible pothoid fossil has also been discovered from the Vale de Agua locality, Portugal, which is more or less contemporaneous with the Vila Verde 2 fossil assemblage, but in this case the periporate pollen has a much finer reticulum and supratectal ornamentation of minute spinules.

Several other undescribed floral fragments, fruits, seeds and stamens from Portuguese Early Cretaceous mesofossil floras may also be related to Araceae, but further studies are needed before these can be placed confidently among the early-diverging lineages of monocots.

Small araceous seeds (Figure 11.14) closely resembling those of certain extant Araceae are known from the Late Cretaceous (Campanian–Maastrichtian) Mira locality, Portugal. Seeds are about 1.5 mm long and about 0.5 mm broad, elongate reniform in shape, with concave ventral and convex dorsal surface. The raphe extends from the hilum to the chalaza in a narrow groove along the ventral surface. The micropyle is seen at one end of the seed as a distinct circular hole in the seed wall. The seed wall has an ornamentation of long, narrow, longitudinal ridges that follow the curvature of the seed. Specimens in which the outer part of the fruit wall is abraded show numerous spherical grooves arranged in longitudinal rows between the ridges. These are interpreted as the remains of large crystal cells. Similar reniform, ribbed seeds are characteristic for many extant members of the subfamily Lasioideae and are also known for some extant members of Monsteroideae (Seubert, 1993).

Albertarum pueri (Figure 11.15) is a silicified araceous infructescence from the Late Cretaceous (Campanian) Horseshoe Canyon Formation, Alberta, Canada (Bogner *et al.*, 2005). The infructescence is stalked and almost spherical in outline with densely packed, bisexual flowers. Flowers have six tepals, the gynoecium is trilocular, and each locule has a single anatropous ovule. Seeds are prominently ribbed with distinctive, sharp, longitudinal ridges. *Albertarum* was referred to the subfamily Orontioideae based on the nature of the stylar region and perianth, as well as supporting ovule and seed characters (Bogner *et al.*, 2005).

11.5.3 Alismataceae

The Alismataceae (including Limnocharitaceae) comprise 15 genera (*c.* 88 species) and have a virtually cosmopolitan distribution. Most species grow in water. Flowers may be

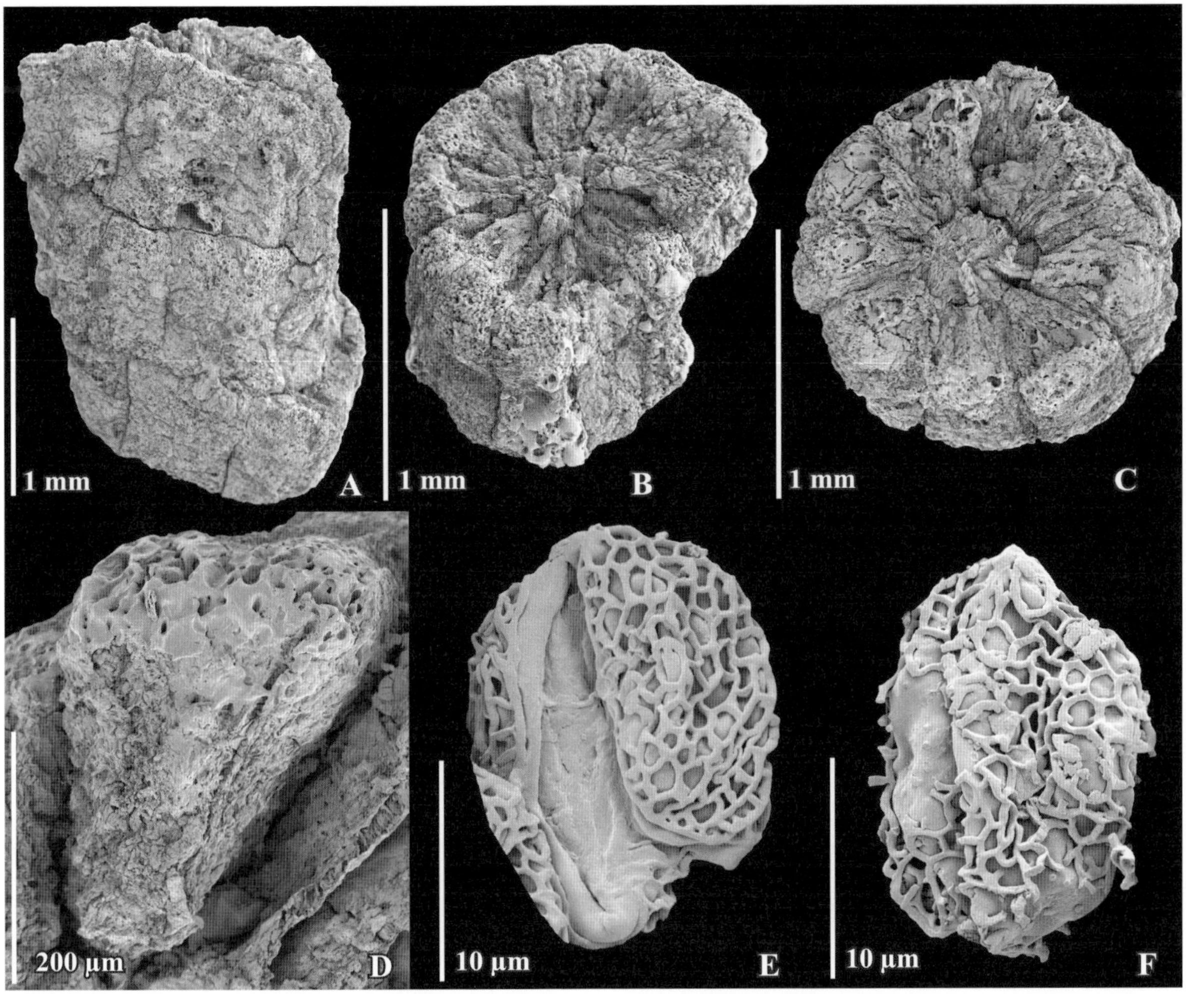

Figure 11.12 Inflorescence and flower fragments of Araceae from the Early Cretaceous (Late Aptian – Early Albian) Vila Verde 2 locality, Portugal. (A) Lateral view of inflorescence fragment showing helically arranged unistaminate flowers. (B) Oblique view of inflorescence fragment showing the arrangement of unistaminate flowers each consisting of a single stamen. (C) Apical view of inflorescence fragment showing ten unistaminate flowers. (D) Detail of stamen showing a pair of pollen sacs separated by a bulky connective with traces of a secretion. (E, F) Pollen grains. SEM micrographs of material in the collections of the Swedish Museum of Natural History, Stockholm.

unisexual or bisexual with two whorls, each of three tepals: the outer sepal-like, the inner petal-like. Stamens range from six to many. Carpels are free and range from three to many. Each carpel contains one to two basal anatropous ovules or numerous anatropous to campylotropous ovules that are attached all over the inner carpel surface (laminar placentation).

Fruits and seeds of Alismataceae are common in the Cenozoic from the latest Eocene – Early Oligocene onwards (Collinson, 1983b; Friis, 1985b; Haggard and Tiffney, 1997). A petrified petiole (*Heleophyton*) from the Middle Eocene Princeton Chert, British Columbia, Canada, shows features of both Alismataceae and Butomaceae (Erwin and Stockey, 1989). The Cretaceous record is sparse, but includes the exquisitely preserved leaves of *Cardstonia tolmanii* (Figure 11.16) described from the Campanian–Maastrichtian near Cardston, Alberta, Canada (Riley and Stockey, 2004).

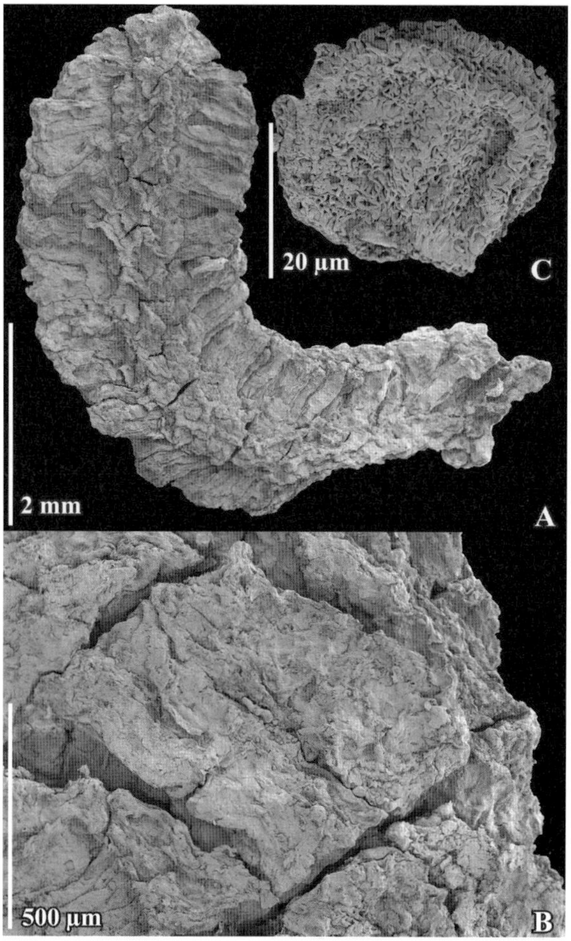

Figure 11.13 Inflorescence of Araceae from the Early Cretaceous (Late Aptian – Early Albian) Vila Verde 2 locality, Portugal. (A) Inflorescence, fractured along the inflorescence axis and seen from the inside. (B) Detail of outer surface of inflorescence showing the outlines of four presumed tepals. (C) Periporate pollen grain. SEM micrographs of material in the collections of the Swedish Museum of Natural History, Stockholm.

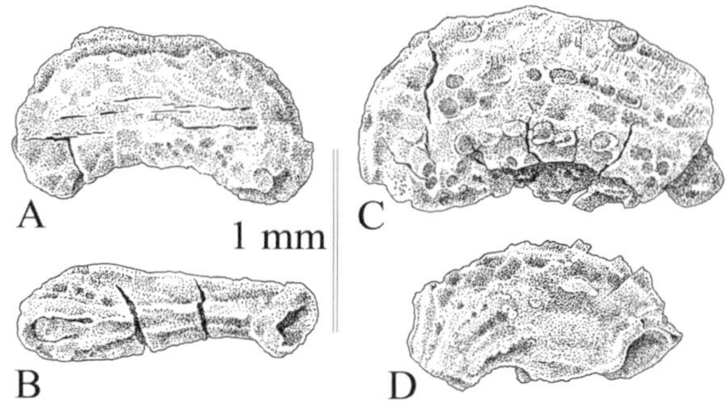

Figure 11.14 Reniform seeds of Araceae from the Late Cretaceous (Campanian–Maastrichtian) Mira locality, Portugal, showing slightly pitted seed wall; seeds seen (A, C, D) in lateral view and (B) towards micropyle and hilum. Drawn from material in the collections of the Swedish Museum of Natural History, Stockholm.

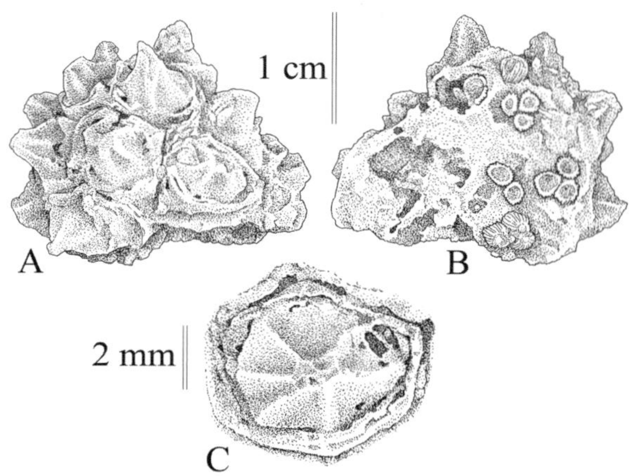

Figure 11.15 Silicified infructescence of *Albertarum pueri* (Araceae) from the Late Cretaceous (Campanian) Horseshoe Canyon Formation, southern Alberta, Canada. (A) Surface view of infructescence fragment showing several fruits. (B) Specimen with prominently ribbed seeds exposed in the locules. (C) Apical view of fruit showing the six carpels and remains of surrounding perianth. Drawn from photographs in Bogner *et al.* (2005).

11.5.4 Butomaceae and Hydrocharitaceae

Butomaceae and Hydrocharitaceae, sometimes referred to as hydrocharitoids, are a group of morphologically diverse water plants. The Butomaceae contain only a single genus. *Butomus umbellatus* is an emergent freshwater aquatic that is widely distributed in temperate Eurasia (Cook, 1998b; Haynes *et al.*, 1998). The Hydrocharitaceae including Najadaceae are more diverse, with 18 genera and about 115 species. Three genera of Hydrocharitaceae (*Enhalus, Thalassia, Halophila*) are marine aquatics. *Najas* includes about 40 species, which grow entirely submerged in fresh or brackish water, and is virtually cosmopolitan in distribution. Flowers of hydrocharitoids may be highly modified, reflecting a variety of different pollination mechanisms. However, it is clear that the flowers are fundamentally trimerous.

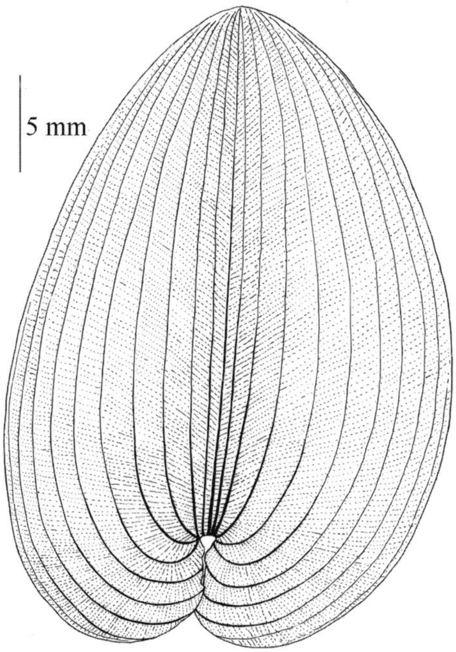

Figure 11.16 Leaf of *Cardstonia tolmanii* from the Late Cretaceous (Campanian–Maastrichtian) near Cardston, Alberta, Canada. Drawn from photograph in Riley and Stockey (2004).

As might be anticipated from their aquatic habit, some hydrocharitoids have an extensive fossil record. However, despite well-documented occurrences from the Cenozoic, there are no reliable Cretaceous records. Seeds similar to those of *Butomus* and *Najas* are known from Oligocene and younger strata in Europe (Mai, 1985a; Collinson *et al.*, 1993). Seeds of *Stratiotes* (Hydrocharitaceae) are also very well represented, based on abundant, well-determined material, from the Late Paleocene onwards in Europe (Collinson, 1983b; Mai, 1985a; Sille *et al.*, 2006).

11.5.5 Potamogetonoids

Potamogetonoids include eight families based on the family circumscriptions in APGIII (2009). All potamogetonoids are water plants and several genera are tolerant of brackish water. Cymodoceaceae, Posidoniaceae and Zosteraceae are marine aquatics. There is great morphological diversity within the group, but systematic diversity is not high. Four families comprise only a single genus: Aponogetonaceae (*Aponogeton*, 45 species), Posidoniaceae (*Posidonia*, nine species), Ruppiaceae (*Ruppia*, one to ten species) and Scheuzeriaceae (*Scheuchzeria*, one species). Cymodoceaceae (five genera, 16 species), Juncaginaceae (four genera, 12 species), Potamogetonaceae (two to three genera, 85–90 species) and Zosteraceae (three genera, 18 species) are also not highly diverse. Floral organisation in the group ranges from regularly trimerous in *Scheuchzeria* to tetramerous in *Potamogeton*. In Zosteraceae both staminate and pistillate flowers are highly modified in association with their unusual underwater pollination.

As in other Alismatales, the aquatic habit of potamogetonoids might be expected to predispose them to good representation in the fossil record, and this is the case for certain groups. However, there are only scattered reports from the Cretaceous. The most informative material includes linear leaves with anatomical preservation known from Late Cretaceous (Maastrichtian) and Early Cenozoic (Danian) marine sediments in The Netherlands. The leaves, assigned to the extinct genus and species *Thalassotaenia debeyi*, closely resemble those of modern seagrasses in Cymodoceaceae and Posidoniaceae (van der Ham *et al.*, 2007). Other linear leaves and stems from the Late Cretaceous of Germany, Belgium and the Netherlands, which are probably related to Cymodoceaceae and Posidoniaceae, are discussed by van der Ham *et al.* (2007). Further Late Cretaceous records include shoots of *Archaeozostera* with attached leaves and fruits. Four species have been recognised from the Late Cretaceous (Maastrichtian) Izumi Group of Shikoku, Honshu, and Hokkaido, Japan (Koriba and Miki, 1958). The material has been compared with Zosteraceae, but additional studies would be helpful to confirm this identification. Fossil seagrasses of uncertain affinities are also known from the Eocene of eastern North America (Lumbert *et al.*, 1984). Today seagrasses are important components of many coastal ecosystems and their establishment in near-shore environments in the latest Cretaceous sea may have had a considerable impact on evolution and ecology of the marine fauna.

In the Cenozoic, fruits of the extinct genus *Limnocarpus*, which are very similar to fruits of both Potamogetonaceae and *Ruppia*, are widespread and abundant from the Late Paleocene onwards in Europe and Asia. Seven genera have been recognised in this complex (Collinson, 1982). Leaves of *Aponogeton* and *Potamogeton* are also reported from the Oligocene (Zhilin, 1974) and Miocene (Kovar-Eder, 1992) respectively, and fruits of

Potamogetonaceae are common and widespread in Cenozoic floras of Europe and Asia (e.g. Friis, 1985b).

11.6 DIOSCOREALES

The Dioscoreales are a well-supported monophyletic group based on molecular data, and include Dioscoreaceae, Burmanniaceae and Nartheciaceae (APGIII, 2009). The Dioscoreaceae comprise three to 20 genera (about 200 species) and the Burmanniaceae comprise 13 genera (about 130 species). Both families are mainly restricted to tropical and warm temperate regions (Huber, 1998; Maas-van de Kamer, 1998). The Nartheciaceae, with four genera and about 40 species, are mainly found in the temperate Northern Hemisphere, but also occur in Venezuela and Guiana. Flowers in all three families are constructed on a basically trimerous plan with two whorls of three tepals, two whorls each of three stamens, and a single tricarpellate ovary. In Burmanniaceae, flowers are bisexual and the ovary is inferior (Maas-van de Kamer, 1998). In Dioscoreaceae, flowers are unisexual and the ovary is inferior (Huber, 1998). In Nartheciaceae flowers are bisexual and the ovary is superior (Tamura, 1998). The number of ovules per locule varies from one to many. Usually ovules are anatropous and bitegmic, but in Nartheciaceae there are also campylotropous and unitegmic ovules (Tamura, 1998).

Burmanniaceae and Nartheciaceae lack any reported fossil record. There are scattered reports of Dioscoreaceae based on fossil leaves such as *Dioscorites cretaceus* from the Late Cretaceous of North America (Berry, 1925), but there are none that are convincing from the Cretaceous and all early records of the family require detailed re-examination (Daghlian, 1981).

11.7 PANDANALES

The Pandanales are a strongly supported monophyletic group based on molecular data and in the APGIII classification comprise Pandanaceae together with Cyclanthaceae, Stemonaceae, Triuridaceae and Velloziaceae. Pandanaceae and Cyclanthaceae have a well-documented fossil record, but the relationships of fossils assigned to the Stemonaceae and Triuridaceae are uncertain.

11.7.1 Pandanaceae

The Pandanaceae comprise 800–900 species in three genera distributed throughout the Old World tropics from West Africa out into the Pacific (Stone *et al.*, 1998). They include trees, shrubs and woody climbers, often with distinctive adventitious roots that are modified in various ways. Reproductive structures of Pandanaceae are extremely condensed (Rudall, 2003) and the limits of individual flowers are sometimes difficult to establish. Flowers are unisexual. Staminate flowers contain variable numbers of tetrasporangiate stamens. Ovulate flowers are composed of one to several carpels. Carpels of different species may contain one to many anatropous, bitegmic ovules.

There is no reliable fossil record of Pandanaceae prior to the latest Cretaceous and these records are themselves in need of further study. Dispersed pollen of *Pandaniidites* was first described by Elsik (1968) from the Early Cenozoic of North America and includes a variety of forms, some of which may be related to Pandanaceae (Nichols, 1992; Hotton *et al.*, 1994). However, pollen similar to *Pandaniidites* has been found *in situ* in flowers of the extinct plant *Limnobiophyllum* that are of araceous affinity and this raises questions about the reliability of previous reports of Pandanaceae from the latest Cretaceous (Maastrichtian) of North America based on fossil pollen (Jarzen, 1983). Other records of possible pandanaceous pollen, for example from the Early Cenozoic of the Old World tropics (Jarzen, 1983), may prove more reliable.

11.7.2 Cyclanthaceae

The Cyclanthaceae are a small family of about 12 genera and 225 species of epiphytes, lianas and terrestrial herbs restricted to Central and South America, where they grow in tropical and montane forests (Heywood *et al.*, 2007). Leaves are large, palm-like and arranged spirally or distichously. Flowers are small, unisexual and aggregated in dense monoecious inflorescences. Female flowers are tetramerous with four tepals, four staminodes and four fused carpels. Male flowers have many tepals in one or two whorls, or tepals are missing. Stamens are usually numerous. In *Cyclanthus*, the female flowers are fused and borne in a series of cycles that form discoidal plates (Smith *et al.*, 2008).

Well-preserved fruits from the Eocene of Germany and England, previously thought to be related to the Cyperaceae, are now convincingly placed in the extant genus *Cyclanthus* (Smith *et al.*, 2008). *Cyclanthus messelensis* from the Middle Eocene Messel flora, Germany, is particularly interesting in showing the distinct discoidal arrangement of the female flowers and seeds with distinctly reticulate surfaces that match those of extant *Cyclanthus*

(Smith *et al.*, 2008). This discovery is also important from a palaeogeographic perspective and adds a further New World element to the Cenozoic floras of Europe. Currently there is no record of Cyclanthaceae from the Cretaceous.

11.7.3 Stemonaceae

The Stemonaceae comprise about 35 species in three genera, mainly distributed in east and southeast Asia and Australia, but with a single species of *Croomia* in southeastern North America (Kubitzki, 1998a). They are creeping or erect perennials with dimerous, usually bisexual flowers with a superior to inferior ovary. Flowers usually have four (occasionally five) tepals in two whorls of two. Stamens are also in two whorls of two. The ovary has a single locule and contains few to many anatropous to hemi-anatropous ovules (Kubitzki, 1998a).

The genus *Spirellea*, established by Knobloch and Mai (1983; 1984) based on small ribbed seeds from the Maastrichtian and Paleocene of Germany, includes several different species that may not be systematically closely related. Mai (1987a) tentatively placed *Spirellea germanica* and *S. walbeckensis* in Stemonaceae. Their precise systematic affinity, however, needs to be documented by more detailed studies (Collinson *et al.*, 1993).

11.7.4 Triuridaceae

Triuridaceae comprise about 45 species in nine genera. They grow in damp, humid, often deep forest habitats and occur from Central and South America to eastern Africa, India and eastern and southeastern Asia to Australia. Triuridaceae are achlorophyllous perennial herbs with reduced scale-like leaves. Flowers are small and usually unisexual, and often have long tepal-like structures, or tepal-like appendages. In one genus (*Lacandonia*) the bisexual flowers/floral structures are very unusual for angiosperms as a whole in having three stamens in the centre surrounded by numerous carpels; they have been described as possible pseudanthia (Rudall, 2003). Pollen is generally inaperturate with a distinctive gemmate surface (Furness *et al.*, 2002).

Fossil flowers (*Mabelia, Nuhliantha*) from the Late Cretaceous (Turonian), Old Crossman locality, New Jersey, USA, have been assigned to Triuridaceae (Gandolfo *et al.*, 1998b, 2002). However, as noted by Rudall (2003) there are differences between some of the features of these fossils and extant Triuridaceae and the pollen is also different. There are also similarities to flowers of eumagnoliids (e.g. Myristicaceae), especially in the form of the staminal column (Chapter 10). The relationships of this fossil material require further investigation (Chapter 9).

11.8 LILIALES

The Liliales are a strongly supported monophyletic group based on molecular data. In the APGIII classification they comprise Liliaceae, Melianthaceae, Smilacaceae and seven other families. The Smilacaceae are well documented from the Cenozoic, and there are a few reports of Liliaceae from the Cretaceous that require reinvestigation, but otherwise none of these families is known reliably from the fossil record.

Pollen assigned to the dispersed pollen genus *Liliacidites* is widespread in Cretaceous sediments, but obviously constitutes a heterogeneous assemblage of species that may represent several different families and orders (section 9.6.4). None of them has been securely assigned to Liliales. The inflorescence described as *Monocotylostrobus bracteatus* from the Deccan Intertrappean Beds, India, is of possible lilialean affinity (Bonde and Kumaran, 1993), but needs critical re-assessment, as does *Cretovarium japonicum* from the Late Cretaceous of Hokkaido, Japan (Chapter 9). Critical re-examination and new material is also needed to support comparisons between leaves of *Acaciaephyllum* and those of extant monocots (Doyle *et al.*, 2008a) (section 11.3.1).

Records of Smilacaceae from the Cenozoic include scattered reports of leaves assigned to *Smilax* (see also Daghlian, 1981; Sun and Dilcher, 1988; Wilde, 1989; Collinson *et al.*, 1993), as well as leaves and pollen assigned to *Ripogonum* from the Miocene of New Zealand (Couper, 1953; Pole, 1993).

11.9 ASPARAGALES

Based on molecular phylogenetics the Asparagales are monophyletic. Definition of the group and relationships within it has been greatly clarified by recent work (Chase *et al.*, 2000). APGIII (2009) recognises a very broadly defined Amaryllidaceae (including Agapanthaceae and Alliaceae) and Asparagaceae (including Agavaceae and six other families). The Amaryllidaceae and Asparagaceae defined in this way are sister taxa. The remaining 12 families of Asparagales form a basal grade within the order, within which various monophyletic groups can be recognised (Chase *et al.*, 2000).

11.9.1 Amaryllidaceae and Asparagaceae

The known fossil record of the Amaryllidaceae and Asparagaceae is extremely poor. There are no reliable records from the Cretaceous and only a single record of petrified *Yucca*-like stems (*Protoyucca shadishii*) from the Middle Miocene of northwestern Nevada, USA (Tidwell and Parker, 1990).

11.9.2 Other Asparagales

The 12 families of Asparagales that make up a basal grade within the order include both Iridaceae and Orchidaceae. The fossil record of all these families, like that of Amaryllidaceae–Asparagaceae, is very sparse. Muller (1981) accepts pollen records of *Astelia* (Asteliaceae, Couper, 1960; Mildenhall, 1980) and *Phormium* (Xanthorrhoeaceae, Couper, 1960) from the Late Eocene as possible records of these groups, but there are no known Cretaceous records. Lack of information on the history of the hyper-diverse Orchidaceae (*c.* 880 genera, 22 000 species, Stevens, 2001 onwards) is especially frustrating. The only exception is a spectacular discovery of an orchid pollinarium (*Meliorchis caribea*) attached to the mesoscutellum of a stingless bee in Oligocene–Miocene amber from the Dominican Republic (Ramirez *et al.*, 2007). Another possible record of the family is a single, minute seed from the Middle Miocene of Jutland, Denmark (Friis, 1985b), but the family is clearly of much greater antiquity than the Neogene.

A permineralised monocot corm, *Eriospermocormus indicus*, from the Maastrichtian Deccan Intertrappean Beds, India (Bonde, 2005), has some anatomical details preserved and has been compared to extant *Eriospermum* (Ruscaceae). However, according to Bonde (2005) the fossil is also similar to other taxa now placed in the Liliales and Asparagales, and its relationship is not fully resolved.

11.10 COMMELINIDS

Commelinids are a major clade of monocots that is well supported based on phylogenetic analyses of molecular data (Chase *et al.*, 2000). In the APGIII (2009) classification, in addition to the Dasypogonaceae, which are of uncertain position within the group, commelinids comprise: Arecales, Poales, Commelinales and Zingiberales. Commelinids have the most extensive Cretaceous fossil record of all monocots, which perhaps reflects both their ecological importance and the relative ease of recognising their fossil remains.

11.10.1 Arecales

The Arecaceae (palms) are the only family of Arecales. They comprise about 190 genera and 2000 species (Dransfield *et al.*, 2008), almost all of which are restricted to the tropics. Palms vary greatly in habit and range from trees and shrubs to climbing and stemless forms, but none are truly herbaceous. Palm wood is distinctive and easily recognised in the fossil record by the large number of scattered vascular bundles often with associated vascular fibres. Fibre bundles composed of non-vascular fibres are also often present. Leaves have sheathing bases, generally with a distinct petiole and a leaf blade that is variously developed on a fundamentally pinnate–palmate plan.

Palm inflorescences are of various kinds and are generally borne in the axil of a leaf. Flowers are unisexual or bisexual, and generally trimerous, with three petals, three sepals and six stamens, but there are also several multistaminate forms (Uhl and Moore, 1980; Dransfield *et al.*, 2008). Palm pollen is diverse both in aperture configuration and exine sculpture (e.g. Halbritter and Hesse, 1993; Harley, 1997; Furness and Rudal, 2003; Harley, 2004). The gynoecium is apocarpous with three carpels, or variously syncarpous with three locules in the ovary. Usually there is a single ovule in each locule, which may be anatropous to orthotropous (Dransfield *et al.*, 2008).

There is a rich fossil record of palms based on dispersed pollen (Muller, 1979; Harley, 2006), but whereas pollen is sometimes readily assigned to the family it is usually more difficult to determine relationships to extant genera. An important exception is the stemless mangrove palm *Nypa*, which has an extensive and well-documented fossil record based on its distinctive pollen and fruits (e.g. Tralau, 1964; Gee, 2001). Fossil fruits comparable to extant *Nypa* are particularly well known from the Eocene (Collinson, 1983a), but have also been reported from the Maastrichtian of India and assigned to *Nypa* or to the fossil genus *Nipadites* (see below). Dispersed pollen is usually referred to the genus *Spinizonocolpites* (see below). *Nypa fruticans* is the only extant member of the genus and is also ecologically distinctive. It is restricted to near coastal mangrove areas from India and Myanmar (Burma) through Thailand, Malaysia, Borneo and Indonesia to the Philippines, Ryukyu Islands, New Guinea, the Solomon Islands, and northern Australia (Tralau, 1964).

Palms have an extensive fossil record through the Cenozoic. In the Paleocene and Eocene there are numerous reports of palm leaves (e.g. Read and Hickey, 1972; Daghlian, 1978), pollen (e.g. Harley *et al.*, 1991; Harley and Morley, 1995; Harley, 1997), fruits (e.g. Reid and Chandler, 1933; Collinson, 1983b; Manchester, 1994) and occasional flowers or inflorescences (e.g. Schaarschmidt and Wilde, 1986). Taken together these records indicate that palms had already undergone substantial diversification prior to the Early Cenozoic.

There are no well-documented records of Arecaceae from the Early Cretaceous although several dispersed and *in situ* pollen types of this age have pollen wall structure that resembles that of some extant palms. The systematic affinity of *Hyphaeneocarpon aegyptiacum*, reported as a palm fruit from the Lower Cretaceous (Aptian) Abu Ballas Formation, Egypt (Vaudois-Miéja and Lejal-Nicol, 1987a), is uncertain according to Pan *et al.* (2006). However, Arecaceae are reliably recorded from the Late Cretaceous. The earliest records are putative palm pollen from the Coniacian/Santonian (Daghlian, 1981), and an apparently reliable record of palm leaves from the Cliffwood Beds, Magothy Formation of New Jersey, USA, which is of probable Early to Late Santonian age (Christopher, 1979).

By the Campanian, putative palm pollen is diverse at low palaeolatitudes (Muller, 1979) and there are also unequivocal records of palm leaves, for example from the Campanian of Montana, USA (Crabtree, 1987), and the Early Campanian Grünbach flora, Austria (Kvaček and Herman, 2004). Palm leaves and probable fruits are also known from the Campanian of Big Bend, Texas, USA (cited in Manchester, 1999). There are also small, multistaminate flowers with a trimerous perianth from the Campanian–Maastrichtian Mira locality, Portugal, that may be related to palms (Friis *et al.*, 2010a).

Evidence of palms from the Maastrichtian is extensive. Palm stems with attached leaves are abundant in the *in situ* Maastrichtian vegetation preserved in the Meteetsee Formation at Big Cedar Ridge, Wyoming, USA (Wing *et al.*, 1993). Palm wood, attributed to *Palmoxylon*, is known from the Maastrichtian Fox Hills Formation, South Dakota, USA (Delevoryas, 1964) and from the Deccan Intertrappean Beds (e.g. Prakash, 1974).

Putative palm pollen is especially diverse in low-palaeolatitude palynofloras from the latest Cretaceous, for example from Africa (Pan *et al.*, 2006). Probable palm pollen may constitute up to 50% of the palynological assemblages in the low-palaeolatitude so-called Palmae Province at this time (Herngreen and Chlonova, 1981; Herngreen *et al.*, 1996). There is considerable potential for further elucidation of the fossil history of palms based on critical studies of dispersed fossil pollen (e.g. Muller, 1979; Harley, 1997, 2006). For example, Harley (2006) enumerated almost 50 different genera of dispersed pollen from the Cretaceous and Cenozoic that have similarities with pollen of extant palms. Most of these genera of dispersed pollen grains have not been studied in detail and their systematic relationships require further investigation. Maastrichtian palm pollen includes genera such as *Longapertites*, *Spinizonocolpites* and *Mauritiidites* (Harley, 2006).

Longapertites is based on pollen first described from the Maastrichtian of Nigeria (van Hoeken-Klinkenberg, 1964) that is common in many low-palaeolatitude palynofloras from the Late Cretaceous. Pollen is monocolpate with a long colpus that extends to the proximal surface. Broadly similar forms include *Monocolpites*, *Proxapertites* and *Quilonipollenites*. Some of these grains, and many *Longapertites* species, are undoubtedly related to palms (Harley, 2006), but these genera may also include pollen of several different families. Most of these fossil grains have been described based only on light microscopy and more detailed research is needed for critical analysis of their relationships (Harley, 1997, 2006).

Spinizonocolpites (Figure 11.17), first described by Muller (1968), is a distinctive element of some Late Cretaceous (Maastrichtian) and Early Cenozoic palynofloras from low and middle palaeolatitudes. From the Late Cretaceous the genus is reported from tropical West Africa, several sites in northern Africa, and also from Borneo (Muller, 1979; Pan *et al.*, 2006). It includes coarsely spiny zonacolpate pollen that often separates into two halves along the encircling aperture. This pollen type has been linked reliably to pollen of extant *Nypa* (Figure 11.17), but include several species indicating that *Nypa* was more diverse in the past than it is today (Harley, 2006). The occurrence of this kind of pollen in coastal and marine sediments indicates that despite its earlier diversity the *Nypa* lineage occupied a similar ecological position in the past to the single living species.

Other distinctive dispersed fossil pollen with spiny, clavate or gemmate exine ornamentation from the latest Cretaceous and Early Cenozoic is also of probable arecaceous affinity. This kind of pollen has been assigned to various genera, of which *Echimonocolpites* and *Mauritiidites* occur in the Maastrichtian (Harley, 1997, 2006). Pollen of *Mauritiidites* was first described from the Maastrichtian of Nigeria (van Hoeken-Klinkenberg, 1964) and is reported in several Late Cretaceous palynofloras from Africa. It is closely similar to pollen of extant *Mauritia*, a genus now restricted to coastal and inland swamps in South America (Muller, 1979).

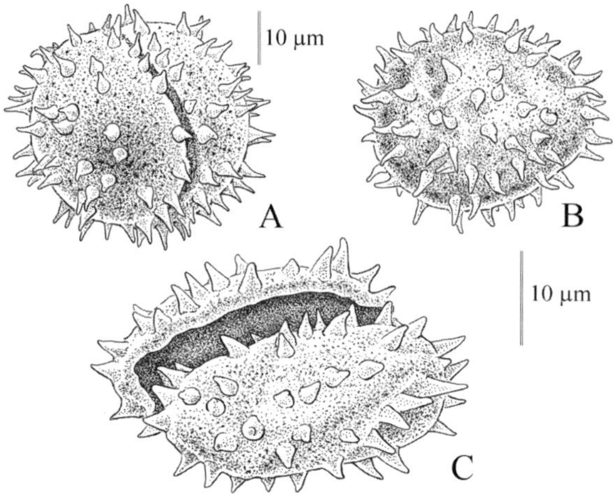

Figure 11.17 (A) Pollen of extant *Nypa fructicans.* (B, C) Fossil pollen of (B) *Spinizonocolpites echinatus* and (C) *S. baculatus* from the Eocene London Clay, southern England. Drawn from SEM images in Harley *et al.* (1991).

Strong support for the diversity of palms in latest Cretaceous low-palaeolatitude vegetation is provided by macrofossils from the Maastrichtian Deccan Intertrappean Beds, India (Agrawal, 1995). Palms are especially well represented and include several species of *Palmoxylon*, along with *Parapalmocaulon* stems, *Sabalophyllum livistonoides* leaves and palm rachilla fragments assigned to *Arecoideostrobus moorei* (for references see Bonde, 1996). Flowers of *Shuklanthus* and fruits of *Viracarpon* from the Deccan Intertrappean Beds (section 11.3.3) (Verma, 1958; Nambudiri and Tidwell, 1978) may also be palms, but could also be related to other monocot families such as Pandanaceae (Nambudiri and Tidwell, 1978).

The Deccan flora also includes palm fruits assigned to *Hyphaeneocarpon* (Bande *et al.*, 1982) and *Nypa* or *Nipadites* (Chitaley, 1960; Prakash, 1960), as well as a *Cocos*-like fruit (Patil and Upadhye, 1984). *Palmocarpon*, established for fossil palm-like fruits from the Cenozoic of North America (Lesquereux, 1878), is represented in the Deccan flora by several species including *P. bracteatum*, *P. compressum*, *P. insigne*, *P. mohgaoense* and *P. takliensis* (e.g. Mahabale, 1950; Prakash, 1954, 1960; Chitaley and Nambudiri, 1995). These fruits are generally ovoid, slightly triangular in cross-section, up to about 6.4 cm long and 4.8 to 2.3 cm broad. Anatomical details are well preserved and show a thin epicarp surrounding a thicker mesocarp that consists of a parenchymatous ground tissue traversed by vascular and fibrous bundles. The endocarp consists mainly of thick walled sclerenchyma.

11.10.2 Poales

The Poales are a well-supported monophyletic group based on phylogenetic analyses of molecular data (Chase *et al.*, 2000). Excluding Hydatellaceae (Chapter 8) they comprise 16 families (APGIII, 2009). Especially important from an ecological standpoint are Poaceae and Cyperaceae, both of which are very diverse. There is a strong support for two large clades within Poales: a cyperoid group and a graminoid group (Bremer, 2002; Chase, 2004). Relationships among the cyperoid and graminoid groups, and the remaining families of Poales are still uncertain (Bremer, 2002).

Cyperoids include three families centred on the Cyperaceae (104 genera, more than 5000 species distributed worldwide), but also including Juncaceae (eight genera, 350 species distributed worldwide) and Thurniaceae (one genus with three species restricted to northern South America). Most cyperoids are perennial or annual grass-like herbs. Flowers are bisexual, or often unisexual, and fundamentally trimerous in their organisation. Typically there are two perianth whorls, each of three tepals, one or two whorls of three stamens, and a single tricarpellate or bicarpellate ovary. In Juncaceae and Thurniaceae the ovary is trilocular with one to many ovules in each locule. In Cyperaceae there is a single locule containing one ovule that is basally attached. Ovules are anatropous and bitegmic.

Cyperoids have no known record from the Cretaceous. However, as would be expected from their frequent occurrence in damp or wet habitats, they are well represented in Cenozoic fossil floras. Fruits of Cyperaceae are recorded throughout the Cenozoic from the Middle Paleocene onwards (e.g. Dorofeev, 1963; Chandler, 1964; Friis, 1985b; Mai and Walther, 1985; Gabel and Bich, 1988; Gabel *et al.*, 1992; Mai, 1995) and are common, even though the Eocene species *Scirpus lakensis* has now been transferred to the genus *Cyclanthus* (Smith *et al.*, 2008) (section 11.7.2). During the Cenozoic the Cyperaceae are also well represented based on pollen (Krutzsch, 1970b). From the Eocene there is a possible record of petrified roots, leaves and stems (Erwin and Stockey, 1992).

The fossil record of Juncaceae is much less extensive than that of Cyperaceae and although seeds have been reported from the Late Eocene–Early Oligocene of southern England (Collinson, 1983b) they are recorded only from the Miocene onwards in other parts of Europe (Friis, 1985b; Mai, 1985a; Collinson *et al.*, 1993).

Graminoids comprise the large family Poaceae with about 10 000 species in more than 650 genera (Stevens, 2001 onwards) and six smaller families (Anarthriaceae, Centrolepidaceae, Ecdeiocoleaceae, Flagellariaceae, Joinvilleaceae, Restionaceae). Most graminoids are small to medium-sized grass-like herbs, but some are very large (e.g. bamboos) or liana-like (e.g. Flagellariaceae). Floral structure is variously modified from a basically trimerous ground plan, with two perianth whorls, each of three tepals, and two whorls of three stamens surrounding a bicarpellate or tricarpellate ovary. Typically there is a single bitegmic ovule in each locule. Ovules vary from anatropous to orthotropous (Dahlgren *et al.*, 1985). Pollen is typically monoporate.

There are no reliable records of graminoids prior to the Maastrichtian. Only Poaceae are recorded from the Late Cretaceous (Maastrichtian). The presence of the family is based mainly on their pollen (Linder, 1987). For interpreting the fossil pollen record of Poales there is also the complexity that pollen of Poaceae, Restionaceae and other graminoids is very similar and difficult to distinguish (Linder and Ferguson, 1985). The various pollen records of Restionaceae reported by Hochuli (1979) from the Cretaceous (Albian and Maastrichtian) and Cenozoic are uncertain (Bonnefille *et al.*, 1990).

The presence of Poaceae in the latest Cretaceous receives support from the discovery of at least five different kinds of grass phytoliths in a Maastrichtian dinosaur coprolite from India (Prasad *et al.*, 2005). Phytoliths are microscopic silica bodies formed inside the cells of grasses and many other plants. They are often systematically diagnostic and have been widely used for studying the history of grasslands (Strömberg, 2005). There is also a record of a grass stem from the Deccan Intertrappean Beds, India (Bonde, 1986), but we have not seen this reference.

In the Paleocene graminoid pollen (referred to Poaceae) is recorded from Brazil, Cameroon and Nigeria (Adegoke *et al.*, 1978). A spikelet with *in situ* pollen is known from the Early Eocene of Tennessee, USA (Crepet and Feldman, 1991), and probable graminoid pollen occurs throughout the tropics during the Eocene (Germeraad *et al.*, 1968). Fruits probably related to Restionaceae have been described from the latest Eocene–Oligocene of Queensland, Australia, as four different species of the extinct genus *Restiocarpus* (Dettmann and Clifford, 2000). They co-occur with pollen of *Milfordia*, a widespread dispersed pollen genus that may also be of graminoid affinity, but that cannot be referred definitively to an extant family (Linder, 1987; Linder *et al.*, 1998).

The development of grass-dominated (and often fire-maintained) ecosystems is detected first in the Oligocene of South America, and then during the Miocene in North America, Eurasia, Africa and Australia (Jacobs *et al.*, 1999; Strömberg, 2005). By the Late Cenozoic, pollen of probable Restionaceae is recorded from all southern continents (Linder *et al.*, 1998).

Other Poales. Families of Poales of uncertain relationship to cyperoids and graminoids are Bromeliaceae, Eriocaulaceae, Mayacaceae, Rapataceae, Typhaceae (including Sparganiaceae) and Xyridaceae. Several of these families have a fossil record from the Early Cenozoic (e.g. Herendeen and Crane, 1995), but there is much less information from the Cretaceous.

Among these families, the fossil record of Typhaceae, which comprise perennial herbs of wet places, is the most extensive and reliable. The family contains two genera, *Sparganium*, with 14 species that are mainly in temperate regions of the Northern Hemisphere, and *Typha*, with about 8–13 species that are widely distributed through both temperate and tropical regions. Inflorescences are globose in *Sparganium* and spike-like in *Typha*. In both genera the perianth is either inconspicuous (*Sparganium*), or more or less absent or reduced to scales or trichomes (*Typha*). Flowers are unisexual. Staminate flowers have one to eight stamens. Pistillate flowers normally have one, but occasionally two or three, carpels (Kubitzki, 1998b). There is a single anatropous, bitegmic ovule in each carpel.

Both *Sparganium* (e.g. Hickey, 1977) and *Typha* (e.g. Grande, 1984) are present in the Eocene and subsequently occur widely in fossil floras from the Cenozoic of Europe (Collinson, 1988; Mai, 1995). *Typha*-like seeds (*Typhaspermum*) are also reported from the latest Eocene–Oligocene of Queensland, Australia (Dettmann and Clifford, 2000). Earlier records are less secure. Pollen assigned to *Sparganium* from the Late Cretaceous (Campanian) of Spain (Medus, 1987) requires more detailed study. More reliable are records of fruits of *Typha* from the Maastrichtian (Knobloch and Mai, 1986), but more detailed comparative studies of these fossils are desirable. Leaves of *Typhacites negevensis* described from the Late Cretaceous (Turonian) of the Negev Desert and assigned to the Typhales (Krassilov *et al.*, 2005) lack critical features for precise systematic assignment (section 11.3.1).

11.10.3 Commelinales

The Commelinales are recognised as monophyletic in the APGIII classification and comprise five families: Commelinaceae, Haemodoraceae, Hanguanaceae, Philydraceae and Pontederiaceae. Commelinales may be the sister group to Zingiberales (Givnish *et al.*, 1999; Chase *et al.*, 2000). Hanguanaceae and Philydraceae are both small families, each with less than five species, that are entirely restricted to tropical southeast Asia, Australia and Micronesia (Bayer *et al.*, 1998; Hamann, 1998). Neither family has any known fossils. The Haemodoraceae, with 13 genera and about 100 species, are more widely distributed, but are also unrecognised in the palaeobotanical record (Simpson, 1998).

The Commelinaceae comprise about 41 genera and 650 species distributed throughout the warm temperate and tropical zones. They are mainly perennial herbs, usually with bisexual, actinomorphic or zygomorphic flowers comprised of three sepals, three petals, six stamens and a superior bi- or trilocular ovary. Two or three of the stamens are often reduced to staminodes. There may be one or many ovules per locule (Faden, 1998).

The Commelinaceae have no known fossil record from the Cretaceous. Leaves and fruits assigned to *Pollia* have been described from the Middle Miocene Ngorora Formation, Kenya (Jacobs and Kabuye, 1989). Extant *Pollia* is widespread through the tropics and warm temperate regions of the Old World (Faden, 1998).

The Pontederiaceae comprise nine genera and about 33 species distributed mainly through the tropics, but with a centre of diversity in the tropics and subtropics of the New World (Cook, 1998a). They are annuals or perennials of freshwater habitats where they may be submerged, floating or emergent. Flowers are bisexual, zygomorphic to almost actinomorphic, and generally with six tepals. The androecium is comprised of six, four or three stamens, or there is a single stamen. The gynoecium consists of three fused carpels that form a superior, trilocular ovary. Ovules are anatropous and borne on axile placentae that may be intrusive into each locule (Cook, 1998a).

The specimen described as *Eichhornia* from the Late Cretaceous (Maastrichtian) Deccan Intertrappean Beds, India (Patil and Singh, 1978), is the earliest record of the family. The fossil is a permineralised axis with attached petioles and roots. Anatomical details are well preserved, but according to Cook (1998a) the fossil may be more similar to extant *Monochoria* (also Pontederiaceae) than it is to *Eichornia*. A leaf fragment from the Eocene Green River Formation, USA, described by Knowlton (1923) as *Pontederites*, resembles *Pontederia* (Cook, 1998a). Seeds resembling those of *Monochoria*, and leaves and seeds resembling those of *Eichhornia*, are recorded from the Late Eocene and through the later Cenozoic (Mai and Walter, 1978, 1985; Wilde, 1989).

11.10.4 Zingiberales

The Zingiberales are a well-supported monophyletic group that was recognised based on morphological and other features by early cladistic studies (Dahlgren *et al.*, 1985). Monophyly of Zingiberales has since been confirmed by molecular phylogenetics (Chase *et al.*, 2000, 2006), which also suggests that Zingiberales and Commelinales may be sister taxa (Givnish *et al.*, 1999; Chase *et al.*, 2000). In the APGIII classification (2009) the Zingiberales comprise Cannaceae, Costaceae, Heliconiaceae, Lowiaceae, Marantaceae, Musaceae, Strelitziaceae and Zingiberaceae. In general, the monophyly of each of these groups is well supported, but relationships among them are much less certain. Among the morphological characters that help define Zingiberales as a whole, are arilate seeds, generally inaperturate pollen, flowers with epigynous ovary, the presence of short-cells in the root hairs, the occurrence of silica bodies, and sieve tube plastids containing starch grains (Dahlgren *et al.*, 1985).

Zingiberales are generally perennial herbs with sympodial rhizomes and herbaceous stems. Leaves are characteristic and generally have a well-defined petiole, an expanded lamina, a distinct midvein, and fine pinnate–parallel secondary veins. Flowers are generally bisexual and fundamentally trimerous. Flowers always have an inferior ovary, and are either zygomorphic or asymmetric. The perianth consists of two whorls, each of three tepals. Stamens are six or fewer. Pollen is generally inaperturate, but variable among the families and may be monocolpate, dicolpate, pantoporate or spiraperturate. The gynoecium consists of three fused carpels with a long style. Generally the ovary has three locules. Ovules are usually anatropous and may be numerous or solitary within the locules (Dahlgren *et al.*, 1985).

The Zingiberales are well represented in the Late Cretaceous and Early Cenozoic fossil record although in many cases it is not certain to which modern family the fossils should be assigned (see below). From the Cenozoic there are numerous, well-documented, records of leaves, as well as fruits and seeds of *Spirematospermum* (Figure 11.18).

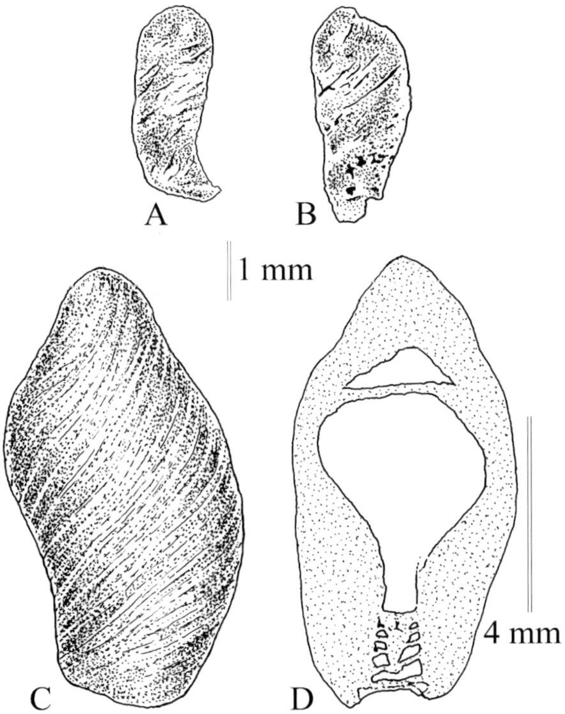

Figure 11.18 Fossil seeds of *Spirematospermum*. (A, B) *Spirematospermum chandlerae* from the Late Cretaceous (Campanian) Neuse River locality, North Carolina, USA. (C, D) *Spirematospermum wetzleri* from the Miocene Fasterholt flora, Denmark, (C) in surface view and (D) in longitudinal section. (A, B) Drawn from SEM images in Friis (1988); (C, D) drawn from photographs in Koch and Friedrich (1971).

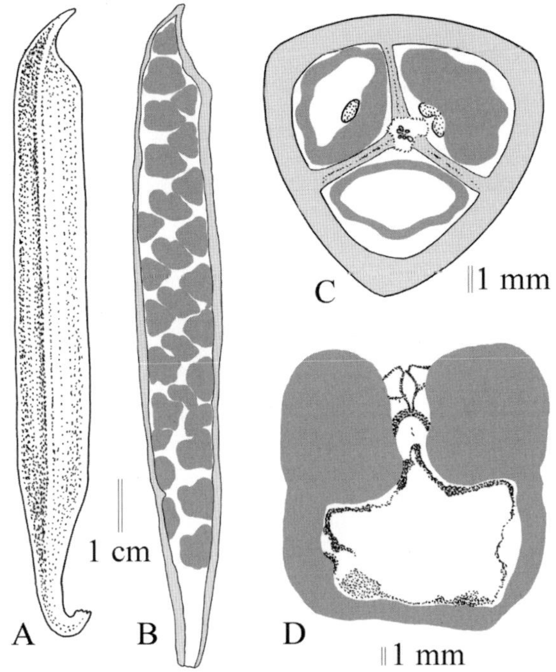

Figure 11.19 Fruits and seeds of *Musa cardiosperma* from the Late Cretaceous (Maastrichtian) Deccan Intertrappean Beds, India. (A) Reconstruction of fruit. (B) Schematic cross-section of fruit. (C) Cross-section of fruit with seeds. (D) Longitudinal sections of seed. Drawn from photographs in Jain (1964).

Fruits of *Spirematospermum* are tricarpellate and trilocular and contain numerous densely packed seeds. Seeds are distinctive and characterised by their spiral striate surface ornamentation, a distinct operculum and the presence of a chalazal chamber (Chandler, 1926, 1964; Friedrich and Koch, 1970; Koch and Friedrich, 1971; Wilde, 1989). *Spirematospermum* has usually been referred to Zingiberaceae (e.g. Koch and Friedrich, 1971), but a closer relationship to Musaceae has been suggested based on the presence of a chalazal chamber, rudimentary micropylar collar and hilar cavity (Rodríguez-de la Rosa and Cevallos-Ferriz, 1994). These features do not occur in extant Zingiberaceae. The precise affinities of *Spirematospermum* are therefore uncertain, but there is no doubt that it belongs to Zingiberales.

There are several records of *Spirematospermum* from the Late Cretaceous of eastern North America and Europe. *Spirematospermum chandlerae* (Figure 11.18) was described based on isolated seeds and groups of seeds from the Late Cretaceous (Early Campanian) Neuse River locality, North Carolina, USA (Friis, 1988). *Spirematospermum* seeds from the Late Cretaceous of Central Europe and Portugal include *Spirematospermum friedrichii* from the Maastrichtian Eisleben locality, Germany (Knobloch and Mai, 1986), *Spirematospermum* sp. from the Campanian, Kössen locality, Austria (Goth, 1986) and *Spirematospermum* sp. from the Campanian–Maastrichtian Mira locality, Portugal (Friis *et al.*, 2010a). The seeds are anatropous and have a characteristic spiral organisation of the tissues of the seed wall that gives the surface its spiral ornamentation. Except for their much smaller size *Spirematospermum chandlerae* and *S. friedrichii*, as well as the two other Cretaceous *Spirematospermum* taxa, are very similar to seeds of *Spirematospermum* from the Cenozoic (Figure 11.18) that are known *in situ* within fruits.

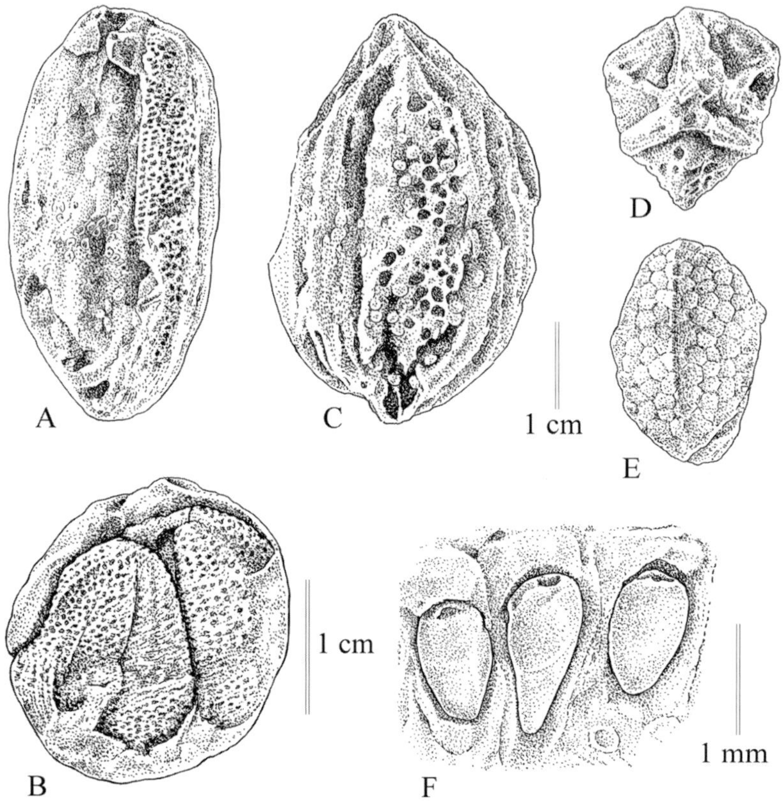

Figure 11.20 Zingiberalean fossils from the Late Cretaceous (Campanian) of southeastern Coahuila, Mexico (A, B) Fruits of *Tricostatocarpon silvapinedae* in (A) lateral and (B) oblique apical view. (C–F) *Striatornata sanantoniensis*: (C) slightly abraded fruit in lateral view; (D) apical view of abraded fruit; (E) group of densely packed seeds; (F) details of seeds showing moulds of embryo sac embedded in remains of seed coat. Drawn from photographs in Rodríguez-de la Rosa and Cevallos-Ferriz (1994).

Musa cardiosperma (Figure 11.19) is based on petrified fruits with *in situ* seeds from the Late Cretaceous (Maastrichtian) Deccan Intertrappean Beds, India (Jain, 1964). Fruits are tricarpellate with seeds borne in a single row in each locule. Manchester and Kress (1993) accept *Musa cardiosperma* as of zingiberalean affinity, but do not consider the attribution to *Musa* to be justified based on differences in details of both the fruits and the seeds.

Other fossil remains related to Zingiberales from the Deccan Intertrappean Beds include the petrified leaves of *Heliconiaites mohgaonensis* (Trivedi and Verma, 1972) and underground rhizomatous stems (*Cyclanthodendron sahnii*) that formed aerial woody pseudostems (*Musocaulon*) and bore strap-like leaves with sheathing bases (*Heliconites*) (Biradar and Bonde, 1990). Although originally assigned to Cyclanthaceae, Biradar and Bonde (1990) concluded that the reconstructed plant was zingiberalean and combined characters of extant Musaceae and Strelitziaceae.

Tricostatocarpon silvapinedae (Figure 11.20) and *Striatornata sanantoniensis* (Figure 11.20) are two genera of fossil zingiberaleans based on permineralised fruits and seeds from the Late Cretaceous (Campanian) of southeastern Coahuila, Mexico (Rodríguez-de la Rosa and Cevallos-Ferriz, 1994). Both genera are based on numerous well-preserved tricarpellate, trilocular fruits with axile placentation that have many, anatropous, operculate seeds in each locule. The two taxa are distinguished by details of their seeds. Seeds of *Striatornata sanantoniensis* have a wide chalazal chamber that takes up almost one-third of the seed volume. They are similar in general organisation to seeds of *Spirematospermum* and may also be related to Musaceae (Rodríguez-de la Rosa and Cevallos-Ferriz, 1994). Seeds of *Tricostatocarpon silvapinedae* are more simple in the structure of the hilum and chalaza.

Zingiberopsis magnifolia is based on fossil leaves with well-preserved fine venation from the Late Cretaceous (? Campanian-Maastrichtian) of Colorado, North and South Dakota, and Wyoming, USA (Hickey and Peterson, 1978; Peppe *et al.*, 2007). The fossils were compared with leaves of extant *Alpinia* (Zingiberaceae) and it seems likely that these leaves are referable to Zingiberales.

12

Fossils of eudicots: early-diverging groups

Eudicots are an important and well-supported monophyletic group of angiosperms. The clade includes almost all dicotyledons; the only groups excluded are eumagnoliids and those dicots at the ANITA grade plus Chloranthaceae and *Ceratophyllum*. Eudicots are therefore broadly equivalent to all dicot lineages except the Magnoliidae *sensu* Takhtajan (1980) (Chapter 7). The term eudicotyledons, or eudicots, was introduced to recognise the monophyletic status of this major group (Doyle and Hotton, 1991), and has been widely accepted in subsequent works. Earlier studies referred to eudicots as non-magnoliid dicots (Walker and Doyle, 1975; Crane, 1989) or tricolpates (Donoghue and Doyle, 1989b; Judd and Olmstead, 2004). The term tricolpates refers to the tricolpate aperture configuration, which is characteristic of the pollen of many early-diverging eudicots. While many eudicots have pollen with other aperture configurations, almost all are based on the triaperturate ground plan (Doyle and Hotton, 1991).

Eudicots are extremely diverse, and exhibit an almost bewildering breadth of morphological and ecological variation. The group contains about three-quarters of all extant angiosperm species (Magallón *et al.*, 1999). The fossil record of eudicots is extensive and informative about their evolutionary history. In this chapter we provide a brief outline of eudicot classification and the early fossil record of the group. We then focus on those lineages that diverged at an early stage from the line that gave rise to the bulk of eudicot species. We emphasise particularly the Cretaceous fossil record, and those taxa that can be recognised based on fossil flowers or other reproductive organs.

12.1 CLASSIFICATION OF EUDICOTS

Most phylogenetic analyses recognise eudicots as a monophyletic group (Chase *et al.*, 1993; Doyle *et al.*, 1994; Hoot *et al.*, 1999; Magallón *et al.*, 1999; Judd and Olmstead, 2004; Soltis and Soltis, 2004). This is well supported by analyses based on molecular data (Chase *et al.*, 1993; Soltis *et al.*, 1997, 1999, 2000a; Judd and Olmstead, 2004; Soltis and Soltis, 2004) and has also been recognised in analyses based on morphology (Donoghue and Doyle, 1989b; Doyle and Endress, 2000), but there are few morphological characters that define the group. So far, only the presence of triaperturate pollen (e.g. tricolpate, tricolporoidate, tricolporate, triporate) has been suggested as a unifying morphological feature (Donoghue and Doyle, 1989a; Nandi *et al.*, 1998). Other morphological features that are typical of most eudicots, such as the presence of a distinct calyx and corolla, do not occur in the earliest-diverging lineages. There are also many groups of eudicots that are nested high up within the clade that lack either calyx or corolla, but in these cases the lack of parts is most probably the result of evolutionary loss.

The developmental modifications that underpin the transition from fundamentally monocolpate to fundamentally triaperturate pollen deserve more detailed study. They appear to reflect significant changes in the behaviour of the spindle during pollen mother cell meiosis. In this context, the apparently convergent tendency of a few magnoliids (Schisandraceae, Chapter 8) and some monocots (e.g. Arecaceae, Harley and Baker, 2001; Harley and Dransfield, 2003) to produce various kinds of triaperturate grain is also of interest, as is the processes through which the triaperturate ground plan has been repeatedly transformed to produce polycolpate or polyporate pollen in some eudicot lineages (e.g. Caryophyllales).

Nelumbo, which is now regarded as having diverged very close to the base of the eudicots, has been considered of particular interest with regard to the origin of triaperturate pollen because it is variable in aperture configuration. Reports of both monoaperturate and tricolpate pollen in *Nelumbo* have also suggested an intermediate phylogenetic position between magnoliids and eudicots (Kuprianova, 1979; Kreunen and Osborn, 1999). However, most pollen grains produced by a single anther in *Nelumbo* are regularly tricolpate, and the small percentage of the grains that are aberrant exhibit a continuum in aperture number, position and form that ranges from monocolpate to tetracolpate

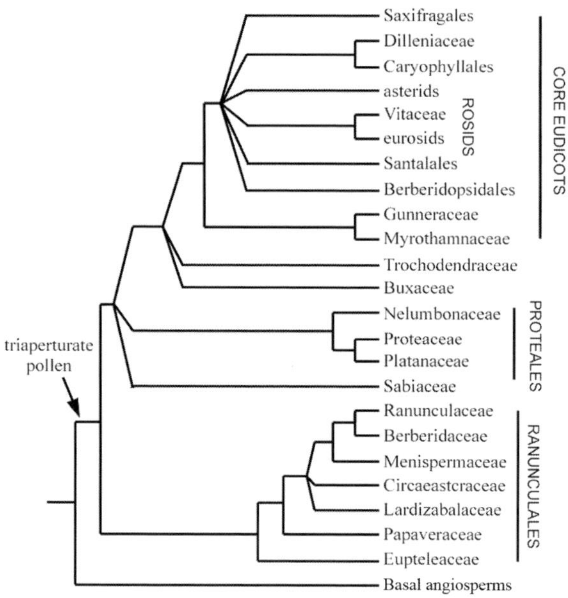

Figure 12.1 Summary of phylogenetic relationships of extant eudicots based on Soltis and Soltis (2004).

forms (Banks *et al.*, 2007) Further, this variation in the pollen of *Nelumbo* is not exceptional among eudicots. It also occurs among other early-diverging lineages such as Platanaceae (Denk and Tekleva, 2006).

The precise phylogenetic position of eudicots with respect to other angiosperms is currently not fully resolved, although recent analyses suggest that *Ceratophyllum* may be the eudicot sister group (Chapter 7). There is also uncertainty about the precise pattern of relationships among basal eudicots. However, some higher-level phylogenetic patterns, and the core constituents of many major monophyletic eudicot groups, have been identified reliably in recent cladistic analyses and are summarised in the APGIII classification (2009), as well as the overview of Magallón *et al.* (1999). According to the APGIII classification there is a basal grade of eudicots comprising several families and orders (Ranunculales, Sabiaceae, Proteales, Buxales, Trochodendrales) that are sister taxa (in various possible configurations) to the core eudicots (Figure 12.1).

Core eudicots comprise two large and strongly supported clades, the rosids (Chapter 14) and the asterids (Chapter 15), together with a small number of lineages whose position in the phylogenetic tree is still poorly resolved (Chapter 7). A four-gene analysis places Gunnerales (Gunneraceae and Myrothamnaceae) as the sister group to all other core eudicots (Soltis *et al.*, 2003).

The grade of early-diverging eudicot lineages recognised in the APGIII classification corresponds approximately to the ranunculids plus the hamamelids I and II of Chase *et al.* (1993). Several traditional groupings, such as the Hamamelididae or Amentiferae, and the Dilleniidae (e.g. Takhtajan, 1969; Cronquist, 1981), are no longer recognised. Analyses based on molecular data have confirmed previous suspicions based on morphological data that these groups are not monophyletic (see papers in Crane and Blackmore, 1989a, b).

12.2 EARLY-DIVERGING EUDICOTS

Eudicot lineages that fall outside the core eudicots include Ranunculales (Berberidaceae, Circaeasteraceae, Eupteleaceae, Lardizabalaceae, Menispermaceae, Papaveraceae, Ranunculaceae), Sabiaceae, Proteales (Nelumbonaceae, Platanaceae, Proteaceae), Buxales (Buxaceae, Didymelaceae) and Trochodendrales (Trochodendraceae including Tetracentraceae) (Figure 12.1). It seems likely that these lineages diverged at an early stage from the main line of eudicot evolution that subsequently produced the bulk of angiosperm species. The diversity of extant species among these early-diverging lineages varies considerably. Some are relatively species-poor, with only one or a few species, whereas others are more diverse. Ranunculaceae and Proteaceae are the most species-rich families with about 1750 and 1350 species, respectively (Figure 12.2). Several families at this level are clearly relictual and have a fossil record that documents greater diversity and a much wider distribution in the past than today (Magallón *et al.*, 1999).

Families of early-diverging eudicots show a great range of vegetative and reproductive characteristics and have often been widely separated in previous classifications. Some families are predominantly herbaceous (e.g. Nelumbonaceae, Ranunculaceae); others include only larger trees (Eupteleaceae, Platanaceae) or climbers (Menispermaceae). Flowers are typically trimerous or dimerous with relatively few parts, an undifferentiated perianth, and helical or opposite–decussate phyllotaxis (Drinnan *et al.*, 1994). Flowers are often small. Larger, showy flowers occur in some taxa, for example in *Nelumbo* or Ranunculaceae, but are less widespread. Many early-diverging eudicots, but not all, have simple tricolpate pollen with a reticulate or foveolate/punctate tectum. Other kinds of aperture configurations occur in Buxaceae (Köhler, 1981), Didymelaceae (Köhler, 1980) and Proteaceae (Dettmann, 1998).

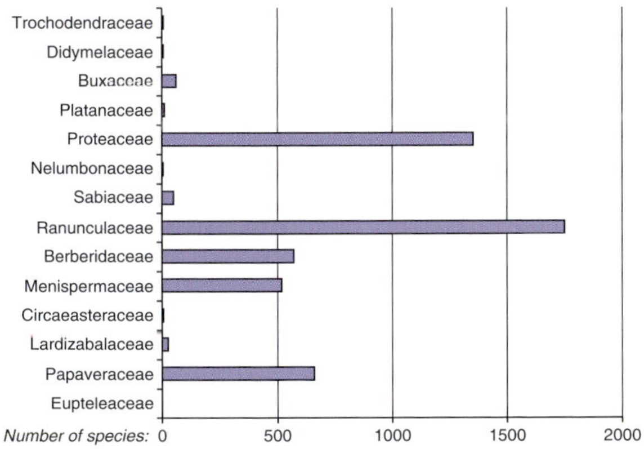

Figure 12.2 Diagram showing great heterogeneity in extant species diversity among early-diverging eudicots. Data from Magallón *et al.* (1999).

12.3 FOSSIL EVIDENCE OF EUDICOT DIVERSIFICATION

Fossils of eudicots are very common in sediments of Cretaceous and Cenozoic age. They often dominate the diversity of Late Cretaceous and Cenozoic plant fossil assemblages. Eudicots also dominate many assemblages in terms of abundance. The group was undoubtedly important in many kinds of vegetation at a worldwide scale from relatively early in the Late Cretaceous.

Eudicots are well represented in the fossil record from the mid- and Late Cretaceous and Cenozoic by dispersed pollen and leaves, fruits, seeds and flowers, as well as wood. Eudicot families that include deciduous woody plants are especially prominent. The record of families that include mainly herbaceous forms is more scattered. However, despite this underlying bias (Chapter 2), palaeobotanical evidence shows that all major eudicot lineages were established by the Early Cenozoic.

The fossil record also documents the sequential nature of eudicot diversification with an initial Early to mid-Cretaceous phase during which the early-diverging eudicot lineages became established, followed by subsequent phases through the Late Cretaceous and Early Cenozoic during which all the major core eudicot lineages are recognised for the first time.

Pollen with generalised features (tricolpate or tricolporate with a reticulate tectum) is produced by many extant eudicots at all levels within the phylogeny of the group. Dispersed fossil pollen of this type is usually referred to taxa such as *Retitricolpites*, *Tricolpites*, *Tricolpopollenites* and *Tricolporopollenites*. In the absence of more distinctive features it can be extremely difficult to assign such dispersed grains to a particular subgroup of extant eudicots. Even when ultrastructural details of the pollen wall are available, the relationships of generalised pollen of this kind from the Cretaceous and Early Cenozoic usually remain uncertain.

There is also an extensive fossil record of probable eudicot leaves that cannot be assigned securely to extant taxa: either because fine details of venation and epidermal features are not available or not diagnostic, or because there is insufficient knowledge of leaf architecture in putative extant counterparts. Many fossil leaves from the Cretaceous and Cenozoic that have been assigned to extant genera of eudicots, especially in early palaeobotanical studies, cannot be taken as reliable records of extant taxa (see Dilcher, 1974). More conservative approaches have sometimes referred fossil dicot leaves to fossil genera such as *Dicotylophyllum* and *Phyllites*. These genera undoubtedly include a mixture of both magnoliid and eudicot leaves.

12.4 FOSSILS OF UNCERTAIN RELATIONSHIPS

12.4.1 Early tricolpate pollen in dispersed palynofloras

The earliest evidence of eudicots is provided by rare occurrences of tricolpate pollen in Late Barremian – Early Aptian palynological assemblages from southern England (Hughes and McDougall, 1990), Egypt (Penny, 1991), Israel (Brenner, 1996), West Africa and North America (Doyle *et al.*, 1977; Doyle, 1992) and from mesofossil assemblages of Portugal (Friis *et al.*, 1999, 2010a, b).

Tricolpate pollen is scarce in the Late Barremian – Early Aptian of southern England (Hughes and McDougall, 1990; Hughes, 1994). So far such grains have been recorded from only two horizons in the Atherfield succession on the Isle of Wight, southern England (Hughes and McDougall, 1990; Hughes, 1994). In contrast, monocolpate pollen, including forms assigned to *Clavatipollenites*, is relatively common through the Atherfield succession and may also be relatively abundant within individual samples (Kemp, 1970). The oldest record of tricolpate pollen from the Isle of Wight is a single grain extracted from a sample collected from Wealden Bed 35. The sample is assigned to Phase 4 in the early angiosperm succession documented by Hughes (1994) (for pollen phases of Hughes, see section 16.2.6). Because only a

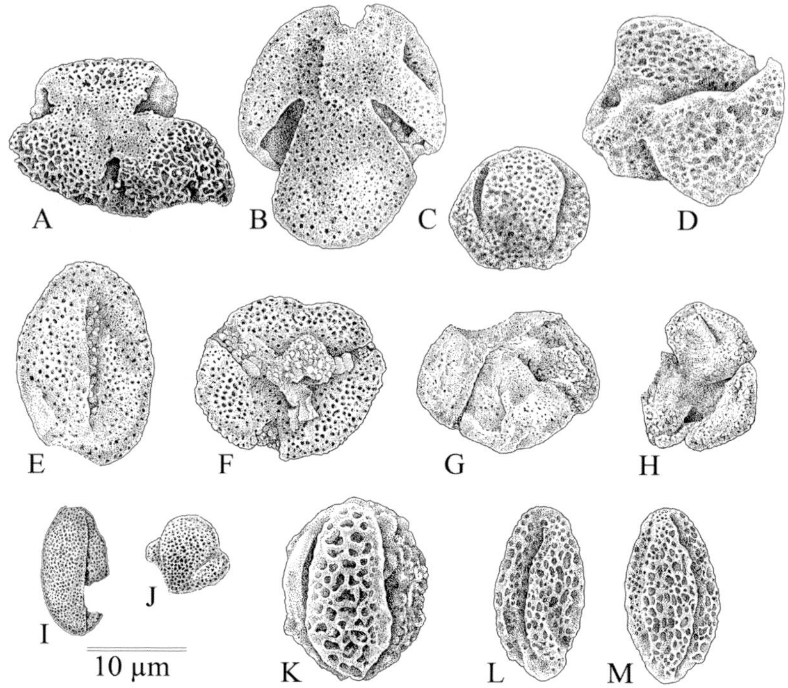

Figure 12.3 Early tricolpate pollen from the Late Barremian–Aptian of England, Egypt, Gabon and Portugal. (A) Pollen with graded reticulum from the Atherfield succession, Isle of Wight, England (Ferruginous Sands, mid-Aptian). (B) Foveolate pollen described as Punctitri-Fineret from the Mersa Matruh 1 Borehole, Egypt (Late Aptian). (C) Finely reticulate pollen described as Retitri-Crotmur from the Mersa Matruh 1 Borehole, Egypt (Late Aptian). (D) Reticulate pollen described as Retitri-Liliret from the Mersa Matruh 1 Borehole, Egypt (mid-Aptian). (E, F) Foveolate to finely reticulate tricolpate pollen found *in situ* in anther from the Torres Vedras locality, Portugal (Late Barremian–Early Aptian), in (E) equatorial and (F) polar view. (G, H) Polar view of foveolate, tricolpate pollen found in coprolite from the Torres Vedras locality, Portugal (Late Barremian – Early Aptian). (I, J) Finely reticulate tricolpate pollen grain from the Atherfield succession, Isle of Wight, England (Barremian) in (I) equatorial and (J) polar view. (K) Coarsely reticulate tricolpate pollen described as Retitri-Lowdown from the Mersa Matruh 1 borehole, Egypt (Late Barremian). (L, M) Coarsely reticulate tricolpate pollen from the Cocobeach sequence, Gabon (Zone C-VIIa; Late Barremian), same pollen grain from two sides. (A, I, J) Drawn from SEM images in Hughes and McDougall (1990); (B–D, K) drawn from SEM images in Penny (1991); (E–H) drawn from material in the collections of the Swedish Museum of Natural History; (L, M) drawn from SEM images in Doyle (1992).

single grain was found, Hughes and McDougall (1990) gave no formal description, but the grain is documented by SEM micrographs. The pollen grain is small, about 16 μm long, with three long colpi and a finely reticulate pollen wall with small rounded to angular lumina and smooth muri (Figure 12.3). The pollen wall is infolded in the colpus areas, which obscures the aperture margin and aperture membrane.

The second record from the Atherfield succession is also based on a single grain, this time extracted from a sample higher in the sequence assigned to Phase 5 of Hughes (1994). The grain is larger, about 25 μm long, also with three long colpi. The pollen wall is reticulate and distinctly columellate with long columellae (Figure 12.3). The reticulum is graded with lumina of irregular shape that decrease in size from the non-apertural regions where the reticulum is coarse towards the apertures and poles where the pollen wall becomes foveolate to almost psilate.

The pollen grain from Phase 5 of Hughes (1994) is very similar to tricolpate specimens from Zone I (Aptian) of the Potomac Group sequence assigned to *Rousea* sp., which are characterised by a similar kind of graded reticulum

(Doyle, 1992). Other Zone I tricolpate–reticulate pollen types from the Potomac Group are aff. *Tricolpites crassimurus*, which is about 35 μm long with a fine reticulum of small, evenly sized lumina, and the slightly smaller *Tricolpites* sp. A, which has a relatively coarse reticulum (Doyle, 1992). In the study of the palynology of two boreholes through the Potomac Group sequence by Doyle and Robbins (1977) two out of the 11 kinds of angiosperm pollen grain recorded from Zone I were tricolpate; the other nine were monocolpate. Already in the early part of Zone II (Late Aptian) of the same sequence tricolpate/triaperturate pollen are more diverse with seven kinds of tricolpate grains among 13 angiosperm taxa in Subzone II-A. In Subzone II-B, 21 out of 32 angiosperm taxa are triaperturate (Doyle and Robbins, 1977).

From equatorial Africa the earliest tricolpate forms from Pollen Zone C-VII (Barremian, Doyle, 1999) of the Cocobeach sequence are assigned to aff. *Tricolpites micromurus* and aff. *Tricolpites crassimurus* (Doyle *et al.*, 1977), or are left unassigned (Doyle, 1992). All of these grains have a reticulate–columellate pollen wall. Grains assigned to aff. *Tricolpites micromurus* are small, about 16 μm long and 12 μm in diameter; grains assigned to aff. *Tricolpites crassimurus* are larger, about 30 μm in diameter (only figured in polar view). The unassigned pollen type is 22 μm long and 12 μm in diameter with colpi that are obliquely and unevenly arranged (Figure 12.3). In Aptian strata from the Cocobeach sequence tricolpate–reticulate pollen is more diverse and includes grains assigned to '*Retitricolpites*' *geranioides*, cf. *Tricolpites georgensis* and *Tricolpites* sp. 2 (Doyle *et al.*, 1977). '*Retitricolpites*' *geranioides* is about 50 μm long, and the largest of the tricolpate grains in this sequence. It is coarsely reticulate with distinct columellae. Tricolpate–striate pollen is reported from the early part of Pollen Zone VIII (Early Aptian) of the Cocobeach sequence (Doyle, 1999).

The earliest tricolpate pollen from Israel is from the Zeweira Formation in the Zohar 1 well (core 6, Barremian or Aptian) and includes several different forms assigned to *Tricolpites*. They are spherical to oblate in shape and distinctly reticulate–columellate (Brenner, 1996).

From the Mersa Matruh 1 Borehole of Egypt the first tricolpate pollen is from the Late Barremian. The record is based on a single grain referred to as Retitri-Lowdown (Figure 12.3). The grain is reticulate–columellate and small, about 15.5 μm long and about 12 μm in diameter. Colpi are long with distinct margins. Muri are smooth, and lumina are rounded or irregularly polygonal in shape (Penny, 1991). Other early tricolpate and reticulate to foveolate pollen from the Mersa Matruh 1 Borehole appears in the mid-Aptian and includes forms described as Punctitri-Fineret, Retitri-Crotmur and Retitri-Liliret (Figure 12.3).

Among the early tricolpate pollen in the Mersa Matruh 1 Borehole, striate forms are an important and distinctive group. These occur first in the earliest Aptian, and at this level they are more diverse and more frequent than the reticulate forms with which they co-occur. Later in the succession (Late Aptian – Early Albian) reticulate pollen types are more diverse and abundant (Penny, 1988a). Early tricolpate pollen grains with striate sculpture are all small with a semitectate, striate and columellate pollen wall (Figure 12.4). Muri are smooth and meander and interweave irregularly over the surface of the grains. The different pollen types are distinguished from each other mainly based on details of the sculpturing of the muri observed by SEM. These features would be very difficult or impossible to observe with light microscopy. Striotri-Smoothmur (Figure 12.4), reported from the earliest Aptian to the latest mid-Aptian, is one of the most common angiosperm pollen types in the early part of the Aptian succession. It accounts for 38.5% of the angiosperm pollen in the earliest parts of its range. It is about 13–19 μm long and about 6.5–17 μm in diameter, and has smooth muri.

Three other tricolpate–striate pollen types occur in the early part of the Aptian succession in the Mersa Matruh 1 Borehole. Striotri-Oval is about 13–19 μm long. Muri are ornamented by faint transverse ridges. Striotri-Segmur is about 14.5–19.4 μm long and has a more distinct ornamentation of transverse ridges over the muri (Figure 12.4). This pollen type has now been found *in situ* in flowers described from the Early Cretaceous of Portugal as *Lusistemon striatus* (Figure 2.18, and below). Pollen described as Strioret-Smooth is about 21.6 μm long with striate sculpture over the polar regions and along the apertures, and striate–reticulate sculpture in the non-apertural regions. Muri are smooth (Figure 12.4).

Similar tricolpate–striate pollen grains are also known from Aptian and younger palynofloras from equatorial Africa (Doyle *et al.*, 1977; Doyle, 1992), Brazil (Regali *et al.*, 1974), and eastern North America (Brenner, 1963; Doyle, 1969; Doyle and Hickey, 1976; Doyle and Robbins, 1977). They are usually referred to the dispersed pollen genera *Striatopollis*, *Reticolpites* or *Retitricolpites*. Based on pollen morphology alone the systematic affinities of these pollen grains would be difficult to assess, but some of the pollen grains are closely comparable to pollen of extant early-diverging eudicots. The parent plants of some of these grains may have been on the stem lineages of extant taxa that

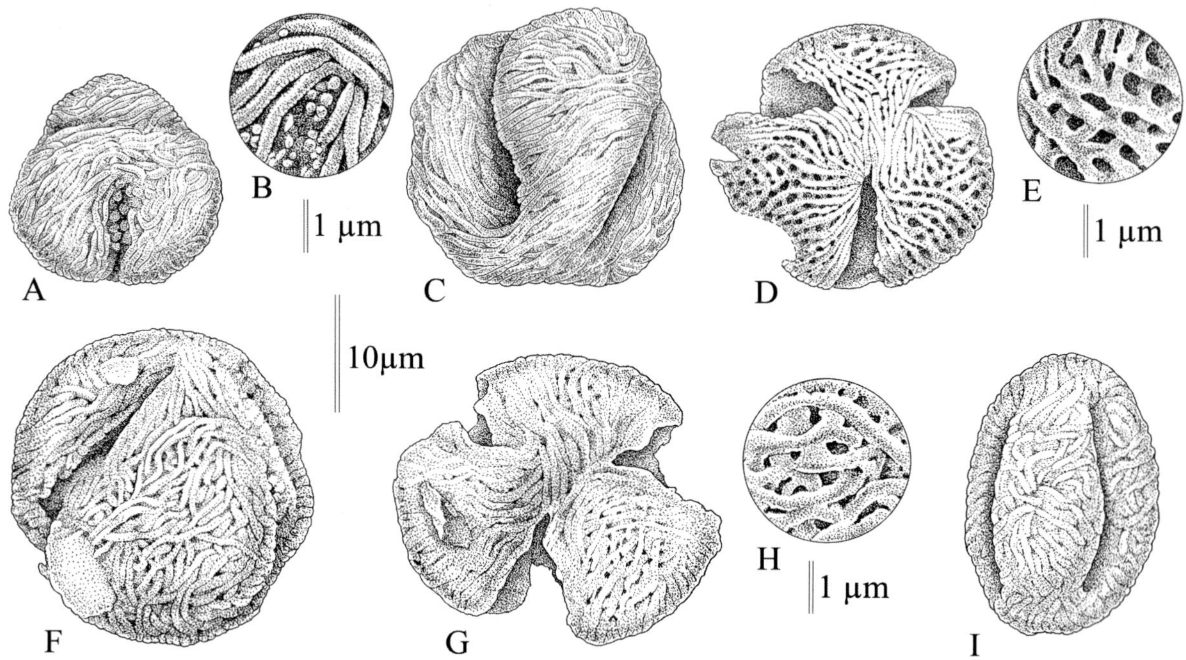

Figure 12.4 Tricolpate–striate pollen from the Early to Late Aptian of the Mersa Matruh 1 Borehole, Egypt. (A–C) Striotri-Segmur pollen with transversely striate muri showing (A) polar view, (B) details of wall ornamentation and (C) equatorial view. (D–F) Strioret-Smooth pollen with smooth muri and striate–reticulate tectum, showing (D) polar view, (E) details of wall ornamentation and (F) oblique equatorial view. (G–I) Striotri-Smoothmur pollen with smooth muri and striate–reticulate tectum showing (G) polar view, (H) details of wall ornamentation and (I) equatorial view. Drawn from SEM images in Penny (1988a).

diverged at an early stage from the main line of eudicot evolution. For example, striate–reticulate pollen described from Egypt as Strioret-Smooth is very similar to grains of extant *Tetracentron* and *Trochodendron* (Trochodendraceae). Other striate pollen is very similar to pollen found *in situ* in Cretaceous flowers that are related to extant Buxaceae. These flowers are *Lusistemon striatus* – *Lusicarpus planatus* from the Late Aptian – Early Albian of Portugal (Pedersen *et al.*, 2007), as well as *Spanomera marylandensis* (Late Albian) and *S. mauldinensis* (Early Cenomanian) from Maryland, USA (Drinnan *et al.*, 1991) (section 12.8). Other striate pollen may have been produced by lineages that are now extinct or that were of diverse relationships among early eudicots.

12.4.2 Early tricolpate pollen *in situ* in reproductive structures

Very few macrofossils or mesofossils with tricolpate pollen *in situ* are known from the earliest phases of angiosperm diversification. From the Late Barremian – Early Aptian mesofossil flora of Torres Vedras, Portugal, two different kinds of tricolpate grain have been discovered, one *in situ* in a stamen and one incorporated in a coprolite (Figure 12.3). Three stamens with tricolpate pollen (Fig. 9 in Friis *et al.*, 1994b) are not from Torres Vedras, but from the Puddledock locality. From the other, younger, Early Cretaceous floras of Portugal, tricolpate pollen is also rare (Catefica, Famalicão, Vale de Agua) or has not been recorded (Buarcos, Vila Verde 2). From the Vale de Agua locality five different kinds of tricolpate grain are known from flowers and dispersed stamens. All are thought to represent eudicot lineages that diverged early in the history of the clade. From Famalicão tricolpate pollen occurs on or near the stigmatic surface of four different kinds of fruits, whereas from the Catefica locality three different tricolpate grains are known, one in a stamen fragment and two in coprolites (Figure 12.5). From the Puddledock flora (Early–Middle Albian), Virginia, USA, *in situ* tricolpate pollen is already much more diverse than that in the older floras, and is known in a variety of dispersed stamens (Figure 12.6).

Figure 12.5 Tricolpate pollen in stamen fragment (A–C) and in coprolites (D–I) from the Early Cretaceous (Late Barremian–Aptian) Catefica locality, Portugal. (A–C) Stamen fragment (A) and (B, C) details of *in situ* finely reticulate pollen. (D–F) Coprolite (D) and (E, F) details of *in situ* pollen. (G–I) Coprolite (G) and (H, I) details of *in situ* pollen. Material in the collections of the Swedish Museum of Natural History, Stockholm.

All *in situ* tricolpate pollen grains recovered so far from the earliest phases of eudicot diversification are prolate and small, ranging from about 10 to 25 µm in length, and have simple, elongate colpi. However, there is more diversity in the structure of the pollen wall, which includes a variety of foveolate, reticulate and striate types (Figures 12.3–12.6). It may also be significant that many of the stamens from Late Aptian–Middle Albian mesofossil floras that

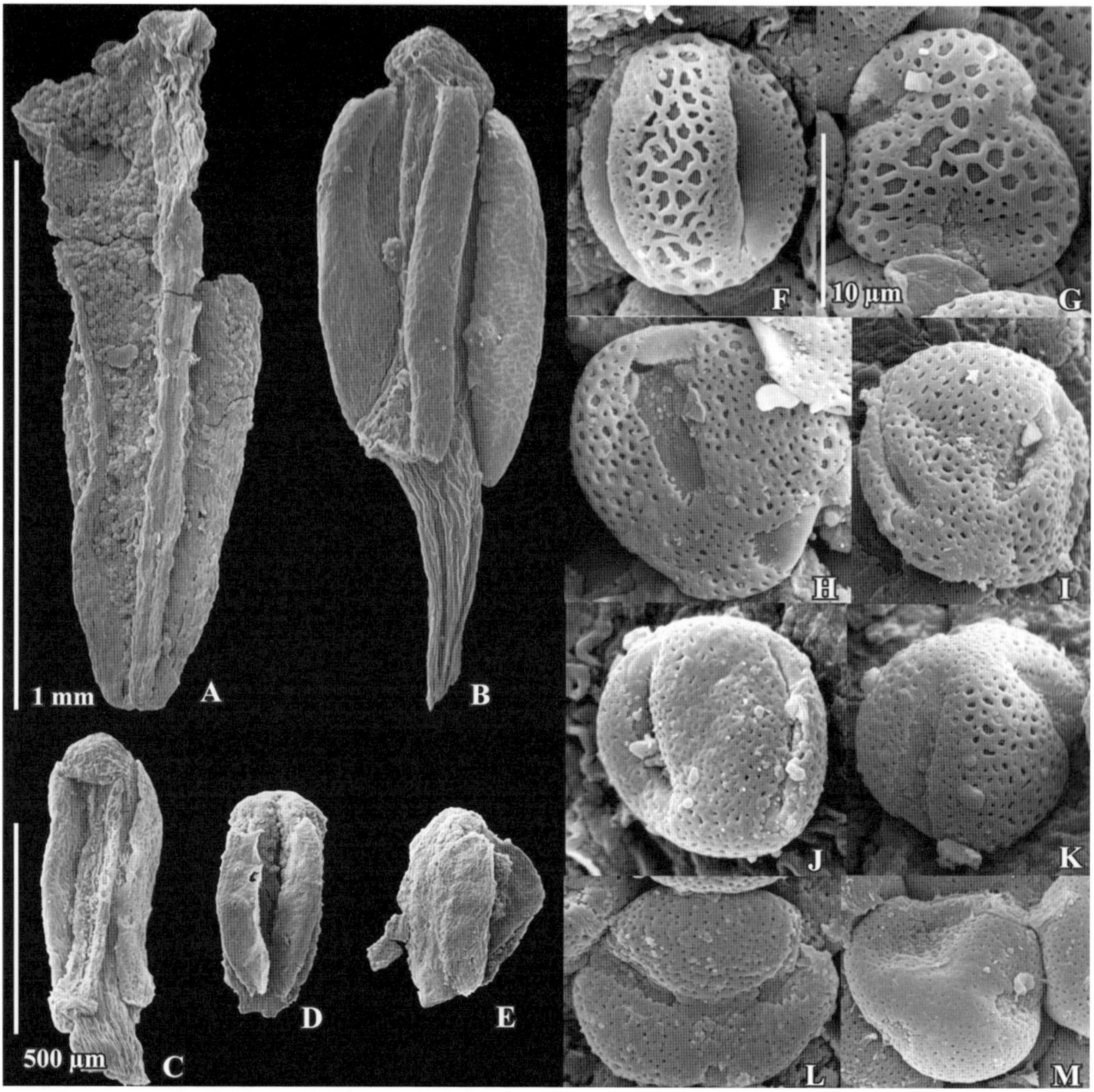

Figure 12.6 Tricolpate pollen from dispersed stamens from the Early Cretaceous (Early–Middle Albian) Puddledock locality, Virginia, USA. (A) Stamen fragment. (B–E) Different stamens, all with valvate dehiscence and a domed apical protrusion of the connective. (F, G) Reticulate–foveolate pollen from stamen in (A). (H–J) Reticulate–foveolate pollen from stamens not figured here. (K) Reticulate–foveolate pollen from stamen in (D). (L) Foveolate–punctate pollen from stamen not figured here. (M) Foveolate–punctate pollen from stamen in (E) Scale bar same for (A, B) for (C–E), and for (F–M). Material housed in the collections of the Field Museum, Chicago.

produced tricolpate pollen have distinct valvate dehiscence and a pronounced apical expansion of the connective (Figure 12.6). Some of these grains, and the floral structures that they are associated with, appear related to lineages of early eudicots that are still extant, such as Ranunculales, Buxales and Platanaceae. Others are clearly eudicots because of their distinctive tricolpate pollen, but their precise systematic position is uncertain.

Figure 12.7 Reconstruction of *Sinocarpus decussatus*, a putative early eudicot from the Early Cretaceous (Early Aptian – early Late Aptian) Yixian Formation of northeastern China, showing fruits with basally fused carpels and associated leaves with chloranthoid teeth. Based on information in Leng and Friis (2003, 2006).

12.4.3 Fossil reproductive structures of probable eudicot relationships

In addition to dispersed tricolpate pollen and dispersed stamens and floral structures that have tricolpate pollen *in situ* that can be assigned to extant groups, there are also several eudicot reproductive structures from the Early Cretaceous that remain to be described and that have not yet been assigned to any extant family or order. There is also a range of material that was probably produced by early eudicots, but that lacks a clear link to associated tricolpate pollen that would provide definitive evidence of this relationship. In these cases the assignment to eudicots is based mostly on features of the inflorescences/ infructescences and sometimes, associated leaves. Fossils of this kind include *Sinocarpus decussatus* from the Early Aptian – early Late Aptian, Yixian Formation, China, *Ranunculaecarpus quinquiecarpellatus* and *Araliaecarpum kolymensis* from the Albian of the Kolyma Basin, East Siberia (Russia), *Hyrcantha karatscheensis* from the Albian of Kazakhstan, *Cathiaria* from the Cenomanian of Siberia (Russia), Kazakhstan and the Czech Republic and the Coniacian of Japan, *Callicrypta chlamydea* and *Freyantha sibirica* from the Cenomanian of Siberia, and a diverse group of follicular fruits from the Late Santonian – Early Campanian of Sweden. Further study is needed to clarify the nature and significance of *Leefructus mirus* recently reported as a fossil eudicot from the Yixion Formation by Sun *et al.* (2011).

Sinocarpus decussatus (Figure 12.7) is based on infructescence fragments preserved as compressions–impressions in a finely laminated siltstone from Dawangzhangzi Village, Liaoning Province, China (Leng and Friis, 2003). The siltstone is from the Dawangzhangzi Beds, middle Yixian Formation and is of Early Aptian – early Late Aptian age (Leng and Friis, 2003). Subsequently an additional specimen was described from a slab that also contained the first unequivocal angiosperm leaves from the Yixian Formation (Leng and Friis, 2006). Other specimens have also been reported by Dilcher *et al.* (2007) from localities in Inner Mongolia and Liaoning.

Fossil infructescences assigned to *Sinocarpus decussatus* are compound, but whether they are ebracteate or bracteate is uncertain. The main axis and lateral branches of the infructescence are slender, with dilated or slightly dilated nodes. Specimens are preserved at fruiting stage and provide details on the structure of the gynoecium, but little information on floral organisation. Lateral reproductive units are typically borne in an opposite–decussate arrangement, or in an arrangement that combines alternate and opposite branching. Each reproductive unit is borne on a long and slender stalk. The perianth is poorly preserved, but perianth parts were apparently free. They are attached below a superior and basally syncarpous gynoecium, composed of four, or more rarely three, carpels arranged in a whorl. The carpels are fused along the ventral side for about half of their length. Mature carpels are elongated, elliptical in outline, apparently with a sessile, or perhaps decurrent, stigma. However, no traces of the stigmatic surface remain. Each carpel has about 20 seeds arranged in two rows along ventral linear placentae that extend for the full length of the carpels. Seeds are anatropous, laterally flattened, and have a smooth surface. They are sometimes embedded in the remains of an amorphous substance that filled the inside of the carpel.

Leaves associated with *Sinocarpus* are simple, minute, and only 1–2 cm long (Figure 12.7). They have pinnate, poorly organised venation and a serrate margin with chloranthoid teeth. Roots described from the base of an inflorescence axis (Dilcher *et al.*, 2007) are not clear in the figured material.

Sinocarpus superficially resembles the inflorescence/infructescence axes of the younger (Middle Albian) *Hyrcantha karatchensis* (see below). However, *Hyrcantha* differs in having an apocarpous gynoecium of three to five free, apparently follicular, carpels borne on a flattened to slightly convex receptacle. Leaves associated with *Hyrcantha* are also compound rather than simple. The transfer of *Sinocarpus decussatus* to *Hyrcantha* suggested by Dilcher *et al.* (2007) required emendation of the diagnosis of *Hyrcantha* to include both syncarpous and apocarpous taxa, which is not followed here.

Sinocarpus is interesting in being the earliest angiosperm with syncarpy. However, the free part of the carpels is relatively large and *Sinocarpus* thus combines a limited degree of syncarpy with a more generalised structure. Comparison with modern plants indicates that *Sinocarpus* is probably related to early-diverging eudicots. It shares several features with extant *Myrothamnus* (Myrothamnaceae; core eudicots) such as decussate, opposite phyllotaxis, 3–4 carpels that are basally fused, and seeds that are embedded in an infilling of the locules. In extant *Myrothamnus* this infilling is a secretion from the ovary wall (Endress and Igersheim, 1999), and it is very likely that the infilling in *Sinocarpus* is the fossilised residue of a similar secretion. However, *Sinocarpus* is distinguished from extant *Myrothamnus* in having seeds present along the full length of the carpels. In *Myrothamnus* seeds are restricted to the basal, syncarpous part. The inflorescence of *Myrothamnus* is also dense and spike-like rather than open as in *Sinocarpus*. *Sinocarpus* also shows similarities with other early-diverging eudicots such as Ranunculaceae and Buxaceae, but cannot be placed with certainty in any extant group based on information currently available.

Hyrcantha karatscheensis (Figure 12.8) from the Middle Albian of Kazakhstan is based on isolated stalked fruits as well as fragmentary inflorescence/infructescence axes with loosely arranged floral structures (Vakhrameev, 1952; Krassilov *et al.*, 1983). The fossils are preserved as impressions/compressions with little organic material remaining. Because of their compressed nature the organisation of the reproductive structures is not fully understood. Individual flowers/fruits are borne on a long stalk. The gynoecium is apocarpous and formed of three to five carpels borne on a flattened to slightly convex receptacle. Carpels are elliptical with a sessile and apical stigma. They are apparently follicular and dehisce towards the centre of the flower along the ventral suture. No ovules or seeds were reported from the gynoecium, but faint transverse grooves in the carpel wall indicate the presence of eight to nine seeds per carpel. In some specimens remains of the androecium and perianth are preserved indicating that the flowers of *Hyrcantha* were probably bisexual. Stamens were apparently numerous, each consisting of a thread-like filament bearing a relatively small anther. Pollen has not been reported from the anthers.

A few small leaves apparently with pinnate venation and compound organisation occur associated with the fruit axes and are referred to as '*Leguminosites*' *karatscheensis* (Krassilov *et al.*, 1983). *Hyrcantha karatscheensis* has been compared to members of extant Ranunculaceae, but has also been compared to extant Paeoniaceae (Krassilov *et al.*, 1983). *Hyrcantha* is similar to *Sinocarpus* in having fruits borne on long slender stalks, but differs in its apocarpous gynoecium. The lack of pollen features and anatomical details of fruits and seeds, as well as the incomplete preservation of floral morphology, precludes thorough comparison with extant plants.

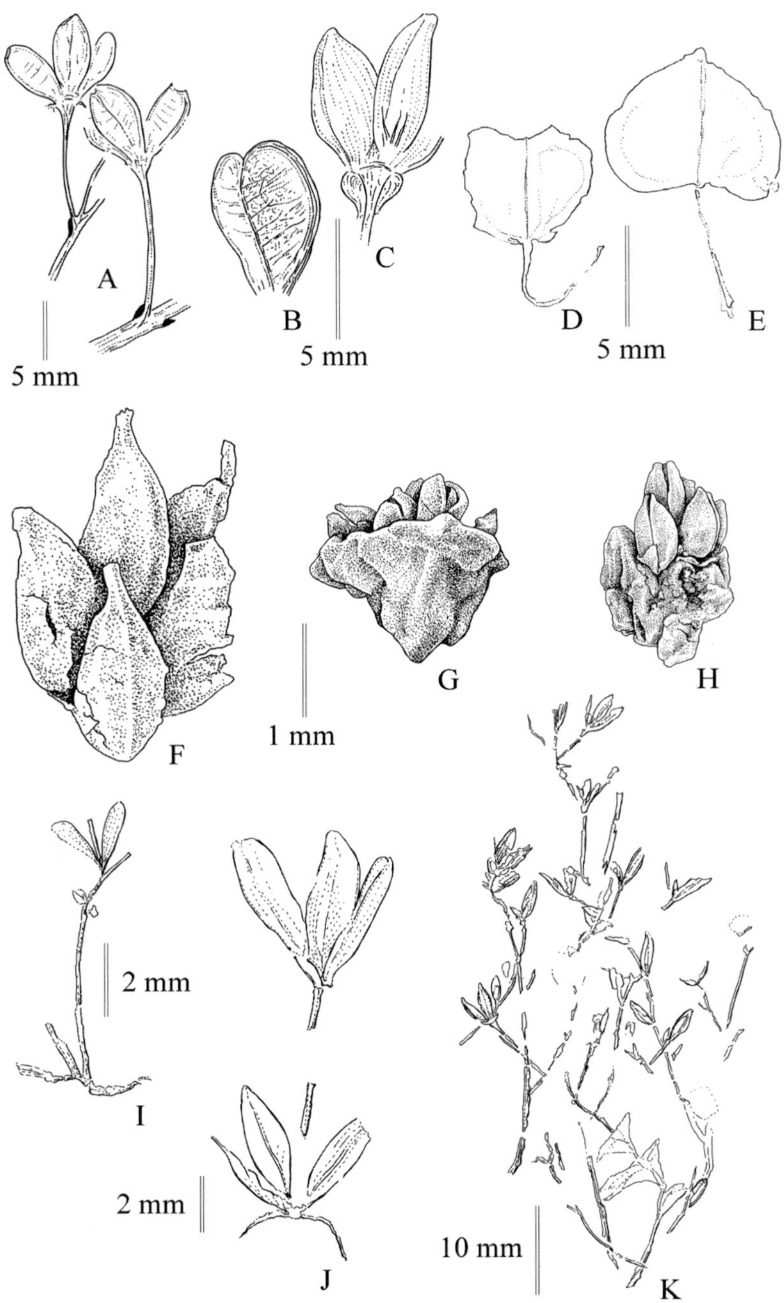

Figure 12.8 Early and mid-Cretaceous reproductive structures of putative basal eudicots. (A–C) *Hyrcantha karatscheensis* from the Early Cretaceous (Albian) of Kazakhstan showing stalked fruits with several free follicular fruitlets. (D, E) *Araliaecarpum kolymensis* from the Early Cretaceous (Albian) Buor-Kemiusskaja locality, Zyrianka River, East Siberia, showing stalked syncarpous gynoecia apparently formed from two carpels. (F–H) *Cathiaria zhilinii* from the Cenomanian–Early Turonian of the Kachar Quarry, northern Kazakhstan, showing (F) dorsal and (H) ventral view of fruits and (G) inflorescence fragment with associated bract. (I–K) *Ternariocarpites floribundus* from the Early–Middle Albian of the Primorye Region, Far East Russia, showing stalked fruits with several free follicles. (A–C) Redrawn from Krassilov *et al.* (1983); (D, E) drawn from photographs in Samylina (1960); (F–H) drawn from SEM images in Hvalj (2001); (I–K) redrawn from Krassilov and Volynets (2008).

Ternariocarpites floribundus (Figure 12.8) is another branched reproductive structure from the Early Cretaceous with apocarpous gynoecia and follicular fruitlets that are very similar to *Hyrcantha* in their general appearance. This fossil was described from the Early–Middle Albian of the Primorye Region in the Russian Far East and comprises slender profusely branched axes with terminal fruits. The fruits are surrounded by five persistent perianth parts and consist of at least three free follicles that are sometimes open along one of the sutures. Each follicle has several seeds arranged in a row. *Ternariocarpites* and *Hyrcantha* are of approximately the same age and their similar structure may indicate that they are closely related.

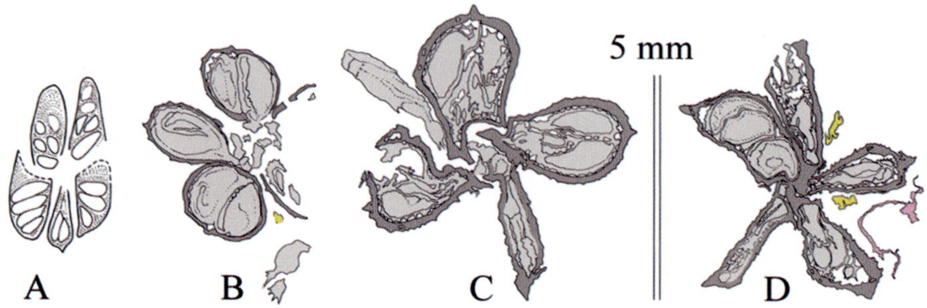

Figure 12.9 *Ranunculaecarpus quinquecarpellatus* from the Early Cretaceous (Albian) Buor-Kemiusskaja locality, Kolyma Basin, East Siberia. (A) Fruit seen on rock surface. (B–D) Sections of fruit from apex (B) towards base (D) showing five free carpels, each containing many reticulate seeds and remains of perianth (pink) and stamens (yellow). (A) redrawn from Samylina (1960); (B–D) drawn from sections prepared by S. Manchester and L. Golovneva.

Ranunculaecarpus quinquecarpellatus (Figure 12.9) from the Albian Buor-Kemiusskaja locality at the Zyrianka River in the Kolyma Basin, East Siberia, Russia (Samylina, 1960) is a small fruiting structure consisting of five free carpels. Carpels are elongate, and apparently follicular, with ventral dehiscence. Each carpel contains many small reticulate seeds. The material is permineralised and new investigations, including longitudinal and transverse sections, have added important details of the morphology and anatomy of carpels and seeds (L.B. Golovneva, S.R. Manchester and E.M. Friis, work in progress). The petrified specimens also preserve remains of tepals and stamens showing that *Ranunculaecarpus* flowers were bisexual. The new information supports assignment of the fossils to the Ranunculales as tentatively suggested by Samylina (1960).

Araliaecarpum kolymensis (Figure 12.8) is another fossil fruit from the Albian Buor-Kemiusskaja locality (Samylina, 1960). The fruit is about 6 mm long and appears to be syncarpous with at least two carpels. Like the specimens of *Ranunculaecarpus quinquiecarpellatus* this material is permineralised. Future study, including sectioning, is likely to provide details of morphology and structure that will facilitate more detailed systematic analysis (L. Golovneva and S. Manchester, personal communication, 2010).

Callicrypta chlamydea is a small flower described from the Cenomanian Timerdyakhskaya Formation at the Tyung River, Vilyuy Basin, eastern Siberia, Russia (Krassilov and Golovneva, 2004). The flower is three-dimensionally preserved, hypogynous, actinomorphic, and apparently with three series of perianth parts and an apocarpous gynoecium of six carpels. The flower was described as pistillate. Androecial remains were interpreted as staminodes (Krassilov and Golovneva, 2004) but could also be remains of stamens from which anthers have been lost. Pollen observed on the flower is tricolpate and reticulate, suggesting a relationship to eudicots. Similar pollen was observed *in situ* in flowers of *Freyantha sibirica* described from the Cenomanian Kemskaya Formation at the Kem River, Tschulymo–Yenisey Basin, West Siberia, Russia (Krassilov and Golovneva, 2001) and it was suggested that these two taxa might represent male and female flowers of the same plant. However, the pollen is of a very generalised type and other evidence linking these two species is needed. *Callicrypta* was compared mainly with extant Ranunculales, but a more precise systematic relationship for the fossils was not suggested (Krassilov and Golovneva, 2004). The relationship of *Callicrypta* and *Freyantha* to eudicots seems secure, but their position within the group is not clear. The interpretation of *Freyantha*, in particular, is uncertain because the flowers are compressed and difficult to study. Also, rather than linking *Freyantha* with *Callicrypta*, Golovneva and Oskolsky (2007) suggested a relationship between *Freyantha* and *Cathiaria* (see below).

Cathiaria was first described and informally named by Hvalj and Golovneva in the doctoral thesis by Hvalj (2001) and was later formally established by Goloneva and Oskolsky (2007) with a single species, *Cathiaria zhilinii*. The type material includes inflorescences and isolated fruits from the Cenomanian – Early Turonian of the Kachar Quarry, northern Kazakhstan (Figure 12.8). Similar material is also known from the Cenomanian of Western Siberia, Russia (Golovneva and Oskolski, 2007) and the Czech Republic (J. Kvaček, personal communication, 2010), as well as the Coniacian of Japan (Takahashi *et al.*, 1999b). Fruits of *Cathiaria* are small, laterally flattened, monocarpellate and with a single seed. They are borne in compound inflorescences consisting of a main axis bearing alternate to subopposite lateral units.

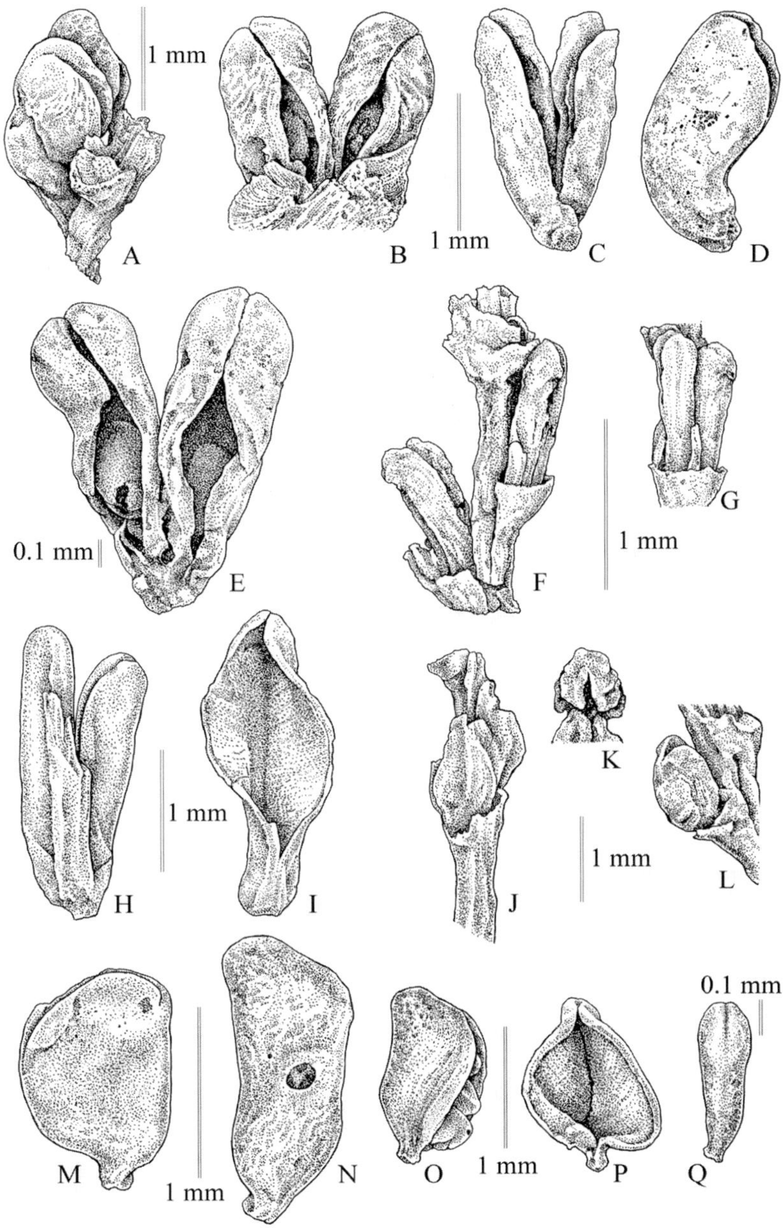

Figure 12.10 Fossil follicles borne in pairs on inflorescences, or isolated, from the Late Cretaceous (Late Santonian – Early Campanian) Åsen locality, Sweden. (A–E) *Maiandrocarpus moirasmenus*, pair of follicles on an inflorescence in (A) lateral and (B) adaxial view, (C, E) pair of isolated follicles and (D) single isolated follicle in lateral view showing sinuous ventral margin. (F, G) *Mitocarpus elegans*, pairs of follicles borne on inflorescence axis showing (F) lateral and (G) abaxial view of follicle pair. (H, I) *Zeugarocarpus adroagathus*, (H) small inflorescence fragment with pair of follicles and (I) isolated follicle. (J–L) *Malliocarpus batrachoides*, inflorescence axis bearing a single follicle in (J) abaxial, (K) apical and (L) lateral view. (M–Q) *Agapitocarpus emisxus*, isolated follicles in (M–O) lateral, (P) adaxial and (Q) abaxial view. (A–D, I–Q) Drawn from SEM-images in Leng *et al.* (2005); (E–H) from Friis *et al.* (2006a).

Although the material is three-dimensionally preserved the organisation of the inflorescence is difficult to interpret. A relationship to eudicots seems most likely, but so far there is no link to triaperturate pollen and the systematic position of *Cathiaria* is uncertain.

A diverse assemblage of follicular fruits (Figure 12.10), which probably represent an extinct complex of early eudicots, has been described from the Late Santonian – Early Campanian Åsen locality, Sweden (Leng *et al.*, 2005). The follicles generally have a distinct ventral slit that extends from the base over the apex to the dorsal side. The stigma is indistinct, sessile and with a double crest at the distal end of the ventral slit. The fruit wall is simple and supplied by a simple vascular system of one dorsal and two ventral bundles. Placentation is marginal and linear with several anatropous, bitegmic ovules. Eight different species in seven genera have been recognised among this complex of follicular fruits (Leng *et al.*, 2005). Infructescence structure is known in part for some of these taxa (*Maiandrocarpus, Malliocarpus, Mitocarpus, Zeugarocarpus*). Others (*Agapitocarpus, Chontrocarpus, Xylocarpus*) are known from dispersed follicles only.

Infructescences consist of a central axis with lateral subunits borne in a helical arrangement. In most taxa the subunits are subtended by a large cup-like bract. In *Malliocarpus* each subunit contains a single follicle, whereas in *Maiandrocarpus, Mitocarpus*, and *Zeugarocarpus* each subunit consists of a pair of follicles that are either free from each other or borne on a common stalk. Each follicle of a pair faces the infructescence axis with its ventral slit. This strongly suggests that the follicle pair is formed by two separate monocarpellate flowers. In a single bicarpellate flower the ventral slits of the two carpels would normally face each other. No scars of perianth parts are observed in any of the taxa. This suggests that the flowers lacked a perianth, which is also consistent with the presence of a large cup-like bract.

The shared features of the fossil follicles and infructescences suggest that they are closely related to each other. Although these fossils cannot be included in any extant family many features of the gynoecium are particularly common among some early-diverging eudicots, including: the plicate carpels that mature into follicles, the fruit wall supplied by three vascular bundles, the presence of a sessile, double-crested stigma, and the linear–marginal placentation with many anatropous, bitegmic ovules (Endress and Igersheim, 1999). A position among early-diverging lineages of eudicots is also supported by the unusual paired arrangement of two monocarpellate flowers subtended by a common bract. In modern plants a similar arrangement is only known in the families Proteaceae and Didymelaceae (Leng *et al.*, 2005).

Maiandrocarpus moirasmenus (Figure 12.10) is based on an infructescence fragment from the Åsen locality with helically arranged subunits, and numerous dispersed follicles (Leng *et al.*, 2005) (see general description above). Each subunit consists of two follicles with a large cup-shaped bract below. Individual follicles are astipitate, but the two follicles of the pair are borne on a common stalk. Follicles are narrowly obovate in outline with a sinuous ventral surface, a rounded apex and a tapering base. There are up to six ovules in each follicle, of which two to three developed into mature seeds. The ovules are smooth, bitegmic and anatropous, with a short flattened funicle and distinct raphe. Mature seeds have a finely wrinkled outer surface.

Mitocarpus elegans (Figure 12.10) is based on a single inflorescence fragment from the Åsen locality (Leng *et al.*, 2005). The fragment shows three lateral subunits in a helical arrangement (see general description above). Each subunit consists of a pair of follicles supported by a cup-shaped median bract and a pair of lateral prophylls. Another organ of unknown nature is present in a median and abaxial position. Individual follicles are astipitate and borne separately. They are narrowly obovoid with almost parallel dorsal and ventral sides. The apex is rounded and beaked, and the base is truncate. Ovules or seeds are not known for this species.

Zeugarocarpus (Figure 12.10) includes two species, *Zeugarocarpus adroagathus* and *Z. leptoagathus*, both from the Åsen locality (Leng *et al.*, 2005). *Zeugarocarpus adroagathus* is known from an inflorescence fragment as well as dispersed follicles and seeds, while *Z. leptoagathus* is known only from dispersed follicles and seeds (see general description above). In *Zeugarocarpus adroagathus* the subunits are composed of a pair of prophylls and a pair of follicles. Individual follicles in both species are astipitate and attached separately. They are narrowly elliptical in outline with almost parallel ventral and dorsal sides, a rounded apex and a truncate base. Mature seeds are anatropous, apparently bitegmic, ellipsoidal in shape and have a spiny surface. In *Zeugarocarpus adroagathus* the spines are large and pronounced, whereas in *Z. leptoagathus* they are smaller and less pronounced.

Malliocarpus batrachoides (Figure 12.10) is based on a single infructescence fragment from the Åsen locality (Leng *et al.*, 2005). The specimen consists of an axis with three lateral subunits in a helical arrangement (see general description above). Each subunit has a large median bract, two lateral prophylls and a single follicle. The follicle is ovate, probably astipitate. The surface of the infructescence and follicle is covered by a dense indumentum of long, uniseriate trichomes. Ovules or seeds are not known.

Agapitocarpus emisxus (Figure 12.10) was described from the Åsen locality based on numerous dispersed follicles (Leng *et al.*, 2005). There is no information on infructescence structure. The follicles are always found separately. Each is stipitate, elongate, elliptical to semicircular and laterally flattened, with rounded apex and slightly pointed base (see general description above). Each carpel has up to eight ovules, of which all, or most, developed into mature seeds. The ovules and seeds are smooth, anatropous and apparently bitegmic. Ovules generally do not fill the ovary cavity.

Xylocarpus rhitidodes is based on a single dispersed follicle from the Åsen locality, Sweden (Leng *et al.*, 2005). The follicle is apparently astipitate, narrowly oblong and slightly flattened laterally with about nine anatropous, ellipsoidal seeds with a reticulate surface.

Chontrocarpus pachytoichus is based on isolated follicles from the Åsen locality, Sweden (Leng *et al.*, 2005). The follicles are astipitate, narrowly obovate to ellipsoidal (see general description above). The base and apex are rounded and the stigma is distinct.

12.5 RANUNCULALES

The Ranunculales, comprising the Berberidaceae, Circaeasteraceae, Eupteleaceae, Lardizabalaceae, Menispermaceae, Papaveraceae and Ranunculaceae, are resolved as sister to all other eudicots in most recent phylogenetic analyses (APGIII, 2009). Extant plants in the order (except Berberidaceae and Eupteleaceae) are predominantly herbs or climbers with low fossilisation potential for their vegetative organs. Perhaps as a result, the macrofossil record of the group is relatively meagre, both from the Cretaceous and from the Cenozoic. Cretaceous leaf fossils such as *Vitiphyllum*, *Cissites* and *Menispermites*, which have gross morphology that may suggest affinity with Ranunculales, are mostly impression fossils and there is the possibility of confusion with the leaves of other angiosperm groups. In most cases venation or epidermal details are not preserved and the systematic affinity of these leaves remains equivocal. In addition, the affinity of the Early Cretaceous fossil plant from Spain first described as *Montsechites ferreri*, and then again later as *Ranunculus aquatilis*, remains uncertain (Chapter 9).

The fossil pollen record of securely identified Ranunculales is also rather poor. This may be due in part to the relatively plesiomorphic features of the pollen in the context of eudicots as a whole; most Ranunculales have tricolpate pollen with reticulate or foveolate exine structure. Tricolporate pollen occurs among Menispermaceae, but otherwise pollen in the order is of rather generalised morphology that occurs in other eudicot groups. Pollen of some Menispermaceae, such as the syncolpate forms, is more characteristic. Distinctive pollen with a spiral aperture occurs in some Berberidaceae. However, while this aperture configuration is unusual it also occurs in other families. Assignment of Cretaceous pollen with spiral apertures to Berberidaceae needs confirmation by detailed ultrastructural studies (Muller, 1981).

Fossil fruits and seeds of Ranunculales are also rare in the Cretaceous. This is somewhat surprising since the fruits and seeds of many extant Ranunculales (e.g. Ranunculaceae) are distinctive with hard tissues in the fruit or seed wall. In addition, many extant Ranunculales grow in environments with good fossilization potential and are also well represented in Cenozoic floras. Currently, the only reproductive organs from the Cretaceous that can be assigned confidently to an extant family of Ranunculales are endocarps assigned to Menispermaceae. The flower of *Teixeiraea lusitanica* is most likely an early representative of the order, but it cannot be placed in an extant family.

Teixeiraea lusitanica (Figure 12.11) from the Early Cretaceous (Late Aptian – Early Albian) Vale de Agua locality, Portugal, is based on a single flower bud apparently fossilised shortly before anthesis (von Balthazar *et al.*, 2005). The flower is unisexual and staminate. There is no evidence of a pistil. All floral parts are free from each other, densely spaced and apparently in a helical arrangement. The bud is not complete and the exact number of parts is unknown. The flower has about 25–30 bracts and perianth parts that exhibit a transition between different organ categories. The outermost organs are obtuse and bract-like. Toward the centre these grade into longer, broadly ligulate, inner organs, which are more sepal-like. In turn these grade internally into sword-shaped, petal-like organs. The androecium consists of about 16 longer, outer stamens, and four shorter, inner stamens. Stamens are dithecate and tetrasporangiate with very long, slightly sagittate anthers topped by a short pointed projection of the connective. Anthers are basifixed, displaced towards the adaxial surface, and apparently attached to the very short filament without a joint. *In situ* pollen is almost spherical, about 20 μm in diameter, tricolpate and tectate–perforate, with an infratectal layer of short, densely spaced columellae, a thick foot layer and an endexine that is thicker under the apertures. Colpi are long, with distinctly delimited margins and granular colpus membranes.

The combined features of pollen and flower suggest relationship with extant Ranunculales, particularly with Berberidaceae, Menispermaceae, Lardizabalaceae and Papaveraceae. However, the fossil flower cannot be placed in any extant family and *Teixeiraea* may represent an extinct lineage among crown-group Ranunculales or perhaps along its stem (von Balthazar *et al.*, 2005).

12.5.1 Berberidaceae

The Berberidaceae comprise about 15 genera and 650 species of herbs or woody shrubs distributed in temperate regions of the Northern Hemisphere and in the Andes (Loconte, 1993). Flowers are hypogynous, typically trimerous, or sometimes tetramerous, and have a single central carpel. Pollen is mostly tricolpate, except for pollen of *Berberis* and *Mahonia*, which

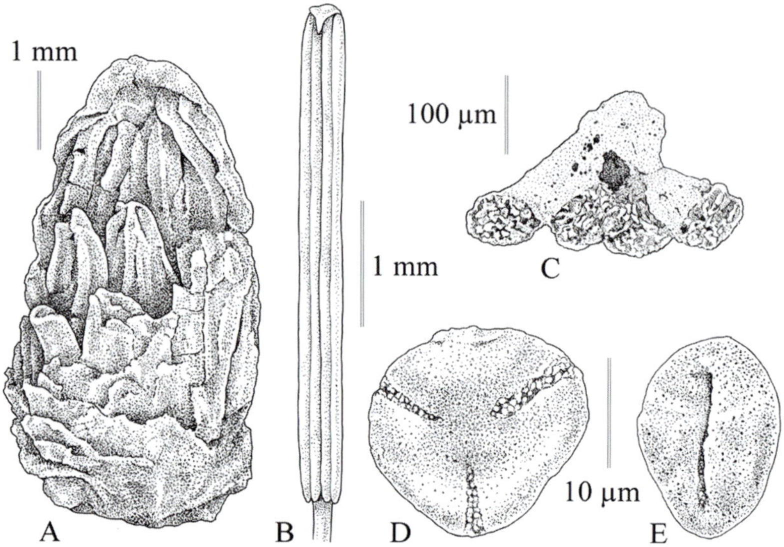

Figure 12.11 *Teixeiraea lusitanica*, an Early Cretaceous (Late Aptian – Early Albian) flower from the Vale de Agua locality, Portugal, assigned to Ranunculales. (A) Flower bud. (B) Reconstruction of stamen in adaxial view. (C) Cross-section of stamen. (D, E) Tricolpate–perforate pollen from stamen in (D) polar view and (E) equatorial view. (A, C–E) Drawn from SEM images in von Balthazar *et al.* 2005.

has irregular or spiral apertures, and *Ranzania*, which has six colpi (Nowicke and Skvarla, 1981).

The fossil record of Berberidaceae is poor and there are no unequivocal records from the Cretaceous. According to Muller (1981) dispersed pollen of *Sigmopollis* with spiral apertures, which has been reported from the Turonian of Canada and compared to pollen of extant *Berberis*, is equivocal. Fossil leaves described as *Winchellia triphylla* from the Late Cretaceous of Yellowstone River, eastern USA, have been assigned to the family (Lesquereux, 1893), but are more likely a platanoid (Crane *et al.*, 1988). From the Cenozoic, Berberidaceae are mainly represented by leaves of *Berberis* (including *Mahonia*) described from the Oligocene and onwards in Europe, Asia and North America (Tanai, 1961; Ernst, 1964; Becker, 1972a; Axelrod, 1987; Mai, 1995; Ramirez and Cevallos-Ferriz, 2000). Many of these fossils undoubtedly represent the extant genus.

12.5.2 Circaeasteraceae

The Circaeasteraceae are a monotypic family with the only species, *Circaeaster agrestis*, restricted to moist and shady forests in southern and southeastern Asia (Wu and Kubitzki, 1993c). *Circaeaster agrestis* is a small herbaceous plant with distinctive leaves that have open dichotomous venation and that are borne in a terminal rosette. The small, simple, bisexual flowers consist of one to three stamens and one to three free carpels. Each carpel matures into a small achene covered with hooked spines.

There is no known fossil record of Circaeasteraceae. Fossil fruits of *Appomattoxia* from the Early Cretaceous of eastern North America (Friis *et al.*, 1995) and Portugal (Friis *et al.*, 2006a) are similar in several features to fruits of *Circaeaster*, but pollen attributed to *Appomattoxia* is monoaperturate and similar to pollen of extant Piperales (Chapter 10).

12.5.3 Eupteleaceae

The Eupteleaceae comprise a single genus, *Euptelea*, with two species of trees in eastern Asia (Endress, 1993f). Flowers of *Euptelea* are bisexual and apetalous with 6–19 long stamens and 8–31 free carpels. The carpels mature into small stipitate samaras. The pollen has three or more short colpi and a finely perforate tectum.

There are no records of Eupteleaceae from the Cretaceous. The Cenozoic record is also sparse, but records of dispersed pollen, leaves and fruits assigned to the family are known from the Paleocene to Miocene of Europe (Krutzsch, 1966a; Mai, 1995), from the Eocene to Miocene of Asia (Zheng and Wang, 1994; Mai, 1995) and from the Eocene–Oligocene of North America (Mai, 1995).

12.5.4 Lardizabalaceae

The Lardizabalaceae (including Sargentodoxaceae) comprise eight genera and about 35 species of woody vines or shrubs. Five of the genera (*Descaisnea, Stauntonia, Akebia, Sinofranchetia, Sargentodoxa*) are restricted to eastern Asia, from the Himalayas to northeast Vietnam and Japan. Two other genera (*Lardizabala, Boquila*) are restricted to temperate regions of South America (Wu and Kubitzki, 1993a). Flowers are usually unisexual, hypogynous and trimerous with a perianth of several cycles. Staminate flowers usually have three to six stamens and pistillate flowers generally have three or sometimes many carpels in whorled or spiral arrangement (Wu and Kubitzki, 1993a). Pollen is tricolpate, usually with a foveolate tectum (Nowicke and Skvarla, 1982).

There are no secure records of the family from the Cretaceous. According to Tiffney (1993) wood described by Page (1970) from the Late Cretaceous of California may be correctly assigned to the family, but the identification needs to be further substantiated. Knobloch and Mai (1986) also mention a Cretaceous record of Lardizabalaceae from Africa, but we have been unable to confirm this occurrence. Fossil seeds assigned to the Asian genera *Decaisnea* and *Akebia* are known from the Miocene of Germany (Mai, 1980, 1995; Mai and Walther, 1991). Fossil seeds assigned to the South American genus *Sargentodoxa* are reported from the Miocene Brandon lignite flora of North America (Tiffney, 1993).

12.5.5 Menispermaceae

The Menispermaceae comprise about 450 species in 71 genera distributed in tropical regions of the world. They are mainly climbers, or sometimes shrubs or small trees, with alternate leaves and are typical of tropical lowlands. Flowers are small, unisexual, with free sepals and tepals in whorls of three (or six), typically borne in racemes, panicles or cymose heads. Pistillate flowers have up to 12 free carpels, each with two anatropous to campylotropous ovules of which only one develops. Fruits are apocarpous and fruitlets usually mature into one-seeded drupes. Endocarps are often distinctly ornamented and curved. Staminate flowers typically have three, six or 12 stamens that are often fused into synandria (Kessler, 1993a). Pollen varies in morphology among the genera and includes tricolpate, tricolporate, and triporate forms, in which the colpi sometimes merge at the pole in a syncolpate arrangement (Thanikaimoni, 1968). In some cases, pollen of Menispermaceae is distinctive and could potentially be recognised in the fossil record.

Fossils assigned to the Menispermaceae from the Late Cretaceous include two different types of endocarp assigned to the extinct genus *Protonomiscium (P. testudinarum* and *P. vangerowii*) from the Turonian to Maastrichtian of Central Europe (Knobloch and Mai, 1986). The endocarps are characterised by a distinct keel that indicates a relationship with extant members of the section Tinosporeae. Another putative record of Menispermaceae from the Cretaceous is *Anamirta pfeifferi* from the Deccan Intertrappean beds of the Nawargoan–Maragsur area, India (Bonde, 1997). This fossil is a fragment of permineralised wood with distinct alternating rings of xylem and phloem. Its characters indicate that the fossil was a liana and Bonde (1997) included it in the Menispermaceae.

Leaves assigned to the extinct genera *Menispermites, Menispermophyllum* and *Cocculophyllum* have been compared to extant Menispermaceae, but their systematic affinity has been questioned (Němejc and Kvaček, 1975) and they require further study. Leaves assigned to the fossil genus *Credneria* have also been referred to Menispermaceae (Rüffle, 1968), but an affinity to Platanaceae is well documented for some species (Tschan *et al.*, 2008) and is more likely for most species of the genus.

Fossil endocarps assigned to Menispermaceae are common and diverse in the Early Cenozoic floras of Europe and North America and have been assigned to a range of extant and extinct genera such as *Anamirta, Atriaecarpum, Bowerbankella, Calycocarpum, Curvitinospora, Davisicarpum, Diploclisia, Eohypserpa, Fritonia, Jatrorrhiza, Menispermicarpum, Odontocaryoidea, Palaeosinomenium, Parabaena, Thanikaimonia, Tinomiscium, Tinomiscoidea, Tinospora* and *Wardensheppeya* (Chandler, 1964; Collinson, 1983a; Manchester, 1994; Mai, 1995; Jacques and De Franceschi, 2005).

12.5.6 Papaveraceae

The Papaveraceae are a family of mostly herbaceous plants that are mainly distributed in temperate regions of the Northern Hemisphere, but with some members also in South Africa and South America. The family comprises about 760 species and 44 genera grouped in two subfamilies, the Papaveroideae and the Fumarioideae (Stevens, 2001 onwards). Flowers are often showy, actinomorphic or monosymmetric, with a basically dimerous organisation. The perianth comprises one to several whorls, the androecium often has numerous stamens and the gynoecium is syncarpous with 2 to many carpels. The ovary is superior with few to numerous ovules. Pollen is very variable ranging from simple tricolpate to polyporate.

The fossil record of Papaveraceae is extremely sparse and currently there are no well-documented occurrences from the Cretaceous. *Palaeoaster*, described from the Late Cretaceous of North America, is based on capsular fruits with numerous seeds. *Palaeoaster* was assigned to the Papaveraceae in a doctoral thesis by Smith (for references see Smith, 2001), but its systematic position is not yet convincingly demonstrated.

Princetonia allenbyensis from the Middle Eocene Princeton Chert of British Columbia, Canada, is a petrified reproductive structure that was originally compared to the Papaveraceae (Stockey, 1987). However, discovery of additional specimens showed that this fossil differs from Papaveraceae and more likely belongs to a currently undetermined group of core eudicots (Stockey and Pigg, 1991). The only other possible fossil record of Papaveraceae is provided by seeds attributed to *Corydalis* (formerly Fumariaceae) from the Late Cenozoic (Miocene) (cited in Collinson *et al.*, 1993).

12.5.7 Ranunculaceae

The Ranunculaceae are a family of herbs, semi-shrubs and lianas with about 2500 species and 59 genera distributed all over the world, particularly at higher latitudes and higher altitudes (Tamura, 1993). Flowers are usually of moderate size, bisexual and with a showy perianth of free sepals and petals. The androecium consists of several to numerous stamens, and the gynoecium usually has many free carpels with one to numerous anatropous ovules. Fruits are usually apocarpous and the individual fruitlets are often dry achenes or follicles. More rarely, fruits are baccate (Tamura, 1993). Pollen shows great variability in aperture configuration, which ranges from tricolpate to pantocolpate and pantoporate. There is also a great variation in pollen sculpture, which includes echinate, or sometimes striate, forms (Tamura, 1993).

There are no unequivocal records of Ranunculaceae from the Cretaceous although there are several leaf fossils that may belong to the family (Johnson, 2002). Seeds described as *Eocaltha zoophilia* from a coprolite from the Campanian of Mexico show some similarity to seeds of extant *Caltha* (Rodriguez-de la Rosa *et al.*, 1998), but the fossil seeds are much more regular, both in morphology of the seed wall and in the shape of the parenchyma cells. The polyporate pollen of *Cretacaeisporites scabratus* from the Cretaceous (probably Late Cenomanian) of Gabon shows strong similarity with pollen of some extant Ranunculaceae (e.g. *Anemone, Coptis, Hepatica*) not only in aperture configuration and tectum ornamentation, but also in features of pollen wall ultrastructure (Ward and Doyle, 1994). The Cenozoic record of Ranunculaceae is more abundant although not as extensive as might be expected considering the widespread occurrence of the family in wetland habitats. Most Cenozoic fossils are fruits and seeds. Fossil leaves of probable Ranunculaceae are very rare (for a review see Pigg and DeVore, 2005).

12.6 PROTEALES

The three families Nelumbonaceae, Platanaceae and Proteaceae, which are now placed together in the Proteales based on molecular data, have been treated as of widely separate relationships in previous classifications. For example, Takhtajan (1969) placed Nelumbonaceae in its own order Nelumbonales (subclass Ranunculidae), Platanaceae in Hamamelidales (subclass Hamamelididae), and Proteaceae in its own order in subclass Rosidae. In other classifications (e.g. Cronquist, 1981) Nelumbonaceae were placed together with water-lilies in Nymphaeales.

In habit and many morphological features Nelumbonaceae, Platanaceae and Proteaceae are very different from each other, and a close relationship among them would not be suspected based on morphology. *Nelumbo* is an aquatic herb with large solitary flowers, whereas *Platanus* is a genus of large trees with small, inconspicuous, simple flowers in dense inflorescences. Proteaceae include shrubs, small trees, or herbs with conspicuous animal-pollinated flowers arranged in dense inflorescences. However, consistent with the predictions from molecular data there is good evidence of the antiquity of all three families. All have a fossil record that extends well back into the Cretaceous.

12.6.1 Nelumbonaceae

The Nelumbonaceae comprise a single extant genus, *Nelumbo*, with two species: *Nelumbo lutea*, restricted to eastern North America, and *Nelumbo nucifera*, from southern Russia, Asia, India and Australia. *Nelumbo* is an aquatic herb with simple leaves and solitary flowers borne on terete petioles and peduncles from horizontal rhizomes. Leaves of the mature plant are large, almost circular, peltate and generally emergent or floating. Flowers are borne above the water surface. They are large, bisexual with many parts in a spiral arrangement. The perianth consists of about two to five sepals and 20–30 petals. The androecium comprises 200–300 tetrasporangiate stamens with long filaments and an elongated apical extension of the connective. Pollen is usually tricolpate, but pollen grains with other aberrant aperture numbers are occasionally produced in the same anther (Banks *et al.*, 2007). The gynoecium consists of 2–30 free carpels, each with a single anatropous ovule. The carpels are sunken in a highly modified receptacle that enlarges to form a distinctive obconical fruiting structure with a truncate upper surface. The mature fruits are embedded in the upper surface with only their tips protruding (Cronquist, 1981).

The fossil history of Nelumbonaceae is mainly based on fossil leaves, but stems and reproductive organs have also been reported. Fossil leaves are usually assigned to the extant genus *Nelumbo* or to the extinct genus *Nelumbites*. They are characterised by their almost circular peltate form, with an entire or crenate leaf margin. The petiole is attached in the basal half of the lamina and primary veins radiate from the point of attachment (Vakhrameev, 1952; Upchurch *et al.*, 1994). Leaves of *Nelumbites* are closely similar to those of extant *Nelumbo*, but differ in details of venation and petiole attachment. Based on these features the two genera have been maintained as distinct (Upchurch *et al.*, 1994).

The earliest records of *Nelumbites* are from the Early Cretaceous of North America where leaves of this type have been reported from two localities in Virginia, USA (Bank near Brooke, Early–Middle Albian, and Quantico, Late Albian, Hickey and Doyle, 1977; Upchurch *et al.*, 1994). Impressions of stems and roots, together with reproductive structures showing similarities to the tepals and fruiting receptacle of *Nelumbo*, occur associated with the *Nelumbites* leaves at the Quantico locality. Although these structures require more detailed study they are sufficiently distinctive to support an affinity of the fossil material to the Nelumbonaceae (Upchurch *et al.*, 1994). Leaves and fruits of Nelumbonaceae have also been recorded from the Late Cretaceous (Campanian–Maastrichtian) of Patagonia, Argentina. The leaves were assigned to the extant genus as *Nelumbo puertae* (Gandolfo and Cuneo, 2005).

There is no unequivocal record of Nelumbonaceae pollen from the Cretaceous. Dettmann (1973) suggested similarity between the dispersed pollen tetrads from the Cenomanian of Australia assigned to *Senectotetradites* and pollen of extant *Nelumbo*. This comparison was questioned by Muller (1981), who suggested that *Senectotetradites amiantopollis* from the Middle to Late Albian Fredericksburg Group of southern Oklahoma, USA (Srivastava, 1977) might represent an early record of the family (Muller, 1981). However, all these reports need more detailed study. Pollen of *Nelumbo* is generalised in its morphology and is hard to distinguish from pollen of many other eudicots. From the Cenozoic, Nelumbonaceae are widely distributed with leaf remains and occasionally remains of rhizomes and fruits recorded from North America, Europe and Asia (for references see Takhtajan, 1974).

12.6.2 Platanaceae

The Platanaceae are a small family containing a single genus with about seven species. All are wind-pollinated trees that have alternate leaves with palmate or rarely pinnate venation. The family has a relictual distribution in temperate to subtropical areas of the Northern Hemisphere with the greatest concentration of species in Mexico. Flowers are small, unisexual, simple, and crowded together into small globose heads that are borne on elongated inflorescence axes. The perianth of both staminate and pistillate flowers is inconspicuous. Staminate flowers typically have three or four tetrasporangiate stamens each with a very short filament. Dehiscence is by longitudinal slits. Pollen is tricolpate and reticulate, but as in some other eudicot families aberrant pollen with irregular aperture configurations have been reported (Denk and Tekleva, 2006). Pistillate flowers sometimes have three or four staminodes. There are three to eight (or more) free carpels. Each has a long linear style and contains a single, semi-orthotropous ovule (Kubitzki, 1993h; von Balthazar and Schönenberger, 2009). Each carpel matures into a single-seeded indehiscent fruit bearing a mass of prominent pappus-like hairs toward the base.

Fossils assigned to Platanaceae are conspicuous in many mid- to Late Cretaceous floras from the Northern Hemisphere, where they include numerous leaf fossils, as well as many reproductive organs. During the Late Cretaceous and Early Cenozoic the family was much more diverse, abundant and widespread than it is today. Platanoid leaves in Cretaceous floras comprise a variety of palmately lobed forms assigned to various genera (e.g. *Araliopsoides, Credneria, Erlingdorfia, Platanus, Tasymia*) (Figures 12.12–12.15). Among the most characteristic and well-known platanoid leaf fossils are forms assigned to *Credneria* (Figure 12.12). This genus was first described from the Santonian of the Quedlinburg area,

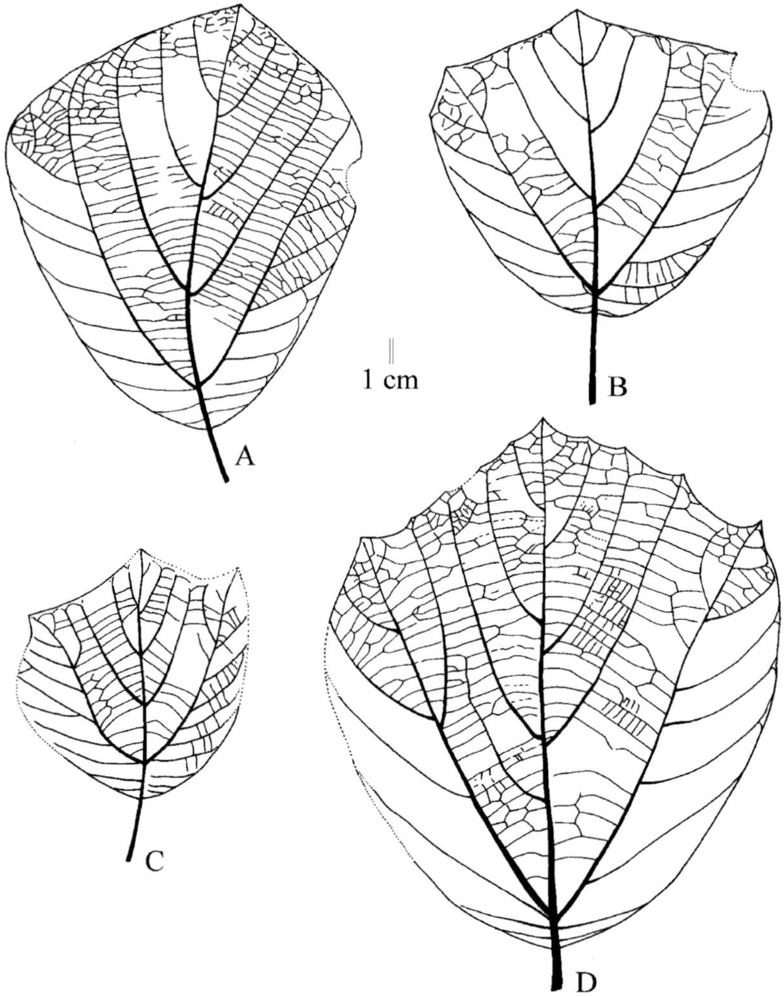

Figure 12.12 Platanoid leaves of *Credneria zenkeri*, an extinct platanoid from the Late Cretaceous (Santonian) Quedlinburg locality, Germany. Redrawn from Richter (1905).

Germany, where it occurs abundantly as impression fossils in well-sorted sandstones and also as compression fossils in clay horizons. Rüffle (1968) placed *Credneria* in the Menispermaceae. However, later studies (Tschan *et al.*, 2008) have shown a clear relationship of at least some species of *Credneria* to Platanaceae.

Leaves assigned to the extant genus *Platanus* are also common in the Late Cretaceous fossil record (Figure 12.13). Leaves of *Platanus* are simple, but often with stipules, which are sometimes very large. Other Cretaceous platanoid leaves are compound (Figure 12.14) as in the Late Cretaceous (Late Maastrichtian) *Erlingdorfia montana* and *Platanites marginata* from the Hell Creek Formation of Montana and North Dakota, USA (Johnson, 1996). In addition to palmately lobed leaves, the widespread pinnately compound and pinnatifid leaves of *Sapindopsis* (Figure 12.15) were apparently also produced by platanaceous plants (Pedersen *et al.*, 1994). Pinnate leaf types are also known in some Early Cenozoic forms, such as *Platanites hebridicus* from the fossil flora of Mull, Scotland (Crane *et al.*, 1988).

There are also numerous platanoid reproductive structures in Cretaceous and Cenozoic floras with the earliest records from the Early – Middle Albian. Flowers are unisexual and crowded in globose staminate and pistillate inflorescences. These heads are found isolated or in strings along elongated axes (Figure 12.16). The heads are typically sessile or sometimes borne on stalks. Both the individual heads and the compound inflorescences are much smaller than those of most extant and Cenozoic Platanaceae (Figure 12.16, Chapter 20).

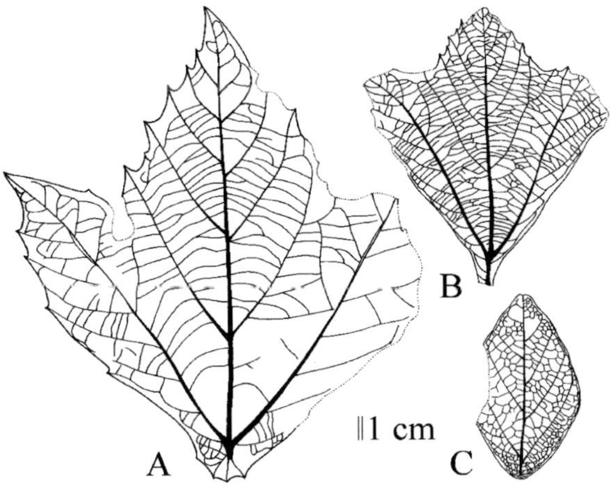

Figure 12.13 Leaves of *Platanus bohemica* from the Late Cretaceous (Cenomanian) Vyšehořovice locality, Czech Republic. Redrawn from Knobloch (1997).

A few of the Cretaceous platanoid reproductive structures have been assigned to the extant genus, for example *Platanus richteri* (Knobloch and Mai, 1986) and *Platanus laevis* (Velenovský, 1889), but most have been placed in a variety of extinct genera (*Archaranthus, Aquia, Chemurnautia, Friisicarpus, Hamatia, Macginicarpa, Oreocarpa, Platananthus, Platanocarpus, Quadriplatanus, Tanyoplatanus, Tricolpopollianthus*) that differ from extant *Platanus* in various respects.

Aquia brookensis (Figure 12.17) includes staminate inflorescences, flowers and dispersed stamens described from the Early–Middle Albian 'Bank near Brooke' locality, Virginia, USA (Crane *et al.*, 1993). Flowers and inflorescences are poorly preserved and details of their organisation not fully understood, but fragments indicate that the flowers were densely arranged in globular inflorescences similar to those of *Platananthus* and extant *Platanus*. Stamens differ from those of other extant or extinct Platanaceae in having much longer filaments and shorter anthers. Anthers are basifixed, tetrasporangiate and dithecate with the two theca separated by a well-developed connective with a distinct peltate apical expansion. Dehiscence is valvate and there are two laterally hinged valves over each theca. Pollen *in situ* in the stamens is small, tricolpate, and tectate–foveolate to reticulate.

Specimens of *Aquia brookensis* are associated with pistillate inflorescences described as *Friisicarpus brookensis* (Crane *et al.*, 1993; see below and Maslova and Herman, 2006, for nomenclature). These comprise elongated infructescence axes bearing numerous sessile globular heads (Figure 12.17), as in some species of the extinct *Platananthus* and extant *Platanus kerrii*. Details of the heads are not well preserved, but individual pistillate flowers have a well-developed perianth and five carpels.

Aquia brookensis and *Friisicarpus brookensis* are the oldest platanoid reproductive structures known so far and their unique stamen features may be plesiomorphic for the group as a whole. The reproductive structures co-occur with abundant pinnatifid leaves of *Sapindopsis* that are thought to have been produced by the same platanoid plant (Crane *et al.*, 1993).

Hamatia elkneckensis (Figure 12.17) is based on staminate inflorescences and flowers from the Early Cenomanian Bull Mountain locality, Maryland, USA (Pedersen *et al.*, 1994). Flowers are densely crowded in small globose heads that are arranged along an elongate axis. Individual flowers are small with a small number of stamens (possibly five) surrounded by prominent tepals. Stamens are basifixed, tetrasporangiate and dithecate with a very short filament and elongated anther. Anthers have valvate dehiscence with two laterally hinged valves over each theca. The most prominent feature of the stamens is the large, and very distinctive, hook-shaped apical extension of the connective that extends down the ventral surface of the anther. Pollen *in situ* in the stamens is small, tricolporate and tectate–reticulate.

Staminate flowers and inflorescences of *Hamatia* closely resemble those of other extinct and extant Platanaceae, but differ in their tricolporate pollen and in the strongly hooked and prominent extension of the connective. Since tricolporate pollen is not known for extant Platanaceae the inclusion of *Hamatia* extends the variation in pollen morphology known in the family (Pedersen *et al.*, 1994). Associated with the staminate flowers are pistillate flowers of *Friisicarpus elkneckensis* with characteristics broadly similar to those of *Friisicarpus brookensis* (see above).

Platananthus (Figure 12.18) is an extinct genus established for fossil staminate flowers and inflorescences closely comparable to those of extant *Platanus* (Manchester, 1986). Flowers are small and densely packed in sessile or stalked globular heads that are borne on elongated axes. The perianth is prominent but undifferentiated, and the androecium is pentamerous. Stamens are tetrasporangiate, dithecate and differentiated into a short filament and an elongated anther with a peltate to conical apical extension. Where dehiscence can be observed it is valvate with two laterally hinged valves over each theca. Pollen *in situ* in the stamens is small, tricolpate and tectate–reticulate.

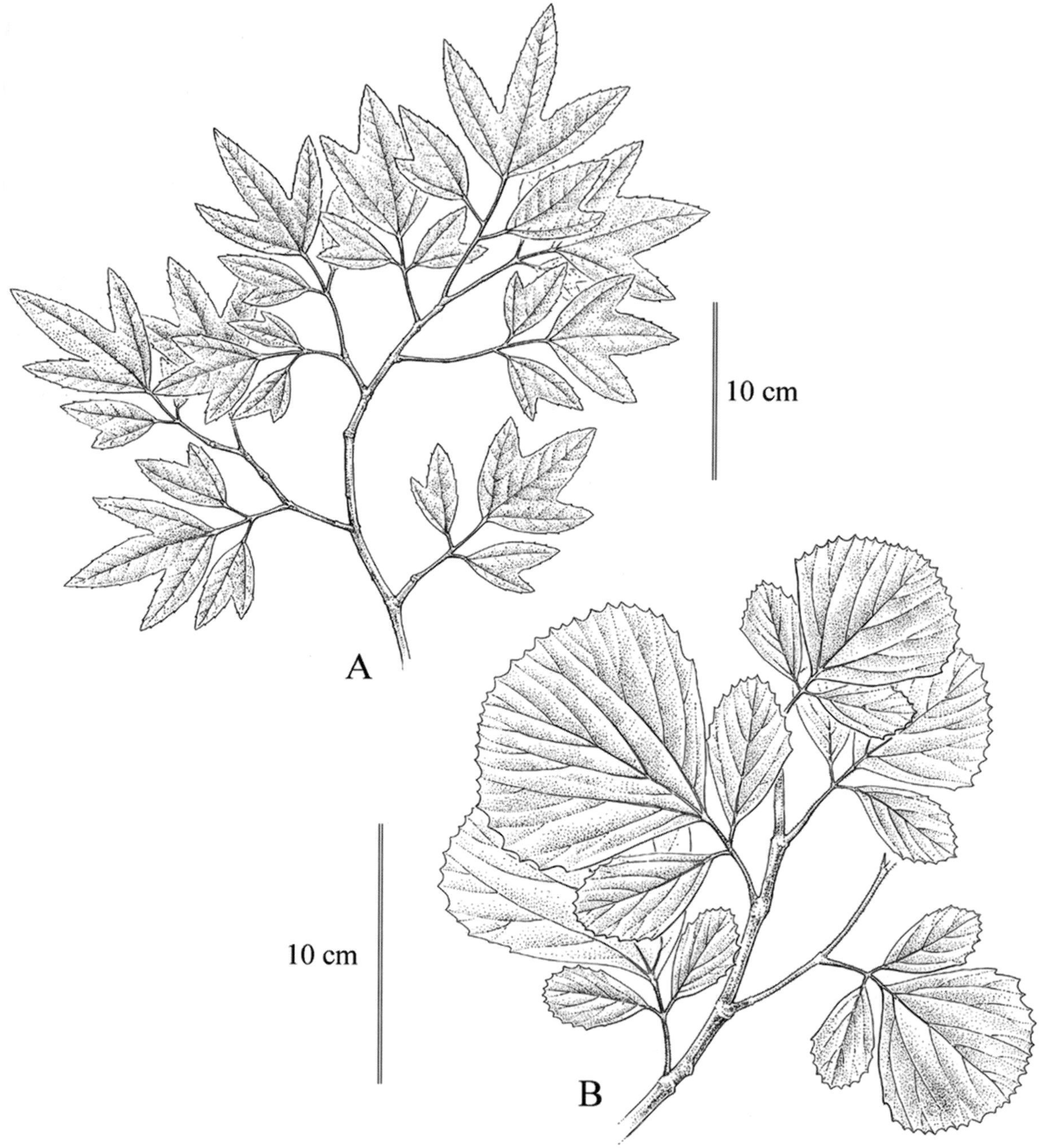

Figure 12.14 Reconstruction of two platanoid fossils with pinnately compound foliage from the Late Cretaceous (Late Maastrichtian) Hell Creek Formation of Montana and North Dakota, USA. (A) *Erlingdorfia montana*. (B) *Platanites marginata*. Redrawn from Johnson (1996).

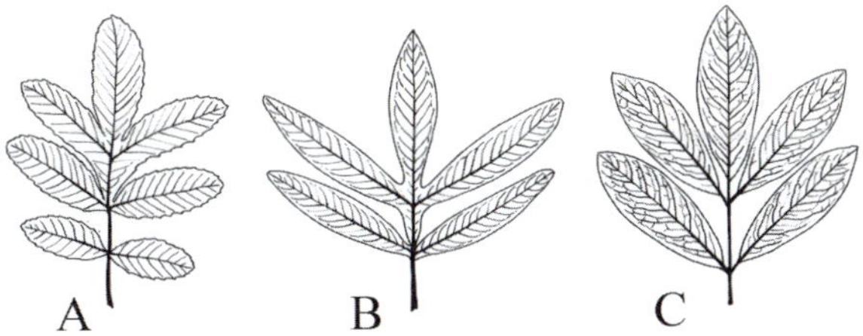

Figure 12.15 Morphological diversity in fossil leaves of *Sapindopsis* from the Cretaceous, apparently produced by platanaceous plants. (A) *Sapindopsis belviderensis.* (B) *Sapindopsis magnifolia.* (C) *Sapindopsis* sp. Redrawn from Crane (1989).

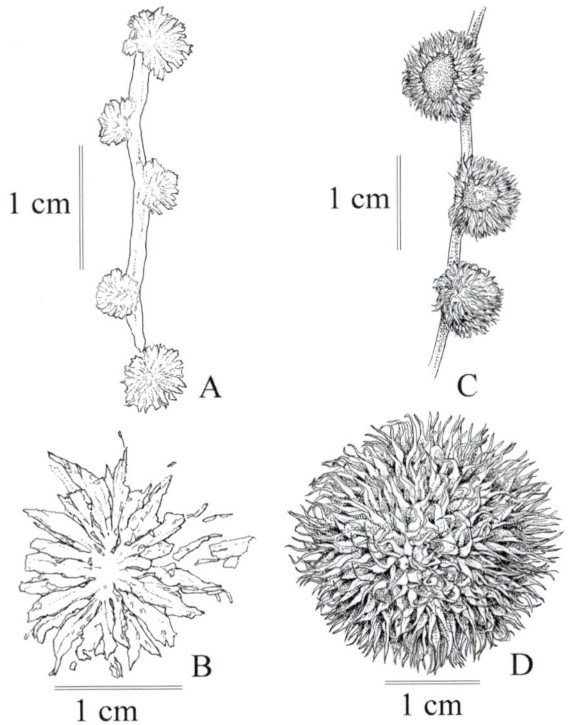

Figure 12.16 Fossil (A, B) and extant (C, D) pistillate inflorescences of Platanaceae. (A) *'Sparganium' aspensis* from the mid-Cretaceous of Wyoming, USA. (B) *Macginicarpa*-like fossil from the Wind River Formation (Middle Eocene), Wyoming, USA. (C) *Platanus kerrii.* (D) *Platanus orientalis.* Drawn from photographs in Crane (1989).

The type species of *Platananthus, Platananthus synandrus*, was described from the Cenozoic of North America, but several species have been described subsequently from mid- and Late Cretaceous floras in North America and Europe. These include *Platananthus potomacensis* from the Late Albian West Brothers locality, Maryland, USA, *P. hueberi* from the Late Santonian – Early Campanian Neuse River locality, Virginia, USA, and *P. scanicus* from the Late Santonian – Early Campanian Åsen locality, Sweden (Friis *et al.*, 1988).

Fossil flowers of *Platananthus* are mainly distinguished from those of extant *Platanus* in having a prominent perianth and a pentamerous androecium, rather than an inconspicuous perianth and a tetramerous androecium. The fossils also differ from extant *Platanus* in their smaller pollen. *Platananthus* is almost certainly an unnatural genus that includes several different types of platanoid plants of diverse relationships with respect to extant and other fossil genera. The Cenozoic species *Platananthus synandrus* is associated with the platanoid flowers and fruiting organs *Macginicarpa*; the Cretaceous species of *Platananthus* are typically associated with pistillate flowers and inflorescences assigned to *Friisicarpus.*

Friisicarpus (Figure 12.17) was instituted by Maslova and Herman (2006) to accommodate fossil pistillate flowers and fruits that were previously assigned to *Platanocarpus* Friis, Crane et Pedersen (1988), a junior homonym of *Platanocarpus* Jarmolenko (1935). *Platanocarpus* (type and only species *P. ovatus*) from the Cretaceous deposits of the Kysyl-zhar locality, northwestern Karatau, Kazakhstan, comprises isolated achenes closely similar to those of extant *Platanus* with a long style and with hairs surrounding the base of the achene (Jarmolenko, 1935). In contrast, flowers of *Friisicarpus* are small and consist of an undifferentiated perianth of prominent tepals, and a gynoecium of several (typically five) free carpels. Carpels have a ventral slit and a poorly developed style. Each carpel is unilocular and contains a single, orthotropous, pendent ovule.

Several species of *Friisicarpus* have been described from the Cretaceous of eastern North America. *Friisicarpus brookensis* from the Early–Middle Albian 'Bank near Brooke' locality, Virginia, co-occurs with staminate fossils of *Aquia brookensis* (Crane *et al.*, 1993), *Friisicarpus elkneckensis* from the Early Cenomanian Bull Mountain locality, Maryland, co-occurs with staminate fossils of *Hamatia elkneckensis* (Pedersen *et al.*, 1994), *Friisicarpus marylandensis* from the Late Albian West Brothers locality, Maryland, co-occurs with *Platananthus potomacensis* (Friis *et al.*, 1988), and *Friisicarpus carolinensis* from the Late Santonian – Early Campanian Neuse River locality, North Carolina, co-occurs with staminate material of *Platananthus hueberi* (Friis *et al.*, 1988). A further undescribed species is known from the Late Santonian – Early Campanian Åsen locality,

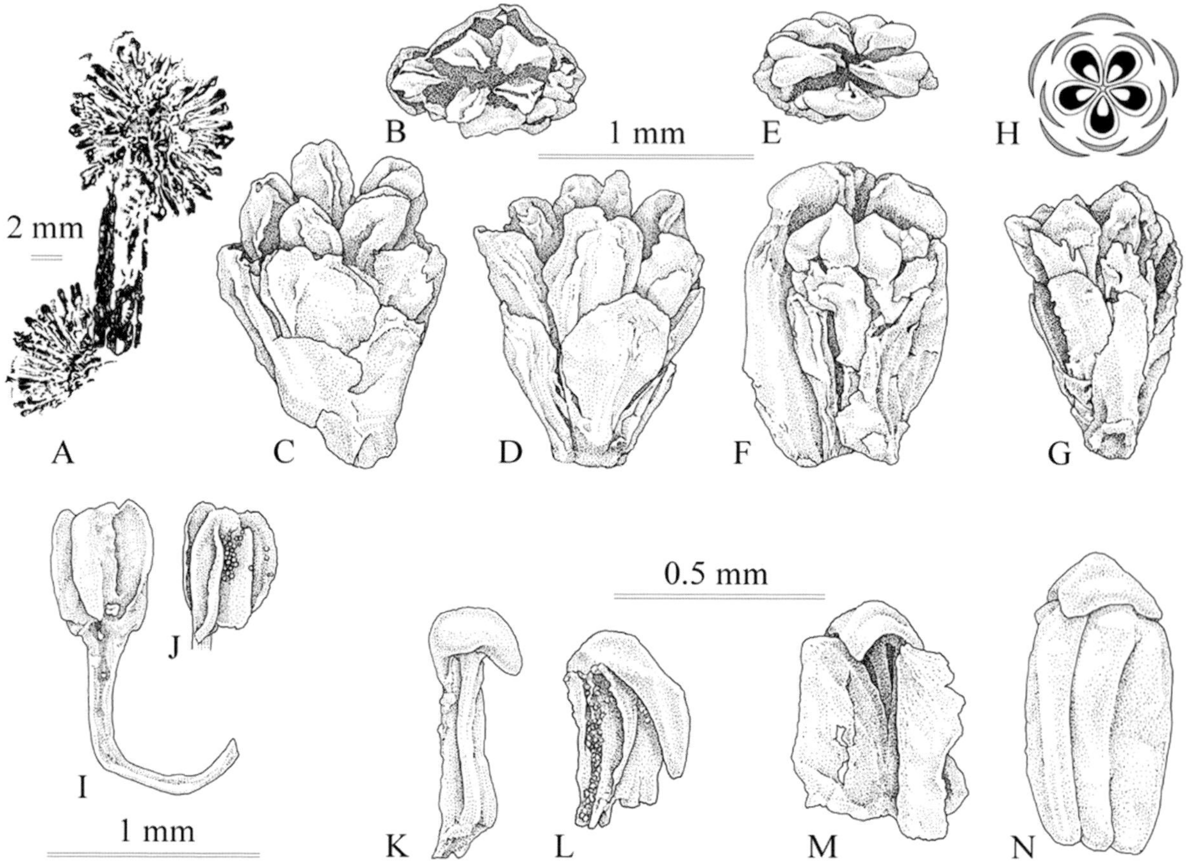

Figure 12.17 Early Cretaceous (Albian) pistillate and staminate platanoid reproductive organs. (A–D) *Friisicarpus brookensis* from the 'Bank near Brooke' locality, Virginia, USA. (A) Inflorescence axis with two sessile, pistillate heads. (B–D) Detached pistillate flower seen in (B) apical and (C, D) lateral view showing well-developed tepals and five free carpels. (E, F) *Friisicarpus elkneckensis* from the Bull Mountain locality, Maryland, USA, detached flower in (E) apical and (F) lateral view showing well-developed tepals, and carpels with prominent apical expansions. (G) *Friisicarpus marylandensis* from the West Brothers locality, Maryland, USA, showing well-developed tepals surrounding the carpels. (H) Floral diagram of *Friisicarpus*; precise arrangement of tepals not fully secure. (I, J) *Aquia brookensis* from the 'Bank near Brooke' locality, Virginia, USA, detached stamens showing (I) long filament and (J) valvate anther dehiscence. (K–N) *Hamatia elkneckensis* from the Bull Mountain locality, Maryland, USA, detached stamens showing the prominent hook-shaped apical extension of the connective and valvate anther dehiscence. (A, I–J) Drawn from photographs and SEM images in Crane *et al.* (1993); (B–D) from Friis *et al.* (2006a); (E–F, K–N) drawn from SEM images in Pedersen *et al.* (1994); (G) drawn from SEM images in Friis *et al.* (1988).

Sweden, and there is another undescribed form from the Cenomanian of Western Siberia, Russia (Maslova and Herman, 2006).

Fossil flowers of *Friisicarpus* are distinguished from pistillate flowers of extant *Platanus*, and also from extinct *Macginicarpa* from the Cenozoic (Manchester, 1986), mainly in the poorly developed style and the lack of trichomes on the ovary and fruit wall. The number of carpels per flower in *Friisicarpus* and in *Macginicarpa* is typically five, whereas in *Platanus* the number is irregular, up to about eight or nine or more. *Friisicarpus* is probably an unnatural taxon that includes several different types of platanoid plant of diverse relationships within the broader platanoid lineage.

Quadriplatanus georgianus (Figure 12.19) includes staminate and pistillate inflorescences described from the

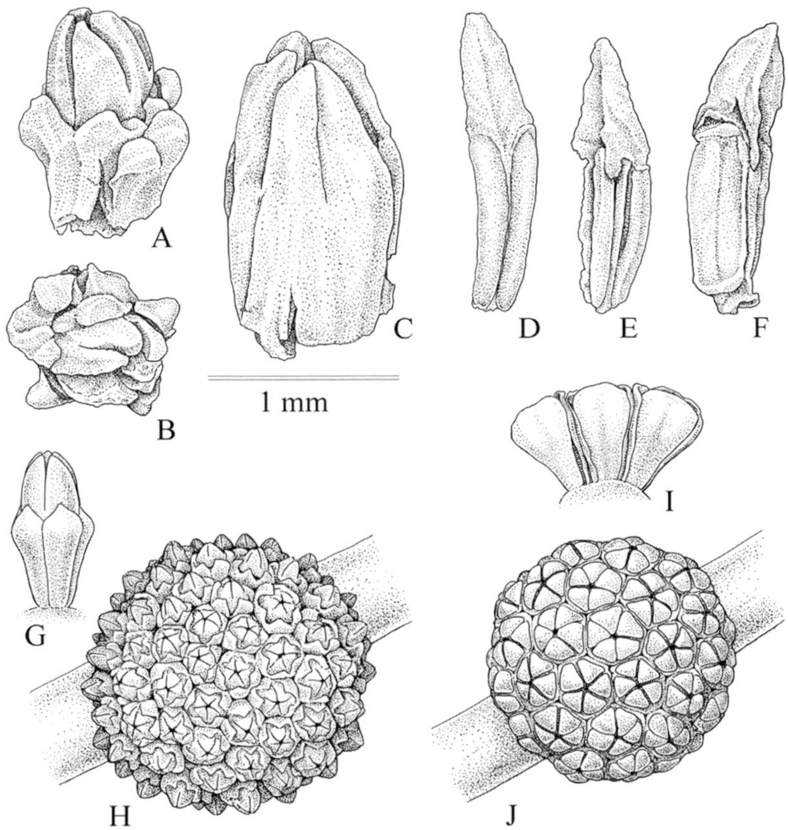

Figure 12.18 Late Cretaceous (Late Santonian–Early Campanian) staminate reproductive organs of (A–H) *Platananthus scanicus* from the Åsen locality, Sweden and (I–J) *Platananthus hueberi* from the Neuse River locality, North Carolina, USA. (A, B) Flowers in (A) lateral and (B) apical view showing apical extensions of the stamen connectives protruding above the perianth parts. (C) Five adhering stamens of a flower in which the perianth has been lost. (D–F) Dispersed stamens showing short filament and elongated apical extension of the connective. (G, H) *Platananthus scanicus*, reconstructions of (G) staminate flower and (H) inflorescence. (I, J) *Platananthus hueberi*, reconstructions of (I) staminate flowers and (J) inflorescence. Drawn from material in the Swedish Museum of Natural History, Stockholm.

Coniacian Upatoi Creek locality, Georgia, USA (Magallón-Puebla *et al.*, 1997). Inflorescences are small, unisexual globular heads, up to about 2.5 mm in diameter, that are helically arranged along an elongate axis. Staminate flowers are actinomorphic and tetramerous with a single whorl of four stamens opposite a single whorl of four tepals. Tepals are free for most of their length, but are fused at the base. Stamens are sessile or with a short filament. Anthers are basifixed, tetrasporangiate and dithecate with a prominent apical prolongation that sometimes forms a hook-shaped extension that extends down the ventral surface of the stamen. Dehiscence is valvate with two laterally hinged valves over each theca. Pollen *in situ* in the stamens is small, tricolpate and tectate–reticulate.

Pistillate flowers of *Quadriplatanus* are also actinomorphic and tetramerous. The perianth consists of two whorls of poorly differentiated tepals. Tepals of the outer whorl are fused to form a narrow floral tube. The gynoecium is apocarpous, consisting of eight carpels in two whorls. Carpels are wedge-shaped, with the ventral suture extending from the flattened apex of the carpels almost to the base. Each carpel contains a single pendent ovule.

Although the pistillate and staminate flowers of *Quadriplatanus* are not known in organic connection their co-occurrence and clear structural similarities suggest that they were probably produced by the same species (Magallón-Puebla *et al.*, 1997). Flowers of *Quadriplatanus* are similar to those of both extant and extinct Platanaceae in most features, but are distinguished by the regular tetramerous/octamerous arrangement of the floral parts in both staminate and pistillate flowers (Magallón-Puebla *et al.*, 1997). In other extinct Platanaceae, such as *Aquia, Hamatia, Platananthus* and *Friisicarpus*, the flowers are typically pentamerous. In extant *Platanus*, staminate flowers are usually tetramerous, while the pistillate flowers have an irregular number of parts.

Platanus quedlinburgensis is a platanoid floral structure described from the Late Cretaceous of Europe based on staminate heads with pollen from the Santonian Quedlinburg locality, Germany (Pacltová, 1982). The heads were

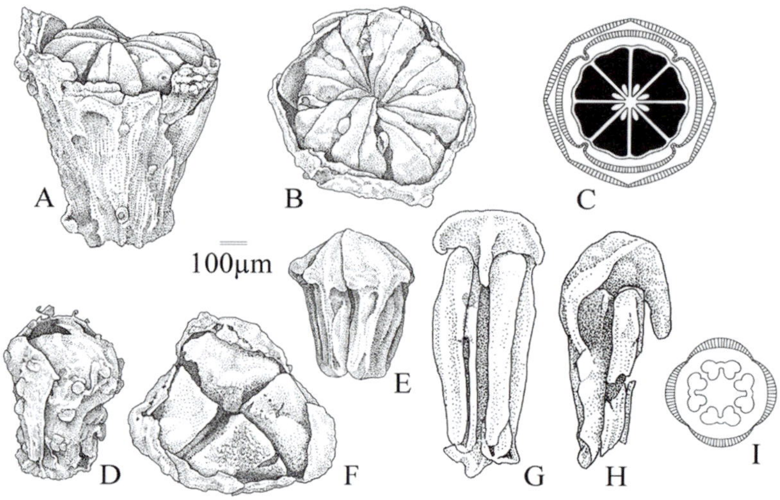

Figure 12.19 Late Cretaceous (Coniacian) pistillate (A–C) and staminate (D–I) flowers of *Quadriplatanus georgianus* from the Upatoi Creek locality, Georgia, USA. (A) Lateral view of pistillate flower. (B) Apical view of pistillate flower showing eight carpels. (C) Floral diagram of pistillate flower. (D) Staminate flower in lateral view showing perianth. (E) Staminate flower with perianth removed. (F) Apical view of staminate flower showing four stamens each with prominent apical extensions of the connective. (G, H) Detached stamens, showing the hook-shaped apical extension of the connective. (I) Floral diagram of staminate flower. Drawn from SEM images and line drawings in Magallón-Puebla *et al.* (1997).

later described as *Platanus richteri* by Knobloch and Mai (1986), who included both staminate and pistillate heads in their new species. Based on reinvestigation of the material Tschan *et al.* (2008) retained the fossils in the extant genus, but details of the floral organisation are not clear and it is difficult to determine with certainty whether they should be assigned to one of the extinct genera. *Platanus laevis* are impression fossils of platanoid pistillate structures (Velenovský, 1882). Details of floral organisation are lacking and it is uncertain whether these fossils belong to the extant genus or to one of the extinct forms.

Archaranthus krassilovii includes an inflorescence axis with small staminate heads and many dispersed stamens from the Maastrichtian–Paleocene of the Amur Region, Russia (Maslova and Kodrul, 2003). Individual heads are about 5 mm in diameter, borne on long stalks, and each consists of about 15 densely packed flowers. Flowers are described as tetramerous and have a well-developed perianth and a single whorl of stamens. Stamens consist of a short stalk and an elongate, massive anther characterised by a distinct triangular apical extension of the connective. Pollen *in situ* is tricolpate, sometimes with an irregular arrangement of the apertures. The colpus margins are clearly delimited. The tectum is finely reticulate and the muri are acutely triangular in profile. In floral organisation, and in anther and pollen morphology, the fossil is very similar to extant Platanaceae. *Archaranthus* is also similar to other staminate heads with a tetramerous floral organisation from the Cretaceous thought to be related to Platanaceae (see, for example, *Quadriplatanus* above). The fossil heads are associated with distinctive platanoid leaves (Maslova and Kodrul, 2003).

There are several other reproductive organs that may be related to the Platanaceae, but that have been assigned to Hamamelidaceae (Chapter 13). For instance, the fruiting head of *Kasicarpa melikianii* from the Turonian Kas flora of western Siberia is associated with the platanoid leaves of *Tasymia pseudoplatanoides* (Maslova *et al.*, 2005; Golovneva, 2008).

The extensive and detailed record of fossil Platanaceae that has emerged from studies of informative fossil material from the Cretaceous and Early Cenozoic over the past few decades highlights several significant differences between most fossil taxa and extant *Platanus* (Chapter 20). Typically, flowers of fossil Platanaceae have a regular number of parts and the perianth parts are more prominent. Stamens that have sufficient details preserved show valvate dehiscence, the pollen is smaller, the styles are shorter, and the stigmatic surface is much less prominent. These structural differences indicate a different reproductive biology in ancient platanoids, probably involving insect pollination rather than wind pollination as in the extant genus (Friis *et al.*, 1988) (Chapter 20). This, is turn, may have had an impact on pollen morphology. Some extinct taxa produced tricolporate pollen in contrast to the uniformly tricolpate grains of extant *Platanus*. There is also much greater variation in the details of the reticulum in fossil taxa than is seen in the extant genus.

In addition to differences in pollination, extant platanoids and many extinct forms also appear to have differed in dispersal biology. Cretaceous platanoids seem to have had fewer flowers per head and fewer carpels per flower. Cretaceous and some Early Cenozoic platanoid fruits also lack the distinctive pappus-like hairs seen in the fruits of extant species (Chapter 20).

12.6.3 Proteaceae

The Proteaceae comprise more than 1700 species in about 80 genera. They are shrubs, or more rarely trees or herbs, with alternate or rarely opposite leaves. The family is widely distributed in tropical and subtropical regions of the Southern Hemisphere. Flowers are usually bisexual, or sometimes unisexual, and are arranged in dense inflorescences. They are typically tetramerous and often with petaloid tepals. The androecium consists of one whorl of stamens with tetrasporangiate, or rarely bisporangiate, anthers with longitudinal dehiscence. Stamens often have an elongated apical expansion of the connective. Pollen is typically triporate, but biporate forms and forms with more than three pores occur in some taxa (e.g. Dettmann, 1998). A few Proteaceae also have tricolporate pollen. The gynoecium consists of a single carpel and the ovary is superior with one or two anatropous to hemiorthropous ovules (McCarthy, 1995; Douglas and Tucker, 1996a, 1996b).

The Proteaceae have an extensive and diverse record of dispersed fossil pollen (Figure 12.20), but the family is also represented in the fossil record by wood, leaves, dispersed cuticles, and reproductive organs. The earliest unequivocal records are from the Late Cretaceous (Santonian) of Australia (Dettmann and Jarzen, 1991). An early putative record is provided by dispersed pollen assigned to *Trioris africaensis* from the Cretaceous (probably Late Cenomanian) of Gabon (Ward and Doyle, 1994). *Trioris africaensis* is similar to pollen of several extant Proteaceae, but could not be placed reliably in any extant subfamily.

Proteaceae-type pollen is particularly abundant in Campanian–Maastrichtian palynofloras of the Otway Basin, southeastern Australia (Dettmann and Jarzen, 1996) and 31 different proteaceous pollen species assigned to genera such as *Beaupreaidites, Cranwellipollis, Lewalanipollis, Propylipollis, Proteacidites* and *Triporopollenites* have been recognised. Among the fossil pollen grains are forms comparable to pollen of extant genera from four different subfamilies of extant Proteaceae, indicating that much of the fundamental lineage diversity of the family was already established by the middle of the Late Cretaceous. Late Cretaceous (Campanian–Maastrichtian) records from New Zealand and the Antarctica Peninsula provide additional evidence of a considerable Late Cretaceous diversification and radiation of Proteaceae (Dettmann and Jarzen, 1996). The Cenozoic fossil record from Australia indicates that this diversification continued through the Paleocene–Eocene (Hill, 1994).

Macrofossil evidence of Proteaceae is less common, but provides strong evidence of the diversity of the family in the Cenozoic in Australia (Hill, 1994). Leaves are assigned to the extinct genera *Banksieaephyllum* and *Banksieaeformis*, and to the extant genera *Banksia* and *Xylomelum*. Of particular interest are infructescences from the Early Eocene of northwestern Australia described as *Banksia archaecarpa* and *Banksia* sp. (McNamara and Scott, 1983), and an inflorescence, *Musgraveinanthus alcoensis* (Figure 12.21), from the Eocene Anglesea deposits that resembles *Musgravea* and *Austromuellera* (Christophel, 1984). These fossils suggest that pollination by birds, or perhaps small mammals, was already underway by the Early Cenozoic. *Banksia archaecarpa* is the earliest well-documented report of the genus and comprises large, elongated infructescences, up to about 10 cm long, with densely spaced flowers, a few of which matured into distinctive follicles. There are also well-documented records of the family in the Cenozoic of New Zealand (e.g. Carpenter *et al.*, 2010).

There are several records of putative Proteaceae fossils from the Northern Hemisphere, but their relationships to the family have not been documented in detail. Leaf fossils described as *Proteophyllum, Grevillea* and *Grevilleophyllum* from the mid-Cretaceous of Bohemia, the Czech Republic, by Velenovský (1883; 1889) are probably all of lauraceous affinity (Kvaček, 1992; Eklund and Kvaček, 1998).

12.7 SABIACEAE

The Sabiaceae are a small family of trees, shrubs, or woody vines comprising about 60 species in three genera (*Meliosma, Ophiocaryon, Sabia*) distributed in southeastern Asia and tropical America. Flowers are bisexual or more rarely unisexual, usually with a pentamerous perianth and androecium. The gynoecium is bi- or tricarpellate, syncarpous and with one or two pendent and hemiorthotropous, unitegmic ovules in each locule (Cronquist, 1981). Pollen is tricolporate.

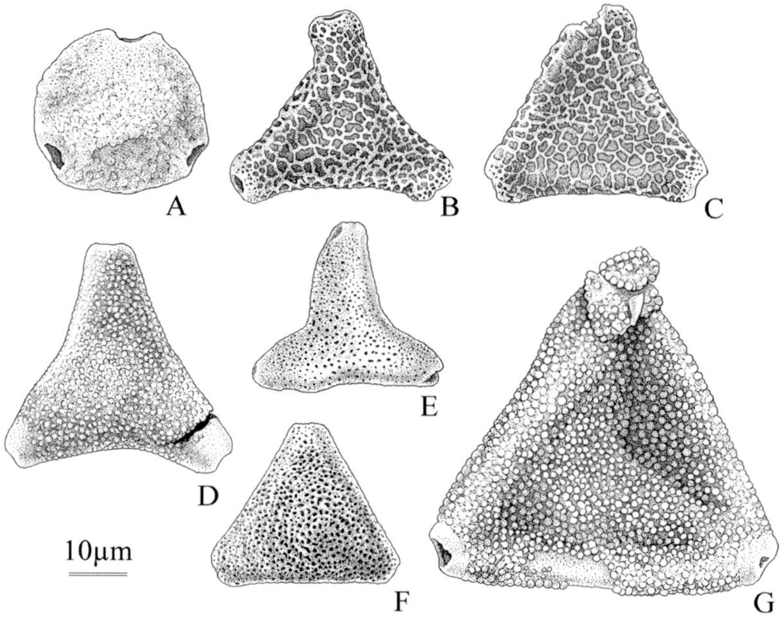

Figure 12.20 Selected dispersed proteaceous pollen from the latest Cretaceous of southeastern Australia showing diversity in shape and pollen wall ornamentation. (A) *Lewalanipollis trycheros*. (B, C) *Proteacidites cooksoniae*. (D) *Proteacidites* sp. cf. *P. fromensis*. (E) *Proteacidites adenanthoides*. (F) *Bleasdalea bleasdalei*. (G) *Propylipollis crotonoides*. Drawn from SEM images in Dettmann and Jarzen (1996).

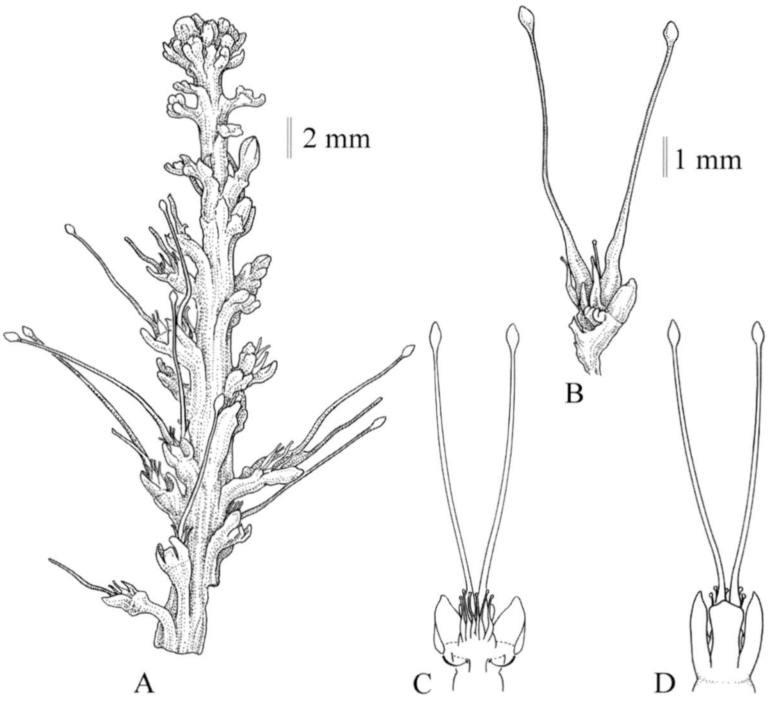

Figure 12.21 Inflorescence of *Musgraveinanthus alcoensis* (Proteaceae) from the Eocene Alcoa Anglesea locality of Victoria, Australia, showing (A) many pairs of flowers along the inflorescence axis, (B) isolated pair of flowers and schematic view of (C) adaxial and (D) abaxial side of flower pair. Redrawn from Christophel (1984).

The family has a sparse fossil record. Currently no fossil floral structures or pollen of the family have been recognised. However, endocarps have been recovered from fossil floras in Europe and North America. The earliest occurrences are from the Late Cretaceous of Central Europe. Fossil endocarps assigned to *Meliosma* (*M. praealba*) have been described from the Maastrichtian of Germany, and fossil endocarps assigned to *Sabia* (*S. menispermoides*, *S. microsperma*, *S. praeovalis*) have been reported from the Turonian–Santonian of the Czech Republic and the Maastrichtian of Germany (Knobloch and Mai, 1986). In both cases the fossil endocarps resemble those of extant Sabiaceae in their distinctive reniform shape and in having a deep funicular invagination of the fruit wall.

Other fossil endocarps from the Late Cretaceous of Central Europe with a possible relationship to Sabiaceae were assigned to the extinct genus *Insitiocarpus* (Knobloch and Mai, 1986). *Insitiocarpus* comprises uni- or bilocular endocarps with a deep funicular cavity that sometimes has a funicular plug. Seeds are semi-anatropous with a thin wall (Knobloch and Mai, 1986). The genus is recorded from the Cenomanian and Maastrichtian, but there are no records from other stages of the Cretaceous. Fossil endocarps and leaves of *Meliosma* and *Sabia* have been reported extensively from Cenozoic floras in Europe and North America (Knobloch and Mai, 1986; Crane *et al.*, 1990; Collinson *et al.*, 1993).

Sabiocaulis sakuraii comprises petrified stems from the Late Cretaceous of Hokkaido, Japan, with wood characters indicating a close relationship with extant *Sabia* (Stopes and Fujii, 1910). Specimens of this species are among the most common angiosperm fossils in the Hokkaido wood assemblage described by Stopes and Fujii (1910). This material is of Late Turonian – Santonian age (Nishida, 1991).

12.8 BUXALES

The Buxales include one family, Buxaceae, that has been expanded to include Didymelaceae (APGIII, 2009). The systematic placement of Buxales has varied considerably in previous classifications (Jarvis, 1989) but in recent phylogenetic analyses it is placed among the early-diverging groups of eudicots, often close to the Trochodendraceae (Chase *et al.*, 1993; von Balthazar *et al.*, 2000; APGIII, 2009).

Several floral structures from the Early and mid-Cretaceous have features that indicate a close relationship with extant Buxaceae. From the Early Cretaceous (Late Aptian – Early Albian) mesofossil floras of Portugal, five different taxa related to Buxaceae have been described (Pedersen *et al.*, 2007). They include staminate flowers and bicarpellate fruits with tricolpate and striate pollen assigned to the fossil species *Lusistemon striatus* and *Lusicarpus planatus*, as well as another bicarpellate fruit, *Silucarpus camptostylus*, and two tricarpellate fruits, *Valecarpus petiolatus* and *Aguacarpus hirsutus* (Pedersen *et al.*, 2007). From slightly younger (Late Albian – Early Cenomanian) floras from the Potomac Group two different species of fossil flower of the extinct genus *Spanomera* also show significant similarities to Buxaceae. However, none of these early flowers can be assigned to the extant family with certainty, mainly because their tricolpate pollen differs from the pollen of their living counterparts. They are treated here as Buxales to emphasise that they are probably part of this lineage, perhaps on the stem group to the modern family.

Spanomera (Figures 12.22, 12.23) includes two species from the mid-Cretaceous of Maryland, USA: *Spanomera mauldinensis* from the Early Cenomanian Mauldin Mountain locality and *S. marylandensis* from the Late Albian West Brothers locality (Drinnan *et al.*, 1991). *Spanomera mauldinensis* is the more completely known of the two species.

The small unisexual flowers of *Spanomera* are borne in inflorescence units that consist of a terminal pistillate flower and two lateral staminate flowers. Inflorescence units and floral parts are borne in an opposite–decussate arrangement. Staminate flowers are short-stalked, with four or five stamens and an equivalent number of tepals. Each stamen is opposite a tepal. Stamens are differentiated into a short filament and an elongated anther. Anthers are dorsifixed, tetrasporangiate and dithecate with a short conical extension of the connective. Dehiscence is by simple longitudinal slits. Pollen *in situ* in the stamens is tricolpate and coarsely reticulate, with muri that form a distinctive striate–rugulate pattern. A prominent, slightly bilobed structure in the centre of the flower is probably a pistillode. The *in situ* pollen of *Spanomera mauldinensis* is similar to dispersed pollen assigned to *Striatopollis paraneus*, which is common in mid-Cretaceous strata of the Potomac Group. The *in situ* pollen of *Spanomera marylandensis* is closely comparable to dispersed grains of *Striatopollis vermimurus* that have been described from the Late Albian of the Potomac Group.

Pistillate flowers consist of four tepals and a bicarpellate gynoecium. The two carpels are follicular, almost free from each other, and fused only along the basal part of the

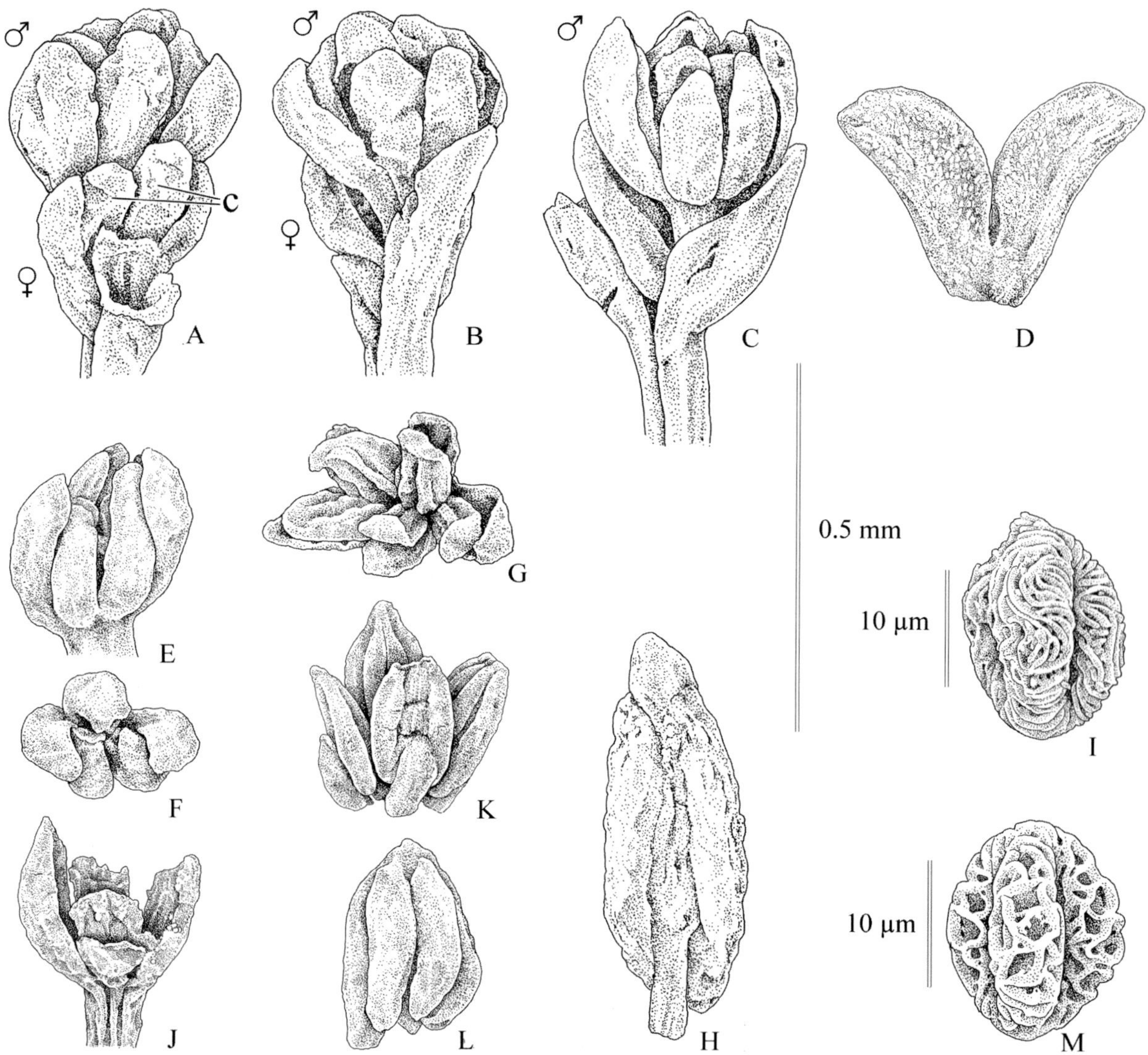

Figure 12.22 Buxalean reproductive organs from Maryland, USA. (A–I) *Spanomera mauldinensis* from the Late Cretaceous (Early Cenomanian) Mauldin Mountain locality. (A, B) Inflorescence unit seen from both sides showing terminal pistillate flower (below) with two carpels (c) and one lateral staminate flower (above). (C) Inflorescence unit showing one staminate flower with five tepals and five stamens. (D) Detached immature gynoecium showing a pair of basally fused carpels. (E, F) Detached staminate flower bud in (E) lateral and (F) apical view showing arrangement of tepals. (G) Detached anthetic staminate flower showing five stamens opposite five tepals. (H) Detached stamen. (I) Tricolpate, striate pollen grains from stamens. (J–M) *Spanomera marylandensis*, from the Early Cretaceous (Late Albian) West Brothers locality. (J) Fragment of staminate flower with pairs of opposite – decussate tepals. (K) Fragmentary staminate flower showing four stamens. (L) Detached stamen. (M) Tricolpate, coarsely reticulate–vermiculate pollen grain from stamen. All drawn from SEM images in Drinnan *et al.* (1991).

ventral margin. The stigma is papillate and decurrent along the crests of the ventral margin, extending almost to the syncarpous region. Abundant pollen grains, of the same kind that are found *in situ* in the stamens, occur on the papillate stigmatic surface. There is no information on ovule or seed characters.

Spanomera flowers are closely similar to those of extant Buxaceae, but differ in their tricolpate pollen and follicular

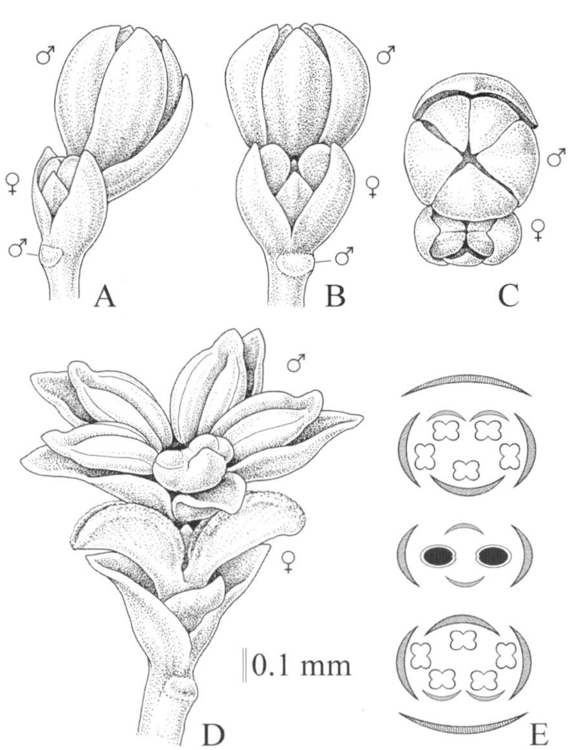

Figure 12.23 Inflorescence reconstruction and floral diagram of the Late Cretaceous buxalean *Spanomera mauldinensis*. (A–C) Inflorescence unit, showing terminal pistillate flower bud, one lateral staminate flower bud, and scar from the other staminate flower (compare with Figure 12.22 (A–C). (D) Inflorescence unit with open flowers. (E) Floral diagrams showing terminal pistillate flower and two lateral staminate flowers. (A–D) From Friis *et al.* (2006a); (E) based on Drinnan *et al.* (1991).

fruits, features that may be plesiomorphic for the group. Flowers of *Spanomera* also share some features with flowers of extant Myrothamnaceae (core eudicots), but pollen of Myrothamnaceae is dispersed in tetrads and the grains have very short tricolpate, almost triporate, apertures.

Lusistemon striatus (Figure 12.24) comprises fragmentary staminate flowers and many isolated stamens recovered from the Early Cretaceous (Late Aptian – Early Albian) Vale de Agua locality, Portugal (Pedersen *et al.*, 2007). The staminate flowers are linked to pistillate organs from the same locality (*Lusicarpus planatus*, see below) by the occurrence of the same characteristic tricolpate and striate pollen *in situ* in the stamens and adhering to the stigmatic surface of the fruits. The flowers are detached and nothing is known about the inflorescence or flower-bearing axis. Since the pistillate and staminate flowers are not found in organic connection they were assigned to separate genera.

Flowers of *Lusistemon striatus* have up to six stamens surrounded by remains of small tepals opposite to the stamens. Stamens are of various lengths, perhaps suggesting a helical arrangement. They consist of a short stalk bearing elongated, narrow anthers, about 0.8–1.2 mm long and 0.2–0.3 mm wide. Anthers are tetrasporangiate, dithecate and apparently dehisced by longitudinal slits. The connective is poorly developed between the thecae, but is extended apically into a small pointed protrusion. Pollen observed in the anthers is tricolpate, striate, circular to slightly elliptical in equatorial view, and about 15–16 μm long and 14–15 μm in diameter (Figure 12.24). Pollen is subangular in polar view. The tectum is striate, with narrow ridges arranged more or less parallel to the polar axis to form a slightly undulating pattern. The ridges dichotomise, and sometimes anastomose or have free endings. They have a rounded to triangular profile and are ornamented by low, transversely oriented ribs. Exine sculpture over the colpus membrane is granular to verrucate, also with finely striate supratectal ornamentation. The pollen wall mainly consists of the tectum; the infratectal layer and foot layer are unusually thin. The infratectal layer is granular and forms an indistinct internal reticulum. The endexine is also granular and is very thin except under the apertures. Pollen of *Lusistemon* is very similar to dispersed pollen described as Striotri-Segmur from the Early Cretaceous (Early–Middle Aptian) of the Mersa Matruh 1 Borehole in Egypt (Penny, 1988a, 1991) (see Figure 12.4). Another closely similar dispersed pollen type was described from the Maastrichtian of Alberta as *Rutihesperipites trochuensis* (Srivastava, 1966).

Lusicarpus planatus (Figure 12.24) is the pistillate structure associated with *Lusistemon striatus* at the Vale de Agua locality. It appears to be preserved at the post-anthetic stage. Pistillate structures are about 1.7 mm long and 1 mm wide, are strongly flattened dorsiventrally, and have a rounded base, with a dimerous, hypogynous and syncarpous gynoecium supported by small tepal-like organs similar to those of the staminate flowers. There are two short, diverging styles, each with a decurrent stigma that extends ventrally for the full length of the free (non-syncarpous) part of the gynoecium. The stigma is covered by the remains of a secretion and was apparently wet in life. Tricolpate striate pollen identical to that found *in situ* in the staminate flowers of *Lusistemon striatus* occurs abundantly on the stigmatic surface, partly or fully embedded in the secretion.

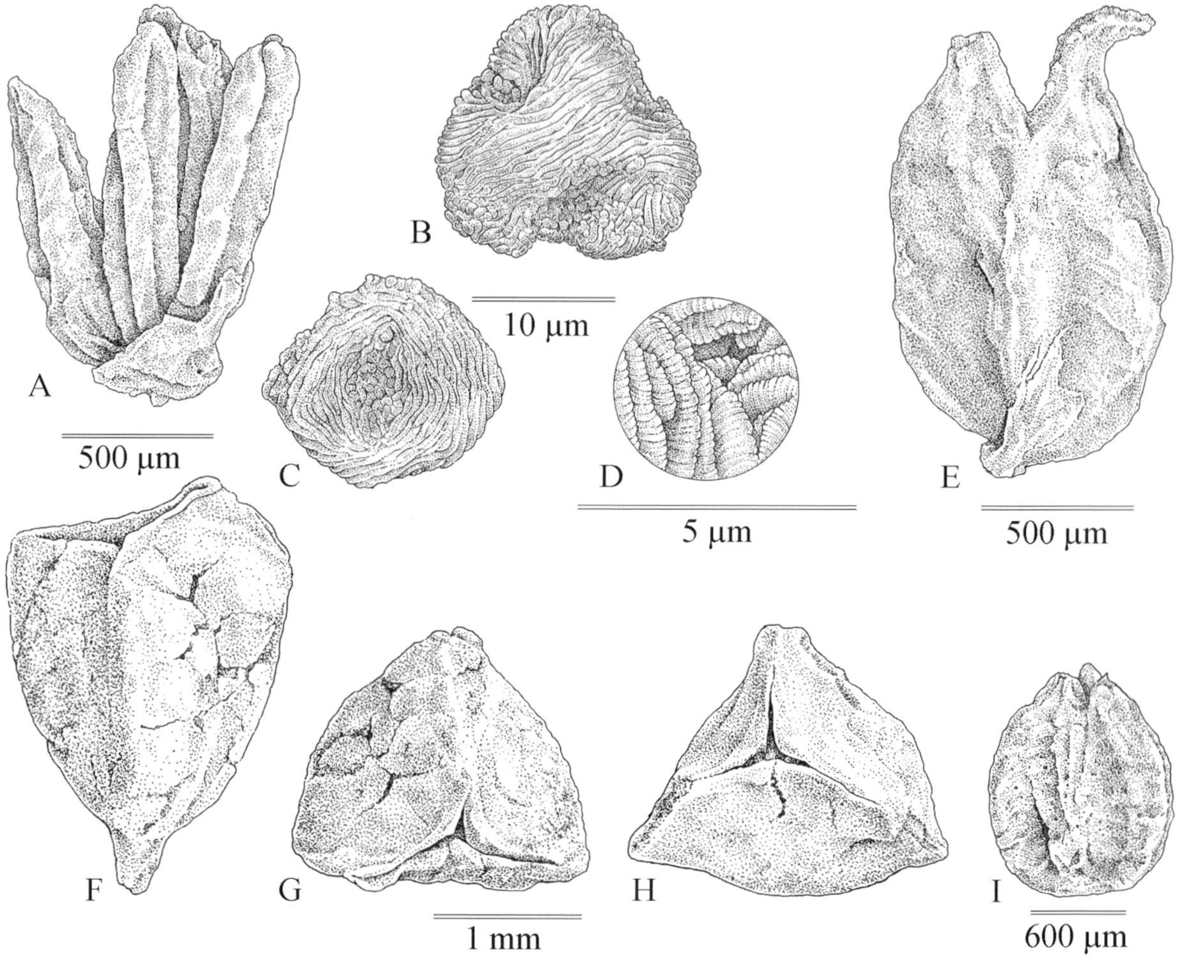

Figure 12.24 Buxalean fossils from the Early Cretaceous (Late Aptian – Early Albian) Vale de Agua locality, Portugal. (A–D) Staminate flower of *Lusistemon striatus* (A) and *in situ* striate, tricolpate pollen grains (B, C), and detail of supratectal ornamentation of pollen wall (D). (E) Pistillate, bicarpellate gynoecium of *Lusicarpus planatus*. (F–H) Pistillate, tricarpellate gynoecium of *Valecarpus petiolatus*. (I) Pistillate, tricarpellate gynoecium of *Aguacarpus hirsutus*. Drawn from SEM images in Pedersen *et al.* (2007).

Staminate *Lusistemon striatus* and pistillate *Lusicarpus planatus* flowers and their pollen share several similarities with flowers of extant Buxaceae. Pistillate flowers are particularly similar to pistillate flowers of *Sarcococca, Pachysandra* and *Styloceras* in their bicarpellate nature and their extended stigmatic surface. Staminate flowers are similar to staminate flowers of extant Buxaceae in the poorly differentiated perianth with stamens opposite the tepals, basifixed anthers and the conical apical extension of the connective (Pedersen *et al.*, 2007). Like the fossil, *Notobuxus* and *Styloceras* also have more than four stamens per flower. The fossils are also similar to *Styloceras* in having very short filaments, but the anthers are proportionally much longer than in the extant genus.

Silucarpus camptostylus is based on a single fragmentary pistillate specimen closely similar to *Lusicarpus*. It differs in having a distinct papillate stigmatic surface. Pollen adhering to the stigmatic surface is reticulate, in contrast to the striate pollen observed on the stigmatic surface of *Lusicarpus* (Pedersen *et al.*, 2007).

Valecarpus pedicellatus (Figure 12.24) is another pistillate structure of possible buxalean affinity from the Early

Cretaceous (Late Aptian – Early Albian) Vale de Agua locality, Portugal (Pedersen *et al.*, 2007). It is preserved at fruiting stage. The only floral organs preserved are the mature carpels. Fruits are tricarpellate and syncarpous, about 3 mm long and 2 mm wide, triangular in cross-section and broadly obovate to obtriangular in lateral view. There are three short styles that are bent outwards. The stigmatic area is lip-like and restricted to the distal part of each style. Fruits are very similar to fruits of extant *Buxus* and *Notobuxus* in their general morphology, particularly the arrangement and shape of styles and stigmas. However, the styles are much shorter in the fossil fruits than in either of the extant taxa. The fossil fruits apparently also lack the interstylar nectaries that characterise extant *Buxus* and *Notobuxus*. No pollen has been observed on the stigmatic surface.

Aguacarpus hirsutus (Figure 12.24) is also based on a tricarpellate gynoecium from the Early Cretaceous (Late Aptian – Early Albian) Vale de Agua locality, Portugal. It is about 1.4 mm long, borne on a short stalk, and has carpels that are united for almost their full length (Pedersen *et al.*, 2007). Styles have distinctly papillate stigmatic surfaces with long, multicellular papillae that are covered with the remains of a secretion. Scattered adhering tricolpate pollen is tectate and striate with transverse ridges on the muri. Grains are about 15 μm long, and similar to those of the *Lusistemon–Lusicarpus* plant described from the same locality. The gynoecium is covered by numerous long trichomes or trichome bases. In addition to having similar kinds of pollen, *Aguacarpus* also resembles *Lusicarpus* in the degree of carpel union and in the nature of the decurrent stigmas, but is tricarpellate. Like *Lusicarpus* it is also similar to some Buxaceae, particularly *Notobuxus*.

12.8.1 Buxaceae

The Buxaceae are a small family of about 70 species in six genera (*Buxus, Didymeles, Notobuxus, Pachysandra, Sarcococca, Styloceras*). In previous classifications *Didymeles* was placed in its own family, which in turn has been variously placed close to the Hamamelidaceae, Buxaceae, Leitneriaceae or Euphorbiaceae (Sutton, 1989). Buxaceae are shrubs or small trees with an almost worldwide distribution. *Didymeles* has two species, both restricted to Madagascar.

Flowers are small, aggregated into spikes or heads, and typically unisexual. Pistillate flowers have a superior ovary supported by many bract-like tepals in a spiral arrangement. The gynoecium in most genera is syncarpous with two or three carpels and two or three locules, each locule with two anatropous ovules. Pistillate flowers of *Didymeles* are unicarpellate and have a unilocular ovary with a single hemiorthotropous ovule (Sutton, 1989; von Balthazar *et al.*, 2003). Staminate flowers typically have tepals and stamens arranged in an opposite–decussate pattern (Cronquist, 1981; Drinnan *et al.*, 1991; von Balthazar and Endress, 2002). Pollen of Buxaceae varies considerably in aperture configuration, which ranges from tricolporate to polycolporate and polyporate. There is also considerable variation in pollen wall ornamentation. The exine is reticulate–rugulate to tectate–foveolate in *Buxus* and *Notobuxus*, but has distinctive crotonoid ornamentation in *Sarcococca* and *Pachysandra* (Gray and Sohma, 1964; Köhler, 1980, 1981; Köhler and Brückner, 1982).

Fossil pollen very similar to that of extant *Pachysandra* and *Sarcococca* is recorded from the Late Cretaceous (Campanian–Maastrichtian) in Siberia, Europe and North America (Srivastava, 1972). Pollen is polyporate with distinctive crotonoid wall ornamentation (Figure 12.25) and is typically assigned to the dispersed pollen genus *Erdtmanipollis*. This genus was established by Krutzsch (1962) for dispersed pollen from the Oligocene that is closely similar to pollen of extant *Pachysandra*. In North America this kind of pollen is known from the Campanian–Maastrichtian of Canada (Srivastava, 1972; Jarzen and Norris, 1975), and from several Campanian and Maastrichtian palynological assemblages in the USA (Gray and Sohma, 1964; Oltz, 1969; Chmura, 1973; Frederiksen, 1987). According to Gray and Sohma (1964) the Californian material includes two distinct polyporate forms, one closely similar to *Pachysandra* and the other closely similar to *Sarcococca*. From Europe *Pachysandra*-like pollen is reported from the Campanian–Maastrichtian of Germany as well as from younger strata (Krutzsch, 1962; Krutzsch and Lenk, 1969).

Dispersed pollen assigned to *Hexaporotricolpites* from the mid-Albian to Cenomanian of West Africa and Brazil (Boltenhagen, 1967; Doyle, 1999) have been assigned to Didymelaceae based on their characteristic tricolporate apertures in which there are two pores placed near each end of the colpi (Boltenhagen, 1967). However, these dispersed grains differs from pollen of extant *Didymeles* in exine sculpture (Muller, 1981) and detailed studies are needed to establish their affinities more securely (Doyle, 1999).

Dispersed pollen assigned to *Schizocolpus marlinensis* from Palaeogene strata of the submerged Ninetyeast Ridge in the eastern Indian Ocean (Kemp and Harris, 1975), from

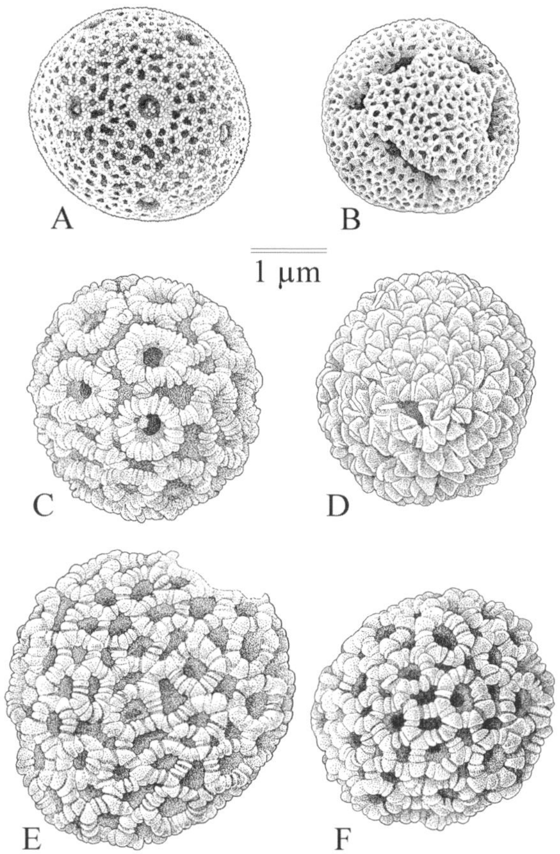

Figure 12.25 Pollen of (A–C) extant and (D–F) fossil Buxaceae. (A) *Styloceras laurifolia*. (B) *Buxus glomerata*. (C) *Sarcococca hookeriana*. (D) *Erdtmanipollis cretacea*. (E, F) *Erdtmanipollis procumbentiformis*. (A–C) Drawn from SEM images in Nowicke and Skvarla (1984); (D–F) drawn from SEM images in Srivastava (1972).

New Zealand (Mildenhall, 1980) and from Australia (Stover and Partridge, 1973) shows the same unusual aperture configuration as *Hexaporotricolpites* and extant *Didymeles* (Stover and Partridge, 1973; Kemp and Harris, 1975). If these grains can be attributed reliably to *Didymeles* or Buxaceae they would further support a much wider distribution for the family in the past than it has today.

Other dispersed pollen of buxaceous affinity includes pollen very similar to extant *Buxus* assigned to the dispersed pollen genus *Buxapollis*, which has been reported from several Cenozoic floras in Europe, Asia and North America. The earliest record of this *Buxus*-like pollen is from the Paleocene of China (Muller, 1981; Song *et al.*, 2004).

12.9 TROCHODENDRALES

12.9.1 Trochodendraceae

The Trochodendraceae (including Tetracentraceae) are a small family comprising two monotypic genera, *Tetracentron* and *Trochodendron*, both restricted to eastern Asia. *Tetracentron* and *Trochodendron* are trees with alternate leaves and vessel-less wood. Leaves have pinnate (*Trochodendron*) or palmate (*Tetracentron*) venation, and a crenate margin with chloranthoid teeth. Flowers are bisexual and borne in elongated inflorescences. Flowers of *Trochodendron* lack a perianth and have up to about 70 stamens and 17 carpels. Flowers of *Tetracentron* are tetramerous with four tepals, four stamens and four carpels. Stamens in both taxa are tetrasporangiate and anthers have valvate dehiscence by two laterally hinged valves. Pollen of *Trochodendron* and *Tetracentron* is very similar: tricolpate and tectate with an irregular, reticulate–striate pattern and longitudinal striations along the colpi. The gynoecium is syncarpous in both genera with few to several anatropous, bitegmic ovules per carpel (Endress, 1986b, 1993b).

The Cretaceous record of Trochodendraceae is currently sparse. Leaves from the Potomac Group sequence previously assigned to *Menispermites potomacensis*, *Populophyllum reniforme* and '*Populus*' *potomacensis* may be related to the Trochodendraceae (Crane, 1989). Similar leaves have also been reported from other mid-Cretaceous localities in North America and Europe (Crane, 1989). All these early leaf records of the group are in need of more detailed study. It will be especially important to link them with reproductive structures in order to document securely their relationship with Trochodendraceae.

In contrast to the Cretaceous record, the Cenozoic record of Trochodendraceae is extensive and documents a much wider geographic distribution than the family has today. Both *Trochodendron* and *Tetracentron* are known from the Eocene of British Columbia, Canada and northwestern USA (Pigg *et al.*, 2007). *Tetracentron* and *Trochodendron* persist in North America at least until the Miocene (Manchester *et al.*, 1991; Manchester and Chen, 2006) and *Tetracentron* has recently been documented also from the Miocene of Iceland (Grímsson *et al.*, 2008).

In addition to fossils referred to extant *Tetracentron* and *Trochodendron*, the fossil history of Trochodendraceae is further expanded by reproductive structures assigned to various species of *Nordenskioeldia* and their associated leaves (Figure 12.26).

Nordenskioeldia was based on woody inflorescence axes with many sessile fruits discovered from the Paleocene of

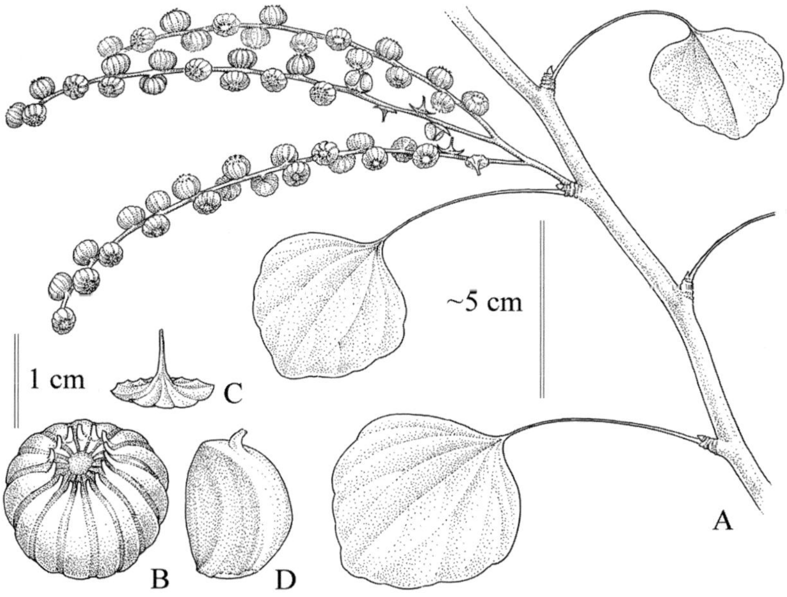

Figure 12.26 Reconstruction of the extinct *Nordenskioeldia* plant (Trochodendraceae) from the Early Cenozoic. (A) Leafy shoot showing short shoots and attached leaves and infructescence. (B) Fruit showing carpels/fruitlets each with a separate, apical, style. (C) Remains of perianth after the fruitlets have been shed. (D) Single fruitlet in lateral view showing apical style. Redrawn from Crane *et al.* (1991).

Spitsbergen (Heer, 1870). The fruits consist of 10 – 20 one-seeded schizocarpous D-shaped fruitlets that are sometimes found dispersed. Associated leaves are petiolate and vary in shape from narrowly to broadly elliptical or obovate to ovate with an entire or crenate margin (Crane *et al.*, 1991). The fruits of *Nordenskioeldia* are mainly distinguished from those of extinct and extant *Trochodendron* in being sessile, in contrast to the stalked flowers and fruits of *Trochodendron*. *Nordenskioeldia* also has a single seed in each fruitlet, and both fruitlets and seeds are shed at maturity.

The *Nordenskioeldia* plant is widespread in Paleocene floras of the Northern Hemisphere ranging from western North America over the Arctic regions to Central Asia and eastern Asia (Crane *et al.*, 1991). In North America the genus persisted into the Miocene before eventually becoming extinct (Manchester *et al.*, 1991). It is interesting that there are no European records of *Nordenskioeldia* even though it is common in Spitsbergen and there are floras of the appropriate age in Europe.

Nordenskioeldia-like fruits also occur in the Late Cretaceous Horseshoe Canyon Formation of Drumheller, Alberta, Canada, preserved as isolated fruitlets and inflorescence fragments (Aulenback, 2009). The infructescences appear shorter, with more densely crowded fruits than typical Cenozoic forms, but have the same distinctive structure (Figs. 543, 544 in Aulenback, 2009). Some of the isolated fruitlets included in the fossil genus (Fig. 541 in Aulenback, 2009) are more similar to fruitlets of *Platanus* and are probably not attributable to *Nordenskioeldia*.

13

Fossils of core eudicots: basal lineages

Core eudicots include the bulk of extant angiosperm species diversity (Magallón *et al.*, 1999) and can be divided into two large clades (rosids, asterids) together with a small number of lineages of uncertain phylogenetic position (Figure 13.1). In this chapter we deal with those lineages of core eudicots that fall outside the species-rich rosids and asterids, but that potentially occupy an important position in the early evolution of the group.

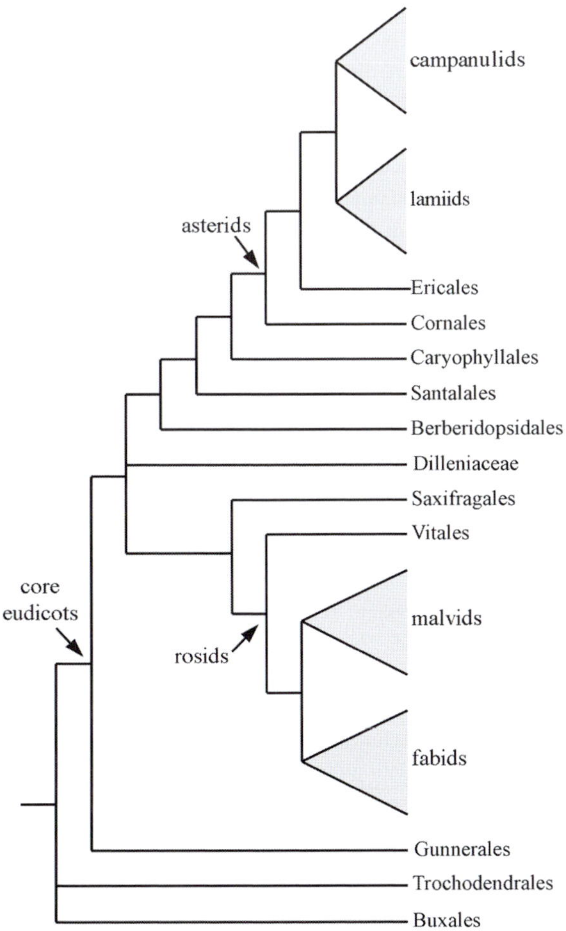

Figure 13.1 Summary of phylogenetic relationships of extant core eudicots based on APGIII (2009).

13.1 CLASSIFICATION OF CORE EUDICOTS

A four-gene phylogenetic analysis places Gunnerales (Gunneraceae and Myrothamnaceae) as the sister group to all other core eudicots (Soltis *et al.*, 2003). In the APGII classification (2003) the precise position of such lineages as Berberidopsidales, Saxifragales, Santalales and Caryophyllales–Dilleniaceae was not yet securely determined, but in the latest APG classification (APGIII, 2009) the positions of some of those taxa have been clarified. Berberidopsidales, Santalales and Caryophyllales are resolved as a grade of successive sister lineages to the asterid clade, while the Saxifragales are placed as sister to the rosid clade. In this chapter we describe the fossil record of these lineages of core eudicots that appear to have diverged before the diversification of both rosids and asterids. Some of these lineages, such as the Saxifragales, have a well-documented fossil history extending back to the Late Cretaceous.

In very general terms core eudicots can be characterised by their flowers, which typically have a perianth composed of a distinct calyx and corolla (heterochlamydous), and a whorled, predominantly pentamerous, organisation. The flowers may have the same number of organs in all whorls (isomerous), or more often they are heteromerous with a pentamerous calyx and corolla, a more variable androecium, and a dimerous or trimerous gynoecium. The relatively fixed floral organisation, and the presence of a well-differentiated perianth, may be coupled to duplications of the floral organ identity genes AP1, AP3 and AG, which occurred at, or near, the base of this important clade of angiosperms (Kramer and Hall, 2005).

13.2 EARLY FOSSIL EVIDENCE OF CORE EUDICOTS

The separation of core eudicots from other, earlier-diverging groups of eudicots most likely took place in the latest part of the Early Cretaceous. The fossil record indicates that the major ecological radiation of core eudicots, and probably also their phylogenetic differentiation, began in the first half of the Late Cretaceous and continued through the later part of the Late Cretaceous and into the Early Cenozoic. Currently there is no clear evidence of core eudicots prior to about the Early–Late Cretaceous boundary.

The first fossils with unequivocal features of core eudicots are from the latest Albian – earliest Cenomanian. They comprise pentamerous, isomerous flowers, with a clearly differentiated calyx and corolla, from the Dakota Formation of Nebraska, USA. These fossils are referred to here as the 'Rose Creek flower' after the locality where they were found (section 14.2). There are also dispersed tetrads of triporoidate pollen recorded from several levels in the mid-Albian to Cenomanian of North America (Doyle, 1969; Ward, 1986) that have been compared with pollen of Myrothamnaceae (Gunnerales, section 13.3). Oblate tricolporate and triporate pollen, which is characteristic of some groups of core eudicots, appears in the Middle and Late Cenomanian (Góczán *et al.*, 1967; Doyle and Robbins, 1977; Ward and Doyle, 1994), indicating that diversification was well under way in certain core eudicot lineages by the early Late Cretaceous (Chapters 14 and 15).

13.3 GUNNERALES

Phylogenetic analyses based on molecular data suggest that the two monogeneric families Gunneraceae and Myrothamnaceae form a monophyletic group that is sister to all other core eudicots. Support for this relationship is not strong, but has increased with the inclusion of evidence from more genes (Soltis *et al.*, 2003, 2005). In earlier classifications *Gunnera* was often placed together with Haloragaceae (Cronquist, 1981) and *Myrothamnus* has been variously placed, often close to Hamamelidaceae (Takhtajan, 1969; Cronquist, 1981).

Gunnera includes about 40 species of herbaceous plants with short upright stems and distinctive ovate to more or less peltate leaves. In some species the leaves may be very large. *Gunnera* is distributed mainly through both the Old and New World tropics as well as in temperate regions of the Southern Hemisphere. Plants may be either monoecious or dioecious, and have small, simple epigynous flowers. Flowers are typically dimerous with four perianth parts in two decussate pairs and two stamens opposite the inner set of perianth parts. The gynoecium consists of an ovary with two terminal styles and a single locule containing a single, pendulous, anatropous ovule. The perianth parts are described as sepals and petals by Wanntorp and Ronse De Craene (2005), and the flowers are interpreted as reduced from a tetramerous ancestral form. Unisexual flowers are even more simple and male flowers may consist of only a single stamen, as in *Gunnera herteri* (Wanntorp and Ronse De Craene, 2005). Pollen is tricolpate with fine reticulate sculpture (Wanntorp *et al.*, 2004b).

Myrothamnus includes two species of shrub: the 'resurrection plants' of southern Africa and Madagascar that grow in extremely arid habitats (Endress, 1989; Kubitzki, 1993b). Plants are dioecious and their simple flowers may either lack, or have up to four, tepals. Pistillate flowers have three to four basally fused carpels that have a sessile, decurrent stigma and many ovules. Staminate flowers have three to eight stamens with an apical extension of the connective (Kubitzki, 1993b). Pollen is triporate and usually shed in tetrads (Zavada and Dilcher, 1986).

There is no macro- or mesofossil record of Gunnerales. However, dispersed tricolpate and finely reticulate pollen assigned to the fossil taxa *Retitricolpites microreticulatus* and *Tricolpites reticulatus*, which show close similarities to pollen of extant *Gunnera*, is reported from the Late Cretaceous of the Southern Hemisphere (Brenner, 1968; Jarzen and Dettmann, 1989; Wanntorp *et al.*, 2004a). The earliest of these records that may be linked reliably to *Gunnera* is from the Turonian of Peru (Jarzen and Dettmann, 1989; Wanntorp *et al.*, 2004a).

Dispersed pollen tetrads from the mid-Albian to Cenomanian of North America, including *Virgo amiantopollis*, show features that appear to indicate a close relationship with *Myrothamnus*. *Virgo amiantopollis* (Figure 13.2) includes tetrahedral tetrads of tricolporoidate to triporoidate pollen. The genus was established by Ward (1986) based on specimens described by Srivastava (1977) as *Senectotetradites amiantopollis*. Individual pollen is oblate to suboblate with very short colpi. Similar pollen tetrads have also been described as *Ajatipollis* sp. *A* from the Middle Albian to Early Cenomanian of the Potomac Group (Doyle and Robbins, 1977), and as

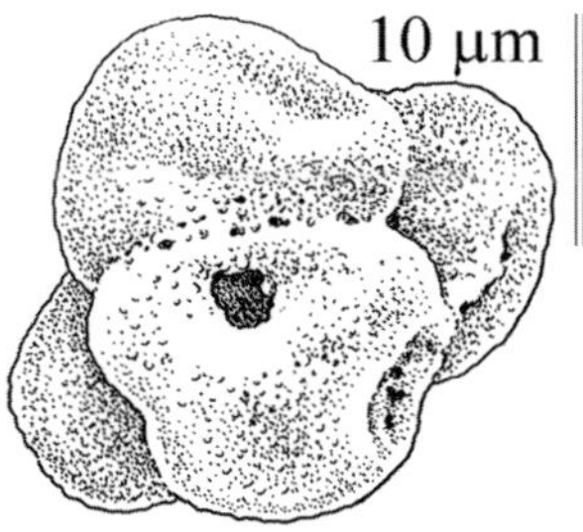

Figure 13.2 Pollen tetrad of *Virgo amiantopollis* from the Early Cretaceous (mid-Albian) of Kansas, USA, similar to extant *Myrothamnus* (Myrothamnaceae); individual grains are tricolporate with very short colpi. Drawn from SEM image in Ward (1986).

Tetradopollenites sp. from the Early Cretaceous of Portugal (Medus and Berthou, 1980). The tetrads from Kansas have been compared to pollen tetrads of extant *Myrothamnus* (Ward and Doyle, 1988) but detailed comparisons have not been published.

13.4 DILLENIACEAE

The Dilleniaceae are an isolated family that are treated as unassigned among core eudicots (APGIII, 2009). The family has sometimes been placed as sister to the Caryophyllales or Vitales. In earlier classifications based on morphology (e.g. Takhtajan, 1969) it was grouped in the superorder Dilleniidae together with Paeoniales, Theales, Violales, Passiflorales, Cucurbitales, Salicales and several other groups that are no longer thought to be closely related. The Dilleniaceae comprise about 12 genera and 300 species in tropical to warm temperate regions with highest species diversity in Australasia (Mabberley, 1997). Flowers of Dilleniaceae are fundamentally pentamerous with an apocarpous gynoecium and numerous stamens, which reflects a secondary increase in number (Judd and Olmstead, 2004). There are no reliable records of Dilleniaceae from the Cretaceous, but seeds that have been compared with those of *Tetracera* and *Hibbertia* are recorded from the Early Eocene of southern England (Chandler, 1964; Collinson *et al.*, 1993).

The calcified fruiting structure from the Late Cretaceous (Coniacian–Santonian) Obirashibe River locality, Hokkaido, Japan, described as *Elsemaria kokubunii* (Figures 2.16, 13.3) was broadly referred to the Dilleniidae (Nishida, 1994b). The single specimen is a capsular fruit with excellent cellular preservation. The fruit is almost spherical in shape, about 13.5 mm in diameter, and borne on a stout stalk. It consists of 10 laterally fused carpels. Dehiscence is loculicidal. Placentae are massive plates that extend into the locules, and each bears more than 40 ovules. The vascularisation of the fruit is well preserved and has been reconstructed in detail, but despite the information available the precise systematic placement of the fossil remains uncertain.

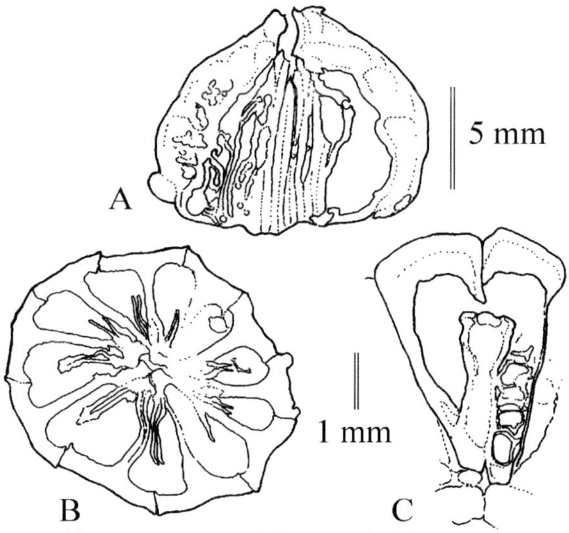

Figure 13.3 *Elsemaria kokubunii*, a calcified fruiting structure from the Late Cretaceous (Coniacian–Santonian) Obirashibe River locality, Hokkaido, Japan. Longitudinal section through centre of fruit (A), transverse section of fruit showing 10 locules (B), and detail of locule in transverse section showing placenta with many ovules (C). Drawn from photographs in Nishida (1994b).

13.5 BERBERIDOPSIDALES

The Berberidopsidales are a well-supported clade comprising two families: Berberidopsidaceae and Aextoxicaceae. The relationship of this group to other core eudicots is not fully resolved (Judd and Olmstead, 2004), but in the APGIII classification it is placed as sister to Santalales, Caryophyllales and all asterids. The Berberidopsidaceae comprise *Berberidopsis*, with one species native to Chile and one species native to eastern Australia, and *Streptothamnus* with a single species, native to eastern Australia. The Aextoxicaceae comprise the genus *Aextoxicon*, also with a single species in Chile (Mabberley, 1997). Neither of these families is known from the fossil record; they are, however, of interest at this level of angiosperm evolution for their very different patterns of floral organisation. *Aextoxicon* has unisexual flowers with well-differentiated sepals and petals, and a drupaceous fruit that develops from a bicarpellate, syncarpous gynoecium. *Berberidopsis* and *Streptothamnus* have bisexual flowers with helically arranged tepals that grade from sepals to petals. The fruit is a berry that develops from a tricarpellate gynoecium (Ronse De Craene, 2004; Soltis *et al.*, 2005).

13.6 SANTALALES

In the APGIII classification the Santalales comprise seven families (Balanophoraceae, Loranthaceae, Misodendraceae, Olacaceae, Opiliaceae, Santalaceae, Schoepfiaceae), which together include about 1000 species (Stevens, 2001 onwards). Like Berberidopsidales, the relationships of Santalales are not fully established, but new sequence data suggest a position as sister group to Caryophyllales plus asterids (APGIII, 2009).

Morphological and ecological diversity within Santalales is extensive, but most are parasitic. Autotrophic woody plants occur only in Olacaceae. Phyllotaxis is typically opposite–decussate. Merosity of flowers is highly variable, ranging from three- to nine-parted. The calyx is often reduced or lacking and the stamens are in one whorl opposite the petals. Pollen also shows considerable variation in aperture number and configuration, pollen wall ornamentation and other features (Figure 13.4) (e.g. Erdtman, 1952; Maguire *et al.*, 1974; Jarzen, 1977).

The fossil record of Santalales includes leaves and reproductive structures from the Cenozoic of Europe, North America and Australia. Leaves assigned to the Olacaceae are known from the Eocene of North America (Taylor, 1990) and fruits similar to those of *Erythropallum* and *Olax* (Olacaceae) are reported from the Eocene of southern England (Chandler, 1964). Leaves assigned to Loranthaceae have been reported from the Eocene of Germany (Mai and Walter, 1978; Wilde, 1989) and Australia (Christophel in Collinson, 1993). It has also been suggested that small shoots with an opposite–decussate arrangement of organs, which have been described as *Ephedrites johnianus* from the Baltic amber (Göppert and Berendt, 1845), are loranthaceous and should be transferred to the extant genus *Patzea* of this family (Conwentz, 1886).

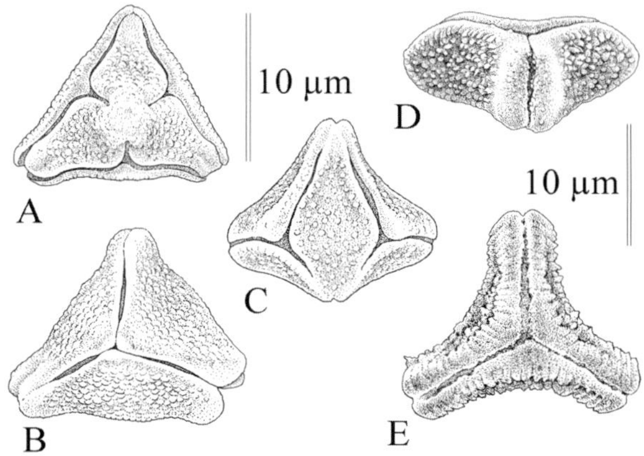

Figure 13.4 Pollen of extant Santalales. (A–C) *Arjona tuberose* (Santalaceae), heteropolar pollen seen (A, B) from both poles and in (C) equatorial view. (D, E) *Nuytsia floribunda* (Loranthaceae), heteropolar pollen seen in (D) equatorial and (E) polar view. (A–C) Drawn from SEM image in Farabee (1991); (D, E) drawn from SEM image in Feuer and Kuijt (1980).

In addition to the macrofossils there are also many Cenozoic records of dispersed pollen ascribed to Santalales. These include such genera as *Gothanipollis*, *Accuratipollis* and *Cranwellia*. *Cranwellia* and *Gothanipollis* both extend back to the Late Cretaceous, where they are widely distributed, both within Laurasia as well as in Gondwana (e.g. Couper, 1953; Srivastava, 1966; Song *et al.*, 2004). It has also been suggested that dispersed pollen assigned to the distinctive and widespread Triprojectacites–*Aquilapollenites* complex is closely related to pollen of extant Santalales, particularly to genera of Loranthaceae (Funkhouser, 1961; Jarzen, 1977; Muller, 1984).

The genus *Aquilapollenites* was established to comprise bilaterally symmetrical pollen from the Maastrichtian of Canada characterised by 'two to four wing-like projections extending from a broadly subrectangular body' (Rouse, 1957). Subsequently about 12 other genera have been described based on dispersed pollen with similar distinctive apertural and equatorial projections (Figure 13.5) (e.g. Farabee, 1991). Collectively these genera are included in the form group Triprojectacites established by Mtchedlishvili (1961). Triprojectacites–*Aquilapollenites* pollen has been reported from many Late Cretaceous floras from high-latitude areas of the Northern Hemisphere in Greenland, North America and Asia. The abundance and diversity of such grains define the so-called

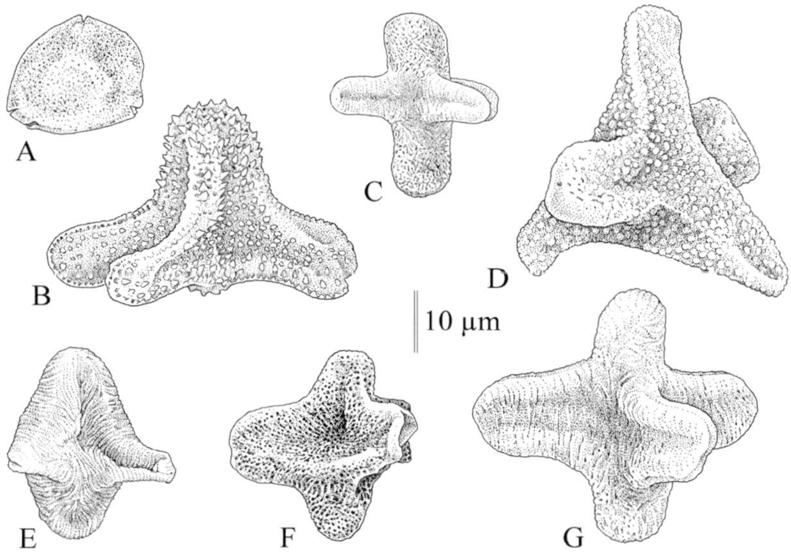

Figure 13.5 Dispersed Triprojectacites–*Aquilapollenites* pollen from the Late Cretaceous and Early Cenozoic, possibly related to Santalales. (A) *Kurtzipites* sp. (B) *Aquilapollenites delicates.* (C) *Integricorpus rigidus.* (D) '*Aquilapollenites*' *quadricretaceus.* (E) *Striatoporites bertillonites.* (F) Unnamed grain. (G) *Bratzevaea* sp. (A–C, E, G) Drawn from SEM image in Farabee (1990); (D, F) drawn from SEM image in Farabee (1991).

Aquilapollenites Province (Chapter 19). Triprojectacites–*Aquilapollenites* pollen first occurs in the Turonian and then attains its maximum diversity and distribution during the Campanian–Maastrichtian (e.g. Herngreen *et al.*, 1996). Pollen of this group is of particular interest because most taxa disappeared at around the K–T boundary. Rare occurrences of the group extend into the Cenozoic (Paleocene to Oligocene) (e.g. Farabee, 1991). Further comparisons of Triprojectacites–*Aquilapollenites* pollen to Santalales and other groups will be of considerable interest because of their importance in Campanian–Maastrichtian palynofloras from high palaeolatitudes in the Northern Hemisphere (Herngreen *et al.*, 1996). However, the group may be systematically heterogeneous. It may include several distinct plant groups with relationships to such diverse families as Apiaceae, Caprifoliaceae, Elaegnaceae, Morinaceae and Simarubaceae (see summary in Farabee, 1991). Until additional information on the *Aquilapollenites* plants is available the relationship of this distinctive pollen to Santalales remains uncertain. The discovery of flowers with Triprojectacites–*Aquilapollenites in situ* would help resolve this issue and would be a major step forward.

13.7 CARYOPHYLLALES

The Caryophyllales comprise 34 families for which details of interrelationships remain unresolved (APGIII, 2009). Judd and Olmstead (2004) recognised two main groups. The first group, centred on Amaranthaceae, Caryophyllaceae, Nyctaginaceae, Phytolaccaceae and Portulacaceae, comprises about 8000 species (Judd *et al.*, 2002) and has no secure fossil record from the Cretaceous. Pollen described as *Retitricolpites multibaculates* from the Late Cretaceous (Late Campanian) of Sakhalin, Russia, is reticulate with distinctive lumina that have isolated bacula extending from the foot layer (Figure 13.6). These grains are similar to pollen of extant *Bougainvillea*, Nyctaginaceae (Takahashi, 1997). Pollen of Nyctaginaceae is also reported from the Eocene of Argentina (Zetter *et al.*, 1999). Seeds similar to those of Caryophyllaceae have been recorded from the Eocene of southern England (Chandler, 1964), and a fossil inflorescence, *Caryophylloflora paleogenica*, with periporate pollen *in situ* was described from northeastern Tasmania, Australia (Jordan and Macphail, 2003). Otherwise, the fossil record of the Caryophyllales is sparse, even from the Cenozoic. Possible seeds of Amaranthaceae from the Late Santonian–Early Campanian Åsen locality, Sweden (Friis in Collinson *et al.*, 1993), are now known to have features that conflict with that potential assignment and their relationship is uncertain.

The second group of Caryophyllales (Polygonales *sensu* Judd *et al.*, 2002; Judd and Olmstead, 2004) can itself be divided into two putative subclades. One subclade, centred on Plumbaginaceae, Polygonaceae and Tamaricaceae, is

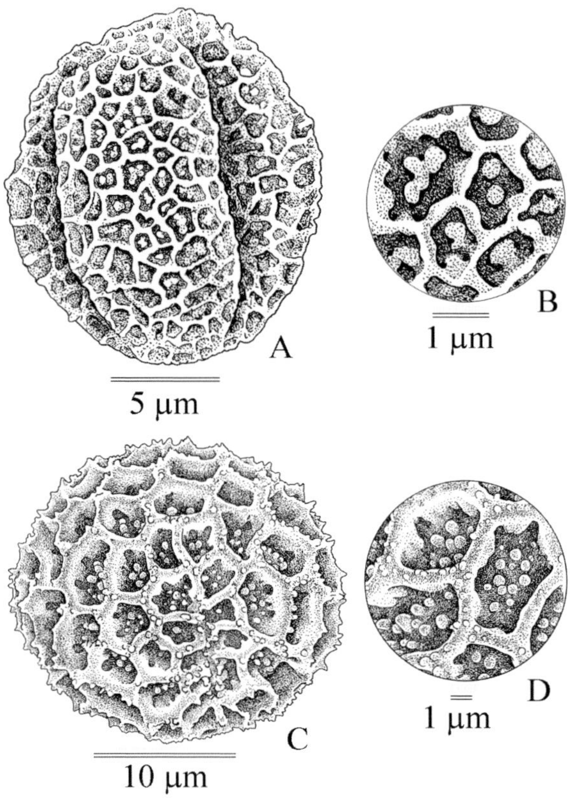

Figure 13.6 (A, B) Pollen of *Retitricolpites multibaculates* from the Late Cretaceous (Late Campanian) of Sakhalin, Russia, (A) in equatorial view and (B) showing detail of pollen wall ornamentation. (C, D) Pollen of extant *Bougainvillea* sp. (Nyctaginaceae) showing (C) morphology and (D) detail of pollen wall ornamentation. (A, B) Drawn from SEM image in Takahashi (1997); (C, D) drawn from SEM image in Halbritter (2000 onwards).

reported from the Paleocene based on dispersed pollen, but these fossils require more detailed study before they can be accepted as reliable records (Collinson, 1993). In younger Cenozoic floras fruits of Polygonaceae are common from Europe and Asia (Friis, 1985b). The other subclade mainly comprises carnivorous plants and includes the families Ancistrocladaceae, Dionocophyllaceae, Droseraceae, Drosophyllaceae and Nepenthaceae.

The Droseraceae are represented in the Cenozoic of Europe and Asia by seeds referable to the extant aquatic carnivorous plant *Aldrovanda* (Chandler, 1964; Mai, 1985a) and by dispersed pollen assigned to the fossil genus *Saxonipollis* (Krutzsch, 1970a). In Asia, seeds of *Aldrovanda* are recorded from several Oligocene and Miocene floras (e.g. Dorofeev, 1963). Seeds of the extinct genus *Palaeoandrovanda*, described from the Late Cretaceous (Turonian–Santonian) Klikov sequence of Bohemia, Czech Republic (Knobloch and Mai, 1984), are fundamentally different from extant *Aldrovanda* and may not even be plant remains (Zuzanna Heřmanová, personal communication, 2010).

13.8 SAXIFRAGALES

Recent phylogenetic analyses based on morphological and molecular data have greatly modified our understanding of relationships among plants traditionally included in Saxifragaceae and Saxifragales. As a result, the circumscription of the group has changed substantially (e.g. Fishbein *et al.*, 2001). In the classification of Takhtajan (1969) the Saxifragales were placed in the Rosidae and included 28 families. Other classifications subsumed many of these same families in the Rosales (Engler, 1930; Cronquist, 1981), and many were considered so closely related that they were sometimes treated merely as subfamilies of Saxifragaceae (Engler, 1930). In the APGIII classification only seven of Takhtajan's 28 families are retained within the modified Saxifragales (Crassulaceae, Grossulariaceae, Iteaceae, Penthoraceae, Pterostemonaceae, Saxifragaceae, Tetracarpaceae). The remainder have very diverse relationships among rosids (e.g. Geraniales, Oxalidales, Myrtales) and asterids (e.g. Cornales, Lamiales, Solanales, Apiales). Other families that are now included in the Saxifragales (Altingiaceae, Aphanopetalaceae, Cercidiphyllaceae, Daphniphyllaceae, Haloragaceae, Hamamelidaceae, Paeoniaceae, Peridiscaceae) were variously placed in the Hamamelididae, Dilleniidae and Rosidae in previous classifications (Takhtajan, 1969; Cronquist, 1981). The systematic position of the newly circumscribed Saxifragales among core eudicots is not yet fully resolved, although a position as sister to all rosids is suggested in several phylogenies based on molecular data (APGIII, 2009).

The Saxifragales have an extensive fossil record, particularly in the Cenozoic where they are represented by numerous leaves, infructescences, fruits and seeds of Cercidiphyllaceae, Altingiaceae, Hamamelidaceae and Haloragaceae (e.g. Mai, 1995; Maslova, 2003). There are also several fossil flowers from the Late Cretaceous that are clearly related to extant Saxifragales, and most prominently to Altingiaceae and Hamamelidaceae. Two fossil

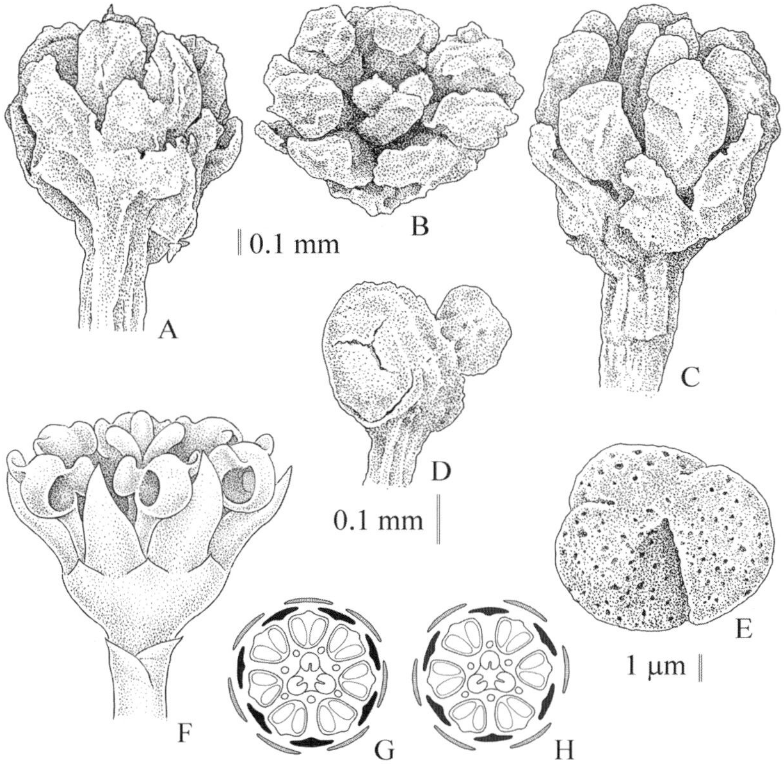

Figure 13.7 *Archamamelis bivalvis* from the Late Cretaceous (Late Santonian – Early Campanian) Åsen locality, Sweden. (A) Lateral view of flower with perianth and androecium in whorls of six. (B, C) Flower in (B) apical, and (C) lateral view, showing perianth and androecium in whorls of seven. (D) Detached stamen showing valvate dehiscence. (E) Tricolpate foveolate pollen. (F) Reconstruction of flower. (G, H) Floral diagrams of flowers with perianth and androecium in whorls of (G) seven and (H) six. (A–E) Drawn from SEM image in Endress and Friis (1991).

inflorescences/infructescences, *Allonia decandra* and *Androdecidua endressii*, can be unequivocally assigned to the Hamamelidaceae, as can the seeds of *Rhodoleia cretacea*. Flowers of *Archamamelis bivalvis* are also closely similar to those of certain Hamamelidaceae, but based on the information currently available they cannot be included in the family with certainty. Several other Cretaceous floral structures have been tentatively referred to the Altingiaceae (*Microaltingia*, *Lindacarpa*, *Viltyungia* and several unnamed inflorescences), but none of these can be reliably assigned to the family. These fossils of uncertain affinity to Hamamelidaceae and Altingiaceae are discussed here as unassigned Saxifragales. Fossil flowers from the Late Cretaceous Åsen locality, Sweden, (*Scandianthus* and *Silvianthemum*) (Friis and Skarby, 1982; Friis, 1990), previously referred to Saxifragales, are related to families that are now placed in the asterid clade (Chapter 15).

The early evolution of the Saxifragales was discussed by Hermsen *et al.* (2006) based on a cladistic analysis that combined molecular and morphological data for 31 extant taxa relevant to Saxifragales in the APGII classification, as well as five putatively related Late Cretaceous reproductive structures.

13.8.1 Unassigned Saxifragales

Archamamelis bivalvis (Figure 13.7) includes small flowers and dispersed stamens from the Late Santonian – Early Campanian Åsen locality, Sweden (Endress and Friis, 1991). Flowers have two whorls of perianth parts and a single whorl of stamens. Each whorl comprises six or seven parts. Stamens are weakly differentiated into a filament and a basifixed anther. Anthers are dithecate and tetrasporangiate with abundant connective tissue between the thecae and at the apex. Each theca dehisced by a single, laterally hinged valve that opened towards the centre of the flower. Pollen *in situ* in the stamens is tricolpate and tectate–foveolate. Internal to the stamens is a whorl of staminodes, and in the centre of the flower there is a trimerous gynoecium with three free styles. Carpels are basally connate and also united with the hypanthium. The fossil flowers are preserved as lignite and many details, including gynoecium structure, are not completely clear.

The combination of floral features observed in *Archamamelis bivalvis* indicates that the fossils are most closely related to extant members of the Hamamelidaceae, but it is uncertain whether they can be placed within the family, or on the Hamamelidaceae stem lineage (Endress and Friis, 1991).

Staminate and pistillate flowers of possible hamamelidaceous affinity reported from the Late Cretaceous (Turonian) Old Crossman locality, New Jersey, USA, have not yet been formally named (Crepet *et al.*, 1992). Inflorescences are small unisexual heads, up to about 2 mm in diameter, each consisting of about 15 densely packed flowers. Staminate flowers consist of a four-lobed sepal cup and four functional stamens positioned opposite to the calyx lobes. Stamens are differentiated into a very short filament and a basifixed, tetrasporangiate, dithecate anther. Dehiscence was apparently valvate, with two laterally hinged valves over each theca. Anthers have prominent triangular apical extensions of the connective with small auricular extensions at the base. Pollen found *in situ* in the stamens is small, tricolpate and tectate–reticulate. A whorl of four short structures, each with a short filament-like stalk and an incurved sagittate head, alternates with the fertile stamens. These structures are interpreted as staminodes and may have had a nectariferous function. The bases of staminodes and stamens are fused to form a short tube. Pistillate flowers are poorly preserved and only incomplete details of their organisation are available. They are bicarpellate and apparently syncarpous with axile placentation.

The fossil flowers share several characters with extant Hamamelidaceae, but they also show similarities to other rosids as well as to Platanaceae. They cannot be accommodated with certainty in the Hamamelidaceae with the information currently available (Crepet *et al.*, 1992).

Divisestylus (Figure 13.8) includes small actinomorphic flowers described from the Old Crossman locality, New Jersey, USA (Hermsen *et al.*, 2003). Flowers have a pentamerous perianth and androecium, a bicarpellate gynoecium and a superior ovary with a smooth intrastaminal nectary disk. They are apparently bisexual, although Hermsen *et al.* (2003) noted that they could perhaps be functionally unisexual. The androecium is haplostemonous with stamens opposite to the sepals. Stamens have short (*Divisestylus brevistamineus*) or long (*D. longistamineus*) filaments. Anthers, which are known only for *D. brevistamineus*, are basifixed, dithecate and tetrasporangiate. Pollen was not observed inside the anthers, which may be immature. However, one kind of prolate and tricolpate pollen with striate exine sculpture was observed on the surface of many specimens and may have been produced by the same kind of flower. The gynoecium is bilocular with axile placentation and contains more than ten ovules. There are two distinct styles that are free basally, but fused at the apex. In *Divisestylus brevistamineus* the styles are shorter and straight; in *D. longistamineus* they are longer and curved. The fruits are capsules with apical dehiscence.

The figured specimens of *Divisestylus brevistamineus* appear to be preserved in a pre-anthetic state, whereas those of *D. longistamineus* are preserved in the fruiting stage. It is possible that the two fossil species are flowers of a single species preserved at different developmental stages. Elongation of filaments and styles at anthesis is common in extant plants and has been recognised in several fossil flowers, such as *Scandianthus costatus* (Chapter 15) and *Normanthus miraensis* (Chapter 14), where many specimens in different developmental stages are available.

Divisestylus shares many features with extant *Itea* and was placed in the Iteaceae–*Pterostemon* clade in the cladistic analysis of Hermsen *et al.* (2006). Iteaceae and *Pterostemon* are distinguished from the fossil species by their distinct diporate pollen. *Pterostemon* is further distinguished from *Divisestylus* by its pentamerous gynoecium and the lack of a nectary disc.

Lindacarpa pubescens is based on compression material collected from Upper Cretaceous (Coniacian) strata at an exposure on the Linda River, a tributary of the Lena River, in eastern Siberia, Russia (Maslova and Goloneva, 2000b). It comprises a single fragmentary inflorescence head, about 20 mm in diameter, composed of about 30 densely packed pistillate flowers that radiate from a globular inflorescence axis. Individual flowers are about 7 mm long, with two perianth whorls. Outer perianth parts are connate for most of their length, forming a short floral tube. The inner perianth parts are not well understood, but they differ from the outer tepals in details of the epidermis. The gynoecium is bicarpellate and basally syncarpous, and the ovary is semi-inferior. There are no details of ovules or seeds.

The fossil was compared to extant Altingiaceae based on the globular inflorescence, the number of flowers per inflorescence and the bicarpellate gynoecium. Flowers of Altingiaceae, however, lack a perianth and the ovary is inferior. It is possible that *Lindacarpa* represents an early

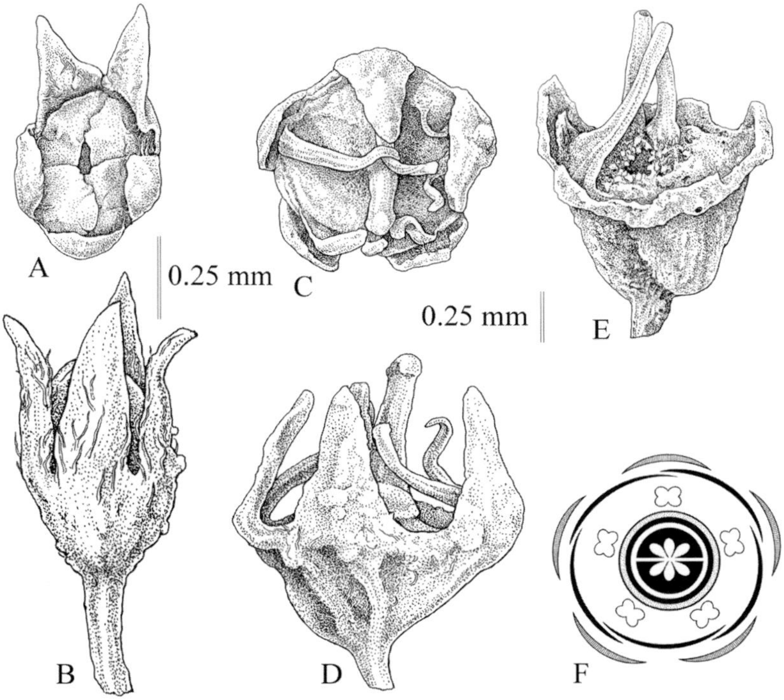

Figure 13.8 Flowers of (A, B) *Divisestylus brevistamineus* and (C–E) *Divisestylus longistamineus* from the Late Cretaceous (Turonian) Old Crossman locality, New Jersey, USA. (A, B) Flower bud in (A) apical and (B) lateral view showing pentamerous calyx and corolla. (C–E) Post-anthetic flowers with corolla abscised, in (C) apical and (D, E) lateral view, showing remains of calyx, filaments and two styles. (F) Floral diagram. (A, C, E) Drawn from SEM image in Hermsen *et al.* (2003); (B, D) from Friis *et al.* (2006a).

member of Altingiaceae, but the information currently available is insufficient for precise systematic assignment.

Microaltingia apocarpela (Figure 13.9) is based on several small, globular inflorescences from the Late Cretaceous (Turonian) Old Crossman locality, New Jersey, USA (Zhou *et al.*, 2001). Inflorescences are pedunculate, about 7 mm in diameter, and consist of 8–12 densely packed pistillate flowers radiating from the inflorescence axis. Individual flowers are sessile and actinomorphic, about 1.2 mm in diameter. The gynoecium is bicarpellate with basally adnate carpels. Styles in young flowers are free and very short and have capitate stigmas with short multicellular papillae. Each carpel contains about ten anatropous ovules/seeds arranged in two rows. The ovary is semi-inferior. Two or three whorls of sterile organs of uncertain nature (termed phyllomes) are borne on top of the hypanthium. Minute tricolpate and reticulate pollen has been observed on the stigmatic surfaces and on the peduncles of the inflorescences.

Zhou *et al.* (2001) assigned *Microaltingia* to the subfamily Altingioideae (Hamamelidaceae). However, in modern Altingiaceae (Altingioideae) there is only a single whorl of sterile organs on the hypanthium and pollen is periporate. *Microaltingia* also shares characters with extant Hamamelidaceae (excluding Altingioidaceae). In the cladistic analyses of Ickert-Bond *et al.* (2007) *Microaltingia* is placed on the stem lineage of the family.

A small unnamed 'capitate inflorescence' of putative altingioid affinity (Figure 13.9) with bicarpellate pistillate flowers was reported from the Late Cretaceous (Late Santonian) Allon locality, Georgia, USA (Herendeen *et al.*, 1999a). The inflorescence is much smaller, about 1.7 mm in diameter, than *Microaltingia* from the Old Crossman locality, and also has more flowers (at least 20) per inflorescence. The phylogenetic relationships of this material remain uncertain pending more detailed description and the discovery of additional specimens (Herendeen *et al.*, 1999a).

Viltyungia eclecta was established based on material collected in the lower part of the Timerdyakhskaya Formation (Cenomanian) exposed at the Tyung River in the Lena–Vilyuy Depression, eastern Siberia, Russia (Maslova and Golovneva, 2000a). The fossil is a fragment of a compound inflorescence with a stout axis bearing sessile, spherical heads consisting of small bisexual and hypogynous flowers crowded on a central receptacle (Maslova and Golovneva, 2000a; Maslova, 2003). The specimen is strongly compressed and details of its organisation and flowers are difficult to establish. The number of flowers per head is estimated to be about 12. The

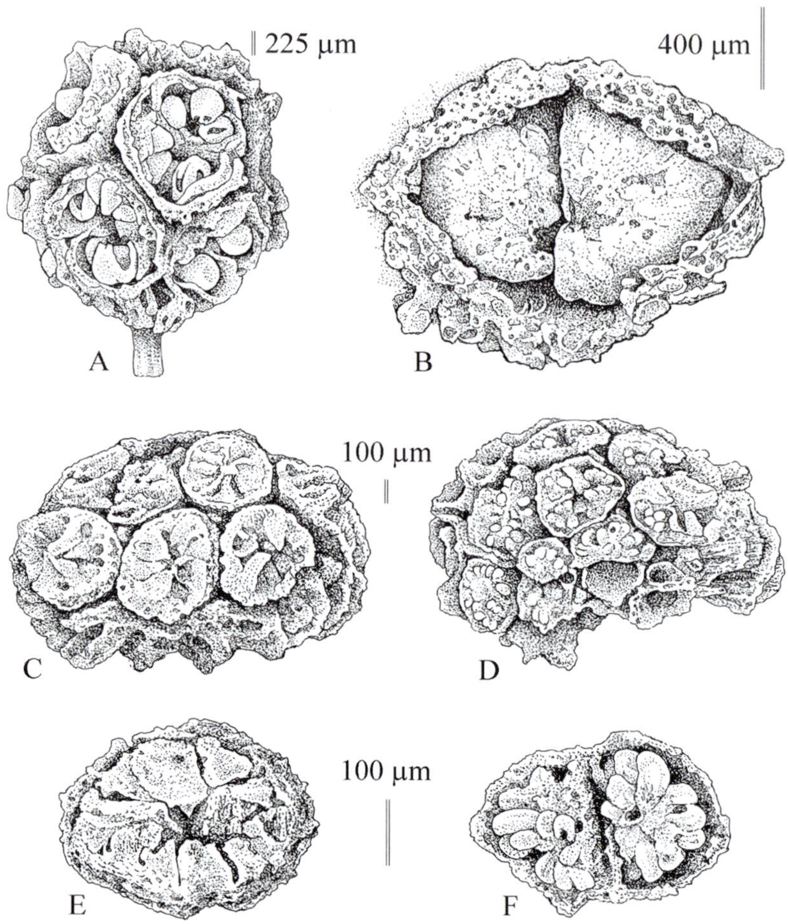

Figure 13.9 Putative altingioid inflorescence/infructescences from the Late Cretaceous of eastern North America. (A, B) *Microaltingia apocarpela* from the Old Crossman locality (Turonian), New Jersey, USA; (A) pistillate head showing fragmentary bicarpellate fruits with seeds and (B) detail of mature gynoecium showing bicarpellate organisation. (C–F) Unnamed 'Capitate inflorescence' from the Allon locality (Late Santonian), Georgia, USA; (C) infructescences showing young fruits and (D) abraded fruits with seeds; (E) details of fruit surface and (F) abraded fruit with two locules. (A, B) Drawn from SEM image in Zhou *et al.* (2001); (C–F) drawn from SEM image in Herendeen *et al.* (1999a).

flowers are about 4 mm long and consist of a massive, bicarpellate gynoecium surrounded by several stamens and staminodes, which in turn are surrounded by a perianth. Anthers are dithecate, tetrasporangiate and open by longitudinal slits. Pollen *in situ* is tricolpate–reticulate. The gynoecium is bilocular and shows imprints of about seven ovules per carpel. In general, the flower and inflorescence structure of *Viltyungia* is closely similar to that of extant Altingiaceae, but flowers in Altingiaceae lack a perianth, are functionally unisexual and have periporate pollen.

13.8.2 Altingiaceae

The Altingiaceae comprise about 16 species in two, or possibly three, genera: *Altingia*, *Liquidambar* and *Semiliquidambar*. The status of *Semiliquidambar* from China is uncertain and the genus is in need of revision (Endress, 1993c). In most earlier classifications the family has been included in the Hamamelidaceae, often as a subfamily Altingioideae (Endress, 1993c), but it is now recognised as a distinct family (APGIII, 2009). The Altingiaceae are trees with alternate, palmately veined, lobed leaves. *Liquidambar* shows a distinct relictual distribution with species in eastern Asia, the eastern Mediterranean region, and southeastern North America and Central America. *Altingia* is restricted to eastern Asia and Malesia. Flowers are small, unisexual, and crowded in dense spherical inflorescences. A perianth is lacking. Staminate flowers have four to ten stamens with bisporangiate anthers that open by longitudinal slits. Pollen is periporate. Pistillate flowers have a bicarpellate, syncarpous ovary with two locules. Each locule contains many ovules.

The extensive fossil record of Altingiaceae during the Cenozoic documents that the family once had a much wider distribution. The extinct genus *Steinhauera* from the Eocene of Europe comprises infructescences very similar to those of extant *Altingia* and the first secure records of *Liquidambar* are also from the Eocene (Mai, 1995). Otherwise most fossil records of the family are from younger floras. *Liquidambar* is particularly common in Miocene and Pliocene floras from the Northern Hemisphere, where there are numerous records of the characteristic palmately lobed leaves and woody infructescences (Mai, 1995; Pigg *et al.*, 2004).

13.8.3 Cercidiphyllaceae

The Cercidiphyllaceae comprise the single genus *Cercidiphyllum* with two species both restricted to China and Japan. Traditionally, the family has been considered close to Trochodendraceae (e.g. Takhtajan, 1969; Endress, 1986b), while cladistic analyses based on morphology suggested a sister group relationship to *Myrothamnus* (Hufford and Crane, 1989; Hufford, 1992). Phylogenetic analyses based on molecular data place the family securely in Saxifragales, but the precise relationships to other taxa of this clade are uncertain. Both species of *Cercidiphyllum* are large trees with opposite to subopposite, cordate and palmately veined leaves with a crenate margin. Flowers are unisexual and naked. The number of stamens in the staminate flowers ranges from 1 to about 13. Pistillate flowers are unicarpellate and borne in clusters of two to eight flowers. Fruits are follicles, each containing numerous small winged seeds (Endress, 1993g).

Leaves of extant *Cercidiphyllum* are broadly similar to leaves of some other eudicots in gross morphology and venation pattern, for example, *Populus* (e.g. Brown, 1939) and *Tetracentron*. Based on leaf impressions alone it may be difficult to reliably assign fossil *Cercidiphyllum*-like leaves to the extant family. There is, however, an extensive fossil record of leaves of this kind associated with *Cercidiphyllum*-like reproductive structures and the reconstructed plants are clearly closely related to the extant family (Brown, 1939; Crane, 1984).

The earliest records of the lineage that includes extant Cercidiphyllaceae are leaves assigned to the form genus *Trochodendroides* from the latest Cretaceous and earliest Cenozoic. These leaves are associated with fruits frequently assigned to the genus *Nyssidium* as well as dispersed winged seeds (Brown, 1939; Crane, 1984). During the Paleocene these extinct *Cercidiphyllum*-like plants were widespread in middle and high latitudes of the Northern Hemisphere. They differed from extant *Cercidiphyllum* in having more robust, branched infructescences and more robust fruits. The fruits also contained a large number of winged seeds with a cellular structure different from that of the seeds of the extant genus. The most completely reconstructed of these *Cercidiphyllum*-like plants is *Joffrea speirsiae* from the Paleocene of Alberta, Canada (Figure 13.10); the occurrence of large numbers of *Joffrea* seedlings demonstrates that this plant was a highly effective early coloniser of bare mud in fluvial settings (Stockey and Crane, 1983; Crane and Stockey, 1985). *Joffrea*-like plants are also known from the Maastrichtian of western North America (Crane and Stockey, 1986). These fossils remain to be studied in detail, but are clearly very similar to extinct *Cercidiphyllum*-like plants from the Early Cenozoic.

Fossil leaves and associated inflorescences that are unequivocally very similar to those of extant *Cercidiphyllum* are first recorded from the Early Oligocene of western North America (Crane and Stockey, 1986). Similar leaves are also known from the Middle Eocene Republic flora (Wehr and Manchester, 1996), but confirmatory evidence from associated fruits and infructescences is not yet available. Reconstruction of one of these Cenozoic cercidiphyllaceous plants has shown that they had staminate flowers with a distinct pentamerous perianth (Kvaček, 2008).

13.8.4 Hamamelidaceae

The Hamamelidaceae comprise about 100 species in 27 genera. Most genera have only one or two species. They are small shrubs or trees, typically with alternate and pinnately veined leaves. Flowers are typically small to medium-sized, bisexual and actinomorphic with a pentamerous, or more rarely tetramerous, perianth and androecium, and a bicarpellate gynoecium. The perianth typically has two whorls or the corolla is lacking. In flowers with a reduced perianth the androecium may be polyandrous (Endress, 1993c). The family shows a widely scattered distribution with several clear Northern Hemisphere disjunctions. The greatest concentration of species is in eastern Asia (Endress, 1993c).

The family has an excellent Cenozoic record based mainly on fossil leaves and seeds (for references see Manchester, 1999). In particular, the characteristic and highly distinctive bony seeds of subfamily Hamamelidoideae, which are

Figure 13.10 Reconstruction of *Joffrea speirsiae*, an extinct member of Cercidiphyllaceae, from the Paleocene of Alberta, Canada, showing short shoots having pistillate inflorescences (above) and infructescences (below). Redrawn from Crane and Stockey (1985).

associated with an explosive seed dispersal mechanism, are well known from the Early Cenozoic, but are not recorded so far from the Cretaceous. The Cretaceous record of Hamamelidaceae is sparse, but the presence of the family in the Late Cretaceous floras is well documented by two fossil flowers, *Allonia decandra* and *Androdecidua endressii*, both from Georgia, USA. Several other flower types from the Late Cretaceous, which may potentially be related to Hamamelidaceae, cannot be included confidently, based on the information currently available (section 13.8.1).

Other possible records of Hamamelidaceae from the Late Cretaceous include dispersed seeds from Central Europe (*Disanthus austriacus*, *Disanthus hercynicus*, *Rhodoleia cretacea*, *Klikovispermum* spp.) (Knobloch and Mai, 1986). In most cases, the absence of information on fruit structure or other floral parts precludes definitive placement of these fossils in Hamamelidaceae, but seeds of *Rhodoleia cretacea* from the Late Cretaceous (Maastrichtian) Walbeck locality, Germany, have been found in clusters that show their original arrangement in the locules. In *R. cretacea* the fruit was clearly bilocular. Seeds are anatropous with the micropyle directed outwards and upwards, and the hilum is a long slit on the sharply angular hilum-raphe side of the seed. The seed coat is thin with a pitted-papillate surface ornamentation, all characters that are also known for seeds of extant *Rhodoleia* (Mai, 2001). In Europe *Rhodoleia* persisted until the Miocene.

Allonia decandra (Figure 13.11), from the Late Cretaceous (Late Santonian) Allon locality, Georgia, USA, is based on a small actinomorphic, staminate flower, together with several flower fragments (Magallón-Puebla *et al.*, 1996). The flower is about 2.75 mm long, pentamerous, with sepals fused for most of their length to form a short floral cup. Remains of the corolla indicate that there were

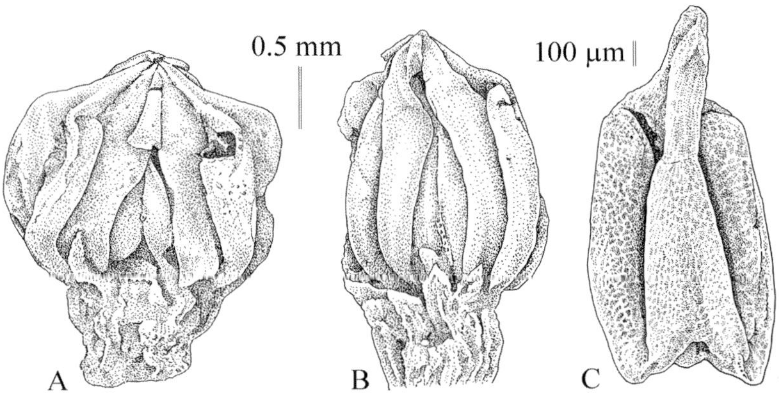

Figure 13.11 *Allonia decandra* from the Late Cretaceous (Late Santonian) Allon locality, Georgia, USA. (A, B) Flower in two different lateral views showing large stamens in two whorls and remains of perianth. (C) Detached anther from outer androecial whorl showing partly dehisced valves and massive apical extension of connective. Drawn from SEM image in Magallón-Puebla *et al.* (1996).

originally five petals. The androecium is composed of ten stamens in two whorls. Each stamen is differentiated into a short filament and a large anther. Anthers are basifixed, dithecate and tetrasporangiate, and have an elongated conical extension of the connective. Dehiscence is valvate with two laterally hinged valves per theca. Pollen *in situ* within the anthers is spherical in outline and tricolpate. Exine sculpture is coarsely reticulate with high, sharply triangular muri and short, densely spaced columellae. Small bodies in the centre of the flower may be the remains of a nectary. No gynoecium was observed.

The fossil flowers are clearly related to extant Hamamelidaceae, and phylogenetic analyses based on morphology place *Allonia* as a sister taxon to the extant genus *Maingaya* in the monophyletic subtribe Loropetalinae of the subfamily Hamamelidoideae. This conclusion is also supported by the cladistic analysis of Hermsen *et al.* (2006). Today the Loropetalinae are restricted to tropical Asia (Magallón-Puebla *et al.*, 1996).

Androdecidua endressii (Figure 13.12) is also from the Late Cretaceous (Late Santonian) Allon locality, Georgia, USA, and is based on a small, actinomorphic, pentamerous flower (Magallón *et al.*, 2001). The flower is preserved at a young stage and is incomplete, but has a well-preserved and very distinctive androecium. The androecium has two whorls of stamens with short filaments and basifixed, tetrasporangiate anthers that open by laterally hinged valves. Apically, the anthers have a distinct triangular extension of the connective. There is no clear evidence of a gynoecium and the arrangement of the perianth is also unclear. Dispersed anthers, similar to those of the flower, have coarsely reticulate and triaperturate pollen. Based on a cladistic analysis the fossil could be placed convincingly in the Hamamelidaceae and the fossil shows a particularly close similarity to flowers in the extant subtribe Loropetalinae (Magallón *et al.*, 2001).

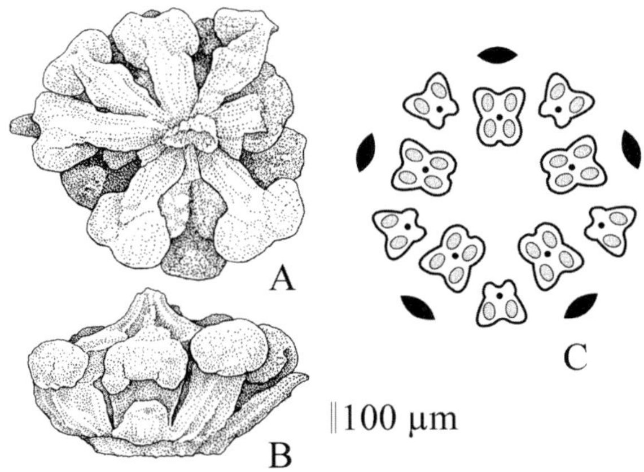

Figure 13.12 *Androdecidua endressii* from the Late Cretaceous (Late Santonian) Allon locality, Georgia, USA. (A) Apical and (B) lateral view of flower. (C) Floral diagram. Drawn from SEM images (A, B) and line drawing (C) in Magallón *et al.* (2001).

13.8.5 Haloragaceae and Crassulaceae

The Haloragaceae and Crassulaceae are resolved as a monophyletic clade that is sister to the Iteaceae–Pterostemonaceae and Grossulariaceae–Saxifragaceae (Fishbein *et al.*, 2001). The Crassulaceae are succulent plants, typically with small pentamerous, isomerous flowers and apocarpous gynoecium. There is no secure fossil record of the family although several leaves and reproductive organs from the Cenozoic have been attributed to the group. Some of these have later been placed in other families (e.g. *Crassulaeceophyllum* was assigned later to Caesalpiniaceae; for references see Friis and Skarby, 1982).

The Haloragaceae comprise nine genera with a total of about 145 species (Mabberley, 1997). The family is almost cosmopolitan in distribution, but most species are concentrated in the Southern Hemisphere. The family includes submerged to emergent aquatics, herbaceous wetland plants, and more rarely shrubs or small trees growing in drier environments (Cronquist, 1981). Flowers are typically small and tetramerous with a syncarpous gynoecium and inferior ovary. Fruits are nut-like or drupaceous, often with remains of perianth preserved at the apex. Flowers appear specialised for wind pollination. Previous classifications that grouped Haloragaceae and Gunneraceae in the same order (Takhtajan, 1969; Cronquist, 1981) are no longer supported based on phylogenetic analyses of molecular data (section 13.3).

The Haloragaceae have an extensive Cenozoic record from both hemispheres, based mainly on dispersed pollen (references in Muller, 1981) as well as their characteristic fruits (e.g. Dorofeev, 1963, 1976; Friis, 1979). However, the only Cretaceous fossils assigned to Haloragaceae are inflorescences/infructescences of *Tarahumara sophiae* and stems of *Obispocaulis myriophylloides* preserved at the Late Cretaceous (Tarahumara Formation, Campanian–Maastrichtian) Huepac Chert locality in northwestern Mexico (Hernández-Castillo and Cevallos-Ferriz, 1999). The reproductive structures are dichasium-like, and composed of a main axis and secondary axes bearing small, unisexual flowers. Pistillate flowers have an inferior ovary formed from four carpels that are fused only at the base. Each carpel has a single, anatropous, pendulous ovule (Hernández-Castillo and Cevallos-Ferriz, 1999). The reproductive organs share several characters with extant Haloragaceae and the authors placed *Tarahumara* in the family. However, they also noted several differences, including the presence in *Tarahumara* of a floral cup and poorly developed syncarpy. Similarly, the fossil stems have distinct aerenchyma and an architecture strongly suggesting an aquatic habit. They are very similar to stems of extant *Myriophyllum*, but they differ from this genus and other extant Haloragaceae in several anatomical features, including their adpressed leaves that almost completely surround the stem (Hernández-Castillo and Cevallos-Ferriz,1999). The placement of *Tarahumara* close to Haloragaceae is supported by the cladistic analysis of Hermsen *et al.* (2006), but whether the fossils should be placed within the family, or in the Haloragaceae stem-group, remains uncertain.

13.8.6 Iteaceae and Pterostemonaceae

The Iteaceae have been regarded as closely related to Grossulariaceae and are sometimes included in that family (e.g. Cronquist, 1981). The family has also sometimes been included in the Pterostemonaceae, but as currently circumscribed the Iteaceae comprise two genera of shrubs and small trees with spirally arranged leaves. The monotypic genus *Choristylis* is from East and South Africa. *Itea* has about 15 species in East Asia and eastern North America. The Pterostemonaceae include a single genus, *Pterostemon*, with two species in tropical Mexico (Mabberley, 1997).

Flowers in Iteaceae and Pterostemonaceae are small, bisexual and semi-inferior with a distinct nectary disc. They are borne in racemose to paniculate inflorescences. Pollen is distinguished from that of other Saxifragales in being diporate. The gynoecium is bicarpellate (*Choristylis*, *Itea*) or with five carpels (*Pterostemon*). The ovary is semi-inferior and matures into a septicidal capsule.

The present distribution of Iteaceae strongly suggests a much wider occurrence in the past. This is supported by the fossil record, which includes *Itea*-like fossils both from Europe and from western North America. In Europe, the Iteaceae are known from the Eocene to Pliocene mostly as dispersed diporate *Itea*-like pollen (Muller, 1981). The fossil flower of *Ademanthemum iteoides* described from Baltic amber (Conwentz, 1886) may also represent Iteaceae. There are several records of fruits, seeds and leaves of Iteaceae from the Cenozoic (e.g. Mai, 1985b; see also Hermsen *et al.*, 2003). In western North America fossil leaves ascribed to *Itea* are known from the Eocene Republic Flora of Washington (Wolfe and Wehr, 1978).

There are no Cretaceous records of Iteaceae, but the small flowers of *Divisestylus* from the Late Cretaceous (Turonian) Old Crossman locality, New Jersey, USA, were described as closely related to the family (section 13.8.1).

13.8.7 Grossulariaceae

The Grossulariaceae are resolved as the sister group to Saxifragaceae. The family includes the single genus, *Ribes*, with about 150 species distributed mainly in temperate regions of the Northern Hemisphere and in the Andes (Mabberley, 1997; Fishbein *et al.*, 2001). Flowers and fruits of *Ribes* show remarkable diversity (Schultheis and Donoghue, 2004), but typically flowers are small, bisexual, and actinomorphic with a tetramerous perianth and androecium, and a bicarpellate, inferior ovary containing many ovules borne on parietal placentae. Pollen is periporate and fruits are berries.

The family has no Cretaceous record. The Cenozoic record is scattered and consists mainly of leaf fossils from North America where *Ribes* extends back to the Eocene (Werker, 1997).

13.8.8 Saxifragaceae

As presently circumscribed Saxifragaceae include about 30 genera and 550 species of herbs, distributed mainly in temperate regions of the Northern Hemisphere, temperate South America and Tasmania, as well as montane areas of the tropics (Fishbein *et al.*, 2001). Flowers are typically small, bisexual, with a free calyx and corolla and two whorls of stamens in an obdiplostemonous arrangement. However, there is considerable variation in the number of parts in the flower, the position of gynoecium and its placentation, as well as pollen morphology and aperture configuration.

There is no record of Saxifragaceae from the Cretaceous and the Cenozoic record is sparse, as is typical for many herbaceous groups. The fossil fruit described as *Saxifragaceaecarpum bifolliculare* from the Miocene of Europe has been reassigned as a species of *Rhodoleia* (Hamamelidaceae) (Mai, 2001).

13.8.9 Paeoniaceae

The Paeoniaceae comprise the single genus *Paeonia* with about 30 species distributed in temperate regions of both the Old and New World (Mabberley, 1997). They are perennial herbs with prominent bisexual flowers. Outer organs of the flower grade from leaves and bracts to showy organs surrounding the polymerous androecium. The gynoecium comprises two to 15 (typically three to five) free carpels.

Relationships of Paeoniaceae have long been problematic and a close relationship has been suggested with a wide array of different groups ranging from Ranunculaceae (e.g. Takhtajan, 1969) to Magnoliaceae (Sawada, 1971) or various groups of Dilleniales (e.g. Cronquist, 1981). Phylogenetic analyses based on molecular sequence data place Paeoniaceae close to the Haloragaceae/Crassulaceae–Saxifragaceae alliance (Pterostemonaceae, Iteaceae, Grossulariaceae, Saxifragaceae), but the precise pattern of relationships at this level remains uncertain (Soltis *et al.*, 2005).

The Paeoniaceae are possibly represented in the Cenozoic by the fruits of *Paeoniaecarpum hungaricum* from the Miocene (Sarmatian) of Hungary (Andreánszky, 1961). The material is preserved as impression and details are not preserved. Andreánszky noted that the fossils could also be related to Magnoliaceae (where he included also Annonaceae). The Paeoniaceae are unknown from the Cretaceous.

14

Fossils of core eudicots: rosids

Rosids are a very large and heterogeneous group of eudicots that exhibit great diversity in both reproductive and vegetative morphology. The group is recognised based on molecular data, but support is currently weak and there are no unique morphological or anatomical features that define the rosids as a group. Among the shared characters noted by Soltis *et al.* (2005) is the presence of a reticulate pollen exine, but this is a common feature in many other groups of angiosperms. Other potential defining characters, such as obdiplostemony, mucilage cells in flowers and/or leaves, simple perforations of vessel end-walls, and alternating inter-vessel pitting, are also found occasionally in other groups and their value as synapomorphies of rosids remains to be determined.

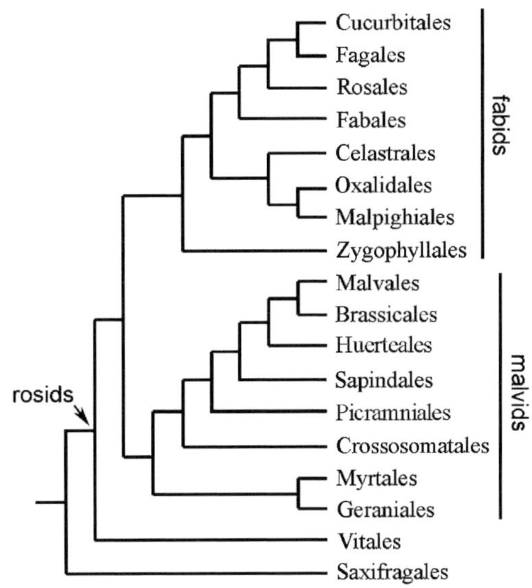

Figure 14.1 Summary of phylogenetic relationships among extant rosids, based on APGIII (2009).

14.1 CLASSIFICATION OF ROSIDS

Core rosids include Vitales, which are sister to two large clades: fabids *sensu* Judd and Olmstead (2004; Eurosids I sensu APGII, 2003) and malvids *sensu* Judd and Olmstead (2004; Eurosids II sensu APGII, 2003) (Figure 14.1). Fabids include Zygophyllales, which are sister to the nitrogen-fixing clade, comprising Cucurbitales, Fabales, Fagales and Rosales, and the COM clade, comprising Celastrales, Oxalidales and Malpighiales. Molecular support for these groupings is not strong (Endress and Matthews, 2006a). Malvids include Geraniales + Myrtales, which are sister to the remaining orders (Crossosomatales, Picramniales, Sapindales, Huerteales, Brassicales, Malvales) (APGIII, 2009).

Studies of floral structure in some lineages of rosids show good correlation between groups defined by molecular data and reproductive features (Endress and Matthews, 2006a), but in other cases there are marked discrepancies. For example, Endress and Matthews (2006a) noted that in floral features members of the COM clade are more similar to malvids than they are to members of the nitrogen-fixing clade.

14.2 FOSSIL EVIDENCE OF ROSIDS

The rosid clade includes about 70 000 species (Magallón *et al.*, 1999). The sheer size, diversity and complexity of the group presents formidable challenges for diagnosing rosid subclades based on morphological characters that are likely to be preserved and recognised in fossil material. In particular, attempting to assign fossil floral material that displays generalised characters to orders and families of rosids is very difficult. Even flowers with strongly specialised floral morphology may be difficult to place because the distinctive features occur in groups that are apparently not closely related. One example of such apparent convergence in floral features among distantly related forms was described by Matthews *et al.* (2001), who showed surprising similarity between some flowers of the Anisophylleaceae (Cucurbitales) and Cunoniaceae (Oxidales)

(Figure 14.2). Both families have trimerous flowers as well as tetra- and pentamerous forms, and the many morphological, histological and anatomical similarities include: whorled and isomerous organisation; valvate sepals; digitate and lobed petals; obdiplostemonous androecium with incurved filaments in the bud; similar, sagittate, dorsifixed and introrse anthers; and an inferior ovary with free styles. *Anisophyllea* (Anisophylleaceae) and *Ceratopetalum* (Cunoniaceae) also share an unusual slit-shaped micropyle, unusual dicolpate pollen and a nectary disc formed by protrusions between and behind each stamen.

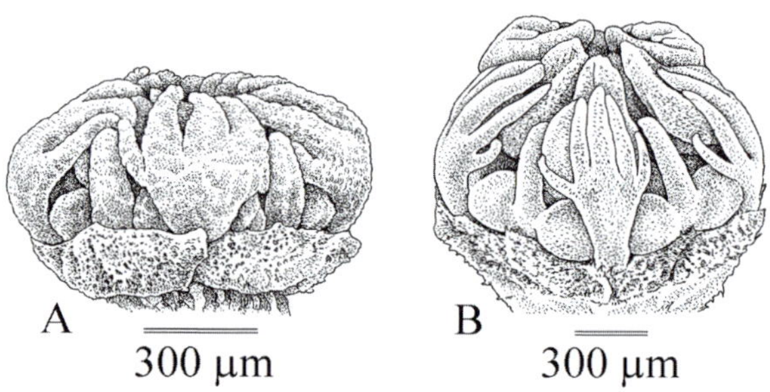

Figure 14.2 Flower buds of (A) extant *Anisophyllea disticha*, Anisophylleaceae and (B) extant *Ceratopetalum gummiferum*, Cunoniaceae, with sepals removed, showing strikingly similar organisation and floral morphology. Note lobed petals and nectary lobes protruding between the stamen filaments. Drawn from SEM images in Matthews *et al.* (2001).

The difficulties of accurately placing even well-understood fossils in the current scheme of rosid phylogeny mean that the present picture of rosid evolution based on Cretaceous fossils is still emerging and somewhat preliminary. However, for some groups (e.g. Fagales) the Cretaceous fossil record is already extensive and informative. Relatively well-known groups such as the Fagales help to provide a solid source of data against which less well documented patterns in the history of other groups can be compared.

The palaeobotanical information available for rosids will increase as more fossil material is investigated and as the morphological synapomorphies of extant groups are clarified by further comparative studies. In this chapter we focus on those rosid clades that have a Cretaceous record. We treat briefly other lineages that are not known from the Cretaceous, but that are well documented in the Cenozoic.

The earliest flowers known to have a distinct calyx and corolla, and that seem likely to have rosid affinities, are from the mid-Cretaceous (latest Albian – earliest Cenomanian) Dakota Formation, Nebraska, USA (Basinger and Dilcher, 1984). These fossil flowers from the Rose Creek locality (Upchurch and Dilcher, 1990) have not been formally named and are referred to as the 'Rose Creek flower'. The flowers are 2–3 cm in diameter, pentamerous and isomerous, apparently with four whorls of floral parts (Figure 14.3). Sepals and petals are free, sepals are robust and persistent, and petals are thin with a narrow base. The androecium is described as haplostemonous with five stamens opposite the petals, but this is not completely clear from the

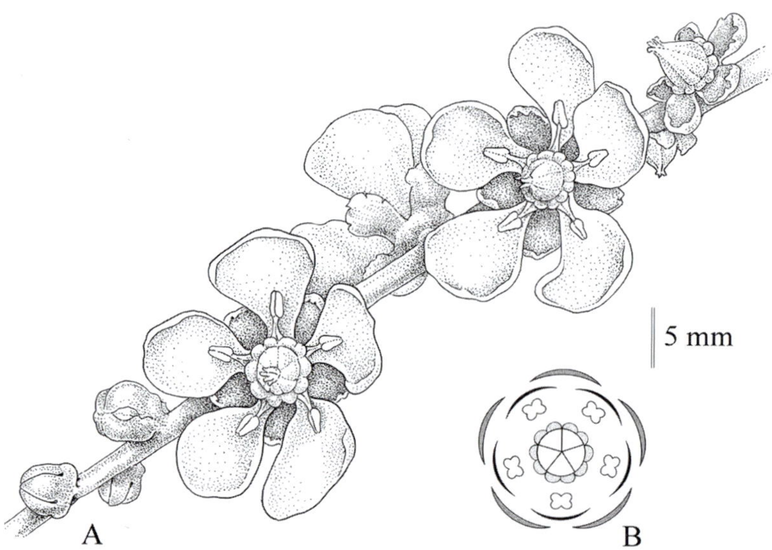

Figure 14.3 Reconstruction (A) and floral diagram (B) of the mid-Cretaceous 'Rose Creek flower' from the Rose Creek locality, Nebraska, USA. (A) Redrawn from Dilcher and Crane (1984b).

illustrations in Basinger and Dilcher (1984). Stamens have well-differentiated filaments and anthers. Anthers are dithecate and tetrasporangiate. Pollen is small, triaperturate, apparently tricolporate, and psilate. The gynoecium is syncarpous, consisting of five carpels with free styles and superior ovary. A distinct nectary disc is described as inserted between the androecium and gynoecium, but this requires clarification.

The 'Rose Creek flower' has been compared to various extant rosids and shows a particularly close similarity to certain extant Rosales and Celastrales (Basinger and Dilcher, 1984; Calvillo-Canadell and Cevallos-Ferriz, 2007), but it has not yet been described in detail and its systematic position requires further critical assessment. The Rose Creek flower may be an early member of Rosales, but the possibility that it belongs to an extinct lineage at a higher level among rosids cannot be excluded.

14.3 VITALES

The Vitales include a single family, Vitaceae (including *Leea*). In several molecular studies the Vitaceae are placed close to the base of core eudicots as the possible sister group to rosids (Soltis *et al.*, 2000a; Judd *et al.*, 2002; APGIII, 2009), but other studies indicate that the family should perhaps be excluded from the rosids (Oxelman *et al.*, 2004) and the phylogenetic position of the order is not fully secure.

The Vitaceae comprise about 12 genera and 770 species of woody climbers, or more rarely herbs, shrubs or small trees, distributed mainly in tropical and subtropical regions (Cronquist, 1981). Flowers are small, bisexual or unisexual, actinomorphic and pentamerous with a bicarpellate gynoecium, superior ovary and distinct nectary disc. Fruits are typically berries with few seeds. Seeds of Vitaceae are very characteristic and easy to recognise. They have a distinctive shape, a hard seed coat, and two grooves delimiting the distinct chalazal area.

Seeds of Vitaceae are extremely common in Cenozoic floras from the Northern Hemisphere, but there are no reports from the Cretaceous. The earliest reliable record of the family is provided by seeds attributed to *Vitis* from the Paleocene. Other extant genera including *Ampelopsis, Parthenocissus* and *Tetrastigma* also have their first occurrence in the Early Cenozoic (e.g. Mai, 1995). Fruits of *Ievlevia dorofeevii* from the Albian of northeastern Siberia, cited by Collinson *et al.* (1993) as a putative early record of Vitaceae, were compared by Samylina (1976) to fruits of extant *Ceratophyllum*. They are spiny and very similar to a complex of spiny fruits known from the Early Cretaceous of East Asia (Chapter 9). However, *Ievlevia* is unlike seeds of Vitaceae. Cretaceous leaves assigned to such genera as *Cissites, Cissophyllum* and *Vitiphyllum* are most commonly compared to members of early-diverging eudicots.

14.4 FABIDS (EUROSIDS I)

The fabid clade includes two families in the Zygophyllales and two major clades: the COM clade (Celastrales, Oxalidales, Malpighiales) and the nitrogen-fixing clade, which comprises Fabales, Rosales, Cucurbitales and Fagales.

Both the COM clade and the nitrogen-fixing clade have a rich fossil record extending back to the Late Cretaceous. In particular, the Fagales are very well represented in the Late Cretaceous by many extinct forms included in the Normapolles complex. In contrast, the Zygophyllales have no confirmed fossil record. Several leaf impressions from the Eocene of North America and Pliocene of South America have been assigned to Zygophyllaceae, but all are in need of reinvestigation (Porter, 1974).

14.5 THE COM CLADE

Molecular support for the COM clade is weak, but there is moderate morphological support for recognition of the group (Endress and Matthews, 2006a). Among rosids, the COM clade may be distinguished by a relatively thin nucellus and the presence of an integumentary endothelium; many members of the clade also have arillate seeds, but this feature also occurs in other rosids as well as in many other angiosperms (Endress and Matthews, 2006a).

14.5.1 Celastrales

In the APGIII classification (2009) the Celastrales comprise the expanded family Celastraceae (including Brexiaceae, Lepuropetalaceae, Hippocrateaceae, Parnassiaceae, Siphonodontaceae, Stackhousiaceae) and Lepidobotryaceae (previously included in Oxalidaceae). There is no support from molecular data for the inclusion of other families traditionally assigned to the Celastrales, such as Aquifoliaceae and Icacinaceae. Most Celastraceae, and all Lepidobotryaceae, are woody plants. An extensive study of floral structure in Celastrales by Matthews and Endress (2005a) shows considerable diversity in floral features both within and among the three families. Flowers of Celastraceae and

Parnassiaceae often have partly fringed sepals and petals, a tricarpellate, completely syncarpous gynoecium, and capsular, loculicidal fruits. A conspicuous nectary disc is present in Celastraceae and Lepidobotryaceae.

Beginning in the Eocene, several fossil leaves and seeds have been assigned to Celastraceae from Europe and North America (Collinson *et al.*, 1993). *Wuyunanthus hexapetalus* is a compressed flower with a hexamerous organisation described from the Paleocene of the Heilongjiang Province, northeastern China, and assigned to Celastraceae (Wang *et al.*, 2001b). However, the hexamerous organisation clearly distinguishes this fossil from extant Celastraceae, and its systematic position is currently uncertain.

14.5.2 Oxalidales

The seven families (Brunelliaceae, Cephalotaceae, Connaraceae, Cunoniaceae, Elaeocarpaceae, Huaceae, Oxalidaceae) now included in Oxalidales were not recognised as closely related in previous classifications based on morphology. The order includes approximately 2000 species, mainly of shrubs and trees that are distributed predominantly in the Southern Hemisphere. Morphological and anatomical support for the clade as a whole is weak, but there is strong support from floral morphology for several of the newly recognised subclades (Matthews and Endress, 2006).

The Elaeocarpaceae are known from only a few Cenozoic floras in the Northern Hemisphere (Mai, 1995), but have a much more extensive record from the Southern Hemisphere based on leaves and pollen, which are known in Australia from the Eocene onwards (O'Dowd *et al.*, 1991; Macphail *et al.*, 1994).

The Cunoniaceae are also well represented in fossil floras from the Southern Hemisphere by a rich diversity of leaves and pollen, as well as several flowers and inflorescences (Carpenter and Buchanan, 1993; Blackburn and Sluiter, 1994; Barnes and Hill, 1999; Barnes and Jordan, 2000). The earliest records are from the Late Cretaceous (Maastrichtian), where the family is recognised in palynofloras from Australia and New Zealand (Macphail *et al.*, 1994). Fossil flowers from the Cenozoic of Australia have been assigned to the extant genus *Ceratopetalum* as two fossil species: *Ceratopetalum wilkinsonii* from the Late Eocene – Early Oligocene, and *C. priscum* from the Middle Miocene (Holmes and Holmes, 1992). There is also an additional undescribed species of flower from the Middle Eocene (Christophel, 1994).

From Cenozoic floras in Europe the Cunoniaceae are represented by fossil wood assigned to the genus *Cunonioxylon*, which is known from the Eocene and Oligocene (Hofmann, 1952; Gottwald, 1992). The Cretaceous record of Cunoniaceae in Europe rests solely on flowers of *Platydiscus peltatus*.

Platydiscus peltatus (Figure 14.4) is based on charcoalified flowers from the Late Cretaceous (Late Santonian – Early Campanian) Åsen locality, Sweden. The fossils are preserved at different developmental stages from small buds to mature, post-anthetic flowers (Schönenberger *et al.*, 2001a). Mature flowers are about 3–4 mm long. Flowers are actinomorphic, tetramerous and isomerous with five whorls of floral parts. The calyx consists of four persistent sepals that are broadly attached to the floral cup, and the corolla consists of four trullate petals with a narrow base. The androecium has two whorls of four stamens in an obdiplostemonous arrangement. Filaments are short in the flower buds, but elongate at anthesis. Pollen is reticulate and triaperturate, apparently tricolporate. The ovary is semi-inferior with the lowermost part four-locular. Above this level the carpels are fused laterally, but not in the centre. Above the floral cup the carpels are completely free. A central canal extends down between the carpels to the fused region. Placentation is axile; the placentae are enlarged and involute and bear many small, reticulate, anatropous ovules. There is a very prominent lobed nectary disc that completely surrounds the bases of the filaments and fills the space between the ovary and the perianth. The flowers are covered with a dense indumentum of stiff hairs and peltate trichomes.

Platydiscus peltatus shares many floral features with members of two apparently unrelated families, the Anisophylleaceae (Cucurbitales) and Cunoniaceae (section 14.2). However, it can be placed unambiguously in the Cunoniaceae based on the involute placenta and the presence of numerous, anatropous ovules (Schönenberger *et al.*, 2001a).

14.5.3 Malpighiales

The Malpighiales comprise about 16 000 species in 35 families including such ecologically and/or economically important groups as Euphorbiaceae, Hypericaceae, Linaceae, Passifloraceae, Rhizophoraceae, Salicaceae and Violaceae. Relationships within Malpighiales are still poorly resolved and there is only weak morphological support for the entire clade. As for several other new constellations in the angiosperm phylogenetic tree, many of the families now included in Malpighiales were not previously recognised as closely related. There are also some lineages that were

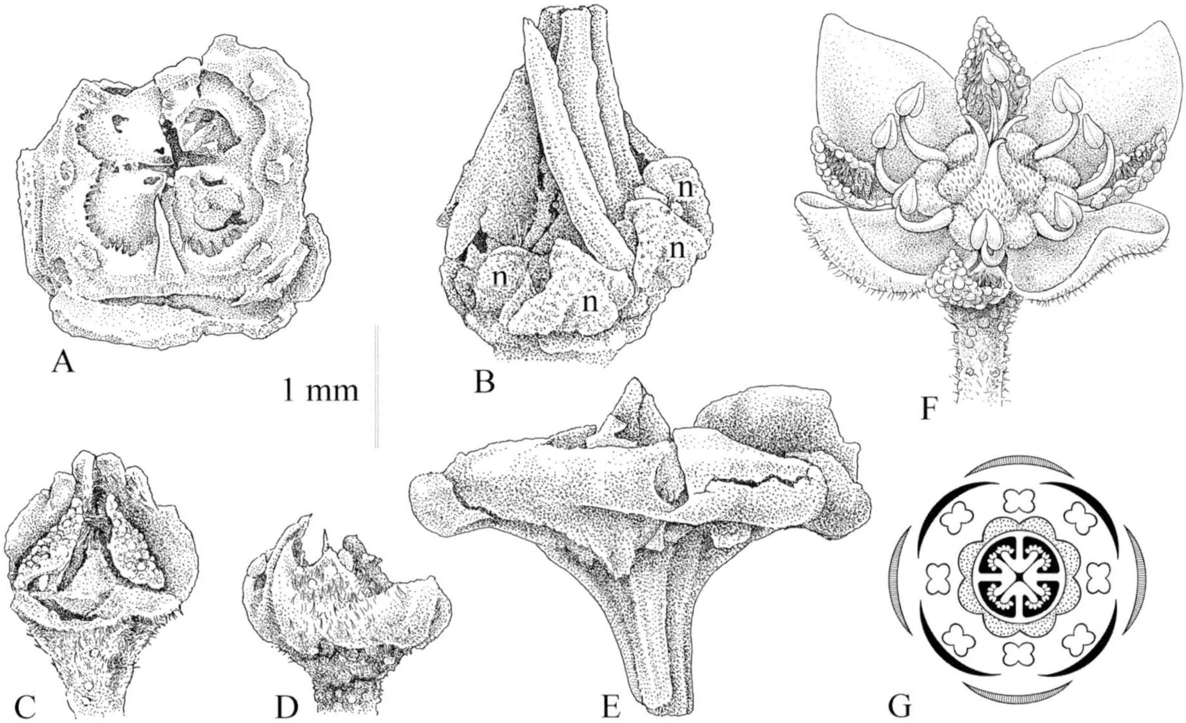

Figure 14.4 *Platydiscus peltatus* from the Late Cretaceous (Late Santonian – Early Campanian) Åsen locality, Sweden. (A) Abraded specimen in apical view showing square outline, tetramerous organisation and remains of stamens; note the prominent nectary disc surrounding and between the stamens. (B) Fragmentary specimen with remains of nectary disc (n), one long filament, and the free apical part of the gynoecium. (C) Flower bud in lateral view with two sepals broken off, showing the petals covered by peltate trichomes on their outer surface. (D) Floral bud in lateral view, showing the broad attachment of a sepal. (E) Anthetic flower in lateral view, showing stalk and sepals. (F) Reconstruction of flower showing four large sepals, four small petals, eight stamens, nectary disc and the gynoecium of four carpels. (G) Floral diagram. (A–F) From Friis *et al.* (2006a); (G) redrawn from Schönenberger *et al.* (2001a).

previous thought to be closely related (Euphorbiaceae, Phyllanthaceae, Picrodendraceae), but that are now separated. Evidence from comparative floral morphology suggests that further changes in the topology of the malpighialean tree are also probable (Sutter *et al.*, 2006). The group is large and heterogeneous, both in floral morphology and habit. It includes small herbs, shrubs, succulents, lianas, and large trees and shows an exceptionally wide ecological range, with mangroves (Rhizophoraceae) at one extreme and cactus-like succulents (some Euphorbiaceae) at the other.

Flowers of *Paleoclusia* (Clusiaceae) provide the only secure record of Malpighiales from the Cretaceous, but the order is well represented by diverse fossil material in the Early Cenozoic. An interesting early record of the order is *Rosenkrantzia picrodendroides* from the earliest Paleocene (Danian) of Nuussuaq, western Greenland, which is based on fossil fruits very similar to those of extant *Picrodendron*, Picrodendraceae (Koch, 1972). Achariaceae may also be present in the Early Cenozoic with fruits assigned to *Oncoba* (previously placed within Flacourtiaceae) known from southern England (Collinson, 1983a). However, the material is not well preserved. It needs to be re-evaluated structurally and also within the new framework for the phylogeny of Malpighiales.

Of the other malpighialean lineages present in the Cenozoic, the Salicaceae have a particularly extensive fossil record with abundant occurrences of fossil leaves as well as some fruits and seeds, in some cases with leaves, flowers and fruits attached (Manchester *et al.*, 1986; Boucher *et al.*, 2003). Other malpighialean lineages reported from the Cenozoic include small foveolate seeds of *Passiflora*

(Passifloraceae) from the Miocene of Europe, and small curved, reticulate to transversely ribbed seeds of *Elatine* (Elatinaceae) from the Pliocene of Europe (Mai, 1995). There are also several reports of fruits assigned to Euphorbiaceae with the oldest records from the Early Eocene London Clay flora of southern England (Collinson, 1983a). The Euphorbiaceae are also present in Early Cenozoic fossil floras from Australia (Macphail *et al.*, 1994), where they are recorded from the Oligocene and onwards (Martin, 1994). Leaves of putative Malpighiaceae (*Banisteriophyllum* and *Malpighiastrum*) originally described as of Cretaceous age (von Ettingshausen, 1895) are now known to be from the Early Cenozoic (Jones, 1926; Rigby and Playford, 1988). The only malpighialean floral structures described from the Early Cenozoic are flowers of *Eoglandulosa warmanensis* from the Eocene of southeastern North America, which is assigned to Malpighiaceae (Taylor and Crepet, 1987).

The fossil history of Rhizophoraceae is of particular interest for understanding the evolution of mangroves. Currently there is no Cretaceous record, but the family is already widespread and diverse by the Eocene based on studies of the dispersed pollen record (Muller and Caratini, 1977). The earliest pollen record is from the Eocene of France based on grains very similar to *Rhizophora* and *Kandelia* (Gruas-Cavagnetto, 1987), as well as the first *Bruguiera*-type pollen (Gruas-Cavagnetto *et al.*, 1988). Grains similar to those of extant *Rhizophora, Cecropia* and *Bruguiera* occur abundantly in dispersed palynofloras from the Cenozoic at low palaeolatitudes. From the Oligocene onwards there are records from Venezuela, Mexico and the Caribbean area, as well as from West Africa (Nigeria) and the Malesian region (Muller and Caratini, 1977). There is also a sparse, but convincing, macrofossil record of Rhizophoraceae from the Eocene of Europe. Two of the best-documented taxa are *Ceriops cantiensis* and *Palaeobruguiera elongata*, both based on distinctive pyritised hypocotyls. They were first described by Chandler from the Eocene of southern England (Chandler, 1964). Subsequent studies have revealed astonishing anatomical details that clearly link these fossils to extant *Ceriops* and *Bruguiera* respectively (Wilkinson, 1981).

Paleoclusia chevalieri (Figure 14.5) comprises small actinomorphic flowers, with many floral details preserved, from the Late Cretaceous (Turonian) Old Crossman locality, New Jersey, USA (Crepet and Nixon, 1998a). The perianth consists of five free sepals and five free petals. The androecium consists of five fascicles of stamens. Filaments vary in length and are sometimes branched. Anthers are dorsifixed, tetrasporangiate and dithecate, and dehisce by longitudinal slits. Pollen was not observed in the stamens, but the anther locules are sometimes filled with an amorphous substance and it was suggested that the stamens were perhaps non-functional. The gynoecium is syncarpous and five-locular, with a prominent five-lobed stigma borne on a very short style. Placentation is axile and there are many anatropous ovules in each locule. Seeds are small and apparently arillate with a distinct reticulate surface pattern.

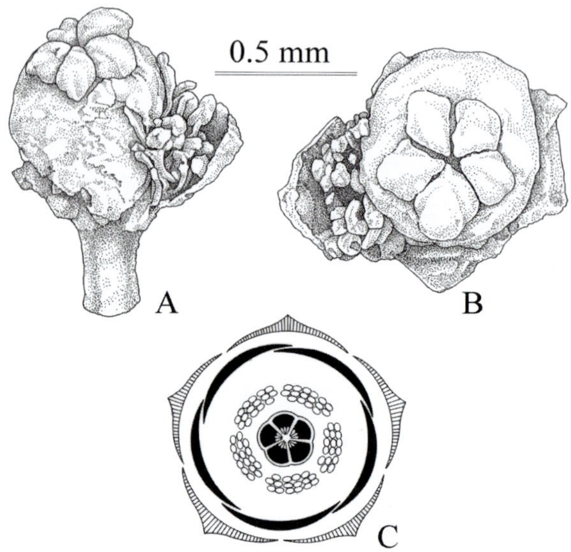

Figure 14.5 *Paleoclusia chevalieri* from the Late Cretaceous (Turonian) Old Crossman locality, New Jersey, USA. (A, B) Flower in (A) lateral and (B) apical view showing a petal with its corresponding cluster of stamens, and the ovary with five stigmatic lobes. (C) Floral diagram. Drawn from SEM images and line drawing in Crepet and Nixon (1998a).

Phylogenetic analysis of the fossil in the context of similar extant taxa places *Paleoclusia chevalieri* in the subfamily Clusioideae of the Clusiaceae (Crepet and Nixon, 1998a). Other fossil flowers from the same locality (currently unpublished) are also closely similar to members of the Clusiaceae, documenting that this family was well established in the Late Cretaceous (Crepet and Nixon, 1998a).

14.6 THE NITROGEN-FIXING CLADE

The four orders of the nitrogen-fixing clade (Cucurbitales, Fabales, Fagales, Rosales) are all species-rich and morphologically very diverse (Figure 14.6). In traditional classifications these groups have not been considered to be closely related. The name of the clade refers to the

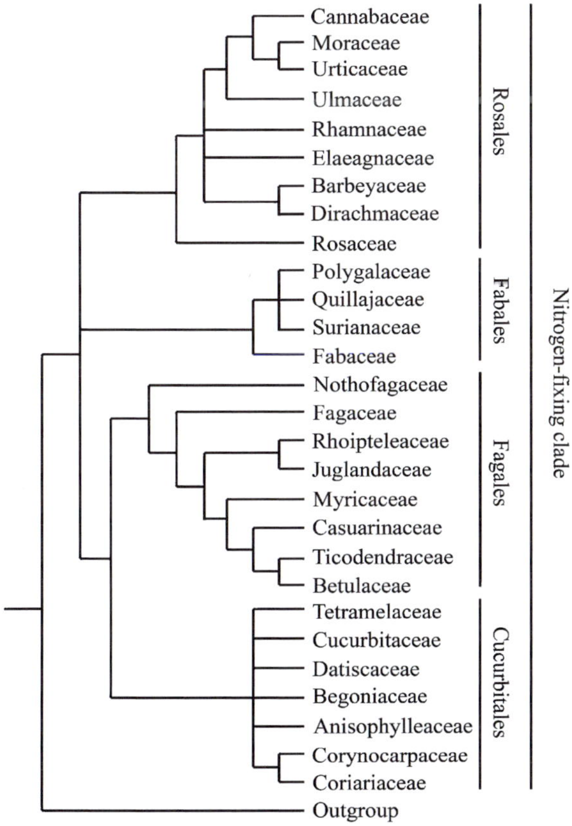

Figure 14.6 Summary of phylogenetic relationships of extant lineages in the nitrogen-fixing clade. Note that the pattern of phylogenetic relationships among families of Fagales differs from that in Figure 14.7 in placing Myricaceae as sister group to Betulaceae, Casuarinaceae and Ticodendraceae, rather than as sister group to Rhoipteleaceae and Juglandaceae. Based on Soltis *et al.* (2005).

symbiotic association with nitrogen-fixing bacteria that occurs in many members of this group, most notably in the subclass Faboideae of the Fabaceae. However, not all members of the clade have this association, nor is the association always with the same kind of bacterium, and the evolution of nitrogen fixation is not yet well understood (e.g. Soltis *et al.*, 2005). Molecular support for the nitrogen-fixing clade is weak; support from morphology is moderate (Endress and Matthews, 2006a).

Flowers in the nitrogen-fixing clade are often apetalous; this condition is most probably related to the common occurrence of wind pollination in some members of the group. Many members of this group also have a single or few ovules per carpel or gynoecium, another character that may be linked to wind pollination (Endress and Matthews, 2006a). Ovules are often unitegmic.

The nitrogen-fixing clade has an extensive fossil record, particularly from the Cenozoic, but also from the Late Cretaceous. Most Late Cretaceous core eudicot fossils that have been described so far either belong to Fagales and Rosales among rosids or to Ericales among asterids (Chapter 15). Within the two rosid orders many of the fossils are difficult to place at the level of extant families or genera, and many, for example those included in the Normapolles complex (section 14.6.6), probably represent once very diverse, but now extinct, stem-lineages, the relationships of which are not yet fully resolved.

14.6.1 Fabales

As currently circumscribed the Fabales include four families: Fabaceae, Polygalaceae, Quillajaceae and Surianaceae. The Fabaceae (Leguminosae) are the largest family of rosids with almost 20 000 species. The family has a worldwide distribution and includes herbs, shrubs, lianas and trees. It is a well-defined group divided into four subfamilies: Cercideae, Caesalpinoideae, Mimosoideae and Faboideae.

The Fabaceae have an extensive fossil record (Herendeen *et al.*, 1992). The family is already diverse and widespread in the Paleocene–Eocene with reports of dispersed pollen from Africa (Mali and Nigeria), North America (Texas), and Europe (southern England), but the most extensive radiation appears to have taken place in the mid-Eocene, by which time the group is well represented by a range of unequivocal legume macrofossils (Herendeen *et al.*, 1992). There are surprisingly few records of Fabaceae from the Cretaceous. Pollen records from the Maastrichtian of North America (Canada), Asia (Siberia) and South America (Colombia), and reports of wood from the Turonian–Santonian of Africa (Sudan) are equivocal and require further examination (Herendeen *et al.*, 1992).

Other lineages of Fabales have a more scattered fossil record, but there are several reports of Polygalaceae. Stephanocolpate pollen assigned to the form genus *Polycolporopollenites* from the Early to Late Cenozoic of the Southern Hemisphere (New Zealand and the Falkland Islands) is now thought to represent Polygalaceae (Macphail and Cantrill, 2006; Raine *et al.*, 2006) and there are several other pollen records from the Southern Hemisphere that have been linked to the family (Forest *et al.*, 2007). From the Northern Hemisphere Crane *et al.* (1990) noted similarities between 'Nonschizocarpic

Samara' fossils from the Paleocene of North Dakota, USA, and fruits of extant *Securidaca*, and the fossils were later assigned to a new fossil genus and species in the family, *Paleosecuridaca curtisii* (Pigg *et al.*, 2008). If correctly identified, wood described as *Suriana inordinate* from the Eocene of Wyoming, USA (Kruse, 1954), is the only record of the Surianaceae and indicates a wider distribution of this family in the past.

14.6.2 Rosales

The Rosales are strongly supported as a monophyletic group by phylogenetic analyses of molecular data, but support from morphology is weak (Soltis *et al.*, 2005). The order now includes nine families (APGIII, 2009) with Rosaceae as the sister to two possible subclades. One well-defined subclade comprises Cannabaceae (including Celtidaceae), Moraceae, Ulmaceae and Urticaceae, which in previous classifications were placed in the subclass Hamamelididae (e.g. Takhtajan, 1969). The other subclade, which may include Barbeyaceae, Dirachmaceae, Elaeagnaceae and Rhamnaceae, is not strongly supported.

Rosaceae. The Rosaceae are a large family that includes about 3000 species of herbs, shrubs and small trees with a worldwide distribution. There is a concentration of species diversity in temperate and subtropical regions of the Northern Hemisphere. The family is varied in floral morphology, but many Rosaceae have bisexual flowers with a well-developed calyx and corolla, a semi-inferior to inferior ovary, and a distinct hypanthium. There are usually two to five united carpels, but sometimes carpels are numerous and free. The Rosaceae also have a great diversity of fruit types, which range from follicles, nuts and drupes to pomes (Cronquist, 1981).

There is no unequivocal Cretaceous record of Rosaceae. Leaves from the earliest Late Cretaceous from the Balkan Peninsula that have been compared to extant *Lyonothamnus* are in need of re-examination (Mai, 1995). However, the family has an extensive fossil record from the Eocene onwards (Manchester, 1999). By the Middle Eocene the family is diverse and widespread with reliable occurrences in North America and Europe. In Europe, the presence of the subfamily Rosoideae is well documented from the Eocene onwards by numerous records of the characteristic endocarps of *Rubus* and several other genera. Two other subfamilies, the Malloideae and the Spiraeoideae, are documented from the Oligocene onwards (Mai, 1995). A recent review documents the diversity of Rosaceae from the Eocene of North America (DeVore and Pigg, 2007). The fossil flower *Paleorosa similkameenensis* from the Middle Eocene Princeton Chert, Canada (Basinger, 1976; Cevallos-Ferriz *et al.*, 1993), is probably the most informative of the early fossil Rosaceae. It is permineralised with excellent cellular preservation and pollen *in situ*. Based on combined floral and pollen features the fossil flowers show a suite of characters intermediate between those of subfamilies Spiraeoideae and Malvoideae (Cevallos-Ferriz *et al.*, 1993).

Cannabaceae. The Cannabaceae, with Celtidoideae now included, is a small family of 11 genera and about 200 species of wind-pollinated herbs, shrubs or trees with simple bisexual or unisexual apetalous flowers. The gynoecium is syncarpous and bicarpellate with a single anatropous ovule. Fruits are nuts, or in Celtidoideae mostly drupes.

The Cannabaceae are well documented in the fossil record with the earliest unequivocal records from the latest Cretaceous (Maastrichtian) of Germany. Endocarps of two extant genera, *Aphananthe* and *Gironniera*, have been reported from the Walbeck (*Aphananthe cretacea*) and Eisleben (*Gironniera gonnensis*) localities, Germany (Knobloch and Mai, 1986). There are also several leaf records of putative Celtidoideae from the Late Cretaceous, such as *Celtidophyllum cretaceum* and *Celtidophyllum praeaustrale* from Moravia, Czech Republic (Krasser, 1896), but none of these leaves have been restudied with modern techniques and their systematic position is unconfirmed.

From the Cenozoic, two fossil flowers with possible affinity to Celtidoideae have been reported based on compression fossils. *Eoceltis dilcheri*, from the Middle Eocene of Texas, USA, comprises unisexual, staminate flowers with four tepals, numerous stamens and *in situ* triporate, scabrate pollen (Zavada and Crepet, 1981). However, several differences between these fossils and flowers of extant Celtidoideae make the relationships of *Eoceltis* uncertain: the fossil flowers are larger than is normal in extant taxa, and also have more, densely crowded, stamens. Pollen is also larger with a distinctly different ultrastructure. The systematic position of the other fossil flower assigned to Celtidoideae, *Celtoidanthus pseudorobustus*, from the Miocene of Germany (Weyland *et al.*, 1958), is also uncertain. This flower is staminate with densely crowded stamens. Anthers are almost sessile and *in situ* pollen is triporate. Similar multistaminate flowers are unknown for extant Celtidoideae or other Cannabaceae. Other Cenozoic fossils of Celtidoideae provide unequivocal records of the family and include abundant leaves, endocarps and pollen from Europe, North America and Asia (Manchester, 1987b).

Moraceae. The Moraceae include about 1000 species, mainly of trees, shrubs and lianas that are widely distributed in tropical and subtropical regions. They are characterised by the production of milky sap in stems and leaves. Flowers are small, unisexual, apetalous, and borne in dense inflorescences. In several genera the inflorescences axis is expanded and invaginated, completely enclosing the flowers as in *Ficus*. Pollen is oblate to spheroidal, porate, with a tectate–scabrate pollen wall. The ovary is superior to inferior and consists of one to three carpels. The wide diversity of fruit types in Moraceae range from small nuts to drupes, sometimes with the fleshy, expanded inflorescence axis functioning in fruit dispersal (Cronquist, 1981). Several different pollination syndromes occur in the family. Most well known is the sophisticated pollination syndrome of *Ficus* involving parasitic wasps, which has been studied in detail from many different perspectives (Rohwer, 1993b). Other taxa are often wind-pollinated (Rohwer, 1993b).

The Moraceae have an extensive Cenozoic record although many of the fossils assigned to the family have been questioned (Collinson, 1989; Martinez-Cabrera and Cevallos-Ferriz, 2006).

The fossil leaves and reproductive structures of *Artocarpus dicksonii* and *Artocarpus* sp. (Figure 2.8) from the Late Cretaceous of Igdlokunguak, Disco Island, Greenland, appear to be correctly assigned, but the fossils have not been restudied since they were first described by Nathorst (1890).

Ulmaceae. Previous classifications generally place Celtidoideae in Ulmaceae, but molecular analyses now place this subfamily in Cannabaceae. The Ulmaceae, as currently defined, are a small family of six genera and about 40 species of trees, or occasionally shrubs, distributed mainly in the Northern Hemisphere, but with two small genera in the Southern Hemisphere. Flowers are small, unisexual or bisexual, apetalous and wind-pollinated. The gynoecium is bicarpellate and syncarpous. Fruits are samaras or drupes.

The family has an excellent Cenozoic record (Manchester, 1989) that includes numerous leaves, fruits and pollen. The earliest records are *Ulmus*-type leaves from the Paleocene of Europe. Ulmaceous fruits are known from the Middle Eocene onwards. Winged fruits of the extinct ulmaceous genus *Cedrelospermum* are reported from Middle to Late Eocene floras of North America and from Late Eocene to Middle Miocene floras of Europe (Manchester, 1987b). The Ulmaceae are particularly well represented in the Miocene of Europe with abundant occurrences of leaves and fruits of *Ulmus* and *Zelkova* (Mai, 1995). There is no unequivocal record of Ulmaceae from the Cretaceous, although dispersed *Ulmus*-type pollen from the Maastrichtian of North America, India, Japan and Brazil may indicate the presence of early representatives of the family (Muller, 1981). Possible ulmaceous leaf fossils are also described from several Late Cretaceous localities in Kazakhstan that range in age from Cenomanian to Santonian–Campanian (Shilin, 1986), but these fossils and their systematic relationships are in need of re-examination.

Urticaceae. The Urticaceae, including also Cecropiaceae (APGII, 2003), comprise about 54 genera and 2600 species of herbs, shrubs, trees and hemi-epiphytes. The family has a worldwide distribution, with a particular concentration of taxa in tropical regions. Flowers are small, unisexual or rarely bisexual, apetalous and wind-pollinated. Stamens are inflexed in bud. The gynoecium is unicarpellate and the ovary is superior with a single ovule. Fruits are typically dry achenes, but the calyx is often persistent, swollen and fleshy, and functions in fruit dispersal.

The fossil record of Urticaceae is scattered and mostly based on dispersed fruits. From the Late Cretaceous of Central Europe Knobloch and Mai (1986) reported 12 species of Urticaceae based on fossil achenes. Most were assigned to extant genera, including *Boehmeria* (5 spp.), *Debregiasia* (1 sp.), *Pouzolzia* (*Memorialis*) (3 spp.). Three species were assigned to the extinct genus *Urticoidea*. Collinson (1989), who re-examined some of the species by SEM, was unable to confirm any of the determinations and noted that only seeds of *Pouzolzia* had cell structure of the perianth characteristic of Urticaceae, but other defining features were lacking.

Rhamnaceae. The Rhamnaceae comprise about 55 genera and 900 species of widely distributed herbs, shrubs and trees. Flowers are small, bisexual or more rarely unisexual, with a well-differentiated calyx and corolla, which are often basally connate to form a floral cup. The androecium is haplostemous with stamens opposite the petals that are often clawed. Flowers are mostly insect-pollinated and have a distinct nectary disc between the androecium and gynoecium. The gynoecium is usually formed from two to three carpels and has a single style. The ovary is superior to inferior with few ovules. Fruits are typically two to three locular drupes, berries or dry fruits with a single, anatropous seed per locule.

Only a single taxon assigned to the Rhamnaceae, *Coahuilanthus belinda*, has been described from the Cretaceous.

Coahuilanthus belinda consists of small actinomorphic and bisexual flowers preserved as compressions/impressions in sediments of Late Campanian age at the El Almácigo locality, Coahuila, Mexico (Calvillo-Canadell and Cevallos-Ferriz, 2007). Flowers are pentamerous with triangular sepals and clawed petals. The five stamens are opposite the petals and the flowers have a distinct nectary. Preservation of the flowers does not allow detailed comparison with extant taxa, but the suite of characters agrees with the features of the family and inclusion of *Coahuilanthus belinda* in Rhamnaceae is probably reliable. The 'Rose Creek flower' from the Early Cenomanian of Nebraska, USA (section 14.2) has been suggested as possibly related to Rhamnaceae (Crepet *et al.*, 2004), but it has five free styles in contrast to the single style in Rhamnaceae. Its systematic position needs further consideration.

The Cenozoic record of Rhamnaceae is extensive and mainly based on leaves and fruits. The characteristic winged fruits of *Paliurus* are particularly common in Europe and Asia with numerous records from the Oligocene onwards (Friis, 1985b). The earliest occurrence of *Paliurus* in North America is fruits from the Eocene (Manchester, 1999).

Elaeagnaceae. The Elaeagnaceae are a small family of three genera and about 50 species of shrubs or small trees, distributed in temperate and subtropical regions of the Northern Hemisphere (Cronquist, 1981). Flowers are small, bisexual, or more rarely unisexual, and apetalous. Sepals are fused to form a tubular or cup-shaped hypanthium and there is a lobed nectary disc. Fruits are drupes, berries or achenes. Pollen is tricolporate and psilate. Most exposed surfaces are covered by a dense indumentum, often of characteristic stellate trichomes.

Except in the latest Cenozoic the fossil record of Elaeagnaceae is based exclusively on dispersed pollen. Song (1989) reported fossil pollen of *Elaeagnacites* from the Late Cretaceous (Senonian) of China. This genus of dispersed pollen is otherwise only known from the Cenozoic. Pollen similar to that of Elaeagnaceae is widespread in the Paleocene. However, none of these putative dispersed eleagnaceous pollen grains have been studied using SEM or TEM and the reliability of the proposed systematic relationships is currently uncertain.

14.6.3 Cucurbitales

Analyses based on molecular data place seven families (Anisophylleaceae, Begoniaceae, Coriariaceae, Corynocarpaceae, Cucurbitaceae, Datiscaceae, Tetramelaceae) in Cucurbitales (APGIII, 2009). Molecular support for the group is not strong and relationships within the order remain to be fully clarified (Matthews and Endress, 2006). The Anisophylleaceae are generally resolved as sister to all other families. Coriariaceae and Corynocarpaceae together are sister to a second subclade comprising the other four families. All families include taxa with apetalous flowers, many are wind-pollinated and there is good support from floral morphology for the recognition of both subclades (Endress and Matthews, 2006a; Matthews and Endress, 2006). Flowers of Anisophylleaceae differ from those of other Cucurbitales and in many features are more similar to flowers of Cunoniaceae (Oxalidales, section 14.2).

The fossil record of Cucurbitales is sparse. Seeds of Cucurbitaceae are known from the Paleocene and Eocene of southern England (Collinson *et al.*, 1993) and dispersed pollen of Cucurbitaceae is recorded from the Oligocene of Africa (Muller, 1981). The Coriariaceae are known from seeds and pollen from the Miocene of Europe (Gregor, 1980; Muller, 1981). Wood of *Tetramelioxylon* (Tetramelaceae) from the Deccan Intertrappean Beds (Maastrichtian) India, may be the earliest record of the order (Lakhanpal, 1970).

14.6.4 Fagales

The families Betulaceae, Casuarinaceae, Fagaceae, Juglandaceae, Myricaceae, Nothofagaceae and Rhoipteleaceae were previously included in the 'Amentiferae' or Hamamelididae (e.g. Takhtajan, 1980; Cronquist, 1981), but are now resolved in all molecular-based phylogenetic analyses as a monophyletic group, the Fagales, within the nitrogen-fixing clade (APGIII, 2009). Also included in Fagales is the recently recognised family Ticodendraceae (Figure 14.7). In the APGIII classification Rhoipteleaceae are included in the Juglandaceae. Here we treat it as a separate family because of its prominent position in discussions of the extinct Normapolles complex.

The Fagales are characterised by their woody habit and their small, typically unisexual, apetalous or sometimes naked, anemophilous flowers. The Nothofagaceae are the sister group to the other families, followed by the Fagaceae, which are sister to the core Fagales. The core Fagales comprise two main clades: the Ticodendraceae–Betulaceae–Casuarinaceae clade and the Rhoipteleaceae–Myricaceae–Juglandaceae clade (Figures 14.7, 14.14). The

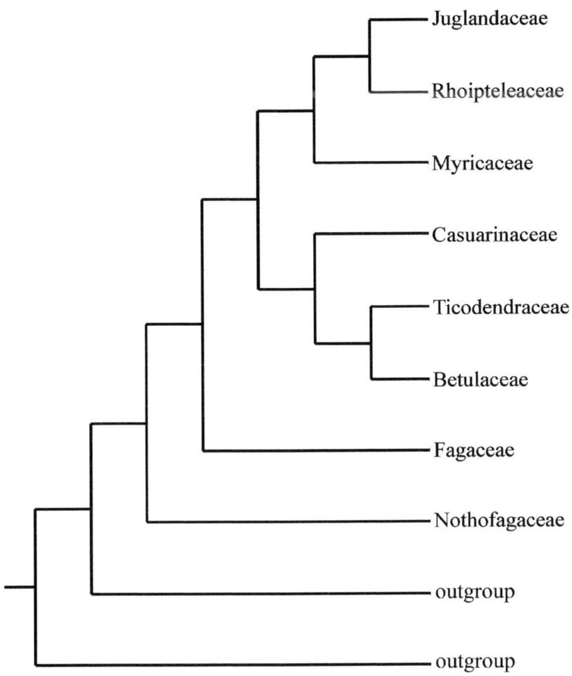

Figure 14.7 Summary of phylogenetic relationships of extant Fagales. Based on Li *et al.* (2004). Note that the pattern of phylogenetic relationships differs from that in Figure 14.6 in placing Myricaceae as sister group to Rhoipteleaceae and Juglandaceae, rather than as sister group to Betulaceae, Casuarinaceae and Ticodendraceae.

Ticodendraceae–Betulaceae–Casuarinaceae clade is characterised by having two ovules per ovary, whereas the ovary in the Rhoipteleaceae–Myricaceae–Juglandaceae clade has a single ovule.

14.6.5 Fossil Fagales of uncertain relationships

The Cenozoic record of Fagales is extensive and includes pollen, flowers, fruits, inflorescences, leaves and wood. There is also a rich Late Cretaceous record, particularly of pollen and reproductive organs, which provides direct evidence of considerable diversity prior to the Cenozoic. The earliest record of probable Fagales is in the Middle Cenomanian and the group diversified strongly through the Late Cretaceous (Chapter 20). As judged by different kinds of dispersed pollen, core Fagales apparently attained their maximum diversification by the Santonian–Campanian. Most of the Cretaceous fagalean fossils identified so far belong to the so-called Normapolles complex (section 14.6.6). These fossils can be broadly assigned to core Fagales, but cannot be referred to a particular extant family. Some Normapolles taxa such as *Caryanthus* and *Budvaricarpus* may belong to Rhoipteleaceae, but the Normapolles complex probably includes mostly extinct lineages that are various kinds of stem groups among core Fagales.

Nothofagaceae and Fagaceae, the two earliest-diverging lineages of the order, are also well documented in Late Cretaceous floras, but perhaps because grains of Fagaceae in particular are harder to recognise they are recorded slightly later. Based on fossil pollen they also appear less diverse than the Normapolles group. The Nothofagaceae are represented in the Late Cretaceous of the Southern Hemisphere by several kinds of dispersed pollen taxa of the *Nothofagus*-type (e.g. Dettmann, 1994). The Fagaceae are represented in the Late Cretaceous of the Northern Hemisphere by fossil flowers of *Protofagacea allonensis* from North America (Herendeen *et al.*, 1995; Sims *et al.*, 1998). Fossil flowers of *Antiquacupula sulcata* and *Archaefagacea futabensis* are also very similar to flowers of extant Fagaceae, but differ in several respects, and are treated here as unassigned Fagales.

Antiquacupula sulcata (Figure 14.8) is a well-characterised fossil fagalean from the Late Cretaceous (Late Santonian) Allon locality, Georgia, USA. It is known from staminate and bisexual flowers, cupules with immature fruits, and isolated mature fruits (Sims *et al.*, 1998). Both staminate and bisexual flowers are actinomorphic, pedicellate and about 2–3 mm in diameter. They have two whorls of three tepals and two whorls of six stamens. Stamens are well differentiated into filament and anther. Anthers are dorsifixed, tetrasporangiate, dithecate and dehisce by longitudinal slits. Pollen observed on the flowers is tricolporate with a finely foveolate tectum. A small vestigial trimerous gynoecium is present in the centre of the staminate flowers. The gynoecium of the bisexual flowers is tricarpellate and the ovary is inferior, trilocular, and with two anatropous ovules in each locule. Mature fruits are distinctly triangular and longitudinally ribbed. Cupules are pedicellate, four-lobed and covered by several orders of bracts that subtend the four lobes of the cupule. Up to six bisexual flowers have been observed in the cupules.

Fossil flowers, fruits and cupules of *Antiquacupula sulcata* are closely similar to extant Fagaceae, but also have similarities to Nothofagaceae and cannot be placed securely in either taxon. They are also similar to the fossil *Protofagacea allonensis* described from the same locality (see below). They differ from *Protofagacea* mainly in pollen morphology and in their pedicellate, rather than sessile, flowers (Sims *et al.*, 1998).

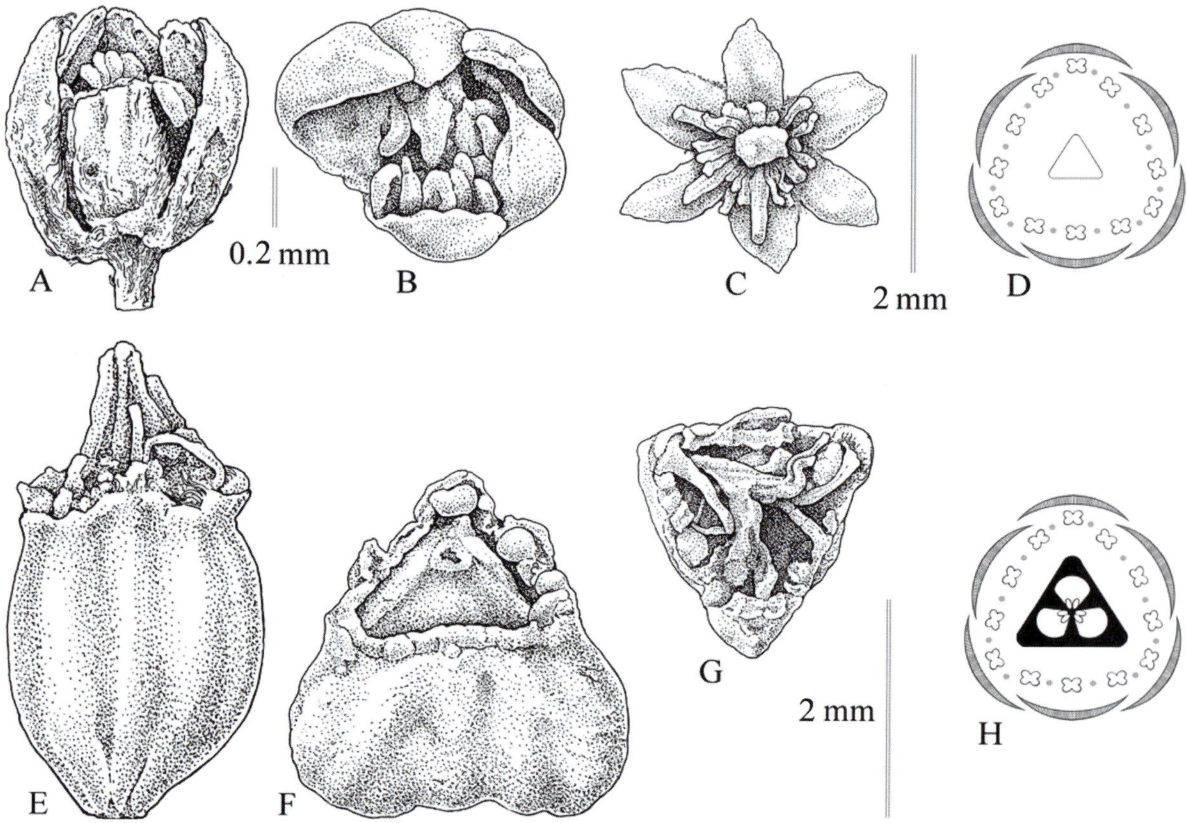

Figure 14.8 *Antiquacupula sulcata* from the Late Cretaceous (Late Santonian) Allon locality, Georgia, USA. (A, B) Staminate flower bud in (A) lateral and (B) apical view. (C) Open staminate flower showing six tepals, persistent stamen filaments separated by prominent nectaries and central pistillode. (D) Floral diagram of staminate flower. (E–G) Bisexual flowers in (E) lateral and (F, G) apical view showing epigynous organisation, stamen filaments and prominent nectaries. (H) Floral diagram of bisexual flower. Drawn from SEM images and line drawings in Sims *et al.* (1998).

Archaefagacea futabensis from the Late Cretaceous (Early Coniacian) Kamikitaba locality, Honshu, Japan (Takahashi *et al.*, 2008a), possibly represents a stem-group lineage of Fagales. The fossil taxon is known from three-dimensionally preserved flowers and fruits (Figure 14.9). Flowers are bisexual and actinomorphic with a tricarpellate gynoecium and inferior ovary. There are six small tepals and six stamens. Pollen found on the flowers is tricolpate with an irregular striate ornamentation. The ovary is trilocular with two ovules in each locule. Fruits are also trilocular, but only a single ovule per carpel develops into a seed.

The characters of *Archaefagacea futabensis* clearly indicate a relationship with Fagales. The general floral structure is similar to several fagalean taxa, including Normapolles flowers such as *Antiquocarya* (section 14.6.6) among core Fagales. Pollen is closely similar to that of some extant Fagaceae, and distinct from that in core Fagales. *Archaefagacea futabensis* is also similar to Fagaceae in having three locules, each with two ovules, only one of which matures into a seed. *Archaefagacea futabensis* cannot be placed in the Fagaceae or other extant families and probably represents an extinct early branch, perhaps on the fagalean stem lineage below the point at which the extant genera of Fagaceae diverged.

14.6.6 Normapolles complex

Among the most conspicuous elements in many Late Cretaceous palynofloras from the Northern Hemisphere

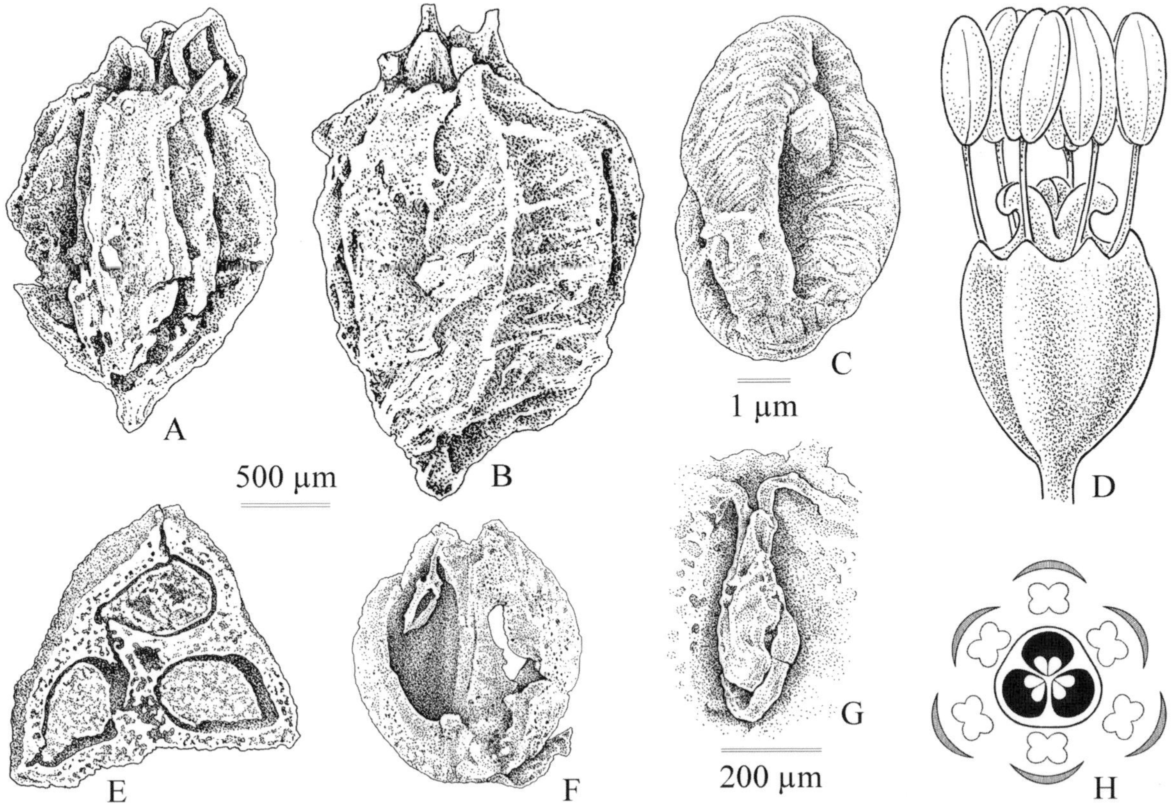

Figure 14.9 *Archaefagacea futabensis* from the Late Cretaceous (Early Coniacian) Kamikitaba locality, Japan. (A) Flower in lateral view showing epigynous organisation and stamen filaments. (B) Fruit in lateral view showing tepals. (C) Pollen, probably tricolporate, with striate ornamentation. (D) Reconstruction of flower. (E) Transverse section of fruit with a single seed in each locule. (F) Longitudinal section of ovary showing apical attachment of ovules. (G) Detail of apex of locule showing two apically attached aborted ovules. (H) Floral diagram. (A–C, E–G) Drawn from SEM images in Takahashi *et al.* (2008a); (D, H) from Takahashi *et al.* (2008a).

are dispersed pollen grains assigned to the so-called Normapolles complex. The term Normapolles (Stemma Normapolles) was suggested by Pflug (1953) for a distinctive group of pollen grains characterised by having a short polar axis and typically three complex apertures with three pores and often three short colpi (Figures 14.10–14.13). More than 100 Normapolles species in more than 80 genera have been described (e.g. Góczán *et al.*, 1967; Skarby, 1968; Batten and Christopher, 1981; Srivastava, 1981; Tschudy, 1981; Zaklinskaya, 1981).

Normapolles pollen grains first appear in the fossil record in the Middle Cenomanian and reach their greatest systematic diversity during the Santonian–Campanian. Normapolles pollen was also abundant and makes up more than 80% of all angiosperm pollen in some European palynofloras from the Late Cretaceous (Pacltová, 1981). During the Late Cretaceous the Normapolles group also reach their maximum geographic distribution extending from eastern North America and Europe to western Siberia in a continuous floristic realm referred to as the Normapolles Province (Batten, 1981; Herngreen and Chlonova, 1981; Zaklinskaya, 1981; Herngreen *et al.*, 1996). The group has another diversity peak in the Eocene perhaps coincident with the initial diversification of extant families that were apparently derived from the Normapolles complex. During the Early Cenozoic the distinctive Normapolles grains seen in the Late Cretaceous decrease in their importance in palynofloras and disappear from the fossil record

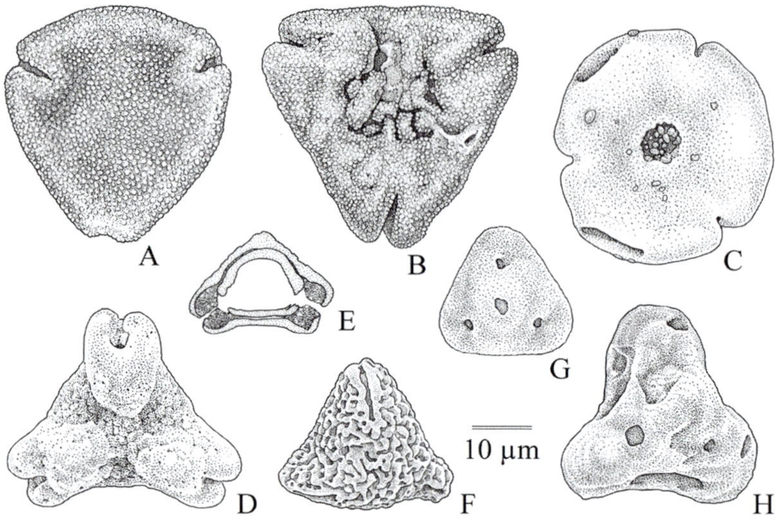

Figure 14.10 Dispersed pollen grains from the Late Cretaceous assigned to the Normapolles complex. (A) *Papillopollis rugulatus.* (B) *Papillopollis* sp. (C) *Vancampopollenites* sp. (D) *Oculopollis* sp. (E) Section through two apertures of an *Oculopollis* grain. (F) *Atlantopollis* sp. (G) *Interporopollenites* sp. (H) *Interporopollenites* sp. (A) Drawn from SEM images in Batten and Morrison (1987); (B–H) drawn from SEM and TEM images in Batten (1986b).

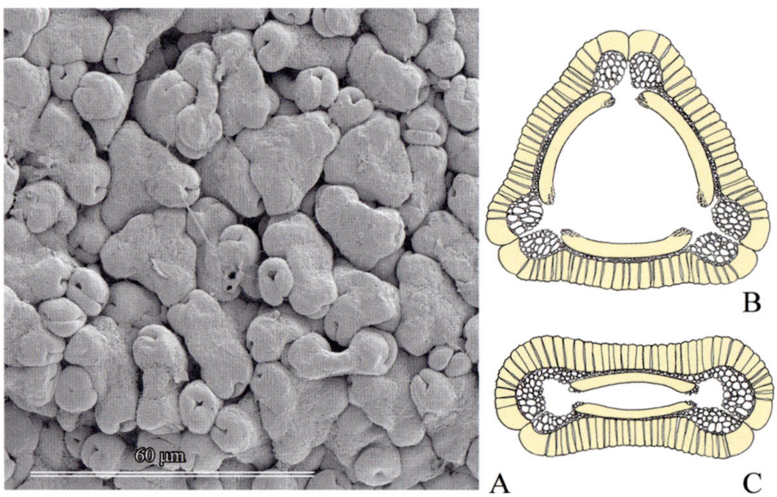

Figure 14.11 *Semioculopollis*-type pollen from a detached anther from the Late Cretaceous (Late Santonian – Early Campanian) Åsen locality, Sweden. (A) Mass of pollen grains in an anther. (B) Section through the equatorial region of a single grain showing thick tectum, granular infratectum and complex structure of the three apertures. (C) Section through two apertures parallel to the polar axis. Material in the collections of the Swedish Museum of Natural History, Stockholm.

by the Oligocene. The only extant plant that produces pollen like some forms included in the Normapolles complex is *Rhoiptelea chiliantha* (Rhoipteleaceae, section 14.6.9): *Rhoiptelea chiliantha* is a rare plant occurring today in southwestern China and northern Vietnam (Sun *et al.*, 2006).

The systematic relationships of dispersed Normapolles pollen have been discussed since they were first described. Many Normapolles forms were compared to pollen of extant Fagales (e.g. Betulaceae, Rhoipteleaceae), but relationships to other families have also been considered. Subsequently, the discovery of a variety of fossil flowers with Normapolles-type pollen *in situ* provided the basis for a more secure and detailed assessment of the systematic relationships of the group.

Currently, about 25 different species of fossil flowers and fruits have been linked to the Normapolles group. This material has been assigned to eight different genera, *Antiquocarya, Bedellia, Calathiocarpus, Caryanthus, Budvaricarpus, Dahlgrenianthus, Manningia* and *Normanthus* (e.g. Friis, 1983; Friis and Crane, 1989; Sims *et al.*, 1999; Schönenberger *et al.*, 2001b; Friis *et al.*, 2003b, 2006b). Pollen found *in situ* is similar to dispersed pollen assigned to *Interporopollenites* (in *Endressianthus*), *Minorpollis* (in *Dahlgrenianthus*), *Plicapollis* (in *Caryanthus* and *Budvaricarpus*), *Pseudopapillopollis* (in *Normanthus*), *Semioculopollis* (in *Antiquocarya*) and *Trudopollis* (in *Calathiocarpus* and *Manningia*) (Figures 14.12, 14.13). Combined evidence from pollen structure and floral

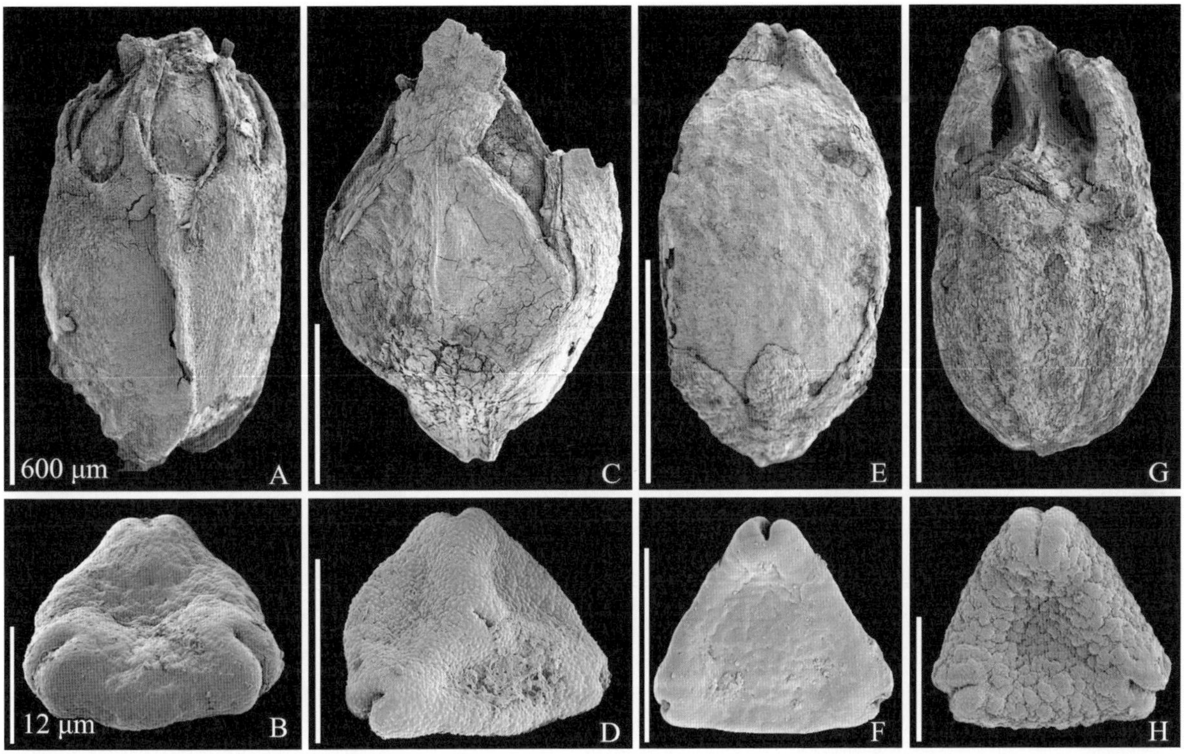

Figure 14.12 SEM images of Normapolles flowers and their associated pollen from the Late Cretaceous (Late Santonian – Early Campanian) Åsen locality, Sweden. (A) *Antiquocarya*, and (B) *Semioculopollis*; (C) *Caryanthus*; and (D) *Plicapollis*; (E) *Dahlgrenianthus*; and (F) *Minorpollis*; (G) *Manningia*; and (H) *Trudopollis*. Material in the collections of the Swedish Museum of Natural History, Stockholm.

morphology suggests that all of the Normapolles pollen types currently known in floral structures were produced by plants related to core Fagales. Within this group, relationships of Normapolles plants were probably diverse, reflecting the presence of several different lineages (Figure 14.14).

The first mesofossils to be described with Normapolles pollen *in situ* were flowers and fruits assigned to the extinct genera *Antiquocarya, Caryanthus* and *Manningia* from the Late Cretaceous (Late Santonian – Early Campanian) Åsen locality, Sweden (Friis, 1983). These were compared to extant Myricaceae, Juglandaceae and Rhoipteleaceae. Many other fossils related to these genera have been reported subsequently from the Late Cretaceous of Central Europe (Knobloch and Mai, 1986) and flowers with Normapolles-type pollen have also been recovered from the Late Cretaceous of Portugal (Schönenberger *et al.*, 2001b; Friis *et al.*, 2003b) and North America (Sims *et al.*, 1999). One of the Portuguese flower types (*Endressianthus*), as well as the North American flower (*Bedellia*) and perhaps also the Portuguese flower *Normanthus*, appears to be related to Betulaceae, while *Caryanthus* and *Budvaricarpus* share many features with Rhoipteleaceae (Figure 14.14).

Most Normapolles flowers are found detached, but in those cases where inflorescence fragments are know (*Endressianthus, Normanthus*) partial inflorescences are dichasia. Normapolles flowers are all bisexual except for *Endressianthus*, which is clearly unisexual. Flowers of *Bedellia* are possibly also unisexual. *Budvaricarpus* is interesting in having both bisexual and unisexual (pistillate) flowers in the same dichasium. Both unisexual and bisexual flowers have inferior ovaries with the exception of *Dahlgrenianthus*, which has a superior ovary (Figure 14.15). The perianth is simple and consists of a single whorl of tepals. Tepals vary in shape and size from distinct and relatively well developed in *Manningia*, to small, thorn-like, and indistinct in *Endressianthus* and *Antiquocarya*. The androecium consists of a single whorl of stamens that are opposite the tepals, except in *Endressianthus* and *Normanthus* where

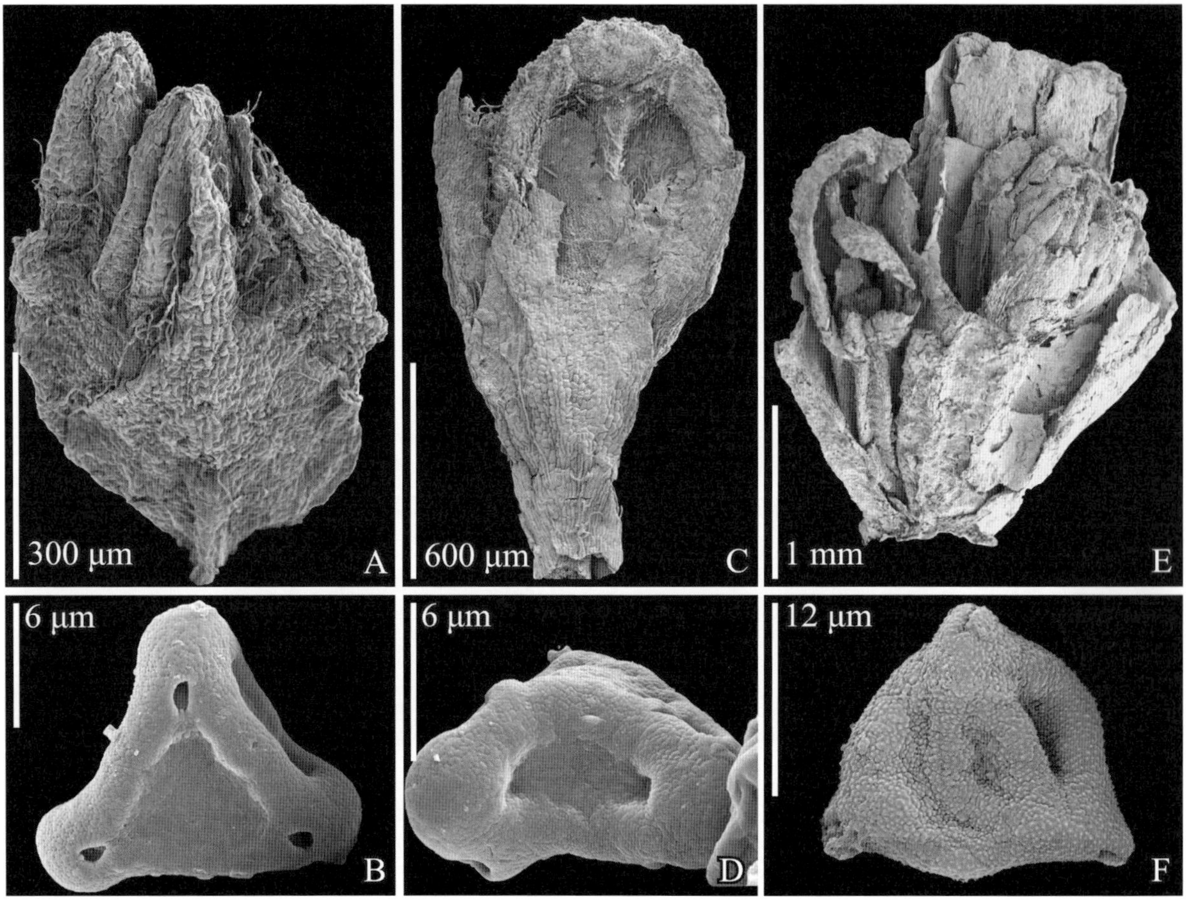

Figure 14.13 SEM images of Normapolles flowers and their associated pollen from the Late Cretaceous (Campanian–Maastrichtian) of Portugal. (A–D) *Endressianthus*, (A) staminate and (C) pistillate flowers and (B, D) *in situ Interporopollenites* pollen; (E) *Normanthus* and (F) *in situ Pseudopapillopollis* pollen. Material from the collections of the Swedish Museum of Natural History, Stockholm.

stamens alternate with the tepals (Figure 14.15), and in *Caryanthus* and *Budvaricarpus* flowers typically have four tepals and six stamens. Stamens have a well-differentiated filament and anther, and filaments were apparently extended considerably at anthesis. Anthers are basifixed, elongate, dithecate and tetrasporangiate and dehisce by longitudinal slits. In several taxa the anthers have stiff trichomes at the apex. The connective is indistinct and there is no apical expansion. Pollen grains are all oblate and triaperturate. In all genera except *Endressianthus* pollen apertures consist of a very short outer colpus and an inner pore. In *Endressianthus* each aperture region has two exopores and one endopore. The pollen wall is tectate with a granular infratectal layer. The gynoecium is bi- or tricarpellate. Mature fruits are nuts, often with a single locule and a single seed.

The following section describes some of the better-known Normapolles flowers. Numerous other flowers and fruits with similar general morphology are reported from the Late Cretaceous of Central Europe, including such genera as *Bicameria* (4 spp.), *Costacarpus* (7 spp.), *Normacarpus* (3 spp.), *Vangerowia* (4 spp.) and *Walbeckia* (10 spp.) (Knobloch and Mai, 1986). Currently these have not been linked definitively to Normapolles pollen, and their affinity to the Normapolles complex, albeit likely, is not fully secure.

Antiquocarya (Figures 14.12, 14.15, 14.16) is based on small, actinomorphic and apparently bisexual flowers from

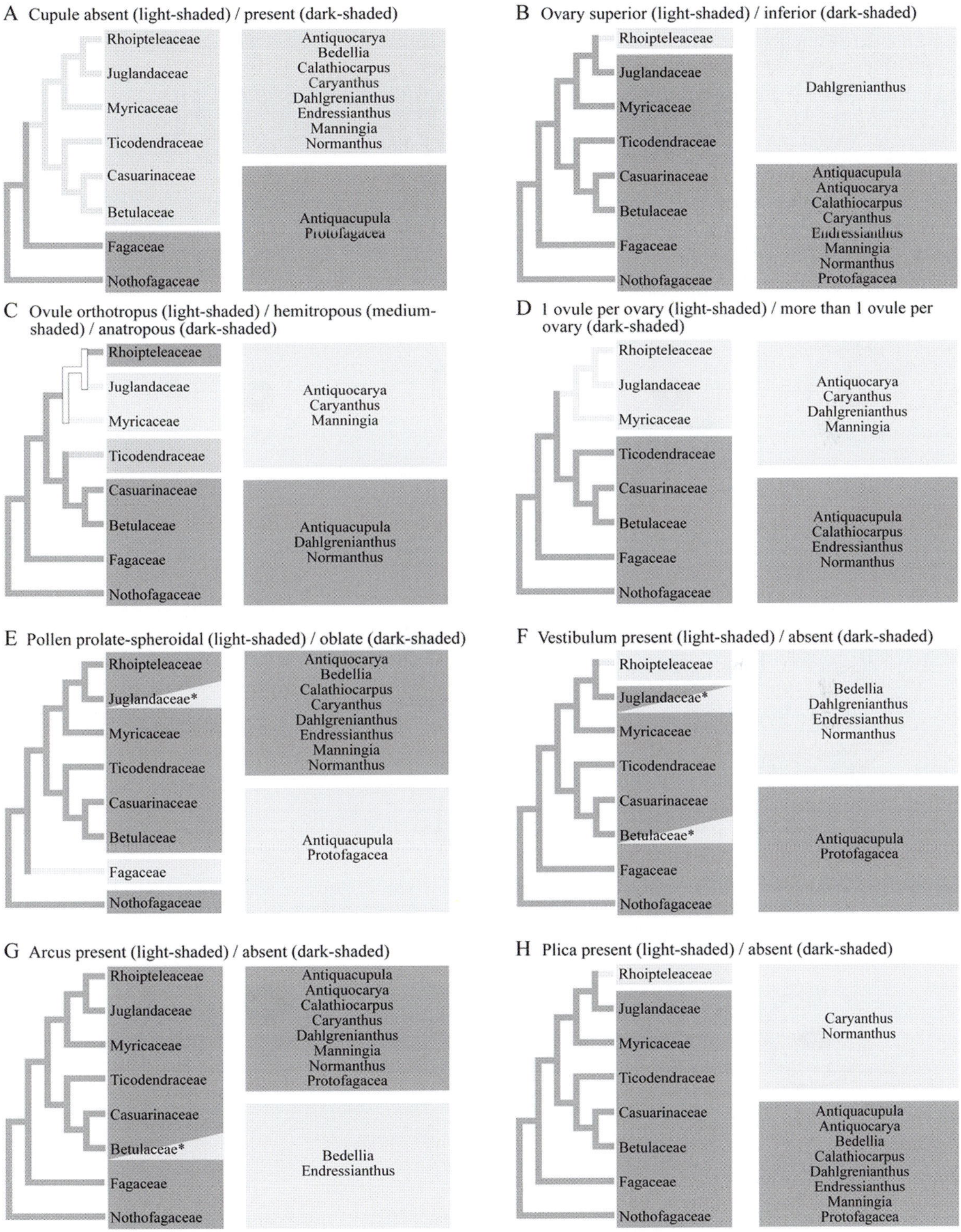

Figure 14.14 Summary of floral and pollen features among extant Fagales and related fossil Normapolles flowers. * indicates that the family is dimorphic for the specific character. From Friis *et al.* (2006b).

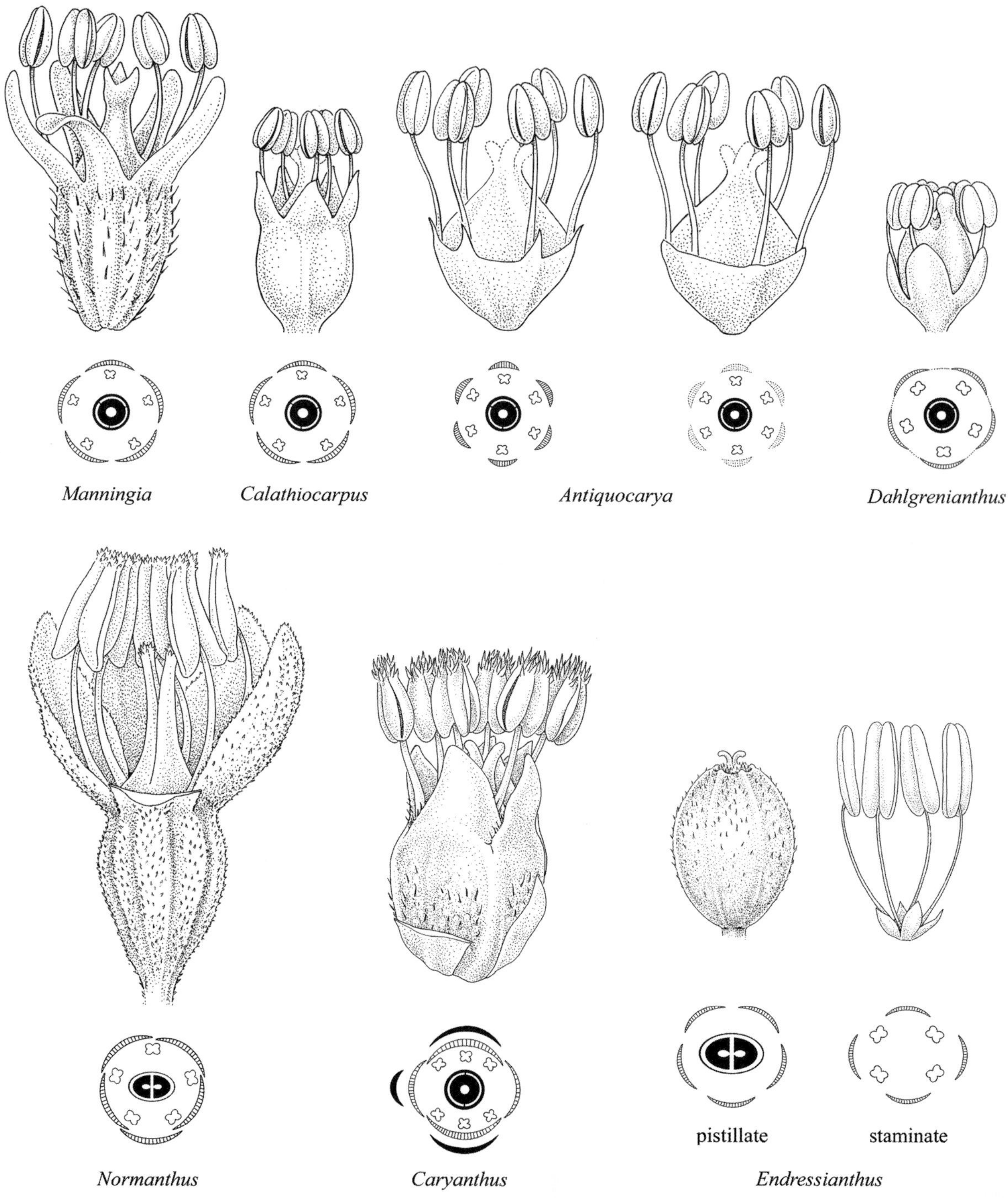

Figure 14.15 Reconstructions and floral diagrams of Normapolles flowers from the Late Cretaceous of Europe. From Friis *et al.* (2006b).

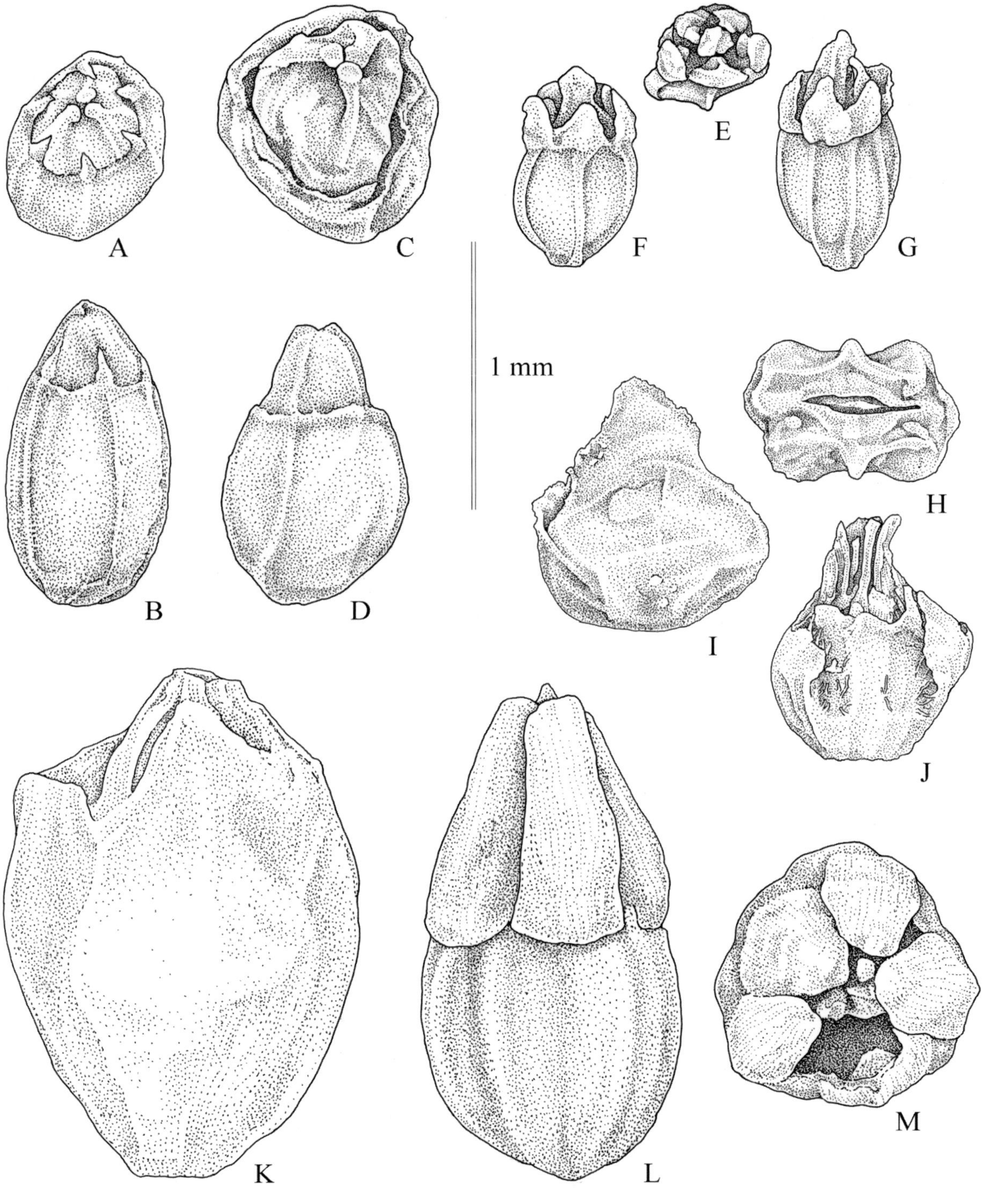

Figure 14.16 Normapolles flowers from the Late Cretaceous of Europe. (A, B) *Antiquocarya verruculosa* in (A) apical and (B) lateral view; Åsen locality (Late Santonian – Early Campanian), Sweden. (C, D) *Antiquocarya nuda* in (C) apical and (D) lateral view; Åsen locality (Late Santonian – Early Campanian), Sweden. (E–G) *Calathiocarpus calyciferus* in (E) apical and (F–G) lateral view; Walbeck locality (Maastrichtian), Germany. (H–J) *Caryanthus knoblochii* in (H) apical and (I, J) lateral view; Åsen locality (Late Santonian – Early Campanian), Sweden. (K) *Caryanthus* sp. in lateral view; Aachen locality (Late Santonian), Germany. (L, M) *Manningia crassa* in (L) lateral and apical view (M); Åsen locality (Late Santonian – Early Campanian), Sweden. (A–D, H–J, L, M) Material in the collections of the Swedish Museum of Natural History, Stockholm; (E–G) material in the collections of the National Museum, Prague; (K) material in the collections of the Geological Institute, Aachen.

the Late Cretaceous (Late Santonian – Early Campanian) Åsen locality, Sweden (Friis, 1983). It includes two species, *Antiquocarya nuda* and *Antiquocarya verruculosa*. Flowers have a single whorl of six short and narrow, almost thorn-like tepals in an epigynous position (*Antiquocarya verruculosa*) or the flowers are naked (*Antiquocarya nuda*). Opposite each tepal (or in an equivalent position in *Antiquocarya nuda*) are the remains of a stamen. Only the bases of the filaments are preserved and anthers have not been observed in any of the specimens. The possibility that the flowers were functionally unisexual cannot be excluded. One specimen of *Antiquocarya nuda* has small flattened, slightly lobed structures in the position of the stamen filaments that may represent aborted stamens. The gynoecium is tricarpellate with a unilocular ovary and three styles. Styles are not complete in any of the specimens studied. There is a single, apparently orthotropous seed with a thin membranous seed coat. Pollen of the *Semioculopollis*-type has been observed on the apical surface of many flowers and fruits.

Bedellia pusilla is based on small staminate flowers with Normapolles-type pollen *in situ* from the Late Cretaceous (Late Santonian) Allon locality, Georgia, USA (Sims *et al.*, 1999). The material is poorly preserved and the organisation of this fossil is not completely clear. There is apparently a perianth consisting of two whorls of five tepals, and an androecium composed of ten stamens. Stamens are differentiated into long filaments and basifixed anthers that dehisce by longitudinal slits. The pollen is peroblate and triangular, and tricolporate with very short outer apertures. Distinct arci are present between the apertures. Based on pollen characters *Bedellia* is similar to extant Betulaceae, but differences in floral morphology preclude assignment of the fossils to this family (Sims *et al.*, 1999).

Budvaricarpus serialis (Figure 14.17) was first described from the Klikov Formation (Coniacian–Santonian), Czech Republic (Knobloch and Mai, 1983, 1984, 1986) as an unusual fruit with three or more carpels fused laterally in a row and forming a series of locules. Reinvestigation of the type material, and examination of new fossils (Heřmanová *et al.*, 2011), shows that the fossil fruiting structures of *Budvaricarpus serialis* are not a single fruit with several locules in a series, but rather three or more fruits, derived from individual flowers, that are laterally fused. The median bisexual flower is flanked by two or more apparently unisexual (pistillate) flowers. The floral structure of the median flower is very similar to that of *Caryanthus* with four tepals in two decussate pairs, six or more stamens, and

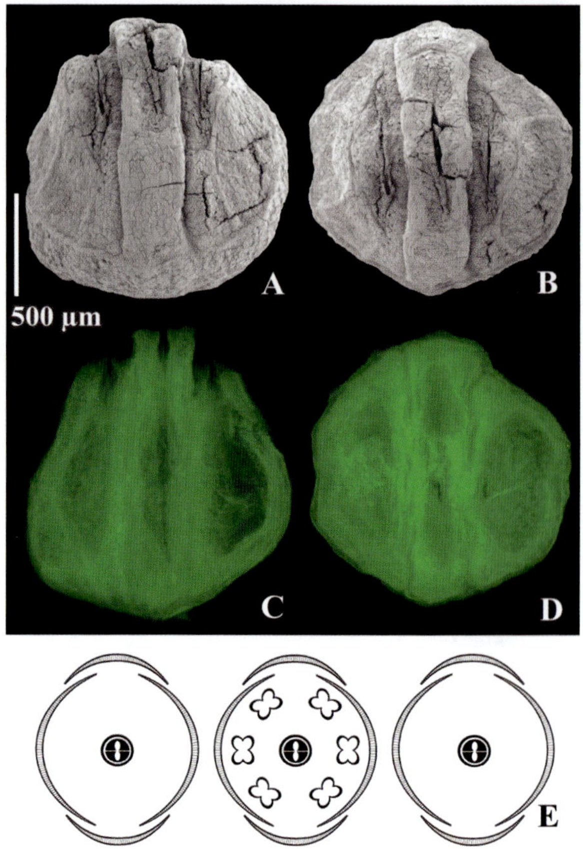

Figure 14.17 *Budvaricarpus serialis* from the Late Cretaceous (Late Turonian–Santonian) of the South Bohemian Basins; specimen consisting of three laterally fused fruits enclosed in a common bract. (A, B) SEM images; (C, D) synchrotron-radiation X-ray tomographic microscopy (SRXTM) images (voltex) and (E) floral diagrams of median, bisexual flower flanked by two lateral, pistillate flowers. Fossil in (A, C) dorsiventral view and (B, D) apical view. Specimen in the collections of the National Museum, Prague (cf. Heřmanová *et al.*, 2011).

a bicarpellate gynoecium forming an inferior ovary. Pollen observed on the apical part of the bisexual flowers is of the *Plicapollis* type. The cluster of fused flowers/fruits is surrounded by fused bracts and the whole unit clearly represents a partial dichasial inflorescence. In the organisation of the flowers and dichasium *Budvaricarpus* is closely similar to extant *Rhoiptelea*. As in *Budvaricarpus*, the typical dichasia of *Rhoiptelea* have a median bisexual flower flanked by two lateral pistillate flowers. Pollen of *Budvaricarpus* is also closely similar to that of *Rhoiptelea* and *Caryanthus* (see below). In *Rhoiptelea* only some of the flowers develop into fruits. In some populations it is the median, bisexual flower; in other populations it is the lateral, pistillate flowers (Sun *et al.*, 2006) (S.Q. Huang, K.R. Pedersen, G. W. Grimm and E.M. Friis, in progress, see also *Caryanthus* below).

Calathiocarpus (Figures 14.15, 14.16) includes four species of flowers and fruits from the Late Cretaceous of Central Europe (Knobloch and Mai, 1986). The type species, *Calathiocarpus calyciferus*, was described from the Maastrichtian of Germany; other species are from Austria, the Czech Republic, the Netherlands and Poland. *Calathiocarpus* ranges from the Turonian to the Maastrichtian. Flowers are up to about 3 mm long, pentamerous and with five triangular tepals in an epigynous position and five stamens opposite the tepals. Only the stamen filaments are preserved, but Normapolles pollen of the *Trudopollis*-type occurs inside the perianth. Pollen is about 15 µm in equatorial diameter, oblate and triaperturate with very short colpi and a tectate–scabrate pollen wall. The gynoecium is tricarpellate with a single solid style. Flowers and the associated pollen of *Calathiocarpus* closely resemble those of *Manningia* and the two genera probably belong to the same lineage of Normapolles plants (Friis *et al.*, 2006b).

Caryanthus (Figures 14.12, 14.15, 14.16) is very common in Late Cretaceous floras from Central and northern Europe. The genus was first described from the Late Santonian – Early Campanian Åsen locality, Sweden, where three species (*Caryanthus elongatus, Caryanthus knoblochii, Caryanthus* sp.) were recognised (Friis, 1983). About ten further species have been described from Central Europe (e.g. Knobloch and Mai, 1986; Friis and Crane, 1989) and the genus may also occur in Late Cretaceous floras from eastern North America (Herendeen *et al.*, 1999a). The material includes numerous fragmentary flowers and fruits. Several thousand specimens have been recovered from the Åsen locality.

Flowers of *Caryanthus* are small, about 1 mm long, bisymmetrical with an inferior ovary, and supported by a small bract and two lateral bracteoles that are fused to the basal part of the ovary. The perianth consists of four tepals arranged in two opposite–decussate pairs. The androecium consists of six to eight stamens. Only the stamen filaments are preserved attached to the flowers. All anthers are detached, but occasional clumps of pollen are observed on the apices of the filaments and the flowers were probably functionally bisexual. Pollen has been found inside the perianth in several species, and is believed to have been produced by the flowers. Pollen is about 18 µm in equatorial diameter, peroblate and triangular with strongly protruding corners. It is slightly heteropolar, and tricolporate with very short outer apertures. There are distinct arci between the apertures. *Caryanthus* pollen is similar to the dispersed Normapolles genus *Plicapollis*. The gynoecium is bicarpellate with an apparently unilocular ovary and two free styles. Fruits are small ribbed nuts including a single seed with a thin, membranous seed coat.

Fossil flowers and fruits of *Caryanthus* are very similar to flowers and fruits of the extant monotypic genus *Rhoiptelea* (Rhoipteleaceae) from China and Vietnam (Figures 14.22, 14.23). All fruits appear to have remains of stamens and were probably developed from bisexual flowers. The flowers are very similar in their organisation to the bisexual median flower of *Budvaricarpus* (see above), but in *Caryanthus* there is no indication of fusion of flowers; pistillate lateral flowers were not observed. It is possible that only bisexual flowers developed, but another possibility is that the lateral flowers aborted very early and that only the bisexual flowers developed into fruits. This is the case for flowers in some *Rhoiptelea* populations.

Dahlgrenianthus (Figures 14.12, 14.15) was established for small Normapolles flowers with a pentamerous perianth and androecium, a trimerous gynoecium and a superior ovary (Friis *et al.*, 2006b). Currently the genus includes three species. Flowers are bisexual and actinomorphic, about 1–1.2 mm long. Tepals are fused basally to form a short floral cup. There are five stamens opposite the tepals. Pollen occurs abundantly inside the perianth and adhering to the stigmas. It is minute, triaperturate and oblate with three very short colpi and circular endopores. In the type species, *Dahlgrenianthus suecicus*, from the Åsen locality, Sweden, pollen is similar to dispersed pollen assigned to the Normapolles genus *Minorpollis*. Another *Dahlgrenianthus* species from the Åsen flora is associated with pollen similar to the Normapolles genus *Semioculopollis*. The ovary is unilocular with a single anatropous ovule in a basal position. Fruits are triangular nuts.

Flowers and fruits of *Dahlgrenianthus* are extremely common at the Åsen locality. They have also been reported from other Late Cretaceous floras from central Europe that range in age from Late Santonian to Maastrichtian (Friis *et al.*, 2006b).

Endressianthus (Figures 14.13, 14.18, 14.19) is currently the best-known Normapolles-producing plant. The genus includes two species, *Endressianthus foveolatus* and *E. miraensis*, both from the Campanian–Maastrichtian of Portugal (Friis *et al.*, 2003b). *Endressianthus miraensis* from the Mira locality is the more complete. It is known from fragmentary pistillate and staminate inflorescences as well as dispersed

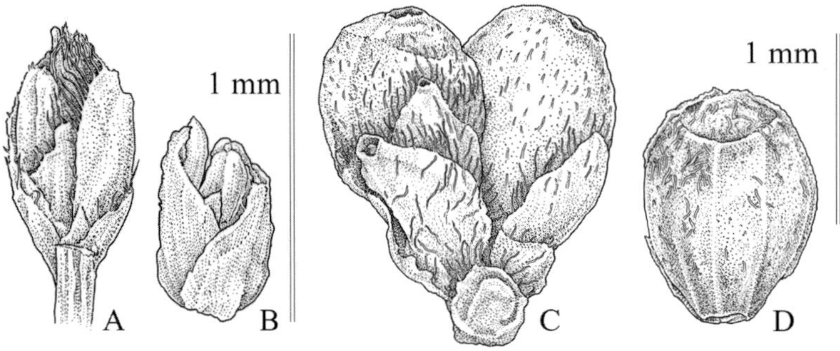

Figure 14.18 *Endressianthus miraensis* from the Late Cretaceous (Campanian–Maastrichtian) Mira locality, Portugal. (A–B) Staminate and (C–D) pistillate inflorescence fragments and flowers. (A, C) From Friis *et al.* (2005); (B, D) drawn from SEM images in Friis *et al.* (2003b).

flowers, stamens, and fruits and pollen *in situ* in staminate flowers. *Endressianthus foveolatus* from the Esgueira locality is currently known only from fruits with pollen adhering to their surface.

Inflorescences of *Endressianthus* are compound cymes with dichasial inflorescence subunits arranged helically along the main inflorescence axis. Each subunit is subtended by a large median bract and two prophylls. These in turn support four unisexual flowers and their prophylls. The surfaces of inflorescence axes, bracts, prophylls and pistillate flowers bear long, unicellular trichomes.

In pistillate subunits one or two flowers develop into mature fruits. Most, perhaps all, pistillate flowers are preserved at fruiting stage. The gynoecium is bicarpellate and syncarpous, and the ovary inferior and bilocular. At maturity one locule is sometimes suppressed. The number of ovules per carpel is unknown, but mature fruits contain a single seed per locule. Small thorn-like structures at the rim of the hypanthium are interpreted as perianth parts. The stigmatic area is not fully understood. No styles have been observed, but in some specimens two circular structures are present in the median plane immediately inside the hypanthium. Between these two structures a distinct slit lined with densely spaced trichomes extends over the apex of the ovary/fruit. Sometimes a secretion completely or partly covers the slit. The fruits are small, up to about 1.3 mm long in *Endressianthus miraensis* and up to 0.75 mm long in *E. foveolatus*. They are indehiscent and apparently dry.

In staminate subunits not all flowers appear to develop mature stamens. Staminate flowers have four stamens, but typically one or more are suppressed. None of the flowers that develop have a full set of mature stamens. Stamens have a very short filament and a narrowly elongated, dithecate, tetrasporangiate anther that dehisces by longitudinal slits.

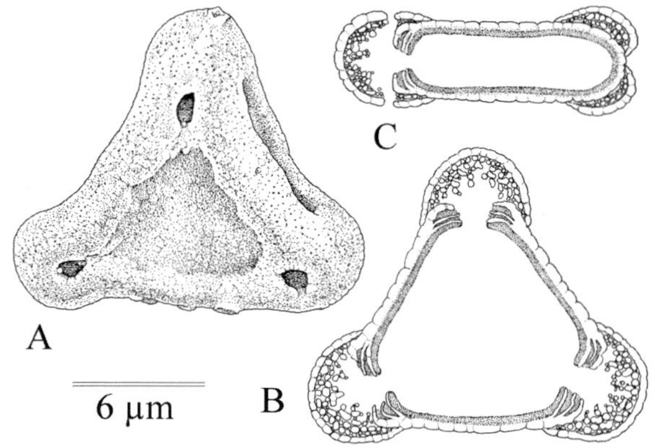

Figure 14.19 *Interporopollenites*-type pollen from the Late Cretaceous (Campanian–Maastrichtian) Normapolles flower *Endressianthus miraensis* from the Mira locality, Portugal. (A) External morphology. (B) Section through equatorial region showing complex wall structure associated with three apertures. (C) Section parallel to polar axis. From Friis *et al.* (2003b).

Pollen *in situ* in the stamens and on the surfaces of the fruits is comparable to the Normapolles genus *Interporopollenites*. It is oblate, about 13–16 μm in diameter, with a triangular equatorial outline and strongly protruding aperture regions. Apertures are distinctive, with three endopores in the equatorial region and six exopores arranged in pairs over the endopore: three on the proximal surface and three on the distal surface. The aperture regions are interconnected by distinct subequatorial crests or arci that extend on both sides of the equator.

Endressianthus shows close similarities to some core Fagales, and particularly extant Betulaceae. It does, however, have several unique characters not known in extant Betulaceae, or other Fagales, such as the presence of six exopores in a subequatorial position and an apical slit on top of the ovary. *Endressianthus* is further distinguished from extant Betulaceae by the larger number of flowers per inflorescence subunit (Friis *et al.*, 2003a).

Manningia (Figures 14.12, 14.15, 14.16) was established based on small flowers and fruits from the Late Cretaceous (Late Santonian – Early Campanian) Åsen locality, Sweden. There is currently only one species, *Manningia crassa*, which has also been reported from localities in Central Europe ranging from Late Turonian to Maastrichtian (Knobloch and Mai, 1986; Friis *et al.*, 2006b). Flowers of *Manningia* are up to about 2 mm long, actinomorphic and bisexual with an inferior ovary. The perianth consists of five ligulate tepals. The androecium consists of five stamens opposite to the tepals with long filaments. Fragments of anthers have been observed in several flowers. Pollen is oblate, triangular and triaperturate with a very short outer aperture. It is very similar to the dispersed Normapolles genus *Trudopollis*. The gynoecium is tricarpellate and the ovary is unilocular. There is a short prominent style and three elongated stigmatic branches. The fruits are ribbed nuts containing a single erect, orthotropous seed with a thin membranous seed coat.

Flowers of *Manningia* are most similar to flowers of extant members of the tribe Engelhardieae (Juglandaceae), but they differ in having a simple undivided locule in the ovary and other features that preclude inclusion in the extant family. The fossil flowers also show some similarity to flowers of certain Myricaceae, but differ in several features of the gynoecium and ovule (Friis, 1983).

Normanthus miraensis (Figures 14.13, 14.20) is described from the Late Cretaceous (Campanian–Maastrichtian) Mira locality, Portugal (Schönenberger *et al.*, 2001b). The flowers are actinomorphic and bisexual with pentamerous perianth and androecium, bicarpellate gynoecium and inferior ovary. Fragments of inflorescences with flowers show a dichasial arrangement of the partial inflorescences. Tepals are distinct and basally fused to form a very short hypanthium cup. The androecium has five stamens alternating with the tepals. Filaments are very short in bud, but elongate at anthesis. Anthers are dithecate and tetrasporangiate and dehisce by longitudinal slits. At the apex, anthers are covered by short, stiff trichomes. Pollen found inside the anthers is oblate, triaperturate and about 17–20 μm in diameter. The protruding apertures consist of a short outer colpus and a circular inner pore with a well-developed vestibulum. Pollen is slightly heteropolar with plica-like thickenings on one surface that extend from the apertures over the pole. The pollen wall is tectate and densely scabrate, with a thick tectum, a thin, granular infratectal layer and a thick foot layer. The grains are closely similar to dispersed pollen assigned to the Normapolles genus *Pseudopapillopollis*. The gynoecium has two long styles that are free for most of their length and have a ventral slit clearly visible along their entire free parts. Stigmatic areas are indistinct and surrounded by short, stiff trichomes. The ovary has two pendent, anatropous ovules of which only one develops into a seed.

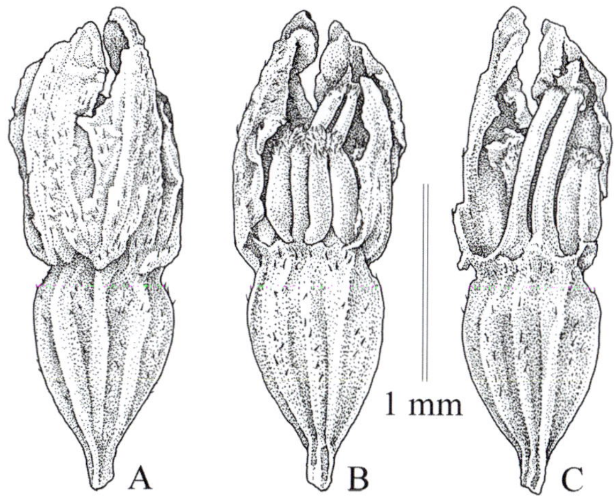

Figure 14.20 *Normanthus miraensis* from the Late Cretaceous (Campanian–Maastrichtian) Mira locality, Portugal. Same flower with (A) perianth intact, (B) one tepal removed and (C) tepals and stamens removed to show the two styles. (A, B) From Friis *et al.* (2005), (C) drawn from SEM image in Schönenberger *et al.* (2001b).

Normanthus is particularly close to extant Betulaceae, but there are also differences, for example in the number of ovules per carpel, that preclude inclusion of *Normanthus* in the extant family (Schönenberger *et al.*, 2001b). *Normanthus* is most likely on the stem group of extant Betulaceae.

14.6.7 Nothofagaceae and Fagaceae

Nothofagaceae. The Nothofagaceae are resolved as sister to all other families of Fagales (Figure 14.7). The single genus, *Nothofagus*, includes about 35 extant species of shrubs or trees with small, unisexual and apetalous flowers that are apparently adapted for wind pollination. Pistillate flowers are borne in woody cupules formed from the subtending bracts. Pollen is typically peroblate with up to 10 short colpi in an equatorial arrangement. The pollen wall is tectate with a granular infratectal layer and spiny to verrucate surface ornamentation. The gynoecium is bi- or tricarpellate and the fruits are nuts. The family is restricted to the Southern Hemisphere, where it shows a classic disjunct distribution between Australasia and South America.

The fossil record of Nothofagaceae is extensive. Currently, all reliable reports are restricted to those regions of the Southern Hemisphere where the genus grows today, as well as Antarctica. There are no reliable records of *Nothofagus* from southern Africa, India or the Northern Hemisphere. In Cretaceous strata the family is represented from the Campanian onwards by pollen of *Nothofagidites* from Australia (Dettmann, 1994) and wood assigned to *Nothofagoxylon* from Antarctica (Poole, 2002). The first record of extant *Nothofagus* is pollen from the Maastrichtian of Australia (Dettmann, 1994). In the Cenozoic, the Nothofagaceae are widely distributed and well documented based on numerous records of pollen, fruits, inflorescence fragments, leaves and wood from the Southern Hemisphere (e.g. Dettmann *et al.*, 1990; Hill and Dettmann, 1996; Poole, 2002).

Fagaceae. The Fagaceae are a family of trees and shrubs, typically with small, simple and unisexual flowers that are wind-pollinated or, in many Quercoideae, insect-pollinated. Flowers are borne in spikes, heads or cymules. Individual female flowers or dichasial clusters of flowers are surrounded by an often woody cupule formed from several series of bracts. Pollen is prolate, tricolporate and tectate with a columellate infratectal layer. The gynoecium is bicarpellate or more commonly tricarpellate, and the ovary is bi- or trilocular. There are two anatropous ovules per carpel. Fruits are nuts (Kubitzki, 1993c).

The Fagaceae are subdivided into two subfamilies. Quercoideae (including Castanoideae) comprises six genera (*Castanea, Castanopsis, Chrysolepis, Lithocarpus, Quercus, Trigonobalanus*) predominantly distributed in temperate and subtropical regions of the Northern Hemisphere, but with some representatives at higher altitudes in the tropics. Fagoideae includes a single genus, *Fagus*, which is sister to the other genera. *Fagus* is predominantly temperate in distribution.

The Fagaceae have an extensive fossil record and there are numerous records of leaves, inflorescence fragments, flowers, pollen and wood in many Cenozoic floras from the Northern Hemisphere. The earliest records of the Fagoideae are leaves and cupules with fruits from the Eocene that are readily assigned to the extant genus *Fagus* (Manchester and Dillhoff, 2004; Denk and Grimm, 2009) but Quercoideae are also diverse in the Eocene (e.g. Mai, 1995). In Late Cretaceous and Paleocene floras there are many different leaves with general fagaceous morphology. These have been assigned to extinct genera such as *Dryophyllum* and *Quercophyllum* (Mai, 1995), but their systematic affinities are in need of thorough re-examination. However, the presence of fossil flowers with distinctive fagaceous characters from Santonian strata indicates that the family was already diverse by the mid-Late Cretaceous. Fossil wood assigned to the Fagaceae is described from the Maastrichtian of Mexico (Estrada-Ruiz *et al.*, 2007).

Protofagacea allonensis (Figure 14.21) from the Late Santonian Allon locality, Georgia, USA, is based on staminate inflorescences and isolated staminate flowers (Herendeen *et al.*, 1995). Inflorescences are pedunculate dichasia typically with seven flowers, or sometimes with three or five. Flowers are sessile and closely spaced in the inflorescence. They are actinomorphic, about 1 mm long, with a perianth consisting of two whorls of three tepals, and an androecium consisting of two whorls of six stamens. Stamens are well differentiated into filament and anther. Anthers are dorsifixed, tetrasporangiate and dithecate, and dehisce by longitudinal slits. Pollen observed on the flowers is small and tricolpate with a foveolate to reticulate tectum.

Associated fruits are linked to the same species by the presence of similar pollen. Fruits are triangular or lenticular in outline. Remains of short tepals at the apex of the gynoecium indicate that the pistillate flowers had an inferior ovary. Triangular fruits are tricarpellate with six tepals. Lenticular fruits are bicarpellate. Associated cupules are pedunculate and four-lobed with scars indicating the presence of a median lenticular fruit and two lateral triangular fruits. Cupules bear several orders of bracts subtending each of the four lobes of the cupule.

The fossils show a suite of characters indicating a close relationship to extant Fagaceae, and particularly extant castaneoids, but they differ from all extant taxa in their pollen morphology and inflorescence structure (Herendeen *et al.*, 1995).

14.6.8 Ticodendraceae, Betulaceae and Casuarinaceae

Ticodendraceae. The Ticodendraceae are a monotypic family comprising the single species *Ticodendron incognitum*, which is today restricted to Central America. *Ticodendron* is a wind-pollinated tree with small unisexual and apetalous flowers aggregated in dense inflorescences. The gynoecium is inferior and bicarpellate with two hemi-orthotropous ovules per carpel. The fruit is a drupe (Gómez-Laurito and Gómez, 1989). Pollen of *Ticodendron*

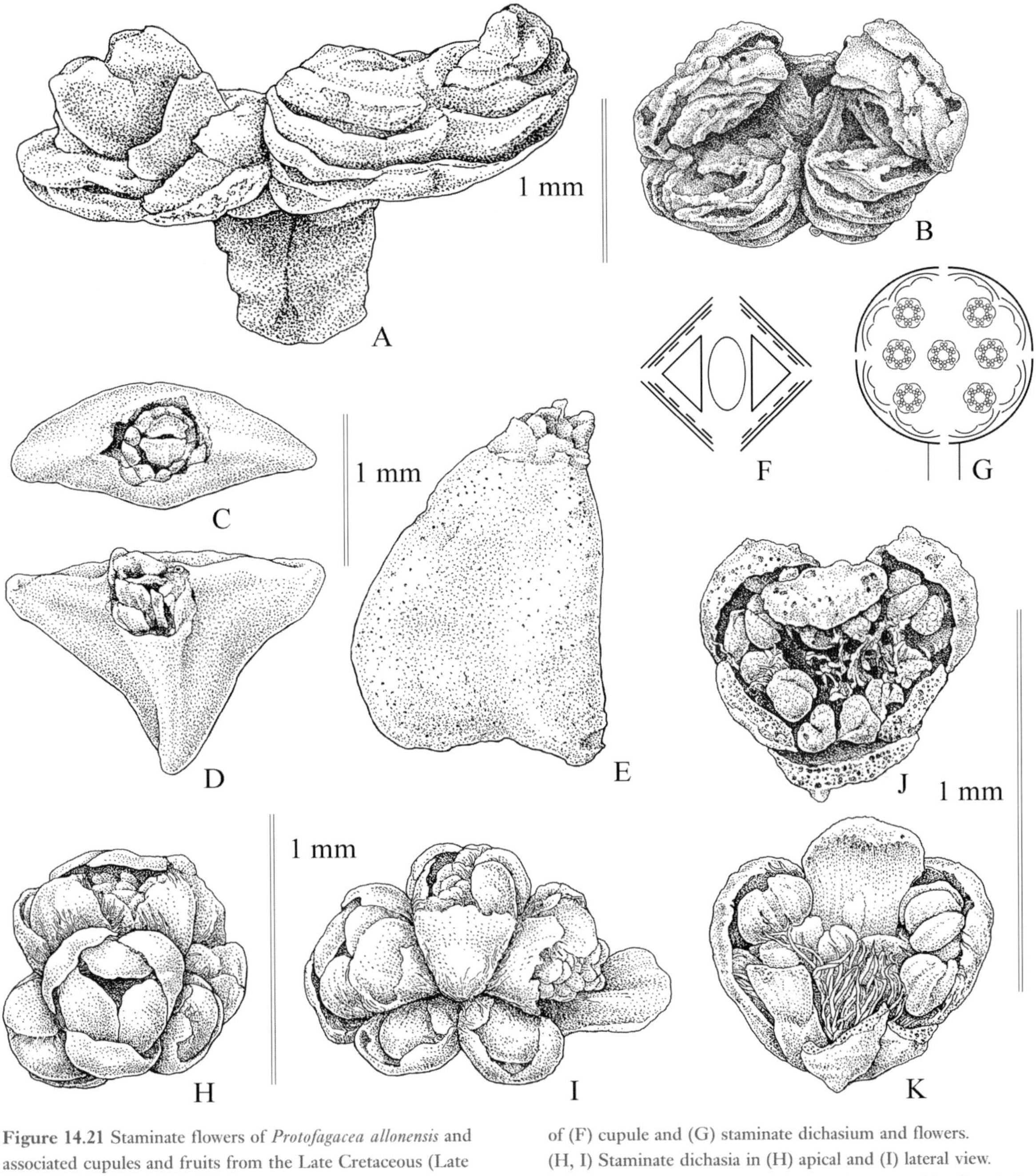

Figure 14.21 Staminate flowers of *Protofagacea allonensis* and associated cupules and fruits from the Late Cretaceous (Late Santonian) Allon locality, Georgia, USA. (A, B) Mature cupules in (A) lateral and (B) apical view. (C, D) Apical view of (C) lenticular and (D) triangular fruits showing epigynous organisation. (E) Lateral view of triangular fruit. (F, G) Inflorescence diagrams of (F) cupule and (G) staminate dichasium and flowers. (H, I) Staminate dichasia in (H) apical and (I) lateral view. (J, K) Staminate flower with three tepals removed in (J) apical and (K) lateral view. Drawn from SEM images and line drawings in Herendeen *et al.* (1995).

is suboblate and triporate with pores slightly displaced towards one pole. The pollen wall is tectate with finely spiny surface ornamentation and a granular infratectal layer (Feuer, 1991).

The Ticodendraceae were recognised only recently (Gómez-Laurito and Gómez, 1989) and have a restricted distribution in Central America. Fossil fruits from the Eocene of Oregon, USA, described as *Ferrignocarpus bivalvis* (Manchester, 1994), and anatomically identical fruits from Early Eocene London Clay of southern England (*Carpolithus* sp. 38, Reid and Chandler, 1933), closely resemble those of extant *Ticodendron* and perhaps indicate a much wider distribution of the family in the past (Manchester and Renner, 2005). This is further supported by the presence of *Ticodendron*-like pollen associated with the *Ferrignocarpus* fruits. However, the fossils also show similarities to Betulaceae and their exact position in the Ticodendraceae–Betulaceae–Casuarinaceae clade is currently under investigation (S. R. Manchester, personal communication, 2010).

Betulaceae. The Betulaceae include six genera and about 150 species of trees and shrubs that are mainly distributed in temperate and cold regions of the Northern Hemisphere. All extant Betulaceae are wind-pollinated with small unisexual flowers borne in lax or upright (*Alnus*) thyrsic inflorescences. Partial inflorescences are unisexual dichasia, usually comprised of two or three pistillate flowers or three staminate flowers. Staminate flowers have two to four tepals, or tepals are lacking. Pistillate flowers are naked or have small tepal lobes. Pollen is oblate and triporate, or in *Alnus* has five pores. The pollen wall is tectate, scabrate to finely rugulate, and the infratectal layer is granular. The gynoecium is bicarpellate and inferior with two pendulous anatropous ovules per carpel. Fruits are nuts, which may have small lateral wings or larger wings formed from the bracts of the dichasia. Fruits of *Corylus* are larger and animal-dispersed.

The Betulaceae have a rich and diverse fossil record in the Northern Hemisphere through the Cenozoic. Some Normapolles flowers from the Late Cretaceous, such as *Bedellia* and *Endressianthus*, show close similarity to extant Betulaceae, but cannot be assigned to the family. They may be on the stem group of the extant family. The earliest unequivocal fossils of Betulaceae are those of *Palaeocarpinus*, first described from the Paleocene of southern England (Crane, 1981) and subsequently reported from several Paleocene and Eocene floras in North America and Europe (Crane *et al.*, 1990). *Palaeocarpinus* plants are now known based on fruits enclosed in spiny involucral bracts and also associated staminate inflorescences and leaves (Crane *et al.*, 1990). There are also unequivocal records of *Alnus, Betula, Corylus* and *Ostrya* in the Early Cenozoic, as well as the extinct genera *Asterocarpinus* and *Cranea*.

Casuarinaceae. The Casuarinaceae are a small family of about 95 species in four genera of trees or shrubs distributed in Southeast Asia, Australia and Melanesia. The morphology is very distinctive with strongly reduced, membranous leaves in a whorled arrangement along slender stems. Staminate flowers are unisexual and borne in dense spikes. Pistillate flowers are naked and borne in globose or elliptical heads. Fruits are bicarpellate with two or four orthotropous ovules. Fruits are winged nuts. At maturity the subtending bracts become lignified and form woody cones. Pollen is oblate and triporate (Johnson and Wilson, 1993).

There are no fossil Casuarinaceae reported from the Cretaceous, but the family was already widespread and diverse in the early part of the Cenozoic. The earliest occurrence of Casuarinaceae is pollen from the Early Paleocene of New Zealand (Campbell and Holden, 1984). Soon after, later in the Paleocene, Casuarinaceae pollen appears in palynofloras from Australia, Argentina and the Indian Ocean. Stem fragments, inflorescences and infructescences are also known from the Early Cenozoic. Taken together these records document a much wider distribution of the family in the past than today.

14.6.9 Myricaceae, Rhoipteleaceae and Juglandaceae

Myricaceae. The Myricaceae include three genera of shrubs or small trees. *Canacomyrica* and *Comptonia* are both monotypic with a very restricted present-day distribution: *Canacomyrica* in New Caledonia and *Comptonia* in eastern North America. *Myrica* has about 50 species and a broader, but also rather patchy, distribution that probably indicates a much wider distribution in the past. *Myrica* is also significant in being the only genus of core Fagales that is widespread in Africa. There is also a single, endemic, probably relictual species in the Macaronesian islands (*Myrica faya*). Flowers are small and unisexual, often without a perianth. Staminate flowers have two to eight stamens, rarely more. Pollen is triporate and peroblate. The gynoecium is syncarpous and bicarpellate with a single orthotropous and erect ovule. Fruits are drupes with a thin exocarp and a small often spherical or ovoid endocarp.

The Myricaceae have a rich fossil record of both leaves and fruits, which indicate that the family was once much more widely distributed. Both *Comptonia* and *Myrica* were

Figure 14.22 *Rhoiptelea chiliantha* (Rhoipteleaceae), the horse-tail tree, from Yunnan, China, with persistent pendulous inflorescences, a survivor of the Normapolles complex; note winged fruits and pinnate leaves.

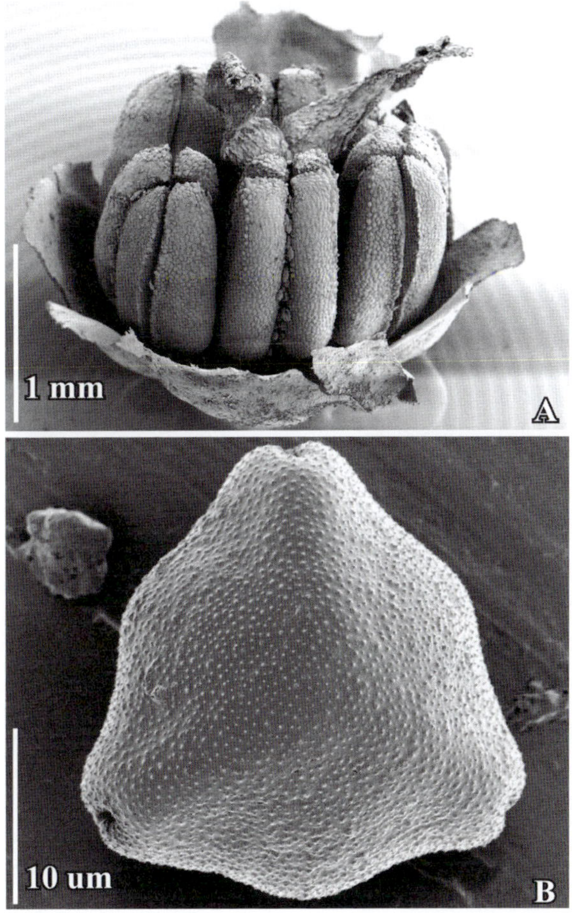

Figure 14.23 *Rhoiptelea chiliantha* (Rhoipteleaceae). (A) Bisexual flower showing tepals, stamens and extended styles. (B) Isolated pollen grain. SEM images of material from Yunnan, China.

diverse and widespread in the swamp floras of Europe and Asia during the later part of the Cenozoic. *Comptonia* is represented by very characteristic narrowly elongate and pinnatifid leaves, as well as slender endocarps. *Myrica* is particularly well represented by its characteristic fruits, which sometimes have their original warty, waxy surface preserved (Friis, 1985b).

Rhoipteleaceae. The Rhoipteleaceae are monotypic with a single species, *Rhoiptelea chiliantha*, restricted to small areas of China and Vietnam. *Rhoiptelea* is a wind-pollinated tree. The Chinese name Maweishu (horse-tail tree) refers to the appearance of the persistent, long, lax, branched inflorescences (Figure 14.22) (Sun *et al.*, 2006). Partial inflorescences are dichasia with three flowers. The median flower is bisexual and the two lateral flowers are pistillate. Studies on floral phenology and sex expression show that *Rhoiptelea* in some populations has functionally gynomonoecious inflorescences (Sun *et al.*, 2006). Flowers have a simple perianth of four tepals in two opposite–decussate pairs in a hypogynous position. In bisexual flowers the androecium consists of six stamens (Figure 14.23). Pollen is slightly heteropolar, and tricolporate with very short outer apertures (Figure 14.23). Grains are peroblate and triangular in shape with protruding corners and with distinct arci between the apertures. The gynoecium is bicarpellate, syncarpous and contains a single anatropous ovule. The fruit is flat and winged. *Rhoiptelea* has an indumentum of simple trichomes as well as peltate glandular trichomes. In populations from Yunnan, Sun *et al.* (2006) observed that the median bisexual flower in the dichasium flowered before the lateral, pistillate flowers, released its pollen and abscised without developing into a fruit. In other populations the median bisexual flower develops into fruit while the pistillate flowers are aborted (S.Q. Huang in progress).

Dispersed pollen grains assigned to the Normapolles genus *Plicapollis*, which is known from the Late Cretaceous and Early Cenozoic, are closely similar to pollen of *Rhoiptelea* (Friis *et al.*, 2006b); it is possible that some of these grains were produced by *Rhoiptelea*-like plants (section 14.6.6). There are also reports of dispersed pollen assigned to *Rhoiptelea*. The earliest of these is from the Maastrichtian of North America (Wolfe, 1973). Fruits assigned to *Rhoiptelea* have been described from Campanian to Maastrichtian floras of Central Europe (Knobloch and Mai, 1986). However, few diagnostic characters are preserved on these fruits, and their relationship to Rhoipteleaceae is not fully secure. In those cases where *Plicapollis* grains have been found *in situ* they are consistently associated with flowers and fruits of *Caryanthus* and *Budvaricarpus* (section 14.6.6). Flowers of *Caryanthus* and *Budvaricarpus* are similar to those of *Rhoiptelea* in having four tepals, six stamens and a bicarpellate, syncarpous gynoecium with a single ovule, as well as similar trichomes and glands, but they have a semi-inferior ovary and they cannot be included in the family as currently circumscribed (Friis *et al.*, 2006b).

Juglandaceae. The Juglandaceae are a well-supported monophyletic group (e.g. Manos *et al.*, 2007). The family comprises eight genera and about 60 species, mostly of trees, or more rarely, large shrubs. They are distributed mainly in temperate regions of the Northern Hemisphere, but with some diversity also in tropical regions (Stone, 1993). All Juglandaceae, except possibly *Platycarya*, are wind-pollinated. Flowers are mostly small, unisexual and apetalous, usually with pistillate and staminate flowers on the same plant. Staminate flowers often have rudiments of a pistil. Pollen is triporate, or more rarely polyporate, peroblate and tectate with a granular infratectal layer (Stone and Broome, 1975). Pistillate flowers have an inferior bicarpellate ovary with a single, orthotropous ovule at the base. The ovary is incompletely partitioned at the base and unilocular towards the apex. Fruits are nuts or samaras (Stone, 1993).

The Juglandaceae have no unequivocal Cretaceous record. However, the Cenozoic record is very extensive (Chapter 20) and many extinct genera from North America, Europe and Asia document a much wider distribution and greater diversity than today, especially in the Early Cenozoic (Manchester, 1987a). The fossil record includes numerous well-preserved fruits, often with elaborate bracts that appear to be adaptations for wind dispersal. Other fruits are larger and clearly modified for animal dispersal. There is also a great diversity of fossil leaves, as well as abundant dispersed pollen and fossil wood. The geological history of the family has been summarised in several important overviews (e.g. Manchester, 1987a) and the position of the extinct genera has been evaluated in a phylogenetic context (Manos *et al.*, 2007).

14.7 MALVIDS (EUROSIDS II)

The malvid clade comprises the Crossosomatales, Picramniales, Sapindales, Huerteales, Brassicales and Malvales (APGIII, 2009). Molecular and morphological support for the clade is strong (Soltis *et al.*, 2005; Endress and Matthews, 2006a). Important shared features include campylotropous ovules with a zigzag micropyle, the presence of gynophores and androphores, a tendency towards gynoecia with more than five carpels, contorted petal aestivation and monosymmetric flowers (Endress and Matthews, 2006a).

14.7.1 Geraniales

The Geraniales are a small order of three families (Geraniaceae, Melianthaceae including Francoaceae, Vivianiaceae including Ledocarpaceae) (APGIII, 2009) comprising about 17 genera and 836 species of herbs, shrubs or trees (Stevens, 2001 onwards). The Geraniaceae are widely distributed in both hemispheres, whereas the Vivianiaceae occur only in restricted parts of South America and the Melianthaceae are restricted to Africa south of the Sahara and South America. Currently the fossil record of Geraniales is extremely sparse and based entirely on scattered records of pollen assigned to *Geranium, Erodium* and *Pelargonium* (Geraniaceae) from the younger Cenozoic (Muller, 1981).

14.7.2 Myrtales

The Myrtales comprise nine families (Alzateaceae, Combretaceae, Crypteroniaceae, Lythraceae, Melastomataceae, Myrtaceae, Onagraceae, Penaeaceae including Olinaceae, Vochysiaceae) and more than 11 000 species in about 380 genera. Several of these families, including Lythraceae, Onagraceae, Myrtaceae and Combretaceae, have an excellent Cenozoic record.

Seeds assigned to extant *Decodon* (Lythraceae), and to extinct genera thought to be closely related to *Decodon* (*Alatospermum, Microdiptera, Mneme*), are common in fossil floras from Europe and North America, with their first occurrences in the Eocene (Friis, 1985b). Recently, fruits and vegetative parts of *Decodon* have also been reported from both the Miocene of Europe (Kvaček and Sakala, 1999) and the Eocene of North America (Little and Stockey, 2003).

The Onagraceae are represented in fossil floras from Europe and Asia by a variety of tiny seeds assigned to the extant genus *Ludwigia*, which are known from the Oligocene onwards (Friis, 1985b). Seeds of possible Onagraceae, assigned to the fossil genus *Palaeeucharidium*, are known from the Eocene London Clay flora of southern England (Reid and Chandler, 1933). The Onagraceae are also present in the fossil record of the Southern Hemisphere with pollen assigned to the extant genus *Fuchsia* reported from the Eocene of Argentina (Zetter *et al.*, 1999).

The Myrtaceae also have an extensive fossil record in Europe and North America and are known from the Paleocene–Eocene onwards (Friis, 1985b; Manchester, 1999). The Myrtaceae are particularly well represented in the Cenozoic floras of the Southern Hemisphere with numerous records of leaves and pollen from Australia (Christophel, 1994; Macphail *et al.*, 1994; Zetter *et al.*, 1999). If correctly assigned, pollen assigned to the extinct genus *Myrtaceidites* from the Maastrichtian of Australia and Paleocene of New Zealand provides the earliest record of the family (Macphail *et al.*, 1994).

The fossil record of Combretaceae consists mostly of leaves and wood from the Cenozoic and there are many records of *Terminalioxylon* wood from tropical regions of the world, including Burma, India, Africa and South America (Mädel-Angeliewa and Müller-Stoll, 1973). There are also several reports of dispersed combretaceous pollen in tropical Cenozoic floras with the earliest records from the Late Eocene of Cameroon (Muller, 1981). The systematic affinity of the several combretaceous reproductive organs reported from the Cenozoic of Europe and North America remains to be fully documented (Friis *et al.*, 1992).

Two genera from the Late Cretaceous, *Esgueiria* and *Gyrocarpusocarpon*, both based on reproductive structures, have been compared to extant Combretaceae. *Gyrocarpusocarpon* is based on a single winged fruit from the Deccan Intertrappean Beds of Mohgaon Kalan, India (Mistri and Kapgate, 1990). The fruit may be combretaceous, but full details of its structure, and comparisons with extant taxa, have not yet been published. *Esgueiria* is known from the Late Cretaceous of Europe and Asia, and potentially similar fossils have also been reported from North America. It shows close similarity to certain extant Combretaceae, but cannot yet be assigned to the family with certainty. We therefore treat it here as unassigned within Myrtales.

Esgueiria (Figure 14.24) is an extinct genus based on numerous well-preserved flowers, fruits and inflorescence fragments from the Late Cretaceous (Campanian–Maastrichtian) Esgueira and Mira localities, Portugal. Two distinct species were described from these Portuguese floras: *Esgueiria adenocarpa* from the Esgueira locality and *E. miraensis* from the Mira locality (Friis *et al.*, 1992). Subsequently an additional species, *E. futabensis*, has been recovered from the Late Cretaceous (Early Coniacian) Kamikitaba locality, Honshu, Japan (Takahashi *et al.*, 1999a). Small bisexual flowers with an inferior ovary, somewhat resembling those of *Esgueiria* have also been described from the Late Santonian Allon locality, Georgia, USA (Flower with spindle-shaped inferior ovary, Herendeen *et al.*, 1999a), but details of perianth and ovary are lacking. There is also no evidence of the peltate multicellular glandular trichomes, and the relationship of these fossils with *Esgueiria* is not secure.

Flowers of *Esgueiria* are densely spaced, and were possibly borne in a racemose inflorescence, but only fragments of inflorescence have been recovered and the organisation is not fully known. Flowers are small, up to about 2.2 mm long, and bisexual with a pentamerous perianth, a basically pentamerous androecium, a trimerous gynoecium with three free styles, and an inferior ovary. The calyx is persistent and consists of five free, imbricate sepals. The corolla consists of five free petals in a contorted arrangement. The organisation of the androecium is difficult to study. In most specimens corolla and androecium have been shed, but the arrangement of stamens has been studied in young specimens and in serial sections of flower buds. Apparently there are two whorls of stamens with three stamens in one whorl and five stamens in the other. Stamens are differentiated into a long filament and a short tetrasporangiate, dithecate anther. Anthers are dorsifixed and sagittate. Pollen observed in the flower (on the stigma and surface of the flower) is small, apparently tricolpate and tectate–foveolate. The ovary is inferior and unilocular with up to six ovules pendent from an apical placenta. Fruits are small, elongate nuts with a single seed. A low ring-shaped disc around the base of the styles is interpreted as a nectary. Outer surfaces

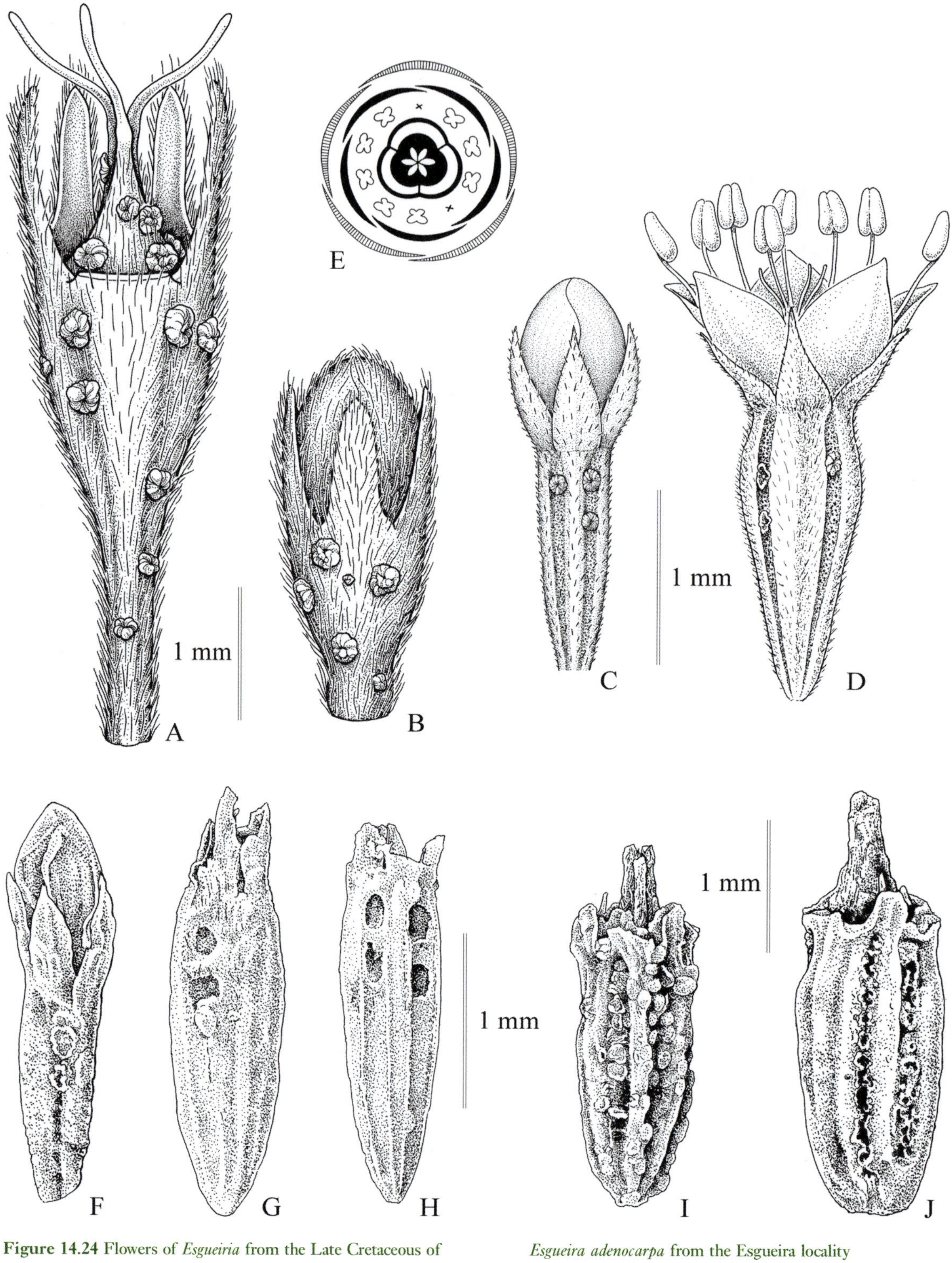

Figure 14.24 Flowers of *Esgueiria* from the Late Cretaceous of Portugal and Japan. (A–D) Reconstructions and (E) floral diagram of (A, B) *Esgueiria adenocarpa* and (C, D) *Esgueiria miraensis* from the Campanian–Maastrichtian of Portugal. (F–H) Flowers of *Esgueira adenocarpa* from the Esgueira locality (Campanian–Maastrichtian), Portugal. (I, J) Flowers of *Esgueiria futabensis* from the Kamikitaba locality (Early Coniacian), Japan. (A–E) redrawn from Friis *et al*. (1992); (F–J) from Friis *et al*. (2006a).

of ovary and perianth parts are covered by simple stiff hairs and distinctive peltate, multicellular, glandular trichomes.

The features observed in the fossil flowers strongly suggest a relationship to extant Myrtales and particularly to Combretaceae. Within Combretaceae the fossil flowers are most similar to those of *Guiera* in the tribe Combreteae. However, the fossils differ from *Guiera* and all other Combretaceae in having three free styles rather than a single style. This and other floral features of *Esgueiria* (e.g. lack of hypanthium cup) may be plesiomorphic in the family (Friis *et al.*, 1992) and perhaps suggest that *Esgueiria* is among stem-group Combretaceae.

14.7.3 Crossosomatales and Picramniales

The Crossosomatales include seven families (Aphloiaceae, Crossosomataceae, Geissolomataceae, Guamatelaceae, Stachyuraceae, Staphyleaceae, Strasburgeriaceae including Ixerbaceae) of shrubs and trees. The order is placed securely within the rosids based on phylogenetic analyses of molecular data, and in the APGIII classification is placed as the sister group to all other malvid orders except the Geraniales and Myrtales. None of the families of Crossosomatales were placed together in older classifications based on morphology, but the new constellation suggested by molecular data is supported by comparative studies of floral morphology (Matthews and Endress, 2005b). All Crossosomatales have solitary flowers, each with a floral cup, a short-stalked gynoecium, and carpels with tips that are postgenitally united to form a compitum. The locules of the ovary taper upwards, and the integuments of the ovules form a zigzag micropyle (Matthews and Endress, 2005b).

The seven extant families of Crossosomatales have a very scattered distribution (e.g. Geissolomataceae restricted to the Cape region, Crossosomataceae to western North America, Stachyuraceae to Southeast Asia except for the included Ixerbaceae that is restricted to New Zealand), which probably indicates that the group is relictual. This is also supported by palaeobotanical evidence although the fossil record of the group is not extensive. Both Stachyuraceae and Staphyleaceae are known from Cenozoic floras in Europe, Asia and North America, where they are represented by seeds of *Stachyurus* (Stachyuraceae) and *Staphylea* and *Turnipia* (Staphyleaceae) (Tiffney, 1979; Mai, 1995). The oldest of these records are from the Oligocene of western Asia and Europe. The Strasburgeriaceae also had a wider distribution during the Cenozoic with pollen of *Bluffopollis scabratus*, which resembles that of *Strasburgeria*, reported from New Zealand as well as from areas outside the present distribution of the genus in Australia, Tasmania and New Caledonia (Jarzen and Pocknall, 1993). Currently there is no Cretaceous record of any member of Crossosomatales.

The Picramniales include a single family, the Picramniaceae, restricted to the New World. To our knowledge the family has no fossil record.

14.7.4 Sapindales

The Sapindales comprise nine families, mainly of trees and shrubs (APGIII, 2009). Several of these families have an extensive and well-documented fossil record from the Early Cenozoic onwards, but representation of Sapindales in the Cretaceous is relatively sparse. Relationships within the order are not yet fully worked out, but Anacardiaceae and Burseraceae appear to be sister taxa, while Meliaceae plus Rutaceae plus Simaroubaceae appear closely related. The Sapindaceae, which includes Aceraceae and Hippocastanaceae, are a further major group within the order.

Anacardiaceae. The Anacardiaceae comprise about 75 genera and almost 1000 species of trees and shrubs distributed in tropical to warm temperate regions of the world. The family is characterised by the presence of resin ducts, as well as distinctive fruits that are often asymmetrical (Stevens, 2001 onwards). The Anacardiaceae are extensively represented by fruits in the Early Eocene of southern England (Chandler, 1964) and North America (Manchester, 1999). There are also several Early Cenozoic records of wood comparable to extant Anacardiaceae. Leaf and pollen records from the Cretaceous are all uncertain, and require critical re-examination (Collinson *et al.*, 1993).

Burseraceae. The Burseraceae are a family of pantropical trees or shrubs, distributed predominantly in Africa and South America. There are about 600 species in around 20 genera (Cronquist, 1981). The family is not known from the Cretaceous, but has a good fossil record, based mainly on fruits, from the Eocene of southern England (Reid and Chandler, 1933; Chandler, 1961, 1964; Collinson, 1983a). Fruits and seeds of Burseraceae are also reported from the Eocene of North America (Taylor, 1990).

Meliaceae. The Meliaceae comprise about 50 genera and 500 species of trees and shrubs (more rarely herbs) with compound leaves. The family is widespread in tropical and subtropical regions, but has no unequivocal Cretaceous record. Fruits similar to those of *Guarea* are reported from the Maastrichtian and Paleocene of North America (Taylor, 1990) but the systematic assignment may be in need of revision (Collinson *et al.*, 1993). Otherwise the family is known from fossil fruits from the Eocene of southern England (Chandler, 1964), as well as fossil fruits and leaves from the Eocene and Oligocene of North America (Manchester and Meyer, 1987; Taylor, 1990).

Rutaceae. The Rutaceae are a large and well-defined family of aromatic trees and shrubs, or more rarely herbs. The family has about 150 genera and 1500 species and is almost cosmopolitan in distribution, with most species in tropical and subtropical regions (Cronquist, 1981). Records of the family from the mid-Cretaceous of North Africa (wood) and North America (leaves) need further documentation (Gregor, 1989). Seeds assigned to the extinct genus *Rutaspermum* (*Rutaspermum biornatum*) from the Late Cretaceous Maastrichtian flora of Walbeck (Knobloch and Mai, 1986) appear to be the earliest record of the family. These seeds are thought to be closely related to extant members of the subfamily Zanthoxyleae, and particularly to seeds of *Zanthoxylon* and *Fagara*. Seeds of *Rutaspermum* were first described from the Early Eocene London Clay flora of southern England (Chandler, 1964; Collinson, 1983a). Fossil seeds of clear rutaceous affinity are common in the younger Cenozoic floras of Europe, North America, and Asia, where several extant genera are well documented (Gregor, 1989; Manchester, 1999). The family is also documented by leaves and wood from the Middle Eocene of Europe (Wilde, 1989) and North America (Taylor, 1990).

Sapindaceae. As currently defined by molecular analyses the Sapindaceae also include both Aceraceae and Hippocastanaceae. Defined in this way the family comprises about 135 extant genera and more than 1500 species. Most are trees and shrubs, but rarely there are also herbaceous vines. The family is distributed world wide, but with most species in tropical and subtropical regions.

Many Sapindaceae have schizocarpous winged fruits, as for example in extant *Acer* and *Dipteronia*. Fruits of this kind have been reported from numerous Cenozoic floras with the earliest record comprising *Acer*-like fruits from the Paleocene of North Dakota (Crane *et al.*, 1990; Manchester, 1999). From the Eocene onwards the family has an extensive record based on twigs, wood and leaves (Chandler, 1964; Collinson, 1983a; Wolfe and Tanai, 1987; Wilkinson, 1988; Mai, 1995). Wood assigned to the form genus *Sapindoxylon* has been recorded from many tropical floras from the Late Cenozoic, for example from Vietnam, Sumatra, Java, Pakistan, Ethiopia, Egypt, Europe and Columbia. *Sapindoxylon* is also known from the Eocene of southern England (Wilkinson, 1988), where there are also reliably determined sapindaceous seeds assigned to extinct genera such as *Palaeallophyllus*, *Palaealectryon* and *Sapindospermum* (Chandler, 1964). Pollen referred to the extant genera *Cupania* and *Serjania* is documented from the Eocene of Argentina (Zetter *et al.*, 1999). The earliest records of Sapindaceae are *Sapindoxylon* from the Late Cretaceous Deccan Intertrappean Beds, India (Wilkinson, 1988), and seeds of *Sapindospermum* from the Late Cretaceous of Central Europe, which range in age from Late Turonian to Maastrichtian (Knobloch and Mai, 1986).

Simaroubaceae. The Simaroubaceae have been radically redefined based on recent phylogenetic analyses using DNA data. As currently circumscribed they comprise about 19 genera and 95 species of mostly tropical trees and shrubs (Stevens, 2001 onwards).

Fruits of some Simaroubaceae are distinctively winged, such as in *Ailanthus*. Similar winged fruits are known from many Cenozoic floras from North America, Asia and Europe (Manchester, 1999). The family is also represented by leaves similar to *Ailanthus* from the Eocene of North America and Europe (Collinson *et al.*, 1993). The only putative records of Simaroubaceae from the Cretaceous are fossil woods of *Ailanthoxylon* and *Simarouboxylon* from the Maastrichtian Deccan Intertrappean flora of India (Prakash, 1965).

14.7.5 Huerteales

According to the APGIII classification the Huerteales includes three families: Dipentodontaceae, Gerrardinaceae and Tapisciaceae. The two small genera, *Tapiscia* and *Huertia*, that currently comprise the Tapisciaceae were placed in Staphyleaceae in previous classifications (Cronquist, 1981). Both genera are trees with compound leaves and small bisexual, flowers with a superior ovary. The family is distinguished from other malvids by their seeds, which have a strongly bullate chalaza and a fibrous exotegmen (Corner, 1976). The disjunct distribution of Tapisciaceae, with one genus in China (*Tapiscia*) and the other in the West Indies and South America (*Huertia*), strongly suggests a much wider distribution in the past, and this is supported by the fossil record. Fruits of *Tapiscia* are known from the Eocene of Europe and North America (Mai, 1995; Manchester, 1999).

14.7.6 Brassicales

The Brassicales are a well-defined monophyletic group based on phylogenetic analyses of molecular, chemical and

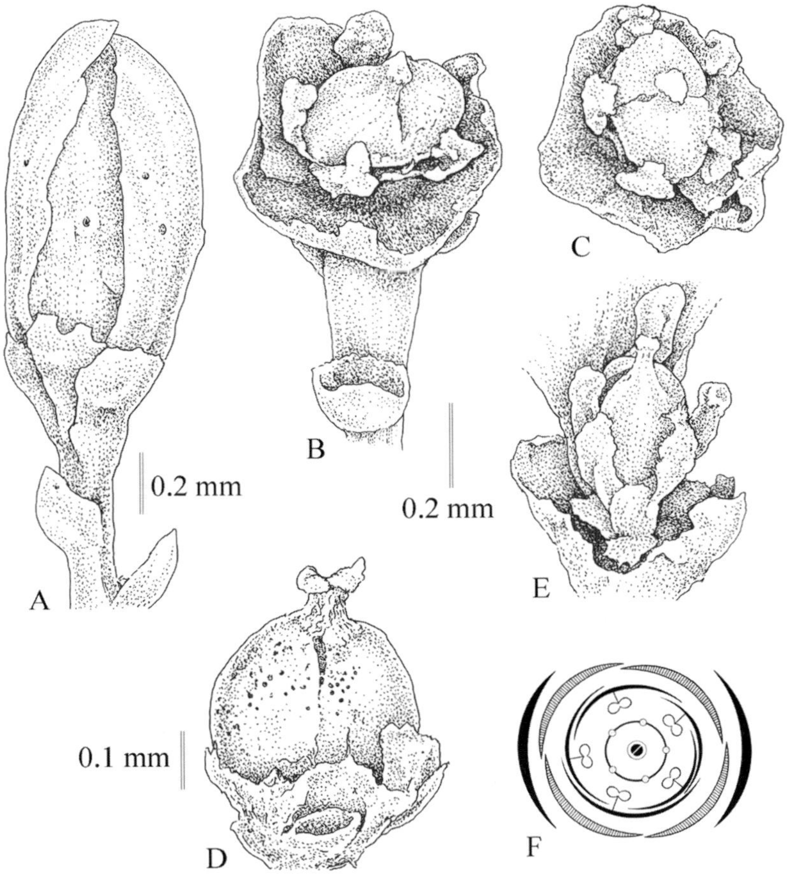

Figure 14.25 *Dressiantha bicarpellata* from the Late Cretaceous (Turonian) Old Crossman locality, New Jersey, USA. (A) Flower with corolla preserved. (B–E) Same flower with corolla removed seen in (B) oblique, (C) apical and (D, E) lateral view showing stamens and bicarpellate ovary. (F) Floral diagram. Drawn from SEM images in Gandolfo *et al.* (1998a).

morphological data (Judd and Olmstead, 2004). The order comprises 17 families (APGIII, 2009) all of which have either no fossil record or very limited representation in the Cenozoic (e.g. Brassicaceae). Against this background the Cretaceous fossil *Dressiantha* is especially remarkable.

Dressiantha bicarpellata (Figure 14.25) from the Late Cretaceous (Turonian) Old Crossman locality, New Jersey, USA (Gandolfo *et al.*, 1998a), includes small pedicellate and bisexual flowers, about 1 mm long, that are described as slightly zygomorphic. The calyx consists of four sepals, apparently in two opposite–decussate pairs, and the corolla consists of five imbricate petals. The androecium consists of one whorl of five stamens opposite the petals, and one alternating whorl of staminodes that are basally fused into a short tube around the base of the gynoecium. Stamens are differentiated into a long filament and a bisporangiate, monothecate, dorsifixed anther. Dehiscence is by a simple longitudinal slit. Pollen observed *in situ* is small, tricolporate and tectate–foveolate. The gynoecium is bicarpellate and syncarpous with the carpels fused only in their basal part of the gynoecium. The ovary is superior, but there is no information on placentation or ovules. The gynoecium is borne on a short broad gynophore.

The fossil flowers show several features not reported for other Late Cretaceous flowers including the presence of a gynophore and monothecate anthers. These and other features of *Dressiantha* indicate a relationship with extant plants included in Brassicales, although the fossils cannot be assigned to any particular family. In a cladistic analysis *Dressiantha* was resolved as sister group to all Brassicales except Gyrostemonaceae (Gandolfo *et al.*, 1998a).

14.7.7 Malvales

The Malvales comprise ten families (APGIII, 2009) that are strongly supported as a monophyletic group based on molecular, chemical and morphological data. However, no floral synapomorphies have yet been identified for the clade as a whole (Schönenberger and von Balthazar, 2006) and relationships within the order are much less certain. In many Malvales the androecium is polystemonous, but there is an enormous variation within the group including apparent multiple origins of polystemony. This, together with the uncertain interfamilial relationships, obscures interpretations of evolutionary pathways in the evolution of the androecium (Schönenberger and von Balthazar, 2006).

The Malvales are not well represented during the Cretaceous, but several families, in particular Dipterocarpaceae, Malvaceae and Thymeleaceae, are represented extensively during the Cenozoic.

Dipterocarpaceae. The Dipterocarpaceae are of considerable importance as major trees of lowland rain forest in Southeast Asia. Members of the family are also known from lowland areas of Africa and South America (Columbia). Similarities between fossil wood of *Woburnia porosa* from southern England and extant Dipterocarpaceae (Stopes, 1912) are accepted by Crawley (2001), but the original age assignment of *Woburnia* to the Early Cretaceous (Aptian) was questioned. Crawley suggests that this wood, like several other fossil woods described by Stopes as Early Cretaceous, is actually Palaeogene in age. Fossil woods of probable dipterocarpaceous affinity are widespread in younger Cenozoic floras from Asia (Java, Sumatra, India, Indo-China) and Africa (Somalia, east Africa) and have been assigned to genera such as *Dipterocarpoxylon, Shoreoxylon, Dryobalanoxylon* and *Grewioxylon* (Ramanujam, 1955; Bande and Prakash, 1986). Pollen resembling that of Dipterocarpaceae is first recorded from the Oligocene of Borneo (Muller, 1970). There are no reliable Cretaceous records of the family based on pollen, mesofossils or leaf fossils.

Malvaceae. The Malvaceae in the sense of APGIII (2009) are broadly circumscribed and include the previously separate Bombacaceae, Malvaceae, Sterculiaceae and Tiliaceae. The Malvaceae include herbs, shrubs and trees, distributed almost worldwide, but with a concentration of species in tropical regions. The Malvaceae exhibit considerable variety in floral form, but often flowers have numerous stamens, as is also the case in many other Malvales (Schönenberger and von Balthazar, 2006). There is also considerable diversity in pollen morphology. Pollen is typically tricolporate, tectate–reticulate or spiny and varies in shape from suboblate to prolate or subprolate.

There is clear evidence of Malvaceae from the Early Cenozoic. Fossil fruits assigned to the extant genus *Craigia*, which is now restricted to southern China, are known from numerous fossil floras in the Northern Hemisphere, including Asia, Europe, Spitsbergen and North America. The first occurrence is in the Eocene. Other well-documented Early Cenozoic records of the family include fossil wood from the Eocene of Oregon, USA, assigned to Sterculiaceae (Manchester, 1980), and fruits comparable to those of extant *Tilia* and *Willisia* (Chandler, 1964) from the Early Eocene London Clay of southern England. There are also many reports of dispersed fossil pollen similar to that of extant Malvaceae ranging from the Paleocene onwards (e.g. Zetter *et al.*, 1999). Cretaceous records of Malvaceae are much less secure. Wood from the Senonian of Egypt (*Hibiscoxylon*, Kräusel, 1939) requires re-examination. A fruit from Australia assigned to the Tiliaceae (*Etheridgea subglobosa*, von Ettingshausen, 1895) is not of Cretaceous age as indicated by von Ettingshausen, but from the Early Cenozoic (Jones, 1926; Rigby and Playford, 1988).

15

Early fossils of eudicots: asterids

The asterids are an extraordinarily diverse group of extant angiosperms that includes more than 80 000 species. The clade also includes two of the most species-rich angiosperm families, the Asteraceae and the Rubiaceae (Bremer *et al.*, 2004) as well as some of the largest angiosperm genera (e.g. *Solanum*). Asterids have an extensive fossil record in the Cenozoic, but their representation in the Cretaceous is more limited and largely confined to Cornales and Ericales.

15.1 CLASSIFICATION OF ASTERIDS

As currently defined by molecular data, asterids include four major clades: Cornales, Ericales, lamiids (Euasterid I), and campanulids (Euasterid II) (Figure 15.1). Cornales are resolved as the earliest-diverging of these four main lineages. They are followed by the Ericales, which are sister to the core asterids, which comprises lamiids and campanulids (APGIII, 2009). Molecular support for the recognition of the asterids as a whole is good, and there is also generally strong support for relationships within the group, although molecular support for the campanulids is weak (Judd and Olmstead, 2004; Soltis *et al.*, 2005).

The families that are now grouped together as asterids have been variously treated in previous classifications based on morphology. The lamiids and campanulids correspond in part to the subclass Sympetalae (Metachlamydeae) and some members of the order Umbelliflorae (subclass Archichlamydeae) in Engler's classification (Melchior, 1964). The core asterids correspond broadly to the subclass Asteridae in the classifications of Takhtajan (1969) and Cronquist (1981), but the group also includes many lineages placed by these earlier authors in the subclass Rosidae.

Morphological support for the group as a whole is weak and there are no unique unifying floral features. Important asterid floral characters noted by Judd and Olmstead (2004) include tenuinucellate, unitegmic ovules and sympetalous corollas, but these features are not shared by all asterids, and unitegmic ovules also occur occasionally outside the group. Several secondary chemical compounds,

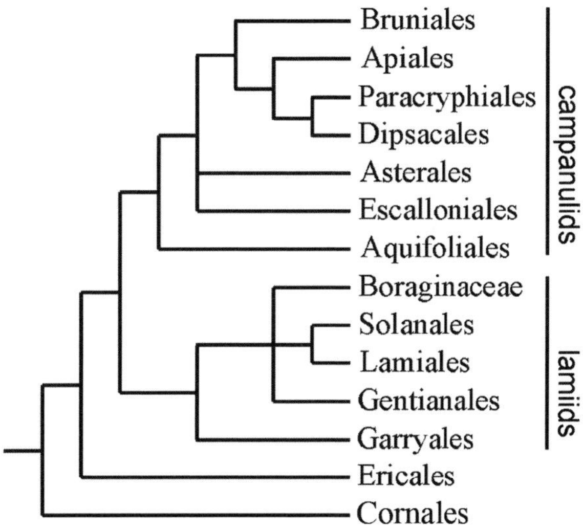

Figure 15.1 Summary of phylogenetic relationships among orders of extant asterids. Based on APGIII (2009).

such as iridoids, tropane alkaloids and caffeic acids, are mainly produced by members of the asterids, but none are universally present and some are also produced by non-asterid angiosperms. Many groups of asterids have an extensive fossil record in the Cenozoic, but in this chapter we focus particularly on those families and orders with a well-documented fossil history that extends back into the Cretaceous.

15.1.1 Asterid fossils of uncertain relationships

Most Cretaceous fossils that can be assigned to the asterids belong to the two lineages that diverged below the level of core asterids: Cornales and Ericales. There are very few Cretaceous fossils of core asterids. It is particularly remarkable that several of the most species-rich lineages of core eudicots, such as Lamiales, Apiales, and Rubiales, have either a very sparse or no reliable Cretaceous record, even though they are well represented in Cenozoic strata.

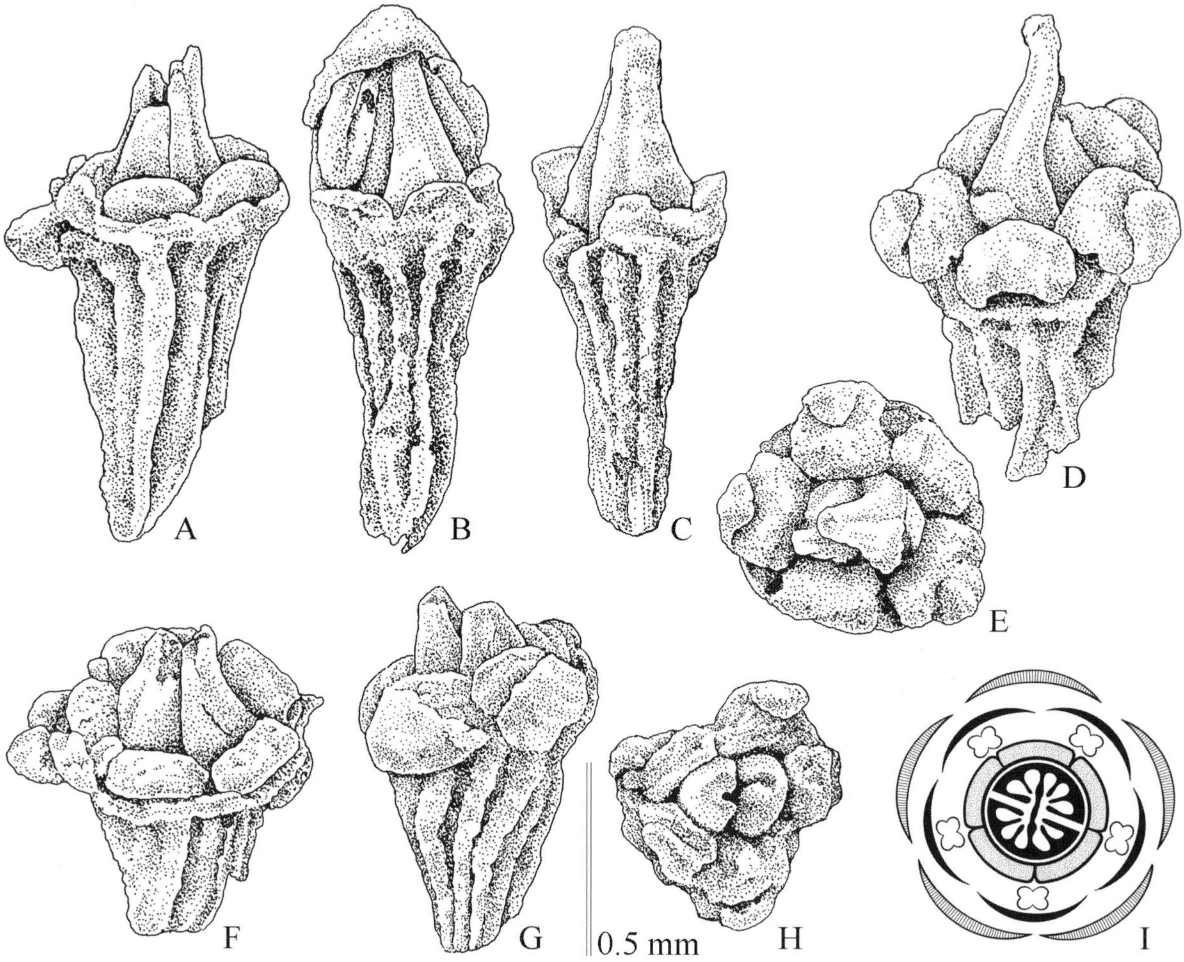

Figure 15.2 Unnamed flower with epigynous ovary from the Late Cretaceous (Late Santonian – Early Campanian) Åsen locality, Sweden. (A) Lateral view of flower showing prominent lobed nectary. (B) Lateral view of flower bud with one petal and two stamens preserved. (C) Lateral view of same flower with floral parts removed. (D) Lateral and (E) apical view of flower showing prominent five-lobed nectary and small calyx lobes. (F–H) Flower showing five-lobed nectary and two robust styles in (F, G) lateral (H) apical view. (I) Floral diagram. (A, C–E, G, H) From material in the collections of the Swedish Museum of Natural History, Stockholm; (B, F) from Friis *et al.* (2006a).

From the Late Cretaceous flora of Åsen, Sweden, several fossil flowers that are still under investigation are probably related to extant asterids, but additional structural information is needed before formal descriptions and systematic evaluations can be completed. One of these taxa includes small flowers with an inferior ovary figured as 'unnamed flower' by Friis and Skarby (1982) and compared to members of the Saxifragales *sensu* Takhtajan (1969). Within that complex of families the fossil flowers are most similar to those of extant families now placed among asterids.

The unnamed flower is about 3 mm long, actinomorphic and bisexual, with a pentamerous perianth and androecium, a bicarpellate gynoecium and an inferior ovary (Figure 15.2). It is known from several specimens. The calyx lobes are small. Petals are larger and

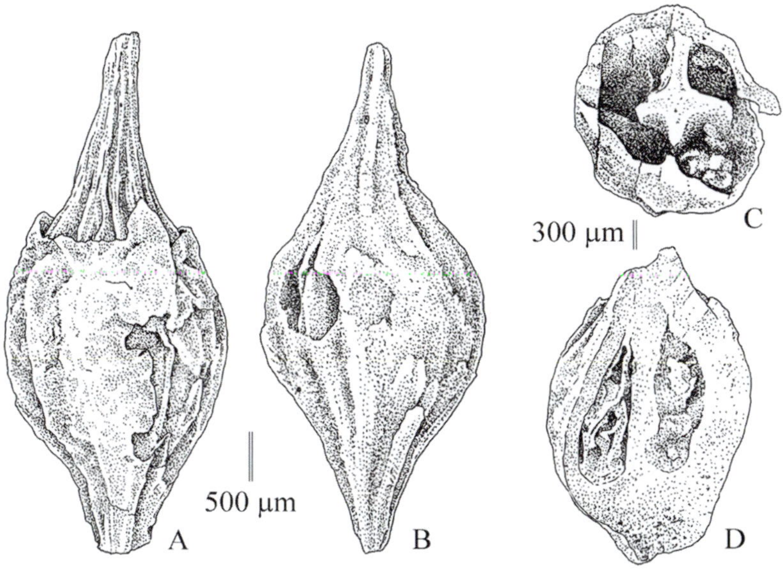

Figure 15.3 *Hironoia fusiformis* from the Late Cretaceous (Early Coniacian) Kamikitaba locality, Japan. (A) Lateral view of fruit showing hypanthium and sepals. (B) Lateral view of abraded fruit. (C) Fragmentary fruit in cross-section showing four locules. (D) Fragmentary fruit in longitudinal section showing locules. Drawn from SEM images in Takahashi *et al.* (2002).

ovate. The androecium consists of one whorl of free stamens. The ovary is bilocular with two stout styles. Each locule contains many ovules. A prominent 5-lobed nectary disc is inserted between the androecium and the styles. The flowers are closely similar to those of fossil *Silvianthemum* and extant *Quintinia* (section 15.19.1) and they probably belong to the same lineage.

15.2 CORNALES

The Cornales are the sister group to all other extant asterids and comprise six or seven families: Cornaceae, Curtisiaceae, Grubbiaceae, Hydrangeaceae, Hydrostachyaceae, Loasaceae and Nyssaceae (sometimes included in Cornaceae). This circumscription deviates substantially from previous classifications and ideas of relationships for the Cornales based on morphology. Interrelationships among the families currently included in Cornales are not fully resolved and morphological support for the order is not strong. Most members of the group have small insect-pollinated flowers with small sepals in an epigynous position and a distinct nectary disc. Fruits are mostly drupes with characteristic endocarps, except for Hydrangeaceae and Loasaceae, which have capsular fruits.

The fossil record of Cornales is extensive, particularly for the Cornaceae and Nyssaceae, which have drupaceous fruits with robust, characteristic endocarps that are easily recognised in the fossil record.

15.2.1 Fossil Cornales of uncertain relationships

The oldest well-documented occurrence of the Cornales is *Hironoia fusiformis* (Takahashi *et al.*, 2002) from the Late Cretaceous (Early Coniacian) Kamikitaba locality, Honshu, Japan. The species is based on small three-dimensionally preserved gynoecia preserved in the fruiting stage (Figure 15.3). Some specimens have a hypanthium and occasional sepals preserved, indicating that the flowers had four or six tepals. The androecium is not preserved. The fruits are fusiform in shape, formed from an inferior ovary, and taper above into a single stout style. The hypanthium is detached from most of the fruits and is only loosely attached to the fruit wall. The fruits are three-loculed or more commonly four-loculed drupes that dehisce by valves. Characters of the endocarp wall and dehiscence valves observed for the fossil occur in extant plants only in the *Nyssa* and *Mastixia* lineages of the Cornales. However, the fossil cannot be placed in any modern genus and there are no other characters that could place the fossil more precisely within the family. It is

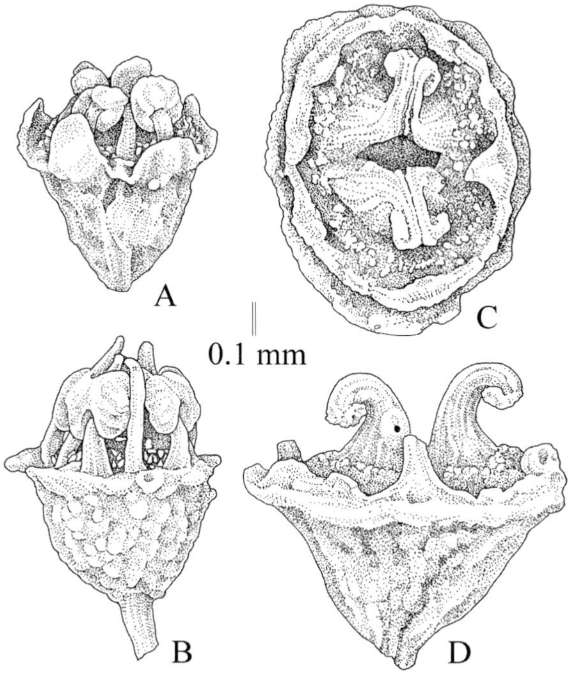

Figure 15.4 *Tylerianthus crossmanensis* from the Late Cretaceous (Turonian) Old Crossman locality, New Jersey, USA. (A) Lateral view of flower showing sepals and stamens. (B) Lateral view of flower with sepals and petals missing showing stamens and staminodes. (C, D) Mature fruit in (C) apical and (D) lateral view, showing apical dehiscence between carpels. (A, B, D) From Friis *et al.* (2006a), (C) drawn from SEM image in Gandolfo *et al.* (1998).

currently uncertain whether *Hironoia* can be placed within crown-group Nyssaceae or belongs to its stem lineage.

Tylerianthus crossmanensis (Figure 15.4) from the Turonian Old Crossman locality, New Jersey, USA (Gandolfo, 1998), is another fossil from the Late Cretaceous assigned to the Cornales. It comprises small bisexual and actinomorphic flowers with a semi-inferior ovary. Flowers are pedicellate, up to about 1.2 mm long, with a calyx of five free sepals. Petals were not observed on any of the fossil specimens and the flowers were described as apetalous. Because all specimens appear to be preserved at anthetic and post-anthetic stages it is possible that the petals had abscised. The androecium consists of one whorl of five shorter stamens alternating with five longer filament-like structures described as staminodes. Stamens are differentiated into a distinct filament and a basifixed, tetrasporangiate and dithecate anther that dehisces by simple longitudinal slits. Pollen found *in situ* in the stamens is minute, tricolporate and tectate–rugulate (described as reticulate). Internal to the androecium is a glandular nectary disc. The gynoecium is bicarpellate and the ovary bilocular with two free styles. Placentation is axile with several anatropous ovules in each locule. The fruit is a capsule that dehisces apically between the styles.

An unnamed flower very similar to *Tylerianthus crossmanensis* was reported from the Late Cretaceous (Late Santonian) Allon locality, Georgia, USA (Herendeen *et al.*, 1999a). The Allon flower appears to differ from *Tylerianthus crossmanensis* in only a few, relatively minor, details and it may belong to the same genus.

Flowers of *Tylerianthus crossmanensis* and the unnamed fossil from the Allon locality show features that indicate a close relationship with extant Hydrangeaceae (asterid), but also with Grossulariaceae (Gandolfo, 1998; Herendeen *et al.*, 1999a) that is now included in the Saxifragales (Chapter 13).

15.2.2 Cornaceae

The Cornaceae comprise two genera, *Cornus* and *Alangium*, with a combined total of about 80 species. Several genera that were previous included in the family (e.g. Cronquist, 1981) have now been assigned elsewhere. The Cornaceae are sister to the Nyssaceae, and the two are sometimes included in a single family (APGIII, 2009).

Cornus and *Alangium* are shrubs, small trees or more rarely woody climbers (species of *Alangium*). *Alangium* is distributed mainly in tropical regions of the Old World, while *Cornus* is distributed mainly in temperate regions of the Northern Hemisphere (Cronquist, 1981). The flowers are small, bisexual or rarely unisexual, and are usually aggregated into capitate or axillary cymes that are sometimes surrounded by petal-like bracts. Flowers have an inferior ovary and a nectary disc inserted between androecium and gynoecium. Pollen is typically colporate or porate, usually with three apertures, but there is a considerable diversity in aperture configuration and pollen wall ornamentation (e.g. Reitsma, 1970). The ovary is typically two- to four-loculed with a single anatropous ovule in each locule. Fruits are drupes with a thick-walled endocarp.

The Cornaceae have an extensive fossil record from the Cenozoic, but there are only few records from the Cretaceous. The fossil record of *Alangium* is based on leaves, fruits, wood and dispersed pollen with the earliest occurrences in the Maastrichtian and Paleocene of western North America (Eyde *et al.*, 1969; Reitsma, 1970; Krutzsch, 1989). According to Manchester (1999) leaves of *Alangium* are strikingly similar to the leaves of some Malvales and many of the leaf fossils assigned to *Alangium* may have been misidentified. Leaves and fruits assignable to *Cornus* are common in the fossil floras of North America and Europe with the oldest record from the Paleocene of North America (Friis, 1985b; Crane *et al.*, 1990; Mai, 1995; Manchester, 1999). Fossil fruits closely resembling those of extant *Cornus* are present in the Late Cretaceous (Late Santonian – Early Campanian) flora from the Åsen locality, Sweden, and are currently under investigation.

15.2.3 Nyssaceae

As currently circumscribed, the Nyssaceae include the Nyssaceae and Mastixiaceae of previous classifications. There are five extant genera (*Camptotheca*, *Davidia*, *Diplopanax*, *Mastixia, Nyssa*) that together comprise about 22 species of trees and shrubs distributed in eastern North America and in East Asia and Indo-Malesia. Flowers are typically small and bisexual, but sometimes staminate and functionally pistillate flowers occur on the same plant. Flowers are borne in a variety of inflorescences sometimes with large petal-like bracts. The perianth may be absent or have one or two whorls of parts. The ovary is inferior, with one or several locules and with a single ovule per carpel. Fruits are drupes with characteristic valvate dehiscence. Valves may extend for the full length of the endocarp (*Mastixia*) or be restricted to the apical part (*Nyssa*). In cross-section endocarps of *Mastixia* are distinguished by a dorsal invagination of the endocarp wall that gives the locule a distinctive U-shaped cross-section.

The Nyssaceae were important in Cenozoic vegetation of Europe and Asia, and the term Mastixioideen-Floren (mastixioid floras) is used for Cenozoic floras of Europe characterised by the presence of *Mastixia* and a variety of extinct forms closely related to *Mastixia* (e.g. *Beckettia*, *Eomastixia*, *Mastixiocarpus*, *Mastixiopsis*, *Retinomastixia*, *Ganitroceras*, *Tectocarya*), which occur together with many tropical–subtropical floristic elements. These Cenozoic mastixioid fossils are known mostly from their characteristic endocarps with valvate dehiscence and a U-shaped cross-section of the locule (Kirchheimer, 1937, 1957; Mai, 1964). They are especially common in the Eocene and Oligocene, but several of these fossil fruits also occur in the Miocene. *Mastixia* has also been recovered in the Eocene of North America, but the group was much less diverse in North America than it was in Europe (Mai, 1995; Manchester, 1999).

Fossils assigned to Mastixioideae are first seen in the Late Cretaceous (Campanian–Maastrichtian) of Central Europe. By the end of the Cretaceous the group was already diverse with five different species assigned to four extinct genera that all continue into the Eocene in Europe (*Beckettia*, *Eomastixia*, *Mastixiocarpus*, *Mastixiopsis*) (Knobloch and Mai, 1986).

Fossils closely related to extant *Nyssa* are also common in the Cenozoic floras of Europe and have also been recorded from both Asia and North America. The earliest records are endocarps from the Eocene of North America and Europe assigned to the extinct genus *Palaeonyssa* and to extant *Nyssa*. In Europe and Asia endocarps of *Nyssa* are particularly abundant in Oligocene and Miocene floras. Their occurrence indicates that *Nyssa* was an important element in the swamp forests that gave rise to the extensive brown coal deposits of Europe (Mai, 1995).

15.3 ERICALES

The Ericales are among those groups of angiosperms that have been re-defined extensively by molecular studies (e.g. Chase *et al.*, 1993). The order now includes families placed previously in the Dilleniidae of Takhtajan (1969) and Cronquist (1981), such as all families of Ericales, Ebenales and Primulales, as well as many families of Theales. The order is well supported by molecular data. Two major subclades can be recognised: core Ericales, comprising Cyrillaceae, Ericaceae, Clethraceae, Sarraceniaceae, Actinidiaceae and Roridulaceae; and primuloids, comprising Maesaceae, Theophrastaceae, Myrsinaceae and Primulaceae. However, the position of many of the other families is poorly resolved (Figure 15.5) and there are only few morphological/chemical features that unite the order (Anderberg *et al.*, 2002; Soltis *et al.*, 2005).

In the overview of Ericales given here (Figure 15.5) we use the framework provided by Soltis *et al.* (2005). The two first branches in the Ericales clade comprise a Marcgraviaceae–Balsaminaceae + Tetrameristaceae clade, and a Fouquieriaceae–Polemoniaceae clade. The fossil record of all these families is poor and to our knowledge there is no Cretaceous record of either group. These groups are not treated further here. Polemoniaceae have been reported from a single Cenozoic macrofossil and a few Late Cenozoic pollen records (Johnson, 2004). We also do not

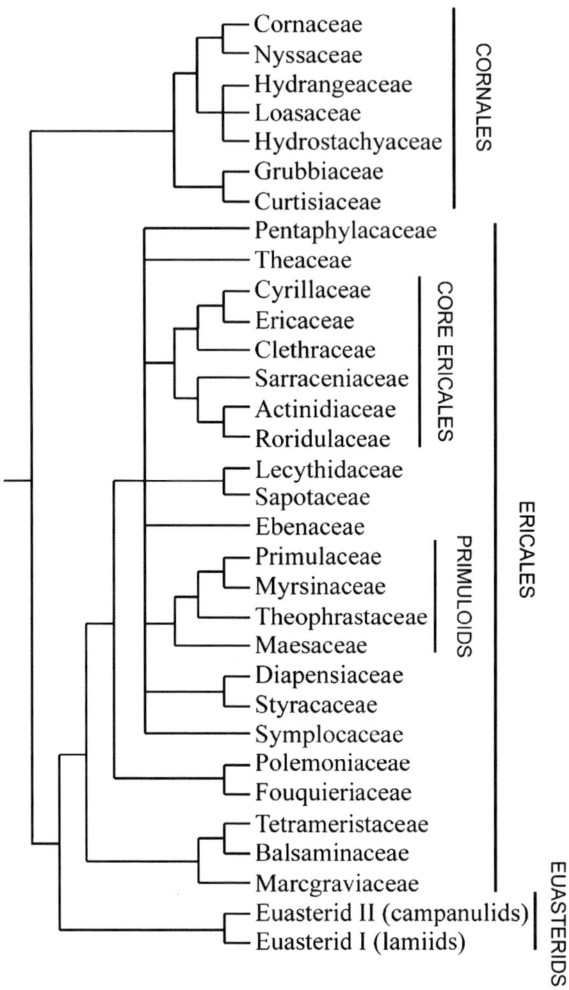

Figure 15.5 Summary of phylogenetic relationships among families of extant Ericales. Based on Soltis *et al.* (2005).

consider Lecythidaceae or Sapotaceae because the fossil record of these families is very sparse.

15.3.1 Ericalean fossils of uncertain relationships

Various fossil reproductive organs of ericalean affinity have been reported from the Late Cretaceous of Europe and North America. These include several different floral structures of which six have been formally named. However, there are still many undescribed flowers and fruits from Late Cretaceous mesofossil floras that are clearly related to Ericales. The group is well represented in late Cretaceous mesofossil floras; for example, in the Old Crossman mesofossil flora of New Jersey, USA, Crepet (1996) estimated that at least 15 different species of Ericales were present. Only one of the Late Cretaceous ericalean flowers that have been formally named can be assigned to a modern family with confidence (*Parasaurauia*, included in the Actinidiaceae). One other flower can be assigned to a subclade of Ericales (unnamed primuloid flower). However, four other flower types (*Actinocalyx bohrii*, *Paleoenkianthus sayrevillensis*, *Paradinandra suecica*, *Pentapetalum trifasciculandricus*) share features with several ericalean families and cannot be assigned to a specific family or subclade with certainty.

Several different types of dispersed fruits and seeds of probable ericalean affinity have been described (Figure 15.6), most of them from Late Cretaceous mesofossil floras from Central Europe (Knobloch and Mai, 1986). According to Knobloch and Mai (1986) these fossils can be related to families such as Actinidiaceae (*Saurauia alenae*, *S. antiqua*), Clethraceae/Cyrillaceae (*Discoclethra maxima*, *D. polysperma*, *D. valvata*, *Epacridicarpum cannelatum*, *E. cretaceum*, *E. rugulatum*, *Puriaeopsis campanulatus*, *Valvaecarpus debeyi*, *V. globulosus*, *V. kerharticensis*, *V. pterocaryaeformis*), Ericaceae (*Leucothoe praecox*, *Viticocarpum minimum*, *Diplycosiopsis walbeckensis*), Pentaphylacaceae (*Allericarpus clausenispermus*, *A. pentaphylacoides*, *Eurya carpatica*, *E. crassitesta*, *E. holyi*, *Pentaphylax protogaea*, *Protovisnea cancellata*, *P. erinacea*, *P. maii*, *P. reticulata*, *P. saxonica*, *P. tetragonalis*, *P. zahajensis*, *Visnea minima*) and Theaceae (*Palaeoschima austriaca*, *P. becvensis*, *P. microvalvata*) (Knobloch and Mai, 1986). However, the amount of structural detail available for these fossils varies greatly, and several are based on capsular fruits without information on floral features or on seeds. In these cases the particular relationship to extant families among Ericales remains to be established with certainty. However, combined evidence from the fruits and seeds indicates that these fossils are very likely to be broadly related to Ericales, and many probably belong to core Ericales. Together with the evidence from fossil flowers they document that Ericales were well established and diverse in Late Cretaceous floras.

Actinocalyx bohrii (Figure 15.7) described from the Late Santonian – Early Campanian Åsen locality, Sweden (Friis, 1985c), is based on small actinomorphic and bisexual flowers, about 1 mm long. These flowers have a hypogynous, pentamerous perianth and androecium, and a trimerous gynoecium. Sepals are imbricate and persistent. In young flowers the sepals are upturned, but in older flowers they become woody and spread into a star-shaped, slightly irregular (often bilateral) form. Petals are fused at the base

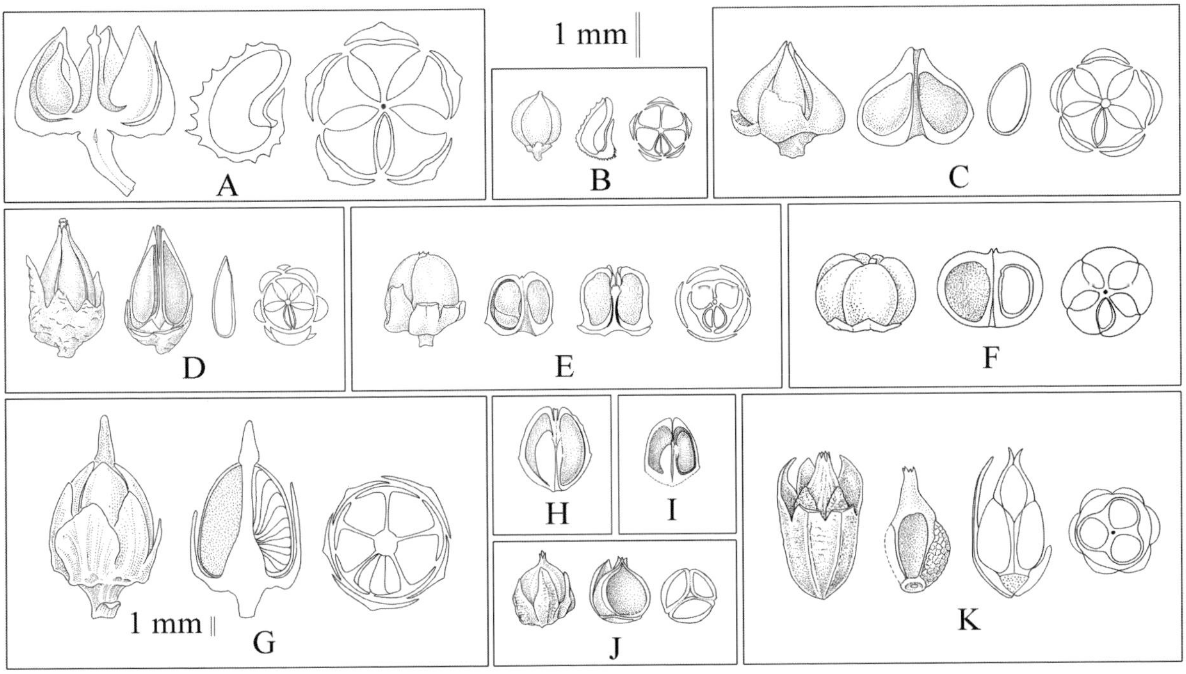

Figure 15.6 Ericalean fruits and seeds from the Late Cretaceous (Maastrichtian) Walbeck (A, C–K) and Eisleben (B) localities, Germany, seen (left to right) in lateral view, schematic longitudinal section (except A and B), sections of seeds (except E–K) and schematic cross-section (except H and I). (A) *Palaeoschima microvalvata*. (B) *Protovisnea saxonica*. (C) *Allericarpus pentaphylacoides*. (D) *Pentaphylax protogaea*. (E) *Discoclethra valvata*. (F) *Viticocarpum minimum*. (G) *Diplycosiopsis walbeckensis*. (H) *Epacridicarpum cannelatum*. (I) *Epacridicarpum cretaceum*. (J) *Valvaecarpus globulosus*. (K) *Purdiaeopsis campanulatus*. Redrawn from Knobloch and Mai (1986).

and form a short corolla tube with five ligulate lobes. The androecium is haplostemonous with five stamens, each differentiated into an elongate filament and a small, slightly lobed, basifixed anther. The bases of the filaments are apparently fused to the corolla while the distal parts are free. Anthers are tetrasporangiate, dithecate and dehisce by longitudinal slits. Pollen *in situ* is minute, up to about 9.5 μm long and about 5 μm in equatorial diameter. Grains are tricolpate and finely foveolate with an almost smooth surface that is often obscured by a pollenkitt-like coating. The gynoecium is syncarpous and the ovary is trilocular with three slender styles, and central placentation. There are many anatropous ovules in each locule. Mature fruits are loculicidal capsules. Seeds are small with a thin seed wall, a reticulate surface and finely pitted cell walls.

The flowers, fruits and seeds of *Actinocalyx bohrii* share many features with various extant Ericales, and particularly with extant Diapensiaceae, but the fossils show several seemingly plesiomorphic features compared to extant members of the family, including more simple anthers and free styles (Friis, 1985c).

Paradinandra suecica (Figure 15.8) is another ericalean fossil from the Late Santonian – Early Campanian Åsen locality, Sweden (Schönenberger and Friis, 2001). Flowers are pentamerous, about 3.5 mm long and 1.2 mm wide, with a hypogynous perianth. The calyx consists of five sepals that are free from each other. Each sepal is thicker in the middle and membraneous laterally. There is a median vein and four or six lateral veins. The corolla is tubular and consists of five petals that are apparently united basally, but free apically. The androecium consists of 15 stamens. Filaments are united basally and also appear to be fused to the base of the corolla. Stamens appear to be in two whorls with an outer whorl of five stamens and an inner whorl of ten. Anthers are basifixed, dithecate and tetrasporangiate with an X-shaped cross-section. Enlarged epidermal cells in the distal part of the theca indicate that anther dehiscence was by apical pores or slits, but anthers

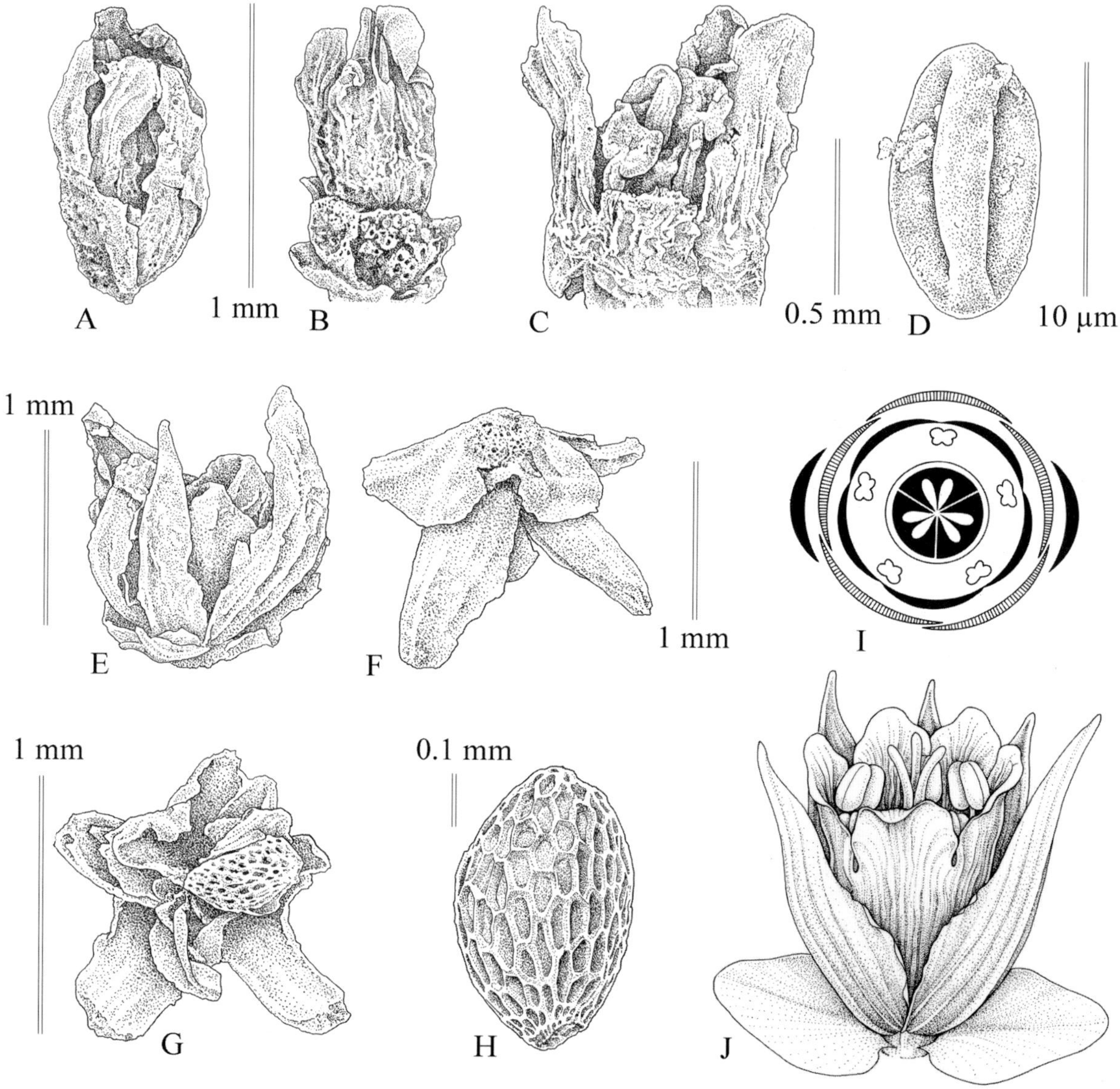

Figure 15.7 *Actinocalyx bohrii* from the Late Cretaceous (Late Santonian – Early Campanian) Åsen locality, Sweden. (A) Flower bud showing calyx and corolla preserved. (B) Flower without calyx showing sympetalous corolla. (C) Detached corolla with stamens attached. (D) Tricolpate–tectate pollen grain. (E) Lateral view of fruit showing persistent calyx. (F) Calyx and pair of bracteoles shown from below. (G) Apical view of open fruit (capsule) showing one seed preserved and persistent, star-shaped calyx. (H) Isolated, reticulate seed. (I) Floral diagram. (J) Reconstruction of flower. (A–H) Drawn from SEM images in Friis (1985c); (I, J) redrawn from Friis (1985c).

at the anthetic stage have not been found. Immature pollen was found *in situ* in a fractured anther in a flower bud. Mature pollen grains were found inside the floral organs of many flowers. Pollen is small, about 10–14 μm, tectate and finely foveolate, with three long colpi. Inside the androecium is a narrow, slightly lobed nectary disc. The gynoecium is tricarpellate and syncarpous with three long styles that are basally united, but free for most of their

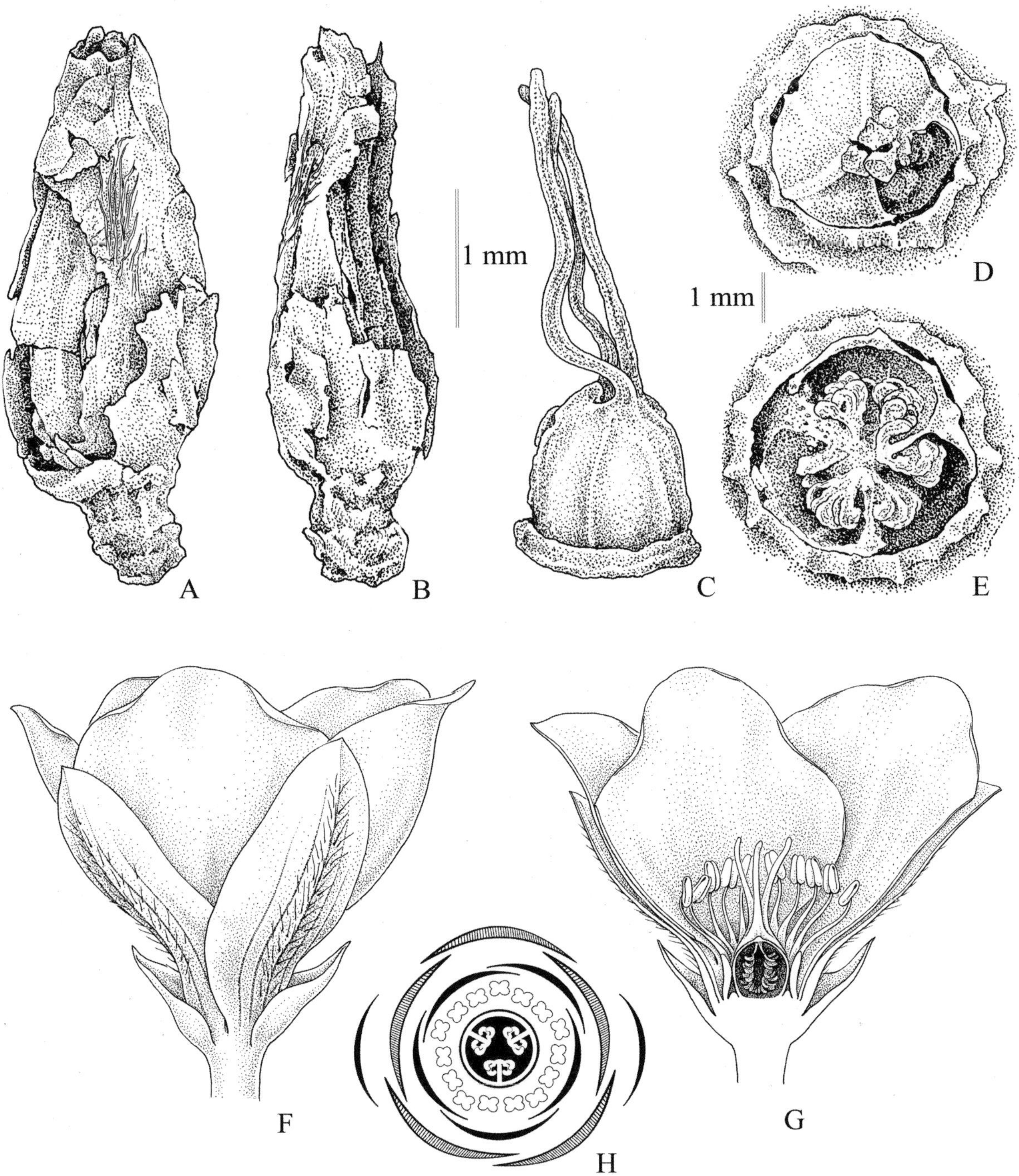

Figure 15.8 *Paradinandra suecica* from the Late Cretaceous (Late Santonian – Early Campanian) Åsen locality, Sweden. (A, B) Flower bud in lateral view showing remains of calyx and corolla. (C) Lateral view of isolated gynoecium showing three free styles and nectary disc. (D) Apical view of gynoecium showing nectary disc. (E) Apical view of abraded gynoecium showing axile placentation with campylotropous ovules. (F, G) Reconstructions of flower. (H) Floral diagram. (A–E) Drawn from SEM images in Schönenberger *et al.* (2001a); (F, G) from Schönenberger (2005) with permission.

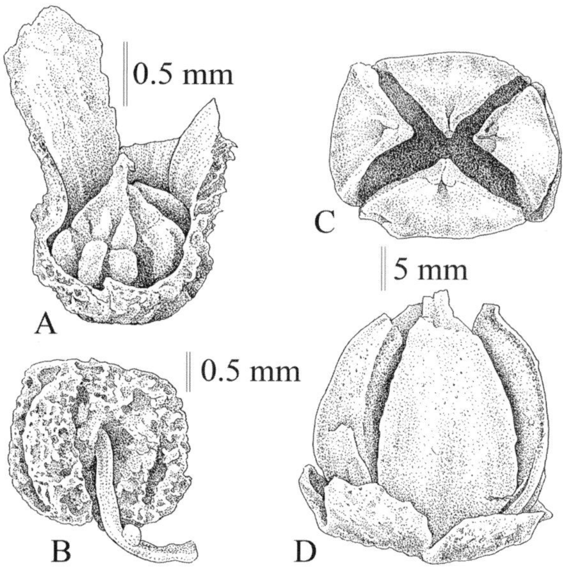

Figure 15.9 *Paleoenkianthus sayrevillensis* from the Late Cretaceous (Turonian) Old Crossman locality, New Jersey, USA. (A) Young fruit showing persistent calyx with some sepal lobes missing. (B) Isolated stamen. (C, D) Mature fruit (capsule) in (C) apical and (D) lateral view. Drawn from SEM images in Nixon and Crepet (1993).

length. Septae are fused at the base, but do not meet in the centre of the ovary and placentae are parietal. Ovules are numerous, campylotropous, and with a reticulate surface. No mature fruits or seeds have been observed.

Paradinandra suecica clearly belongs to the Ericales and the flowers are particularly similar to those of certain Pentaphylacaceae, Theaceae and Actinidiaceae. As indicated by the name, *Paradinandra* was thought to be very similar to *Adinandra* of the Pentaphylacaceae (formerly included in the Ternstroemiaceae), but some features of the gynoecium in the fossil have not been described among extant Ericales and the precise phylogenetic position of *Paradinandra* remains uncertain.

Paleoenkianthus sayrevillensis (Figure 15.9) comprises flowers, fruits and isolated stamens described from the Late Cretaceous (Turonian) Old Crossman locality, New Jersey, USA (Nixon and Crepet, 1993). Flowers are small, about 2 mm long and 1 mm in diameter, actinomorphic and bisexual with a pentamerous, hypogynous perianth. The perianth consists of one whorl of imbricate sepals and one whorl of valvate petals that are connate at the base. The androecium has eight stamens, apparently arranged in a single whorl, and the gynoecium is tetramerous. Stamens have a well-differentiated filament and anther. Anthers are basifixed, inverted, tetrasporangiate and deeply lobed, with two or three short awns near their base. How the anthers dehisced is uncertain. Tricolporate and reticulate pollen occurs on the surface of the stigma and other parts of the flowers, but was not observed *in situ*. Thread-like structures are scattered over the surface of the pollen. These structures were interpreted as viscin threads (Nixon and Crepet, 1993) although they are more like the fungal hyphae that are common in some fossil material. The gynoecium is syncarpous with a four-loculed ovary. The ovary is lobed, with eight distinct ridges that are particularly pronounced in young specimens. There are four free styles that are slightly unequal and curved, occasionally with the stigmatic area preserved. The fruits are capsules that dehisced both loculicidally and septicidally.

Flowers of *Paleoenkianthus sayrevillensis* show a suite of floral features that clearly indicate a relationship with core Ericales. Nixon and Crepet (1993) suggested a particular affinity with extant *Enkianthus* (Ericaceae) as indicated by the name. Because details of placentation, as well as ovules and seeds, are not known, the precise systematic placement of this fossil within Ericales is uncertain.

Pentapetalum trifasciculandricus (Figure 15.10) is another ericalean taxon from the Late Cretaceous (Turonian) Old Crossman locality, New Jersey, USA, based on fossil flowers (Martínez-Millán *et al.*, 2009). Flowers are actinomorphic and bisexual with five sepals, five petals, numerous stamens, a tricarpellate gynoecium and three free styles. Sepals are arranged quincuncially; petals are imbricate. Stamens are grouped in three clusters. Stamens have slender thread-like filaments and tetrasporangiate, basifixed anthers. Pollen is triaperturate, probably tricolporoidate. The ovary is trilocular and superior with many small ovules. Seeds appear thin-walled with a reticulate surface pattern.

Martínez-Millán *et al.* (2009) performed several cladistic analyses to access the likely relationships of *Pentapetalum*, some with morphological data alone, some combining morphology and molecular sequence data. Extant taxa included in the analyses were identified using morphological keys and literature studies. Conflicting results were obtained from the various analyses, but all point to a relationship of *Pentapetalum* with early-branching lineages of Ericales, probably close to the Pentaphylacaceae.

In addition to *Paleoenkianthus* and *Pentapetalum*, mentioned above, three other flowers were briefly described and illustrated from the Late Cretaceous (Turonian) Old

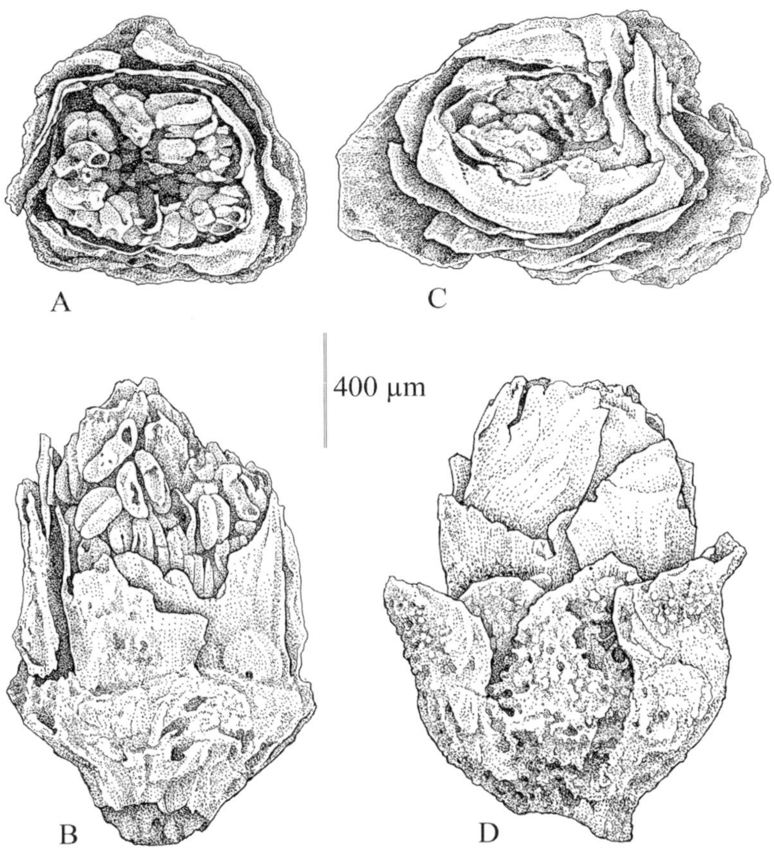

Figure 15.10 *Pentapetalum trifasciculandricus* from the Late Cretaceous (Turonian) Old Crossman locality, New Jersey, USA. (A, B) Flower bud in (A) apical and (B) lateral view, showing many stamens. (C) Apical view of flower with well-preserved perianth. (D) Lateral view of flower, showing calyx and corolla. Drawn from SEM images in Martinez-Millan *et al.* (2009).

Crossman locality in New Jersey, USA (Crepet, 1996). One of these flower types 'with some characters in common with modern Pyrolaceae' (Plates IV–V in Crepet, 1996) is similar to flowers reported as 'Taxon 26' from the Late Santonian Allon locality, Georgia, USA (Herendeen *et al.*, 1999a). These flowers are small, about 2–3 mm long, bisexual, pentamerous–isomerous and actinomorphic with a superior ovary (Figure 15.11). The sepals are free and in specimens from the Old Crossman locality the sepals have large glands along the abaxial margins. Petals are clawed. The androecium is haplostemonous with stamens opposite the sepals that alternate with five nectary lobes. Stamens have flattened filaments and basifixed anthers that dehisce by longitudinal slits. The gynoecium consists of five carpels with a single stout style and a spherical five-lobed stigma. Fruits are five-loculed capsules with axile placentation bearing many seeds.

Among the dispersed ericalean fruits and seeds described from the Late Cretaceous of Central Europe three species thought to belong to the Theaceae (as currently circumscribed) are assigned to the extinct genus *Palaeoschima*. *Palaeoschima austriaca* is from the Coniacian–Santonian of Austria, *P. becvensis* from the Campanian–Maastrichtian of Austria, the Czech Republic and Germany, and *P. microvalvata* from the Maastrichtian of Germany (Knobloch and Mai, 1984, 1986). *Palaeoschima* is characterised by having a five-loculed capsule with a single seed in each locule (Figure 15.6). At dehiscence the valves split from the central column. The seeds released are slightly curved, sometimes with a dorsal crest, and have a pitted, spiny surface. Knobloch and Mai (1986) also compare the seeds with several members of the Pentaphylacaceae (e.g. *Eurya*, *Ternstroemia*). Although these fossils clearly belong to the Ericales their more precise position within the order is not yet fully established.

Pentaphylax protogaea (Figure 15.6) from the Maastrichtian Walbeck locality, Germany (Knobloch and Mai, 1986), comprises five-valved capsular fruits that have a

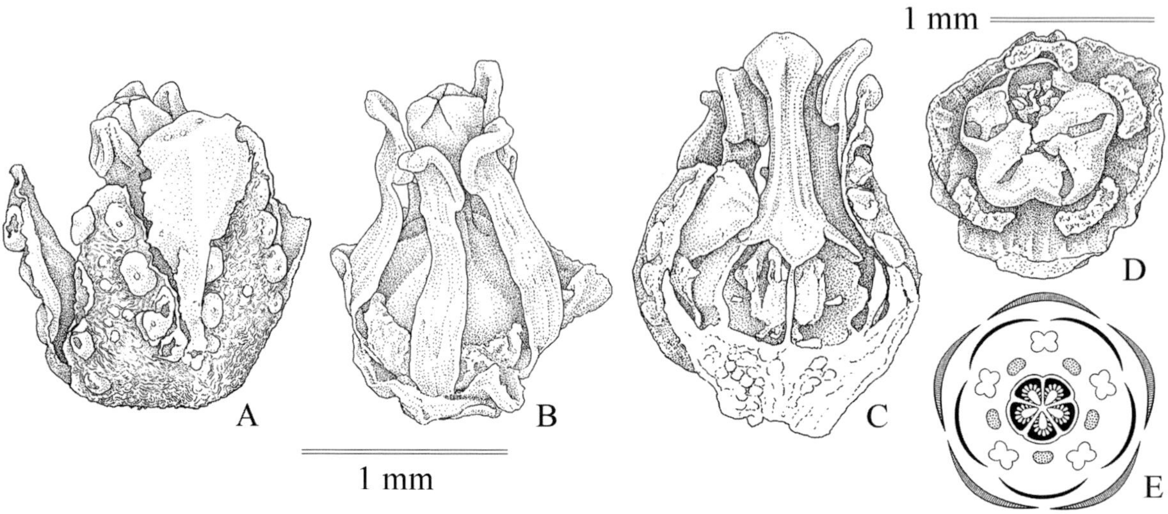

Figure 15.11 Unnamed ericalean flowers from the Late Cretaceous (Turonian) Old Crossman locality, New Jersey, USA. (A) Flower showing glandular sepals and clawed petals. (B) Flower with perianth missing, showing stamens alternating with nectary lobes, and the spherical five-lobed apex of the style. (C) Fragmentary flower, showing broken gynoecium with placentae and style. (D) Apical view of flower, showing the pentamerous organisation. (E) Floral diagram. (A–D) Drawn from SEM images in Crepet (1996).

thick, woody, fruit wall and a persistent, leathery calyx. The calyx has five lobes of unequal length that are fused basally to a shallow cup. Fruit dehiscence is loculicidal and leaves a central column. There are two anatropous seeds per locule. *Pentaphylax protogaea* was compared to the extant *P. euryoides*, but the fruits of modern *Pentaphylax* are much bigger, the fruit wall is thinner, and the calyx lobes are of equal length. In addition, seeds of the extant species are campylotropous. The inclusion of the fossil material in the extant genus needs further documentation. Other taxa from the Walbeck locality assigned to the extinct genera *Allericarpus, A. clausenispermus* and *A. pentaphylacoides* (Figure 15.6), were also compared to extant *Pentaphylax* (Knobloch and Mai, 1986).

15.3.2 Pentaphylacaceae and Theaceae

The phylogenetic position of Pentaphylacaceae and Theaceae among Ericales is uncertain. The two families are treated together here because many extant taxa now included in the Pentaphylacaceae were previously assigned to the Theaceae, and many fossil records assigned to the Theaceae should now be assigned to the Pentaphylacaceae.

The Pentaphylacaceae were formerly a monotypic family with a single species, *Pentaphylax euryoides*, distributed in Southeast Asia, Malesia and Indonesia. However, as currently circumscribed, the family includes taxa previously assigned to the Ternstroemiaceae (*Adinandra*, *Anneslea*, *Archboldiodendron*, *Balthasaria*, *Cleyera*, *Eurya*, *Freziera*, *Killipiodendron*, *Paranneslea*, *Symplococarpon*, *Ternstroemia*, *Ternstroemiopsis*, *Visnea*) and Sladeniaceae (*Sladenia*, *Ficalhoa*); two families that in many earlier classifications were included in the Theaceae (Cronquist, 1981). Extant members of the broadly defined Pentaphylacaceae are shrubs and trees with many species in Southeast Asia, Malesia and Central to South America, and a few representatives in Central to East Africa and the Canary Islands. Flowers are usually small to medium-sized. Fruits are berries with characteristic, often campylotropous, seeds.

The Pentaphylacaceae have an extensive fossil record and are particularly well represented by fruits and seeds from the Cenozoic, but there are also several different kinds of fruits and seeds from Late Cretaceous strata that may be assigned to the family. Other Late Cretaceous fossils assigned to Pentaphylacaceae are more problematic, including fossil fruits described as an extinct member of the modern genus *Pentaphylax* (*Pentaphylax protogaea*, section 15.3.1). Fossil flowers of *Paradinandra suecica* from the Late Cretaceous of Sweden are very similar to flowers of extant Pentaphylacaceae, but differ in details of the gynoecium (section 15.3.1).

Seeds of *Eurya* are particularly common in Cenozoic deposits. They are easily recognised by their strongly curved shape (campylotropous) with a horseshoe-shaped embryo cavity and a thick sclerotic testa with strongly thickened and finely pitted anticlinal cells walls (Friis, 1985b). Seeds assignable to extant *Eurya* are also recorded from the Late Cretaceous with the earliest record from the Santonian (Knobloch and Mai, 1986). Three species have been described so far. Of these, seeds of *Eurya crassitesta* reported from several localities in central Europe ranging in age from Campanian to Paleocene (Knobloch, 1975) are particularly similar to seeds of extant *Eurya*, while *Eurya carpatica* (Knobloch and Mai, 1983) from the Campanian–Maastrichtian of Moravia (Czech Republic) and Austria (Knobloch and Mai, 1986) differs in having a verrucate surface, a character not observed in seeds of the extant genus. *Eurya holyi* (Knobloch, 1977) has a very thick seed wall that is also unlike that of extant *Eurya*.

In addition to seeds of *Eurya*, Knobloch and Mai (1983, 1984, 1986) also described several different kinds of seeds that were compared to those of extant *Visnea* (Figure 15.6). These include an extinct species, *V. minima*, from the Maastrichtian of Walbeck, Germany, and several species assigned to the extinct genus *Protovisnea* (*P. cancellata*, *P. erinacea*, *P. maii*, *P. reticulata*, *P. saxonica*, *P. tetragonalis*, *P. zahajensis*) that range in age from Late Turonian to Maastrichtian. Some of these species are based on fruits with seeds; others are based solely on seeds. The fruits are dry capsules with five valves borne on a persistent five-parted calyx. Characters of the fossils indicate a close relationship to Pentaphylacaceae, but Knobloch and Mai (1986) noted that many of the characters of the fossils also occur in other ericalean families.

The Theaceae, as previously defined, included a heterogeneous assemblage of shrubs and trees that were sometimes grouped into four subfamilies: Asteropeioideae, Bonnetiodeae, Theoideae and Ternstroemioideae (including *Sladenia*) (e.g. Cronquist, 1981). However, molecular phylogenetics has shown that these groups are only remotely related. The Asteropeioideae are now recognised as a family of Caryophyllales, and Bonnetiodeae as a family of Malpighiales. Ternstroemioideae are included in Pentaphylacaceae. The remaining taxa (Theeae, Gordonieae, Stewartieae) are shrubs and trees with leathery leaves with serrate margins, usually capsular fruits and anatropous seeds with a straight embryo cavity. The Theaceae are distributed in Southeast Asia and Malesia, as well as in the New World where the family extends from eastern North America to South America.

The fossil record of Theaceae, as the family is currently circumscribed, is less extensive than for Pentaphylacaceae, and most records are from Cenozoic strata. Several fruits and seeds from the Late Cretaceous may indicate the presence of the family. These are assigned to the fossil genus *Palaeoschima*, and thought to be closely related to extant *Schima*, tribe Gordonieae (Knobloch and Mai, 1984, 1986). However, there are also similarities in seed structure to members of the Pentaphylacaceae and the precise systematic position of *Palaeoschima* remains to be established (section 15.3.1).

15.3.3 Core Ericales: Clethraceae, Cyrillaceae and Ericaceae

The extant Cyrillaceae occur in eastern North America and northern South America and include two genera, *Cyrilla* and *Cliftonia*, each with a single species. The genus *Purdiaea*, which was included in the family in earlier classifications, is now transferred to Clethraceae (Anderberg and Zhang, 2002). Flowers of Cyrillaceae are small, actinomorphic and bisexual with perianth and stamens in whorls of five and a superior ovary of two to five carpels. Fruits are capsules with a single seed in each locule.

The fossil record of Cyrillaceae is poorly understood. The extinct genus *Epacridicarpum*, known from small five-loculed fruits, is common in many Cenozoic floras from Europe (Friis, 1985b) and has also been reported from the Late Cretaceous of Central Europe (Knobloch and Mai, 1986). The genus was assigned to the Cyrillaceae by Mai (1976), but this assignment was later questioned by Friis (1985b). Knobloch and Mai (1986) also compared *Epacridicarpum* to extant *Purdiaea* (now Clethraceae), but the phylogenetic position of *Epacridicarpum* is still uncertain. Several other taxa from the Late Cretaceous of Central Europe assigned to Cyrillaceae were mostly compared to extant *Purdiaea* (Knobloch and Mai, 1986) and are discussed under Clethraceae.

The Ericaceae are a large cosmopolitan family with more than 4000 species of shrubs or more rarely herbs, climbers, or small trees (Heywood *et al.*, 2007). Flowers are typically actinomorphic and bisexual with a superior ovary, but there are also zygomorphic flowers and flowers with an inferior ovary. The number of floral parts is usually four or five, but there is a considerable variation within the family in this feature. There is also great variation in the number of ovules. Stamen dehiscence is usually by apical pores and pollen is typically shed in permanent tetrads. Fruits are usually small capsules, berries or drupes, and seeds are small.

The Ericaceae have an extensive fossil record, particularly from the Cenozoic, and the family was clearly once much more diverse in Europe than it is today. Genera such as *Eubotrys*, *Leucothoe*, *Lyonia* and *Zenobia* that are today restricted to eastern North America and/or eastern Asia were present in the Cenozoic of Europe (Mai, 1964; Friis, 1985b). The Ericaceae have also been reported from the Late Cretaceous of Central Europe with one species assigned to the modern genus *Leucothoe (L. praecox)* and two species assigned to extinct genera. *Leucothoe praecox* comprises small five-loculed capsules from the Maastrichtian of Eisleben. No seeds were observed (Knobloch and Mai, 1986). *Viticocarpum minimum* from the Maastrichtian of Walbeck (Figure 15.6) comprises small five-loculed drupes with a single seed in each locule and was compared to extant *Arctostaphylos*. *Diplycosiopsis walbeckensis*, also from the Maastrichtian of Walbeck (Figure 15.6), comprises five-loculed capsules with many seeds in each locule and was compared to extant *Diplycosia* and *Pernettya*. Although all these fruits were most likely produced by ericalean plants, the information currently available precludes more precise systematic placement.

The Clethraceae include small evergreen or deciduous trees and shrubs. Until recently the family included a single genus, *Clethra*, with about 120 species, but the genus *Purdiaea*, formerly included in Cyrillaceae, is now also included (Anderberg and Zhang, 2002). The Clethraceae have a disjunct distribution in tropical and subtropical Asia and America, but with a single species also in Madeira. Flowers are small, actinomorphic with pentamerous perianth and androecium, and a superior ovary of three to five carpels. Anther dehiscence is by apical pores. Fruits of *Clethra* are three-loculed with many, typically winged, seeds. Fruits of *Purdiaea* are three- to five-loculed with a single seed per locule.

The Clethraceae clearly had a much wider geographic distribution during the Cenozoic, and two distinct species have been identified from the Miocene of Europe (Friis, 1985b). *Clethra cimbrica* from the Middle Miocene of Denmark is based on well-preserved capsular fruits containing characteristic winged seeds, very similar to those of extant *Clethra arborea* from Madeira, and an undescribed fossil seed from the Miocene of Poland closely resembles *Clethra alnifolia* from eastern North America (Friis, 1985b). The relationships of other fossils from Europe assigned to Clethraceae are equivocal (Friis, 1985b).

Knobloch and Mai (1986) reported several possible fossil Clethraceae from the Late Cretaceous (Maastrichtian) mesofossil flora of Walbeck, Germany, and other localities in Central Europe. The extinct genus *Discoclethra* (Figure 15.6) includes three species (*D. maxima*, *D. polysperma*, *D. valvata*) of three-loculed, capsular fruits with many seeds. *Discoclethra* was assigned to Clethraceae by Knobloch and Mai (1986), but the authors also indicated similarities to Celastraceae, Hamamelidaceae, Cunoniaceae and Ericaceae. The extinct genus *Puriaeopsis*, with a single species *P. campanulatus*, is based on three- or rarely four-loculed fruits with a single seed in each locule (Figure 15.6). The fruits were described as capsules, but with indehiscent locules. *Puriaeopsis* was compared to extant *Purdiaea*, but also shows some similarity to extant *Cyrilla* (Cyrillaceae) (Knobloch and Mai, 1986). The extinct genus *Valvaecarpus* (Figure 15.6) includes four species (*V. debeyi*, *V. globulosus*, *V. kerharticensis*, *V. pterocaryaeformis*) based on capsular fruits with three to five single-seeded locules. *Valvaecarpus* is similar to fruits of extant *Purdiaea*, but the seeds are fused to the pericarp as in some extant Cyrillaceae (Knobloch and Mai, 1986).

15.3.4 Core Ericales: Sarraceniaceae, Actinidiaceae and Roridulaceae

The Actinidiaceae are currently the only family of the Sarraceniaceae–Actinidiaceae–Roridulaceae clade with a well-documented fossil record. No fossils have been assigned to the Roridulaceae, a small South Africa family of semicarnivorous plants. The only fossil assigned to the Sarraceniaceae, *Archaeamphora longiservia* from the Lower Cretaceous Yixian Formation of Liaoning, northeastern China, is of very uncertain affinity and may not be an angiosperm. It comprises shoots with small linear leaves in a helical arrangement along a central stalk (Li, 2005). Pitcher-like structures along the linear leaves were compared to the pitchers of extant *Heliamphora* (Sarraceniaceae). The material is, however, strongly compressed and there are very few details preserved, which precludes more detailed assessment of structure and relationships. The pitcher-like structures could also be deformations caused by galls or other insect attacks, and the linear leaves are very similar in size and shape to those of certain conifers from the Yixian Formation.

The Actinidiaceae include about 355 species of shrubs, small trees and climbers in three genera, *Actinidia*, *Clematoclethra* and *Saurauia*. Flowers are bisexual or unisexual and actinomorphic, usually with a pentamerous and hypogynous perianth and numerous, or rarely only ten, stamens. Stamens are versatile, invert at anthesis and open by apical pores or slits. Pollen is tricolporate. The gynoecium is formed from three, or sometimes many, carpels. Fruits are berries or sometimes capsules, with three to many locules. Each locule contains many anatropous ovules (Cronquist, 1981).

The Actinidiaceae are today restricted to Southeast Asia, Malesia and Central and western South America, but the family previously had a much wider distribution. Fossil remains of Actinidiaceae occur in numerous Cenozoic floras from Europe and Asia and most are represented by their characteristic seeds. Seeds of *Actinidia* are known from the Eocene and onwards (Friis, 1985b). *Saurauia* is known from the Paleocene of Europe, but also extends back into the Late Cretaceous.

The presence of Actinidiaceae in the Late Cretaceous (Santonian) of North America is documented by the small, well-preserved flowers of *Parasaurauia allonensis* and two fossil species of *Saurauia* from the Late Cretaceous of Europe are based on dispersed seeds (Knobloch and Mai, 1983, 1986). The seeds are small, angular, up to about 2 mm long, and characterised by a reticulate surface. Both species are reported from several localities ranging from Late Turonian to Maastrichtian in age. *Saurauia alenae* is reported from the Klikov sequence of the Czech Republic, and from the Hergenrather beds and Eisleben localities of Germany (Knobloch and Mai, 1983, 1986). *Saurauia antiqua*, also from the Klikov sequence and the Eisleben locality as well as from Quedlinburg, Germany, ranges from the Late Turonian to the Maastrichtian (Knobloch and Mai, 1986).

Parasaurauia allonensis (Figure 15.12) was established for fossil flowers of actinidiaceous affinity from the Late Cretaceous (Late Santonian) Allon locality, Georgia, USA (Keller *et al.*, 1996). Flowers are small, about 0.7–1.2 mm long and 0.5–0.8 mm in diameter. They are actinomorphic and bisexual, with a pentamerous perianth and androecium, and trimerous gynoecium. The ovary is superior. Sepals and petals are imbricate and free. The outer surface of the sepals is covered by large multicellular trichomes that are particularly prominent along the sepal margins. Petals are glabrous. The androecium consists of ten stamens apparently arranged in two whorls. In flower buds stamens opposite the sepals are larger and with longer filaments than those opposite the petals indicating that they were initiated earlier and belong to an outer whorl. Stamens have a well-differentiated filament and anther. Anthers are basifixed, tetrasporangiate and deeply lobed at the base. Pollen was not observed in any of the anthers or on other floral parts. The gynoecium is syncarpous with a trilocular ovary with three free styles. Numerous ovules in each locule are borne on axile placentae.

Floral characters of *Parasaurauia allonensis* indicate close relationship with extant *Saurauia*, *Actinidia*, and *Clematoclethra* in the Actinidiaceae, as well as flowers of other ericalean families such as Sarraceniaceae and Clethraceae. Phylogenetic analysis including the fossil flower and representatives of all extant ericalean families placed *Parasaurauia* within the Actinidiaceae as a sister taxon to *Saurauia* and *Actinidia* (Keller *et al.*, 1996). However, the fossil differs from the two extant genera in having ten stamens. This is a potentially plesiomorphic feature shared with *Clematoclethra*, which is the sister group to *Actinidia* and *Saurauia*.

15.3.5 Ebenaceae and primuloids

The position of Ebenaceae within Ericales is uncertain, but in some molecular analyses the family is placed as sister to the primuloids (Schönenberger *et al.*, 2005), a monophyletic group of four families: Maesaceae, Theophrastaceae, Myrsinaceae and Primulaceae. The Maesaceae are a new family separated from Myrsinaceae (Anderberg *et al.*, 2000, 2002), while Theophrastaceae, Myrsinaceae and Primulaceae, which were formerly placed in the separate order Primulales, are now resolved as a monophyletic group within Ericales.

The Ebenaceae, Maesaceae, Theophrastaceae and Myrsinaceae are shrubs and small trees mainly distributed in tropical and subtropical regions, while the Primulaceae are herbs that occur predominantly in northern temperate regions. Flowers of Ebenaceae are actinomorphic and usually unisexual, with a sympetalous corolla. Staminate flowers have two whorls of stamens, each with up to four times as many parts as in the perianth whorls. Pistillate flowers usually have a single whorl of staminodes. The ovary is superior and matures into a berry that often has a persistent leathery calyx. Flowers of extant primuloids are typically actinomorphic, bisexual, with a pentamerous perianth and androecium. They have characteristic five-carpellate, unilocular ovaries, with a central stalked placenta.

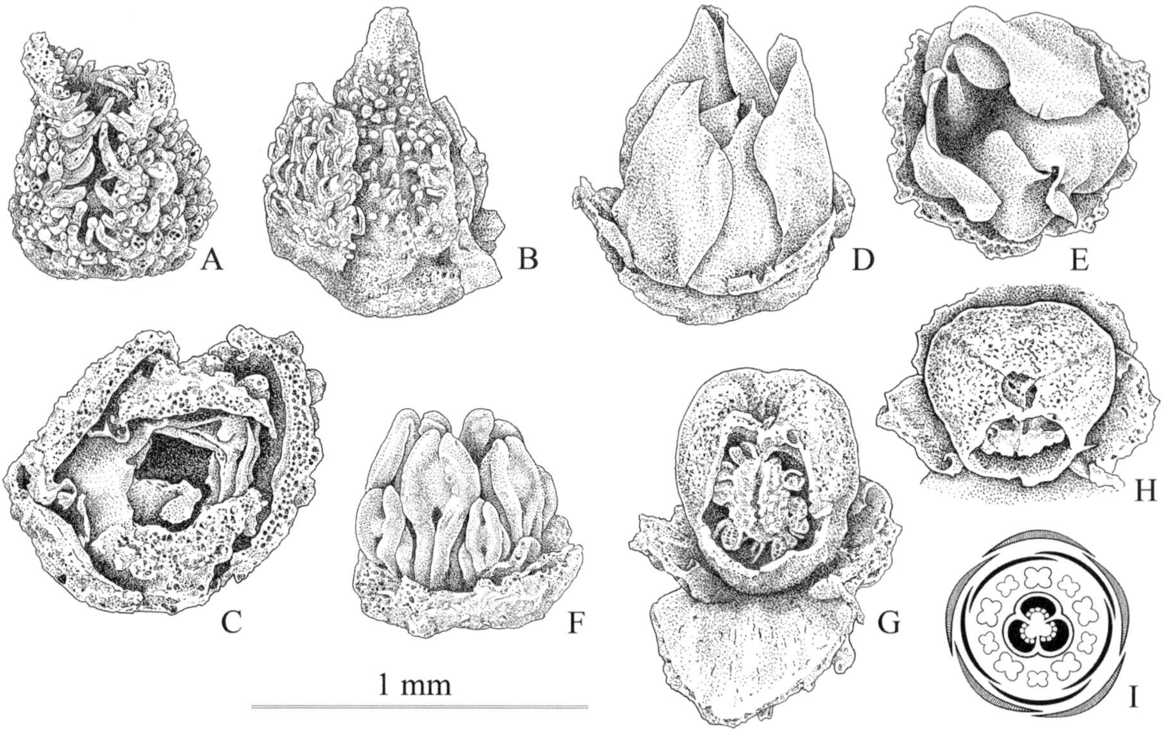

Figure 15.12 *Parasaurauia allonensis* from the Late Cretaceous (Late Santonian) Allon locality, Georgia, USA. Flower showing pentamerous perianth and androecium, and trimerous gynoecium. (A, B) Flower buds with calyx preserved, in lateral view. (C) Flower bud in apical view, showing five tepals. (D, E) Flower with calyx removed, showing pentamerous corolla in (D) lateral and (E) apical view. (F) Flower bud with perianth removed, showing pentamerous androecium composed of larger and smaller stamens. (G, H) Partly broken gynoecium in (G) lateral and (H) apical view, showing three carpels and minute seeds in the locules. (I) Floral diagram. Drawn from SEM images and line drawing in Keller *et al.* (1996).

Ovules are numerous and may be embedded in, or free from, the placenta (e.g. Caris *et al.*, 2000).

The Ebenaceae are well documented from the Early Cenozoic onwards by fossil leaves, flowers and fruits, although the record is not extensive. *Austrodiospyros cryptostoma* (Figure 15.13) from the Eocene of southern Australia (Basinger and Christophel, 1985) comprises staminate flowers with *in situ* pollen and associated leaves. There are also unequivocal records of the family from the Eocene of Europe (Mai, 1995). Primuloids have a scattered fossil record with a few reports of *Myrsine* (Myrsinaceae) and *Lysimachia* from the later Cenozoic of Europe (Friis, 1985b; Mai, 1995) and one species of distinctive primuloid flowers from the Late Cretaceous of Portugal.

The primuloid flowers from Portugal (Figure 15.14) are preserved as charcoal from the Late Cretaceous (Campanian–Maastrichtian) Mira locality (Friis *et al.*, 2010a). All flowers are apparently preserved after anthesis and the corolla and androecium have abscised. Flowers are borne on a long, slender pedicel with no bracts or scars from bracts. Flowers are actinomorphic, pentamerous and isomerous, about 0.8 mm long without the stalk and about 0.8 mm in diameter. The calyx is persistent, tightly enclosing the ovary, and consists of broadly ovate lobes that taper to long and narrow tips. The ovary is superior to slightly semi-inferior, unilocular and with a long and slender style that is five-angled in cross-section. Placentation is free and central. The placenta is dome-shaped, almost globose, and borne on a central stalk. Ovules are numerous, densely packed, anatropous and with a reticulate wall. Schizogenous secretory cavities are present in the outer tissues of all floral organs except for the placenta and seeds. The cavities are irregularly scattered and sometimes burst onto the surface of the different floral organs.

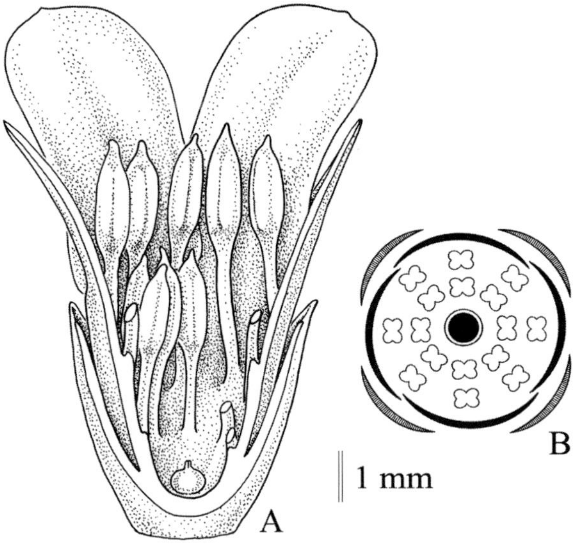

Figure 15.13 *Austrodiospyros cryptostoma* from the Early Cenozoic (Eocene) of southern Australia. (A) Reconstruction of longitudinal section through the flower. (B) Floral diagram. (A) Redrawn from Basinger and Christophel (1985); (B) redrawn from Christophel and Basinger (1982).

Although these small pedicellate fossil flowers are not complete, their pentamerous and isomerous organisation, together with the free, central and dome-shaped placenta bearing numerous anatropous ovules, unequivocally place the fossils in the primuloid clade.

15.3.6 Symplocaceae, Diapensiaceae and Styracaceae

Diapensiaceae together with Styracaceae are sometimes resolved as a monophyletic group with the Symplocaceae as their closest relative, but support for this topology from both molecular and structural data is relatively weak (Soltis *et al.*, 2005).

The Symplocaceae include a single genus, *Symplocos*, with over 300 species of trees and shrubs. *Symplocos* has a disjunct distribution and occurs in eastern North America and South America as well as India, southeastern Asia and northeastern Australia. Flowers are small, actinomorphic and bisexual. The androecium sometimes has numerous stamens that are fused or clustered. The ovary is inferior or semi-inferior, and formed from three to five carpels. Fruits are drupaceous with distinct apical pores (Heywood *et al.*, 2007).

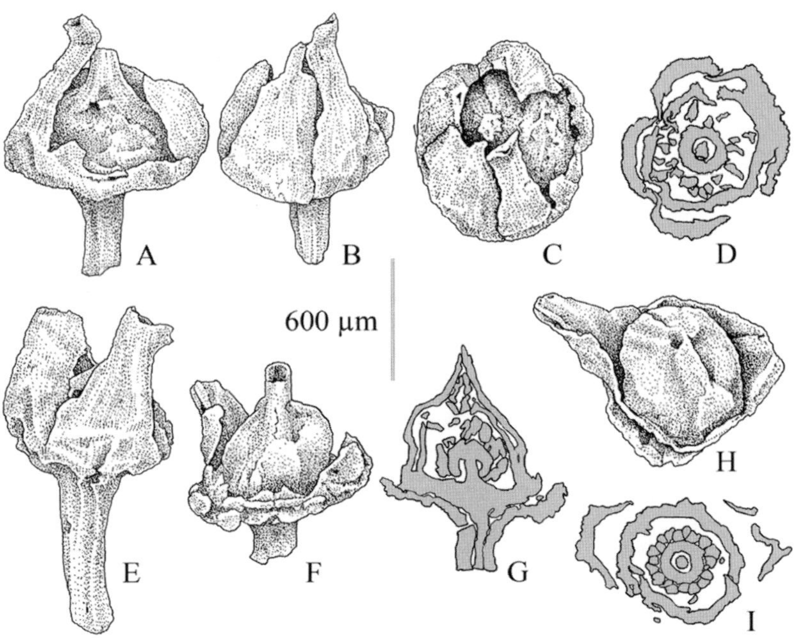

Figure 15.14 Unnamed primuloid flower from the Late Cretaceous (Campanian–Maastrichtian) Mira locality, Portugal. (A–D) Flower in (A, B) two different lateral views, (C) apical view, and (D) transverse section. (E) Flower in lateral view. (F–I) Flower in (F) lateral view, (G) longitudinal section, (H) apical view and (I) transverse section. (D, G, I) Sections from SXRTM studies, showing placentation and ovules. Material in the collections of the Swedish Museum of Natural History, Stockholm.

The current distribution of Symplocaceae suggest a wider geographic range in the past, and the family has an extensive record in the Cenozoic of Europe consisting mainly of the characteristic endocarps of *Symplocos*, as well as endocarps assigned to two extinct genera, *Pallioporia* and *Sphenotheca* (Kirchheimer, 1950; Mai, 1970, 1995). There are also records of leaves of *Symplocos* and numerous records of dispersed pollen assigned to Symplocaceae from the Cenozoic of Europe (Krutzsch, 1989). The presence of Symplocaceae in the Cretaceous is not yet fully documented although two species of pollen, assigned to the dispersed pollen genus *Triporopollenites* (*T. andersonii* and *T. scabroporis*) from the Maastrichtian of California (Chmura, 1973), were accepted by Muller (1981) as possible records of the family. Muller (1981), however, questioned the assignment to Symplocaceae of the pollen of *Symplocoipollenites vestibulum* described by Jarzen and Norris (1975) from the Campanian of western Canada.

The Styracaceae include 11 genera and about 150 species of trees and shrubs. The family also shows a disjunct distribution extending from southeastern North America and South America, to the Mediterranean region, southeastern Asia and New Guinea (Heywood *et al.*, 2007). Flowers of Styracaceae are small, actinomorphic and bisexual with a tubular calyx and short corolla lobes. The ovary is superior, formed by three to five carpels, and the fruits are drupes or capsules with one or few seeds (Heywood *et al.*, 2007). The distribution of Styracaceae also suggests a much wider distribution in the past, but currently the family has no known fossil record.

The Diapensiaceae include small herbaceous plants mainly distributed in montane and cold regions of the Northern Hemisphere. Flowers are actinomorphic and bisexual with pentamerous perianth and androecium, and a tricarpellate superior ovary. Fruits are capsules with few to many seeds (Heywood *et al.*, 2007). The Diapensiaceae have no documented fossil record. The fossil flower *Actinocalyx bohrii*, described from the Late Cretaceous of Sweden, closely resembles flowers of Diapensiaceae, but differs from them in certain respects and cannot be included in the family with certainty (section 15.3.1).

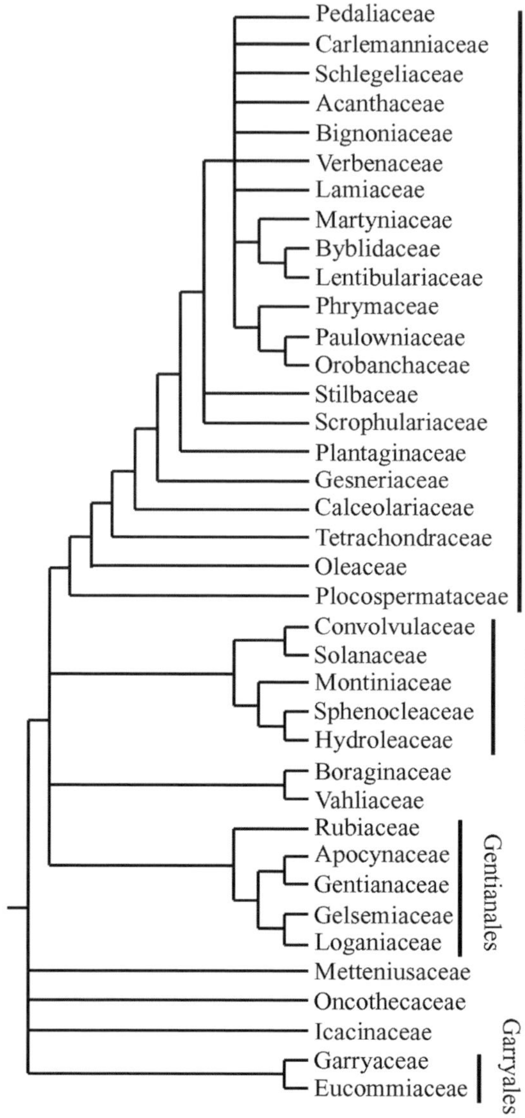

Figure 15.15 Summary of phylogenetic relationships among families of the lamiid clade. Based on Soltis *et al.* (2005).

15.4 LAMIIDS (EUASTERIDS I)

The lamiid clade (Bremer *et al.*, 2002), or Euasterids I (Chase *et al.*, 1993), is very species-rich. In some analyses Boraginaceae, including Hoplestigmataceae, together with Icacinaceae, Metteniusaceae, Oncothecaceae and Vahliaceae are resolved as the earliest branching lineage within the group, but their precise position remains to be established. In other analyses Garryales are sister to all other lamiids (Figure 15.15). In addition, there are three well-supported clades: Gentianales, Solanales and Lamiales (APGIII, 2009). Interrelationships among the major clades of lamiids are still unresolved, but relationships within each group are generally more or less well understood and several distinct subclades have been identified (Figure 15.15).

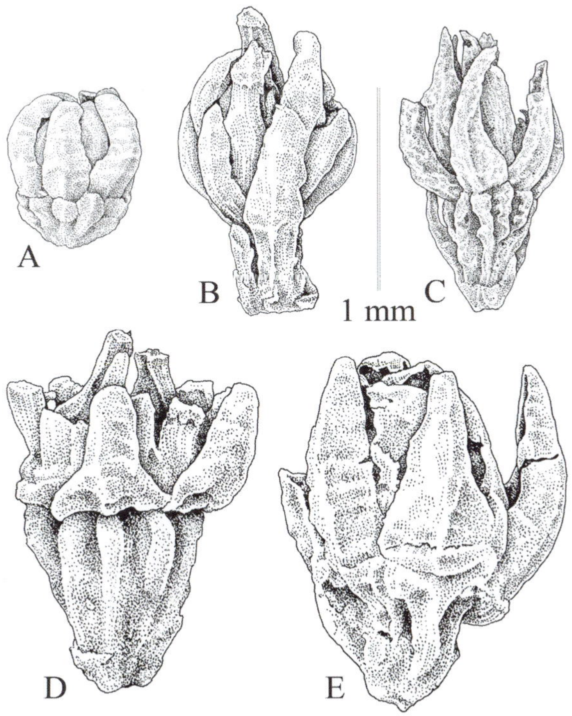

Figure 15.16 Flowers of *Scandianthus* from the Late Cretaceous (Late Santonian – Early Campanian) Åsen locality, Sweden. (A–D) *Scandianthus costatus*, specimens in different developmental stages; (A) flower bud, (B, C) anthetic flowers with corolla preserved, (D) fruiting stage with corolla missing. (E) *Scandianthus major*, flower in anthetic stage with corolla preserved. (A, C, E) From Friis *et al.* (2006); (B, D) from material in the collections of the Swedish Museum of Natural History, Stockholm.

Compared to Cornales and Ericales the macro- and mesofossil record of lamiids is poor and mainly restricted to families with woody members such as the Garryales (Eucommiaceae), Icacinaceae and Rubiaceae, although there are some reports of herbaceous lamiids from younger Cenozoic strata. The record of lamiids in the Late Cretaceous is particularly poor and currently the only extant family that is well documented prior to the Cenozoic is the Icacinaceae. The pollen record of lamiids is more extensive (Muller, 1981), but is also often difficult to evaluate. It is referred to here only where identifications are unequivocal.

15.4.1 Lamiid fossils of uncertain relationships

There are currently very few fossils of Cretaceous age that are of possible lamiid affinity. However, fossil flowers of *Scandianthus* from the Late Cretaceous of Sweden may be attributable to this group.

Scandianthus (Figures 15.16, 15.17) comprises two species, *Scandianthus costatus* and *S. major*, both described from the Late Santonian – Early Campanian, Åsen locality, Sweden (Friis and Skarby, 1982). Flowers are small, up to about 2.5 mm long, bisexual and actinomorphic with a pentamerous perianth and androecium and a bicarpellate gynoecium. The perianth consists of five free sepals that are apparently imbricate, and five free petals with open aestivation. The androecium is diplostemonous with the outer whorl of stamens opposite the sepals and the inner whorl opposite the petals. Stamens are differentiated into a band-like filament and a short dorsifixed anther. Anthers are tetrasporangiate and dithecate, and dehiscence is by simple longitudinal slits. Pollen observed *in situ* in the flowers is small, tricolporate and tectate – finely foveolate. The gynoecium is syncarpous with two free and stout styles, and the ovary is inferior and unilocular. There are two lateral, pendent placentae with numerous ovules. Fruits are capsules that open by apical dehiscence between the styles. Seeds are small and anatropous with a thin seed coat. Internal to the androecium there is a prominent lobed nectary disc bearing five bilobed structures arranged opposite to the sepals.

Fossil flowers of *Scandianthus* were originally referred to the Saxifragales *sensu* Takhtajan (1969) by Friis and Skarby (1982). That group of families is now recognised as phylogenetically very heterogeneous, but the fossil flowers show particular similarity to Hydrangeaceae and Vahliaceae, as well as Escalloniaceae and Saxifragaceae. According to new phylogenetic assessments based on molecular data these families are now placed separately in modern systems of classification with the Vahliaceae, Hydrangeaceae and Escalloniaceae, placed among asterids, and Saxifragaceae, retained among rosids (Chapter 13). It is likely that *Scandianthus* is an early asterid, but a detailed re-evaluation of its position is needed before its systematic affinity can be established with certainty.

15.5 BORAGINACEAE, ICACINACEAE AND VAHLIACEAE

The position of the Boraginaceae among the asterids is not completely clear. In many classifications based on morphology the family is placed among or close to the Solanaceae, but molecular analyses indicate a more isolated

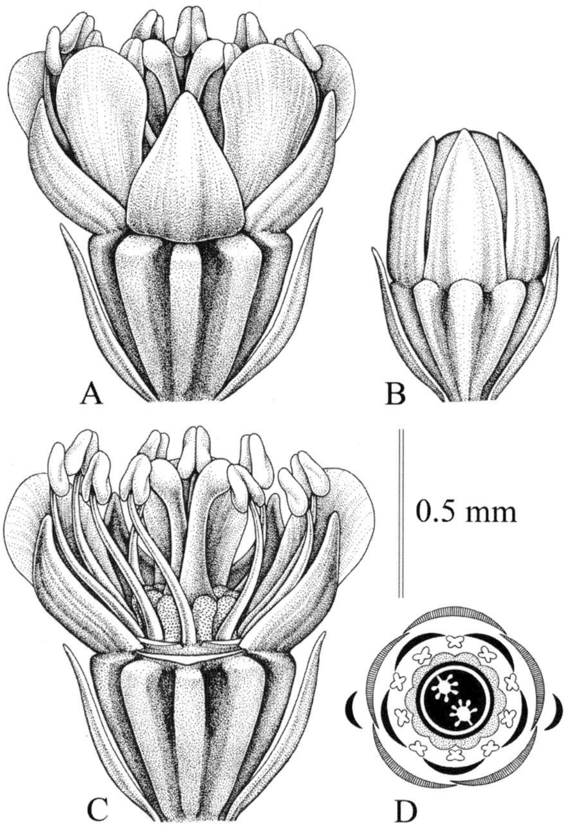

Figure 15.17 Reconstructions of *Scandianthus costatus* from the Late Cretaceous (Late Santonian – Early Campanian) Åsen locality, Sweden; (A) anthetic flower, (B) floral bud, (C) anthetic flower with one sepal and two petals removed, showing nectary disc and androecium, and (D) floral diagram. Based on material in the collections of the Swedish Museum of Natural History, Stockholm.

position (Soltis *et al.*, 2005). As currently circumscribed the Boraginaceae are very diverse and range from large tropical trees to small temperate annual herbs (Heywood *et al.*, 2007).

The Icacinaceae are a pantropical family of about 35 genera and 300 species of shrubs, trees and lianas. A few taxa extend into subtropical and warm temperate regions. Flowers are small, actinomorphic and bisexual or unisexual. The perianth and androecium is tetramerous or pentamerous, while the gynoecium is unicarpellate or bicarpellate (Heywood *et al.*, 2007). Fruits are drupes, often with a flattened and distinctively grooved or ribbed endocarp surface that sometimes has a papillate inner lining.

The Vahliaceae are a small family that includes the single genus, *Vahlia* with five extant species of shrubs or herbs, mainly distributed in Africa, Madagascar and from Iraq to India (Heywood *et al.*, 2007). The Vahliaceae have traditionally been thought to be closely related to Saxifragaceae and in some classifications were included in that family or in a broad concept of Saxifragales (e.g. Takhtajan, 1969). However, molecular studies now show that the family should be placed among asterids, even though its exact position within the group is currently uncertain. In the analyses of Bremer *et al.* (2002) Vahliaceae are placed as sister to the Boraginaceae.

The fossil record of both Boraginaceae and Vahliaceae is poor. The Boraginaceae are known from dispersed pollen from the Oligocene and Miocene (Muller, 1981) and there are several records of fruits or fruitlets. In Europe woody members of the Boraginaceae are known from the Eocene to the Pliocene represented by endocarps of *Ehretia*, a genus now occurring in tropical regions of Asia, Africa and America (Gottschling *et al.*, 2002). From North America the Boraginaceae are represented by small silicified nutlets from the Late Miocene of South Dakota with well-preserved morphology and cellular details that place the fossils in the extant genus *Lithospermum* (Gabel, 1987).

From the Late Cretaceous the small flowers of *Scandianthus costatus* and *Scandianthus major* are very similar to those of extant Vahliaceae (Friis and Skarby, 1982), but the information presently available for the fossils does not allow an assignment of *Scandianthus* to the Vahliaceae (section 15.4.1).

In contrast, the fossil record of Icacinaceae is extensive and although mainly based on the characteristic endocarps, there are also well-documented records of leaves, wood and pollen (Mai, 1995). The family is particularly common in tropical–subtropical floras from the Early Cenozoic of Europe, such as the Eocene London Clay, Geiseltal and Messel floras (Reid and Chandler, 1933; Collinson, 1983a; Mai, 1995). In addition to fossils that can be assigned to extant genera of Icacinaceae these floras also include a number of extinct genera such as *Icacinicarya* and *Palaeophytocrene* that were first described from southern England (Reid and Chandler, 1933). The Icacinaceae are also well represented in Early Cenozoic floras from Eastern North America where the family includes several genera known from Europe as well as a number of extinct genera reported only for North America (Manchester, 1999). *Stizocaryopsis* is an extinct genus described from the Paleocene of Egypt (Gregor and Hagn, 1982).

There are also fossil endocarps from the Late Cretaceous that are referable to Icacinaceae. Endocarps assigned to an extinct species of *Iodes* as *Iodes germanica* are known from the Maastrichtian of Eisleben, Germany. The endocarps are flattened, unilocular and about 4–5 mm long, with a grooved surface and a finely papillate inner lining (Knobloch and Mai, 1986). Two other species of icacinaceous affinity from the Late Cretaceous of Central Europe were assigned to the extinct genus *Icacinicarya*. They are based on similar flattened and unilocular, grooved endocarps, with an indistinct inner papillate lining. *Icacinicarya budvarensis* is known from the Late Turonian to the Santonian and *Icacinicarya papillaris* co-occurs with *Iodes germanica* in the Eisleben flora. According to Knobloch and Mai (1986) *Icacinicarya* extends from the Late Turonian to the Santonian.

15.6 GARRYALES

Both families of Garryales (Eucommiaceae, Garryaceae) contain only a single genus. *Garrya* comprises 14 species of shrub and small tree restricted to western and central North America. *Eucommia* includes a single species, *Eucommia ulmoides*, a small tree restricted to southern and central China. *Garrya* and *Eucommia* both have small, simple, unisexual flowers that may be either naked (*Eucommia*) or with a single perianth whorl (*Garrya*). The gynoecium is bicarpellate or, in *Garrya*, sometimes tricarpellate. The fruit is a samara (*Eucommia*) or a dry berry with few seeds (*Garrya*). *Eucommia* is characterised by the presence of latex in both vegetative and reproductive parts (Heywood *et al.*, 2007).

The Garryaceae have no documented fossil record, but *Eucommia* is common in Cenozoic floras. The genus is known from fossil leaves and winged fruits from eastern Asia, Europe and North America. It clearly had a much wider distribution in former geological periods. The earliest records of *Eucommia* are from the Eocene of eastern North America (Call and Dilcher, 1997) and Japan (Huzioka, 1961; Manchester, 1999). In several cases the distinctive latex strands seen in the extant genus are well preserved in both fossil leaves and fruits, providing further security in the identification. The genus persists in central and northern America through the Oligocene to the Miocene (Manchester, 1999). In Europe the genus is known from the Oligocene, is particularly common in the Late Miocene and Pliocene and persisted into the Pleistocene, where its presence is documented by distinctive tricolpate pollen (Mai, 1995).

Dispersed pollen from the Mesozoic assigned to *Eucommiidites* is not relevant to the evolution of the asterids. The genus was originally established to accommodate dispersed pollen thought to be related to *Eucommia* (Erdtman, 1948), but this pollen type has long been known to belong to a group of gymnosperms, now recognised as the Erdtmanithecales (Chapter 5).

15.7 GENTIANALES

The Gentianales are a group of five families in which Rubiaceae are resolved as sister to two subclades: the Gentianaceae–Apocynaceae clade and the Loganiaceae–Gelsemiaceae clade. Among Gentianales, as far as we are aware, only Rubiaceae have a well-documented fossil record. The family is one of the most species-rich of all angiosperms with more than 13 000 species in more than 600 genera. The family is cosmopolitan in its distribution, but is particularly diverse in tropical to subtropical regions. It exhibits enormous diversity of form from small herbaceous plants with diminutive flowers to large trees with large, showy flowers. Flowers are typically actinomorphic and bisexual with a tetramerous or pentamerous perianth and androecium, and a bicarpellate, inferior ovary. The sympetalous corolla is often extended into a long narrow tube. There is also considerable variation in fruit and seed type within the family.

The pollen record of Rubiaceae is extensive in the Cenozoic from the Late Eocene onwards (Muller, 1981; Graham, 2009). However, according to Muller, the pollen record is difficult to evaluate because pollen of extant Rubiaceae is currently not well known. Fossil leaves of Rubiaceae are reported from the Eocene of southeastern North America (Roth and Dilcher, 1979). There are also records of fossil fruits, which show that some genera with relatively restricted distributions today were once more widespread. For example, *Cephalanthus*, now mainly occuring in North America, is well documented based on fossil fruits from Cenozoic floras in Europe (Friis, 1985b). Similarly, fossil fruits of *Emmenopterys*, which is now restricted to southern China, are known in Eocene floras in North America (Manchester, 1999).

15.8 SOLANALES AND LAMIALES

The Solanales include two clades of which the larger comprises Solanaceae and Convolvulaceae. The Convolvulaceae consist of about 1600 species of mostly herbaceous vines, but also with some trees. Flowers are pentamerous with free calyx lobes and a sympetalous corolla. The gynoecium is syncarpous and bicarpellate. The family has an almost worldwide distribution. The Solanaceae include about 2600 species in about 100 genera of herbs, shrubs or more rarely small trees. The family is also widely distributed, but with a concentration of species in South America. Flowers are usually bisexual, actinomorphic to zygomorphic, with a pentamerous or sometimes tetra- or hexamerous perianth and androecium. Sepals are typically free and the petals fused. A nectary disc is usually present. The gynoecium is bicarpellate or sometimes tricarpellate and the fruit is often a berry.

The Solanaceae–Convolvulaceae clade has only a scattered fossil record. Muller (1981) accepted fossil pollen from the Eocene of Cameroon, Nigeria and Brazil as representing the group, but to our knowledge there are no meso- or macrofossils that can be attributed to the family. Fossil floral structures and winged fruits from the Cenozoic of Europe, North America and Asia previously assigned to the extant genus *Porana* (Convolvulaceae) have now been transferred to an extinct genus, *Chaneya*, more likely related to members of the Rutales (Wang and Manchester, 2000; Teodoridis and Kvaček, 2005). The Solanaceae are represented in the fossil record by a single seed described as *Cantisolanum daturoides* from the Eocene of southern England (Reid and Chandler, 1933; Collinson *et al.*, 1993) and by scattered records of seeds of *Solanum* and *Physalis* from the Late Cenozoic of Siberia and Europe (e.g. Dorofeev, 1963; Collinson *et al.*, 1993).

The other clade of Solanales comprises three small families, the Montiniaceae, Sphenocleaceae and Hydroleaceae. The Montiniaceae are small trees and shrubs distributed in Africa and Madagascar, whereas Sphenocleaceae and Hydroleaceae are herbaceous to shrubby plants distributed in warm temperate to tropical regions. None of these families are known from the fossil record.

The Lamiales are a large and diverse group with about 1100 genera and 18 000 species placed in 23 families (APGIII, 2009). The order is strongly supported based on phylogenetic analyses of molecular sequence data (Bremer *et al.*, 2002), but relationships among the constituent families are poorly resolved (Soltis *et al.*, 2005). There is a basal grade of six families (Plocospermataceae, Oleaceae, Tetrachondraceae, Calceolariaceae, Gesneriaceae, Plantaginaceae) and an unresolved core group including large families such as Lamiaceae (Labiatae) with more than 7000 species, Acanthaceae with about 4000 species, Bignoniaceae with about 800 species, Orobanchaceae with more than 2000 species, Scrophulariaceae with about 1700 species and Verbenaceae with almost 1200 species. Certain chemical characters appear to unite the group and there are also a few potential morphological synapomorphies including diacytic stomata, details of trichome anatomy and reduction of stamen number to less than five (e.g. Soltis *et al.*, 2005). Many Lamiales also have opposite leaves.

There are no Cretaceous fossils that can be assigned to Lamiales and the Cenozoic meso- and macrofossil record is also meagre. Fruits assigned to the Lamiaceae are known from Europe, Asia and North America (e.g. Dorofeev, 1963; Friis, 1985b; Taylor, 1990; Mai, 1995) with the earliest record in the Eocene (Collinson *et al.*, 1993). The verbenaceous affinity cited by Collinson *et al.* (1993) for the leaf fossil *Premnophyllum trigonum* from the Late Cretaceous of Bohemia, Czech Republic, has not been confirmed by later studies (J. Kvaček, personal communication, 2009). The pollen record of Lamiales is also sparse and represented by a few records of Verbenaceae and Lamiaceae, all from the Late Cenozoic (Muller, 1981). Other lamialean families with a Cenozoic record include Oleaceae and Bignoniaceae, both known from winged seeds. Oleaceae is known from Europe, Asia and North America with the earliest occurrence in the Eocene of North America, while Bignoniaceae is known from Oligocene strata of Europe and North America (Manchester, 1999).

15.9 CAMPANULIDS (EUASTERIDS II)

The campanulids (Bremer *et al.*, 2002), or Euasterids II (Chase *et al.*, 1993), contain seven orders: Aquifoliales, Escalloniales, Asterales, Bruniales, Apiales, Paracryphiales and Dipsacales (APGIII, 2009). The campanulids are mainly united as a clade based on molecular sequence data, and there are few morphological features that appear to define the group. Soltis *et al.* (2005) mention alternate leaves, but leaves in many Dipsacales are opposite. The Aquifoliales are resolved as sister to the other lineages, but the relationship among the other clades is not fully resolved (Figure 15.18).

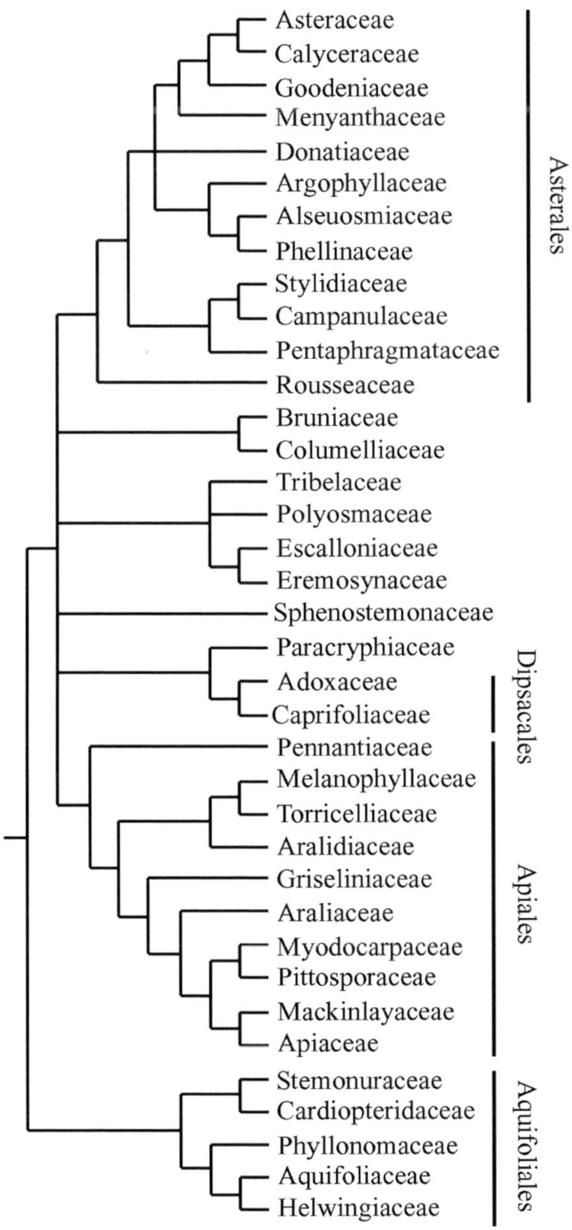

Figure 15.18 Summary of phylogenetic relationships among families of the campanulid clade. Based on Soltis *et al.* (2005).

The fossil record of campanulids is more extensive than that of the lamiids, but compared to the size of the group the fossil record is still surprisingly sparse. This may be explained partly by the herbaceous habit of the most species-rich lineages, such as Asterales and Apiales, but it is also possible that the diversification of some of these lineages was indeed relatively recent and related to the climatic cooling that occurred during the mid- and Late Cenozoic.

15.9.1 Campanulid fossils of uncertain relationships

Silvianthemum (Figure 15.19) from the Late Cretaceous (Late Santonian – Early Campanian) Åsen locality, Sweden, is known from small, well-preserved flower buds as well as more mature specimens (Friis, 1990). Flowers are bisexual and actinomorphic, with a pentamerous and epigynous perianth and androecium, and a tricarpellate gynoecium. The calyx consists of five short lanceolate sepals that do not cover the petals in the bud. The corolla consists of five ovate and imbricate petals. The organisation of the androecium is not fully understood, but there are at least eight or nine stamens in one, or more likely two, whorls. Stamens are differentiated into a filament and a tetrasporangiate, dithecate anther that opens by simple slits. Pollen *in situ* in the flowers is small, tricolpate and tectate–foveolate. The gynoecium is syncarpous and formed from three carpels. The ovary is unilocular with three free, stout styles. There are three large, stalked, parietal placentae bearing numerous ovules. Seeds are small, and anatropous with a thin seed coat. The surface of the ovary and perianth is covered by distinct trichomes, which include both simple and multicellular, peltate types.

Features of *Silvianthemum* strongly suggest a relationship with the extant genus *Quintinia*, formerly included in the Escalloniaceae (Friis, 1990). This proposed relationship has been supported by a morphologically based cladistic analysis of extant Dipsacales and related taxa that also included the fossil *Silvianthemum* (Backlund, 1996). In the most recent analyses *Quintinia* is placed in the Paracryphiaceae, which are sister to Dipsacales but ovules in *Quintinia* are bitegmic (Endress, 2010a, personal communication), which make a position of *Quintinia* (and *Silvianthemum*) among campanulids unlikely.

15.10 AQUIFOLIALES, ESCALLONIALES AND ASTERALES

The Aquifoliales include two subclades: Phyllonomaceae–Aquifoliaceae–Helwingiaceae and Cardiopteridaceae–Stemonuraceae. Based on morphological analyses these families were previously placed in separate orders, but

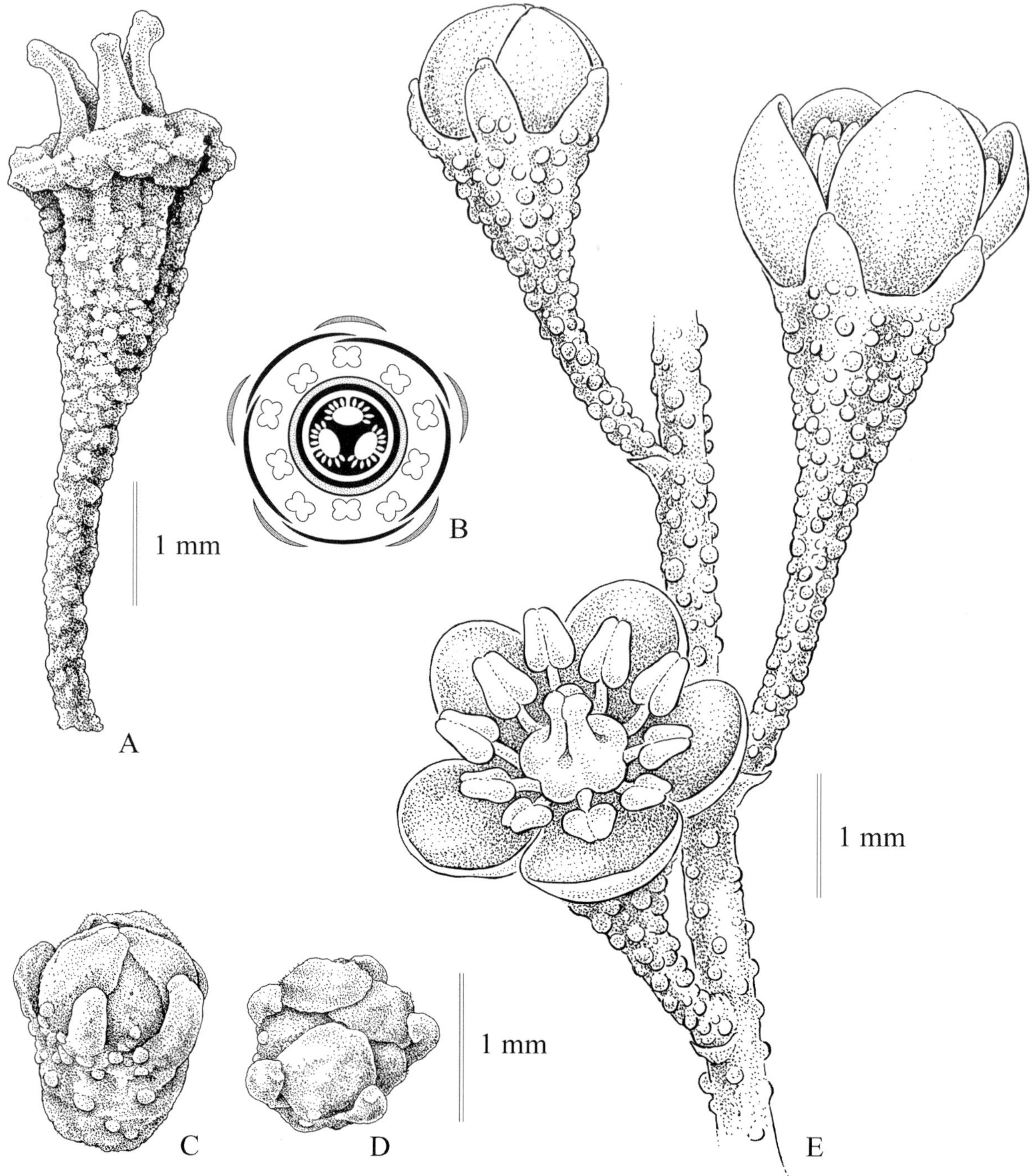

Figure 15.19 *Silvianthemum suecicum* from the Late Cretaceous (Late Santonian – Early Campanian) Åsen locality, Sweden. (A) Flower at fruiting stage, showing three persistent styles. (B) Floral diagram. Note diagram indicates presence of 10 stamens, but observed numbers are eight and nine. (C, D) Flower bud in (C) lateral and (D) apical view. (E) Reconstruction of inflorescence bearing flower and flower buds. (A, C, D) From Friis *et al.* (2006a).

molecular analyses provide strong support for their close relationship (e.g. Soltis *et al.*, 2005). However, there are only a few morphological features that unite the group. These include unisexual flowers and fruits that are drupes or berries (e.g. Soltis *et al.*, 2005).

The Aquifoliaceae include a single genus, *Ilex*, with more than 400 species of shrubs and trees mainly distributed in tropical and subtropical regions, but with some species extending to temperate areas. The flowers are small, actinomorphic and unisexual with an inconspicuous perianth (Heywood *et al.*, 2007). Pollen is tricolpate and distinctively verrucate. The ovary is superior and formed from two to five carpels. Fruits are drupes with several small flattened endocarps that are semi-circular in outline and often have distinct longitudinal ribs.

The Escalloniales include a single family, Escalloniaceae. The order is a heterogeneous and poorly defined group that is also of uncertain position within the campanulids, but is probably one of the earliest-diverging lineages in the clade (APGIII, 2009).

The Asterales include 12 families and are a strongly supported monophyletic group based on molecular sequence data (Bremer *et al.*, 2002). In earlier classifications based on morphology several of these families were widely separated in different orders. The Asterales are the largest of the eudicot orders with about 26 300 species, about 1720 genera and most of the diversity in the two families Campanulaceae and Asteraceae (Compositae) (Bremer, 1994; Kadereit, 2007). The Asteraceae are by far the largest family in the order, with almost 24 000 species. The Asterales are mainly herbaceous or small shrubby plants with flowers often borne in densely packed inflorescences, as in the capitulate heads of Asteraceae. The extensive species diversity within the Asteraceae may be of very recent origin (Magallón *et al.*, 1999).

The fossil record of Aquifoliales, Escalloniales and Asterales is provided mainly by records of Aquifoliaceae. This family is represented in the fossil record by distinctive ribbed endocarps of *Ilex* and verrucate pollen similar to that of extant *Ilex*. The Cenozoic pollen record is extensive (Muller, 1981) but there are also several reports from the Late Cretaceous. According to Muller (1981) the earliest records are species of *Ilexpollenites* from the Turonian and Coniacian of southeastern Australia, Gabon and Egypt, and further pollen records from younger Cretaceous strata (Maastrichtian) are reported from California (Chmura, 1973). All these early records require re-examination. Endocarps assigned to *Ilex* are common in the European Cenozoic floras and are also reported from the Late Cretaceous of Europe including a single species, *Ilex antiqua*, from the Maastrichtian of Eisleben (Knobloch and Mai, 1986).

The Asteraceae are mainly known from fossil pollen with unequivocal occurrences in the Oligocene (Muller, 1981). The earliest occurrence of Campanulaceae is also pollen from Oligocene strata (Muller, 1981).

15.11 BRUNIALES, APIALES, PARACRYPHIALES AND DIPSACALES

According to the APGIII (2009) classification Bruniales are sister to Apiales, Paracryphiales and Dipsacales. Bruniales consists of two small families, Bruniaceae and Columelliaceae, while the Paracryphiales contain a single family Paracryphiaceae including Quintiniaceae. Apiales and Dipsacales are much larger groups.

The current circumscription of Apiales with its seven families has only been recognised recently based on molecular systematics. In previous classifications based on morphological comparisons most of these families were included in the rosids (Cronquist, 1988). Relationships among the families of Apiales are not fully resolved, but the present molecular-based phylogenetic analyses indicate that simple leaves and few carpels are plesiomorphic features for the group while compound leaves and higher carpel numbers are derived (e.g. Soltis *et al.*, 2005). Flowers of Apiales are typically small, bisexual and with free perianth parts, although taxa with sympetalous corollas are also present in the order.

The largest family is Apiaceae (Umbelliferae), with almost 3800 species of mostly herbaceous plants and an almost worldwide distribution. The highest species concentration is in temperate regions. The Araliaceae comprise almost 1500 species of mostly woody, tropical plants. Other families included in the order are typically small and several have restricted distributions in southeastern Asia, Australia and Madagascar.

Only the Araliaceae have a substantial meso- and macrofossil record that extends to the Late Cretaceous. From the Cenozoic there is also a single fossil occurrence of Toricelliaceae as well as few scattered reports of Apiaceae. More than 20 species of Araliaceae are recorded from the Cenozoic floras of Eurasia mainly based on endocarps, but there are also some leaves (Mai, 1995) and pollen (Muller, 1981). Endocarps of Araliaceae are known from North America (Manchester, 1999). Manchester (1999) also reported a fossil fruit from the Eocene of North America assigned to the extant genus *Torricellia*. At that time *Torricellia* was included in Araliaceae, but is now separated into its own family that is restricted today to southeastern Asia, Malesia and Madagascar. The presence of *Torricellia* in the Eocene of North America is a striking example of the difficulties in inferring past distributions from present distribution patterns.

From the Late Cretaceous of Central Europe Knobloch and Mai (1986) reported four species of *Acanthopanax* (*A. friedrichii*, *A. gigantocarpus*, *A. mansfeldensis*, *A. obliquocostatus*) and one species of *Aralia* (*A. antiqua*), all from the Maastrichtian of Germany (Eisleben and Walbeck localities). They all appear to be sound determinations, but more detailed studies of fruit wall and seed organisations may be needed for fully secure placement of these fossils.

The Dipsacales include two families, Caprifoliaceae and Adoxaceae, or the Caprifoliaceae are sometimes subdivided into several smaller families (Soltis *et al.*, 2005). *Sambucus* and *Viburnum* were previously included in the Caprifoliaceae, but are now transferred to Adoxaceae. The group has been recognised earlier based on morphological evidence, but as suggested by the exclusion of *Sambucus* and *Viburnum* from the Caprifoliaceae morphological support for the group is not strong. Among uniting morphological features for the whole group, Soltis *et al.* (2005) noted the presence of pericyclic cork and a gynoecium of three or more carpels. The relationship of Dipsacales to other campanulids is still uncertain. Currently no fossils of Dipsacales have been reported from the Cretaceous. The order includes about 100 species in about 40 genera, and have a well-documented and diverse fossil record in the Early and Late Cenozoic.

The Caprifoliaceae in its present circumscription includes genera that were previously assigned to Diervillaceae, Linnaeaceae, Morinaceae, Dipsacaceae and Valerianaceae. The family has an almost worldwide distribution, but with different lineages showing different geographic distributions. Caprifoliaceae and Adoxaceae are mainly herbs or shrubs. Flowers of Caprifoliaceae are zygomorphic and have a long style with a capitate stigma, while in Adoxaceae the flowers are actinomorphic and have a short style with lobed stigma. The two families can also be distinguished by their pollen ornamentation, which is spiny in Caprifoliaceae and reticulate in Adoxaceae (Soltis *et al.*, 2005).

Fossils of Caprifoliaceae include species assigned to extinct as well as extant genera and document a wider distribution in the past than in the present. In North America the family is known from the Eocene and onwards represented by winged fruits of the extinct genus *Diplodipelta* (Manchester and Donoghue, 1995), which is closely related to extant *Dipelta* that today is restricted to Asia. During the Cenozoic *Dipelta* was also present in Europe, where it is known from the Eocene (Manchester and Donoghue, 1995). *Weigela* is another genus now restricted to Asia that had a wider distribution during the Cenozoic with records from the Miocene of Siberia (Dorofeev, 1963) and from the Miocene–Pliocene of Europe (Łańcucka-Środoniowa, 1967; Friis, 1985b). In the Cenozoic floras of Europe Adoxaceae are represented by about 20 species of *Sambucus* known based on their characteristic seeds from the Eocene onwards. Endocarps of *Viburnum* are recorded from the Miocene and onwards (Mai, 1995). Both Adoxaceae and Caprifoliaceae also have a well-documented Cenozoic pollen record (Muller, 1981), but neither is currently known from the Cretaceous.

16

Patterns of structural diversification in angiosperm reproductive organs

This chapter provides an overview of the main patterns of structural diversification in angiosperm flowers as inferred from the occurrence of floral features in the fossil record in the context of phylogenetic analyses of angiosperm relationships (Chapter 7). The systematic chapters (Chapters 8–15) provide descriptions of the individual fossils and cite the relevant literature. Here we focus on broader patterns, the appearance of major structural novelties and, where possible, what these imply about floral innovation among angiosperms as a whole. In part these large-scale patterns in the evolution of floral structure reflect the successive diversification of different lineages of angiosperms at different times in the past. The consequences of floral innovation for angiosperm pollination and dispersal biology are considered in Chapters 17 and 18. Aspects of the diversification of reproductive structure in specific angiosperm lineages are addressed in Chapter 20.

Building on earlier work by Brenner (1963), Muller (1970) and others, studies in the 1970s on early angiosperms from the Potomac Group of eastern North America were the first to demonstrate a coordinated increase in the diversity and 'advancement' of pollen and leaves through the Early and mid-Cretaceous (Doyle and Hickey, 1976; Hickey and Doyle, 1977; Hickey, 1978). This work confirmed that monoaperturate angiosperm pollen appeared first in the fossil record, an observation that was (and remains) consistent with phylogenetic concepts based on extant plants. A parallel increase in the structural complexity of fossil angiosperm leaves was also broadly consistent with interpretations of the evolution of leaf architecture among extant taxa (Hickey and Wolfe, 1975; Doyle and Hickey, 1976; Hickey and Doyle, 1977).

The first attempt to present a similar overview of the stratigraphic appearance of floral features followed about a decade after these pioneering studies (Friis and Crepet, 1987). But in the mid-1980s investigations of mesofossil flowers from the Cretaceous had only just begun and information on the floral structure of angiosperms from the Early Cretaceous was especially limited. Since then numerous new mesofossil floras containing rich assemblages of angiosperm reproductive organs have been discovered in many different parts of the world (Chapter 4). Taken together these floras now cover most of the Cretaceous period, from the Late Barremian – Early Aptian to the Maastrichtian, and have yielded a very wide range of fossil angiosperm flowers (Chapters 8–15). These continuing discoveries provide an increasingly secure basis for reconstructing the timing of major events in floral evolution, including through the critical early phases of angiosperm diversification.

An important point, which has become especially clear over the last several decades, is that the rich fossil record of flowers, fruits and seeds from the Cretaceous is strikingly different from that from the Cenozoic (Figure 16.1). Although mesofossil floras with three-dimensional, coalified fossils are very common in Cenozoic sediments, angiosperm reproductive material of this age mainly consists of fruits and seeds. Flowers are rare in these Cenozoic samples, in strong contrast to the situation in the Cretaceous. This most likely reflects fundamental changes in the nature of the vegetation from which Cretaceous and Cenozoic mesofossil assemblages were derived, which in turn will have been influenced both by evolutionary change and the changing nature of Cretaceous and Cenozoic environments and climates.

It is also important that angiosperm mesofossils from the Cretaceous are often charcoalified. Charcoal is present, but much less common, in Cenozoic sediments. Most of the three-dimensionally preserved angiosperm material from the Cenozoic is lignitic. Cretaceous vegetation, at least at mid-palaeolatitudes where angiosperm-rich mesofossil floras are common, was probably rather open, and grew under warm conditions with strong seasonality. It was evidently exposed to frequent natural fires. In contrast, during the Early Cenozoic dense subtropical–tropical forests with closed canopy became much more extensively developed at mid-palaeolatitudes. These new kinds of environments were evidently much less subject to fire and

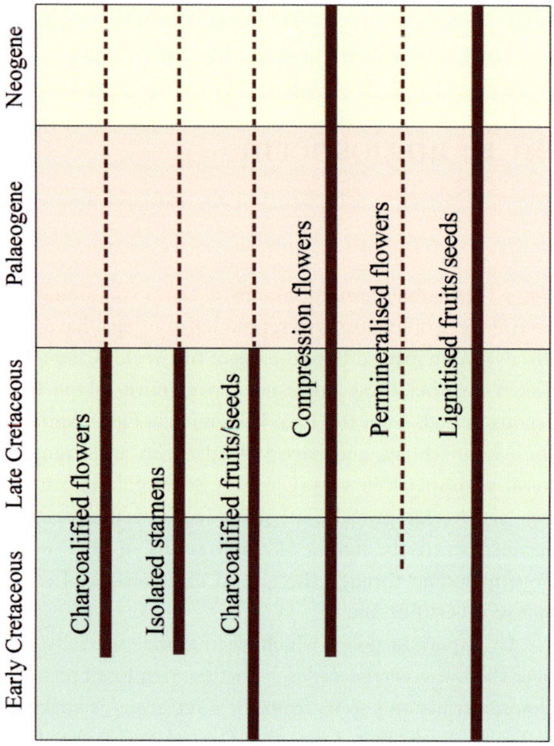

Figure 16.1 Preservation of angiosperm flowers and other reproductive organs through the Cretaceous and Cenozoic. Thick lines indicate common/abundant occurrences; dashed lines indicate occasional occurrences.

much less conducive to the preservation of angiosperm floral structures as charcoalified mesofossils. The details of this vegetational change, and the extent to which it reflects an accurate, or partly distorted, insight into global and regional environmental change, remains to be determined by future research.

16.1 INFLORESCENCE STRUCTURE

Phylogenetic analyses of extant plants infer that the basic condition in angiosperms is to have flowers borne in inflorescences. Whether the initial inflorescence type was monotelic with a terminal flower, as in botryoid, paniculate and thyrsoid inflorescences, or whether it was polytelic, open and without a terminal flower, as in racemes, spikes and thyrses, remains equivocal (Endress and Doyle, 2009). Solitary flowers, however, are thought to be derived and have probably evolved several times from different kinds of inflorescences (Endress and Doyle, 2009). This interpretation contrasts with earlier ideas that solitary *Magnolia*-like flowers reflect the basic condition in angiosperms (e.g. Takhtajan, 1969).

Early Cretaceous inflorescences. The earliest angiosperm floral structures are from the Late Barremian – Early Aptian. Most of the mesofossil flowers and other reproductive structures known from this very early phase of angiosperm evolution are from the Torres Vedras locality, Portugal. Some of these flowers are found in inflorescences, but generally they are detached and usually provide no information on how the flowers were arranged on the parent plant. Other early fossils, including those from the Yixian Formation of northeastern China, are preserved as compressions or impressions and provide more information on inflorescence structure in early angiosperms.

At this very early phase of angiosperm evolution, where evidence of floral attachment is available, the flowers are arranged in inflorescences. There are no clear examples of solitary angiosperm flowers from the Late Barremian – Early Aptian. Both simple and compound inflorescences are known. Simple inflorescences include the small *Hedyosmum*-like staminate inflorescences from Torres Vedras and other Early Cretaceous localities in Portugal (Catefica, Vale de Agua). They are spikes with densely packed whorls of unisexual, monostaminate and naked flowers (Figure 16.2). According to our interpretation simple inflorescences also occur in *Archaefructus* from the Yixian Formation, which consists of an elongate axis bearing staminate flowers below and pistillate flowers above (Chapter 9). *Sinocarpus*, also from the Yixian Formation, has flowers borne on long pedicels in a compound indeterminate inflorescence that has opposite branching, at least in the partial inflorescences.

From slightly younger floras there is a similar range of inflorescence morphology to that seen at older localities. There are several simple spikes bearing pistillate or staminate organs in the Late Aptian – Early Albian floras from Portugal. Among these are the different inflorescences of Araceae from the Vila Verde 2 locality, which have densely packed flowers in a spiral arrangement (Figure 16.2). The inflorescences assignable to the subfamily Aroideae present a similar situation to the *Hedyosmum*-like material. They consist of a main axis bearing densely packed, monostaminate and naked flowers that are not supported by bracts. The inflorescence related to the subfamily Pothoideae bears bisexual flowers. In the Late Aptian – Early Albian Crato Formation there are several angiosperms such as

Figure 16.2 SEM images of Early Cretaceous inflorescences from Portugal showing many small flowers in helical or whorled arrangement; (A, D, C) from the Catefica locality (Late Barremian – Aptian); (B) from the Vale de Agua locality (Late Aptian – Early Albian), (E–H) from the Vila Verde 2 locality (Late Aptian – Early Albian). (A) Pistillate inflorescence of uncertain relationship. (B–D) *Hedyosmum*-like staminate inflorescences. (E, F) Staminate inflorescences of Araceae, showing stamens with almost square outline. (G) Staminate inflorescence of Araceae with distinct sterile, apical extensions of the stamens. (H) Bisexual inflorescence of Araceae. Material in the collections of the Swedish Museum of Natural History, Stockholm.

Araripia florifera and a number of undescribed fossils (e.g. Figs. 3c, 4b, 4c in Mohr and Friis, 2000) that have indeterminate inflorescences.

Other Aptian–Albian inflorescences that are preserved as compressions/impressions and that also have densely crowded flowers include *Xingxueina heilongjiangensis* from the Aptian Chengzihe Formation of Heilongjiang Province, China, the unnamed Koonwarra fossil from Australia, which is also of Aptian age, and *Caspiocarpus paniculiger* from the Albian of Kazakhstan. *Xingxueina* is apparently a simple inflorescence axis that bears sessile flowers, while the Koonwarra and *Caspiocarpus* inflorescences are apparently compound. Densely crowded, spherical, and presumed compound inflorescences are also known for the pistillate *Friisicarpus brookensis* and staminate *Aquia brookensis*, both fossils of platanaceous affinity from the Albian of the Potomac Group. The Albian inflorescences of *Hyrcantha karatscheensis* from Kazakhstan and *Ternariocarpites floribundus* from the Primorye Region of the Russian Far East also have flowers gathered into inflorescences, but these are open with few flowers in each.

There are a few flowers from the Aptian – Middle Albian of Portugal that may have been solitary, such as the *Amborella*-like flower from the Catefica flora and the *Monetianthus* flower from Vale de Agua. In both cases, however, this is an indirect inference based only on extrapolation from the likely living relatives of the fossil material. *Virginianthus* (Calycanthaceae) from the Early–Middle Albian Puddledock flora presents a similar situation. Direct evidence for solitary flowers of this age is provided by the magnoliid fossil *Endressinia brasiliana*, *Pluricarpellatia peltata* and several other undescribed angiosperms from the Crato Formation (Mohr and Friis, 2000; Mohr *et al.*, 2007; but see Doyle and Endress, 2010). These fossils are preserved as whole plants with root systems attached to stems that bear leaves and flowers.

Mid-Cretaceous inflorescences. As in earlier forms, angiosperm inflorescences from the mid-Cretaceous (Late Albian – Cenomanian) are mainly simple or compound elongated spikes with small and densely crowded floral units. Spike-like inflorescences from the mid-Cretaceous include several fossils of uncertain affinity such as *Caloda delevoryana* from Kansas, USA, species of *Zlatkocarpus* and *Myricanthium* from the Czech Republic, and the *Chloranthistemon* plant (Chloranthaceae) from the German Karst deposits. There are also a variety of small, spherical, and presumably compound, pistillate and staminate inflorescences most of which are related to early eudicots, and especially extant Platanaceae. Compound inflorescences also occur among early eudicots in *Spanomera*, from the mid-Cretaceous Potomac Group, in which pistillate and staminate flowers are borne close together. The terminal, more or less sessile, pistillate flower is flanked by two lateral, short-stalked, staminate flowers.

A very distinctive group of elongated compound inflorescences that occurs for the first time in the mid-Cretaceous includes the inflorescences of *Mauldinia* (and *Prisca*), from several localities in the USA, the Czech Republic, Germany and Kazakhstan, and *Pragocladus* from the Czech Republic. These compound inflorescences were all produced by plants related to extant Lauraceae. As seen most clearly in *Mauldinia*, the long and catkin-like inflorescences bear lateral, bilobed, cladode-like partial inflorescences that are interpreted as homocladic thyrses with flowers and scales arranged in a strongly condensed dichasial–monochasial branch system.

Solitary flowers appear to be more common in mid-Cretaceous floras than in earlier angiosperm assemblages. The ‘Rose Creek flower’ from Nebraska may have been borne singly, but this is clearly the case in *Archaeanthus* (Magnoliales) from Kansas, USA, *Lovellea* (Laurales) from Australia and *Triplicarpus* from the Czech Republic. Three fruiting structures with apocarpous gynoecia from North America (*Lesqueria elocata*, *Palaeanthus problematicus*, *‘Williamsonia’ recentior*) were probably also solitary. It may be significant that these apparently solitary flowers are generally larger than other angiosperm flowers known from the Late Albian – Early Cenomanian interval (see below).

Late Cretaceous inflorescences. From the Late Cretaceous the diversity of isolated fossil flowers is much greater than from previous time intervals. Many of these Late Cretaceous flowers have not yet been studied in detail and how they were arranged on their parent plants is usually unknown. However, as in older floras, some were apparently borne in small, densely packed inflorescences. This is inferred either from comparison with their extant relatives or from direct evidence of inflorescence fragments that have attached flowers or fruits. Late Cretaceous inflorescences of this kind are elongated spikes of the *Chloranthistemon* type (Chloranthaceae), *Maiandrocarpus*, *Malliocarpus*, *Mitocarpus* and *Zeugarocarpus* (early-diverging eudicots), as well as diverse small spherical heads related to Platanaceae (e.g. *Archaranthus*, *Friisicarpus*, *Platananthus*, *Platanus*, *Quadriplatanus*).

During the Late Cretaceous there are well-documented records of compound inflorescences with dichasial partial inflorescences, mainly from fossils related to extant Fagales. Especially well understood are the elongated, compound inflorescences with cymose branching and dichasial subunits, of the Normapolles genus *Endressianthus* (Figure 16.3). *Endressianthus* is closely related to extant Betulaceae and was probably wind-pollinated. Its staminate inflorescences may have been pendulous as in its extant relatives. Other genera of Normapolles flowers, such as *Caryanthus* and *Budvaricarpus*, are more closely related to Rhoipteleaceae and Juglandaceae, but were probably also borne in lax, pendulous inflorescences. There are several other well-preserved fagalean fossils such as *Antiquacupula* and *Protofagacea* from Georgia, USA, that have flowers that were clearly borne in dichasial partial inflorescences. In *Protofagacea* the dichasial system of bracts and bracteoles is fused to form a condensed woody cup (‘cupule’) as in extant Fagaceae and Nothofagaceae.

Indirect evidence of solitary flowers from the Turonian–Maastrichtian comes from the systematic affinity of flowers

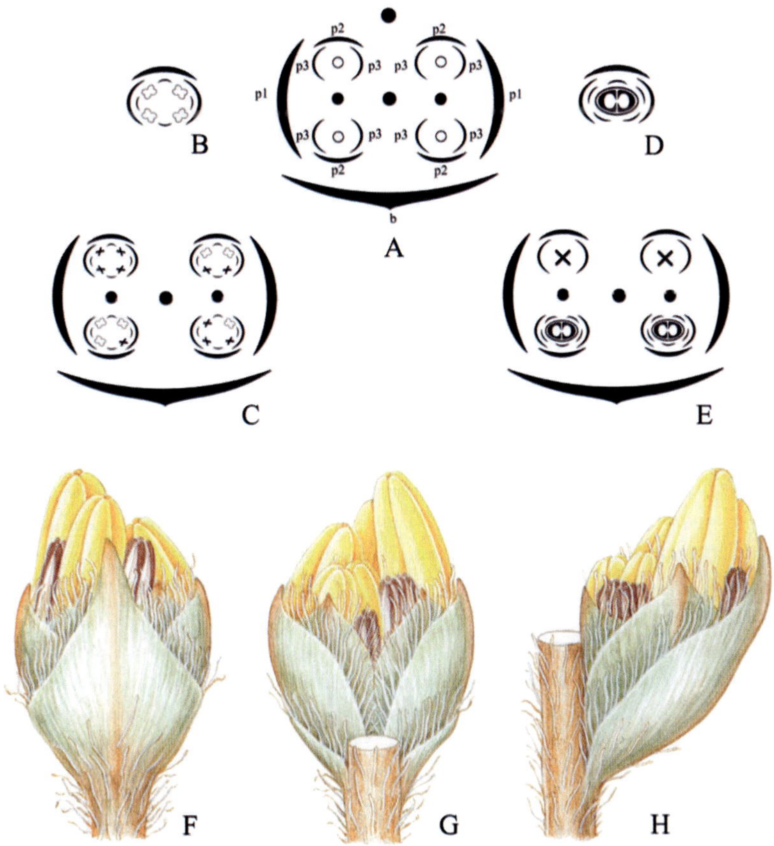

Figure 16.3 Floral diagrams (A–E) and reconstructions (F–H) of inflorescence and flowers in the Normapolles genus *Endressianthus* from the Late Cretaceous (Campanian–Maastrichtian) of Portugal. (A) Diagram of partial inflorescence with four flowers. (B) Floral diagram of individual staminate flower with perianth parts and supporting bracts. (C) Diagram of partial inflorescence with staminate flowers; none of the flowers has a full set of stamens. (D) Floral diagram of individual pistillate flower with perianth parts and supporting bracts. (E) Diagram of partial inflorescence with pistillate flowers; only two flowers are fully developed. (F–H) Reconstructions of partial inflorescences with staminate flowers in (F) abaxial, (G) adaxial and (H) lateral view. From Friis *et al.* (2003b).

such as *Futabanthus* (Annonaceae). Two other flowers, also from Japan, *Protomonimia* and *Hidakanthus*, are terminal on stout stalks and were probably also solitary. *Protomonimia* and *Hidakanthus* are preserved at the fruiting stage, but both probably developed from relatively large solitary flowers.

16.2 FLORAL ORGANISATION

16.2.1 Sex of flowers

Flowers of extant *Amborella* are functionally unisexual, which is regarded as a probable derived feature (Endress and Doyle, 2009). *Amborella* is generally regarded as organisationally bisexual because sterile stamens are present in the pistillate flowers (Endress and Doyle, 2009). Bisexual flowers also occur in those lineages (Nymphaeales and Austrobaileyales) that diverge from all other angiosperms immediately above the level of *Amborella*. This provides further support for the standard conclusion that bisexual flowers are basic for angiosperms as a whole. However, the reproductive structures of most non-angiosperm seed plants are unisexual and the transfer of Hydatellaceae with its small unisexual flowers to Nymphaeales (Saarela *et al.*, 2007) further complicates the picture. Unisexual flowers also occur in Chloranthaceae and *Ceratophyllum*, as well as early-diverging eudicots and many eumagnoliids. Phylogenetic trees based on extant taxa therefore do not provide an unambiguous resolution of the ancestral form of sex expression in angiosperm flowers.

Sex distribution in Early Cretaceous flowers. The fossil record is equivocal on the relative antiquity of unisexual versus bisexual flowers in angiosperms. There is no information about floral organisation or about individual floral organs before the Late Barremian – Early Aptian, and when angiosperm floral structures are first encountered there is already some structural diversity. The oldest assemblage of angiosperm flowers from Portugal (Torres Vedras, Late Barremian – Early Aptian) includes both unisexual (e.g. *Hedyosmum*-like) and bisexual flowers (e.g. small, hypogynous and trimerous flower). *Archaefructus* from the

Yixian Formation has unisexual staminate and pistillate flowers on the same axis, at least under the inflorescence interpretation (Chapter 9).

Slightly younger floras from the Aptian – Early Albian of Portugal (Buarcos, Catefica, Famalicão, Juncal, Vale de Agua, Vila Verde 2) and from the Aptian – Middle Albian of the Potomac Group sequence of eastern North America (mainly the Puddledock flora) also include both unisexual and bisexual flowers. Unisexual flowers are seemingly slightly more common, but in some cases it is difficult to establish floral sex expression securely in fossil material. Stamens are often shed at an early stage in floral development and do not always leave traces in developing fruits.

Aptian – Middle Albian angiosperms with unisexual flowers include several eudicots, such as the ranunculalean flower *Teixeiraea*, the buxalean flowers *Lusistemon/Lusicarpus*, and the platanoid flowers *Friisicarpus brookensis* and *Aquia brookensis*. Other early buxalean fossils, such as *Aguacarpus*, *Silucarpus* and *Valecarpus*, which are related to the *Lusistemon/Lusicarpus* plant, were probably also unisexual. There are also unisexual flowers of Chloranthaceae (*Hedyosmum*), and among monocots there are unisexual flowers of Araceae. Fossil flowers of this age that are structurally bisexual include the ANITA-grade and eumagnoliid forms *Canrightia Carpestella*, Laurales B, *Endressinia*, *Monetianthus*, *Potomacanthus* and several undescribed fossils. There are also bisexual flowers among the early fossil Araceae.

Sex distribution in mid-Cretaceous flowers. In fossil floras from the mid-Cretaceous the proportion of unisexual to bisexual flowers appears to be more or less equal, but mesofossil floras from this time interval remain to be studied in detail. Direct evidence of bisexual flowers is provided by the 'Rose Creek flower', the various species of *Mauldinia* and the *Chloranthistemon* flowers from Germany. *Archaeanthus* was also a bisexual flower based on the small, probable stamen scars that occur between the carpels and the larger scars of the perianth parts. Unisexual flowers of mid-Cretaceous age include various platanoid taxa from the Potomac Group, such as *Hamatia elkneckensis – Friisicarpus elkneckensis* from the Mauldin Mountain locality and *Platananthus potomacensis – Friisicarpus marylandensis* from the West Brothers locality, as well as species of *Spanomera* (Buxales) also from the Mauldin Mountain and West Brothers localities. From the Dakota Formation there is *Caloda delevoryana* and several unnamed fossils. In many flowers of this age (e.g. *Lesqueria*) sex expression is not known securely.

Sex distribution in Late Cretaceous flowers. Bisexual flowers, which predominate among extant angiosperms, only become dominant in fossil floras during the Late Cretaceous. At this time bisexual flowers were also common in some lineages that today usually have unisexual flowers. Most notably, bisexual flowers predominate in the extinct Normapolles complex, while members of extant core Fagales typically have unisexual flowers. Among the Normapolles flowers so far recognised only the two species of *Endressianthus*, and perhaps *Bedellia*, were clearly unisexual. All others were apparently bisexual, or may have had both bisexual and unisexual flowers in the same inflorescences as in *Budvaricarpus*.

Typical specimens of *Budvaricarpus* have three flowers that are laterally fused and probably reflect a condensed partial inflorescence. The median flower, which is often the largest, has both carpels and stamens, while the two lateral flowers are pistillate. *Antiquocarya* is another Normapolles plant that probably has both bisexual and functionally unisexual flowers. In several specimens of *Antiquocarya* there are remains of filaments in flowers and fruits that also have a well-developed gynoecium. However, in other flowers small irregular structures that apparently represent aborted stamens are seen in the equivalent positions. A similar situation occurs among flowers of Late Cretaceous Fagales that fall outside the core (Normapolles) group. *Antiquacupula* has both staminate and bisexual flowers.

Unisexual flowers known from the Late Cretaceous include various platanoids such as species of *Friisicarpus*, *Platananthus* and *Quadriplatanus*, and probably also species of *Maiandrocarpus*, *Malliocarpus*, *Mitocarpus*, and *Zeugarocarpus* (presumed early-diverging eudicots, Chapter 12).

16.2.2 Receptacle, floral phyllotaxis and merism

An elongated receptacle with spiral floral phyllotaxis, similar to the situation in extant *Magnolia*, has often been regarded as the basic condition in angiosperms (Arber and Parkin, 1907; Takhtajan, 1969; Cronquist, 1981). However, elongated receptacles are relatively rare among extant angiosperms and based on modern phylogenetic interpretations appear to have evolved independently in several lineages from flowers with a short receptacle (e.g. *Myosurus* in Ranunculaceae, Magnoliaceae in Magnoliales, *Schisandra* in Schisandraceae). Similarly, floral phyllotaxis among early-diverging lineages of angiosperms is often not truly spiral. Floral phyllotaxis is also variable across lineages and also within clades (Endress, 2001). Because of this variability among early-diverging lineages of angiosperms the basic condition among angiosperms as a whole remains uncertain (Endress and Doyle, 2009).

In some members of the ANITA grade, the arrangement of parts is spiral (*Amborella*, *Austrobaileya*, Trimeniaceae, Schisandraceae), but in Nymphaeaceae the arrangement is predominantly whorled. In Nymphaeaceae floral phyllotaxis may also be mixed with some organs arranged in spirals and others arranged in whorls. Similar plasticity is also expressed in other lineages at or near this level of angiosperm evolution. For example, in *Ceratophyllum*, even in a single species, flowers may have either spiral or whorled phyllotaxis (Endress, 2001).

As in Nymphaeaceae, careful studies of floral phyllotaxis among eumagnoliids demonstrate that in many cases parts often thought to be in a spiral are actually in whorls. Spiral arrangements are neither common nor plesiomorphic among eumagnoliids (Doyle and Endress, 2000; Endress and Doyle, 2009). In monocots, as well as most eudicots, floral phyllotaxis is typically whorled (Ronse De Craene *et al.*, 2003; Endress and Doyle, 2007).

As with floral phyllotaxis, the number of floral organs is often variable in flowers of early-diverging extant angiosperm lineages. In part, this may be linked to variable floral phyllotaxis, as well as the shape and size of the perianth parts in relation to the size of the flower. The extent of variability makes it impossible to establish securely a clear model of perianth evolution among early angiosperms. The plesiomorphic condition of the perianth among angiosperms as a whole is uncertain (Endress, 2001).

Flowers with whorled phyllotaxis generally have a fixed number of parts in each whorl and the number of parts in the perianth and androecial whorls, as well as the number of carpels, may be the same. Flowers of this kind are described as isomerous. More often, carpel number is lower, or more rarely higher, than the number of parts in the androecial whorl. Stamen number may also be higher than the number of tepals or petals. Such flowers are described as heteromerous.

Many eumagnoliids with whorled phyllotaxis are heteromerous with perianth and androecium on a trimerous plan and a monomerous gynoecium. Monocots generally have trimerous and isomerous or heteromerous flowers, while among core eudicots flowers are typically pentamerous and heteromerous. There is, of course, much variation in these features among the species of these very species-rich clades.

Receptacle, floral phyllotaxis and merism in Early Cretaceous flowers. Phyllotaxis is not always easy to assess in flowers of extant angiosperms and may be still more difficult to establish with certainty in fossil matrial. However, both whorled and spiral phyllotaxis appears to be present among Late Barremian – Early Aptian mesofossils and there is no obvious dominance of one type over the other. The flowers from the Yixian Formation of China, and also from the Torres Vedras locality, Portugal, are already diverse in phyllotaxis and in number of floral parts. Generally, however, the floral organs are borne on a short, indistinct receptacle. *Sinocarpus* from the Yixian Formation apparently had whorled phyllotaxis and in the Torres Vedras flora whorled phyllotaxis occurs in the pistillate *Hedyosmum*-like flowers as well as an undescribed trimerous flower. Spiral phyllotaxis occurs in two undescribed multicarpellate structures from Torres Vedras (Figure 16.4).

In slightly younger flowers, spiral floral phyllotaxis is known for *Carpestella*, *Virginianthus*, and many unnamed taxa, while whorled arrangements occurs in the lauraceous flower *Potomacanthus*, and several unnamed flowers. The phyllotaxis of *Monetianthus* is perhaps mixed with spirally arranged tepals, irregularly arranged stamens and whorled carpels. At this level there are several fossils that have elongated receptacles, such as the floral structures associated with *Anacostia* (Figure 16.5). *Monetianthus* and *Carpestella* are unique in having an extension of the floral axis to the apex, as also occurs in extant Nymphaeaceae and Schisandraceae. Most flowers at this level have relatively few parts, such as the staminate flowers of the fossil aroids inflorescence, which consist only of a single organ, but there are also several taxa with multiparted flowers such as *Virginianthus*, *Potomacanthus* and *Monetianthus*.

Receptacle, floral phyllotaxis and merism in mid-Cretaceous flowers. Flowers from the mid-Cretaceous show similar features of floral phyllotaxis and merism as in earlier flowers, although the proportion of flowers with whorled phyllotaxis and a relatively low number of floral parts appears to be somewhat higher, probably reflecting the continuing diversification of eudicots. However, in the mid-Cretaceous floral structures with strongly extended receptacles and numerous floral parts, such as *Archaeanthus*, *Lesqueria*, and similar forms, are also prominent. Even though these fossils are preserved at fruiting stage it seems likely that the anthetic flowers also had an elongated receptacle. *Archaeanthus* had nine to twelve tepals apparently in three or four whorls of three tepals each. It is not known whether the estimated 50–60 stamens and 100–130 carpels were arranged in a spiral, in whorls or were

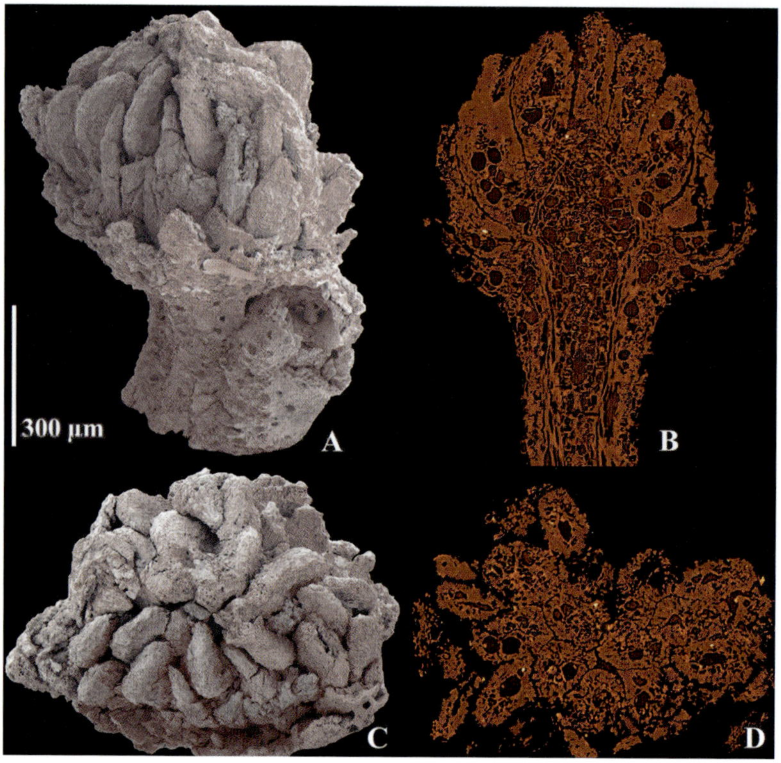

Figure 16.4 (A, C) SEM and (B, D) SXRTM images of a multicarpellate, apocarpous flower from the Torres Vedras locality, Portugal (Late Barremian – Early Aptian) in (A, B) lateral and (C, D) apical view, showing many carpels in a helical arrangement on a slightly elongated dome-shaped receptacle. Material in the collections of the Swedish Museum of Natural History, Stockholm.

irregular, but a spiral arrangement appears most likely based on the situation in extant *Magnolia*. *Lovellea* has a receptacular floral cup (hypanthium) with two series of tepals and two series of stamens of about eight parts each, as well as many carpels in a spiral arrangement. Other flowers from the mid-Cretaceous, such as the various species of *Mauldinia*, the 'Rose Creek flower' and several undescribed taxa, have a simple unexpanded receptacle, whorled phyllotaxis and few floral parts.

Receptacle, floral phyllotaxis and merism in Late Cretaceous flowers. In Late Cretaceous floras, reflecting the rapidly expanding diversity of core eudicots, most fossil flowers have a short indistinct receptacle bearing whorled floral organs (Figure 16.6). The perianth and androecium are often on a pentamerous plan, while the gynoecium is typically dimerous or trimerous. Only few flowers with spiral phyllotaxis and many floral parts have been recorded from Late Cretaceous floras. Among these are flowers with various types of expanded receptacle that include the probable eumagnoliids, *Detrusandra*, *Cronquistiflora* and *Microvictoria*, which have shallow to deep receptacular floral cups, as well as the annonaceous *Futabanthus* from Japan. Indirect evidence for the presence of elongated floral axes in the later part of the Cretaceous is provided by the many different species of *Liriodendron*-like seeds and leaves (Magnoliaceae). *Litocarpon beardii* (probably Magnoliales) from the Late Cretaceous (Late Santonian – Early Campanian) of British Columbia, Canada, provides direct evidence of the presence of flowers with an elongated, probably *Magnolia*-like, receptacles at this stage of angiosperm evolution.

16.2.3 Position of floral organs

Hypogyny, the condition where the gynoecium is placed above the other floral organs on the floral axis, has generally been assumed to be the ancestral condition in angiosperms. Conversely, congenital fusion of the surrounding floral organs into a hypanthium that forms a floral cup, which can either surround the gynoecium or be fused to it to various degrees, has generally been considered a derived feature (Endress and Doyle, 2009). Phylogenetic trees based on molecular data do not give a clear signal regarding the likely pattern of evolution of this feature although

Figure 16.5 SEM images of Early Cretaceous multicarpellate flowers/fruits with strongly elongate receptacle and many carpels in a spiral arrangement. (A) Specimen from the Buarcos locality (Late Aptian – Early Albian), Portugal. (B) Specimen from the Puddledock locality (Early–Middle Albian), Virginia, USA. Material in the collections of (A) the Swedish Museum of Natural History, Stockholm, and (B) the Field Museum, Chicago.

hypogyny predominates at the ANITA grade, and the floral cup of *Amborella* and the perigynous to epigynous condition seen in some Nymphaeales probably reflects secondary modification (Endress and Doyle, 2009).

Position of floral organs in Early Cretaceous flowers. Most of the angiosperm flowers from the Early Cretaceous have a hypogynous organisation, but there are also some that are perigynous–epigynous (Figures 16.7, 16.8), such as *Carpestella*, *Monetianthus* and several unnamed flowers, in which the hypanthium is fused to the gynoecium. Other early flowers have a hypanthium forming a floral cup that may be deep (e.g. *Virginianthus*) or shallow (e.g. *Potomacanthus*).

Position of floral organs in mid-Cretaceous flowers. Flowers in mid-Cretaceous floras mostly have a hypogynous organisation. There are only a few exceptions such as the flowers of *Mauldinia* that have a shallow floral cup.

Position of floral organs in Late Cretaceous flowers. Whereas the flowers known from the Early Cretaceous are predominately hypogynous in their organisation, flowers that are epigynous become much more common in Late Cretaceous floras (Figures 16.8–16.9). Although there are no mesofossil floras from the Late Cretaceous for which all the taxa present have been documented in detail, preliminary surveys have been presented for the flora at the Allon locality Georgia, USA (Herendeen *et al.*, 1999a), and for the flora at the Kamikitaba locality, Japan (Takahashi *et al.*, 1999b). Complete lists of taxa have also been published for several mesofossil floras of central Europe (Knobloch and Mai, 1986). A rough estimate of the ratio of hypogynous to epigynous flowers in these floras, as well as other floras that we have collected and investigated, indicates that almost half of the flowers recorded are epigynous or semi-epigynous.

In the Allon flora of Georgia, USA, 25 taxa show the position of the ovary. Of these 11 have an epigynous organisation and 14 are hypogynous. In the Kamikitaba flora of Japan the equivalent numbers are eight and five. Out of the 26 angiosperm taxa that have been formally described so far from the Åsen flora of southern Sweden, 14 show hypogynous organisation with 12 epigynous, but a rough survey suggests that the epigynous organisation is the most common type in the flora. In the Campanian–Maastrichtian floras of Mira and Esgueira from Portugal epigyny is also very common and apparently dominates over hypogyny. The proportion of flowers with an epigynous organisation in these Late Cretaceous floras is remarkably high compared to surveys of extant angiosperms in which the ratio of hypogynous to epigynous flowers is about 3:1. It has been suggested that protection of the developing ovules during pollination may be a key factor promoting epigyny (Grant, 1950). The higher proportion of flowers with epigynous organisation in Late Cretaceous floras may reflect initial strategies for protection of the ovules in the face of increased biotic pollination, as well as the relative prominence of particular groups of core eudicots at this stage in angiosperm evolution.

16.2.4 Perianth

Phylogenetic analyses of extant angiosperms suggest that the presence of a perianth is plesiomorphic for angiosperm flowers and that its absence, even in groups that diverged rather early from the main line of angiosperm evolution, such as *Hydatella*, *Ceratophyllum*, Chloranthaceae (except pistillate *Hedyosmum*) and some eumagnoliids, is a result of

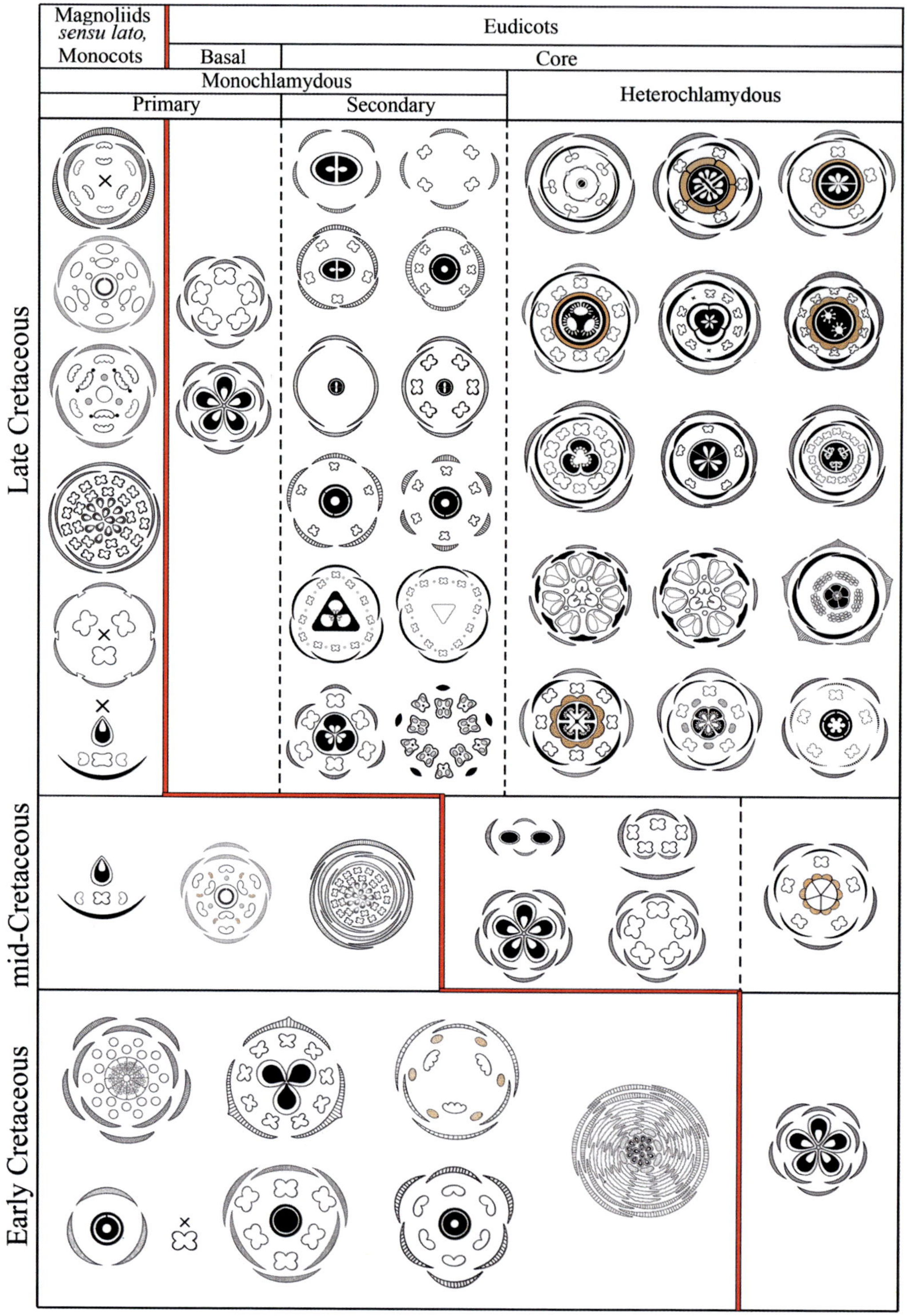

Figure 16.6 Compilations of floral diagrams of Cretaceous flowers from the Early, mid- and Late Cretaceous. Note that Early and mid-Cretaceous flowers typically have a simple perianth of one kind of organ (tepals). Heterochlamydous flowers with differentiated calyx and corolla are first recorded in the mid-Cretaceous, but are dominant in the Late Cretaceous. A nectary disc (orange) is also characteristic of many Late Cretaceous flowers. The red line separates eudicot angiosperms from other angiosperms.

Figure 16.7 SEM images of Early Cretaceous flowers with an inferior ovary. (A, B) Undescribed flowers from the Puddledock locality (Early–Middle Albian), Virginia, USA. (C) *Hedyosmum*-like pistillate flower from Torres Vedras locality (Late Barremian – Early Aptian), Portugal. (D) Undescribed trimerous flower from the Catefica locality (Late Barremian–Aptian), Portugal. (E) *Virginianthus calycanthoides*, from the Puddledock locality. (F) *Monetianthus mirus* from the Vale de Agua locality (Late Aptian – Early Albian), Portugal. (G) Undescribed flower with trimerous organisation from the Vale de Agua locality, Portugal. Material in the collections of the Swedish Museum of Natural History, Stockholm (C, D, F, G), and the Field Museum, Chicago (A, B, E).

secondary loss (Endress and Doyle, 2009). Among the lineages of the ANITA grade, as well as among eumagnoliids and monocots, the perianth consists typically of a single kind of organ (tepals) that may be sepaloid, as in *Amborella*, or petaloid, as in *Cabomba* and *Brasenia*. True petals, defined by more differentiated anatomy and delayed development, are rare in these groups and are regarded as only present in *Nuphar* among plants of the ANITA grade (Endress and Doyle, 2009). By comparison, petals are characteristic of most groups among core eudicots (Endress and Matthews, 2006b), and in most eudicot flowers the perianth is clearly differentiated into sepals and petals (heterochlamydous), which are arranged in whorls.

Perianth in Early Cretaceous flowers. In fossil flowers it is sometimes difficult to distinguish bracts from sepals or tepals, but in most cases the perianth and its organisation is straightforward to interpret. In the earliest fossil flowers from the Late Barremian – Early Aptian, and also in the

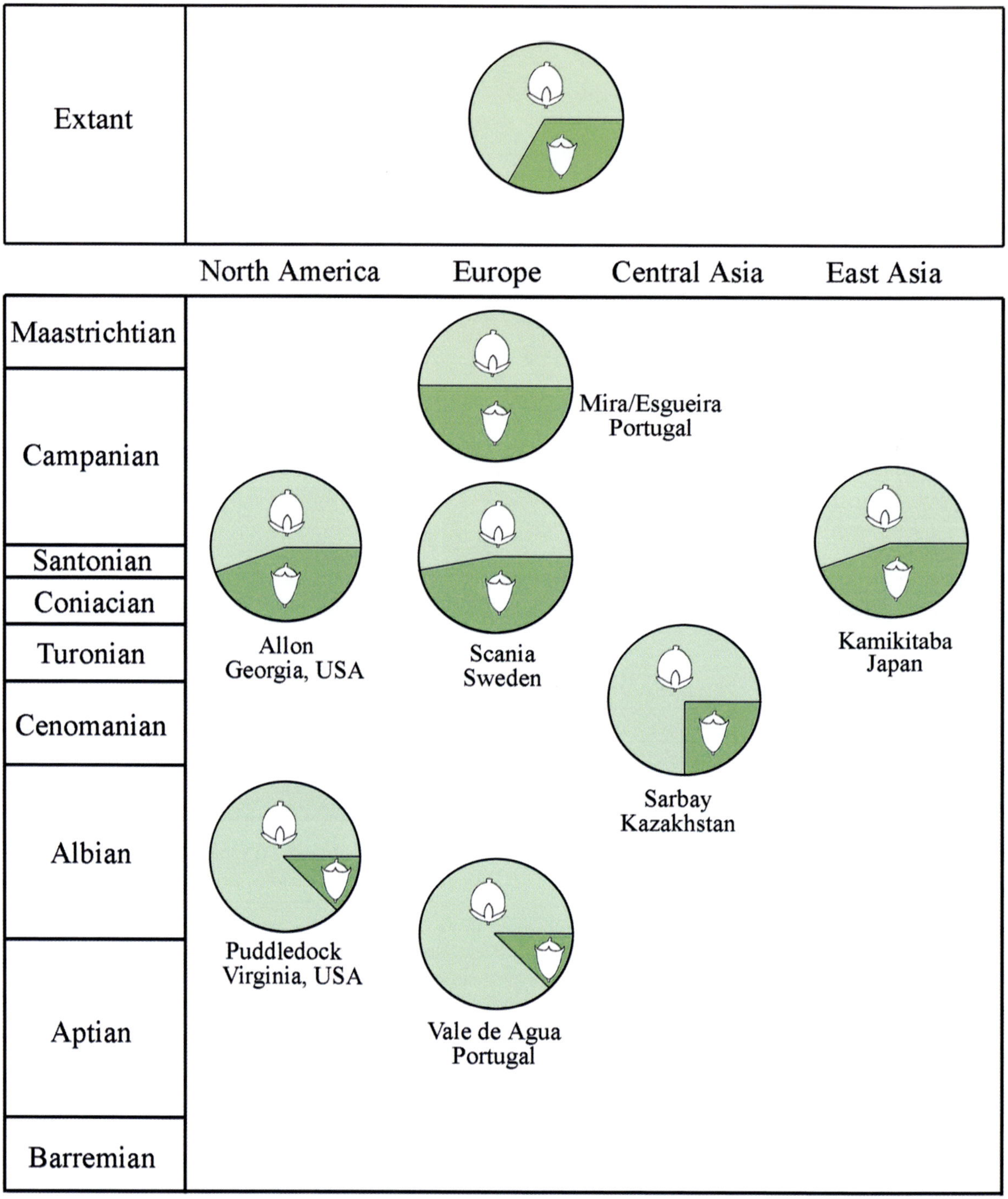

Figure 16.8 Changes in the proportion of flowers with an inferior (darker green) and superior (lighter green) ovary in mesofossil floras through the Cretaceous and compared to their distribution in extant angiosperms. Note the prominence of inferior ovaries in Late Cretaceous mesofossil floras.

Figure 16.9 SEM-images of flowers from the Late Cretaceous (Campanian–Maastrichtian) of Mira and Esgueira, Portugal, with inferior ovaries. Material in the collections of the Swedish Museum of Natural History, Stockholm.

more diverse flower assemblages from the Late Aptian – Middle Albian, the perianth consists only of tepals. For example, in the *Amborella*-like flower, *Friisicarpus*, *Teixeiraea*, *Virginianthus*, the pistillate *Hedyosmum*-like flowers and the bisexual flowers of Araceae there is no evidence of differentiation into sepals and petals. In the staminate *Hedyosmum*-like flowers and the unisexual aroid flowers tepals are lacking entirely. In *Teixeiraea* and *Virginianthus* the tepals appear to grade from sepaloid to petaloid, whereas in the *Amborella*-like flower, the *Hedyosmum*-like pistillate flowers, the various lauraceous flowers such as *Potomacanthus* and *Cohongarootonia*, as well as in several undescribed forms, the tepals appear to be sepaloid (Figures 16.6, 16.7).

Perianth in mid-Cretaceous flowers. Most fossil flowers known from the mid-Cretaceous have a perianth that is more or less undifferentiated, as is the case in most of the more ancient fossil material. An undifferentiated perianth is present, for example, in *Spanomera* and among early platanoids known from this interval. A more or less undifferentiated perianth is also present in *Mauldinia*, although there is some variation in the size of the outer and inner tepals in the various species. In *Mauldinia mirabilis* the outer tepals are much smaller than the inner. Similarly, in the reconstructed flower of *Archaeanthus* the outer three tepals are leathery with relatively broad attachment, while the six or nine inner tepals are thinner with narrower attachments.

An important innovation in the form of the perianth, which appears for the first time in the mid-Cretaceous, is the presence of flowers with a distinct heterochlamydous perianth. These flowers have an outer whorl of sepals (calyx) and an inner whorl of petals (corolla). This is seen most clearly in the 'Rose Creek flower' where the sepals are thick and broadly attached, whereas the petals are thin and more membranous, and have a narrow claw-shaped base.

Perianth in Late Cretaceous flowers. In Late Cretaceous flowers the heterochlamydous condition, which is characteristic of core eudicots, predominates. Most flowers of this age have a distinct calyx and corolla (Figures 16.6, 17.10). However, they are generally small and there is relatively little variation in the size and shape of sepals and petals. Most petals are elliptical to spathulate and membranous, while the sepals are often leathery or woody. In these flowers the perianth parts are also regularly arranged in whorls with consistent symmetry in which the number of parts is generally fixed rather than variable. All of these features are characteristic of core eudicots; most fossil flowers of this kind are clearly attributable to this major group.

The increasing complexity of floral form seen among fossil flowers from the Late Cretaceous is paralleled in one key group of angiosperms by a reverse trend toward simplification. In several mesofossil assemblages from the Late Cretaceous of Europe and eastern North America a large proportion of the flowers appear to be secondarily simple. These flowers have only a single whorl of perianth parts, which are often not well developed, or they lack a perianth entirely (Figure 16.6). This is the case for all the known flowers of the Normapolles group and also occurs in other fagalean flowers from the Late Cretaceous, such as *Antiquacupula* and *Protofagacea*. The Normapolles flowers are thought to be wind-pollinated and the poorly developed perianth perhaps reflects loss of petals (or loss of both petals and sepals) in association with this mode of pollen transfer.

The prominence of flowers with fixed, whorled, floral phyllotaxis and a regular number of parts in the Late Cretaceous is a prerequisite for increased integration of the different parts of the flower (section 16.3.3). Stabilisation of floral structure creates new possibilities for fusion between adjacent organs of the same floral whorl, as well as for fusion among the organs of different floral whorls (Endress, 1994b). An early indication of the new possibilities for synorganisation resulting from increased regularity in floral phyllotaxis is seen in *Actinocalyx* from the Late Santonian – Early Campanian Åsen locality, Sweden, which provides the earliest record of sympetaly. Congenital or postgenital fusion of adjacent petals to form a corolla tube had important implications for subsequent directions of floral evolution in many species-rich groups of angiosperms. With the exception of *Actinocalyx* sympetaly is otherwise known only from the Cenozoic fossil record.

16.2.5 Androecium

Extrapolations from extant taxa and their relationships (Endress and Doyle, 2009) suggest that the basic condition of the androecium in angiosperm flowers involved several fertile stamens arranged in more than two whorls, or in a spiral. The unistaminate flowers of Chloranthaceae, *Ceratophyllum* and Hydatellaceae, all key early-diverging lineages, are considered derived in this respect. The presence of inner staminodes, which has sometimes been viewed as a possible plesiomorphic condition, is also considered derived (Endress and Doyle, 2009). Stabilisation of the number of stamens, and the arrangement of the stamens in two whorls, is characteristic of the flowers of monocots and most eudicots.

The basic condition for angiosperm stamen morphology is much less clear. Several taxa that diverge near the base of the angiosperm phylogenetic tree, including *Amborella*, Nymphaeaceae and *Austrobaileya*, have stamens in which there is no clearly differentiated filament. The lower part of the stamen is long and broad, and anther dehiscence is introrse. However, other taxa at this level of angiosperm evolution, such as Hydatellaceae and Cabombaceae, have well-differentiated long narrow filaments and anther dehiscence is latrorse or extrorse.

Androecium in Early Cretaceous flowers. There are very few complete flowers from Early Cretaceous mesofossil floras, but there are a surprisingly large number of dispersed stamens with pollen *in situ*. The abundance of dispersed stamens with *in situ* pollen in early mesofossil floras may indicate that stamens were important for protection and attraction in the biology of early flowers. Stamens may also have abscised rather early in floral development, even during the male phase of anthesis, as known for several extant Magnoliales. Dispersed stamens of certain extant Magnoliales are still visited by pollinators (Endress, 2010b).

With regard to organisation of the androecium, even at this early stage in angiosperm evolution there is already considerable diversity. Staminate flowers of the *Hedyosmum*-like inflorescences consist of a single stamen and if the fertile zone of *Archaefructus* is correctly interpreted as an inflorescence then individual staminate flowers consist of two or sometimes three stamens. Similarly, two different aroid inflorescences from Vila Verde 2 have naked staminate flowers with one or two stamens. Several early flowers, such as the trimerous hypogynous form from Torres Vedras, have nine stamens probably in three whorls. Other flowers of Aptian – Middle Albian age, such as the *Amborella*-like flower, *Endressinia*, *Monetianthus*, *Teixeiraea* and *Virginianthus*, have many stamens in an irregular or spiral arrangement.

In the earliest floras that contain angiosperm macrofossils (e.g. Yixian flora) and mesofossils (Torres Vedras) the stamens have simple anthers with only modest development of the connective between the pollen sacs (Figure 16.10). The filament is not well developed and the stamen base is short. Anthers, on average, are smaller than those from younger floras and they vary in shape and size from sagittate and short to narrow and elongate. At this level, anthers are dithecate, but with the four pollen sacs more or less equally distributed. Sometimes there is an apical expansion of the connective, but typically it is small as in *Archaefructus*. In the Torres Vedras flora only the *Hedyosmum*-like stamens have an apical extension. Such an extension is lacking or indistinct in all other stamens from the earliest phases of the angiosperm fossil record.

In slightly younger, Aptian – Middle Albian, mesofossil floras from Portugal (Buarcos, Catefica, Famalicão, Vale de Agua) and North America (Puddledock) there is a much greater variation in anther shape and size than in earlier floras (Figures 16.10, 16.11). Many of the anthers have an extensive development of the connective. The apical sterile extension may be strongly developed and in some cases may be as long as the pollen sacs. The lower part of the stamen may be short or long, and developed into a more or less distinct filament. In many cases it is flattened, and it always grades into the anther without a distinct joint. Pollen sacs are often very small compared with the total size of the stamen (Figures 16.10, 16.11).

Flattened stamens typically have pollen sacs borne on one side. They may bulge over the connective, as in several undescribed stamens from Puddledock, or they may be embedded in the stamen surface, as in *Virginianthus*. In isolated stamens of this kind it is difficult to establish whether the pollen sacs are adaxial or abaxial. In *Virginianthus*, where the flattened stamens are still attached to the floral cup, the pollen sacs are abaxial and dehiscence is extrorse. Dehiscence is also extrorse in the *Amborella*-like flower from the Catefica locality, Portugal.

Almost all early anthers appear to be tetrasporangiate. The only exceptions encountered so far are the bisporangiate anthers known from two lauralean flowers from the Early–Middle Albian Puddledock flora. Extant Laurales show considerable variation in this feature, and the fossil record shows that this variation occurred very early in the evolution of the group.

Anther dehiscence in the earliest angiosperm stamens from the Late Barremian – Early Aptian was probably by longitudinal slits, but the mode of dehiscence is not very clear. In slightly younger mesofossil floras from the Aptian – Middle Albian there are a greater variety of anthers, many of them with valvate dehiscence. This mode of dehiscence remains common throughout the Early Cretaceous. Most anthers with valvate dehiscence have two laterally hinged valves over each theca. The two pollen sacs of each theca are then exposed as the pair of valves open like two shutters. This is seen clearly in the platanoid *Aquia brookensis* from the 'Bank near Brooke' locality, and also in many stamens from the Puddledock flora that occur dispersed or in flowers. A few stamens of this age show an interesting variant of valvate dehiscence that is especially common in extant Laurales. In these cases there are apically hinged valves over each theca, or over individual pollen sacs. This is seen for example in the various lauralean floral structures that occur in the Puddledock flora.

Staminodes are known from early in the Cretaceous with the first distinct types described from the Late Aptian – Early Albian of Brazil (*Endressinia*) and from the Early–Middle Albian Puddledock flora, where they occur in *Virginianthus* and together with staminal appendages in *Cohongarootonia* and other lauralean flowers.

Androecium in mid-Cretaceous flowers. In mid-Cretaceous flowers there is a clear shift in stamen morphology with the first appearance of stamens with very distinct filaments. In stamens of this kind the lower part of the stamen is elongated, band- or thread-like, and there is a distinct joint between filament and anther. This is seen in stamens of the 'Rose Creek flower' as well as in several undescribed flowers from the Mauldin Mountain flora. However, stamens with massive bases that grade broadly into the anthers also persist into the mid-Cretaceous and in some extant taxa. One prominent example is the fossil *Chloranthistemon* from the karst deposits of Germany,

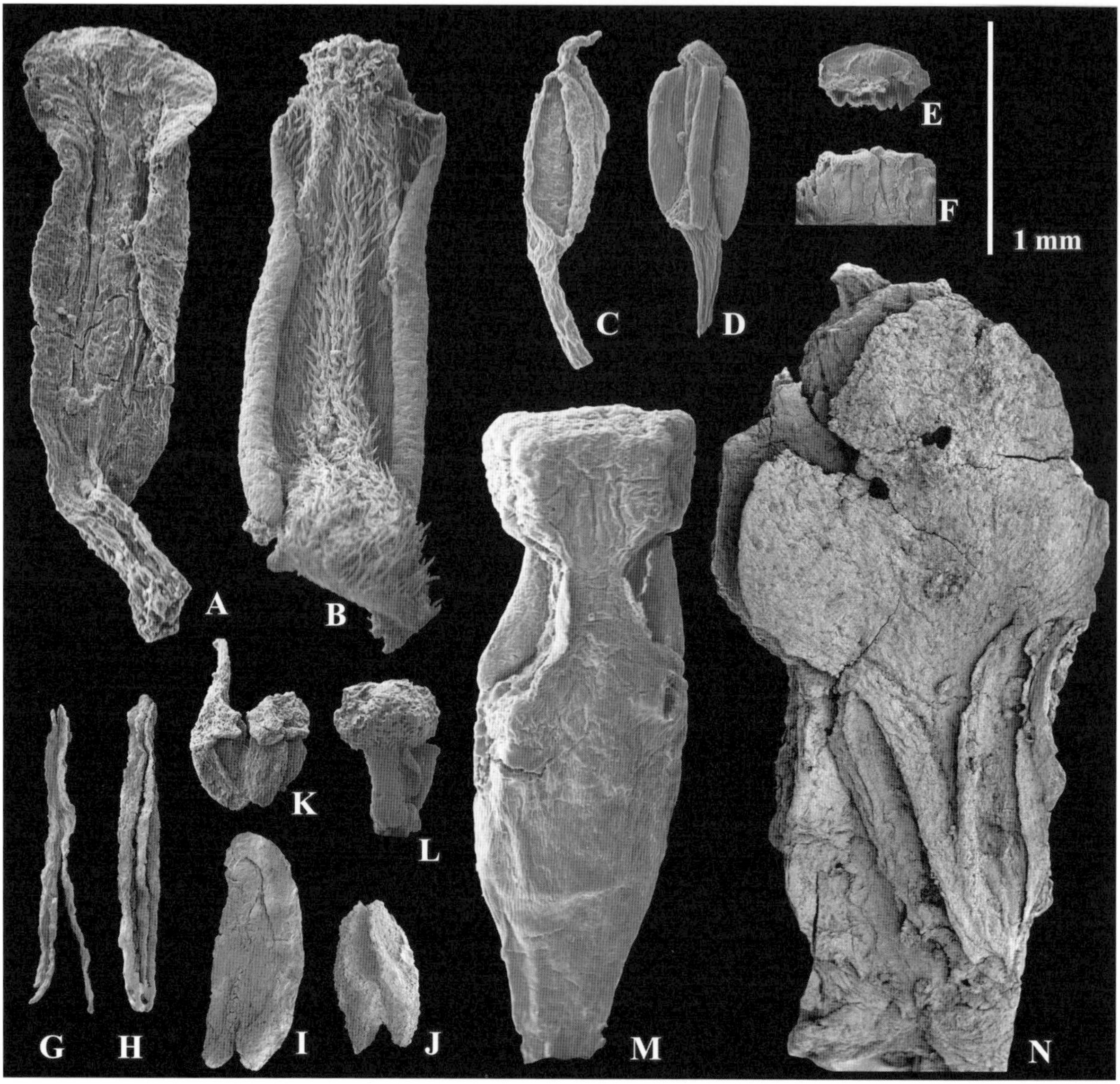

Figure 16.10 SEM images of isolated stamens from the Early Cretaceous. (A–F) Puddledock locality (Early–Middle Albian), USA. (G–J) Torres Vedras (Late Barremian – Early Aptian), Portugal. (K, L) Catefica (Late Barremian – Aptian), Portugal. (M) Vale de Agua (Late Aptian – Early Albian), Portugal. (N) Vila Verde 2, Portugal (Late Aptian – Early Albian). Note that the very simple stamens (G–J) are Late Barremian – Aptian, but the larger, more bulky stamens are Late Aptian – Middle Albian. Material in the collections of the Field Museum, Chicago (A-F) and in the collections of the Swedish Museum of Natural History, Stockholm (G-N).

although in this case the broad base may represent fusion of three stamens. Other examples occur in the various species of *Mauldinia*, including those from the same karst deposits.

An intriguing feature of mid-Cretaceous angiosperm stamens is the common occurrence of anthers that have a very extensive sterile apex. Such stamens occur, for example in flowers of various platanoids, such as *Hamatia*

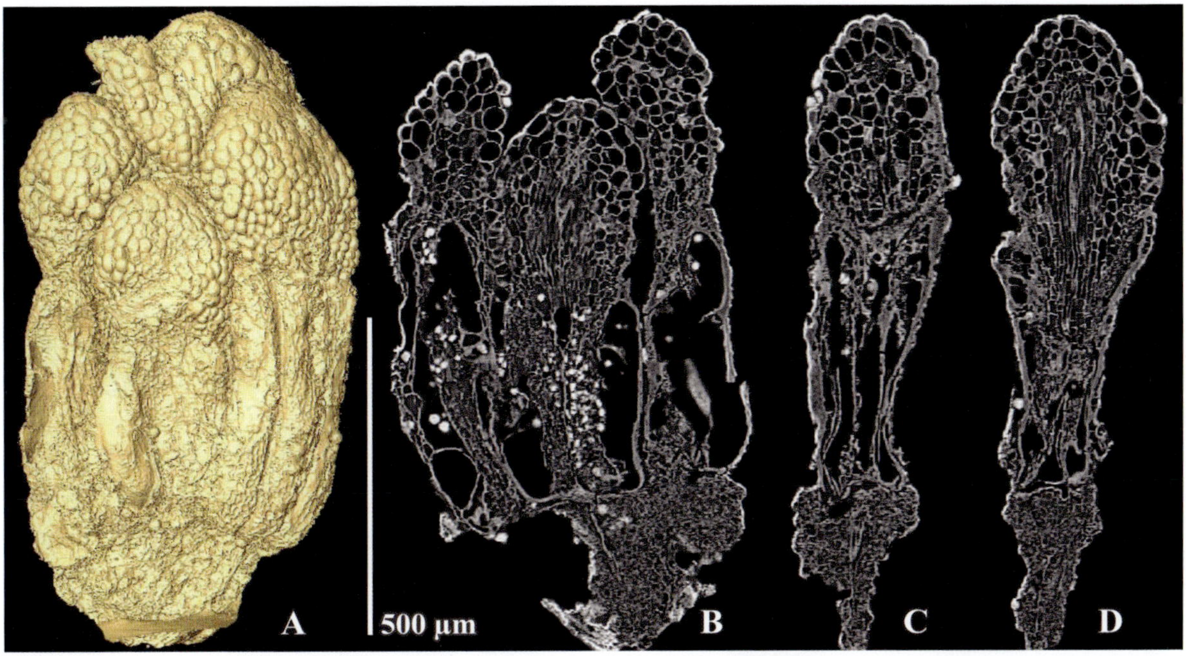

Figure 16.11 Surface rendering and longitudinal slices of staminate flower from the Puddledock locality (Early–Middle Aptian), Virginia, USA, based on synchrotron radiation X-ray tomographic microscopy (SRXTM). Note very small pollen sacs and bulky apical extension of the connective. Material in the collections of the Field Museum, Chicago.

elkneckensis and *Platananthus potomacensis*. At this level of angiosperm evolution a sterile anther extension may function both for attraction, as in extant *Nelumbo*, in which the apical staminal appendage is thermogenic (Meeuse and Raskin, 1988), and for protection, as in extant *Asimina* (Annonaceae), in which the heads of the stamens form a protective shield (Endress, 1985).

All mid-Cretaceous anthers are dithecate; most are tetrasporangiate, as in earlier floras, but bisporangiate anthers are also known from this time interval in various species of *Mauldinia* (Lauraceae). *Mauldinia* flowers also have distinct staminodes and staminal appendages.

Androecium in Late Cretaceous flowers. The common form of the androecium in flowers from the Late Cretaceous mesofossil floras is markedly different from that in Early Cretaceous flowers. It is also distinctive compared with flowers from the mid-Cretaceous. Stamen phyllotaxis is mainly whorled and the number of androecial whorls is typically one or two (Figure 16.6), but rarely more. Anthers are mostly dithecate with a distinct band- or thread-shaped filament that is separated from the anther by a distinct joint. Androecia of this kind are characteristic of extant core eudicots.

A few stamens from the Late Cretaceous have a massive connective often combined with extensive sterile apical extensions. This is seen clearly in the various species of *Chloranthistemon*, but also occurs in *Detrusandra* and *Cronquistiflora*. Valvate dehiscence is typical of stamens of this type. Valvate dehiscence is also characteristic of the various species of *Platananthus* (Platanaceae), fossil species related to the Hamamelidaceae (*Allonia*, *Androdecidua*, *Archamamelis*), and flowers of Lauraceae. In several of these groups stamens also have very prominent apical extensions of the connective. In general, however, the prevailing mode of dehiscence in Late Cretaceous anthers is by longitudinal slits.

Innovations in anther form are also seen in stamens from the Late Cretaceous. From the Turonian Old Crossman locality anthers with awns occur in *Paleoenkianthus sayrevillensis*, monothecate anthers were described for *Dressiantha*, and dithecate–monosporangiate anthers are perhaps present in the two species of *Mabelia*.

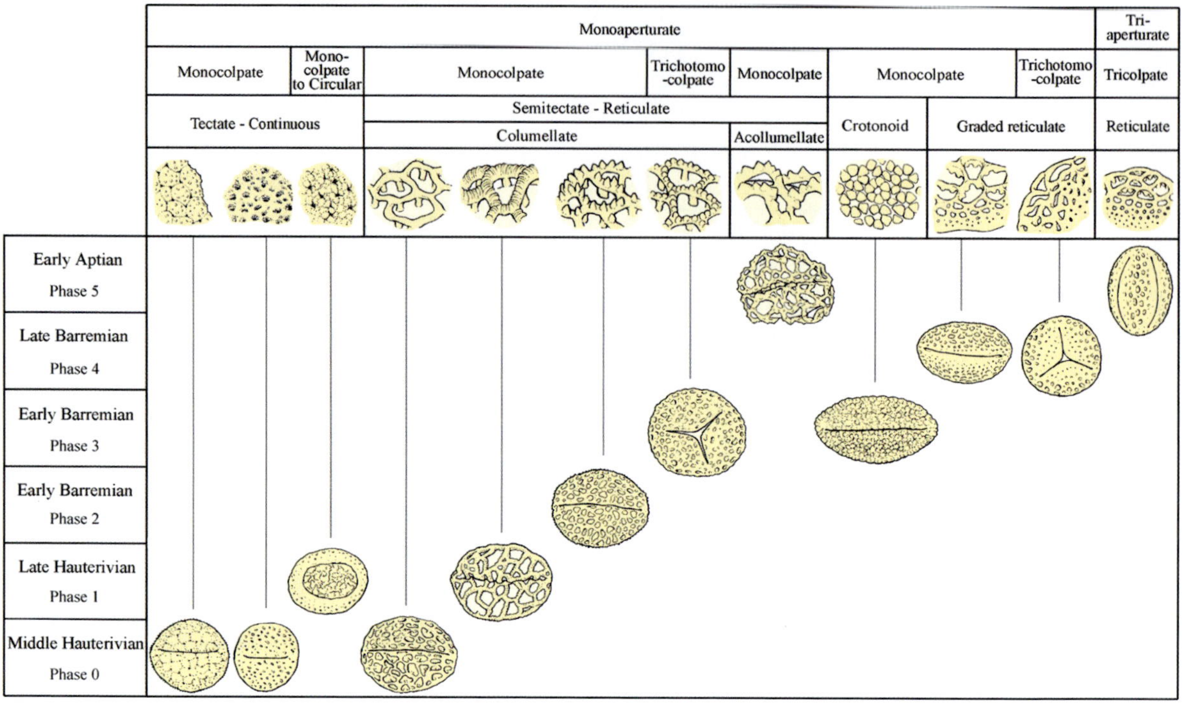

Figure 16.12 Stratigraphic appearance of major angiosperm pollen types in the Early Cretaceous (Middle Hauterivian – Early Aptian) of England showing aperture configuration as well as tectum structure and ornamentation in the six pollen phases of Hughes (1994). Compiled from data in Hughes (1994).

16.2.6 Pollen

There is general consensus that the triaperturate pollen that characterises eudicots is a derived feature at the level of angiosperms as a whole and this is also consistent with the relatively late stratigraphic appearance of these grains. The probable ancestral angiosperm pollen morphology has been hypothesised to be globose grains with a monocolpate aperture. The hypothesised basic pollen wall had a continuous tectum, a columellate infratectal layer and a thin endexine (Doyle, 2005). According to these ideas, which are based mainly on extrapolation from the phylogenetic distribution of pollen features among extant taxa, probable derived features include the presence of a granular infratectal layer, and the presence of more than one pollen aperture. However, the granular infratectal layer that occurs in several potential angiosperm outgroups (Chapter 5) indicates the importance of including fossils in these considerations. Whether the earliest angiosperm pollen was inaperturate or monoaperturate is also not completely clear.

Early Cretaceous pollen. The earliest pollen grains that have been attributed to angiosperms are of possible Late Valanginian – Early Hauterivian age from Israel (Brenner, 1996). They are circular in outline, apparently inaperturate, tectate–columellate and finely reticulate. They range in size from about 15 μm to about 26 μm (Brenner, 1996). So far these grains have been illustrated only by light micrographs (Brenner, 1996), which reveal only few structural details. Without these details, including features of the pollen wall, the angiospermous affinity of these grains remains to be securely established. There is also a single report of Valanginian *Clavatipollenites*-like pollen from Portugal (Trincão, 1990).

More secure evidence of early angiosperm pollen comes from very detailed studies of palynofloras from southern England (Figures 16.12, 16.13). From the Middle Hauterivian – Early Barremian, Hughes (1994) distinguished three phases; phase 0 dated as Middle Hauterivian, phase 1 dated as Late Hauterivian, and phase 2 dated as Early Barremian. The probable early angiosperm pollen grains from these three phases vary considerably in size. With the exception of a few grains, which have a continuous tectum (Monosulc group of Hughes), they are all

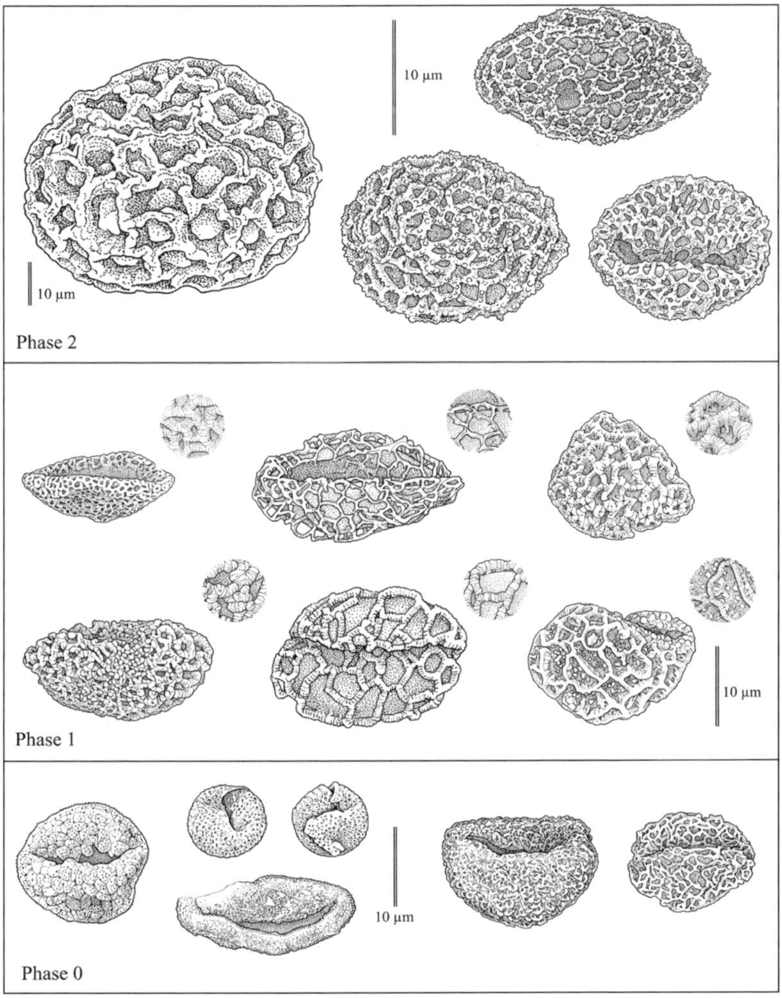

Figure 16.13 Angiosperm pollen from the Hauterivian to Early Barremian (pollen phases 0–2) of England, showing the diversity of tectum structure and ornamentation in monoaperturate pollen in the earliest phases of angiosperm diversification. Drawn from SEM images in Hughes (1994).

reticulate (Retisulc group of Hughes). The smallest grains are about 9 µm in diameter and the largest grains about 29 µm in diameter. They are mostly monocolpate with a long straight colpus. More rarely the aperture is large and more or less circular.

Even at this very early stage in angiosperm history these pollen grains exhibit considerable diversity in wall structure and ornamentation. From phase 0 Hughes (1994) reported two different pollen types with a continuous tectum, one type with a foveolate tectum, and two further types with a semi-tectate reticulate pollen wall. The muri of the reticulate pollen types from phase 0 are smooth. They lack supratectal ornamentation and are supported by distinct columellae. From phase 1 there is a much larger diversity of reticulate pollen grains, including the first record of reticulate grains that have a supratectal ornamentation of fine transverse striations on the muri. This kind of ornamentation is very common in later Barremian and Aptian angiosperm pollen. Monocolpate and reticulate pollen with spinulose supratectal ornamentation first appears in phase 2 and also remains common in angiosperm pollen from later in the Barremian and Aptian.

During the Late Barremian, and then through the Aptian and into the Albian, the diversity of angiosperm pollen expands considerably, in terms of both aperture configuration and pollen wall sculpture. However, at this level angiosperm pollen continues to be small. In the Early Barremian – Early Aptian of southern England (phases 3–5 of Hughes, 1994) angiosperm pollen is more abundant and diverse than in phases 0–2. In particular there is a greater diversity of reticulate pollen with beaded/spinulose supratectal ornamentation of the muri. This is also seen in other dispersed palynofloras and *in situ* in pollen from stamens and flowers (Figure 16.14). Pollen grains with crotonoid wall ornamentation also occur for the first time in phase 3 and among monoaperturate pollen there are the first trichotomocolpate types. During phase 4 the first pollen grains with a graded reticulum enter the fossil record. A graded reticulum is characteristic of many monocots (Chapter 11), but may also occur in other angiosperms.

The first appearance of tricolpate pollen, based on a single dispersed grain examined in SEM, is in phase 4 of Hughes (1994), which is thought to date from approximately the Barremian–Aptian boundary (Hughes, 1994). A further single tricolpate grain is recorded from phase 5, which is dated as Early Aptian (Hughes, 1994). At these levels monocolpate angiosperm pollen is relatively common, but records of tricolpate pollen are rare and sporadic. Scattered records of tricolpate pollen types are also known from more or less contemporaneous pollen assemblages from Israel (Brenner, 1996), Equatorial Africa (Doyle, 1992), Egypt (Penny, 1988a) and North America (Doyle, 1992) as well as from the Torres Vedras mesofossil assemblage in Portugal.

The earliest triaperturate pollen grains are all tricolpate and prolate with simple, elongate colpi. Most are small, foveolate or reticulate, but there are also pollen types that are larger and some that have other kinds of sculpture. In particular, there are a variety of grains with striate ornamentation (Chapter 12). Tricolpate pollen increased rapidly in diversity and abundance through the Aptian and Albian, and a rough estimate based on *in situ* triaperturate pollen in flowers and stamens from the Puddledock mesofossil flora indicates that about one-third of the angiosperm taxa at this level are eudicots. All triaperturate pollen types from the Puddledock flora are tricolpate with simple, elongate colpi and reticulate or foveolate–punctate tectum. Triaperturate grains with modifications of the apertures occur only after the appearance of tricolpate forms with simple colpi. Tricolporoidate and tricolporate pollen enter the fossil record during the Middle – Late Albian (Doyle and Hickey, 1976; Doyle and Robbins, 1977).

Periporate pollen was also present early in angiosperm history. The systematic distribution of periporate pollen in extant angiosperms indicates that this aperture configuration evolved independently in many different groups. Such grains occur sporadically in magnoliids (*sensu lato*), monocots as well as eudicots. Periporate pollen in Early Cretaceous angiosperms shows some variation in shape and form, and exine sculpture varies from echinate to reticulate. Some of these grains were probably produced by monocots, but the systematic affinity of all of the forms has not yet been established.

Mid-Cretaceous pollen. Pollen grains from the mid-Cretaceous are predominately triaperturate, with an increased number of tricolporate forms. Already in the Late Albian triaperturate pollen may account for about two-thirds of all angiosperm species in some pollen floras. Further innovation is seen from the Middle Cenomanian onwards (Doyle and Hickey, 1976), which is when triporate grains are recorded for the first time. Most notably the earliest breviaxal grains assignable to the Normapolles complex (core Fagales) with elaborate tricolporate apertures appear at this time (e.g. Pacltová, 1971).

Late Cretaceous pollen. In Late Cretaceous mesofossil floras triaperturate pollen types increase further in diversity. In many palynofloras from Europe and eastern North America they also become the most abundant forms. Nevertheless, throughout the Late Cretaceous angiosperm pollen found *in situ* within flowers remains generally small. Sometimes pollen grains are aggregated together in the anthers or on the stigmas by a pollenkitt-like substance. Simple tricolpate as well as tricolporate forms further diversify through the Late Cretaceous and are common in dispersed palynological assemblages as well as *in situ* in fossil flowers.

In Late Cretaceous pollen floras a great range of new and distinctive pollen types appear, which include periporate and pericolpate forms. Some of these distinctive grains can be attributed to extant taxa (e.g. Buxaceae) while the systematic affinities of others remain uncertain.

The most marked palynological change during the Late Cretaceous is the very extensive diversification of triporate/tri-brevicolporate pollen attributed to the Normapolles group. These pollen grains, many of which are securely attributable to core Fagales, are especially diverse and abundant at middle latitudes in western Europe and eastern North America (Chapter 20). At higher latitudes through the Late Cretaceous there is also the first appearance and radiation of distinctive pollen of the Triprojectacites complex. Both Triprojectacites and Normapolles pollen types are characterised by their elaborate and often highly distinctive pollen apertures.

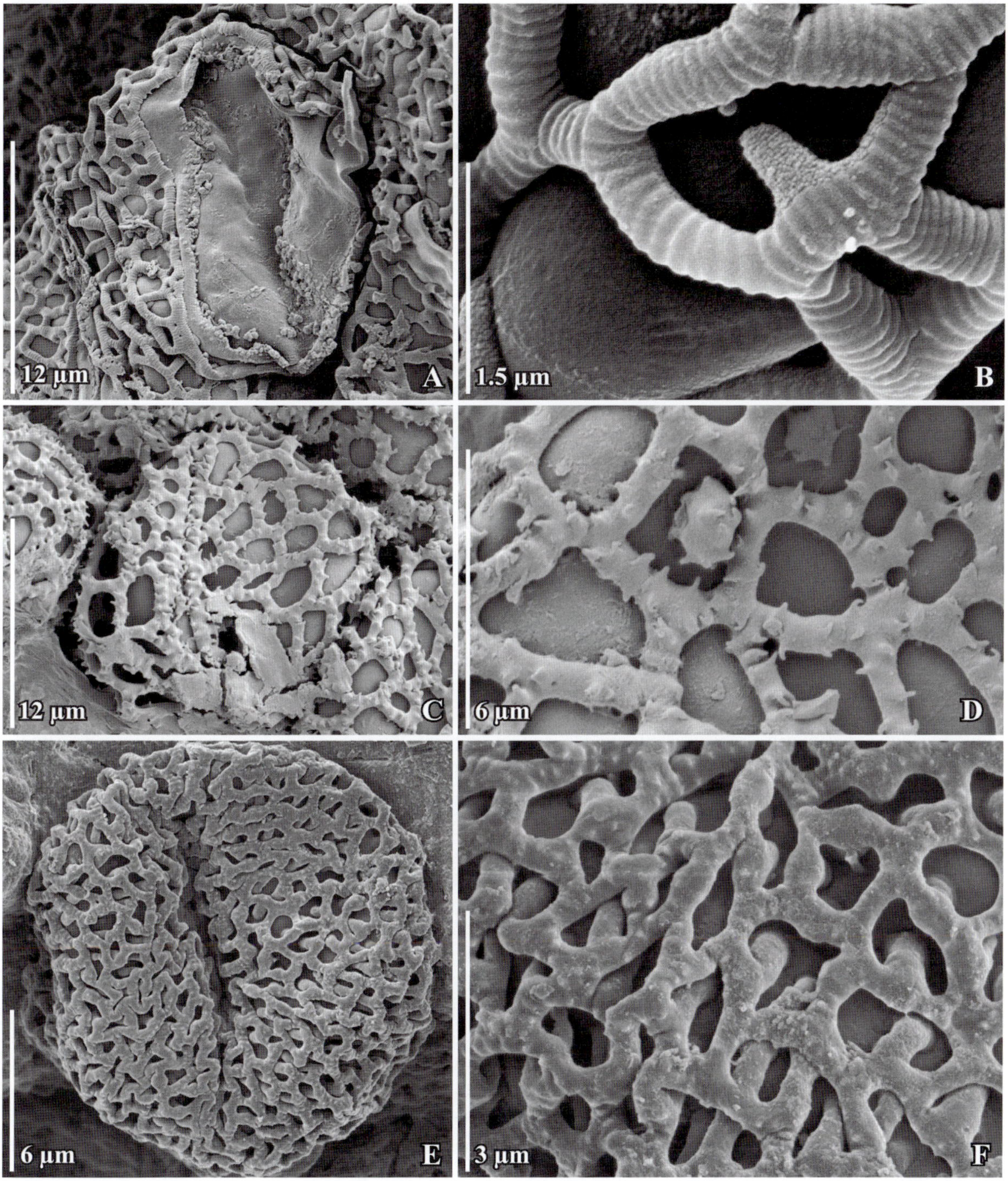

Figure 16.14 SEM images of Early Cretaceous *in situ* monocolpate angiosperm pollen showing variation in tectum ornamentation. All grains from the Torres Vedras locality (Late Barremian – Early Aptian), Portugal. Pollen grains with beaded, striate or spiny muri are very common in the earliest phases of angiosperm diversification from the Late Hauterivian and onwards. (A, B) *Retimonocolpites*-type pollen with (A) loosely attached reticulum and (B) detail of reticulum. (C, D) *Pennipollis* sp., grain with acolumellate pollen wall and spiny tectum ornamentation. (E, F) *Clavatipollenites* sp., grain with long columellae and finely spiny muri. Material in the collections of the Swedish Museum of Natural History, Stockholm.

Most Normapolles grains are in the size range of about 10–30 μm and have a relatively thick, continuous, psilate or finely spinulose tectum, characters that are consistent with adaptation to wind pollination. Triprojectacites-type pollen is typically larger, about 20–60 μm, and exhibits a wider range of wall ornamentation. This could indicate insect pollination, although the abundant occurrence of Triprojectacites grains in some palynological assemblages from the latest Cretaceous perhaps points to wind pollination.

16.2.7 Gynoecium

The characteristic angiosperm ovary, containing one or more ovules, ultimately matures into a fruit containing seeds. Here we mainly focus on the structure of the carpels/ovaries at anthesis. Patterns in the evolution of angiosperm fruits and seeds are considered in relation to dispersal in Chapter 18.

With the exception of extant Nymphaeaceae, apocarpy is widespread among early-diverging lineages of angiosperms, suggesting that the presence of several free carpels per flower is the basic condition in angiosperms. This is also in accordance with the Cronquist/Takhtajan formulation of the Euanthial model of angiosperm floral evolution. Apocarpy also predominates among eumagnoliids and among early-diverging groups of eudicots, while among early-diverging monocots both *Acorus* and Araceae have syncarpous ovaries. Unicarpellate flowers, which occur for example in Chloranthaceae, Hydatellaceae and Ceratophyllaceae, as well as many eumagnoliids, are relatively derived based on optimisation of gynoecial characters onto the pattern of angiosperm relationships inferred from phylogenetic studies of extant taxa (Endress and Doyle, 2009).

Among core eudicots, and also among most groups of monocots, carpels are typically fused (syncarpous). However, there are scattered occurrences of syncarpous ovaries among predominantly apocarpous lineages and this feature has apparently evolved repeatedly. Endress and Doyle (2009) distinguish between eusyncarpous gynoecia, where carpels are fused at the centre to form a locular ovary usually with axile placentation, and parasyncarpous gynoecia, where the carpels are fused to form a unilocular ovary with parietal placentation.

In the Cronquist/Takhtajan concept of gynoecial evolution plicate carpels, which often mature into follicular fruits, were considered basic and widespread among 'primitive' angiosperms. However, more detailed studies (Endress and Igersheim, 1997; Igersheim and Endress, 1997; Endress and Igersheim, 2000a, b) have demonstrated that plicate carpels are actually rare among early-diverging angiosperm lineages. In contrast, ascidiate carpels are widespread and occur in most of the ANITA lineages. Carpels are closed in these early lineages by mucilaginous secretions rather than by 'fusion' of the carpel walls (Figure 1.10). Optimisation of this feature onto the molecular phylogeny of extant taxa indicates that this is most likely the basic condition among angiosperm flowers (Endress and Doyle, 2009). Carpels referred to as intermediate ascidiate (Endress and Doyle, 2009), which have both an ascidiate and a plicate zone, are rare and mainly characterise Laurales. In other eumagnoliids and monocots, as well as in more derived eudicots, carpels are plicate and closure is by postgenital fusion.

Carpels and ovaries in Early Cretaceous flowers. The distinction between plicate and ascidiate carpels is sometimes difficult to make even in extant angiosperms. Certainty with regard to this distinction often requires developmental studies. For fossils the difficulties are obvious, especially when it comes to establishing whether tissues are truly fused, or only tightly adhering because of maturation of the organ, compression during fossilisation, or changes during diagenesis. *Archaefructus* provides a case in point. It was described as having follicular fruits with the implication that these were derived from plicate carpels. However, none of the fossil specimens of *Archaefructus* studied so far show fruits that have split and none of the specimens show a distinct ventral slit. Among modern angiosperms *Brasenia* has somewhat similar fruits that also superficially appear follicular. However, they do not open along a ventral slit and developmental studies show that the carpels are ascidiate.

Both apocarpous and syncarpous gynoecia occur very early in angiosperm history and in the Yixian Formation of China flowers with both apocarpous (*Archaefructus*) and syncarpous (*Sinocarpus*) gynoecia are present. Both types of gynoecia also occur at the Torres Vedras locality. In the Torres Vedras flora monocarpellate *Hedyosmum*-like flowers (Figure 16.7) co-occur with flowers that have more than one carpel (Figure 16.4). These include both an apocarpous, multicarpellate flower and a trimerous syncarpous flower. Carpels of the *Hedyosmum*-like flower were most likely ascidiate as in the extant genus, while carpels of the apocarpous multicarpellate flower have a distinct ventral suture and may have been plicate. Carpels of the slightly younger *Monetianthus* appear to be synascidiate

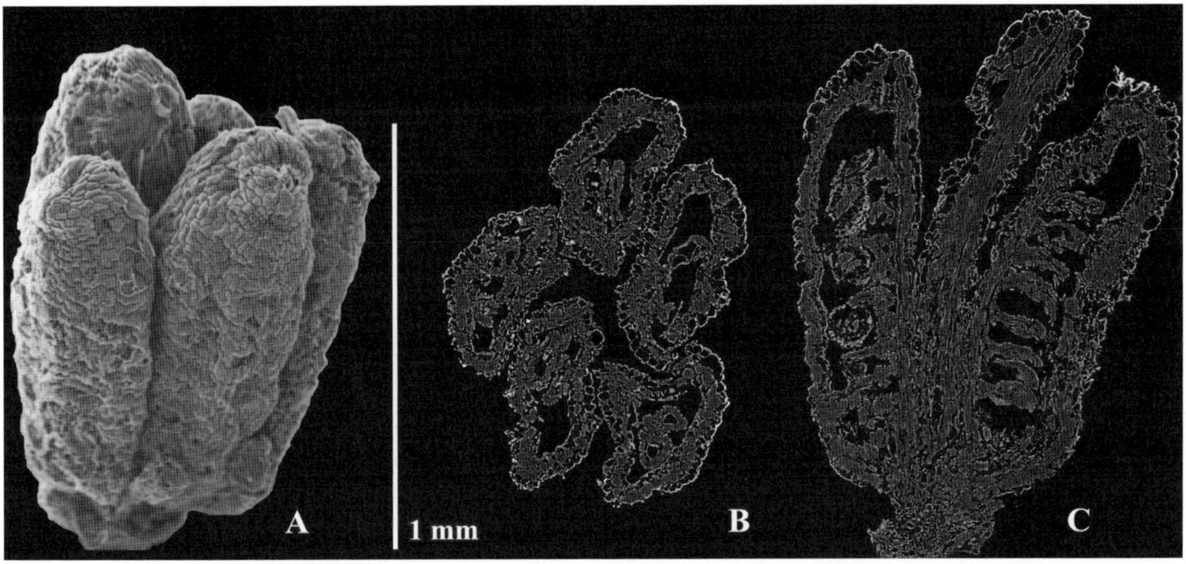

Figure 16.15 SEM and SXRTM images of apocarpous flower/fruit from the Early Cretaceous (Early–Middle Albian) Puddledock locality, Virginia, USA, showing five free carpels in lateral view (A), cross-section through all five carpels (B) and longitudinal section (C) through three carpels. The carpels are follicular with two rows of seeds borne along the ventral suture. Material in the collections of the Field Museum, Chicago.

below and plicate above, with postgenital or incomplete fusion of the distal carpel margins.

An interesting feature of the early record of fossil flowers is that gynoecia from this time interval typically lack a distinct style and the stigma is not elevated. In some key fossils the stigmatic surface is often poorly preserved; for example, in both *Archaefructus* and *Sinocarpus* the nature of the stigma is unknown. However, in mesofossil material the stigma can be exceptionally well preserved and the most common stigma type in this early phase of angiosperm evolution appears to be extended. An extended stigma occurs, for example, in the pistillate *Hedyosmum*-like flowers.

The Late Barremian – Early Aptian Torres Vedras mesofossil flora includes several small monocarpellate, usually uniovulate, fruiting units. Similar fruiting units of various kinds are present in slightly younger mesofossil floras from the Aptian – Middle Albian. More complete specimens show that some of these are fruits derived from monocarpellate flowers, while others may be isolated fruitlets from apocarpous, multicarpellate structures. In the younger assemblages flowers with monocarpellate (e.g. *Potomacanthus*, Laurales B) or apocarpous (e.g. *Endressinia, Friisicarpus, Pluricarpellatia, Virginianthus*) gynoecia predominate, but there are also syncarpous flowers such as *Monetianthus* and the probable nymphaealean flower *Carpestella*, the four genera, *Aguacarpus, Lusicarpus, Silucarpus* and *Valecarpus*, assigned to Buxales and *Canrightia*. In one of the undescribed apocarpous flowers/fruits from the Puddledock locality the individual fruitlets are follicle-like and there are many seeds in two rows along the ventral margin (Figure 16.15).

Carpels and ovaries in mid-Cretaceous flowers. Through the mid-Cretaceous the proportion of angiosperm flowers with either monocarpellate (e.g. *Mauldinia, Myricanthium, Pragocladus, Zlatkocarpus*) or apocarpous, multicarpellate (e.g. *Archaeanthus, Friisicarpus, Lesqueria, Lovellea*) gynoecia continues to be significant. As in earlier floras there are also a variety of flowers with at least partially syncarpous gynoecia, such as *Spanomera* and the 'Rose Creek flower'. *Spanomera* presents an especially common situation among early eudicots in which there are two carpels that are fused to varying extents.

The 'Rose Creek flower' is the first flower with styles that are distinctly raised above the ovary, and in which the stigmatic areas are apparently restricted. Other flowers with restricted (or almost insignificant) stigmas include *Mauldinia* and *Zlatkocarpus*. An extended, sessile, stigmatic area was probably also present in *Archaeanthus* and is especially prominent on the long, free adaxial carpel margins in *Spanomera*.

Carpels and ovaries in Late Cretaceous flowers. In the Late Cretaceous there is a clear dominance of flowers with syncarpous gynoecia (Figure 16.6). Flowers with a syncarpous, unilocular ovary also became more abundant in the Late Cretaceous. One example is the primuloid flower from Mira, Portugal, which has a unilocular ovary formed from five carpels and a free central placenta. There are also many trilocular forms that are of probable ericalean affinity, as well as syncarpous ovaries in flowers related to other groups such as Cornales. Syncarpous gynoecia also occur in flowers of presumed early-diverging Fagales (e.g. *Antiquacupula*). The unilocular ovaries of Normapolles are also syncarpous and in several cases bilocular or trilocular in their construction. Two locules, each with one ovule, occur in several taxa such as *Endressianthus*, *Normanthus* and *Budvaricarpus*. They all have bicarpellate flowers and typically only one seed matures suppressing the other locule and its ovule during maturation.

Floral structures with an apocarpous gynoecium are relatively rare during the Late Cretaceous. They include species of eumagnoliids such as *Cronquistiflora*, *Detrusandra*, *Jerseyanthus* and *Futabanthus*, as well as the putative eumagnoliids *Hidakanthus*, *Keraocarpon* and *Protomonimia*. The various species of *Liriodendroidea* probably also had apocarpous gynoecia based on extrapolations from their magnoliaceous living relatives. Several species related to early-diverging eudicots, such as the platanoid *Friisicarpus*, also have apocarpous gynoecia. In the Late Cretaceous, for the first time, there is also a clear dominance of flowers that have distinct, and often very stout, styles.

16.2.8 Ovules and seeds

Angiosperm ovules are typically bitegmic. Unitegmic ovules are less common and are scattered among many different groups of extant angiosperms. In the broader context of angiosperm phylogeny according to both traditional and more recent interpretations, unitegmic ovules are a derived feature.

With regard to the number of ovules per carpel, current and traditional interpretations come to different conclusions about the basic condition in angiosperms. In the Takhtajan/Cronquist formulation of the Euanthial model carpels with several or many ovules are considered ancestral, while uniovulate carpels are considered derived. However, recent phylogenetic interpretations suggest that a single, probably pendent, anatropous ovule in each carpel is plesiomorphic for angiosperms as a whole (Endress and Doyle, 2009). Although several early-diverging angiosperm lineages, such as Nymphaeales and *Austrobaileya*, have carpels with many ovules, these seem likely to be secondary modifications (Endress and Doyle, 2009).

Ovules and seeds in Early Cretaceous flowers. In Early Cretaceous floras flowers with uniovulate carpels are overwhelmingly dominant. They include fossil *Hedyosmum*-like flowers, *Anacostia*, *Appomattoxia*, *Friisicarpus*, Laurales B *Pennicarpus*, *Potomacanthus* and many undescribed taxa. In most cases the ovules in these taxa are also anatropous. The position of the ovules in the locule has only been established for a few taxa. *Canrightia* has a unilocular ovary formed from two to five carpels, each with a single, orthotropous and pendent ovule.

Among the earliest angiosperms there are also flowers with many ovules per carpel. For example, *Archaefructus* and *Sinocarpus* from the Yixian Formation both have carpels with several seeds. In the Torres Vedras flora there is also a multicarpellate flower apparently with many seeds and several other undescribed taxa have more than one ovule per carpel. Slightly younger floras also include a variety of taxa with several to many ovules, most notably *Monetianthus* from the Vale de Agua flora, *Pluricarpellatia* from the Crato flora, and several flowers from the Puddledock flora.

Although the number of integuments may be difficult to establish in fossil seeds, by comparison with extant taxa most Early Cretaceous seeds appear to have been derived from bitegmic ovules. Seeds are typically exotestal with the resistant 'mechanical' layer in the outer seed wall (outer epidermis). The tegmen, derived from the inner integument, is thin. More rarely, seeds are endotestal as in *Canrightia*. Exotestal seeds typically have a smooth outer wall, often with a distinct digitate pattern reflecting the outline of the sclerenchymatous epidermal cells. There are also a variety of exotestal seeds with a finely spiny sculpture (Figure 16.16).

Ovules and seeds in mid-Cretaceous flowers. Angiosperm flowers from the mid-Cretaceous show more or less the same distribution in ovule number per carpel as in earlier flowers. Although there are many unilocular carpels containing a single, anatropous, pendent ovule (e.g. *Couperites*) there are also forms in which the number of ovules is much larger, although they are also apparently anatropous and pendent. *Archaeanthus*, for example, is estimated to have had 10–18 seeds in each fruitlet. The ovules appear to be suspended from the placentae that run on either side of the ventral suture. This is seen even more clearly in an

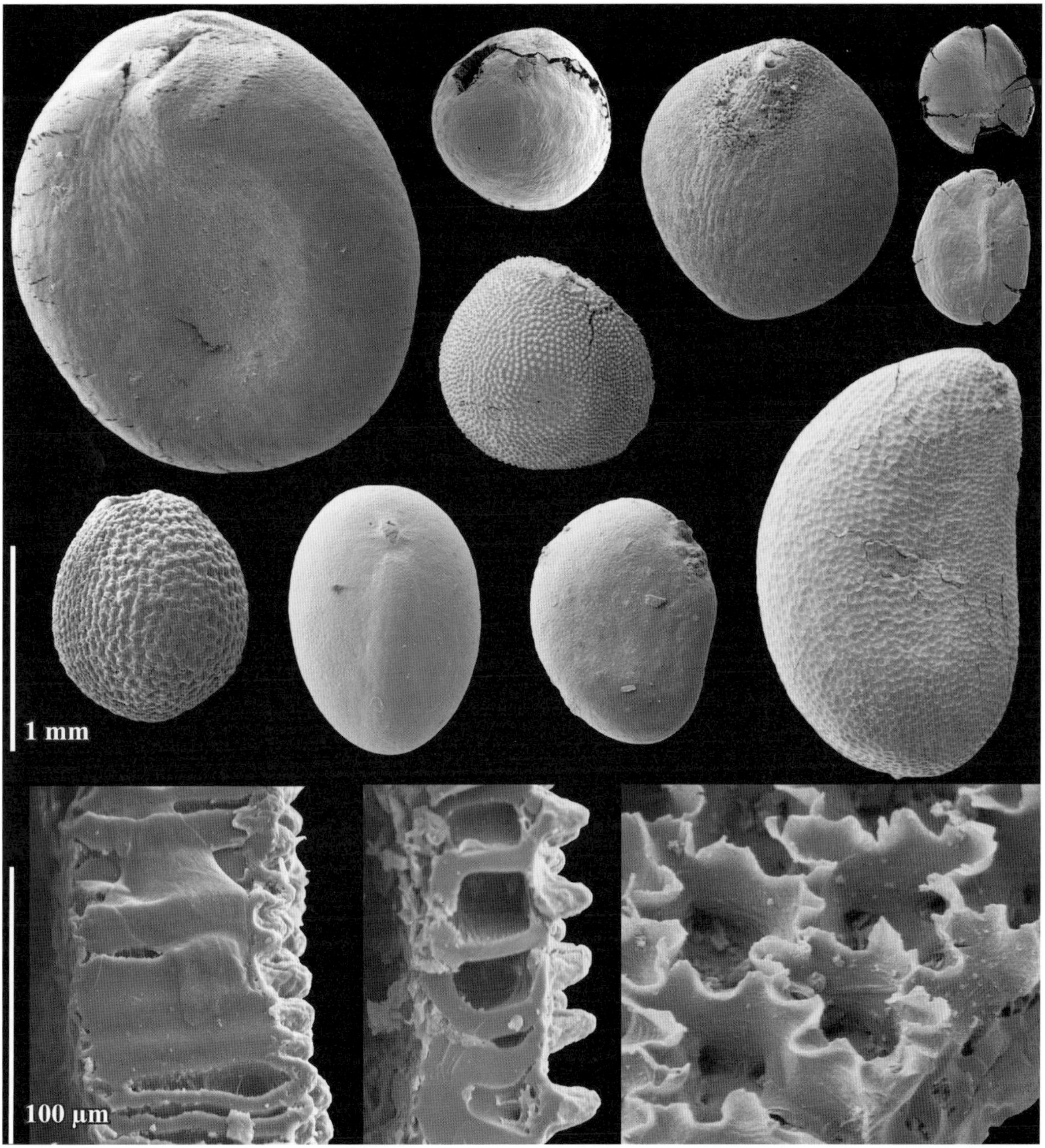

Figure 16.16 SEM images of exotestal seeds from the Early Cretaceous of Portugal showing differences in size and general morphology, as well as sections through the palisade cells of the exotesta. Material in the collections of the Swedish Museum of Natural History, Stockholm.

impression of a large follicle from the Dakota Sandstone flora in which there are the scars of about 40 ovules or seeds, 20 on either side of the ventral suture (Hollick, 1903).

Ovules and seeds in Late Cretaceous flowers. Angiosperm flowers from the Late Cretaceous show considerable variation in ovule number. However, at this stage of angiosperm evolution the situation is the reverse of that in earlier floras. There is a clear dominance of taxa with many ovules per carpel. Most of these taxa are closely related to groups of extant core eudicots and include genera such as *Divisestylus*, *Dressiantha*, *Microaltingia*, *Paleoclusia* and *Platydiscus* that are of probable rosid affinity, and *Scandianthus* and *Silvianthemum* that are of uncertain relationship among asterids. Among asterids there are also numerous genera assigned to the Ericales with many ovules, such as *Actinocalyx*, *Paradinandra*, *Pentapetalum*, the unnamed primuloid flower and many others.

In the broader context of Late Cretaceous mesofossil floras, flowers of the Normapolles complex are unusual. In some taxa, such as species of *Normanthus* and *Budvaricarpus*, there is apparently only one ovule per carpel, and in *Caryanthus* there is perhaps only a single ovule in each bicarpellate (but unilocular) ovary. Other Normapolles taxa, such as *Endressianthus*, may have had two ovules per carpel, but typically only one ovule per ovary developed into a seed. The low number of ovules/seeds in these taxa is most probably a derived feature linked to wind pollination.

16.2.9 Nectaries

Staminal appendages observed in Early Cretaceous (Early–Middle Albian) lauralean flowers from Puddledock may have functioned as nectar-producers, but the first floral nectaries are otherwise reported from the mid-Cretaceous 'Rose Creek flower'. In Late Cretaceous flowers from mesofossil floras, nectary discs are common and often very prominent (Figures 16.6, 17.10). They occur, for example, in *Platydiscus*, *Silvianthemum*, *Scandianthus* and a variety of unnamed taxa of core eudicots from several different fossil floras. The evolution of nectaries is an important innovation among the Late Cretaceous flowers that reflects the evolution of more efficient biotic pollination, especially among rosids and early asterids.

16.2.10 Floral symmetry

Fossil flowers from the Early Cretaceous, as well as most of those known from the Late Cretaceous, are actinomorphic with radial floral symmetry. In Santonian–Campanian mesofossil floras several fossil Normapolles flowers document the first occurrence of forms with two planes of symmetry (bisymmetric). Other than the unique bisexual *Chloranthus*-like flowers from the mid-Cretaceous of Germany, the first direct evidence of zygomorphic flowers (with a single plane of symmetry) is *Raoanthus* from the Late Maastrichtian Deccan Intertrappean Beds. *Raoanthus* exhibits weakly developed monosymmetry. There is currently no direct evidence of zygomorphic flowers older than the Maastrichtian although such flowers might be expected from the presence of ancient Zingiberales that produced *Spirematospermum* seeds in the Campanian–Maastrichtian.

More pronounced zygomorphy is clearly present in the Early Cenozoic, for example in unequivocal papilionoid flowers from the Late Paleocene – Early Eocene of southeastern North America (Crepet and Taylor, 1985). These early papilionoid flowers have clear differentiation of the petals into a standard, a keel and two wing petals.

16.3 OTHER ASPECTS OF FLORAL CONSTRUCTION

16.3.1 Flower size

Flower size is to a large extent linked to the nature of the pollination syndrome (Chapter 17). Flower size may also vary significantly among relatively closely related taxa (e.g. Davis *et al.*, 2007) often overwhelming any phylogenetic component in this feature. Nevertheless flower size has often been discussed in a phylogenetic context. According to the Euanthial model elaborated by Cronquist and Takhtajan, ancestral flowers were thought to be large, much like the flowers of modern *Magnolia* or related groups. A different picture emerges from the phylogenetic framework now available based on molecular data from extant taxa.

Among extant angiosperms, many of the lineages that diverge at an early stage from the main line of angiosperm evolution have flowers that are more commonly of small to medium size. *Amborella*, *Cabomba*, *Hydatella*, Trimeniaceae, Chloranthaceae and Ceratophyllaceae all have flowers that are less than 1 cm in diameter. Small flowers are also typical among early-branching lineages of eudicots and monocots. Flowers of medium size occur in Schisandraceae and many Nymphaeaceae. Large flowers occur only in certain Nymphaeaceae (e.g. *Victoria*), which clearly reflects secondary gigantism (Endress, 2001). This general trend of increasing flower size in angiosperm phylogeny is also supported by observations of both compressed and three-dimensionally preserved fossil flowers.

Flower size in the Early Cretaceous. All of the early angiosperm flowers known from the Late Barremian – Middle Albian are of small to medium size. Individual floral units in *Xingxueina heilongjiangensis* from northeastern China are only few millimetres in diameter. Other early compression/impression fossils of angiosperm floral structures include *Archaefructus* and *Sinocarpus* from the Early–Late Aptian Yixian Formation. Under the inflorescence interpretation, flowers of *Archaefructus* are also small and simple. Most *Archaefructus* specimens are preserved in the fruiting stage, but there are also a few specimens with younger floral structures. Male flowers, borne proximally on the inflorescence axes, are about 2 mm long; the female flowers, borne distally, are about 5 mm long. *Sinocarpus* appears to be preserved in the fruiting stage or at least after anthesis. The fruits are up to about 12 mm long; smaller reproductive structures that may represent the anthetic stage are about 5 mm long and 5 mm in diameter.

Flowers from Early Cretaceous mesofossil floras range from about 0.5 mm up to about 5 mm. These charcoalified fossils have been subject to some alteration during charring; based on experiments with extant taxa Lupia *et al.* (1995) indicated that overall shrinkage may vary from 14% to 41% depending on the nature of the organs (Chapter 2). An estimate based on the fragmentary *Monetianthus* flower (Nymphaeaceae) from the Vale de Agua locality suggests that at anthesis, with its tepals intact, the flower would have been about 1 cm in diameter. *Teixeiraea*, a probable ranunculalean flower, also from the Vale de Agua locality, was probably also about 1 cm in diameter in life. The angiosperm flowers preserved as compressions in the Late Aptian – Early Albian Crato flora of Brazil (Mohr and Friis, 2000; Mohr and Eklund, 2003) are also small to medium-sized. The largest is *Endressinia*, which is about 4–5 mm long and 6–7 mm in diameter.

Flower size in the mid-Cretaceous. During the mid-Cretaceous smaller flowers, including those preserved as mesofossils (e.g. *Mauldinia*, *Spanomera*) and compressions (e.g. *Zlatkocarpus*, *Myricanthium*, *Pragocladus*, *Mauldinia*), continue to dominate fossil floras. The 'Rose Creek flower' is slightly larger, about 2–3 cm in diameter, and there are also several forms that are even larger such as *Archaeanthus*, *Lesqueria* and *Triplicarpus*. Even though they are all preserved at the fruiting stage it is likely that the flowers were also large. The probable outer tepals of *Archaeanthus* flowers (*Archaepetala beekeri*) are 70–80 mm long and up to 40 mm wide. The more membranous inner tepals (*Archaepetala obscura*) are estimated to have been about 80 mm long. At anthesis with the tepals still partly closed, as in a modern *Magnolia*, these flowers may have been about 8–10 cm in diameter. Judging from the probable stipular bud scales (*Kalymmanthus walkeri*) the flower buds of *Archaeanthus* may have been up to about 65 mm long. Other tepals known from the Dakota Formation, which are very similar to those apparently produced by *Archaeanthus*, range in length from about 28 mm (*'Ficus' neurocarpa*) to about 15 cm (*'Magnolia' palaeopetala*) (Hollick, 1903). Flowers of these plants would have been relatively large.

Flower size in the Late Cretaceous. Flowers of small size continue to dominate fossil assemblages from the Late Cretaceous. Records of larger floral structures are relatively sparse and mostly preserved at fruiting stage as permineralisations. For example, *Elsemaria*, *Hidakanthus* and *Protomonimia* are all calcified fruiting structures from Hokkaido with a diameter of about 4 cm. It is interesting that all larger flowers from the mid- and Late Cretaceous were probably solitary, while the smaller flowers were probably borne in more or less dense inflorescences (section 16.1).

16.3.2 Protection in floral bud

In almost all flowers the anthers, in which the pollen grains are produced, and the ovary, which contains the ovules, are protected during development. This is achieved in different ways at different levels of angiosperm evolution. The typical situation, as occurs in most eudicots and many monocots, is that protection is afforded by the outer whorl of floral organs (sepals) and sometimes the floral bracts. At anthesis the calyx opens to expose the corolla, androecium and gynoecium. However, among eumagnoliids, and also at the ANITA grade, there is much more variation in how the anthers and ovary/ovules are protected. In the solitary flowers of *Liriodendron* and *Magnolia* protection in the bud occurs by a modified stipular bud-scale, while in *Eupomatia* there is a cap-like calyptra formed by one (possibly two) sheathing bracts (Endress, 1994b). In other eumagnoliid flowers protection is often by the tepals, which commonly show little differentiation for attraction. Among extant eumagnoliids with numerous stamens, such as many Annonaceae, the stamens often have prominent extensions of the connectives that fit together to form a protective shield over the pollen sacs.

In Early Cretaceous flowers, stamens with a prominent apical extension of sterile tissue are common, and many have well-developed connectives. In these flowers stamens may have been important for protecting the developing pollen sacs and ovules. In Early Cretaceous flowers tepals clearly also played a role in protection, but from the mid-Cretaceous and onwards the most common mode of protection was probably by the calyx or by floral bracts. In Late Cretaceous mesofossil floras more normal heterochlamydous flowers with a clearly differentiated calyx and corolla predominate.

16.3.3 Synorganisation of floral parts

The concept of floral synorganisation relates to the spatial and functional integration of individual floral parts to form a whole functional flower (Endress, 1994b). All but the very simplest flowers show some level of synorganisation, but the extent to which it is developed varies considerably among the flowers of different angiosperm groups. At one end of the spectrum are the relatively simple flowers of *Cabomba* (Nymphaeales) or *Alisma* (Alismatales); at the other end are the complex highly synorganised flowers of Asclepidaceae, Stylidiaceae or Orchidaceae. One manifestation of high levels of synorganisation is fusion of different floral organs one to another and among themselves, as for example in sympetalous flowers in which the stamens are fused to the corolla tube (section 16.2.4). In general, higher levels of floral synorganisation require stable patterns of floral morphogenesis, which is typical of flowers with relatively few parts that are organised in a stable whorled floral phyllotaxis.

In the fossil record there is a clear temporal trend from flowers with relatively weak levels of floral synorganisation to more complex, synorganised flowers, which are encountered in the Late Cretaceous and Cenozoic. The first occurrence of sympetaly in the Late Cretaceous is in the flowers of *Actinocalyx bohrii* from the Åsen locality, Sweden. These flowers are very small and thus the corolla tube is short. At least among eudicots, complex flowers with high levels of synorganisation appear to be largely a Cenozoic phenomenon.

17

History and evolution of pollination in angiosperms

Pollination, the successful transfer of pollen from the pollen sacs (microsporangia) into proximity with the ovule, is the essential precursor to fertilisation and therefore to sexual reproduction in seed plants. Pollination has been studied most intensively in angiosperms, although few species have been examined in detail compared with the great variety of flowers within the group (e.g. Proctor *et al.*, 1996; Thien *et al.*, 2009). Pollination in extant non-angiosperm seed plants has received less attention, but studies over the past few decades now provide a more complete context within which pollination in angiosperms can be evaluated and studied (e.g. Owens *et al.*, 1998).

Interpretation of pollination in extinct plants faces significant difficulties. Only rarely is there relatively direct evidence of flower–pollinator interactions (e.g. insect gut contents, coprolites, insects preserved within flowers, insects carrying pollen) and interpretations of pollination in extinct plants therefore depend heavily on extrapolations to extant taxa based on structural similarities. This often leads to plausible interpretations, but it may also be constraining. There is no *a priori* reason why the spectrum of plant–pollinator interactions existing today should also include all of those that existed in the past, and inferring floral function from floral structure, even in extant plants, can sometimes be difficult. It is therefore especially challenging to infer pollination in extinct seed plants (e.g. *Caytonia*, Bennettitales) that have no clear close living relatives.

The enormous diversity in modes of pollination among living plants can be broadly divided into pollination involving either animals (biotic) or wind/water (abiotic). In this chapter we review knowledge of pollination among living seed plants, focusing in particular on biotic pollination in angiosperms. We also provide an outline of the relationships and fossil history of the major groups of animal pollinators. We then summarise what can be inferred from the fossil record about the likely history of pollination in angiosperms and potential patterns of co-evolution with pollinators.

17.1 POLLINATION IN EXTANT NON-ANGIOSPERM SEED PLANTS

17.1.1 Pollination in *Ginkgo* and cycads

Pollen transfer in *Ginkgo* is by wind (Del Tredici, 2007). Large quantities of pollen are produced from lax cones on male plants over a short period (just a few days) in the spring. At this stage the ovules are exposed and secrete a small pollination drop in which the pollen is captured. Experience from commercial production of *Ginkgo* seeds shows that wind pollination can be highly effective and pollination of a large number of ovules on many female trees can be achieved from a small number of widely scattered male trees.

It was long assumed that extant cycads, like *Ginkgo*, were wind-pollinated, based in part on the production of large amounts of pollen and the lack of obvious adaptations for attracting pollinators. However, although wind may be important for pollen transfer in some cycad species (e.g. *Dioon edule*) those cycads that have been studied in detail also have some involvement of insects (Norstog and Nicholls, 1997). The occurrence of insect pollination in cycads is also consistent with the fact that species often grow as scattered individuals, and rarely in dense monospecific stands. Many cycads also grow in the understory of closed forest where wind pollination is generally less effective. In most species of cycads the ovules are also tightly enclosed by the cone scales at the time of pollen release (Jones, 2002).

Interactions of beetles with the pollen and ovulate cones of *Encephalartos* have been known for many years (Rattray, 1913), but only relatively recently has the widespread involvement of insects in pollination in *Encephalartos* and other cycads been recognised and studied in detail (Norstog and Nicholls, 1997). Experimental aerodynamic studies of *Cycas*, *Dioon* and *Zamia* suggest that wind may not be sufficient for effective pollen transport from plant to plant (Niklas and Norstog, 1984). Pollen collects on one or more whorls of sterile basal cone scales, but then it is transferred to the ovules by insects. Secondary transport within a single cone may also occur by water. Pollination of cycads in *ex situ* living collections is often effective when viable pollen is introduced into a ripe female cone as a slurry in water (Crane, personal observations at Royal Botanic Gardens, Kew).

In many cycads, however, pollination appears to be affected solely by insects, especially by beetles, and predominantly by weevils. In *Zamia furfuracea* (Norstog *et al.*, 1986; Norstog and Fawcett, 1989) and *Zamia pumila* (Tang, 1987) different species of the weevil *Rhopalotria* have larvae and adults that feed in the pollen cones, which are rich in starch. The adults become covered in pollen and are attracted to ovulate cones, perhaps by a combination of heat and odour production (e.g. Tang *et al.*, 1987; Jones, 2002). In *Zamia pumila* a second species of beetle (*Pharaxontha zamiae*, a langurid beetle rather than a weevil) is also involved in pollination (Tang, 1987). In *Macrozamia communis* (Chadwick, 1993) and *Lepidozamia peroffskyana* (Kennedy, 1991; Hall *et al.*, 2004) pollination is affected by different species of the weevil genus *Tranes*. In *Cycas media* a species of small bee (*Trigona* sp.) is known to collect pollen from pollen cones (Jones, 2002).

17.1.2 Pollination in conifers

Pollination in all extant conifers is thought to be by wind (e.g. Owens *et al.*, 1998). However, there is considerable variation in details of the pollination mechanism within the general conifer wind-pollinated syndrome. The basic pollination mechanism in conifers appears to involve saccate pollen and the presence of a pollination drop. The pollination drop may capture pollen directly from the air, or it may scavenge grains that have been trapped on or near the ovules. A pollination drop is secondarily absent in certain Pinaceae, and secondary loss is also presumed to have occurred in all Araucariaceae. In *Larix* and *Pseudotsuga* (Pinaceae) non-saccate pollen is trapped on an extension of the integument (Doyle and O'Leary, 1935) and brought into the ovule by integumentary ingrowth. In Araucariaceae and *Tsuga* (Pinaceae) pollen is inaperturate and lands on the cone scale. Gametes are delivered to the archegonia in the ovule by long pollen tubes (Tomlinson, 1994; Tomlinson *et al.*, 1997), a situation somewhat similar to that in angiosperms.

Among conifers that produce a pollination drop, Podocarpaceae and most Pinaceae (e.g. *Picea*, *Pinus*) have saccate pollen. These grains float in the pollination drop (Owens *et al.*, 1998) and enter the micropyle because the ovules are inverted. In this way saccate pollen is captured preferentially. In some Pinaceae pollen may be transferred to the nucellus by reabsorption of the pollination drop, which is apparently an active process stimulated by the presence of pollen. This implies a chemical signal from the pollen that is detected by the ovule and surrounding tissues. In some conifers pollen scavenging may be enhanced if the pollination drop is enlarged by rainwater. In this way pollen that is lodged on the ovuliferous scales may be captured and once this occurs the production of the pollination drop ceases. In most Podocarpaceae, pollen scavenging occurs by extension of the pollination drop onto a wettable surface of the bract and cone axis (Tomlinson, 1991). Pollen is transferred to the nucellus as the pollination drop evaporates and in some species pollination drops may continue to be produced even after pollen is captured.

In other conifers that have a pollination drop the pollen is non-saccate or has only vestigial sacci. In these conifers pollen sinks in the drop and the ovules are in no preferred orientation. In Cephalotaxaceae, Taxaceae and Cupressaceae the pollen bursts in water to release the protoplast, whereas in *Phyllocladus* pollen grains remain intact and are oriented with the germination furrow facing the nucellus (Singh, 1978).

It is interesting that the diversity of pollination mechanisms in the seemingly uniformly anemophilous conifers incorporates a variety of processes (e.g. trapping of pollen on structures external to the ovule, growth of long pollen tubes, involvement of secretions in various ways in the pollination process) that also occur in angiosperms.

17.1.3 Pollination in Gnetales

There is evidence of the involvement of insects, as well as pollen transfer by wind, in the pollination of Gnetales. It is interesting that despite the morphological diversity of the three extant genera there is often an association between the pollen-producing organs and so-called 'non-functional' ovules that do not develop into seeds.

In *Ephedra*, some species (e.g. *Ephedra trifurca*) are thought to be wind-pollinated (Niklas *et al.*, 1986), whereas other species (e.g. *Ephedra aphylla*) have 'non-functional' ovules closely associated with pollen-producing flowers. These ovules produce pollination drops that are visited by insects (Bino *et al.*, 1984a, b; Meeuse *et al.*, 1990). The pollination drop has high sugar content and it has been inferred that insects (flies) attracted to the pollination drop are also involved in pollen transfer.

In *Welwitschia* the pollen-producing structures on male plants have a central 'non-functional' ovule (Figure 5.4). They are therefore structurally, but not functionally, bisexual. The ovule has a pronounced, funnel-shaped micropyle that produces a large pollination drop at anthesis. This may play a role in the attraction of insects, in a way analogous to that inferred for *Ephedra*. The flowers also have an odour similar to that of fly-pollinated angiosperm flowers, and visits by flies, ants and wasps are reported (references in Henschel and Seely, 2000).

Inflorescences of *Gnetum* have long been known to be visited by a variety of insects (e.g. van der Pijl, 1953), and in some species (e.g. *G. gnemon*) a whorl of 'non-functional' ovules occurs regularly above the whorls of pollen-producing structures. These ovules produce pollination drops and are visited by nocturnal moths, which transfer pollen from male to female plants (Kato and Inoue, 1994). Insect pollination is also consistent with the habit of *Gnetum*, the scattered distribution of individual plants, and the environment in which it grows. Wind pollination is uncommon among understory plants or lianas of rainforest environments.

17.2 POLLINATION IN EXTANT ANGIOSPERMS

Given that transfer of pollen from one plant to another is crucial to the generation and maintenance of intraspecific variation in plant populations, it is perhaps surprising that most angiosperm flowers are bisexual. The explanation of this apparent paradox relates to the lack of mobility of plants, the different timing of pollen and seed maturation, and the economy of producing one, rather than two, sets of attractants in species that have biotic pollinators (Charnov *et al.*, 1976). By the same reasoning unisexual flowers are more common in plants (both dioecious, monoecious) with abiotic pollination, in which attractants are typically not as well developed as in plants with biotic pollination.

In practice, in most hermaphrodite flowers, pollen production and stigma receptivity are slightly separated in time (dichogamy, sequential hermaphroditism) and anthers and stigma(s) within a single flower may also be separated spatially, or kept apart by carefully choreographed movements of the different floral organs (Endress, 2010b). These features help prevent self-pollination and self-fertilisation and there are also a range of incompatibility mechanisms that are mediated in various ways by genetically controlled signals between pollen and carpellary tissues. In many flowers there are also elegant, and sometimes very elaborate, structural modifications that control the interaction of flowers and pollinators and prevent self-pollination (Endress, 2010b). These diverse attributes, combined with the range of attractants that have evolved in angiosperms, contribute significantly to the structural and functional complexity of angiosperm flowers. Attractants include those that provide a direct energetic reward to pollinators (nectar, pollen, oil) and those that help to draw attention to the flower (colour, odour, temperature, motion). Some flowers are also visited because they provide brood sites in which the young of pollinators can develop.

17.2.1 Abiotic versus biotic pollination in early angiosperms

Most extant angiosperms are insect-pollinated. Insect pollination predominates in monocots and eudicots, as well as among extant eumagnoliids. In almost all cases, abiotic pollination in angiosperms is clearly a secondary specialisation. However, among angiosperms at or around the ANITA grade generalist pollination modes, which often combine abiotic and biotic pollination, occur in *Amborella*, which is functionally dioecious, as well as in species of *Illicium*, other Schisandraceae and *Trimenia*. Abiotic pollination occurs in Nymphaeales (wind pollination in *Brasenia*, water pollination in *Euryale*) (Thien *et al.*, 2009). In Chloranthaceae wind pollination characterises *Ascarina* and *Hedyosmum*, while *Chloranthus* and *Sarcandra* are insect-pollinated (Endress, 1994b; von Balthazar and Endress, 1999). Based on the current rooting of the angiosperm tree it is therefore difficult to establish the ancestral pollination mode for angiosperms. Uncertainties over the rooting, as well as the possibility of significant extinction at this level of angiosperm evolution, further complicate the picture. These uncertainties suggest caution in developing generalisations about pollination and other aspects of the biology of the earliest angiosperms (sections 17.5, 17.6).

17.2.2 Abiotic pollination

Wind pollination (anemophily) is far more widespread among extant angiosperms than water pollination (hydrophily) and most species of aquatic angiosperms produce flowers that project above the water surface and are pollinated by wind or animals. Some water plants, however, have specialised pollination that clearly reflects adaptation to an aquatic habitat. In some cases, staminate flowers, anthers or pollen grains may float across the water surface to pistillate flowers (e.g. *Vallisneria*). More rarely, pollen is transported to the stigma beneath the water surface. Among those lineages that were established very early in angiosperm evolution, water pollination occurs in extant Nymphaeales (*Euryale*) and *Ceratophyllum*.

Although wind pollination has a scattered phylogenetic distribution among angiosperms (e.g. grasses, Betulaceae, Polygonaceae, Ulmaceae, *Fraxinus*, *Populus*) it is especially characteristic of certain lineages and is also more common in certain habitats. Typically, anemophily occurs in open vegetation (e.g. grasslands, savannahs, around open water) or in places where animal pollinators are rare (e.g. salt marshes, some semi-arid habitats). It is also common among the dominant trees of temperate regions, many of which are concentrated in particular lineages (e.g. families of Fagales). In temperate regions wind pollination often occurs before trees and shrubs are fully leafed out and is often characteristic of plants that grow in large, relatively dense populations in which pollen shedding and ovule production is well synchronised. Wind pollination is relatively rare in tropical rain forests and among plants that flower in the forest understory, but it does occur in trees that emerge above an otherwise dense canopy or that grow in more open habitats.

Wind-pollinated flowers typically have high pollen production, reduced perianth parts, and generally only one or few ovules per carpel. Frequently there are separate staminate and pistillate flowers (dicliny). Increased pollen production is often reflected in increased size of the anthers, increased numbers of stamens in the flowers, or an increased number of staminate flowers on the plant (Endress, 1977, 1986a). Pollen grains typically have a more or less smooth surface, do not adhere together in clumps, and are generally between 20 and 40 μm in diameter (Whitehead, 1969). The stigmas of wind-pollinated plants are often dry and divided (e.g. grasses). Friedman and Barrett (2008) investigated correlations between wind pollination and several ecological and floral features and noted a particularly strong correlation between sexual system (unisexual), floral size (small), showiness (plain), and nectar (absent).

The occasional occurrence of wind-pollinated taxa in otherwise insect-pollinated groups (e.g. *Acer saccharum* in *Acer*, *Fraxinus* in Oleaceae), and the occurrence of insect-pollinated taxa in otherwise wind-pollinated groups (e.g. some grasses of the rainforest understory) (e.g. Soderstrom and Calderon, 1971), shows that the switch between biotic and abiotic pollination can be relatively labile. On the other hand, it is also clear that wind pollination, although secondary in a broader phylogenetic context, predominates and has remained relatively fixed for long periods of time in some large and ecologically important lineages such as certain Fagales (e.g. Betulaceae, Nothofagaceae) and many commelinid monocots (e.g. grasses).

17.2.3 Biotic pollination

Biotic pollination occurs in a very wide range of ecosystems and is particularly common in closed vegetation. A study of pollination in trees from tropical lowland rain forests estimated animal pollination for between 98% and 99% of all angiosperm species (Bawa, 1990). A great variety of animals participate in pollination in angiosperms (Fægri and Van der Pijl, 1980; Proctor *et al.*, 1996). Insects, such as butterflies, flies and moths, are particularly important. Birds and bats are also significant pollinators, especially in the tropics. Pollination mechanisms involving non-flying mammals are less common and have only been recognised relatively recently. Among animal-pollinated (zoophilous) flowers there is a complete gradation from forms that are specialists, strictly pollinated by only one or a few pollinators (monophilic or oligophilic), to flowers that are generalists and visited by a great range of potential pollinators (polyphilic). Similarly, among pollinators there is every intermediate between animals that faithfully and very effectively visit flowers of only one or a few species, and animals that visit flowers of many species and are 'opportunistic' pollinators. Other animal visitors to flowers may be nectar or pollen thieves that contribute nothing in terms of pollination.

Flowers of different taxa that are pollinated by similar animal pollinators often show similar specialisation in structure and other features (e.g. mechanisms of attraction). These similarities have been used to recognise 'syndromes' among the diversity of angiosperm floral form that are characteristic of certain kinds of flowers visited by pollinators (Fægri and Van der Pijl, 1980; Proctor *et al.*, 1996). These syndromes have predictive value in inferring pollination in fossil and extant species for which direct evidence is lacking, but they can also be misleading. In many angiosperm species the relationship to pollinators is diffuse rather than highly specific, and in many groups there is a great variation in pollination mode among closely related species. It is also important that the behaviour of most insect species has never been studied directly. Inferences about the involvement of various insect groups in pollination, therefore, involve a degree of extrapolation.

17.3 INSECTS AS POLLINATORS

True insects (Ectognatha) are an ancient group of arthropods first recognised in the Early Devonian (Engel and Grimaldi, 2004) soon after the initial development of macrophytic vegetation on land, but their extraordinary diversity at the species level arose largely through the Mesozoic and Cenozoic by exploiting the more complex environments and new food sources created by the escalating diversification of land plants (Grimaldi and Engel, 2005). Angiosperm diversification in particular seems to have created new opportunities that helped drive the diversification of certain insect groups, especially herbivores and pollinators, to new levels (Grimaldi, 1999). At the same time, enhanced opportunities for reproductive isolation that were made possible by insect pollination have long been considered a key factor in generating angiosperm diversity (e.g. Saporta and Marion, 1873). There is no doubt that co-evolution of angiosperm flowers and insect pollinators over a long period, and with various degrees of fidelity and specificity, have had a profound effect on structural and systematic diversity in both groups. Our treatment of the phylogeny and fossil history of insects is based largely on the reviews of Grimaldi (1999) and Grimaldi and Engel (2005), and we refer to these studies for more detailed information and literature.

The earliest true insect (*Rhyniognatha hirsti*), and the earliest evidence of the probable insect sister group (springtails/collembolans; *Rhyniella praecursor*), is from the Early Devonian Rhynie Chert (Engel and Grimaldi, 2004). The earliest insects from the Devonian were probably detritivores (Zherikhin, 2002) although some may also have fed on spores. Large-scale insect herbivory was well established by the Carboniferous, and damage caused to plants and plant tissues is well documented. There is no evidence that any of these early insects were pollinators. The major orders of insects that are important pollinators today (Thysanoptera, Coleoptera, Hymenoptera, Diptera, Lepidoptera) first appeared in the Permian or through the Mesozoic (Labandeira, 1998) and several important insect pollinators were already present in the Early Cretaceous during the early phases of angiosperm diversification (Figure 17.1).

17.3.1 Thrip pollination and the fossil history of Thysanoptera

Thrips (Thysanoptera) are an important group among the paraneopteran insects (Grimaldi and Engel, 2005), which also include the Psocodea (barklice, booklice, true lice) as well as the true bugs (Hemiptera). Most of the *c.* 5500 species of thrips are divided between two major families that may be sister taxa. The Phlaeothripidae (*c.* 3200 species) include species that feed on leaves, flowers and fungal spores. They may also form galls, and in some cases are predatory. The Thripidae (*c.* 2500 species) are mainly specialised flower feeders. Outside these two families the diversity of thrips is relatively low.

Pollen feeding among thrips appears to have evolved at least three times (Heming, 1993). Pollen is punctured and drained individually by a feeding tube, which is typically 1–2 μm, sometimes 5–10 μm, in diameter (Kirk, 1984; Grimaldi and Engel, 2005). Thrips may feed on nectar or on special food bodies on the stamens or perianth parts (Endress, 1994b). In addition, thrips may also be attracted to flowers for shelter. Thrip-pollinated flowers are often white with narrow passages (funnels or slits) that may provide protection. Typically, scent is the primary attractant (Endress, 1994b). Pollination by thrips is known in several groups of derived eudicots (e.g. Dipterocarpaceae, Verbenaceae). It is known also in some eumagnoliids (e.g. Monimiaceae, Myristicaceae, Winteraceae) (Endress, 1994b), but has not been reported for ANITA-grade angiosperms (Thien *et al.*, 2009).

The history of thrips may extend back to the Late Palaeozoic based on the presence of its possible sister group in the Permian (Grimaldi, 1999). The oldest unequivocal Thysanoptera are from the Triassic, but exactly how these fossils relate to extant groups is uncertain. Fossil thrips that can be referred unequivocally to modern groups, and that could have been potential pollinators, are first recorded in Lebanese amber from the Early Cretaceous. Thrips are also diverse in most Late Cretaceous ambers, as well as in the Early Cenozoic Baltic amber, in which most of the fossil thrips are referable to modern families (Larsson, 1978; Grimaldi, 1999).

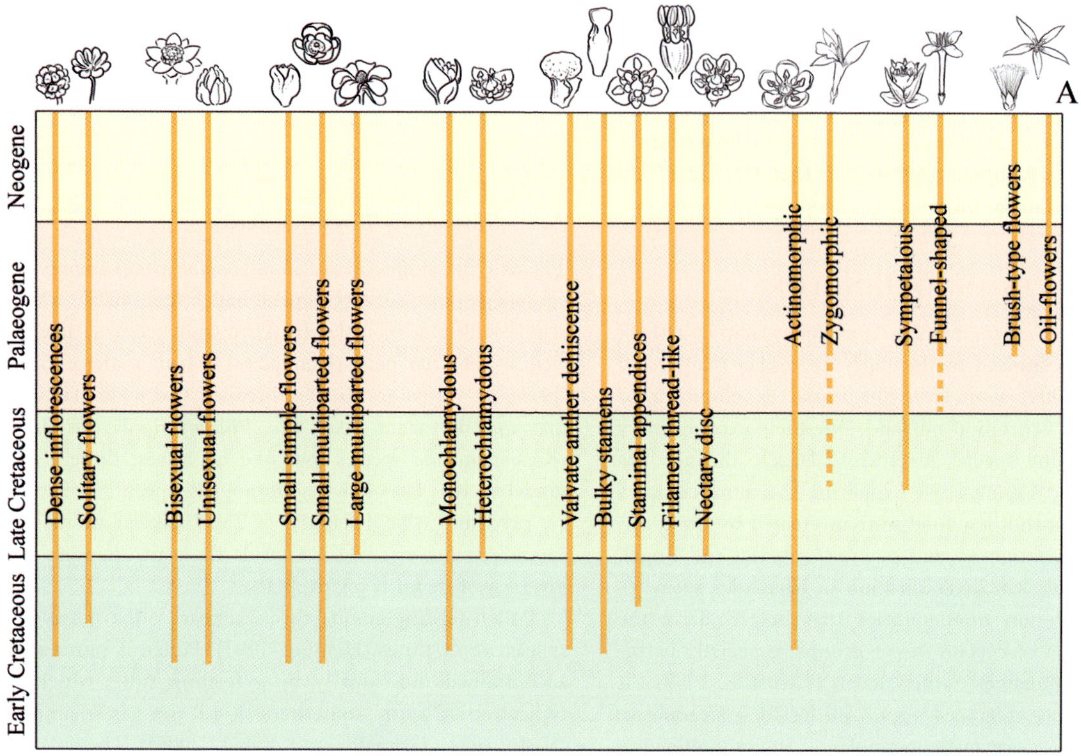

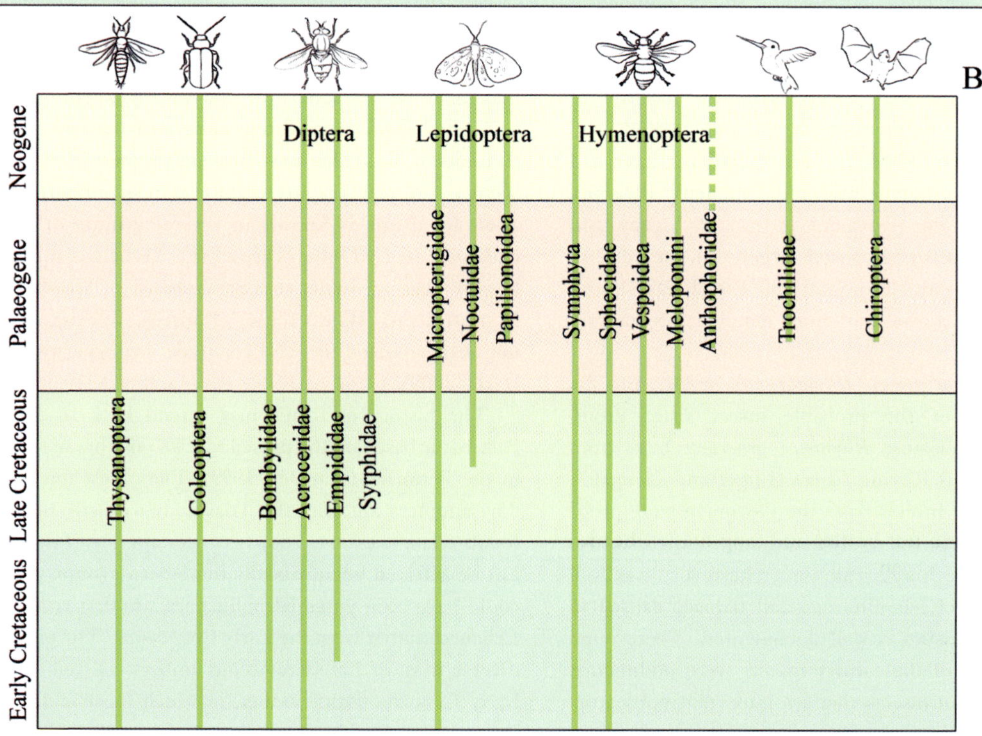

Figure 17.1 Cretaceous and Cenozoic occurrences of (A) major floral features relevant for pollination and (B) major groups of pollinators of angiosperm flowers. Solid lines indicate direct fossil evidence for first occurrence of feature or group; dashed lines indicate indirect fossil evidence of feature or group.

17.3.2 Beetle pollination and the fossil history of Coleoptera

Beetles (Coleoptera) include *c.* 350 000 named species, which together account for 40% of all insects (Grimaldi and Engel, 2005). In general, beetles are well adapted to a cryptic lifestyle, often in tightly confined spaces, but they also retain the ability to fly using their membranous hind wings, which are crossed and folded under the protective elytra (modified front wings) while the animal is on the ground. Several major groups of beetles are specialised for feeding on plants, but beetles occupy a great variety of ecological niches. Phylogenetic analyses suggest that many different groups of beetles have independently made the transition to herbivorous, predatory, fungivorous and aquatic lifestyles (Hunt *et al.*, 2007).

Extant Coleoptera can be divided into four suborders: Archostemata, Myxophaga, Adephaga and Polyphaga. Archostemata comprise *c.* 35–40 species that are mainly specialised wood borers, although some are specialised pollen feeders. Myxophaga comprise about 65 species of minute, specialised aquatic or semi-aquatic beetles that typically graze on cyanobacteria and green algae in wet sand. Adephaga comprise *c.* 45 000 species, and phylogenetic analyses (Hunt *et al.*, 2007) support the traditional division into water beetles (Hydradephaga) and ground beetles (Geadephaga). Hydradephaga include the diving beetles (Dytiscidae) and whirligig beetles (Gyrinidae). Geadephaga include the very diverse carabid beetles (*c.* 40 000 species).

The Polyphaga account for about 90% of all beetle species. They display extraordinary diversity in feeding and other aspects of their ecology and among them there are several very large clades (e.g. Scarabaeoidea, *c.* 35 000 species; Staphylinidae, *c.* 48 000 species; Buprestoidea, *c.* 14 000 species; most Elateroidea, *c.* 23 000 species). Most important in terms of their interactions with plants are the hyper-diverse Cucujiformia (*c.* 190 000 species in 90 families), which make up more than half of all beetle species. Within Cucujiformia there are two important groups of plant feeders: weevils and bark beetles (Curculionoidea, *c.* 70 000 species) and leaf beetles and long-horned beetles (Chrysomeloidea, *c.* 58 000 species).

Flowers that are pollinated by beetles (cantharophilous flowers) are often large, dull in colour (cream, greenish, brownish), produce relatively large amounts of pollen and sometimes have special nutritive tissues in the perianth parts or stamens (Endress, 1984). Scent is the primary attractant, and pollen, or special food bodies in the perianth parts or stamens, are the primary reward. In general beetles are messy pollinators that transfer pollen in the process of feeding on both pollen and floral organs. Beetles (Polyphaga) involved commonly in pollination include Chrysomeloidea, Staphylinidae, Cantharidae (Elateroidea), Curculionoidea and Scarabaeoidea (Endress, 1994b).

Beetles contribute to pollination in many ANITA-grade angiosperms, including *Amborella*, Nymphaeaceae and Austrobaileyales (Thien *et al.*, 2009), and are also prominent pollinators of certain eumagnoliids (e.g. Annonaceae, Eupomatiaceae, Degeneriaceae, Myristicaceae, Gottsberger, 1977, 1988). However, while beetles are often considered as the archetypical pollinators of archaic angiosperms, they are also important in a range of other groups (e.g. Araceae, Cyclanthaceae, Hydnoraceae, palms, Gottsberger, 1977). Equally, there are many groups of eumagnoliids that are not beetle-pollinated.

Beetles have the most ancient fossil record of all insects with complete metamorphosis (holometabolous insects), with probable stem-group Coleoptera (Protocoleoptera) and possible stem-group Archostemata known from the Early and Late Permian, respectively (Grimaldi and Engel, 2005). By the Late Triassic there is unequivocal evidence of true beetles and some of these Triassic fossils have been assigned to crown-group Archostemata and Polyphaga as well as possible crown-group Adephaga. The earliest Polyphaga are specimens assigned to the Staphylinidae from the Late Triassic of Virginia; in the Late Jurassic Karatau fossil insect assemblages from Kazakhstan, Polyphaga account for 80% of the beetle specimens collected. Among them are unequivocal weevils (Curculionidea) that provide the earliest definitive evidence of Cucujiformia. Taken together the fossil record of beetles provides clear evidence that the group was diverse well before the initial Early Cretaceous diversification of angiosperms (Grimaldi, 1999) and that beetles could potentially have been important pollinators from the Triassic onwards.

17.3.3 Bee/wasp pollination and the fossil history of Hymenoptera

Many new species of Hymenoptera continue to be discovered, adding to the more than 120 000 species already described. Ultimately, the diversity of Hymenoptera may rival or eclipse that of beetles (Grimaldi and Engel, 2005). Xyelidae (Archihymenoptera) appears to be the sister group to all other Hymenoptera (Neohymenoptera) within which there is a basal grade of other sawfly and wood wasp lineages

('Symphyta'). The bulk of hymenopteran diversity occurs in a single clade, the Euhymenoptera.

The larvae of many xyelids feed on conifers and at this level of hymenopteran diversification most lineages are fundamentally phytophagous. Sawflies are important pollinators for some groups of angiosperms in the temperate regions (e.g. Apiaceae, Proctor *et al.*, 1996). They visit flowers to feed on nectar and pollen, as well as stamens and petals. Like beetles, they may inflict damage in the process.

Among Euhymenoptera, the Apocrita, defined by the narrow 'wasp-waist', include several lineages important for pollination and plant–insect interrelationships. The Chalcidoidea (*c.* 20 000 species) include mostly parasitoids, but there are secondary reversions to phytophagy and the group includes the fig wasps, in which there is a particularly tight obligate relationship between the plants and their pollinators (Ramirez, 1994). Wasp larvae develop in specialised flowers. The Cynipoidea include the gall wasps that are especially important gall-formers on Fagaceae. Especially diverse among the Apocrita are the Ichneumonoidea, with perhaps 100 000 species, which are mainly sophisticated endoparasitoids.

The key group of Apocrita in terms of pollinator interactions with angiosperms is Aculeata, which includes ants, bees, and wasps, and in which the ovipositor is modified into a sting. Of the three superfamilies, the Chrysidoidea are exclusively parasitoids whereas the Vespoidea, which includes the wasps and ants, are mainly parasitoids or predatory. The Apoidea are also mainly parasitoids or predatory, except for the bees, which are specialised for feeding their larvae on pollen, floral oils and nectar (Grimaldi and Engel, 2005).

Among wasps the most important group of pollinators are the pollen wasps (Masarinae), which construct simple burrows in which the larvae are fed pollen and nectar (Grimaldi and Engel, 2005). In general, flowers pollinated by wasps have short corollas with open nectar (Fægri and Van der Pijl, 1980). Ants are less significant pollinators, but they are important plant dispersers (Chapter 18) and plant consumers. They have also evolved various close and sophisticated symbiotic interactions with many groups of angiosperms. Frequently ants that occur in flowers are nectar robbers. However, in a few plants, particularly in hot, dry habitats, ants may be involved in pollination. Ant-pollinated plants are typically prostrate or low-growing with their flowers close to the ground (Proctor *et al.*, 1996).

The Apoidea comprise *c.* 30 000 described species known as sphecid wasps, a grade of four lineages (Heterogynidae, Ampulicidae, Sphecidae, Crabronidae) that are successively more closely related to the bees (Anthophila, Grimaldi and Engel, 2005). Sphecids are mainly predatory and larvae are fed mainly on small arthropods. Adults, however, may feed on nectar and pollen and may be important pollinators. At least one genus of Crabronidae (sister group to bees) feeds its larvae on pollen and nectar.

Bees are a well-defined natural group. They are almost exclusively vegetarian and comprise about 20 000 species. Bees can be divided into short-tongued bees (e.g. Colletidae, Halictidae, Andrenidae, Melittidae) and long-tongued bees (Megachilidae, Apidae). Most bees are solitary. The eusocial forms, which account for only a few per cent of living species, occur mainly among a subgroup of Apidae (corbiculate apines: Euglossini, Bombini, Apini, Meliponini) (Grimaldi and Engel, 2005).

Bees are probably the group of pollinators that have coevolved most closely with angiosperms and this relationship may account for much of the diversity of angiosperm flowers. Flowers that are pollinated by bees (melittophilous flowers) are often blue, yellow or ultraviolet with nectar guides, and are open in the early morning. Frequently, flowers are zygomorphic, which forces the visiting bee into a position where it makes contact with stigma or pollen. Attraction is mainly visual and the reward is generally nectar and/or pollen. Certain groups of bees also collect oil, scent or resin (Fægri and Van der Pijl, 1980; Renner and Schaefer, 2010). Among the important groups of bees of the family Apidae are members of the tribes Apini (*Apis*), Bombini (*Bombus*), certain Euglossini (especially *Eufriesea*, *Euglossa*, *Eulaema*), Meliponini (*Melipona*, *Trigona*) and Xylocopini (a group of non-corbiculate apines that include the carpenter bees, *Xylocopa*, in the tropics). Although there are several plants pollinated by Hymenoptera among ANITA-grade angiosperms, particularly in the Nymphaeaceae (Thien *et al.*, 2009), pollination by bees is most common among groups that are nested at much higher levels in angiosperm phylogeny.

The Hymenoptera probably began to differentiate from around the mid- to the Late Triassic (Grimaldi, 1999). The earliest forms may have had phytophagous larvae similar to extant taxa at the 'Symphyta' grade. Consistent with the phylogenetic pattern, the earliest fossil Hymenoptera are putative xyelids from the Late Triassic of Australia, central Asia and southern Africa. Euhymenoptera, with their predominantly parasitoid habit, probably arose around the

Triassic–Jurassic boundary based on probable early Proctotrupomorpha from the Early Jurassic of Asia. The Cynipoidea, which include the gall wasps, and the Chalcidoidea, which include the fig wasps, are not recorded until the Early Cretaceous.

The earliest ichneumonoids, and the earliest evidence of their sister group, the Aculeata, is from the Early Cretaceous of Asia and Australia. The initial diversification of both groups probably occurred around the Jurassic–Cretaceous boundary. The earliest fossil aculeate Hymenoptera are from the Late Jurassic of Kazakhstan and are assigned to the extinct family Bethylonymidae. Sphecid wasps are recorded from the earliest Cretaceous onwards and many groups of aculeates are also known from the Early Cretaceous. Among these, ants are first recorded in Burmese amber and also in amber from France. These records are from around the Early–Late Cretaceous boundary (Grimaldi and Engel, 2005). Even though the age of the Burmese amber is controversial, subsequent fossils from the Late Cretaceous clearly establish that the radiation of ants was well under way by this time.

The initial diversification of bees probably followed a trajectory similar to that of ants. The earliest evidence of their sister group (Crabronidae) comes from around the Early–Late Cretaceous boundary, suggesting that bees had already diverged by this time. However, the earliest direct fossil evidence comes from later in the Cretaceous based on a single specimen, *Cretotrigona prisca* (Meliponini, Figure 17.2), preserved in amber from the very latest Cretaceous (Maastrichtian) of New Jersey, USA (Michener and Grimaldi, 1988; Engel, 2000). The controversy over the age of this fossil (estimated at *c.* 65 Myr) is reviewed by Grimaldi (1999) and is important because this fossil can be attributed to a group of bees (Apidae) that is nested well within the Aculeata. Its presence in the latest Cretaceous implies that the origin of other lineages was earlier in the Cretaceous, an inference also consistent with the presence of possible nests of halictine bees in wood from the mid-Cretaceous Dakota Formation of Arizona, USA (Elliott and Nations, 1998), and also from the Maastrichtian (Grimaldi and Engel, 2005). From the Eocene–Oligocene onwards the fossil history of bees is more extensive. There are unequivocal fossil bees from the Eocene–Oligocene Baltic amber and Florissant Beds, Colorado, USA, as well as from the Miocene of Spain and the Oligocene–Miocene Dominican amber (Grimaldi, 1999). There is also a spectacular find of stingless bee from the Dominican amber with a single pollinarium, comprising the two complete pollinia, attached to its mesoscutellum (Ramirez *et al.*, 2007). This rare, direct evidence of plant–pollinator interaction in the past also provides the most reliable example of an orchid in the fossil record (Chapter 11). Indirect evidence of oil-collecting bees in the Eocene is from the fossil malpighiaceous flower, *Eoglandulosa warmanensis*, which has paired elaiophores, a feature characteristic for extant flowers visited by anthohorid bees.

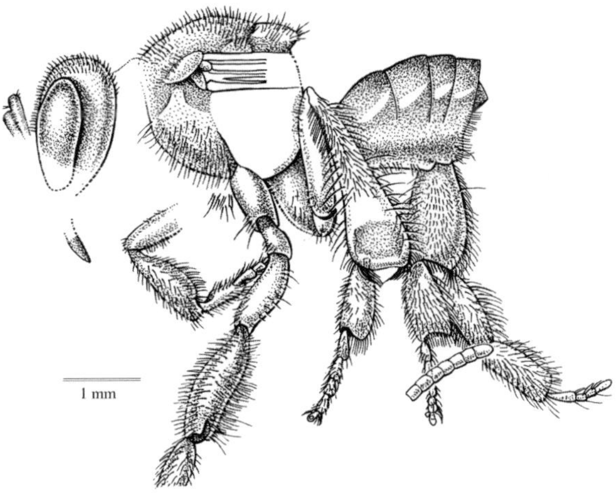

Figure 17.2 *Cretotrigona prisca*, the earliest fossil bee discovered in Cretaceous (presumed Maastrichtian) amber from New Jersey, USA. Redrawn from Grimaldi (1999).

17.3.4 Fly pollination and the fossil history of Diptera

True flies (Diptera) are a large group of about 120 000 described species (Grimaldi, 1999). They are almost all liquid feeders, but display extraordinary diversity in their ecology. Adults include predators, parasitoids and pollen feeders. Based on the presence of long proboscides it seems likely that flower visiting is scattered through Diptera and that obligate flower visitors may be more common than is recognised generally. Many flies are fast fliers and excellent hoverers, both features that enhance their ability to forage effectively among scattered flowers. However, inferring possible involvement in pollination of fossil Diptera requires inferences based on lengths of proboscides combined with extrapolation to those few living relatives for which behaviour is known (Grimaldi, 1999).

The Diptera can be divided into six major groups of which the Brachycera are the most significant for pollination. The Brachycera comprise 13 major clades, about half of which contain important pollinators. However, among

Calyptrata, which has about 50 000 species and is the most diverse of these clades, pollinators are relatively sparse. Key groups of extant brachyceran pollinators include the Acroceridae, Apioceridae, Bombylidae, Empididae, Mydidae, Nemestrinidae, Scenopinidae, Stratiomyidae and Syrphidae (Grimaldi, 1999).

The Acroceridae (small-headed flies) include some of the best-documented flower visitors. They have long proboscides and are good hoverers. The Syrphidae (hoverflies or flower flies) are also a very important, cosmopolitan and diverse group of pollinators with about 6000 species and 180 genera. Many species in the two largest subfamilies (Syrphinae, Eristalinae) are anthophilic, feeding on nectar and pollen, but none have especially long proboscides. The Bombylidae (bee flies) are well-known and important pollinators, especially in xeric environments. Many have a long proboscis and exhibit flower-constant foraging behaviour. Similarly, both the Apioceridae (flower-loving flies) and Mydidae (mydas flies) include flies that are good fliers, some of which have long proboscides and are known to be flower feeders. The Empididae (dagger flies and balloon flies) are sometimes regarded as important pollinators, but Grimaldi (1999) notes that only a small number of empidids appears to be frequent or obligate pollen feeders. The Nemestrinidae (tangle-veined flies) mainly have a long slender proboscis, and are also superb fliers.

Fly pollination (myophily) of unspecialised, open flowers is important in many angiosperm families. These flowers are typically small and simple with exposed nectar and pale colours. However, flowers in some groups of angiosperms appear highly specialised for fly pollination. Flowers pollinated by flies (e.g. Calliphoridae, Drosophilidae, Muscidae, Sarcophagidae) that oviposit in carrion or dung (sapromyophily) mimic the normal sites of oviposition in their musky or putrid smell, their greenish, brownish or purple perianth that often has dark speckles, and sometimes floral organs with thread-like extensions (Endress, 1994b). Sapromyophilous flowers are often large and deep with traps of various kinds (e.g. *Rafflesia arnoldii*, *Aristolochia grandiflora*, *Stapelia*, *Sapranthus*) and they do not produce nectar. Flowers mimicking mushrooms (fungus gnat flowers) occur in *Asarum* and *Aristolochia* (Aristolochiaceae), *Arisarum* and *Arisaema* (Araceae) and *Dracula* (Orchidaceae) (Endress, 1994b). Many ANITA-grade angiosperms are also fly-pollinated (Thien *et al.*, 2009), and fly pollination also occurs in many groups of monocots and eudicots, including families such as Araceae, Asclepidaceae and Orchidaceae (Endress, 1994b).

The fossil record of anthophilous fly lineages extends from the Middle Jurassic and onwards (Grimaldi, 1999). The oldest fossil bombylid, *Paleoplatypygus zaitsevi*, is from the Middle Jurassic of Siberia and is related to taxa with small rudimentary mouthparts (Evenhuis, 1994 cited in Grimaldi, 1999), as is also the case for the only other Mesozoic bombylids, which are from the Late Cretaceous (Santonian) amber of Tamyr, Siberia. It seems unlikely that these Jurassic and Cretaceous forms were flower visitors. Fossil bombylids with features more suggestive of pollinator behaviour are not known until the Eocene and Oligocene. They are present in the Baltic amber as well as classic Cenozoic compression insect faunas such as those from the Green River and Florissant shales. Based on the frequent occurrence of various Brachycera in Cretaceous insect assemblages Grimaldi (1999) concludes that the absence of the more derived, and commonly flower-pollinating, groups of bee flies in the Cretaceous reflects real absence. The major diversification of bombylids probably occurred in the Late Cretaceous and Early Cenozoic.

A similar pattern pertains for fossil acrocerids and empidids (Grimaldi, 1999). The earliest fossil acrocerids, from the Late Jurassic of Karatau, Kazakhstan (Nartshuk, 1996), have short proboscides, were probably not good hoverers and may not have been pollinators (Grimaldi, 1999). There is no Cretaceous record of the group, but fossils from the Eocene – Oligocene Baltic amber are very closely related to extant genera. Empidids are very well represented in the fossil record beginning in the Late Jurassic and extending through the Cretaceous. However, forms with a long proboscis do not appear until the Cenozoic.

The Mesozoic fossil record of other groups of anthophilous flies is relatively sparse (Grimaldi, 1999). The earliest unequivocal syrphid fossil is from Late Cretaceous (Santonian) amber from Taymyr, Siberia, and Early Cenozoic flower fly faunas are essentially modern in aspect. The Scenopinidae (window flies) include some species with a long proboscis that feed on flowers, but the earliest fossils from the Late Cretaceous (Turonian) of New Jersey have a short proboscis, suggesting that they were probably not regular pollinators. For Apioceridae and Mydidae (mydas flies) biogeographic and phylogenetic arguments also suggest that the long proboscis and pollen feeding evolved during the Late Cretaceous.

The Nemestrinidae are first recorded from the Late Jurassic of Karatau, and one of these fossils (*Protonemestrius*) appears to have a long proboscis (Grimaldi, 1999). If this interpretation is substantiated it would represent the earliest occurrence of this specialised feeding structure in the fossil record. Other nemestrinids from the Early Cretaceous of China may include some of earliest obligate pollinators (Ren, 1998). Grimaldi (1999) suggested that the Nemestrinidae are 'the best candidate for the earliest obligate pollinator of angiosperms'.

17.3.5 Butterfly/moth pollination and the fossil history of Lepidoptera

Butterflies and moths (Lepidoptera) are second only to beetles in species diversity with *c.* 150 000 named species (Grimaldi, 1999). However, in contrast to beetles, Lepidoptera are almost uniformly plant feeders and they are very strongly linked with angiosperms. Larval development in Lepidoptera frequently depends on angiosperm food plants, and adults are prominent pollinators of many angiosperm flowers (Ehrlich and Raven, 1964; Grimaldi, 1999). Because Lepidoptera can fly long distances they can be efficient pollinators when individual plants, plant populations, or both, are widely scattered.

The Lepidoptera are a clearly defined monophyletic group (Kristensen, 1984; Kristensen *et al.*, 2007) in which the earliest-diverging extant clades, Micropterigidae, Agathiphagidae and Heterobathmiidae, are all species-poor. The Micropterigidae comprise about 220 species and have larvae that mainly feed among leaf litter. Adults feed on pollen, including the pollen of some ancient lineages of angiosperms. The Agathiphagidae (two species) have larvae that feed on the seeds of the conifer *Agathis*. The Heterobathmiidae (10 species) have leaf-mining larvae that are restricted to *Nothofagus*. Adults of all three lineages are mandibulate with chewing mouthparts.

Micropterigidae, Agathiphagidae and Heterobathmiidae are successive sisters to the much more species-rich Glossata, which are defined by the presence of the characteristic, slender, usually long and coiled, lepidopteran proboscis that is especially adept at sucking up fluids. Within Glossata, the Ditrysia account for about 98% of all Lepidoptera and include about 28 major groups (Grimaldi and Engel, 2005). The groups of Glossata that fall outside Ditrysia are all relatively species-poor; most have leaf-mining larvae, and many have restricted distributions in the Southern Hemisphere.

The Macrolepidoptera comprise a putative monophyletic group within Ditrysia and include several prominent clades that are of particular interest for angiosperm pollination: the Bombycoidea (including saturniids, sphingids/hawkmoths and silkworm moths), Pyraloidea, Geometroidea, Noctuoidea (gypsy moth and tiger moths) and the Rhopalocera (true butterflies, skippers and American butterfly moths).

Flowers pollinated by Lepidoptera (lepidopterophilous flowers) are often actinomorphic and have nectaries concealed in long narrow tubes or nectar spurs that can be accessed only by the long butterfly or moth proboscis (Endress, 1994b).

Noctuid moths and hawkmoths are mainly active at dusk or at night and are attracted primarily by scent. Flowers pollinated by these insects tend to be light, presumably aiding recognition in the dark, but are not generally vividly coloured. Nectar is the primary reward. Hawkmoths have extremely long proboscides and the corresponding flowers (sphingophilous flowers) often produce nectar at the bases of long tubes or spurs. The flowers are often delicate because hawkmoths generally hover while feeding. The proboscides of noctuid moths are generally shorter and the corresponding flowers (phalaenophilous flowers) have relatively short tubes (Endress, 1994b). Groups of angiosperms with long spurs that are pollinated by hawkmoths include many orchids, as well as species of *Bauhinia* (Fabaceae) and *Pelargonium* (Geraniaceae). Angiosperms with long corolla tubes that are pollinated by hawkmoths include many asterids (e.g. Apocynaceae, Rubiaceae, Solanaceae, Campanulaceae) as well as certain Liliaceae, Caricaceae, Combretaceae, Nyctaginaceae and Pittosporaceae (Endress, 1994b).

True butterflies and their relatives (Rhopalocera) comprise three main groups: the Hedylidae (American butterfly moths, *c.* 35 species) sister to the Hesperiidae (skippers, *c.* 3500 species) and Papilionoidea (true butterflies with clubbed antennae, *c.* 14 500 species). They are mainly active during the day and have shorter proboscides than hawkmoths. Butterfly-pollinated flowers (psychophilous flowers) have relatively short tubes or spurs, may or may not be scented, and are generally brightly coloured (often orange, red or pink) (Endress, 1994b). Because butterflies land when feeding, individual flowers are either large and provide a landing place (e.g. certain Fabaceae, Liliaceae, Vochysiaceae, Zingiberaceae), or are small and grouped together to form a landing platform (e.g. certain Boraginaceae, certain Fabaceae, Nyctaginaceae, Verbenaceae) (Endress, 1994b).

Despite their seemingly ephemeral adult structure, the fossil record of Lepidoptera is surprisingly extensive (Grimaldi, 1999). But it is also strikingly concentrated in the Cenozoic. The most important records come either from amber or from exquisite preservation in fine-grained lake sediments (e.g. Florissant beds, Colorado, USA). The sparse earlier fossil record has been interpreted in dramatically different ways by different authors (e.g. Whalley, 1987; Labandeira *et al.*, 1994; Grimaldi, 1999). Grimaldi (1999) and Grimaldi and Engel (2005) excluded all Triassic records of the group as unreliable, and viewed the Jurassic records as of uncertain relationship among the glossate Lepidoptera.

The earliest evidence of Lepidoptera is *Archaeolepis manae* from the Early Jurassic (Sinemurian) of southern England, which is securely attributed to the group, based on both wing venation and the presence of scales on the wings (Grimaldi and Engel, 2005). There are also slightly younger records based on wing venation from the Early Jurassic (Toarcian) of Germany, and five species of Lepidoptera are known from the diverse Late Jurassic insect faunas of Karatau, Kazakhstan (Grimaldi and Engel, 2005). The relationships of these early records of Lepidoptera to living groups are uncertain. None can be assigned to extant clades and therefore they do not provide unequivocal evidence of crown-group Lepidoptera. The relationships of probable Lepidoptera from the Early Cretaceous from the Crato Formation of Brazil (Grimaldi and Engel, 2005; Bechly in Martill *et al.*, 2007) are similarly uncertain.

The only records of crown-group Lepidoptera from the Early Cretaceous are a micropterygid moth (Whalley, 1986) and a larva attributed to Glossata (Grimaldi, 1999), both from Lebanese amber. Subsequently, through the Late Cretaceous Glossata are represented by five records of non-ditrysian taxa. Micropterygids and undetermined Glossata are represented in putative mid-Cretaceous Burmese amber. The earliest record of possible Ditrysia is based on leaf mines attributed to gracillarid larvae from the mid-Cretaceous Dakota Formation of Kansas, USA (Labandeira *et al.*, 1994). The earliest records of Macrolepidoptera are provided by fossils attributed to Noctuoidea and Hesperiidae from the Mo Clay (Fur Formation, Early Eocene) of Denmark. By the Eocene–Oligocene there is good representation of Lepidoptera outside the Macrolepidoptera in Baltic amber. Macrolepidoptera are also well represented in the Florissant beds, Colorado, USA, by several groups (Geometroidea, Papilionoidea).

17.4 VERTEBRATES AS POLLINATORS

The two main groups of vertebrates that participate in the pollination of extant angiosperms are birds and mammals (principally bats). Early in angiosperm evolution it is possible that other groups of vertebrates could also have played a role, but clear evidence for this is currently lacking.

17.4.1 Bird pollination and the fossil history of birds

Phylogenetic relationships among the c. 10 000 species of modern birds (Neornithes) have been greatly clarified by a genome-scale study based on a sample of 169 species representing all of the major living groups (Hackett *et al.*, 2008). This analysis confirms previous ideas of two major groups of birds: Paleognathae (ostriches, rheas, tinamous, cassowaries, emu, kiwis) and Neognathae (all other birds). Previous hypotheses of a basal split within neognathes between Galloanserae (pheasants, quail, ducks, geese, chickens and relatives) and Neaves (most modern birds) were also strongly supported. Relationships within Neaves, which comprises the bulk of bird diversity, have been more controversial, but three major groups can be recognised (Hackett *et al.*, 2008). The group comprising waterbirds and cranes does not include important pollinators, whereas the other two groups contain the hummingbirds and sunbirds.

Hummingbirds (Trochilidae) belong to the Apodiformes in a clade that also includes swifts and nightjars. In the New World, hummingbirds are important pollinators, and because they hover during feeding the corresponding flowers (trochilophilous flowers) are often horizontal or pendent and lack landing platforms (Endress, 1994b).

Sunbirds (Nectariniidae) belong to the perching birds (Passeriformes, passerines) that are part of the large group comprising shorebirds and land birds. Other important passerine pollinators are honeyeaters (Meliphagidae), honeycreepers (Drepaniidae) and orioles (Icteridae). Other important pollinators among land birds are the lorikeets, which belong to the true parrots (Psittacidae). Both the passerines and the lorikeets perch while feeding and the corresponding flowers are often oriented so that they are accessible from that position.

Birds are especially important pollinators in the tropics. Flowers that are pollinated by birds (ornithophilous flowers) open during the day. They are often red in colour and lack scent. The primary reward is generally large amounts of watery nectar, and bird-pollinated flowers tend to have prominent nectaries. Typically they also have a well-protected, inferior ovary and a tubular construction, often with the stamens or stigmas protruding. Frequently, tissues at the periphery of the floral organs or in positions that could potentially be damaged by beaks or feet are robust and sclerified (Endress, 1994b). Proctor *et al.* (1996) list five general types of bird-pollinated flowers: gullet flowers, tubular flowers, brush-flowers, capitula flowers and spurred flowers. Bird-pollinated flowers occur in a wide range of angiosperms (e.g. Acanthaceae, Ericaceae, Gesneriaceae, Myrtaceae, Orchidaceae, Proteaceae, Rubiaceae), but bird pollination is not reported for any magnoliid angiosperm (Thien *et al.*, 2009; Endress, 2010b).

Molecular clock estimates for the divergence times of bird lineages suggest that modern orders of birds were already diverse by the Late Cretaceous (Pereira and Baker, 2006), but this is so far not supported by the fossil record. The earliest unequivocal bird is *Archaeopteryx lithographica* from the Late Jurassic Solnhofen Limestone of Germany. Subsequently, various fossil birds are known from the Cretaceous, and many of the most informative fossils have only been described relatively recently. Especially important have been recent discoveries from the Early Cretaceous of northeastern China, which has yielded a wealth of exquisitely preserved early birds (e.g. Zhou and F. Zhang, 2002; Zhou *et al.*, 2003; Zhou and Zhang, 2003, 2005). Currently more than 20 species are known from northeastern China, but all belong to a grade of extinct lineages below the true birds and none of the fossils can be placed in the Neornithes (Zhou and Zhang, 2005). Fossils from the Late Cretaceous resembling modern birds remain in question and according to Feduccia (2003) there are no modern birds in the Cretaceous.

In contrast to the Cretaceous, there are several bird fossils from the Cenozoic that can be assigned unequivocally to modern groups. By the Eocene almost all of the approximately 35 extant orders of birds have been recognised, with the possible exception of the most species-rich modern group, the Passeriformes, which may appear slightly later. Hummingbirds, one of the most important groups of bird pollinators, are first recorded from the Middle Eocene (Mayr, 2004). Surprisingly, this earliest record and several subsequent Palaeogene records are from Germany, France and Caucasus in the Old World (Louchart *et al.*, 2008) rather than from the New World, to which hummingbirds are restricted today.

Taken as a whole, current palaeontological data indicate an explosive diversification of bird lineages through the Early Cenozoic (Feduccia, 2003). The implication is that birds were probably not significant pollinators, or dispersal agents (Chapter 18), until after about 65 million years ago. In this context, and in the context of its scattered systematic distribution, bird pollination in angiosperm flowers is clearly a secondary specialisation that has arisen independently, presumably most often from insect-pollinated precursors, in many different lineages during the Cenozoic.

17.4.2 Mammal pollination and the fossil history of mammals

The traditional classification of extant mammals into three major groups is well supported by modern molecular phylogenetics. Monotremata (echidna, platypus; five species) are sister group to therian mammals, which comprise Marsupialia (marsupials, *c.* 334 species) and Eutheria (placental mammals; *c.* 5050 species). The most important group of mammalian pollinators are bats (Chiroptera), which are placental mammals.

Both the Old World fruit bats (megachiropteran bats) and the echolocating bats (microchiropteran bats) are important pollinators throughout the tropics. Bat pollination (chiropterophily) has been reported from more than 60 angiosperm families. It is mainly restricted to core eudicots and some families of monocots. Bat pollination is particularly common in the Agavaceae, Bignoniaceae, Bromeliaceae, Cactaceae, Campanulaceae, Fabaceae, Gesneriaceae, Malvaceae, Marcgraviaceae and Myrtaceae (Fleming *et al.*, 2009). Among magnoliids bat pollination is only recorded for one species in Annonaceae and one genus in Lauraceae (Bawa, 1990). Bat-pollinated flowers are visited at night. The primary attractant is the smell of rotting or fermenting fruit, and the rewards are mainly nectar and pollen. Bat-pollinated flowers are typically exposed and free from foliage or branches. Most frequently they are produced by trees, although some are produced by giant herbs, epiphytes or large cacti (Endress, 1994b). Like hummingbirds, some bats hover while feeding. Other bats land to feed and in these cases the flowers are large and robust. Bat-pollinated flowers are frequently white, creamy, dull greenish or brownish in colour, but the variability in morphology and colour is considerable (Bawa, 1990). Bat-pollinated flowers often open at dusk or at night and produce large quantities of pollen and nectar (Proctor *et al.*, 1996).

Pollination by mammals other than bats has only been confirmed relatively recently, but is now known to occur sporadically in several different groups of angiosperms (e.g. Acanthaceae, Clusiaceae, Combretaceae, Fabaceae, Malvaceae, Myrtaceae, Proteaceae, Strelitziaceae). These unusual pollination mechanisms, which are clearly a secondary specialisation for both the flowers and for the mammals, involve diverse animals including rodents, marsupials, lemurs, and monkeys (Endress, 1994b; Proctor *et al.*, 1996). Some flowers are visited by both birds and mammals, or by bats and non-flying mammals. These flowers are typically exposed and similar in shape and structure to other bird- or bat-pollinated flowers. Other flowers are fully adapted to pollination by non-flying mammals. Flowers are often sessile, gathered in robust inflorescences that are placed away from tips of the branches. Rodent-pollinated flowers are often placed close to the ground and are more or less concealed (Proctor *et al.*, 1996).

Stem-group mammals are first known from the Late Triassic and diversified through the Jurassic. The clade that encompasses extant Monotremata was already differentiated by the Early Cretaceous. The number and diversity of fossil mammals increases through the Cretaceous and the available information has expanded dramatically in recent years. However, most of these Cretaceous fossils either can be assigned to stem lineages of marsupials or placentals, or lack sufficient diagnostic characters for confident placement in the therian crown group (Novacek, 1999).

Diet and mode of locomotion was already diverse among the Late Jurassic – Early Cretaceous mammals, and a variety of terrestrial, arboreal and gliding forms are known from the Daohugou and Yixian formations, China (Zhou, 2004; Meng *et al.*, 2006). From the Yixian Formation Zhou (2004) recorded eight different kinds of mammal, six of them adapted to life on the ground and two adapted for a scansorial and arboreal existance.

The most abundant and diverse group of Mesozoic mammals are the multituberculates. This group of rodent-like herbivores, which are also known from the Yixian Formation, have teeth suitable for fruit and seed eating (Novacek, 1999). Multituberculates extend into the Early Cenozoic, but their phylogenetic position appears to be outside the therian clade. Although they could have been involved as pollinators in the Cretaceous, it seems more likely that multituberculates interacted with angiosperms as dispersal vectors (Chapter 18).

The overall pattern of therian evolution is therefore similar to that of birds, but the fossil record is better understood. Again, estimates based on molecular clock techniques indicate the divergence of extant orders in the mid-Cretaceous or earlier, but supporting paleontological evidence is so far lacking. The traditional view of an intense pulse of diversification of modern placental and marsupial subclasses during the Early Cenozoic apparently still holds. More specifically, the earliest records of the main lineages of mammals involved in the pollination of angiosperm flowers (bats, primates, rodents, modern groups of marsupials) are unknown until the Early Cenozoic. The rapid radiation of bats starting in the Early Eocene and onwards has been linked to increased plant and insect diversity (Teeling *et al.*, 2005). At this time there is also secure early evidence for dense tropical canopy forests, the habitat where bat pollination is most common today.

17.5 HISTORY OF POLLINATION IN ANGIOSPERMS

The history of pollination in angiosperms can be inferred independently from both the fossil record of potential animal pollinators as well as the fossil record of plants. The extent to which these inferences suggest a broadly similar picture increases confidence that the overall pattern reflects evolutionary change and perhaps broad co-evolutionary interaction. In this section we consider both lines of evidence to infer pollination in pre-angiosperm vegetation and changes in pollination biology through the Cretaceous and into the Cenozoic.

17.5.1 Pollination in pre-angiosperm vegetation

Wind pollination almost certainly predominated among those seed plants that were abundant in the pre-angiosperm Mesozoic vegetation. Based on the presence of non-saccate and saccate pollen grains Mesozoic conifers probably exhibited a similar range of pollination mechanisms to those seen in extant groups. Mesozoic *Ginkgo* was probably also wind-pollinated, and to judge from the lax, presumed pendulous, catkin-like pollen cones of *Androstrobus*, pollination in many Mesozoic cycads may also have been predominantly by wind (Harris, 1961). Some fossil cycads, like their living relatives, may have been insect-pollinated. Based on extrapolation from extant taxa and the massive occurrence on some bedding planes of ginkgoalean or cycadalean fossils, it seems likely that some Mesozoic cycads and *Ginkgo* probably grew in sufficiently large populations for pollen transfer by wind to be effective.

For those groups of Mesozoic plants that lack an obvious modern analogue it is more difficult to infer pollination. In *Caytonia* and corystosperms the presence of saccate pollen indicates the probable involvement of a pollination drop mechanism. This is also consistent with the recurved cupules in which ovules would most likely have been inverted. It is interesting that in *Caytonia* the lip and mouth of the cupule is distinctly papillate. The papillae were perhaps involved in secreting the pollination drop, or holding it in place. The presence of pollen in the micropyle of *Caytonia* ovules indicates that the pollen grains initially floated upwards in the pollination drop and were perhaps then drawn more deeply into the cupule and ultimately to the micropyles of the seeds as the pollination drop dried or was resorbed. *Caytonia* pollen is small (about 25 μm) compared with that of corystosperms and other saccate grains. It seems most likely that it was transferred by wind, but some involvement of insects cannot be ruled out.

The large pollen grains of the Elaterates (e.g. *Elaterosporites*, *Galeacornea*), which often have elaborate projections and that may have been produced by plants related to Gnetales (Chapter 5), seem unlikely to have been wind-dispersed. Early Cretaceous ephedroids, as well as the Erdtmanithecales, have smaller pollen grains and may have been insect-pollinated, as are some of their living relatives. Aggregation of presumed pollen-producing and ovule-producing organs into a bisexual structure, as occurs in the enigmatic Jurassic plant *Schweitzeria* (*Irania*) *hermaphroditica* (Chapter 6), may also imply insect pollination.

There is evidence of insects feeding on pollen from as early as the Permian (Krassilov *et al.*, 2007) and the fossil record of pollen-feeding insects is particularly rich from the Late Jurassic and Early Cretaceous. Three different species of sawfly (Symphyta, Xyelidae) were described from the Early Cretaceous Baissa locality of eastern Siberia, with masses of pollen grains in their guts (Krassilov *et al.*, 2003, 2007). Pollen grains were all non-angiospermous and included saccate forms of probable conifer affinity. All specimens studied had only a single kind of pollen in their gut, which suggests selective pollen feeding. However, one species of sawfly was represented by two specimens, each with different grains in their gut contents. Coprolites consisting almost exclusively of pollen are also common in mesofossil floras throughout the Early Cretaceous (Figure 17.3) including mesofossil floras from pre-angiosperm assemblages. Most of the coprolites from pre-angiosperm assemblages include saccate pollen of conifers and *Classopollis* (Cheirolepidiaceae). Some of these pollen types are found in great quantities in dispersed palynofloras and were most likely dispersed by wind. The insects feeding on them may merely have been exploiting a temporary food source in an opportunistic way.

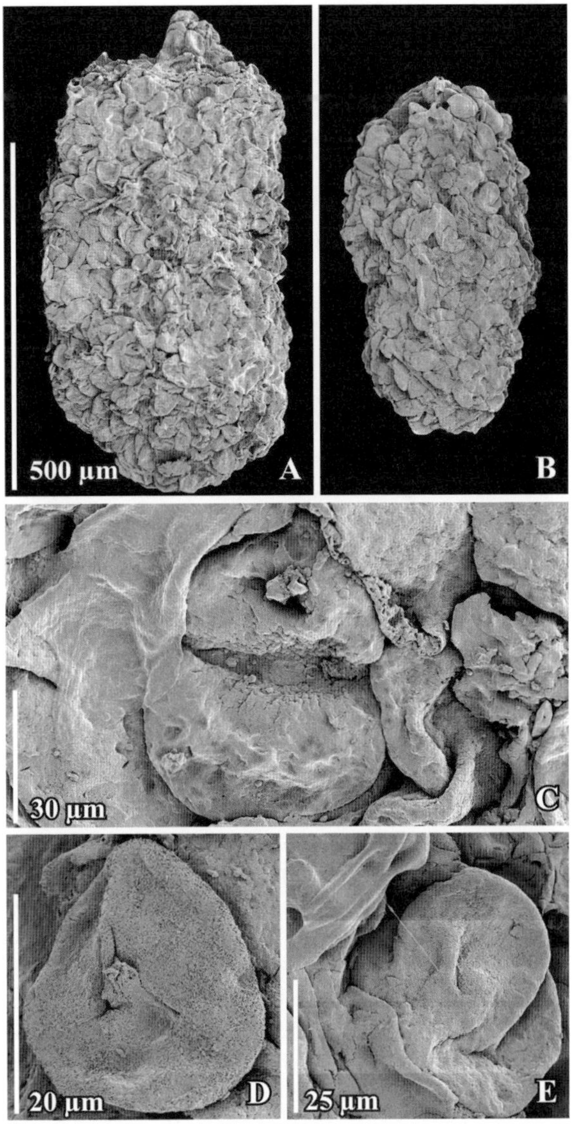

Figure 17.3 Two coprolites from the earliest Cretaceous (Valanginian) of Portugal consisting almost exclusively of gymnosperm pollen. (A) Coprolite with (C) bisaccate pollen. (B) Coprolite with (D) *Classopollis* pollen and (E) various bisaccate grains. (A) and (B): same scale bar. Material in the collections of the Swedish Museum of Natural History, Stockholm.

Insect pollination among extinct Mesozoic seed plants is suggested for the flower-like reproductive structures of the Bennettitales (Chapter 5). Different kinds of flower, probably with different kinds of pollination biology occur in the group. In Williamsoniaceae flowers were borne exposed sometimes on slender, often branched, stems. In Cycadeoidaceae flowers were borne embedded in the persistent leaf bases on stems that were generally unbranched and pachycaul. Bisexual flowers are known from both families.

Most flowers of Williamsoniaceae from the Late Triassic to the Cretaceous are unisexual. The pollen-producing structures of *Weltrichia* from the Jurassic are generally large and robust, consisting of a whorl of spreading bracts that may be free or fused into a cup for about one-third to one-half of their length. Typically they have a thick fibrous texture, suggesting that their primary function may have been protection of the developing pollen sacs, but perhaps they also played some role in attraction. In most Bennettitales the pollen sacs are borne inside bivalved synangia. This may also reflect specialisation for protection of the developing pollen grains. In *Weltrichia sol* (Figure 17.4) from the Middle Jurassic of Yorkshire the internal surface of the shallow cup is covered with numerous small sacs, each of which contains a small resinous sphere interpreted as possible remains of nectaries (Harris, 1969).

Williamsonia, the corresponding ovulate structures from the Jurassic and Cretaceous, were also large and similarly robust with thick fibrous bracts, which formed a perianth-like structure around the ovules in the centre. Considerable variation in the size, the shape of the bracts, and the form of the ovulate structure in species of *Williamsonia* may imply different pollination mechanisms. This variation is particularly marked among those Bennettitales from the Early Cretaceous Wealden flora of southern England (Figure 17.5) that have been studied by Watson and Sincock (1992). The largest of these Wealden flowers, *Williamsonia margotiana*, had a perianth-like involucre composed of twelve broadly spathulate bracts, each 8–10 cm long. When fully open, the flower was saucer-like and 16 cm or more in diameter. *Williamsonia carruthersii* and *W. cynthiae* were smaller, globular to ovoid when not fully open, and about 10–12 cm in diameter at maturity. Smaller still are the flowers of *Bennetticarpus antoinetteae*. These are cup-shaped, about 2 cm long and 3 cm in diameter, with rounded bracts apparently arranged in a dense spiral. The bennettitalean affinity of this taxon

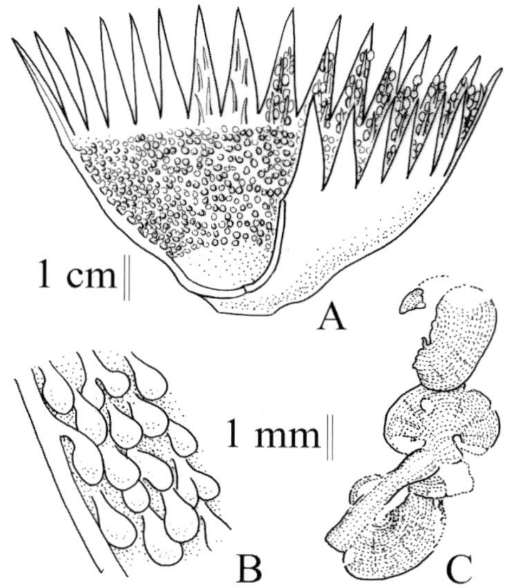

Figure 17.4 Pollen-producing reproductive structure of *Weltrichia sol* from the Middle Jurassic flora of Yorkshire, England. (A) Reconstruction of reproductive structure with fused bracts; (B) inner surface of the shallow cup, lined with small sacs containing resinous bodies that have been interpreted as the remains of nectaries; (C) a row of apparently paired synangia. Redrawn from Harris (1969).

is uncertain, but this small, robust, open flower suggests biotic pollination.

How the large flowers of Williamsoniaceae were pollinated is unknown, but specimens of *Williamsonia bryonyae* preserved at different developmental stages provide some insight into the possible floral biology of one of the Wealden species. The young flowers (Figure 17.5) are up to about 8 cm long and consist of a central ovule-bearing structure surrounded by at least ten spathulate bracts apparently arranged in a spiral. The flowers are funnel-shaped with the proximal portions of the bracts forming a narrow tube, but distally they expand to form a broader cup. Pollen is present in the micropyles of the ovules, clearly documenting that pollination occurred at this young stage when the tube was present (Watson and Sincock, 1992). Later, the flowers open and the narrow bracts were separated and widely spread (Figure 17.5).

It is interesting to speculate on how the reproductive structures of Williamsoniaceae may have been pollinated.

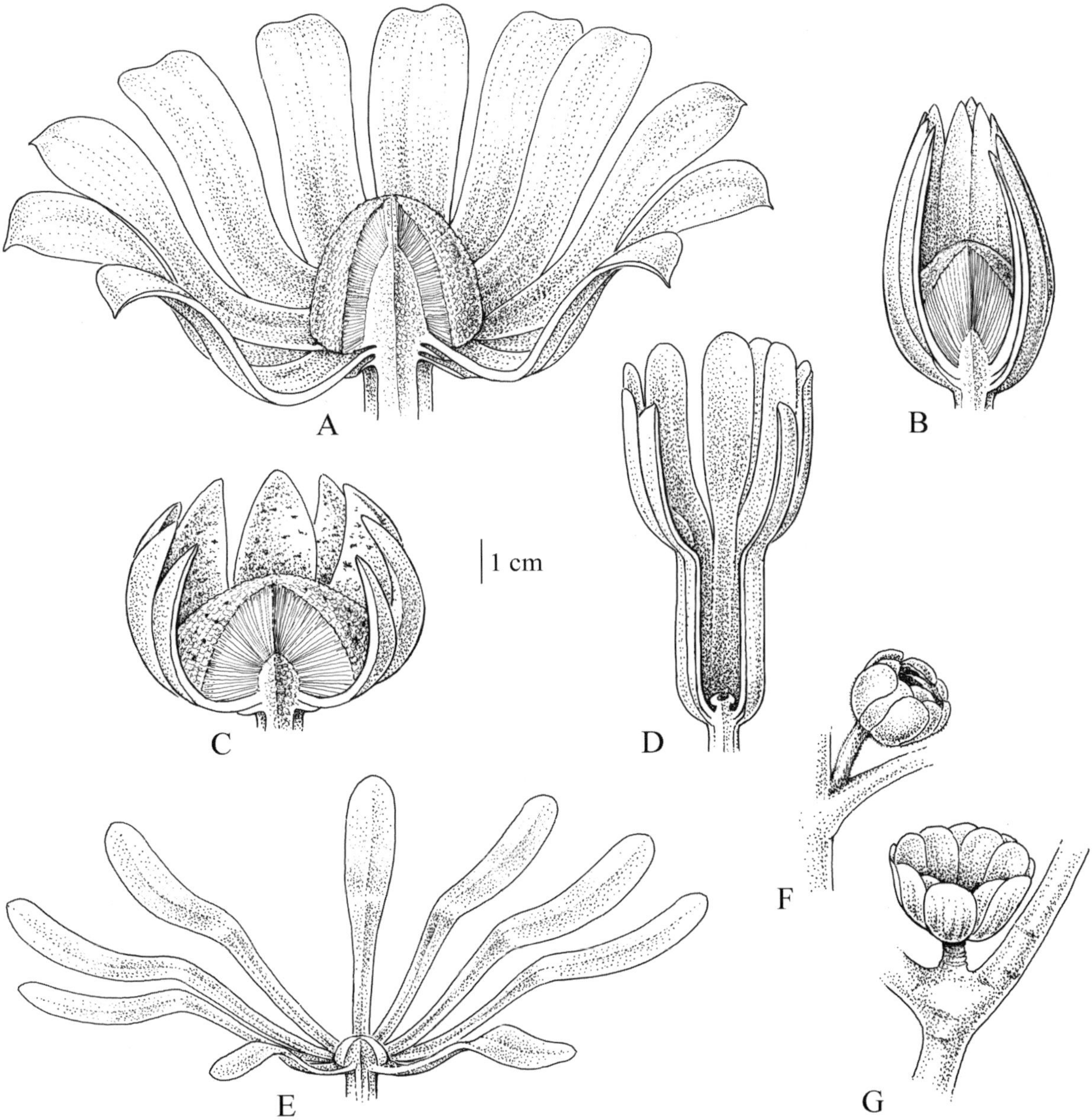

Figure 17.5 Reconstruction of reproductive structures of Bennettitales from the Wealden of southern England. (A) *Williamsonia margotiana*. (B) *Williamsonia carruthersii*. (C) *Williamsonia cynthiae*. (D, E) *Williamsonia bryonyae*, (D) early in development and (E) open. (F, G) *Bennetticarpus antoinetteae*, (F) bud and (G) open. Redrawn from Watson and Sincock (1992).

One possibility is that *Williamsonia bryonyae* and other flowers of *Williamsonia* were visually attractive and pollinated during the day. Pollination may have been by small insects that found shelter in the tube and that were perhaps also attracted by some reward or perhaps warmth generated by the ovule-bearing structures. Pollinators may have entered before the bracts had opened fully, while they still provided protection for the insects foraging inside. This may also explain the protection of the ovules beneath the resistant surface formed by the heads of the interseminal scales. This interpretation makes these flowers functionally similar to the flowers of many *Magnolia* species. As in *Magnolia*, flowers of Williamsoniaceae may have functioned as temporary traps by slight opening then closing of the bracts. An alternative possibility, especially for larger *Williamsonia* flowers, is that they were pollinated while wide open, perhaps at night, by small vertebrates (e.g. multituberculates) that visited for resin and/or pollen. This interpretation makes these flowers functionally similar to the robust inflorescences of *Protea* that are pollinated today by small mammals.

Compared with most *Williamsonia* reproductive structures, the bisexual structures of *Williamsoniella* are small, about 2 cm long, and stalked, borne exposed on slender branching stems in the axils of leaves (Zimmerman, 1933). Despite structural similarities the biology of *Williamsoniella* seems to differ from that of a typical angiosperm flower. On the whole, development of the reproductive structure and the maturation of the different organs seems less integrated (Harris, 1944; Crane and Herendeen, 2009) and the pollination biology of *Williamsoniella* is more difficult to interpret than appears at first sight. It is possible that the bracts functioned more for protection during the earliest stages of development, than for attraction at anthesis. The fleshy pollen-producing structures, which tightly enclose the developing ovulate structure in some species, may have been more important for attracting pollinators by colour or odour.

The reproductive structures of at least some Cycadeoidaceae were also clearly bisexual. They appear to have been protandrous in their development. In specimens with pollen organs preserved, the ovules and interseminal scales are not well developed. In other cases the pollen-producing structures are not well preserved or lacking, and in such cases it is difficult to be sure whether such flowers were bisexual or unisexual. They are assigned to several genera (*Amarjolia*, *Cycadeoidea*, *Monanthesia*) known mainly from permineralised material (Chapter 5). Typically they are about 2–4 cm in diameter and have distinct, densely hairy, bracts surrounding fleshy, pinnate pollen-producing organs that in some species are fused into a short cup at the base.

Given that the reproductive structures of Cycadeoidaceae were borne embedded in the leaf bases current interpretations suggest they probably never opened and that the reconstruction by Wieland (1906), showing a fully open bisexual structure with bracts and pollen organs widely spread, may not be correct. It has been suggested subsequently that the reproductive structures remained closed and were mainly self-pollinated (Delevoryas, 1963, 1965; Crepet, 1972, 1974). Permineralised specimens of *Cycadeoidea* sometimes show distinct feeding channels that may contain frass, which have been taken as evidence of burrowing insects that may have acted as pollinators, although pollen was never observed in the channels.

Monanthesia appears to have had reproductive structures produced in the axil of every leaf. All flowers matured at the same time. One spectacular specimen from the Cretaceous of Utah is almost 2.1 m tall and about 0.3 m in diameter (Watson and Sincock, 1992). Except for a zone at the base of the trunk, the entire specimen is covered with hundreds of closely packed flowers. They evidently flowered synchronously in the axils of the relatively small petiole bases. In life, such massed flowers would have been a magnet for insect or perhaps small mammal pollinators.

In summary, evidence from the pre-angiosperm fossil record indicates that although wind pollination was almost certainly widespread among earliest Cretaceous seed plants, different kinds of insect pollination were already established in the Late Triassic, Jurassic and Early Cretaceous, at least in the Bennettitales, and probably also in other groups, for example Gnetales and their relatives as well as most likely in some cycads. This is a significant difference from the situation in the Palaeozoic in which there is no clear evidence of insect pollination. Purported insect pollination in Palaeozoic medullosans, which has mainly been inferred from the large size of *Monoletes* pollen grains, is not well supported and has not yet been investigated in detail.

Likely pollinators of Jurassic and Early Cretaceous plants (Figure 17.1) could have included a wide range of beetles. For example, the Late Jurassic Karatau insect fauna from Kazakhstan contains a diverse assemblage of beetles that includes many groups of potential pollinators. Among Diptera, sawflies may also have been important given direct evidence of pollen feeding in Mesozoic fossils of the group. Various groups of Brachycera are also present at this time and could also have visited flowers of Bennettitales and the reproductive structures of other groups.

17.5.2 Pollination in Early Cretaceous angiosperms

Given the sparse evidence of angiosperms in the earliest phase of angiosperm diversification (Valanginian – Early Barremian) there is little direct evidence from which to infer their mode of pollination. No unequivocal angiosperm floral structures of this age have been described, nor

Figure 17.6 Coprolites from the Early Cretaceous of Portugal (Late Barremian – Early Aptian) containing angiosperm pollen. (A, B, F) Coprolites containing only one kind of pollen (*Pennipollis*-type). (C, H) Coprolite/pollen clump with several different kinds of pollen including *Retimonocolpites*-type, *Pennipollis*-type and *Transitoripollis*/*Tucanopollis*-type. (D, G) Coprolite containing two different kinds of pollen (*Transitoripollis*/*Tucanopollis*-type and unnamed polyporate, spiny pollen). (E, I) Coprolite/pollen clump with plant fragments and many different kinds of pollen. (A–E) Same scale bar. Material in the collections of the Swedish Museum of Natural History, Stockholm.

Figure 17.7 Watercolour illustrations of extant *Nymphaea* sp. and a reconstruction of Early Cretaceous *Monetianthus mirus* at natural size, to illustrate the difference in size between flowers of extant and fossil Nymphaeaceae.

is there evidence of coprolites containing angiosperm pollen. The dispersed angiosperm pollen grains of this age are generally small, tectate–psilate or semitectate–reticulate. The reticulum is sometimes very coarse and muri often have a fine sculpture of spines or ridges, but none of these features definitively indicates biotic or abiotic pollination.

The fossil record of angiosperm flowers and isolated floral organs, beginning around the Late Barremian – Early Aptian, provides a firmer basis from which to infer possible modes of pollination. The presence of numerous coprolites in Early Cretaceous mesofossil floras (Friis *et al.*, 2000a, 2006a) also provides some evidence of animal pollination early in angiosperm history. These coprolites are all very small, often angular in shape, and are thought to be from insects. Some have only one kind of pollen, but others have a mixture of two to many pollen types, indicating that the visiting insects were not specialising on a particular plant (Figure 17.6).

Most angiosperm flowers of this age are small. Their very small size is particularly obvious when the *Hedyosmum*-like flowers from the Early Cretaceous of Portugal are compared with those of extant *Hedyosmum* species, or *Monetianthus* is compared with flowers of extant Nymphaeaceae (Figure 17.7). This small size contrasts markedly with the much larger structures of many Jurassic and Early Cretaceous Bennettitales and suggests a different kind of pollination mechanism. Early angiosperm flowers are more similar in size to the pollen organs of extinct Erdtmanithecales and probably ephedroid plants with which early angiosperms co-occurred.

In the earliest flowers from the Late Barremian – Early Aptian, stamens typically have poorly developed sterile tissue. Several of these stamens contain pollen grains similar to those described from dispersed palynological assemblages, including species of *Asteropollis*, *Clavatipollenites*, *Pennipollis*, *Transitoripollis*/*Tucanopollis* and *Retimonocolpites*. The regular occurrence of these pollen types in depositional basins

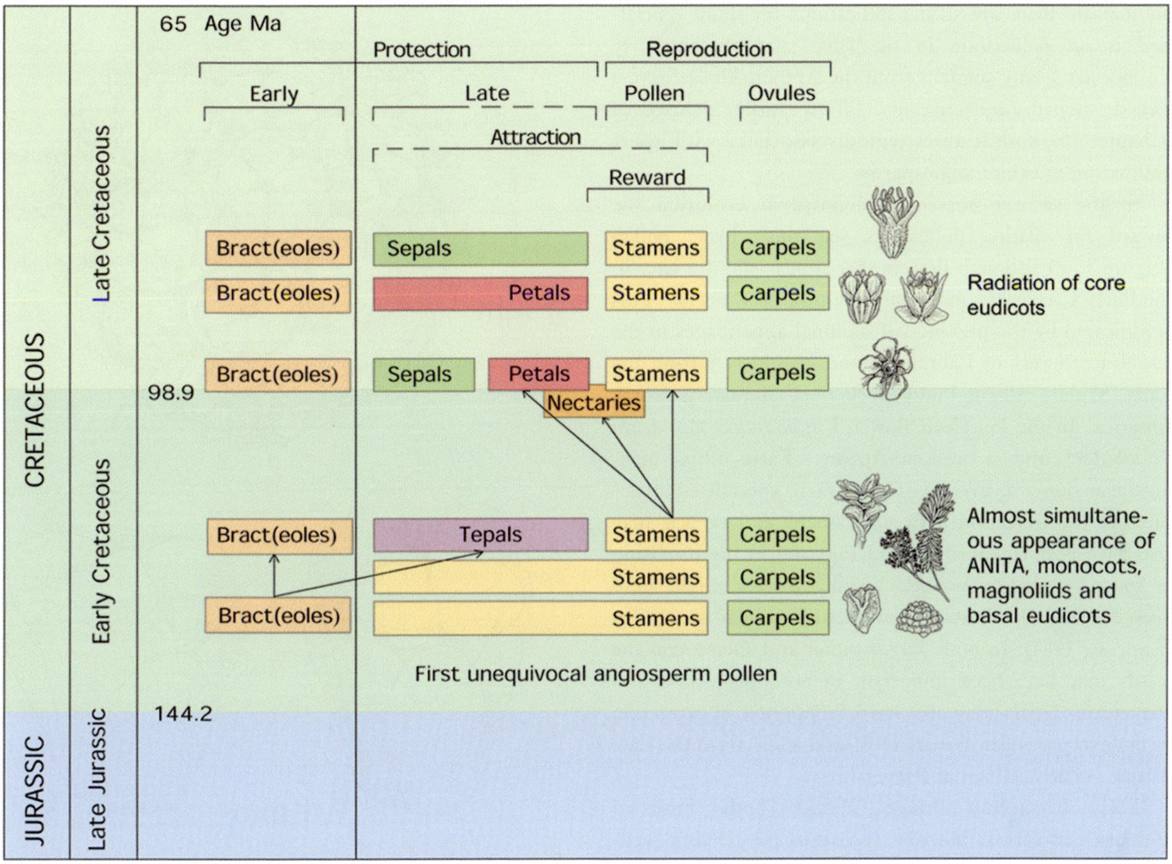

Figure 17.8 Appearance and function of angiosperm floral organs through the Cretaceous, illustrating the distinct shift, probably linked to the diversification of core eudicots, around the Early–Late Cretaceous boundary. The perianth in Early Cretaceous flowers was typically monochlamydous with only one kind of perianth parts, whereas the perianth of most Late Cretaceous flowers was heterochlamydous with clearly differentiated calyx (sepals) and corolla (petals). Some Late Cretaceous flowers, such as the flowers of the Normapolles complex also had an undifferentiated perianth resulting from loss of either calyx or corolla as an adaptation to wind pollination. From Friis *et al.* (2006a).

indicates high pollen production, perhaps associated with wind pollination. Indications of wind pollination for the *Asteropollis*-producing plants come also from pistillate and staminate flowers with *Asteropollis in situ*. Floral structures and pollen are very similar to those of extant *Hedyosmum*, which is thought to be exclusively wind-pollinated. The large extended stigmatic surface of the *Hedyosmum*-like fossils is also similar to that of modern forms and is consistent with wind pollination.

However, notwithstanding likely wind pollination in plants that produced *Asteropollis* pollen, *Clavatipollenites*, *Pennipollis*, *Transitoripollis/Tucanopollis* and *Retimonocolpites*, as well as *Asteropollis*, have also been found in coprolites, indicating that the flowers producing the pollen were occasionally visited by insects, and that insects potentially contributed to their pollination. Indirect evidence of insect pollination in Late Barremian – Early Aptian angiosperms is also provided by the finely striate pollen of the aroid fossil *Mayoa* from Torres Vedras. In extant Araceae this kind of pollen is generally associated with unspecialised pollination (Grayum, 1990, 1992), and extant Araceae with very similar pollen (*Spathiphyllum*) are visited by an array of different insects including beetles, flies and bees (Gibernau, 2003).

Although Late Barremian – Early Aptian angiosperms may have been ambophilous or with unspecialised insect pollination, there are strong indications for more specialised insect pollination in the later part of the Early Cretaceous. Many stamens from the Aptian–Albian have a well-developed connective and valvate anther dehiscence (Chapter 16), both features typically associated with insect pollination in extant angiosperms.

In the earliest phases of angiosperm evolution the reward for visiting pollinators was most likely pollen (Figure 17.8) although flowers that appear slightly later in the Early Cretaceous may also have produced nectar. This is indicated by the presence of staminal appendages in the lauralean flowers of Laurales B and Laurales A from the Early–Middle Albian Puddledock flora of eastern North America. In the lauralean flower *Virginianthus* also from Puddledock, and in the Late Aptian – Early Albian magnolialean flower *Endressinia* from Brazil, specialised structures at the transition between stamens and carpels may also have been involved in attraction and in the provision of food for visiting insects. Visiting insects might have been beetles, as in extant flowers with similar structures (Endress, 1984). In both *Virginianthus* and *Endressinia* the tepals may have been important in both protection and attraction. Tepals may also have functioned in attraction in the nymphaealean flowers of *Monetianthus* from the Late Aptian – Early Albian of Portugal.

Early Cretaceous flowers include both unisexual and bisexual forms, but few appear to have had a well-differentiated perianth. In Late Barremian – Early Aptian flowers protection and attraction were probably mostly by the associated floral bracts. In the slightly younger flowers with extensive sterile tissue in the anthers, stamens probably played an important role both in protection and in attraction (Figure 17.8). In extant taxa with stamens of this kind the expanded apical heads above the pollen sacs appear to function in both protection and attraction. The massed heads of the stamens form a protective shield over the developing pollen sacs. The expanded connectives may also be modified to attract insects by colour or odour. In some cases they even produce nectar (Endress, 1994b). We think it likely that similar mechanisms of attraction operated in many early angiosperm flowers from the Aptian–Albian. There is also evidence that the small flowers of early angiosperms were often massed together into a dense inflorescence, where they would have made a stronger visual impact (Chapter 16).

In plants with unisexual flowers and presumed insect pollination, such as the various early platanoids described from the Albian of North America, pollinators were probably attracted to the staminate flowers by the extended apical shields of the stamens and the pollen. However, no nectar or food bodies are known for the staminate or pistillate flowers and pollinators may have been attracted to the pistillate flowers by deceit. In these platanoids in particular pistillate and staminate flowers show striking similarity in gross morphology, with apical swellings of the carpels resembling the apical shields of the stamens (Figure 17.9).

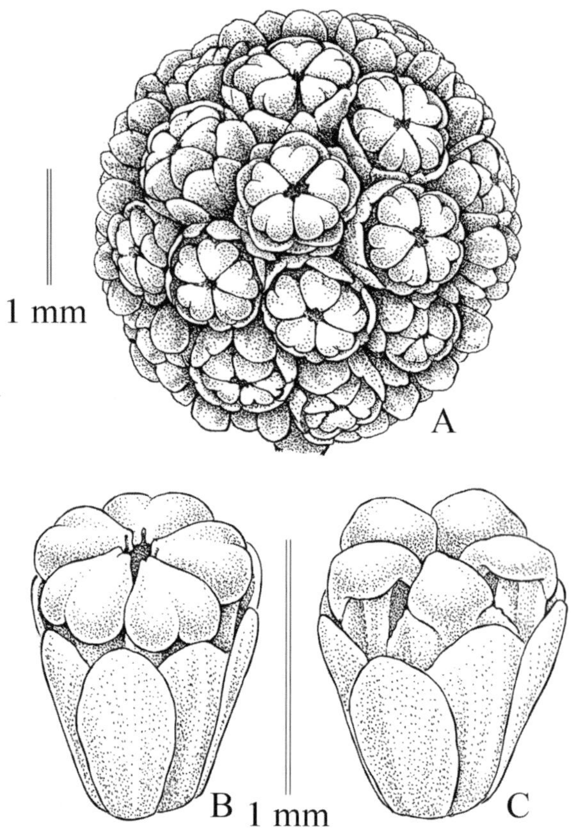

Figure 17.9 Possible mimicry of floral organs in an Early Cretaceous eudicot with unisexual flowers. Staminate flowers of *Hamatia elkneckensis* (C) probably offered pollen as reward and the swollen apical extension of the connectives may have functioned in attraction. Pistillate flowers of *Friisicarpus elkneckensis* (A, B) resemble the staminate flowers in the distinctively swollen apex of each carpel.

Aggregation of pollen into larger units, such as tetrads, polyads or pollinia or clusters adhering by pollenkitt or viscin threads, is thought to promote pollination efficiency (Harder and Johnson, 2008) and has a scattered distribution among angiosperms. It is often, but not always, linked to insect pollination. Aggregation of pollen was established early in angiosperm history as evidenced by permanent tetrads of *Walkeripollis* (possibly Winteraceae) from the Late Barremian – Early Aptian of Gabon and the Late Aptian – Early Albian of Israel, as well as several different kinds of permanent tetrads with tricolpate (eudicot) grains. Pollen clumps also occur frequently in palynological assemblages and in the mesofossil floras indicating a tendency for the pollen grains of early angiosperms to adhere together.

Potential insect pollinators during the Early Cretaceous would have included the beetles and flies that were already diverse by the Late Jurassic and earliest Cretaceous (Figure 17.1) and that may have already been involved in pollination interactions with other kinds of seed plants. In addition, other important groups of pollinators were also beginning to diversify at this time. There is direct evidence from the Early Cretaceous of further differentiation among several groups of Diptera that may include potential pollinators, for example Nemestrinidae (section 17.3.4). Among Hymenoptera, sphecid wasps and other groups of aculeates are also recognised for the first time in the Early Cretaceous (Grimaldi and Engel, 2005) and a fossil xyelid (Hymenoptera) from the Crato Formation of Brazil of Late Aptian – Early Albian age has been reported with masses of *Afropollis* in its gut (Caldas *et al.*, 1989). Early Lepidoptera with chewing mouthparts were also present and together with other fossils from the Jurassic suggest that diversification of these important pollinators was already underway (section 17.3.5).

Given the occurrence of water plants early in angiosperm history (e.g. *Archaefructus*, *Monetianthus*, *Pluricarpellatia*) the possible early occurrence of specialised pollination associated with the aquatic habit needs to be considered. In most extant aquatic angiosperms flowers are held at, or above, the water surface and this was probably also the case for *Monetianthus*. However, underwater pollination does occur in *Ceratophyllum* (Endress, 1994a) and several aquatic monocots with submerged flowering and is generally associated with a very simple floral construction. Water pollination could potentially account for the simple floral structure of *Archaefructus* (Friis *et al.*, 2003a).

17.5.3 Pollination in mid-Cretaceous angiosperms

During the latest Albian – earliest Cenomanian the fossil record of angiosperm flowers is more extensive than in the earlier phases of angiosperm radiation and provides clear evidence of more specialised insect pollination. Compression floras, for example from the Dakota Formation, provide evidence of flowers with much larger petaloid organs of attraction. Fossil tepals of *Archaeanthus* and other probable Magnoliales are relatively large. These large flowers are more comparable in size to flowers of Bennettitales and fossils of the two groups have sometimes been confused. For example, *Williamsonia problematica*, *W. recentior* and *W. elocata* have all been reinterpreted as angiosperms (Crane and Dilcher, 1984). This similarity in size and form may also reflect similarities in pollination mechanisms.

The 'Rose Creek flower', also from the Dakota Formation, is relatively large compared to earlier angiosperm flowers. It shows for the first time a perianth differentiated into a distinct calyx, presumably with a protective function during flower development, and a distinct corolla, which presumably functioned in attracting pollinators. A similar separation of functions is also seen earlier in the angiosperm fossil record, for instance in *Virginianthus* from the Puddledock flora, but here the organs are all tepals and the differentiation is much less marked. The 'Rose Creek flower' also provides evidence of nectar as a reward for insect visitors and is the earliest fossil with a probable nectary disc (Figure 17.10).

In addition to the nectaries of the 'Rose Creek flower', several other mid-Cretaceous flowers show early evidence of floral organs specialised for provision of rewards to insect visitors. In *Mauldinia* (Lauraceae) staminal glands probably functioned in a similar way to those of extant Lauraceae in the production of nectar. It is also interesting that these glands are well concealed inside the space formed by the inner tepals.

In general, by the mid-Cretaceous, the variety of angiosperm flower types had increased and the diversity of likely insect pollinators was probably also correspondingly greater (Figure 17.1). However, at this stage of angiosperm evolution there is no direct evidence of either sympetaly or zygomorphy. Nor is there indirect evidence based on the likely relationship of fossil taxa to extant forms with zygomorphic flowers. Similarly, from the insect fossil record, although there is evidence by the mid-Cretaceous for a range of potential pollinators among Coleoptera, Diptera, Hymenoptera and Thysanoptera, other pollinators that are important today are missing. There is no direct evidence of highly specialised pollinators, such as bees or glossate Lepidoptera. The only evidence of Lepidoptera with a well-developed proboscis is indirect, based on mines in leaves from the Dakota Formation at around the Lower–Upper Cretaceous boundary (Labandeira *et al.*, 1994).

Figure 17.10 Reconstructions of mid- and Late Cretaceous flowers from North America and Europe with adaptation for insect pollination. (A) *Mauldinia mirabilis* from the Mauldin Mountain locality, Maryland, USA; note staminodes and prominent nectariferous staminal appendages. (B) 'The Rose Creek flower' from Nebraska, USA; note membraneous petals and lobed nectary (C) *Scandianthus costatus* from the Åsen locality, Sweden; note prominent ten-lobed nectaries around the gynoecium. (D) Unnamed asterid flower from the Åsen locality, Sweden; note prominent five-lobed nectaries around the gynoecium. (E) *Silvianthemum suecicum* from the Åsen locality, Sweden; note slightly raised nectariferous zone in the gynoecium wall. (F) *Platydiscus peltatus* from the Åsen locality, Sweden; note prominent four-lobed nectaries around the gynoecium. (G, H) Unnamed primuloid flower from the Mira locality, Portugal with (G) and without (H) corolla; note indistinct nectariferous zone in the gynoecium wall. (I) *Divisestylus longistamineus* from the Old Crossman locality, New Jersey, USA; note indistinct nectariferous zone in the gynoecium wall.

17.5.4 Pollination in Late Cretaceous angiosperms

In the Late Cretaceous, consistent with the increasing modernisation and rapid diversification of eudicots, monocots and eumagnoliids, there is strong evidence for increasingly sophisticated pollination mechanisms, mediated by insect pollinators that were important earlier in the Cretaceous as well as by new groups of insect pollinators that were perhaps also diversifying rapidly at this time. The characteristic flowers of this Late Cretaceous diversification are generally small, radially symmetrical, and often show clear differentiation of the perianth into calyx and corolla (heterochlamydous). These flowers are mostly bisexual (Figure 17.10).

Flowers of this kind strongly imply insect pollination, and some may be similar to flowers pollinated today by relatively unspecialised flies and wasps (see also Crepet and Friis, 1987; Proctor *et al.*, 1996). Among these flowers there is also a high proportion of forms with inferior ovaries and robust styles, perhaps suggesting that relatively destructive pollinators, such as beetles, may still have been important (Crepet and Friis, 1987). Although not as common as in earlier floras, some flowers from the Late Cretaceous also have relatively robust stamens with prominent apical extensions above the pollen sacs and valvate anther dehiscence. This suggests that the importance of stamens in protection and attraction seen earlier in the Cretaceous may also have been retained in some lineages. Many Late Cretaceous flowers had well-developed and conspicuous nectaries (Figure 17.10) and nectar was clearly an important reward for Late Cretaceous pollinators. The presence in some mesofossil floras of abundant coprolites containing pollen (e.g. Lupia *et al.*, 2002) suggests that pollen continued to be an important reward for insect visitors during this time interval, but perhaps not as prominently in earlier floras.

Two key innovations that are seen first in Late Cretaceous angiosperm flowers are a sympetalous corolla and the presence of a common style, which is usually combined with syncarpy. A single style provides a common tract for pollen tube growth that may result in more efficient fertilisation of ovules in all carpels. It also provides a new arena for pollen competition. Syncarpous flowers are often insect-pollinated. Typically they also have many ovules per carpel and fruits with many seeds.

The sympetalous corolla, which is characteristic of many of the more derived groups of monocots and eudicots, is first seen in *Actinocalyx bohrii* from the Late Santonian – Early Campanian Åsen flora, Sweden. Sympetaly is characteristic of flowers with more specialised pollination by nectar-feeding insects such as butterflies and bees. The flowers of *Actinocalyx* are small and the corolla tube short, but the tube is constricted close to the opening, perhaps restricting access to only those more specialised insects that could reach the nectar at the base of the tube.

Other features indicating insect pollination among Late Cretaceous heterochlamydous flowers are the predominance of minute pollen and the frequent occurrence of pollenkitt on the surface of the pollen grains. This substance glues the grains together in clusters, securing the transport of many pollen grains from one flower to the other in a single pollination event. Pollenkitt-covered clusters of fossil pollen have been observed both *in situ* in anthers and also on the stigmatic surface of flowers.

Fossil flowers from the Deccan Intertrappean beds complement the picture that emerges from mesofossil floras by emphasising the presence of small flowers also at lower palaeolatitudes. The palm flowers preserved in the Intertrappean beds were almost certainly insect-pollinated, as were the flowers of probable Myrtales that occur with them. Several of the myrtalean flowers (*Sahnianthus parijai*, *S. dinectrianum*) have nectaries preserved and *Raoanthus* provides the first direct indication of zygomorphy in the fossil record. Zygomorphy is the essential prelude to certain kinds of structural modification in angiosperm flowers associated with many of the more specialised insect pollination mechanisms seen in both monocots and eudicots. Indirect evidence of zygomorphy in Late Cretaceous flowers also comes from the presence of seeds related to Zingiberales from Upper Santonian – Maastrichtian strata of Europe and North America.

In addition to the presence of numerous, heterochlamydous, probably insect-pollinated flowers, Late Cretaceous floras are also characterised by a great diversity of floral structures with features indicating wind pollination. Most of these belong to the Normapolles complex, a diverse group of flowers with Normapolles-type pollen *in situ*. They are all placed among the core Fagales, some close to Rhoipteleaceae and Juglandaceae, others closer to Betulaceae. These flowers typically have a reduced perianth consisting of a single whorl of sepals or petals that are minute, or the flowers may be completely naked. Most of these flowers are bisexual in contrast to their living relatives that are mostly unisexual, although unisexual and perhaps functional unisexual fossil flowers have also been identified. The stigmatic areas of these flowers are extended and the syncarpous gynoecium has one or two ovules per carpel. Fruits are typically one-seeded. The pollen grains are of small to medium size, typically oblate to peroblate, triaperturate with continuous tectum that is psilate or finely rugulate to spiny (Chapter 16). Both pollen characters and floral features indicate wind pollination. Wind pollination is also supported by the abundant occurrence of Normapolles-type pollen in many dispersed palynofloras from this time interval. The occurrence of wind pollination in the Normapolles complex and other core Fagales reported from the Late Cretaceous and Early Cenozoic is clearly secondary.

Increased evidence of a greater variety of insect-pollinated flowers in the Late Cretaceous is also consistent with an increased range of insect pollinators known from the fossil record (Figure 17.1). The earliest syrphid flies are of this age and other groups of flies probably also diversified at this time. By the end of the Cretaceous there is also the earliest evidence of bees and shortly afterwards in the Early Cenozoic there is an extensive record of glossate Lepidoptera.

17.5.5 Pollination in Cenozoic angiosperms

The variety of angiosperm families that are convincingly represented in diverse and well-studied fruit and seed floras from the Eocene, for example from the London Clay of southern England, and the Clarno Chert of Oregon, USA, provides compelling indirect evidence that a considerable range of flowers and modes of pollination had already evolved in the Early Cenozoic. However, direct fossil evidence of floral form and likely modes of pollination during the Early Cenozoic is necessary to test these assumptions.

Most fossil flowers from the Cenozoic are known from a relatively small number of important occurrences. Flowers from the Eocene–Oligocene Baltic amber have been known for more than a hundred years (e.g. Conwentz, 1886) and discoveries of amber from other Cenozoic sites such as the Oligocene–Miocene Dominican and Mexican amber have also produced well-preserved and informative material (e.g. Dilcher *et al.*, 1992; Poinar, 2002; Castañeda-Posadas and Cevallos-Ferriz, 2007). Another important source of Cenozoic flowers is the Eocene Princeton chert of British Columbia, Canada (e.g. Stockey and Pigg, 1991).

Insights into floral form in Cenozoic angiosperms also come from floras preserved as compressions or impressions where the flowers are usually preserved together with leaf fossils. Among the important macrofossil floras with well-preserved flowers from North America are rich assemblages from the Mississippi Embayment in Kentucky and Tennessee, USA (e.g. Crepet *et al.*, 1974, 1975; Crepet and Dilcher, 1977; Crepet, 1979; Crepet and Taylor, 1986), and well-preserved flowers from the Eocene Green River Formation of Wyoming, USA (e.g. MacGinitie, 1969; Crepet, 1984). From Europe the most informative Early Cenozoic floral fossils are from the Middle Eocene of Messel, Germany (Schaarschmidt, 1984; Schaarschmidt and Wilde, 1986).

Cenozoic flowers show a much greater range in form than is seen in the Cretaceous. Smaller flowers are often arranged in dense inflorescences and many of the larger flowers are solitary. Prominent in the Early Cenozoic assemblages are flowers that are obviously adapted for specialised insect pollination. These include forms with narrow funnel-shaped sympetallous corolla tubes, which may be related to extant Rubiaceae, as well as several kinds of flowers in which the corolla tube is shorter and wider (Friis and Crepet, 1987). Also present in these Early Cenozoic assemblages are zygomorphic flowers with pronounced bilateral symmetry, including one that is clearly referable to extant Fabaceae (Papilionoideae), and that has sculpted wing petals, which are associated with bee pollination (Friis and Crepet, 1987; Crepet and Herendeen, 1992). Flowers possibly related to extant Zingiberaceae provide further evidence of bilateral symmetry; other papilionoid flowers show connate stamen filaments, which are often associated with zygomorphy in legumes (Crepet and Friis, 1987; Friis and Crepet, 1987).

Evidence of zygomorphy, the presence of well-developed corolla tubes and connation of parts all provide evidence of increased floral synorganisation. They suggest new levels of specialisation for insect pollination beyond what is known from the Cretaceous, most probably by bees, butterflies and moths. Brush blossom inflorescences, placed securely in extant Fabaceae (Mimosoideae, *Eomimosoidea plumosa*) (Daghlian *et al.*, 1980) are also present. These inflorescences, composed of many small flowers with sympetalous corollas, strongly exserted stamens, and tubular stigmas, are also consistent with pollination by bees and Lepidoptera (Crepet and Friis, 1987).

Fossil inflorescences of Euphorbiaceae, known from the Eocene of southeastern North America, provide further indirect evidence of probable bee pollination (Crepet and Friis, 1987). Staminate inflorescences referable to tribe Hippomaneae (Crepet and Daghlian, 1982) as well as pseudanthia referable to tribe Euphorbiaceae (Crepet and Friis, 1987) are both suggestive of bee pollination, which is widespread in extant Euphorbiaceae.

Bee pollination is also inferred for the fossil flowers of *Eoglandulosa warmanensis* assigned to Malpighiaceae from the Early Cenozoic (Taylor and Crepet, 1987). These flowers have five sepals, five clawed petals and ten stamens in two groups of five. The sepals have large, paired abaxial structures that appear to be the remains of oil glands (elaiophores, Taylor and Crepet, 1987). Flowers of extant Malpighiaceae have a close relationship with three tribes of anthophorid bees (Crepet and Friis, 1987). These bees collect oil from the elaiophores by reaching around the clawed bases of the petals to access the sepal glands (Crepet and Friis, 1987). This pollination syndrome is widespread in most Malpighiaceae, which, like their anthophorid bee pollinators, are confined mainly to the New World tropics. In the few Old World Malpighiaceae, in the absence of the appropriate pollinators, the sepal glands are reduced or apparently absent (Crepet and Friis, 1987).

Taken together, the picture that emerges from the rich representation of extant families of angiosperms known from Eocene fossil floras, as well as from the direct evidence of fossil flowers, is that by the Early Cenozoic, interactions between angiosperms and their pollinators had increased substantially in their variety and sophistication over that seen in the Cretaceous. These interactions were becoming fundamentally more modern. Over the succeeding 50–60 million years, effectively the second half of angiosperm evolutionary history, that process of modernisation undoubtedly continued, sometimes involving tight or diffuse co-evolution, but ultimately resulting in the vast diversity of pollination interactions that occur in angiosperms today.

17.6 LARGE-SCALE TRENDS IN THE HISTORY OF ANGIOSPERM POLLINATION

Optimisation of pollination modes on the phylogeny of extant angiosperm based on molecular data (Hu *et al.*, 2008a) suggests that insect pollination is basic among angiosperms. However, the situation may be more complex than it appears at first sight. Pollination among extant ANITA-grade angiosperms is diverse. Insect pollination predominates, but some taxa, including *Amborella*, are generalists that are pollinated by both wind and insects. This is also recorded for *Brasenia* (Cabombaceae), and some species of *Trimenia* (Trimeniaceae) (Thien *et al.*, 2009). Therefore, in attempting to infer how the earliest angiosperms may have been pollinated, focusing on a strict dichotomy between insect and wind pollination may be misleading.

Pollination among extant groups of angiosperms at the ANITA grade may be mediated by insects, wind or water. Insect pollinators are mainly Coleoptera and Diptera with pollination by Hymenoptera also reported. Bees are major pollinators for several Nymphaeaceae (Thien *et al.*, 2009). Thrip pollination is reported for Chloranthaceae (Luo and Li, 1999) with wind pollination occuring in *Ascarina* and *Hedyosmum* (Chloranthaceae). *Ceratophyllum* and *Euryale* stand out in being water-pollinated.

At the next level of angiosperm evolution, the earliest-diverging lineages of monocots (e.g. Alismatales) and eudicots (e.g. *Nelumbo*, *Platanus*, *Euptelea*) also exhibit great diversity in their modes of pollination, but above this level insect pollination predominates. In these groups entomophily is widespread and variously developed, but there are also some large and important clades in which wind pollination is the norm.

Insect pollination predominates among extant eumagnoliids. Wind pollination is reported only in Lactoridaceae and Saururaceae. Pollination by beetles is especially widespread in this group, with Diptera, Hymenoptera and Thysanoptera also being important. The apparent absence of Lepidoptera from among the pollinators of ANITA-grade angiosperms and eumagnoliids is especially notable. Winteraceae are the only exception, but in this case pollination is by micropterygid moths with chewing mouthparts (Thien *et al.*, 1985). Their involvement with pollination in extant Winteraceae, a lineage that has a fossil history that extends back about 125 million years, may provide a remarkable example of the constancy of flower-pollinator interactions over long periods of time. Whether such long-term fidelity of flower–pollinator interactions is the norm or the exception, in different groups of angiosperms, would be interesting to investigate.

The earliest fossil floras that provide direct evidence of the flowers of early angiosperms suggest that both insect

(Nymphaeales, Araceae) and wind (*Hedyosmum*, Chloranthaceae) pollination developed very early in the evolution of the group. However, this early insect pollination appears relatively unspecialised. It is especially interesting that there is evidence of wind pollination in the earliest floras and it is possible that at least some of the earliest angiosperms were generalists (ambophilous), predominately wind-pollinated, but occasionally visited and pollinated by beetles and flies. There may also have been water pollination, for example in *Archaefructus* very early in angiosperm evolution.

It is also noteworthy that the predominant insect pollinators of ANITA-grade extant angiosperms are beetles and flies, two groups of holometabolous insects that have a well-established fossil history that extends back to the early Mesozoic, well before the first appearance of angiosperms in the Early Cretaceous. Given the antiquity of these insect groups, as well as the antiquity of subgroups of beetles and flies that are important pollinators today, it seems very likely that insect pollination developed first in non-angiosperm seed plants such as Bennettitales, Gnetales and perhaps some cycads. In addition to developing new interactions with insect pollinators early angiosperms probably also co-opted pollinators from pre-existing pollinator relationships that were already well established.

Striking features of the fossil record of angiosperm floral structures around the Barremian–Aptian boundary are their very small size, the relatively poor development of the perianth, and the very small stamens with only little sterile tissue between and above the pollen sacs. Later in the Early Cretaceous (Aptian–Albian) many stamens have relatively robust anther connectives and often prominent sterile extension above the pollen sacs. These features suggest that some protection, and perhaps some increase in visual prominence, may have been achieved by the aggregation of these features into larger units that perhaps consist of numerous tightly packed flowers. Direct evidence of pollen feeding is also provided by the numerous coprolites that occur in mesofossil floras of this age. Features suggesting insect pollination, such as valvate pollen dehiscence and well-developed connectives, which are common in Late Aptian – Middle Albian flowers, also occur in slightly younger angiosperm flowers, and beetle pollination is indicated by the glandular staminodes of *Endressinia*.

Flowers generally remained small through the mid-Cretaceous, but in fossil floras from the mid-Cretaceous larger flowers with a more prominently developed perianth are encountered for the first time. In addition to the prominent thin petals seen in the 'Rose Creek flower', which were presumably attractive; the sepals were also well developed and leathery suggesting a protective function. Also in the mid-Cretaceous the first appearance of new structural types among angiosperm flowers may indicate new kinds of relationships with pollinators, including some perhaps mimicking pollination in Bennettitales. For example, the large solitary flowers of *Archaeanthus* are similar to large-flowered Bennettitales and there is direct evidence that the flower was protected in bud, as in extant Magnoliaceae by a leathery floral bud-scale, probably of foliar origin. These large magnolialean flowers raise the possibility of direct competition for pollinators between ancient Bennettitales and a newly evolved group of eumagnoliid angiosperms.

During the Late Cretaceous there is further diversification in floral form accompanying the rapid diversification of core eudicots and perhaps also core monocots. Most flowers are still small, perhaps implying aggregation into compact inflorescences, but there is also increased evidence of specialisation for pollination. Especially important is the first appearance of sympetaly in *Actinocalyx bohrii*. This marks the beginning of a new kind of pollination mechanism in which access to the flower is restricted in a new way. Further development of the corolla tube was a key step towards flowers in which physical positioning of insects within the flower is more tightly controlled. This in turn created opportunities for more precise pollen deposition, possibilities for new kinds of floral mechanism (e.g. heterostyly) and a basis for isolating mechanisms that could have promoted speciation. The opportunities created by sympetaly have been exploited most fully by flowers that are pollinated by groups of insects that are nested relatively high up within insect phylogeny and that also appear later in the fossil record: most notably, bees, butterflies and moths. The same is also true of those groups of angiosperms with which they are most closely associated. In the context of angiosperm evolution as whole specialised interactions with these kinds of pollinators, which have undoubtedly contributed to the diversification of many groups of angiosperms, are clearly secondary.

During the Cenozoic there is direct palaeobotanical evidence for more pronounced zygomorphy, as well as clear evidence of specialised wind pollination in both monocots (e.g. grasses, Crepet and Feldman, 1991) and eudicots (e.g. Betulaceae, Crane, 1981; Juglandaceae, Manchester, 1987a). There is also evidence of more specialised entomophily, for example in *Eoglandulosa*, which shows specialisation for pollination by anthophorid bees. The more specialised flowers that are seen first in the Early Cenozoic coincide broadly with the first substantial evidence for Lepidoptera and bees in the paleontological record. It seems very likely that the close interactions between bees, butterflies and moths and the flowers of angiosperms are primarily phenomena that developed during the Cenozoic.

The same conclusion probably also applies to pollination by vertebrates. Among birds the important groups of pollinators, hummingbirds and sunbirds, both occupy relatively derived positions in bird phylogeny. Hummingbirds are first recorded in the fossil record in the Middle Eocene (Mayr, 2004). Passerine birds, the group to which sunbirds and several other important pollinators belong, are first recorded from the Oligocene of Europe (Mayr and Manegold, 2004). In mammals, groups that have been implicated as potential pollinators of angiosperms occupy scattered positions in mammal phylogeny and the same appears true of the plants involved. In all cases, these unusual pollination interactions seem likely to reflect sporadic opportunism that sometimes resulted in secondary adaptation for both the animals and plants involved. In terms of major patterns of floral specialisation in angiosperm vertebrate pollination is of relatively minor significance, but it nevertheless demonstrates the great breadth of plant–animal interactions that appear to have been advantageous in facilitating this crucial phase in the angiosperm lifecycle.

18

History and evolution of dispersal in angiosperms

In addition to great diversity in floral structure and modes of pollination, extant angiosperms also exhibit astonishing variety in fruit and seed morphology. This variety reflects a great range of different modes of dispersal as well as other aspects of dispersal biology. Dispersal and establishment is a key phase in the plant life cycle. Dispersal allows new populations to be established and new habitats to be colonised. It therefore has an important influence on the structure of plant populations. The evolution of angiosperm dispersal modes may have had important consequences for large-scale patterns of angiosperm evolution.

As with pollination, dispersal in angiosperms and other seed plants may be abiotic or may involve interactions with animals. However, in contrast to pollination, where pollen is carried from flower to flower, in dispersal the destination for the propagule is generally much less tightly constrained (Wheelwright and Orians, 1982). The potential reward for the disperser is also available only at the outset, rather than also on completion. The evolutionary dynamic is therefore different. As a result, co-evolution between plants and their dispersers may be much less specific than is often the case for pollination (Wheelwright and Orians, 1982). Co-evolution is more likely to be 'loose' rather than 'tight' (Herrera, 1985; Fleming, 1991).

In this chapter we provide an overview of the fossil history of angiosperm fruits and seeds in the context of dispersal in extant seed plants and the fossil history of extant dispersal vectors. We also consider the potential connections between large-scale changes in angiosperm dispersal biology and possible changes in the nature of Cretaceous and Early Cenozoic vegetation.

18.1 DISPERSAL IN EXTANT NON-ANGIOSPERM SEED PLANTS

Dispersal of vegetative (asexually produced) propagules, which results in the formation of clones and the spread of genetically identical individuals, is common in many groups of land plants from liverworts to angiosperms. However, from the standpoint of establishing new, genetically distinct and genetically variable populations, which comprise the raw material of evolution, the dispersal of propagules that result from sexual reproduction is more important.

In seed plants, the gametophyte phase of the life cycle is reduced and telescoped to be almost totally reliant on the sporophyte. The ovule is an integumented, indehiscent sporangium containing a single functional megaspore within which the megagametophyte develops. Pollination, at least as seen in extant seed plants, reduces reliance on water for fertilisation, and if fertilisation and embryogenesis are successful the mature ovule (seed) contains an embryo that will ultimately germinate to produce a new sporophyte. Usually the seed is retained on the parent plant until embryogenesis is complete. As a result, the developing gametophyte and embryo can be provisioned in large part from maternal resources. Dispersal usually occurs when the embryo is mature. The unit of dispersal may be just the seed, or a seed with accessory structures, such as an aril or surrounding fleshy bracts. In angiosperms, the unit of dispersal is frequently a fruit, with one or several seeds inside it.

18.1.1 Dispersal in cycads

In addition to haphazard dispersal, for example through chance physical disturbance or gravity, extant cycads are dispersed by animals and by water. Animals are attracted to the mature seeds by the brightly coloured, fleshy sarcotesta (often glossy red, orange or yellow), which also contains starch-rich cells that provide a food reward (Jones, 2002). In some cycads (*Bowenia*) the sarcotesta has a distinctive (unpleasant) odour. In several species prominent coloration of the seeds is supplemented by brightly coloured cones (e.g. some species of *Encephalartos*).

A variety of animals feed on the ovulate cones and seeds of extant cycads, and in the process distribute the seeds. Generally only the sarcotesta is consumed and the hard sclerotesta, containing the rest of the seed and the embryo, is discarded. Mammals known to feed on cycad seeds include rodents, peccaries, bats, bears, monkeys and, in Australia, kangaroos and wallabies. Large birds, such as parrots, cockatoos, crows, mockingbirds and hornbills are also attracted to cycad seeds (Jones, 2002). Less commonly, whole cycad seeds may be consumed and are voided with the sarcotesta having been digested away. Elephants have been recorded to eat the mature ovulate cones of *Encephalartos poggei*; crows, emus and cassowaries are also known to swallow cycad seeds whole (Jones, 2002).

Three species of *Cycas* (*C. circinalis*, *C. rumphii*, *C. thouarsii*), which occur mainly in coastal or near-coastal forests, appear to be dispersed by flotation in seawater (Jones, 2002). *Cycas rumphii* often grows on stabilised dunes (Jones, 2002). The seeds of these three species float when viable, apparently because of a layer of spongy tissue inside the sarcotesta. Viable seeds of the other species sink. Seeds of *C. circinalis*, *C. rumphii* and *C. thouarsii* are also larger than those of other *Cycas* species and, unlike some other species of *Cycas*, are generally not brightly coloured.

18.1.2 Dispersal in *Ginkgo*

Seeds of *Ginkgo* are similar to those of cycads in their basic construction. There is a soft fleshy sarcotesta, which has a strong, unpleasant odour at maturity, and a hard sclerotesta that protects the gametophyte and developing embryo. In China, dispersal of *Ginkgo* seeds by masked palm civets (*Paguma larvata*) and the leopard cat (*Felis bengalensis*) has been reported; in Japan, raccoon dogs (*Nyctereutes procyonoides*) have been known to consume *Ginkgo* seeds (Rothwell and Holt, 1997; Del Tredici, 2007). Seeds that have passed through these animals lack the sarcotesta, and germinate in the droppings to produce a clump of *Ginkgo* seedlings (Rothwell and Holt, 1997). Experiments have shown that the removal of the sarcotesta enhances germination and that, in addition to its likely involvement in dispersal, the sarcotesta may also contribute to seed dormancy (Rothwell and Holt, 1997).

Unfortunately, nothing is known about dispersal of *Ginkgo* in its likely native habitat in China. Tiffney (1984) has suggested that the 'fleshy and odoriferous seed seems well adapted to attracting reptiles.' He also suggested that the subsequent loss of dispersal vectors impaired the ability of *Ginkgo* to re-expand its range after it was restricted by the glacial episodes of the Pleistocene (Tiffney, 1984). Del Tredici (1989) has suggested that extinct carnivorous families (Viverravidae, related to cats, and Miacidae, related to dogs) were perhaps also involved in *Ginkgo* dispersal during the Early Cenozoic.

18.1.3 Dispersal in conifers

Conifers are sometimes regarded as 'one of the world's migrationally sedentary groups' (Page, 1990) perhaps because animal dispersal occurs in relatively few species. The suggested correlation between fleshy propagules and dioecy in conifers (Givnish, 1980) may to a large extent reflect phylogenetic contingency rather than a functional relationship (Donoghue, 1989).

Most conifers have wind-dispersed seeds that are shed separately from the ovuliferous scale. Often the seeds have well-developed wings (e.g. *Agathis*, most Pinaceae), but in many genera the wings are narrow and poorly developed (e.g. *Sciadopitys*, many Cupressaceae). In some species of *Pinus* the cones only open after they have been burnt. In *Araucaria* there is one seed per cone scale, and both cone scale and seed are shed as a single unit at maturity.

Animal dispersal in conifers occurs in Cephalotaxaceae, *Juniperus* (Cupressaceae), most Podocarpaceae and Taxaceae as well as some Pinaceae. Fleshy propagules are formed by modification of different structures around the seeds. In Taxaceae the mature (poisonous) seed is surrounded wholly or in part by a sweet-tasting fleshy aril. Taxaceae are presumed to be bird-dispersed. In *Pseudotaxus* the aril is greenish-white at maturity, in *Taxus* it is bright red, and in *Austrotaxus* and *Torreya* the aril is purple-green or blue-black (Page, 1990). *Amentotaxus* (Cephalotaxaceae) also has a blue-black aril that envelops the seed almost completely. In *Cephalotaxus* (Cephalotaxaceae) there is no aril, but the seed has a fleshy sarcotesta, whereas in *Juniperus* (Cupressaceae) the cone scales surrounding the seeds are fleshy and coalescent. *Juniperus* is presumed to be bird-dispersed.

The Podocarpaceae exhibit a variety of specialisations in which the seeds are associated with fleshy and often brightly coloured parts. Based on phylogenetic analyses of combined molecular and morphological data the presence of a large fleshy epimatium (false aril, modified cone scale) surrounding the seed is apparently basic within the family, although this has been lost or reduced in *Dacrydium*, *Lagarostrobos* and *Microstrobos* (Kelch, 1998). An epimatium is also lacking in *Phyllocladus* although in this case the evolutionary interpretation is equivocal. In several genera of Podocarpaceae (e.g. *Podocarpus*), in addition to the epimatium, the 'receptacle' below the seed, together with the attached bract bases, is swollen into a fleshy, warty, structure, which is often brightly coloured. In *Phyllocladus* there is no epimatium, but the seed is black and surrounded at the base by a white aril. All of these modifications are presumed to be associated with bird dispersal (Wardle, 1969), especially in New Zealand, where mammals are not part of the native fauna.

18.1.4 Dispersal in Gnetales

Modifications for both wind and animal dispersal occur in extant Gnetales. Seeds of *Ephedra* have two layers around the nucellus: a thin inner integument and a thicker, outer envelope that includes a hard, sclerenchymatous layer that is functionally equivalent to a sclerotesta. In some species of *Ephedra* the bracts associated with the seed, outside the envelope, are swollen, fleshy and often brightly coloured (red or yellow). The seeds, together with their associated bracts, are typically dispersed by birds (Kubitzki, 1990b; Hollander and Vander Wall, 2009). In other species of *Ephedra* the bracts form membranous, keeled wings and these are mostly dispersed by wind (Kubitzki, 1990b; Hollander and Vander Wall, 2009). Hollander and Vander Wall (2009) observed rodent dispersal for both types of *Ephedra* seeds in North America. Seeds of *Welwitschia* show similar but even more marked specialisation for wind dispersal (Kubitzki, 1990c). *Welwitschia* has broad membranous wings formed by a pair of inner bracts that enclose the seed. These bracts are equivalent to those that form the envelope in seeds of *Ephedra* (Chapter 5).

Seeds of *Gnetum* are modified for animal dispersal, with the outer of the three layers that surrounds the nucellus becoming fleshy and brightly coloured (red, pink, yellow). The middle layer, equivalent to the seed envelope of *Ephedra*, forms a hard sclerotesta-like layer. Possible dispersers of *Gnetum* seeds in the New World include toucans and monkeys; those from Asia include larger birds and civets (Kubitzki, 1990a). Water dispersal, and dispersal by fish, has been observed for *Gnetum* species that grow along rivers; seeds of these species are less brightly coloured (Ridley, 1930; Kubitzki, 1990a).

18.2 DISPERSAL IN EXTANT ANGIOSPERMS

Consistent enclosure of seeds by the carpel seems to have permitted structural diversification of the dispersal unit among angiosperms beyond what has been possible in most non-angiosperm seed plants. However, despite the great structural diversity of angiosperm fruits and seeds, their modes of dispersal can be broadly classified into six groups (van der Pijl, 1982; Willson *et al.*, 1990). Wind-dispersed (anemochorous) fruits and seeds generally have wings, plumes or hairs that help slow their rate of fall, thereby increasing the probability of dispersal by air currents. Vertebrate-dispersed fruits and seeds that are consumed by the animal (endozoochorous), either entirely or in part, generally have a fleshy, often brightly coloured pulp that attracts vertebrate frugivores. Ant-dispersed (myrmecochorous) seeds generally have a food body (elaiosome) that is associated with consumption by ants. Ballistic-dispersed (ballochorous) fruits and seeds are generally expelled explosively from a specialised fruit. Externally dispersed (epizoochorous) fruits and seeds generally have hooks, burrs, hairs or other devices that aid attachment to the fur or feathers of vertebrates. Spines and hooks also characterise some fruits and seeds of aquatic plants that are mainly water-dispersed. In these cases the spines and hooks may act as anchors (Collinson and Van Bergen, 2004). The sixth category includes those fruits and seeds that appear to have no special structural modification associated with a particular mode of dispersal (i.e., unassisted). Modes of dispersal that fall outside these six main groups (e.g. water-dispersed) are relatively rare.

The two most common modes of dispersal of angiosperms are by wind and by vertebrates (Willson *et al.*, 1990). Fruits and seeds that are apparently unspecialised with regard to their mode of dispersal are also common. In some cases this category has been used in a way that may include other known dispersal modes (e.g. scatterhoarding by vertebrates), but the mode of dispersal for many angiosperms is uncertain and cannot be inferred simply from morphology (Willson *et al.*, 1990). Some fruits and seeds may also have more than one mode of dispersal. In some cases different modes of dispersal may also operate sequentially (e.g. ballistic dispersal followed by ant dispersal, Willson *et al.*, 1990).

The frequency of dispersal by vertebrates generally increases as vegetation becomes more diverse and structurally complex (i.e. as herbaceous vegetation changes to tree-dominated communities). The percentage of species that apparently have no specialised dispersal mechanism decreases along the same gradient (Willson *et al.*, 1990). These trends may also apply to the progression through a successional sequence (Foster *et al.*, 1986; Willson *et al.*, 1990). Similarly, ballistically dispersed and externally

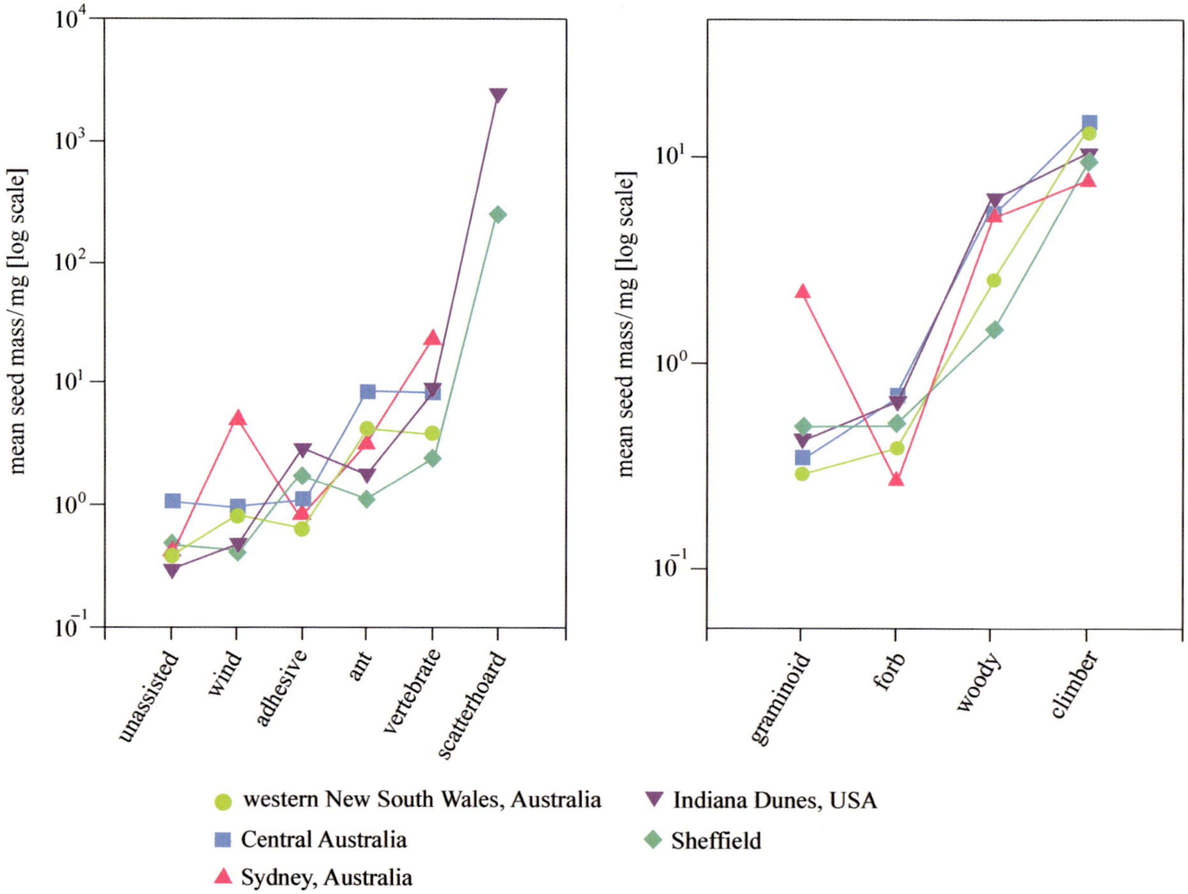

Figure 18.1 The relationship between seed mass, dispersal systems and growth form in extant floras from the temperate region in Australia, Europe and North America. Based on Westoby *et al.* (1996).

transported fruits and seeds occur more frequently in plants of small stature in the ground layer of the vegetation.

A key factor linked to these large-scale trends appears to be seed size (Figure 18.1). In general, seeds with a mass greater than 100 mg tend to be adapted to dispersal by vertebrates, whereas seeds with a mass of less than 0.1 mg tend to have an unspecialised (unassisted) mode of dispersal. Between 0.1 and 100 mg all dispersal modes are possible (Hughes *et al.*, 1994; Westoby *et al.*, 1996). Similarly, woody plants and climbers have an average seed mass about one order of magnitude larger than the seeds of graminoids and other herbaceous plants (forbs) (Leishman *et al.*, 1995; Westoby *et al.*, 1996).

Experiments show that seed size affects survivorship of seedlings subjected to deep shade, drought, physical damage and competition from other plants in the vegetation (Westoby *et al.*, 1996). This advantage may derive from a 'reserve effect', whereby during germination and early growth larger-seeded species have more uncommitted seed reserves that are available to support respiration or to repair damage. Seed size is also strongly correlated with a combined measure of seed number and the vegetative mass supporting seed production (Shipley and Dion, 1992).

18.3 ANIMAL DISPERSERS

As with pollination, both insects and vertebrates participate in the dispersal mechanisms of angiosperms, but at this phase of the angiosperm life cycle it is vertebrates rather than insects that are the more important.

18.3.1 Insects

Although a great range of insects are important in the pollination of angiosperms (Chapter 17) their involvement in the dispersal of fruits and seeds is much more limited. The only group of insects to play a substantial role in dispersal in angiosperms is ants. Ants are commonly involved in the transport of seeds from a wide range of angiosperm herbs. Ant dispersal (myrmecochory) is especially well developed in the temperate and Mediterranean climates where many herbaceous plants have an elaiosome (oil body) on the exterior of the seed that appears to enhance ant dispersal (Forest *et al.*, 2007).

In the fossil record, the earliest ants are known from the Late Cretaceous (Turonian) of New Jersey (Agosti *et al.*, 1998). This early fossil is referred to a group with well-developed social behaviour. Ants may therefore have been involved in the dispersal of angiosperm propagules for as much as the past 90 million years.

18.3.2 Vertebrates

A variety of vertebrates, ranging from fish to primates, contribute to the dispersal of angiosperm fruits and seeds. Birds and mammals are especially important and both play a much more significant role in angiosperm dispersal than in angiosperm pollination (Chapter 17). Dispersal of angiosperm propagules by fish (ichthyochory) and by reptiles (saurochory) also occurs, but is much less common.

Fish are known to be important dispersal vectors in the nutrient-poor water of seasonally inundated rain forest in the Amazon Basin (e.g. Gottsberger, 1978; Goulding, 1983). There are also instances of fish dispersal reported from temperate regions (Pollux *et al.*, 2007). It is possible that fish have served as relatively generalised dispersal agents of swamp and riparian plants since the Palaeozoic (Tiffney, 1984). However, the sarcopterygians (lobe-finned fishes) that predominated in the Palaeozoic today comprise only a small number of species-poor lineages that are not known to be involved in fruit or seed dispersal. The fish that are known to be involved in angiosperm dispersal today are bony fishes (osteichthyes), and specifically diverse lineages of ray-finned fishes (actinopterygians). Actinopterygians diversified through the Mesozoic, giving rise to a range of modern groups by the Cretaceous.

Dispersal of propagules by reptiles, although not widespread, does occur in some groups of angiosperms (van der Pijl, 1982). Reptile-dispersed fruits or seeds tend to be brightly coloured, often with strong odours, and are borne near the ground. Among extant reptiles, turtles and tortoises are the most significant dispersers. Turtles (testudines) first appear in the Triassic fossil record, but diversify during the Mesozoic. Many extant lineages are known from the Cretaceous. The oldest squamates (lizards and snakes) are known from the Late Triassic, but fossils that can be referred confidently to modern groups do not appear until the Cretaceous (Novacek, 1999). Non-avian dinosaurs may have been involved in the inadvertent dispersal of angiosperm propagules, as part of a general syndrome of large-scale herbivory.

Birds are very important dispersers of angiosperm fruits and seeds in both temperate and tropical plant communities today. Dispersal by birds (ornithochory) may be by consumption and passage through the gut (internal), or by direct transport in the beak or adhering to the body (external). In either case, the distance that propagules can be moved may be considerable. Bird-dispersed propagules are generally brightly coloured, conspicuous on the plant, and lacking an odour. Typically they are fleshy fruits. Some have one or a few seeds that are protected by the hard endocarp of the fruit (drupes). Others contain many seeds that are protected by the hard seed wall (berries). Modern groups of birds appear to have diversified during the Early Cenozoic (Chapter 17), and by the Eocene most of the orders of extant birds, with the possible exception of the passerines, can be recognised (Ericson *et al.*, 2003). Analyses at the family level of the changing composition of the avifauna during the Cenozoic indicate a steady increase in the proportion of omnivorous or plant-eating birds from relatively low levels in the Paleocene–Eocene, to much higher levels in the Pliocene to Recent (Tiffney, 1984).

Bats are very important dispersal agents of angiosperm propagules in subtropical and tropical regions (Fleming and Heithaus, 1981; van der Pijl, 1982). Bat-dispersed (chiropterochory) fruits are generally large, fleshy and with a strong odour. There may be one to many seeds that have a hard inner layer. Similarly, bat-dispersed seeds generally have a fleshy aril or sarcotesta (van der Pijl, 1957). Chiropterochory may not have been established until very late in the history of angiosperms. Bats first appear in the Eocene and diversified rapidly through the Cenozoic. All Eocene bats are interpreted as insectivorous (Simmons *et al.*, 2008) and the report of a putative frugivorous bat from the Oligocene is controversial (e.g. Schutt and Simmons, 1998).

In addition to bats, a variety of other mammals, ranging from rodents to primates, are also involved in the dispersal of angiosperm fruits and seeds. Typically, mammal-dispersed fruits or seeds have a distinctive odour. Small seeds appear to be adapted to passage through the masticatory apparatus and gut; larger seeds generally have more obvious mechanical protection. In the case of many mammal-dispersed fruits or seeds the presence of a hard protective layer may also encourage scatterhoarding. Living groups of therian mammals appear to have undergone their major initial radiation during the Cenozoic (Chapter 17). Based on this, and inferences about the evolution of dispersal mechanisms in different angiosperm lineages, Tiffney (1984, p. 568) concluded that 'mammals became important as dispersal agents only during the Tertiary.' However, eutherian mammals were already present in the Early Cretaceous (Ji *et al.*, 2002) and the Mesozoic fossil record of multituberculate mammals (an extinct group outside the therian clade) has been greatly expanded in recent years. These small, rodent-sized herbivores are known from the Late Jurassic and were both diverse and abundant during the Cretaceous (Kielan-Jaworowska *et al.*, 1987; Novacek, 1999). Their importance has been compared to the high diversity of rodent species in many recent habitats (Novacek, 1999). The dentition of multituberculates has been interpreted as suited for fruit and seed eating (Novacek, 1999) and it is possible that multituberculates played a significant role in the dispersal of angiosperm fruits and seeds in the Cretaceous (Eriksson *et al.*, 2000b).

18.4 HISTORY OF DISPERSAL IN ANGIOSPERMS

The history of dispersal in angiosperms, like that of pollination (Chapter 17), can be inferred only indirectly, either from the fossil record of angiosperm fruits and seeds or, in the case of animal-dispersed propagules, from the fossil record of potential dispersers. In this section we briefly consider these two lines of evidence through different stages of angiosperm evolutionary history. We begin with pre-angiosperm vegetation and follow the history of dispersal in angiosperms to the Cenozoic.

18.4.1 Dispersal in pre-angiosperm vegetation

Jurassic and earliest Cretaceous floras provide some evidence of modes of seed dispersal among the plants that comprised pre-angiosperm vegetation. Jurassic and Cretaceous conifers appear to show a similar range of dispersal systems to those occurring among conifers today. Cone scales of *Araucarites phillipsii* from the Middle Jurassic of Yorkshire, England, were relatively large (*c.* 16 mm × 13 mm) and the seeds (*c.* 12 mm × 8 mm) are among the largest known in this flora. As in extant *Araucaria*, each cone scale contained a single, large embedded seed. Seeds were also shed as part of a cone scale complex in many Cheirolepidiaceae (e.g. *Hirmeriella*), as well as in other enigmatic conifers such as *Schizolepis* (Harris, 1979). In contrast, seeds of probable Cupressaceae (e.g. *Elatides williamsonii*) were small (2.0 mm × 1.0 – 1.4 mm), presumably shed separately from the cone scale, and probably dispersed abiotically. Evidence for animal dispersal in pre-angiosperm conifers is provided by fossil Taxaceae (e.g. *Marskea*, Harris, 1979), which had relatively large seeds with a thin aril as in some extant taxa.

Seeds of extinct *Ginkgo* and cycads from the Middle Jurassic flora of Yorkshire have similar anatomy to those of their living relatives, with a fleshy sarcotesta surrounding a hard sclerotesta. They were presumably animal-dispersed. Seeds attributed to *Ginkgo huttonii* were relatively large in comparison to other propagules in the Yorkshire flora (*c.* 16 mm × 9 mm), but they are smaller than those typically produced by extant *Ginkgo biloba*. Seeds of *Beania gracilis* were 16 mm long and 13 mm wide (Harris, 1964) and are among the largest seeds from the flora. Seeds of *Beania mamayi* were much smaller (4 mm in diameter). From the Early Cretaceous Yixian Formation seeds found in the stomach of the fossil bird *Jeholornis prima* (Zhou and F. Zhang, 2002) are of unknown affinity, but show some resemblance in size and shape to seeds of living *Ginkgo*. The large seed of *Onoana* from the Early Cretaceous of California, USA, originally described as an angiosperm (Chapter 9), shows some resemblance to seeds of cycads and may have been water-dispersed.

Two important extinct groups in pre-angiosperm vegetation that both show evidence of animal dispersal are Bennettitales and Caytoniales. In both cases, the ovulate structure appears fleshy.

In Bennettitales the seeds are embedded in a mass of interseminal scales, the cutinised heads of which form a covering to the ovulate structure. Only the micropylar tips of the seeds protrude from the surface between the interseminal scales (Chapter 5). In *Vardekloeftia* from the Triassic of Greenland, and *Bennetticarpus wettsteinii* from the Triassic Lunz flora, Austria, there are relatively few large ovules/seeds in the ovulate structures. In most Bennettitales, seeds are smaller and more numerous. The heads of the interseminal scales often have a thick cuticle. Together, they form a firm, perhaps leathery, covering to the ovulate structure. Nathorst (1909b) suggested that the ovulate structures, for example in *Williamsonia leckenbyi*, were succulent. Similarly, Harris (1969) suggested that ovulate structure comprising seeds and fleshy interseminal scales were edible 'and that nearly all our specimens are fragments of the useless parts, rind or core left by some animal after its meal.' This may account for the paucity of bennettitalean seeds recovered from bedding planes or isolated through bulk maceration. However, the occurrence on bedding planes of some whole bennettitalean reproductive structures, such as *Williamsoniella* (Harris, 1944; Crane and Herendeen, 2009) suggest that the structures were shed at various stages of maturity. This is also consistent with the well-defined abscission layer across the base of the ovulate structures that has been recognised in several species of *Cycadeoidea* (Watson and Sincock, 1992). Whole ovulate structures appear to have been shed. This may have been linked to animal dispersal, but it is also possible that gynoecia and the seeds that they included were retained within the leaf bases until they were released by fire, in a manner comparable to the serotinous cones of some living pines (Watson and Sincock, 1992). In terms of the protection of the seeds and the probable mode of animal dispersal, angiosperm fruits or inflorescences like the cherimoya (*Annona muricata*), the Osage apple (*Maclura pomifera*) or the breadfruit (*Artocarpus edulis*) are crude analogues to the situation envisaged in extinct Bennettitales. However, compared to these angiosperms, bennettitalean ovulate structures and the seeds they contain, are generally much smaller (ovulate structures typically 10 mm – 50 mm in diameter).

In Caytoniales animal dispersal probably occurred through consumption of the fleshy cupules, each of which contained several small seeds. The cupules were typically 3–4 mm in diameter containing small flattened ellipsoidal seeds 1.5 mm long and 1.1 mm wide. Direct evidence that the *Caytonia* cupules were edible, and perhaps fleshy and berry-like, is provided by coprolites about 4 mm in diameter that contain remains of the *Caytonia* plant. One of these coprolites from the Gristhorpe Bed consisted largely of two species of *Caytonia* and their well-preserved seeds (Harris, 1964).

Propagules inferred to have been animal-dispersed were thus produced in several groups of plants during the Jurassic and Cretaceous, including cycads, Ginkgoales, Bennettitales, Caytoniales and even some groups of conifers. This indicates that a variety of animal dispersal systems were already in effect prior to the major diversification of birds and mammals. The beginnings of the adaptations for animal dispersal seen in some living seed plants (e.g. cycads, *Ginkgo*, Taxaceae) were probably already in place in the ancient pre-angiosperm ecosystems of the Mesozoic.

The various spiny, hooked and hairy propagules of unknown affinity from the Late Jurassic and Early Cretaceous (Chapters 5, 9) also provide some insight into dispersal in pre- or early angiosperm vegetation. Some of these fossils, such as *Lappacarpus*, *Ievlevia*, *Donlesia* and *Beipiaoa*, superficially resemble the angiosperm fruits of *Ceratophyllum* and *Trapa*. They may have been water-dispersed with the spines acting as anchors. Alternatively, they may have been become attached to early birds foraging in the lakes.

18.4.2 Dispersal in Early Cretaceous angiosperms

Fossil angiosperm fruits and seeds are unknown from the very earliest (Valanginian – Early Barremian) phases of angiosperm evolution. The oldest fossil assemblages that contain abundant angiosperm reproductive structures are from the Late Barremian – Early Albian of Portugal, and the most intensively studied of these assemblages is the rich mesofossil flora from the Famalicão locality. Fossil fruits and seeds from Famalicão provide the most complete basis for evaluating the propagules of early angiosperms and their likely modes of dispersal (Eriksson *et al.*, 2000b).

From the total of 106 angiosperm taxa recognised at the Famalicão locality only three (*Anacostia portugallica*, *A. teixeirae*, Friis *et al.*, 1997a; *Canrightia resinifera*, Friis and Pedersen, 2011) are currently formally described and named. Remains of non-eudicot angiosperms dominate the mesofossil flora. Their systematic relationships appear to be largely with ANITA-grade angiosperms, some eumagnoliids and early monocots. With the exception of one periporate pollen type all other pollen grains recovered *in situ* within stamens or attached to carpels/fruits are monoaperturate (Friis *et al.*, 1999).

Among the 106 early angiosperms recognised from Famalicão seed size could be estimated for 64 taxa, and fruit size could be estimated for 37 taxa. In five cases both fruit and seed size could be assessed for the same taxon. Within taxon variation in seed size was also estimated for the 13 most abundant taxa. Average seed volume in the Famalicão mesofossil flora is less than 6.78 mm^3, with an estimated weight in the range of 0.1–1.0 mg

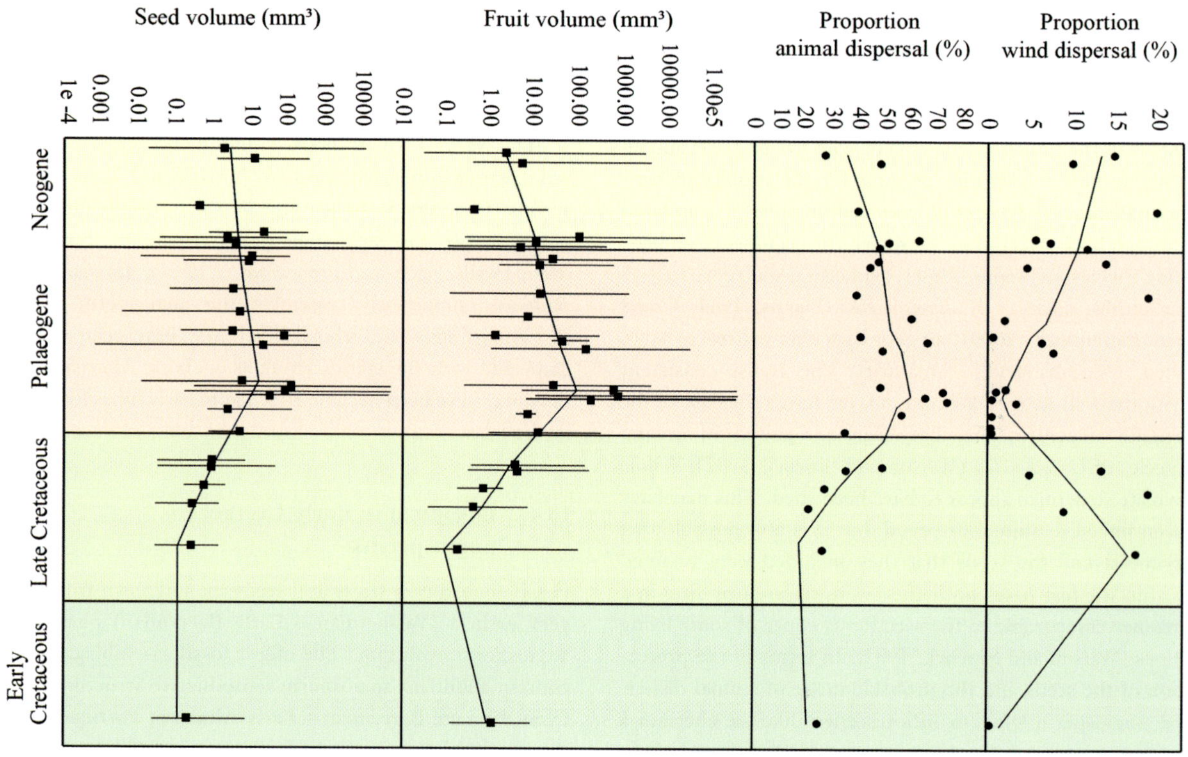

Figure 18.2 Seed and fruit volumes in 25 fossil floras ranging from the Early Cretaceous (Aptian) to the Neogene (Pliocene) and the proportion of animal- and wind-dispersed (or non-assisted) taxa inferred for each flora. The mean value in seed and fruit size is indicated by a square and the full size range indicated by a solid line. Note the increase in size and proportion of animal-dispersed fruits and seeds from the Cretaceous into the early part of the Palaeogene (Eocene) and the decline to the late Neogene (Pliocene), which is similar to the present situation. Based on Eriksson *et al.* (2000a).

(Eriksson *et al.*, 2000b). Variation in volume ranges from 0.02 to 6.86 mm^3. Average fruit volume is 2.06 mm^3 (range 0.12–8.34 mm^3) and for those fruits in which the number of seeds could be counted, the maximum number of seeds per fruit is six. Most species (12 of 18) have single-seeded fruits (Eriksson *et al.*, 2000b).

Average seed volume in the Famalicão assemblage is not too different from that of many modern floras (Leishman *et al.*, 1995). For those taxa in which it can be assessed, within-taxon variation in seed volume is also comparable to that in many living plants (coefficient of variation = 28%). However, it is striking that large fruits and seeds are missing, even though the presence of large fossils (e.g. twigs of cheirolepidiaceous conifers) indicates that larger fruits and seeds of angiosperms could potentially have been preserved (Eriksson *et al.*, 2000b). Whereas seed size assessed for recent vegetation often varies across five orders of magnitude (Leishman *et al.*, 1995), seed size at Famalicão only ranges across two orders of magnitude and seeds in the upper part of the modern size range are missing.

Observations from the Famalicão assemblage also hold for other Early Cretaceous mesofossil floras (Figure 18.2). In Portugal, the angiosperm fruit and seed assemblages from Buarcos, Catefica, Torres Vedras, Vale de Agua and several other floras are all consistently small in size. Larger fruits and seeds are not present and there is also no evidence of them in contemporary macrofossil floras (Saporta, 1894; Teixeira, 1948, 1950). Similarly, at Drewry's Bluff, Dutch Gap and Puddledock in the Potomac Group of eastern North America, angiosperm fruits and seeds are all consistently small.

Another consistent feature among mesofossil floras from the Early Cretaceous is the kind of fruits preserved. Drupes, berries, nuts and small follicles are all common, but there is no evidence of large follicles. At Famalicão, of the 37 taxa for which fruit type could be assessed, about 15% were drupes, 9% were berries (Figures 18.3, 18.4) and the remainder (75%) were either follicles or nuts. On the assumption that drupes and berries were animal-dispersed, approximately a quarter of the Famalicão taxa were biotically dispersed. The remaining three quarters may have been abiotically dispersed.

In several cases, there is evidence that the individual fruits from Famalicão and from other Early Cretaceous mesofossil floras were aggregated into larger, multiple-fruited units, consistent with the evidence for more or less dense inflorescences in many early angiosperms (Chapter 16). Aggregated in this way the fruits may have been more attractive to potential dispersers than their individual small size would suggest. Similar aggregations of small fruits also occur in some of the earliest-diverging lineages of extant angiosperms, such as *Amborella*, *Chloranthus*, *Piper* or *Trimenia*.

The small size of the angiosperm seeds in the Famalicão assemblage is interesting in relation to the size of seeds in some non-angiosperm seed plants. For example, co-occurring with early angiosperm seeds in the Early Cretaceous mesofossil floras from Portugal and eastern North America are seeds of Cretaceous ephedroids assigned to *Ephedra* and *Ephedrispermum* (Rydin *et al.*, 2006a) and related genera such as *Buarcospermum*, *Lignierispermum*, *Lobospermum* and *Rugonella* (Friis *et al.*, 2009a). These seeds are all of similar small size to those of early angiosperms recorded from Famalicão and other Late Barremian – Middle Albian mesofossil assemblages. They also have a well-developed sclerenchyma layer in the envelope that surrounds the nucellus and integument, and in some of these seeds there are indications of a fleshy outer layer.

There are even stronger similarities between the berries of early angiosperms from Famalicão and seed-bearing cupules of *Caytonia* from the Middle Jurassic of Yorkshire. The fleshy cupules of *Caytonia* are of similar size to some of the larger angiosperm fruits from Famalicão. Seeds of *Caytonia* are also of similar size to many Famalicão seeds and have a similar hard seed coat.

The possible dispersers for the drupes and berries recovered from Famalicão are unknown, but they could have been animals (presumed tetrapods, perhaps multituberculates) similar to those that made the coprolites containing *Caytonia* seeds and cupules in the Middle Jurassic. There is also the possibility that the fruits and seeds of early angiosperms may have been attractive to early birds or pterosaurs (Fleming and Lips, 1991).

In addition to endozoochory, epizoochory may also have been established very early in angiosperm evolution. Small uniovulate fruits assigned to the fossil genus *Appomattoxia* are known from the Early–Middle Albian Puddledock locality, Virginia, USA, and from the Late Barremian – Early Aptian flora of Torres Vedras, Portugal. The fruit surface is covered by long stiff hairs, each of which terminates in a distinct hook similar to that of extant *Circaeaster*. Like *Circaeaster*, the *Appomattoxia* plant may have been a herb or small shrub with fruits exposed closed to the ground where they could adhere to the fur of small mammals (Friis *et al.*, 1995). Alternatively, *Appomattoxia* may have been water- or wind-dispersed with the hooks acting as anchors to the substrate or to other plants (Collinson and Van Bergen, 2004).

Fruits slightly larger than those known from mesofossils occur in some of the compression/impression fossils from the Crato Formation (Late Aptian – Early Albian). *Pluricarpellatia*, for example, has apocarpous fruits with mature carpels up to about 1 cm in length. However, the seeds are tiny, about 1 mm long. Only few seeds were observed in mature carpels, indicating that the seeds were the dispersal units and had been shed. *Pluricarpellatia* was interpreted as an aquatic plant and the seeds were most likely dispersed in water.

Archaefructus from the Late Barremian – Early Aptian Yixian Formation is also interpreted as an aquatic plant. The fruiting units are up to about 2 cm long and the seeds 1–2 mm long. Like *Pluricarpellatia* the propagules of *Archaefructus* were probably dispersed in water, but no open carpels have been observed and seeds appear to be intact in the carpels. The dispersal units in *Archaefructus* may have been each carpel with its enclosed seeds.

18.4.3 Dispersal in mid-Cretaceous angiosperms

Data comparable to that of the Famalicão flora have not been compiled for mesofossil floras from the mid-Cretaceous, but the same generalisations seem to apply. Average seed size is small, and larger seeds are not recovered. Drupes, berries and nuts are the fruits most commonly encountered, and as in earlier floras, presumed animal-dispersed fruits are common. *Couperites mauldinensis* from the earliest Cenomanian mesofossil flora at Mauldin Mountain, Maryland, USA, and the two species of *Zlatkocarpus* from the Cenomanian of the Czech Republic, are good examples of berries from this level of angiosperm evolution. As in the case of berries (or drupes) from earlier

Figure 18.3 Early Cretaceous fruits from Portugal showing small size and typical one-seeded dispersal units. (A, B, G) Single-seeded berries. (C, H–K) Single-seeded drupes. (D) Spiny fruit, probably animal- or perhaps water-dispersed. (E, F, M) Dry fruits, probably unassisted dispersal. (L) Berry containing several seeds. Material in the collections of the Swedish Museum of Natural History, Stockholm.

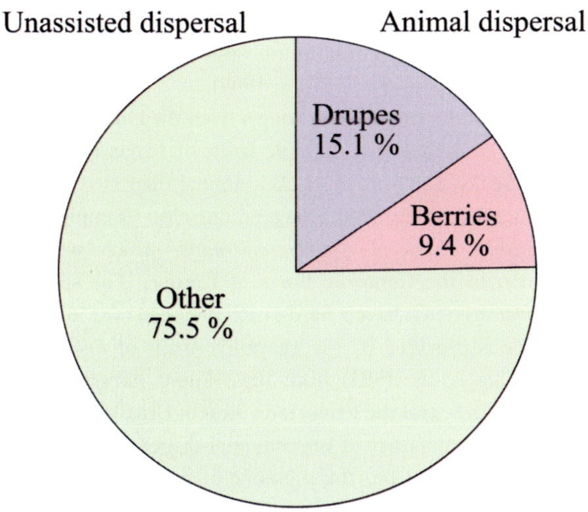

Figure 18.4 The proportion of angiosperm propagules from the Famalicão mesofossil flora, Portugal (Late Aptian), Portugal, interpreted as animal-dispersed (drupes and berries) and unassisted (other fruits). Based on Eriksson *et al.* (2000b).

mesofossil floras such as Famalicão, the outer fruit wall is thin and parenchymatous and the protective sclerenchymatous layer is in the seed coat (or endocarp), indicating animal dispersal.

In some fossil floras from this time interval there are indications of larger fruits, which also corresponds with the first appearance of larger flowers (Chapters 16 and 17). For example, compression fossils of *Archaeanthus* have larger follicles, up to 38 mm long, which probably contained up to 20 seeds. However, in all cases, the seeds produced in these larger fruits were still in the range of those in the Famalicão flora (up to *c.* 2–3 mm) or perhaps slightly larger. Dispersal was probably unspecialised or was perhaps by wind.

18.4.4 Dispersal in Late Cretaceous angiosperms

More data are available on angiosperm seed size in Late Cretaceous floras than for older plant assemblages. Compilations of data by Tiffney (1984) and by Eriksson *et al.* (2000a) both provide evidence that angiosperm seed size remained small through the Cretaceous with only a slight increase through the Late Cretaceous (Figure 18.2). In the mesofossil flora from the Campanian of Gay Head, Massachusetts, USA, average seed volume is 5.7 mm^3 with the largest estimated at 55 mm^3 (Tiffney, 1984). Although explicit analyses have not been carried out for other Late Cretaceous mesofossil floras from North America, those from Allon and Upatoi Creek, Georgia, USA, which are of Santonian and Coniacian age, respectively (Herendeen *et al.*, 1999a), and from the Old Crossman locality, New Jersey, USA, which is Turonian in age, appear to conform to this pattern.

The Åsen mesofossil flora from Sweden, which is of Santonian–Campanian age, has the same characteristics seen in Late Cretaceous mesofossil floras from North America. Other Late Cretaceous floras from Europe that have been measured show a slightly more narrow size range and smaller propagules (Tiffney, 1984; Eriksson *et al.*, 2000a). For instance in the Santonian – Early Campanian flora from Staré Hamry in the Czech Republic, and in the Campanian floras from Aachen, Germany, and Petrovice, Czech Republic, average seed volumes are 1.4 mm^3, 0.2 mm^3 and 0.3 mm^2, respectively (Tiffney, 1984). The corresponding maximum seed sizes are 3 mm, 0.5 mm and 0.73 mm. Minute fruits and seeds also characterise the Budurone mesofossil flora from the latest Cretaceous (Maastrichtian) of Romania (Lindfors *et al.*, 2010). Fruits are typically indehiscent nuts and drupes, and both fruits and seeds apparently lack special ornamentation or modification for dispersal. Fruits and seeds range from 0.6 to 2.6 mm in length and from 0.5 to 1.9 mm in breadth. Estimated fruit volume is in the range 0.08–3.59 mm^3 and seed volume ranges between 0.08 and 0.79 mm^3.

Other Late Cretaceous mesofossil floras from Asia (Kazakhstan, Japan) and from the Antarctic Peninsula also have small fruits and seeds (Frumin and Friis, 1996, 1999; Takahashi *et al.*, 1999b; Hvalj, 2001; Eklund, 2003; Eklund *et al.*, 2004a). This was apparently the general condition at mid- to high palaeolatitudes during the Late Cretaceous.

It may be significant that larger fruits and seeds of Late Cretaceous age have been reported from only a single locality (Paki, Senegal) from the palaeoequatorial region (Monteillet and Lappartient, 1981). The fossils occur in a marine deposit of Late Campanian-Maastrichtian age and are apparently preserved as moulds. The fruits and seeds range from about 1.6 mm to 106 mm in length and Tiffney (1984) estimated an average volume for propagules as 51 950 mm^3.

As in the mid-Cretaceous, larger angiosperm fruits are also known from mid- and high palaeolatitudes in the Late Cretaceous (e.g. *Elsemaria, Hidakanthus*, *Keraocarpon*, *Protomonimia*), but most appear to be follicles containing smaller seeds. For example, *Hidakanthus* and *Protomonimia* (Chapter 9) from the Coniacian to Santonian of Hokkaido, Japan, are probably eumagnoliid fruiting structures up to about 40 mm in diameter bearing numerous follicles. Each follicle is about 15–20 mm long. No seeds were observed in the follicles of *Hidakanthus*, but in *Protomonimia* each carpel contains about 15 small seeds, each about 5 mm long. Similar seeds occur isolated in the Early Coniacian Kamikitaba flora, also from Japan, as part of a typical Late Cretaceous mesofossil assemblage of small seeds (Takahashi *et al.*, 1999b).

18.4.5 Dispersal in Cenozoic angiosperms

Based on several independent compilations of data from a variety of different fossil floras, there is a pronounced increase in seed size in the Early Cenozoic (Figure 18.2). This pattern was first recognised by Tiffney (1984), and has since been confirmed in subsequent analyses by Wing and Boucher (1998) and Eriksson *et al.* (2000a, b). The pattern is pronounced and apparently unambiguous. For example, compilations by Tiffney (1984) of seed size in the Early Eocene fossil flora of the London Clay, southern England, yield an average seed size of 1957 mm^3, three to four orders of magnitude greater than the typical Cretaceous floras. Analyses of the Middle Eocene fossil flora from the Clarno Nut Beds yield an even larger value of 3729 mm^3. Maximum seed sizes recorded for these floras are 61 318 mm^3 and 59 150 mm^3, respectively (Tiffney, 1984).

With regard to potential taphonomic biases and their potential influence on these results it is important that these two Cenozoic fossil floras are from quite different depositional environments. In the case of the London Clay, the propagules had drifted out to sea and are preserved in marine sediments. In the case of the Clarno Nut Beds, the flora records the vegetation of a volcanic terrain and the fruits and seeds are preserved in a mud flow. Because both of these floras are collected by hand, rather than by sieving and panning, it is possible that some very small seeds may have been overlooked, but in both cases the minimum seed size is 0.25mm^3 (Tiffney, 1984). The Geiseltal flora, from Germany, which is of Middle Eocene age and is collected and studied in a similar way to the Late Cretaceous mesofossil floras (Mai, 1976), confirms the pattern. It has a mean seed size of 308 mm^3 with the largest and smallest seeds of 3182 and 2.1 mm^3, respectively, still substantially larger than the norm in the Late Cretaceous (Tiffney, 1984).

The pattern of large seed size established in angiosperms from the Early Cenozoic continues in fossil floras up to the Recent (Tiffney, 1984; Eriksson *et al.*, 2000a). For example, the Pliocene fossil fruit and seed assemblage from Bergheim, Germany, has an average seed size of 100 mm^3 with maximum and minimum values of 1400 mm^3 and 0.9 mm^3, respectively (Tiffney, 1984).

Many of the larger fruits known from the Early Cenozoic are nuts or drupes, for example fruits of Juglandaceae and other Fagales, and were most likely animal-dispersed. Berries are much less prominent. However, one clear example is the banana-like fruits of *Spirematospermum*, which are very common in the Cenozoic floras of Europe. The seeds of *Spirematospermum* have a hard protective seed coat. Another example is provided by the guava-like fruits of *Paleomyrtinaea* (Pigg *et al.*, 1993) from the Almont flora of North Dakota, USA, and the Princeton Chert of British Columbia, Canada. The presence of larger animal-dispersed berries can also be deduced from the presence of many small isolated seeds assignable to such extant genera as *Passiflora*, *Actinidia*, *Vitis*, *Ampelopsis* and *Myrtus*.

The dramatic increase in the abundance and size of berries, nuts and drupes in the Early Cenozoic would have had important consequences for the maintenance and support of populations of birds and mammals, and may also have been a key factor that promoted aspects of their diversification.

18.5 LARGE-SCALE TRENDS IN THE HISTORY OF ANGIOSPERM DISPERSAL

The recognition of significant correlations between seed size and various ecological factors (Harper *et al.*, 1970) has stimulated several studies of large-scale ecological trends in this parameter in Cretaceous and Cenozoic angiosperms. In his classic paper Tiffney (1984) plotted the range and mean volume for fruits and seeds in 27 Late Cretaceous and Cenozoic floras. He noted a significant increase in size following the Cretaceous–Cenozoic boundary after which maximum fruit and seed size did not increase greatly. Similarly, the minimum size of angiosperm fruits and seeds remained approximately constant since the first appearance of the group in the Cretaceous.

Tiffney (1984) interpreted the marked temporal trend in seed size in terms of the relationship between seed size, dispersal mechanism and successional status in living angiosperms. He concluded that 'Cretaceous angiosperms were primarily small-seeded, abiotically dispersed shrubs or opportunistic trees perhaps occupying marginal or open habitats in the gymnosperm dominated vegetation, but probably not forming a closed-canopy climax community.'

He also inferred that the relative paucity of appropriate bird and mammal dispersal agents limited the success of large angiosperm diaspores until the rapid diversification of birds, bats and terrestrial mammals in the latest Cretaceous and Palaeogene. He concluded that the advent of large, animal-dispersed propagules 'allowed the establishment of angiosperm seedlings in areas of low light intensity and led to the development of stable, closed-canopy, climax communities', which were similar, for the first time, to those of the Recent. Tiffney's data were expanded to a larger sample of 49 fossil floras by Wing and Boucher (1998), who also increased the number of floras sampled from the Cretaceous. They demonstrated a similar temporal trend in diaspore volume and concluded that the Early Cenozoic 'was the time when animal dispersal and shade-tolerant life history strategies became common among flowering plants' (Wing and Boucher, 1998, p 403).

Analyses similar to those of Tiffney (1984) and Wing and Boucher (1998), based on a critical sampling of 25 floras from the Barremian–Aptian to the Neogene, were carried out by Eriksson *et al.* (2000a). In this study temporal trends in fruit and seed size were examined separately to avoid confounding effect such as large fruits containing numerous small seeds, and sampling of Cretaceous plant assemblages was further increased. Changes in the relative proportions of animal-dispersed and wind-dispersed propagules (as judged by comparison to extant analogues) were also evaluated (Figure 18.4).

In the earliest fossil assemblage examined (the Famalicão mesofossil flora), which is Late Aptian in age, angiosperm seeds and fruits are generally very small (average seed volume 0.78 mm^3). Seeds and fruits from other Cretaceous floras are generally also small with some indication of an increase in size before the end of the Cretaceous, as noted also by Tiffney (1984). In the Paleocene and Eocene mean seed size increases by three orders of magnitude. Subsequently there is a slight decline before seed size stabilises in the Oligocene to Pliocene at one to two orders of magnitude greater than Cretaceous levels. There is also some indication that for the floras sampled the percentage of wind-dispersed propagules was higher in the Late Cretaceous and in the Oligocene–Miocene, but relatively low in the Paleocene–Eocene (Figure 18.2). Both seed and fruit size appear to decrease after the Early Eocene with some indication of a transitory increase during the Miocene at about 30–20 million years ago. A decrease in mean angiosperm propagule size after the Eocene was also evident in the analysis of Tiffney (1984).

Although larger fruits and seeds do occur among Cretaceous angiosperms it is remarkable that the vast majority of angiosperm reproductive organs discovered from numerous floras ranging from Antarctica to Asia, Europe and North America are minute. This contrasts markedly to the much wider size range and larger size of fruits and seeds observed from Cenozoic strata. The Cretaceous and Cenozoic floras are extracted from the same kind of sediments, deposited in the same kind of lacustrine and fluviatile environments, and many of the floras are extracted using the same techniques. The pattern therefore cannot be explained by taphonomic differences, but must be related to basic differences in reproductive biology and/or ecological conditions, more so because taphonomic processes tend to discriminate against smaller fruits and seeds (Sims and Cassara, 2009).

Based on these data, and a review of evidence of vegetation and climate change, as well as the evolution of potential mammalian and avian dispersers, Eriksson *et al.* (2000a) concluded that the most likely explanation of the pattern of change in angiosperm seed and fruit features was the climate-driven development of closed forest vegetation during the early Cenozoic. Under this interpretation the prevalence of semi-open, dry, and probably herbivore-disturbed (dinosaur-disturbed) vegetation during the Cretaceous was a key factor in influencing angiosperm propagule size. There was no strong selective pressure to stimulate the evolution of larger angiosperm seeds and fruits. Furthermore, because small seed size is strongly influenced by the size of the plant body (Shipley and Dion, 1992) propagule size may also have been limited by the herbaceous to shrubby stature inferred for many Cretaceous angiosperms. Cretaceous angiosperms may have been dispersed both by animals with specific feeding strategies (e.g. early multituberculates and perhaps early birds) and by the indiscriminate mass feeding of larger animals in which the fruits were consumed with the foliage (e.g. herbivorous dinosaurs, Janzen, 1983).

Toward the end of the Cretaceous, climatic changes resulting from new configurations of the continental landmasses, perhaps combined with the demise of large dinosaur herbivores around the Cretaceous–Palaeogene boundary, promoted the development of closed multistratal forest with the concomitant increase in selection pressure for larger seed size. Under this interpretation the evolution of specialised frugivores in the Early Cenozoic, such as bats, rodent and birds, would be a secondary phenomenon resulting from the increased availability of large angiosperm fruits and seeds (Eriksson *et al.*, 2000a). In the mid- and high latitude regions that currently provide most of the data on fossil floras the climate cooled around the end of the Eocene, perhaps resulting in relatively more open habitats (Eriksson *et al.*, 2000a). This trend was further intensified in the Miocene with the widespread development of grasslands. After the Eocene, in certain environments, selection for large propagule size was thereby relaxed.

Eriksson *et al.* (2000a) termed their interpretation the 'recruitment hypothesis', which they contrasted with the 'disperser hypothesis' of Tiffney (1984) in which the evolution of large propagule size is envisaged as driven largely by the latest Cretaceous or Early Cenozoic diversification of vertebrate dispersal vectors. Under the 'disperser hypothesis' ('coevolution hypothesis' *sensu* Eriksson, 2008) large diaspores were disadvantageous in the Cretaceous owing to the paucity of potential dispersers. However, data from the Jurassic and Early Cretaceous indicate that animal dispersal was already common in pre-angiosperm communities and remained significant throughout the Cretaceous (Eriksson *et al.*, 2000a). In all Cretaceous mesofossil floras surveyed from the Late Barremian – Early Aptian and onwards the percentage of propagules with some indication of animal dispersal ranges from *c.* 20% to *c.* 50%. In the Cenozoic the percentage of angiosperm propagules inferred to be animal-dispersed ranges from *c.* 30% to *c.* 75%. These data imply that animal dispersal was not an innovation that first appeared during the early Cenozoic with the availability of new kinds of dispersers.

Evidence from pre-angiosperm fossil floras, and from Cretaceous angiosperms themselves, therefore suggests that animal dispersal of propagules has been important from the earliest phases of angiosperm evolution. As with pollination, early angiosperms may have co-opted dispersal vectors that were already engaged in dispersal interactions in earlier Mesozoic vegetation. Perhaps toward the end of the Cretaceous, but certainly, and more markedly, during the Early Cenozoic, seed size increased. However, the precise factors driving an increase in seed size in the Early Cenozoic are difficult to disentangle (Eriksson, 2008). Whereas the 'recruitment hypothesis' suggests that the development of closed multistratal forest drove selection for larger seed size, Moles *et al.* (2005) suggested a third, but related, alternative, 'the life form hypothesis'. This suggests that large seeds and biotic dispersal evolved as 'co-adapted traits' (Eriksson, 2008) as a consequence of the evolution of large plant life forms.

It is clear that all of the factors recognised in the disperser (co-evolution), recruitment and life form hypotheses are linked. Larger plant life forms themselves create conditions that favour recruitment from larger seeds, and at the same time may create plants with improved capacity (through larger size and longer lifespan) for the production of larger propagules. Furthermore, the herbivorous fauna itself may have been an active participant in those long-term interactions. Wing and Tiffney (1987b) proposed that the extinction of large herbivorous dinosaurs may have been a key driver. They suggest that intense herbivory during the Cretaceous may have favoured small plants with small seeds that could rapidly exploit disturbed, open habitats. However, in the Early Cenozoic, with reduction in the intensity of herbivory, competition among plants resulted in more dense vegetation, which in turn favoured larger seeds. The result was a 'coevolutionary spiral' in which more abundant, larger fruits and seeds favoured frugivory, which in turn favoured still larger diaspores (Wing and Tiffney, 1987b). Changing climatic conditions, especially changing patterns of rainfall, may also have favoured the development of dense multistratal forests and a better-developed angiosperm tree flora. The larger propagules that are advantageous in such vegetation types would also have provided a new food source for diversifying avian and mammalian frugivores. After the Eocene, changing climatic conditions in many areas favoured more open conditions and propagule size decreased, although it still remained above typical values for the Cretaceous.

The predominant mode of animal dispersal through the Cretaceous and through the Cenozoic was endozoochory, in which propagules either pass through the gut or are discarded as their digestible portions are consumed. Clear specialisations for external dispersal by animals (epizoochory) were not common, but may have been established early in the Cretaceous, if dispersal in *Appomattoxia* involved animals rather than water. Other specialisations, for dispersal by ants or for ballistic dispersal, are not well represented in the fossil record and may be predominantly a phenomenon of the later Cenozoic associated with the diversification of herbaceous lineages of angiosperms.

In the context of the history of angiosperm dispersal mechanisms, and of turnover in the vertebrate fauna, it is very likely that most dispersal modes of extant non-angiosperm seed plants are of relatively recent, rather than ancient, origin. More recent extinctions (e.g. Pleistocene extinctions) may well have resulted in some extant plants for which the specialised dispersers are now extinct (Janzen and Martin, 1982). Tiffney (1984) has suggested that this is the case for extant *Ginkgo*. Climatic and other environmental changes, together with different patterns of local and regional dispersal and extinction, may also lead to situations in which fruits and seeds with specialised dispersal modes become spatially separated from their corresponding dispersal agents (e.g. elaiosome-bearing seeds in southern New Zealand where there are no ants, Willson *et al.*, 1990).

Relationships between fleshy-fruited plants and their dispersers are generally diffuse (e.g. Herrera, 1986; Wheelwright, 1988; Jordano, 1995), but in some lineages (notably within Juglandaceae, Chapter 20) there is evidence of long-established relationships (e.g. *Carya* and squirrels) as well as probable selection to improve the efficiency of animal dispersal (e.g. disincentives for immediate consumption, incentives for scatterhoarding). Several angiosperm lineages show evidence of increased specialisation for animal dispersal during the Early Cenozoic. Interestingly, in many of these same lineages (e.g. Betulaceae, Juglandaceae) there is also evidence of increased specialisation for wind dispersal. This may perhaps be explained in part by the increased stature of woody angiosperms, which facilitates wind dispersal. Because the production of large numbers of small seeds with unspecialised dispersal mode is itself successful in many lineages (Willson, 1993; Clark *et al.*, 1998) any reduction in seed production associated with increased reserves leads to increased selection for dispersal attributes (Eriksson and Jakobsson, 1999). The marked structural diversification associated with wind dispersal, which occurs independently in several different lineages of angiosperms in the Early Cenozoic, may thus simply be a by-product of selection for fewer larger propagules.

19

Vegetational context of early angiosperm diversification

Overwhelming palaeobotanical evidence from many geographical areas indicates that angiosperms first attained ecological prominence during the mid-Cretaceous and that this led ultimately to profound changes in the composition of terrestrial plant communities and ecosystems. Knowledge of how these changes were manifested in changes in ecosystem structure and function, and how this affected, and was influenced by, larger-scale changes in the global environment is still at an early stage. A major difficulty is the relatively poor stratigraphic resolution for many Cretaceous terrestrial deposits as well as difficulties of correlating floras from different environmental settings, different latitudes and different continents. An exhaustive treatment of these issues is beyond the scope of this book, but here we provide an overview of vegetational change through the Cretaceous to place the evolutionary changes discussed in previous chapters in a broader context.

19.1 TRANSITION TO ANGIOSPERM-DOMINATED VEGETATION

The transition to angiosperm-dominated vegetation did not occur simultaneously or uniformly in all parts of the world, and our understanding of how this great transition unfolded geographically is still rudimentary. Nevertheless it is possible to distinguish some broad patterns, and estimates of the timing and magnitude of angiosperm diversification through the Cretaceous have been made at both regional (e.g. Samylina, 1976; Lupia *et al.*, 1999; Cantrill and Poole, 2005) and global scales (e.g. Niklas *et al.*, 1980; Crane, 1987; Crane and Lidgard, 1990; Lidgard and Crane, 1990). Data on the transition to angiosperm-dominated vegetation are most complete from middle- and high-palaeolatitude regions of Laurasia.

Angiosperms first appear in mid-palaeolatitude regions of Laurasia in the Valanginian–Hauterivian with isolated records from Portugal, Israel and Italy. Angiosperms are extremely rare in these early assemblages. For instance, in the Late Valanginian – Early Hauterivian Helez Formation of Israel they account for less than 0.2% of the palynomorphs examined (Brenner, 1996). By the Middle Hauterivian – Early Barremian angiosperms were still restricted in their distribution with scattered occurrences of pollen in mid-palaeolatitudes and palaeoequatorial regions. However, by that time pollen already shows some morphological diversity (Chapter 16), although exclusively at the magnoliid and monocot level. By the Late Barremian – Early Aptian angiosperms were established widely at low to mid-palaeolatitudes and also include the first eudicots. By around the Albian–Cenomanian boundary the geographic expansion of flowering plants into high latitudes, at least in Laurasia, was essentially complete. From the Late Barremian – Early Aptian onwards, in addition to pollen, the fossil record of angiosperms also includes meso- and macrofossil remains with many flowers, fruits, seeds and stamens, as well as leaf fossils.

By the mid-Cretaceous angiosperm leaves, reproductive structures, and pollen are much more diverse, both in form and in their systematic relationships, than earlier in the Cretaceous. Angiosperms producing monoaperturate pollen (monocots and magnoliids) continue to be important, but the proportion of eudicots in the floras is markedly increased. Core eudicots are present for the first time, although they are still rare.

From the Turonian onwards angiosperms predominate in most areas and core eudicots are the most conspicuous angiosperms in all fossil assemblages. However, an interesting feature of the mid- and Late Cretaceous fossil record is that levels of diversity assessed from dispersed palynofloras are often different from values obtained from macrofossil floras. For example, in many mid-Cretaceous floras angiosperms are often more diverse as macrofossils than in coeval palynofloras (Pierce, 1961; Lidgard and Crane, 1990). Even at macrofossil localities that are dominated by angiosperm leaves, the relative diversity of angiosperm pollen extracted from the same sediments may only be about 25% and this pattern persists into the latest Cretaceous (Lidgard and Crane, 1990).

One explanation for these differences is that angiosperms were disproportionately well represented in those environments that are most frequently sampled in macrofossil assemblages, but were perhaps less important in interfluvial areas away from those usual sites of deposition. In the Potomac Group, for example, it has been suggested that conifers remained dominant in backswamp areas throughout the Albian, while angiosperms flourished in riparian settings (Hickey and Doyle, 1977). Another explanation is that angiosperms are underrepresented in many palynofloras because of the predominance of insect-pollinated taxa. Insect-pollinated plants typically produce small quantities of poorly dispersed pollen. Often such pollen from Cretaceous insect-pollinated plants is also small, less than about 10 µm in length, and is easily overlooked or lost during standard palynological preparation procedures. Difficulties of discriminating different species among the small pollen grains of many early angiosperms, especially using standard light microscopy, may also contribute to seemingly low levels of angiosperm pollen diversity.

19.2 COMPONENTS OF EARLY CRETACEOUS VEGETATION

Despite the complexities of interpreting large-scale floristic patterns it is clear that the expansion in the relative importance of angiosperms through the Cretaceous is mirrored by a marked decline in the diversity of other kinds of plants (Figures 19.1, 19.2).

19.2.1 Free-sporing plants

Ferns are abundant and diverse in earliest Cretaceous macrofossil floras. They are mainly assignable to families such as Cyatheaceae, Dipteridaceae, Gleicheniaceae, Marattiaceae, Matoniaceae, Osmundaceae or Schizaeaceae, rather than the 'higher' polypodioid groups that account for most of the diversity of extant ferns (Pryer *et al.*, 1995). Species from these different families of ferns often grew together; Schizaeaceae and Gleicheniaceae are particularly common in floras from mid- and high palaeolatitudes.

Onychiopsis (Dicksoniaceae) and *Ruffordia* (Schizaeaceae) are among the most characteristic Berriasian–Valanginian ferns, and *Onychiopsis* was sufficiently abundant in some more or less brackish back-barrier flats to form local layers of peat (Friis and Pedersen, 1990). Both *Onychiopsis* and *Ruffordia* have thick, leathery pinnules probably indicative of both physiological and climatic water stress. Xeromorphy is also very marked in the reproductive structures of *Onychiopsis*. The sporangia are tightly enclosed in a leathery, apparently indehiscent envelope (Figure 3.11).

Also conspicuous from the Early Cretaceous is the extinct fern genus *Weichselia* (Matoniaceae), which ranges from the Jurassic to the Cenomanian and has an almost worldwide distribution. *Weichselia* is recorded from Europe, North America, northeastern Asia, Tibet, northern Africa, the Middle East, India and South America (e.g. Saiki and Wang, 2003; Diez *et al.*, 2005). It was also widespread in Europe during the Hauterivian–Barremian (e.g. Alvin, 1968; Daber, 1968; Barale *et al.*, 1984; Diéguez and Meléndez, 2000; Diéguez *et al.*, 2010) and is especially common in marginal marine habitats. *Weichselia* exhibits a range of xeromorphic features (Alvin, 1968, 1971, 1974; Daber, 1968) including a thick maceration-resistant cuticle, sunken stomata and tough, cone-like aggregations of sori. Pinnules on either side of the same leaf were often held more or less upright, perhaps to reduce direct insolation (Watson and Alvin, 1996).

There is strong evidence that during the Hauterivian–Barremian ferns formed extensive stands in prairie or savannah-like habitats that would be dominated today by grasses or other herbaceous angiosperms (Harris, 1981). Probable gleicheniaceous and schizaeaceous spores are especially abundant in Early Cretaceous palynofloras, and extant species in these families are often characteristic of open habitats. Most likely ferns were also important in the ground cover of more closed forests, and some may have been epiphytes as suggested by the *Lindsaea*-type roots growing among the root mass of the characteristic Early Cretaceous tree fern *Tempskya* (Schneider and Kenrick, 2001).

Equisetum is also ubiquitous in earliest Cretaceous macrofossil floras (e.g. Schenk, 1871; Watson, 1983). It was evidently important in similar habitats to where it occurs today, around ponds, lakes, and in other wet places. Significant stands of *Equisetites lyelli* are known *in situ* from the Berriasian–Valanginian of southern England and are thought to have to grown in swamps with up to half a metre of standing water (e.g. Watson, 1983).

Isoetes-like and *Selaginella*-like megaspores are abundant throughout the Cretaceous (Batten and Kovach, 1990). Together with occasional macrofossils of *Nathorstiana*, which is particularly common in the Hauterivian–Barremian deposits of Quedlinburg (Mägdefrau, 1968), this suggests that heterosporous lycopsids were prominent in partly

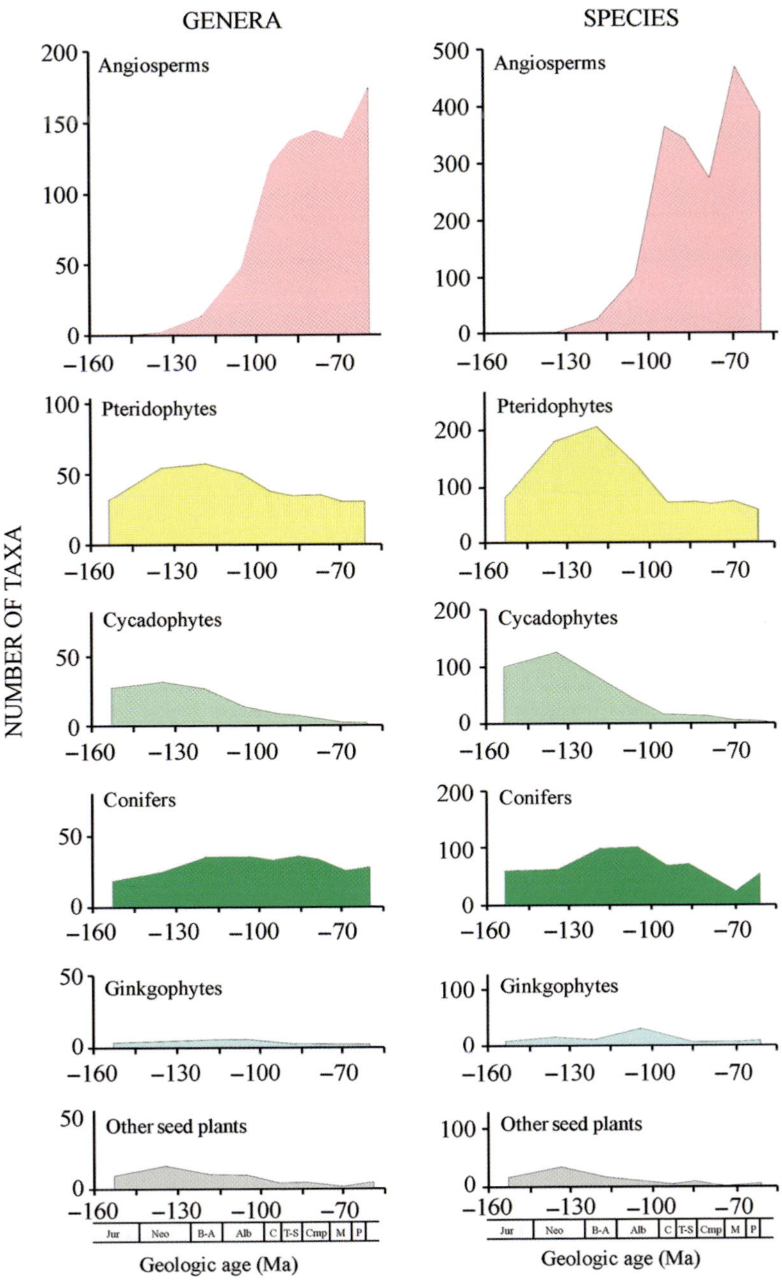

Figure 19.1 Changes in summed diversity patterns for major groups of land plants from the Late Jurassic to the Early Cenozoic (Paleocene), primarily from the Northern Hemisphere, based on fossil leaves. For details of analysis see Lupia *et al.* (2000). Pteridophytes include lycopsids, sphenopsids and ferns. Cycadophytes include cycads and Bennettitales and some presumed seed ferns with pinnate leaves. Ginkgophytes include *Ginkgo*, fossil ginkgoalean plants as well as Czekanowskiales. Other seed plants mostly include other Mesozoic seed plants with distinctive leaves, such as Caytoniales. Jur, Late Jurassic; Neo, Neocomian; B–A, Barremian–Aptian; Alb, Albian; C, Cenomanian; T–S, Turonian–Santonian; Cmp, Campanian; M, Maastrichtian; P, Paleocene. Ma, million years. Based on Lupia *et al.* (2000).

flooded areas and perhaps other open habitats. Megaspores of probable heterosporous water ferns (e.g. *Arcellites*, Marsileaceae) are also commonly encountered in the Early Cretaceous and these plants were probably also important colonisers of wet places (e.g. Batten and Kovach, 1990).

19.2.2 Bennettitales and Cycadales

Pinnate leaves referable to both Bennettitales and Cycadales are characteristic of Berriasian–Valanginian floras. In the Wealden of northern Europe both macrofossils and

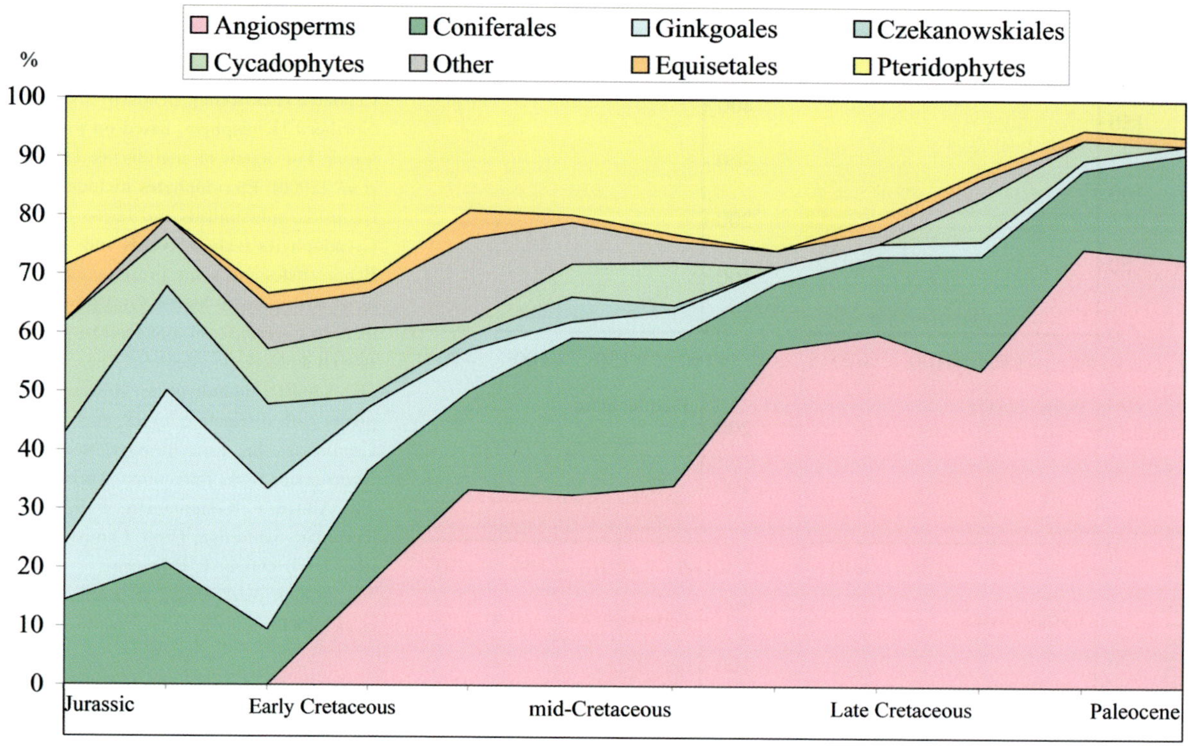

Figure 19.2 Changes in diversity patterns for major land plant groups from the Late Jurassic to the Early Cenozoic (Paleocene) of Northeastern Russia based on leaf fossils from 12 stratofloras distinguished during this interval. The first appearance of angiosperms is in the Early Albian. Pteridophytes are mainly ferns in this case and exclude Equisetales. Data from Samylina (1974) and Goloneva (1998).

distinctive dispersed cuticles document that the Bennettitales were much more abundant than cycads. Cuticles of some of the larger bennettitalean leaves show circular cells that are heavily cutinised on all walls and also have a circular pore. Sincock and Watson (1988) suggested that these cells may be the remains of salt glands: an adaption to growing in soils that are saline, either because of aridity or because of brackish groundwater.

Many Early Cretaceous Bennettitales were short, perhaps stunted, pachycaul plants most likely of open savannah-like habitats, but others had more slender stems, 2 m high or taller (Chapter 5). As in the Jurassic some Bennettitales from the Early Cretaceous may have been sumach-like shrubs.

In the most characteristic Early Cretaceous Bennettitales (e.g. *Cycadeoidea*) the flowers were concealed and embedded among the woody petiole bases. Large quantities of resin appear to have been secreted between the leaf bases and ramenta. Watson and Alvin (1996) draw an analogy with 'serotinous' species of *Pinus* in which the seeds are retained by resin secreted among the cone scales. In these pines, and perhaps in cycadeoids, the cones open and the seeds are released only when the resin has been melted by the heat of fire. In Bennettitales with exposed reproductive structures (e.g. *Williamsonia*) possible adaptation to fire is again manifested in the thick-textured, hairy, artichoke-like bracts. Frequently, these reproductive structures, which are sometimes quite large, were shed with the bracts still attached.

Bennettitalean seeds were small and in at least some cases they were evidently produced in large numbers. For one *Monanthesia* plant, about 2 m high, Watson and Alvin (1996) estimate a total production of about 1.3 million seeds from about 4600 reproductive structures. The small size of most bennettitalean seeds is more consistent with germination in open, rather than shaded, habitats. Permineralised seeds contain well-preserved dicotyledonous embryos (Wieland, 1906) showing that there was control of dormancy and that embryogenesis was not followed immediately by germination. Features of reproductive biology, combined with stem and leaf morphology, support the inference that Bennettitales were predominantly plants of open habitats. Some may have been early-successional shrubs.

19.2.3 Gnetales, Erdtmanithecales and Elaterates

A distinctive feature of palynofloras from mid- and low palaeolatitudes, especially from the later part of the Early Cretaceous, is the abundance of a great variety of *Ephedra*-like pollen. Pollen of the *Ephedrites/Gnetaceaepollenites* complex may reach levels of 50%–70% in palynofloras from around the Barremian–Aptian boundary (Herngreen *et al.*, 1996). In low-palaeolatitude areas of northern Gondwana, the diversification of *Ephedra*-like pollen occurs more or less simultaneously with that of angiosperms (Crane and Lidgard, 1989, 1990). During the Aptian there is also evidence that fluctuations in the abundance of ephedroid pollen seem to follow similar fluctuations in the abundance of angiosperm grains (Doyle *et al.*, 1982).

The Erdtmanithecales are thought to be closely related to Gnetales and Bennettitales (Chapter 5) and may also have had similar reproductive biology. Currently there is no information on the vegetative parts of these plants, but in the Berriasian–Valanginian mesofossil floras of Bornholm, Denmark, abundant *Eucommiidites* seeds occur associated with abundant *Onychiopsis* and other plants with distinctive xeromorphic features (Pedersen *et al.*, 1989b). It is possible that Erdtmanithecales, like Early Cretaceous Gnetales, may have been plants of open habitats.

Also part of the BEG complex (Chapter 5) are a great variety of dispersed seeds (e.g. *Buarcospermum*, *Lignierispermum*, *Lobospermum*, *Rugonella*), which are abundant in some mid-Cretaceous mesofossil floras. Further information is needed on the leaves and pollen organs of these plants for a better understanding of their biology and ecology, but again, like Early Cretaceous Gnetales, the plants that produced them may have been important, perhaps weedy, colonisers of open, disturbed habitats.

The Elaterates are known only from dispersed polyplicate pollen with elaborate appendages (Chapter 5) that are recorded from Peru, through Brazil and West Africa to East Africa and New Guinea. They are characteristic of these areas during the mid-Cretaceous and give their name to the Elaterates Province (*sensu* Herngreen *et al.*, 1996). The size and complexity of these grains probably indicates that their parent plants were insect pollinated (Chapter 17). Like that of more conventional ephedroid pollen, the diversity and abundance of elaterate grains increases dramatically in palaeoequatorial regions in the Early Albian (Figure 19.3), but these grains are rare in post-Cenomanian sediments and disappear from the fossil record around the Coniacian.

19.2.4 Cheirolepidiaceae

In most Early Cretaceous floras conifers are both abundant and diverse. The extinct family Cheirolepidiaceae is especially conspicuous in the Early and mid-Cretaceous, but declines in importance through the Late Cretaceous. The youngest macrofossil occurrence of the family is from the Campanian–Maastrichtian of Portugal (Pons and Broutin, 1978). The Cheirolepidiaceae are known mainly from leaves and shoots, but are also characterised by distinctive and unusual pollen assigned to the genus *Classopollis* (Figure 3.8). These grains have a unique structure that is not comparable to pollen of any other conifer or seed plant. The ovulate structures of Cheirolepidiaceae are poorly understood, but also have features that may suggest an unusual reproductive biology.

Leafy shoots of Cheirolepidiaceae appear segmented because the leaves have elaborate, sheathing leaf bases with only a short free tip (Watson, 1988). In some species (e.g. *Pseudofrenelopsis parceramosa*) the shoots may have been deciduous (Watson and Alvin, 1996). Leaves are reduced, more or less xeromorphic, and often strongly so (Figure 3.8). Marginal hairs on the leaves of many species have been compared with similar hairs on the leaves of the extant Mediterranean conifer *Tetraclinis articulata*, where the hairs play a role in condensing moisture from the air during the night (Pons, 1979). Cuticles are often very thick, stomata are sunken, and there are well-developed papillae over the stomatal pits.

The Cheirolepidiaceae appear to have included large forest trees (Alvin, 1982, 1983) as well as smaller plants that may have been fleshy (Watson, 1977). They probably occupied a variety of habitats, especially in areas that were dry for much of the year or that had saline groundwater (Upchurch and Doyle, 1981; Alvin, 1983; Watson, 1988). Both physiological and climatic stress may have influenced their morphology. The Cheirolepidiaceae often appear to have formed stands that were almost monospecific (Francis, 1983, 1984; Watson and Alvin, 1996).

Reconstruction of the palaeoecology of a single species of Cheirolepidiaceae from the Purbeck Formation (latest Jurassic) of southern England (Francis, 1983, 1984) illustrates features that may have been characteristic of many

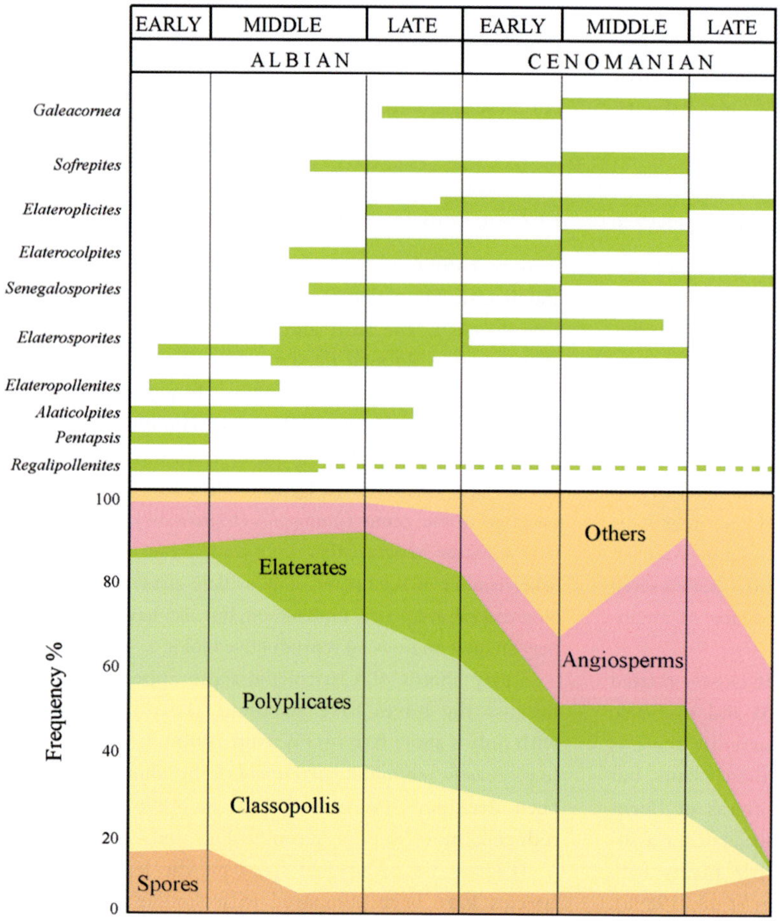

Figure 19.3 Stratigraphic range of genera of Elaterates pollen and changes in diversity patterns for the Elaterates and other major groups of pollen and spores in the Albian–Cenomanian of the Elaterates Province of Brazil. Note the prominence of the Elaterates (possible Gnetales), polyplicates (ephedroid) and *Classopollis* (Cheirolepidiaceae), all indicators for xeric conditions during the Albian and early part of the Cenomanian. All three groups declined drastically in the Late Cenomanian, while angiosperms increased in diversity during the Middle Cenomanian. Based on Dino *et al.* (1999).

members of the group. The trees were rooted in a well-drained calcareous soil and are known from associated wood (*Protocupressinoxylon purbeckensis*), leafy shoots (*Cupressinocladus valdensis*) and small pollen cones containing *Classopollis*. Associated plant fossils are sparse, consisting only of wood fragments (two other species of conifers), occasional stumps of *Cycadeoidea* and spores (lycopsids and ferns). Wood of the *Cupressinocladus valdensis* plant has narrow and highly irregular growth rings. The climate in which it was formed is interpreted as strongly seasonal, similar to Mediterranean regions of southwestern Australia. Summers were interpreted as very arid and separated by warm, wet winters during which tree growth took place. Occasional interruption of growth in the spring by hot dry spells is thought to be responsible for sporadic additional growth rings. Associated sediments also reflect a strongly seasonal climate. They include calcretes, as well as muds that contain both evaporites and freshwater fossils. Charcoal in the soil provides direct evidence of occasional forest fire that was not sufficiently intense to kill the large trees (Francis, 1983, 1984).

The Cheirolepidiaceae are particularly abundant at mid- and low palaeolatitudes and the pollen of *Dicheiropollis*, which characterises many Early Cretaceous palynofloras in palaeoequatorial regions, is also thought to be of cheirolepidiaceous affinity. *Classopollis* is frequently very abundant and may comprise as much as 80%–90% of some Early Cretaceous palynofloras. *Classopollis* pollen is relatively thick-walled, but wind pollination and high pollen production has generally been inferred. *Classopollis* may be relatively overrepresented in many dispersed palynofloras.

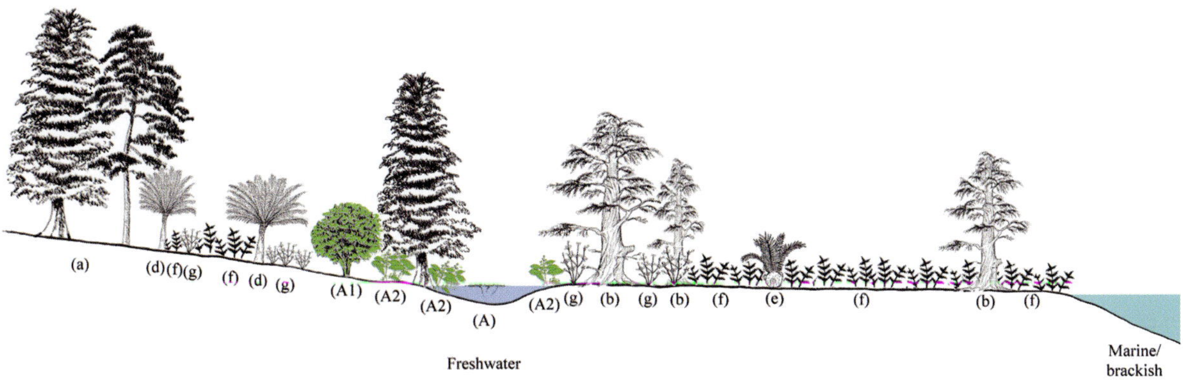

Figure 19.4 Reconstruction of Hauterivian–Barremian vegetation from southern Laurasia illustrating the vegetational context of earliest angiosperm diversification. Angiosperms were small understorey herbs or shrubs of patchy distribution; the aquatic habit was probably also established early among angiosperms. Angiosperms in green; for legend see Figure 19.5.

19.2.5 Other conifers

Shoots with spirally arranged, more or less scale-like leaves, generally assigned to *Brachyphyllum* or *Pagiophyllum*, are conifers of probably varied relationships. Some may be related to Cheirolepidiaceae. Others may be assignable to the extant family Cupressaceae, which also includes the extinct *Elatides* and *Sphenolepis* as well as extant *Sciadopitys*. Conifer shoots of this kind occur abundantly in some floras from the earliest Cretaceous. Some of the parent plants may have occupied habitats that were waterlogged for most of the year, similar to those occupied today by extant *Taxodium* and *Glyptostrobus*. Like their modern counterparts, Early Cretaceous conifers were almost certainly wind-pollinated and produced large quantities of wind-dispersed pollen grains. The group may therefore be relatively overrepresented in Cretaceous palynofloras. In southern Gondwana the occurrence of saccate conifer pollen different from that seen in Laurasia most likely indicates the presence of Podocarpaceae.

19.2.6 Ginkgoales and Czekanowskiales

Other seed plants that are present in earliest Cretaceous vegetation include Czekanowskiales (*Czekanowskia*, *Phoenicopsis*, *Solenites*) and Ginkgoales (e.g. *Ginkgo*). Both groups have an almost global distribution during the Early Cretaceous, but are much more abundant at higher northern palaeolatitudes (Figure 19.2). They are the predominant seed plants in many floras from the Arctic region (Samylina, 1974; Vakhrameev *et al.*, 1978).

The structure and biology of the Czekanowskiales, and especially their reproductive structures, are poorly understood (Chapter 5). Their most distinctive vegetative feature is the presence of deciduous short shoots bearing linear leaves, and their abundance at higher northern palaeolatitudes suggests that this feature may be related to climatic seasonality. In the north, growth may have come to a halt during the winter darkness, and the shoots may have been shed.

Early Cretaceous *Ginkgo* seems to have been similar to extant *G. biloba*, but with smaller seeds (Zhou and Zheng, 2003). Fossil *Ginkgo* was probably wind-pollinated and animal-dispersed as in extant *Ginkgo biloba*. Like Czekanowskiales, during the Early Cretaceous, *Ginkgo* is especially common in fossil floras from high northern palaeolatitudes (Samylina, 1974). It is also conspicuous in Early Cretaceous macrofossil floras from high latitudes in southern Gondwana (e.g. Drinnan and Chambers, 1986).

19.3 VEGETATION DURING THE EARLY DIVERSIFICATION OF ANGIOSPERMS

Fossil floras from the earliest Cretaceous (Berriasian–Barremian) provide information on the kinds of vegetation and landscapes in which angiosperms first evolved (Figure 19.4). The best-known and most intensively studied earliest Cretaceous macrofossil floras are from Europe. Fossil floras from the English Wealden (Chapter 4) include both palynofloras (e.g. Couper, 1958; Hughes and Moody-Stuart, 1967; Batten, 1973; Hughes, 1975, 1994) and diverse macrofossil assemblages (e.g. Seward, 1894, 1895; Watson, 1969; Watson and Sincock,

1992; Watson and Alvin, 1996; Watson *et al.*, 2001; Watson and Cusack, 2005, and Chapter 4). The Wealden is the old lithostratigraphic term for a mainly terrestrial sequence of pre-Aptian sediments from southern England that comprise the Hastings Beds (Berriasian–Valanginian) and the Weald Clay (Hauterivian–Barremian).

Floras from other parts of Europe, of about the same age as those from the Wealden, are very similar to the English floras in general composition. Plant assemblages of Berriasian–Valanginian age, corresponding to the Hastings Beds in southern England, include the classic Wealden-coal flora from Bückeburg, Germany (Osterwald and Obernkirchen members of the Bückeburg Formation, Pelzer and Wilde, 1987), and other floras from northwestern Germany studied by Dunker (1846), Schenk (1871), Benda (1961), Riegel *et al.* (1986) (see also Mägdefrau, 1968). Assemblages from Europe, corresponding in age to those of the Weald Clay (Hauterivian–Barremian), include the Quedlinburg flora first described by Richter (1906; 1909) and later by Daber (1968, 1990; see also Mägdefrau, 1968), the flora of Bernissart, Belgium (e.g. Dejax *et al.*, 2007) and several floras from northern Spain including the floras from the Las Hoyas Sub-Basin in Serranía de Cuenca (e.g. Diéguez *et al.*, 1995; Diéguez and Meléndez, 2000) and from the Sierra del Montsec (e.g. Barale *et al.*, 1984; Blanc-Louvel, 1984; Diéguez *et al.*, 2010).

Typically Berriasian–Valanginian floras in northwestern Europe are much more diverse than the Hauterivian–Barremian floras, with rich assemblages of ferns (e.g. *Onychiopsis* and *Ruffordia*), Bennettitales and conifers. The climate in northwest Europe during the Berriasian–Barremian has been interpreted as warm and seasonally dry (Riegel *et al.*, 1986; Watson and Alvin, 1996) with a short interval of more humid conditions in the later part of the Valanginian (Gale, 2000). Xeromorphic features of many of the fossil plants of this age clearly support interpretations of seasonal aridity and charcoal is common. Watson and Alvin (1996) suggest that periods of aridity were probably severe and often accompanied by lightning-induced fires. Plant assemblages from the Hastings Beds were interpreted as deposited on near-coastal mudplains with lagoons and sandy watercourses of variable salinity (e.g. Allen, 1976, 1981; Watson and Alvin, 1996). Increased fluvial activity culminated periodically in coalescing alluvial fans. Large flats extending over the alluvial plain were bare of trees and bushes, but areas where deposition and erosion were temporarily halted were colonised by rich growths of ferns and other pteridophytes (Watson and Alvin, 1996). Floodplain environments with meandering rivers, levees, back swamps and marsh lakes were also suggested for the contemporaneous Bückeburg Formation (Riegel *et al.*, 1986) and a similar near-coastal environment was also suggested for the contemporaneous plant assemblages of the Jydegård Formation of Bornholm, Denmark (Friis and Pedersen, 1990).

The depositional environment of the Hauterivian–Barremian Weald Clay of southern England reflects the transition from more active fluvial systems in the Hasting Beds, through reduced river flow and the development of extensive (often saline) mudswamps, into the marine conditions of the succeeding Greensand. Macrofossils from the Weald Clay are sparse and those plant assemblages that do occur are generally dominated by a single species. Charcoalified leaf fragments of *Weichselia reticulata* are common in lenses within the Weald Clay, but these rarely contain other plant fossils. Contemporaneous assemblages from Quedlinburg are also characterised by low diversity (Daber, 1968). Watson and Alvin (1996) suggest that the single species assemblages with *Weichselia* may reflect even-aged plant communities formed by 'stand replacement' fires that periodically destroyed entire plant communities.

The abundance of charcoalified matoniaceous and gleicheniaceous leaves at some localities, combined with an almost complete absence of woody seed plants, suggests a fern-prairie (Figure 19.4). Conifers and other woody plants were perhaps excluded by periodic fire under a climatic regime with a long dry season and brief, but stormy, pluvial intervals (Batten, 1975; Harris, 1981). Following burning, ferns would have sprouted from underground rhizomes. Numerous spores and seeds may also have been available for germination after having been released from their tough enclosing structures.

Much less information is available for the earliest Cretaceous vegetation at low palaeolatitudes. Palynofloras from the Berriasian and Hauterivian of Brazil and West Africa contain common pteridophyte spores and bisaccate conifer pollen that have been used to suggest humid conditions. Subsequently, a trend toward drier climates has been inferred from the increase in the diversity and abundance of ephedroid pollen during the Barremian (Herngreen *et al.*, 1996). In the Barremian of Gabon, sediments of the developing mid-Atlantic rift valley (Cocobeach Sequence), which contain some of the first diverse assemblages of monosulcate angiosperm grains, also have a high proportion of ephedroid pollen relative to pollen of cheirolepidiaceous conifers (Doyle *et al.*, 1982). This has been used to infer that early angiosperms initially colonised habitats in which ephedroid plants were common. The palynological association between angiosperm and ephedroid pollen is also consistent with evidence from the macrofossil assemblage in the Drewry's Bluff Leaf Bed (Potomac Group) that at least some early Cretaceous Gnetales and angiosperms grew in similar habitats (Chapter 5; Crane and Upchurch, 1987). It is also interesting that in the Hauterivian–Barremian of southern England angiosperm and ephedroid pollen make their first appearance at the same stratigraphic level (Hughes, 1994).

Angiosperms are unknown from high palaeolatitudes during the Berriasian–Barremian. Macrofossil floras of this age from the Arctic are characterised by low species diversity and are dominated by ferns such as species of *Coniopteris*. Many of the fern genera known further south, such as *Onychiopsis*, *Ruffordia* and *Matonidium*, are either rare or lacking. Palynofloras are also characterised by low species diversity and are mainly composed of fern spores and bisaccate pollen of conifers. High-palaeolatitude palynofloras from southern Gondwana show the same general composition and in both hemispheres Cheirolepidiaceae and Gnetales are not prominent in high-palaeolatitude floras.

19.4 EARLY ANGIOSPERMS: DIVERSITY IN OBSCURITY

From around the Late Barremian – Early Aptian angiosperms are reported from numerous low- and mid-palaeolatitude palynological assemblages and are observed in mesofossil floras for the first time (Figures 19.4, 19.5). The earliest angiosperm leaves are apparently also of this age although some have been described as Hauterivian. In palynofloras angiosperm abundance and diversity is typically very low, but mesofossil floras show that angiosperm diversification was already well under way by the Late Barremian–Aptian, and so far, the mesofossil floras containing the most diverse angiosperm assemblages are those from Portugal.

From the Late Barremian – Early Aptian flora of Torres Vedras, about 50 different taxa of angiosperms are distinguished based on flowers, fruits, seeds and dispersed stamens (Friis *et al.*, 2010b). The slightly younger flora of Famalicão from the Late Aptian includes more than 100 different kinds of angiosperms, mostly based on fruits and seeds (Eriksson *et al.*, 2000b). The diversity of angiosperms in other mesofossil floras from the later part of the Early Cretaceous is also high. It is significant that the angiosperm diversity observed in the mesofossil floras is not matched by a similar diversity of angiosperm pollen in palynofloras, or by fossil leaves in macrofossil floras. It is also interesting that mesofossil assemblages collected at different points along the same horizon are often very different in terms of their composition, suggesting that the individual assemblages represents local, perhaps patchy, populations of different kinds of angiosperms.

Studies of the Late Barremian – Early Albian mesofossil assemblages suggest that many Early Cretaceous angiosperms were herbs or small shrubs. The systematic relationships of many of these early angiosperms are uncertain, but many appear to be related to ANITA-type angiosperms, Chloranthaceae and probably early monocots (Chapter 20). Eumagnoliids are present, but apparently increased in diversity later in the Cretaceous and are not prominent at this time. Eudicots appeared around the Late Barremian – Early Aptian boundary. The earliest forms were probably herbaceous or small shrubs related to basal grade eudicots such as Ranunculales and Buxales. Eudicots did not rise to dominance until the mid-Cretaceous (Chapter 20).

There are several indications that the Late Barremian – Albian floras, at least at mid-palaeolatitudes, grew in more humid climates than prevailed earlier in the Cretaceous, but the presence of Cheirolepidiaceae and Gnetales, as well as charcoal in some horizons, indicates that climate continued to be seasonally dry and that the vegetation was probably relatively open. In palynofloras of this age Schizaeaceae and Gleicheniaceae are conspicuous, and among conifers Pinaceae and probable Podocarpaceae, are especially well represented. Pollen of *Classopollis* (Cheirolepidiaceae) is also locally abundant.

The habitats occupied by early angiosperms are unknown, but based on the common occurrence of delicate flowers and other floral parts it is very likely that angiosperms grew close to depositional environments and therefore probably in relatively humid environments. This is also supported by extrapolation from the closest relatives of those fossil taxa that can be assigned to modern families or orders (e.g. Chloranthaceae, Araceae) and the texture of early angiosperm leaves that frequently have an herbaceous appearance (e.g. Barale and Ouaja, 2002). Based on extrapolation from the ecology of extant angiosperms and their phylogenetic relationships it has also been suggested that early angiosperms grew in dark, disturbed and damp environments (e.g. Feild *et al.*, 2004). The presence of

Figure 19.5 Summary of vegetational changes through the Cretaceous and Cenozoic in southern Laurasia for selected intervals from the Berriasian to Eocene showing angiosperm expansion. (a) conifers; (b) cheirolepidiaceous conifers; (c) ginkgoaleans; (d) cycads; (e) bennettitaleans; (f) ferns; (g) ephedroids; (A1) angiosperm shrubs; (A2) angiosperm herbs; (A3) angiosperm trees, insect-pollinated; (A4) angiosperm trees, wind-pollinated; (A5) angiosperm trees, canopy; (A: angiosperm aquatics, shown only in Figure 19.4) Angiosperms in green. Modified from Crane (1987).

aquatic angiosperms early in angiosperm history is also securely documented by the presence of Nymphaeales (Chapter 20), *Archaefructus* and angiosperm leaves that indicate an aquatic habitat (Barale and Ouaja, 2002; Bravi *et al.*, 2010).

At low palaeolatitudes in Brazil many of the angiosperms preserved in the Crato Formation were also clearly herbaceous and several were probably aquatic. However, the non-angiospermous component of the Crato flora shows structural features consistent with arid environments (fibrous leaves, sunken stomata, well-developed indumentum, possible deciduous branches).

The fossil data thus suggest that angiosperms were diverse in the Early Cretaceous at least from the Late Barremian – Early Aptian onwards, but were not especially prominent in the vegetation. They occurred somewhat in obscurity in small scattered populations (Friis *et al.*, 2010b). Their diversity was at the level of present day ANITA and similar taxa, and in habit they were low stature plants. They grew in the more wet environments in an otherwise open and semi-arid forest vegetation dominated by various conifer trees including *Brachyphyllum*-like plants, Cheirolepidiaceae and probable Araucariaceae. They were sometimes part of understory or weedy, colonising plant communities, together with ferns, various ephedroids and related (BEG) plants.

19.5 MID-CRETACEOUS VEGETATION

By the mid-Cretaceous the geographic expansion of flowering plants into high latitudes, at least in Laurasia, was essentially complete (Figure 19.2). Angiosperm leaves, reproductive structures, and pollen from the mid-Cretaceous are more diverse in their systematic relationships than in older floras. Angiosperms producing monoaperturate pollen (monocots and magnoliids *sensu lato*) continue to be important, but the proportion of eudicots in fossil floras has increased markedly. Core eudicots are present for the first time, although they are still rare. In general the physiognomic diversity among mid-Cretaceous leaf macrofossils is also much greater than in earlier macrofossil floras (Doyle and Hickey, 1976; Hickey and Doyle, 1977; Upchurch and Wolfe, 1987). According to Upchurch and Wolfe (1987) the new foliar physiognomic types that appear at this level all have features of leaves of early successional plants, whereas other kinds of leaves, for example leaves with drip-tips, or vine-like forms, are absent or rare. The non-angiosperm component in mid-Cretaceous floras consists mainly of various lycopsids, ferns and conifers. Fern families reported for the Early Cretaceous, such as Schizaeaceae, Gleicheniaceae, Osmundaceae and Dicksoniaceae, continue to be important.

Eudicots are important elements in mid-Cretaceous floras and for the first time they may be the dominant angiosperm component (Figure 19.5). However, the early-diverging lineages that comprise the basal grade of extant eudicots are the most prominent. Triporate pollen, probably produced by core eudicots, appears for the first time in the Middle Cenomanian of Europe and eastern North America (e.g. Pacltová, 1971; Doyle and Hickey, 1976) and begins to diversify rapidly soon afterwards, foreshadowing the later diversification of the Normapolles group.

Especially important in mid-Cretaceous floras from mid-palaeolatitudes in Laurasia are the leaves of the platanoid–*Sapindopsis* complex, interpreted as leaves of plants of early-successional thickets. These leaves were probably all produced by plants closely related to extant Platanaceae. This platanoid–*Sapindopsis* complex is not present in the earlier floras from Portugal, but is common, both in the Potomac Group, and elsewhere, during the mid-Cretaceous (Chapter 20). Many of these platanoid fossils are preserved in presumed levee sands (e.g. Retallack and Dilcher, 1981a), and the abundance of platanoid leaves in such environments recalls the streamside habit of extant *Platanus*. Platanoid leaves from the Potomac Group sediments were inferred to have been produced by plants that grew in a range of successional habitats (Doyle and Hickey, 1976; Hickey and Doyle, 1977) and in several places platanoid leaves occur in probable near-coastal settings (e.g. Mägdefrau, 1968; Golovneva, 2007).

Angiosperms also increase in importance in low palaeolatitude regions through the mid-Cretaceous. Typically, by the Cenomanian, angiosperm pollen may account for 70% of the pollen and spores in some palynofloras (Herngreen *et al.*, 1996) and often accounts for 50% of the species. Among the characteristic and often abundant angiosperm pollen grains are a great variety of tricolpate and tricolporate species.

An important component of low-palaeolatitude vegetation during the mid-Cretaceous were plants producing Elaterates and ephedroid (polyplicate) pollen (Figure 19.3). The Elaterates (Chapter 5) are restricted to the low palaeolatitude so-called Elaterates Province, while ephedroids extend further south and north, even though they are more diverse and abundant in low palaeolatitudes. The climate at low palaeolatitudes in the mid-Cretaceous is thought to have been semi-arid or drier, perhaps with more humid conditions persisting in northwestern South America. The climate evidently changed at the end of the Cenomanian as indicated by the abrupt decline of *Classopollis* and ephedroids and the loss of Elaterates (Figure 19.3). These events on land also correspond to a significant extinction among marine invertebrates, and may be related to the development of a permanent connection between the North and South Atlantic (Herngreen *et al.*, 1996).

At high palaeolatitudes in Laurasia Ginkgoales and Czekanowskiales continue to be of some importance (Figure 19.2) and all of the woody species are interpreted as deciduous. Woods from the Late Cenomanian of the North Slope of Alaska have large and distinct growth rings with practically no latewood (Spicer and Parrish, 1986), indicating abundant precipitation through the growing season and the rapid onset of winter dormancy (Upchurch and Wolfe, 1987). At high palaeolatitudes in southern Gondwana araucariaceous and podocarp conifers are especially significant components of the vegetation.

19.6 LATE CRETACEOUS VEGETATION AND FLORISTIC PROVINCES

Angiosperms continue to proliferate through the Late Cretaceous and eudicots become the dominant angiosperms in almost all floras (Figure 19.5). During the Late Cretaceous floristic provinciality, as assessed from palynofloras, reached its maximum for the whole of the Cretaceous, probably reflecting wider separation of the landmasses as a result of continued breakup of Laurasia and Gondwana, together with the extensive transgressions and generally higher sea levels.

At mid-palaeolatitudes in Europe and eastern North America the Normapolles Province is especially distinctive. It is bounded in the west by the Western Interior Seaway and in the east by the Turgai Straight. The province takes its name from the abundant occurrence of various kinds of Normapolles pollen and more than 80 Normapolles genera have been recognised (Batten and Christopher, 1981). The maximum diversity of Normapolles pollen occurs in Europe. A variety of Normapolles grains have now been found in flowers and their relationship to core Fagales is well established (Chapters 14, 20). In general, the group is thought to have been wind-pollinated and associated with rather dry climates. The thick pollen wall and complex apertures of most Normapolles grains may have been an adaptation to such conditions. Limited leaf physiognomic data indicate that more arid conditions persisted in Europe than in eastern North America at this time (Wolfe and Upchurch, 1987).

At high palaeolatitudes in Laurasia different kinds of pollen predominate in the *Aquilapollenites* Province. *Aquilapollenites* palynofloras had a circumboreal distribution down to about 50°N, except in western North America and eastern Asia where they extend further south. Palynological assemblages from the *Aquilapollenites* Province are characterised by abundant and diverse pollen taxa belonging to the *Aquilapollenites*/Triprojectate Group, such as *Aquilapollenites, Manicorpus* and *Triprojectus* possibly related to Santalales (Chapter 13).

At mid-palaeolatitudes in the Late Cretaceous of western North America Wing and Tiffney (1987b, 1987a) envisaged an open landscape consisting of herbaceous angiosperms and ferns, with scattered low angiosperm shrubs. Based on analogies with recent vegetation associated with large herbivores (e.g. East Africa) they suggested intensive grazing by herbivorous dinosaurs as a major factor contributing to the maintenance of an open structure in these Late Cretaceous plant communities. Fire may also have been an important factor, although perhaps to a lesser extent than in the Normapolles Province.

Based on the leaf physiognomy of Campanian and Maastrichtian floras from the Northern Hemisphere, Wolfe and Upchurch (1987) suggested that almost all plants were exposed to full light intensity, but concluded that open conditions were controlled in large part by climatic factors rather than grazing. As possible analogues they pointed to the low-seasonality climate of New Caledonia, and stunted, open, savannah-like vegetation growing on nutrient-poor podsols in the tropics. Climate, large herbivore disturbance, poor soils, and perhaps most importantly fire, may all have been important features of the ecosystems that sustained the rich Late Cretaceous dinosaur faunas of both the *Aquilapollenites* Province and the Normapolles Province in the Late Cretaceous.

At low palaeolatitudes, in the later part of the Late Cretaceous, a distinctive palynological assemblage (the Palm Province) developed across tropical South America, tropical Africa and India presumably reflecting increased moisture and declining aridity. This is manifested in an increased diversity and abundance of palm-like monocolpate pollen that may constitute up to 10%–15% of the pollen flora (Herngreen *et al.*, 1996). In the Maastrichtian Deccan Intertrappean Beds of India there are diverse flowers and fruits of probable palms (e.g. *Deccananthus,*

Nypa, Tricoccites, Palmocarpon, Viracarpon). These occur along with fruits of banana-like monocots (Zingiberales) and a variety of reproductive structures suggestive of warm, wet and perhaps tropical rainforest conditions. Other macrofossil floras are rare from the Palm Province.

At high palaeolatitudes in the Southern Hemisphere, floras are broadly similar and assigned to the *Proteacidites/Nothofagidites* Province (Srivastava, 1978; Herngreen *et al.*, 1996). Southern Africa, Madagascar and India, however, are distinctly different. *Nothofagus* appears never to have been part of the vegetation in these areas. The *Nothofagidites* Province is characterised by the first appearance of *Nothofagidites* pollen (referable to Nothofagaceae) at around the Santonian–Campanian boundary (Herngreen *et al.*, 1996). Subsequently, pollen types attributed to extant subgroups of *Nothofagus* begin to appear in the Campanian and Maastrichtian.

20

The accumulation of angiosperm diversity

The palaeontological information summarised in earlier chapters provides direct historical evidence of the pattern of angiosperm evolution through time. Such information is an important complement to contemporary approaches that use data from living plants to look back into evolutionary history. At the same time, living plants are the essential points of reference for any meaningful interpretation of the palaeontological record. Only from studies of the modern world can we understand the lives of plants and the roles they play in ecological systems. For these reasons the emphasis in this book has been on the integration of information from living and fossil plants, which we see as vital for a full understanding of plant evolution.

In this chapter we provide a brief integrated overview of the major patterns of angiosperm evolution as revealed by studies of living plants and the fossil record. We also provide more detailed consideration of selected groups of angiosperms with a particularly interesting or informative palaeontological history. Finally, we conclude with a brief consideration of angiosperm evolution through the Cenozoic to make the temporal connection between our primary focus on the Cretaceous fossil record, and the diversity of angiosperms that exists today.

20.1 LARGE-SCALE PATTERNS IN ANGIOSPERM DIVERSIFICATION

Angiosperm fossils that are about 135 million years old are known from around the Valanginian–Hauterivian boundary in the Early Cretaceous. These earliest fossils are dispersed pollen grains for which parent plants are unknown. However, by about 10 million years later, around the Barremian–Aptian boundary, there are rich assemblages of angiosperm reproductive structures from several localities, as well as unequivocal angiosperm leaves. In most cases these early angiosperm fossils comprise individual isolated organs, but many instances of flowers with pollen *in situ* confirm that dispersed pollen grains of this age, which were previously inferred to reflect the presence of angiosperms, were indeed produced by extinct members of the group. In a few instances, flowers and leaves attached to the same fossil specimen confirm that characteristic angiosperm flowers were produced on plants that also bore characteristic angiosperm leaves.

In Barremian–Aptian palynofloras angiosperm pollen grains are represented mainly by monocolpate grains, or by grains with apertures that are clearly modifications of a basically monocolpate form. Most of these early angiosperm fossils probably indicate the presence of the angiosperm crown group. This is also consistent with the relationships of those fossils for which the systematic affinities are clear. Relationships are with groups of extant angiosperms at or around the ANITA grade (e.g. Chloranthaceae, Nymphaeales), or with a few lineages that diverged early in the history of eumagnoliids, eudicots and monocots. However, the diversity of all three major groups of extant angiosperms is very restricted in the earliest phases of the angiosperm fossil record. Among eumagnoliids, there is evidence of possible stem-group Winteraceae (Canellales). Among monocotyledons, so far, only Araceae have been recognised unequivocally. In the case of Araceae and Nymphaeaceae, and perhaps also Chloranthaceae, the likely phylogenetic position of these early fossils is in the crown group of the extant family. There is no evidence in Barremian–Aptian fossil assemblages of diverse eumagnoliids, or of later-diverging lineages of angiosperms, such as commelinid monocots, or rosid or asterid eudicots. Eudicots are especially sparse at this level. Insofar as the relationships of these earliest eudicots to extant lineages are known, they are related to early-diverging lineages within the clade, such as Buxales. In early mesofossil floras there are a very large number of well-preserved angiosperm fossils that are of uncertain relationship in terms of extant families and orders, and that appear to represent lineages of early crown-group angiosperms that today are extinct.

During this early phase of angiosperm evolution angiosperm pollen is generally much less abundant than that of other groups of seed plants. Only occasionally do angiosperm flowers, fruits and seeds dominate mesofossil floras. To judge from the systematic affinities of the angiosperms recognised so far, and also from more direct evidence provided by leaves and wood, the habit of Barremian–Aptian angiosperms included small shrubby, herbaceous and aquatic forms.

By around the Early–Late Cretaceous boundary about 100 million years ago, angiosperm diversity, and the number of extant taxa that can be recognised, had increased considerably. Diverse eumagnoliids are known by this time. They include Calycanthaceae and Lauraceae (Laurales), Winteraceae (Canellales) and Magnoliaceae (Magnoliales). The monocot record remains poor, but this may be explained in part by the difficulty of recognising early monocots unless they have very distinctive features characteristic of an extant order or family. It may be significant that fully convincing evidence of monocots from the Early Cretaceous did not come until the discovery of the very distinctive pollen of *Mayoa* and the highly characteristic inflorescences of fossil Araceae. The pollen grains or stamens from the inflorescences would not have been recognised as having been produced by monocots if they were only known as isolated fossils.

Among eudicots, by around the Early–Late Cretaceous boundary, there is clear evidence of several lineages (e.g. Buxales, Proteales) that appear to have diverged at a relatively early stage from the main line of eudicot evolution based on phylogenetic analyses of data from extant plants. The 'Rose Creek flower' also provides reasonably secure, early documentation of core eudicots, but otherwise, in the latest Albian and earliest Cenomanian, evidence of core eudicots is relatively sparse. As in the Barremian–Aptian, there is no evidence of later-diverging lineages of angiosperms, such as commelinid monocots, core rosids or the earliest-diverging lineages of asterid eudicots, all of which are well represented in the Late Cretaceous.

By around the Early–Late Cretaceous boundary it is clear that, at least in certain kinds of vegetation and in some geographic locations, angiosperms were ecologically dominant. This is reflected in the common occurrence of angiosperm-dominated macro- and mesofossil floras. Angiosperms also dominate some palynofloras. Increased ecological diversity is also reflected in an increased range of leaf types as well as occasional larger fragments of angiosperm wood. However, in some ecological settings, as judged by palynofloras, and also by some macrofossil assemblages that are preserved more or less *in situ*, angiosperms apparently remained subordinate to other elements in Cretaceous vegetation. This may suggest that, at least in the initial phases of angiosperm diversification, phylogenetic and ecological diversification may have been partly decoupled, as appears to occur, for example, for Poaceae in the Cenozoic. The family diversified in the Early Cenozoic, but underwent its most dramatic ecological expansion only later (Strömberg, 2005).

Through the early part of the Late Cretaceous, and continuing through to the Cretaceous–Palaeogene boundary, angiosperm diversification continues rapidly. Lineages such as Chloranthaceae evidently remained important and the presence of very distinctive *Chloranthus*-like plants in three different mesofossil floras from the Late Cretaceous is striking in the context of the diminutive role that extant *Chloranthus* and *Sarcandra* play in modern vegetation.

Eumagnoliids, especially Lauraceae, remain important in many mesofossil floras through the Late Cretaceous, and new forms that can be linked to extant taxa (e.g. Annonaceae) appear for the first time. However, in Late Cretaceous floras there are also unequivocal eumagnoliids that are not readily accommodated in extant taxa (Chapter 10). These extinct forms appear to fall outside the crown groups of the relatively few families of eumagnoliids that exist today.

Even though the fossil record of monocots remains relatively sparse through the Late Cretaceous, the unequivocal first appearance of palms and Zingiberales implies the presence of many other monocot lineages (Chapter 11). There is also evidence of monocot diversification at low palaeolatitudes later in the Cretaceous. The diversity of probable palm pollen in South America and Africa during the later part of the Late Cretaceous provides an important body of evidence on monocot diversification that still remains to be studied in detail. Further information on the monocots preserved in the Deccan Intertrappean Beds will also be important to develop a more detailed perspective on the diversification of the group prior to the Early Cenozoic, when many extant taxa at the level of families and genera are recognised for the first time.

Most prominent in the Late Cretaceous diversification of angiosperms is a massive and rapid increase in the diversity of core eudicots. This is reflected in the increasing diversity and abundance of triaperturate pollen as well as the first appearance of extant lineages of both rosids (e.g. Fagales) and asterids (e.g. Cornales, Ericales).

As a consequence of the Late Cretaceous diversification of eudicots and monocots, by around the Cretaceous–Palaeogene boundary about 65 million years ago angiosperms were diverse at all latitudes and in most ecological settings. As in the mid-Cretaceous, angiosperm abundance in some situations may have been lower than might be implied by their diversity, but at many other localities angiosperms were clearly the dominant floristic element. By the Maastrichtian there is also increased evidence of large trees, suggesting that the nature of angiosperm-dominated forest was also beginning to change.

The factors that led to the diversification and rise to dominance of angiosperms through the Cretaceous have been widely debated. However, in our view, palaeobotanical and phylogenetic patterns provide no strong evidence in support of a single causal factor. It seems more likely that the success of angiosperms reflects an ongoing cascade of evolutionary innovation resulting in a great variety of adaptive types. Many different factors must have underpinned repeated local ecological success as environments changed. Ongoing geological changes that had profound effects on palaeogeography, palaeoenvironment and climate may also have helped create the conditions under which angiosperms were successful. The massive accumulation of species diversity most likely resulted from both high speciation rates, perhaps stimulated by several different aspects of angiosperm reproductive biology, and low extinction rates. Ultimately angiosperm species overwhelmed the pre-existing vegetation from the earliest Cretaceous and latest Jurassic in almost all habitats.

An interesting question is whether the tempo of angiosperm diversification varied through the Cretaceous. A large-scale quantitative analysis of floristic patterns for the whole of the Phanerozoic using summed diversity for different groups of land plants per interval of geological time (e.g. Knoll *et al.*, 1979; Niklas *et al.*, 1980; Tiffney, 1981; Tiffney and Niklas, 1990) shows a gradual rise in angiosperm species diversity, beginning in the Barremian and extending through the mid- and Late Cretaceous, such that, by the end of the Cretaceous, angiosperms only account for about 40% of all land plant species. However, analyses based on summed diversity using leaf fossils suggest a much more rapid and dramatic rise to systematic, floristic and vegetational dominance (e.g. Crane, 1987). This is also more consistent with independent analyses of changes in within-flora diversity based on both fossil pollen and leaves (Lidgard and Crane, 1990). Other approaches have sought to assess rates of evolutionary innovation in the evolution of key characters (Doyle and Hickey, 1976; Hickey and Doyle, 1977; Lupia *et al.*, 1999) and also rates of modernisation in terms of the recognition of extant families (Muller, 1981).

Based on these studies, and also on the patterns documented in previous chapters, the initial phase of angiosperm diversification during the Early Cretaceous may have been somewhat gradual, but from the mid-Cretaceous onwards the accumulation of angiosperm species seems likely to have been rapid. It is also clear that angiosperm diversification was by no means complete by the end of the Cretaceous. Much of the diversity of living angiosperms seems to have accumulated relatively recently, and hence relatively rapidly, as is also implied by inferences using molecular data from extant taxa. Angiosperm species diversity may have more than doubled since the earliest Cenozoic.

Securing a deeper understanding of how angiosperm species diversity accumulated is difficult. Estimating speciation and extinction rates directly from the appearance and duration of species in the fossil record, as has been done, for example, for various groups of marine invertebrates, is complicated in land plants by several factors. One concern is that estimates based on 'species' of single organs, such as pollen grains or leaves, will potentially mask changes occurring in other organs that would be recognised if the whole plant were available. This is a particular problem with pollen species, which often have long stratigraphic durations. Another concern is that the reverse is true with the angiosperm leaf record. Compilation of the stratigraphic ranges of fossil leaves though the Cretaceous show that most leaf species occur in only one fossil flora (P. Crane and S. Lidgard, unpublished data). In effect they therefore have no stratigraphic range, and rates of speciation and extinction are thus impossible to calculate in any meaningful way. It will be some time before the angiosperm fossil record is sufficiently well known that it can provide useful estimates of speciation and extinction rates for different angiosperm subgroups.

In the absence of appropriate palaeobotanical data, alternative attempts to examine rates of speciation and extinction in different groups of angiosperms rely on a variety of modelling approaches. These approaches have no direct way to account for the potential effect of extinct diversity on both speciation and extinction rates. For relatively young clades that are rich in species diversity today, this may not be a major problem. The living sample of all species that have ever existed within the clade may be relatively complete. But for older clades, especially those with limited extant diversity, there is a much greater likelihood of significant extinction, and hence that extant taxa are a poor sample of the full diversity of that group over time. This may also have important consequences for models of phylogenetic relationships. Ineffective sampling of extant taxa in phylogenetic studies can have an important impact on phylogenetic results. This will also be the case when sampling is constrained by extinction.

20.2 PATTERNS OF ANGIOSPERM DIVERSIFICATION: EARLY LINEAGES

A striking feature of the Early Cretaceous record of angiosperm flowers, fruits and seeds is that many of the fossil taxa, even those that are relatively well understood, are not readily assigned to extant taxa. Examples such as the *Hedyosmum*-like staminate and pistillate inflorescences (section 20.2.2) are the exception rather than the rule. The palaeobotanical evidence of considerable extinction at this level of angiosperm evolution is incontrovertible. Consequently, while the extant Amborellaceae, Nymphaeales, Austrobaileyales, Chloranthaceae and Ceratophyllaceae are the only lineages from this early phase of angiosperm evolution that have survived until the Recent, they are unlikely to be fully representative of the diversity that once existed early in angiosperm history. In the case of Nymphaeales there has also been considerable more recent diversification. Extrapolations from extant taxa about the biology, ecology, and perhaps even phylogenetic interrelationships of the plants involved in the earliest phases of angiosperm diversification therefore need to be made with caution.

As the record of early flowers becomes better known, and as the morphological features that define extant groups become better understood, it seems likely that many fossils from early mesofossil assemblages will be accommodated in various positions close to the base of the angiosperm phylogenetic tree. Some probably belong to the crown group of extant families (e.g. *Monetianthus*) or perhaps even to extant genera (e.g. *Hedyosmum*-like fossils). Others may be on the stem group to more diverse extant clades, such as eudicots, monocots and eumagnoliids, or may fit between clades comprised of one or more of these extant taxa (e.g. *Canrightia*). However, a further intriguing possibility, as has been suggested already for *Archaefructus* (section 9.2.1), is that some of these taxa may belong on the stem group to angiosperms as a whole.

In our view, so far, the evidence for placing Early Cretaceous angiosperm-like fossils outside the angiosperm crown group is weak. A more interesting and more likely possibility, as considered in Chapter 6, is that one or more of the groups of extinct seed plants may ultimately be recognised as having some, but not all, of the characteristic features of angiosperms. Although such plants may be much less angiosperm-like than Early Cretaceous plants such as *Archaefructus*, they could be much more significant for understanding angiosperm evolution. Fossil plants such as *Caytonia* and corystosperms could belong on the angiosperm stem group, especially if the basic anatropous condition of the angiosperm ovule is fundamentally homologous with their recurved ovule-bearing structures, and therefore if the angiosperm outer integument is homologous to the cupule. It would be of great interest if that homology also extended to the envelope seen in the seeds of the BEG group.

If the long-discussed idea of homology between the anatropous ovule of angiosperms and the cupule of *Caytonia* and corystosperms ultimately finds strong support it would also imply that early angiosperm pollination, with stigmatic germination but a fluid-filled carpel, evolved from a situation where the ovule was inverted. This, in turn, implies that angiosperm ancestors possessed a pollination drop system involving the upward flotation of saccate pollen. Enclosure of the ovule to form a carpel, and the evolution of the characteristic angiosperm stamen, would both be subsequent innovations on the lineage leading to angiosperms.

20.2.1 Nymphaeales

The fossil record of Nymphaeales is based largely on their characteristic, hard-walled seeds. Fossil leaves of Nymphaeales are much less common. As in many other water plants the leaves are not shed, the leaf tissue is soft, and the leaves usually decompose while still attached to the plant. Leaves of Nymphaeales are likely to fossilise only under special conditions.

Until relatively recently, a striking feature of the fossil history of angiosperms has been the very sparse record of Nymphaeales during the Cretaceous, despite their extensive representation based on seeds during the Cenozoic. However, as predicted by phylogenetic analyses of extant taxa, new palaeobotanical data confirm that the Nymphaeales were indeed present at a very early stage in the evolutionary history of angiosperms. Evidence of Nymphaeales in the Early Cretaceous formerly rested entirely on leaves similar to those of extant forms, but is now supported by several records of probable nymphaealean seeds and a few probable nymphaealean floral structures.

For most of these fossils their precise position within Nymphaeales is difficult to evaluate, but the fossil flower *Monetianthus mirus* from the Early Cretaceous of Vale de Agua is especially significant because it can be placed with reasonable confidence as sister to extant *Barclaya* and Nymphaeoideae in crown-group Nymphaeaceae. In turn, this implies the presence of other lineages of Nymphaeales, including Cabombaceae and Hydatellaceae. While the presence of the Nymphaeales in the Early Cretaceous is unsurprising given its phylogenetic position, the recognition of very early diversity among crown-group Nymphaeaceae is an unanticipated result. It implies a much earlier diversification into the lineages of several modern genera than had previously been suspected.

While *Monetianthus* provides secure evidence of the nymphaealean lineage at a very early stage in angiosperm evolution, more direct evidence of the aquatic habit in early

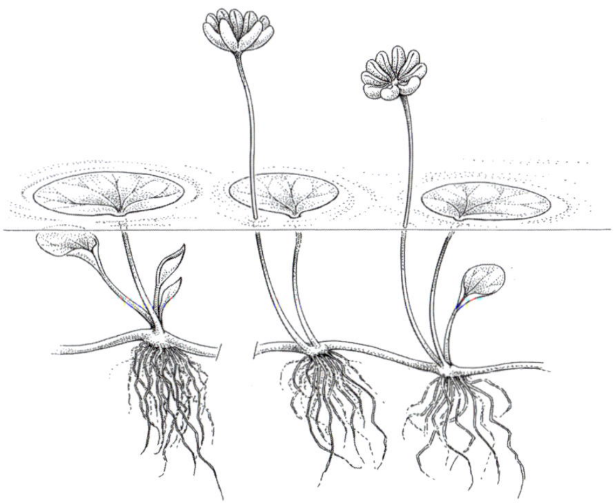

Figure 20.1 Reconstruction of *Pluricarpellatia peltata*, an aquatic plant from the Early Cretaceous (Late Aptian – Early Albian) Crato Formation, Brazil, probably related to Cabombaceae. The plant had creeping rhizomes that gave rise to slender stalks bearing floating leaves and flowers/fruits that probably emerged above the water surface. Based on Mohr *et al.* (2008).

members of the group comes from the similarity between leaves of extant taxa and putative nymphaealean fossils from the Early Cretaceous. These include *Braseniopsis venulosa* from Portugal and *Scutifolium jordanicum* from Jordan, as well as *Menispermites 'tenuinervis'* and perhaps *Populophyllum reniforme* from the Potomac Group of eastern North America. Further evidence is now provided by similarities in both leaves and general habit between extant Nymphaeales and Early Cretaceous *Pluricarpellatia peltata* (Figure 20.1), which is preserved as a whole plant. Early Cretaceous forms probably inhabited aquatic environments similar to those of extant Nymphaeales and were also bottom-rooted water plants. *Archaefructus* may also be an early member of this clade of early angiosperm aquatics (Chapter 9).

Ideas on the evolution of floral structure and biology in Nymphaeales come mainly from studies of phylogenetic relationships among extant taxa. Based on morphological analyses and molecular sequence data, the earliest Nymphaeales probably had smaller, simpler flowers than extant *Nymphaea*, *Euryale* and *Victoria*. They probably also lacked the kinds of sophisticated trap pollination mechanisms seen in these genera. The small size of *Monetianthus* is consistent with the size of other Early Cretaceous flowers, but is much smaller than the flowers of most extant taxa. Only *Brasenia*, *Cabomba* and *Barclaya* have flowers that may be of comparable size. The placement of *Hydatella* as the probable sister group to all other Nymphaeales also raises interesting questions about patterns of floral evolution within the group.

Recent studies have expanded information on Early Cretaceous Nymphaeales, but during the Late Cretaceous the fossil record of Nymphaeales remains sparse. There are occasional records of dispersed nymphaealean seeds, but these are not common in the Late Cretaceous mesofossil floras that we have studied, especially compared to the situation in the Cenozoic where seeds of Nymphaeales are abundant in many fossil assemblages (e.g. Collinson, 1980; Mai, 1995). From the Early Cenozoic onwards Nymphaeales were widespread and common in certain kinds of aquatic habitats. Furthermore, the occurrence of seeds very similar to those of several extant genera in the Early Cenozoic, as well as seeds that are intermediate between those of extant genera (e.g. *Sabrenia*, Collinson, 1980), may indicate that active diversification was still under way. Some extant genera may already have differentiated, but they were also accompanied by plants with intermediate morphological features. The diversification that produced the more than 40 extant species of *Nymphaea* almost certainly occurred in the Cenozoic and may well have been an even more recent event that occurred mainly in the Neogene (Figure 20.2). Similarly, the very large flowers of *Victoria*, and the very large leaves of both *Victoria* and *Euryale*, which make these genera so distinctive, are almost certainly secondary, and perhaps quite recent, developments.

An important issue is whether the apparently rare occurrence of Nymphaeales in the Late Cretaceous fossil record reflects a bias in the kinds of sediments preserved, or so far examined, or whether it reflects a more fundamental difference between Late Cretaceous freshwater systems and those of earlier and later times. Vast marine transgressions during the Late Cretaceous may have reduced land surfaces, and freshwater lakes may also have been more restricted in their extent and distribution. This is also consistent with the more restricted occurrences of freshwater fishes from the Late Cretaceous compared to the Early Cretaceous and the Cenozoic (Chang Meemann, personal communication, 2010) and the similar scarcity in Late Cretaceous floras of many characteristic wetland and aquatic plants from the Early Cenozoic. Modern kinds of freshwater ecosystems may only have developed when sea levels fell and when more mesic, rather than more arid, conditions were established at low to mid-palaeolatitudes in the Early Cenozoic. The explanation for the relatively sparse record of Nymphaeales in the Late Cretaceous, and the trigger for much of the secondary Cenozoic radiation of Nymphaeales, especially core Nymphaeaceae, may well have been environmental.

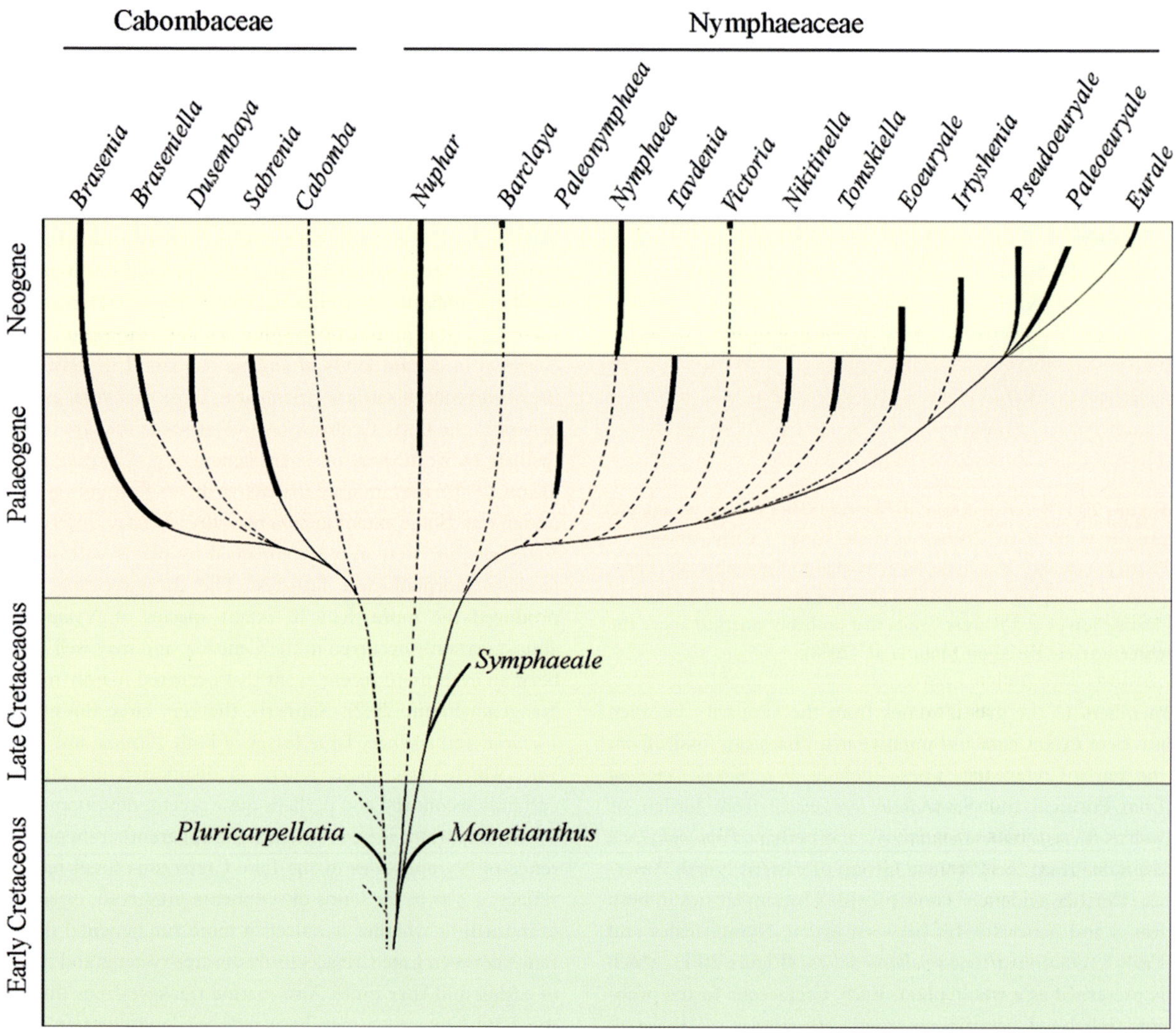

Figure 20.2 Stratigraphic occurrence of Nymphaeales. Based on seeds and floral structures the diversification of Nymphaeales began very early in the evolutionary history of angiosperms, but may have slowed in the Late Cretaceous before a more dramatic and extensive second phase of diversification in the Early Cenozoic. Note the many extinct taxa particularly in the Palaeogene. Thick lines indicate the stratigraphic occurrence of the genera. Patterns of relationships based on Borsch *et al.* (2008); stratigraphic ranges of nymphaealean taxa based on information in this book as well as Dorofeev (1973), Collinson (1980) and Mai (1995).

20.2.2 Chloranthaceae

The Chloranthaceae have come into focus as a key group in angiosperm evolution, mainly because of their unusually simple flowers, their extensive fossil record from both the Early and Late Cretaceous, and their interesting, perhaps pivotal, position in angiosperm phylogeny at a level above the ANITA grade, close to the point of divergence of monocots, eudicots and eumagnoliids. Among the four extant genera morphological and molecular data indicate that *Hedyosmum* is sister to *Ascarina* and *Chloranthus* (Qiu *et al.*, 1999; Doyle *et al.*, 2003; Eklund *et al.*, 2004b). *Sarcandra* may be either sister to *Chloranthus* or nested within it.

In Chloranthaceae, as in Nymphaeales, there are good indications that the lineages leading to some modern genera diverged in the earliest phases of angiosperm evolution, but in this case the similarities of early fossils to extant genera are clearer and even stronger. Stamens, staminate inflorescences, pollen and fruits that are very similar to those of extant *Hedyosmum* are among the first fossil angiosperms encountered during the Early Cretaceous. The relationship between these fossils and the extant genus is secure and only incomplete information on other parts of the plants makes it uncertain whether the fossils should be placed among extant species in the *Hedyosmum*-crown group, or as sister to the extant taxa in the *Hedyosmum*-stem group (Figure 20.3).

The early occurrence of *Ascarina*-like plants also seems likely based on similarities between dispersed pollen assigned to the fossil genus *Clavatipollenites* and pollen of extant *Ascarina* (Couper, 1960; Walker and Walker, 1984). However, so far no clear *Ascarina*-like flowers or inflorescences have been identified from the fossil record. Although there is great similarity between extant and fossil pollen grains in aperture configuration, details of sculpture and pollen wall ultrastructure, it is important to be cautious because this general pollen type also occurs in other extant lineages. Unfortunately, dispersed fossil stamens with *Ascarina*-like pollen *in situ* are not sufficiently diagnostic to support secure assignment to the extant genus.

Floral structures closely similar to those of extant *Chloranthus* are first recorded from the mid-Cretaceous of Germany. They consist of small, bisexual flowers and inflorescence fragments assigned to *Chloranthistemon*, and the same fossil genus is also known from the Late Cretaceous of southern Sweden and eastern North America. Fossil *Chloranthistemon* and extant *Chloranthus* both have a very distinctive, peculiar, tripartite androecium that consists of a fully developed (tetrasporangiate) median stamen and two bisporangiate lateral stamens. In *Chloranthistemon* from the mid-Cretaceous of Germany the pollen sacs on the bisporangiate lateral stamens are borne on the inner margin, whereas in the Late Cretaceous fossils and in extant *Chloranthus* the pollen sacs are borne on the outer margin. This may suggest that the lateral stamens of the ancestral form were tetrasporangiate and that the outer pollen sacs were lost in mid-Cretaceous *Chloranthistemon*, but the inner pollen sacs were lost in the Late Cretaceous taxa and the modern genus.

In addition to fossil material with clear relationship to extant Chloranthaceae there is also other material from the Early Cretaceous that may have chloranthaceous affinities, but for which relationships are currently uncertain. Most importantly, pollen grains generally assigned to *Clavatipollenites* based on LM and SEM studies are diverse and may have been produced by a variety of plants that are not systematically closely related. Some may have been close to Chloranthaceae; others may be related to other early angiosperm lineages. *Clavatipollenites*-type pollen observed on the stigmas of the mid-Cretaceous *Couperites* fruits is a case in point. These grains have a more distinctly delimited colpus margin than pollen of extant *Ascarina*, and the *Couperites* fruits on which they occur have anatropous, exotestal rather than orthotropous, endotestal seeds. Although *Couperites* fruits and their associated pollen were probably produced by plants close to Chloranthaceae and the ANITA grade, their precise phylogenetic position remains uncertain.

In addition to documenting the early appearance of fossil Chloranthaceae, palaeobotanical evidence also suggests that these early fossils were similar to their extant counterparts in their mode of pollination. Early Cretaceous *Hedyosmum*-like plants seem likely to have been wind-pollinated. The fossil staminate inflorescences and stamens are much smaller than those of extant species, but they show the same kind of simple floral construction and produced very similar pollen grains. The pollen grains of the fossil material and extant *Hedyosmum* are almost identical in size, shape and exine ornamentation. It also seems likely that mid- and Late Cretaceous *Chloranthistemon* were insect-pollinated, as in extant *Chloranthus*. Both living and fossil forms have extensive, probably osmophoric, stamens and very small pollen sacs that open by lateral valves.

Despite the availability of well-understood early fossils, and a clear hypothesis of relationships among the living genera, interpretation of the pattern of floral evolution in Chloranthaceae is complicated by the uncertain phylogenetic position of the family and currently equivocal patterns of floral evolution among the early-diverging lineages of angiosperms as a whole. It is not clear whether floral evolution in Chloranthaceae proceeded from forms with simple unisexual flowers, as implied by the earlier stratigraphic appearance of *Hedyosmum*-like flowers, or whether the origin of chloranthaceous flowers was by reduction from bisexual, probably trimerous or tetramerous, forms as inferred from the closely related extinct genus *Canrightia*. Which of these hypotheses (or other intermediate hypotheses) is regarded as more plausible also has implications for ideas about the evolution of pollination biology. If evolution proceeded from simple-flowered, perhaps wind-pollinated, progenitors then the occurrence of well-developed insect pollination in *Sarcandra* and *Chloranthus* is most likely a secondary development. However, if evolution proceeded from bisexual flowers to the unisexual condition in *Hedyosmum* and *Ascarina*, then the pattern of

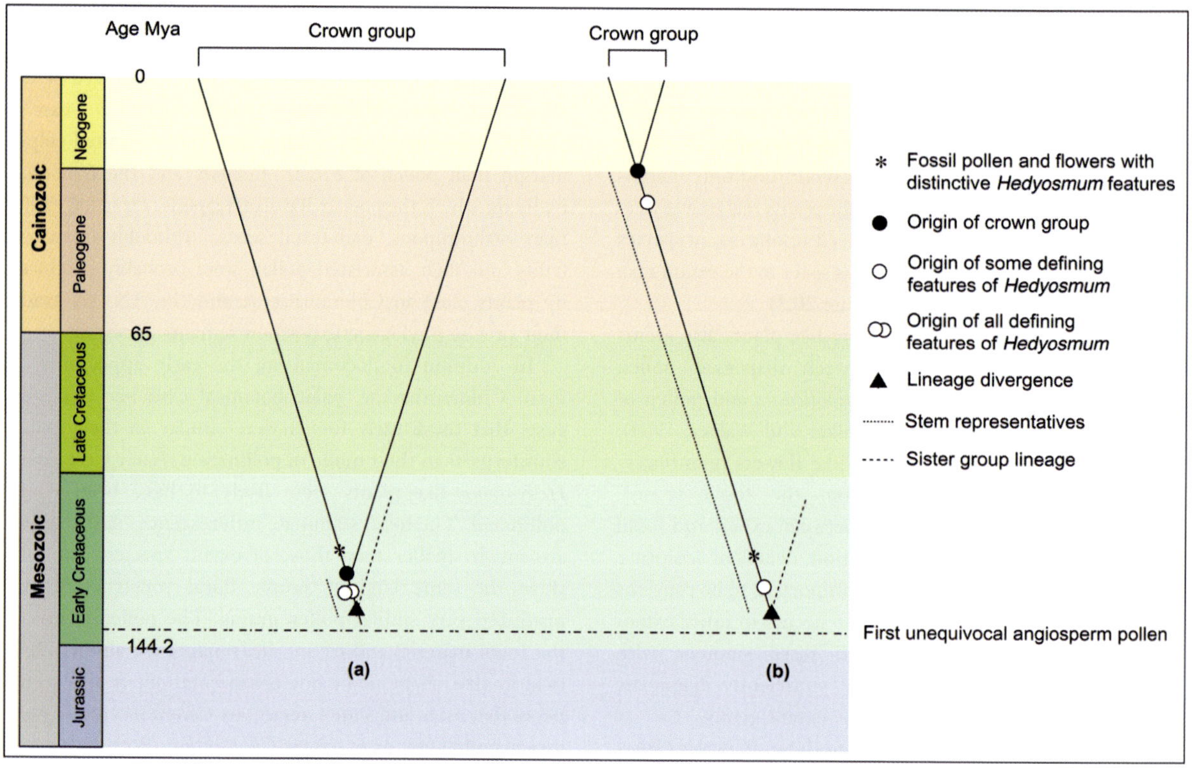

Figure 20.3 Two alternative models for the age of crown-group *Hedyosmum*. In the first model (a) Early Cretaceous *Hedyosmum*-like fossils possess all defining features of extant *Hedyosmum* and are included in the crown group of the extant genus, implying that *Hedyosmum* is at least 125 million years old. In the second model (b) the Early Cretaceous fossils lack some of the defining features of extant *Hedyosmum*, which are acquired much later. The age of the extant genus is therefore much younger (Cenozoic). Which of these two models is accepted requires assumptions about characters not preserved in the fossil material. From Friis *et al.* (2005).

relationships among extant genera in the family implies that the transition to wind pollination occurred at least twice. Resolution of these issues could also become of more general importance if Chloranthaceae eventually come to occupy a position among the ANITA lineages or even closer to the base of the angiosperm phylogenetic tree.

A further interesting feature of extant Chloranthaceae is its currently restricted distribution. The centre of diversity of extant *Hedyosmum* is in South America, and present species richness in this area includes about 45 species of tree. This diversity most likely reflects a secondary development of relatively recent origin, perhaps connected to the Cenozoic uplift of the Andes (Todzia, 1993). One species, *Hedyosmum orientale*, is herbaceous and occurs in East Asia.

In contrast to this disjunct and apparently relictual distribution of the extant genus, Early to mid-Cretaceous fossils related to *Hedyosmum* had an almost global distribution, at least to judge from dispersed *Asteropollis* pollen. *Hedyosmum*-like fossils are first reported at around the Barremian–Aptian boundary and through Aptian and Albian they occur in the palaeoequatorial regions of Africa and South America, as well as at mid-palaeolatitudes from North and Central America, Europe and Asia. From higher palaeolatitudes (Antarctica, Greenland) and from Australia the earliest reports are from the Albian. The importance of *Hedyosmum*-like plants clearly decreased after the Turonian, and there are no reports of *Asteropollis*/*Hedyosmum* pollen from later in the Cretaceous. It is only much later, in South America during the Cenozoic, that these grains reappear and proliferate (Muller, 1981).

The occurrence of *Clavatipollenites* pollen and putative *Ascarina* fossils shows a somewhat similar pattern. Extant *Ascarina* has a Southern Hemisphere distribution, occuring on Pacific islands and from New Zealand to Madagascar (Todzia, 1993). As in *Hedyosmum*, there is an almost world wide distribution in the Early Cretaceous from the Barremian through the Albian at low to mid palaeolatitudes in Africa, South America, North and Central America, Europe and Asia. At high palaeolatitudes (Antarctica, Greenland), and also in Australia, these grains appear only later, during the Albian. In most areas, *Ascarina*-like pollen becomes markedly less common during the mid-Cretaceous. Late Cretaceous and Cenozoic records are scarce and mainly from the Campanian–Maastrichtian of South America, Europe, Asia and Antarctica. Only in Australia is there an almost continuous record from the Albian to the Eocene (Muller, 1970). Cenozoic records of *Ascarina* from New Zealand, Australia, Africa and the Indian Ocean (Ninetyeast Ridge) document that the present restricted distribution of *Ascarina* is of relative recent origin (Muller, 1981; Eklund, 1999).

Evidence on the past distribution of the *Chloranthus–Sarcandra* lineage is much less extensive than for *Hedyosmum* and *Ascarina*, mainly because knowledge of the fossil record of this group is entirely reliant on evidence from mesofossils rather than fossil pollen. Grains of *Hammenia fredericksburgensis* from the Albian–Cenomanian of North America may be related to Chloranthaceae, and were tentatively compared to both *Chloranthus* and *Hedyosmum* (Ward, 1986), but were not assigned to either extant genus. Currently there are no dispersed pollen records that can be attributed securely to the *Chloranthus–Sarcandra* lineage.

The paucity of *Chloranthus*-like fossils in the Cretaceous compared with fossils related to *Ascarina* and *Hedyosmum* may reflect the different pollination biology of the two lineages. Extant *Chloranthus* and *Sarcandra* are insect-pollinated and the structure of fossil *Chloranthistemon* strongly indicates that this was also true of the Cretaceous members of the group. Pollen sacs were small and pollen production was probably very low as in extant *Chloranthus*. Interestingly, although the two species of *Chloranthistemon* from the Åsen locality, Sweden, have very distinct pollen, similar pollen grains were not identified in the Åsen palynological assemblages. This again emphasises the under-representation of pollen from insect-pollinated plants in the fossil record, which is partly alleviated by the study of mesofossils. To a great extent, palynological assemblages record most faithfully woody, wind-pollinated plants that produce pollen in large quantities.

Notwithstanding the absence of evidence from fossil pollen, the presence of *Chloranthistemon* in the mid- and Late Cretaceous of Europe and the Turonian of eastern North America is sufficient to show that the past distribution of this group was different from its exclusively Asian distribution today. There are no fossils from the latest Cretaceous or from the Cenozoic that could be assigned to the *Chloranthus–Sarcandra* lineage and there is thus a considerable gap in the fossil record from the mid- and Late Cretaceous records in Europe and North America to the present Asian populations. Whether the extant diversity of *Chloranthus* is also of relatively recent origin, as suggested for *Hedyosmum*, remains an open question.

For *Ascarina* and *Hedyosmum*, and for the *Chloranthus–Sarcandra* lineage, what caused the change from being widely distributed in the Cretaceous to being relictual today is uncertain. Muller (1970) suggested that the mid-Cretaceous decrease of the *Clavatipollenites/Ascarina* complex in the palaeoequatorial regions of Central Africa and South America might be explained by increased aridity in that region, as indicated by an increase in the abundance of ephedroid pollen. In mid-palaeolatitude regions increased restriction may similarly be related to increased seasonal aridity during the Late Cretaceous. However, in contrast to Nymphaeales and other angiosperm groups such as Lauraceae (section 20.3.2), the Chloranthaceae do not appear to have diversified strongly during the Early Cenozoic. Although the warm and humid conditions that the group appears to favour apparently prevailed in many low and mid-palaeolatitude regions during the Palaeogene, the most recent phase in the diversification of the family probably only started in the later Cenozoic.

20.3 PATTERNS OF ANGIOSPERM DIVERSIFICATION: EUMAGNOLIIDS

The fossil record provides clear evidence for the differentiation of three of the four main lineages of eumagnoliids by around the Early–Late Cretaceous boundary *c.* 100 million years ago. However, there is also evidence of much more recent diversification. For example, much of the diversity of extant Lauraceae, an unusually species-rich group among eumagnoliids, seems likely to be of relatively recent origin. This is also consistent with the relatively short molecular branch lengths that separate many genera of extant Lauraceae in recent phylogenetic analyses and the relatively minor morphological features by which they are distinguished. Similarly, in Magnoliales, although *Archaeanthus* documents the presence of stem-group Magnoliaceae by around the mid-Cretaceous, it is clear that much of the diversification in extant *Magnolia* occurred much later during the Cenozoic and was then followed by complex patterns of regional extinction.

20.3.1 Canellales–Piperales

The antiquity of the Canellales–Piperales clade is documented by the presence of *Walkeripollis*, a distinctive pollen tetrad very similar to tetrads of extant Winteraceae, in the Late Barremian – Early Aptian of Gabon and Late Aptian – Albian of Israel (Chapter 10). Nothing else is known about the plants that produced this pollen, and even though *Walkeripollis* was most likely produced by stem-group Winteraceae, rather than the extant family (Doyle *et al.*, 1990b), it sets a minimum age for the divergence of Canellaceae and Winteraceae. *Walkeripollis* also places the divergence of Canellales from Piperales as a very early event in angiosperm history.

The occurrence of *Walkeripollis* in West Africa and Israel during the Cretaceous (Doyle *et al.*, 1990a), and the occurrence of distinct Winteraceae pollen in the Miocene of South Africa (Coetzee and Praglowski, 1987), all records outside the range of extant Winteraceae, emphasises the extent to which the current distribution of the family is relictual. This is also reinforced by the presence of the endemic *Takhtajania* on Madagascar.

The fossil record of Piperales is scarce. Possible Early Cretaceous Piperales include the fruits of *Appomattoxia* and their associated pollen as well as similar dispersed pollen from geographically widespread areas assigned to species of *Tucanopollis* and *Transitoripollis* (Chapter 9). The systematic assignment of these fossils, however, remains to be established with certainty.

20.3.2 Laurales

Compared with the relatively sparse fossil history of Canellales and Piperales, the record of the Laurales–Magnoliales group is much more extensive. Neither Laurales nor Magnoliales have been recognised in the earliest mesofossil floras from the Barremian–Aptian, but Laurales in particular are well documented from the Early–Middle Albian (Puddledock flora) onwards.

The Laurales exemplify a case of marked imbalance in species diversity among the extant families included in the order (Renner, 2005) with the Lauraceae including between 2500 and 3000 extant species and the six other families comprising between one species (Gomortegaceae) and 195 species (Monimiaceae) (Renner, 2005). The Lauraceae are one of the most species-rich of all eumagnoliid lineages and are especially well represented in the tropics of Southeast Asia and the New World. They are also present in mainland Asia. Based on phylogenetic and geographic patterns, it seems likely that much of the diversity in these regions may reflect relatively recent radiations during the later Cenozoic (Renner, 2005).

Not surprisingly, the Lauraceae account for most of the extinct diversity known for Laurales, but despite the meagre fossil record of the other six families the early presence of Lauraceae, combined with its phylogenetic position, makes it likely that all the major lineages of the Laurales were established in the Early Cretaceous. This is also supported by the presence of calycanthoid flowers in the Early–Middle Albian Puddledock flora of Virginia, USA (*Virginianthus calycanthoides*), and the Late Cretaceous of New Jersey, USA (*Jerseyanthus calycanthoides*), as well as fossil wood convincingly attributed to Atherospermataceae from the Late Cretaceous of Antarctica. Although scattered, the fossil record also indicates a different geographic distribution for both Laurales and Lauraceae than today.

In addition to *Virginianthus calycanthoides* the Puddledock flora also includes a number of other fossils related to Lauraceae or to core Laurales (*Cohongarootonia, Potomacanthus, Powhatania* and several unnamed taxa). There is also a possible lauralean flower from the Late Albian – Early Cenomanian of Western Australia (*Lovellea*). The Lauraceae are also common in many early Late Cretaceous meso- and macrofossil floras, where they are represented by both flowers and fruits. Especially prominent is the extinct genus *Mauldinia* that includes at least five different species occurring in mid-continent and eastern North America as well as in Europe (Czech Republic, Germany) and Kazakhstan (Figure 20.4). Other flowers of Lauraceae in the Late Cretaceous assemblages include *Perseanthus crossmanensis* from the Turonian of New Jersey, USA, *Neusenia tetrasporangiata* and several unnamed flowers from the Campanian of North Carolina, USA, and *Lauranthus futabensis* from the Coniacian of Japan.

It is interesting that so far Laurales have not been recognised from any of the Portuguese mesofossil floras, from either the Early or the Late Cretaceous. The Lauraceae are, however, diverse in the Cenomanian of the Czech Republic, where the family is represented by several kinds of leaf fossils and a variety of inflorescences and flowers.

In post-Cenomanian floras from Europe, Laurales appear to be rare in both meso- and macrofossil floras. So far they have not been observed in the Åsen flora, the most diverse of these European Late Cretaceous floras. This is intriguing in the light of the rich diversity of Lauraceae in the Early Cenozoic floras of Europe, particularly in the Palaeogene where the family is represented extensively by

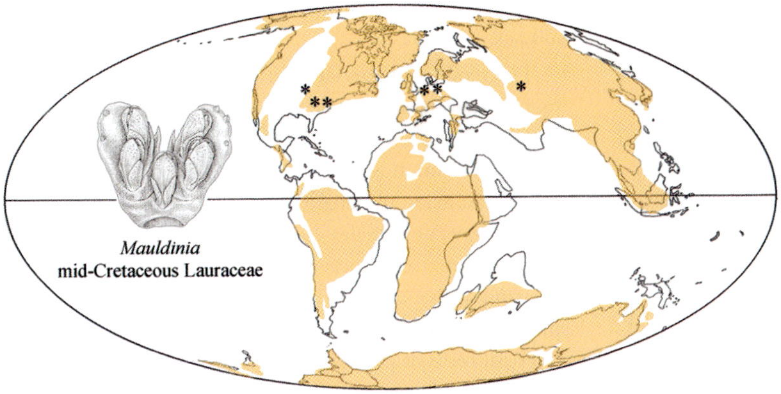

Figure 20.4 Distribution of the genus *Mauldinia*, an extinct member of Lauraceae, that was prominent in many mid-Cretaceous floras perhaps in near-coastal environments. Map showing position of continents and emergent land surface during the Cenomanian based on Smith *et al.* (1994).

fruit and leaf fossils, as well as occasional flowers, such as the beautiful flowers of *Cinnamomum felixii* and *Cinnamomum prototypum* preserved in Baltic amber (Conwentz, 1886). Laurales are particularly diverse in tropical lowland rain forests and their expansion in the Early Cenozoic of Europe most likely reflects a change in environment and community structure, from warm, seasonally dry climates and open vegetation in the Late Cretaceous, to warm, humid climates and closed stratified forest in the Early Cenozoic.

During the cooling of the Neogene most of the Palaeogene Lauraceae in Europe became victims of regional extinction. Only the genus *Laurus* persisted in mainland Europe, although there is greater relictual diversity on the Canary Islands and Madeira. In contrast, Lauraceae and other Laurales may have diversified in other regions during the Neogene. Renner (2005) suggested that most of the South American diversity of Laurales, particularly in Lauraceae and Siparunaceae, is relatively young and associated with the uplift of the Andes, a situation recalling that proposed for the diversity of extant *Hedyosmum* (section 20.2.2).

20.3.3 Magnoliales

The fossil history of Magnoliales, like that of Laurales, is extensive. The oldest unequivocal evidence of Magnoliales is *Archaeanthus* from the mid-Cretaceous, which implies that the lineage leading to Myristicaceae had also diverged by that time. So far, there are no unequivocal Myristicaceae from the Cretaceous. The earliest record of the family is fossil wood from the earliest Palaeogene, the systematic assignment of which needs to be clarified (Collinson *et al.*, 1993). Univocal seeds of Myristicaceae are known from the Eocene London Clay flora (Doyle *et al.*, 2008b). It would be interesting if the fossil flowers from the Late Cretaceous Old Crossman locality, which have been assigned to Triuridaceae (monocots), are recognised, ultimately, as an earlier record of the Myristicaceae lineage (Chapter 9).

Traditionally, extant Magnoliaceae have been divided among several genera, but phylogenetic analyses based on both morphological and molecular data demonstrate conclusively that there are two clades in the family; *Liriodendron* and *Magnolia sensu lato*. *Magnolia sensu lato* encompasses all other extant genera that have been recognised previously in Magnoliaceae.

All extant Magnoliaceae are woody shrubs or trees, and all have flower buds that are protected during development by bud-scales that are clearly homologous to the stipules of vegetative leaves. *Archaeanthus linnenbergeri* from the mid-Cretaceous of central Kansas, USA, shows conclusively that woody plants with this form of bud protection had already evolved by around the Early–Late Cretaceous boundary. Leaves of *Liriophyllum* from the mid-Cretaceous of Colorado, USA, also show well-developed stipules fused to the base of the petioles. In the context of related families of Magnoliales, the presence of stipules and their characteristic modification into a calyptra that protects the floral buds is the defining feature of extant Magnoliaceae.

The generalised form of leaves seen among species of *Magnolia sensu lato* makes them difficult to trace through the Late Cretaceous and into the Early Cenozoic, but this is much more straightforward for the very distinctive leaves of the *Liriodendron* clade. Early leaves of the group assigned to *Liriophyllum* from around the Early–Late Cretaceous boundary are very deeply lobed and at the point where the midrib reaches the sinus, it branches into two strong secondary veins. These two veins form the edge of the sinus before entering each of the two lobes. In contrast, in extant *Liriodendron*, the midrib terminates at the base of the sinus. It does not produce a strong pair of apical secondary veins and the sinus margin is formed by the lamina rather than two veins. Leaf form is very similar in the two extant species, but it would be interesting to quantify whether they are more variable in the North American *Liriodendron tulipifera*, a species with a wide distribution and high levels of genetic variability, than in the Chinese species, *L. chinense*, which is geographically more restricted and genetically much less variable.

It is difficult to generalise about the ecology of extinct Magnoliaceae, but it is interesting that in the mid-Cretaceous of Kansas, leaves of *Liriophyllum* occur in two of the compression fossil localities, but as far as we are aware have never been recorded among the coeval and very extensively collected leaf floras of the Dakota Sandstone. This perhaps suggests that these ancient liriodendroid species were not plants of riparian environments and instead occurred more commonly in environments away from the main fluvial channels.

In both extant *Liriodendron* and *Magnolia sensu lato*, pollination is by insects. The large flowers with robust tepals facilitate pollination by trapping, and then releasing, a variety of pollinators, which range from thrips to beetles. The trapping–release sequence is synchronised with stigma receptivity followed by shedding of pollen. *Archaeanthus* may have had a similar pollination mechanism, based on inferences from the scars of the floral organs, which suggest very similar floral structure to extant taxa.

In contrast to the relative uniformity of the pollination mechanism in extant Magnoliaceae, there is a clear distinction between dispersal in *Liriodendron* and in *Magnolia sensu lato*. In *Liriodendron* one or two seeds develop in each carpel, they have a well-developed sclerotic (endotestal) layer, the exotesta is thin, and the carpel itself develops into an indehiscent samara with a long, flattened, distal wing. Each samara is a single unit of dispersal, which is shed from the receptacle at maturity. The long wing, and the position of the seed(s), results in autorotational flight, which slows the rate of descent of the propagule and increases the likelihood of wind dispersal. In *Magnolia*, and also in the other genera now subsumed within it, there are typically one or two to eight seeds per carpel. The carpel develops into a follicle. Dehiscence may occur along both the ventral and the dorsal suture, or there may be more irregular shedding of the outer parts of the carpels. The seeds hang from the dehisced follicle on a thread derived from the funicle. The epidermis of the seeds is developed into a fleshy sarcotesta that is generally red or pink, and the seeds are presumed to be dispersed mainly by birds.

The mode of dispersal inferred for the mid-Cretaceous fossil *Archaeanthus* differs from that of all extant Magnoliaceae, and in several respects appears more generalised. There are many tiny seeds in each follicle. As in *Liriodendron* the follicles are shed from the receptacle, but unlike *Liriodendron* the follicles also dehisce to release the seeds. *Archaeanthus* thus appears to lack both the specialisation for wind dispersal seen in extant *Liriodendron*, and the specialisation for bird dispersal seen in extant *Magnolia*.

Fossil seeds of *Liriodendroidea* from the Late Cretaceous of Central Asia, Europe and eastern North America are very similar in general structure to seeds of extant *Liriodendron*, but in contrast to the samaras of *Liriodendron*, fruits of *Liriodendroidea* have several winged seeds per fruit and the small, winged seeds, rather than the fruits, appear to have been the dispersal unit.

Both *Liriodendron* and *Magnolia* have an extensive Cenozoic fossil record. Both were evidently important components of Cenozoic vegetation in Eurasia particularly in the later Cenozoic where they are often found associated with Neogene lignite. They are represented mainly by their characteristic seeds. Both genera appear to have had a circumboreal distribution in the Cenozoic with records also from the Neogene of Iceland (Mai, 1995; Grímsson *et al.*, 2007). *Liriodendron* and *Magnolia* persisted in Europe until the Pliocene, and like many other taxa they were extinguished from Europe by the glaciations of the Late Cenozoic.

20.4 PATTERNS OF ANGIOSPERM DIVERSIFICATION: MONOCOTS

The fossil record provides clear evidence of Chloranthaceae, Nymphaeales, eumagnoliids and eudicots in the Early Cretaceous and the phylogenetic position of monocots among these lineages implies that they also differentiated at an early phase in angiosperm evolution. An early occurrence of fossil monocots has long been suggested based on dispersed pollen (Walker and Walker, 1984, 1986) and is now confirmed by the discovery of floral structures of alismatalean affinity from the Early Cretaceous (Chapter 11). However, fossils of monocots are scarce through the Cretaceous and are not common in fossil floras until the Cenozoic, where they occur abundantly in fruit and seed floras.

The paucity of monocot fossils during the earliest phases of angiosperm diversification has been emphasised in several studies (Doyle, 1973; Herendeen and Crane 1995; Gandolfo *et al.*, 2000; Doyle *et al.*, 2008a). Gandolfo *et al.* (2000) highlighted the importance of fossil flowers assigned to Triuridaceae, from the Turonian of New Jersey, USA, as the earliest unequivocal record of the group. They suggested that the lack of clear monocot fossils until the Late Cretaceous indicated that monocots were of low diversity, and were ecologically restricted until the Late Cretaceous. In the later Cretaceous, and especially during the Campanian–Maastrichtian, there is unequivocal evidence of several groups of monocots (e.g. palms, Zingiberales), and by the Eocene the presence of many different families of monocots is well documented.

The diversity of small monocot floral structures recently discovered in Lower Cretaceous strata now indicates that monocots were probably more prominent in Early Cretaceous angiosperm vegetation than has previously appeared to be the case (Friis *et al.* 2010b). These Early Cretaceous monocots include araceous flowers as well as other floral structures of uncertain affinity that are probably also referable to the Alismatales. Currently, at least three subfamilies of Araceae appear to be present already in the Early Cretaceous (Figure 20.5) based on flowers, and several of these have pollen grains *in situ* that can be linked to dispersed pollen types. Further studies of these dispersed pollen grains may provide a broader geographic and stratigraphic framework for reconstructing patterns of early monocot diversification in time and space.

Pollen grains found *in situ* in one of the aroid flowers are similar to some of the earliest dispersed fossil angiosperm pollen, such as grains from the mid-Hauterivian described as 'CfB Retisulc-newling'. Similar dispersed grains are also known from Barremian–Albian palynofloras and are often assigned to the pollen genera *Retimonocolpites* or *Liliacidites*. Dispersed pollen of *Pennipollis* is another conspicuous and widespread pollen type in Early Cretaceous palynofloras, which may also be an early monocot. It is known *in situ* in the staminate flowers of *Pennistemon* and on the associated pistillate structures of *Pennicarpus*. Among extant angiosperms, as far as we are aware, the characteristic granular infratectal layer observed in the pollen wall of *Pennipollis* is known only in monocots. We therefore think it very likely that *Pennistemon*/*Pennipollis* plants are related to monocots rather than to Chloranthaceae as suggested by other authors (e.g. Doyle and Endress, 2010). There are also a remarkable diversity of other Early Cretaceous monoaperturate pollen grains with features suggestive of a relationship with monocots (Walker and Walker, 1984, 1986), but currently these

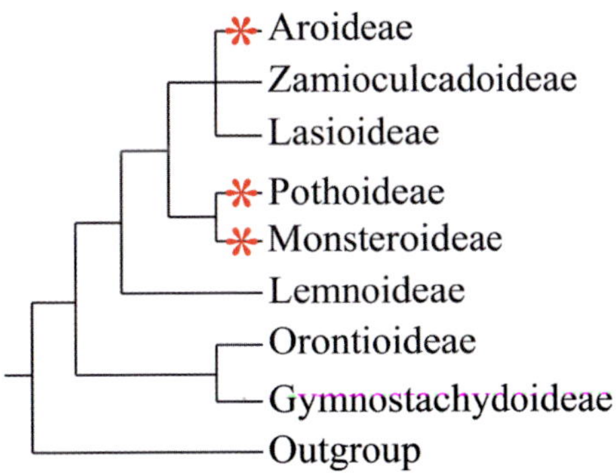

Figure 20.5 The monocot family Araceae was established very early in angiosperm history. One subfamily (Monsteroideae) is recorded from the Torres Vedras locality (Late Barremian – Early Aptian), Portugal. Two other subfamilies (Pothoideae, Aroideae) are present in the slightly younger flora from the Vila Verde 2 locality (Late Aptian – Early Albian), Portugal. Based on the *Retimonocolpites*-type pollen associated with one of the fossil aroid flowers at the Vila Verde 2 locality, the Araceae were probably diverse and widespread during the early phases of angiosperm evolution in the Early Cretaceous. Lineages known from the Early Cretaceous indicated by *. Patterns of relationship based on Stevens (2001 onwards).

grains are known only from dispersed palynofloras and cannot be assigned with certainty to any extant group.

The recognition of Alismatales as an early component of Early Cretaceous vegetation is consistent with the most recent phylogenetic analyses that resolve the small monocot genus *Acorus* (Acorales), and then Alismatales, as successive sister lineages to all other monocots. The fossil record of *Acorus* is meagre, but several of the 14 families of Alismatales are well represented in the fossil record from the Cenozoic.

During the Late Cretaceous, monocots may have been more ecologically restricted than in the Aptian–Albian. Pollen data indicate that palms, and perhaps other monocots, were widespread in some palaeoequatorial regions, but at mid-palaeolatitudes monocot fossils are scarce. It is particularly striking that many monocot families that are characteristic of wetland environments, such as Cyperaceae, Typhaceae, Sparganiaceae and Potamogetonaceae, have not yet been identified with certainty from the Late Cretaceous. In contrast, these groups are all very well represented in the Cenozoic fossil record from the Palaeogene onwards. This is similar to the pattern observed for Nymphaeales (section 20.2.1) and could also indicate more restricted freshwater and wetland habitats during the Late

Cretaceous than in the Early Cretaceous and Early Cenozoic. High sea levels, combined with xeric conditions, may have resulted in a more restricted distribution of freshwater lakes, at least at mid-palaeolatitudes, which have yielded most of the currently known meso- and macrofossil floras.

Despite the extensive Cenozoic record of monocots there are still several highly diverse groups of monocots that have a very poor fossil record, or for which a fossil record is entirely lacking. In most cases these absences can be explained by some combination of herbaceous habit, pollination biology, ecology and the probably poor resistance of various organs to depositional processes. Orchids provide one such example. Orchids are a huge group of monocots that is represented by only a few exceptional fossils. The vast majority of orchids are epiphytes in wet tropical forests and it is very unlikely that leaves or whole plants would ever reach depositional basins. Orchids have a highly specialised mode of biotic pollination. The seeds are also typically dust seeds with a thin seed coat. The most reliable record of a fossil orchid is a result of exceptional preservation in Dominican amber (Oligocene–Miocene) where an insect carrying pollinia became trapped in resin, perhaps far from the orchid that it had visited.

20.5 PATTERNS OF ANGIOSPERM DIVERSIFICATION: EUDICOTS

Based on the presence of their distinctive triaperturate pollen, eudicots are recognised at an early stage in angiosperm evolution, about 10–15 million years after the first appearance of monoaperturate angiosperm pollen. Where the systematic relationships of early eudicot fossils are clear, they are with those extant taxa that diverged relatively early from the main line of eudicot evolution (Ranunculales, Buxales, Proteales). These lineages tend to have low generic diversity and the differentiation of many of the extant species in these groups probably occurred relatively recently. Among early eudicots there are also fossils that are not clearly referable to any extant group.

Core eudicots are recognised in the mid-Cretaceous, and successive groups then appear through the Late Cretaceous and into the Early Cenozoic. In some cases, Late Cretaceous taxa appear to fall within the crown group of extant families (*Platydiscus*: Cunoniaceae); in other cases they appear to fall within the stem group (*Archamamelis*: Hamamelidaceae). An interesting feature of the Late Cretaceous record is that, although some major groups of rosids and asterids are very well represented, others, very conspicuously, are not. This is especially the case among rosids, where there is also no obvious correlation with the pattern of relationships among the various groups revealed by molecular phylogenetics. For example, the palaeobotanical evidence is very clear that Fagales began to diversify early in the Late Cretaceous. They have a very extensive and well-studied fossil record that extends from the Late Cenomanian through the Late Cretaceous and into the Early Cenozoic (section 20.5.2). However, the phylogenetic position of Fagales, nested relatively high up within the fabid clade, implies the existence of many other lineages (e.g. Fabales) for which there is currently little or no Late Cretaceous record. Fabales, which are much more diverse today, are not known in the Cretaceous and their major radiation appears to have taken place in the Cenozoic.

A similar situation occurs among malvids: the apparently secure record of Brassicales, which is nested high up in the clade, implies the Late Cretaceous presence of other lineages, such as Sapindales, for which there is currently no secure evidence from the Late Cretaceous. Many of the most diverse extant lineages in the rosid clade (e.g. Euphorbiaceae, Fabaceae) are well documented in the Early Cenozoic, but not yet recognised in the Late Cretaceous. As more is learnt about the rosid fossil record, especially from low palaeolatitudes, it will be interesting to see if this seeming mismatch between the stratigraphic and phylogenetic pattern can be resolved.

For asterids the congruence between the stratigraphic and phylogenetic pattern is much clearer. The bulk of Late Cretaceous asterid diversity is accounted for by the two lineages (Cornales, Ericales) that are the successive sisters to the lamiids and campanulids. The Ericales in particular are very diverse in Late Cretaceous floras. The first record of Ericales (Turonian, Nixon and Crepet, 1993) provides the earliest evidence of asterids. Subsequently, flowers and fruits of this order are abundant and diverse in many Late Cretaceous mesofossil floras. There is also early evidence of Cornales at around the same time (Early Coniacian, Takahashi *et al.*, 2002).

The Cornales are the sister group to all other asterids and therefore stand in a key position for understanding the early evolution of the most diverse clade of angiosperms. Within Cornales there have also been several analyses of relationships among extant genera based on both morphological and molecular data, which provide a basis for understanding aspects of the evolution of the biology of the group (e.g. Fan and Xiang, 2003; Feng *et al.*, 2009). The Cornales are diverse in many aspects of their structure, but fruits of the group are distinctive in being drupaceous and syncarpous, with two to four locules. In some taxa there is a characteristic valve for each locule that opens at germination. Distinctive fruits of this kind are not known from the Early Cretaceous, nor have they been recovered, so far, from the earliest stages of the Late Cretaceous. However, *Hironoia fusiformis* from the Early Coniacian of Japan has fruits with distinctive cornalean features, and the Turonian flowers of *Tylerianthus crossmanensis* may also represent early Cornales (Chapter 15). The sparse Cretaceous occurrences of Cornales are in strong contrast to the extensive Cenozoic record, which includes a diverse and abundant complex of extinct taxa that characterise certain distinctive vegetational types (Mastixioidean Flora) during the Eocene–Miocene.

The Ericales are the sister group to all asterids except Cornales. Fossil flowers, fruits and seeds of probable ericalean affinity are very well represented through most of the Late Cretaceous in Turonian and younger mesofossil floras, and provide clear evidence of the Late Cretaceous diversification of the group. The morphology and phylogenetic relationships of several Cretaceous Ericales have been assessed in detail and the fossil evidence is already sufficient to indicate that Ericales were phylogenetically diverse, and also geographically widespread in Turonian and younger rocks. It is possible that the Late Cretaceous diversification of the group may be linked to increased aridity at mid-palaeolatitudes or perhaps the expansion of nutrient-poor environments.

In addition to Cornales and Ericales, there are also certain forms among the great variety of 'saxifragalean flowers' known from the Coniacian onwards that may eventually be shown to be early asterids when their relationships are understood in more detail. For example, *Scandianthus* is similar in several respects to extant Vahliaceae, while *Silvianthemum* shares many features with extant *Quintinia*. The current position of *Quintinia* in the Paracryphiaceae is uncertain, and *Quintinia* and related fossils may represent a more basal lineage. As new information accumulates on phylogenetic relationships, and also on the morphological features that can be used to define some of the key asterid groups, it will be important to re-evaluate the relationships of these fossils (Chapter 15).

There is currently little secure evidence of lamiids or campanulids from the Late Cretaceous. However, in both cases, there are a few records that appear reliable and these involve lineages that split at an early stage from the main line of evolution in the group. Among lamiids, it is interesting that Icacinaceae, one of the few asterid families outside of Cornales and Ericales that have a reliable fossil record in the later part of the Cretaceous, are also resolved as one of the earliest-diverging lineages in phylogenetic analyses based on molecular data. Fossil fruits of Icacinaceae (endocarps) are well represented in fossil floras from the Early Cenozoic, but smaller fruits, with the distinctive pitted endocarps characteristic of the family, have also been described from the Late Cretaceous of Central Europe. Similarly, among campanulids, the extensive fossil history of Aquifoliaceae during the Cenozoic also extends back into the Cretaceous. Fossil fruits assigned to *Ilex* are known from the Campanian and Maastrichtian of Europe (Knobloch and Mai, 1986). In both lamiids and campanulids it is striking that many of the most diverse families (e.g. Asteraceae) and orders (e.g. Solanales–Lamiales) are not represented currently by fossils from the Cretaceous.

20.5.1 Proteales

Recognition of the Proteales, which includes Nelumbonaceae, Platanaceae and Proteales, as a monophyletic group is one of the less intuitive results from the application of molecular data to angiosperm phylogenetics. There are no clear morphological synapomorphies for the clade and in general appearance and reproductive biology members of the three families are very distinct. As a result, their association is interesting from the standpoint of biogeography, habitat, diversification rates and diversity patterns. Extant Proteaceae have a mainly extratropical Gondwanan distribution, while the distribution of extant Nelumbonaceae and Platanaceae is mainly in the Northern Hemisphere. There is also an imbalance of species diversity in the Proteales, similar to that in Laurales (section 20.3.2). Nelumbonaceae and Platanaceae each consist of a single extant genus with only a few species. In contrast, the Proteaceae comprise more than 1700 extant species in about 80 genera. The Proteaceae are comparable in species richness to families of more derived eudicots (Magallón and Sanderson, 2001). Extant Nelumbonaceae are aquatic herbs and the fossil record indicates that this life form was established in the family already in the Early Cretaceous.

All three families of Proteales have a fossil record that can be traced back to the mid-Cretaceous (Chapter 12). Cretaceous Nelumbonaceae fossils are mainly leaves recorded from the Northern Hemisphere, but the presence of *Nelumbo* leaves and *Nelumbo*-like fruiting structures from the Campanian–Maastrichtian of Argentina (Gandolfo and Cuneo, 2005) indicates that this family had an amphitropical distribution in the Cretaceous. The Cretaceous record of Platanaceae is especially extensive and well-documented, based on leaves, wood, inflorescences, flowers, pollen and fruits from many Northern Hemisphere localities, and the family was clearly much more diverse in the Cretaceous, and perhaps also in the Early Cenozoic. In contrast, while the Proteaceae were already present and had a distinct Gondwanan distribution in the Late Cretaceous they appear to have continued to diversify rapidly through the Cenozoic (Sauquet *et al.*, 2009).

The Cretaceous history of Nelumbonaceae is perhaps comparable to that of Nymphaeales. Early Cretaceous forms were aquatic plants similar to their extant descendants, and the humid phases of the Albian may have offered lake systems suitable for *Nelumbo*, Nymphaeaceae and a variety of other aquatic plants. In the Late Cretaceous extensive transgressions and increasing aridity may have reduced the environments available for Nelumbonaceae and similar plants of shallow lakes. However, unlike Nymphaeaceae, Nelumbonaceae did apparently not diversify markedly through the Cenozoic although the genus appears to have been widespread. Today the single living genus has a relictual distribution with one species in eastern North America and one species in Asia extending to northern Australia.

Extant *Platanus* includes about ten species. One species, *Platanus kerrii* from Southeast Asia, is distinguished from the others by its simple pinnate leaves, its evergreen life form, and other features. *Platanus kerrii* is placed in its own subgenus *Castaneophyllum*. Species of subgenus *Platanus* are deciduous plants typically with palmate leaves. All extant *Platanus* have unisexual, wind-pollinated flowers that are crowded into small balls borne along an elongated inflorescence axis. Pistillate flowers have an apocarpous gynoecium and a single orthotropous ovule in each carpel. Fruits are achenes with an obconical shape and numerous trichomes at the base, which are most likely involved in dispersal. At the apex of the fruit there is a distinct swelling formed from spongy tissue, which may help the fruits to float in water. The style is long, with an extended stigmatic area that is typical of wind-pollinated plants. In extant *Platanus* there are up to nine or ten fruits per flower and there are many flowers, densely crowded in each inflorescence ball. Each compound inflorescence axis typically bears only a few balls. *Platanus kerrii* is different. It has many balls that are sessile on an elongated axis.

The earliest platanoids are pistillate inflorescences and flowers of *Friisicarpus brookensis* and the associated staminate structures described as *Aquia brookensis* from the Middle Albian. They have many small balls that are sessile on a compound inflorescence axis as in *Platanus kerrii*. Each ball contains relatively few (*c*. 50) flowers and fruits (*c*. 250). Early Cretaceous, and most Late Cretaceous, platanoids have small actinomorphic and unisexual flowers with an undifferentiated but well-developed perianth. Staminate flowers have five stamens. Anther dehiscence is by laterally hinged valves and the connective is apically expanded into a conspicuous connective extension. The pistillate flowers have five carpels arranged in a star-shaped pattern. The stigmatic area is inconspicuous and there is no extended style. Valvate dehiscence is typical of insect-pollinated flowers, together with the actinomorphic symmetry, the open form of the flowers, and the presence of pollenkitt on the pollen. This combination of features suggests that these fossils were most likely insect-pollinated.

The details of Cretaceous platanoid flowers contrast strongly with those of wind-pollinated extant *Platanus*. Extant *Platanus* has irregular flowers with a variable number of parts, anther dehiscence by longitudinal slits and pistillate flowers that have carpels with long expanded styles. In fossil platanoid flowers the connective expansion may have served as protection early in flower development and as attractant at anthesis. The reward for pollinators was most likely pollen. Carpels are typically swollen apically and superficially look like stamens with expanded apical connectives. They may have attracted insects by mimicking the staminate flowers. This swelling may also have had a protective function in the pre-anthetic flower and could also have had a function in dispersal. It is also interesting that the number of flowers per inflorescence is smaller in the fossil platanoids than in the extant wind-pollinated *Platanus*.

Although the dominant mode of pollination in Cretaceous platanoids was probably by insects, there are also indications of wind pollination in Late Cretaceous platanoids. Pistillate flowers described as *Platanus quedlinburgensis* from the Santonian Quedlinburg locality have long extended styles and inconspicuous perianth (Tschan *et al.*, 2008). The material is, however, strongly compressed and poorly preserved and details of floral organisation are unclear.

In the Late Cretaceous, the diversity and abundance of Platanaceae appears to increase; platanoid fossils are among the most abundant macrofossils in mid- and Late Cretaceous floras from Siberia, Central Asia and Northeastern Russia. In western Siberia abundant *Sapindopsis* and *Platanus* are recorded for floras deposited along the margins of the Turgai Strait (e.g. Golovneva, 2007).

Extant Platanaceae are tall, fast-growing, riparian trees that may be drought-tolerant once they are established. Leaves that can be assigned to Platanaceae from the

Cretaceous (e.g. *Sapindopsis*, *Platanus*, *Credneria*, *'Aralia'*) are often large and tough, and indicate that the Cretaceous platanoids were probably woody plants as their extant descendants. Perhaps drought tolerance was established in certain Platanaceae under the arid, and sometimes coastal, conditions that were widespread in the Late Cretaceous mid-palaeolatitude regions. Such an environment is envisaged for instance for *Credneria*, which in the Quedlinburg area occurs in coarse, well-sorted sand deposits interpreted as coastal dunes.

Extant species of Proteaceae are mainly distributed in the Southern Hemisphere and are present on most of the now separated fragments of the former Gondwanan continent (Australia, South Africa, Central and South America, New Zealand, Madagascar) extending into Southeast Asia. The highest species diversity is in Southwest Australia and in the Cape Floristic Province of southern Africa, but the highest generic diversity is in tropical rain forests of Queensland and Malesia (Sauquet *et al.*, 2009). The diversity of Proteaceae in Australia is mainly within the subfamily Grevillioideae (e.g. *Banksia*, *Dryandra*, *Grevillea*, *Hakea*), with one or two diverse genera also in the subfamilies Proteoideae (e.g. *Conospermum*, *Petrophile*) and Persoonioideae (e.g. *Persoonia*). The African diversity is within the subfamily Proteoideae (e.g. *Protea*, *Leucadendron*). Occurrences in Southeast Asia and Mexico may be the result of geologically more recent dispersal events.

The Cretaceous record of Proteaceae consists mainly of dispersed pollen, but the Cenozoic record comprises many leaf fossils, inflorescences and flowers in addition to a much more diverse pollen record. The scleromorphic nature of many Cenozoic Proteaceae leaves, together with the woody character of many inflorescences, may have enhanced their fossilisation potential.

Patterns of diversity among genera of Proteaceae in many respects parallel the patterns seen in other eudicots, with the early-diverging lineages being generally species-poor, while species-rich genera such as *Grevillea*, *Hakea*, *Protea* and *Leucadendron* are typically nested higher up in the phylogeny. There is also an ecological dimension to diversity patterns within the family, with the early-diverging lineages occurring in tropical rain forests, while the more derived taxa are typically scleromorphic, occur in more nutrient-poor environments, and often grow in seasonally dry Mediterranean climates (Sauquet *et al.*, 2009).

Evidence from Australia indicates that scleromorphic features in Proteaceae had already evolved in the Early Cenozoic in plants that grew in wet environments (Hill, 1998). The fossil pollen record also indicates that extant lineages of scleromorphic Proteaceae are older than the mid- to Late Cenozoic establishment of Mediterranean climate (Sauquet and Cantrill, 2007; Sauquet *et al.*, 2009). This suggests that scleromorphy in this largely Southern Hemisphere family may have evolved primarily as a response to nutrient-poor soil rather than xeric conditions. This adaptation may then have favoured these plants with the subsequently spread of seasonally dry climates. The diversification and radiation of the Proteaceae may have been a three-fold event, an initial appearance and diversification in the Late Cretaceous, followed by diversification onto oligotrophic substrates (Paleocene–Eocene), and finally by renewed diversification associated with the shift from permanently moist conditions to arid and Mediterranean climates in the Oligocene/Miocene.

20.5.2 Fagales

The Fagales are represented extensively by a wide range of fossils from both the Late Cretaceous and the Cenozoic, and the systematics, fossil history and evolution of the group has been the focus of intensive neobotanical and palaeobotanical study. The order comprises eight families with Nothofagaceae sister to Fagaceae plus core Fagales. Within core Fagales there are two subclades: Myricaceae are sister to Rhoipteleaceae plus Juglandaceae, and Ticodendraceae are sister to Betulaceae plus Casuarinaceae (Chapter 14). Flowers are typically simple and unisexual with features characteristic of anemophilous plants; except for a few insect-pollinated taxa, especially in Fagaceae, most Fagales are wind-pollinated.

The extensive fossil history of Fagales provides excellent documentation of key evolutionary transitions that would be much less evident from the study of living material alone. Similar to the situation in Proteales, the Fagales have a mainly amphitropical distribution that was already established by the end of the Cretaceous. However, in contrast to Proteales most of the species diversity of Fagales is in the Northern Hemisphere.

Fossil pollen attributable to Nothofagaceae (*Nothofagites*) is first recorded from the Campanian (Late Cretaceous) of Australia and at the same time there are also records of fossil wood that can be assigned to the family. The earliest record of pollen that can be assigned to the extant genus is from the Maastrichtian of Australia. The presence of *Nothofagites*/*Nothofagus* in fossil pollen floras from the Late Cretaceous of southern Australia (including Tasmania), New Zealand, South America and Antarctica suggests that it was, from the beginning, a circum-austral plant of relatively high latitudes and therefore of relatively cool climates. Despite occasional reports from the Northern Hemisphere there is no reliable evidence that *Nothofagus* ever existed outside the Southern Hemisphere. There are also no reliable records of *Nothofagus* from southern Africa. This may reflect the relatively early separation of Africa from the rest of Gondwana, and perhaps warmer climates resulting from the more northerly position of this continent. With the exception of wood, the earliest macrofossils of *Nothofagus* appear in the Early Cenozoic. Subsequently, the genus is well represented in the Cenozoic of the Southern Hemisphere by fossil leaves, cupules, wood and pollen.

In the Northern Hemisphere, there is also clear evidence of Fagales in the Late Cretaceous. The earliest record of plants that may be in a relatively basal position in the phylogenetic tree of extant taxa is from the Coniacian of Japan (*Archaefagacea futabensis*), but there are also well-preserved pistillate and staminate flowers, fruits and cupules from the Santonian of southeastern North America (*Protofagacea allonensis*, *Antiquacupula sulcata*) that are similar to extant Fagaceae. However, although pollen grains of *Nothofagus* are diagnostic of subgroups in the genus, and therefore provide unequivocal evidence of crown group *Nothofagus* in the latest Cretaceous, the systematic attribution of the more completely understood fossil material from the Northern Hemisphere is much less certain. Potentially their phylogenetic position could be on the stem group to extant Fagaceae, or on the stem group to core Fagales, the stem group of Fagaceae plus core Fagales, or perhaps the stem group to *Nothofagus* itself. These early fossils could also potentially fall in the stem group for Fagales as a whole. The impediment to more precise phylogenetic placement is a lack of clear morphological synapomorphies that would define these different groupings of extant taxa.

Whatever the precise phylogenetic position of *Antiquacupula*, *Archaefagacea* and *Protofagacea*, they conclusively establish the presence of Fagales in the Northern Hemisphere by the middle of the Late Cretaceous. They also suggest that the fundamental vicariant event in the family that established its currently disjunct northern–southern distribution has very deep roots that go back into the mid-Cretaceous. A mid-Cretaceous appearance of Fagales as a whole is also suggested by the presence in the Cenomanian of triporate pollen characteristic of core Fagales, and the earliest records of *Caryanthus* fossil flowers, which are clearly related to core Fagales. Given the absence of evidence of core eudicots in the Early Cretaceous the current evidence suggests that the lineages leading to extant Nothofagaceae and Fagaceae diverged from the lineage that gave rise to core Fagales very quickly around the Early Cretaceous – Late Cretaceous boundary (Figures 20.6, 20.7).

The core Fagales are mainly represented in the Late Cretaceous by dispersed pollen assigned to the Normapolles complex and flowers with associated Normapolles-type pollen. In some Late Cretaceous palynological assemblages from the so-called Normapolles floristic province of Europe and North America (Chapter 19) Normapolles pollen may constitute up to 80% of all angiosperm pollen recorded. The Normapolles group was also enormously diverse: more than 80 different genera and more than 100 species are currently recognised (Chapter 14).

Flowers with Normapolles-type pollen are also common in some of the mesofossil floras from the Normapolles Province. Currently more than 25 species have been identified and it has been possible to link some of the dispersed taxa to floral structures (Figure 20.6). Undoubtedly, Normapolles-producing plants were important in the Late Cretaceous vegetation of the Normapolles Province. Floral and pollen morphology strongly suggest that these plants were wind-pollinated.

Analyses of systematic relationships of currently known Normapolles flowers show that they are all Fagales, and probably all core Fagales, but it is also clear that the Normapolles-producing plants constitute a heterogeneous complex. Some are most closely related to the Ticodendraceae–Betulaceae–Casuarinaceae subclade and some most closely related to the Myricaceae–Rhoipteleaceae–Juglandaceae subclade.

The Ticodendraceae–Betulaceae–Casuarinaceae group includes 11 extant genera: one in Ticodendraceae, four in Casuarinaceae and six in Betulaceae. Genera of the Betulaceae are resolved into two tribes: the Betuleae (*Alnus*, *Betula*) and the Coryleae (*Corylus*, *Ostryopsis*, *Carpinus*, *Ostrya*). They are all wind-pollinated. Fruits in *Alnus* and *Betula* are small and in many species (especially in *Betula*) a pair of membranous wings indicates clear specialisation for wind-dispersal. In contrast, fruits in the Coryleae are larger. In *Ostryopsis* there is apparently little specialisation of the fruit for dispersal. In *Carpinus* and *Ostrya* fruits are dispersed attached to the involucre, which is specialised for

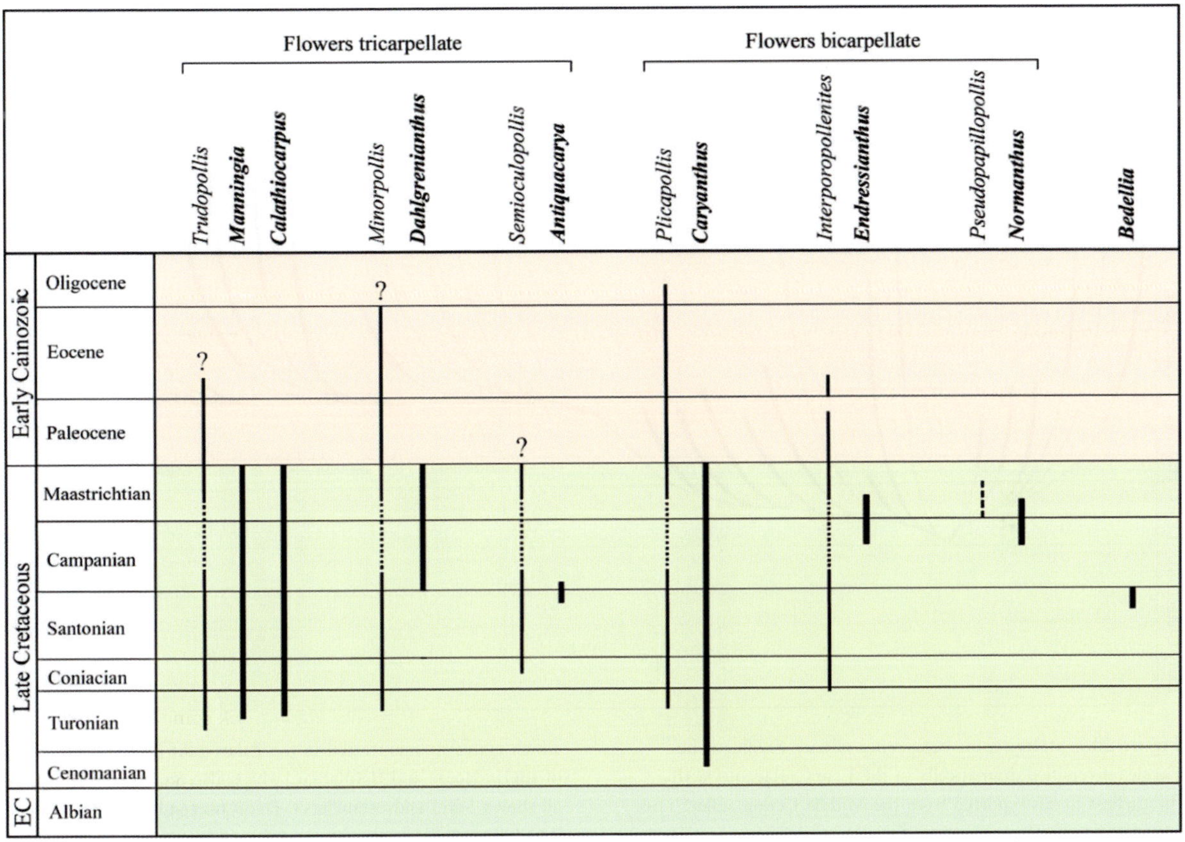

Figure 20.6 Stratigraphic ranges of Normapolles genera known from fossil flowers (bold italics) and pollen types associated with the Normapolles flowers (regular italics). From Friis *et al.* (2006b). EC, Early Cretaceous.

wind-dispersal. In *Corylus* the fruit is relatively large and animal-dispersed (probably by scatterhoarding). All of these dispersal specialisations, including several others that are no longer represented among extant taxa, were present by the Eocene–Oligocene, although there is some indication of an increase in size of *Corylus* fruits through the Cenozoic.

There are no confirmed records of Betulaceae in the Cretaceous. However, the Late Cretaceous (Campanian–Maastrichtian) flowers and fruits of *Endressianthus* described from the mesofossil floras of Mira and Esgueira in Portugal are closely related to extant Betulaceae. Pollen found *in situ* in the *Endressianthus* flowers are identical to the dispersed Normapolles genus *Interporopollenites* that occurs commonly in the Mediterranean region during the Late Cretaceous, and that ranges in age from the latest Turonian into the Early Cenozoic (Figure 20.6). This pollen type, with its paired exopores, is unique and distinctive, but in other characters the pollen grains show strong similarities with those of extant Betulaceae. Gynoecium and ovule features also support a placement of *Endressianthus* as an early member of the Betulaceae–Ticodendraceae–Casuarinaceae subclade. *Endressianthus* fruits are small and lack the specialisations for wind or animal dispersal seen in Early Cenozoic members of core Fagales and extant Betulaceae. Pollen morphology and size, as well as the abundant occurrence of *Interporopollenites* in the dispersed palynological assemblages, strongly suggest wind pollination, as has also been inferred for other plants broadly referred to the Normapolles complex. *Endressianthus* is the only one of the Normapolles plants discovered so far that has only unisexual flowers.

Most Normapolles taxa have bisexual flowers; rarely, they have both bisexual and unisexual flowers as in *Budvaricarpus*. *Budvaricarpus*, and the closely related genus *Caryanthus*, are most likely early members of the

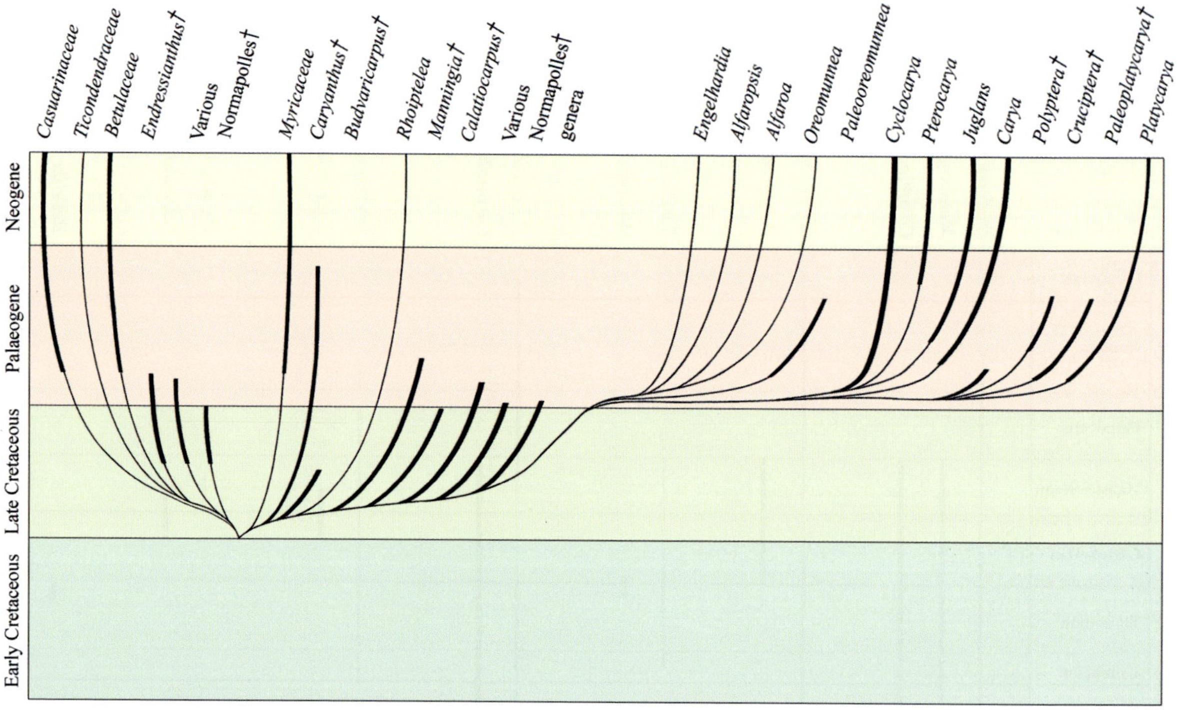

Figure 20.7 Stratigraphic occurrence of core Fagales. The Fagales diversified dramatically in the Late Cretaceous with the earliest fossils reported from the Middle Cenomanian. The Late Cretaceous diversification is mainly within the core Fagales related to the Normapolles complex. The extinct plants placed within the Normapolles complex comprises a heterogeneous group of core Fagales with some taxa closely related to *Rhoiptelea* and Juglandaceae, and others more closely related to the Ticodendraceae–Betulaceae–Casuarinaceae lineage. Core Fagales had a second extensive radiation in the Early Cenozoic, which is mainly manifested in the Palaeogene diversification of Juglandaceae and Betulaceae, which also included extinct (not all shown here) and extant taxa. Fossil taxa indicated by †. Thick lines indicate the stratigraphic occurrence of the genera. Patterns of relationship based on Li *et al.* (2004) and Manos and Stone (2001); stratigraphic ranges of Normapolles taxa based on Friis *et al.* (2006b); stratigraphic ranges of Juglandaceae taxa based mainly on Manchester (1987a). Normapolles taxa are here compound, including floral structures and associated pollen types.

Myricaceae–Rhoipteleaceae–Juglandaceae subclade, very closely similar to extant *Rhoiptelea* in floral organisation, gynoecium characters and the associated *Plicapollis*-type pollen. Extant *Rhoiptelea* is restricted today to small populations in southern China and northern Vietnam. It is interesting in having partial inflorescences with three flowers, a median bisexual flower and two lateral, pistillate flowers. Late Cretaceous *Budvaricarpus* also has the same floral organisation with four tepals in two pairs and six stamens in the bisexual flower. *Caryanthus*, another Late Cretaceous flower, also has very similar floral organisation and is probably also very closely related to *Rhoiptelea*. Both *Caryanthus* and *Budvaricarpus* have pollen that is identical to that of *Rhoiptelea*, but when found isolated such grains are assigned to the Normapolles genus *Plicapollis*. Cretaceous *Rhoiptelea*-like fossils differ mainly from the extant genus in having small unspecialised nuts. Fruits of *Rhoiptelea* are larger, winged, and obviously adapted for wind dispersal.

In addition to *Budvaricarpus* and *Caryanthus* there are several other Normapolles plants that appear to be closely related to the Myricaceae–Rhoipteleaceae–Juglandaceae subclade (e.g. *Manningia*, *Antiquocarya*), but none of these can be placed in any of the existing families. Future work may show them to be on the stem to the subclade, or one or more of the extant families.

Extant Juglandaceae comprise eight extant genera distributed in two tribes; Platycaryoideae (*Platycarya, Engelhardia*, *Oreomunnea*, *Alfaroa*) and Juglandoideae (*Carya*,

Pterocarya, *Cyclocarya*, *Juglans*). There is considerable variation in fruit form and dispersal biology in the family. In Platycaryoideae, fruits of *Platycarya* (tribe Platycaryeae) are small with two weakly developed lateral wings, and are dispersed without any attached bracts whereas in the tribe Engelhardieae fruits of *Engelhardia* and *Oreomunnea* are larger and are dispersed with prominent trilobed wings formed by the surrounding bracts. In *Alfaroa* fruit size is larger and the associated bracts are greatly reduced.

Similar variation in fruits occurs in the Juglandoideae. In *Carya* (tribe Hicoreae) and *Juglans* (tribe Juglandeae) the fruits are relatively large, surrounded by the fleshy husk (dehiscent in *Carya*). There is clear specialisation for animal dispersal (probably by scatterhoarding). In *Pterocarya* and *Cyclocarya* (Juglandeae) the fruit has a well-developed wing or wings formed from the bract and bracteoles and there is clear specialisation for wind dispersal.

The Cretaceous fossil record of the Juglandaceae is based almost exclusively on dispersed pollen, which grades morphologically into that produced by the Normapolles complex. All fossil fruits with associated Normapolles pollen that have been described from the Late Cretaceous are small and show no obvious specialisation for either wind or animal dispersal.

As in Betualaceae, specialisation for dispersal in Juglandaceae appears to be a phenomenon of the Early Cenozoic. In addition to this overall trend, within *Carya* and *Juglans* there are also indications of increased specialisation for animal dispersal. In the Hicoreae, of which *Carya* is the only extant representative, the earliest probable members of the group (e.g. *Juglandicarya simplicarpa* from the Paleocene of North America and *Juglandicarya depressa* from the Early Eocene London Clay flora) have a relatively simple locule morphology from which the seed could probably be easily removed. In contrast, the subsequent development of a secondary septum and inner ribs in Oligocene and Miocene *Carya* species effectively locks the seed in the hard fruit and may have discouraged the direct removal of seeds from freshly fallen nuts (Manchester, 1987a). Scatterhoarding of such fruits may promote decay and softening of the endocarp. Direct early evidence of nut storage, presumably by squirrels, is provided by a fossilised cache of *Carya washingtonensis* nuts discovered within a silicified *Platanus* stump from the Miocene of Washington (Manchester, 1987a).

A pattern of increasing locule complexity similar to that seen in the Hicoreae also occurs in the walnuts (*Juglans*). The earliest walnuts (*Juglans clarnensis*) from the Middle Eocene of Oregon are similar to modern black walnuts (*Juglans* sect. *Rhysocaryon*) but have a simpler seed and locule morphology. Later (Oligocene and Miocene) species of *Juglans* sect. *Rhysocaryon*, like fossil *Carya* species of this age, have thicker fruit walls and more complex locule and septal configurations. Interestingly, in *Juglans* sect. *Cardiocaryon* a simple locule with easily removed seeds was retained but in association with an endocarp that has a reticulum of blade-like ridges that seems to deter gnawing by rodents (Miki, 1955; Manchester, 1987a).

In summary, the initial diversification of core Fagales appears to have taken place associated with wind pollination, which was perhaps an adaptation to open vegetation in a seasonally dry environment. Following the dramatic diversification of the Normapolles complex, a second burst of evolution within core Fagales may have been associated with adaptation to more specialised modes of fruit dispersal: development of elaborated wings for more efficient wind dispersal in some groups, and the development of much larger fruits for animal dispersal in the more dense forest types that expanded in the Early Cenozoic (Chapter 18).

20.6 ANGIOSPERM EVOLUTION AND GLOBAL CHANGE THROUGH THE CENOZOIC

Although the focus of this book is on angiosperm fossils from the Cretaceous, this long interval of geologic time comprises only about the first half of angiosperm evolutionary history. Angiosperm evolution, including substantial speciation, innovation and extinction, continued through the Cenozoic, as is clearly seen in the fossil history of Nymphaeales, Magnoliales, monocots, Proteales and Fagales reviewed in previous sections. Diversification may even have accelerated in many groups. Therefore, even though angiosperms become steadily more similar to those of today through the Cenozoic, studies of Palaeogene and Neogene plant fossils still have much to contribute to a more complete understanding of angiosperm evolution.

For this reason, in the Cenozoic, as in the Cretaceous, it is important to take a critical view of the assignment of plant fossils to extant taxa. Many of the assignments of Early Cenozoic leaf fossils to extant genera in the older palaeobotanical literature are unreliable (Dilcher, 1974) and genera that are extinct are known from as late as the Miocene (e.g. *Nordenskioeldia*, *Pseudofagus*). There is also evidence of extinct species from the Quaternary (Jackson and Weng, 1999). Even where there are clear morphological similarities between fossil and extant taxa, detailed studies or plant reconstructions often reveal significant differences, either in the characters themselves, or in the combinations in which they occur. The evolution of angiosperms as we know them today was certainly not complete by the end of the Cretaceous.

Just as many Cenozoic plants are not identical to extant genera and species, Palaeogene, Neogene and even Quaternary vegetation also differed in species composition from that of extant plant communities. Fossil floras from the Cenozoic often contain mixtures of taxa that seem anomalous compared with plant communities of today. Studies of Quaternary vegetational history show that, in general, plant species respond individualistically to climatic and other environmental change. Plant communities generally fail to retain their integrity in the face of climatic perturbation, and ecological tolerances of species and genera may also change over time.

20.6.1 Angiosperm evolution and the Cretaceous–Palaeogene (K–T) extinction

In many groups of organisms the transition from the Cretaceous to the Palaeogene (the Cretaceous–Palaeogene, or K–T, boundary) is marked by levels of extinction second only to those at the end of the Permian in their global impact. These and other changes around the K–T boundary modified the selective landscape of angiosperm evolution, probably with profound indirect impacts on vegetational structure. A possible extraterrestrial cause of extinctions at the K–T boundary is widely accepted, as is the Chixulub structure in the Yucatán Peninsula, Mexico, as the likely associated impact crater (Hildebrand *et al.*, 1991), but there may be more complex alternative explanations for the floristic changes that took place from the Cretaceous to the Cenozoic. Extensive Deccan Trap volcanism around the K–T boundary, together with marine regressions and cooling at around the same time, must also have had profound impacts on terrestrial ecosystems.

Because plants provide the energy fundamental to most terrestrial ecosystems, the nature of vegetational change at the end of the Cretaceous and its likely impact on angiosperm macroevolution is of great interest. Initially, plants were thought to have been little affected by events at the K–T boundary (Hickey, 1981), but recent work, especially on the most extensive series of boundary sections in continental deposits in western North America, has identified a more complex pattern that includes significant levels of species loss. In some sections there is palynological evidence for a mass kill of the vegetation followed by near-immediate recolonisation of the landscape by ferns (Tschudy *et al.*, 1984; Nichols *et al.*, 1986).

The most detailed study of plant extinction at the K–T boundary, based on macrofossils, is from the Williston Basin of southwestern North Dakota, USA (Wilf and Johnson, 2004). Across the K–T boundary almost all the previously dominant species are lost. Only a few species continued from the Cretaceous into the Paleocene. Most of these were relatively unimportant in the Cretaceous, but come to dominate macrofossil floras in the Paleocene. Species richness declined significantly; a low-diversity flora characterises a large area through Colorado, Wyoming, Montana and North and South Dakota for almost a million years (Wilf and Johnson, 2004).

Careful analysis of the macrofossil record in the Williston Basin shows that many species disappeared well below the K–T boundary itself, but nevertheless there are sufficient losses close to the boundary to suggest a maximum estimate of 57% regional extinction at the species level (Wilf and Johnson, 2004). Estimates based on palynological data from the same area point to a lower level of about 30% extinction (e.g. Nichols and Johnson, 2002).

The macroevolutionary consequences of the regional extinctions and vegetational changes at the K–T boundary, which may also have been exacerbated by significant changes in climate (Wilf *et al.*, 2003; Nichols and Johnson, 2008), are not yet clear. Certainly many of the extant lineages of angiosperms reviewed in Chapters 8 and 10–15 were already present in the Late Cretaceous and evidently survived the perturbations at the K–T boundary, but the extent to which extinction played a direct or indirect role in winnowing out certain Late Cretaceous groups, or stimulating the diversification of those taxa that radiated in the Early Cenozoic, is unknown. For example, although the diversity of Normapolles pollen clearly declined from the Late Cretaceous into the Early Cenozoic, several taxa continued at least into the Eocene and there was substantial renewed diversification of both subclades of core Fagales.

20.6.2 Palaeogene vegetation

Following the K–T boundary the diversity of angiosperms continued to increase. The earliest unequivocal record of neotropical rain forests is from the Late Paleocene of Colombia and appears to be of lower diversity than its modern counterparts (Wing *et al.*, 2009). There is also evidence of transitory rapid global warming around the Paleocene–Eocene boundary (Wing *et al.*, 2005).

The Eocene was the warmest interval of the Cenozoic. Eocene vegetation is well known based on a large number of macrofossil floras, some of which (e.g. London Clay flora) have been collected and studied for well over a century. Many Eocene floras provide clear indications of extraordinary warmth at latitudes that today are much cooler. During the Eocene, Wolfe (1985) estimates that full tropical multistratal rain forests occurred as far north as 30°N, while subtropical vegetation may have extended to 60°N. For example, the fruit and seed flora from the London Clay of southern England comprises drifted plant material preserved in marine sediments and permineralised in iron pyrite (e.g. Reid and Chandler, 1933; Chandler, 1961; Collinson, 1983a). The source vegetation for the London Clay flora is interpreted as a diverse, paratropical rain forest (Wing and Sues, 1992). About 140 genera (both extinct and extant) can be assigned to living families and the present-day distribution of these taxa is strongly skewed towards the tropics. Many of the components of the London Clay flora are clearly related to plants now living in southeastern Asia (e.g. *Nypa*, Icacinaceae, Annonaceae), supporting the interpretation of a tropical climate. However, the presence of characteristic subtropical and temperate genera in the flora indicates that the London Clay vegetation was not a direct analogue of any recent tropical forest.

Eocene floras from North America provide similar indications of extreme warmth. Especially thoroughly studied is the flora from the Clarno Nut Beds, Oregon, USA, from which more than 20 000 permineralised fossils of fruits, seeds and wood, as well as leaf fossils, have been collected (Manchester, 1994). More than 170 species assigned to 145 genera have been recognised based on fruits and seeds. The vegetation of the Clarno Nut Beds contains a variety of taxa that were most likely frost-intolerant and that evidently grew under warm conditions, perhaps similar to those that today support paratropical rain forest. However, it is interesting that the flora also contains elements for which the closest living relatives today grow in temperate regions. It also contains many species, including some very common elements that cannot yet be assigned to any extant group.

Elsewhere in the USA, considerable warmth is indicated by the classic Middle Eocene flora from the Green River Formation in southwest Wyoming, northeast Utah and northwest Colorado (MacGinitie, 1969), as well as Early and Middle Eocene floras from western Tennessee and Kentucky. In the Green River Formation overwhelming evidence from both plants (e.g. large leaves of palms and aroids) and animals (e.g. crocodiles, turtles) indicates warm tropical to subtropical conditions (Grande, 1998). In Tennessee and Kentucky securely identified eudicots include papilionoid, caesalpinioid and mimosoid legumes and other fabids as well as various Fagales, Euphorbiaceae, Oleaceae, Malpighiaceae and several Ericales (e.g. Crepet *et al.*, 1975; Crepet and Dilcher, 1977). The vegetation has been interpreted as deciduous forest with a distinct dry season (Upchurch and Wolfe, 1987).

Eocene fossil floras from Asia and from the Southern Hemisphere also provide clear indications of general climatic warmth, including high-diversity vegetation at relatively high latitudes. Eocene floras from Argentina indicate subtropical to tropical forest (Romero, 1986); palynofloras and macrofloras from Australia, which was still situated at very high latitude, provide evidence that climates in the region were warmer, and probably wetter, than at any other time during the Cenozoic (Macphail *et al.*, 1994).

At around the Eocene–Oligocene transition in many different parts of the world there are clear indications of a colder and drier climate. The trigger for the onset of a cooler global climate in the Early Oligocene seems to have been the combined effect of various tectonic factors on oceanic and atmospheric circulation, which also included the development of a circum-Antarctic deep thermohaline current. This may have suppressed oceanic heat transfer to the Antarctic, resulting in the build-up of snow and ice (Pickering, 2000). Ice sheets may have been present in Antarctica by the mid-Eocene, but were certainly in place by the Eocene–Oligocene boundary (DeConto and Pollard, 2003). Major polar ice sheets were present on the continent by about the mid-Miocene.

20.6.3 Neogene vegetation and the origin of modern biomes

The dominant trend through the Neogene, initiated around the Eocene–Oligocene transition, was towards cooler and drier climates, driven largely by continuing tectonic changes that had direct impacts on possibilities for overland migration, ocean currents and movements of associated airmasses. As climates changed, and as diversification of the plant and animal groups that were already well established in the Palaeogene continued, Neogene ecosystems, and the processes by which they were regulated, took on a more modern appearance. New kinds of evergreen tropical rain forest developed at low latitudes. Desert and semi-desert developed in the zones of high pressure on either side of the equatorial belt, while more modern kinds of temperate and boreal forest formed further north and further south. Mediterranean sclerophyll woodland developed independently from temperate and boreal forests in five different parts of the world. At a global level the result was a reduction in total forest area and an increase in more open, lower-biomass, vegetation such as savannah, grasslands, deserts and tundra.

Modernisation of the Neogene environment was accompanied by regional extinction and further angiosperm diversification. The number of extant angiosperm families recorded, and the ease with which they can be recognised, increases substantially through the Neogene. It is during the Neogene that much of the generic and species-level diversity within extant families of angiosperms seems to have evolved. Most strikingly, more open vegetation created new opportunities for the diversification and spread of herbaceous species. Grasses, for example, undergo dramatic expansion from the Miocene onwards (e.g. Strömberg, 2005; Strömberg *et al.*, 2007). Asteraceae show a similar explosion in the abundance of fossil pollen through the same interval, which probably also reflects a dramatic expansion in their diversity and ecological importance. Other herbaceous groups that appear to show a similar pattern include the Lamiaceae, Apiaceae, Brassicaceae and most Caryophyllales. Together these groups account for a significant proportion of the species richness of extant angiosperms.

The rapid accumulation of species in both dominantly woody and dominantly herbaceous groups during the Neogene is especially suggestive of high speciation rates and low extinction rates. It is also interesting that such groups frequently appear to be nested in a more inclusive clade that includes tropical, although not necessarily tropical rainforest, members. Dominantly herbaceous groups attain their greatest diversity outside the lowland tropics.

Angiosperm diversification continued through the Neogene up to the Recent. Extinction, especially as a result of Pleistocene glaciations, also played an important role in shaping modern plant communities. In Europe there is clear evidence of major regional extinction. Increased aridity in Africa may also have been a potent force. But at the same time these climatic pertubations stimulated further speciation and microevolutionary change. Such changes, as well as extinction, also continue today in response to ongoing environmental changes, including those induced by human activity.

20.7 PROSPECTS

The fossil material described in this book illustrates the remarkable increase in the knowledge of angiosperm fossils that has occurred over the past 50 years. This knowledge, almost all of which has been based on new fossil discoveries in many different parts of the world, has greatly clarified the pattern of angiosperm evolution, especially through the Cretaceous, and has illuminated many aspects of Darwin's 'abominable mystery' (Darwin, in a letter to Hooker 1879; see Friedman, 2009; Friis *et al.*, 2010b). Most notably, with regard to 'the rapid appearance' of flowering plants, it is now clear that a wide range of essentially modern, and highly diverse, angiosperms did not appear suddenly in the mid-Cretaceous. Instead, the early history of angiosperms extends through a long period of more than 35 million years, most of the Early Cretaceous, during which there is a clear and orderly pattern of increasing phylogenetic diversity, increasing structural complexity, and increasing abundance. This is now clearly documented by the fossil record of angiosperm leaves, pollen and flowers, fruits and seeds. Furthermore, the diversification of angiosperms continues through the mid-Cretaceous, Late Cretaceous and Cenozoic, and again in broad conformity with the patterns implicit in recent phylogenetic analyses of extant angiosperms based on independent molecular data. The patterns of diversity seen among extant angiosperms, the dominance of different groups of angiosperms in different kinds of present-day vegetation, and the restriction of different groups of angiosperms to different parts of the world, all reflect the contingent nature of this process of diversification, together with extinction, in the context of long-term, substantial and ongoing environmental change.

Much still remains to be done to further elucidate the large-scale patterns of angiosperm evolution and also to clarify the nature of the processes that have been played out over tens of million years both within the group as a whole, and also within individual clades. A large amount of fossil material from both the Early and Late Cretaceous in existing collections still remains to be described, and renewed field work to find new localities will surely be rewarded with new discoveries and new surprises. However, the overall pattern, as it is currently known, appears robust. Further work on specific fossils and specific groups will add to the picture. It will also be greatly aided by increasingly integrated studies of well-preserved fossil material and careful morphological investigations of extant taxa. However, another, and perhaps even more important goal, on which only rudimentary progress has been made so far, will be to better understand how the patterns and processes that shaped angiosperm evolution over the past 135 million years can be linked to the massive geological and environmental changes that occurred through the same interval. Together, the interactions among these biological and geological processes created the ecology of the living planet of which we are part.

References

Adegoke, O. S., Jan du Chêne, R. E., Agumanu, A. E. & Ajayi, P. O. 1978. Palynology and the age of the Kerri–Kerri Formation, Nigeria. *Revista Española de Micropaleontología*, **10**, 267–83.

Adendorff, R., McLoughlin, S. & Bamford, M. K. 2002. A new genus of ovuliferous glossopterid fruits from South Africa. *Palaeontologia Africana*, **38**, 1–17.

Agosti, D., Grimaldi, D. A. & Carpenter, J. M. 1998. Oldest known ant fossils discovered. *Nature*, **391**, 447.

Agrawal, S. K. 1995. *Palmoxylon mathuri* Sahni from Kachchh (Gujarat, W. India) and its stratigraphic bearing. In *Global Environment and Diversification of Plants through Geological Time* (ed. D. D. Pant) Allahabad, India: Society of Indian Plant Taxonomists, pp. 23–6.

Albert, V. A., Backlund, A., Bremer, K. *et al.* 1994. Functional constraints and *rbc*L evidence for land plant phylogeny. *Annals of the Missouri Botanical Garden*, **81**, 534–67.

Allègre, C. J., Birck, J. L., Capmas, F. & Courtillot, V. 1999. Age of the Deccan traps using ^{187}Re–^{187}Os systematics. *Earth and Planetary Science Letters*, **170**, 197–204.

Allen, P. 1976. Wealden of the Weald: a new model. *Proceedings of the Geologists' Association*, **86**, 389–437.

Allen, P. 1981. Pursuit of Wealden models. *Journal of the Geological Society, London*, **138**, 375–405.

Allen, P. 1990. Wealden research – ways ahead. *Proceedings of the Geologists' Association*, **100**, 529–64.

Alvarez, L. W., Alvarez, W., Sarao, F. & Michel, V. H. 1980. Extraterrestrial cause for the Cretaceous-Tertiary extinction. *Science*, **208**, 1095–108.

Alvin, K. L. 1968. The spore-bearing organs of the Cretaceous fern *Weichselia*. *The Journal of the Linnean Society (Botany)*, **61**, 87–92.

Alvin, K. L. 1971. *Weichselia reticulata* (Stokes et Webb) Fontaine from the Wealden of Belgium. *Memoires de l'Institut Royal des Sciences Naturelles de Belgique*, **166**, 1–33.

Alvin, K. L. 1974. Leaf anatomy of *Weichselia* based on fusainized material. *Palaeontology*, **17**, 587–98.

Alvin, K. L., Fraser, C. J. & Spicer, R. A. 1981. Anatomy and palaeoecology of *Pseudofrenelopsis* and associated conifers in the English Wealden. *Palaeontology*, **24**, 759–78.

Alvin, K. L. 1982. Cheirolepidiaceae: biology, structure and palaeoecology. *Review of Palaeobotany and Palynology*, **37**, 71–98.

Alvin, K. L. 1983. Reconstructions of a Lower Cretaceous conifer. *Botanical Journal of the Linnean Society*, **86**, 169–76.

Anderberg, A. A., Ståhl, B. & Källersjö, M. 2000. Maesaceae, a new primuloid family in the order Ericales s.l. *Taxon*, **49**, 183–7.

Anderberg, A. A., Rydin, C. & Kjällersjö, M. 2002. Phylogenetic relationships in the order Ericales s.l.: analysis of molecular data of five genes from the plastid and mitochondrial genomes. *American Journal of Botany*, **89**, 677–87.

Anderberg, A. A. & Zhang, X. P. 2002. Phylogenetic relationships of Cyrillaceae and Clethraceae (Ericales) with special emphasis on the genus *Purdiaea* Planch. *Organisms Diversity & Evolution*, **2**, 127–37.

Anderson, C. N., Bremer, K. & Friis, E. M. 2005. Dating phylogenetically basal eudicots using *rbc*L sequences and multiple fossil reference points. *American Journal of Botany*, **92**, 1737–48.

Anderson, J. M. & Anderson, H. M. 1983. *Palaeoflora of Southern Africa: Molteno Formation (Triassic)*. Volume 1: Part 1. *Introduction;* Part 2. *Dicroidium*. Rotterdam: A.A. Balkema.

Anderson, J. M. & Anderson, H. M. 1985. *Palaeoflora of Southern Africa. Prodromus of South African Megafloras. Devonian to Cretaceous.* Rotterdam: A.A. Balkema.

Anderson, J. M. & Anderson, H. M. 1989. *Palaeoflora of Southern Africa: Molteno Formation (Triassic)*. Volume 2: Part 3. *Gymnosperms (excluding Dicroidium)*. Rotterdam: A.A. Balkema.

Anderson, J. M. & Anderson, H. M. 2003. *Heyday of the Gymnosperms: Systematics and Biodiversity of the Late Triassic Molteno Fructifications.* Pretoria: National Botanical Institute.

Ando, H., Seishi, M., Oshima, M. & Matsumaru, T. 1995. Fluvial shallow marine deposited systems of the Futaba Group (Upper Cretaceous): depositional facies and sequence. *Journal of Geography (Tokyo)*, **104**, 284–303.

Andreánzky, G. 1961. Ergänzungen zur Kenntnis der sarmatischen Flora Ungarns I. *Annales Historico-Naturales Musei Nationalis Hungarici. Mineralogica et Palaeontologica*, **53**, 13–33.

Andrews, H. N. 1961. *Paleobotany*. New York, London, Sydney: John Wiley & Sons.

APG 1998. An ordinal classification for the families of flowering plants. *Annals of the Missouri Botanical Garden*, **85**, 531–53.

APGII 2003. An update of the Angiosperm Phylogeny Group classification for the orders and families of flowering plants. *Botanical Journal of the Linnean Society*, **141**, 399–436.

APGIII 2009. An update of the Angiosperm Phylogeny Group classification for the orders and families of flowering plants: APG III. *Botanical Journal of the Linnean Society*, **161**, 105–21.

Arber, E. A. N. & Parkin, J. 1907. On the origin of angiosperms. *Journal of the Linnean Society, Botany*, **38**, 29–80.

Arber, E. A. N. & Parkin, J. 1908. Studies on the evolution of the angiosperms. *Annals of Botany*, **12**, 489–515.

Archangelsky, A., Taylor, T. N. & Kurmann, M. H. 1986. Ultrastructural studies of fossil plant cuticles: *Tico harrisii* from the early Cretaceous of Argentina. *Botanical Journal of the Linnean Society*, **92**, 101–16.

Archangelsky, A., Andreis, R. R., Archangelsky, S. & Artabe, A. 1995. Cuticular characters adapted to volcanic stress in a new Cretaceous cycad leaf from Patagonia, Argentina. Considerations on the stratigraphy and depositional history of the Baqueró Formation. *Review of Palaeobotany and Palynology*, **89**, 213–33.

Archangelsky, A. 2001. The Tíco Flora (Patagonia) and the Aptian Extinction Event. *Acta Palaeobotanica*, **41**, 115–22.

Archangelsky, S. & Brett, D. W. 1961. Studies on Triassic fossil plants from Argentina. I. *Rhexoxylon* from the Ischigualasto Formation. *Philosophical Transactions of the Royal Society of London* B, **244**, 1–19.

Archangelsky, S. 1963. A new Mesozoic flora from Ticó, Santa Cruz Province, Argentina. *Bulletin of the British Museum (Natural History) Geology*, **8**, 45–92.

Archangelsky, S. 1967. Estudio de la Formación Baqueró Cretácico Inferior de Santa Cruz, Argentina. *Revista del Museo de la Plata (Nueva Serie), Paleontología*, **5**, 63–171.

Archangelsky, S. 1968. Studies on Triassic fossil plants from Argentina. IV. The leaf genus *Dicroidium* and its possible relation to *Rhexoxylon* stems. *Palaeontology*, **11**, 500–12.

Archangelsky, S. 1980. Palynology of the Lower Cretaceous in Argentina. *Proceedings of the Fourth Palynological Conference, Lucknow (1976–77)*, **2**, 425–8.

Archangelsky, S. & Taylor, T. N. 1986. Ultrastructural studies of fossil plant cuticles. II. *Tarphyderma* gen. n., a Cretaceous conifer from Argentina. *American Journal of Botany*, **73**, 1577–87.

Archangelsky, S. & Taylor, T. N. 1993. The ultrastructure of *in situ Clavatipollenites* pollen from the Early Cretaceous of Patagonia. *American Journal of Botany*, **80**, 879–85.

Archangelsky, S., Barreda, V., Passalia, M. G. *et al.* 2009. Early angiosperm diversification: evidence from southern South America. *Cretaceous Research*, **30**, 1073–82.

Archibald, J. D. 1996. *Dinosaur Extinction and the End of an Era: What the Fossils Say*. New York: Columbia University Press.

Arnold, C. A. 1953. Origin and relationships of the cycads. *Phytomorphology*, **3**, 51–65.

Artabe, A. E., Spalletti, L. A., Brea, M. *et al.* 2007. Structure of a corystosperm fossil forest from the Late Triassic of Argentina. *Palaeogeography, Palaeoclimatology, Palaeoecology*, **243**, 451–70.

Ash, S. R. 1972. *Marcouia* gen. nov., a problematical plant from the Late Triassic of the southwestern U.S.A. *Palaeontology*, **15**, 423–9.

Ash, S. R. 1976a. The systematic position of *Eoginkgoites*. *American Journal of Botany*, **63**, 1327–31.

Ash, S. R. 1976b. Occurrence of the controversial plant fossil *Sanmiguelia* in the Upper Triassic of Texas. *Journal of Paleontology*, **50**, 799–804.

Ash, S. R. 1977. An unusual bennettitalean leaf from the Upper Triassic of the south-western United States. *Palaeontology*, **20**, 641–59.

Askin, R. A. 1990. Campanian to Paleocene spore and pollen assemblages of Seymour Island, Antarctica. *Review of Palaeobotany and Palynology*, **65**, 105–13.

Atta-Peters, D. & Salami, M. B. 2006. Aptian-Maastrichtian palynomorphs from the offshore Tano Basin, western Ghana. *Journal of African Earth Sciences*, **46**, 379–94.

Aulenback, K. R. 2009. *Identification Guide to the Fossil Plants of the Horseshoe Canyon Formation of Drumheller, Alberta*. Calgary: University of Calgary Press.

Axelrod, D. I. 1952. A theory of angiosperm evolution. *Evolution*, **6**, 29–60.

Axelrod, D. I. 1960. The evolution of flowering plants. In *The Evolution of Life* (ed. S. Tax) Chicago: University of Chicago Press, pp. 227–305.

Axelrod, D. I. 1970. Mesozoic paleogeography and early angiosperm history. *The Botanical Review*, **36**, 277–319.

Axelrod, D. I. 1987. The late Oligocene Creede flora, Colorado. *University of California Publications in Geological Sciences*, **130**, 1–235.

Axsmith, B. J., Taylor, E. L., Thomas, T. N. & Cuneo, N. R. 2000. New perspectives on the Mesozoic seed fern order Corystospermales based on attached organs from the Triassic of Antarctica. *American Journal of Botany*, **87**, 757–68.

Backlund, A. 1996. Phylogeny of the Dipsacales. Uppsala: PhD Thesis, Faculty of Science and Technology, University of Uppsala, pp. 1–34.

Bailey, I. W. & Sinnott, E. W. 1916. The climatic distribution of certain types of angiosperm leaves. *American Journal of Botany*, **3**, 24–39.

Bailey, I. W. 1944. The development of vessels in angiosperms and its significance in morphological research. *American Journal of Botany*, **31**, 421–8.

Bande, M. B., Prakash, U. & Ambwani, K. 1982. A fossil palm fruit *Hyphaeneocarpon indicum* gen. et sp. nov. from the Deccan Intertrappean beds, India. *The Palaeobotanist*, **30**, 303–9.

Bande, M. B. & Prakash, U. 1986. The Tertiary flora of Southeast Asia with remarks on its palaeoenvironment and phytogeography of the Indo-Malayan region. *Review of Palaeobotany and Palynology*, **49**, 203–33.

Banks, H., Stafford, P. & Crane, P. R. 2007. Aperture variation in the pollen of *Nelumbo* (Nelumbonaceae). *Grana*, **46**, 157–62.

Barale, G., Blanc-Louvel, C., Buffetaut, E. *et al.* 1984. Les gisements de calcaires lithographiques du Crétacé Inférieur du Montsech (Province de Lérida, Espagne). Considérations Paléoécologiques. *Geobios, Mémoirs spécial*, **8**, 275–83.

Barale, G., Philippe, M., Tayech-Mannai, B. & Zarbout, M. 1997. Découverte d'une flore à Ptéridophytes et Gymnospermes dans le Crétacé inférieur de la région de Tataouine (Sud Tunisien). *Comptes Rendus de l'Académie des Sciences, Paris, Sciences de la terre et des planètes*, **325**, 221–4.

Barale, G., Zarbout, M. & Philippe, M. 1998. Niveaux à végétaux fossiles en environnement fluviatile à marin proximal dans le Dahar (Bathonien à Albien – Sud Tunisien). *Bulletin de la Société géologique de France*, **169**, 811–19.

Barale, G. & Ouaja, M. 2001. Découverte de nouvelles flores avec des restes à affinités angiospermiennes dans le Crétacé inférieur du Sud Tunisien. *Cretaceous Research*, **22**, 131–43.

Barale, G. & Ouaja, M. 2002. La biodiversité végétale des gisements d'âge Jurassique supérieur-Crétacé inférieur de Merbah El Asfer (Sud-Tunisien). *Cretaceous Research*, **23**, 707–37.

Baranova, M. 1972. Systematic anatomy of some leaf epidermis in the Magnoliaceae and some related families. *Taxon*, **21**, 447–69.

Barbacka, M. & Bóka, K. 2000. A new Early Liassic fructification of the Caytoniales from Hungary. *Acta Palaeobotanica*, **40**, 83–109.

Barbosa, B. P. 1981. *Carta Geológica de Portugal na Escala de 1/50000. Notícia Explicativa da Folha 16-C Vagos.* Lisbon: Serviços Geológicos de Portugal.

Barclay, R. S., Johnson, K. R., Betterton, W. J. & Dilcher, D. L. 2003. Stratigraphy and megaflora of a K-T boundary section in the eastern Denver Basin, Colorado. *Rocky Mountain Geology*, **38**, 45–71.

Barnes, R. W. & Hill, R. S. 1999. Macrofossils of *Callicoma* and *Codia* (Cunoniaceae) from Australian Cainozoic sediments. *Australian Systematic Botany*, **12**, 647–70.

Barnes, R. W. & Jordan, G. J. 2000. *Eucryphia* (Cunoniaceae) reproductive and leaf macrofossils from Australian Cainozoic sediments. *Australian Systematic Botany*, **13**, 373–94.

Barreda, V. 1997. Palynomorph assemblage of the Chenque Formation, late Oligocene?-Miocene from Golfo San Jorge basin, Patagonia, Argentina. Part 4. Polyporate and porate pollen. *Ameghiniana*, **34**, 145–54.

Barrett, P. M. 2000. Evolutionary consequences of dating the Yixian Formation. *Trends in Ecology and Evolution*, **15**, 99–103.

Barron, E. J., Thompson, S. L. & Schneider, S. H. 1981. An Ice-Free Cretaceous? Results from Climate Model Simulations. *Science*, **212**, 501–8.

Barron, E. J., Fawcett, P. J., Pollard, D. & Thompson, S. 1994. Model simulations of Cretaceous climates: the role of geography and carbon dioxide. In *Palaeoclimates and their Modelling* (eds J. R. L. Allen, B. J. Hoskin, B. W. Sellwood, R. A. Spicer & P. J. Valdes) London, Glasgow, New York, Tokyo, Melbourne, Madras: Chapman & Hall, pp. 99–108.

Basinger, J. F. 1976. *Paleorosa similkameenensis*, gen. et sp. nov., permineralized flowers (Rosaceae) from the Eocene of British Columbia. *Canadian Journal of Botany*, **54**, 2293–305.

Basinger, J. F. & Dilcher, D. L. 1984. Ancient bisexual flowers. *Science*, **224**, 511–13.

Basinger, J. F. & Christophel, D. C. 1985. Fossil flowers and leaves of the Ebenaceae from the Eocene of southern Australia. *Canadian Journal of Botany*, **63**, 1825–43.

Batten, D. J. 1973. Palynology of early Cretaceous soil beds and associated strata. *Palaeontology*, **16**, 399–424.

Batten, D. J. 1975. Wealden palaeoecology from the distribution of plant fossils. *Proceedings of the Geologists' Association*, **85**, 433–58.

Batten, D. J. 1981. Stratigraphy, palaeogeography and evolutionary significance of Late Cretaceous and Early Tertiary Normapolles pollen. *Review of Palaeobotany and Palynology*, **35**, 125–37.

Batten, D. J. & Christopher, R. A. 1981. Key to recognition of Normapolles and some morphologically similar genera. *Review of Palaeobotany and Palynology*, **35**, 359–83.

Batten, D. J. & Maclennan, A. M. 1984. The paleoenvironmental significance of the conifer family Cheirolepidiaceae in the Cretaceous of Portugal. In *Third Symposium on Mesozoic Terrestrial Ecosystems, Short Papers* (eds W.-E. Reif & F. Westphal) Tübingen: Attempto Verlag, pp. 7–12.

Batten, D. J. 1986a. The Cretaceous Normapolles pollen genus *Vancampopollenites*: occurrence, form, and function. *Palaeontology, Special Papers*, **35**, 21–39.

Batten, D. J. 1986b. Possible functional implications of exine sculpture and architecture in some Late Cretaceous Normapolles pollen. In *Pollen and Spores. Form and Function* (eds S. Blackmore & I. K. Ferguson) London: Academic Press, pp. 219–32.

Batten, D. J. & Morrison, L. 1987. Morphology and occurrence of the Normapolles pollen genus *Papillopollis* in the Cretaceous of Portugal. *Palynology*, **11**, 133–54.

Batten, D. J. 1988. A revision of S.J. Diikstra's Late Cretaceous megaspores and other plant microfossils from Limburg, The Netherlands. *Mededelingen Rijks Geologische Dienst*, **41–3**, 1–55.

Batten, D. J., Dupagne-Kievits, J. & Lister, J. K. 1988. Palynology of the Upper Cretaceous Aachen Formation of northeastern Belgium. In *The Chalk District of the Euregio Meuse-Rhine* (eds M. Streel & M. J. M. Bless) Liège: Nat. Mus. Lab. Paleontol. Univ. Liège, pp. 95–103.

Batten, D. J. & Kovach, W. L. 1990. Catalog of Mesozoic and Tertiary megaspores. *American Association of Stratigraphic Palynologists, Contributions Series*, **24**, 1–227.

Batten, D. J. & Dutta, R. J. 1997. Ultrastructure of exine of gymnospermous pollen grains from Jurassic and basal Cretaceous deposits in northwest Europe and implications for botanical relationships. *Review of Palaeobotany and Palynology*, **99**, 25–54.

Batten, D. J. 1998. Palaeoenvironmental implication of plant, insect and other organic-walled microfossils in the Weald Clay Formation (Lower Cretaceous) of southeast England. *Cretaceous Research*, **19**, 279–315.

Batten, D. J., Collinson, M. E. & Brain, A. P. R. 1998. Ultrastructural interpretation of the Late Cretaceous megaspore *Glomerisporites pupus* and its associated microspores. *American Journal of Botany*, **85**, 724–35.

Batten, D. J. 2007. Spores and pollen from the Crato Formation: biostratigraphic and palaeoenvironmental implications. In *The Crato Fossil Beds of Brazil: Window into an Ancient World* (eds D. M. Martill, G. Bechly & R. F. Loveridge) Cambridge: Cambridge University Press, pp. 566–73.

Bawa, K. S. 1990. Plant-pollinator interactions in tropical rain forests. *Annual Review of Ecology and Systematics*, **21**, 399–422.

Bayer, C., Appel, O. & Rudall, P. J. 1998. Hanguanaceae. In *The Families and Genera of Vascular Plants.* IV. *Flowering Plants – Monocotyledones. Alismatanae and Commelinanae (except Gramineae)*. Volume I (ed. K. Kubitzki) Berlin, Heidelberg, New York: Springer-Verlag, pp. 223–5.

Beauvais, M., Berthou, P.-Y. & Lauverjat, J. 1975. Le gisement Campanien de Mira (Beira Litorale, Portugal): sédimentologie, micropaléonotologie, révision des Madréporaires. *Comunicações dos Serviços Geológicos de Portugal*, **59**, 37–58.

Beck, C. B. 1966. On the origin of gymnosperms. *Taxon*, **15**, 337–9.

Beck, C. B. 1970. The appearance of gymnospermous structure. *Biological Reviews*, **45**, 379–400.

Beck, C. B. 1971. On the anatomy and morphology of lateral branch systems of *Archaeopteris*. *American Journal of Botany*, **58**, 758–84.

Beck, C. B. 1981. *Archaeopteris* and its role in vascular plant evolution. In *Paleobotany, Paleoecology, and Evolution* (ed. K. J. Niklas) New York: Praeger, pp. 193–230.

Becker, H. F. 1972a. The Metzel Ranch flora of the Upper Ruby River Basin, southwestern Montana. *Palaeontographica* B, **141**, 1–61.

Becker, H. F. 1972b. *Sanmiguelia*, an enigma compounded. *Palaeontographica* B, **138**, 181–5.

Behrensmeyer, A. K. & Hook, R. H. 1992. Paleoenvironmental contexts and taphonomic modes. In *Terrestrial Ecosystems through Time. Evolutionary Paleoecology of Terrestrial Plants and Animals* (eds A. K. Behrensmeyer, J. D. Damuth, W. A. DiMichele, R. Potts, H.-D. Sues & S. L. Wing) Chicago, IL: The University of Chicago Press, pp. 15–136.

Behrensmeyer, A. K., Kidwell, S. M. & Gastaldo, R. A. 2000. Taphonomy and paleobiology. *Paleobiology*, **26**, 103–47.

Béland, P. & Russell, D. A. 1978. Paleoecology of Dinosaur Provincial Park (Cretaceous), Alberta, interpreted from the distribution of articulated vertebrate remains. *Canadian Journal of Earth Science*, **15**, 1012–24.

Bell, C. D., Soltis, D. E. & Soltis, P. S. 2005. The age of the angiosperms: a molecular timescale without a clock. *Evolution*, **59**, 1245–58.

Benda, L. 1961. Beiträge zur Flora des nordwestdeutschen Wealden. I. "Blätterkohlen" aus dem Hils und Osterwald. *Geologische Jahrbuch*, **78**, 621–52.

Bender, F. & Mädler, K. 1969. Die sandige Schichtenfolge der Kreide mit einer Angiospermen-Flora in Südjordanien. *Beihefte zum Geologischen Jahrbuch*, **81**, 35–92.

Benson, R. N. 2006. *Internal Stratigraphic Correlation of the Subsurface Potomac Formation, New Castle County, Delaware, and Adjacent Areas in Maryland and New Jersey*. Newark, Delaware: State of Delaware, Delaware Geological Survey.

Benton, M. J., Wills, M. A. & Hitchin, R. 2000. Quality of the fossil record through time. *Nature*, **403**, 534–7.

Benton, M. J. 2001. Finding the tree of life: matching phylogenetic trees to the fossil record through the 20th century. *Proceedings of the Royal Society of London* B, **268**, 2123–30.

Berner, R. A. 1991. A model for atmospheric CO_2 over Phanerozoic time. *American Journal of Science*, **291**, 339–76.

Berridge, E. M. 1911. On some points of resemblance between Gnetalean and Bennettitean seeds. *New Phytologist*, **10**, 140–4.

Berry, E. W. 1907. Contribution to the Mesozoic flora of the Atlantic coastal plain. II. North Carolina. *Bulletin of the Torrey Botanical Club*, **34**, 185–206.

Berry, E. W. 1909. Contribution to the Mesozoic flora of the Atlantic coastal plain. III. New Jersey. *Bulletin of the Torrey Botanical Club*, **36**, 245–64.

Berry, E. W. 1910a. Contribution to the Mesozoic flora of the Atlantic coastal plain. V. North Carolina. *Bulletin of the Torrey Botanical Club*, **37**, 181–200.

Berry, E. W. 1910b. Contribution to the Mesozoic flora of the Atlantic coastal plain. VI. Georgia. *Bulletin of the Torrey Botanical Club*, **37**, 503–11.

Berry, E. W. 1911. Systematic paleontology, Lower Cretaceous: Pteridophyta, Cycadophytae, Gymnospermae, Monocotyledonae, Dicotyledonae. In *Lower Cretaceous* (eds W. B. Clark, A. B. Bibbins & E. W. Berry) Baltimore: The Johns Hopkins Press, pp. 214–596.

Berry, E. W. 1912. Contributions to the Mesozoic flora of the Atlantic coastal plain. VIII. Texas. *Bulletin of the Torrey Botanical Club*, **39**, 387–406.

Berry, E. W. 1919. Upper Cretaceous floras of the eastern Gulf Region in Tennessee, Mississippi, Alabama, and Georgia. *US Geological Survey Professional Paper*, **112**, 1–177.

Berry, E. W. 1922. The flora of the Woodbine Sand at Arthurs Bluff, Texas. *US Geolological Survey Professional Paper*, **129-G**, 153–81.

Berry, E. W. 1925. The flora of the Ripley Formation. *US Geological Survey Professional Paper*, **129**, 199–226.

Berry, E. W. 1937. *Gyrocarpus* and other fossil plants from the Cumarebo field in Venezuela. *Journal of the Washington Academy of Sciences*, **27**, 501–6.

Berry, E. W. 1939. Fossil plants from the Cretaceous of Minnesota. *Journal of the Washington Academy of Sciences*, **29**, 331–6.

Berthou, P. Y. 1990. Le Bassin d'Araripe et les petits bassins intracontinentaux voisins (N.E. du Brésil): formation et évolution dans le cadre de l'ouverture de l'Atlantique équatorial. Comparaison avec les bassins ouest-Africains situés dans le même contexte. In *Atas do I Simposio sobre a Bacia do Araripe e Bacias Interiores do Nordeste, Crato, 14–16 de Junho de 1990* (eds D. de A. Campos, M. S. S. Viana, P. M. Brito & G. Beurlen), pp. 113–34.

Bessey, C. E. 1894. Evolution and classification. *Proceedings of the American Association for the Advancement of Science*, **42**, 237–51.

Bessey, C. E. 1896. The point of divergence of monocotyledons and dicotyledons. *Botanical Gazette*, **22**, 229–32.

Bessey, C. E. 1897. Phylogeny and taxonomy of the angiosperms. *Botanical Gazette*, **24**, 145–78.

Bessey, C. E. 1915. The phylogenetic taxonomy of flowering plants. *Annals of the Missouri Botanical Garden*, **2**, 109–64.

Bice, K. L., Huber, B. T. & Norris, R. D. 2003. Extreme polar warmth during the Cretaceous greenhouse? Paradox of the late Turonian $\delta^{18}O$ record at Deep Sea Drilling Project Site 511. *Paleoceanography*, **18**, 10.1029/2002PA000848

Bino, R. J., Dafni, A. & Meeuse, A. D. J. 1984a. Entomophily in the dioecious gymnosperm *Ephedra aphylla* Forsk. (=*E. alte* C.A. Mey.), with some notes on *E. campylopoda* C.A. Mey. I. Aspects of the entomophilous syndrome. *Proceedings of the Koninklijke Nederlandse Academie van Wetenschappen. Series C: Biological and Medical Sciences*, **87**, 1–13.

Bino, R. J., Devente, N. & Meeuse, A. D. J. 1984b. Entomophily in the dioecious gymnosperm *Ephedra aphylla* Forsk. (=*E. alte* C.A. Mey.), with some notes on *E. campylopoda* C.A. Mey. II. Pollination droplets, nectaries, and nectarial secretion in *Ephedra*. *Proceedings of the Koninklijke Nederlandse Akademie van Wetenschappen. Series C: Biological and Medical Sciences*, **87**, 15–24.

Biradar, N. V. & Bonde, S. D. 1990. The genus *Cyclanthodendron* and its affinities. In *Proceedings of the 3rd International Organization of Palaeobotany Conference* (eds J. G. Douglas & D. C. Christophel) Melbourne, pp. 51–7.

Blackburn, D. T. & Sluiter, I. R. K. 1994. The Oligo-Miocene coal floras of southeastern Australia. In *History of the Australian Vegetation: Cretaceous to Recent* (ed. R. S. Hill) Cambridge: Cambridge University Press, pp. 328–67.

Blanc-Louvel, C. 1984. Le genre "*Ranunculus* L." dans le Berriasien (Crétace inf.) de la province de Lérida (Espagne). *Instituto de Estudios Ilerdenses de la Diputación Provincial de Lleida*, **45**, 83–92.

Bocherens, H., Friis, E. M., Mariotti, A. & Pedersen, K. R. 1993. Carbon isotopic abundances in Mesozoic and Cenozoic fossil plants: palaeoecological implications. *Lethaia*, **26**, 347–58.

Bogner, J. 1976. Die systematische Stellung von *Acoropsis* Conwentz, einer fossilen Aracee aus dem Bernstein. *Mitteilungen Bayerischen Staatssammlungen, Paläontologie und historischen Geologie*, **16**, 95–8.

Bogner, J. & Mayo, S. J. 1998. Acoraceae. In *The Families and Genera of Vascular Plants*. IV. *Flowering Plants – Monocotyledons. Alismatanae and Commelinanae (except Gramineae)* (ed. K. Kubitzki) Berlin, Heidelberg, New York: Springer-Verlag, pp. 7–11.

Bogner, J. 2001. What is *Acorus brachystachys* Heer? *Aroideana*, **24**, 100–1.

Bogner, J., Hoffman, G. L. & Aulenback, K. R. 2005. A fossilized aroid inflorescence, *Albertarum pueri* gen. nov. et sp. nov., of Late Cretaceous (Late Campanian) age from the Horseshoe Canyon Formation of southern Alberta, Canada. *Canadian Journal of Botany*, **83**, 591–8.

Bogner, J., Johnson, K. R., Kvaček, Z. & Upchurch, G. R. 2007. New fossil leaves of Araceae from the Late Cretaceous and Paleogene of western North America. *Zitteliana*, **A47**, 133–47.

Boltenhagen, E. 1967. Spores et pollen du Crétacé supérieur du Gabon. *Pollen et Spores*, **9**, 335–55.

Bond, W. J. & van Wilgen, B. W. 1996. *Fire and Plants. Population and Community Biology Series*. London, Glasgow, Weinheim, New York, Tokyo, Melbourne, Madras: Chapman & Hall.

Bonde, S. D. 1986. A new gramineous stem from the Deccan Intertrappean beds of Nawargaon in Wardha District, Maharashtra, India. *Biovigyanam*, **12**, 39–43.

Bonde, S. D. & Kumaran, K. P. N. 1993. A liliaceous inflorescence from the Deccan Intertrappean beds of India. *Current Science*, **65**, 776–8.

Bonde, S. D. 1995. A palm peduncle and fruit from the Deccan Intertrappean beds of India. In *Global Environment and Diversification of Plants through Geological Time* (ed. D. D. Pant) Allahabad, India: Society of Indian Plant Taxonomists, pp. 63–9.

Bonde, S. D. 1996. *Arecoideostrobous moorei* gen. et sp. nov. a palm rachilla from the Deccan Intertrappean beds of India. *The Palaeobotanist*, **43**, 102–9.

Bonde, S. D. 1997. Fossil dicotyledonous liana *Anamirta pfeifferi* sp. nov. (Menispermaceae) from the Deccan Intertrappean beds of India. *The Palaeobotanist*, **47**, 89–94.

Bonde, S. D. 2000. *Rhodospathodendron tomlinsonii* gen. et sp. nov., an araceous viny axis from the Nawargaon intertrappean beds of India. *The Palaeobotanist*, **49**, 85–92.

Bonde, S. D. 2005. *Eriospermocormus indicus* gen. et sp. nov. (Liliales: Eriospermaceae): first record of a monocotyledonous corm from the Deccan Intertrappean beds of India. *Cretaceous Research*, **26**, 197–205.

Bonnefille, R., Hamilton, A. C., Linder, H. P. & Riollet, G. 1990. 30,000-Year-old fossil Restionaceae pollen from central equatorial Africa and its biogeographical significance. *Journal of Biogeography*, **17**, 307–314.

Bornemann, A., Norris, R. D., Friedrich, O. *et al.* 2008. Isotopic evidence for glaciation during the Cretaceous supergreenhouse. *Science*, **319**, 189–92.

Borsch, T., Löhne, C. & Wiersema, J. 2008. Phylogeny and evolutionary patterns in Nymphaeales: integrating genes, genomes and morphology. *Taxon*, **57**, 1052–81.

Bose, M. N. 1968. A new species of *Williamsonia* from the Rajmahal Hills, India. *The Journal of the Linnean Society (Botany)*, **61**, 121–7.

Bose, M. N. & Banerji, J. 1984. The fossil flora of Kachchh, I – Mesozoic megafossil flora. *The Palaeobotanist*, **33**, 1–189.

Bose, M. N., Banerji, J. & Pal, P. K. 1984a. *Amarjolia dactylota* (Bose) comb. nov., a bennettitalean bisexual flower from the Rajmahal Hills, India. *The Palaeobotanist*, **32**, 217–29.

Bose, M. N., Pal, P. K. & Harris, T. M. 1984b. *Carnoconites rajmahalensis* (Wieland) comb. nov. from the Jurassic of Rajmahal Hills, India. *The Palaeobotanist*, **32**, 368–9.

Bose, M. N., Pal, P. K. & Harris, T. M. 1985. The *Pentoxylon* plant. *Philosophical Transactions of the Royal Society of London* B, **310**, 77–108.

Boucher, L. D., Manchester, S. R. & Judd, W. S. 2003. An extinct genus of Salicaceae based on twigs with attached flowers, fruits, and foliage from the Eocene Green River Formation of Utah and Colorado, USA. *American Journal of Botany*, **90**, 1389–99.

Boureau, E. 1950. Étude paléoxylologique du Sahara (IX). Sur un *Myristicoxylon princeps* n. gen., nov. sp., du Danien d'Asselar. *Bulletin du Museum national d'Histoire naturelle, Paris, II*, **22**, 523–8.

Bowe, L. G., Coat, G. & DePamphilis, C. W. 2000. Phylogeny of seed plants based on all three genomic compartments: extant gymnosperms are monophyletic and Gnetales' closest relatives are conifers. *Proceedings of the National Academy of Sciences, USA*, **97**, 4092–97.

Boyd, A. 1992. Revision of the Late Cretaceous Pautût flora from West Greenland: Gymnospermopsida (Cycadales, Cycadeoidales, Caytoniales, Ginkgoales, Coniferales). *Palaeontographica* B, **225**, 105–72.

Braukmann, T. W. A., Kuzmina, M. & Stefanović, S. 2009. Loss of all plastid *ndh* genes in Gnetales and conifers: extent and evolutionary significance for the seed plant phylogeny. *Current Genetics*, **55**, 323–37.

Bravi, S., Civile, D., Martino, C., Barone Lumaga, M. R. & Nardi, G. 2004. Geological and paleontological observations on a fossil plant horizon in the Cenomanian of Mount Chianello (Southern Apennines). *Bollettino della Società Geologica Italiana*, **123**, 19–38.

Bravi, S., Barone Lumaga, M. R. & Mickle, J. E. 2010. *Sagaria cilentana* gen. et sp. nov. – a new angiosperm fructification from the Middle Albian of Southern Italy. *Cretaceous Research*, **31**, 285–90.

Bremer, B., Bremer, K., Heidari, N. *et al.* 2002. Phylogenetics of asterids based on 3 coding and 3 non-coding chloroplast DNA markers and the utility of non-coding DNA at higher taxonomic levels. *Molecular Biology and Evolution*, **24**, 274–301.

Bremer, K. & Wanntorp, H. E. 1981. The cladistic approach to plant classification. In *Advances in Cladistics: Proceedings of the First Meeting of the Willi Henning Society* (eds V. A. Funk & D. R. Brooks) New York: New York Botanical Garden, pp. 87–94.

Bremer, K. 1994. *Asteraceae: Cladistics and Classification*. Portland, OR: Timber Press.

Bremer, K. 2000. Early Cretaceous lineages of monocot flowering plants. *Proceedings of the National Academy of Sciences, USA*, **97**, 4707–11.

Bremer, K. 2002. Gondwanan evolution of the grass alliance of families (Poales). *Evolution*, **56**, 1374–87.

Bremer, K., Friis, E. M. & Bremer, B. 2004. Molecular phylogenetic dating of asterid flowering plants shows Early Cretaceous diversification. *Systematic Biology*, **53**, 496–505.

Brenner, G. J. 1963. The spores and pollen of the Potomac Group of Maryland. *Maryland Department of Geology, Mines and Water Resources, Bulletin*, **27**, 1–215.

Brenner, G. J. 1967. The gymnospermous affinity of *Eucommiidites* Erdtman, 1948. *Review of Palaeobotany and Palynology*, **5**, 123–7.

Brenner, G. J. 1968. Middle Cretaceous spores and pollen from northeastern Peru. *Pollen et Spores*, **10**, 341–83.

Brenner, G. J. 1976. Middle Cretaceous floral provinces and early migration of angiosperms. In *Origin and Early Evolution of Angiosperms* (ed. C. B. Beck) New York: Columbia University Press, pp. 23–47.

Brenner, G. J. & Bickoff, I. S. 1992. Palynology and the age of the Lower Cretaceous basal Kurnub Group from the coastal plain to the northern Negev of Israel. *Palynology*, **16**, 137–85.

Brenner, G. J. 1996. Evidence for the earliest stage of angiosperm pollen evolution: a paleoequatorial section from Israel. In *Flowering Plant Origin, Evolution and Phylogeny* (eds D. W. Taylor & L. J. Hickey) New York: Chapman & Hall, pp. 91–115.

Brenner, R. L., Ludvigson, G. A., Witzke, B. J. *et al.* 2000. Late Albian Kiowa–Skull Creek marine transgression, Lower Dakota Formation, eastern margin of Western Interior Seaway, U.S.A. *Journal of Sedimentary Research*, **70**, 868–78.

Brown, R. W. 1939. Fossil leaves, fruits, and seeds of *Cercidiphyllum*. *Journal of Paleontology*, **13**, 485–99.

Brown, R. W. 1956. Palmlike plants from the Dolores Formation (Triassic) southwestern Colorada. *US Geological Survey Professional Papers*, **274-H**, 205–9.

Bugdaeva, E. V. 1984. Flora and correlation of Turgian strata in Transbaikalia. *Geologiya i Geofizika*, **11**, 22–7 (in Russian).

Burger, D. 1980. Palynological studies in the Lower Cretaceous of the Surat Basin, Australia. *Bureau of Mineral Resources, Geology and Geophysics (Canberra), Bulletin*, **189**, 1–106.

Burger, D. 1990. Early Cretaceous angiosperms from Queensland, Australia. *Review of Palaeobotany and Palynology*, **65**, 153–63.

Burger, W. C. 1977. The Piperales and the monocots; alternative hypotheses for the orgin of monocotyledonous flowers. *The Botanical Review*, **43**, 345–93.

Burleigh, J. G. & Mathews, S. 2004. Phylogenetic signal in nucleotide data from seed plants: implications for resolving the seed plant tree of life. *American Journal of Botany*, **91**, 1599–613.

Burnham, R. J. 1989. Relationships between standing vegetation and leaf litter in a paratropical forest: implications for paleobotany. *Review of Palaeobotany and Palynology*, **58**, 5–32.

Cabrera, L. I., Salazar, G. A., Chase, M. W. *et al.* 2008. Phylogenetic relationships of aroids and duckweeds (Araceae) inferred from coding and noncoding plastid DNA. *American Journal of Botany*, **95**, 1153–65.

Caldas, E. B., Martins Neto, R. G. & Lima Filho, F. P. 1989. *Afropollis* sp. (polén) no trato intestinal de vespa (Hymenoptera: Apocrita: Xyelidae) no Cretáceo da Bacia do Araripe. *Abstracts II Simpósio Nacional de Estudos Tectônicos,* Sociedade Brasileira de Geología, **11**, 195–6.

Call, V. B. & Dilcher, D. L. 1997. The fossil record of *Eucommia* (Eucommiaceae) in North America. *American Journal of Botany*, **84**, 798–814.

Calvillo-Canadell, L. & Cevallos-Ferriz, S. R. S. 2007. Reproductive structures of Rhamnaceae from the Cerro del Pueblo (Late Cretaceous, Coahuila) and Coatzingo (Oligocene, Puebla) Formations, Mexico. *American Journal of Botany*, **94**, 1658–69.

Campbell, J. D. & Holden, A. M. 1984. Miocene casuarinacean fossils from Southland and Central Otago, New Zealand. *New Zealand Journal of Botany*, **22**, 159–67.

Canright, J. E. 1952. The comparative morphology and relationships of the Magnoliaceae. I. Trends of specialization in the stamens. *American Journal of Botany*, **39**, 484–97.

Cantrill, D. J. 1995. The occurrence of the fern *Hausmannia* Dunker (Dipteridaceae) in the Cretaceous of Alexander Island, Antarctica. *Alcheringa*, **19**, 243–54.

Cantrill, D. J. 1996. Fern thickets from the Cretaceous of Alexander Island, Antarctica containing *Alamatus bifurcatus* Douglas and *Aculea acicularis* sp. nov. *Cretaceous Research*, **17**, 169–82.

Cantrill, D. J. & Nichols, G. J. 1996. Taxonomy and palaeoecology of Early Cretaceous (Late Albian) angiosperm leaves from Alexander Island, Antarctica. *Review of Palaeobotany and Palynology*, **92**, 1–28.

Cantrill, D. J. 2000. A Cretaceous macroflora from a freshwater lake deposit, President Head, Snow Island, Antarctica. *Palaeontographica* B, **253**, 153–91.

Cantrill, D. J. & Poole, I. P. 2002. Cretaceous to Tertiary patterns of diversity change in the Antarctic Peninsula. *Geological Society of London, Special Publication*, **194**, 141–52.

Cantrill, D. J. & Poole, I. 2005. Taxonomic turnover and abundance in Cretaceous to Tertiary wood floras of Antarctica: implications for changes in forest ecology. *Palaeogeography, Palaeoclimatology, Palaeoecology*, **215**, 205–19.

Cantrill, D. J., Wanntorp, L. & Drinnan, A. N. 2011. Mesofossil flora from the Late Cretaceous of New Zealand. *Cretaceous Research*, **32**, 164–73.

Cao, Z., Wu, S., Zhang, P. & Li, J. 1997. Discovery of fossil monocotyledons from Yixian Formation, western Liaoning. *Chinese Science Bulletin*, **16**, 1764–6 (in Chinese).

Cao, Z., Wu, S., Zhang, P. & Li, J. 1998. Discovery of fossil monocotyledons from Yixian Formation, western Liaoning. *Chinese Science Bulletin*, **43**, 230–3.

Caris, P., Ronse Decraene, L. P., Smets, E. & Clinckemaillie, D. 2000. Floral development of three *Maesa* species, with special emphasis on the position of the genus within Primulales. *Annals of Botany*, **86**, 87–97.

Carlquist, S. 1983. Wood anatomy of *Bubbia* (Winteraceae), with comments on origin of vessels in dicotyledons. *American Journal of Botany*, **70**, 578–90.

Carlquist, S. 1988. *Comparative Wood Anatomy. Springer Series in Wood Science*. Berlin, Heidelberg, New York, London, Paris, Tokyo: Springer-Verlag.

Carlquist, S. 1996a. Wood anatomy of primitive angiosperms: new perspectives and syntheses. In *Flowering Plant Origin, Evolution and Phylogeny* (eds D. W. Taylor & L. J. Hickey) New York: Chapman & Hall, pp. 68–90.

Carlquist, S. 1996b. Wood, bark, and stem anatomy of Gnetales: a summary. *International Journal of Plant Sciences*, **157** (6 Suppl.), S58–S76.

Carlquist, S. & Schneider, E. L. 2002. The tracheid-vessel element transition in angiosperms involves multiple independent features: cladistic consequences. *American Journal of Botany*, **89**, 185–95.

Carpenter, F. M. 1992. *Treatise of Invertebrate Paleontology (R), Arthropoda 4*. Parts 3 and 4. *Hexapoda*. Lawrence, KS: Geological Society of America, University of Kansas Press.

Carpenter, R. J. & Buchanan, A. M. 1993. Oligocene leaves, fruits and flowers of the Cunoniaceae from Cethana, Tasmania. *Australian Systematic Botany*, **6**, 91–109.

Carpenter, R. J., Hill, R. S. & Jordan, G. J. 1994. Cenozoic vegetation in Tasmania: macrofossil evidence. In *History of the Australian Vegetation* (ed. R. S. Hill) Cambridge: Cambridge University Press, pp. 276–98.

Carpenter, R. J., Bannister, J. M., Jordan, G. J. & Lee, D. E. 2010. Leaf fossils of Proteaceae tribe Persoonieae from the Late Oligocene-Early Miocene of New Zealand. *Australian Systematic Botany*, **23**, 1–15.

Castañeda-Posadas, C. & Cevallos-Ferriz, S. R. S. 2007. *Swietenia* (Meliaceae) flower in Late Oligocene-Early Miocene amber from Simojovel de Allende, Chiapas, Mexico. *American Journal of Botany*, **94**, 1821–7.

Césari, S. N., Marenssi, S. A. & Santillana, S. N. 2001. Conifers from the Upper Cretaceous of Cape Lamb, Vega Island, Antarctica. *Cretaceous Research*, **22**, 309–19.

Cevallos-Ferriz, S. R. S. & Stockey, R. A. 1988. Permineralized fruits and seeds from the Princeton chert (Middle Eocene) of British Columbia: Araceae. *American Journal of Botany*, **75**, 1099–113.

Cevallos-Ferriz, S. R. S. & Stockey, R. A. 1990. Vegetative remains of the Magnoliaceae from the Princeton chert (Middle Eocene) of British Columbia. *Canadian Journal of Botany*, **68**, 1327–39.

Cevallos-Ferriz, S. R. S., Erwin, D. M. & Stockey, R. A. 1993. Further observations on *Paleorosa similkameenensis* (Rosaceae) from the Middle Eocene Princeton chert of British Columbia, Canada. *Review of Palaeobotany and Palynology*, **78**, 277–91.

Chadwick, C. E. 1993. The roles of *Tranes lyterioides* and *T. sparsus* Boh (Col., Curculionidae) in the pollination of *Macrozamia communis* (Zamiaceae). In *Proceedings of CYCAD 90, Second International Conference on Cycad Biology* (eds D. W. Stevenson & N. J. Norstog) Townsville, Queensland, Australia.: Palm and Cycad Societies of Australia, pp. 77–80.

Chaloner, W. G. 1969. Triassic spores and pollen. In *Aspects of Palynology* (eds R. H. Tschudy & R. A. Scott) New York: Wiley Interscience, pp. 291–309.

Chamberlain, C. J. 1913. *Macrozamia morei*, a connecting link between living and fossil cycads. *Botanical Gazette*, **55**, 141–54.

Chamberlain, C. J. 1920. The living cycads and the phylogeny of seed plants. *American Journal of Botany*, **7**, 125–45.

Chamberlain, C. J. 1935. *Gymnosperms: Structure and Evolution*. Chicago, IL: University of Chicago Press.

Chandler, M. E. J. 1926. The Upper Eocene flora of Hordle, Hants. *Palæontographical Society Monograph*, **77**, 1–52.

Chandler, M. E. J. 1954. Some Upper Cretaceous and Eocene fruits from Egypt. *Bulletin of the British Museum (Natural History) Geology*, **2**, 147–87.

Chandler, M. E. J. 1961. *The Lower Tertiary Floras of Southern England*. Volume 1. *Paleocene Floras, London Clay Flora (Suppl.), Text and Atlas*. London: British Museum (Natural History).

Chandler, M. E. J. & Axelrod, D. I. 1961. An Early Cretaceous (Hauterivian) angiosperm fruit from California. *American Journal of Science*, **259**, 441–6.

Chandler, M. E. J. 1964. *The Lower Tertiary Floras of Southern England*. Volume IV. *A Summary and Survey of Findings in Light of Recent Botanical Observations*. London: British Museum (Natural History).

Chang, M.-M., Chen, P.-J., Wang, Y.-Q., Wang, Y. & Mioa, D.-S. (eds) 2003. *The Jehol Biota*. Shanghai: Shanghai Scientific & Technical Publishers.

Chapman, J. L. & Smellie, J. L. 1992. Cretaceous fossil wood and palynomorphs from Williams Point, Livingston Island, Antarctic Peninsula. *Review of Palaeobotany and Palynology*, **74**, 163–92.

Charnov, E. L., Maynard-Smith, J. & Bull, J. J. 1976. Why be an hermaphrodite? *Nature*, **263**, 125–6.

Chase, M. W., Soltis, D. E., Olmstead, R. G. *et al.* 1993. Phylogenetics of seed plants: an analysis of nucleotide sequences from the plastid gene *rbc*L. *Annals of the Missouri Botanical Garden*, **80**, 528–80.

Chase, M. W., Soltis, D. E., Soltis, P. S. *et al.* 2000. Higher-level systematics of the monocotyledons: an assessment of current knowledge and a new classification. In *Monocots. Systematics and Evolution* (eds K. L. Wilson & D. A. Morrison) Melbourne: CSIRO, pp. 3–16.

Chase, M. W. 2004. Monocot relationships: an overview. *American Journal of Botany*, **91**, 1645–55.

Chase, M. W., Fay, M. F., Devey, D. S. *et al.* 2006. Multigene analyses of monocot relationships: a summary. *Aliso*, **22**, 63–75.

Chaw, S.-M., Zharkikh, A., Sung, H.-M. & Lau, T.-C. 1997. Molecular phylogeny of extant gymnosperms and seed plant evolution: analysis of nuclear 18S rRNA sequences. *Molecular Bioloy and Evolution*, **14**, 56–68.

Chaw, S.-M., Parkinson, C. L., Cheng, Y., Vincent, T. M. & Palmer, J. D. 2000. Seed plant phylogeny inferred from all three plant genomes: monophyly of extant gymnosperms and origin of Gnetales and conifers. *Proceedings of the National Academy of Sciences, USA*, **97**, 4086–91.

Chen, P.-J., Dong, Z.-M. & Zhen, S.-N. 1998. An exceptionally well-preserved theropod dinosaur from the Yixian Formation of China. *Nature*, **391**, 147–52.

Chen, P.-J. 2000. Palaeoenvironmental changes during the Cretaceous in eastern China. In *Cretaceous Environments of Asia*, Volume 17 (eds H. Okada & N. J. Mateer) Amsterdam: Elsevier, pp. 81–90.

Chen, P.-J. 2003. Cretaceous biostratigraphy of China. In *Biostratigraphy of China* (eds W. Zhang, P. Chen & A. R. Palmer) Beijing: Science Press, pp. 465–523.

Chesters, K. I. M. 1955. Some plant remains from the Upper Cretaceous and Tertiary of West Africa. *The Annals and Magazine of Natural History*, Ser. 12, 8, 498–503.

Chitaley, S. D. 1954. On a fructification from the Intertrappean flora of the Madhya Pradesh, India. *The Palaeobotanist*, **3**, 9–17.

Chitaley, S. D. 1956. On the fructification of *Tricoccites trigonum* Rode from the Deccan Intertrappean Series of India. *The Palaeobotanist*, **5**, 56–63.

Chitaley, S. D. 1958. Seeds of *Viracarpon hexaspermum* Sahni from the Intertrappean beds of Mohgaon Kalan, India. *The Journal of the Indian Botanical Society*, **37**, 408–11.

Chitaley, S. D. 1960. *Nipa* fruits from the Deccan Intertrappeans of India. *The Bulletin of the Botanical Society*, **1**, 31–5.

Chitaley, S. D. 1968. On *Aerorhizos harrisii* gen. et sp. nov. from India. *The Journal of the Indian Botanical Society*, **67**, 7–12.

Chitaley, S. D., Shallom, L. J. & Mehta, N. V. 1969. *Viracarpon sahnii*, sp. nov. from the Deccan Intertrappean beds of Mahurzari, India. In *J. Sen Memorial Volume* (eds H. Santapau, A. K. Ghosh, S. K. Roy, S. Chanda & S. K. Chaudhuri) Calcutta: J. Sen Memorial Committee and Botanical Society of Bengal, pp. 331–4.

Chitaley, S. D. & Patil, G. V. 1970. A petrified leaf from the Deccan Intertrappean beds of India. *Journal of Biological Sciences*, **13**, 30–6.

Chitaley, S. D. & Patil, G. V. 1971. Reinvestigation of *Shuklanthus superbum* Verma. *The Botanique*, **2**, 41–9.

Chitaley, S. D. & Sheikh, M. T. 1973. *Harrisostrobus intertrappea* gen. et sp. nov., a petrified gymnospermous cone from the Deccan Intertrappean beds of India. *Palaeontographica* B, **144**, 25–30.

Chitaley, S. D. & Kate, U. R. 1974. On a new petrified flower *Deccananthus savrii* gen. et sp. nov. from the Deccan Intertrappean beds of India. *The Palaeobotanist*, **21**, 317–20.

Chitaley, S. D. & Patel, M. Z. 1975. *Raoanthus intertrappea*, a new petrified flower from India. *Palaeontographica* B, **153**, 141–9.

Chitaley, S. D. & Nambudiri, E. M. V. 1995. Anatomy of *Nypa* fruits reviewed from new specimens from the Deccan Intertrappean flora of India. In *Global Environment and Diversification of Plants through Geological Times* (ed. D. D. Pant) Allahabad, India: Society of Indian Plant Taxonomists, pp. 83–94.

Chmura, C. A. 1973. Upper Cretaceous (Campanian-Maastrichtian) angiosperm pollen from the Western San Joaquin Valley, California, U.S.A. *Palaeontographica* B, **141**, 89–171.

Christensen, W. K. 1975. Upper Cretaceous belemnites from the Kristianstad area in Scania. *Fossils and Strata*, **7**, 1–69.

Christophel, D. C. & Basinger, J. F. 1982. Earliest floral evidence for the Ebenaceae in Australia. *Nature*, **296**, 439–41.

Christophel, D. C. 1984. Early Tertiary Proteaceae: the first floral evidence for the Musgraveinae. *Australian Journal of Botany*, **32**, 177–86.

Christophel, D. C. 1994. The early Tertiary macrofloras of continental Australia. In *History of the Australian Vegetation: Cretaceous to Recent* (ed. R. S. Hill) Cambridge: Cambridge University Press, pp. 262–75.

Christopher, R. A. 1979. Normapolles and triporate pollen assemblages from the Raritan and Magothy Formations (Upper Cretaceous) of New Jersey. *Palynology*, **3**, 73–121.

Christopher, R. A. 1982. Palynostratigraphy of the basal Cretaceous units of the eastern Gulf and southern Atlantic Coastal Plains. In *Second Symposium on the Geology of the Southeastern Coastal Plain*, Volume 53 (eds D. D. Arden, B. F. Beck & E. Morrow) Atlanta, GA: Georgia Geological Survey Circular, pp. 10–23.

Clark, J. S., Macklin, E. & Wood, L. 1998. Stages and spatial scales of recruitment limitation in southern Appalachian forests. *Ecological Monographs*, **68**, 213–35.

Coen, E. S. & Meyerowitz, E. M. 1991. The war of the whorls: genetic interactions controlling flower development. *Nature*, **353**, 31–7.

Coetzee, J. A. & Muller, J. 1984. The phytogeographic significance of some extinct Gondwana pollen types from the Tertiary of the southwestern Cape (South Africa). *Annals of the Missouri Botanical Garden*, **71**, 1088–99.

Coetzee, J. A. & Praglowski, J. 1987. Winteraceae pollen from the Miocene of the southwesten Cape (South Africa). *Grana*, **27**, 27–37.

Collinson, M. E. 1980. Recent and Tertiary seeds of the Nymphaeaceae *sensu lato* with a revision of *Brasenia ovula* (Brong.) Reid and Chandler. *Annals of Botany*, **46**, 603–32.

Collinson, M. E. 1982. A reassessment of fossil Potamogetoneae fruits with description of new material from Saudi Arabia. *Tertiary Research*, **4**, 83–104.

Collinson, M. E. 1983a. *Fossil Plants of the London Clay. Palaeontological Association Field Guides to Fossils no. 1.* London: The Palaeontological Association.

Collinson, M. E. 1983b. Palaeofloristic assemblages and palaeoecology of the Lower Oligocene Bembridge Marls, Hamstead ledge, Isle of Wight. *Botanical Journal of the Linnean Society*, **86**, 177–225.

Collinson, M. E. & Hooker, J. J. 1987. Vegetational and mammalian faunal changes in the Early Tertiary of southern England. In *The Origins of Angiosperms and their Biological Consequences* (eds E. M. Friis, W. G. Chaloner & P. R. Crane) Cambridge: Cambridge University Press, pp. 259–304.

Collinson, M. E. 1988. Freshwater macrophytes in palaeolimnology. *Palaeogeography, Palaeoclimatology, Palaeoecology*, **62**, 317–42.

Collinson, M. E. 1989. The fossil history of the Moraceae, Urticaceae (including Cecropiaceae), and Cannabaceae. In *The Evolution, Systematics and Fossil History of the Hamamelidae.* Volume 2. *'Higher' Hamamelidae* (eds P. R. Crane & S. Blackmore) Oxford: Clarendon Press, pp. 319–39.

Collinson, M. E. 1993. Taphonomy and fruiting biology of Recent and fossil *Nypa*. *Special Papers in Palaeontology*, **49**, 165–80.

Collinson, M. E., Boulter, M. C. & Holmes, P. L. 1993. Magnoliophyta ('Angiospermae'). In *The Fossil Record 2* (ed. M. J. Benton) London: Chapman & Hall, pp. 809–41.

Collinson, M. E., Kvaček, Z. & Zastawniak, E. 2002. The aquatic plants *Salvinia* (Salviniales) and *Limnobiophyllum* (Arales) from the Late Miocene flora of Sosnica (Poland). *Acta Palaeobotanica*, **41**, 253–82 (2001).

Collinson, M. E. & Van Bergen, P. F. 2004. Evolution of angiosperm fruit and seed dispersal biology and ecophysiology: morphological, anatomical and chemical evidence from fossils. In *The Evolution of Plant Physiology* (eds A. R. Hemsley & I. Poole) London: Linnean Society of London, pp. 343–77.

Columbus, J. T., Friar, E. A., Porter, J. M., Prince, L. M. & Simpson, M. G. (eds) 2006. *Monocots: Comparative Biology and Evolution (excluding Poales)*. Claremont, CA.: Rancho Santa Ana Botanic Garden.

Conwentz, H. 1886. *Die Flora des Bernsteins.* Volume 2. *Die Angiospermen des Bernsteins.* Leipzig: Wilhelm Engelmann.

Cook, C. D. K. 1998a. Pontederiaceae. In *The Families and Genera of Vascular Plants.* Volume IV. *Flowering Plants – Monocotyledones. Alismatanae and Commelinanae (except Gramineae)* (ed. K. Kubitzki) Berlin, Heidelberg, New York: Springer-Verlag, pp. 395–403.

Cook, C. D. K. 1998b. Butomaceae. In *The Families and Genera of Vascular Plants.* Volume IV. *Flowering Plants – Monocotyledones. Alismatanae and Commelinanae (except Gramineae)* (ed. K. Kubitzki) Berlin, Heidelberg, New York: Springer-Verlag, pp. 100–2.

Corner, E. J. H. 1976. *The Seeds of Dicotyledons.* Cambridge: Cambridge University Press.

Cornet, B. 1986. The leaf venation and reproductive structures of a Late Triassic angiosperm, *Sanmiguelia lewisii*. *Evolutionary Theory*, **7**, 231–309.

Cornet, B. 1989a. Late Triassic angiosperm-like pollen from the Richmond Rift Basin of Virginia, U.S.A. *Palaeontographica* B, **213**, 37–87.

Cornet, B. 1989b. The reproductive morphology and biology of *Sanmiguelia lewisii*, and its bearing on angiosperm evolution in the Late Triassic. *Evolutionary Trends in Plants*, **3**, http://bcornet.tripod.com/evtrend/Sanmig2.htm.

Cornet, B. & Habib, D. 1992. Angiosperm-like pollen from the ammonite-dated Oxfordian (Upper Jurassic) of France. *Review of Palaeobotany and Palynology*, **71**, 269–94.

Coulter, J. M. & Chamberlain, C. J. 1917. *Morphology of Gymnosperms.* Chicago, IL: University of Chicago Press.

Couper, R. A. 1953. Upper Mesozoic and Cainozoic spores and pollen from New Zealand. *New Zealand Geological Survey, Palaeontological Bulletin*, **22**, 1–77.

Couper, R. A. 1956. Evidence for a possible gymnospermous affinity for *Tricolpites troedssonii* Erdtman. *New Phytologist*, **55**, 280–5.

Couper, R. A. 1958. British Mesozoic microspores and pollen grains. A systematic and stratigraphic study. *Palaeontographica* B, **103**, 75–179.

Couper, R. A. 1960. New Zealand Mesozoic and Cainozoic plant microfossils. *New Zealand Geological Survey, Palaeontological Bulletin*, **32**, 1–87.

Crabtree, D. R. 1987. Angiosperms of the northern Rocky Mountains: Albian to Campanian (Cretaceous) megafossil floras. *Annals of the Missouri Botanical Garden*, **74**, 707–47.

Crame, J. A., Francis, J. E., Cantrill, D. J. & Pirrie, D. 2004. Maastrichtian stratigraphy of Antarctica. *Cretaceous Research*, **25**, 411–23.

Crane, P. R. 1981. Betulaceous leaves and fruits from the British Upper Paleocene. *Botanical Journal of the Linnean Society*, **83**, 103–36.

Crane, P. R. & Manchester, S. R. 1982. An extinct juglandaceous fruit from the Upper Palaeocene of southern England. *Botanical Journal of the Linnean Society*, **85**, 89–101.

Crane, P. R. 1984. A re-evaluation of *Cercidiphyllum*-like plant fossils from the British Lower Tertiary. *Botanical Journal of the Linnean Society*, **89**, 199–230.

Crane, P. R. & Dilcher, D. L. 1984. *Lesqueria:* an early angiosperm fruiting axis from the mid-Cretaceous. *Annals of the Missouri Botanical Garden*, **71**, 384–402.

Crane, P. R. 1985a. Phylogenetic analysis of seed plants and the origin of angiosperms. *Annals of the Missouri Botanical Garden*, **72**, 716–93.

Crane, P. R. 1985b. Phylogenetic relationships in seed plants. *Cladistics*, **1**, 329–48.

Crane, P. R. & Stockey, R. A. 1985. Growth and reproductive biology of *Joffrea speirsii* gen. et sp. nov., a *Cercidiphyllum*-like plant from the Late Paleocene of Alberta, Canada. *Canadian Journal of Botany*, **63**, 340–64.

Crane, P. R. 1986. Form and function in wind dispersed pollen. In *Pollen and Spores: Form and Function* (eds S. Blackmore & I. K. Ferguson) London: Academic Press, pp. 179–202.

Crane, P. R., Friis, E. M. & Pedersen, K. R. 1986. Lower Cretaceous angiosperm flowers: fossil evidence on early radiation of dicotyledons. *Science*, **232**, 852–4.

Crane, P. R. & Stockey, R. A. 1986. Morphology and development of pistillate inflorescences in extant and fossil Cercidiphyllaceae. *Annals of the Missouri Botanical Garden*, **73**, 382–93.

Crane, P. R. 1987. Vegetational consequences of the angiosperm diversification. In *The Origins of Angiosperms and their Biological Consequences* (eds E. M. Friis, W. G. Chaloner & P. R. Crane) Cambridge: Cambridge University Press, pp. 107–44.

Crane, P. R. & Upchurch, G. R. 1987. *Drewria potomacensis* gen. et sp. nov., an early Cretaceous member of Gnetales from the Potomac Group of Virginia. *American Journal of Botany*, **74**, 1722–36.

Crane, P. R. 1988. Major clades and relationships in the "higher" gymnosperms. In *Origin and Evolution of Gymnosperms* (ed. C. B. Beck) New York: Columbia University Press, pp. 218–72.

Crane, P. R., Manchester, S. R. & Dilcher, D. L. 1988. Morphology and phylogenetic significance of the angiosperm *Platanites hebridicus* from the Palaeocene of Scotland. *Palaeontology*, **31**, 503–17.

Crane, P. R. 1989. Paleobotanical evidence on the early radiation of nonmagnoliid dicotyledons. *Plant Systematics and Evolution*, **162**, 165–91.

Crane, P. R. & Blackmore, S. (eds) 1989a. *The Evolution, Systematics and Fossil History of the Hamamelidae*. Volume 1. *Introduction and 'Lower' Hamamelidae*. Oxford: Oxford University Press.

Crane, P. R. & Blackmore, S. (eds) 1989b. *The Evolution, Systematics and Fossil History of the Hamamelidae*. Volume 2. *'Higher' Hamamelidae*. Oxford: Oxford University Press.

Crane, P. R., Doyle, J. A., Donoghue, M. J. & Friis, E. M. 1989a. Angiosperm origins. *Nature*, **342**, 131.

Crane, P. R., Friis, E. M. & Pedersen, K. R. 1989b. Reproductive structure and function in Cretaceous Chloranthaceae. *Plant Systematics and Evolution*, **165**, 211–26.

Crane, P. R. & Lidgard, S. 1989. Angiosperm diversification and palaeolatitudinal gradients in Cretaceous floristic diversity. *Science*, **246**, 675–8.

Crane, P. R. & Lidgard, S. 1990. Angiosperm radiation and patterns of Cretaceous palynological diversity. In *Evolutionary Radiations* (eds P. D. Taylor & G. P. Larwood) Oxford: Oxford University Press, pp. 377–407.

Crane, P. R., Manchester, S. R. & Dilcher, D. L. 1990. A preliminary survey of fossil leaves and well-preserved reproductive structures from the Sentinel Butte Formation (Paleocene) near Almont, North Dakota. *Fieldiana, Geology New Series*, **20**, 1–63.

Crane, P. R., Manchester, S. R. & Dilcher, D. L. 1991. Reproductive and vegetative structure of *Nordenskioldia* (Trochodendraceae), a vesselless dicotyledon from the Early Tertiary of the Northern Hemisphere. *American Journal of Botany*, **78**, 1311–34.

Crane, P. R., Pedersen, K. R., Friis, E. M. & Drinnan, A. N. 1993. Early Cretaceous (Early to Middle Albian) platanoid inflorescences associated with *Sapindopsis* leaves from the Potomac Group of Eastern North America. *Systematic Botany*, **18**, 328–44.

Crane, P. R., Friis, E. M. & Pedersen, K. R. 1994. Paleobotanical evidence on the early radiation of magnoliid angiosperms. *Plant Systematics and Evolution* (Suppl.), **8**, 51–72.

Crane, P. R. 1996. The fossil history of Gnetales. *International Journal of Plant Sciences*, **157** (6 Suppl.), S50–S57.

Crane, P. R. & Herendeen, P. S. 1996. Cretaceous floras containing angiosperm flowers and fruits from eastern North America. *Review of Palaeobotany and Palynology*, **90**, 319–37.

Crane, P. R. & Kenrick, P. 1997. Diverted development of reproductive organs: a source of morphological innovation in land plants. *Plant Systematics and Evolution*, **206**, 161–74.

Crane, P. R., Herendeen, P. S. & Friis, E. M. 2004. Fossils and plant phylogeny. *American Journal of Botany*, **91**, 1683–99.

Crane, P. R. & Herendeen, P. S. 2009. Bennettitales from the Grisethorpe Bed (Middle Jurassic) at Cayton Bay, Yorkshire, UK. *American Journal of Botany*, **96**, 284–95.

Cranwell, L. M. 1959. Fossil pollen from Seymour Island, Antarctica. *Nature*, **184**, 1782–5.

Crawley, M. 2001. Angiosperm woods from British Lower Cretaceous and Palaeogene deposits. *Special Papers in Palaeontology*, **66**, 1–100.

Creber, G. T. & Chaloner, W. G. 1985. Tree growth in the Mesozoic and Early Tertiary and the reconstruction of palaeoclimates. *Palaeogeography, Palaeoclimatology, Palaeoecology*, **52**, 35–50.

Crepet, W. L. 1972. Investigations of North America cycadeoids: pollination mechanisms in *Cycadeoidea*. *American Journal of Botany*, **59**, 1048–56.

Crepet, W. L. & Delevoryas, T. 1972. Investigations of North American cycadeoids: early ovule ontogeny. *American Journal of Botany*, **59**, 209–15.

Crepet, W. L. 1974. Investigations of North American cycadeoids: the reproductive biology of *Cycadeoidea*. *Palaeontographica* B, **148**, 144–59.

Crepet, W. L., Dilcher, D. L. & Potter, F. W. 1974. Eocene angiosperm flowers. *Science*, **185**, 781–2.

Crepet, W. L., Dilcher, D. L. & Potter, F. W. 1975. Investigations of angiosperms from the Eocene of southeastern North America: a catkin with juglandaceous affinities. *American Journal of Botany*, **62**, 813–23.

Crepet, W. L. & Dilcher, D. L. 1977. Investigations of angiosperms from the Eocene of North America: a mimosoid inflorescence. *American Journal of Botany*, **64**, 714–25.

Crepet, W. L. 1978. Investigations of angiosperms from the Eocene of North America: an aroid inflorescence. *Review of Palaeobotany and Palynology*, **25**, 241–52.

Crepet, W. L. 1979. Insect pollination: a paleontological perspective. *BioScience*, **29**, 102–8.

Crepet, W. L. & Daghlian, C. P. 1982. Euphorbioid inflorescences from the Middle Eocene Clairborne Formation. *American Journal of Botany*, **69**, 258–66.

Crepet, W. L. 1984. Advanced (constant) insect pollination mechanisms: pattern of evolution and implications vis-à-vis angiosperm diversity. *Annals of the Missouri Botanical Garden*, **71**, 607–30.

Crepet, W. L. & Taylor, D. W. 1985. The diversification of the Leguminosae: first fossil evidence of the Mimosoideae and Papilionoideae. *Science*, **228**, 1087–9.

Crepet, W. L. & Taylor, D. W. 1986. Primitive mimosoid flowers from the Paleocene-Eocene and their systematic and evolutionary implications. *American Journal of Botany*, **73**, 548–63.

Crepet, W. L. & Friis, E. M. 1987. The evolution of insect pollination in angiosperms. In *The Origins of Angiosperms and their Biological Consequences* (eds E. M. Friis, W. G. Chaloner & P. R. Crane) Cambridge: Cambridge University Press, pp. 181–201.

Crepet, W. L. & Feldman, G. D. 1991. The earliest remains of grasses in the fossil record. *American Journal of Botany*, **78**, 1010–14.

Crepet, W. L. & Herendeen, P. S. 1992. Papilionoid flowers from the early Eocene of southeastern North America. In *Advances in Legume Systematics*. Part 4. *The Fossil Record* (eds P. S. Herendeen & D. L. Dilcher) Kew: Royal Botanic Gardens, pp. 43–55.

Crepet, W. L., Nixon, K. C., Friis, E. M. & Freudenstein, J. V. 1992. The oldest flowers of hamamelidaceous affinity, from the Upper Cretaceous of New Jersey, North America. *Proceedings of the National Academy of Sciences, USA*, **89**, 8986–9.

Crepet, W. L. & Nixon, K. C. 1994. Flowers of Turonian Magnoliidae and their implications. *Plant Systematics and Evolution* (Suppl.), **8**, 73–91.

Crepet, W. L. 1996. Timing in the evolution of derived floral characters: Upper Cretaceous (Turonian) taxa with tricolpate and tricolpate-derived pollen. *Review of Palaeobotany and Palynology*, **90**, 339–59.

Crepet, W. L. & Nixon, K. C. 1998a. Fossil Clusiaceae from the Late Cretaceous (Turonian) of New Jersey and implications regarding the history of bee pollination. *American Journal of Botany*, **85**, 1122–33.

Crepet, W. L. & Nixon, K. C. 1998b. Two new fossil flowers of magnoliid affinity from the Late Cretaceous of New Jersey. *American Journal of Botany*, **85**, 1273–88.

Crepet, W. L., Nixon, K. C. & Gandolfo, M. A. 2004. Fossil evidence and phylogeny: the age of major angiosperm clades based on mesofossil and macrofossil evidence from Cretaceous deposits. *American Journal of Botany*, **91**, 1666–82.

Crepet, W. L., Nixon, K. C. & Gandolfo, M. A. 2005. An extinct calycanthoid taxon, *Jerseyanthus calycanthoides*, from the Late Cretaceous of New Jersey. *American Journal of Botany*, **92**, 1475–85.

Cridland, A. A. 1957. *Williamsoniella papillosa*, a new species of bennettitalean flower. *Annals and Magazine of Natural History, ser. 12*, **10**, 383–8.

Cronquist, A. 1968. *The Evolution and Classification of Flowering Plants*. London: Nelson.

Cronquist, A. 1981. *An Integrated System of Classification of Flowering Plants*. New York: Columbia University Press.

Cronquist, A. 1988. *The Evolution and Classification of Flowering Plants*. 2nd ed. New York: New York Botanical Garden, Bronx.

Cruickshank, R. D. & Ko, K. 2003. Geology of an amber locality in the Hukawng Valley, northern Myanmar. *Journal of Asian Earth Sciences*, **21**, 441–55.

Czeczott, H. 1951. Srodkowo-miocenska flora Zalesiec kolo Wisniowca. 1. *Acta Geologica Polonica*, **2**, 349–445.

Daber, R. 1968. A *Weichselia-Stiehleria*-Matoniaceae community within the Quedlinburg Estuary of Lower Cretaceous age. *The Journal of the Linnean Society (Botany)*, **61**, 75–85.

Daber, R. 1990. Zur Paläogeographie und Biologie von *Weichselia reticulata* im Hauterive-Barrême (Unterkreide). *Feddes Repertorium*, **101**, 319–31.

Daghlian, C. P. 1978. Coryphoid palms from the Lower and Middle Eocene of southeastern North America. *Palaeontographica* B, **166**, 44–82.

Daghlian, C. P., Crepet, W. L. & Delevoryas, T. 1980. Investigations of Tertiary angiosperms: a new flora including *Eomimosoidea plumosa* from the Oligocene of Eastern Texas. *American Journal of Botany*, **67**, 309–20.

Daghlian, C. P. 1981. A review of the fossil record of monocotyledons. *The Botanical Review*, **47**, 517–55.

Dahlgren, R. M. T. 1980. A revised system of classification of the angiosperms. *Botanical Journal of the Linnean Society*, **80**, 91–124.

Dahlgren, R. M. T. 1983. General aspects of angiosperm evolution and macrosystematics. *Nordic Journal of Botany*, **3**, 119–49.

Dahlgren, R. M. T., Clifford, H. T. & Yeo, P. F. 1985. *The Families of the Monocotyledons: Structure, Evolution, and Taxonomy.* Berlin: Springer-Verlag.

Dam, G., Nøhr-Hansen, H., Pedersen, G. K. & Sønderholm, M. 2000. Sedimentary and structural evidence of a new early Campanian rift phase in the Nuussuaq Basin, West Greenland. *Cretaceous Research*, **21**, 127–54.

Dam, G., Pedersen, G. K., Sønderholm, M. *et al.* 2009. Lithostratigraphy of the Cretaceous–Paleocene Nuussuaq Group, Nuussuaq Basin, West Greenland. *Geological Survey of Denmark and Greenland Bulletin*, **19**, 1–171.

Davis, C. C., Latvis, M., Nickrent, D. L., Wurdack, K. J. & Baum, D. A. 2007. Floral gigantism in Rafflesiaceae. *Science*, **315**, 1812.

Dawson, J. W. 1886. On the Mesozoic floras of the Rocky Mountain Region of Canada. *Proceedings and Transactions of the Royal Society of Canada*, **3**, 1–22.

Debey, M. H. 1848. Uebersicht der urweltlichen Pflanzen des Kreidegebirges überhaupt und der Aachener Kreideschichten insbesondere. *Verhandlungen des Naturhistorischen Vereines des Preussischen Rheinlandes*, **5**, 113–25.

DeConto, R. M., Brady, E. C., Bergengren, J. & Hay, W. H. 2000a. Late Cretaceous climate, vegetation, and ocean interaction. In *Warm Climates in Earth History* (eds B. T. Huber, K. G. MacLeod & S. L. Wing) Cambridge: Cambridge University Press, pp. 275–96.

DeConto, R. M., Thompson, S. L. & Pollard, D. 2000b. Recent advances in paleoclimate modeling: toward better simulations of warm paleoclimates. In *Warm Climates in Earth History* (eds B. T. Huber, K. G. MacLeod & S. L. Wing) Cambridge: Cambridge University Press, pp. 21–49.

DeConto, R. M. & Pollard, D. 2003. Rapid Cenozoic glaciation of Antarctica induced by declining atmospheric CO_2. *Nature*, **421**, 245–9.

Dejax, J. 1987. Sur la présence de grains de pollen à sculpture crotonoide dans le Crétacé inférieur du Congo. *Mémoires et Travaux de l'Institut de Montpellier, Ecole Pratique des Hautes Etudes*, **17**, 253–71.

Dejax, J. & Brunet, M. 1996. Les flores fossiles du bassin D'Hama-Koussou, crétacé inférieur du nord-Cameroun: Correlations biochronologiques avec le fossé de la Bénoué, implication paléogéographiques. Géologie de l'Afrique et de l'Atlantique Sud: Actes Colloques Angers, 1994. *Bulletin Centres Recherches Exploration-Production Elf-Aquitaine*, **16**, 145–73.

Dejax, J., Pons, D. & Yans, J. 2007. Palynology of the dinosaur-bearing Wealden facies in the natural pit of Bernissart (Belgium). *Review of Palaeobotany and Palynology*, **144**, 25–38.

Del Tredici, P. 1989. Ginkgos and multituberculates: evolutionary interactions in the Tertiary. *Biosystems*, **22**, 327–39.

Del Tredici, P. 2007. The phenology of sexual reproduction in *Ginkgo biloba*: ecological and evolutionary implications. *The Botanical Review*, **73**, 267–78.

Delevoryas, T. 1959. Investigations of North American cycadeoids: *Monanthesia*. *American Journal of Botany*, **46**, 657–66.

Delevoryas, T. 1963. Investigations of North American cycadeoids: cones of *Cycadeoidea*. *American Journal of Botany*, **50**, 45–52.

Delevoryas, T. 1964. Two petrified angiosperms from the Upper Cretaceous of South Dakota. *Journal of Paleontology*, **38**, 584–6.

Delevoryas, T. 1965. Investigations of North American cycadeoids: microsporangiate structures and phylogenetic implications. *The Palaeobotanist*, **14**, 89–93.

Delevoryas, T. 1968. Some aspects of cycadeoid evolution. *Botanical Journal of the Linnean Society*, **61**, 137–46.

Delevoryas, T. & Hope, R. C. 1976. More evidence for a slender growth habit in Mesozoic cycadophytes. *Review of Palaeobotany and Palynology*, **21**, 93–100.

Delevoryas, T. & Mickle, J. E. 1995. Upper Cretaceous magnoliaceous fruit from British Columbia. *American Journal of Botany*, **82**, 763–8.

Denk, T. & Dillhoff, R. M. 2005. *Ulmus* leaves and fruits from the Early–Middle Eocene of northwestern North America: systematics and implications for character evolution within Ulmaceae. *Canadian Journal of Botany*, **83**, 1663–81.

Denk, T. & Oh, I.-C. 2005. Phylogeny of Schisandraceae based on morphological data: evidence from modern plants and the fossil record. *Plant Systematics and Evolution*, **256**, 113–45.

Denk, T. & Tekleva, M. V. 2006. Comparative pollen morphology and ultrastructure of *Platanus*: implications for phylogeny and evaluation of the fossil record. *Grana*, **45**, 195–221.

Denk, T. & Grimm, G. W. 2009. The biogeographic history of beech trees. *Review of Palaeobotany and Palynology*, **158**, 83–100.

Dettmann, M. E. 1973. Angiospermous pollen from Albian to Turonian sediments of eastern Australia. *Geological Society of Australia Special Publication*, **4**, 3–34.

Dettmann, M. E. 1986. Early Cretaceous palynoflora of subsurface strata correlative with the Koonwarra Fossil Bed, Victoria. In *Plants and Invertebrates from the Lower Cretaceous Koonwarra Fossil Beds, South Gippsland, Victoria* (eds P. A. Jell & J. Roberts) Sydney: Association of Australasian Palaeontologists, pp. 79–110.

Dettmann, M. E. & Thomson, M. R. A. 1987. Cretaceous palynomorphs from James Ross Island area – a pilot study. *British Antarctic Survey Bulletin*, **77**, 13–59.

Dettmann, M. E., Pocknall, D. T., Romero, E. J. & del Carmen Zamaloa, M. 1990. *Nothofagites* Erdtman ex Potonié, 1960; a catalogue of species with notes on the paleogeographic distribution of *Nothofagus* Bl. (Southern Beech). *New Zealand Geological Survey, Palaeontological Bulletin*, **60**, 1–79.

Dettmann, M. E. & Jarzen, D. M. 1991. Pollen evidence for Late Cretaceous differentiation of Proteaceae in southern polar forests. *Canadian Journal of Botany*, **69**, 901–6.

Dettmann, M. E. 1994. Cretaceous vegetation: the microfossil record. In *History of the Australian Vegetation: Cretaceous to Recent* (ed. R. S. Hill) Cambridge: Cambridge University Press, pp. 143–70.

Dettmann, M. E. & Jarzen, D. M. 1996. Pollen of proteaceous-type from latest Cretaceous sediments, southeastern Australia. *Alcheringa*, **20**, 103–60.

Dettmann, M. E. 1998. Pollen morphology of Eidotheoideae: implications for phylogeny in the Proteaceae. *Australian Systematic Botany*, **11**, 605–12.

Dettmann, M. E. & Clifford, H. T. 2000. Monocotyledon fruits and seeds, and an associated palynoflora from Eocene-Oligocene sediments of coastal central Quensland, Australia. *Review of Palaeobotany and Palynology*, **110**, 141–73.

Dettmann, M. E., Clifford, H. T. & Peters, M. 2009. *Lovellea wintonensis* gen. et sp. nov. – Early Cretaceous (late Albian), anatomically preserved, angiosperm flowers and fruits from the Winton Formation, western Queensland. *Cretaceous Research*, **30**, 339–55.

DeVore, M. L. & Pigg, K. B. 2007. A brief review of the fossil history of the family Rosaceae with a focus on the Eocene Okanogan Highlands of eastern Washington State, USA, and British Columbia, Canada. *Plant Systematics and Evolution*, **266**, 45–57.

Dickison, W. C. 1975. The bases of angiosperm phylogeny: vegetative anatomy. *Annals of the Missouri Botanical Garden*, **62**, 590–620.

Diéguez, C., Martín-Closas, C., Trincão, P. & Lopéz-Morón, N. 1995. IV. Paleontology. 1 Flora. In *Las Hoyas, a Lacustrine Konservat-Lagerstätte, Cuenca, Spain. II International Symposium on Lithographic Limestones. Field Trip Guide Book* (ed. M. N. Meléndez) Madrid: Universidad Complutense de Madrid, pp. 29–32.

Diéguez, C. & Meléndez, N. 2000. Early Cretaceous ferns from lacustrine limestones at Las Hoyas, Cuenca Province, Spain. *Palaeontology*, **43**, 1113–41.

Diéguez, C., Peyrot, D. & Barrón, E. 2010. Floristic and vegetational changes in the Iberian Peninsula during Jurassic and Cretaceous. *Review of Palaeobotany and Palynology*, **162**, 325–40.

Diez, J. B., Sender, L. M., Villanueva-Amadoz, U., Ferrer, J. & Rubio, C. 2005. New data regarding *Weichselia reticulata*: soral clusters and the spore developmental process. *Review of Palaeobotany and Palynology*, **135**, 99–107.

Dijkstra, S. J. 1949. Megaspores and some other fossils from the Aachenian (Senonian) in south Limburg, Netherlands. *Mededeelingen van de geologische Stichting, Nieuwe Serie*, **3**, 19–32.

Dilcher, D. L. 1973. A paleoclimatic interpretation of the Eocene floras of southeastern North America. In *Vegetation and Vegetational History of Northern Latin America* (ed. A. Graham) Amsterdam, London, New York: Elsevier, pp. 39–59.

Dilcher, D. L. 1974. Approaches to the identification of angiosperm leaf remains. *The Botanical Review*, **40**, 1–157.

Dilcher, D. L. & Daghlian, C. P. 1977. Investigations of angiosperms from the Eocene of south-eastern North America: *Philodendron* leaf remains. *American Journal of Botany*, **64**, 526–34.

Dilcher, D. L. 1979. Early angiosperm reproduction: an introductory report. *Review of Palaeobotany and Palynology*, **27**, 291–328.

Dilcher, D. L. & Crane, P. R. 1984a. *Archaeanthus*: an early angiosperm from the Cenomanian of the Western Interior of North America. *Annals of the Missouri Botanical Garden*, **71**, 351–83.

Dilcher, D. L. & Crane, P. R. 1984b. In pursuit of the first flower. *Natural History Magazine*, **93**, 56–61.

Dilcher, D. L. & Kovach, W. L. 1986. Early angiosperm reproduction: *Caloda delevoryana* gen. et sp. nov., a new fructification from the Dakota Formation (Cenomanian) of Kansas. *American Journal of Botany*, **73**, 1230–7.

Dilcher, D. L. 1989. The occurrence of fruits with affinity to Ceratophyllaceae in Lower and mid-Cretaceous sediments. *American Journal of Botany*, **76**, 162.

Dilcher, D. L. & Basson, P. W. 1990. Mid-Cretaceous angiosperm leaves from a new fossil locality in Lebanon. *Botanical Gazette*, **151**, 538–47.

Dilcher, D. L., Herendeen, P. S. & Hueber, F. M. 1992. Fossil *Acacia* flowers with attached anther glands from Dominican Republic amber. In *Advances in Legume Systematics*. Part 4. *The Fossil Record* (eds P. S. Herendeen & D. L. Dilcher) Kew: Royal Botanic Gardens, pp. 34–42.

Dilcher, D. L., Bernardes-De-Oliveira, M. E., Pons, D. & Lott, T. A. 2005. Welwitschiaceae from the Lower Cretaceous of northeastern Brazil. *American Journal of Botany*, **92**, 1294–310.

Dilcher, D. L., Sun, G., Ji, Q. & Li, H. 2007. An early infructescence *Hyrcantha decussata* (comb. nov.) from the Yixian Formation in northeastern China. *Proceedings of the National Academy of Sciences, USA*, **104**, 9370–4.

Dilcher, D. L. & Wang, H. 2009. An Early Cretaceous fruit with affinities to Ceratophyllaceae. *American Journal of Botany*, **96**, 2256–69.

Dinis, J. L. & Trincão, P. 1995. Recognition and stratigraphical significance of the Aptian unconformity in the Lusitanian Basin, Portugal. *Cretaceous Research*, **16**, 171–86.

Dinis, J. L. 1999. Estratigraphia sedimentologia da formação de Figueira da Foz. Aptiano a Cenomaniano do sector norte da Bacia Lusitânica. Coimbra: PhD Thesis, Departemento de Ciências da Terra, University of Coimbra, pp. 1–381.

Dinis, J. L. 2001. Definicão da Formacão da Figueira da Foz – Aptiano a Cenomaniano do sector central da margem oeste ibérica / Definition of the Figueira da Foz Formation – Aptian to Cenomanian of the central sector of the western Iberian margin. *Comunicações Instituto Geológico e Mineiro*, **88**, 127–60.

Dinis, J. L., Rey, J. & Graciansky, P.-C. de 2002. Le bassin lusitanien (Portugal) à l'Aptien supérieur-Albien: organisation séquentielle, proposition de corrélations, évolution. *Comptes Rendus Geoscience*, **334**, 757–64.

Diniz, F. 1967. Spores a flagelles nouvelles pour le Crétacé du Portugal. *Pollen et Spores*, **9**, 569–77.

Diniz, F., Kedves, M. & Simoncsics, P. 1974. Les sporomorphes principaux de sédiments crétacés de Vila Flor et de Carrajão, Portugal. *Comunicações dos Serviços Geológicos de Portugal*, **48**, 161–78.

Dino, R., Pocknall, D. T. & Dettmann, M. E. 1999. Morphology and ultrastructure of elater-bearing pollen from the Albian to Cenomanian of Brazil and Ecuador: implications for botanical affinity. *Review of Palaeobotany and Palynology*, **105**, 201–35.

Dischinger, J. B. 1987. Late Mesozoic and Cenozoic stratigraphic and structural framework near Hopewell, Virginia. *US Geological Survey Bulletin*, **1567**, 1–48.

Dobruskina, I. A. 1996. Connections of Israeli Upper Cretaceous flora with coeval floras of adjacent regions. *Rheedea*, **6**, 43–58.

Dobruskina, I. A. 1997. Turonian plants from the southern Negev, Israel. *Cretaceous Research*, **18**, 87–107.

Doludenko, M. P. & Orlovskaya, E. R. 1976. *Jurassic Flora of the Karatau*. Moscow: Nauka (in Russian).

Donoghue, M. J. 1989. Phylogenies and the analysis of evolutionary sequences, with examples from seed plants. *Evolution*, **43**, 1137–56.

Donoghue, M. J. & Doyle, J. A. 1989a. Phylogenetic studies of seed plants and angiosperms based on morphological characters. In *The Hierarchy of Life* (eds B. Fernholm, K. Bremer & H. Jörnwall) Amsterdam, New York, Oxford: Excerpta Medica, pp. 181–93.

Donoghue, M. J. & Doyle, J. A. 1989b. Phylogenetic analysis of angiosperms and the relationships of Hamamelidae. In *Evolution, Systematics, and Fossil History of the Hamamelidae*. Volume 1. *Introduction and 'Lower' Hamamelidae* (eds P. R. Crane & S. Blackmore) Oxford: Clarendon Press, pp. 17–45.

Dorofeev, P. I. 1963. *Treticnye Flory Zapadnoj Sibiri*. Moscow and Leningrad: Akademia Nauk SSSR (in Russian).

Dorofeev, P. I. 1970. *Treticnye Flory Urala*. Leningrad: Nauka (in Russian).

Dorofeev, P. I. 1973. Systematics of ancestral forms of *Brasenia*. *Paleontological Journal*, **7**, 219–27 (in Russian).

Dorofeev, P. I. 1974a. *Nymphaeales*. In *Iskopaemye Cvetkovye Rastenija SSSR*, Volume **1** (ed. A. Takhtajan) Leningrad: Nauka, pp. 52–85 (in Russian).

Dorofeev, P. I. 1974b. *Liriodendron*. In *Iskopaemye Cvetkovye Rastenija SSSR*, Volume **1** (ed. A. Takhtajan) Leningrad: Nauka, pp. 18–20 (in Russian).

Dorofeev, P. I. 1976. K sistematike neogenovych *Proserpinaca* Belorusii. *Doklady Akademia Nauk SSSR*, **20**, 1036–8 (in Russian).

Dorofeev, P. I. 1983. Dva novych vida *Liriodendron* iz treticnych otlozenii SSSR. *Botanicheskij Zhurnal*, **68**, 1401–8 (in Russian).

Dorofeev, P. I. 1988. *Miocenovye Flory Tambovskoi Oblasti.* Leningrad: Nauka (in Russian).

Douglas, A. W. & Tucker, S. C. 1996a. The developmental basis of diverse carpel orientations in Grevilleoideae (Proteaceae). *International Journal of Plant Sciences*, **157**, 373–97.

Douglas, A. W. & Tucker, S. C. 1996b. Comparative floral ontogenies among Persoonioideae including *Bellendena* (Proteaceae). *American Journal of Botany*, **83**, 1528–55.

Douglas, J. G. 1969. The Mesozoic floras of Victoria. Parts I and 2. *Geological Survey of Victoria, Memoir*, **28**, 1–310.

Douglas, J. G. 1986. The Cretaceous vegetation, and palaeoenvironment of Otway Group sediments. *Second South-Eastern Australia Oil Exploration Symposium*, pp. 233–40.

Douglas, J. G. 1994. Cretaceous vegetation: the macrofossil record. In *History of the Australian Vegetation* (ed. R. S. Hill) Cambridge: Cambridge University Press, pp. 171–88.

Doyle, J. & O'Leary, M. 1935. Pollination in *Tsuga*, *Cedrus*, *Pseudotsuga*, and *Larix*. *Scientific Proceedings of the Royal Dublin Society*, **21**, 191–204.

Doyle, J. A. 1969. Cretaceous angiosperm pollen of the Atlantic Coastal Plain and its evolutionary significance. *Journal of the Arnold Arboretum*, **50**, 1–35.

Doyle, J. A. 1973. Fossil evidence on early evolution of the monocotyledons. *Quarterly Review of Biology*, **48**, 399–413.

Doyle, J. A., Van Campo, M. & Lugardon, B. 1975. Observations on exine structure of *Eucommiidites* and Lower Cretaceous angiosperm pollen. *Pollen et Spores*, **17**, 429–86.

Doyle, J. A. & Hickey, L. J. 1976. Pollen and leaves from the mid-Cretaceous Potomac Group and their bearing on early angiosperm evolution. In *Origin and Early Evolution of Angiosperms* (ed. C. B. Beck) New York: Columbia University Press, pp. 139–206.

Doyle, J. A. 1977. Patterns of evolution in early angiosperms. In *Patterns of Evolution* (ed. A. Hallam) Amsterdam: Elsevier, pp. 501–46.

Doyle, J. A., Biens, P., Doerenkamp, A. & Jardiné, S. 1977. Angiosperm pollen from the pre-Albian Lower Cretaceous of Equatorial Africa. *Bulletin Centres Recherches Exploration-Production Elf-Aquitaine*, **1**, 451–73.

Doyle, J. A. & Robbins, E. I. 1977. Angiosperm pollen zonation of the continental Cretaceous of the Atlantic Coastal Plain and its application to deep wells in the Salisbury Embayment. *Palynology*, **1**, 43–78.

Doyle, J. A., Jardiné, S. & Doerenkamp, A. 1982. *Afropollis*, a new genus of early angiosperm pollen, with notes on the Cretaceous palynostratigraphy and paleoenvironments of northern Gondwana. *Bulletin Centres Recherches Exploration-Production Elf-Aquitaine*, **6**, 39–117.

Doyle, J. A. & Donoghue, M. J. 1985. Relationships of angiosperms and Gnetales: a numerical cladistic analysis. In *Systematic and Taxonomic Approaches in Palaeobotany.* Systematics Association Special Volume No. **31** (eds R. A. Spicer & B. A. Thomas) Oxford: Clarendon Press, pp. 177–98.

Doyle, J. A. & Donoghue, M. J. 1986. Seed plant phylogeny and the origin of angiosperms: an experimental cladistic approach. *The Botanical Review*, **52**, 321–431.

Doyle, J. A. & Donoghue, M. J. 1987. The origin of angiosperms: a cladistic approach. In *The Origins of Angiosperms and their Biological Consequences* (eds E. M. Friis, W. G. Chaloner & P. R. Crane) Cambridge: Cambridge University Press, pp. 17–49.

Doyle, J. A., Hotton, C. L. & Ward, J. V. 1990a. Early Cretaceous tetrads, zonasulculate pollen, and Winteraceae. I. Taxonomy, morphology, and ultrastructure. *American Journal of Botany*, **77**, 1544–57.

Doyle, J. A., Hotton, C. L. & Ward, J. V. 1990b. Early Cretaceous tetrads, zonasulculate pollen, and Winteraceae. II. Cladistic analysis and implications. *American Journal of Botany*, **77**, 1558–68.

Doyle, J. A. & Hotton, C. L. 1991. Diversification of early angiosperm pollen in a cladistic context. In *Pollen and Spores, Patterns of Diversity* (eds S. Blackmore & S. H. Barnes) Oxford: Clarendon Press, pp. 169–95.

Doyle, J. A. 1992. Revised palynological correlations of the lower Potomac Group (USA) and the Cocobeach sequence of Gabon (Barremian-Aptian). *Cretaceous Research*, **13**, 337–49.

Doyle, J. A. & Donoghue, M. J. 1992. Fossils and seed plant phylogeny reanalyzed. *Brittonia*, **44**, 89–106.

Doyle, J. A. & Donoghue, M. J. 1993. Phylogenies and angiosperm diversification. *Paleobiology*, **19**, 141–67.

Doyle, J. A., Donoghue, M. J. & Zimmer, E. A. 1994. Integration of morphological and ribosomal RNA data on the origin of angiosperms. *Annals of the Missouri Botanical Garden*, **81**, 419–50.

Doyle, J. A. & Le Thomas, A. 1994. Cladistic analysis and pollen evolution in Annonaceae. *Acta Botanica Gallica*, **141**, 149–70.

Doyle, J. A. 1996. Seed plant phylogeny and the relationships of Gnetales. *International Journal of Plant Sciences* (6 Suppl.), **157**, S3–S39.

Doyle, J. A. & Le Thomas, A. 1996. Phylogenetic analysis and character evolution in Annonaceae. *Bulletin du Muséum National d'Histoire Naturelle, Paris, 4e Série, Section B, Adansonia*, **3–4**, 279–334.

Doyle, J. A. 1999. The rise of angiosperms as seen in the African Cretaceous pollen record. In *Proceedings of the 3rd Conference on African Palynology, Johannesburg 14–19 September 1997. Palaeoecology of Africa and the Surrounding Islands* (ed. K. Heine) Rotterdam: A.A. Balkema, pp. 3–29.

Doyle, J. A. 2000. Paleobotany, relationships, and geographic history of Winteraceae. *Annals of the Missouri Botanical Garden*, **87**, 303–16.

Doyle, J. A. & Endress, P. K. 2000. Morphological phylogenetic analysis of basal angiosperms: comparison and combination with molecular data. *International Journal of Plant Sciences*, **161** (6 Suppl.), S121–S153.

Doyle, J. A., Eklund, H. & Herendeen, P. S. 2003. Floral evolution in Chloranthaceae: implications of a morphological phylogenetic analysis. *International Journal of Plant Sciences*, **164** (6 Suppl.), S365–S382.

Doyle, J. A. 2005. Early evolution of angiosperm pollen as inferred from molecular and morphological phylogenetic analyses. *Grana*, **44**, 227–51.

Doyle, J. A. 2006. Seed ferns and the origin of angiosperms. *Journal of the Torrey Botanical Society*, **133**, 169–209.

Doyle, J. A., Endress, P. K. & Upchurch, G. R. 2008a. Early Cretaceous monocots: a phylogenetic evaluation. *Sborník Národního Muzea v Praze, B*, **64**, 59–87.

Doyle, J. A., Manchester, S. R. & Sauquet, H. 2008b. A seed related to Myristicaceae in the Early Eocene of Southern England. *Systematic Botany*, **33**, 636–46.

Doyle, J. A. & Endress, P. K. 2010. Integrating Early Cretaceous fossils into the phylogeny of living angiosperms: Magnoliidae and eudicots. *Journal of Systematics and Evolution*, **48**, 1–35.

Dransfield, N. W., Uhl, N. W., Asmussen, C. B. *et al.* 2008. *Genera Palmarum: The Evolution and Classification of Palms.* Kew: Royal Botanic Gardens.

Drinnan, A. N. & Chambers, T. C. 1985. A reassessment of *Taeniopteris daintreei* from the Victorian Early Cretaceous: a member of the Pentoxylales and a significant Gondwanaland plant. *Australian Journal of Botany*, **33**, 89–100.

Drinnan, A. N. & Chambers, T. C. 1986. Flora of the Lower Cretaceous Koonwarra Fossil Bed (Korumburra Group), South Gippsland, Victoria. *Memoir of the Association of Australasian Palaeontologists*, **3**, 1–77.

Drinnan, A. N. & Crane, P. R. 1989. Cretaceous paleobotany and its bearing on the biostratigraphy of Austral angiosperms. In *Antarctic Paleobiology* (eds T. N. Taylor & E. L. Taylor) New York: Springer-Verlag, pp. 192–219.

Drinnan, A. N., Crane, P. R., Friis, E. M. & Pedersen, K. R. 1990. Lauraceous flowers from the Potomac Group (mid-Cretaceous) of eastern North America. *Botanical Gazette*, **151**, 370–84.

Drinnan, A. N., Crane, P. R., Pedersen, K. R. & Friis, E. M. 1991. Angiosperm flowers and tricolpate pollen of buxaceous affinity from the Potomac Group (mid-Cretaceous) of eastern North America. *American Journal of Botany*, **78**, 153–76.

Drinnan, A. N., Crane, P. R. & Hoot, S. B. 1994. Patterns of floral evolution in the early diversification of non-magnoliid dicotyledons (eudicots). *Plant Systematics and Evolution* (Suppl.), **8**, 93–122.

Drozdzewski, G., Hartkopf-Fröder, C., Lange, F.-G. *et al.* 1998. Vorläufige Mitteilung über unterkretazischen Tiefenkarst im Wülfrather Massenkalk (Rheinisches Schiefergebirge). *Mitteilungen des Verband der deutscher Höhlen- und Karstforscher*, **44**, 54–64.

Duan, S. 1998. The oldest angiosperm – a tricarpous female reproductive fossil from western Liaoning Province, NE China. *Science in China*, **41**, 14–20.

Duarte, L. & Santos, R. S. 1993. Plant and fish megafossils of the Codó Formation, Parnaíba Basin, NE Brazil. *Cretaceous Research*, **14**, 735–46.

Dunker, W. 1846. *Monographie der Norddeutschen Wealdenbildung. Ein Beitrag zur Geognosie und Naturgeschichte der Vorwelt.* Braunschweig: Oehme und Müller.

Dusén, P. 1908. Über die tertiäre Flora der Seymour-Insel. *Wissenschaftliche Ergebnisse der Schwedischen Südpolar-Expedition, 1901–1903*, **3** (3), 1–27.

Dutra, T. L. & Batten, D. J. 2000. Upper Cretaceous floras of King George Island, West Antarctica, and their palaeoenvironmental and phytogeographic implications. *Cretaceous Research*, **21**, 181–209.

Eames, A. J. 1952. Relationships of Ephedrales. *Phytomorphology*, **2**, 79–100.

Eames, A. J. 1961. *Morphology of the Angiosperms.* New York: McGraw-Hill.

Ehrendorfer, F. 1976. Evolutionary significance of chromosome differentiation patterns in gymnosperms and primitive angiosperms. In *Origin and Early Evolution of Angiosperms* (ed. C. B. Beck) New York: Columbia University Press, pp. 220–40.

Ehrlich, P. R. & Raven, P. H. 1964. Butterflies and plants: a study in coevolution. *Evolution*, **18**, 586–608.

Eklund, H., Friis, E. M. & Pedersen, K. R. 1997. Chloranthaceous floral structures from the Late Cretaceous of Sweden. *Plant Systematics and Evolution*, **207**, 13–42.

Eklund, H. & Kvaček, J. 1998. Lauraceous inflorescences and flowers from the Cenomanian of Bohemia (Czech Republic, Central Europe). *International Journal of Plant Sciences*, **159**, 668–86.

Eklund, H. 1999. Big survivors with small flowers. Fossil history and evolution of Laurales and Chloranthaceae. Uppsala: PhD Thesis, Faculty of Science and Technology, University of Uppsala, pp. 1–52.

Eklund, H. 2000. Lauraceous flowers from the Late Cretaceous of North Carolina, USA. *Botanical Journal of the Linnean Society*, **132**, 397–428.

Eklund, H. 2003. First Cretaceous flowers from Antarctica. *Review of Palaeobotany and Palynology*, **127**, 187–217.

Eklund, H., Cantrill, D. J. & Francis, J. E. 2004a. Late Cretaceous mesofossil assemblage from Table Nunatak, Antarctica. *Cretaceous Research*, **26**, 211–28.

Eklund, H., Doyle, J. A. & Herendeen, P. S. 2004b. Morphological phylogenetic analysis of living and fossil Chloranthaceae. *International Journal of Plant Sciences*, **165**, 107–51.

El-Ghazaly, G. A. & Chaudhary, R. 1993. Pollen morphology of some species of the genus *Euphorbia* L. *Review of Palaeobotany and Palynology*, **78**, 293–319.

El-Ghazaly, G. & Rowley, J. R. 1997. Pollen wall of *Ephedra foliata*. *Palynology*, **21**, 7–18.

Elliott, D., K. & Nations, D. 1998. Bee burrows in the Late Cretaceous (Late Cenomanian) Dakota Formation, northeastern Arizona. *Ichnos*, **5**, 243–53.

Elsik, W. C. 1968. Palynology of a Paleocene Rockdale lignite, Milam County, Texas. II. Morphology and taxonomy. *Pollen et Spores*, **10**, 599–664.

Endress, P. K. 1977. Evolutionary trends in the Hamamelidales-Fagales-group. *Plant Systematics and Evolution*, (Suppl.) **1**, 321–47.

Endress, P. K. & Honegger, R. 1980. The pollen of the Austrobaileyaceae and its phylogenetic significance. *Grana*, **19**, 177–82.

Endress, P. K. 1982. Syncarpy and alternative modes of escaping disadvantages of apocarpy in primitive angiosperms. *Taxon*, **31**, 48–52.

Endress, P. K. & Lorence, D. H. 1983. Diversity and evolutionary trends in the floral structure of *Tambourissa* (Monimiaceae). *Plant Systematics and Evolution*, **143**, 53–81.

Endress, P. K. & Sampson, F. B. 1983. Floral structure and relationships of the Trimeniaceae (Laurales). *Journal of the Arnold Arboretum*, **64**, 447–73.

Endress, P. K. 1984. The flowering process in the Eupomatiaceae (Magnoliales). *Botanische Jahrbücher für Systematik*, **104**, 297–319.

Endress, P. K. 1985. Stamenabzission und Pollenpräsentation bei Annonaceae. *Flora*, **176**, 95–8.

Endress, P. K. 1986a. An entomophily syndrome in Juglandaceae: *Platycarya strobilacea*. *Veröffentlichungen des Geobotanischen Instituts der ETH, Stiftung Rübel, Zürich*, **87**, 100–11.

Endress, P. K. 1986b. Floral structure, systematics, and phylogeny in Trochodendrales. *Annals of the Missouri Botanical Garden*, **73**, 297–324.

Endress, P. K. 1987. The Chloranthaceae: reproductive structures and phylogenetic position. *Botanische Jahrbücher für Systematik, Pflanzengeschichte und Pflanzengeographie*, **109**, 153–226.

Endress, P. K. 1989. The systematic position of the Myrothamnaceae. In *Evolution, Systematics, and Fossil History of Hamamelidae*. Volume 1. *Introduction and 'Lower' Hamamelidae* (eds P. R. Crane & S. Blackmore) Oxford: Clarendon Press, pp. 193–200.

Endress, P. K. & Hufford, L. D. 1989. The diversity of stamen structures and dehiscence patterns among Magnoliidae. *Botanical Journal of the Linnean Society*, **100**, 45–85.

Endress, P. K. 1990. Evolution of reproductive structures and functions in primitive angiosperms (Magnoliidae). *Memoirs of the New York Botanical Garden*, **55**, 5–34.

Endress, P. K. & Friis, E. M. 1991. *Archamamelis*, hamamelidalean flowers from the Upper Cretaceous of Sweden. *Plant Systematics and Evolution*, **175**, 101–14.

Endress, P. K. & Stumpf, S. 1991. The diversity of stamen structures in 'Lower' Rosidae (Rosales, Fabales, Proteales, Sapindales). *Botanical Journal of the Linnean Society*, **107**, 217–93.

Endress, P. K. 1993a. Austrobaileyaceae. In *The Families and Genera of Vascular Plants*. Volume II. *Flowering Plants – Dicotyledons. Magnoliid, Hamamelid and Caryophyllid Families* (eds K. Kubitzki, J. G. Rohwer & V. Bittrich) Berlin, Heidelberg, New York: Springer-Verlag, pp. 138–40.

Endress, P. K. 1993b. Trochodendraceae. In *The Families and Genera of Vascular Plants*. Volume II. *Flowering Plants – Dicotyledons. Magnoliid, Hamamelid and Caryophyllid Families* (eds K. Kubitzki, J. G. Rohwer & V. Bittrich) Berlin, Heidelberg, New York: Springer-Verlag, pp. 599–602.

Endress, P. K. 1993c. Hamamelidaceae. In *The Families and Genera of Vascular Plants.* Volume II. *Flowering Plants – Dicotyledons. Magnoliid, Hamamelid and Caryophyllid Families* (eds K. Kubitzki, J. G. Rohwer & V. Bittrich) Berlin, Heidelberg, New York: Springer-Verlag, pp. 322–31.

Endress, P. K. 1993d. Himantandraceae. In *The Families and Genera of Vascular Plants.* Volume II. *Flowering Plants – Dicotyledons. Magnoliid, Hamamelid and Caryophyllid Families* (eds K. Kubitzki, J. G. Rohwer & V. Bittrich) Berlin, Heidelberg, New York: Springer-Verlag, pp. 338–41.

Endress, P. K. 1993e. Eupomatiaceae. In *The Families and Genera of Vascular Plants.* Volume II. *Flowering Plants – Dicotyledons. Magnoliid, Hamamelid and Caryophyllid Families* (eds K. Kubitzki, J. G. Rohwer & V. Bittrich) Berlin, Heidelberg, New York: Springer-Verlag, pp. 296–8.

Endress, P. K. 1993f. Eupteleaceae. In *The Families and Genera of Vascular Plants.* Volume II. *Flowering Plants – Dicotyledons. Magnoliid, Hamamelid and Caryophyllid Families* (eds K. Kubitzki, J. G. Rohwer & V. Bittrich) Berlin, Heidelberg, New York: Springer-Verlag, pp. 299–301.

Endress, P. K. 1993g. Cercidiphyllaceae. In *The Families and Genera of Vascular Plants.* Volume II. *Flowering Plants – Dicotyledons. Magnoliid, Hamamelid and Caryophyllid Families* (eds K. Kubitzki, J. G. Rohwer & V. Bittrich) Berlin, Heidelberg, New York: Springer-Verlag, pp. 250–2.

Endress, P. K. 1994a. Evolutionary aspects of the floral structure in *Ceratophyllum. Plant Systematics and Evolution* (Suppl.), **8**, 175–83.

Endress, P. K. 1994b. *Diversity and Evolutionary Biology of Tropical Flowers. Cambridge Tropical Biology Series.* Cambridge: Cambridge University Press.

Endress, P. K. 1996a. Structure and function of female and bisexual organ complexes in Gnetales. *International Journal of Plant Sciences*, **157** (6 Suppl.), S113–S125.

Endress, P. K. 1996b. Diversity and evolutionary trends in angiosperm anthers. In *The Anther: Form, Function and Phylogeny* (eds W. G. D'Arcy & R. C. Keating) Cambridge: Cambridge University Press, pp. 92–110.

Endress, P. K. 1996c. Evolutionary aspects of fruits in basal flowering plants. *Det Norske Videnskaps-Akademi. I. Matematisk Naturvetenskaplig Klasses Avhandlinger Ny serie*, **18**, 21–32.

Endress, P. K. & Igersheim, A. 1997. Gynoecium diversity and systematics of the Laurales. *Botanical Journal of the Linnean Society*, **125**, 93–168.

Endress, P. K. & Igersheim, A. 1999. Gynoecium diversity and systematics of the basal eudicots. *Botanical Journal of the Linnean Society*, **130**, 305–93.

Endress, P. K. & Igersheim, A. 2000a. Gynoecium structure and evolution in basal angiosperms. *International Journal of Plant Sciences*, **161** (6 Suppl.), S211–S223.

Endress, P. K. & Igersheim, A. 2000b. The reproductive structures of the basal angiosperm *Amborella trichopoda* (Amborellaceae). *International Journal of Plant Sciences*, **161** (6 Suppl.), S237–S248.

Endress, P. K. 2001. The flowers in extant basal angiosperms and inferences on ancestral flowers. *International Journal of Plant Sciences*, **162**, 1111–40.

Endress, P. K. & Matthews, M. L. 2006a. First step towards a floral structural characterization of the major rosid subclades. *Plant Systematics and Evolution*, **260**, 223–51.

Endress, P. K. & Matthews, M. L. 2006b. Elaborate petals and staminodes in eudicots: diversity, function, and evolution. *Organisms Diversity & Evolution*, **6**, 257–93.

Endress, P. K. & Doyle, J. A. 2007. Floral phyllotaxis in basal angiosperms: development and evolution. *Current Opinion in Plant Biology*, **10**, 52–7.

Endress, P. K. 2008. Perianth biology in the basal grade of extant angiosperms. *International Journal of Plant Sciences*, **169**, 844–62.

Endress, P. K. & Doyle, J. A. 2009. Reconstructing the ancestral angiosperm flower and its initial specializations. *American Journal of Botany*, **96**, 22–66.

Endress, P. K. 2010a. Flower structure and trends of evolution in eudicots and their major subclades. *Annals of the Missouri Botanical Garden*, **97**, 541–83.

Endress, P. K. 2010b. The evolution of floral biology in basal angiosperms. *Philosophical Transactions of the Royal Society* B, **365**, 411–21.

Engel, M. S. 2000. A new interpretation of the oldest fossil bee (Hymenoptera: Apidae). *American Museum Novitates*, **3296**, 1–11.

Engel, M. S. & Grimaldi, D. A. 2004. New light shed on the oldest insect. *Nature*, **427**, 627–30.

Engler, A. 1930. Saxifragaceae. In *Die Natürlichen Pflanzenfamilien*, Volume 18a (ed. A. E. H. Prant) Leipzig: W. Engelman, pp. 74–226.

Erdtman, G. 1948. Did dicotyledonous plants exist in early Jurassic time? *Geologiska Föreningens i Stockholm Förhandlingar*, **70**, 265–71.

Erdtman, G. 1952. *Pollen Morphology and Plant Taxonomy. Angiosperms.* Stockholm: Almquist & Wiksell.

Ericson, P. G. P., Irestedt, M. & Johansson, U. S. 2003. Evolution, biogeography, and patterns of diversification in passerine birds. *Journal of Avian Biology*, **34**, 3–15.

Eriksson, O. & Jakobsson, A. 1999. Recruitment trade-offs and the evolution of dispersal mechanisms in plants. *Evolutionary Ecology*, **13**, 411–23.

Eriksson, O., Friis, E. M. & Löfgren, P. 2000a. Seed size, fruit size and dispersal spectra in angiosperms from the Early Cretaceous to the Late Tertiary. *American Naturalist*, **156**, 47–58.

Eriksson, O., Friis, E. M., Pedersen, K. R. & Crane, P. R. 2000b. Seed size and dispersal systems of Early Cretaceous angiosperms from Famalicão, Portugal. *International Journal of Plant Sciences*, **161**, 319–29.

Eriksson, O. 2008. Evolution of seed size and biotic seed dispersal in angiosperms: paleoecological and neoecological evidence. *International Journal of Plant Sciences*, **169**, 863–70.

Ernst, W. R. 1964. The genera of Berberidaceae, Lardizabalaceae, and Menispermaceae in the southeastern United States. *Journal of the Arnold Arboretum*, **45**, 1–35.

Erwin, D. M. & Stockey, R. A. 1989. Permineralized monocotyledons from the Middle Eocene Princeton chert (Allenby Formation) of British Columbia: Alismataceae. *Canadian Journal of Botany*, **67**, 2636–45.

Erwin, D. M. & Stockey, R. A. 1992. Vegetative body of a permineralized monocotyledon from the Middle Eocene Princeton chert of British Columbia. *Courier Forschungsinstitut Senckenberg*, **147**, 309–27.

Erwin, D. M. & Stockey, R. A. 1994. Permineralized monocotyledons from the Middle Eocene Princeton chert: Arecaceae. *Palaeontographica* B, **234**, 19–40.

Esau, K. 1977. *Anatomy of Seed Plants.* New York, Santa Barbara, London, Sydney, Toronto: John Wiley and Sons.

Estrada-Ruiz, E., Martinez-Cabrera, H. I. & Cevallos-Ferriz, S. R. S. 2007. Fossil woods from the late Campanian-early Maastrichtian Olmos Formation, Coahuila, Mexico. *Review of Palaeobotany and Palynology*, **145**, 123–33.

Eyde, R. H., Bartlett, A. & Barghoorn, E. S. 1969. Fossil record of *Alangium*. *Bulletin of the Torrey Botanical Club*, **96**, 288–314.

Faden, R. B. 1998. Commelinaceae. In *The Families and Genera of Vascular Plants.* Volume IV. *Flowering Plants – Monocotyledones. Alismatanae and Commelinanae (except Gramineae)* (ed. K. Kubitzki) Berlin, Heidelberg, New York: Springer-Verlag, pp. 109–28.

Fægri, K. & Iversen, J. 1964. *Textbook of Pollen Analysis.* Copenhagen: Munksgaard.

Fægri, K. & Van der Pijl, L. 1980. *The Principles of Pollination Ecology.* Oxford: Pergamon Press.

Fagerlind, F. 1947. Strobilus und Blüte von *Gnetum* und die Möglichkeit, aus ihrer Struktur den Blütenbau der Angiospermen zu deuten. *Arkiv för Botanik*, **33a**, 1–57.

Fairon-Demaret, M. & Scheckler, S. E. 1987. Typification and redescription of *Moresnetia zalesskyi* Stockmans, 1948, an early seed plant from the Upper Famennian of Belgium. *Bulletin de l'Institut Royal des Sciences Naturelles de Belgique, Science de la Terre*, **57**, 183–99.

Fan, C. & Xiang, Q.-Y. 2003. Phylogenetic analyses of Cornales based on 26S rRNA and combined 26S rDNA-MAT*K*-RBC*L* sequence data. *American Journal of Botany*, **90**, 1357–72.

Farabee, M. J. 1990. Triprojectate fossil genera pollen. *Review of Palaeobotany and Palynology*, **65**, 341–7.

Farabee, M. J. 1991. Botanical affinities of some Triprojectacites fossil pollen. *American Journal of Botany*, **78**, 1172–81.

Feduccia, A. 2003. 'Big bang' for Tertiary birds? *Trends in Ecology and Evolution*, **18**, 172–6.

Feild, T. S., Arens, N. C., Doyle, J. A., Dawson, T. E. & Donoghue, M. J. 2004. Dark and disturbed: a new image of early angiosperm ecology. *Paleobiology*, **30**, 82–107.

Feistmantel, O. 1876. Jurassic (Oolitic) flora of Kach. *Memoirs of the Geological Survey of India, Palaeontologia Indica*, Series 11, **4**, 192–224.

Feng, C.-M., Manchester, S. R. & Xiang, Q.-Y. 2009. Phylogeny and biogeography of Alangiaceae (Cornales) inferred from DNA sequences, morphology, and fossils. *Molecular Phylogenetics and Evolution*, **51**, 201–14.

Feuer, S. 1991. Pollen morphology and the systematic relationships of *Ticodendron incognitum*. *Annals of the Missouri Botanical Garden*, **78**, 143–51.

Feuer, S. M. & Kuijt, J. 1980. Fine structure of Mistletoe pollen. III. Large-flowered neotropical Loranthaceae and their Australian relatives. *American Journal of Botany*, **67**, 34–50.

Firbas, F. 1947. Spermatophyta, Samenpflanzen. In *Lehrbuch der Botanik für Hochschulen* (ed. E. Strasburger) Jena: Gustav Fischer, pp. 454–567.

Fishbein, M., Hibsch-Jetter, C., Soltis, D. E. & Hufford, L. 2001. Phylogeny of Saxifragales (angiosperms, eudicots): analysis of a rapid, ancient radiation. *Systematic Biology*, **50**, 817–47.

Fleming, T. H. & Heithaus, E. R. 1981. Frugivorous bats, seed shadows, and the structure of tropical forests. *Biotropica*, **13**, 45–53.

Fleming, T. H. 1991. Fruiting plant-frugivore mutualism: The evolutionary theater and the ecological play. In *Plant–Animal Interactions: Evolutionary Ecology in Tropical and Temperate Regions* (eds P. W. Price, T. M. Lewinsohn, G. W. Fernandes & W. W. Benson) New York: John Wiley and Sons, pp. 119–44.

Fleming, T. H. & Lips, K. R. 1991. Angiosperm endozoochory: were pterosaurs Cretaceous seed dispersers? *American Naturalist*, **138**, 1058–65.

Fleming, T. H., Geiselman, C. & Kress, W. J. 2009. The evolution of bat pollination: a phylogenetic perspective. *Annals of Botany*, **104**, 1017–43.

Florin, R. 1933. Studien über die Cycadales des Mesozoicums. Nebst Erörterungen über die Spaltöffnungsapparate der Bennettitales. *Kungliga Svenska Vetenskapsakademiens Handlingar, 3. Serie*, **12**, 1–134.

Flynn, J. J. & Wyss, A. R. 1998. Recent advances in South American mammalian paleontology. *Trends in Ecology and Evolution*, **13**, 449–54.

Fontaine, W. M. 1889. The Potomac or younger Mesozoic flora. *US Geological Survey Monograph*, **15**, 1–377, plates **1**–180.

Fontaine, W. M. 1893. Notes on some fossil plants from the Trinity Division of the Comanche Series of Texas. *Proceedings of the US National Museum*, **16**, 261–82.

Foreman, D. B. & Sampson, F. B. 1987. Pollen of *Palmeria scandens* and *Wilkiea huegeliana* (Monimiaceae). *Grana*, **26**, 127–33.

Forest, F., Chase, M. W., Persson, C., Crane, P. R. & Hawkins, J. A. 2007. The role of biotic and abiotic factors in evolution of ant dispersal in the Milkwort family (Polygalaceae). *Evolution*, **61**, 1675–94.

Forster, A., Schouten, S., Baas, M. & Damsté, J. S. S. 2007. Mid-Cretaceous (Albian-Santonian) sea surface temperature record of the tropical Atlantic Ocean. *Geology*, **35**, 919–22.

Foster, R. B., Arce, B. J. & Wachter, T. S. 1986. Dispersal and the sequential plant communities in Amazonian Peru floodplain. In *Frugivores and Seed Dispersal* (eds A. Estrada & T. H. Fleming) Dordrecht: Dr W. Junk Publishers, pp. 357–69.

França, J. C. & Zbyszewski, G. 1963. *Carta Geológica de Portugal na Escala de 1/50 000. Notícia Explicativa da Folha 26-B Alcobaça*. Lisbon: Serviços Geológicos de Portugal.

Francis, J. E. 1983. Reconstruction of the dominant conifer of the "Fossil Forests" of the Lower Purbeck Formation (Upper Jurassic) Dorset, southern England. *Palaeontology*, **26**, 277–94.

Francis, J. E. 1984. The seasonal environment of the Purbeck (Upper Jurassic) fossil forests. *Palaeogeography, Palaeoclimatology, Palaeoecology*, **48**, 285–307.

Francis, J. E. 1986. Growth rings in Cretaceous and Tertiary wood from Antarctica and their palaeoclimatic implications. *Palaeontology*, **29**, 665–84.

Frederiksen, N. O. 1987. Tectonic and paleogeographic setting of a new latest Cretaceous floristic province in North America. *Palaios*, **2**, 533–42.

Fregenal, M. A. & Meléndez, N. 1995. Geology: stratigraphy, basin evolution and geochemistry. I. Geological setting. In *Las Hoyas, a Lacustrine Konservat-Lagerstätte, Cuenca, Spain. II International Symposium on Lithographic Limestones. Field Trip Guide Book* (ed. M. N. Meléndez) Madrid: Universidad Complutense de Madrid, pp. 1–10.

Friedman, J. & Barrett, S. C. H. 2008. A phylogenetic analysis of the evolution of wind pollination in the angiosperms. *International Journal of Plant Sciences*, **169**, 49–58.

Friedman, W. E. 1990. Double fertilization in *Ephedra*, a nonflowering seed plant: its bearing on the origin of angiosperms. *Science*, **247**, 951–4.

Friedman, W. E. 1992. Evidence of a pre-angiosperm origin of endosperm: implications for the evolution of flowering plants. *Science*, **255**, 336–9.

Friedman, W. E. 1993. The evolutionary history of the seed plant male gametophyte. *Trends in Ecology and Evolution*, **8**, 15–21.

Friedman, W. E. & Carmichael, J. S. 1996. Double fertilization in Gnetales: implications for understanding reproductive diversification among seed plants. *International Journal of Plant Sciences*, **157** (6 Suppl.), S77–S94.

Friedman, W. E. 2001. Comparative embryology of basal angiosperms. *Current Opinion in Plant Biology*, **4**, 14–20.

Friedman, W. E. 2009. The meaning of Darwin's 'abominable mystery'. *American Journal of Botany*, **96**, 5–21.

Friedman, W. E. & Ryerson, K. C. 2009. Reconstructing the ancestral female gametophyte of angiosperms: insights from *Amborella* and other ancient lineages of flowering plants. *American Journal of Botany*, **96**, 129–43.

Friedrich, W. L. & Koch, B. E. 1970. Comparison of fruits and seeds of fossil *Spirematospermum* with those of living *Cenolophon*. *Bulletin of the Geological Society of Denmark*, **20**, 192–5.

Friis, E. M. 1979. The Damgaard flora: a new Middle Miocene flora from Denmark. *Bulletin of the Geological Society of Denmark*, **27**, 117–42.

Friis, E. M. & Skarby, A. 1981. Structurally preserved angiosperm flowers from the Upper Cretaceous of southern Sweden. *Nature*, **291**, 485–6.

Friis, E. M. & Skarby, A. 1982. *Scandianthus* gen. nov., angiosperm flowers of saxifragalean affinity from the Upper Cretaceous of southern Sweden. *Annals of Botany*, **50**, 569–83.

Friis, E. M. 1983. Upper Cretaceous (Senonian) floral structures of juglandalean affinity containing Normapolles pollen. *Review of Palaeobotany and Palynology*, **39**, 161–88.

Friis, E. M. 1984. Preliminary report on Upper Cretaceous angiosperm reproductive organs from Sweden and their level of organization. *Annals of the Missouri Botanical Garden*, **71**, 403–18.

Friis, E. M. 1985a. Structure and function in Late Cretaceous angiosperm flowers. *Biologiske Skrifter, Det Kongelige Danske Videnskabernes Selskab*, **25**, 1–37.

Friis, E. M. 1985b. Angiosperm fruits and seeds from the Middle Miocene of Jutland (Denmark). *Biologiske Skrifter, Det Kongelige Danske Videnskabernes Selskab*, **24**, 1–165.

Friis, E. M. 1985c. *Actinocalyx* gen. nov., sympetalous angiosperm flowers from the Upper Cretaceous of southern Sweden. *Review of Palaeobotany and Palynology*, **45**, 171–83.

Friis, E. M., Crane, P. R. & Pedersen, K. R. 1986. Floral evidence for Cretaceous chloranthoid angiosperms. *Nature*, **320**, 163–4.

Friis, E. M. & Crepet, W. L. 1987. Time of appearance of floral features. In *The Origins of Angiosperms and their Biological Consequences* (eds E. M. Friis, W. G. Chaloner & P. R. Crane) Cambridge: Cambridge University Press, pp. 145–79.

Friis, E. M. 1988. *Spirematospermum chandlerae* sp. nov., an extinct species of Zingiberaceae from the North American Cretaceous. *Tertiary Research*, **9**, 7–12.

Friis, E. M., Crane, P. R. & Pedersen, K. R. 1988. Reproductive structures of Cretaceous Platanaceae. *Biologiske Skrifter, Det Kongelige Danske Videnskabernes Selskab*, **31**, 1–55.

Friis, E. M. & Crane, P. R. 1989. Reproductive structures of Cretaceous Hamamelidae. In *Evolution, Systematics, and Fossil History of the Hamamelidae*. Volume 1. *Introduction and 'Lower' Hamamelidae* (eds P. R. Crane & S. Blackmore) Oxford: Clarendon Press, pp. 155–74.

Friis, E. M. 1990. *Silvianthemum suecicum* gen. et sp. nov., a new saxifragalean flower from the Late Cretaceous of Sweden. *Biologiske Skrifter, Det Kongelige Danske Videnskabernes Selskab*, **36**, 1–35.

Friis, E. M. & Pedersen, K. R. 1990. Structure of the Lower Cretaceous fern *Onychiopsis psilotoides* from Bornholm, Denmark. *Review of Palaeobotany and Palynology*, **66**, 47–63.

Friis, E. M., Pedersen, K. R. & Crane, P. R. 1992. *Esgueiria* gen. nov., fossil flowers with combretaceous features from the Late Cretaceous of Portugal. *Biologiske Skrifter, Det Kongelige Danske Videnskabernes Selskab*, **41**, 1–45.

Friis, E. M., Eklund, H., Pedersen, K. R. & Crane, P. R. 1994a. *Virginianthus calycanthoides* gen. et sp. nov. – a calycanthaceous flower from the Potomac Group (Early Cretaceous) of eastern North America. *International Journal of Plant Sciences*, **155**, 772–85.

Friis, E. M., Pedersen, K. R. & Crane, P. R. 1994b. Angiosperm floral structures from the Early Cretaceous of Portugal. *Plant Systematics and Evolution* (Suppl.), **8**, 31–49.

Friis, E. M., Pedersen, K. R. & Crane, P. R. 1995. *Appomattoxia ancistrophora* gen. et sp. nov., a new Early Cretaceous plant with similarities to *Circaeaster* and extant Magnoliidae. *American Journal of Botany*, **82**, 933–43.

Friis, E. M. & Pedersen, K. R. 1996. *Eucommiitheca*, a new pollen organ with *Eucommiidites* pollen from the Early Cretaceous of Portugal. *Grana*, **35**, 104–12.

Friis, E. M., Crane, P. R. & Pedersen, K. R. 1997a. *Anacostia*, a new basal angiosperm from the Early Cretaceous of North America and Portugal with monocolpate/trichotomocolpate pollen. *Grana*, **36**, 225–44.

Friis, E. M., Crane, P. R. & Pedersen, K. R. 1997b. Fossil history of magnoliid angiosperms. In *Evolution and Diversification of Land Plants* (eds K. Iwatsuki & P. H. Raven) Tokyo, Berlin, Heidelberg, New York: Springer-Verlag, pp. 121–56.

Friis, E. M., Pedersen, K. R. & Crane, P. R. 1999. Early angiosperm diversification: the diversity of pollen associated with angiosperm reproductive structures in Early Cretaceous floras from Portugal. *Annals of the Missouri Botanical Garden*, **86**, 259–6.

Friis, E. M. & Pedersen, K. R. 2000. Die fossilen Blüten von Åsen in Schonen, Süd-Schweden. In *Europäische Fossillagerstätten* (ed. D. Meischner) Heidelberg: Springer-Verlag, pp. 151–4.

Friis, E. M., Pedersen, K. R. & Crane, P. R. 2000a. Fossil floral structures of a basal angiosperm with monocolpate, reticulate-acolumellate pollen from the Early Cretaceous of Portugal. *Grana*, **39**, 226–45.

Friis, E. M., Pedersen, K. R. & Crane, P. R. 2000b. Reproductive structure and organization of basal angiosperms from the Early Cretaceous (Barremian or Aptian) of Western Portugal. *International Journal of Plant Sciences*, **161** (6 Suppl.), S169–S182.

Friis, E. M., Pedersen, K. R. & Crane, P. R. 2001. Fossil evidence of water lilies (Nymphaeales) in the Early Cretaceous. *Nature*, **410**, 357–60.

Friis, E. M., Doyle, J. A., Endress, P. K. & Leng, Q. 2003a. *Archaefructus* – angiosperm precursor or specialized early angiosperm? *Trends in Plant Science*, **8**, 369–373.

Friis, E. M., Pedersen, K. R. & Schönenberger, J. 2003b. *Endressianthus*, a new Normapolles producing plant genus of fagalean affinity from the Late Cretaceous of Portugal. *International Journal of Plant Sciences*, **164** (5 Suppl.), S201–S223.

Friis, E. M., Pedersen, K. R. & Crane, P. R. 2004. Araceae from the Early Cretaceous of Portugal: Evidence on the emergence of monocotyledons. *Proceedings of the National Academy of Sciences, USA*, **101**, 16 565–70.

Friis, E. M., Pedersen, K. R. & Crane, P. R. 2005. When Earth started blooming: insights from the fossil record. *Current Opinion in Plant Biology*, **8**, 1–8.

Friis, E. M., Pedersen, K. R. & Crane, P. R. 2006a. Cretaceous angiosperm flowers: Innovation and evolution in plant reproduction. *Palaeogeography, Palaeoclimatology, Palaeoecology*, **232**, 251–93.

Friis, E. M., Pedersen, K. R. & Schönenberger, J. 2006b. Normapolles plants: a complex of extinct fagalean lineages. *Plant Systematics and Evolution*, **260**, 107–40.

Friis, E. M., Crane, P. R., Pedersen, K. R. *et al.* 2007. Phase contrast enhanced synchrotron-radiation X-ray analyses of Cretaceous seeds link Gnetales to extinct Bennettitales. *Nature*, **450**, 549–52.

Friis, E. M., Pedersen, K. R. & Crane, P. R. 2009a. Early Cretaceous mesofossils from Portugal and eastern North America related to the Bennettitales-Erdtmanithecales-Gnetales group. *American Journal of Botany*, **96**, 252–83.

Friis, E. M., Pedersen, K. R., von Balthazar, M., Grimm, G. W. & Crane, P. R. 2009b. *Monetianthus mirus* gen. et sp. nov., a nymphaealean flower from the Early Cretaceous of Portugal. *International Journal of Plant Sciences*, **170**, 1086–101.

Friis, E. M., Pedersen, K. R. & Crane, P. R. 2010a. Cretaceous diversification of angiosperms in the western part of the Iberian Peninsula. *Review of Palaeobotany and Palynology*, **162**, 341–61.

Friis, E. M., Pedersen, K. R. & Crane, P. R. 2010b. Diversity in obscurity: fossil flowers and the early history of angiosperms. *Philosophical Transactions of the Royal Society* B, **365**, 369–82.

Friis, E. M. & Pedersen, K. R. 2011. *Canrightia resinifera*, a new Early Cretaceous fruit of magnolialean affinity from Portugal. *Grana*, **50**, 3–29.

Frumin, S. & Friis, E. M. 1996. Liriodendroid seeds from the Late Cretaceous of Kazakhstan and North Carolina, USA. *Review of Palaeobotany and Palynology*, **94**, 39–55.

Frumin, S. & Friis, E. M. 1999. Magnoliid reproductive organs from the Cenomanian-Turonian of north-western Kazakhstan: Magnoliaceae and Illiciaceae. *Plant Systematics and Evolution*, **216**, 265–88.

Frumin, S., Eklund, H. & Friis, E. M. 2004. *Mauldinia hirsuta* sp. nov., a new member of the extinct genus *Mauldinia* (Lauraceae) from the Late Cretaceous (Cenomanian-Turonian) of Kazakhstan. *International Journal of Plant Sciences*, **165**, 883–95.

Frumina, S. I., Zhilin, S. G. & Korchagina, I. A. 1995. *Alapaja* (Taxodiaceae) seeds from the Cenomanian-Turonian of Northern Kazakhstan. *Paleontological Journal*, **29**, 194–202.

Fryns-Claessens, E. & Van Cotthem, W. R. 1973. A new classification of the ontogenetic types of stomata. *The Botanical Review*, **39**, 71–138.

Funkhouser, J. W. 1961. Pollen of the genus *Aquilapollenites*. *Micropaleontology*, **7**, 193–8.

Furness, C. A., Rudall, P. J. & Eastman, A. 2002. Contribution of pollen and tapetal characters to the systematics of Triuridaceae. *Plant Systematics and Evolution*, **235**, 209–18.

Furness, C. A. & Rudall, P. J. 2003. Apertures with lids: distribution and significance of operculate pollen in monocotyledons. *International Journal of Plant Sciences*, **164**, 835–54.

Gabel, M. L. 1987. A fossil *Lithospermum* (Boraginaceae) from the Tertiary of South Dakota. *American Journal of Botany*, **71**, 1690–3.

Gabel, M. L. & Bich, H. 1988. A range extension for the Tertiary fossil *Eleofimbris* (Cyperaceae). *The Southwestern Naturalist*, **33**, 110–12.

Gabel, M. L., Backlund, D. C. & Haffner, J. 1992. Sedge (Cyperaceae) achenes from the Late Barstovian of Nebraska. *Journal of Paleontology*, **66**, 525–9.

Gale, A. S. 2000. The Cretaceous world. In *Biotic Response to Global Change* (eds S. J. Culver & P. R. Rawson) Cambridge: Cambridge University Press, pp. 4–19.

Gandolfo, M. A., Nixon, K. C., Crepet, W. L. & Ratcliffe, G. E. 1997. A new fossil fern assignable to Gleicheniaceae from Late Cretaceous sediments of New Jersey. *American Journal of Botany*, **84**, 483–93.

Gandolfo, M. A. 1998. *Tylerianthus crossmanensis* gen. et sp. nov. (aff. Hydrangeaceae) from the Upper Cretaceous of New Jersey. *American Journal of Botany*, **85**, 376–86.

Gandolfo, M. A., Nixon, K. C. & Crepet, W. L. 1998a. A new fossil flower from the Turonian of New Jersey: *Dressiantha bicarpellata* gen. et sp. nov. (Capparales). *American Journal of Botany*, **85**, 964–74.

Gandolfo, M. A., Nixon, K. C., Crepet, W. L. & Friis, E. M. 1998b. Oldest known fossils of monocotyledons. *Nature*, **394**, 532–3.

Gandolfo, M. A., Nixon, K. C. & Crepet, W. L. 2000. Monocotyledons: a review of their Early Cretaceous record.

In *Monocots: Systematics and Evolution* (eds K. I. Wilson & D. A. Morrison) Melbourne: CSIRO, pp. 44–51.

Gandolfo, M. A., Nixon, K. C. & Crepet, W. L. 2002. Triuridaceae fossil flowers from the Upper Cretaceous of New Jersey. *American Journal of Botany*, **89**, 1940–57.

Gandolfo, M. A., Nixon, K. C. & Crepet, W. L. 2004. Cretaceous flowers of Nymphaeaceae and implications for complex insect entrapment pollination mechanisms in early angiosperms. *Proceedings of the National Academy of Sciences, USA*, **101**, 8056–60.

Gandolfo, M. A. & Cuneo, R. N. 2005. Fossil Nelumbonaceae from the La Colonia Formation (Campanian-Maastrichtian, Upper Cretaceous), Chubut, Patagonia, Argentina. *Review of Palaeobotany and Palynology*, **133**, 169–78.

Gasser, C. S., Broadhvest, J. & Hauser, B. A. 1998. Genetic analysis of ovule development. *Annual Review of Plant Physiology and Plant Molecular Biology*, **49**, 1–24.

Gastaldo, R. A. 1989. Preliminary observations on phytotaphonomic assemblages in a subtropical/temperate Holocene bayhead delta: Mobile Delta, Gulf Coastal Plain, Alabama. *Review of Palaeobotany and Palynology*, **58**, 61–83.

Gaussen, H. 1946. Les Gymnospermes, actuelles et fossiles. *Travaux du Laboratoire Forestier de Toulouse. Tome II*, **1**, 1–130.

Gee, C. T. 2001. The mangrove palm *Nypa* in the geologic past of the New World. *Wetlands Ecology and Management*, **9**, 181–94.

Germeraad, J. H., Hopping, C. A. & Muller, J. 1968. Palynology of Tertiary sediments from tropical areas. *Review of Palaeobotany and Palynology*, **6**, 189–348.

Gibernau, M. 2003. Pollinators and visitors of aroid inflorescences. *Aroideana*, **26**, 66–83.

Gierlowski-Kordesch, E. & Janofske, D. 1989. Paleoenvironmental reconstruction of the Weald around Uña (Serrania de Cuenca, Cuenca Province, Spain). In *Proceedings of the 3rd International Cretaceous Symposium* (ed. J. Wiedmann) Tübingen: E. Schweizerbart'sche Verlagsbuchhandlung (Nägele u. Obermiller), pp. 239–64.

Givnish, T. J. 1980. Ecological constraints on the evolution of breeding systems in seed plants: dioecy and dispersal in gymnosperms. *Evolution*, **34**, 959–72.

Givnish, T. J., Evans, T. M., Pires, J. C. & Sytsma, K. J. 1999. Polyphyly and convergent morphological evolution in Commelinales and Commelinidae: evidence from *rbc*L sequence data. *Molecular Phylogenetics and Evolution*, **12**, 360–85.

Glaser, J. D. 1969. Petrology and origin of Potomac and Magothy (Cretaceous) sediments, middle Atlantic Coastal Plain. *Maryland Geological Survey Report of Investigations*, **11**, 1–101.

Glasspool, I. J., Hilton, J., Collinson, M. E., Wang, S.-J. & Li, C.-S. 2004. Foliar physiognomy in Cathaysian gigantopterids and the potential to track Palaeozoic climates using an extinct plant group. *Palaeogeography, Palaeoclimatology, Palaeoecology*, **250**, 69–110.

Góczán, F., Groot, J. J., Krutzsch, W. & Pacltová, B. 1967. Die Gattungen des "Stemma Normapolles Pflug 1953b" (Angiospermae) – Neubeschreibungen und Revision europäischer Formen (Oberkreide bis Eozän). *Paläontologische Abhandlungen* B, **2**, 427–633.

Góczán, F. & Juhász, M. 1984. Monosulcate pollen grains of angiosperms from Hungarian Albian sediments I. *Acta Botanica Hungarica*, **30**, 289–319.

Golovneva, L. B. 1998. Cretaceous floral evolution in northeastern Russia. *Paleontological Journal*, **32**, 633–41.

Golovneva, L. B. 2007. Occurrence of *Sapindopsis* (Platanaceae) in the Cretaceous of Eurasia. *Paleontological Journal*, **41**, 1077–90.

Golovneva, L. B. & Oskolski, A. A. 2007. Infructescences of *Cathiaria* gen. n. from the late Cretaceous of North Kazakhstan and Siberia (Russia). *Acta Palaeobotanica*, **47**, 57–87.

Golovneva, L. B. 2008. A new platanaceous genus *Tasymia* (angiosperms) from the Turonian of Siberia. *Paleontological Journal*, **42**, 192–202.

Gómez-Laurito, J. & Gómez, L. D. 1989. *Ticodendron*: a new tree from Central America. *Annals of the Missouri Botanical Garden*, **76**, 1148–51.

Gomez, B., Martín-Closas, C., Méon, H., Thévenard, F. & Barale, G. 2001. Plant taphonomy and palaeoecology in the lacustrine Uña delta (Late Barremian, Iberian Range, Spain). *Palaeogeography, Palaeoclimatology, Palaeoecology*, **170**, 133–48.

Gomez, B., Martínez-Delclòs, X., Bamford, M. & Philippe, M. 2002. Taphonomy and palaeoecology of plant remains from the oldest African Early Cretaceous amber locality. *Lethaia*, **35**, 300–8.

Goremykin, V., Bobrova, V., Pahnke, J. *et al.* 1996. Noncoding sequences from the slowly evolving chloroplast inverted repeat in addition to *rbc*L data do not support gnetalean affinities of angiosperms. *Molecular Biology and Evolution*, **13**, 383–96.

Goth, K. 1986. Erster Nachweis von *Spirematospermum*-Samen aus der Oberkreide von Kössen in Tirol. *Courier Forschungsinstitut Senckenberg*, **86**, 171–5.

Gothan, W. 1908. Die fossilen Hölzer von der Seymour- und Snow Hill-Insel. *Wissenschaftliche Ergebnisse der*

Schwedischen Südpolar-Expedition, 1901–1903, **3** (8), 1–33.

Gottsberger, G. 1977. Some aspects of beetle pollination in the evolution of flowering plants. *Plant Systematics and Evolution* (Suppl.), **1**, 211–26.

Gottsberger, G. 1978. Seed dispersal by fish in the inundated regions of Humaita, Amazonia. *Biotropica*, **10**, 170–83.

Gottsberger, G. 1988. The reproductive biology of primitive angiosperms. *Taxon*, **37**, 630–43.

Gottschling, M., Mai, D. H. & Hilger, H. H. 2002. The systematic position of *Ehretia* fossils (Ehretiaceae, Boraginales) from the European Tertiary and implications for character evolution. *Review of Palaeobotany and Palynology*, **121**, 149–56.

Gottwald, H. 1992. Hölzer aus marinen Sanden des oberen Eozän von Helmstedt (Niedersachsen). *Palaeontographica* B, **225**, 27–103.

Gould, R. E. & Delevoryas, T. 1977. The biology of *Glossopteris*: evidence from petrified seed-bearing and pollen-bearing organs. *Alcheringa*, **1**, 387–99.

Goulding, M. 1983. The role of fishes in seed dispersal and plant distribution in Amazonian floodplain ecosystems. *Sonderbände des Naturwissenschaftlichen Vereins in Hamburg*, **7**, 271–83.

Gradstein, F. M., Ogg, J. G. & Smith, A. G. (eds) 2004. *A Geologic Time Scale*. Cambridge: Cambridge University Press.

Graham, A. & Jarzen, D. M. 1969. Studies in neotropical paleobotany. I. The Oligocene communities of Puerto Rico. *Annals of the Missouri Botanical Garden*, **56**, 308–57.

Graham, A. 2009. Fossil record of the Rubiaceae. *Annals of the Missouri Botanical Garden*, **96**, 90–108.

Grande, L. 1984. Paleontology of the Green River Formation, with a review of the fish fauna. 2nd ed. *Geological Survey of Wyoming, Bulletin*, **63**, 1–333.

Grande, L. 1998. Locked in stone: an extinct 50-million-year-old lake system. *Natural History Magazine*, **107**, 66–9.

Grant, V. 1950. The protection of the ovule in flowering plants. *Evolution*, **4**, 179–201.

Gray, J. & Sohma, K. 1964. Fossil *Pachysandra* from Western America with a comparative study of pollen in *Pachysandra* and *Sarcococca*. *American Journal of Science*, **262**, 1159–97.

Grayum, M. H. 1990. Evolution and phylogeny of the Araceae. *Annals of the Missouri Botanical Garden*, **77**, 628–97.

Grayum, M. H. 1992. Comparative External Pollen Ultrastructure of the Araceae and Putative Related Taxa. *Monographs in Systematic Botany from the Missouri Botanical Garden*, **43**, 1–167.

Greenwood, D. R. & Conran, J. G. 2000. The Australian Cretaceous and Tertiary monocot fossil record. In *Monocots: Systematics and Evolution* (eds K. I. Wilson & D. A. Morrison) Melbourne: CSIRO, pp. 52–9.

Gregor, H.-J. 1980. Seeds of the genus *Coriaria* Linne (Coriariaceae) in the European Neogene. *Tertiary Research*, **3**, 61–9.

Gregor, H.-J. 1981. *Schisandra geissertii* nova spec. – ein exotisches Element in Elsässer Pliozän (Sessenheim, Brunssumien). *Mitteilungen des Badischen Landesvereins für Naturkunde und Naturschutz*, **12**, 241–7.

Gregor, H.-J. & Hagn, H. 1982. Fossil fructifications from the Cretaceous-Palaeocene-boundary of SW-Egypt (Danian, Bir Abu Munqar). *Tertiary Research*, **4**, 121–47.

Gregor, H.-J. 1989. Aspects of the fossil record and phylogeny of the family Rutaceae (Zanthoxyleae, Toddalioideae). *Plant Systematics and Evolution*, **162**, 251–65.

Gregor, H. J. 1977. Subtropische Elemente im europäischen Tertiär. II (Fruktifikationen) *Paläontologische Zeitschrift*, **51**, 199–226.

Grigorescu, D. 1992. Nonmarine Cretaceous formations of Romania. In *Aspects of Nonmarine Cretaceous Geology* (eds N. J. Mateer & P. J. Chen) Beijing: China Ocean Press, pp. 142–64.

Grimaldi, D. A. 1996. *Amber: Window to the Past*. New York: Harry N. Abrams, Inc. and The American Museum of Natural History.

Grimaldi, D. A. 1999. The co-radiations of pollinating insects and angiosperms in the Cretaceous. *Annals of the Missouri Botanical Garden*, **86**, 373–406.

Grimaldi, D. A. (ed.). 2000. *Studies on Fossils in Amber, with Particular Reference to the Cretaceous of New Jersey*. Leiden: Backhuys.

Grimaldi, D. A. & Engel, M. H. 2005. *Evolution of the Insects*. Cambridge: Cambridge University Press.

Grímsson, F., Denk, T. & Símonarson, L. A. 2007. Middle Miocene floras of Iceland – the early colonization of an island? *Review of Palaeobotany and Palynology*, **144**, 181–219.

Grímsson, F., Denk, T. & Zetter, R. 2008. Pollen, fruits, and leaves of *Tetracentron* (Trochodendraceae) from the Cainozoic of Iceland and western North America and their palaeobiogeographic implications. *Grana*, **47**, 1–14.

Groot, J. J. & Groot, C. R. 1962. Plant microfossils from Aptian, Albian and Cenomanian deposits of Portugal. *Comunicações dos Serviços Geológicos de Portugal*, **46**, 133–76.

Gruas-Cavagnetto, C. 1987. Nouveaux éléments mégathermes dans la palynoflore éocène du Bassin Parisien. *Mémoires et*

Travaux de l'Institut de Montpellier, Ecole Pratique des Hautes Etudes, **17**, 207–33.

Gruas-Cavagnetto, C., Tambareau, Y. & Villatte, J. 1988. Données paléoécologiques nouvelles sur le Thanétien et l'Ilerdien de l'avant–pays pyrénéen et de la Montagne Noire. *Institut français de Pondichéry. Traveaux de la Section Scientifique et Technique*, **25**, 219–35.

Grönwall, K. A. 1915. Nordöstra Skånes kaolin- och kritbildningar samt deras praktiska användning. *Sveriges Geologiska Undersökning Ser. C*, **261**, 1–185.

Guo, S.-X. & Wu, X.-W. 2000. *Ephedrites* from latest Jurassic Yixian Formation in western Liaoning, Northeast China. *Acta Palaeontologica Sinica*, **39**, 81–91.

Gutierrez, G. & Lauverjat, J. 1978. Les charophytes du sénonian supérieur de la Beira Littorale (Portugal). *103e Congrès national des sociétés savantes. Sciences II*, 105–17.

Gübeli, A. A., Hochuli, P. A. & Wildi, W. 1984. Lower Cretaceous turbiditic sediment from the Rif chain (Northern Marocco) – palynology, stratigraphy and palaeogeographic setting. *Geologische Rundschau*, **73**, 1081–114.

Göppert, H. R. & Berendt, G. C. 1845. *Der Bernstein und die in ihm befindlichen Pflanzenreste der Vorwelt*. Berlin: Nicolaischen Buchhandlung.

Hackett, S. J., Kimball, R. T., Reddy, S. *et al.* 2008. A phylogenomic study of birds reveals their evolutionary history. *Science*, **320**, 1763–8.

Hagerup, O. 1934. Zur Abstammung einiger Angiospermen durch Gnetales und Coniferæ, I. *Biologiske Meddelelser, Kongelige Danske Videnskabernes Selskab*, **11**, 1–83.

Haggard, C. & Tiffney, B. H. 1997. The flora of the early Miocene Brandon Lignite, Vermont, USA, VIII. *Caldesia* (Alismataceae). *American Journal of Botany*, **84**, 239–52.

Halbritter, H. & Hesse, H. 1993. Sulcus morphology in some monocot families. *Grana*, **32**, 87–99.

Halbritter, H. 2000 onwards. *Bougainvillea* sp. In *PalDat – a palynological database: Descriptions, illustrations, identification, and information retrieval*. http://www.paldat.org/ (eds R. Buchner & M. Weber).

Hall, J. A., Walter, G. H., Bergstrom, D. M. & Machin, P. 2004. Pollination ecology of the Australian cycad *Lepidozamia peroffskyana* (Zamiaceae). *Australian Journal of Botany*, **52**, 333–43.

Hallam, A. 1984. Continental humid and arid zones during the Jurassic and Cretaceous. *Palaeogeography, Palaeoclimatology, Palaeoecology*, **47**, 195–223.

Hallam, A. 1985. A review of Mesozoic climates. *Journal of the Geological Society of London*, **142**, 433–45.

Hallam, A. 1992. *Phanerozoic Sea-Level Changes. Perspectives in Paleobiology and Earth History Series*. New York: Columbia University Press.

Hallam, A. 1994. *An Outline of Phanerozoic Biogeography. Oxford Biogeography Series*. Oxford, New York, Tokyo: Oxford University Press.

Hallier, H. 1900. Ueber Kautschuklianen und andere Apocyneen, nebst Bemerkungen über *Hevea* und einem Versuch zur Lösung der Nomenklaturfrage. *Jahrbuch der Hamburgischen Wissenschaftlichen Anstalten, Beiheft 3*, **17**, 17–216.

Hallier, H. 1901. Ueber die Verwandtschaftsverhältnisse der Tubifloren und Ebenalen, den polyphyletischen Ursprung der Sympetalen und Apetalen und die Anordnung der Angiospermen überhaupt. Vorstudien zum Entwurf eines Stammbaums der Blümenpflanzen. *Abhandlungen aus dem Gebiete der Naturwissenschaften in Hamburg*, **16**, 1–112.

Hallier, H. 1902. Beiträge zur Morphogenie der Sporophylle und des Trophophylls in Beiziehung zur Phylogenie der Kormophyten. *Jahrbuch der Hamburgischen Wissenschaftlichen Anstalten*, **19**, 1–110.

Hallier, H. 1912. L'origine et le système phylétique des angiospermes exposés à l'aide de leur arbre généalogique. *Archives Néerlandaises des Sciences Exactes et Naturelles, Séries IIIB*, **1**, 146–234.

Hamann, U. 1998. Philydraceae. In *The Families and Genera of Vascular Plants*. Volume IV. *Flowering Plants – Monocotyledons. Alismatanae and Commelinanae (except Gramineae)* (ed. K. Kubitzki) Berlin, Heidelberg, New York: Springer-Verlag, pp. 389–94.

Hamby, R. K. & Zimmer, E. A. 1992. Ribosomal RNA as a phylogenetic tool in plant systematics. In *Molecular Systematics of Plants* (eds P. S. Soltis, D. E. Soltis & J. J. Doyle) New York: Chapman and Hall, pp. 50–91.

Hansen, A., Hansmann, S., Samigullin, T., Antonov, A. & Martin, W. 1999. *Gnetum* and the angiosperms: molecular evidence that their shared morphological characters are convergent rather than homologous. *Molecular Biology and Evolution*, **16**, 1006–9.

Haq, B. U., Hardenbol, J. & Vail, P. R. 1987. Chronology and fluctuating sea levels since the Triassic. *Science*, **235**, 1156–66.

Harder, L. D. & Johnson, S. D. 2008. Function and evolution of aggregated pollen in angiosperms. *International Journal of Plant Sciences*, **169**, 59–78.

Harley, M. M., Kurmann, M. H. & Ferguson, I. K. 1991. Systematic implication of comparative morphology in selected fossil and extant pollen from the Palmae and the Sapotaceae. In *Pollen and Spores: Patterns of Diversification* (eds S. Blackmore & S. Barnes) Oxford: Clarendon Press, pp. 225–38.

Harley, M. M. & Morley, R. J. 1995. Ultrastructural studies of some fossil and extant palm pollen, and the reconstruction of the biogeographical history of subtribes Iguanurinae and Calaminae. *Review of Palaeobotany and Palynology*, **85**, 153–83.

Harley, M. M. 1997. Palm pollen and the fossil record. London: PhD Thesis, University of East London in collaboration with The Royal Botanic Gardens, Kew.

Harley, M. M. & Baker, W. J. 2001. Pollen aperture morphology in Arecaceae: application within phylogenetic analyses, and a summary of the fossil record of palms. *Grana*, **40**, 45–77.

Harley, M. M. & Dransfield, J. 2003. Triporate pollen in the Arecaceae. *Grana*, **42**, 3–19.

Harley, M. M. 2004. Triaperturate pollen in the monocotyledons: configurations and conjectures. *Plant Systematics and Evolution*, **247**, 75–122.

Harley, M. M. 2006. A summary of fossil records for Arecaceae. *Botanical Journal of the Linnean Society*, **151**, 39–67.

Harper, J. L., Lovell, P. H. & Moore, K. G. 1970. The shapes and sizes of seeds. *Annual Review of Ecology and Systematics*, **1**, 327–56.

Harris, T. M. 1932a. The fossil flora of Scoresby Sound East Greenland. Part 3: Caytoniales and Bennettitales. *Meddelelser om Grønland*, **85**(5), 1–133.

Harris, T. M. 1932b. The fossil flora of Scoresby Sound East Greenland. Part 2: Description of seed plants *incertae sedis* together with a discussion of certain cycadophyte cuticles. *Meddelelser om Grønland*, **85**(3), 1–114.

Harris, T. M. 1933. A new member of the Caytoniales. *New Phytologist*, **32**, 97–114.

Harris, T. M. 1937. The fossil flora of Scoresby Sound, East Greenland. Part 5: Stratigraphic relations of the plant beds. *Meddelelser om Grønland*, **112**, 1–114.

Harris, T. M. 1940a. On some Jurassic specimens of *Sagenopteris*. *Annals of Magazine of Natural History, Ser. 11*, **6**, 249–65.

Harris, T. M. 1940b. On *Caytonia* Thomas. *Annals of Botany*, **4**, 713–34.

Harris, T. M. 1941. *Caytonanthus*, the microsporophyll of *Caytonia*. *Annals of Botany*, **5**, 47–58.

Harris, T. M. 1944. A revision of *Williamsoniella*. *Philosophical Transactions of the Royal Society of London* B, **231**, 313–28.

Harris, T. M. 1945. On a coprolite of *Caytonia* pollen. *Annals of Magazine of Natural History, Ser. 11*, **12**, 357–78.

Harris, T. M. 1951. The relationships of the Caytoniales. *Phytomorphology*, **1**, 29–39.

Harris, T. M. 1954. Mesozoic seed cuticles. *Svensk Botanisk Tidskrift*, **48**, 281–91.

Harris, T. M. 1956. The mystery of flowering plants. *The Listener*, **26** April, 514–16.

Harris, T. M. 1958. The seed of *Caytonia*. *The Palaeobotanist*, **7**, 93–106.

Harris, T. M. 1961. The fossil cycads. *Palaeontology*, **4**, 313–23.

Harris, T. M. 1962. The occurrence of the fructification *Carnoconites* in New Zealand. *Transactions of the Royal Society of New Zealand, Geology*, **1**, 17–27.

Harris, T. M. 1964. *The Yorkshire Jurassic Flora*. Volume II. *Caytoniales, Cycadales & Pteridosperms*. London: British Museum (Natural History).

Harris, T. M. 1969. *The Yorkshire Jurassic Flora*. Volume III. *Bennettitales*. London: British Museum (Natural History).

Harris, T. M. 1971. The stem of *Caytonia*. *Geophytology*, **1**, 23–9.

Harris, T. M. 1974. *Williamsoniella lignieri*: its pollen and the compression of spherical pollen grains. *Palaeontology*, **17**, 125–48.

Harris, T. M., Millington, W. & Miller, J. 1974. *The Yorkshire Jurassic Flora* Volume IV. Part 1. *Ginkgoales*. Part 2. *Czekanowskiales*. London: British Museum (Natural History).

Harris, T. M. 1976. The Mesozoic gymnosperms. *Review of Palaeobotany and Palynology*, **21**, 119–35.

Harris, T. M. 1979. *The Yorkshire Jurassic Flora*. Volume V. *Coniferales*. London: British Museum (Natural History).

Harris, T. M. 1981. Burnt ferns from the English Wealden. *Proceedings of the Geologists' Association*, **92**, 47–58.

Harris, T. M. 1983a. The stem of *Pachypteris papillosa* (Thomas & Bose) Harris. *Botanical Journal of the Linnean Society*, **86**, 149–59.

Harris, T. M. 1983b. Fossils from New Zealand ascribed to the *Pentoxylon* plant. *Phyta, D.D. Pant Commemorative Volume*, 695–721.

Hasebe, M., Kofuji, R., Ito, M., Iwatsuki, K. & Ueda, K. 1992. Phylogeny of gymnosperms inferred from *rbc*L gene sequences. *Botanical Magazine, Tokyo*, **105**, 673–9.

Hasenboehler, B. 1981. Étude paléobotanique et palynologique de l'Albien et du Cenomanien du "Bassin occidental Portugais" au sud de l'accident de Nazare (Province d'Estremadure, Portugal). Paris: PhD Thesis, Université Pierre et Marie Curie, pp. 1–319.

Hayes, P. A., Francis, J. E., Cantrill, D. J. & Crame, J. A. 2006. Palaeoclimate analysis of Late Cretaceous angiosperm leaf floras, James Ross Island, Antarctica. In *Cretaceous–Tertiary High-Latitude Palaeoenvironments, James Ross Basin, Antarctica*, vol. **258** (eds J. E. Francis, D. Pirrie & J. A. Crame) London: The Geological Society of London, Special Publications, pp. 49–62.

Haynes, R. R., Holm-Nielsen, L. B. & Les, D. H. 1998. Najadaceae. In *The Families and Genera of Vascular Plants*. Volume IV. *Flowering Plants – Monocotyledons. Alismatanae and Commelinanae (except Gramineae)* (ed. K. Kubitzki) Berlin, Heidelberg, New York: Springer-Verlag, pp. 301–6.

He, H. Y., Wang, X. L., Zhou, Z. H. *et al.* 2004. Timing of the Jiufotang Formation (Jehol Group) in Liaoning, northeastern China and its implications. *Geophysical Research Letters*, **31**, L12605, doi:10.1029/2004GL019790.

He, H. Y., Wang, X. L., Zhou, Z. H. *et al.* 2006. $^{40}Ar/^{39}Ar$ dating of Lujiatun Bed (Jehol Group) in Liaoning, northeastern China. *Geophysical Review Letters*, **33**, L04303, doi:10.1029/2005GL025274.

Hedlund, R. W. & Norris, G. 1968. Spores and pollen grains from Fredericksburgian (Albian) strata, Marshall County, Oklahoma. *Pollen et Spores*, **10**, 129–59.

Heer, O. 1868. *Flora Fossilis Arctica*. Volume I. *Die fossile Flora der Polarländer enthaltend die in Nordgrönland, auf der Melville-Insel, im Banksland, am Mackenzie, in Island und in Spitzbergen entdeckten fossilen Pflanzen*. Zurich: F. Schulthess.

Heer, O. 1870. Die Miocene Flora und Fauna Spitzbergens *Kongliga Svenska Vetenskaps-Akademiens Handlingar*, **8**, 1–98.

Heer, O. 1876. Beiträge zur Jura-Flora Ostsibiriens und des Amurlandes. *Mémoires de l'Académie Impériale des Sciences de St.-Pétersbourg, VII. Série*, **22**, 1–122.

Heer, O. 1882. *Flora Fossilis Grönlandica. Die Fossile Flora Grönlands*. Part I. *Flora Fossilis Arctica. Die Fossile Flora der Polarländer*. Zurich: J. Wurster & Comp.

Heer, O. 1883a. *Flora Fossilis Grönlandica. Die Fossile Flora Grönlands*. Part II. *Flora Fossilis Arctica. Die Fossile Flora der Polarländer*. Zurich: J. Wurster & Comp.

Heer, O. 1883b. Oversigt over Grønlands fossile flora. *Meddelelser om Grønland*, **5**, 81–202.

Heer, O. 1883c. Flora fossilis Grønlandica. Afbildninger af Grønlands fossile flora. *Meddelelser om Grønland* (Suppl.), **5**, 1–27.

Heimhofer, U., Hochuli, P. A., Burla, S., Dinis, J. M. L. & Weissert, H. 2005. Timing of Early Cretaceous angiosperm diversification and possible links to major paleoenvironmental change. *Geology*, **33**, 141–4.

Heimhofer, U., Hochuli, P. A., Burla, S. & Weissert, H. 2007. New records of Early Cretaceous angiosperm pollen from Portuguese coastal deposits: implications for the timing of the early angiosperm radiation. *Review of Palaeobotany and Palynology*, **144**, 39–76.

Heming, B. S. 1993. Structure, function, ontogeny, and evolution of feeding in thrips (Thysanoptera). In *Functional Morphology of Insect Feeding* (eds C. W. Shaefer & R. A. B. Leschen) Lanham, MD: Thomas Say Publications in Entomology, Entomological Society of America, pp. 3–41.

Hennig, W. 1965. Phylogenetic systematics. *Annual Review of Entomology*, **10**, 97–116.

Henschel, J. R. & Seely, M. K. 2000. Long-term growth patterns of *Welwitschia mirabilis*, a long-lived plant of the Namib Desert (including a bibliography). *Journal of Plant Ecology*, **150**, 7–26.

Herendeen, P. S., Les, D. H. & Dilcher, D. L. 1990. Fossil *Ceratophyllum* (Ceratophyllaceae) from the Tertiary of North America. *American Journal of Botany*, **77**, 7–16.

Herendeen, P. S. 1991a. Charcoalified angiosperm wood from the Cretaceous of eastern North America and Europe. *Review of Palaeobotany and Palynology*, **70**, 225–39.

Herendeen, P. S. 1991b. Lauraceous wood from the mid-Cretaceous Potomac Group of eastern North America – *Paraphyllanthoxylon marylandense* sp. nov. *Review of Palaeobotany and Palynology*, **69**, 277–90.

Herendeen, P. S., Crepet, W. L. & Dilcher, D. L. 1992. The fossil history of the Leguminosae: phylogenetic and biogeographic implications. In *Advances in Legume Systematics*. Part 4. *The Fossil Record* (eds P. S. Herendeen & D. L. Dilcher) Kew: The Royal Botanic Gardens, pp. 303–16.

Herendeen, P. S., Crepet, W. L. & Nixon, K. C. 1993. *Chloranthus*-like stamens from the Upper Cretaceous of New Jersey. *American Journal of Botany*, **80**, 865–71.

Herendeen, P. S., Crepet, W., L. & Nixon, K. C. 1994. Fossil flowers and pollen of Lauraceae from the Upper Cretaceous of New Jersey. *Plant Systematics and Evolution*, **189**, 29–40.

Herendeen, P. S., Crane, P. R. & Drinnan, A. N. 1995. Fagaceous flowers, fruits, and cupules from the Campanian

(Late Cretaceous) of Central Georgia, USA. *International Journal of Plant Sciences*, **156**, 93–116.

Herendeen, P. S., Magallón-Puebla, S., Lupia, R., Crane, P. R. & Kobylinska, J. 1999a. A preliminary conspectus of the Allon flora from the Late Cretaceous (Late Santonian) of central Georgia, USA. *Annals of the Missouri Botanical Garden*, **86**, 407–71.

Herendeen, P. S., Wheeler, E. A. & Baas, P. 1999b. Angiosperm wood evolution and the potential contribution of paleontological data. *The Botanical Review*, **65**, 278–300.

Herendeen, P. S., Crane, P. R. 1995. The fossil history of the monocotyledons. In *Monocotyledons: Systematics and Evolution* (eds P. J. Rudall, P. J. Cribb, D. F. Cutler & C. J. Humphries) Kew: Royal Botanic Gardens, pp. 1–21.

Herman, A. 1990. Late Cretaceous floras and climate of the Anadyr-Koryakian subregion (North-East USSR). *Proceedings of the Symposium Paleofloristic and Paleoclimatic Changes in the Cretaceous and Tertiary*, ed. E. Knobloch and Z. Kvaček, pp. 73–9. Prague: Geological Survey of Czechoslovakia.

Herman, A. B. & Lebedev, E. L. 1991. *Stratigraphy and Flora of the Cretaceous Deposits of North-West Kamtchatka*. Moscow: Academia Nauk SSSR.

Herman, A. B. 1994. A review of Late Cretaceous floras and climates of Arctic Russia. In *Cenozoic Plants and Climates of the Arctic* (eds M. C. Boulter & H. C. Fischer) Heidelberg: Springer-Verlag, pp. 127–49.

Herman, A. B. & Spicer, R. A. 1995. Latest Cretaceous flora of northeastern Russia and the "terminal Cretaceous event" in the Arctic. *Palaeontological Journal*, **29**, 22–35.

Herman, A. B. & Kvaček, J. 2002. Campanian Grünbach Flora of Lower Austria: preliminary floristics and palaeoclimatology. *Annalen des Naturhistorischen Museums in Wien*, **103A**, 1–21.

Heřmanová, Z., Kvaček, J. & Friis, E. M. 2011. *Budvaricarpus serialis* Knobloch & Mai, an unusual new member of Normapolles complex from the Late Cretaceous of the Czech Republic. *International Journal of Plant Sciences*, **172**, 285–93.

Hermsen, E. J., Gandolfo, M. A., Nixon, K. C. & Crepet, W. L. 2003. *Divisestylus* gen. nov. (aff. Iteaceae), a fossil saxifrage from the Late Cretaceous of New Jersey, USA. *American Journal of Botany*, **90**, 1373–88.

Hermsen, E. J., Nixon, K. C. & Crepet, W. L. 2006. The impact of extinct taxa on understanding the early evolution of angiosperm clades: an example incorporating fossil reproductive structures of Saxifragales. *Plant Systematics and Evolution*, **260**, 141–69.

Hernández-Castillo, G. R. & Cevallos-Ferriz, S. R. S. 1999. Reproductive and vegetative organs with affinities to Haloragaceae from the Upper Cretaceous Huepac Chert Locality of Sonora, Mexico. *American Journal of Botany*, **86**, 1717–34.

Herngreen, G. F. W. 1973. Palynology of Albian-Cenomanian strata of borehole 1-QS-1-MA, State of Maranhão, Brazil. *Pollen et Spores*, **15**, 515–55.

Herngreen, G. F. W. 1975. Palynology of Middle and Upper Cretaceous strata in Brazil. *Mededelingen Rijks Geologische Dienst, Nieuwe Serie*, **26**, 39–116.

Herngreen, G. F. W. & Chlonova, A. F. 1981. Cretaceous microfloral provinces. *Pollen et Spores*, **23**, 441–555.

Herngreen, G. F. W., Kedves, M., Rovnina, L. V. & Smirnova, S. B. 1996. Cretaceous palynofloral provinces: a review. In *Palynology: Principles and Applications*, Volume 3 (eds J. Jansonius & D. C. McGregor) Salt Lake City: American Association of Stratigraphic Palynologists Foundation, pp. 1157–88.

Herrera, C. M. 1985. Determinants of plant-animal coevolution: the case of mutualistic dispersal of seeds by vertebrates. *Oikos*, **44**, 132–41.

Herrera, C. M. 1986. Vertebrate dispersed plants: why they don't behave the way they should. In *Frugivores and Seed Dispersal* (eds A. Estrada & T. H. Fleming) Dordrecht: Dr W. Junk Publications, pp. 5–18.

Heslop-Harrison, J. 1983. Self-incompatibility: phenomenology and physiology. *Proceedings of the Royal Society of London* B, **202**, 73–92.

Hesse, M., Weber, M. & Halbritter, H. 2000. A comparative study of the polyplicate pollen types in Arales, Laurales, Zingeberales and Gnetales. In *Pollen and Spores: Morphology and Biology* (eds M. M. Harley, C. M. Morton & S. Blackmore) Kew: Royal Botanic Gardens, pp. 227–39.

Hesse, M. & Zetter, R. 2007. The fossil pollen record of Araceae. *Plant Systematics and Evolution*, **263**, 93–115.

Heywood, V. H., Brummitt, R. K., Culham, A. & Seberg, O. 2007. *Flowering Plant Families of the World*. Kew: Royal Botanic Gardens.

Hickey, L. J. & Wolfe, J. A. 1975. The bases of angiosperm phylogeny: vegetative morphology. *Annals of the Missouri Botanical Garden*, **62**, 538–89.

Hickey, L. J. 1977. Stratigraphy and paleobotany of the Golden Valley Formation (Early Tertiary) of western North Dakota. *Memoir of the Geological Society of America, Memoir*, **150**, 1–183.

Hickey, L. J. & Doyle, J. A. 1977. Early Cretaceous fossil evidence for angiosperm evolution. *The Botanical Review*, **43**, 2–104.

Hickey, L. J. 1978. Origin of the major features of angiospermous leaf architecture in the fossil record. *Courier Forschungsinstitut Senckenberg*, **30**, 27–34.

Hickey, L. J. & Peterson, R. K. 1978. *Zingiberopsis*, a fossil genus of the ginger family from Late Cretaceous to early Eocene sediments of Western Interior North America. *Canadian Journal of Botany*, **56**, 1136–52.

Hickey, L. J. 1981. Land plant evidence compatible with gradual, not catastrophic, change at the end of the Cretaceous. *Nature*, **292**, 529–31.

Hickey, L. J. 1984. Northeast of Washington, D.C. to Brooke, Virginia. In *Cretaceous and Tertiary Stratigraphy, Paleontology and Structure, Southwestern Maryland and Northwestern Virginia* (eds N. O. Frederiksen & K. Kraff) Reston, VA: US Geological Survey, pp. 193–209.

Hicks, J. F., Johnson, K. R., Obradovich, J. D., Miggins, D. P. & Tauxe, L. 2003. Magnetostratigraphy of Upper Cretaceous (Maastrichtian) to lower Eocene strata of the Denver Basin, Colorado. *Rocky Mountain Geology*, **38**, 1–27.

Hildebrand, A. R., Penfield, G. T., Kring, D. A. *et al.* 1991. Chicxulub Crater: a possible Cretaceous/Tertiary boundary impact crater on the Yucatan Peninsula, Mexico. *Geology*, **19**, 867–71.

Hill, C. R. & Crane, P. R. 1982. Evolutionary cladistics and the origin of angiosperms. In *Problems of Phylogenetic Reconstruction* (eds K. A. Joysey & A. E. Friday) New York: Academic Press, pp. 269–361.

Hill, C. R. 1996. A plant with flower-like organs from the Wealden of the Weald (Lower Cretaceous), southern England. *Cretaceous Research*, **17**, 27–38.

Hill, R. S. 1994. The history of selected Australian taxa. In *History of the Australian Vegetation* (ed. R. S. Hill) Cambridge: Cambridge University Press, pp. 390–419.

Hill, R. S. & Dettmann, M. E. 1996. Origin and diversification of the genus *Nothofagus*. In *The Ecology and Biodiversity of Nothofagus Forest* (eds T. T. Veblen, R. S. Hill & J. Read) New Haven, CT: Yale University Press, pp. 11–23.

Hill, R. S. 1998. Fossil evidence for the onset of xeromorphy and scleromorphy in Australian Proteaceae. *Australian Systematic Botany*, **11**, 391–400.

Hilton, J. & Bateman, R. M. 2006. Pteridosperms are the backbone of seed-plant phylogeny. *Journal of the Torrey Botanical Society*, **133**, 119–68.

Hjelmroos, M. 1991. Evidence of long-distance transport of *Betula* pollen. *Grana*, **30**, 215–28.

Hochuli, P. A. 1979. Ursprung und Verbreitung der Restionaceae. *Vierteljahrschrift der Naturforschenden Gesellschaft in Zürich*, **124**, 109–31.

Hofmann, C.-C. & Zetter, R. 2007. Upper Cretaceous pollen flora from the Vilui Basin, Siberia: circumpolar and endemic *Aquilapollenites*, *Manicorpus*, and *Azonia* species. *Grana*, **46**, 227–49.

Hofmann, C., Féraud, G. & Courtillot, V. 2000. ^{40}Ar/^{39}Ar dating of mineral separates and whole rocks from the Western Ghats lava pile: further constraints on duration and age of the Deccan traps. *Earth and Planetary Science Letters*, **180**, 13–27.

Hofmann, E. 1952. Pflanzenreste aus dem phosphoritvorkommen von Prambachkirchen in Oberösterreich. II. *Palaeontographica* B, **92**, 122–83.

Hollander, J. L. & Vander Wall, S. B. 2009. Dispersal syndromes in North American *Ephedra*. *International Journal of Plant Sciences*, **170**, 323–30.

Hollick, A. 1894. Observations on the geology and botany of Martha's Vineyard. *Transactions of the New York Academy of Sciences*, **13**, 8–22.

Hollick, A. 1902. Geological and botanical notes: Cape Cod and Chappaquidick Island, Mass. *Bulletin of the New York Botanical Garden*, **2**, 381–407.

Hollick, A. 1903. A fossil petal and a fossil fruit from the Cretaceous (Dakota Group) of Kansas. *Bulletin of the Torrey Botanical Club*, **30**, 102–5.

Hollick, A. 1906. *The Cretaceous Flora of Southern New York and New England. Monographs of the US Geological Survey.* Washington: United States Geological Survey.

Hollick, A. 1912. Additions to the paleobotany of the Cretaceous Formation on Long Island. No. III. *Bulletin of the Torrey Botanical Club*, **8**, 154–70.

Hollick, A. 1930. The Upper Cretaceous floras of Alaska. *US Geological Survey Professional Paper*, **159**, 1–219.

Holmes, W. B. K. & Ash, S. R. 1979. An Early Triassic megafossil flora from the Lorne Basin, New South Wales. *Proceedings of the Linnean Society of New South Wales*, **103**, 47–70.

Holmes, W. B. K. 1987. New corystosperm ovulate fructifications from the Middle Triassic of eastern Australia. *Alcheringa*, **11**, 165–73.

Holmes, W. B. K. & Holmes, F. M. 1992. Fossil flowers of *Ceratopetalum* Sm. (family Cunoniaceae) from the Tertiary of Australia. *Proceedings of the Linnean Society of New South Wales*, **113**, 265–70.

Hoot, S. B., Magallón, S. & Crane, P. R. 1999. Phylogeny of basal eudicots based on three molecular data sets: *atp*B, *rbc*L, and 18S nuclear ribosomal DNA sequences. *Annals of the Missouri Botanical Garden*, **86**, 1–32.

Hotton, C. L. K., Leffingwell, H. A. & Skvarla, J. 1994. Pollen ultrastructure of Pandanaceae and the fossil genus *Pandaniidites*. In *Ultrastructure of Fossil Spores and Pollen. Its Bearing on Relationships among Fossil and Living Groups* (eds M. H. Kurmann & J. A. Doyle) Kew: The Royal Botanic Gardens, pp. 173–91.

Hou, L., Martin, L. D., Zhou, Z. & Feduccia, A. 1996. Early adaptive radiation of birds: evidence from fossils from northeastern China. *Science*, **274**, 1164–7.

Howe, J. & Cantrill, D. J. 2001. Palaeoecology and taxonomy of Pentoxylales from the Albian of Antarctica. *Cretaceous Research*, **22**, 779–93.

Hu, S., Dilcher, D. L., Jarzen, D. M. & Taylor, D. W. 2008a. Early steps of angiosperm-pollinator coevolution. *Proceedings of the National Academy of Sciences, USA*, **105**, 240–5.

Hu, S., Jarzen, D. M. & Dilcher, D. L. 2008b. New species of angiosperm pollen from the Dakota Formation (Cenomanian, Upper Cretaceous) of Minnesota, U.S.A. *Palynology*, **32**, 17–26.

Huber, B. T. & Watkins, D. K. 1992. Biogeography of Campanian-Maastrichtian calcareous plankton in the region of the Southern Ocean: paleogeographic and paleoclimatic implications. *Antarctic Research Series*, **56**, 31–60.

Huber, H. 1993. Aristolochiaceae. In *The Families and Genera of Vascular Plants.* Volume II. *Flowering Plants – Dicotyledons. Magnoliid, Hamamelid and Caryophyllid Families* (eds K. Kubitzki, J. G. Rohwer & V. Bittrich) Berlin, Heidelberg, New York: Springer-Verlag, pp. 129–37.

Huber, H. 1998. Dioscoreaceae. In *The Families and Genera of Vascular Plants.* Volume III. *Flowering Plants – Monocotyledons. Lilianae (except Orchidaceae)* (ed. K. Kubitzki) Berlin, Heidelberg, New York: Springer-Verlag, pp. 216–35.

Huddlestun, P. F. & Hetrick, J. H. 1991. *The Stratigraphic Framework of the Fort Valley Plateau and the Central Georgia Kaolin District. Georgia Geological Society Guidebooks.* Carrollton, GA: Georgia Geological Society.

Hueber, F. M. & Watson, J. 1988. The unusual Upper Cretaceous conifer *Androvettia* from eastern U.S.A. *Botanical Journal of the Linnean Society*, **98**, 117–33.

Huelsenbeck, J. P., Larget, B. & Swofford, D. L. 2000. A compound poisson process for relaxing the molecular clock. *Genetics*, **154**, 1879–92.

Hufford, L. D. & Crane, P. R. 1989. A preliminary phylogenetic analysis of 'lower' Hamamelidae. In *Evolution, Systematics and Fossil History of the Hamamelidae.* Volume 1. *Introduction and 'Lower' Hamamelidae* (eds P. R. Crane & S. Blackmore) Oxford: Clarendon Press, pp. 175–92.

Hufford, L. D. & Endress, P. K. 1989. The diversity of stamen structures and dehiscence patterns among Hamamelididae. *Botanical Journal of the Linnean Society*, **99**, 301–46.

Hufford, L. D. 1992. Rosidae and their relationships to other nonmagnoliid dicotyledons: a phylogenetic analysis using morphological and chemical data. *Annals of the Missouri Botanical Garden*, **79**, 218–48.

Hufford, L. D. 1996. The origin and early evolution of angiosperm stamens. In *The Anther. Form, Function and Phylogeny* (eds W. G. D'Arcy & R. C. Keating) Cambridge: Cambridge University Press, pp. 58–91.

Hughes, L., Dunlop, M., French, K. *et al.* 1994. Predicting dispersal spectra; a minimum set of hypotheses based on plant attributes. *Journal of Ecology*, **82**, 933–50.

Hughes, N. F. 1961a. Fossil evidence and angiosperm ancestry. *Science Progress*, **49**, 84–102.

Hughes, N. F. 1961b. Further interpretation of *Eucommiidites* Erdtman 1948. *Palaeontology*, **4**, 292–9.

Hughes, N. F. & Moody-Stuart, J. C. 1967. Palynological facies and correlation in the English Wealden. *Review of Palaeobotany and Palynology*, **1**, 259–68.

Hughes, N. F. 1975. Plant succession in the English Wealden strata. *Proceedings of the Geologists' Association*, **86**, 439–55.

Hughes, N. F. 1976. *Palaeobiology of Angiosperm Origins.* Cambridge: Cambridge University Press.

Hughes, N. F., Drewry, G. & Laing, J. F. 1979. Barremian earliest angiosperm pollen. *Palaeontology*, **22**, 513–36.

Hughes, N. F. 1986. The problems of data-handling for early angiosperm-like pollen. In *Systematic and Taxonomic Approaches in Palaeobotany* (eds R. A. Spicer & B. A. Thomas) Oxford: Clarendon Press, pp. 233–51.

Hughes, N. F. & McDougall, A. B. 1987. Records of angiospermid pollen entry into the English Early Cretaceous succession. *Review of Palaeobotany and Palynology*, **50**, 255–72.

Hughes, N. F. & McDougall, A. B. 1989. New Wealden correlation for the Wessex Basin. *Proceedings of the Geologists' Association*, **101**, 85–90.

Hughes, N. F. & McDougall, A. B. 1990. Barremian-Aptian angiospermid pollen records from southern England. *Review of Palaeobotany and Palynology*, **65**, 145–51.

Hughes, N. F., McDougall, A. B. & Chapman, J. L. 1991. Exceptional new record of Cretaceous Hauterivian angiospermid pollen from southern England. *Journal of Micropalaeontology*, **10**, 75–82.

Hughes, N. F. 1994. *The Enigma of Angiosperm Origins.* Cambridge: Cambridge University Press.

Hunt, T., Bergsten, J., Levkanicova, Z. *et al.* 2007. A comprehensive phylogeny of beetles reveals the evolutionary origins of a superradiation. *Science*, **318**, 1913–16.

Hutchinson, J. 1959. *The Families of Flowering Plants.* Oxford: Clarendon Press.

Huzioka, K. 1961. A new Paleogene species of the genus *Eucommia* from Hokkaido, Japan. *Transactions and Proceedings of Palaeontological Society of Japan, New Series*, **41**, 9–12.

Hvalj, A. V. 2001. Karpologija srednemelovykh khvojnykh i tsvetkovykh Kachara (Jugo-Vostochnoe Zaural'e). Sankt Petersburg: PhD Thesis, Komarov Botanical Institute, Russian Academy of Sciences, pp. 1–279 (in Russian).

Ibrahim, M. & Schrank, E. 1996. Palynological studies on the Late Jurassic – Early Cretaceous of the Kahraman -1 well, northern Western Desert, Egypt. *Géologie de l'Afrique et de l'Atlantique Sud: Actes Colloques Angers* 1994, **611**–29.

Ibrahim, M. I. A. 2002. New angiosperm pollen from the Upper Barremian–Aptian of the Western Desert, Egypt. *Palynology*, **26**, 107–33.

Ickert-Bond, S. M. & Wojciechowski, M. F. 2004. Phylogenetic relationships in *Ephedra* (Gnetales): Evidence from nuclear and chloroplast DNA sequence data. *Systematic Botany*, **29**, 834–49.

Ickert-Bond, S. M., Pigg, K. B. & Wen, J. 2007. Comparative infructescence morphology in *Altingia* (Altingiaceae) and discordance between morphological and molecular phylogenies. *American Journal of Botany*, **94**, 1094–115.

Igersheim, A. & Endress, P. K. 1997. Gynoecium diversity and systematics of the Magnoliales and winteroids. *Botanical Journal of the Linnean Society*, **124**, 213–71.

Igersheim, A. & Endress, P. K. 1998. Gynoecium diversity and systematics of the paleoherbs. *Botanical Journal of the Linnean Society*, **127**, 289–370.

Jackson, S. T. & Weng, C. 1999. Late Quaternary extinction of a tree species in eastern North America. *Proceedings of the National Academy of Sciences, USA*, **96**, 13 847–52.

Jacobs, B. F. & Kabuye, C. H. S. 1989. An extinct species of *Pollia* Thunberg (Commelinaceae) from the Miocene Ngorora Formation, Kenya. *Review of Palaeobotany and Palynology*, **59**, 67–76.

Jacobs, B. F., Kingston, J. D. & Jacobs, L. L. 1999. The origin of grass-dominated ecosystems. *Annals of the Missouri Botanical Garden*, **86**, 590–643.

Jacques, F. M. B. & De Franceschi, D. 2005. Endocarps of Menispermaceae from Le Quesnoy outcrop (Sparnacian facies, Lower Eocene, Paris Basin). *Review of Palaeobotany and Palynology*, **135**, 61–70.

Jaeger, J.-J., Courtillot, V. & Tapponnier, P. 1989. Paleontological view of the ages of the Deccan Traps, the Cretaceous/Tertiary boundary, and the India-Asia collision. *Geology*, **17**, 316–19.

Jain, R. K. 1964. Studies in Musaceae – 1. *Musa cardiosperma* sp. nov., a fossil banana fruit from the Deccan Intertrappean Series, India. *The Palaeobotanist* **12**, 45–58 (1963).

Jan du Chêne, R. E., Adegoke, O. S., Adediran, S. A. & Petters, S. W. 1978a. Palynology and foraminifera of the Lokoja Sandstone (Maastrichtian), Bida Basin, Nigeria. *Revista Español de Micropaleontología*, **10**, 379–93.

Jan du Chêne, R. E., Klasz, I. de & Archibong, E. E. 1978b. Biostratigraphic study of the borehole OJO-1, SW Nigeria, with special emphasis on the Cretaceous microflora. *Revue de Micropaléontologie*, **21**, 123–39.

Janchen, E. 1950. Die Herkunft der Angiospermen-Blüte und die systematische Stellung der Apetalen. *Österreichische Botanische Zeitschrift*, **97**, 129–67.

Jansen, R. K., Cai, Z., Raubeson, L. A. *et al.* 2007. Analysis of 81 genes from 64 plastid genomes resolves relationships in angiosperms and identifies genome-scale evolutionary patterns. *Proceedings of the National Academy of Sciences, USA*, **104**, 19 369–74.

Janzen, D. H. & Martin, P. S. 1982. Neotropical anachronisms: the fruits the gomphotheres ate. *Science*, **215**, 19–27.

Janzen, D. H. 1983. Dispersal of seeds in vertebrate guts. In *Coevolution* (eds D. J. Futuyma & M. Slatkin) Sunderland, MA: Sinauer, pp. 232–62.

Jardiné, S. & Magloire, L. 1965. Palynologie et stratigraphie du Crétacé des bassins du Sénegal et de Côte d'Ivoire. *Mémoires du Bureau de Recherches Géologiques et Minières*, **32**, 187–245.

Jardiné, S. 1967. Spores à expansions en forme d'élatères du Crétacé moyen d'Afrique occidentale. *Review of Palaeobotany and Palynology*, **1**, 235–58.

Jardiné, S., Doerenkamp, A. & Biens, P. 1974. *Dicheiropollis etruscus*, un pollen caractéristique de Crétacé inférieur afro-sudaméricain. Conséquences pour l'évaluation des unités climatiques et implications dans la dérive des continents. *Bulletin des sciences géologiques de Strasbourg*, **27**, 69–85.

Jarmolenko, A. V. 1935. The Upper Cretaceous flora of the North-Western Kara-Tau. *Acta Universitatis Asiae Mediae. Series Viiib. Botanica*, **28**, 1–36.

Jarvis, C. E. 1989. A review of the family Buxaceae Dumortier. In *Evolution, Systematics and Fossil History of the Hamamelidae. I: Introduction and 'Lower' Hamamelidae* (eds P. R. Crane & S. Blackmore) Oxford: Clarendon Press, pp. 273–8.

Jarzembowski, E. A. 1991. The Weald Clay of the Weald: report of 1988/1989 field meetings. *Proceedings of the Geologists' Association*, **102**, 83–92.

Jarzen, D. M. & Norris, G. 1975. Evolutionary significance and botanical relationships of Cretaceous angiosperm pollen in the western Canadian interior. *Geoscience and Man*, **11**, 47–60.

Jarzen, D. M. 1977. *Aquilapollenites* and some Santalalean genera. A botanical comparison. *Grana*, **16**, 29–39.

Jarzen, D. M. 1982. Palynology of Dinosaur Provincial Park (Campanian) Alberta. *Syllogeus*, **38**, 1–69.

Jarzen, D. M. 1983. The fossil pollen record of the Pandanaceae. *Gardens' Bulletin*, **36**, 163–75.

Jarzen, D. M. & Dettmann, M. E. 1989. Taxonomic revision of *Tricolpites reticulatus* Cookson ex Couper, 1953 with notes on the biogeography of *Gunnera* L. *Pollen et Spores*, **31**, 97–112.

Jarzen, D. M. & Pocknall, D. T. 1993. Tertiary *Bluffopollis scabratus* (Couper) Pocknall & Mildenhall, 1984 and modern *Strasburgeria* pollen: a botanical comparison. *New Zealand Journal of Botany*, **31**, 185–92.

Ji, Q., Luo, Z.-X., Yuan, C.-X. *et al.* 2002. The earliest known eutherian mammal. *Nature*, **416**, 816–22.

Ji, Q., Li, H., Bowe, L. M., Liu, Y. & Taylor, D. W. 2004. Early Cretaceous *Archaefructus eoflora* sp. nov., with bisexual flowers from Beipiao, Western Liaoning, China. *Acta Geologica Sinica*, **78**, 883–96.

Johnson, K. R. 1996. Description of seven common plant megafossils from the Hell Creek Formation (Late Cretaceous: late Maastrichtian), North Dakota, South Dakota, and Montana. *Proceedings of the Denver Museum of Nature & Science*, **3**, 1–47.

Johnson, K. R. 2002. Megaflora of the Hell Creek and lower Fort Union Formations in the western Dakotas: vegetational response to climate change, the Cretaceous-Tertiary boundary event, and rapid marine transgression. *Geological Society of America, Special Paper*, **361**, 329–91.

Johnson, K. R., Reynolds, M. L., Werth, K. W. & Thomasson, J. R. 2003. Overview of the Late Cretaceous, early Paleocene, and early Eocene megafloras of the Denver Basin, Colorado. *Rocky Mountain Geology*, **38**, 101–20.

Johnson, L. 2004. Polemoniaceae. Phlox Family. Version 25 April 2004, http://tolweb.org/Polemoniaceae/20812/ 2004.04.25 in The Tree of Life Web Project, http://tolweb.org/

Johnson, L. A. S. & Wilson, K. L. 1990. General traits of the Cycadales. In *The Families and Genera of Vascular Plants.* Volume I. *Pteridophytes and Gymnosperms* (eds K. Kubitzki, K. U. Kramer & P. S. Green) Berlin, Heidelberg, New York: Springer-Verlag, pp. 363–77.

Johnson, L. A. S. & Wilson, K. L. 1993. Casuarinaceae. In *The Families and Genera of Vascular Plants.* Volume II. *Flowering Plants – Dicotyledons. Magnoliid, Hamamelid and Caryophyllid Families* (eds K. Kubitzki, J. G. Rohwer & V. Bittrich) Berlin, Heidelberg, New York: Springer-Verlag, pp. 237–42.

Jones, D. L. 2002. *Cycads of the World.* Washington, D.C.: Smithsonian Institution Press.

Jones, O. A. 1926. The Tertiary deposits of the Moreton district. *Proceedings of the Royal Society of Queensland*, **38**, 23–46.

Jordan, G. J. & Macphail, M. K. 2003. A Middle-Late Eocene inflorescence of Caryophyllaceae from Tasmania, Australia. *American Journal of Botany*, **90**, 761–8.

Jordan, R. R. 1983. Stratigraphic nomenclature of nonmarine Cretaceous rocks of inner margin of coastal plain in Delaware and adjacent states. *Delaware Geological Survey Report of Investigation*, **37**, 1–43.

Jordano, P. 1995. Angiosperm fleshy fruits and seed dispersers: a comparative analysis of adaptation and constraints on plant-animal interactions. *American Naturalist*, **145**, 163–91.

Judd, W. S., Campbell, C. S., Kellogg, E. A., Stevens, P. F. & Donoghue, M. J. 2002. *Plant Systematics: A Phylogenetic Approach*. Sunderland, MA: Sinauer Associates.

Judd, W. S. & Olmstead, R. G. 2004. A survey of tricolpate (eudicot) phylogenetic relationships. *American Journal of Botany*, **91**, 1627–44.

Juhász, M. & Góczán, F. 1985. Comparative study of Albian monosulcate angiosperm pollen grains. *Acta Biologica Szegediensis*, **31**, 147–72.

Jähnichen, H. 1965. Beiträge zur Tertiärflora der Lausitz. Inkohlte Blätter und Epidermisstrukturen. *Monatsberichte der deutschen Akademie der Wissenschaften zu Berlin*, **7**, 664–70.

Kadereit, J. W. 2007. Asterales: introduction and conspectus. In *The Families and Genera of Vascular Plants.* Volume VIII. *Flowering Plants – Eudicots. Asterales* (ed. K. Kubitzki) Berlin, Heidelberg, New York: Springer-Verlag, pp. 1–6.

Karsten, G. 1918. Zur Phylogenie der Angiospermen. *Zeitschrift für Botanik*, **10**, 369–88.

Kato, M. & Inoue, T. 1994. Origin of insect pollination. *Nature*, **368**, 195.

Kedves, M. & Diniz, F. 1967. Quelque type sporomorphes de sediments crétacé d'Aveiro, Portugal. *Comunicações dos Serviços Geológicos de Portugal*, **52**, 17–26.

Kedves, M. & Pittau, P. 1979. Contribution à la connaissance des pollens des Normapolles du Crétacé Supérieur du Portugal. *Pollen et Spores*, **21**, 169–209.

Kedves, M. & Diniz, F. 1981. Probrevaxones, a new pollen group for the first Brevaxones form-genera from the Upper Cenomanian of Portugal. *Acta Botanica Academiae Scientiarum Hungaricae*, **27**, 383–402.

Kelber, K.-P., Franz, L., Stachel, T., Lorenz, V. & Okrusch, M. 1992/93. Plant fossils from Gross Brukkaros (Namibia) and their biostratigraphical significance. *Communications of the Geological Survey of Namibia*, **8**, 57–66.

Kelch, D. G. 1998. Phylogeny of Podocarpaceae: comparison of evidence from morphology and 18S rDNA. *American Journal of Botany*, **85**, 986–96.

Keller, J. A., Herendeen, P. S. & Crane, P. R. 1996. Fossil flowers and fruits of the Actinidiaceae from the Campanian (Late Cretaceous) of Georgia. *American Journal of Botany*, **83**, 528–41.

Kelley, D. R. & Gasser, C. S. 2009. Ovule development: genetic trends and evolutionary considerations. *Sexual Plant Reproduction*, **22**, 229–34.

Kemp, E. M. 1968. Probable angiosperm pollen from British Barremian to Albian strata. *Palaeontology*, **11**, 421–34.

Kemp, E. M. 1970. Aptian and Albian miospores from southern England. *Palaeontographica* B, **131**, 73–143.

Kemp, E. M. & Harris, W. K. 1975. The vegetation of Tertiary islands on the Ninetyeast Ridge. *Nature*, **258**, 303–7.

Keng, H. 1993. Illiciaceae. In *The Families and Genera of Vascular Plants*. Volume II. *Flowering Plants – Dicotyledons. Magnoliid, Hamamelid and Caryophyllid Families* (eds K. Kubitzki, J. G. Rohwer & V. Bittrich) Berlin, Heidelberg, New York: Springer-Verlag, pp. 344–7.

Kennedy, E. M., Spicer, R. A. & Rees, P. M. 2002. Quantitative palaeoclimate estimates from Late Cretaceous and Paleocene leaf floras in the northwest of the South Island, New Zealand. *Palaeogeography, Palaeoclimatology, Palaeoecology*, **184**, 321–45.

Kennedy, E. M., Lovis, J. D. & Daniel, J. L. 2003. Discovery of a Cretaceous angiosperm reproductive structure from New Zealand. *New Zealand Journal of Geology and Geophysics Abstracts*, **46**, 519–22.

Kennedy, P. 1991. Cycad-insect relationships. *Encephalartos*, **27**, 22–6.

Kennedy, W. J., Crame, J. A., Bengtson, P. & Thomson, M. R. A. 2007. Coniacian ammonites from James Ross Island, Antarctica. *Cretaceous Research*, **28**, 509–31.

Kenrick, P. & Crane, P. R. 1997. *The Origin and Early Diversification of Land Plants. A Cladistic Study.* Smithsonian Series in Comparative Evolutionary Biology. Washington, D.C.: Smithsonian Institution Press.

Kerp, J. H. F. 1988. Aspects of Permian palaeobotany and palynology. X. The west and central European species of the genus *Autunia* Krasser emend. Kerp (Peltaspermaceae) and the form-genus *Rachiphyllum* Kerp (callipterid foliage). *Review of Palaeobotany and Palynology*, **54**, 249–360.

Kerr, R. A. 1984. How to make a warm Cretaceous climate. *Science*, **222**, 677–8.

Kessler, P. J. A. 1993a. Menispermaceae. In *The Families and Genera of Vascular Plants*. Volume II. *Flowering Plants – Dicotyledons. Magnoliid, Hamamelid and Caryophyllid Families* (eds K. Kubitzki, J. G. Rohwer & V. Bittrich) Berlin, Heidelberg, New York: Springer-Verlag, pp. 402–18.

Kessler, P. J. A. 1993b. Annonaceae. In *The Families and Genera of Vascular Plants*. Volume II. *Flowering Plants – Dicotyledons. Magnoliid, Hamamelid and Caryophyllid Families* (eds K. Kubitzki, J. G. Rohwer & V. Bittrich) Berlin, Heidelberg, New York: Springer-Verlag, pp. 93–129.

Khand, Y., Badamgarav, D., Ariunchimeg, Y. & Barsbold, R. 2000. Cretaceous system in Mongolia and its depositional environments. In *Cretaceous Environments of Asia* (eds H. Okada & N. J. Mateer) Amsterdam: Elsevier, pp. 49–79.

Kielan-Jaworowska, Z., Dashzeveg, D. & Trofimov, B. A. 1987. Early Cretaceous multituberculates from Mongolia and a comparison with Late Jurassic forms. *Acta Palaeontologica Polonica*, **32**, 3–47.

Kim, S., Park, C.-W., Kim, Y.-D. & Suh, Y. 2001. Phylogenetic relationships in family Magnoliaceae inferred from *ndhF* sequences. *American Journal of Botany*, **88**, 717–28.

Kimura, T. & Ohana, T. 1992. Cretaceous palaeobotany and phytogeography in eastern Eurasia. *Paleontological Society of Korea, Special Publication*, **1**, 27–34.

Kimura, T. 2000. Early Cretaceous climatic provinces in Japan and adjacent regions on the basis of fossil land plants. In *Cretaceous Environments of Asia* (eds H. Okada & N. J. Mateer) Amsterdam: Elsevier, pp. 155–61.

Kirchheimer, F. 1937. Beiträge zur Kenntnis der Tertiärflora. Früchte und Samen aus dem deutschen Tertiär. *Palaeontographica* B, **82**, 73–141.

Kirchheimer, F. 1950. Die Symplocaceen der erdgeschichtlichen Vergangenheit. *Palaeontographica* B, **90**, 1–52.

Kirchheimer, F. 1957. *Die Laubgewächse der Braunkohlenzeit*. Halle: Wilhelm Knapp Verlag.

Kirillova, G. L., Markevitch, V. S. & Belyi, V. F. 2000. Cretaceous environmental changes of East Russia. In *Cretaceous Environments of Asia* (eds H. Okada & N. J. Mateer) Amsterdam: Elsevier, pp. 1–47.

Kirk, W. D. J. 1984. Pollen-feeding in thrips (Insecta: Thysanoptera). *Journal of Zoology*, **204**, 107–17.

Klavins, S. D., Taylor, T. N. & Taylor, E. L. 2002. Anatomy of *Umkomasia* (Corystospermales) from the Triassic of Antarctica. *American Journal of Botany*, **89**, 664–76.

Knappe, H. & Rüffle, L. 1975. Neue Monimiaceen-Blätter im Santon des Subherzyn und ihre phytogeographischen Beziehungen zur Flora des ehemaligen Gondwana-Kontinents. *Wissenschaftliche Zeitschrift der Humboldt-Universität zu Berlin, Mathematisch-Naturwissenschaftliche Reihe*, **24**, 493–9.

Knobloch, E. 1964. Neue Pflanzenfunde aus dem südböhmischen Senon. *Jahrbuch des Staatlichen Museums für Mineralogie und Geologie zu Dresden*, **1964**, 133–201.

Knobloch, E. 1971. Fossile Früchte und Samen aus der Flyschzone der mährischen Karpaten. *Sborník geologických Ved, Paleontologie*, **13**, 7–46.

Knobloch, E. 1975. Früchte und Samen aus der Gosauformation von Kössen in Österreich. *Věstnik Ústředního ústavu geologického*, **50**, 83–91.

Knobloch, E. 1977. Paläokarpologische Characteristik der Flyschzone der mährischen Karpaten. *Sborník geologických Ved, Paleontologie*, **19**, 79–137.

Knobloch, E. & Mai, D.-H. 1983. Carbonized seeds and fruits from the Cretaceous of Bohemia and Moravia and their stratigraphical significance. *Knihovnicka Zemního plynu a nafty*, **4**, 305–32.

Knobloch, E. & Mai, D. H. 1984. Neue Gattungen nach Früchten und Samen aus dem Cenoman bis Maastricht (Kreide) von Mitteleuropa. *Feddes Repertorium*, **95**, 3–41.

Knobloch, E. 1985. Paläobotanisch-biostratigraphische Characteristik der Klikov-Schichtenfolge (Oberturon-Santon) in Südböhmen. *Sborník geologických Ved, Geologie*, **40**, 101–45.

Knobloch, E. & Mai, D. H. 1986. Monographie der Früchte und Samen in der Kreide von Mitteleuropa. *Rozpravy ústredního ústavu geologickénho, Praha*, **47**, 1–219.

Knobloch, E. 1997. *"Credneria" bohemica* Velenovský – eine altertümliche Platane. *Palaeontographica* B, **242**, 127–48.

Knoll, A. H., Niklas, K. J. & Tiffney, B. H. 1979. Phanerozoic land-plant diversity in North America. *Science*, **206**, 1400–2.

Knowlton, F. H. 1901. Reports by Professor F. H. Knowlton on fossil plants collected by T. Wayland Vaughan in Lamar County; by G.H. Ragsdale in Cooke County; by T.V. Munson in Rhamey Hill, Denison, Texas. *US Geological Survey 21st Annual Report 1899–1900, pt. 7*, 314–18.

Knowlton, F. H. 1923. Revision of the flora of the Green River Formation. *US Geolological Survey, Professional Paper*, **131**, 133–82.

Knowlton, F. H. 1930. The flora of the Denver and associated formations of Colorado. *US Geological Survey, Professional Paper*, **155**, 1–142.

Koch, B. E. 1964. Review of fossil floras and nonmarine deposits of West Greenland. *Geological Society of America Bulletin*, **75**, 535–48.

Koch, B. E. & Friedrich, W. L. 1971. Früchte und Samen von *Spirematospermum* aus der miozänen Fasterholt-Flora in Dänemark. *Palaeontographica* B, **136**, 1–46.

Koch, B. E. 1972. Fossil *Picrodendron* fruit from the Upper Danian of Nûgssuaq, West Greenland. *Meddelelser om Grønland*, **193**, 1–32.

Kolakovsky, A. A. 1957. Pervoe dopolnenie k kodorskoj flore (Meore-Atara). *Trudy Sukhumskago Botanicheskago sada*, **10**, 237–318 (in Russian).

Kolakovsky, A. A. 1964. Pliotsenovaja flora Pitsundy. *Trudy Sukhumskago Botanicheskago sada*, **14**, 1–209 (in Russian).

Konopka, A. S., Herendeen, P. S., Merrill, G. L. S. & Crane, P. R. 1997. Sporophytes and gametophytes of Polytrichaceae from the Campanian (Late Cretaceous) of Georgia. *International Journal of Plant Sciences*, **158**, 489–99.

Konopka, A. S., Herendeen, P. S. & Crane, P. R. 1998. Sporophytes and gametophytes of Dicranaceae from the Santonian (Late Cretaceous) of Georgia, USA. *American Journal of Botany*, **85**, 714–23.

Koponen, T. 1968. Generic revision of Mniaceae Mitt. (Bryophyta). *Annales Botanici Fennici*, **5**, 117–51.

Koppelhus, E. B. & Batten, D. J. 1989. Late Cretaceous megaspores from southern Sweden: morphology and palaeoenvironmental significance. *Palynology*, **13**, 91–120.

Koriba, K. & Miki, S. 1958. *Archeozostera*, a new genus from Upper Cretaceous in Japan. *The Palaeobotanist*, **7**, 107–10.

Kovar-Eder, J. 1992. A remarkable preservation state of fossil leaves recognized in *Potamogeton*. *Courier Forschungsinstitut Senckenberg*, **147**, 393–7.

Kramer, E. M. & Hall, J. C. 2005. Evolutionary dynamics of genes controlling floral development. *Current Opinion in Plant Biology*, **8**, 13–18.

Kramer, E. M. & Hodges, S. A. 2010. *Aquilegia* as a model system for the evolution and ecology of petals. *Philosophical Transactions of the Royal Society* B, **365**, 477–90.

Krasser, F. 1896. Beiträge zur Kenntniss der fossilen Kreideflora von Kunstadt in Mähren. *Beiträge zur Paläontologie Österreich-Ungarns und des Orientes*, **10**, 113–52.

Krassilov, V. A. 1967. *Early Cretaceous Flora of South Primorye and its Stratigraphic Significance*. Moscow: Nauka (in Russian).

Krassilov, V. A. 1973. The Jurassic disseminules with pappus and their bearing on the problem of angiosperm ancestry. *Geophytology*, **3**, 1–4.

Krassilov, V. A. 1975a. Dirhopalostachyaceae – a new family of proangiosperms and its bearing on the problem of angiosperm ancestry. *Palaeontographica* B, **153**, 100–10.

Krassilov, V. A. 1975b. *Paleoecology of Terrestrial Plants. Basic Principles and Techniques*. New York, Toronto: John Wiley & Sons.

Krassilov, V. A. 1977a. The origin of angiosperms. *The Botanical Review*, **43**, 143–76.

Krassilov, V. A. 1977b. Contribution to the knowledge of the Caytoniales. *Review of Palaeobotany and Palynology*, **24**, 155–78.

Krassilov, V. A. 1982. Early Cretaceous flora of Mongolia. *Palaeontographica* B, **181**, 1–43.

Krassilov, V. A. & Bugdaeva 1982. Achene-like fossils from the Lower Cretaceous of the Lake Baikal area. *Review of Palaeobotany and Palynology*, **36**, 279–95.

Krassilov, V. A., Shilin, P. V. & Vachrameev, V. A. 1983. Cretaceous flowers from Kazakhstan. *Review of Palaeobotany and Palynology*, **40**, 91–113.

Krassilov, V. A. 1984. New paleobotanical data on origin and early evolution of angiospermy. *Annals of the Missouri Botanical Garden*, **71**, 577–92.

Krassilov, V. A. 1986. New floral structures from the Lower Cretaceous of Lake Baikal area. *Review of Palaeobotany and Palynology*, **47**, 9–16.

Krassilov, V. A. & Bugdaeva, E. V. 1988a. Gnetalean plants from the Jurassic of Ust-Balej, East Siberia. *Review of Palaeobotany and Palynology*, **53**, 359–74.

Krassilov, V. A. & Bugdaeva, E. V. 1988b. Protocycadopsid pteridosperms from the Lower Cretaceous of Transbaikalie and the origin of cycads. *Palaeontographica* B, **208**, 27–32.

Krassilov, V. A. & Dobruskina, I. A. 1995. Angiosperm fruit from the Lower Cretaceous of Israel and orgins in rift valleys. *Paleontological Journal*, **29**, 110–15.

Krassilov, V. A. & Shilin, P. V. 1995. New platanoid staminate heads from the mid-Cretaceous of Kazakhstan. *Review of Palaeobotany and Palynology*, **85**, 207–11.

Krassilov, V. A. 1997. *Angiosperm Origins: Morphological and Ecological Aspects*. Sofia: Pensoft.

Krassilov, V. A., Dilcher, D. L. & Douglas, J. G. 1998. New ephedroid plant from the Lower Cretaceous Koonwarra Fossil Bed, Victoria, Australia. *Alcheringa*, **22**, 123–33.

Krassilov, V. A. & Dobruskina, I. A. 1998. A graminoid plant from the Cretaceous of the Middle East. *Paleontological Journal*, **32**, 429–34.

Krassilov, V. A. & Bacchia, F. 2000. Cenomanian florule of Nammoura, Lebanon. *Cretaceous Research*, **21**, 785–99.

Krassilov, V. A. & Bugdaeva, E. V. 2000. Gnetophyte assemblage from the Early Cretaceous. *Palaeontographica* B, **253**, 139–51.

Krassilov, V. A. & Golovneva, L. B. 2001. Inflorescence with tricolpate pollen grains from the Cenomanian of Tschulymo-Yenisey Basin, West Siberia. *Review of Palaeobotany and Palynology*, **40**, 91–113.

Krassilov, V. A., Tekleva, M., Meyer-Melikyan, N. & Rasnitsyn, A. P. 2003. New pollen morphotype from the gut compression of a Cretaceous insect, and bearing on palynomorphological evolution and paleoecology. *Cretaceous Research*, **24**, 149–56.

Krassilov, V. A. 2004. Cretaceous floral structures from Negev, Israel as evidence of angiosperm radiation in the Gondwana realm. *Acta Palaeobotanica*, **44**, 37–53.

Krassilov, V. A. & Golovneva, L. B. 2004. A minute mid-Cretaceous flower from Siberia and implications for the problem of basal angiosperms. *Geodiversitas*, **26**, 5–15.

Krassilov, V. A., Lewy, Z., Nevo, E. & Silantieva, N. 2005. *Late Cretaceous (Turonian) Flora of Southern Negev, Israel*. Sofia-Moscow: Pensoft.

Krassilov, V. A., Rasnitsyn, A. P. & Afonin, S. A. 2007. Pollen eaters and pollen morphology: co-evolution through the Permian and Mesozoic. *African Invertebrates*, **48**, 3–11.

Krassilov, V. A. & Volynets, Y. 2008. Weedy Albian angiosperms. *Acta Palaeobotanica*, **48**, 151–69.

Krause, D. W. & Hartman, J. H. 1996. Late Cretaceous fossils from Madagascar and their implications for biogeographic relationships with the Indian subcontinent. *Memoirs of the Geological Survey of India, Palaeontologia Indica*, **37**, 135–54.

Kreunen, S. S. & Osborn, J. M. 1999. Pollen and anther development in *Nelumbo* (Nelumbonaceae). *American Journal of Botany*, **86**, 1662–76.

Kristensen, N. P. 1984. Studies on the morphology and systematics of primitive Lepidoptera (Insecta). *Steenstrupia*, **10**, 141–91.

Kristensen, N. P., Scoble, M. J. & Karsthol, O. 2007. Lepidoptera phylogeny and systematics: the state of inventorying moth and butterfly diversity. *Zootaxa*, **1668**, 699–747.

Kruse, H. O. 1954. Some Eocene dicotyledonous woods from Eden Valley, Wyoming. *The Ohio Journal of Science*, **54**, 243–68.

Krutzsch, W. 1962. Stratigraphisch bzw. botanisch wichtige neue Sporen- und Pollenformen aus dem deutschen Tertiär. *Geologie*, **11**, 265–307.

Krutzsch, W. 1966a. Zur Kenntnis der präquartären periporaten Pollenformen im nördlichen Mitteleuropa. *Geologie*, **15**, 16–71.

Krutzsch, W. 1966b. Die sporenstratigraphische Gliederung der Oberkreide im nördlichen Mitteleuropa – Methodische Grundlagen und gegenwärtiger Stand der Untersuchungen. *Abhandlungen des Zentralen geologischen Instituts*, **5**, 111–37.

Krutzsch, W. & Lenk, G. 1969. Über 5 neue Pollenarten aus dem Maastricht der Bohrung Colbitz 10 (Calvoorder Scholle, D.D.R.). *Monatsberichte der Deutschen Akademie der Wissenchaften zu Berlin*, **11**, 938–45.

Krutzsch, W. 1970a. Zur Kenntnis fossiler disperser Tetradenpollen. *Paläontologische Abhandlungen Abteilung B, Paläobotanik*, **3**, 399–433.

Krutzsch, W. 1970b. *Atlas der Mittel- und Jungtertiaren Dispersen Sporen und Pollen – Sowie der Mikroplanktonformen des Nordlichen Mitteleuropas*, Lieferung VII. *Monoporate, Monocolpate, Longicolpate, Dicolpate und Ephedroide (Polyplicate) Pollenformen*. Jena: Gustav Fischer Verlag.

Krutzsch, W. 1989. Paleogeography and historical phytogeography (paleochorology) in the Neophyticum. *Plant Systematics and Evolution*, **162**, 5–61.

Kryshtofovich, A. N. 1929. Discovery of the oldest dicotyledons of Asia in the equivalents of the Potomac Group in Suchan, Ussuriland, Siberia. *Bulletins du Comité Géologique*, **48**, 1357–90 (in Russian with extended summary in English).

Kräusel, R. 1922. Beiträge zur Kenntnis der Kreideflora. I. Über einige Kreidepflanzen von Swalmen (Niederlande). *Mededelingen Rijks Geologischen Dienst, Serie A*, **2**, 1–40.

Kräusel, R. 1923. Über pflanzenführende Kreideschichten aus der Umgebung von Heerlen (Holländ. Limburg) und die Verbreitung des Aachener Sandes in den südlichen Niederlanden. *Senckenbergiana*, **5**, 145–54.

Kräusel, R. 1939. Ergebnisse der Forschungsreisen Prof. E. Strömer's in der Wüsten Ägyptens. IV. Die fossilen Floren Ägyptens. *Abhandlungen der Bayerischen Akademie der Wissenschaften, Mathematisch-naturwissenschaftlige Abteilung. Neue Folge*, **47**, 1–140.

Kräusel, R. 1948. *Sturiella langeri* nov. gen. nov. sp., eine Bennettitee aus der Trias von Lunz (Nieder-Österreich). *Senckenbergiana*, **29**, 141–9.

Kräusel, R. & Weyland, H. 1950. Kritische Untersuchungen zur Kutikularanalyse tertiärer Blätter. I. *Palaeontographica* B, **91**, 7–92.

Kräusel, R. & Schaarschmidt, F. 1968. *Scoresbya* Harris (Dipteridaceae) aus dem Unteren Jura von Sassendorf. *Palaeontographica* B, **123**, 124–31.

Kubitzki, K. 1990a. Gnetaceae. In *The Families and Genera of Vascular Plants.* Volume I. *Pteridophytes and Gymnosperms* (ed. K. Kubitzki) Berlin, Heidelberg, New York: Springer-Verlag, pp. 383–6.

Kubitzki, K. 1990b. Ephedraceae. In *The Families and Genera of Vascular Plants.* Volume I. *Pteridophytes and Gymnosperms* (ed. K. Kubitzki) Berlin, Heidelberg, New York: Springer-Verlag, pp. 379–82.

Kubitzki, K. 1990c. Welwitschiaceae. In *The Families and Genera of Vascular Plants.* Volume I. *Pteridophytes and Gymnosperms* (ed. K. Kubitzki) Berlin, Heidelberg, New York: Springer-Verlag, pp. 387–91.

Kubitzki, K. 1993a. Calycanthaceae. In *The Families and Genera of Vascular Plants.* Volume II. *Flowering Plants – Dicotyledons. Magnoliid, Hamamelid and Caryophyllid Families* (eds K. Kubitzki, J. G. Rohwer & V. Bittrich) Berlin, Heidelberg, New York: Springer-Verlag, pp. 197–200.

Kubitzki, K. 1993b. Myrothamnaceae. In *The Families and Genera of Vascular Plants.* Volume II. *Flowering Plants – Dicotyledons. Magnoliid, Hamamelid and Caryophyllid Families* (eds K. Kubitzki, J. G. Rohwer & V. Bittrich) Berlin, Heidelberg, New York: Springer-Verlag, pp. 468–9.

Kubitzki, K. 1993c. Fagaceae. In *The Families and Genera of Vascular Plants.* Volume II. *Flowering Plants – Dicotyledons. Magnoliid, Hamamelid and Caryophyllid Families* (eds K. Kubitzki, J. G. Rohwer & V. Bittrich) Berlin, Heidelberg, New York: Springer-Verlag, pp. 301–9.

Kubitzki, K. 1993d. Hernandiaceae. In *The Families and Genera of Vascular Plants.* Volume II. *Flowering Plants – Dicotyledons. Magnoliid, Hamamelid and Caryophyllid Families* (eds K. Kubitzki, J. G. Rohwer & V. Bittrich) Berlin, Heidelberg, New York: Springer-Verlag, pp. 457–67.

Kubitzki, K. 1993e. Canellaceae. In *The Families and Genera of Vascular Plants.* Volume II. *Flowering Plants – Dicotyledons. Magnoliid, Hamamelid and Caryophyllid Families* (eds K. Kubitzki, J. G. Rohwer & V. Bittrich) Berlin, Heidelberg, New York: Springer-Verlag, pp. 200–3.

Kubitzki, K. 1993f. Gomortegaceae. In *The Families and Genera of Vascular Plants.* Volume II. *Flowering Plants – Dicotyledons. Magnoliid, Hamamelid and Caryophyllid Families* (eds K. Kubitzki, J. G. Rohwer & V. Bittrich) Berlin, Heidelberg, New York: Springer-Verlag, pp. 318–20.

Kubitzki, K. 1993g. Lactoridaceae. In *The Families and Genera of Vascular Plants.* Volume II. *Flowering Plants – Dicotyledons. Magnoliid, Hamamelid and Caryophyllid Families* (eds K. Kubitzki, J. G. Rohwer & V. Bittrich) Berlin, Heidelberg, New York: Springer-Verlag, pp. 359–61.

Kubitzki, K. 1993h. Platanaceae. In *The Families and Genera of Vascular Plants.* Volume II. *Flowering Plants – Dicotyledons. Magnoliid, Hamamelid and Caryophyllid Families* (eds K. Kubitzki, J. G. Rohwer & V. Bittrich) Berlin, Heidelberg, New York: Springer-Verlag, pp. 521–2.

Kubitzki, K. 1998a. Stemonaceae. In *The Families and Genera of Vascular Plants.* Volume III. *Flowering Plants – Monocotyledons. Lilianae (except Orchidaceae)* (ed. K. Kubitzki) Berlin, Heidelberg, New York: Springer-Verlag, pp. 422–5.

Kubitzki, K. 1998b. Typhaceae. In *The Families and Genera of Vascular Plants.* Volume IV. *Flowering Plants – Monocotyledons. Alismatanae and Commelinanae (except Gramineae)* (ed. K. Kubitzki) Berlin, Heidelberg, New York: Springer-Verlag, pp. 457–60.

Kulkarni, A. R. & Patil, K. S. 1977. *Aristolochioxylon prakashii* from the Deccan Intertrappean beds of Wardha district, Maharashtra. *Geophytology*, **7**, 44–9.

Kuprianova, L. A. 1979. On the possibility of the development of tricolpate pollen from monosulcate. *Grana*, **18**, 1–4.

Kuyl, O. S., Muller, J. & Waterbolk, H. T. 1955. The application to palynology of oil geology with special reference of western Venezuela. *Geologie en Mijnbouw*, **17**, 49–75.

Kvaček, J. 1998. Cuticle analysis of gymnosperms of the Bohemian Cenomanian. Prague: PhD Thesis, Department of Palaeontology, Charles University, pp. 1–229.

Kvaček, J. 1999. New data and revision of three gymnosperms from the Cenomanian of Bohemia – *Sagenopteris variabilis* (Velenovský) Velenovský, *Mesenea bohemica* (Corda) comb. n. and *Eretmophyllum obtusum* (Velenovský) comb. n. *Acta Musei Nationalis Pragae, Series B, Historia Naturalis*, **55**, 15–24.

Kvaček, J. & Pacltová, B. 2001. *Bayeritheca hughesii* gen. et sp. nov., a new *Eucommiidites*- bearing pollen organ from the Cenomanian of Bohemia. *Cretaceous Research*, **22**, 695–704.

Kvaček, J. & Eklund, H. 2003. A report on newly recovered reproductive structures from the Cenomanian of Bohemia (Central Europe). *International Journal of Plant Sciences*, **164**, 1021–39.

Kvaček, J. & Herman, A. B. 2004. Monocotyledons from the Early Campanian (Cretaceous) of Grünbach, Lower Austria. *Review of Palaeobotany and Palynology*, **128**, 323–53.

Kvaček, J. & Friis, E. M. 2010. *Zlatkocarpus* gen. nov., a new angiosperm reproductive structure with monocolpate-reticulate pollen from the Late Cretaceous (Cenomanian) of the Czech Republic. *Grana*, **49**, 115–27.

Kvaček, Z. 1992. Lauralean angiosperms in the Cretaceous. *Courier Forschungsinstitut Senckenberg*, **147**, 345–67.

Kvaček, Z., Manum, S. B. & Boulter, M. C. 1994. Angiosperms from the Paleocene of Spitsbergen, including an unfinished work by A.G. Nathorst. *Palaeontographica* B, **232**, 103–28.

Kvaček, Z. 1995. *Limnobiophyllum* Krassilov – a fossil link between the Araceae and the Lemnaceae. *Aquatic Botany*, **50**, 49–61.

Kvaček, Z. & Sakala, J. 1999. Twig with attached leaves, fruits and seeds of *Decodon* (Lythraceae) from the Lower Miocene of northern Bohemia, and implications for the identification of detached leaves and seeds. *Review of Palaeobotany and Palynology*, **107**, 201–22.

Kvaček, Z. 2008. Whole-plant reconstructions in fossil angiosperm research. *International Journal of Plant Sciences*, **169**, 918–27.

Köhler, E. 1980. Zur Pollenmorphologie und systematischen Stellung der Didymelaceae Leandri. *Feddes Repertorium*, **91**, 581–91.

Köhler, E. 1981. Pollen morphology of the West Indian-Central American species of the genus *Buxus* L. (Buxaceae). *Pollen et Spores*, **32**, 37–91.

Köhler, E. & Brückner, P. 1982. Die Pollenmorphologie der afrikanischen *Buxus*- und *Notobuxus*-arten (Buxaceae) und ihre systematische Bedeutung. *Grana*, **21**, 71–82.

Labandeira, C. C. & Sepkoski, J. J. 1993. Insect diversity in the fossil record. *Science*, **261**, 310–15.

Labandeira, C. C., Dilcher, D. L., Davis, D. R. & Wagner, D. L. 1994. Ninety-seven million years of angiosperm-insect association: paleobiological insights into the meaning of coevolution. *Proceedings of the National Academy of Sciences, USA*, **91**, 12 278–82.

Labandeira, C. C. 1998. The role of insects in Late Jurassic to Middle Cretaceous Ecosystems. In *Lower and Middle Cretaceous Terrestrial Ecosystems*, (eds G. S. Lucas, J. I. Kirkland & J. W. Estep) *New Mexico Museum of Natural History and Science Bulletin*, **14**, 105–24.

Lacey, W. S. 1976. Further observations on the Molteno flora of Rhodesia. *Arnoldia*, **36**, 1–14.

Lakhanpal, R. N. 1970. Tertiary floras of India and their bearing on the historical geology of the region. *Taxon*, **19**, 675–94.

Lam, H. J. 1950. Stachyospory and phyllospory as factors in the natural system of Cormophyta. *Svensk Botanisk Tidskrift*, **44**, 517–34.

Łańcucka-Środoniowa, M. 1967. Two new genera: *Hemiptelea* Planch. and *Weigela* Thunb. in the younger Tertiary of Poland. *Acta Palaeobotanica*, **8**, 1–19.

Larson, R. L. 1991. Latest pulse of the Earth: evidence for a mid-Cretaceous super plume. *Geology*, **19**, 547–50.

Larsson, S. G. 1978. *Baltic Amber – a Palaeobiological Study*. Klampenborg: Scandinavian Science Press Ltd.

Lauverjat, J. & Pons, D. 1978. Le gisement sénonien d'Esgueira (Portugal): stratigraphie et flore fossile. In *103e Congrès National des Sociétés Savantes*, vol. sci. II. *Nancy 1978*, pp. 119–37.

Le Thomas, A., Suárez-Cervera, M. & Goldblatt, P. 1996. Deux types polliniques originaux dans le genre *Aristea* (Iridaceae-Nivenioideae): implications phylogéniques. *Grana*, **35**, 87–96.

Leishman, M. R., Westoby, M. & Jurado, E. 1995. Correlates of seed size variation: a comparison among five temperate floras. *Journal of Ecology*, **83**, 517–30.

Leng, Q. & Friis, E. M. 2003. *Sinocarpus decussatus* gen. et sp. nov, a new angiosperm with syncarpous fruits from the Yixian Formation of Northeast China. *Plant Systematics and Evolution*, **241**, 77–88.

Leng, Q., Wu, S. & Friis, E. M. 2003. Angiosperms. In *The Jehol Biota* (eds M. Chang, P. Chen, Y. Wang, Y. Wang & D. Miao) Shanghai: Shanghai Scientific & Technical Publishers, pp. 178–85.

Leng, Q. & Yang, H. 2003. Pyrite framboids associated with the Mesozoic Jehol Biota in northeastern China: implications for microenvironment during early fossilization. *Progress in Natural Science*, **13**, 206–12.

Leng, Q., Schönenberger, J. & Friis, E. M. 2005. Late Cretaceous follicular fruits from southern Sweden with systematic affinities to early diverging eudicots. *Botanical Journal of the Linnean Society*, **148**, 377–407.

Leng, Q. & Friis, E. M. 2006. Angiosperm leaves associated with *Sinocarpus* Leng et Friis infructescences from the Yixian Formation (mid-Early Cretaceous) of NE China. *Plant Systematics and Evolution*, **262**, 173–87.

Les, D. H. 1993. Ceratophyllaceae. In *The Families and Genera of Vascular Plants.* Volume II. *Flowering Plants – Dicotyledons. Magnoliid, Hamamelid and Caryophyllid Families* (eds K. Kubitzki, J. G. Rohwer & V. Bittrich) Berlin, Heidelberg, New York: Springer-Verlag, pp. 246–50.

Les, D. H., Cleland, M. A. & Waycott, M. 1997. Phylogenetic studies in Alismatidae, II: evolution of marine angiosperms ('seagrasses') and hydrophily. *Systematic Botany*, **22**, 443–63.

Lesquereux, L. 1868. On some Cretaceous fossil plants from Nebraska. *American Journal of Science*, **46**, 91–105.

Lesquereux, L. 1874. Contributions to the fossil flora of the Western Territories. Part I. The Cretaceous flora. *Report of the United States Geological Survey of the Territories*, **6**, 1–136.

Lesquereux, L. 1878. Contributions to the fossil flora of the Western Territories. Part II. The Tertiary flora. *Report of the United States Geological Survey of the Territories*, **7**, 1–336.

Lesquereux, L. 1883. Contributions to the fossil flora of Western Territories. Part III. The Cretaceous and Tertiary floras. *Report of the United States Geological Survey of the Territories*, **8**, 1–283.

Lesquereux, L. 1892. Flora of the Dakota Group. *Monographs of the United States Geological Survey*, **17**, 1–400.

Lesquereux, L. 1893. The genus *Winchellia*. *American Geologist*, **12**, 209–13.

Levina, A. P., Zhelezko, V. I., Leiptsig, A. B. *et al.* 1990. The Sokolov and Sarbay ironmine quarries. In *Upper Cretaceous Deposits of the Southern Transural (The Region of the Upper Current of the Tobol River)* (eds G. N. Papulov, V. I. Zhelezko & A. P. Levina) Sverdlovsk: Akademia Nauk SSSR, pp. 46–58 (in Russian).

Li, H., Taylor, E. L. & Taylor, T. N. 1996. Permian vessel elements. *Science*, **271**, 188–9.

Li, H. 2003. Lower Cretaceous angiosperm leaf from Wuhe in Anhui, China. *Chinese Science Bulletin*, **48**, 611–14.

Li, H.-Q. 2005. Early Cretaceous sarraceniacean-like pitcher plants from China. *Acta Botanica Gallica*, **152**, 227–34.

Li, R.-Q., Chen, Z.-D., Lu, A.-M. *et al.* 2004. Phylogenetic relationships in Fagales based on DNA sequences from three genomes. *International Journal of Plant Sciences*, **165**, 311–24.

Li, W. & Liu, Z. 1994. The Cretaceous palynofloras and their bearing on stratigraphic correlation in China. *Cretaceous Research*, **15**, 333–65.

Li, X. (ed.) 1995. *Fossil Floras of China through the Geological Ages*. Guangzhou: Guangdong Science and Technology Press.

Lidgard, S. & Crane, P. R. 1988. Quantitative analyses of the early angiosperm radiation. *Nature*, **331**, 344–6.

Lidgard, S. & Crane, P. R. 1990. Angiosperm diversification and Cretaceous floristic trends: a comparison of palynofloras and leaf macrofloras. *Paleobiology*, **16**, 77–93.

Lignier, O. 1894. *Structure et Affinités du Bennettites morierei Sap. & Mar. (sp.)*. Caen: E. Lanier.

Lignier, O. 1911. Le *Bennettites morierei* (Sap. et Mar.) Lignier se reproduisait probablement par parthénogénèse. *Bulletin de la Société Botanique de France*, **11**, 224–7.

Lima, M. R. de 1978. Palinologia da Formação Santana (Cretáceo do Nordeste do Brasil). Introdução geologica e descrição sistemática dos polens da subturma Azonotriletes. *Ameghiniana*, **15**, 333–65.

Lima, M. R. de 1979. Palinologia da Formação Santana (Cretáceo do Nordeste do Brasil). II. Descrição sistemática dos esporos da subturma Zonotriletes e turma Monoletes, e dos polens das turmas Saccites e Aletes. *Ameghiniana*, **16**, 27–63.

Lima, M. R. de 1980. Palinologia da Formação Santana (Cretáceo do Nordeste do Brasil). III. Descrição sistemática dos polens da turma Plicates (subturma Costates). *Ameghiniana*, **17**, 15–47.

Lima, W. de 1900. Notícia sobre alguns vegetaes fósseis da flora senoniana (sensu lato) do solo portuguez. *Communicações da Direcção dos Trabalhos Geologicos de Portugal*, **4**, 1–12.

Linder, H. P. & Ferguson, I. K. 1985. On the pollen morphology and phylogeny of the Restionales and Poales. *Grana*, **24**, 65–76.

Linder, H. P. 1987. The evolutionary history of the Poales/Restionales – a hypothesis. *Kew Bulletin*, **42**, 297–318.

Linder, H. P., Briggs, B. G. & Johnson, L. A. S. 1998. Restionaceae. In *The Families and Genera of Vascular Plants*. Volume IV. *Flowering Plants – Monocotyledons. Alismatanae and Commelinanae (except Gramineae)* (ed. K. Kubitzki) Berlin, Heidelberg, New York: Springer-Verlag, pp. 425–45.

Lindfors, S. M., Csikic, Z., Grigorescu, D. & Friis, E. M. 2010. Preliminary account of plant mesofossils from the Maastrichtian Budurone microvertebrate site of the Haţeg Basin, Romania. *Palaeogeography, Palaeoclimatology and Palaeoecology*, **293**, 353–9.

Linnaeus, C. 1735. *Systema Naturae*. Leiden: Lugduni batavorum.

Linnaeus, C. 1758. *Systema Naturae*. Stockholm: Laurentii Salvii.

Lipka, T. R., Therrien, F., Weishampel, D. B. *et al.* 2006. A new turtle from the Arundel Clay facies (Potomac Formation, Early Cretaceous) of Maryland, U.S.A. *Journal of Vertebrate Paleontology*, **26**, 300–7.

Little, S. A. & Stockey, R. A. 2003. Vegetative growth of *Decodon allenbyensis* (Lythraceae) from the Middle Eocene Princeton Chert with anatomical comparisons to *Decodon verticillatus*. *International Journal of Plant Sciences*, **164**, 453–69.

Liu, H.-M., Ferguson, D. K., Hueber, F. M., Li, C.-S. & Wang, Y.-F. 2008. Taxonomy and systematics of *Ephedrites cheniae* and *Alloephedra xingxuei* (Ephedraceae) *Taxon*, **57**, 577–82.

Loconte, H. & Stevenson, D. W. 1990. Cladistics of the Spermatophyta. *Brittonia*, **42**, 197–211.

Loconte, H. 1993. Berberidaceae. In *The Families and Genera of Vascular Plants*. Volume II. *Flowering Plants – Dicotyledons. Magnoliid, Hamamelid and Caryophyllid Families* (eds K. Kubitzki, J. G. Rohwer & V. Bittrich) Berlin, Heidelberg, New York: Springer-Verlag, pp. 147–52.

Louchart, A., Tourment, N., Carrier, J., Roux, T. & Mourer-Chauviré, C. 2008. Hummingbird with modern feathering: an exceptionally well-preserved Oligocene fossil from southern France. *Naturwissenschaften*, **95**, 171–5.

Ludvigsen, R. & Beard, G. 1997. *West Coast Fossils*. Madeira Park, British Columbia: Harbour Publishing.

Lumbert, S. H., den Hartog, C., Phillips, R. C. & Olson, F. S. 1984. The occurrence of fossil seagrasses in the Avon Park Formation (late Middle Eocene), Levy County, Florida (U.S.A.). *Aquatic Botany*, **20**, 121–9.

Lundegren, A. 1931. De kretaceiska ler- och sandförekomsterna N om Ivösjön (Mit einer Zusammenfassung in deutscher Sprache). *Geologiska Föreningens i Stockholm Förhandlingar*, **53**, 298–320.

Lundegren, A. 1934. Kristianstadsområdets kritbildningar (Mit einer Zusammenfassung in deutscher Sprache). *Geologiska Föreningens i Stockholm Förhandlingar*, **56**, 125–313.

Luo, Y. B. & Li, Z. Y. 1999. Pollination ecology of *Chloranthus serratus* (Thunb.) Roem. et Schult. and *Ch. fortunei* (A. Gray) Solms-Laub. (Chloranthaceae). *Annals of Botany*, **83**, 489–99.

Lupia, R. 1995. Paleobotanical data from fossil charcoal: an actualistic study of seed plant reproductive structures. *Palaios*, **10**, 465–77.

Lupia, R. 1999. Discordant morphological disparity and taxonomic diversity during the Cretaceous angiosperm radiation: North American pollen record. *Paleobiology*, **25**, 1–28.

Lupia, R., Lidgard, S. & Crane, P. R. 1999. Comparing palynological abundance and diversity: implications for biotic replacement during the Cretaceous angiosperm radiation. *Paleobiology*, **25**, 305–40.

Lupia, R., Crane, P. R. & Lidgard, S. L. 2000. Angiosperm diversification and mid-Cretaceous environmental change. In *Biotic Responses to Global Change: The Last 145 Million Years* (eds S. J. Culver & P. F. Rawson) Cambridge: Cambridge University Press, pp. 207–22.

Lupia, R., Herendeen, P. S. & Keller, J. A. 2002. A new fossil flower and associated coprolites: evidence for angiosperm-insect interactions in the Santonian (Late Cretaceous) of Georgia, U.S.A. *International Journal of Plant Sciences*, **163**, 675–86.

Maas-van de Kamer, H. 1998. Burmanniaceae. In *The Families and Genera of Vascular Plants.* Volume III. *Flowering Plants – Monocotyledons. Lilianae (except Orchidaceae)* (ed. K. Kubitzki) Berlin, Heidelberg, New York: Springer-Verlag, pp. 154–64.

Mabberley, D. J. 1997. *The Plant-Book. A Portable Dictionary of the Higher Plants.* Cambridge: Cambridge University Press.

MacGinitie, H. D. 1953. Fossil plants of the Florrissant Beds, Colorado. *Publications of the Carnegie Institution, Washington*, **599**, 1–197.

MacGinitie, H. D. 1969. The Eocene Green River flora of northwestern Colorado and northeastern Utah. *University of California Publications in Geological Sciences*, **83**, 1–140.

MacGinitie, H. D. 1974. An early middle Eocene flora from the Yellowstone-Absaroka Volcanic province, northwestern Wind River Basin, Wyoming. *University of California Publications in Geological Sciences*, **108**, 1–103.

MacLeod, K. G., Huber, B. T. & Ducharme, M. L. 2000. Paleontological and geochemical constraints on the deep ocean during the Cretaceous greenhouse interval. In *Warm Climates in Earth History* (eds B. T. Huber, K. G. MacLeod & S. L. Wing) Cambridge: Cambridge University Press, pp. 241–74.

MacNeal, D. L. 1958. The flora of the Upper Cretaceous Woodbine Sand in Denton County, Texas. *Monographs of the Academy of Natural Sciences of Philadelphia*, **10**, 1–152.

Macphail, M. K., Alley, N. F., Truswell, E. M. & Sluiter, I. R. K. 1994. Early Tertiary vegetation: evidence from spores and pollen. In *History of the Australian Vegetation: Cretaceous to Recent* (ed. R. S. Hill) Cambridge: Cambridge University Press, pp. 189–261.

Macphail, M. K. & Cantrill, D. J. 2006. Age and implications of the Forest Bed, Falkland Islands, southwest Atlantic: evidence from fossil pollen and spores. *Palaeogeography, Palaeoclimatology, Palaeoecology*, **240**, 602–29.

Magallón-Puebla, S., Herendeen, P. S. & Endress, P. K. 1996. *Allonia decandra*: floral remains of the tribe Hamamelideae (Hamamelidaceae) from Campanian strata of southeastern USA. *Plant Systematics and Evolution*, **202**, 177–98.

Magallón-Puebla, S., Herendeen, P. S. & Crane, P. R. 1997. *Quadriplatanus georgianus* gen. et sp. nov.: staminate and pistillate platanaceous flowers from the Late Cretaceous (Coniacian-Santonian) of Georgia, USA. *International Journal of Plant Sciences*, **158**, 373–94.

Magallón, S., Crane, P. R. & Herendeen, P. S. 1999. Phylogenetic pattern, diversity and diversification of eudicots. *Annals of the Missouri Botanical Garden*, **86**, 297–372.

Magallón, S., Herendeen, P. S. & Crane, P. R. 2001. *Androdecidua endressii* gen. et sp. nov., from the Late Cretaceous of Georgia (United States): Further floral diversity in Hamamelidoideae (Hamamelidaceae) *International Journal of Plant Sciences*, **162**, 963–83.

Magallón, S. & Sanderson, M. J. 2001. Absolute diversification rates in angiosperm clades. *Evolution*, **55**, 1762–80.

Maguire, B., Wurdack, J. J. & Huang, Y.-C. 1974. Pollen grains of some American Olacaceae. *Grana Palynologica*, **14**, 26–38.

Mahabale, T. S. 1950. Central Provinces Mohgaon Kalan, (Chhindawa District). Palaeobotany in India, 7. *Journal of the Indian Botanical Society*, **29**, 31–3.

Mai, D. H. 1964. Die Mastixioideen Floren im Tertiär der Oberlauzitz. *Paläontologische Abhandlungen Abteilung B, Paläobotanik*, **2**, 1–192.

Mai, D. H. 1970. Subtropische Elemente im europäischen Tertiär I. Die Gattungen *Gironniera*, *Sarcococca*, *Illicium*, *Evodia*, *Ilex*, *Mastixia*, *Alangium*, *Symplocos* und *Rehderodendron*. *Paläontologische Abhandlungen Abteilung B, Paläobotanik*, **3**, 441–503.

Mai, D. H. 1976. Fossile Früchte und Samen aus dem Mitteleozän des Geiseltales. *Abhandlungen des Zentralen geologischen Instituts*, **26**, 93–149.

Mai, D. H. & Walther, H. 1978. Die Floren der Haselbacher Serie im Weisselster-Becken (Bezirk Leipzig, DDR). *Abhandlungen des Staatlichen Museums für Mineralogie und Geologie zu Dresden*, **28**, 1–200.

Mai, D. H. 1980. Zur Bedeutung von Relikten in der Florengeschichte. *100 Jahre Arboretum (1879–1979)*, 281–307.

Mai, D. H. 1985a. Entwicklung der Wasser- und Sumpfpflanzen-Gesellschaften Europas von der Kreide bis ins Quartär. *Flora*, **176**, 449–511.

Mai, D. H. 1985b. Beiträge zur Geschichte einiger holziger Saxifragales-Gattungen. *Gleditschia*, **13**, 75–88.

Mai, D. H. & Walther, H. 1985. Die obereozänen Floren des Weisselster-Beckens (Bezirk Leipzig, DDR). *Abhandlungen des Staatlichen Museums für Mineralogie und Geologie zu Dresden*, **33**, 1–220.

Mai, D. H. 1987a. Neue Früchte und Samen aus paläozänen Ablagerungen Mitteleuropas. *Feddes Repertorium*, **98**, 197–229.

Mai, D. H. 1987b. Neue Arten nach Früchten und Samen aus dem Tertiär von Nordwestsachsen und der Lausitz. *Feddes Repertorium*, **98**, 105–26.

Mai, D. H. & Walther, H. 1991. Die oligozänen und untermiozänen Floren NW-Sachsens und des Bitterfelder Raumes. *Abhandlungen des Staatlichen Museums für Mineralogie und Geologie zu Dresden*, **38**, 1–230.

Mai, D. H. 1995. *Tertiäre Vegetationsgeschichte Europas*. Jena, Stuttgart, New York: Gustav Fischer Verlag.

Mai, D. H. 2001. The fossils of *Rhodoleia* Champion (Hamamelidaceae) in Europe. *Acta Palaeobotanica*, **41**, 161–75.

Maisey, J. G. 1991. *Santana Fossils: An Illustrated Atlas*. Neptune City, NJ: TFH.

Manchester, S. R. 1980. *Chattawaya* (Sterculiaceae): a new genus of wood from the Eocene of Oregon and its implication for the xylem evolution of the extant genus *Pterospermum*. *American Journal of Botany*, **67**, 59–67.

Manchester, S. R. 1986. Vegetative and reproductive morphology of an extinct plane tree (Platanaceae) from the Eocene of Western North America. *Botanical Gazette*, **147**, 200–26.

Manchester, S. R., Dilcher, D. L. & Tidwell, W. D. 1986. Interconnected reproductive and vegetative remains of *Populus* (Salicaceae) from the Middle Eocene Green River Formation, northeastern Utah. *American Journal of Botany*, **73**, 156–60.

Manchester, S. R. 1987a. The fossil history of the Juglandales. *Monographs in Systematic Botany*, **21**, 1–137.

Manchester, S. R. 1987b. Extinct ulmaceous fruits from the Tertiary of Europe and western North America. *Review of Palaeobotany and Palynology*, **52**, 119–29.

Manchester, S. R. & Meyer, H. W. 1987. Oligocene fossil plants of the John Day Formation. Fossil, Oregon. *Oregon Geology*, **49**, 115–26.

Manchester, S. R. 1989. Systematics and fossil history of the Ulmaceae. In *Evolution, Systematics and Fossil History of the Hamamelidae*. Volume 2. *'Higher' Hamamelidae* (eds P. R. Crane & S. Blackmore) Oxford: Clarendon Press, pp. 221–51.

Manchester, S. R., Crane, P. R. & Dilcher, D. L. 1991. *Nordenskioldia* and *Trochodendron* (Trochodendraceae) from the Miocene of Northwestern North America. *Botanical Gazette*, **152**, 357–68.

Manchester, S. R. & Kress, W. J. 1993. Fossil bananas (Musaceae): *Ensete oregonense* sp. nov. from the Eocene of western North America and its phytogeographic significance. *American Journal of Botany*, **80**, 1264–72.

Manchester, S. R. 1994. Fruits and seeds of the Middle Eocene Nut Beds Flora, Clarno Formation, Oregon. *Palaeontographica Americana*, **58**, 1–205.

Manchester, S. R. & Donoghue, M. J. 1995. Winged fruits of Linnaeeae (Caprifoliaceae) in the Tertiary of Western North America: *Diplodipelta* gen. nov. *International Journal of Plant Sciences*, **156**, 709–22.

Manchester, S. R. 1999. Biogeographical relationships of North American Tertiary floras. *Annals of the Missouri Botanical Garden*, **86**, 472–522.

Manchester, S. R. & Dillhoff, R. M. 2004. *Fagus* (Fagaceae) fruits, foliage, and pollen from the Middle Eocene of Pacific Northwestern North America. *Canadian Journal of Botany*, **82**, 1509–17.

Manchester, S. R. & Renner, S. S. 2005. Paleobotany and former range of the Central American relict family Ticodendraceae. In *XVII International Botanical Congress (Abstracts)*. Vienna, Austria, p. 135.

Manchester, S. R. & Chen, I. 2006. *Tetracentron* fruits from the Miocene of Western North America. *International Journal of Plant Sciences*, **167**, 601–5.

Manos, P. S. & Stone, D. E. 2001. Evolution, phylogeny, and systematics of the Juglandaceae. *Annals of the Missouri Botanical Garden*, **88**, 231–69.

Manos, P. S., Soltis, P. S., Soltis, D. E. *et al.* 2007. Phylogeny of extant and fossil Juglandaceae inferred from the integration of molecular and morphological data sets. *Systematic Biology*, **56**, 412–30.

Mapes, G. & Rothwell, G. W. 1984. Permineralized ovulate cones of *Lebachia* from the Late Paleozoic limestones of Kansas. *Palaeontology*, **27**, 69–94.

Markgraf, F. 1926. Ephedraceae. In *Die Natürlichen Pflanzenfamilien*, Volume **13** (eds H. G. A. Engler *et al.*) Leipzig and Berlin: W. Engelmann, pp. 409–19.

Marshall, L. G. 1980. Marsupial paleobiography. In *Aspects of Vertebrate History: Essays in Honor of Edwin Harris Colbert* (ed. L. L. Jacobs) Flagstaff, A2: Museum of Northern Arizona Press, pp. 345–86.

Martens, P. 1971. *Les Gnétophytes. Handbuch der Pflanzenanatomie*. Berlin, Stuttgart: Gebrüder Borntraeger.

Martill, D. M., Brito, P. M., Wenz, S. & Wilby, P. R. 1993. Fossils of the Santana and Crato Formations, Brazil.

In *Palaeontological Association Field Guides to Fossils Series, 5* (ed. E. A. Jarzembowski) London: The Palaeontological Association, pp. 1–159.

Martill, D. M. 2007. The geology of the Crato Formation. In *The Crato Fossil Beds of Brazil: Window into an Ancient World* (eds D. M. Martill, G. Bechly & R. F. Loveridge) Cambridge: Cambridge University Press, pp. 8–24.

Martill, D. M., Bechly, G. & Loveridge, R. F. (eds) 2007. *The Crato Fossil Beds of Brazil: Window into an Ancient World.* Cambridge: Cambridge University Press.

Martin, H. A. 1994. Australian Tertiary phytogeography: evidence from palynology. In *History of the Australian Vegetation. Cretaceous to Recent* (ed. R. S. Hill) Cambridge: Cambridge University Press, pp. 104–42.

Martin, W., Gieri, A. & Saedler, H. 1989. Molecular evidence for pre-Cretaceous angiosperm origins. *Nature*, **339**, 46–8.

Martin, W., Lydiate, D., Brinkmann, H. *et al.* 1993. Molecular phylogenies in angiosperm evolution. *Molecular Biology and Evolution*, **10**, 140–62.

Martinez-Cabrera, H. I. & Cevallos-Ferriz, S. R. S. 2006. *Maclura* (Moraceae) wood from the Miocene of the Baja California Peninsula, Mexico: fossil and biogeographic history of its closer allies. *Review of Palaeobotany and Palynology*, **140**, 113–22.

Martínez-Millán, M., Crepet, W. L. & Nixon, K. C. 2009. *Pentapetalum trifasciculandricus* gen. et sp. nov., a thealean fossil flower from the Raritan Formation, New Jersey, USA (Turonian, Late Cretaceous). *American Journal of Botany*, **96**, 933–49.

Maslova, N. P. & Golovneva, L. B. 2000a. A hamamelid inflorescence with *in situ* pollen grains from the Cenomanian of eastern Siberia. *Paleontological Journal*, **34**, (Suppl.) 1, S40–S49.

Maslova, N. P. & Golovneva, L. B. 2000b. *Lindacarpa* gen. nov., a new hamamelid fructification from the Upper Cretaceous of eastern Siberia. *Paleontological Journal*, **34**, 462–8.

Maslova, N. P. 2003. Extinct and extant Platanaceae and Hamamelidaceae: morphology, systematics, and phylogeny. *Paleontological Journal*, **37**, (Suppl.) 5, S467–S590.

Maslova, N. P. & Kodrul, T. M. 2003. New platanaceous inflorescence *Archaranthus* gen. nov. from the Maastrichtian-Paleocene of the Amur Region. *Paleontological Journal*, **37**, 89–98.

Maslova, N. P. & Herman, A. 2004. New finds of fossil hamamelids and data on the phylogenetic relationships between the Platanaceae and Hamamelidaceae. *Paleontological Journal*, **38**, 563–75.

Maslova, N. P., Golovneva, L. B. & Tekleva, M. V. 2005. Infructescences of *Kasicarpa* gen. nov. (Hamamelidales) from the Late Cretaceous (Turonian) of the Chulym-Enisey Depression, Western Siberia, Russia. *Acta Palaeobotanica*, **45**, 121–37.

Maslova, N. P. & Herman, A. B. 2006. Infructescences of *Friisicarpus* nom. nov. (Platanaceae) and associated foliage of the platanoid type from the Cenomanian of Western Siberia. *Palaeontological Journal*, **40**, 109–13.

Mathews, S. 2009. Phylogenetic relationships among seed plants: persistent questions and the limits of molecular data. *American Journal of Botany*, **96**, 228–36.

Mathews, S. 2010. A duplicate gene rooting of seed plants and the phylogenetic position of flowering plants. *Philosophical Transactions of the Royal Society* B, **365**, 383–95.

Matthews, M. L., Endress, P. K., Schönenberger, J. & Friis, E. M. 2001. A comparison of floral structures of Anisophylleaceae and Cunoniaceae and the problem of their systematic position. *Annals of Botany*, **88**, 439–55.

Matthews, M. L. & Endress, P. K. 2005a. Comparative floral structure and systematics in Celastrales (Celastraceae, Parnassiaceae, Lepidobotryaceae). *Botanical Journal of the Linnean Society*, **149**, 129–94.

Matthews, M. L. & Endress, P. K. 2005b. Comparative floral structure and systematics in Crossosomatales (Crossosomataceae, Stachyuraceae, Staphyleaceae, Aphloiaceae, Geissolomataceae, Ixerbaceae, and Strasburgeriaceae). *Botanical Journal of the Linnean Society*, **147**, 1–46.

Matthews, M. L. & Endress, P. K. 2006. Floral structure and systematics in four orders of rosids, including a broad survey of floral mucilage cells. *Plant Systematics and Evolution*, **260**, 199–221.

May, F. E. 1975. *Dichastopollenites reticulatus* gen. et sp. nov. – potential Cenomanian guide fossil from southern Utah and northeastern Arizona. *Journal of Paleontology*, **49**, 528–33.

Mayo, S. J., Bogner, J. & Boyce, P. C. 1997. *The Genera of the Araceae*. Kew: Royal Botanic Gardens.

Mayr, G. 2004. Old World fossil record of modern-type hummingbirds. *Science*, **304**, 861–4.

Mayr, G. & Manegold, A. 2004. The oldest European fossil songbird from the early Oligocene of Germany. *Naturwissenschaften*, **90**, 173–7.

McCarthy, P. (ed.) 1995. *Flora of Australia*. Volume 16. *Elaeagnaceae, Proteaceae 1*. Melbourne: CSIRO Australia.

McIntyre, D. J. 1968. Further new pollen species from New Zealand Tertiary and uppermost Cretaceous deposits. *New Zealand Journal of Botany*, **6**, 177–204.

McLachlan, I. R. & Pieterse, A. 1978. Preliminary palynological results: Site 361, Leg 40, Deep Sea Drilling Project. *Initial Reports Deep Sea Drilling Project*, **40**, 857–81.

McLoughlin, S. 1990. Some Permian glossopterid fructifications and leaves from the Bowen Basin, Queensland, Australia. *Review of Palaeobotany and Palynology*, **62**, 11–40.

McLoughlin, S. & Drinnan, A. N. 1995. A Middle Jurassic flora from the Walloon Coal Measures, Mutdapilly, Queensland, Australia. *Memoirs of the Queensland Museum*, **38**, 257–72.

McLoughlin, S., Drinnan, A. N. & Rozefelds, A. C. 1995. A Cenomanian flora from the Winton Formation, Eromanga Basin, Queensland. *Memoirs of the Queensland Museum*, **38**, 273–313.

McLoughlin, S. 1996. Early Cretaceous macrofloras of Western Australia. *Records of the Western Australian Museum*, **18**, 19–65.

McLoughlin, S., Tosolini, A.-M., Nagalingum, N. & Drinnan, A. N. 2002. The Early Cretaceous (Neocomian) flora and fauna of the lower Strzelecki Group, Gippsland Basin, Victoria. *Memoirs of the Association of Australasian Palaeontologists*, **26**, 1–144.

McLoughlin, S., Pott, C. & Elliott, D. 2010. The Winton Formation flora (Albian–Cenomanian, Eromanga Basin): implications for vascular plant diversification and decline in the Australian Cretaceous. *Alcheringa*, **34**, 303–23.

McNamara, K. J. & Scott, J. K. 1983. A new species of *Banksia* (Proteaceae) from the Eocene Merlinleigh Sandstone of the Kennedy Range, Western Australia. *Alcheringa*, **7**, 185–93.

McQueen, D. C. 1956. Leaves of Middle and Upper Cretaceous pteridophytes and cycads from New Zealand. *Transactions of the Royal Society of New Zealand*, **83**, 673–85.

Medus, J. & Berthou, P. Y. 1980. Palynoflores dans la coupe de l'Albien de Foz do Folcão (Portugal). *Geobios*, **13**, 263–9.

Medus, J. 1981. Pollens Normapolles de coupes stratotypiques du Crétacé supérieur des Charentes et du Sénonien du Portugal. *Comunicações dos Serviços Geológicos de Portugal*, **67**, 19–28.

Medus, J. 1987. Analyse quantitative des palynoflores du Campanien de Sedano, Espagne. *Review of Palaeobotany and Palynology*, **51**, 309–26.

Meeuse, A. D. J. 1972a. Facts and fiction in floral morphology with special reference to the Polycarpicae 2. Interpretation of the floral morphology of various taxonomic groups. *Acta Botanica Neerlandica*, **21**, 235–52.

Meeuse, A. D. J. 1972b. Facts and fiction in floral morphology with special reference to the Polycarpicae 1. A general survey. *Acta Botanica Neerlandica*, **21**, 113–27.

Meeuse, A. D. J. 1972c. Facts and fiction in floral morphology with special reference to the Polycarpicae 3. Consequences and various additional aspects of the anthocorm theory. *Acta Botanica Neerlandica*, **21**, 351–65.

Meeuse, A. D. J. 1986. *Anatomy of Morphology.* Leiden: Brill.

Meeuse, A. D. J. 1987. *All About Angiosperms.* Delft: Eburon, CIP-Gegevens Koninklijke Bibliotheek, Den Haag.

Meeuse, A. D. J., De Meijer, A. H., Mohr, O. W. P. & Wellinga, S. M. 1990. Entomophily in the dioecious gymnosperm *Ephedra aphylla* Forsk. (= *E. alte* C.A. Mey.), with some notes on *Ephedra compylopoda* C.A. Mey. III. Further anthecological studies and relative importance of entomophily. *Israel Journal of Botany*, **39**, 113–23.

Meeuse, B. J. D. & Raskin, I. 1988. Sexual reproduction in the arum lily family, with emphasis on thermogenicity. *Sexual Plant Reproduction*, **1**, 3–15.

Meijer, J. J. F. 2000. Fossil woods from the Late Cretaceous Aachen Formation. *Review of Palaeobotany and Palynology*, **112**, 297–336.

Melchior, H. 1964. *A. Engler's Syllabus der Pflanzenfamilien mit besonderer Berücksichtigung der Nutzpflanzen nebst einer Übersicht über die Florenreiche und Florengebiete der Erde. II. Angiospermen. Übersicht über die Florengebiete der Erde.* Berlin-Nikolassee: Gebrüder Borntraeger.

Meléndez, M. N. (ed.) 1995. *Las Hoyas, a Lacustrine Konservat-Lagerstätte, Cuenca, Spain.* Madrid: Universidad Complutense de Madrid.

Melville, R. 1962. A new theory of the angiosperm flower. I. The gynoecium. *Kew Bulletin*, **16**, 1–50.

Melville, R. 1963. A new theory of the angiosperm flower. II. The androecium. *Kew Bulletin*, **17**, 1–63.

Melville, R. 1983a. Glossopteridae, Angiospermidae and the evidence for angiosperm origin. *Botanical Journal of the Linnean Society*, **86**, 279–323.

Melville, R. 1983b. Two new genera of Glossopteridae. *Botanical Journal of the Linnean Society*, **86**, 275–7.

Mendes, M. M., Friis, E. M. & Pais, J. 2008a. *Erdtmanispermum juncalense* sp. nov., a new species of the extinct order Erdtmanithecales from the Early Cretaceous (Berriasian) of Portugal. *Review of Palaeobotany and Palynology*, **149**, 50–6.

Mendes, M. M., Pais, J. & Friis, E. M. 2008b. *Raunsgaardispermum lusitanicum* gen. et sp. nov., a new seed with *in situ* pollen from the Early Cretaceous (probably Berriasian) of Portugal: further support for the Bennettitales-Erdtmanithecales-Gnetales link. *Grana*, **47**, 211–19.

Mendes, M. M., Pais, J., Pedersen, K. R. & Friis, E. M. 2010. *Erdtmanitheca portucalensis*, a new pollen organ from the Early Cretaceous (Aptian-Albian) of Portugal with *Eucommiidites*-type pollen. *Grana*, **49**, 26–36.

Meng, J., Hu, Y., Wang, Y., Wang, X. & Li, C. 2006. A Mesozoic gliding mammal from northeastern China. *Nature*, **444**, 889–93.

Meyen, S. V. 1984. Basic features of gymnosperm systematics and phylogeny as evidenced by the fossil record. *The Botanical Review*, **50**, 1–111.

Meyen, S. V. 1988. Gymnosperms of the Angara flora. In *Origin and Evolution of Gymnosperms* (ed. C. B. Beck) New York: Columbia University Press, pp. 338–81.

Michener, C. D. & Grimaldi, D. A. 1988. A *Trigona* from Late Cretaceous amber of New Jersey (Hymenoptera: Apidae: Meliponinae). *American Museum Novitates*, **2917**, 1–10.

Mickle, J. E. 1996. *Grexlupus carolinensis*, a new probable lauraceous fruit from the Late Cretaceous of North Carolina. *Journal of the Elisha Mitchell Scientific Society*, **112**, 1–6.

Miki, S. 1955. Nut remains of Juglandaceae in Japan. *Journal of the Institute of Polytechnics, Osaka City University, Series D*, **6**, 131–44.

Miki, S. 1964. Mesozoic flora of *Lycoptera* beds in South Manchuria. *Bulletin of Mukogawa Women's University*, **12**, 13–22.

Mildenhall, D. C. & Crosbie, Y. M. 1979. Some porate pollen from the Upper Tertiary of New Zealand. *New Zealand Journal of Geology and Geophysics*, **22**, 499–508.

Mildenhall, D. C. 1980. New Zealand Late Cretaceous and Cenozoic plant biography: a contribution. *Palaeogeography, Palaeoclimatology, Palaeoecology*, **31**, 197–233.

Mistri, P. B. & Kapgate, D. K. 1990. Report of a winged fruit of the family Combretaceae from the Deccan Intertrappean Beds of Mohgaonkalan (M.P.) India. In *3 IOP Conference. Melbourne*, pp. 93–6.

Mohr, B. A. R. & Friis, E. M. 2000. Early angiosperms from the Aptian Crato Formation (Brazil), a preliminary report. *International Journal of Plant Sciences*, **161** (6 Suppl.), S155–S167.

Mohr, B. A. R. & Rydin, C. 2002. *Trifurcatia flabellata* n. gen. n. sp., a putative monocotyledon angiosperm from the Lower Cretaceous Crato Formation (Brazil). *Mitteilungen des Museums für Naturkunde Berlin, Geowissenschaftliche Reihe*, **5**, 335–44.

Mohr, B. A. R. & Eklund, H. 2003. *Araripia florifera*, a magnoliid angiosperm from the Lower Cretaceous Crato Formation (Brazil). *Review of Palaeobotany and Palynology*, **126**, 279–92.

Mohr, B. A. R. & Bernardes-de-Oliveira, M. E. C. 2004. *Endressinia brasiliana*, a magnolialean angiosperm from the Lower Cretaceous Crato Formation (Brazil). *International Journal of Plant Sciences*, **165**, 1121–33.

Mohr, B. A. R., Bernardes-de-Oliveira, M. E. C., Barale, G. & Ouaja, M. 2006. Palaeogeographic distribution and ecology of *Klitzschophyllites*, an early Cretaceous angiosperm in southern Laurasia and northern Gondwana. *Cretaceous Research*, **27**, 464–72.

Mohr, B. A. R., Bernardes-de-Oliveira, M. E. C. & Loveridge, R. F. 2007. The macrophyte flora of the Crato Formation. In *The Crato Fossil Beds of Brazil: Window into an Ancient World* (eds D. M. Martill, G. Bechly & R. F. Loveridge) Cambridge: Cambridge University Press, pp. 537–65.

Mohr, B. A. R., Bernardes-de-Oliveira, M. E. C. & Taylor, D. W. 2008. *Pluricarpellatia*, a nymphaealean angiosperm from the Lower Cretaceous of northern Gondwana (Crato Formation, Brazil). *Taxon*, **57**, 1147–58.

Moles, A. T., Ackerly, D. D., Webb, C. O. *et al.* 2005. A brief history of seed size. *Science*, **307**, 576–80.

Moncrieff, A. C. M. & Kelly, S. R. A. 1993. Lithostratigraphy of the uppermost Fossil Bluff Group (Early Cretaceous) of Alexander Island, Antarctica: history of an Albian regression. *Cretaceous Research*, **14**, 1–15.

Monteillet, J. & Lappartient, J.-R. 1981. Fruits et graines du Crétacé supérior des carrières de Paki (Senegal). *Review of Palaeobotany and Palynology*, **34**, 331–44.

Moore, M. J., Bell, C. D., Soltis, P. S. & Soltis, D. E. 2007. Using plastid genome-scale data to resolve enigmatic relationships among basal angiosperms. *Proceedings of the National Academy of Sciences, USA*, **104**, 19 363–8.

Mtchedlishvili, N. D. 1961. Triprojectacites. In *Pollen and Spores from Western Siberia, Jurassic-Paleocene*, Volume 177 (ed. S. R. Samoilovitch) Leningrad: Trudy Vses. Neft. Geofiz. Rass. Issled., Akademia Nauk, pp. 203–99.

Mtchedlishvili, N. D. & Shakhmoundes, V. A. 1973. Occurrence of Araceae pollen in the Lower Cretaceous sediments. In *Palynology of Mesophytes* (ed. A. F. Chlonova) Moscow: Nauka, pp. 137–42 (in Russian with English summary).

Mueller, F. 1883. *Observations on New Vegetable Fossils of the Auriferous Drifts.* Second Decade. Melbourne: Geological Survey of Victoria, John Ferres Government Printer.

Muhammad, A. F. & Sattler, R. 1982. Vessel structure of *Gnetum* and the origin of angiosperms. *American Journal of Botany*, **69**, 1004–21.

Mulcahy, D. L. 1979. The rise of the angiosperms: a genecological factor. *Science*, **206**, 20–3.

Muller, J. 1968. Palynology of the Pedawan and Plateau sandstone formations (Cretaceous – Eocene) in Sarawak, Malaysia. *Micropaleontology*, **14**, 1–37.

Muller, J. 1970. Palynological evidence on early differentiation of angiosperms. *Biological Reviews*, **45**, 417–50.

Muller, J. & Caratini, C. 1977. Pollen of *Rhizophora* (Rhizophoraceae) as a guide fossil. *Pollen et Spores*, **19**, 361–89.

Muller, J. 1979. Reflections on fossil palm pollen. *Proceedings of the IV International Palynological Conference, Lucknow (1976–77)*, **1**, 568–78.

Muller, J. 1981. Fossil pollen records of extant angiosperms. *The Botanical Review*, **47**, 1–142.

Muller, J. 1984. Significance of fossil pollen for angiosperm history. *Annals of the Missouri Botanical Garden*, **71**, 419–43.

Mundry, M. & Stützel, T. 2004. Morphogenesis of the reproductive shoots of *Welwitschia mirabilis* and *Ephedra distachya* (Gnetales), and its evolutionary implications. *Organisms, Diversity and Evolution*, **4**, 91–108.

Mädel-Angeliewa, E. & Müller-Stoll, W. R. 1973. Kritische Studien über fossile Combretaceen-Hölzer: Über Hölzer vom Typus *Terminalioxylon* G. Schönfeld mit einer Revision der bisher zu *Evodioxylon* Chiarugi gestellten Arten. *Palaeontographica* B, **142**, 117–36.

Mädel, E. 1960. Monimiacaeen-Hölzer aus der oberkretazischen Umzamba-Schichten von Ost-Pondoland (S-Afrika). *Senckenbergiana Lethaea*, **41**, 331–91.

Mägdefrau, K. 1968. *Paläobiologie der Pflanzen*. Jena: VEB Gustav Fischer Verlag.

Nambudiri, E. M. V. & Tidwell, W. D. 1978. On probable affinities of *Viracarpon* Sahni from the Deccan Intertrappean flora of India. *Palaeontographica* B, **166**, 30–43.

Nandi, O. I., Chase, M. W. & Endress, P. K. 1998. A combined cladistic analysis of angiosperms using *rbc*L and non-molecular data sets. *Annals of the Missouri Botanical Garden*, **85**, 137–212.

Nartshuk, E. P. 1996. A new fossil acrocerid fly from the Jurassic beds of Kazakhstan (Diptera: Acroceridae). *Zoosystematica Rossica*, **4**, 313–15.

Nathorst, A. G. 1890. Ueber die Reste eines Brotfruchtbaums, *Artocarpus dicksoni* n.sp., aus den cenomanen Kreideablagerungen Grönlands. *Kongliga Svenska Ventenskaps-Akademiens Handlingar*, **24**, 1–10.

Nathorst, A. G. 1909a. Über die Gattung *Nilssonia* Brogn. mit besonderer Berücksichtigung schwedisher Arten. *Kungliga Svenska Vetenskabsakademiens Handlingar*, **43**, 1–40.

Nathorst, A. G. 1909b. Paläobotanische Mitteilungen. 8. Über *Williamsonia*, *Wielandia*, *Cycadocephalus* und *Weltrichia*. *Kungliga Svenska Vetenskapsakademiens Handlingar*, **45**, 1–37.

Nathorst, A. G. 1911. Paläobotanische Mitteilungen. No. 9. *Kungliga Svenska Vetenskabsakademiens Handlingar*, **46**, 1–33.

Němejc, F. 1961. Fossil plants from Klikov in S. Bohemia (Senonian). *Rozpravy Československé Věd*, **71**, 1–42.

Němejc, F. & Kvaček, Z. 1975. *Senonian Plant Macrofossils from the Region of Zliv and Hluboká (near Ceské Budejovice) in South Bohemia*. Praha: Univerzita Karlova.

Neumeyer, H. 1924. Die Geschichte der Blüte. *Abhandlungen der Zoologische-Botanischen Gesellschaft in Wien*, **14**, 1–112.

Newberry, J. S. 1868. Notes on the later extinct floras of North America. *American Journal of Science*, **46**, 401–7.

Newberry, J. S. 1895. The flora of the Amboy Clays: edited by A. Hollick. *Monographs of the United States Geological Survey*, **26**, 1–260.

Nichols, D. 1992. Plants at the K/T boundary. *Nature*, **356**, 295.

Nichols, D. J., Jarzen, D. M., Orth, C. J. & Oliver, P. Q. 1986. Palynological and Ir anomalies at Cretaceous-Tertiary boundary, South-Central Saskatchewan. *Science*, **231**, 714–17.

Nichols, D. J., Watabe, M., Ichinnorov, N. & Ariunchimeg, Y. 1997. Preliminary report on the palynology of the Cretaceous of the Gobi Desert, Mongolia. *Proceedings of the Ninth International Palynological Congress, American Association of Stratigraphic Palynologists, Dallas.*

Nichols, D. J. & Johnson, K. R. 2002. Palynology and microstratigraphy of the Hell Creek Formation in North Dakota: a microfossil record of plants at the end of Cretaceous time. In *The Hell Creek Formation and the Cretaceous-Tertiary Boundary in the northern Great Plains: an Integrated Continental Record of the End of the Cretaceous*, Volume **361** (eds J. H. Hartman, K. R. Johnson & D. J. Nichols) Boulder, Colorado: Geological Society Special Paper, pp. 95–143.

Nichols, D. J., Matsukawa, M. & Ito, M. 2006. Palynology and age of some Cretaceous nonmarine deposits in Mongolia and China. *Cretaceous Research*, **27**, 241–51.

Nichols, D. J. & Johnson, K. R. 2008. *Plants and the K-T Boundary*. Cambridge: Cambridge University Press.

Niklas, K. J., Tiffney, B. H. & Knoll, A. H. 1980. Apparent changes in the diversity of fossil plants. In *Evolutionary Biology*, Volume **12** (eds W. C. Steere, M. K. Hecht & B. Wallace) New York: Plenum Publishers, pp. 1–89.

Niklas, K. J. & Norstog, K. 1984. Aerodynamics and pollen grain depositional patterns on cycad megastrobili: implications on the reproduction of three cycad genera (*Cycas*, *Dioon*, and *Zamia*). *Botanical Gazette*, **145**, 92–104.

Niklas, K. J., Buchmann, S. L. & Kerchner, V. 1986. Aerodynamics of *Ephedra trifurca*. I. Pollen grain velocity fields around stems bearing ovules. *American Journal of Botany*, **73**, 966–99.

Nishida, H. 1985. A structurally preserved magnolialean fructification from the mid-Cretaceous of Japan. *Nature*, **318**, 58–9.

Nishida, H. & Nishida, M. 1988. *Protomonimia kasai-nakajhongii* gen. et sp. nov.: a permineralized magnolialean fructification from the mid-Cretaceous of Japan. *Botanical Magazine, Tokyo*, **101**, 397–437.

Nishida, H. 1991. Diversity and significance of Late Cretaceous permineralized plant remains from Hokkaido, Japan. *Botanical Magazine, Tokyo*, **104**, 253–73.

Nishida, H. 1994a. Morphology and the evolution of Cycadeoidales. *Journal of Plant Research*, **107**, 479–92.

Nishida, H. 1994b. *Elsemaria*, a Late Cretaceous angiosperm fructification from Hokkaido, Japan. *Plant Systematics and Evolution* (Suppl.), **8**, 123–35.

Nishida, H., Pigg, K. B. & Rigby, J. F. 2003. Swimming sperm in an extinct Gondwanan plant. *Nature*, **422**, 396–7.

Nishida, M. 1984. The anatomy and the affinities of the petrified plants from the Tertiary of Chile. IV. Dicotyledonous wood from Quiriquina Island, near Concepción. In *Contribution to the Botany of the Andes* (ed. M. Nishida) Tokyo: Academia Science Books, pp. 111–21.

Nishida, M., Nishida, H. & Nasa, T. 1988. Anatomy and affinities of the petrified plants from the Tertiary of Chile. V. *Botanical Magazine, Tokyo*, **101**, 293–309.

Nishida, M., Nishida, H. & Sugiyama, R. 1993. Structure and affinities of the petrified plants from the Cretaceous of Northern Japan and Saghalien. XIV. Coniferous woods from the Upper Cretaceous of Taneichi, Iwate Prefecture. *Research Institute of Evolutionary Biology, Scientific Report*, **7**, 69–86.

Nishida, M., Ohsawa, T., Nishida, H., Yoshida, A. & Kanie, Y. 1996. A permineralized magnolialean fructification from the Upper Cretaceous of Hokkaido, Japan. *Research Institute of Evolutionary Biology, Scientific Report*, **8**, 19–30.

Nixon, K. C. & Crepet, W. L. 1993. Late Cretaceous fossil flowers of ericalean affinity. *American Journal of Botany*, **80**, 616–23.

Nixon, K. C., Crepet, W. L., Stevenson, D. & Friis, E. M. 1994. A reevaluation of seed plant phylogeny. *Annals of the Missouri Botanical Garden*, **81**, 484–533.

Nooteboom, H. P. 1993. Magnoliaceae. In *The Families and Genera of Vascular Plants.* Volume II. *Flowering Plants – Dicotyledons. Magnoliid, Hamamelid and Caryophyllid Families* (eds K. Kubitzki, J. G. Rohwer & V. Bittrich) Berlin, Heidelberg, New York: Springer-Verlag, pp. 391–401.

Norstog, K. & Fawcett, P. K. S. 1989. Insect–cycad symbiosis and its relation to the pollination of *Zamia furfuracea* (Zamiaceae) by *Rhopalotria mollis* (Curculionidae). *American Journal of Botany*, **76**, 1380–94.

Norstog, K. J., Stevenson, D. W. & Niklas, K. J. 1986. The role of beetles in the pollination of *Zamia furfuracea* L. fil. (Zamiaceae). *Biotropica*, **18**, 300–6.

Norstog, K. J. & Nicholls, T. J. 1997. *The Biology of the Cycads.* Ithaca, NY: Cornell University Press.

Novacek, M. J. 1999. 100 million years of land vertebrate evolution: The Cretaceous – Early Tertiary transition. *Annals of the Missouri Botanical Garden*, **86**, 230–58.

Nowicke, J. W. & Skvarla, J. J. 1981. Pollen morphology and phylogenetic relationships of the Berberidaceae. *Smithsonian Contributions to Botany*, **50**, 1–83.

Nowicke, J. W. & Skvarla, J. J. 1982. Pollen morphology and the relationships of *Circaeaster*, of *Kingdonia*, and of *Sargentodoxa* to the Ranunculaceae. *American Journal of Botany*, **69**, 990–8.

Nowicke, J. W. & Skvarla, J. J. 1984. Pollen morphology and the relationships of *Simmondsia chinensis* to the order Euphorbiales. *American Journal of Botany*, **71**, 210–15.

Nykvist, N. 1957. Kretaceiska vedrester vid Åsen i Skåne. *Svenska Skogsvårdsföreningens Tidskrift*, **55**, 477–81.

O'Dowd, D. J., Brew, C. R., Christophel, D. C. & Norton, R. A. 1991. Mite-plant associations from the Eocene of southern Australia. *Science*, **252**, 99–101.

Oh, I.-C., Denk, T. & Friis, E. M. 2003. Evolution of *Illicium* (Illiciaceae): mapping morphological characters on the molecular tree. *Plant Systematics and Evolution*, **240**, 175–209.

Ohana, T. & Kimura, T. 1995. Late Mesozoic phytogeography in eastern Eurasia, with special reference to the origin of angiosperms in time and site. *Proceedings of the 15th International Symposium of Kyungpook National University*, 293–328.

Ohana, T., Kimura, T. & Chitaley, S. 1999. *Keraocarpon* gen. nov., magnolialean fruits from the Upper Cretaceous of Hokkaido, Japan. *Paleontological Research*, **3**, 294–302.

Okada, H. & Sakai, T. 2000. The Cretaceous system of the Japanese islands and its physical environment. In *Cretaceous Environments of Asia* (eds H. Okada & N. J. Mateer) Amsterdam: Elsevier, pp. 113–44.

Oldham, T. & Morris, J. 1863. The fossil flora of the Rajmahal Series, the Rajmahal Hills, Bengal. *Memoirs of the Geological Survey of India, Palaeontologia Indica, Series 2*, **1**, 1–52.

Oltz, D. F. 1969. Numerical analyses of palynological data from Cretaceous and Early Tertiary sediments in east central Montana. *Palaeontographica* B, **128**, 90–166.

Orth, C. J., Gilmore, J. S., Knight, J. D. *et al.* 1981. An iridium abundance anomaly at the palynological Cretaceous-Tertiary boundary in northern New Mexico. *Science*, **214**, 1341–3.

Osborn, J. M., Taylor, T. N. & Crane, P. R. 1991. The ultrastructure of *Sahnia* pollen (Pentoxylales). *American Journal of Botany*, **78**, 1560–9.

Osborn, J. M., Taylor, T. N. & Lima, M. R. de 1993. The ultrastructure of fossil ephedroid pollen with gnetalean affinities from the Lower Cretaceous of Brazil. *Review of Palaeobotany and Palynology*, **77**, 171–84.

Osborn, J. M. & Taylor, T. N. 1994. Comparative ultrastructure of fossil gymnosperm pollen and its phylogenetic implications. In *Ultrastructure of Fossil Spores and Pollen. Its Bearing on Relationships among Fossil and Living Groups* (eds M. H. Kurmann & J. A. Doyle) Kew: The Royal Botanic Gardens, pp. 99–121.

Owens, J. N., Takaso, T. & Runions, C. J. 1998. Pollination in conifers. *Trends in Plant Science*, **3**, 479–85.

Owens, J. P. & Sohl, N. F. 1989. *Campanian and Maastrichtian Depositional Systems of the Black Creek Group of the Carolinas. Field Trip Guidebook*. Raleigh, NC: Carolina Geological Society.

Oxelman, B., Yoshikawa, N., McConaughy, B. L. *et al.* 2004. RPB2 gene phylogeny in flowering plants, with particular emphasis on asterids. *Molecular Phylogenetics and Evolution*, **32**, 462–79.

Pacltová, B. 1971. Palynological study of Angiospermae from the Peruc Formation (?Albian-Lower Cenomanian) of Bohemia. *Sborník geologických ved, Paleontologie*, **13**, 105–41.

Pacltová, B. 1981. The evolution and distribution of Normapolles pollen during the Cenophytic. *Review of Palaeobotany and Palynology*, **35**, 175–208.

Pacltová, B. 1982. Some pollen of recent and fossil species of the genus *Platanus* L. *Acta Universitatis Carolinae – Geologica*, **4**, 367–91.

Page, C. N. 1990. General traits of Conifers. In *The Families and Genera of Vascular Plants*. Volume I. *Pteridophytes and Gymnosperms* (eds K. Kubitzki, K. U. Kramer & P. S. Green) Berlin, Heidelberg, New York: Springer-Verlag, pp. 290–2.

Page, V. M. 1968. Angiosperm wood from the Upper Cretaceous of central California, Part I. *American Journal of Botany*, **55**, 168–72.

Page, V. M. 1970. Angiosperm wood from the Upper Cretaceous of central California. III. *American Journal of Botany*, **69**, 990–8.

Page, V. M. 1979. Dicotyledonous wood from the Upper Cretaceous of Central California. *Journal of the Arnold Arboretum*, **60**, 323–49.

Pais, J. & Reyre, Y. 1981. Problèmes posés par la population sporo-pollinique d'un niveau à plantes de la série de Buarcos (Portugal). *Boletim da Sociedade Geológica de Portugal*, **22**, 35–40.

Pan, A. D., Jacobs, B. F., Dransfield, J. & Baker, W. J. 2006. The fossil history of palms (Arecaceae) in Africa and new records from the Late Oligocene (28–27 Mya) of north-western Ethiopia. *Botanical Journal of the Linnean Society*, **151**, 69–81.

Pant, D. D. & Nautiyal, D. D. 1960. Some seeds and sporangia of *Glossopteris* flora from Raniganj coalfield, India. *Palaeontographica* B, **107**, 41–64.

Pant, D. D. & Basu, N. 1973. *Pteruchus indicus* sp. nov. from the Triassic of India. *Palaeontographica* B, **144**, 11–24.

Pant, D. D. 1977. The plant of *Glossopteris*. *Journal of the Indian Botanical Society*, **56**, 1–23.

Pant, D. D. & Basu, N. 1979. Some further remains of fructifications from the Triassic of Nidpur, India. *Palaeontographica* B, **168**, 129–46.

Papulov, G. N. 1990. Biostratigrafiya verchnemelobych otlozhenij rajona. In *Verkhnemelovye Otlozhenniya Juzhnogo Zauralya (Rajon Verkhnego Pritobolya)* (eds G. N. Papulov, V. I. Zhelezko & A. P. Levina) Sverdlovsk: Akademiya nauk SSSR – Uralskoe otdelenie, pp. 154–73 (in Russian).

Papulov, G. N., Zhelezko, V. I. & Levina, A. P. (eds) 1990. *Verkhnemelovye Otlozhenniya Juzhnogo Zauralya (Rajon Verkhnego Pritobolya)*. Sverdlovsk: Akademiya nauk SSSR – Uralskoe otdelenie (in Russian).

Parenti, L. R. 1980. A phylogenetic analysis of the land plants. *Biological Journal of the Linnean Society*, **13**, 225–42.

Parrish, J. T., Ziegler, A. M. & Scotese, C. R. 1982. Rainfall patterns and the distribution of coals and evaporites in the Mesozoic and Cenozoic. *Palaeogeography, Palaeoclimatology, Palaeoecology*, **40**, 67–101.

Parrish, J. T. 1987. Global palaeogeography and palaeoclimate of the Late Cretaceous and Early Tertiary. In *The Origins of Angiosperms and their Biological Consequences* (eds E. M. Friis, W. G. Chaloner & P. R. Crane) Cambridge: Cambridge University Press, pp. 51–73.

Passalia, M. G., Romero, E. J. & Panza, J. L. 2001. Improntas foliares del Cretácico de la provincia de Santa Cruz, Argentina. *Ameghiniana*, **38**, 73–84.

Patil, G. V. 1972. *Viracarpon chitaleyi*, sp. nov., from the Deccan Intertrappean beds of Mohgaon Kalan, India. *Botanique*, **3**, 21–6.

Patil, G. V. & Singh, R. B. 1978. Fossil *Eichhornia* from the Eocene Deccan Intertrappean beds, India. *Palaeontographica* B, **167**, 1–7.

Patil, G. V. & Upadhye, E. V. 1984. *Cocos*-like fruit from Mohgaonkalan and its significance towards the stratigraphy of Mohgaonkalan Intertrappean beds. In *Proceedings of the Symposium on Evolutionary Botany and Biostratigraphy, University of Calcutta 1979* (eds A. K. Sharma, G. C. Mitra & M. Banerjee) New Delhi: Today & Tomorrow's Printers & Publishers, pp. 541–54.

Pearson, D. A., Schaefer, T., Johnson, K. R. & Nichols, D. J. 2001. Palynological calibrated vertebrate record from North Dakota consistent with abrupt dinosaur extinction at the Cretaceous-Tertiary boundary. *Geology*, **29**, 39–42.

Pearson, H. H. W. 1929. *Gnetales*. Cambridge: Cambridge University Press.

Pedersen, G. K. & Pulvertaft, T. C. R. 1992. The nonmarine Cretaceous of the West Greenland Basin, onshore West Greenland. *Cretaceous Research*, **13**, 263–72.

Pedersen, K. R. 1976. Fossil floras of Greenland. In *Geology of Greenland* (eds A. Escher & W. S. Watt) Copenhagen: The Geological Survey of Greenland, pp. 519–35.

Pedersen, K. R. & Friis, E. M. 1986. *Caytonanthus* pollen from the Lower and Middle Jurassic. *Geoskrifter*, **24**, 255–67.

Pedersen, K. R., Crane, P. R. & Friis, E. M. 1989a. Morphology and phylogenetic significance of *Vardekloeftia* Harris (Bennettitales). *Review of Palaeobotany and Palynology*, **60**, 7–24.

Pedersen, K. R., Crane, P. R. & Friis, E. M. 1989b. Pollen organs and seeds with *Eucommiidites* pollen. *Grana*, **28**, 279–94.

Pedersen, K. R., Crane, P. R., Drinnan, A. N. & Friis, E. M. 1991. Fruits from the mid-Cretaceous of North America with pollen grains of the *Clavatipollenites* type. *Grana*, **30**, 577–90.

Pedersen, K. R., Friis, E. M. & Crane, P. R. 1993. Pollen organs and seeds with *Decussosporites* Brenner from Lower Cretaceous Potomac Group sediments of eastern USA. *Grana*, **32**, 273–89.

Pedersen, K. R., Friis, E. M., Crane, P. R. & Drinnan, A. N. 1994. Reproductive structures of an extinct platanoid from the Early Cretaceous (latest Albian) of eastern North America. *Review of Palaeobotany and Palynology*, **80**, 291–303.

Pedersen, K. R., von Balthazar, M., Crane, P. R. & Friis, E. M. 2007. Early Cretaceous floral structures and *in situ* tricolpate-striate pollen: New early eudicots from Portugal. *Grana*, **46**, 176–96.

Pelzer, G. & Wilde, V. 1987. Klimatische Tendenzen während der Ablagerung der Wealden-Fazies in Nordwesteuropa. *Geologisches Jahrbuch Reihe A*, **96**, 239–63.

Penny, J. H. J. 1986. An Early Cretaceous angiosperm pollen assemblage from Egypt. *Special Papers in Palaeontology*, **35**, 121–34.

Penny, J. H. J. 1988a. Early Cretaceous striate pollen from the Borehole Mersa Matruh 1, North West Desert, Egypt. *Journal of Micropalaeontology*, **7**, 201–15.

Penny, J. H. J. 1988b. Early Cretaceous acolumellate semitectate pollen from Egypt. *Palaeontology*, **31**, 373–418.

Penny, J. H. J. 1989. New Early Cretaceous forms of the angiosperm pollen genus *Afropollis* from England and Egypt. *Review of Palaeobotany and Palynology*, **58**, 289–99.

Penny, J. H. J. 1991. Early Cretaceous angiosperm pollen from the borehole Mersa Matruh 1, North West Desert, Egypt. *Palaeontographica* B, **222**, 31–88.

Penny, J. H. J. 1992. The relevance of the Early Cretaceous angiosperm palynology of Egypt to biostratigraphy and reconstruction of angiosperm palaeolatitude migrations. *Cretaceous Research*, **13**, 369–78.

Peppe, D. J., Erickson, J. & Hickey, J. 2007. Fossil leaf species from the Fox Hills Formation (Upper Cretaceous: North Dakota, USA) and their paleogeographic significance. *Journal of Paleontology*, **81**, 550–67.

Pereira, S. L. & Baker, A. J. 2006. A mitogenomic timescale for birds detects variable phylogenetic rates of molecular evolution and refutes the standard molecular clock. *Molecular Biology and Evolution*, **23**, 1731–40.

Petriella, B. 1978. La reconstrucción de *Dicroidium* (Pteridospermopsida, Corystospermaceae). *Obra del Centenario del Museo de La Plata*, **5**, 107–10.

Petriella, B. 1981. Sistematica y vinculaciones de las Corystospermaceae H. Thomas. *Ameghiniana*, **18**, 221–34.

Pflug, H. D. 1953. Zur Entstehung und Entwicklung des angiospermiden Pollen in der Erdgeschichte. *Palaeontographica* B, **95**, 60–171.

Philipson, W. R. 1993a. Monimiaceae. In *The Families and Genera of Vascular Plants.* Volume II. *Flowering Plants – Dicotyledons. Magnoliid, Hamamelid and Caryophyllid Families* (eds K. Kubitzki, J. G. Rohwer & V. Bittrich) Berlin, Heidelberg, New York: Springer-Verlag, pp. 426–37.

Philipson, W. R. 1993b. Amborellaceae. In *The Families and Genera of Vascular Plants.* Volume II. *Flowering Plants – Dicotyledons. Magnoliid, Hamamelid and Caryophyllid Families* (eds K. Kubitzki, J. G. Rohwer & V. Bittrich) Berlin, Heidelberg, New York: Springer-Verlag, pp. 92–3.

Philipson, W. R. 1993c. Trimeniaceae. In *The Families and Genera of Vascular Plants.* Volume II. *Flowering Plants – Dicotyledons. Magnoliid, Hamamelid and Caryophyllid Families* (eds K. Kubitzki, J. G. Rohwer & V. Bittrich) Berlin, Heidelberg, New York: Springer-Verlag, pp. 596–9.

Phillips, P. P. & Felix, C. J. 1971. A study of Lower and Middle Cretaceous spores and pollen from the southeastern United States. II. Pollen. *Pollen et Spores*, **13**, 447–73.

Pickering, K. 2000. The Cenozoic world. In *Biotic Response to Global Change* (eds S. J. Culver & P. R. Rawson) Cambridge: Cambridge University Press, pp. 20–34.

Pierce, R. L. 1961. Lower Upper Cretaceous plant microfossils from Minnesota. *Minnesota Geological Survey Bulletin*, **42**, 1–86.

Pigg, K. B. 1990. Anatomically preserved *Glossopteris* foliage from the central Transantarctic Mountains. *Review of Palaeobotany and Palynology*, **66**, 105–27.

Pigg, K. B., Stockey, R. A. & Maxwell, S. L. 1993. *Paleomyrtinaea princetonensis* gen. et sp. nov., permineralized myrtaceous fruits and seeds from the Princeton chert and related Myrtaceae from Almont, North Dakota. *Canadian Journal of Botany*, **71**, 1–9.

Pigg, K. B. & Taylor, T. N. 1993. Anatomically preserved *Glossopteris* stems with attached leaves from the Central Transantarctic Mountains, Antarctica. *American Journal of Botany*, **80**, 500–16.

Pigg, K. B. & Trivett, M. L. 1994. Evolution of the glossopterid gymnosperms from Permian Gondwana. *Journal of Plant Research*, **107**, 461–77.

Pigg, K. B., Ickert-Bond, S. M. & Wen, J. 2004. Anatomically preserved *Liquidambar* (Altingiaceae) from the middle Miocene of Yakima Canyon, Washington State, USA, and its biogeographic implications. *American Journal of Botany*, **91**, 499–509.

Pigg, K. B. & DeVore, M. L. 2005. *Paleoactaea* gen. nov. (Ranunculaceae) fruits from the Paleogene of North Dakota and the London Clay. *American Journal of Botany*, **92**, 1650–9.

Pigg, K. B., Dillhoff, R. M., DeVore, M. L. & Wehr, W. C. 2007. New diversity among the Trochodendraceae from the Early/Middle Eocene Okanogan Highlands of British Columbia, Canada, and Northeastern Washington State, United States. *International Journal of Plant Sciences*, **168**, 521–32.

Pigg, K. B., DeVore, M. L. & Wojciechowski, M. F. 2008. *Paleosecuridaca curtisii* gen. et sp. nov., *Securidaca*-like samaras (Polygalaceae) from the Late Paleocene of North Dakota and their significance to the divergence of families within the Fabales. *International Journal of Plant Sciences*, **169**, 1304–13.

Pignal, M., Lugardon, B., Jérémie, J. & Le Thomas, A. 1999. Morphologie et ultrastructure du pollen des Siparunaceae (Laurales). *Grana*, **38**, 210–17.

Pimenova, N. V. 1954. Sarmatskaya Flora Amvrosievki. *Akademiya Nauk Ukrainskoj SSR. Trudy Instituta Geologicheskikh Nauk Serie Startigrafii i Paleontologii*, **8**, 1–95.

Poinar, G. 2002. Fossil palm flowers in Dominican and Mexican amber. *Botanical Journal of the Linnean Society*, **138**, 57–61.

Poinar, G. & Chambers, K. L. 2005. *Palaeoanthella huangii* gen. et sp. nov., an early Cretaceous flower (Angiospermae) in Burmese amber. *Sida*, **21**, 2087–92.

Poinar, G., Chambers, K. L. & Buckley, R. 2007. *Eoepigynia burmensis* gen. et sp. nov., an Early Cretaceous eudicot flower (Angiospermae) in Burmese amber. *Journal of the Botanical Research Institute of Texas*, **1**, 91–6.

Poinar, G. O. 1992. *Life in Amber.* Stanford, CA: Stanford University Press.

Poinar, G. O. & Chambers, K. L. 2008. An Early Cretaceous angiosperm fossil of possible significance in rosid floral diversification. *Journal of the Botanical Research Institute Texas*, **2**, 1183–92.

Pole, M. 1998. The Proteaceae record in New Zealand. *Australian Systematic Botany*, **11**, 343–72.

Pole, M. 2000. Dicotyledonous leaf macrofossils from the latest Albian-earliest Cenomanian of the Eromanga Basin, Queensland, Australia. *Paleontological Research*, **4**, 39–52.

Pole, M. & Vajda, V. 2009. A new terrestrial Cretaceous-Paleogene site in New Zealand – turnover in macroflora confirmed by palynology. *Cretaceous Research*, **30**, 917–38.

Pole, M. & Philippe, M. 2010. Cretaceous plant fossils of Pitt Island, the Chatham group, New Zealand. *Alcheringa*, **34**, 231–63.

Pole, M. S. 1993. Early Miocene flora of the Manuherikia Group, New Zealand. 5. Smilacaceae, Polygonaceae, and Elaeocarpaceae. *Journal of the Royal Society of New Zealand*, **23**, 289–302.

Pole, M. S. & Douglas, B. J. 1999. Plant macrofossils of the Upper Cretaceous Kaitangata Coalfield, New Zealand. *Australian Systematic Botany*, **12**, 331–64.

Pollux, B. J. A., Ouborg, N. J., Van Groenendael, J. M. & Klaassen, M. 2007. Consequences of intraspecific seed-size variation in *Sparganium emersum* for dispersal by fish. *Functional Ecology*, **21**, 1084–91.

Pons, D. & Broutin, J. 1978. Les organes reproducteurs de *Frenelopsis oligostomata* (Crétacé, Portugal). *Comptes Rendus 103e Congrès National des Sociétés Savantes, Nancy 1978*, pp. 139–59.

Pons, D. 1979. Les organes reproducteurs de *Frenelopsis alata* (K. Feistm) Knobloch, Cheirolepidiaceae du Cenomanien de l'Anjou, France. *Comptes Rendus 104e Congrès National des Sociétés Savantes, Bordeaux 1979*, pp. 209–31.

Pons, D., Lauverjat, J. & Broutin, J. 1980. Paléoclimatologie comparée de deux gisements du Crétacé supérieur d'Europe occidentale. *Mémoires de la Société Géologique de France, nouvelle série*, **139**, 151–8.

Pons, D., Berthou, P. Y. & Almeida-Campos, D. 1990. Upper Aptian and Albian palynology of the Araripe Basin (Brazil): ecological and floristical characteristics. In *International Conference of Late Palaeozoic and Mesozoic Floristic Change*. Córdoba, Spain, p. 21.

Poole, I. & Francis, J. E. 1999. The first record of fossil atherospermataceous wood from Antarctica. *Review of Palaeobotany and Palynology*, **107**, 97–107.

Poole, I., Cantrill, D. J., Hayes, P. & Francis, J. E. 2000a. The fossil record of Cunoniaceae: new evidence from Late Cretaceous wood of Antarctica. *Review of Palaeobotany and Palynology*, **111**, 127–44.

Poole, I. & Francis, J. E. 2000. The first record of fossil wood of Winteraceae from the Upper Cretaceous of Antarctica. *Annals of Botany*, **85**, 307–15.

Poole, I., Gottwald, H. & Francis, J. E. 2000b. *Illicioxylon*, an element of Gondwanan Polar forests? Late Cretaceous and early Tertiary woods of Antarctica. *Annals of Botany*, **86**, 421–32.

Poole, I., Richter, H. G. & Francis, J. E. 2000c. Evidence for Gondwanan origins for *Sassafras* (Lauraceae)? Late Cretaceous fossil wood of Antarctica. *IAWA Journal*, **21**, 463–75.

Poole, I. & Gottwald, H. 2001. Monimiaceae *sensu lato*, an element of Gondwanan polar forests: evidence from the Late Cretaceous-Early Tertiary wood flora of Antarctica. *Australian Systematic Botany*, **14**, 207–30.

Poole, I. 2002. Systematics of Cretaceous and Tertiary *Nothofagoxylon*: implications for Southern Hemisphere biogeography and evolution of the Nothofagaceae. *Australian Systematic Botany*, **15**, 247–76.

Poole, I. & Cantrill, D. J. 2006. Cretaceous and Tertiary vegetation of Antarctica implications from the fossil wood record. In *Cretaceous – Tertiary High-Latitude Palaeoenvironments, James Ross Basin, Antarctica*, (eds J. E. Francis, D. Pirrie & J. A. Crame): London Geological Society of London, Special Publication, **258**, pp. 63–81.

Porter, D. M. 1974. Disjunct distributions in the New World Zygophyllaceae. *Taxon*, **23**, 339–46.

Potonié, R. 1956. Synopsis der Gattungen der Sporae dispersae. I. Sporites. *Beihefte zum Geologischen Jahrbuch*, **23**, 1–103.

Pott, C., Krings, M., Kerp, H. & Friis, E. M. 2010. Reconstruction of a bennettitalean flower from the Carnian (Upper Triassic) of Lunz, Lower Austria. *Review of Palaeobotany and Palynology*, **159**, 94–111.

Praglowski, J. 1976. Schisandraceae Bl. *World Pollen and Spore Flora*, **5**, 1–32.

Praglowski, J. 1979. Winteraceae Lindl. *World Pollen and Spore Flora*, **8**, 1–36.

Prakash, U. 1954. *Palmocarpon mohgaoense* sp. nov., a palm fruit from the Deccan Intertrappean series, India. *The Palaeobotanist*, **3**, 91–6.

Prakash, U. 1960. A survey of the Deccan Intertrappean flora of India. *Journal of Paleontology*, **34**, 1027–40.

Prakash, U. 1965. A survey of the fossil dicotyledonous woods from India and the Far East. *Journal of Paleontology*, **39**, 815–27.

Prakash, U. 1974. Palaeogene angiospermous woods. In *Aspects and Appraisal of Indian Palaeobotany* (eds K.R. Surange, R. N. Lakhanpal & D. C. Bharadwaj) Lucknow: Birbal Sahni Institute of Palaeobotany, pp. 306–20.

Prasad, V., Strömberg, C. A. E., Alimohammadian, H. & Sahni, A. 2005. Dinosaur coprolites and the early evolution of grasses and grazers. *Science*, **310**, 1177–80.

Prevec, R., McLoughlin, S. & Bamford, M. K. 2008. Novel double wing morphology revealed in a South African ovuliferous glossopterid fructification: *Bifariala intermittens* (Plumstead 1958) comb. nov. *Review of Palaeobotany and Palynology*, **150**, 22–36.

Price, R. A. 1996. Systematics of the Gnetales: a review of morphological and molecular evidence. *International Journal of Plant Sciences*, **157** (6 Suppl.), S40–S49.

Proctor, M., Yeo, P. & Lack, A. 1996. *The Natural History of Pollination. The New Naturalist Library – A Survey of British Natural History.* London: Harper Collins.

Pryer, K. M., Smith, A. R. & Skog, J. E. 1995. Phylogenetic relationships of extant ferns based on evidence from morphology and *rbc*L sequences. *American Fern Journal*, **85**, 205–82.

Pryer, K. M., Schneider, H., Smith, A. R. *et al.* 2001. Horsetails and ferns are a monophyletic group and the closest living relatives to seed plants. *Nature*, **409**, 618–22.

Pryer, K. M., Schuettpelz, E., Wolf, P. G. *et al.* 2004. Phylogeny and evolution of ferns (monilophytes) with a focus on the early leptosporangiate divergences. *American Journal of Botany*, **91**, 1582–98.

Qiu, Y.-L., Chase, M. W., Les, D. H. & Park, C. R. 1993. Molecular phylogenetics of the Magnoliidae: cladistic analyses of nucleotide sequences of the plastid gene *rbc*L. *Annals of the Missouri Botanical Garden*, **80**, 587–606.

Qiu, Y.-L., Lee, J., Bernasconi-Quadroni, F. *et al.* 1999. The earliest angiosperms: evidence from mitochondrial, plastid and nuclear genomes. *Nature*, **402**, 404–7.

Qiu, Y.-L. & Palmer, J. D. 1999. Phylogeny of early land plants: insights from genes and genomes. *Trends in Plant Science*, **4**, 26–30.

Qiu, Y.-L., Lee, J., Bernasconi-Quadroni, F. *et al.* 2000. Phylogeny of basal angiosperms: analyses of five genes from three genomes. *International Journal of Plant Science*, **161** (6 Suppl.), S3–S27.

Rai, H. S., Reeves, P. A., Peakall, R., Olmstead, R. G. & Graham, S. W. 2008. Inference of higher-order conifer relationships from a multi-locus plastid data set. *Botany*, **86**, 658–69.

Raine, J. I. 1984. Outline of a palynological zonation of Cretaceous to Paleogene terrestrial sediments in west coast region South Island, New Zealand. *Report of the New Zealand Geological Survey*, **109**, 1–81.

Raine, J. I., Mildenhall, D. C. & Kennedy, E. M. 2006. *New Zealand Fossil Spores and Pollen: an Illustrated Catalogue.* 2nd ed. GNS Science miscellaneous series no. 4. http://www.gns.cri.nz/what/earthhist/fossils/spore_pollen/catalog/index.htm.

Ramanujam, C. G. K. 1955. Fossil wood of Dipterocarpaceae from the Tertiary of South Arcot district, India. *The Palaeobotanist*, **4**, 45–56.

Ramirez, B. W. 1994. Coevolution of *Ficus* and Agaonidae. *Annals of the Missouri Botanical Garden*, **61**, 770–80.

Ramirez, J. L. & Cevallos-Ferriz, S. R. S. 2000. Leaves of Berberidaceae (*Berberis* and *Mahonia*) from Oligocene sediments, near Tepexi de Rodríguez, Puebla. *Review of Palaeobotany and Palynology*, **110**, 247–57.

Ramirez, S. R., Gravendeel, B., Singer, R. B., Marshall, C. R. & Pierce, N. E. 2007. Dating the origin of the Orchidaceae from a fossil orchid with its pollinator. *Nature*, **448**, 1042–5.

Ramshaw, J. A. M., Richardson, D. L., Meatyard, B. T. *et al.* 1972. The time of origin of the flowering plants determined by using amino acid sequence data of cytochrome *c*. *New Phytologist*, **71**, 773–9.

Rao, A. R. 1976. Problems in the Pentoxyleae. *The Palaeobotanist*, **25**, 393–6.

Rao, A. R. 1981. The affinities of the Pentoxylon plant. *The Palaeobotanist*, **28–29**, 207–9.

Rasmussen, E. S., Lomholt, S., Andersen, C. & Vejbæk, O. V. 1998. Aspects of the structural evolution of the Lusitanian Basin in Portugal and the shelf and slope area offshore Portugal. *Tectonophysics*, **300**, 199–225.

Rat, P. 1989. The Iberian Cretaceous: Climatic interpretations. In *Proceedings of the 3rd International Cretaceous Symposium* (ed. J. Wiedmann). Tübingen: E. Schweizerbart'sche Verlagsbuchhandlung (Nägele u. Obermiller), pp. 17–25.

Rattray, G. 1913. Notes on the pollination of some South African cycads. *Transactions of the Royal Society of South Africa*, **3**, 623–7.

Raubeson, L. A. & Gensel, P. G. 1991. Upper Cretaceous conifer leaf fossils from the Black Creek Formation with an assessment of affinities using principal components analysis. *Botanical Gazette*, **152**, 380–91.

Ray, J. 1703. *Methodus Plantarum Emendata et Aucta*. London: Impensis Samuelis Smith & Benjamini Walford. Et veneunt Amstelædami apud Janssonio – Waasbergios.

Rayner, R. J. 1993. The fossils from the Orapa Diamond mine: a review. *Botswana Notes and Records*, **25**, 1–17.

Raynolds, R. G. & Johnson, K. R. 2003. Synopsis of the stratigraphy and paleontology of the uppermost Cretaceous and lower Tertiary strata in the Denver Basin, Colorado. *Rocky Mountain Geology*, **38**, 171–81.

Read, R. W. & Hickey, L. J. 1972. A revised classification of fossil palm and palm-like leaves. *Taxon*, **21**, 129–37.

Rees, P. M. & Smellie, J. L. 1989. Cretaceous angiosperms from an allegedly Triassic flora at Williams Point, Livingston Island, South Shetland Islands. *Antarctic Science*, **1**, 239–48.

Rees, P. M. 1993. Caytoniales in Early Jurassic floras from Antarctica. *Geobios*, **26**, 33–42.

Regali, M. S. P., Uesugui, N. & Santos, A. S. 1974. Palinologia dos sedimentos meso-cenozóicos do Brazil. *Boletim Tecnico da Petrobrás*, **17**, 117–91.

Regali, M. S. P. 1989. *Tucanopollis*, um gênero novo das angiospermas primitivas. *Boletim de Geociências da Petrobrás*, **3**, 395–402.

Regali, M. S. P. & Viana, C. F. 1989. *Late Jurassic-Early Cretaceous in Brazilian Sedimentary Basins: Correlation with the International Standard Scale*. Rio de Janeiro: Petrobrás.

Reid, E. M. & Chandler, M. E. J. 1933. *The Flora of the London Clay*. London: British Museum (Natural History).

Reitsma, T. 1970. Pollen morphology of Alangiaceae. *Review of Palaeobotany and Palynology*, **10**, 249–332.

Remizowa, M. V., Sokoloff, D. D., Macfarlane, T. D. *et al.* 2008. Comparative pollen morphology in the early-divergent angiosperm family Hydatellaceae reveals variation at the infraspecific level. *Grana*, **47**, 81–100.

Ren, D. 1998. Flower-associated Brachycera flies as fossil evidence for Jurassic angiosperm origins. *Science*, **280**, 85–8.

Renner, S. S. 1998. Phylogenetic affinities of Monimiaceae based on cpDNA gene and spacer sequences. *Perspectives in Plant Ecology, Evolution and Systematics*, **1**, 61–77.

Renner, S. S. 1999. Circumscription and phylogeny of Laurales: evidence from molecular and morphological data. *American Journal of Botany*, **86**, 1301–15.

Renner, S. S. & Chanderbali, A. S. 2000. What is the relationship among Hernandiaceae, Lauraceae, and Monimiaceae, and why is this question so difficult to answer? *International Journal of Plant Sciences*, **161** (6 Suppl.), S109–S119.

Renner, S. S., Foreman, D. B. & Murray, D. 2000. Timing transarctic disjunctions in the Atherospermataceae (Laurales): evidence from coding and noncoding chloroplast sequences. *Systematic Botany*, **49**, 579–91.

Renner, S. S. 2005. Variation in diversity among Laurales, Early Cretaceous to Present. *Biologiske Skrifter, Kongelige Danske Videnskabernes Selskab*, **55**, 441–58.

Renner, S. S. & Schaefer, H. 2010. The evolution and loss of oil-offering flowers: new insights from dated phylogenies for angiosperms and bees. *Philosophical Transactions of the Royal Society* B, **365**, 423–35.

Retallack, G. & Dilcher, D. L. 1981a. A coastal hypothesis for the dispersal and rise to dominance of flowering plants. In *Palaeobotany, Palaeoecology and Evolution* (ed. K. J. Niklas) New York: Praeger Publishers, pp. 27–77.

Retallack, G. & Dilcher, D. L. 1981b. Arguments for a glossopterid ancestry of angiosperms. *Paleobiology*, **7**, 54–67.

Retallack, G. & Dilcher, D. L. 1981c. Early angiosperm reproduction: *Prisca reynoldsii* gen. et sp. nov. from mid-Cretaceous coastal deposits, Kansas, USA. *Palaeontographica* B, **179**, 103–37.

Rey, J. 1972. Recherches géologiques sur le Crétacé inférieur de l'Estremadura (Portugal). *Serviços Geológicos de Portugal, Memórias (Nova Série) 3*, **21**, 1–477.

Rey, J. 1979. Le Crétacé inférieur de la marge atlantique portugaise: biostratigraphie, organisation séquentielle, évolution paléogéographique. *Ciências da Terra (UNL)*, **5**, 97–120.

Rey, J. 1982. Dynamique et paléoenvironnements du bassin mésozoique d'Estremadura (Portugal), au Crétacé inférieur. *Cretaceous Research*, **3**, 103–11.

Rey, J. 1983. Le Crétacé de l'Algarve: essai de synthèse. *Comunicações dos Serviços Geológicos de Portugal*, **69**, 87–101.

Rey, J. 1992. Les unités lithostratigraphiques du Crétacé inférieur de la région de Lisbonne. *Comunicações dos Serviços Geológicos de Portugal*, **78**, 103–24.

Rey, J. 1993. Les unités lithostratigraphiques du groupe de Torres Vedras (Estremadura, Portugal). *Comunicações Instituto Geológico e Mineiro*, **79**, 75–85.

Rey, J., Dinis, J. L., Callapez, P. & Cunha, P. P. 2006. *Da Rotura Continental à Margem Passiva. Composição e Evolução do Cretácico de Portugal. Cadernos de Geologia de Portugal*. Lisbon: Ministério da Economia e da Inovação.

Reymanówna, M. 1960. A cycadeoidean stem from the western carpathians. *Acta Palaeobotanica*, **1**, 1–28.

Reymanówna, M. 1968. On seeds containing *Eucommiidites troedssonii* pollen from the Jurassic of Grojec, Poland. *Botanical Journal of the Linnean Society*, **61**, 147–52.

Reymanówna, M. 1973. The Jurassic flora from Grojec near Kraków in Poland. Part II. Caytoniales and anatomy of *Caytonia*. *Acta Palaeobotanica*, **14**, 45–87.

Reyre, Y. 1973. Palynologie du Mésozoïque saharien. *Mémoires du Muséum National d'Histoire Naturelle, Ser. C*, **27**, 1–284.

Ribeiro, A., Antunes, M. T., Ferreira, M. P. *et al.* 1980. *Introduction à la géologie générale du Portugal*. Lisbon: Serviços Geológicos de Portugal.

Richter, P. B. 1905. *Beiträge zur Flora der oberen Kreide Quedlinburgs und seiner Umgebung. I. Die Gattung Credneria und einige seltnere Pflanzenreste*. Leipzig: Wilhelm Engelmann.

Richter, P. B. 1906. *Beiträge zur Flora der unteren Kreide Quedlinburgs. I Die Gattung Hausmannia Dunker und einige seltenere Pflanzenreste*. Leipzig: Wilhelm Engelmann.

Richter, P. B. 1909. *Beiträge zur Flora der unteren Kreide Quedlinburgs. II. Die Gattung Nathorstiana P. Richter und Cylindrites spongioides Goeppert*. Leipzig: Wilhelm Engelmann.

Ridley, H. N. 1930. *The Dispersal of Plants Throughout the World*. Ashford, United Kingdom: Reeve.

Riegel, W., Wilde, V. & Pelzer, G. 1986. Erste Ergebnisse einer paläobotanischen Grabung in der fluviatilen Wealden-Fazies des Osterwaldes bei Hannover. *Courier Forschungsinstitut Senckenberg*, **86**, 137–170.

Rigby, J. F. & Playford, G. 1988. Upper Triassic and Lower Tertiary megafossil floras of the Ipswich area, Southeast Queensland; selected localities. *Excursion Guide, 7 International Palynological Congress, Brisbane*, **SA4**, 1–9.

Riley, M. G. & Stockey, R. A. 2004. *Cardstonia tolmanii* gen. et sp. nov. (Limnocharitaceae) from the Upper Cretaceous of Alberta, Canada. *International Journal of Plant Sciences*, **165**, 897–916.

Rioult, M. 1966. Sur l'age Albien de *Cycadeoidea micromyela* Moriere (Bennettitinée). *Bulletin de la Société linnéenne de Normandie, Caen, 10è série*, **7**, 9–18.

Rocha, R., Manuppella, G., Mouteride, R., Ruget, C. & Zbyszewski, G. 1981. *Carta Geológica de Portugal na Escala de 1/50 000. Notícia Explicativa da Folha 19-C Figueira da Foz*. Lisbon: Serviços Geológicos de Portugal.

Rode, K. P. 1933. A note on fossil angiospermous fruits from the Deccan Intertrappean beds of Central Provinces. *Current Science*, **2**, 171–2.

Rodin, R. J. 1953. Distribution of *Welwitschia mirabilis*. *American Journal of Botany*, **40**, 280–5.

Rodríguez-de la Rosa, R. A. & Cevallos-Ferriz, S. R. S. 1994. Upper Cretaceous Zingiberalean fruits with in situ seeds from southeastern Coahuila, Mexico. *International Journal of Plant Sciences*, **155**, 786–805.

Rodríguez-de la Rosa, R. A., Cevallos-Ferriz, S. R. S. & Silva-Pineda, A. 1998. Paleobiological implications of Campanian coprolites. *Palaeogeography, Palaeoclimatology, Palaeoecology*, **142**, 231–54.

Rohwer, J. G. 1993a. Lauraceae. In *The Families and Genera of Vascular Plants*. Volume II. *Flowering Plants – Dicotyledons. Magnoliid, Hamamelid and Caryophyllid Families* (eds K. Kubitzki, J. G. Rohwer & V. Bittrich) Berlin, Heidelberg, New York: Springer-Verlag, pp. 366–91.

Rohwer, J. G. 1993b. Moraceae. In *The Families and Genera of Vascular Plants*. Volume II *Flowering Plants – Dicotyledons. Magnoliid, Hamamelid and Caryophyllid Families* (eds K. Kubitzki, J. G. Rohwer & V. Bittrich) Berlin, Heidelberg, New York: Springer-Verlag, pp. 438–52.

Rohwer, J. G. 2000. Towards a phylogenetic classification of the Lauraceae: Evidence from *mat*K sequences. *Systematic Botany*, **25**, 60–71.

Romero, E. J. 1986. Paleogene phytogeography and climatology of South America. *Annals of the Missouri Botanical Garden*, **73**, 449–61.

Romero, E. J. & Archangelsky, S. 1986. Early Cretaceous angiosperm leaves from southern South America. *Science*, **234**, 1580–2.

Ronse De Craene, L. P., Soltis, P. S. & Soltis, D. E. 2003. Evolution of floral structures in basal angiosperms. *International Journal of Plant Sciences*, **164** (5 Suppl.), S329–S363.

Ronse De Craene, L. P. 2004. Floral development of *Berberidopsis corallina*: a crucial link in the evolution of flowers in the core eudicots. *Annals of Botany*, **94**, 741–51.

Ross, N. E. 1949. On a Cretaceous pollen and spore bearing clay of Scania. *Bulletin of the Geological Institution of Uppsala*, **34**, 25–43.

Roth, J. L. & Dilcher, D. L. 1979. Investigations of angiosperms from the Eocene of North America: Stipulate leaves of the Rubiaceae including a probable polyploid population. *American Journal of Botany*, **66**, 1194–207.

Rothwell, G. W. 1981. The Callistophytaceae (Pteridospermopsida), reproductively sophisticated gymnosperms. *Review of Palaeobotany and Palynology*, **32**, 103–21.

Rothwell, G. W. 1982. New interpretations of the earliest conifers. *Review of Palaeobotany and Palynology*, **37**, 7–28.

Rothwell, G. W., Scheckler, S. E. & Gillespie, W. H. 1989. *Elkinsia* gen. nov., a late Devonian gymnosperm with cupulate ovules. *Botanical Gazette*, **150**, 170–89.

Rothwell, G. W. & Serbet, R. 1994. Lignophyte phylogeny and the evolution of spermatophytes: A numerical cladistic analysis. *Systematic Botany*, **19**, 443–82.

Rothwell, G. W. & Holt, B. F. 1997. Fossils and phenology in the evolution of *Ginkgo biloba*. In *Ginkgo biloba – a Global Treasure* (eds T. Hori, R. W. Ridge, W. Tulecke, P. Del Tredici, J. Tremouillaux-Guiller & H. Tobe) Tokyo: Springer-Verlag, pp. 223–30.

Rothwell, G. W. & Stockey, R. A. 2002. Anatomically preserved *Cycadeoidea* (Cycadeoidaceae), with a reevaluation of systematic characters for the seed cones of Bennettitales. *American Journal of Botany*, **89**, 1447–58.

Rothwell, G. W., Crepet, W. L. & Stockey, R. A. 2009. Is the anthophyte hypothesis alive and well? New evidence from the reproductive structures of Bennettitales. *American Journal of Botany*, **96**, 296–322.

Rouse, G. E. 1957. The application of a new nomenclatural approach to Upper Cretaceous plant microfossils from Western Canada. *Canadian Journal of Botany*, **35**, 349–75.

Rudall, P. J., Cribb, P. J., Cutler, D. F. & Humphries, C. J. 1995. *Monocotyledons: Systematics and Evolution*. Kew: Royal Botanic Gardens.

Rudall, P. J. 2003. Monocot pseudanthia revisited: floral structure of the mycoheterotrophic family Triuridaceae. *International Journal of Plant Sciences*, **164** (5 Suppl.), S307–S320.

Rudall, P. J., Sokoloff, D. D., Remizowa, M. V. *et al.* 2007. Morphology of Hydatellaceae, an anomalous aquatic family recently recognized as an early-divergent angiosperm lineage. *American Journal of Botany*, **94**, 1073–92.

Rudall, P. J. & Bateman, R. M. 2010. Defining the limits of flowers: the challenge of distinguishing between the evolutionary products of simple versus compound strobili. *Philosophical Transactions of the Royal Society* B, **365**, 397–409.

Ruffell, A. H. & Batten, D. J. 1990. The Barremian-Aptian arid phase in western Europe. *Palaeogeography, Palaeoclimatology, Palaeoecology*, **80**, 197–212.

Russell, D. A. 1989. *An Odyssey in Time; The Dinosaurs of North America*. Toronto: National Museum of Natural Sciences, Canada, University of Toronto Press.

Russell, L. S. 1983. Evidence for an unconformity at the Scollard-Battle contact, Upper Cretaceous strata, Alberta. *Canadian Journal of Earth Science*, **20**, 1219–31.

Rydin, C., Källersjö, M. & Friis, E. M. 2002. Seed plant relationships and the systematic position of Gnetales based on nuclear and chloroplast DNA: conflicting data, rooting problems and the monophyly of conifers. *International Journal of Plant Sciences*, **163**, 197–214.

Rydin, C., Mohr, B. & Friis, E. M. 2003. *Cratonia cotyledon* gen. et sp. nov.: a unique Cretaceous seedling related to *Welwitschia*. *Proceedings of the Royal Society of London* B (Suppl.) *Biology Letters*, **270**, S29–S32.

Rydin, C., Pedersen, K. R. & Friis, E. M. 2004. On the evolutionary history of *Ephedra*; Cretaceous fossils and extant molecules. *Proceedings of the National Academy of Sciences, USA*, **101**, 16 571–6.

Rydin, C., Pedersen, K. R., Crane, P. R. & Friis, E. M. 2006a. Former diversity of *Ephedra* (Gnetales): evidence from Early Cretaceous seeds from Portugal and North America. *Annals of Botany*, **98**, 123–40.

Rydin, C., Wu, S. Q. & Friis, E. M. 2006b. *Liaoxia* (Gnetales): ephedroids from the Early Cretaceous Yixian Formation in China. *Plant Systematics and Evolution*, **262**, 239–65.

Rydin, C. & Friis, E. M. 2010. *Siphonospermum simplex* gen. et sp. nov. from the Yixian Formation of northeast China. *BML Evolutionary Biology*, **10**, 183. http://www.biomedcentral.com/1471-2148/10/183.

Rüffle, L. 1965. Monimiaceen-Blätter im älteren Senon von Mitteleuropa. *Geologie*, **14**, 78–89.

Rüffle, L. 1968. Merkmalskomplexe bei älteren Angiospermen-Blättern und die Kutikula von *Credneria* Zenker (Menispermaceae). *Palaeontographica* B, **123**, 132–43.

Rüffle, L. 1995. Some artificial genera (Fagaceae, Platanaceae, Araliaceae) of Upper Cretaceous of the Northern Hemisphere and heterophylly in some modern hybrids. *The Palaeobotanist*, **44**, 225–37.

Rüffle, L., Knappe, H. 1988. Ökologische und paläogeographische Bedeutung der Oberkreideflora von Quedlinburg, besonders einiger Loranthaceae und Monimiaceae. *Hallesches Jahrbuch für Geowissenschaften*, **13**, 49–65.

Saarela, J. M., Rai, H. S., Doyle, J. A. *et al.* 2007. Hydatellaceae identified as a new branch near the base of the angiosperm phylogenetic tree. *Nature*, **446**, 312–15.

Sahni, B. 1932. A petrified *Williamsonia* (*W. sewardiana*, sp. nov.) from the Rajmahal Hills, India. *Memoirs of the Geological Survey of India, Palaeontologia Indica, New Series*, **10**, 1–19.

Sahni, B. 1934. The silicified flora of the Deccan Intertrappean Series, Part II. Gymnospermous and angiospermous fruits. *Proceedings of the 21st Indian Science Congress, Section V, Botany*, **317**–18.

Sahni, B. & Rode, K. P. 1937. Fossil plants from the Intertrappean beds of the Mohgaon Kalan, in the Deccan, with a sketch of the geology of Chhindwara district. *Proceedings of the National Academy of Sciences, India*, **7**, 165–74.

Sahni, B. 1943. Indian silicified plants. 2. *Enigmocarpon Parijai*, a silicified fruit from the Deccan, with a review of the fossil history of the Lythraceae. *The Proceedings of the Indian Academy of Sciences* B, **17**, 59–93.

Sahni, B. 1944. Takli near Nagpur. Genus *Viracarpon* Sahni. Palaeobotany in India V. *Proceedings of the National Academy of Sciences, India*, **14**, 80–2.

Sahni, B. 1948. The Pentoxyleae: a new group of Jurassic gymnosperms from the Rajmahal Hills of India. *Botanical Gazette*, **110**, 47–80.

Saiki, K. & Wang, Y.-D. 2003. Preliminary analysis of climate indicator plant distribution in the Early Cretaceous of China. *Journal of Asian Earth Sciences*, **21**, 813–22.

Sampson, F. B. 1975. Aperture orientation in *Laurelia* pollen (Atherospermataceae syn. subfamily Atherospermoideae of Monimiaceae). *Grana*, **15**, 153–7.

Sampson, F. B. & Endress, P. K. 1984. Pollen morphology in the Trimeniaceae. *Grana*, **23**, 129–37.

Sampson, F. B. 1993. Pollen morphology of the Amborellaceae and Hortoniaceae (Hortonioideae: Monimiaceae). *Grana*, **32**, 154–62.

Sampson, F. B. 2000a. The pollen of *Takhtajania perrieri* (Winteraceae). *Annals of the Missouri Botanical Garden*, **87**, 380–8.

Sampson, F. B. 2000b. Pollen diversity in some modern magnoliids. *International Journal of Plant Sciences*, **161** (6 Suppl.), S193–S210.

Samylina, V. A. 1960. Angiosperms from the Lower Cretaceous of the Kolyma Basin. *Botanicheskij Zhurnal*, **45**, 335–52 (in Russian).

Samylina, V. A. 1961. New data on the Lower Cretaceous flora of the southern part of the Maritime Territory of the R.F.S.R. *Botanicheskij Zhurnal*, **46**, 634–45 (in Russian).

Samylina, V. A. 1968. Early Cretaceous angiosperms of the Soviet Union based on leaf and fruit remains. *The Journal of the Linnean Society (Botany)*, **61**, 207–18.

Samylina, V. A. 1974. Early Cretcaeous Flora of Northeastern USSR. *Komarovskie chtniya*, **27**, 1–56 (in Russian).

Samylina, V. A. 1976. *The Cretaceous Flora of Omsukchan (Magadan District)*. Leningrad: Nauka.

Sanderson, M. J. 1997. A nonparametric approach to estimating divergence times in the absence of rate constancy. *Molecular Biology and Evolution*, **14**, 1218–31.

Sanderson, M. J. & Doyle, J. A. 2001. Sources of error and confidence intervals in estimating the age of angiosperms from *rbc*L and 18S rDNA data. *American Journal of Botany*, **88**, 1499–16.

Saporta, G. de & Marion, A.-F. 1873. Essai sur l'état de la végétation à l'époque des Marnes Heersiennes de Gelinden. *Memoires couronnes et Memories des Savants etrangers, publies par l'Academie royale des Sciences des Lettres et Beaux-Arts de Belgique*, **37**.

Saporta, G. de 1877. L'ancienne végétation polaire d'après les travaux de M. le professeur Heer et les dernières découvertes des explorateurs suédois. In *Comptes-rendues du Congrès International des Sciences Géographiques, Paris*, pp. 1–47.

Saporta, G. de 1894. *Flore Fossile du Portugal. Nouvelles Contributions à la Flore Mésozoique. Accompagnées d'une Notice Stratigraphique par Paul Choffat*. Lisbon: Imprimerie de l'Academie Royale des Sciences.

Sauquet, H. 2003. Androecium diversity and evolution in Myristicaceae (Magnoliales), with a description of a new Malagasy genus, *Doyleanthus* gen. nov. *American Journal of Botany*, **90**, 1293–305.

Sauquet, H., Doyle, J. A., Scharaschkin, T. *et al.* 2003. Phylogenetic analysis of Magnoliales and Myristicaceae based on multiple data sets: implications for character evolution. *Botanical Journal of the Linnean Society*, **142**, 125–86.

Sauquet, H. & Cantrill, D. J. 2007. Pollen diversity and evolution in Proteoideae (Proteales: Proteaceae). *Systematic Botany*, **32**, 271–316.

Sauquet, H., Weston, P. H., Anderson, C. L. *et al.* 2009. Contrasted patterns of hyperdiversification in Mediterranean hotspots. *Proceedings of the National Academy of Sciences, USA*, **106**, 221–5.

Sawada, M. 1971. Floral vasculation of *Paeonia japonica* with some consideration on systematic position of the Paeoniaceae. *Botanical Magazine, Tokyo*, **84**, 51–60.

Schaarschmidt, F. 1984. Flowers from the Eocene oil-shale of Messel: a preliminary report. *Annals of the Missouri Botanical Garden*, **71**, 599–606.

Schaarschmidt, F. & Wilde, V. 1986. Palmenblüten und blätter aus dem Eozän von Messel. *Courier Forschungsinstitut Senckenberg*, **86**, 177–202.

Schenk, A. 1871. *Die Fossile Flora der Nordwestdeutschen Wealdenformation*. Kassel: Theodor Fischer.

Schneider, E. l. & Williamson, P. S. 1993. Nymphaeaceae. In *The Families and Genera of Vascular Plants.* Volume II. *Flowering Plants – Dicotyledons. Magnoliid, Hamamelid and Caryophyllid Families* (eds K. Kubitzki, J. G. Rohwer & V. Bittrich) Berlin, Heidelberg, New York: Springer-Verlag, pp. 486–93.

Schneider, H. & Kenrick, P. 2001. An Early Cretaceous root-climbing epiphyte (Lindsaeaceae) and its significance for calibrating the diversification of polypodiaceous ferns. *Review of Palaeobotany and Palynology*, **115**, 33–41.

Schrank, E. 1982. Kretazische Pollen und Sporen aus dem "Nubischen Sandstein" des Dakhla-Beckens (Ägypten). *Berliner Geowissenschaftliche Abhandlungen A*, **40**, 87–109.

Schrank, E. 1983. Scanning electron and light microscopic investigations of angiosperm pollen from the Lower Cretaceous of Egypt. *Pollen et Spores*, **25**, 213–42.

Schrank, E. 1987. Paleozoic and Mesozoic palynomorphs from Northeast Africa (Egypt and Sudan) with special reference to Late Cretaceous pollen and dinoflagellates. *Berliner Geowissenschaftliche Abhandlungen A*, **75**, 249–310.

Schrank, E. 1992. Nonmarine Cretaceous correlations in Egypt and northern Sudan: palynological and palaeobotanical evidence. *Cretaceous Research*, **13**, 351–68.

Schrank, E. & Nesterova, E. V. 1993. Palynofloristic changes and Cretaceous climates in northern Gondwana (NE Africa) and southern Laurasia (Kazakhstan). In *Geoscientific Research in Northeast Africa* (ed. U. Thonweihe & H. Schandelmeier) Rotterdam: A.A. Balkema, pp. 381–90.

Schrank, E. 1994a. Palynology of the Yesomma Formation in Northern Somalia: A study of pollen, spores and associated phytoplankton from the Late Cretaceous. *Palaeontographica* B, **231**, 63–112.

Schrank, E. 1994b. Nonmarine Cretaceous palynology of northern Kordofan, Sudan, with notes on fossil Salviniales (water ferns). *Geologische Rundschau*, **83**, 773–86.

Schrank, E. & Ibrahim, M. I. A. 1995. Cretaceous (Aptian-Maastrichtian) palynology of foraminifera-dated wells (KRM-1, AG-18) in northwestern Egypt. *Berliner Geowissenschaftliche Abhandlungen A*, **177**, 1–44.

Schrank, E. & Mahmoud, M. S. 1998. Palynology (pollen, spores and dinoflagellates) and Cretaceous stratigraphy of the Dakhla Oasis, central Egypt. *Journal of African Earth Sciences*, **26**, 167–93.

Schrank, E. 1999. Mesozoic Floren aus Nordost-Afrika und ihre Beziehungen zum Klima am Paläoäquator. *Deutsche Forschungsgemeinschaft*, **1999**, 137–66.

Schrank, E. & Rüffle, L. 2003. The Late Cretaceous leaf flora from Jebel Mudaha, Sudan. *Courier Forschungsinstitut Senckenberg*, **241**, 119–29.

Schultheis, L. M. & Donoghue, M. J. 2004. Molecular phylogeny and biogeography of *Ribes* (Grossulariaceae), with an emphasis on gooseberries (subg. *Grossularia*). *Systematic Botany*, **9**, 77–96.

Schutt, W. A. & Simmons, N. B. 1998. Morphology and homology of the chiropteran calcar, with comments on the phylogenetic relationships of *Archaeopteropus*. *Journal of Mammalian Evolution*, **5**, 1–32.

Schweitzer, H.-J. 1977. Die räto-jurassischen Floren des Iran und Afghanistans. 4. Die rätische Zwitterblüte *Irania hermaphroditica* nov. spec. und ihre Bedeutung für die phylogenie der Angiospermen. *Palaeontographica* B, **161**, 98–145.

Schönenberger, J. & Friis, E. M. 2001. Fossil flowers of ericalean s.l. affinity from the Late Cretaceous of southern Sweden. *American Journal of Botany*, **88**, 467–80.

Schönenberger, J., Friis, E. M., Matthews, M. L. & Endress, P. K. 2001a. Cunoniaceae in the Cretaceous of Europe: evidence from fossil flowers. *Annals of Botany*, **88**, 423–37.

Schönenberger, J., Pedersen, K. R. & Friis, E. M. 2001b. Normapolles flowers of fagalean affinity from the Late Cretaceous of Portugal. *Plant Systematics and Evolution*, **226**, 205–30.

Schönenberger, J. 2005. Rise from the ashes – the reconstruction of charcoalified fossil flowers. *Trends in Plant Science*, **10**, 436–43.

Schönenberger, J., Anderberg, A. A. & Sytsma, K. J. 2005. Molecular phylogenetics and patterns of floral evolution in the Ericales. *International Journal of Plant Sciences*, **166**, 265–88.

Schönenberger, J. & von Balthazar, M. 2006. Reproductive structures and phylogenetic framework of the rosids – progress and prospects. *Plant Systematics and Evolution*, **260**, 87–106.

Schöning, M. & Bandel, K. 2004. A diverse assemblage of fossil hardwood from the Upper Tertiary (Miocene?) of the Arauco Peninsula, Chile. *Journal of South American Earth Sciences*, **17**, 59–71.

Scotese, C. R., Gahagan, L. M. & Larson, R. L. 1988. Plate tectonic reconstructions of the Cretaceous and Cenozoic ocean basins. *Tectonophysics*, **155**, 27–48.

Scott, A. C. & Chaloner, W. G. 1983. The earliest fossil conifer from the Westphalian B of Yorkshire. *Proceedings of the Royal Society of London* B, **220**, 163–82.

Scott, A. C. & Jones, T. P. 1991. Microscopical observations of recent and fossil charcoal. *Microscopy and Analysis*, July, **1991**, 13–15.

Scott, D. H. 1923. *Studies in Fossil Botany.* Volume II. *Spermophyta*. London: A. & C. Black, Ltd.

Scott, R. A., Barghoorn, E. S. & Leopold, E. 1960. How old are the angiosperms? *American Journal of Science*, **258-A**, 284–99.

Scott, R. A., Williams, P. L., Craig, L. C. *et al.* 1972. “Pre-Cretaceous” angiosperms from Utah: evidence for Tertiary age of the palm woods and roots. *American Journal of Botany*, **59**, 886–96.

Scott, R. A. & Wheeler, E. 1982. Fossil wood from the Eocene Clarno Formation of Oregon. *IAWA Bulletin, n. s.*, **3**, 135–54.

Sellwood, B. W. & Price, G. D. 1994. Sedimentary facies as indicators of Mesozoic palaeoclimate. *Philosophical Transactions of the Royal Society* B, **341**, 225–33.

Sender, L. M., Diez, J. B., Ferrer, J., Pons, D. & Rubio, C. 2005. Preliminary data on a new Albian flora from the Valle del Río Martín, Teruel, Spain. *Cretaceous Research*, **26**, 898–905.

Serbet, R. & Stockey, R. A. 1991. Taxodiaceous pollen cones from the Upper Cretaceous (Horseshoe Canyon Formation) of Drumheller, Alberta, Canada. *Review of Palaeobotany and Palynology*, **70**, 67–76.

Seubert, E. 1993. *Die Samenmerkmale der Araceen und ihre Bedeutung für die Gliederung der Familie*. Koenigstein: Koeltz Scientific Books.

Seward, A. C. 1894. *Catalogue of the Mesozoic Plants in the Department of Geology, British Museum (Natural History).*

The Wealden Flora. Part I. *Thallophyta-Pteridophyta*. London: British Museum (Natural History).

Seward, A. C. 1895. *Catalogue of the Mesozoic Plants in the Department of Geology, British Museum (Natural History). The Wealden Flora*. Part II. *Gymnospermae*. London: British Museum (Natural History).

Seward, A. C. 1904. *Catalogue of the Mesozoic Plants in the British Museum (Natural History). The Jurassic Flora. 2.* London: British Museum (Natural History).

Seward, A. C. 1912. A petrified *Williamsonia* from Scotland. *Philosophical Transactions of the Royal Society of London* B, **203**, 101–26.

Seward, A. C. 1917. *Fossil Plants*. Volume III. *Pteridospermae, Cycadofilices, Cordaitales, Cycadophyta*. Cambridge: Cambridge University Press.

Seward, A. C. 1926. The Cretaceous plant-bearing rocks of western Greenland. *Philosophical Transactions of the Royal Society of London* B, **215**, 57–175.

Seward, A. C. & Conway, V. M. 1935. Additional Cretaceous plants from western Greenland. *Kungliga Svenska Vetenskapsakademiens Handlingar*, **15**, 1–51.

Seward, A. C. & Conway, V. M. 1939. Fossil plants from Kingigtoq and Kangdlunguaq, West Greenland. *Meddelelser om Grønland*, **93**, 1–41.

Sha, J., Matsukawa, M., Cai, H. *et al.* 2003. The Upper Jurassic-Lower Cretaceous of eastern Heilongjiang, Northeast China: stratigraphy and regional basin history. *Cretaceous Research*, **24**, 715–28.

Sharma, B. D. 1969. Further observations on *Williamsonia santalensis* Sitholey and Bose with the description of a new species. *Palaeontographica* B, **125**, 93–103.

Shilin, P. V. 1986. *Pozdnemelovye flory Kazakhstana. Systematicheskii sostav, istoriya razvitiya, stratigraficheskoe znachenie*. Alma-Ata: Nauka.

Shipley, B. & Dion, J. 1992. The allometry of seed production in herbaceous angiosperms. *American Naturalist*, **139**, 467–83.

Shipunov, A. B. & Sokoloff, D. D. 2003. *Schweitzeria*, a new name for *Irania* Schweitzer (fossil Gymnospermae). *Bulletin of Moscow Society of Naturalists*, **108**, 89–90.

Shukla, V. B. 1944. On *Sahnianthus*, a new genus of petrified flowers from the Intertrappean Beds at Mohgaon Kalan in the Deccan and its relation with the fruit *Enigmocarpon Parjai* Sahni from the same locality. *Proceedings of the National Academy of Sciences, India*, Section B, **14**, 1–39.

Sille, N. P., Collinson, M. E., Kucera, M. & Hooker, J. J. 2006. Morphological evolution of *Stratiotes* through the Paleogene in Europe. *Palaois*, **21**, 272–88.

Simmons, N. B., Seymour, K. L., Habersetzer, J. & Gunnell, G. F. 2008. Primitive Early Eocene bat from Wyoming and the evolution of flight and echolocation. *Nature*, **451**, 818–21.

Simpson, M. G. 1998. Haemodoraceae. In *The Families and Genera of Vascular Plants*. Volume IV. *Flowering Plants – Monocotyledons. Alismatanae and Commelinanae (except Gramineae)* (ed. K. Kubitzki) Berlin, Heidelberg, New York: Springer-Verlag, pp. 212–22.

Sims, H. J., Herendeen, P. S. & Crane, P. R. 1998. A new genus of fossil Fagaceae from the Santonian (Late Cretaceous) of central Georgia, USA. *International Journal of Plant Sciences*, **159**, 391–404.

Sims, H. J., Herendeen, P. S., Lupia, R., Christopher, R. A. & Crane, P. R. 1999. Fossil flowers with Normapolles pollen from the Late Cretaceous of southeastern North America. *Review of Palaeobotany and Palynology*, **106**, 131–51.

Sims, H. J. & Cassara, J. A. 2009. The taphonomic fidelity of seed size in fossil assemblages: a live-dead case study. *Palaios*, **24**, 387–93.

Sincock, C. A. & Watson, J. 1988. Terminology used in the description of bennettitalean cuticle characters. *Botanical Journal of the Linnean Society*, **97**, 179–87.

Singh, H. 1978. *Embryology of Gymnosperms. Encyclopedia of Plant Anatomy.* Berlin, Stuttgart: Gebrüder Borntraeger.

Sitholey, R. V. & Bose, M. N. 1953. *Williamsonia santalensis* sp. nov. – a male fructification from the Rajmahal series, with remarks on the structure of *Ontheanthus polyandra* Ganju. *The Palaeobotanist*, **2**, 29–39.

Sitholey, R. V. & Bose, M. N. 1971. *Weltrichia santalensis* (Sitholey & Bose) and other bennettitalean male fructifications from India. *Palaeontographica* B, **131**, 151–9.

Skarby, A. 1964. Revision of *Gleicheniidites senonicus* Ross. *Stockholm Contributions in Geology*, **11**, 59–77.

Skarby, A. 1968. *Extratriporopollenites* (Pflug) emend. from the Upper Cretaceous of Scania, Sweden. *Stockholm Contributions in Geology*, **16**, 1–60.

Skarby, A. 1986. Normapolles anthers from the Upper Cretaceous of southern Sweden. *Review of Palaeobotany and Palynology*, **46**, 235–56.

Skarby, A., Rowley, J. R. & Nilsson, L. 1990. Exine structure of Upper Cretaceous Normapolles grains from anthers (northeastern Scania, Sweden). *Palynology*, **14**, 145–73.

Skelton, P. W. 2003a. Fluctuating sea-level. In *The Cretaceous World* (ed. P. W. Skelton) The Open University. Cambridge: Cambridge University Press, pp. 67–83.

Skelton, P. W. (ed.) 2003b. *The Cretaceous World*. The Open University. Cambridge: Cambridge University Press.

Skog, J. E. & Dilcher, D. L. 1992. A new species of *Marsilea* from the Dakota Formation in central Kansas. *American Journal of Botany*, **79**, 982–8.

Smiley, C. J. 1969. Cretaceous floras of Chandler-Coville region, Alaska. Stratigraphy and preliminary floristics. *The American Association of Petroleum Geologists Bulletin*, **53**, 482–502.

Smith, A. G., Hurley, A. M. & Briden, J. C. 1981. *Phanerozoic Paleocontinental World Maps. Cambridge Earth Science Series.* Cambridge: Cambridge University Press.

Smith, A. G., Smith, D. G. & Funnel, B. M. 1994. *Atlas of Mesozoic and Cenozoic Coastlines.* Cambridge: Cambridge University Press.

Smith, P. E., Evensen, N. M., York, D. *et al.* 1995. Dates and rates in ancient lakes: 40Ar-39Ar evidence for an Early Cretaceous age for the Jehol Group, northeast China. *Canadian Journal of Earth Science*, **32**, 1426–31.

Smith, S. A. & Donoghue, M. J. 2008. Rates of molecular evolution are linked to life history in flowering plants. *Science*, **322**, 86–9.

Smith, S. Y. & Stockey, R. A. 2003. Aroid seeds from the Middle Eocene Princeton Chert (*Keratosperma allenbyense*, Araceae): comparisons with extant Lasioideae. *International Journal of Plant Sciences*, **164**, 239–50.

Smith, S. Y. & Stockey, R. A. 2007a. Pollen morphology and ultrastructure of Saururaceae. *Grana*, **46**, 250–67.

Smith, S. Y. & Stockey, R. A. 2007b. Establishing a fossil record for the perianthless Piperales: *Saururus tuckerae* sp. nov. (Saururaceae) from the Middle Eocene Princeton Chert. *American Journal of Botany*, **94**, 1642–57.

Smith, S. Y., Collinson, M. E. & Rudall, P. J. 2008. Fossil *Cyclanthus* (Cyclanthaceae, Pandanales) from the Eocene of Germany and England. *American Journal of Botany*, **95**, 688–99.

Smith, U. R. 2001. Revision of the Cretaceous fossil genus *Palaeoaster* (Papaveraceae) and clarification of pertinent species of *Eriocaulon*, *Palaeoaster*, and *Sterculiocarpus*. *Novon*, **11**, 258–60.

Soderstrom, T. R. & Calderon, C. E. 1971. Insect pollination in tropical rainforest grasses. *Biotropica*, **3**, 1–16.

Sokoloff, D. D., Remizowa, M. V., Macfarlane, T. D. & Rudall, P. J. 2008. Classification of the early-divergent angiosperm family Hydatellaceae: one genus instead of two, four new species and sexual dimorphism in dioecious taxa. *Taxon*, **57**, 179–200.

Sole de Porta, N. 1971. Algunos géneros nuevos de polen procedentes de la Formación Guaduas (Maastrichtiense-Paleoceno) de Colombia. *Studia Geologica (Salamanca)*, **2**, 133–43.

Solms-Laubach, H. 1891. On the fructification of *Bennettites gibsonianus*, Carr. *Annals of Botany*, **5**, 419–54.

Soltis, D. E., Soltis, P. S., Nickrent, D. L. *et al.* 1997. Angiosperm phylogeny inferred from 18S ribosomal DNA sequences. *Annals of the Missouri Botanical Garden*, **84**, 1–49.

Soltis, D. E., Soltis, P. S., Chase, M. W. *et al.* 2000a. Angiosperm phylogeny inferred from 18S rDNA, *rbc*L, and *atp*B sequences. *Botanical Journal of the Linnean Society*, **133**, 381–461.

Soltis, D. E., Soltis, P. S. & Zanis, M. J. 2002. Phylogeny of seed plants based on evidence from eight genes. *American Journal of Botany*, **89**, 1670–81.

Soltis, D. E., Senters, A. E., Zanis, M. J. *et al.* 2003. Gunnerales are sister to other core eudicots: implications for the evolution of pentamery. *American Journal of Botany*, **90**, 461–70.

Soltis, D. E., Soltis, P. S., Endress, P. K. & Chase, M. W. 2005. *Phylogeny and Evolution of Angiosperms.* Sunderland, MA: Sinauer Associates.

Soltis, P. S., Soltis, D. E. & Chase, M. W. 1999. Angiosperm phylogeny inferred from multiple genes as a tool for comparative biology. *Nature*, **402**, 402–4.

Soltis, P. S., Soltis, D. E., Zanis, M. J. & Kim, S. 2000b. Basal lineages of angiosperms: relationships and implications for floral evolution. *International Journal of Plant Sciences*, **161** (6 Suppl.), S97–S107.

Soltis, P. S. & Soltis, D. E. 2004. The origin and diversification of angiosperms. *American Journal of Botany*, **91**, 1614–26.

Song, Z. C. 1989. General aspects of the floristic regions on Late Cretaceous and Early Tertiary of China. *Acta Palynologica*, **1**, 1–8.

Song, Z. C., Wang, W. M. & Huang, F. 2004. Fossil pollen records of extant angiosperms in China. *The Botanical Review*, **70**, 425–58.

Spackman, W. 1948. A dicotyledonous wood found associated with the Idaho Tempskyas. *Annals of the Missouri Botanical Garden*, **35**, 107–15.

Spicer, R. A. & Parrish, J. T. 1986. Paleobotanical evidence for cool North Polar climates in the mid-Cretaceous (Albian-Cenomanian). *Geology*, **14**, 703–6.

Spicer, R. A. 1987. The significance of the Cretaceous flora of northern Alaska for the reconstruction of the climate of the Cretaceous. *Geologisches Jahrbuch Reihe A*, **96**, 265–91.

Spicer, R. A., Davies, K. S. & Herman, A. B. 1994a. Circum-Arctic plant fossils and the Cretaceous-Tertiary transition. In *Cenozoic Plants and Climates of the Arctic* (eds M. C. Boulter & H. C. Fischer) Berlin, Heidelberg: Springer-Verlag, pp. 161–74.

Spicer, R. A., Rees, P. M. & Chapman, J. L. 1994b. Cretaceous phytogeography and climate signals. In *Palaeoclimates and their Modelling* (eds J. R. L. Allen, B. J. Hoskin, B. W. Sellwood, R. A. Spicer & P. J. Valdes) London, Glasgow, New York, Tokyo, Melbourne, Madras: Chapman & Hall, pp. 69–78.

Spix, J. B. & Martius, C. F. P. 1823–1831. *Reise in Brasilien.* Munich: Lindauer.

Sporne, K. R. 1974. *The Morphology of Angiosperms.* London: Hutchinson University Library.

Srinivasan, V. & Friis, E. M. 1989. Taxodiaceous conifers from the Upper Cretaceous of Sweden. *Biologiske Skrifter, Det Kongelige Danske Videnskabernes Selskab*, **35**, 1–57.

Srinivasan, V. 1992. Two new species of the conifer *Glenrosa* from the Lower Cretaceous of North America. *Review of Palaeobotany and Palynology*, **72**, 245–55.

Srivastava, S. C. 1974. Pteridospermic remains from the Triassic of Nidpur, Madhya Pradesh, India. *Geophytology*, **4**, 54–9.

Srivastava, S. K. 1966. Upper Cretaceous microflora (Maestrichtian) from Scollard, Alberta, Canada. *Pollen et Spores*, **8**, 497–552.

Srivastava, S. K. 1969. Assorted angiosperm pollen from the Edmonton Formation (Maastrichtian), Alberta, Canada. *Canadian Journal of Botany*, **47**, 975–89.

Srivastava, S. K. 1972. Pollen genus *Erdtmanipollis* Krutzsch 1962. *Pollen et Spores*, **14**, 309–22.

Srivastava, S. K. 1977. Microspores from the Fredericksburg Group (Albian) of the southern United States. *Palébiologie Continentale*, **6**, 1–119.

Srivastava, S. K. 1978. Cretaceous spore-pollen floras: a global evaluation. *Biological Memoires*, **3**, 1–130.

Srivastava, S. K. 1981. Evolution of Upper Cretaceous phytogeoprovinces and their pollen flora. *Review of Palaeobotany and Palynology*, **35**, 155–73.

Stafford, P. J. 1995. The Northwest European Pollen Flora, 53. Ulmaceae. *Review of Palaeobotany and Palynology*, **88**, 25–46.

Stapf, O. 1889. Die Arten der Gattung *Ephedra*. *Denkschriften der Kaiserlichen Akademie der Wissenschaften in Wien. Mathematisch-Naturwissenschaftliche Klasse*, **56**, 1–112.

Stebbins, G. L. 1974. *Flowering Plants. Evolution Above the Species Level.* Cambridge, MA: Harvard University Press.

Stephanović, S., Jager, M., Deutsch, J., Broutin, J. & Masselot, M. 1998. Phylogenetic relationships of conifers inferred from partial 28S rRNA gene sequences. *American Journal of Botany*, **85**, 688–97.

Stein, W. S. & Beck, C. B. 1987. Paraphyletic groups in phylogenetic analysis: Progymnospermopsida and Préphanérogames in alternative views of seed plant relationships. *Bulletin de la Société botanique de France*, **134**, 107–19.

Stewart, W. N. 1983. *Paleobotany and the Evolution of Plants.* Cambridge: Cambridge University Press.

Stevens, P. F. 2001 onwards. Angiosperm Phylogeny Website. Version 7, May 2006 [and more or less continuously updated] http://www.mobot.org/MOBOT/research/APweb/

Stevenson, D. W. 1992. A formal classification of the extant cycads. *Brittonia*, **44**, 220–3.

Stevenson, D. W., Davis, J., Freudenstein, J. *et al.* 2000. A phylogenetic analysis of the monocotyledons based on morphological and molecular character sets, with comments on the placement of *Acorus* and Hydatellaceae. In *Monocots: Systematics and Evolution* (eds K. Wilson & D. Morrison) Melbourne: CSIRO Publishing, pp. 17–24.

Stockey, R. A. & Crane, P. R. 1983. In situ *Cercidiphyllum*-like seedlings from the Paleocene of Alberta, Canada. *American Journal of Botany*, **70**, 1564–8.

Stockey, R. A. 1987. A permineralized flower from the Middle Eocene of British Columbia. *American Journal of Botany*, **74**, 1878–87.

Stockey, R. A. & Pigg, K. B. 1991. Flowers and fruits of *Princetonia allenbyensis* (Magnoliopsida; family indet.) from the Middle Eocene Princeton chert of British Columbia. *Review of Palaeobotany and Palynology*, **70**, 163–72.

Stockey, R. A., Hoffman, G. L. & Rothwell, G. W. 1997. The fossil monocot *Limnobiophyllum scutatum*: resolving the phylogeny of Lemnaceae. *American Journal of Botany*, **84**, 355–68.

Stockey, R. A. & Rothwell, G. W. 2003. Anatomically preserved *Williamsonia* (Williamsoniaceae): evidence for bennettitalean reproduction in the Late Cretaceous of Western North America. *International Journal of Plant Sciences*, **164**, 251–62.

Stockey, R. A. 2006. The fossil record of basal monocots. *Aliso*, **22**, 91–106.

Stockey, R. A., Rothwell, G. W. & Johnson, K. R. 2007. *Cobbania corrugata* gen. et comb. nov. (Araceae): a floating aquatic monocot from the Upper Cretaceous of western North America. *American Journal of Botany*, **94**, 609–24.

Stone, B. C., Huynh, K.-L. & Poppendieck, H.-H. 1998. Pandanaceae. In *The Families and Genera of Vascular Plants.* Volume III. *Flowering Plants – Monocotyledons. Lilianae (except Orchidaceae)* (ed K. Kubitzki) Berlin, Heidelberg, New York: Springer-Verlag, pp. 397–404.

Stone, D. E. & Broome, C. R. 1975. Juglandaceae A Rich. ex Kunth. *World Pollen and Spore Flora*, **4**, 1–35.

Stone, D. E. 1993. Juglandaceae. In *The Families and Genera of Vascular Plants.* Volume II. *Flowering Plants – Dicotyledons. Magnoliid, Hamamelid and Caryophyllid Families* (eds K. Kubitzki, J. G. Rohwer & V. Bittrich) Berlin, Heidelberg, New York: Springer-Verlag, pp. 348–59.

Stopes, M. C. & Fujii, K. 1910. Studies on the structure and affinities of Cretaceous plants. *Philosophical Transactions of the Royal Society of London* B, **201**, 1–90.

Stopes, M. C. 1912. Petrifactions of the earliest European angiosperms. *Philosophical Transactions of the Royal Society of London* B, **203**, 75–100.

Stopes, M. C. 1915. *Catalogue of the Mesozoic Plants in the British Museum (Natural History). The Cretaceous Flora.* Part II. *Lower Greensand (Aptian) Plants of Britain*. London: British Museum (Natural History).

Stopes, M. C. 1918. New bennettitean cones from the British Cretaceous. *Philosophical Transactions of the Royal Society of London* B, **208**, 389–440.

Stover, L. E. 1964. Cretaceous ephedroid pollen from West Africa. *Micropaleontology*, **10**, 145–56.

Stover, L. E. & Partridge, A. D. 1973. Tertiary and Late Cretaceous spores and pollen from the Gippsland Basin, Southeastern Australia. *Proceedings of the Royal Society of Victoria*, **85**, 237–86.

Strömberg, C. A. E. 2005. Decoupled taxonomic radiation and ecological expansion of open-habitat grasses in the Cenozoic of North America. *Proceedings of the National Academy of Sciences, USA*, **102**, 11 980–4.

Strömberg, C. A. E., Werdelin, L., Friis, E. M. & Saraç, G. 2007. The spread of grass-dominated habitats in the Eastern Mediterranean during the Cainozoic: phytolith evidence. *Palaeogeography, Palaeoclimatology, Palaeoecology*, **250**, 18–49.

Stuessy, T. F. 2004. A transitional-combinational theory for the origin of angiosperms. *Taxon*, **53**, 3–16.

Sun, G., Guo, S. X., Zheng, S. L., Piao, T. Y. & Sun, X. K. 1993. First discovery of the earliest angiospermous megafossils in the world. *Science in China*, **36**, 249–56.

Sun, G., Cao, Z., Li, H. & Wang, X. 1995. Cretaceous floras. In *Fossil Floras of China Through the Geological Ages* (ed. X. Li) Guangzhou: Guangdong Science and Technology Press, pp. 411–52.

Sun, G. & Dilcher, D. L. 1997. Discovery of the oldest known angiosperm inflorescence in the world from Lower Cretaceous of Jixi, China. *Acta Palaeontologica Sinica*, **36**, 135–42.

Sun, G., Dilcher, D. L., Zheng, S. & Zhou, Z. 1998. In search of the first flower: a Jurassic angiosperm, *Archaefructus*, from northeast China. *Science*, **282**, 1692–5.

Sun, G., Zheng, S. L. & Mei, S. W. 2000. Discovery of *Liaoningocladus* gen. nov. from the lower part of the Yixian Formation (Upper Jurassic) in western Liaoning, China. *Acta Palaeontologica Sinica*, **39** (Suppl.), 200–8.

Sun, G., Zheng, S., Dilcher, D. L., Wang, Y. & Mei, S. 2001. *Early Angiosperms and their Associated Plants from Western Liaoning, China*. Shanghai: Shanghai Scientific and Technological Education Publishing House.

Sun, G. & Dilcher, D. L. 2002. Early angiosperms from the Lower Cretaceous of Jixi, eastern Heilongjiang, China. *Review of Palaeobotany and Palynology*, **121**, 91–112.

Sun, G., Ji, Q., Dilcher, D. L. *et al.* 2002. Archaefructaceae, a new basal angiosperm family. *Science*, **296**, 899–904.

Sun, G., Dilcher, D. L., Wang, H. & Chen, Z. 2011. A eudicot from the Early Cretaceous of China. *Nature*, **471**, 625–8.

Sun, S.-G., Lu, Y. & Huang, S.-Q. 2006. Floral phenology and sex expression in functionally monoecious *Rhoiptelea chiliantha* (Rhoipteleaceae). *Botanical Journal of the Linnean Society*, **152**, 145–51.

Sun, Z. & Dilcher, D. L. 1988. Fossil *Smilax* from Eocene sediments in western Tennessee. *American Journal of Botany*, **75**, 118.

Surange, K. R. & Chandra, S. 1971. *Dekania indica* gen. et sp. nov. – a Glossopteridean fructification from the Lower Gondwana of India. *The Palaeobotanist*, **20**, 264–70.

Surange, K. R. & Chandra, S. 1974a. Some male fructifications of Glossopteridales. *The Palaeobotanist*, **21**, 255–66.

Surange, K. R. & Chandra, S. 1974b. *Lidgettonia mucronata* sp. nov. a female fructification from the Lower Gondwana of India. *The Palaeobotanist*, **21**, 121–6.

Surange, K. R. & Chandra, S. 1974c. Further observations on *Glossotheca* Surange & Maheshwari: a male fructification of Glossopteridales. *The Palaeobotanist*, **21**, 248–54.

Surange, K. R. & Chandra, S. 1975. Morphology of the gymnospermous fructifications of the *Glossopteris* flora and their relationships. *Palaeontographica* B, **149**, 153–80.

Surange, K. R. & Chandra, S. 1976. Morphology and affinities of *Glossopteris*. *The Palaeobotanist*, **25**, 509–24.

Sutter, D., Forster, P. & Endress, P. K. 2006. Female flowers and systematic position of Picrodendraceae (Euphorbiaceae s.l., Malpighiales). *Plant Systematics and Evolution*, **261**, 187–215.

Sutton, D. A. 1989. The Didymelales: a systematic review. In *Evolution, Systematics, and Fossil History of the*

Hamamelidae. Volume 1. *Introduction and 'Lower' Hamamelidae* (eds P. R. Crane & S. Blackmore) Oxford: Clarendon Press, pp. 279–84.

Swisher, C. C., Wang, Y.-Q., Wang, X.-I., Xu, X. & Wang, Y. 1999. Cretaceous age for the feathered dinosaurs of Liaoning, China. *Nature*, **400**, 58–61.

Süss, H. 1960. Ein Monimiaceen-Holz aus der oberen Kreide Deutschlands, *Hedycaryoxylon subaffine* (Vater) nov. comb. *Senckenbergiana Lethaea*, **41**, 317–30.

Takahashi, M. 1997. Fossil spores and pollen grains of Cretaceous (Upper Campanian) from Sakhalin, Russia. *Journal of Plant Research*, **110**, 283–98.

Takahashi, M., Crane, P. R. & Ando, H. 1999a. *Esgueiria futabensis* sp. nov.; a new angiosperm flower from the Upper Cretaceous (lower Coniacian) of northeastern Honshu, Japan. *Paleontological Research*, **3**, 81–7.

Takahashi, M., Crane, P. R. & Ando, H. 1999b. Fossil flowers and associated plant fossils from the Kamikitaba locality (Ashizawa Formation, Futuba Group, lower Coniacian, Upper Cretaceous) of Northeast Japan. *Journal of Plant Research*, **112**, 187–206.

Takahashi, M., Herendeen, P. S. & Crane, P. R. 2001. Lauraceous flowers from the Kamikitaba locality (Lower Coniacian; Upper Cretaceous) of Northeast Japan. *Journal of Plant Research*, **114**, 429–34.

Takahashi, M., Crane, P. R. & Manchester, S. R. 2002. *Hironoia fusiformis* gen. et sp. nov.: a cornalean fruit from the Kamikitaba locality (Upper Cretaceous, Lower Coniacian) in northeastern Japan. *Journal of Plant Research*, **115**, 463–73.

Takahashi, K. & Suzuki, M. 2003. Dicotyledonous fossil wood flora and early evolution of wood characters in the Cretaceous of Hokkaido, Japan. *IAWA Journal*, **24**, 269–309.

Takahashi, M., Crane, P. R. & Friis, E. M. 2007. Fossil seeds of Nymphaeales from the Tamayama Formation (Futaba Group), Upper Cretaceous (Early Santonian) of northeastern Honshu, Japan. *International Journal of Plant Sciences*, **168**, 341–50.

Takahashi, M., Friis, E. M., Herendeen, P. S. & Crane, P. R. 2008a. Fossil flowers of Fagales from the Kamikitaba Locality (Early Coniacian; Late Cretaceous) of Northeastern Japan. *International Journal of Plant Sciences*, **169**, 899–907.

Takahashi, M., Friis, E. M., Uesugi, K., Suzuki, Y. & Crane, P. R. 2008b. Floral evidence of Annonaceae from the Late Cretaceous of Japan. *International Journal of Plant Sciences*, **169**, 890–8.

Takhtajan, A. L. 1969. *Flowering Plants. Origin and Dispersal.* Edinburgh: Oliver & Boyd.

Takhtajan, A. L. 1980. Outline of the classification of flowering plants (Magnoliophyta). *The Botanical Review*, **46**, 225–359.

Takhtajan, A. L. 1997. *Diversity and Classification of Flowering Plants*. New York: Columbia University Press.

Takhtajan, A. L. (ed.) 1974. *Magnoliaceae-Eucommiaceae. Magnoliophyta Fossilia URSS.* Leningrad: Nauka (in Russian).

Tamura, M. 1993. Ranunculaceae. In *The Families and Genera of Vascular Plants.* Volume II. *Flowering Plants – Dicotyledons. Magnoliid, Hamamelid and Caryophyllid Families* (eds K. Kubitzki, J. G. Rohwer & V. Bittrich) Berlin, Heidelberg, New York: Springer-Verlag, pp. 563–83.

Tamura, M. N. 1998. Nartheciaceae. In *The Families and Genera of Vascular Plants.* Volume III. *Flowering Plants – Monocotyledons. Lilianae (except Orchidaceae)* (ed. K. Kubitzki) Berlin, Heidelberg, New York: Springer-Verlag, pp. 381–92.

Tanai, T. 1961. Neogene floral change in Japan. *Journal of the Faculty of Sciences. Hokkaido University Ser. 4*, **11**, 119–398.

Tang, M., Sternberg, L. & Price, D. 1987. Metabolic aspects of thermogenesis in male cones of five cycad species. *American Journal of Botany*, **74**, 1555–9.

Tang, W. 1987. Insect pollination in the cycad *Zamia pumila* (Zamiaceae). *American Journal of Botany*, **74**, 90–9.

Tao, J.-R. & Zhang, C.-B. 1992. Two angiosperm reproductive organs from the Early Cretaceous of China. *Acta Phytotaxonomica Sinica*, **30**, 423–6.

Tao, J.-R. & Yang, Y. 2003. *Alloephedra xingxuei* gen. et sp. nov., an Early Cretaceous member of Ephedraceae from Dalazi Formation in Yanji Basin, Jilin Province of China. *Acta Palaeontologica Sinica*, **42**, 208–15.

Taylor, D. W. & Crepet, W. L. 1987. Fossil floral evidence of Malpighiaceae and an early plant-pollinator relationship. *American Journal of Botany*, **74**, 274–86.

Taylor, D. W. 1988. Eocene floral evidence of Lauraceae: corroboration of the North American megafossil record. *American Journal of Botany*, **75**, 948–57.

Taylor, D. W. 1990. Paleobiogeographic relationships of angiosperms from the Cretaceous and Early Tertiary of the North American area. *The Botanical Review*, **56**, 279–420.

Taylor, D. W. & Hickey, L. J. 1990. An Aptian plant with attached leaves and flowers: implications for angiosperm origin. *Science*, **247**, 702–4.

Taylor, D. W., Brenner, G. J. & Basha, S. H. 2008. *Scutifolium jordanicum* gen. et sp. nov. (Cabombaceae), an aquatic fossil plant from the Lower Cretaceous of Jordan, and the relationships of related leaf fossils to living genera. *American Journal of Botany*, **95**, 340–52.

Taylor, E. L. & Taylor, T. N. 1992. Reproductive biology of the Permian Glossopteridales and their suggested relationship to flowering plants. *Proceedings of the National Academy of Sciences, USA*, **89**, 11 495–7.

Taylor, E. L. 1996. Enigmatic gymnosperms? Structurally preserved Permian and Triassic seed ferns from Antarctica. *Review of Palaeobotany and Palynology*, **90**, 303–18.

Taylor, E. L. & Taylor, T. N. 2009. Seed ferns from the late Paleozoic and Mesozoic: any angiosperm ancestors lurking there? *American Journal of Botany*, **96**, 237–51.

Taylor, T. N. 1973. A consideration of the morphology, ultrastructure and multicellular microgametophyte of *Cycadeoidea dacotensis* pollen. *Review of Palaeobotany and Palynology*, **16**, 157–64.

Taylor, T. N., Cichan, M. A. & Baldoni, A. M. 1984. The ultrastructure of Mesozoic pollen: *Pteruchus dubius* (Thomas) Townrow. *Review of Palaeobotany and Palynology*, **41**, 319–27.

Taylor, T. N., Del Fueyo, G. M. & Taylor, E. L. 1994. Permineralized seed fern cupules from the Triassic of Antarctica: implications for cupule and carpel evolution. *American Journal of Botany*, **81**, 666–77.

Tebbs, M. C. 1993. Piperaceae. In *The Families and Genera of Vascular Plants.* Volume II. *Flowering Plants – Dicotyledons. Magnoliid, Hamamelid and Caryophyllid Families* (eds K. Kubitzki, J. G. Rohwer & V. Bittrich) Berlin, Heidelberg, New York: Springer-Verlag, pp. 516–20.

Teeling, E. C., Springer, M. S., Madsen, O. *et al.* 2005. A molecular phylogeny for bats illuminates biogeography and the fossil record. *Science*, **307**, 580–4.

Teixeira, C. 1945. *Nymphéacées Fossiles du Portugal.* Lisbon: Serviços Geológicos de Portugal.

Teixeira, C. 1946. Flora cretácica de Esgueira (Aveiro). *Portugaliae Acta Biologica*, **1**, 235–42.

Teixeira, C. 1947. Nouvelles recherches et revision de la flore de Cercal. *Brotéria, Série Trimestral Ciências Naturais*, **16**, 5–15.

Teixeira, C. 1948. *Flora Mesozóica Portuguesa.* Part I. Lisbon: Serviços Geológicos de Portugal.

Teixeira, C. 1950. *Flora Mesozóica Portuguesa.* Part II. Lisbon: Serviços Geológicos de Portugal.

Teixeira, C. 1952. Notes sur quelques gisements de végétaux fossiles du Cretacé des environs de Leiria. *Revista da Faculdade de Ciências de Lisboa. 2.ª Serie, C*, **2**, 133–54.

Teixeira, C. 1954. La flore fossile des calcaires lithographique de Santa Maria de Mayá (Lérida, Espagne). *Boletim da Sociedade Geológica de Portugal*, **12**, 139–52.

Tekleva, M. V. & Krassilov, V. A. 2009. Comparative pollen morphology and ultrastructure of modern and fossil gnetophytes. *Review of Palaeobotany and Palynology*, **156**, 130–8.

Teodoridis, V. & Kvaček, Z. 2005. The extinct genus *Chaneya* Wang et Manchester in the Tertiary of Europe – a revision of *Porana*-like fruit remains from Öhningen and Bohemia. *Review of Palaeobotany and Palynology*, **134**, 85–103.

Thanikaimoni, G. 1968. *Morphologie des Pollens des Ménispermacées.* Pondichéry: Institut Français de Pondichéry. Traveaux de la Section Scientifique et Technique.

Thayn, G. F., Tidwell, W. D. & Stokes, W. L. 1985. Flora of the Lower Cretaceous Cedar Mountain Formation of Utah and Colorado. Part III: *Icacinoxylon pittiensis* n. sp. *American Journal of Botany*, **72**, 175–80.

The Board of Trustees of the Royal Botanic Gardens, Kew. 2008. *World Checklist of Monocotyledons.* http://apps.kew.org/wcsp/monocots/

Thien, L. B., Bernhardt, P., Gibbs, G. W. *et al.* 1985. The pollination of *Zygogynum* (Winteraceae) by a moth, *Sabatinca* (Micropterygidae): an ancient association? *Science*, **227**, 540–3.

Thien, L. B., Bernhardt, P., Devall, M. S. *et al.* 2009. Pollination biology of basal angiosperms (ANITA grade). *American Journal of Botany*, **96**, 166–82.

Thoday, M. G. 1911. The female inflorescence and ovules of *Gnetum africanum*, with notes on *Gnetum scandens*. *Annals of Botany*, **25**, 1101–35.

Thomas, H. H. 1925. The Caytoniales, a new group of angiospermous plants from the Jurassic rocks of Yorkshire. *Philosophical Transactions of the Royal Society of London* B, **213**, 299–363.

Thomas, H. H. 1933. On some pteridospermous plants from the Mesozoic rocks of South Africa. *Philosophical Transactions of the Royal Society of London* B, **222**, 193–265.

Thomas, H. H. 1958. *Lidgettonia*, a new type of fertile *Glossopteris*. *Bulletin of the British Museum (Natural History) Geology*, **3**, 179–89.

Thompson, W. P. 1918. Independent evolution of vessels in Gnetales and angiosperms. *Botanical Gazette*, **65**, 83–90.

Thorne, R. F. 1976. A phylogenetic classification of the Angiospermae. *Evolutionary Biology*, **9**, 35–106.

Tidwell, W. D., Rushforth, S. R., Reveal, J. L. & Behunin, H. 1970. *Palmoxylon simperi* and *Palmoxylon pristina*: two pre-Cretaceous angiosperms from Utah. *Science*, **168**, 835–40.

Tidwell, W. D., Simper, A. D. & Thayn, G. F. 1977. Additional information concerning the controversial Triassic plant: *Sanmiguelia*. *Palaeontographica* B, **163**, 143–51.

Tidwell, W. D., Ash, S. R. & Parker, L. R. 1981. Cretaceous and Tertiary floras of the San Juan Basin. In *Advances in San Juan Basin Paleontology* (eds. S. G. Lucas, J. K. Rigby & B. S. Kues) Albuquerque, NM: University of New Mexico Press.

Tidwell, W. D. & Parker, L. R. 1990. *Protoyucca shadishii* gen. et sp. nov., an arborescent monocotyledon with secondary growth from the Middle Miocene of northwestern Nevada, USA. *Review of Palaeobotany and Palynology*, **62**, 79–95.

Tiffney, B. H. 1977a. Fruits and seeds of the Brandon Lignite: Magnoliaceae. *Botanical Journal of the Linnean Society*, **75**, 299–393.

Tiffney, B. H. 1977b. Dicotyledonous angiosperm flower from the Upper Cretaceous of Martha's Vineyard. Massachusetts. *Nature*, **265**, 136–7.

Tiffney, B. H. 1979. Fruits and seeds of the Brandon Lignite III. *Turpinia* (Staphyleaceae). *Brittonia*, **31**, 39–51.

Tiffney, B. H. & Barghoorn, E. S. 1979. Fruits and seeds of the Brandon Lignite IV. Illiciaceae. *American Journal of Botany*, **66**, 321–9.

Tiffney, B. H. 1981. Diversity and major events in the evolution of land plants. In *Paleobotany, Paleoecology and Evolution* (ed. K. J. Niklas) New York: Praeger Press, pp. 193–230.

Tiffney, B. H. 1984. Seed size, dispersal syndromes, and the rise of the angiosperms: evidence and hypothesis. *Annals of the Missouri Botanical Garden*, **71**, 551–76.

Tiffney, B. H. & McClammer, J. U. A. 1988. Seed of the Anonaceae from the Palaeocene of Pakistan. *Tertiary Research*, **9**, 13–20.

Tiffney, B. H. & Niklas, K. J. 1990. Continental area, dispersion, latitudinal distribution and topographic variety: a test of correlation with terrestrial plant diversity. In *Causes of Evolution* (eds R. M. Ross & W. D. Allmon) Chicago, IL: University of Chicago Press, pp. 76–102.

Tiffney, B. H. 1993. Fruits and seeds of the Tertiary Brandon Lignite. VII. *Sargentodoxa* (Sargentodoxaceae). *American Journal of Botany*, **80**, 517–23.

Todzia, C. A. 1993. Chloranthaceae. In *The Families and Genera of Vascular Plants.* Volume II. *Flowering Plants – Dicotyledons. Magnoliid, Hamamelid and Caryophyllid Families* (eds K. Kubitzki, J. G. Rohwer & V. Bittrich) Berlin, Heidelberg, New York: Springer-Verlag, pp. 281–9.

Tomlinson, P. B. 1974. Development of the stomatal complex as a taxonomic character in the monocotyledons. *Taxon*, **23**, 109–28.

Tomlinson, P. B. 1991. Pollen scavenging. *National Geographic Research and Exploration*, **7**, 188–95.

Tomlinson, P. B. 1994. Functional morphology of saccate pollen in conifers with special reference to Podocarpaceae. *International Journal of Plant Sciences*, **155**, 699–715.

Tomlinson, P. B., Braggins, J. E. & Rattenbury, J. A. 1997. Contrasted pollen capture mechanisms in Phyllocladaceae and certain Podocarpaceae (Coniferales). *American Journal of Botany*, **84**, 214–23.

Townrow, J. A. 1957. On *Dicroidium*, probably a pteridospermous leaf, and other leaves now removed from this genus. *Transactions of the Geological Society of South Africa*, **60**, 21–56.

Townrow, J. A. 1962. On *Pteruchus* a microsporophyll of the Corystospermaceae. *Bulletin of the British Museum (Natural History) Geology*, **6**, 289–320.

Townrow, J. A. 1965. A new member of the Corystospermaceae Thomas. *Annals of Botany*, **29**, 495–511.

Tralau, H. 1964. The genus *Nypa* van Wurmb. *Kungliga Svenska Vetenskapsakademiens Handlingar 4th Serie*, **10**, 1–29.

Tralau, H. 1967. The phytogeographic evolution of the genus *Ginkgo*. *Botaniska Notiser*, **120**, 409–22.

Traverse, A. 1988. *Paleopalynology*. Boston, MA: Unwin Hyman.

Trevisan, L. 1988. Angiospermous pollen (monosulcate-trichotomosulcate phase) from very early Lower Cretaceous of Southern Tuscany (Italy): some aspects. *Seventh International Palynological Conference, Brisbane, Abstracts*, **165**.

Trincão, P. R. P. 1990. Esporos e pólenes do Cretácio inferior (Berriasiano-Aptiano) de Portugal: paleontologia e biostratigrafia. Lisbon: PhD Thesis, Universidade Nova de Lisboa, pp. 1–312, 47 plates.

Trivedi, B. S. & Verma, C. L. 1972. Occurrence of *Heliconiaites mohgaonensis* gen. et sp. nov. from the Early Eocene of Deccan Intertrappean series, M. P., India. *Palaeontographica* B, **139**, 73–82.

Troitsky, A. V., Melekhovets, Y. F., Rakhimova, G. M. *et al.* 1991. Angiosperm origin and early stages of seed plant evolution deduced from rRNA sequence comparisons. *Journal of Molecular Evolution*, **32**, 253–61.

Truswell, E. M. 1991. Antarctica: a history of terrestrial vegetation. In *The Geology of Antarctica* (ed. R. J. Tingey) Oxford: Clarendon, pp. 499–537.

Tschan, G. F., Denk, T. & von Balthazar, M. 2008. *Credneria* and *Platanus* (Platanaceae) from the Late Cretaceous

(Santonian) of Quedlinburg, Germany. *Review of Palaeobotany and Palynology*, **152**, 211–36.

Tschudy, R. H. 1981. Geographic distribution and dispersal of Normapolles genera in North America. *Review of Palaeobotany and Palynology*, **35**, 283–314.

Tschudy, R. H., Pillmore, C. L., Orth, C. J., Gilmore, J. S. & Knight, J. D. 1984. Disruption of the terrestrial plant ecosystem at the Cretaceous–Tertiary boundary, Western Interior. *Science*, **225**, 1030–4.

Turutanova-Ketova, A. 1930. Jurassic flora of the chain Kara-Tau (Tian-Shan). *Travaux du Musée Geologique près l'Académie des sciences de l'URSS*, **6**, 131–72.

Uhl, N. W. & Moore, H. E. 1980. Androecial development in six polyandrous genera representing five major groups of palms. *Annals of Botany*, **45**, 57–75.

Upchurch, G. R. & Doyle, J. A. 1981. Paleoecology of the conifers *Frenelopsis* and *Pseudofrenelopsis* (Cheirolepidiaceae) from the Cretaceous Potomac Group of Maryland and Virginia. In *Geobotany*, Volume **2** (ed. R. C. Romans) New York: Plenum Publishing Corporation, pp. 167–202.

Upchurch, G. R. 1984. Cuticle evolution in Early Cretaceous angiosperms from the Potomac Group of Virginia and Maryland. *Annals of the Missouri Botanical Garden*, **71**, 522–50.

Upchurch, G. R. & Wolfe, J. A. 1987. Mid-Cretaceous to Early Tertiary vegetation and climate: evidence from fossil leaves and woods. In *The Origins of Angiosperms and their Biological Consequences* (eds E. M. Friis, W. G. Chaloner & P. R. Crane) Cambridge: Cambridge University Press, pp. 75–105.

Upchurch, G. R. & Dilcher, D. L. 1990. Cenomanian angiosperm leaf megafossils, Dakota Formation, Rose Creek Locality, Jefferson County, southeastern Nebraska. *US Geological Survey Bulletin*, **1915**, 1–55.

Upchurch, G. R., Crane, P. R. & Drinnan, A. N. 1994. *The Megaflora from the Quantico Locality (Upper Albian), Lower Cretaceous Potomac Group of Virginia*. Martinsville, VA: Virginia Museum of Natural History.

Vajda, V., Raine, J. I. & Hollis, C. J. 2001. Indication of global deforestation at the Cretaceous-Tertiary boundary by New Zealand fern spike. *Science*, **294**, 1700–2.

Vakhrameev, V. A. 1952. *Stratigraphy and Fossil Flora of the Cretaceous Deposits in the Western Kazakhstan*. Moscow, Leningrad: Akademia Nauk SSSR.

Vakhrameev, V. A. 1958. *Stratigraphy and Fossil Flora of the Jurassic and Cretaceous Deposits in the Vilyuy Depression and the Adjacent Part of the Priverhoyansk Trough*. Moscow: USSR Academy of Science Press.

Vakhrameev, V. A. & Kotova, I. Z. 1977. Ancient angiosperms and accompanying plants from the Lower Cretaceous of Transbailkalia. *Paleontological Journal*, **1977**, 487–95.

Vakhrameev, V. A., Dobruskina, I. A., Meyen, S. V. & Zaklinskaja, E. D. 1978. *Paläozoische und Mesozoische Floren Eurasiens und die Phytogeographie dieser Zeit*. Jena: VEB Gustav Fischer Verlag.

Vakhrameev, V. A. & Krassilov, V. A. 1979. Reproductive structures of angiosperms from the Albian of Kazakhstan. *Paleontological Journal*, **1979**, 112–18.

Vakhrameev, V. A. 1991. *Jurassic and Cretaceous Floras and Climates of the Earth*. Cambridge: Cambridge University Press.

van der Ham, R. W. J. M., van Konijnenburg-van Cittert, J. H. A. & Indeherberge, L. 2007. Seagrass foliage from the Maastrichtian type area (Maastrichtian, Danian, NE Belgium, SE Netherlands). *Review of Palaeobotany and Palynology*, **144**, 301–21.

van der Pijl, L. 1953. On the flower biology of some plants from Java with general remarks on fly-traps (species of *Annona*, *Artocarpus*, *Typhonium*, *Gnetum*, *Arisaema* and *Abroma*). *Annales Bogorienses*, **1**, 77–99.

van der Pijl, L. 1957. The dispersal of plants by bats (chiropterochory). *Acta botanica Neerlandica*, **6**, 291–315.

van der Pijl, L. 1982. *Principles of Dispersal in Higher Plants*. New York: Springer-Verlag.

van Hoeken-Klinkenberg, P. M. J. 1964. A palynological investigation of some Upper Cretaceous sediments in Nigeria. *Pollen et Spores*, **6**, 209–31.

Van Itterbeeck, J., Markevich, V. S. & Horne, D. J. 2004. The age of the dinosaur-bearing Cretaceous sediments at Dashuiguo, Inner Mongolia, P.R. China based on charophytes, ostracods and palynomorphs. *Cretaceous Research*, **25**, 391–409.

Vangerow, E. F. 1954. Megasporen und andere pflanzliche Mikrofossilien aus der Aachener Kreide. *Palaeontographica* B, **96**, 24–38.

Vaudois-Miéja, N. & Lejal-Nicol, A. 1987a. Paléocarpologie africaine: apparition dès l'Aptien en Égypte d'un Palmier (*Hyphaeneocarpon aegypticum* n.sp.). *Comptes Rendus de l'Académie des Sciences Paris, Série II*, **304**, 233–8.

Vaudois-Miéja, N. & Lejal-Nicol, A. 1987b. Paléocarpologie africaine: *Cercidiphyllocarpon prasadii* n. gen. n. sp., un nouveau fruit du Nubien de l'Afrique orientale. *Comptes Rendus de l'Académie des Sciences Paris, Série II*, **305**, 149–53.

Vaudois-Miéja, N. & Lejal-Nicol, A. 1988. Paléocarpologie africaine: *Sapindaceaecarpum koelreuterioides* n.sp., un nouveau fruit du Nubien de l'Afrique orientale. *Comptes*

Rendus de l'Académie des Sciences Paris, Série II, **307**, 855–62.

Velenovský, J. 1882. Die Flora der Böhmischen Kreideformation. *Beiträge zur Paläontologie Österreich-Ungarns und des Orientes*, **2**, 8–32.

Velenovský, J. 1883. Die Flora der Böhmischen Kreideformation. *Beiträge zur Paläontologie Österreich-Ungarns und des Orientes*, **3**, 1–22.

Velenovský, J. 1884. Die Flora der Böhmischen Kreideformation. *Beiträge zur Paläontologie Österreich-Ungarns und des Orientes*, **4**, 1–14.

Velenovský, J. 1885. *Die Gymnospermen der Böhmischen Kreideformation*. Praha: E. Greger.

Velenovský, J. 1889. Květena Českého Cenomanu. *Rozpravy Královské České Společnosti Nauk*, **7**, 1–75.

Velenovský, J. & Viniklář, L. 1926. Flora Cretacea Bohemiae. I. *Rozpravy Státního Geologického Ústavu Ceskoslovenské republiky*, **1**, 1–57.

Velenovský, J. & Viniklář, L. 1927. Flora Cretacea Bohemiae. II. *Rozpravy Státního Geologického Ústavu Ceskoslovenské republiky*, **2**, 1–54.

Velenovský, J. & Viniklář, L. 1929. Flora Cretacea Bohemiae. III. *Rozpravy Státního Geologického Ústavu Ceskoslovenské republiky*, **3**, 1–33.

Verma, J. K. 1958. On an inflorescence of a new petrified monocot flower, *Shuklanthus superbum* gen. et sp. nov. from the Deccan Intertrappean series of Madhya Pradesh, India. *Journal of the Palaeontological Society of India*, **3**, 185–200.

Viehofen, A., Hartkopf-Fröder, C. & Friis, E. M. 2008. Inflorescences and flowers of *Mauldinia angustiloba* sp. nov. (Lauraceae) from mid-Cretaceous karst infillings in the Rhenish Massif, Germany. *International Journal of Plant Sciences*, **169**, 871–89.

Villanueva-Amadoz, U., Pons, D., Diez, J. B., Ferrer, J. & Sender, L. M. 2010. Angiosperm pollen grains of San Just site (Escucha Formation) from the Albian of the Iberian Range (north-eastern Spain). *Review of Palaeobotany and Palynology*, **162**, 362–81.

Vink, W. 1993. Winteraceae. In *The Families and Genera of Vascular Plants*. Volume II. *Flowering Plants – Dicotyledons. Magnoliid, Hamamelid and Caryophyllid Families* (eds K. Kubitzki, J. G. Rohwer & V. Bittrich) Berlin, Heidelberg, New York: Springer-Verlag, pp. 630–8.

Vishnu-Mittre 1953. A male flower of the Pentoxyleae with remarks on the structure of the females cones of the group. *The Palaeobotanist*, **2**, 75–84.

Volkheimer, W. & Salas, A. 1975. Die älteste Angiospermen-Palynoflora Argentiniens von der Typuslokalität der unterkretazischen Huitrín-Folge des Neuquén-Beckens. Mikrofloristische Assoziation und biostratigraphische Bedeutung. *Neues Jahrbuch für Geologie und Paläontologie Monatsheft*, **7**, 424–36.

von Balthazar, M. & Endress, P. K. 1999. Floral bract function, flowering process and breeding systems of *Sarcandra* and *Chloranthus* (Chloranthaceae). *Plant Systematics and Evolution*, **218**, 161–78.

von Balthazar, M., Endress, P. K. & Qiu, Y.-L. 2000. Phylogenetic relationships in Buxaceae based on nuclear internal transcribed spacers and plastid *ndh*F sequences. *International Journal of Plant Sciences*, **161**, 785–92.

von Balthazar, M. & Endress, P. K. 2002. Development of inflorescences and flowers in Buxaceae and the problem of perianth interpretation. *International Journal of Plant Sciences*, **163**, 847–76.

von Balthazar, M., Schatz, G. E. & Endress, P. K. 2003. Female flowers and inflorescences of Didymelaceae. *Plant Systematics and Evolution*, **237**, 199–208.

von Balthazar, M., Pedersen, K. R. & Friis, E. M. 2005. *Teixeiraea lusitanica* gen. et sp. nov., a ranunculalean flower from the Early Cretaceous of Portugal. *Plant Systematics and Evolution*, **255**, 55–75.

von Balthazar, M., Pedersen, K. R., Crane, P. R. & Friis, E. M. 2007. *Potomacanthus lobatus* gen. et sp. nov., a new flower of probable Lauraceae from the Early Cretaceous (Early to Middle Albian) of eastern North America. *American Journal of Botany*, **94**, 2041–53.

von Balthazar, M., Pedersen, K. R., Crane, P. R. & Friis, E. M. 2008. *Carpestella lacunata* gen. et sp. nov., a new basal angiosperm flower from the Early Cretaceous (Early to Middle Albian) of eastern North America. *International Journal of Plant Sciences*, **169**, 890–8.

von Balthazar, M. & Schönenberger, J. 2009. Floral structure and organization in Platanaceae. *International Journal of Plant Sciences*, **170**, 210–25.

von Balthazar, M., Crane, P. R., Pedersen, K. R. & Friis, E. M. 2011. New flowers of Laurales from the Early Cretaceous (Early to Middle Albian) of eastern North America. In *Flowers on the Tree of Life* (eds L. Wanntorp & L.P. Ronse De Craene) Systematic Association Special Volume Series Cambridge University Press, pp. (In press.)

von Ettingshausen, C. 1895. Beiträge zur Kenntnis der Kreideflora Australiens. *Denkschriften der Kaiserlichen Akademie der Wissenschaften in Wien. Mathematisch-Naturwissenschaftliche Classe*, **62**, 1–56.

Walker, J. W. 1974a. Aperture evolution in the pollen of primitive angiosperms. *American Journal of Botany*, **61**, 1112–37.

Walker, J. W. 1974b. Evolution of the exine structure in the pollen of primitive angiosperms. *American Journal of Botany*, **61**, 891–902.

Walker, J. W. & Doyle, J. A. 1975. The bases of angiosperm phylogeny. *Annals of the Missouri Botanical Garden*, **62**, 664–723.

Walker, J. W. 1976. Comparative pollen morphology and phylogeny of the ranalean complex. In *Origin and Early Evolution of Angiosperms* (ed. C. B. Beck) New York: Columbia University Press, pp. 241–99.

Walker, J. W., Brenner, G. J. & Walker, A. G. 1983. Winteraceous pollen in the Lower Cretaceous of Israel: early evidence of a magnolialean angiosperm family. *Science*, **220**, 1273–5.

Walker, J. W. & Walker, A. G. 1984. Ultrastructure of Lower Cretaceous angiosperm pollen and the origin and early evolution of flowering plants. *Annals of the Missouri Botanical Garden*, **71**, 464–521.

Walker, J. W. & Walker, A. G. 1986. Ultrastructure of Early Cretaceous angiosperm pollen and its evolutionary implications. In *Pollen and Spores: Form and Function* (eds S. Blackmore & I. K. Ferguson) London: Academic Press, pp. 203–17.

Wang, H. & Dilcher, D. L. 2006. Aquatic angiosperms from the Dakota Formation (Albian, Lower Cretaceous), Hoisington III locality, Kansas, USA. *International Journal of Plant Sciences*, **167**, 385–401.

Wang, X., Zheng, S. & Jin, J. 2010. Structure and relationships of *Problematospermum*, an enigmatic seeds from the Jurassic of China. *International Journal of Plant Sciences*, **171**, 447–56.

Wang, X.-L., Zhou, Z., Swisher, L. C. C. *et al.* 2001a. A Cretaceous age for the Jehol Biota: evidence of new $^{40}Ar/^{39}Ar$ dates and fossil vertebrates. In *An International Symposium on Exploring the History of Life on the Earth: Paleontology in China during the Last 15 Years.* Abstracts of Papers. Beijing, China: National Natural Science Foundation of China, pp. 50–51.

Wang, X. L., Wang, Y. Q., Xu, X. *et al.* 1999. Record of the Sihetun vertebrate mass mortality events, western Liaoning, China: caused by volcanic eruptions. *Geological Review*, **45** (Suppl.), 458–67.

Wang, Y.-F. & Manchester, S. R. 2000. *Chaneya*, a new genus of winged fruit from the Tertiary of North America and Eastern Asia. *International Journal of Plant Sciences*, **161**, 167–78.

Wang, Y.-F., Li, C.-S. & Li, Z.-Y. 2001b. *Wuyunanthus* gen. nov., a flower of Celastraceae from the Palaeocene of north-east China. *Botanical Journal of the Linnean Society*, **136**, 323–7.

Wanntorp, L., Dettmann, M. E. & Jarzen, D. M. 2004a. Tracking the Mesozoic distribution of *Gunnera*: comparison with the fossil pollen species *Tricolpites reticulatus* Cookson. *Review of Palaeobotany and Palynology*, **132**, 163–74.

Wanntorp, L., Praglowski, J. & Grafström, E. 2004b. New insight into the pollen morphology of *Gunnera* (Gunneraceae). *Grana*, **43**, 15–21.

Wanntorp, L. & Ronse De Craene, L. P. 2005. The *Gunnera* flower: key to eudicot diversification or response to pollination mode? *International Journal of Plant Sciences*, **166**, 945–53.

Ward, J. V. 1986. Early Cretaceous angiosperm pollen from the Cheyenne and Kiowa Formations (Albian) of Kansas, USA. *Palaeontographica* B, **202**, 1–81.

Ward, J. V. & Doyle, J. A. 1988. Possible affinities of two triporoidate tetrads from the mid-Cretaceous of Laurasia. In *Abstract Seventh International Palynological Congress, Brisbane*, p. 179.

Ward, J. V., Doyle, J. A. & Hotton, C. L. 1989. Probable granular magnoliid angiosperm pollen from the Early Cretaceous. *Pollen et Spores*, **31**, 113–32.

Ward, J. V. & Doyle, J. A. 1994. Ultrastructure and relationships of mid-Cretaceous polyforate and triporate pollen from northern Gondwana. In *Ultrastructure of Fossil Spores and Pollen. Its Bearing on Relationships among Fossil and Living Groups* (eds M. H. Kurmann & J. A. Doyle) Kew: The Royal Botanic Gardens, pp. 161–72.

Ward, L. F. 1888. Evidence of the fossil plants as to the age of the Potomac Formation. *American Journal of Science 3rd ser.*, **36**.

Ward, L. F. 1895. The Potomac Formation. *US Geological Survey Annual Report*, **15**, 307–97.

Ward, L. F. 1905. Status of the Mesozoic floras of the United States. Second Paper. *Monographs of the United States Geological Survey*, **48**, 1–616.

Wardle, P. 1969. Biological flora of New Zealand. 4. *Phyllocladus alpinus* Hook F. (Podocarpaceae) Mountain Toatoa, Celery Pine. *New Zealand Journal of Botany*, **7**, 76–95.

Watson, J. 1969. A revision of the English Wealden flora, I. Charales-Ginkgoales. *Bulletin of the British Museum (Natural History) Geology*, **17**, 209–54.

Watson, J. 1977. Some Lower Cretaceous conifers of the Cheirolepidiaceae from the U.S.A. and England. *Palaeontology*, **20**, 715–49.

Watson, J. 1983. Two Wealden species of *Equisetum* found *in situ*. *Acta Palaeobotanica*, **28**, 265–9.

Watson, J. & Fischer, H. L. 1984. A new conifer genus from the Lower Cretaceous Glen Rose Formation, Texas. *Palaeontology*, **27**, 719–27.

Watson, J. 1988. The Cheirolepidiaceae. In *Origin and Evolution of Gymnosperms* (ed. C. B. Beck) New York: Columbia University Press, pp. 382–447.

Watson, J. & Sincock, C. A. 1992. *Bennettitales of the English Wealden*. London: The Palaeontographical Society.

Watson, J. & Alvin, K. L. 1996. An English Wealden floral list, with comments on the possible environmental indicators. *Cretaceous Research*, **17**, 5–26.

Watson, J., Lydon, S. J. & Harrison, N. A. 2001. A revision of the English Wealden Flora, III: Czekanowskiales, Ginkgoales & allied Coniferales. *Bulletin of the British Museum (Natural History) Geology*, **57**, 29–82.

Watson, J. & Lydon, S. J. 2004. The bennettitalean trunk genera *Cycadeoidea* and *Monanthesia* in the Purbeck, Wealden and Lower Greensand of southern England: a reassessment. *Cretaceous Research*, **25**, 1–26.

Watson, J. & Cusack, H. A. 2005. *Cycadales of the English Wealden*. London: The Palaeontographical Society.

Weber, R. 1995. A new species of *Scoresbya* Harris and *Sonoraphyllum* gen. nov. (Plantae incertae sedis) from the Late Triassic of Sonora, Mexico. *Revista Mexicana de Ciencias Geológicas*, **12**, 94–107.

Wehr, W. C. & Manchester, S. R. 1996. Paleobotanical significance of Eocene flowers, fruits and seeds from Republic, Washington. *Washington Geology*, **24**, 25–7.

Werker, E. 1997. *Seed Anatomy. Handbuch der Pflanzenanatomie*. Berlin, Stuttgart: Gebrüder Borntraeger.

Westoby, M., Leishman, M. R. & Lord, J. M. 1996. Comparative ecology of seed size and dispersal. *Philosophical Transactions of the Royal Society of London* B, **351**, 1309–18.

Wettstein, R. R. von 1907. *Handbuch der Systematischen Botanik 2*. Leipzig: Franz Deuticke.

Weyland, H. 1938. Beiträge zur Kenntnis der rheinischen Tertiärflora. III. *Palaeontographica* B, **83**, 123–71.

Weyland, H., Kilpper, K. & Berendt, W. 1967. Kritische Untersuchungen zur Kutikularanalyse tertiärer Blätter VII. *Palaeontographica* B, **120**, 151–68.

Weyland, W., Pflug, H. D. & Jähnischen, H. 1958. *Celtoidanthus pseudorobustus* n. gen., n. sp. eine Ulmaceen-Blüte aus Braunkohle der Niederlausitz. *Palaeontographica* B, **105**, 67–74.

Whalley, P. 1986. A review of current fossil evidence of Lepidoptera in the Mesozoic. *Biological Journal of the Linnean Society*, **28**, 253–72.

Whalley, P. 1987. Insects and Cretaceous mass extinctions. *Nature*, **327**, 562.

Wheeler, E., Scott, R. A. & Barghoorn, E. S. 1977. Fossil dicotyledonous wood from Yellowstone National Park. *Journal of the Arnold Arboretum*, **58**, 280–306.

Wheeler, E. F., Lee, M. & Matten, L. C. 1987. Dicotyledonous woods from the Upper Cretaceous of southern Illinois. *Botanical Journal of the Linnean Society*, **95**, 77–100.

Wheelwright, N. T. & Orians, G. H. 1982. Seed dispersal by animals: contrasts with pollen dispersal, problems of terminology, and constraints on coevolution. *American Naturalist*, **119**, 402–13.

Wheelwright, N. T. 1988. Fruit-eating birds and bird-dispersed plants in the tropics and the temperate zone. *Trends in Ecology and Evolution*, **3**, 270–4.

White, M. E. 1981. Revision of the Talbragar Fish Bed flora (Jurassic) of New South Wales. *Records of the Australian Museum*, **33**, 695–721.

Whitehead, D. R. 1969. Wind pollination in the angiosperms: evolutionary and environmental considerations. *Evolution*, **23**, 28–35.

Whitehead, D. R. 1983. Wind pollination: some ecological and evolutionary perpespectives. In *Pollination Biology* (ed. L. Real) London: Academic Press, pp. 97–108.

Wieland, G. R. 1906. *American Fossil Cycads*. Washington, D.C.: Carnegie Institution of Washington.

Wieland, G. R. 1911. On the *Williamsonia* Tribe. *The American Journal of Science*, **32**, 433–76.

Wieland, G. R. 1916. *American Fossil Cycads*. Volume II. *Taxonomy*. Washington, D.C.: Carnegie Institution of Washington.

Wieland, G. R. 1934. Fossil cycads, with special reference to *Raumeria Reichenbachiana* Goeppert sp. of the Zwinger of Dresden. *Palaeontographica* B, **79**, 85–130.

Wikström, N., Savolainen, V. & Chase, M. W. 2001. Evolution of the angiosperms. Calibrating the tree. *Proceedings of the Royal Society of London* B, **268**, 2211–20.

Wilde, V. 1989. Untersuchungen zur Systematik der Blattreste aus dem Mitteleozän der Grube Messel bei Darmstadt (Hessen, Bundesrepublik Deutschland). *Courier Forschungsinstitut Senckenberg*, **115**, 1–213.

Wilf, P., Johnson, K. R. & Huber, B. T. 2003. Correlated terrestrial and marine evidence for global climate changes before mass extinction at the Cretaceous-Paleogene boundary. *Proceedings of the National Academy of Sciences, USA*, **100**, 599–604.

Wilf, P. & Johnson, K. R. 2004. Land plant extinction at the end of the Cretaceous: a quantitative analysis of the North Dakota megafloral record. *Paleobiology*, **30**, 347–68.

Wilford, G. E. & Brown, P. J. 1994. Maps of late Mesozoic-Cenozoic Gondwana break-up: some palaeogeographical implications. In *History of the Australian Vegetation* (ed. R. S. Hill) Cambridge: Cambridge University Press, pp. 5–13.

Wilkinson, H. P. 1981. The anatomy of the hypocotyls of *Ceriops*. *Botanical Journal of the Linnean Society*, **82**, 139–64.

Wilkinson, H. P. 1988. Sapindaceous pyritised twigs from the Eocene of Sheppey, England. *Tertiary Research*, **9**, 81–6.

Williamson, P. S. & Schneider, E. I. 1993. Cabombaceae. In *The Families and Genera of Vascular Plants.* Volume II. *Flowering Plants – Dicotyledons. Magnoliid, Hamamelid and Caryophyllid Families* (eds K. Kubitzki, J. G. Rohwer & V. Bittrich) Berlin, Heidelberg, New York: Springer-Verlag, pp. 157–61.

Willson, M. F., Rice, B. L. & Westoby, M. 1990. Seed dispersal spectra: a comparison of temperate plant communities. *Journal of Vegetation Science*, **1**, 547–62.

Willson, M. F. 1993. Mammals as seed-dispersal mutualists in North America. *Oikos*, **67**, 59–76.

Wilson, K. I. & Morrison, D. A. (eds) 2000. *Monocots: Systematics and Evolution*. Melbourne: CSIRO.

Wilson, L. R. 1962. Permian plant microfossils from the Flowerpot Formation, Greer County, Oklahoma. *Oklahoma Geological Survey, Circular* **49**, 5–50.

Wing, S. L. & Tiffney, B. H. 1987a. The reciprocal interaction of angiosperm evolution and tetrapod herbivory. *Review of Palaeobotany and Palynology*, **50**, 179–210.

Wing, S. L. & Tiffney, B. H. 1987b. Interactions of angiosperms and herbivorous tetrapods through time. In *The Origins of Angiosperms and their Biological Consequences* (eds E. M. Friis, W. G. Chaloner & P. R. Crane) Cambridge: Cambridge University Press, pp. 203–24.

Wing, S. L. & Sues, H. D. 1992. Mesozoic and early Cenozoic terrestrial systems. In *Terrestrial Ecosystems Through Time* (ed. A. K. Behrensmeyer) Chicago, IL: Chicago University Press, pp. 327–416.

Wing, S. L., Hickey, L. J. & Swisher, C. C. 1993. Implications of an exceptional fossil flora for Late Cretaceous vegetation. *Nature*, **363**, 342–4.

Wing, S. L. & Boucher, L. D. 1998. Ecological aspects of the Cretaceous flowering plant radiation. *Annual Review of Earth and Planetary Science*, **26**, 379–421.

Wing, S. L., Harrington, G. J., Smith, F. A. *et al.* 2005. Transient floral change and rapid global warming at the Paleocene-Eocene boundary. *Science*, **310**, 993–6.

Wing, S. L., Herrera, F, Jaramillo, C. A. *et al.* 2009. Late Paleocene fossils from the Cerrejon Formation, Colombia, are the earliest record of Neotropical rainforest. *Proceedings of the National Academy of Sciences, USA*, **106**, 18 627–32.

Winter, K.-U., Becker, A., Münster, T. *et al.* 1999. MADS-box genes reveal that gnetophytes are more closely related to conifers than to flowering plants. *Proceedings of the National Academy of Sciences, USA*, **96**, 7342–7.

Wolfe, J. A. & Pakiser, H. M. 1971. Stratigraphic interpretations of some Cretaceous microfossil floras of the Middle Atlantic States. *US Geological Survey Professional Papers, **750-B***, B35–B47.

Wolfe, J. A. 1973. Fossil forms of Amentiferae. *Brittonia*, **25**, 334–55.

Wolfe, J. A. & Wehr, W. 1978. Middle Eocene dicotyledonous plants from Republic, northeastern Washington. *US Geological Survey Bulletin*, **1597**, 1–25.

Wolfe, J. A. 1979. Temperature parameters of humid to mesic forests of eastern Asia and their relation to forests of other areas of the Northern Hemisphere and Australasia. *US Geological Survey Professional Papers*, **1106**, 1–37.

Wolfe, J. A. 1985. Distribution of major vegetational types during the Tertiary. In *The Carbon Cycle and Atmospheric CO_2: Natural Variations Archean to Present* (eds W. S. Broeckera & E. T. Sundquist). Washington, D.C.: American Geophysical Union, Geophysical Monograph, **32**, pp. 357–75.

Wolfe, J. A. 1987. Late Cretaceous-Cenozoic history of deciduousness and the terminal Cretaceous event. *Paleobiology*, **13**, 215–26.

Wolfe, J. A. & Tanai, T. 1987. Systematics, phylogeny, and distribution of *Acer* (maples) in the Cenozoic of western North America. *Journal of the Faculty of Sciences. Hokkaido University Ser. 4*, **22**, 1–246.

Wolfe, J. A. & Upchurch, G. R. 1987. North American nonmarine climates and vegetation during the Late Cretaceous. *Palaeogeography, Palaeoclimatology, Palaeoecology*, **61**, 33–77.

Wolfe, J. A. 1990. Estimates of Pliocene precipitation and temperature based on multivariate analysis of leaf physiognomy. In *Pliocene Climates – Scenario for Global Warming*, vol. **90–94** (eds L. B. Gosnell & R. Z. Poore): US Geological Survey Open-File Report, pp. 39–42.

Wolfe, J. A. 1991. Palaeobotanical evidence for a June 'impact' at the Cretaceous/Tertiary boundary. *Nature*, **352**, 420–3.

Wolfe, J. A. 1993. A method of obtaining climatic parameters from leaf assemblages. *US Geological Survey Bulletin*, **2040**, 1–71.

Wolfe, J. A. & Spicer, R. A. 1999. Fossil leaf character states: multivariate analyses. In *Fossil Plants and Spores: Modern Techniques* (eds T. P. Jones & N. P. Rowe) London: Geological Society, pp. 233–9.

Wolfe, K. H., Gouy, M., Yang, Y.-W., Sharp, P. M. & Li, W.-H. 1989. Date of the monocot-dicot divergence estimated from chloroplast DNA sequence data. *Proceedings of the National Academy of Sciences, USA*, **86**, 6201–5.

Won, H. & Renner, S. S. 2006. Dating dispersal and radiation in the gymnosperm *Gnetum* (Gnetales) – clock calibration when outgroup relationships are uncertain. *Systematic Biology*, **55**, 610–22.

Woodhouse, R. P. 1935. *Pollen Grains*. New York: McGraw-Hill Book Co.

Wu, C.-Y. & Kubitzki, K. 1993a. Lardizabalaceae. In *The Families and Genera of Vascular Plants*. Volume II. *Flowering Plants – Dicotyledons. Magnoliid, Hamamelid and Caryophyllid Families* (eds K. Kubitzki, J. G. Rohwer & V. Bittrich) Berlin, Heidelberg, New York: Springer-Verlag, pp. 361–5.

Wu, C.-Y. & Kubitzki, K. 1993b. Saururaceae. In *The Families and Genera of Vascular Plants*. Volume II. *Flowering Plants – Dicotyledons. Magnoliid, Hamamelid and Caryophyllid Families* (eds K. Kubitzki, J. G. Rohwer & V. Bittrich) Berlin, Heidelberg, New York: Springer-Verlag, pp. 586–8.

Wu, C.-Y. & Kubitzki, K. 1993c. Circaeasteraceae. In *The Families and Genera of Vascular Plants*. Volume II. *Flowering Plants – Dicotyledons. Magnoliid, Hamamelid and Caryophyllid Families* (eds K. Kubitzki, J. G. Rohwer & V. Bittrich) Berlin, Heidelberg, New York: Springer-Verlag, pp. 288–9.

Wu, S.-Q. 1999. A preliminary study of the Jehol flora from western Liaoning. *Palaeoworld*, **11**, 7–37 (in Chinese with English summary).

Wu, S.-Q. 2003. Land plants. In *The Jehol Biota* (eds M. Chang & P. J. Chen) Shanghai: Shanghai Scientific & Technical Publishers, pp. 167–77.

Yabe, H. & Endo, S. 1935. *Potamogeton* remains from the Lower Cretaceous? *Lycoptera* bed of Jehol. *Proceedings of the Imperial Academy, Tokyo*, **11**, 274–6.

Yamada, T. & Kato, M. 2008. *Stopesia alveolata*, a fossil seed of Trimeniaceae from the Lower Cretaceous (Albian) of Hokkaido, Northern Japan. *Acta Phytotaxonomica et Geobotanica*, **59**, 228–30.

Yang, X. J. 2003. New material of fossil plants from the Early Cretaceous Muling Formation of the Jixi Basin, Eastern Heilongjiang Province, China. *Acta Palaeontologica Sinica*, **42**, 561–84.

Yang, Y., Geng, B.-Y., Dilcher, D. L., Chen, Z.-D. & Lott, T. A. 2005. Morphology and affinities of an Early Cretaceous *Ephedra* (Ephedraceae) from China. *American Journal of Botany*, **92**, 231–41.

Yang, Z., Cheng, Y. & Wang, H. 1986. *The Geology of China. Oxford Monographs on Geology and Geophysics*. Oxford: Clarendon Press.

Yao, X., Taylor, T. N. & Taylor, E. L. 1995. The corystosperm pollen organ *Pteruchus* from the Triassic of Antarctica. *American Journal of Botany*, **82**, 535–46.

Yao, Y.-F., Xi, Y.-Z., Geng, B.-Y. & Li, C.-S. 2004. The exine ultrastructure of pollen grains in *Gnetum* (Gnetaceae) from China and its bearing on the relationship with the ANITA Group. *Botanical Journal of the Linnean Society*, **146**, 415–25.

Yoder, A. D. & Yang, Z. 2000. Estimation of primate speciation dates using local molecular clocks. *Molecular Biology and Evolution*, **17**, 1081–90.

Zaklinskaya, E. D. 1981. Phylogeny and classification of the Normapolles. *Review of Palaeobotany and Palynology*, **35**, 139–47.

Zanis, M. J., Soltis, D. E., Soltis, P. S., Mathews, S. & Donoghue, M. J. 2002. The root of the angiosperms revisited. *Proceedings of the National Academy of Sciences, USA*, **99**, 6848–53.

Zanis, M. J., Soltis, D. E., Soltis, P. S., Qiu, Y.-L. & Zimmer, E. A. 2003. Phylogenetic analyses and perianth evolution in basal angiosperms. *Annals of the Missouri Botanical Garden*, **90**, 129–50.

Zastawniak, E. 1994. Upper Cretaceous leaf flora from the Blaszyk Moraine (Zamek Formation), King George Island, South Shetland Islands, West Antarctica. *Acta Palaeobotanica*, **34**, 119–63.

Zavada, M. S. & Crepet, W. L. 1981. Investigations of angiosperms from the Middle Eocene of North America: Flowers of the Celtidoideae. *American Journal of Botany*, **68**, 924–33.

Zavada, M. S. & Crepet, W. L. 1986. Pollen wall ultrastructure of the type material of *Pteruchus africanus*, *Pteruchus dubius* and *Pteruchus papillatus*. *Pollen et Spores*, **27**, 271–6.

Zavada, M. S. & Dilcher, D. L. 1986. Comparative pollen morphology and its relationship to phylogeny of pollen in the Hamamelidae. *Annals of the Missouri Botanical Garden*, **73**, 348–81.

Zavada, M. S. & Taylor, T. N. 1986. Pollen morphology of Lactoridaceae. *Plant Systematics and Evolution*, **154**, 31–9.

Zavada, M. S. & Benzon, J. M. 1987. First fossil evidence for the primitive angiosperm family Lactoridaceae. *American Journal of Botany*, **74**, 1590–4.

Zavada, M. S. 2003. The ultrastructure of angiosperm pollen from the Lower Cenomanian of the Morondova Basin, Madagascar. *Grana*, **42**, 20–32.

Zavialova, N., van Konijnenburg-van Cittert, J. & Zavada, M. 2009. The pollen ultrastructure of *Williamsoniella coronata* Thomas (Bennettitales) from the Bajocian of Yorkshire. *International Journal of Plant Sciences*, **170**, 1195–200.

Zbyszewski, G., Moitinho d'Almeida, F. & Torre de Assunção, C. 1955. *Carta Geológica de Portugal na Escala de 1/50 000. Notícia Explicativa da Folha 30-C Torres Vedras.* Lisbon: Serviços Geológicos de Portugal.

Zeba-Bano, Maheshawari, H. K. & Bose, M. N. 1979. Some plant remains from Pathargama, Rajmahal Hills, Bihar. *The Palaeobotanist*, **26**, 144–56.

Zetter, R., Hofmann, C.-C., Draxler, I. *et al.* 1999. A rich Middle Eocene microflora at Arroyo de los Mineros, near Canãdón Beta, NE Tierra del Fuego Province, Argentina. *Abhandlungen der Geologischen Bundesanstalt*, **56**, 439–60.

Zetter, R., Hesse, M. & Huber, K. H. 2002. Combined LM, SEM and TEM studies of Late Cretaceous pollen and spores from Gmünd, Lower Austria. *Stapfia*, **80**, 201–30.

Zhang, Y.-Y. 1999. The evolutionary succession of Cretaceous angiosperm pollen in China. *Acta Palaeontologica Sinica*, **38**, 435–53 (in Chinese with English abstract).

Zheng, Y. & Wang, W. 1994. Sequence of Miocene Fotan Group in SE Fujian and its palyno-assemblages. *Acta Palaeontologica Sinica*, **33**, 200–216 (in Chinese with English abstract).

Zherikhin, V. V. 2002. Ecological history of the terrestrial insects. In *History of Insects* (eds A. P. Rasnytsyn & D. L. J. Quicke) Dordrecht: Kluwer Academic Press, pp. 331–88.

Zhilin, S. G. 1974. *The Tertiary Floras of the Plateau Ustjurt (Transcaspia)*. Leningrad: Nauka.

Zhou, Z.-H. & Zhang, F. 2002. A long-tailed, seed-eating bird from the Early Cretaceous of China. *Nature*, **418**, 405–9.

Zhou, Z.-H., Barrett, P. M. & Hilton, J. 2003. An exceptionally preserved Lower Cretaceous ecosystem. *Nature*, **421**, 807–14.

Zhou, Z.-H. & Zhang, F. 2003. Anatomy of the primitive bird *Sapeornis chaoyangensis* from the Early Cretaceous of Liaoning, China. *Canadian Journal of Earth Science*, **40**, 731–47.

Zhou, Z.-H. 2004. Vertebrate radiations of the Jehol Biota and their environmental background. *Chinese Science Bulletin*, **49**, 754–6.

Zhou, Z.-H. & Zhang, F. 2005. Discovery of an ornithurine bird and its implication for Early Cretaceous avian radiation. *Proceedings of the National Academy of Sciences, USA*, **102**, 18 998–9002.

Zhou, Z.-H. 2006. Evolutionary radiation of the Jehol Biota: chronological and ecological perspectives. *Geological Journal*, **41**, 377–93.

Zhou, Z.-K., Crepet, W. L. & Nixon, K. C. 2001. The earliest fossil evidence of the Hamamelidaceae: Late Cretaceous (Turonian) inflorescences and fruits of Altingioideae. *American Journal of Botany*, **88**, 753–66.

Zhou, Z.-Y., Li, H.-M. & Cao, Z. Y. 1990. Some Cretaceous plants from Pingzhou (Ping Chau) Island, Hong Kong. *Acta Palaeontologica Sinica*, **29**, 415–26.

Zhou, Z.-Y. & Li, H. 1994. Some Late Cretaceous plants from King George Island, Antarctica. In *Stratigraphy and Palaeontology of Fildes Peninsula, King Georges Island, Antarctica, State Antarctic Committee. Monograph 3* (ed. Y. Shen) Beijing: Science Press, pp. 85–95.

Zhou, Z.-Y. 1997. Mesozoic Ginkgoalean megafossils: a systematic review. In *Ginkgo biloba, A Global Treasure* (eds T. Hori, R. W. Ridge, W. Tulecke, P. Del Tredici, J. Trémouillaux-Guiller & H. Tobe) Tokyo: Springer-Verlag, pp. 183–206.

Zhou, Z.-Y. & Zheng, S. 2003. The missing link in *Ginkgo* evolution. *Nature*, **423**, 821–2.

Ziegler, A. M., Raymond, A. L., Gierlowski, T. C. *et al.* 1987. Coal, climate and terrestrial productivity: the present and early Cretaceous compared. In *Coal and Coal-bearing Strata: Recent Advances*, Volume **32** (ed. A. C. Scott): Geological Society Special Publication, pp. 25–49.

Zimmerman, W. 1933. Paläobotanische und phylogenetische Beiträge I-V. *Palaeobiologica*, **5**, 321–48.

Index